PARALLELOGRAM

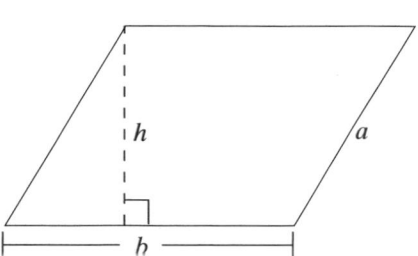

Perimeter: $P = 2a + 2b$
Area: $A = bh$

CIRCLE

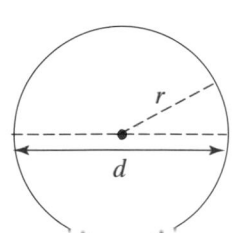

Circumference: $C = \pi d$
$C = 2\pi r$
Area: $A = \pi r^2$

RECTANGULAR SOLID

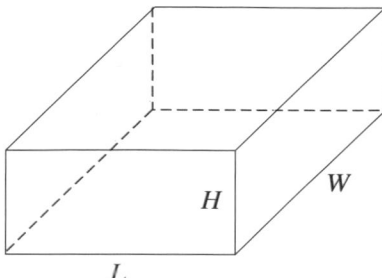

Volume: $V = LWH$
Surface Area: $A = 2HW + 2LW + 2LH$

673 - 7181 김영 cha
073 - 7185 kir munks

CUBE

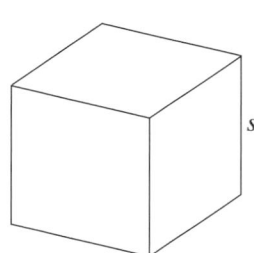

Volume: $V = s^3$
Surface Area: $A = 6s^2$

CONE

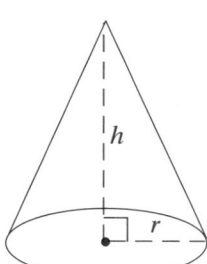

Volume: $V = \frac{1}{3}\pi r^2 h$
Lateral Surface Area: $A = \pi r \sqrt{r^2 + h^2}$

RIGHT CIRCULAR CYLINDER

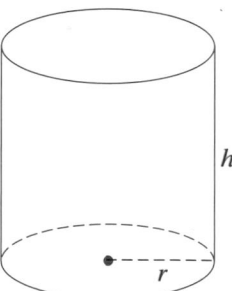

Volume: $V = \pi r^2 h$
Surface Area:
$A = 2\pi rh + 2\pi r^2$

D0575743

OTHER FORMULAS

Distance: $d = rt$ (r = rate, t = time)
Temperature: $F = \frac{9}{5}C + 32$ $C = \frac{5}{9}(F - 32)$
Simple Interest: $I = Prt$
  (P = principal, r = annual interest rate, t = time in years)
Compound Interest: $A = P\left(1 + \frac{r}{n}\right)^{nt}$
  (P = principal, r = annual interest rate, t = time in years, n = number of compoundings per year)

Companion Website
http://www.prenhall.com/martin-gay

BEGINNING AND INTERMEDIATE ALGEBRA

BEGINNING AND INTERMEDIATE ALGEBRA

Second Edition

K. ELAYN MARTIN-GAY

University of New Orleans

Prentice Hall

Upper Saddle River, New Jersey 07458

Library of Congress Cataloging-in-Publication Data

Martin-Gay, K. Elayn
    Beginning and intermediate algebra/K. Elayn Martin-Gay.—2nd ed.
      p. cm.
    Includes index.
     1. Algebra. I. Title.
QA152.2.M3674 2001   00-065268
512.9—dc21   CIP

Executive Acquisition Editor: Karin E. Wagner
Editor in Chief: Christine Hoag
Project Managers: Mary Beckwith/Anne Marie Jones
Vice President/Director of Production and Manufacturing: David W. Riccardi
Executive Managing Editor: Kathleen Schiaparelli
Senior Managing Editor: Linda Mihatov Behrens
Project Management: Elm Street Publishing Services, Inc.
Manufacturing Buyer: Alan Fischer
Manufacturing Manager: Trudy Pisciotti
Senior Marketing Manager: Eilish Collins Main
Marketing Assistant: Dan Auld
Director of Marketing: John Tweeddale
Editor in Chief, Development: Carol Trueheart
Development Editors: Tony Palermino/Emily Keaton
Associate Editor, Mathematics/Statistics Media: Audra J. Walsh
Editorial Assistant: Glenda Pinto
Art Director: Maureen Eide
Assistant to the Art Director: John Christiana
Interior Designer: Donna Wickes
Cover Designer: Joseph Sengotta
Art Editor: Grace Hazeldine
Director of Creative Services: Paul Belfanti
Photo Researcher: Kathy Ringrose
Photo Editor: Beth Boyd
Cover Photo: Alex Demyan/Demyan Photographs
Art Studio: Academy Artworks
Compositor: Preparé Inc., Italy

 © 2001, 1996 by Prentice-Hall, Inc.
Upper Saddle River, NJ 07458

The first edition of this book was published in 1996 under the title *Introductory and Intermediate Algebra*.

Printed in the United States of America
10 9 8 7 6 5

ISBN 0-13-016636-7

Prentice-Hall International (UK) Limited, *London*
Prentice-Hall of Australia Pty. Limited, *Sydney*
Prentice-Hall Canada, Inc., *Toronto*
Prentice-Hall Hispanoamericana, S.A., *Mexico*
Prentice-Hall of India Private Limited, *New Delhi*
Prentice-Hall of Japan, Inc., *Tokyo*
Pearson Education Asia, Pte. Ltd.
Editora Prentice-Hall do Brasil, Ltda., *Rio de Janeiro*

To Joseph C. Buccaran

Lives of great men all remind us
We can make our lives sublime,
And, departing, leave behind us
Footprints on the sands of time.
                    —*Henry Wadsworth Longfellow*

From your years of dedication to excellent
education, I think of all the footprints you
have already left and the many yet to come.

CONTENTS

3 GRAPHS AND FUNCTIONS   150

4 SOLVING SYSTEMS OF LINEAR EQUATIONS   238

5 EXPONENTS AND POLYNOMIALS   278

6 FACTORING POLYNOMIALS   340

7 RATIONAL EXPRESSIONS   408

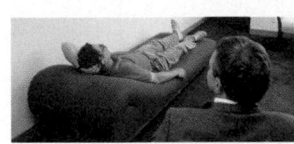

PREFACE

ABOUT THIS BOOK

Beginning and Intermediate Algebra, Second Edition, was written to provide a **solid foundation in algebra** as well as to develop students' problem-solving skills. Specific care has been taken to ensure that students have the most **up-to-date and relevant** text preparation for their next mathematics course, as well as to help students succeed in nonmathematical courses that require a grasp of algebraic fundamentals. I have tried to achieve this by writing a user-friendly text that is keyed to objectives and contains many worked-out examples. The basic concepts of graphs and functions are introduced early, and problem solving techniques, real-life and real-data applications, data interpretation, appropriate use of technology, mental mathematics, number sense, critical thinking, decision-making, and geometric concepts are emphasized and integrated throughout the book.

The many factors that contributed to the success of the first edition have been retained. In preparing this edition, I considered the comments and suggestions of colleagues throughout the country, students, and many users of the prior edition. The AMATYC Crossroads in Mathematics: Standards for Introductory College Mathematics before Calculus and the MAA and NCTM standards (plus Addenda), together with advances in technology, also influenced the writing of this text.

Beginning and Intermediate Algebra, Second Edition, is **part of a series of texts** that can include *Basic College Mathematics* and *Prealgebra, Third Edition.* Also available are *Beginning Algebra, Third Edition, Intermediate Algebra, Third Edition,* and *Intermediate Algebra: A Graphing Approach, Second Edition.* Throughout the series, pedagogical features are designed to develop student proficiency in algebra and problem solving, and to prepare students for future courses.

KEY PEDAGOGICAL FEATURES IN THE SECOND EDITION

Readability and Connections   I have tried to make the writing style as clear as possible while still retaining the mathematical integrity of the content. When a new topic is presented, an effort has been made to **relate the new ideas to those that students**

XIII

may already know. Constant reinforcement and connections within problem solving strategies, data interpretation, geometry, patterns, graphs, and situations from everyday life can help students gradually master both new and old information.

Problem-Solving Process   This is formally introduced in Chapter 2 with a **new four-step process that is integrated throughout the text**. The four steps are Understand, Translate, Solve, and Interpret. The repeated use of these steps throughout the text in a variety of examples shows their wide applicability. Reinforcing the steps can increase students' confidence in beginning problems.

Applications and Connections   Every effort was made to include as many accessible, interesting and relevant real-life applications as possible throughout the text in both worked-out examples and exercise sets. The applications **strengthen students' understanding of mathematics in the real world** and help to motivate students. They show connections to a wide range of fields including agriculture, astronomy, automotive ownership, business, chemistry, communication, computer technology, construction, consumer affairs, demographics, earth science, education, entertainment, environmental issues, finance and economics, food service, geography, government, hobbies, labor and career issues, life science, medicine, music, nutrition, physics, political science, population, recreation, sports, technology, transportation, travel, weather, and important related mathematical areas such as geometry and statistics. (See the Index of Applications on page xxiv.) Many of the applications are based on **recent and interesting real-life data**. Sources for data include newspapers, magazines, government publications, publicly held companies, special interest groups, research organizations, and reference books. Opportunities for obtaining your own real data with and without using the internet are also included.

Helpful Hints   Helpful Hints contain practical advice on applying mathematical concepts. These are found throughout the text and **strategically placed** where students are most likely to need immediate reinforcement. They are highlighted in a box for quick reference and, as appropriate, an indicator line is used to precisely identify the particular part of a problem or concept being discussed. For instance, see pages 90 and 365.

Visual Reinforcement of Concepts   The text contains numerous graphics, models, and illustrations to visually clarify and reinforce concepts. These include **new and updated** bar graphs and circle graphs in two and three dimensions, line graphs, calculator screens, application illustrations, photographs, and geometric figures. There are now **approximately 1,000 figures**.

Real World Chapter Openers   The new two-page chapter opener focuses on how math is used in a specific career, provides links to the World Wide Web, and references a "Spotlight on Decision Making" feature within the chapter for further exploration of the **career and the relevance of algebra**. For example, look at the opener for Chapter 4. The opening pages also contain a list of section titles, and an introduction to the mathematics to be studied together with mathematical connections to previous chapters in the text.

Student Resource Icons   At the beginning of each section, videotape, tutorial software CD Rom, Student Solutions Manual, and Study Guide icons are displayed. These icons help remind students that these learning aids are available should they choose to use them to review concepts and skills at their own pace. These items have **direct correlation to the text** and emphasize the text's methods of solution.

Chapter Highlights   Found at the end of each chapter, the Chapter Highlights contain key definitions, concepts, *and* examples to **help students understand and retain** what they have learned.

Chapter Project   This feature occurs at the end of each chapter, often serving as a chapter wrap-up. For **individual or group completion**, the multi-part Chapter Project, usually hands-on or data based, allows students to problem solve, make interpretations, and to think and write about algebra.

In addition, a reference to alternative or additional Real World Activities is given. This **internet option** invites students to find and retrieve real data for use in solving problems. Visit the Real World Activities Website by going to http://www.prenhall.com/martin-gay.

Functional Use of Color and New Design   Elements of this text are highlighted with color or design to make it easier for students to read and study. Special care has been taken to use color within solutions to examples or in the art to **help clarify, distinguish, or connect concepts**. For example, look at page 301 in Section 5.3.

EXERCISE SETS

Each text section ends with an exercise set, usually divided into two parts. Both parts contain graded exercises. The **first part is carefully keyed** to at least one worked example in the text. Once a student has gained confidence in a skill, the **second part contains exercises not keyed to examples**. Exercises and examples marked with a video icon (◆) have been worked out step-by-step by the author in the videos that accompany this text.

Throughout the text exercises there is an emphasis on **data and graphical interpretation** via tables, charts, and graphs. The ability to interpret data and read and create a variety of types of graphs is developed gradually so students become comfortable with it. Similarly, throughout the text there is integration of **geometric concepts**, such as perimeter and area. Exercises and examples marked with a geometry icon (△) have been identified for convenience.

Each exercise set contains one or more of the following features.

Spotlight on Decision Making   These unique **new, specially designed applications** help students develop their decision-making and problem-solving abilities, skills useful in mathematics and in life. Appropriately placed before an exercise set begins, students have an opportunity to immediately practice and reinforce basic algebraic concepts found in the accompanying section in relevant, accessible contexts. There is an emphasis on workplace or job-related career situations (such as the decisions of a Meteorologist in Section 3.1, a phychologist in Section 9.6, or a Webmaster in Section 11.4) as well as decision making in general (such as choosing a credit card in Section 6.5 or deciding between two job offers in Section 4.3).

Mental Mathematics   These problems are found at the beginning of many exercise sets. They are mental warm-ups that **reinforce concepts** found in the accompanying section and increase students' confidence before they tackle an exercise set. By relying on their own mental skills, students increase not only their confidence in themselves but also their number sense and estimation ability.

Writing Exercises   These exercises now found in almost every exercise set are marked with the icon (✎). They require students to **assimilate information** and provide a written response to explain concepts or justify their thinking. Guidelines recommended by the American Mathematical Association of Two Year Colleges (AMATYC) and other professional groups recommend incorporating writing in mathematics courses to reinforce concepts. Writing opportunities also occur within features such as Spotlight on Decision Making and Chapter Projects.

Data and Graphical Interpretation   Throughout the text there is an emphasis on data interpretation in exercises via tables, bar charts, line graphs, or circle graphs. The ability to interpret data and read and create a variety of graphs is **developed gradually** so students become comfortable with it.

Calculator Explorations and Exercises   These optional explorations offer guided instruction, through examples and exercises, on the proper use of **scientific and graphing calculators or computer graphing utilities as tools in the mathematical problem-solving process**. Placed appropriately throughout the text, these explorations reinforce concepts or motivate discovery learning.

Additional exercises building on the skills developed in the Explorations may be found in exercise sets throughout the text and are marked with the icon 📟 for scientific calculator use or with the icon 📱 for graphing calculator use.

Review Exercises   These exercises occur in each exercise set (except for those in Chapter 1). These problems are **keyed to earlier sections** and review concepts learned earlier in the text that are needed in the next section or in the next chapter. These exercises show the **links between earlier topics and later material**.

A Look Ahead   These exercises occur at the end of some exercise sets. This section contains examples and problems similar to those found in a subsequent algebra course. "A Look Ahead" is presented as **a natural extension of the material** and contains an example followed by advanced exercises.

In addition to the approximately 7000 exercises within sections, exercises may also be found in the Vocabulary Checks, Chapter Reviews, Chapter Tests, and Cumulative Reviews.

Vocabulary Checks   Vocabulary checks, **new to this edition**, provide an opportunity for students to become more familiar with the use of mathematical terms as they strengthen their verbal skills.

Chapter Review and Chapter Test   The end of each chapter contains a review of topics introduced in the chapter. The review problems are keyed to sections. The chapter test is not keyed to sections.

Cumulative Review   Each chapter after the first contains a **cumulative review of all chapters beginning with the first** up through the chapter at hand. Each problem contained in the cumulative review is actually an earlier worked example in the text that is referenced in the back of the book along with the answer. Students who need to see a complete worked-out solution, with explanation, can do so by turning to the appropriate example in the text.

KEY CONTENT FEATURES IN THE SECOND EDITION

Overview   This new edition retains many of the factors that have contributed to its success. Even so, **every section of the text was carefully re-examined**. Throughout the new edition you will find numerous new applications, examples, and many real-life applications and exercises. Some sections have internal re-organization to better clarify and enhance the presentation.

Table of Content Changes in the Second Edition   The second edition includes a **new Chapter 8, Transitions to Intermediate Algebra**. Although intermediate algebra topics are woven into earlier chapters where appropriate, the purpose of this chapter is to help students make the transition from beginning algebra to intermediate algebra. For example, Chapter 8 contains types of equations and inequalities normally found in intermediate algebra, such as absolute value equations and inequalities, system of equations in three variables as well as matrices and determinants.

By moving these intermediate algebra topics to Chapter 8, **Chapters 2 and 3 were combined to form a new Chapter 2, Equations, Inequalities, and Problem Solving**. As a result, **graphing is now covered in Chapter 3, Graphs and Functions**. A new Section 3.1 is devoted to introducing the rectangular coordinate system and creating scatter diagrams from real data. Functions are introduced in Section 3.3 and continually revisited to help students fully understand and see the importance of this topic. For example, see Sections 3.4, 5.3, 6.8, and 7.1 just to name a few.

Increased Integration of Geometry Concepts   In addition to the traditional topics in beginning algebra courses, this text contains a strong emphasis on problem solving, and geometric concepts are integrated throughout. The geometry concepts presented are those most important to a students' understanding of algebra, and I have included **many applications and exercises** devoted to this topic. These are marked with the icon $\triangle$. Also, geometric figures, a review of angles, lines, and special triangles, are covered in the appendices. The inside front cover provides a quick reference of geometric formulas.

Real Numbers and Algebraic Expressions   Chapter 1 now begins with Tips for Success in Mathematics (Section 1.1). Chapter 1 has been streamlined and refreshed for **greater efficiency and relevance**. New applications and real data enhance the chapter.

Early and Intuitive Introduction to Graphs and Functions   As bar and line graphs are gradually introduced in Chapters 1 and 2, an emphasis are placed on the notion of paired data. This leads naturally to the concepts of ordered pair and the rectangular coordinate system introduced in Chapter 3. This edition offers more real data and conceptual type applications and further strengthens the introduction to slope.

Once students are comfortable with graphing equations, functions are introduced in Chapter 3. The concept of function is illustrated in numerous ways to ensure student understanding: by listing ordered pairs of data, showing rectangular coordinate system graphs, visually representing set correspondences, and including numerous real-data and conceptual examples. **The importance of a function is continuously reinforced** by not treating it as a single, stand-alone topic but by constantly integrating functions in appropriate sections of this text.

Increased Attention to Problem Solving   Building on the strengths of the prior edition, a special emphasis and strong commitment are given to contemporary, accessible, and practical applications of algebra. **Real data** was drawn from a variety of sources including internet sources, magazines, newspapers, government publications, and reference books. **Unique Spotlight on Decision Making exercises and a new four-step problem-solving process are incorporated throughout** to focus on helping to build students problem-solving skills.

Increased Opportunities for Using Technology   Optional explorations for a calculator or graphing calculator (or graphing utility such as Texas Instruments Interactive), are integrated appropriately **throughout the text** in Calculator Explorations features and in exercises marked with a calculator icon. The Martin-Gay **Companion Website** includes links to internet sites to allow opportunities for finding data and using it for problem solving such as with the accompanying on-line Real World Activities.

The Website also includes links to search potential mathematically related careers branching from the chapter openers. Instructors may also choose from a variety of **distance learning or on-line delivery options** including Blackboard or Web CT.

New Examples   Detailed step-by-step examples were added, deleted, replaced, or updated as needed. Many of these reflect real life. **Examples are used in two ways.** Often there are numbered, formal examples, and occasionally an example or application is used to introduce a topic or informally discuss the topic.

New Exercises   A significant amount of time was spent on the exercise sets. New exercises and examples **help address a wide range of student learning styles and abilities**. The text now includes the following types of exercises: spotlight on decision making exercises, mental math, computational exercises, real-life applications, writing exercises, multi-part exercises, review exercises, a look ahead exercises, optional calculator or graphing calculator exercises, data analysis from tables and graphs, vocabulary checks, and projects for individual or group assignment. Also available are new on-line Real World Activities accessed via this textbook's companion website, and a selection of group activities in a worksheet ready, easy to use format, found in the Instructor's Resource Manual with Tests.

Enhanced Supplements Package   The new Second Edition is supported by a wealth of supplements designed for **added effectiveness and efficiency**. New items include the MathPro 4.0 Explorer tutorial software together with a unique video clip feature, a new computerized testing system TestGenEQ, and an expanded and improved Martin-Gay companion website. Some highlights in print materials include the addition of teaching tips in the Annotated Instructor's Edition, and an expanded Instructor's Resource Manual with Tests including additional exercises and short group activities in a ready-to use-format. Please see the list of supplements for descriptions.

OPTIONS FOR ON-LINE AND DISTANCE LEARNING

For maximum convenience, Prentice Hall offers on-line interactivity and delivery options for a variety of distance learning needs. Instructors may access or adopt these in conjunction with this text, *Beginning and Intermediate Algebra*.

Companion Website

Visit *http://www.prenhall.com/martin-gay*
The companion Website includes basic distance learning access to provide links to the text's Real World Activities, career-related sites referenced in the chapter opening pages and a selection of on-line self quizzes. E-mail is available. For quick reference, the inside front cover of this text also lists the companion Website URL.

WebCT

WebCT includes distance learning access to content found in the Martin-Gay Companion Website plus more. WebCT provides tools to create, manage, and use on-line course materials. Save time and take advantage of items such as on-line help, communication tools, and access to instructor and student manuals. Your college may already have WebCT's software installed on their server or you may choose to download it. Contact your local Prentice Hall sales representative for details.

Blackboard

Visit *http://www.prenhall.com/demo*
For distance learning access to content and features from the Martin-Gay Companion Website plus more, Blackboard provides simple templates and tools to create,

manage, and use on-line course materials. Save time and take advantage of items such as on-line help, course management tools, communication tools, and access to instructor and student manuals. No technical experience required. Contact your local Prentice Hall sales representative for details.

For a *complete* computer-based internet course ...
Prentice Hall Interactive Math
Visit *http://www.prenhall.com/interactive_math*

Prentice Hall Interactive Math is an exciting, proven choice to help students succeed in math. Created for a computer-based course, it provides the effective teaching philosophy of K. Elayn Martin-Gay in an Internet-based course format. Interactive Math, Introductory and Intermediate Algebra, takes advantage of state-of-the-art technology to provide highly flexible and user-friendly course management tools and an engaging, highly interactive student learning program that easily accommodates the variety of learning styles and broad spectrum of students presented by the typical beginning and intermediate algebra class. Personalized learning includes reading, writing, watching video clips, and exploring concepts through interactive questions and activities. Contact your local Prentice Hall sales representative for details.

SUPPLEMENTS FOR THE INSTRUCTOR

Printed Supplements

Annotated Instructor's Edition (ISBN 0-13-016637-5)
- Answers to exercises on the same text page or in Graphing Answer Section
- Graphing Answer Section contains answers to exercises requiring graphical solutions, chapter projects, and Spotlight on Decision Making exercises.
- Teaching Tips throughout the text placed at key points in the margin where students historically need extra help. These tips provide ideas on how to help students through these concepts, as well as ideas for expanding upon a certain concept or ideas for classroom activities

Instructor's Solutions Manual (ISBN 0-13-017339-8)
- Detailed step-by-step solutions to even-numbered section exercises
- Solutions to every Spotlight on Decision Making exercise
- Solutions to every Calculator Exploration exercise
- Solutions to every Chapter Test and Chapter Review exercise
- Solution methods reflect those emphasized in the textbook

Instructor's Resource Manual with Tests (ISBN 0-13-017330-4)
- Notes to the Instructor that includes an introduction to Interactive Learning, Interpreting Graphs and Data, Alternative Assessment, Using Technology, and Helping Students Succeed
- Eight Chapter Tests per chapter (5 free response, 3 multiple choice)
- Two Cumulative Review Tests (one free response, one multiple choice) following every two chapters
- Eight Final Exams (4 free response, 4 multiple choice)
- Twenty additional exercises per section for added test exercises or worksheets, if needed
- Group Activities by Bettie A. Truitt, Ph.D. (on average of two per chapter; providing short group activities in a convenient ready-to-use handout format)
- Answers to all items

Media Supplements

TestGen EQ CD-ROM (Windows/Macintosh) (ISBN 0-13-018591-4)

- Algorithmically driven, text specific testing program
- Networkable for administering tests and capturing grades on-line
- Edit or add your own questions to create a nearly unlimited number of tests and worksheets
- Use the new "Function Plotter" to create graphs
- Tests can be easily exported to HTML so they can be posted to the Web for student practice

Computerized Tutorial Software Course Management Tools
MathPro 4.0 Explorer Network CD-ROM (ISBN 0-13-018593-0)

- Enables instructors to create either customized or algorithinically generated practice tests from any section of a chapter, or a test of random items
- Includes an e-mail function for network users, enabling instructors to send a message to a specific student or to an entire group
- Network based reports and summaries for a class or student and for cumulative or selected scores are available

Companion Website: *http://www.prenhall.com/martin-gay*

- Create a customized online syllabus with Syllabus Manager
- Assign Internet-based Real World Activities, wherein students find and retrieve real data for use in guided problem solving
- Assign quizzes or monitor student self quizzes by having students e-mail results, such as true/false reading quizzes or vocabulary check quizzes
- Destination links provide additional opportunities to explore related sites

SUPPLEMENTS FOR THE STUDENT

Printed Supplements

Student Solutions Manual (ISBN 0-13-017338-X)

- Detailed step-by-step solutions to odd-numbered section exercises
- Solutions to every (odd and even) Mental Math exercise
- Solutions to odd-numbered Calculator Exploration exercises
- Solutions to every (odd and even) exercise found in the Chapter Reviews and Chapter Tests
- Solution methods reflect those emphasized in the textbook
- Ask your bookstore about ordering

Student Study Guide (ISBN 0-13-017341-X)

- Additional step-by-step worked out examples and exercises
- Practice tests and final examination
- Includes Study Skills and Note-taking suggestions
- Includes Hints and Warnings section
- Solutions to all exercises, tests, and final examination
- Solution methods reflect those emphasized in the text
- Ask your bookstore about ordering

How to Study Mathematics

- Have your instructor contact the local Prentice Hall sales representative

Math on the Internet: A Student's Guide

- Have your instructor contact the local Prentice Hall sales representative

Prentice Hall/New York Times, Theme of the Times Newspaper Supplement

- Have your instructor contact the local Prentice Hall sales representative

Media Supplements

Computerized Tutorial Software
MathPro 4.0 Explorer Network CD-Rom (ISBN 0-13-018593-0)
MathPro 4.0 Explorer Student CD-Rom (ISBN 0-13-018594-9)

- Keyed to each section of the text for text-specific tutorial exercises and instruction
- Warm-up exercises and graded Practice Problems
- Video clips, providing a problem (similar to the one being attempted) being explained and worked out on the board
- Explorations, allowing explorations of concepts associated with objectives in more detail
- Algorithmically generated exercises, and includes bookmark, on-line help, glossary, and summary of scores for the exercises tried
- Interactive feedback
- Have your instructor contact the local Prentice Hall sales representative— also available for home use

Videotape Series (ISBN 0-13-018598-1)

- Written and presented by textbook author K. Elayn Martin-Gay
- Keyed to each section of the text
- Presentation and step-by-step solutions to exercises from each section of the text.
- Examples or exercises taken directly from the text are marked with a video icon 📹.
- Key concepts are explained

Companion Website: *www.prenhall.com/martin-gay*

- Offers Warm-ups, Real World Activities, True/False Reading Quizzes, Chapter Quizzes, and Vocabulary Check Quizzes
- Includes a link to the Real World Activities referenced in each chapter of this text
- Option to e-mail results to your instructor
- Destination links provide additional opportunities to explore other related sites, such as those mentioned in this text's chapter opening pages

ACKNOWLEDGMENTS

First, as usual, I would like to thank my husband, Clayton, for his constant encouragement. I would also like to thank my children, Eric and Bryan, for continuing to eat my burnt meals. Thankfully, they have started to cook a little themselves.

I would also like to thank my extended family for their invaluable help and wonderful sense of humor. Their contributions are too numerous to list. They are Rod and Karen Pasch; Peter, Michael, Christopher, Matthew, and Jessica Callac; Stuart, Earline, Melissa, Mandy, and Bailey Martin; Mark, Sabrina, and Madison Martin; Leo, Barbara, Aaron, and Andrea Miller; and Jewett Gay.

A special thank you to all the users of the first edition of this text and for their suggestions for improvements that were incorporated into the second edition. I would also like to thank the following reviewers for their input and suggestions:

Rick Armstrong, *St. Louis Community College–Florissant Valley*
Arlene Bakner, *Collin County Community College*
Patrick Cross, *University of Oklahoma*
Joanne Kendall, *College of the Mainland*
Donald Kinney, *University of Minnesota*
Judith Marwick, *Prairie State College*
Sharon North, *St. Louis Community College–Florissant Valley*
Karen Stewart, *College of the Mainland*

Previous Edition Reviewers:

Carol Achs, *Mesa Community College*
Gabrielle Andries, *University of Wisconsin–Milwaukee*
Jan Archibald, *Ventura College*
Carol Atnip, *University of Louisville*
Sandra Beken, *Horry-Georgetown Technical College*
Nancy J. Bray, *San Diego Mesa College*
Helen Burrier, *Kirkwood Community College*
Celeste Carter, *Richland College*
Deann Christianson, *The University of the Pacific*
John Coburn, *St. Louis Community College*
Iris DeLoach-Johnson, *Miami University*
Omar L. DeWitt, *University of New Mexico*
Catherine Folio, *Brookdale Community College*
Robert W. Gesell, *Cleary College*
Dauhrice Gibson, *Gulf Coast Community College*
Marian Glasby, *Anne Arundel Community College*
Margaret (Peg) Greene, *Florida Community College at Jacksonville*
Frank Gunnip, *Oakland Community College*
Doug Jones, *Tallahassee University*
Mike Mears, *Manatee Community College*
James W. Newsom, *Tidewater Community College*
Randy Pittman, *J. Sargeant Reynolds Community College*
Mary Kay Schippers, *Fort Hays State University*
Mary Lee Seitz, *Erie Community College–City Campus*
Ken Seydel, *Skyline College*
Edith Silver, *Mercer County Community College*
Ventura Simmons, *Medgar Evers College*
Bonnie Simon, *Naugatuck Valley Community Technical College*
Debbie Singleton, *Lexington Community College*
Ronald Smith, *Edison Community College*
Richard Spangler, *Tacoma Community College*
Lauren Syda, *Yuba Community College*

Diane Trimble, *Collin County Community College*
Patrick C. Ward, *Illinois Central College*
John C. Wenger, *City College of Chicago–Harold Washington College*
Jerry Wilkerson, *Missouri Western State College*

There were many people who helped me develop this text and I will attempt to thank some of them here. Richard Semmler was invaluable for contributing to the overall accuracy of this text. Emily Keaton was also invaluable for her many suggestions and contributions during the development of the manuscript, and a special thank you to Mary Beckwith and Ann Marie Jones, my project managers. They kept us all organized and on task. Ingrid Mount at Elm Street Publishing Services provided guidance throughout the production process. I very much appreciated the writers, formatters, and accuracy checkers of the supplements. I thank Bettie A. Truitt for the selection of group activities. I thank Terri Bittner, Trisha Bergthold, Cindy Trimble, Jeff Rector, and Teri Lovelace at Laurel Technical Services for all their work on some of the supplements, and providing a thorough accuracy check. Lastly, a special thank you to my editor Karin Wagner for her support and assistance throughout the development and production of this text and to all the staff at Prentice Hall: Chris Hoag, Linda Behrens, Alan Fischer, Maureen Eide, Grace Hazeldine, Gus Vibal, Audra Walsh, Eilish Main, Daniel Auld, Elise Schneider, Stephanie Szolusha, John Tweedale, Paul Corey, and Tim Bozik.

K. Elayn Martin-Gay

ABOUT THE AUTHOR

K. Elayn Martin-Gay has taught mathematics at the University of New Orleans for more than 20 years. Her numerous teaching awards include the local University Alumni Association's Award for Excellence in Teaching, and Outstanding Developmental Educator at University of New Orleans, presented by the Louisiana Association of Developmental Educators.

Prior to writing textbooks, K. Elayn Martin-Gay developed an acclaimed series of lecture videos to support developmental mathematics students in their quest for success. These highly successful videos originally served as the foundation material for her texts. Today the tapes specifically support each book in the Martin-Gay series.

Elayn is the author of nine published textbooks as well as multimedia interactive mathematics, all specializing in developmental mathematics courses such as basic mathematics, prealgebra, beginning and intermediate algebra. She has provided author participation across the broadest range of materials: textbook, videos, tutorial software, and Interactive Math courseware. All the components are designed to work together. This offers an opportunity of various combinations for an integrated, consistent teaching and learning package.

APPLICATIONS INDEX

Highlights of *Beginning and Intermediate Algebra,* Second Edition

Beginning and Intermediate Algebra, Second Edition, has been written and designed to help you succeed in this course. Specific care has been taken to ensure you have the most up-to-date and relevant text features to provide you with a solid foundation in algebra, as many accessible real-world applications as possible, and to prepare you for future courses.

Get Motivated!

pages 676, 677

◀ REAL-WORLD CHAPTER OPENERS

New Real-World Chapter Openers focus on how algebraic concepts relate to the world around you so that you see the relevance and practical applications of algebra in daily life.

They also provide links to the World Wide Web and reference a *Spotlight on Decision Making* feature within the chapter for further exploration.

SPOTLIGHT ON ▶ DECISION MAKING

These unique new applications encourage you to develop your decision-making and problem-solving abilities, and develop life skills, primarily using workplace or career-related situations.

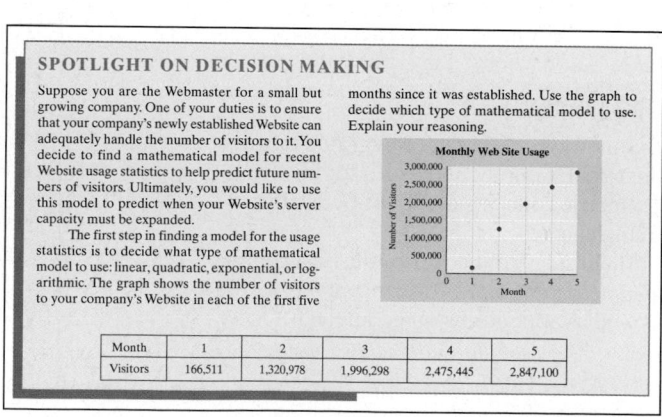

page 707

XXVII

Become a Confident Problem Solver!

A goal of this text is to help you develop problem-solving abilities.

Example 1 **FINDING TIME GIVEN RATE AND DISTANCE**

A glacier is a giant mass of rocks and ice that flows downhill like a river. Portage Glacier in Alaska is about 6 miles, or 31,680 *feet*, long and moves 400 *feet* per year. Icebergs are created when the front end of the glacier flows into Portage Lake. How long does it take for ice at the head (beginning) of the glacier to reach the lake?

Solution

1. UNDERSTAND. Read and reread the problem. The appropriate formula needed to solve this problem is the distance formula, $d = rt$. To become familiar with this formula, let's find the distance that ice traveling at a rate of 400 feet per year travels in 100 years. To do so, we let time t be 100 years and rate r be the given 400 feet per year, and substitute these values into the formula $d = rt$. We then have that distance $d = 400(100) = 40,000$ feet. Since we are interested in finding how long it takes ice to travel 31,680 feet, we now know that it is less than 100 years.

   Since we are using the formula $d = rt$, we let

   t = the time in years for ice to reach the lake
   r = rate or speed of ice
   d = distance from beginning of glacier to lake

2. TRANSLATE. To translate to an equation, we use the formula $d = rt$ and let distance $d = 31,680$ feet and rate $r = 400$ feet per year.

$$d = r \cdot t$$
$$31,680 = 400 \cdot t \quad \text{Let } d = 31,680 \text{ and } r = 400.$$

3. SOLVE. Solve the equation for t. To solve for t, divide both sides by 400.

$$\frac{31,680}{400} = \frac{400 \cdot t}{400} \quad \text{Divide both sides by 400.}$$
$$79.2 = t \quad \text{Simplify.}$$

> **HELPFUL HINT**
> Don't forget to include units, if appropriate.

4. INTERPRET.

   Check: To check, substitute 79.2 for t and 400 for r in the distance formula and check to see that the distance is 31,680 feet.

   State: It takes 79.2 years for the ice at the head of Portage Glacier to reach the lake.

page 114

◀ **GENERAL STRATEGY FOR PROBLEM SOLVING**

Save time by having a plan. This text's organization can help you. Note the outlined problem-solving steps, *Understand, Translate, Solve,* and *Interpret.*

Problem solving is introduced early and emphasized and integrated throughout the book. The author provides patient explanations and illustrates how to apply the problem-solving procedure to the in-text examples.

60. Stalactites join stalagmites to form columns. A column found at Natural Bridge Caverns near San Antonio, Texas, rises 15 feet and has a *diameter* of only 2 inches. Find the volume of this column in cubic inches. (*Hint:* Use the formula for volume of a cylinder and use a calculator approximation for π.) Round to the nearest tenth of a cubic inch.

page 121

GEOMETRY ▶

Geometric concepts are integrated throughout the text. Examples and exercises involving geometric concepts are identified with a triangle icon.

The inside front cover of this text contains *Geometric Formulas* for convenient reference, as well as appendices on geometry.

Get Involved!

Real-world applications in this textbook will help to reinforce your problem-solving skills and show you how algebra is connected to a wide range of fields like consumer affairs, sports, and business. You will be asked to evaluate and interpret real data in graphs, tables, and in context to solve applications. See also the Index of Applications located on page xxiv for a quick way to locate those in your areas of interest.

---

90. In 1996, the number of U.S. paging subscribers (in millions) was 42. The number of subscribers in 1999 (in millions) was 58. Let y be the number of subscribers (in millions) in the year x, where $x = 0$ represents 1996. (*Source:* Strategis Group for Personal Communications Asso.)

a. Write a linear equation that models the number of U.S. paging subscribers (in millions) in terms of the year x. [*Hint:* Write 2 ordered pairs of the form (years past 1996, number of subscribers).]

b. Use this equation to predict the number of U.S. paging subscribers in the year 2007. (Round to the nearest million.)

page 235

INTERESTING, RELEVANT, AND PRACTICAL REAL-WORLD APPLICATIONS

Accessible applications reinforce concepts needed for success in this course and relate those concepts to everyday life.

INTERESTING REAL DATA ▶

Real-world applications include those based on real data.

23. One of Japan's superconducting "bullet" trains is researched and tested at the Yamanashi Maglev Test Line near Otsuki City. The steepest section of the track has a rise of 2580 for a horizontal distance of 6450 meters. What is the grade for this section of track? (*Source:* Japan Railways Central Co.)

2580 meters

6450 meters

page 206

Example 2 The three busiest airports in the United States are in the cities of Chicago, Atlanta, and Dallas/Ft. Worth. The airport in Atlanta has 7.7 million more arrivals and departures than the Dallas/Ft. Worth airport. The Chicago airport has 9.8 million more arrivals and departures than the Dallas/Ft. Worth airport. Write the sum of the arrivals and departures from these three cities as a simplified algebraic expression. Let x be the number of arrivals and departures at the Dallas/Ft. Worth airport.

...rrivals and departures at the Dallas/Ft. Worth airport, then ...of arrivals and departures at the Atlanta airport and ...of arrivals and departures at the Chicago airport ...sum, we have

page 485

◀ CHAPTER PROJECT

New *Chapter Projects* for individuals or groups provide chapter wrap-up and extend the chapter's concepts in a multi-part application.

In addition, references to *Real World Activities* are given. This **internet option** invites you to find and retrieve real data for use in solving problems.

11

For additional Chapter Projects, visit the Real World Activities Website by going to http://www.prenhall.com/martin-gay.

CHAPTER PROJECT

Modeling Temperature

When a cold object is placed in a warm room, the object's temperature gradually rises until it becomes, or nearly becomes, room temperature. Similarly, if a hot object is placed in a cooler room, the object's temperature gradually falls to room temperature. The way in which a cold or hot object warms up or cools off is modeled by an exact mathematical relationship, known as Newton's law of cooling. This law relates the temperature of an object to the time elapsed since its warming or cooling began. In this project, you will have the opportunity to investigate this model of cooling and warming. This project may be completed by working in groups or individually.

To investigate Newton's law of cooling in this project, you will collect experimental data in one of two methods: Method 1, using a stopwatch and thermometer, or Method 2, using Texas Instruments' Calculator-Based Laboratory (CBL™) or Second Generation Calculator-Based Laboratory (CBL 2™).

Method 1 Materials
- Container of either cold or hot liquid
- Thermometer
- Stopwatch
- Graphing calculator with regression capabilities

Method 2 Materials
- Container of either cold or hot liquid
- A TI-82, TI-83, or TI-85 graphing calculator with unit-to-unit link cable
- CBL™ or CBL 2™ unit with temperature probe

Steps for Collecting Data with Method 1:

a. Insert the thermometer into the liquid and allow a thermometer reading to register. Take a temperature reading T as you start the stopwatch (at $t = 0$) and record it in the accompanying data table.

b. Continue taking temperature readings at uniform intervals anywhere between 5 and 10 minutes long. At each reading use the stopwatch to measure the length of time that has elapsed since the temperature readings started with your first reading at $t = 0$. Record your time t and liquid temperature T in the data table. Gather data for six to twelve readings.

c. Plot the data from the data table. Plot t on the horizontal axis and T on the vertical axis.

Steps for Collecting Data with Method 2:

a. Enter the HEAT program appropriate for your calculator.

b. Prepare the CBL or CBL 2 and the graphing calculator. Insert the temperature probe into the liquid.

c. Start the HEAT program on the graphing calculator and follow its instructions to begin collecting data. The program will collect 36 temperature readings in degrees Celsius and plot them in real time with t on the horizontal axis and T on the vertical axis.

1. Which of the following mathematical models best fits the data you collected? Explain your reasoning. (Assume $a > 0$.)

page 727

Be Confident!

Several features of this text can be helpful in building your confidence and mathematical competence. As you study, also notice the connections the author makes to relate new material to ideas that you may already know.

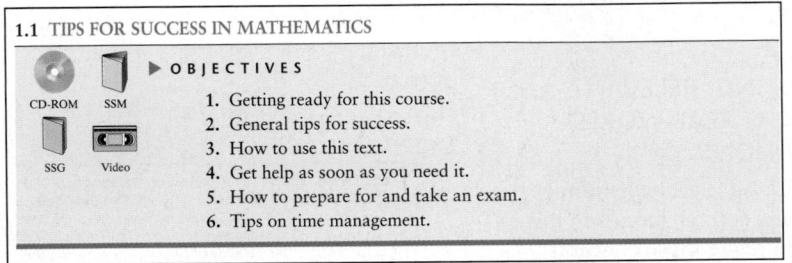

page 4

◄ **TIPS FOR SUCCESS**

New coverage of study skills in Section 1.1 reinforces this important component to success in this course.

MENTAL MATH ►

Mental Math warm-up exercises reinforce concepts found in the accompanying section and can increase your confidence before beginning an exercise set.

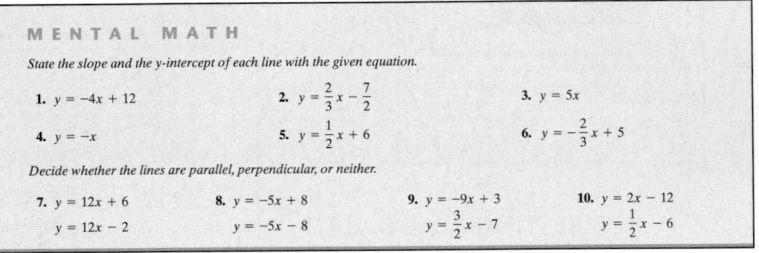

page 217

HELPFUL HINT
Note that $f(x)$ is a special symbol in mathematics used to denote a function. The symbol $f(x)$ is read "f of x." It does *not* mean $f \cdot x$ (f times x).

page 180

◄ **HELPFUL HINTS**

Found throughout the text, these contain practical advice on applying mathematical concepts. They are strategically placed where you are most likely to need immediate reinforcement.

CHAPTER 3 VOCABULARY CHECK

Fill in each blank with one of the words or phrases listed below.

relation	line	function	standard	slope	domain
slope–intercept	x	y	range	parallel	linear function
point–slope	perpendicular	linear inequality			

1. A _____ is a set of ordered pairs.
2. The graph of every linear equation in two variables is a _____.
3. The statement $-x + 2y > 0$ is called a _____ in two variables.
4. _____ form of linear equation in two variables is $Ax + By = C$.
5. The _____ of a relation is the set of all second components of the ordered pairs of the relation.
6. _____ lines have the same slope and different y-intercepts.
7. _____ form of a linear equation in two variables is $y = mx + b$.
8. A _____ is a relation in which each first component in the ordered pairs corresponds to exactly one second component.
9. In the equation $y = 4x - 2$, the coefficient of x is the _____ of its corresponding graph.
10. Two lines are _____ if the product of their slopes is −1.
11. To find the x-intercept of a linear equation, let _____ = 0 and solve for the other variable.
12. The _____ of a relation is the set of all first components of the ordered pairs of the relation.
13. A _____ is a function that can be written in the form $f(x) = mx + b$.
14. To find the y-intercept of a linear equation, let _____ = 0 and solve for the other variable.
15. The equation $y - 8 = -5(x + 1)$ is written in _____ form.

VOCABULARY CHECKS ►

New *Vocabulary Checks* allow you to write your answers to questions about chapter content and strengthen verbal skills.

Visualize It!

The Second Edition increases emphasis on visualization. Graphing is introduced early and intuitively. Knowing how to read and use graphs is a valuable skill in the workplace as well as in this and other courses.

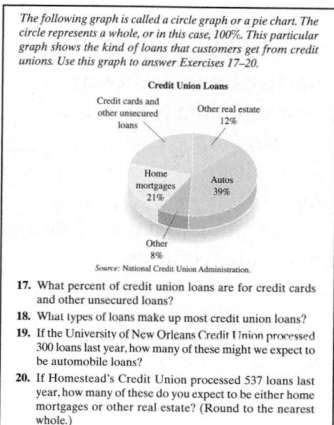

The following graph is called a circle graph or a pie chart. The circle represents a whole, or in this case, 100%. This particular graph shows the kind of loans that customers get from credit unions. Use this graph to answer Exercises 17–20.

Credit Union Loans

Credit cards and other unsecured loans
Other real estate 12%
Home mortgages 21%
Autos 39%
Other 8%

Source: National Credit Union Administration.

17. What percent of credit union loans are for credit cards and other unsecured loans?
18. What types of loans make up most credit union loans?
19. If the University of New Orleans Credit Union processed 300 loans last year, how many of these might we expect to be automobile loans?
20. If Homestead's Credit Union processed 537 loans last year, how many of these do you expect to be either home mortgages or other real estate? (Round to the nearest whole.)

page 492

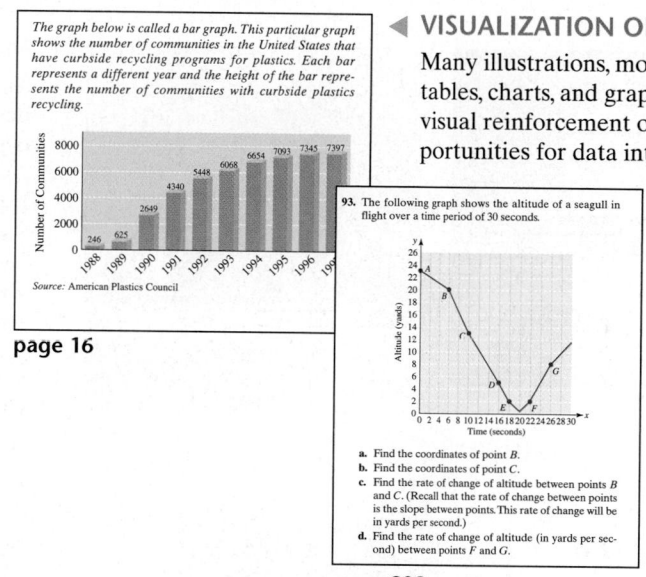

The graph below is called a bar graph. This particular graph shows the number of communities in the United States that have curbside recycling programs for plastics. Each bar represents a different year and the height of the bar represents the number of communities with curbside plastics recycling.

Number of Communities: 246, 625, 2649, 4340, 5448, 6068, 6654, 7093, 7345, 7397
(years 1988–199_)

Source: American Plastics Council

page 16

◄ VISUALIZATION OF TOPICS

Many illustrations, models, photographs, tables, charts, and graphs provide visual reinforcement of concepts and opportunities for data interpretation.

93. The following graph shows the altitude of a seagull in flight over a time period of 30 seconds.

(graph with points A, B, C, D, E, F, G; Altitude (yards) vs Time (seconds))

a. Find the coordinates of point *B*.
b. Find the coordinates of point *C*.
c. Find the rate of change of altitude between points *B* and *C*. (Recall that the rate of change between points is the slope between points. This rate of change will be in yards per second.)
d. Find the rate of change of altitude (in yards per second) between points *F* and *G*.

page 209

SCIENTIFIC AND GRAPHING CALCULATOR EXPLORATIONS ►

Enhanced *Explorations* contain examples and exercises to **reinforce** concepts, help **interpret** graphs, or **motivate** discovery learning. Scientific and graphing utility exercises can also be found in exercise sets.

▼

81. The number of people employed in the United States as medical assistants was 225 thousand in 1996. By the year 2006, this number is expected to rise to 391 thousand. Let *y* be the number of medical assistants (in thousands) employed in the United States in the year *x*, where *x* = 0 represents 1996. (*Source:* Bureau of Labor Statistics)
a. Write a linear equation that models the number of people (in thousands) employed as medical assistants in the year *x*. (See hint for Exercise 79. a.)
b. Use this equation to estimate the number of people who will be employed as medical assistants in the year 2004.

page 219

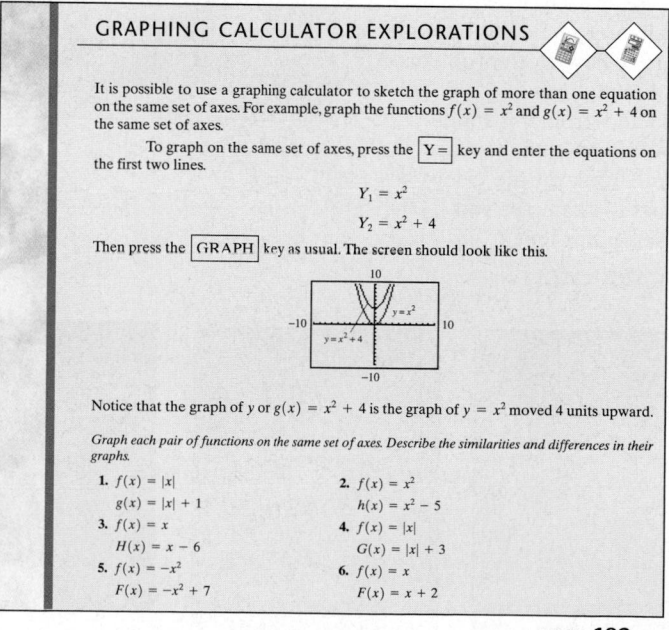

GRAPHING CALCULATOR EXPLORATIONS

It is possible to use a graphing calculator to sketch the graph of more than one equation on the same set of axes. For example, graph the functions $f(x) = x^2$ and $g(x) = x^2 + 4$ on the same set of axes.

To graph on the same set of axes, press the $\boxed{Y=}$ key and enter the equations on the first two lines.

$$Y_1 = x^2$$
$$Y_2 = x^2 + 4$$

Then press the $\boxed{GRAPH}$ key as usual. The screen should look like this.

(graphing calculator screen showing $y = x^2$ and $y = x^2 + 4$, window −10 to 10)

Notice that the graph of *y* or $g(x) = x^2 + 4$ is the graph of $y = x^2$ moved 4 units upward.

Graph each pair of functions on the same set of axes. Describe the similarities and differences in their graphs.

1. $f(x) = |x|$
$g(x) = |x| + 1$
2. $f(x) = x^2$
$h(x) = x^2 - 5$
3. $f(x) = x$
$H(x) = x - 6$
4. $f(x) = |x|$
$G(x) = |x| + 3$
5. $f(x) = -x^2$
$F(x) = -x^2 + 7$
6. $f(x) = x$
$F(x) = x + 2$

page 182

Discover the Best Supplemental Resource for You!
Integrated Learning Program

All of the components of the Martin-Gay supplemental resources fit together to help you learn and understand algebra. Use these student resources based on your personal learning style to enhance what you learn from your instructor and textbook.

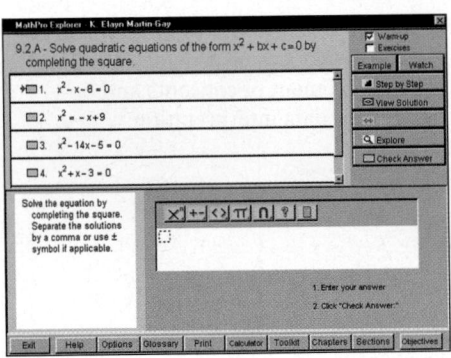 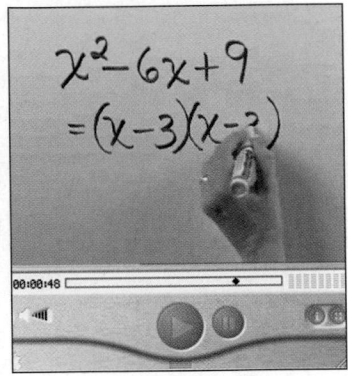

◄ **MATHPRO EXPLORER 4.0**

This **interactive** tutorial software is developed around the content and concepts of *Beginning and Intermediate Algebra*.
It provides:
- virtually unlimited practice problems with immediate feedback
- video clips
- step-by-step solutions
- exploratory activities
- on-line help
- summary of progress

Available on CD-ROM

LECTURE VIDEO SERIES BY K. ELAYN MARTIN-GAY ▶

Hosted by the award-winning teacher and author of *Beginning and Intermediate Algebra*, these videos cover each objective in every chapter section as a supplementary review. Both problems taken directly from the text and problems similar to those found in the text are included. Look for the video icon next to examples and exercises in this text.

WWW.PRENHALL.COM/MARTIN-GAY COMPANION WEBSITE!

The website offers warm-ups, real world activities, reading quizzes, vocabulary quizzes, and chapter quizzes, that you can e-mail to your instructor, and links to explore related sites such as those noted in this text's chapter-opening pages.

ALSO AVAILABLE

The New York Times/ Themes of the Times
Newspaper-format supplement

BEGINNING AND INTERMEDIATE ALGEBRA

A Demanded Technical Position

Computers are everywhere: banks, real estate offices, churches, post offices, factories, corporations, grocery stores, libraries, homes, and schools. With so many computers in use in today's society, it has become necessary to network, or connect together, computers to share resources such as printers or file storage. A popular computer network for linking computers in the same building is the Local Area Network (LAN). According to the Unicom Institute of Technology, LAN administrators are among the technical workers most in demand in today's job market.

LAN administrators are generally responsible for designing and configuring the network, adding printers or other devices to the network as necessary, installing networked application software, maintaining system files, ensuring network security, and troubleshooting problems on the network. LAN administrators enjoy solving problems, communicate effectively, and read and understand technical material.

 For more information about the field of computer networking, or a number of other information technology (IT) professions, visit the Planet IT Website by going to www.prenhall.com/martin-gay.

In the Spotlight on Decision Making feature on page 32, you will have the opportunity to make a decision concerning the feasibility of a proposed computer network as a LAN administrator.

REVIEW OF REAL NUMBERS

1

The power of mathematics is its flexibility. We apply numbers to almost every aspect of our lives, from an ordinary trip to the grocery store to a rocket launched into space. The power of algebra is its generality. Using letters to represent numbers, we tie together the trip to the grocery store and the launched rocket.

In this chapter we review the basic symbols and words—the language—of arithmetic and introduce using variables in place of numbers. This is our starting place in the study of algebra.

1.1 TIPS FOR SUCCESS IN MATHEMATICS

CD-ROM   SSM

SSG   Video

▶ **OBJECTIVES**

1. Getting ready for this course.
2. General tips for success.
3. How to use this text.
4. Get help as soon as you need it.
5. How to prepare for and take an exam.
6. Tips on time management.

Before reading this section, remember that your instructor is your best source for information. Please see your instructor for any additional help or information.

1   Now that you have decided to take this course, remember that a positive attitude will make all the difference in the world. Your belief that you can succeed is just as important as your commitment to this course. Make sure that you are ready for this course by having the time and positive attitude that it takes to succeed.

Next, make sure that you have scheduled your math course at a time that will give you the best chance for success. For example, if you are also working, you may want to check with your employer to make sure that your work hours will not conflict with your course schedule.

Now you are ready for your first class period. Double-check your schedule and allow yourself extra time to arrive in case of traffic or in case you have trouble locating your classroom. Make sure that you bring at least your textbook, paper, and a writing instrument with you. Are you required to have a lab manual, graph paper, calculator, or some other supplies besides this text? If so, bring this material with you also.

2   Below are some general tips that will increase your chance for success in a mathematics class. Many of these tips will also help you in other courses you may be registered for.

Exchange names and phone numbers with at least one other person in class. This contact person can be a great help in case you miss the class assignment or want to discuss math concepts or exercises that you find difficult.

Choose to attend all class periods. If possible, sit near the front of the classroom. This way, you will see and hear the presentation better. It may also be easier for you to participate in classroom activities.

Do your homework. You've probably heard the phrase "practice makes perfect" in relation to music and sports. It also applies to mathematics. You will find that the more time you spend solving mathematics problems, the easier the process becomes. Be sure to block enough time to complete your assignments before the next class period.

Check your work. Review the steps you made while working a problem. Learn to check your answers in the original problems. You may also compare your answers to the answers to selected exercises listed in the back of the book. If you have made a mistake, figure out what went wrong. Then correct your mistake. If you can't find your mistake, don't erase your work or throw it away. Bring your work to your instructor, a tutor in a math lab, or a classmate. Someone can help you find where you had trouble if they have your work to look at.

Learn from your mistakes. Everyone, even your instructor, makes mistakes. (That definitely includes me—Elayn Martin-Gay. You usually don't see my mistakes because many other people doublecheck my work in this text. If I make a mistake on a videotape, it is edited out so that you are not confused by it.) Use your mistakes to learn and to become a better math student. The key is finding and understanding your mistakes. Was your mistake a careless mistake or did you make it because you can't read your own "math" writing? If so, try to work more slowly or write more neatly and make a conscious effort to carefully check your work. Did you make a mistake because you don't understand a concept? Take the time to review the concept or ask questions to better understand the concept.

Know how to get help if you need it. It's OK to ask for help. In fact, it's a good idea to ask for help whenever there is something that you don't understand. Make sure you know when your instructor has office hours and how to find his or her office. Find out if math tutoring services are available on your campus. Check out the hours, location, and requirements of the tutoring service. Know whether videotapes or software are available and how to access those resources.

Organize your class materials, including homework assignments, graded quizzes and tests, and notes from your class or lab. All of these items will make valuable references throughout your course and as you study for upcoming tests and your final exam. Make sure that you can locate any of these materials when you need them.

Read your textbook before class. Reading a mathematics textbook is unlike entertainment reading such as reading a newspaper. Your pace will be much slower. It is helpful to have a pencil and paper with you when you read. Try to work out examples on your own as you encounter them in your text. You may also write down any questions that you want to ask in class. I know that when you read a mathematics textbook, sometimes some of the information in a section will still be unclear. But once you hear a lecture or watch a video on that section, you will understand it much more easily than if you had not read your text.

Don't be afraid to ask questions. From experience, I can tell you that you are not the only person in class with questions. Other students are normally grateful that someone has spoken up.

Hand in assignments on time. This way you can be sure that you will not lose points needlessly for being late. Show every step of a solution and be neat and organized. Also be sure that you understand which problems are assigned for homework. You can always doublecheck this assignment with another student in your class.

3 There are many helpful resources that are available to you in this text. It is important that you become familiar with and use these resources. This should increase your chances for success in this course.

For example:

- If you need help in a particular section, check at the beginning of the section to see what videotapes or software are available. These resources are usually available to you in a tutorial lab, resource center, or library.
- Many of the exercises in this text are referenced by an example(s). Use this referencing in case you have trouble completing an assignment from the exercise set.

- Make sure that you understand the meaning of the icons that are beside many exercises. The video icon ✎ tells you that the corresponding exercise may be viewed on the videotape that corresponds with that section. The pencil icon ✏ tells you that this exercise is a writing exercise in which you should answer in complete sentences. The calculator icons are placed by exercises that can be worked more efficiently with the use of a scientific calculator ▦ or a graphing calculator. ▦ The geometry icon △ indicates you are working with a geometry problem.

- There are many opportunities at the end of each chapter to help you understand the concepts of the chapter.

  Vocabulary Checks provide a vocabulary self-check to make sure that you know the vocabulary in that chapter.

  Highlights contain chapter summaries with examples.

  Chapter Review contains additional exercises that are keyed to sections of the chapter.

  Chapter Test is a sample test to help you prepare for an exam.

  Cumulative Review is a review consisting of material from the beginning of the book to the end of the particular chapter.

4 If you have trouble completing assignments or understanding the mathematics, get help as soon as you need it! This tip is presented as an objective on its own because it is *so* important. In mathematics, usually the material presented in one section builds on your understanding of the previous section. What does this mean? It means that if you don't understand the concepts covered during a class period, there is a good chance that you will not understand the concepts covered during the next class period. If this happens to you, get help as soon as you can.

Where can you get help? Many suggestions have been made in this section on where to get help and now it is up to you to do it. Try your instructor, a tutor center, or math lab, or you may want to form a study group with fellow classmates. If you do decide to see your instructor or go to a tutor center, make sure that you have a neat notebook and be ready with your questions.

5 Make sure that you allow yourself plenty of time to prepare for a test. If you think that you are a little math anxious, it may be that you are not preparing for a test in a way that will ensure success. The way that you prepare for a test in mathematics is important.

To prepare for a test,

1. Review your previous homework assignments.
2. Review any notes from class and section level quizzes you may have taken. (If this is a final exam, review chapter tests you have taken also.)
3. Review concepts and definitions by reading the Highlights at the end of each chapter.
4. Practice working exercises by completing the Chapter Review found at the end of each chapter. (If this is a final exam, work a Cumulative Review. There is one found at the end of each chapter (except Chapter 1). Choose the review found at the end of the latest chapter that you have covered in your course.)

Don't stop here!

5. It is important that you place yourself in conditions similar to test conditions to see how you will perform. In other words, once you feel that you know the material, get out a few blank sheets of paper and take a sample test. There is a Chapter Test available at the end of each chapter, or you can work selected

problems from the Chapter Review, or your instructor may provide you with a review sheet. During this sample test, do not use your notes or your textbook. Then check your sample test. If you are not satisfied with the results, study the areas that you are weak in and try again.

6. On the day of the test, allow yourself plenty of time to arrive where you will be taking your exam.

When taking your test,

1. Read the directions on the test carefully.
2. Read each problem carefully as you take your test. Make sure that you answer the question asked.
3. Watch your time and pace yourself so that you may attempt each problem on your test.
4. If you have time, check your work and answers.
5. Do not turn your test in early. If you have extra time, spend it double-checking your work.

6 As a college student, you know the demands that classes, homework, work, and family place on your time. Some days you probably wonder how you'll ever get everything done. One key to managing your time is developing a schedule. Here are some hints for making a schedule:

1. Make a list of all of your weekly commitments for the term. Include classes, work, regular meetings, extracurricular activities, etc. You may also find it helpful to list such things as doing laundry, regular workouts, grocery shopping, etc.
2. Next, estimate the time needed for each item on the list. Also make a note of how often you will need to do each item. Don't forget to include time estimates for reading, studying, and homework you do outside of your classes. You may want to ask your instructor for help estimating the time needed for this item.
3. In the exercise set below, you are asked to block out a typical week on the schedule grid given. Start with items with fixed time slots, like classes and work.
4. Next, include the items on your list with flexible time slots. Think carefully about how best to schedule some items such as study time.
5. Don't fill up every time slot on the schedule. Remember that you need to allow time for eating, sleeping, and relaxing! You should also allow a little extra time in case things take longer than planned.
6. If you find that your weekly schedule is too full for you to handle, you may need to make some changes in your workload, class load, or in other areas of your life. You may want to talk to your advisor, manager or supervisor at work, or someone in your college's academic counseling center for help with such decisions.

Exercise Set 1.1

1. What is your instructor's name?
2. What are your instructor's office location and office hours?
3. What is the best way to contact your instructor?
4. What does this icon ↘ mean?
5. What does this icon ◈ mean?

6. Do you have the name and contact information of at least one other student in class?
7. Will your instructor allow you to use a calculator in this class?
8. Are videotapes and/or tutorial software available to you?

9. Is there a tutoring service available? If so, what are its hours?

10. Have you attempted this course before? If so write down ways that you may improve your chances of success during this attempt.

11. List some steps that you may take in case you begin having trouble understanding the material or completing an assignment.

12. Read or reread objective ⁶ and fill out the schedule grid below.

	Monday	Tuesday	Wednesday	Thursday	Friday	Saturday	Sunday
7:00 A.M.							
8:00 A.M.							
9:00 A.M.							
10:00 A.M.							
11:00 A.M.							
12:00 P.M.							
1:00 P.M.							
2:00 P.M.							
3:00 P.M.							
4:00 P.M.							
5:00 P.M.							
6:00 P.M.							
7:00 P.M.							
8:00 P.M.							
9:00 P.M.							

1.2 SYMBOLS AND SETS OF NUMBERS

CD-ROM   SSM

SSG   Video

▶ OBJECTIVES

1. Use the number line to order numbers.
2. Translate sentences into mathematical statements.
3. Identify natural numbers, whole numbers, integers, rational numbers, irrational numbers, and real numbers.
4. Find the absolute value of a real number.

1

We begin with a review of the set of natural numbers and the set of whole numbers and how we use symbols to compare these numbers. A **set** is a collection of objects, each of which is called a **member** or **element** of the set. A pair of brace symbols { } encloses the list of elements and is translated as "the set of" or "the set containing."

NATURAL NUMBERS

The set of **natural numbers** is $\{1, 2, 3, 4, 5, 6, \ldots\}$.

WHOLE NUMBERS

The set of **whole numbers** is $\{0, 1, 2, 3, 4, \dots\}$.

The three dots (an ellipsis) at the end of the list of elements of a set means that the list continues in the same manner indefinitely.

These numbers can be pictured on a **number line**. We will use the number line often to help us visualize distance and relationships between numbers. Visualizing mathematical concepts is an important skill and tool, and later we will develop and explore other visualizing tools.

To draw a number line, first draw a line. Choose a point on the line and label it 0. To the right of 0, label any other point 1. Being careful to use the same distance as from 0 to 1, mark off equally spaced distances. Label these points 2, 3, 4, 5, and so on. Since the whole numbers continue indefinitely, it is not possible to show every whole number on the number line. The arrow at the right end of the line indicates that the pattern continues indefinitely.

Picturing whole numbers on a number line helps us to see the order of the numbers. Symbols can be used to describe concisely in writing the order that we see.

The **equal symbol** $=$ means "is equal to."
The symbol $\neq$ means "is not equal to."

These symbols may be used to form a **mathematical statement**. The statement might be true or it might be false. The two statements below are both true.

$2 = 2$ states that "two is equal to two"
$2 \neq 6$ states that "two is not equal to six"

If two numbers are not equal, then one number is larger than the other. The symbol $>$ means "is greater than." The symbol $<$ means "is less than." For example,

$2 > 0$ states that "two is greater than zero"
$3 < 5$ states that "three is less than five"

On the number line, we see that a number **to the right of** another number is **larger**. Similarly, a number **to the left of** another number is smaller. For example, 3 is to the left of 5 on the number line, which means that 3 is less than 5, or $3 < 5$. Similarly, 2 is to the right of 0 on the number line, which means 2 is greater than 0, or $2 > 0$. Since 0 is to the left of 2, we can also say that 0 is less than 2, or $0 < 2$.

The symbols $\neq$, $<$, and $>$ are called **inequality symbols**.

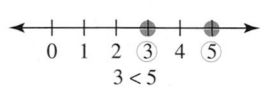

$2 > 0$ or $0 < 2$

$3 < 5$

> **HELPFUL HINT**
> Notice that $2 > 0$ has exactly the same meaning as $0 < 2$. Switching the order of the numbers and reversing the "direction of the inequality symbol" does not change the meaning of the statement.
>
> $5 > 3$ has the same meaning as $3 < 5$.
>
> Also notice that, when the statement is true, the inequality arrow points to the smaller number.

Example 1   Insert $<$, $>$, or $=$ in the space between the paired numbers to make each statement true.

     a. $2 \angle 3$ **b.** $7 \big\rangle 4$ **c.** $72 \big\rangle 27$

Solution   **a.** $2 < 3$ since 2 is to the left of 3 on the number line.
         b. $7 > 4$ since 7 is to the right of 4 on the number line.
         c. $72 > 27$ since 72 is to the right of 27 on the number line.

Two other symbols are used to compare numbers. The symbol $\leq$ means "is less than or equal to." The symbol $\geq$ means "is greater than or equal to." For example,

$$7 \leq 10 \text{ states that "seven is less than or equal to ten"}$$

This statement is true since $7 < 10$ is true. If either $7 < 10$ or $7 = 10$ is true, then $7 \leq 10$ is true.

$$3 \geq 3 \text{ states that "three is greater than or equal to three"}$$

This statement is true since $3 = 3$ is true. If either $3 > 3$ or $3 = 3$ is true, then $3 \geq 3$ is true.

The statement $6 \geq 10$ is false since neither $6 > 10$ nor $6 = 10$ is true.

The symbols $\leq$ and $\geq$ are also called **inequality symbols**.

Example 2   Tell whether each statement is true or false.

     a. $8 \geq 8$ **b.** $8 \leq 8$ **c.** $23 \leq 0$ **d.** $23 \geq 0$

Solution   **a.** True, since $8 = 8$ is true.   **b.** True, since $8 = 8$ is true.
         c. False, since neither $23 < 0$ nor $23 = 0$ is true.   **d.** True, since $23 > 0$ is true.

2   Now, let's use the symbols discussed above to translate sentences into mathematical statements.

Example 3   Translate each sentence into a mathematical statement.

     a. Nine is less than or equal to eleven.
     b. Eight is greater than one.
     c. Three is not equal to four.

Solution

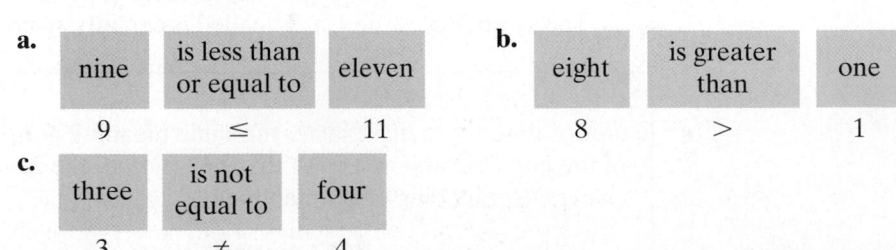

3   Whole numbers are not sufficient to describe many situations in the real world. For example, quantities smaller than zero must sometimes be represented, such as temperatures less than 0 degrees.

We can picture numbers less than zero on the number line as follows:

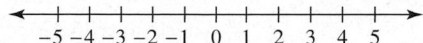

Numbers less than 0 are to the left of 0 and are labeled −1, −2, −3, and so on. A − sign, such as the one in −1, tells us that the number is to the left of 0 on the number line. In words, −1 is read "negative one." A + sign or no sign tells us that a number lies to the right of 0 on the number line. For example, 3 and +3 both mean positive three.

The numbers we have pictured are called the set of **integers**. Integers to the left of 0 are called **negative integers**; integers to the right of 0 are called **positive integers**. The integer 0 is neither positive nor negative.

INTEGERS

The set of **integers** is $\{\dots, -3, -2, -1, 0, 1, 2, 3, \dots\}$.

Notice the ellipses (three dots) to the left and to the right of the list for the integers. This indicates that the positive integers and the negative integers continue indefinitely.

Example 4 Use an integer to express the number in the following. "Pole of Inaccessibility, Antarctica, is the coldest location in the world, with an average annual temperature of 72 degrees below zero." (*Source: The Guinness Book of Records*)

Solution The integer −72 represents 72 degrees below zero.

A problem with integers in real-life settings arises when quantities are smaller than some integer but greater than the next smallest integer. On the number line, these quantities may be visualized by points between integers. Some of these quantities between integers can be represented as a quotient of integers. For example,

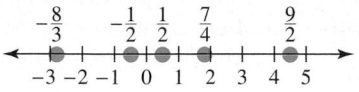

The point on the number line halfway between 0 and 1 can be represented by $\frac{1}{2}$, a quotient of integers.

The point on the number line halfway between 0 and −1 can be represented by $-\frac{1}{2}$. Other quotients of integers and their graphs are shown.

The set numbers, each of which can be represented as a quotient of integers, is called the set of **rational numbers**. Notice that every integer is also a rational number since each integer can be expressed as a quotient of integers. For example, the integer 5 is also a rational number since $5 = \frac{5}{1}$.

RATIONAL NUMBERS

The set of **rational numbers** is the set of all numbers that can be expressed as a quotient of integers.

The number line also contains points that cannot be expressed as quotients of integers. These numbers are called **irrational numbers** because they cannot be represented by rational numbers. For example, $\sqrt{2}$ and π are irrational numbers.

IRRATIONAL NUMBERS

The set of **irrational numbers** is the set of all numbers that correspond to points on the number line but that are not rational numbers. That is, an irrational number is a number that cannot be expressed as a quotient of integers.

Rational numbers and irrational numbers can be written as decimal numbers. The decimal equivalent of a rational number will either terminate or repeat in a pattern. For example, upon dividing we find that

$$\frac{3}{4} = 0.75 \text{ (decimal number terminates or ends) and}$$

$$\frac{2}{3} = 0.66666\dots \text{ (decimal number repeats in a pattern)}$$

The decimal representation of an irrational number will neither terminate nor repeat. (For further review of decimals, see the Appendix.)

The set of numbers, each of which corresponds to a point on the number line, is called the set of **real numbers**. One and only one point on the number line corresponds to each real number.

REAL NUMBERS

The set of **real numbers** is the set of all numbers each of which corresponds to a point on the number line.

On the following number line, we see that real numbers can be positive, negative, or 0. Numbers to the left of 0 are called **negative numbers**; numbers to the right of 0 are called **positive numbers**. Positive and negative numbers are also called **signed numbers**.

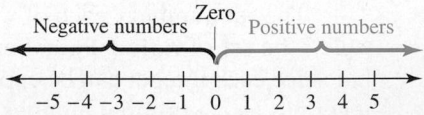

Several different sets of numbers have been discussed in this section. The following diagram shows the relationships among these sets of real numbers.

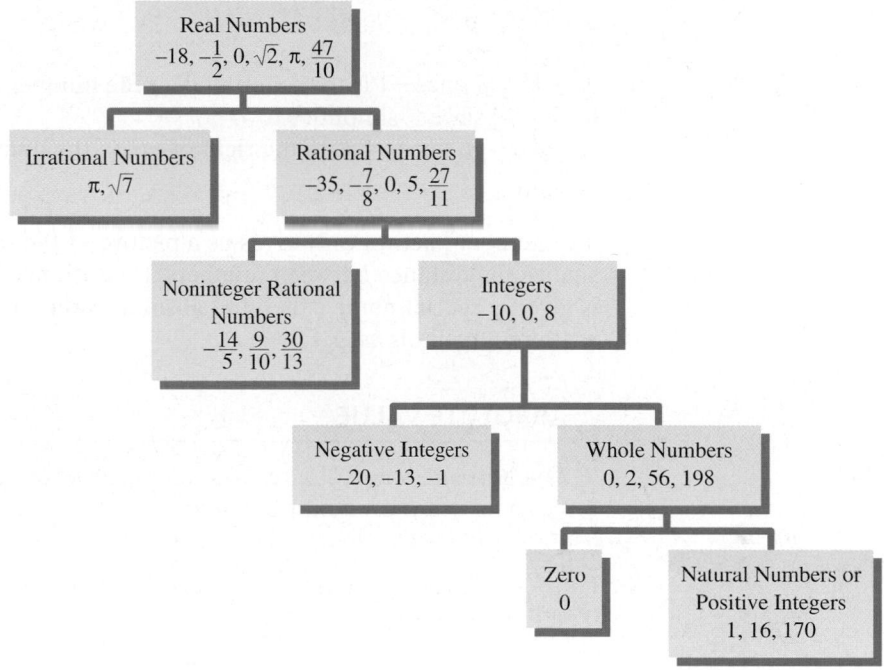

Common Sets of Numbers

Example 5   Given the set $\left\{-2, 0, \frac{1}{4}, 112, -3, 11, \sqrt{2}\right\}$, list the numbers in this set that belong to the set of:

   a. Natural numbers   **b.** Whole numbers   **c.** Integers
   d. Rational numbers   **e.** Irrational numbers   **f.** Real numbers

Solution
   a. The natural numbers are 11 and 112.
   b. The whole numbers are 0, 11, and 112.
   c. The integers are $-3, -2, 0, 11$, and 112.
   d. Recall that integers are rational numbers also. The rational numbers are $-3, -2$, $0, \frac{1}{4}, 11$, and 112.
   e. The irrational number is $\sqrt{2}$.
   f. The real numbers are all numbers in the given set.

We can now extend the meaning and use of inequality symbols such as $<$ and $>$ to apply to all real numbers.

ORDER PROPERTY FOR REAL NUMBERS

Given any two real numbers a and b, $a < b$ if a is to the left of b on the number line. Similarly, $a > b$ if a is to the right of b on the number line.

Example 6 Insert $<, >,$ or $=$ in the appropriate space to make the statement true.

    a. -1 0 **b.** 7 $\frac{14}{2}$ **c.** -5 -6

Solution   **a.** $-1 < 0$ since -1 is to the left of 0 on the number line.
    b. $7 = \frac{14}{2}$ since $\frac{14}{2}$ simplifies to 7.
    c. $-5 > -6$ since -5 is to the right of -6 on the number line.

4   The number line not only gives us a picture of the real numbers, it also helps us visualize the distance between numbers. The distance between a real number a and 0 is given a special name called the **absolute value** of a. "The absolute value of a" is written in symbols as $|a|$.

ABSOLUTE VALUE

The absolute value of a real number a, denoted by $|a|$, is the distance between a and 0 on a number line.

For example, $|3| = 3$ and $|-3| = 3$ since both 3 and -3 are a distance of 3 units from 0 on the number line.

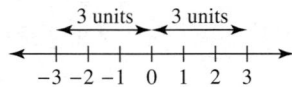

> **HELPFUL HINT**
> Since $|a|$ is a distance, $|a|$ is always either positive or 0, never negative. That is, **for any real number a, $|a| \geq 0$.**

Example 7 Find the absolute value of each number.

    a. $|4|$ **b.** $|-5|$ **c.** $|0|$

Solution   **a.** $|4| = 4$ since 4 is 4 units from 0 on the number line.
    b. $|-5| = 5$ since -5 is 5 units from 0 on the number line.
    c. $|0| = 0$ since 0 is 0 units from 0 on the number line.

Example 8 Insert $<, >,$ or $=$ in the appropriate space to make the statement true.

    a. $|0|$ 2 **b.** $|-5|$ 5 **c.** $|-3|$ $|-2|$ **d.** $|5|$ $|6|$ **e.** $|-7|$ $|6|$

Solution   **a.** $|0| < 2$ since $|0| = 0$ and $0 < 2$.
    b. $|-5| = 5$.
    c. $|-3| > |-2|$ since $3 > 2$.
    d. $|5| < |6|$ since $5 < 6$.
    e. $|-7| > |6|$ since $7 > 6$.

SPOTLIGHT ON DECISION MAKING

Suppose you are a quality control engineering technician in a factory that makes machine screws. You have just helped to install programmable machinery on the production line that measures the length of each screw. If a screw's length is greater than 4.05 centimeters or less than or equal to 3.98 centimeters, the machinery is programmed to discard the screw. To check that the machinery works properly, you test six screws with known lengths. The results of the test are displayed. Is the new machinery working properly? Explain.

TEST RESULTS

Test Screw	Actual Length of Test Screw (cm)	Machine Action on Test Screw
A	4.03	Accept
B	3.96	Reject
C	4.05	Accept
D	4.08	Reject
E	3.98	Reject
F	4.01	Accept

Exercise Set 1.2

Insert <, >, or = in the appropriate space to make the statement true. See Example 1.

1. 4   10
2. 8   5
3. 7   3
4. 9   15
5. 6.26   6.26
6. 2.13   1.13
7. 0   7
8. 20   0

9. The freezing point of water is 32° Fahrenheit. The boiling point of water is 212° Fahrenheit. Write an inequality statement using < or > comparing the numbers 32 and 212.

10. The freezing point of water is 0° Celsius. The boiling point of water is 100° Celsius. Write an inequality statement using < or > comparing the numbers 0 and 100.

11. The average salary in the United States for an experienced registered nurse is $44,300. The average salary for a drafter is $34,611. Write an inequality statement using < or > comparing the numbers 44,300 and 34,611. (*Source*: U.S. Department of Labor)

12. The state of New York is home to 312 institutions of higher learning. California claims a total of 384 colleges and universities. Write an inequality statement using < or > comparing the numbers 312 and 384. (*Source*: U.S. Department of Education)

Are the following statements true or false? See Example 2.

13. $11 \le 11$
14. $4 \ge 7$
15. $10 > 11$
16. $17 > 16$
17. $3 + 8 \ge 3(8)$
18. $8 \cdot 8 \le 8 \cdot 7$
19. $7 > 0$
20. $4 < 7$

△ **21.** An angle measuring 30° is shown and an angle measuring 45° is shown. Use the inequality symbol ≤ or ≥ to write a statement comparing the numbers 30 and 45.

△ **22.** The sum of the measures of the angles of a triangle is 180°. The sum of the measures of the angles of a parallelogram is 360°. Use the inequality symbol ≤ or ≥ to write a statement comparing the numbers 360 and 180.

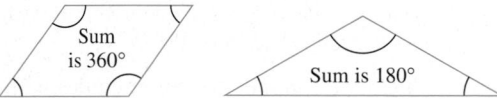

Write each sentence as a mathematical statement. See Example 3.

23. Eight is less than twelve.
24. Fifteen is greater than five.
25. Five is greater than or equal to four.
26. Negative ten is less than or equal to thirty-seven.
27. Fifteen is not equal to negative two.
28. Negative seven is not equal to seven.

Use integers to represent the values in each statement. See Example 4.

29. Driskill Mountain, in Louisiana, has an altitude of 535 feet. New Orleans, Louisiana lies 8 feet below sea level. (*Source*: U.S. Geological Survey)

30. During a Green Bay Packers football game, the team gained 23 yards and then lost 12 yards on consecutive plays.

31. From 1997 to 1998, Chicago Cubs home attendance decreased by 433,853. (*Source:* The Baseball Archive)

32. From 1997 to 1998, Boston Red Sox home attendance increased by 108,559. (*Source:* The Baseball Archive)

33. Aaron Miller deposited $350 in his savings account. He later withdrew $126.

34. Aris Peña was deep-sea diving. During her dive, she ascended 30 feet and later descended 50 feet.

The graph below is called a bar graph. This particular graph shows the number of communities in the United States that have curbside recycling programs for plastics. Each bar represents a different year and the height of the bar represents the number of communities with curbside plastics recycling.

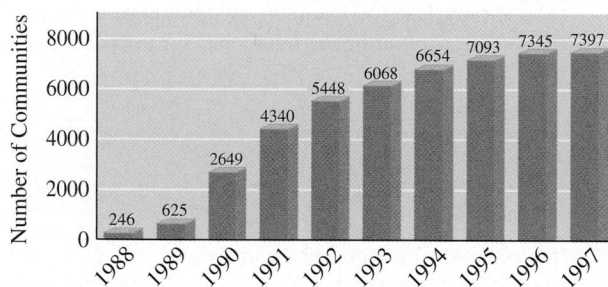

Source: American Plastics Council

35. In which year was the number of communities with curbside plastics recycling the lowest?

36. What is the highest number of communities with curbside plastics recycling?

37. In what years was the number of curbside plastic recycling programs less than 5000?

38. In what years was the number of curbside plastic recycling programs greater than 7000?

39. Write an inequality statement using ≤ or ≥ comparing the number of communities with curbside plastic recycling for 1993 and 1990.

40. Do you notice any trends shown by this bar graph?

Tell which set or sets each number belongs to: natural numbers, whole numbers, integers, rational numbers, irrational numbers, and real numbers. See Example 5.

41. 0

42. $\frac{1}{4}$

43. −2

44. $-\frac{1}{2}$

45. 6

46. 5

47. $\frac{2}{3}$

48. $\sqrt{3}$

49. $-\sqrt{5}$

50. $-1\frac{5}{6}$

Tell whether each statement is true or false.

51. Every rational number is also an integer.

52. Every negative number is also a rational number.

53. Every natural number is positive.

54. Every rational number is also a real number.

55. 0 is a real number.

56. Every real number is also a rational number.

57. Every whole number is an integer.

58. $\frac{1}{2}$ is an integer.

59. A number can be both rational and irrational.

60. Every whole number is positive.

Insert <, >, or = in the appropriate space to make a true statement. See Examples 6 through 8.

61. −10   −100

62. −200   −20

63. 32   5.2

64. 7   −7

65. $\frac{18}{3}$ $\frac{24}{3}$

66. $\frac{8}{2}$ $\frac{12}{3}$

67. −51   −50

68. |−20| −200

69. |−5| −4

70. 0 |0|

71. |−1| |1|

72. $\left|\frac{2}{5}\right|$ $\left|-\frac{2}{5}\right|$

73. |−2| |−3|

74. −500 |−50|

75. |0| |−8|

76. |−12| $\frac{24}{2}$

The apparent magnitude of a star is the measure of its brightness as seen by someone on Earth. The smaller the apparent magnitude, the brighter the star. Use the apparent magnitudes in the table to answer Exercises 77–82.

Star	Apparent Magnitude	Star	Apparent Magnitude
Arcturus	−0.04	Spica	0.98
Sirius	−1.46	Rigel	0.12
Vega	0.03	Regulus	1.35
Antares	0.96	Canopus	−0.72
Sun	−26.7	Hadar	0.61

Source: Norton's 2000.0: Star Atlas and Reference Handbook, 18th ed., Longman Group, UK, 1989

77. The apparent magnitude of the sun is −26.7. The apparent magnitude of the star Arcturus is −0.04. Write an inequality statement comparing the numbers −0.04 and −26.7.

78. The apparent magnitude of Antares is 0.96. The apparent magnitude of Spica is 0.98. Write an inequality statement comparing the numbers 0.96 and 0.98.

79. Which is brighter, the sun or Arcturus?

80. Which is dimmer, Antares or Spica?

81. Which star listed is the brightest?

82. Which star listed is the dimmest?

Rewrite the following inequalities so that the inequality symbol points in the opposite direction and the resulting statement has the same meaning as the given one.

83. $25 \geq 20$ **84.** $-13 \leq 13$

85. $0 < 6$ **86.** $5 > 3$

87. $-10 > -12$ **88.** $-4 < -2$

89. In your own words, explain how to find the absolute value of a number.

90. Give an example of a real-life situation that can be described with integers but not with whole numbers.

1.3 FRACTIONS

CD-ROM SSM SSG Video

▶ **OBJECTIVES**

1. Write fractions in simplest form.
2. Multiply and divide fractions.
3. Add and subtract fractions.

1 A quotient of two numbers such as $\frac{2}{9}$ is called a **fraction**. In the fraction $\frac{2}{9}$, the top number, 2, is called the **numerator** and the bottom number, 9, is called the **denominator**.

A fraction may be used to refer to part of a whole. For example, $\frac{2}{9}$ of the circle below is shaded. The denominator 9 tells us how many equal parts the whole circle is divided into and the numerator 2 tells us how many equal parts are shaded.

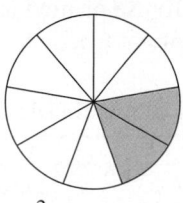

$\frac{2}{9}$ of the circle is shaded.

To simplify fractions, we can factor the numerator and the denominator. In the statement $3 \cdot 5 = 15$, 3 and 5 are called **factors** and 15 is the **product**. (The raised dot symbol indicates multiplication.)

$$3 \qquad \cdot \qquad 5 \qquad = \qquad 15$$
$$\uparrow \qquad\qquad \uparrow \qquad\qquad\qquad \uparrow$$
$$\text{factor} \qquad \text{factor} \qquad\qquad \text{product}$$

To **factor** 15 means to write it as a product. The number 15 can be factored as $3 \cdot 5$ or as $1 \cdot 15$.

A fraction is said to be **simplified** or in **lowest terms** when the numerator and the denominator have no factors in common other than 1. For example, the fraction $\frac{5}{11}$ is in lowest terms since 5 and 11 have no common factors other than 1.

To help us simplify fractions, we write the numerator and the denominator as a product of **prime numbers**.

PRIME NUMBER

A prime number is a whole number, other than 1, whose only factors are 1 and itself. The first few prime numbers are

$$2, 3, 5, 7, 11, 13, 17, 19, 23, 29, \text{ and so on.}$$

Example 1   Write each of the following numbers as a product of primes.

   a. 40   **b.** 63

Solution   **a.** First, write 40 as the product of any two whole numbers, other than 1.

$$40 = 4 \cdot 10$$

Next, factor each of these numbers. Continue this process until all of the factors are prime numbers.

$$40 = \quad 4 \quad \cdot \quad 10$$
$$= 2 \cdot 2 \cdot 2 \cdot 5$$

All the factors are now prime numbers. Then 40 written as a product of primes is

$$40 = 2 \cdot 2 \cdot 2 \cdot 5$$

b. $63 = \quad 9 \cdot 7$
$$= 3 \cdot 3 \cdot 7$$

To use prime factors to write a fraction in lowest terms, apply the fundamental principle of fractions.

FUNDAMENTAL PRINCIPLE OF FRACTIONS

If $\frac{a}{b}$ is a fraction and c is a nonzero real number, then

$$\frac{a \cdot c}{b \cdot c} = \frac{a}{b}$$

Example 2   Write each fraction in lowest terms.

a. $\dfrac{42}{49}$ **b.** $\dfrac{11}{27}$ **c.** $\dfrac{88}{20}$

Solution   **a.** Write the numerator and the denominator as products of primes; then apply the fundamental principle to the common factor 7.

$$\frac{42}{49} = \frac{2 \cdot 3 \cdot 7}{7 \cdot 7} = \frac{2 \cdot 3}{7} = \frac{6}{7}$$

b. $\dfrac{11}{27} = \dfrac{11}{3 \cdot 3 \cdot 3}$

There are no common factors other than 1, so $\frac{11}{27}$ is already in lowest terms.

c. $\dfrac{88}{20} = \dfrac{2 \cdot 2 \cdot 2 \cdot 11}{2 \cdot 2 \cdot 5} = \dfrac{22}{5}$

2   To multiply two fractions, multiply numerator times numerator to obtain the numerator of the product; multiply denominator times denominator to obtain the denominator of the product.

MULTIPLYING FRACTIONS

$$\frac{a}{b} \cdot \frac{c}{d} = \frac{a \cdot c}{b \cdot d}, \qquad \text{if } b \neq 0 \text{ and } d \neq 0$$

Example 3   Find the product of $\dfrac{2}{15}$ and $\dfrac{5}{13}$. Write the product in lowest terms.

Solution $\dfrac{2}{15} \cdot \dfrac{5}{13} = \dfrac{2 \cdot 5}{15 \cdot 13}$ Multiply numerators. Multiply denominators.

Next, simplify the product by dividing the numerator and the denominator by any common factors.

$$= \frac{2 \cdot 5}{3 \cdot 5 \cdot 13}$$

$$= \frac{2}{39}$$

Before dividing fractions, we first define **reciprocals**. Two fractions are reciprocals of each other if their product is 1. For example $\frac{2}{3}$ and $\frac{3}{2}$ are reciprocals since $\frac{2}{3} \cdot \frac{3}{2} = 1$. Also, the reciprocal of 5 is $\frac{1}{5}$ since $5 \cdot \frac{1}{5} = \frac{5}{1} \cdot \frac{1}{5} = 1$.

To divide fractions, multiply the first fraction by the reciprocal of the second fraction.

DIVIDING FRACTIONS

$$\frac{a}{b} \div \frac{c}{d} = \frac{a}{b} \cdot \frac{d}{c}, \qquad \text{if } b \neq 0, d \neq 0, \text{ and } c \neq 0$$

Example 4 Find each quotient. Write all answers in lowest terms.

a. $\dfrac{4}{5} \div \dfrac{5}{16}$
    b. $\dfrac{7}{10} \div 14$
    c. $\dfrac{3}{8} \div \dfrac{3}{10}$

Solution **a.** $\dfrac{4}{5} \div \dfrac{5}{16} = \dfrac{4}{5} \cdot \dfrac{16}{5} = \dfrac{4 \cdot 16}{5 \cdot 5} = \dfrac{64}{25}$

b. $\dfrac{7}{10} \div 14 = \dfrac{7}{10} \div \dfrac{14}{1} = \dfrac{7}{10} \cdot \dfrac{1}{14} = \dfrac{7 \cdot 1}{2 \cdot 5 \cdot 2 \cdot 7} = \dfrac{1}{20}$

c. $\dfrac{3}{8} \div \dfrac{3}{10} = \dfrac{3}{8} \cdot \dfrac{10}{3} = \dfrac{3 \cdot 2 \cdot 5}{2 \cdot 2 \cdot 2 \cdot 3} = \dfrac{5}{4}$

3 To add or subtract fractions with the same denominator, combine numerators and place the sum or difference over the common denominator.

ADDING AND SUBTRACTING FRACTIONS WITH THE SAME DENOMINATOR

$$\frac{a}{b} + \frac{c}{b} = \frac{a + c}{b}, \qquad \text{if } b \neq 0$$

$$\frac{a}{b} - \frac{c}{b} = \frac{a - c}{b}, \qquad \text{if } b \neq 0$$

Example 5 Add or subtract as indicated. Write each result in lowest terms.

a. $\dfrac{2}{7} + \dfrac{4}{7}$
    b. $\dfrac{3}{10} + \dfrac{2}{10}$
    c. $\dfrac{9}{7} - \dfrac{2}{7}$
    d. $\dfrac{5}{3} - \dfrac{1}{3}$

Solution **a.** $\dfrac{2}{7} + \dfrac{4}{7} = \dfrac{2 + 4}{7} = \dfrac{6}{7}$

b. $\dfrac{3}{10} + \dfrac{2}{10} = \dfrac{3 + 2}{10} = \dfrac{5}{10} = \dfrac{5}{2 \cdot 5} = \dfrac{1}{2}$

c. $\dfrac{9}{7} - \dfrac{2}{7} = \dfrac{9 - 2}{7} = \dfrac{7}{7} = 1$

d. $\dfrac{5}{3} - \dfrac{1}{3} = \dfrac{5 - 1}{3} = \dfrac{4}{3}$

To add or subtract fractions without the same denominator, first write the fractions as **equivalent fractions** with a common denominator. Equivalent fractions are

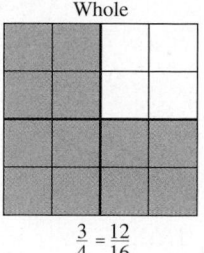

Whole

$\dfrac{3}{4} = \dfrac{12}{16}$

fractions that represent the same quantity. For example, $\frac{3}{4}$ and $\frac{12}{16}$ are equivalent fractions since they represent the same portion of a whole, as the diagram shows. Count the larger squares and the shaded portion is $\frac{3}{4}$. Count the smaller squares and the shaded portion is $\frac{12}{16}$. Thus, $\frac{3}{4} = \frac{12}{16}$.

We can write equivalent fractions by multiplying a given fraction by 1, as shown in the next example. Multiplying a fraction by 1 does not change the value of the fraction.

Example 6   Write $\dfrac{2}{5}$ as an equivalent fraction with a denominator of 20.

Solution   Since $5 \cdot 4 = 20$, multiply the fraction by $\dfrac{4}{4}$. Multiplying by $\dfrac{4}{4} = 1$ does not change the value of the fraction.

Multiply by $\dfrac{4}{4}$ or 1.

$$\frac{2}{5} = \frac{2}{5} \cdot \frac{4}{4} = \frac{2 \cdot 4}{5 \cdot 4} = \frac{8}{20}$$

Example 7   Add or subtract as indicated. Write each answer in lowest terms.

a. $\dfrac{2}{5} + \dfrac{1}{4}$ **b.** $\dfrac{1}{2} + \dfrac{17}{22} - \dfrac{2}{11}$ **c.** $3\dfrac{1}{6} - 1\dfrac{11}{12}$

Solution   **a.**   Fractions must have a common denominator before they can be added or subtracted. Since 20 is the smallest number that both 5 and 4 divide into evenly, 20 is the **least common denominator**. Write both fractions as equivalent fractions with denominators of 20. Since

$$\frac{2}{5} \cdot \frac{4}{4} = \frac{2 \cdot 4}{5 \cdot 4} = \frac{8}{20} \quad \text{and} \quad \frac{1}{4} \cdot \frac{5}{5} = \frac{1 \cdot 5}{4 \cdot 5} = \frac{5}{20}$$

then

$$\frac{2}{5} + \frac{1}{4} = \frac{8}{20} + \frac{5}{20} = \frac{13}{20}$$

b.   The least common denominator for denominators 2, 22, and 11 is 22. First, write each fraction as an equivalent fraction with a denominator of 22. Then add or subtract from left to right.

$$\frac{1}{2} = \frac{1}{2} \cdot \frac{11}{11} = \frac{11}{22}, \qquad \frac{17}{22} = \frac{17}{22}, \qquad \text{and} \qquad \frac{2}{11} = \frac{2}{11} \cdot \frac{2}{2} = \frac{4}{22}$$

Then

$$\frac{1}{2} + \frac{17}{22} - \frac{2}{11} = \frac{11}{22} + \frac{17}{22} - \frac{4}{22} = \frac{24}{22} = \frac{12}{11}$$

c. To find $3\frac{1}{6} - 1\frac{11}{12}$, first rewrite each mixed number as follows:

$$3\frac{1}{6} = 3 + \frac{1}{6} = \frac{18}{6} + \frac{1}{6} = \frac{19}{6}$$

$$1\frac{11}{12} = 1 + \frac{11}{12} = \frac{12}{12} + \frac{11}{12} = \frac{23}{12}$$

Then

$$3\frac{1}{6} - 1\frac{11}{12} = \frac{19}{6} - \frac{23}{12} = \frac{38}{12} - \frac{23}{12} = \frac{15}{12} = \frac{5}{4} \text{ or } 1\frac{1}{4}$$

HELPFUL HINT

Notice that we wrote the mixed number $3\frac{1}{6}$ as $\frac{19}{6}$ in Example 7c. Recall a short-cut process for writing a mixed number as an improper fraction:

$$3\frac{1}{6} = \frac{6 \cdot 3 + 1}{6} = \frac{19}{6}$$

SPOTLIGHT ON DECISION MAKING

Suppose you are fishing on a freshwater lake in Canada. You catch a whitefish weighing $14\frac{5}{32}$ pounds. According to the International Game Fish Association, the world's record for largest lake whitefish ever caught is $14\frac{3}{8}$ pounds. Did you set a new world's record? Explain. By how much did you beat or miss the existing world record?

MENTAL MATH

Represent the shaded part of each geometric figure by a fraction.

1.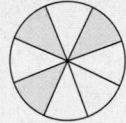

2.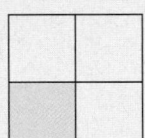

3.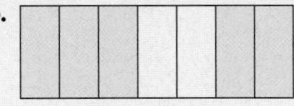

4.

For Exercises 5 and 6, fill in the blank.

5. In the fraction $\frac{3}{5}$, 3 is called the _____ and 5 is called the _____.

6. The reciprocal of $\frac{7}{11}$ is _____.

Exercise Set 1.3

Write each number as a product of primes. See Example 1.

1. 33

2. 60

3. 98

4. 27

5. 20

6. 56

7. 75

8. 32

9. 45

10. 24

Write the fraction in lowest terms. See Example 2.

11. $\dfrac{2}{4}$ **12.** $\dfrac{3}{6}$

13. $\dfrac{10}{15}$ **14.** $\dfrac{15}{20}$

15. $\dfrac{3}{7}$ **16.** $\dfrac{5}{9}$

17. $\dfrac{18}{30}$ **18.** $\dfrac{42}{45}$

Multiply or divide as indicated. Write the answer in lowest terms. See Examples 3 and 4.

19. $\dfrac{1}{2} \cdot \dfrac{3}{4}$ **20.** $\dfrac{10}{6} \cdot \dfrac{3}{5}$

21. $\dfrac{2}{3} \cdot \dfrac{3}{4}$ **22.** $\dfrac{7}{8} \cdot \dfrac{3}{21}$

23. $\dfrac{1}{2} \div \dfrac{7}{12}$ **24.** $\dfrac{7}{12} \div \dfrac{1}{2}$

25. $\dfrac{3}{4} \div \dfrac{1}{20}$ **26.** $\dfrac{3}{5} \div \dfrac{9}{10}$

27. $\dfrac{7}{10} \cdot \dfrac{5}{21}$ **28.** $\dfrac{3}{35} \cdot \dfrac{10}{63}$

29. $2\dfrac{7}{9} \cdot \dfrac{1}{3}$ **30.** $\dfrac{1}{4} \cdot 5\dfrac{5}{6}$

The area of a plane figure is a measure of the amount of surface of the figure. Find the area of each figure below. (The area of a rectangle is the product of its length and width. The area of a triangle is $\frac{1}{2}$ the product of its base and height.)

△ **31.**
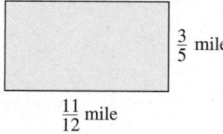
$\dfrac{3}{5}$ mile
$\dfrac{11}{12}$ mile

△ **32.**
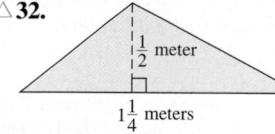
$\dfrac{1}{2}$ meter
$1\dfrac{1}{4}$ meters

Add or subtract as indicated. Write the answer in lowest terms. See Example 5.

33. $\dfrac{4}{5} - \dfrac{1}{5}$ **34.** $\dfrac{6}{7} - \dfrac{1}{7}$

35. $\dfrac{4}{5} + \dfrac{1}{5}$ **36.** $\dfrac{6}{7} + \dfrac{1}{7}$

37. $\dfrac{17}{21} - \dfrac{10}{21}$ **38.** $\dfrac{18}{35} - \dfrac{11}{35}$

39. $\dfrac{23}{105} + \dfrac{4}{105}$ **40.** $\dfrac{13}{132} + \dfrac{35}{132}$

Write each fraction as an equivalent fraction with the given denominator. See Example 6.

41. $\dfrac{7}{10}$ with a denominator of 30

42. $\dfrac{2}{3}$ with a denominator of 9

43. $\dfrac{2}{9}$ with a denominator of 18

44. $\dfrac{8}{7}$ with a denominator of 56

45. $\dfrac{4}{5}$ with a denominator of 20

46. $\dfrac{4}{5}$ with a denominator of 25

Add or subtract as indicated. Write the answer in lowest terms. See Example 7.

47. $\dfrac{2}{3} + \dfrac{3}{7}$ **48.** $\dfrac{3}{4} + \dfrac{1}{6}$

49. $2\dfrac{13}{15} - 1\dfrac{1}{5}$ **50.** $5\dfrac{2}{9} - 3\dfrac{1}{6}$

51. $\dfrac{5}{22} - \dfrac{5}{33}$ **52.** $\dfrac{7}{10} - \dfrac{8}{15}$

53. $\dfrac{12}{5} - 1$ **54.** $2 - \dfrac{3}{8}$

Each circle below represents a whole, or 1. Determine the unknown part of the circle.

55.
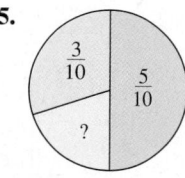

56.
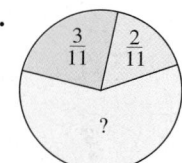

57.
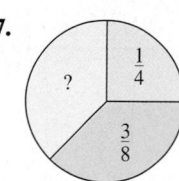

58.
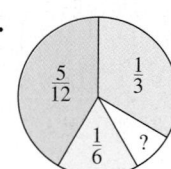

59.
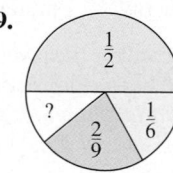

60.
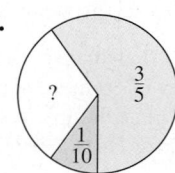

Perform the following operations. Write answers in lowest terms.

61. $\dfrac{10}{21} + \dfrac{5}{21}$ **62.** $\dfrac{11}{35} + \dfrac{3}{35}$

63. $\dfrac{10}{3} - \dfrac{5}{21}$ **64.** $\dfrac{11}{7} - \dfrac{3}{35}$

65. $\dfrac{2}{3} \cdot \dfrac{3}{5}$

66. $\dfrac{2}{3} \div \dfrac{3}{4}$

67. $\dfrac{3}{4} \div \dfrac{7}{12}$

68. $\dfrac{3}{5} + \dfrac{2}{3}$

69. $\dfrac{5}{12} + \dfrac{4}{12}$

70. $\dfrac{2}{7} + \dfrac{4}{7}$

71. $5 + \dfrac{2}{3}$

72. $7 + \dfrac{1}{10}$

73. $\dfrac{7}{8} \div 3\dfrac{1}{4}$

74. $3 \div \dfrac{3}{4}$

75. $\dfrac{7}{18} \div \dfrac{14}{36}$

76. $4\dfrac{3}{7} \div \dfrac{31}{7}$

77. $\dfrac{23}{105} - \dfrac{2}{105}$

78. $\dfrac{57}{132} - \dfrac{13}{132}$

79. $1\dfrac{1}{2} + 3\dfrac{2}{3}$

80. $2\dfrac{3}{5} + 4\dfrac{7}{10}$

81. $\dfrac{2}{3} - \dfrac{5}{9} + \dfrac{5}{6}$

82. $\dfrac{8}{11} - \dfrac{1}{4} + \dfrac{1}{2}$

The perimeter of a plane figure is the total distance around the figure. Find the perimeter of each figure below.

△ **83.**

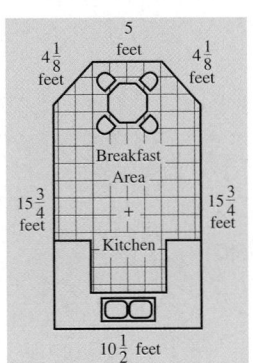

△ **84.**

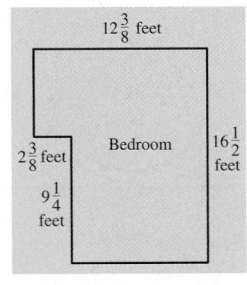

85. In 1988, Petra Felke of East Germany set an Olympic record in the women's javelin throw of $74\dfrac{17}{25}$ meters. Later that year she set a new world record for the javelin throw of 80 meters. By how much did her world record beat her Olympic record? (*Source: World Almanac and Book of Facts, 1999*)

86. In March 1999, a proposal to increase the size of rectangular escape vents in lobster traps was brought before the Maine state legislature. The proposed change would add $\dfrac{1}{16}$ of an inch to the current vent height of $1\dfrac{7}{8}$ inches. What would be the new vent height under the proposal? (*Source: The Boston Sunday Globe*, April 4, 1999)

87. In your own words, explain how to add two fractions with different denominators.

88. In your own words, explain how to multiply two fractions.

The following trail chart is given to visitors at the Lakeview Forest Preserve.

Trail Name	Distance (miles)
Robin Path	$3\dfrac{1}{2}$
Red Falls	$5\dfrac{1}{2}$
Green Way	$2\dfrac{1}{8}$
Autumn Walk	$1\dfrac{3}{4}$

89. How much longer is Red Falls Trail than Green Way Trail?

90. Find the total distance traveled by someone who hiked along all four trails.

Most of the water on Earth is in the form of oceans. Only a small part is fresh water. The graph below is called a circle graph or pie chart. This particular circle graph shows the distribution of fresh water.

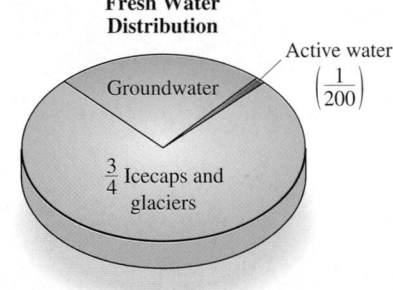

91. What fractional part of fresh water is icecaps and glaciers?

92. What fractional part of fresh water is active water?

93. What fractional part of fresh water is groundwater?

94. What fractional part of fresh water is groundwater or icecaps and glaciers?

1.4 INTRODUCTION TO VARIABLE EXPRESSIONS AND EQUATIONS

CD-ROM   SSM

SSG   Video

▶ **OBJECTIVES**

1. Define and use exponents and the order of operations.
2. Evaluate algebraic expressions, given replacement values for variables.
3. Determine whether a number is a solution of a given equation.
4. Translate phrases into expressions and sentences into equations.

1

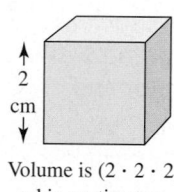

Volume is $(2 \cdot 2 \cdot 2)$
cubic centimeters.

Frequently in algebra, products occur that contain repeated multiplication of the same factor. For example, the volume of a cube whose sides each measure 2 centimeters is $(2 \cdot 2 \cdot 2)$ cubic centimeters. We may use **exponential notation** to write such products in a more compact form. For example,

$$2 \cdot 2 \cdot 2 \quad \textit{may be written as} \quad 2^3.$$

The 2 in 2^3 is called the **base**; it is the repeated factor. The 3 in 2^3 is called the **exponent** and is the number of times the base is used as a factor. The expression 2^3 is called an **exponential expression**.

$$\overset{\text{exponent}}{2^3} = 2 \cdot 2 \cdot 2 = 8$$
$$\underset{\text{base}}{\qquad} \text{2 is a factor 3 times}$$

Example 1   Evaluate the following:

    a. 3^2 [read as "3 squared" or as "3 to second power"]
    b. 5^3 [read as "5 cubed" or as "5 to the third power"]
    c. 2^4 [read as "2 to the fourth power"]
    d. 7^1 **e.** $\left(\dfrac{3}{7}\right)^2$

Solution   **a.** $3^2 = 3 \cdot 3 = 9$
    b. $5^3 = 5 \cdot 5 \cdot 5 = 125$
    c. $2^4 = 2 \cdot 2 \cdot 2 \cdot 2 = 16$
    d. $7^1 = 7$
    e. $\left(\dfrac{3}{7}\right)^2 = \left(\dfrac{3}{7}\right)\left(\dfrac{3}{7}\right) = \dfrac{9}{49}$

> **HELPFUL HINT**
> $2^3 \neq 2 \cdot 3$ since 2^3 indicates repeated **multiplication** of the same factor.
> $$2^3 = 2 \cdot 2 \cdot 2 = 8, \text{whereas } 2 \cdot 3 = 6.$$

Using symbols for mathematical operations is a great convenience. The more operation symbols presented in an expression, the more careful we must be when performing the indicated operation. For example, in the expression $2 + 3 \cdot 7$, do we add first or multiply first? To eliminate confusion, **grouping symbols** are used.

Examples of grouping symbols are parentheses ( ), brackets [ ], braces { }, and the fraction bar. If we wish $2 + 3 \cdot 7$ to be simplified by adding first, we enclose $2 + 3$ in parentheses.

$$(2 + 3) \cdot 7 = 5 \cdot 7 = 35$$

If we wish to multiply first, $3 \cdot 7$ may be enclosed in parentheses.

$$2 + (3 \cdot 7) = 2 + 21 = 23$$

To eliminate confusion when no grouping symbols are present, use the following agreed upon order of operations.

ORDER OF OPERATIONS

Simplify expressions using the following order. If grouping symbols such as parentheses are present, simplify expressions within those first, starting with the innermost set. If fraction bars are present, simplify the numerator and the denominator separately.

1. Evaluate exponential expressions.
2. Perform multiplications or divisions in order from left to right.
3. Perform additions or subtractions in order from left to right.

Now simplify $2 + 3 \cdot 7$. There are no grouping symbols and no exponents, so we multiply and then add.

$$2 + 3 \cdot 7 = 2 + 21 \qquad \text{Multiply.}$$
$$= 23 \qquad \text{Add.}$$

Example 2   Simplify each expression.

a. $6 \div 3 + 5^2$ **b.** $\dfrac{2(12 + 3)}{|-15|}$ **c.** $3 \cdot 10 - 7 \div 7$ **d.** $3 \cdot 4^2$ **e.** $\dfrac{3}{2} \cdot \dfrac{1}{2} - \dfrac{1}{2}$

Solution   **a.** Evaluate 5^2 first.

$$6 \div 3 + 5^2 = 6 \div 3 + 25$$

Next divide, then add.

$$6 \div 3 + 25 = 2 + 25 \qquad \text{Divide.}$$
$$= 27 \qquad \text{Add.}$$

b. First, simplify the numerator and the denominator separately.

$$\frac{2(12 + 3)}{|-15|} = \frac{2(15)}{15} \qquad \text{Simplify numerator and denominator separately.}$$
$$= \frac{30}{15}$$
$$= 2 \qquad \text{Simplify.}$$

c. Multiply and divide from left to right. Then subtract.

$$3 \cdot 10 - 7 \div 7 = 30 - 1$$
$$= 29 \qquad \text{Subtract.}$$

d. In this example, only the 4 is squared. The factor of 3 is not part of the base because no grouping symbol includes it as part of the base.

$$3 \cdot 4^2 = 3 \cdot 16 \qquad \text{Evaluate the exponential expression.}$$
$$= 48 \qquad \text{Multiply.}$$

e. The order of operations applies to operations with fractions in exactly the same way as it applies to operations with whole numbers.

$$\frac{3}{2} \cdot \frac{1}{2} - \frac{1}{2} = \frac{3}{4} - \frac{1}{2} \qquad \text{Multiply.}$$

$$= \frac{3}{4} - \frac{2}{4} \qquad \text{The least common denominator is 4.}$$

$$= \frac{1}{4} \qquad \text{Subtract.}$$

HELPFUL HINT

Be careful when evaluating an exponential expression. In $3 \cdot 4^2$, the exponent 2 applies only to the base 4. In $(3 \cdot 4)^2$, we multiply first because of parentheses, so the exponent 2 applies to the product $3 \cdot 4$.

$$3 \cdot 4^2 = 3 \cdot 16 = 48 \qquad (3 \cdot 4)^2 = (12)^2 = 144$$

Expressions that include many grouping symbols can be confusing. When simplifying these expressions, keep in mind that grouping symbols separate the expression into distinct parts. Each is then simplified separately.

Example 3 Simplify $\dfrac{3 + |4 - 3| + 2^2}{6 - 3}$.

Solution The fraction bar serves as a grouping symbol and separates the numerator and denominator. Simplify each separately. Also, the absolute value bars here serve as a grouping symbol. We begin in the numerator by simplifying within the absolute value bars.

$$\frac{3 + |4 - 3| + 2^2}{6 - 3} = \frac{3 + |1| + 2^2}{6 - 3} \qquad \begin{array}{l}\text{Simplify the expression} \\ \text{inside the absolute value} \\ \text{bars.}\end{array}$$

$$= \frac{3 + 1 + 2^2}{3} \qquad \begin{array}{l}\text{Find the absolute value} \\ \text{and simplify the denomi-} \\ \text{nator.}\end{array}$$

$$= \frac{3 + 1 + 4}{3} \qquad \begin{array}{l}\text{Evaluate the exponential} \\ \text{expression.}\end{array}$$

$$= \frac{8}{3} \qquad \text{Simplify the numerator.}$$

Example 4 Simplify $3[4(5 + 2) - 10]$.

Solution Notice that both parentheses and brackets are used as grouping symbols. Start with the innermost set of grouping symbols.

$$
\begin{aligned}
3[4(5 + 2) - 10] &= 3[4(7) - 10] & \text{Simplify the expression in parentheses.}\\
&= 3[28 - 10] & \text{Multiply 4 and 7.}\\
&= 3[18] & \text{Subtract inside the brackets.}\\
&= 54 & \text{Multiply.}
\end{aligned}
$$

Example 5 Simplify $\dfrac{8 + 2 \cdot 3}{2^2 - 1}$.

Solution $\dfrac{8 + 2 \cdot 3}{2^2 - 1} = \dfrac{8 + 6}{4 - 1} = \dfrac{14}{3}$

2 In algebra, we use symbols, usually letters such as x, y, or z, to represent unknown numbers. A symbol that is used to represent a number is called a **variable**. An **algebraic expression** is a collection of numbers, variables, operation symbols, and grouping symbols. For example,

$$
2x, \qquad -3, \qquad 2x + 10, \qquad 5(p^2 + 1), \qquad \text{and} \qquad \frac{3y^2 - 6y + 1}{5}
$$

are algebraic expressions. The expression $2x$ means $2 \cdot x$. Also, $5(p^2 + 1)$ means $5 \cdot (p^2 + 1)$ and $3y^2$ means $3 \cdot y^2$. If we give a specific value to a variable, we can **evaluate an algebraic expression**. To evaluate an algebraic expression means to find its numerical value once we know the values of the variables.

Algebraic expressions often occur during problem solving. For example, the expression

$$16t^2$$

gives the distance in feet (neglecting air resistance) that an object will fall in t seconds. (See Exercise 63 in this section.)

Example 6 Evaluate each expression if $x = 3$ and $y = 2$.

a. $2x - y$ **b.** $\dfrac{3x}{2y}$ **c.** $\dfrac{x}{y} + \dfrac{y}{2}$ **d.** $x^2 - y^2$

Solution **a.** Replace x with 3 and y with 2.

$$
\begin{aligned}
2x - y &= 2(3) - 2 & \text{Let } x = 3 \text{ and } y = 2.\\
&= 6 - 2 & \text{Multiply.}\\
&= 4 & \text{Subtract.}
\end{aligned}
$$

b. $\dfrac{3x}{2y} = \dfrac{3 \cdot 3}{2 \cdot 2} = \dfrac{9}{4}$ Let $x = 3$ and $y = 2$.

c. Replace x with 3 and y with 2. Then simplify.

$$\frac{x}{y} + \frac{y}{2} = \frac{3}{2} + \frac{2}{2} = \frac{5}{2}$$

d. Replace x with 3 and y with 2.

$$x^2 - y^2 = 3^2 - 2^2 = 9 - 4 = 5$$

3   Many times a problem-solving situation is modeled by an equation. An **equation** is a mathematical statement that two expressions have equal value. The equal symbol "=" is used to equate the two expressions. For example, $3 + 2 = 5$, $7x = 35$, $\dfrac{2(x - 1)}{3} = 0$, and $I = PRT$ are all equations.

HELPFUL HINT

An equation contains the equal symbol "=". An algebraic expression does not.

When an equation contains a variable, deciding which values of the variable make an equation a true statement is called **solving** an equation for the variable. A **solution** of an equation is a value for the variable that makes the equation true. For example, 3 is a solution of the equation $x + 4 = 7$, because if x is replaced with 3 the statement is true.

$$x + 4 = 7$$
$$\downarrow$$
$$3 + 4 = 7 \qquad \text{Replace } x \text{ with 3.}$$
$$7 = 7 \qquad \text{True.}$$

Similarly, 1 is not a solution of the equation $x + 4 = 7$, because $1 + 4 = 7$ is **not** a true statement.

Example 7   Decide whether 2 is a solution of $3x + 10 = 8x$.

Solution   Replace x with 2 and see if a true statement results.

$$3x + 10 = 8x \qquad \text{Original equation}$$
$$3(2) + 10 \stackrel{?}{=} 8(2) \qquad \text{Replace } x \text{ with 2.}$$
$$6 + 10 \stackrel{?}{=} 16 \qquad \text{Simplify each side.}$$
$$16 = 16 \qquad \text{True.}$$

Since we arrived at a true statement after replacing x with 2 and simplifying both sides of the equation, 2 is a solution of the equation.

4 Now that we know how to represent an unknown number by a variable, let's practice translating phrases into algebraic expressions and sentences into equations. Often-times solving problems involves the ability to translate word phrases and sentences into symbols. Below is a list of some key words and phrases to help us translate.

Addition (+)	*Subtraction* (−)	*Multiplication* (·)	*Division* (÷)	*Equality* (=)
Sum	Difference of	Product	Quotient	Equals
Plus	Minus	Times	Divide	Gives
Added to	Subtracted from	Multiply	Into	Is/was/ should be
More than	Less than	Twice	Ratio	Yields
Increased by	Decreased by	Of	Divided by	Amounts to
Total	Less			Represents Is the same as

Example 8 Write an algebraic expression that represents each phrase. Let the variable x represent the unknown number.

   a. The sum of a number and 3
   b. The product of 3 and a number
   c. Twice a number
   d. 10 decreased by a number
   e. 5 times a number increased by 7

Solution   **a.** $x + 3$ since "sum" means to add
   b. $3 \cdot x$ and $3x$ are both ways to denote the product of 3 and x
   c. $2 \cdot x$ or $2x$
   d. $10 - x$ because "decreased by" means to subtract
   e. $\underbrace{5x}_{\substack{5 \text{ times} \\ \text{a number}}} + 7$

> **HELPFUL HINT**
> Make sure you understand the difference when translating phrases containing "decreased by," "subtracted from," and "less than."
>
> | Phrase | Translation |
> |---|---|
> | A number decreased by 10 | $x - 10$ |
> | A number subtracted from 10 | $10 - x$ |
> | 10 less than a number | $x - 10$ |
>
> Notice the order.

Now let's practice translating sentences into equations.

Example 9 Write each sentence as an equation. Let x represent the unknown number.

   a. The quotient of 15 and a number is 4.
   b. Three subtracted from 12 is a number.
   c. Four times a number added to 17 is 21.

Solution   **a.** In words:

the quotient of 15 and a number	is	4
↓	↓	↓

Translate: $\dfrac{15}{x}$ $=$ 4

b. In words:

three subtracted **from** 12	is	a number
↓	↓	↓

Translate: $12 - 3$ $=$ x

Care must be taken when the operation is subtraction. The expression $3 - 12$ would be incorrect. Notice that $3 - 12 \neq 12 - 3$.

c. In words:

four times a number	added to	17	is	21
↓	↓	↓	↓	↓

Translate: $4x$ $+$ 17 $=$ 21

CALCULATOR EXPLORATIONS

Exponents

To evaluate exponential expressions on a scientific calculator, find the key marked $\boxed{y^x}$ or $\boxed{\wedge}$. To evaluate, for example, 3^5, press the following keys: $\boxed{3}$ $\boxed{y^x}$ (or $\boxed{\wedge}$) $\boxed{5}$ $\boxed{=}$ (or $\boxed{\text{ENTER}}$). The display should read $\boxed{243}$ or $\boxed{\begin{array}{l} 3\wedge5 \\ \qquad 243 \end{array}}$.

Order of Operations

Although most calculators follow the order of operations, parentheses must sometimes be inserted when evaluating an expression. For example, to find the value of $\frac{5}{12 \ - \ 7}$, press the keys

$\boxed{5}$ $\boxed{\div}$ $\boxed{(}$ $\boxed{1}$ $\boxed{2}$ $\boxed{-}$ $\boxed{7}$ $\boxed{)}$ $\boxed{=}$ (or $\boxed{\text{ENTER}}$).

The display should read $\boxed{1}$ or $\boxed{\begin{array}{l} 5/(12 - 7) \\ \qquad\qquad 1 \end{array}}$.

Use a calculator to evaluate each expression.

1. 5^3 **2.** 7^4

3. 9^5 **4.** 8^6

5. $2(20 - 5)$ **6.** $3(14 - 7) + 21$

7. $24(862 - 455) + 89$ **8.** $99 + (401 + 962)$

SPOTLIGHT ON DECISION MAKING

Suppose you are a local area network (LAN) administrator for a small college and you are configuring a new LAN for the mathematics department. The department would like a network of 20 computers so that each user can transmit data over the network at a speed of 0.25 megabits per second. The collective speed for a LAN is given by the expression rn, where r is the data transmission speed needed by each of the n computers on the LAN. You know that the network will drastically lose its efficiency if the collective speed of the network exceeds 8 megabits per second. Decide whether the LAN requested by the math department will operate efficiently. Explain your reasoning.

MENTAL MATH

Fill in the blank with add, subtract, multiply, or divide.

1. To simplify the expression $1 + 3 \cdot 6$, first _____ .
2. To simplify the expression $(1 + 3) \cdot 6$, first _____ .
3. To simplify the expression $(20 - 4) \cdot 2$, first _____ .
4. To simplify the expression $20 - 4 \div 2$, first _____ .

Exercise Set 1.4

Evaluate. See Example 1.

1. 3^5
2. 5^3
3. 3^3
4. 4^4
5. 1^5
6. 1^8
7. 5^1
8. 8^1
9. $\left(\dfrac{1}{5}\right)^3$
10. $\left(\dfrac{6}{11}\right)^2$
11. $\left(\dfrac{2}{3}\right)^4$
12. $\left(\dfrac{1}{2}\right)^5$
13. 7^2
14. 9^2
15. 4^2
16. 2^4
17. $(1.2)^2$
18. $(0.07)^2$

Simplify each expression. See Examples 2 through 5.

19. $5 + 6 \cdot 2$
20. $8 + 5 \cdot 3$
21. $4 \cdot 8 - 6 \cdot 2$
22. $12 \cdot 5 - 3 \cdot 6$
23. $2(8 - 3)$
24. $5(6 - 2)$
25. $2 + (5 - 2) + 4^2$
26. $6 - 2 \cdot 2 + 2^5$
27. $5 \cdot 3^2$
28. $2 \cdot 5^2$

29. $\dfrac{1}{4} \cdot \dfrac{2}{3} - \dfrac{1}{6}$
30. $\dfrac{3}{4} \cdot \dfrac{1}{2} + \dfrac{2}{3}$
31. $\dfrac{6 - 4}{9 - 2}$
32. $\dfrac{8 - 5}{24 - 20}$
33. $2[5 + 2(8 - 3)]$
34. $3[4 + 3(6 - 4)]$
35. $\dfrac{19 - 3 \cdot 5}{6 - 4}$
36. $\dfrac{4 \cdot 3 + 2}{4 + 3 \cdot 2}$
37. $\dfrac{|6 - 2| + 3}{8 + 2 \cdot 5}$
38. $\dfrac{15 - |3 - 1|}{12 - 3 \cdot 2}$
39. $\dfrac{3 + 3(5 + 3)}{3^2 + 1}$
40. $\dfrac{3 + 6(8 - 5)}{4^2 + 2}$
41. $\dfrac{6 + |8 - 2| + 3^2}{18 - 3}$
42. $\dfrac{16 + |13 - 5| + 4^2}{17 - 5}$

43. Are parentheses necessary in the expression $2 + (3 \cdot 5)$? Explain your answer.

44. Are parentheses necessary in the expression $(2 + 3) \cdot 5$? Explain your answer.

For Exercises 45 and 46, match each expression in the first column with its value in the second column.

45.
a. $(6 + 2) \cdot (5 + 3)$	19
b. $(6 + 2) \cdot 5 + 3$	22
c. $6 + 2 \cdot 5 + 3$	64
d. $6 + 2 \cdot (5 + 3)$	43

46.
a. $(1 + 4) \cdot 6 - 3$	15
b. $1 + 4 \cdot (6 - 3)$	13
c. $1 + 4 \cdot 6 - 3$	27
d. $(1 + 4) \cdot (6 - 3)$	22

Evaluate each expression when $x = 1$, $y = 3$, and $z = 5$. See Example 6.

47. $3y$

48. $4x$

49. $\dfrac{z}{5x}$

50. $\dfrac{y}{2z}$

51. $3x - 2$

52. $6y - 8$

53. $|2x + 3y|$

54. $|5z - 2y|$

55. $5y^2$

56. $2z^2$

Evaluate each expression if $x = 12$, $y = 8$, and $z = 4$. See Example 6.

57. $\dfrac{x}{z} + 3y$

58. $\dfrac{y}{z} + 8x$

59. $x^2 - 3y + x$

60. $y^2 - 3x + y$

61. $\dfrac{x^2 + z}{y^2 + 2z}$

62. $\dfrac{y^2 + x}{x^2 + 3y}$

Neglecting air resistance, the expression $16t^2$ gives the distance in feet an object will fall in t seconds.

63. Complete the chart below. To evaluate $16t^2$, remember to first find t^2, then multiply by 16.

Time t (in seconds)	Distance $16t^2$ (in feet)
1	
2	
3	
4	

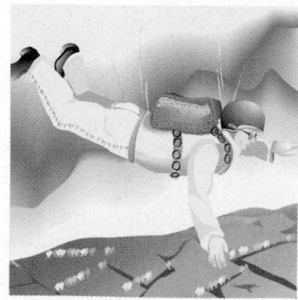

64. Does an object fall the same distance *during* each second? Why or why not? (See Exercise 63.)

Decide whether the given number is a solution of the given equation. See Example 7.

65. Is 5 a solution of $3x - 6 = 9$?

66. Is 6 a solution of $2x + 7 = 3x$?

67. Is 0 a solution of $2x + 6 = 5x - 1$?

68. Is 2 a solution of $4x + 2 = x + 8$?

69. Is 8 a solution of $2x - 5 = 5$?

70. Is 6 a solution of $3x - 10 = 8$?

71. Is 2 a solution of $x + 6 = x + 6$?

72. Is 10 a solution of $x + 6 = x + 6$?

73. Is 0 a solution of $x = 5x + 15$?

74. Is 1 a solution of $4 = 1 - x$?

Write each phrase as an algebraic expression. Let x represent the unknown number. See Example 8.

75. Fifteen more than a number

76. One-half times a number

77. Five subtracted from a number

78. The quotient of a number and 9

79. Three times a number increased by 22

80. The product of 8 and a number

Write each sentence as an equation. Use x to represent any unknown number. See Example 9.

81. One increased by two equals the quotient of nine and three.

82. Four subtracted from eight is equal to two squared.

83. Three is not equal to four divided by two.

84. The difference of sixteen and four is greater than ten.

85. The sum of 5 and a number is 20.

86. Twice a number is 17.

87. Thirteen minus three times a number is 13.

88. Seven subtracted from a number is 0.

89. The quotient of 12 and a number is $\dfrac{1}{2}$.

90. The sum of 8 and twice a number is 42.

91. In your own words, explain the difference between an expression and an equation.

92. Determine whether each is an expression or an equation.
   a. $3x^2 - 26$
   b. $3x^2 - 26 = 1$
   c. $2x - 5 = 7x - 5$
   d. $9y + x - 8$

Solve the following.

△ **93.** The perimeter of a figure is the distance around the figure. The expression $2l + 2w$ represents the perimeter of a rectangle when l is its length and w is its width. Find the perimeter of the following rectangle by substituting 8 for l and 6 for w.

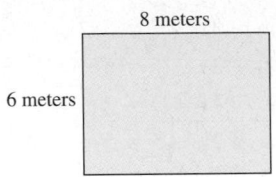

8 meters

6 meters

△ **94.** The expression $a + b + c$ represents the perimeter of a triangle when a, b, and c are the lengths of its sides. Find the perimeter of the following triangle.

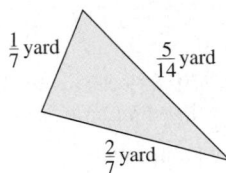

$\frac{1}{7}$ yard $\frac{5}{14}$ yard

$\frac{2}{7}$ yard

△ **95.** The area of a figure is the total enclosed surface of the figure. Area is measured in square units. The expression lw represents the area of a rectangle when l is its length and w is its width. Find the area of the following rectangular-shaped lot.

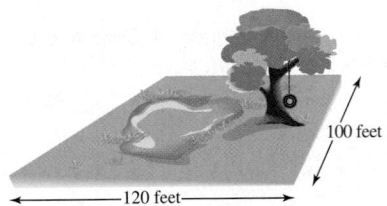

100 feet

120 feet

△ **96.** A trapezoid is a four-sided figure with exactly one pair of parallel sides. The expression $\frac{1}{2}h(B + b)$ represents its area, when B and b are the lengths of the two parallel sides and h is the height between these sides. Find the area if $B = 15$ inches, $b = 7$ inches, and $h = 5$ inches.

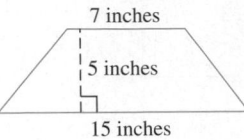

7 inches

5 inches

15 inches

97. The expression $\dfrac{I}{PT}$ represents the rate of interest being charged if a loan of P dollars for T years required I dollars in interest to be paid. Find the interest rate if a \$650 loan for 3 years to buy a used IBM personal computer requires \$126.75 in interest to be paid.

98. The expression $\dfrac{d}{t}$ represents the average speed r in miles per hour if a distance of d miles is traveled in t hours. Find the rate to the nearest whole number if the distance between Dallas, Texas, and Kaw City, Oklahoma, is 432 miles, and it takes Peter Callac 8.5 hours to drive the distance.

99. Sprint Communications Company offers a long-distance telephone plan called Sprint Sense AnyTime that charges \$4.95 per month and \$0. 10 per minute of calling. The expression $4.95 + 0.10m$ represents the monthly long-distance bill for a customer who makes m minutes of long-distance calling on this plan. Find the monthly bill for a customer who makes 228 minutes of long-distance calls on the Sprint Sense AnyTime plan.

100. In forensics, the density of a substance is used to help identify it. The expression $\dfrac{M}{V}$ represents the density of an object with a mass of M grams and a volume of V milliliters. Find the density of an object having a mass of 29.76 grams and a volume of 12 milliliters.

1.5 ADDING REAL NUMBERS

CD-ROM   SSM

SSG   Video

▶ **OBJECTIVES**

1. Add real numbers with the same sign.
2. Add real numbers with unlike signs.
3. Solve problems that involve addition of real numbers.
4. Find the opposite of a number.

1 Real numbers can be added, subtracted, multiplied, divided, and raised to powers, just as whole numbers can. We use the number line to help picture the addition of real numbers.

Example 1   Add: $3 + 2$

Solution   We start at 0 on a number line, and draw an arrow representing 3. This arrow is three units long and points to the right since 3 is positive. From the tip of this arrow, we draw another arrow representing 2. The number below the tip of this arrow is the sum, 5.

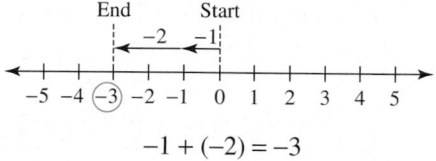

$$3 + 2 = 5$$

Example 2   Add: $-1 + (-2)$

Solution   We start at 0 on a number line, and draw an arrow representing -1. This arrow is one unit long and points to the left since -1 is negative. From the tip of this arrow, we draw another arrow representing -2. The number below the tip of this arrow is the sum, -3.

$$-1 + (-2) = -3$$

Thinking of signed numbers as money earned or lost might help make addition more meaningful. Earnings can be thought of as positive numbers. If \$1 is earned and later another \$3 is earned, the total amount earned is \$4. In other words, $1 + 3 = 4$.

On the other hand, losses can be thought of as negative numbers. If \$1 is lost and later another \$3 is lost, a total of \$4 is lost. In other words, $(-1) + (-3) = -4$.

Using a number line each time we add two numbers can be time consuming. Instead, we can notice patterns in the previous examples and write rules for adding signed numbers. When adding two numbers with the same sign, notice that the sign of the sum is the same as the sign of the addends.

ADDING TWO NUMBERS WITH THE SAME SIGN

Add their absolute values. Use their common sign as the sign of the sum.

Example 3   Add.

    a. $-3 + (-7)$ **b.** $-1 + (-20)$ **c.** $-2 + (-10)$

Solution   Notice that each time, we are adding numbers with the same sign.

    a. $-3 + (-7) = -10$ ← Add their absolute values: $3 + 7 = 10$.
                              Use their common sign.

b. $-1 + (-20) = -21$ ← Add their absolute values: $1 + 20 = 21$.
                   └——— Common sign.

c. $-2 + (-10) = -12$ ← Add their absolute values.
                   └———Common sign.

2   Adding numbers whose signs are not the same can also be pictured on a number line.

Example 4   Add: $-4 + 6$

Solution

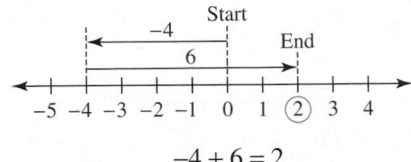

$$-4 + 6 = 2$$

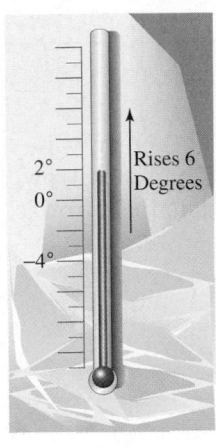

Using temperature as an example, if the thermometer registers 4 degrees below 0 degrees and then rises 6 degrees, the new temperature is 2 degrees above 0 degrees. Thus, it is reasonable that $-4 + 6 = 2$.

Once again, we can observe a pattern: when adding two numbers with different signs, the sign of the sum is the same as the sign of the addend whose absolute value is larger.

ADDING TWO NUMBERS WITH DIFFERENT SIGNS

Subtract the smaller absolute value from the larger absolute value. Use the sign of the number whose absolute value is larger as the sign of the sum.

Example 5   Add.

    a. $3 + (-7)$ **b.** $-2 + 10$ **c.** $0.2 + (-0.5)$

Solution   Notice that each time, we are adding numbers with the different signs.

    a. $3 + (-7) = -4$ ← Subtract their absolute values: $7 - 3 = 4$.
                 └——— The negative number, -7, has the larger absolute value so the sum is negative.

    b. $-2 + 10 = 8$ ← Subtract their absolute values: $10 - 2 = 8$.
                   └——The positive number, 10, has the larger absolute value so the sum is understood positive.

    c. $0.2 + (-0.5) = -0.3$ ← Subtract their absolute values: $0.5 - 0.2 = 0.3$.
                     └——The negative number, -0.5, has the larger absolute value so the sum is negative.

Example 6   Add.

    a. $-8 + (-11)$ **b.** $-5 + 35$ **c.** $0.6 + (-1.1)$

    d. $-\dfrac{7}{10} + \left(-\dfrac{1}{10}\right)$ **e.** $11.4 + (-4.7)$ **f.** $-\dfrac{3}{8} + \dfrac{2}{5}$

Solution
a. $-8 + (-11) = -19$ Same sign. Add absolute values and use the common sign.
b. $-5 + 35 = 30$ Different signs. Subtract absolute values and use the
sign of the number with the larger absolute value.

c. $0.6 + (-1.1) = -0.5$ Different signs.
d. $-\dfrac{7}{10} + \left(-\dfrac{1}{10}\right) = -\dfrac{8}{10} = -\dfrac{4}{5}$ Same sign.
e. $11.4 + (-4.7) = 6.7$
f. $-\dfrac{3}{8} + \dfrac{2}{5} = -\dfrac{15}{40} + \dfrac{16}{40} = \dfrac{1}{40}$

Example 7   Add.

a. $3 + (-7) + (-8)$ **b.** $\left[7 + (-10)\right] + \left[-2 + (-4)\right]$

Solution   **a.** Perform the additions from left to right.

$$3 + (-7) + (-8) = -4 + (-8)$$ Adding numbers with different signs.
$$= -12$$ Adding numbers with like signs.

b. Simplify inside brackets first.

$$\left[7 + (-10)\right] + \left[-2 + (-4)\right] = \left[-3\right] + \left[-6\right]$$
$$= -9$$ Add.

3   Positive and negative numbers are often used in everyday life. Stock market returns
show gains and losses as positive and negative numbers. Temperatures in cold cli-
mates often dip into the negative range, commonly referred to as "below zero" temperatures. Bank statements report deposits and withdrawals as positive and negative
numbers.

Example 8   **FINDING THE GAIN OR LOSS OF A STOCK**

During a three-day period, a share of Electronic's International stock recorded the
following gains and losses:

Monday	**Tuesday**	**Wednesday**
a gain of $2	a loss of $1	a loss of $3

Find the overall gain or loss for the stock for the three days.

Solution   Gains can be represented by positive numbers. Losses can be represented by negative numbers. The overall gain or loss is the sum of the gains and losses.

In words: | gain | plus | loss | plus | loss |

Translate: $2 \quad + \quad (-1) \quad + \quad (-3) = -2$

The overall loss is $2.

$\dfrac{4}{\ }$ To help us subtract real numbers in the next section, we first review the concept of opposites. The graph of 4 and −4 is shown on the number line below.

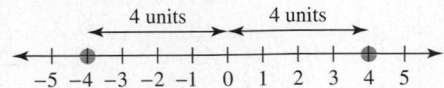

Notice that 4 and −4 lie on opposite sides of 0, and each is 4 units away from 0.

This relationship between −4 and +4 is an important one. Such numbers are known as **opposites** or **additive inverses** of each other.

OPPOSITES OR ADDITIVE INVERSES

Two numbers that are the same distance from 0 but lie on opposite sides of 0 are called opposites or additive inverses of each other.

Example 9   Find the opposite or additive inverse of each number.

a. 5   **b.** −6   **c.** $\dfrac{1}{2}$ **d.** −4.5

Solution   **a.** The opposite of 5 is −5. Notice that 5 and −5 are on opposite sides of 0 when plotted on a number line and are equal distances away.

b. The opposite of −6 is 6.

c. The opposite of $\dfrac{1}{2}$ is $-\dfrac{1}{2}$.

d. The opposite of −4.5 is 4.5.

We use the symbol "−" to represent the phrase "the opposite of" or "the additive inverse of." In general, if a is a number, we write the opposite or additive inverse of a as $-a$. We know that the opposite of −3 is 3. Notice that this translates as

the opposite of	−3	is	3
↓	↓	↓	↓
−	(−3)	=	3

This is true in general.

If a is a number, then $-(-a) = a$.

Example 10   Simplify each expression.

a. $-(-10)$ **b.** $-\left(-\dfrac{1}{2}\right)$ **c.** $-(-2x)$ **d.** $-|-6|$

Solution   **a.** $-(-10) = 10$ **b.** $-\left(-\dfrac{1}{2}\right) = \dfrac{1}{2}$ **c.** $-(-2x) = 2x$

d. Since $|-6| = 6$, then $-|-6| = -6$.

Let's discover another characteristic about opposites. Notice that the sum of a number and its opposite is 0.

$$10 + (-10) = 0$$
$$-3 + 3 = 0$$
$$\frac{1}{2} + \left(-\frac{1}{2}\right) = 0$$

In general, we can write the following:

> The sum of a number a and its opposite $-a$ is 0.
> $$a + (-a) = 0$$

Notice that this means that the opposite of 0 is then 0 since $0 + 0 = 0$.

SPOTLIGHT ON DECISION MAKING

Suppose you own stock in XYZ Corp and like to follow the stock's price. You see the following information on a stock ticker indicating the current price of XYZ stock followed by the change in price from its previous closing price. You decide that you would like to make a trade. Should you sell some of the XYZ stock you already own or should you buy additional XYZ stock? Explain your reasoning. What other factors would you want to consider?

```
XYZ   63.625   -7.125     PCX   46.875   +3.375     JCP   5
```

MENTAL MATH

Tell whether the sum is a positive number, a negative number, or 0. Do not actually find the sum.

1. $-80 + (-127)$ **2.** $-162 + 164$ **3.** $-162 + 162$

4. $-1.26 + (-8.3)$ **5.** $-3.68 + 0.27$ **6.** $-\frac{2}{3} + \frac{2}{3}$

Exercise Set 1.5

Add. See Examples 1 through 7.

1. $6 + 3$ **2.** $9 + (-12)$ **17.** $5 + (-7)$ **18.** $3 + (-6)$

3. $-6 + (-8)$ **4.** $-6 + (-14)$ **19.** $-16 + 16$ **20.** $23 + (-23)$

5. $8 + (-7)$ **6.** $6 + (-4)$ **21.** $27 + (-46)$ **22.** $53 + (-37)$

7. $-14 + 2$ **8.** $-10 + 5$ **23.** $-18 + 49$ **24.** $-26 + 14$

9. $-2 + (-3)$ **10.** $-7 + (-4)$ **25.** $-33 + (-14)$ **26.** $-18 + (-26)$

11. $-9 + (-3)$ **12.** $7 + (-5)$ **27.** $6.3 + (-8.4)$ **28.** $9.2 + (-11.4)$

13. $-7 + 3$ **14.** $-5 + 9$ **29.** $|-8| + (-16)$ **30.** $|-6| + (-61)$

15. $10 + (-3)$ **16.** $8 + (-6)$ **31.** $117 + (-79)$ **32.** $144 + (-88)$

33. $-9.6 + (-3.5)$

34. $-6.7 + (-7.6)$

35. $-\dfrac{3}{8} + \dfrac{5}{8}$

36. $-\dfrac{5}{12} + \dfrac{7}{12}$

37. $-\dfrac{7}{16} + \dfrac{1}{4}$

38. $-\dfrac{5}{9} + \dfrac{1}{3}$

39. $-\dfrac{7}{10} + \left(-\dfrac{3}{5}\right)$

40. $-\dfrac{5}{6} + \left(-\dfrac{2}{3}\right)$

41. $-15 + 9 + (-2)$

42. $-9 + 15 + (-5)$

43. $-21 + (-16) + (-22)$

44. $-18 + (-6) + (-40)$

45. $-23 + 16 + (-2)$

46. $-14 + (-3) + 11$

47. $|5 + (-10)|$

48. $|7 + (-17)|$

49. $6 + (-4) + 9$

50. $8 + (-2) + 7$

51. $[-17 + (-4)] + [-12 + 15]$

52. $[-2 + (-7)] + [-11 + 22]$

53. $|9 + (-12)| + |-16|$

54. $|43 + (-73)| + |-20|$

55. $-1.3 + [0.5 + (-0.3) + 0.4]$

56. $-3.7 + [0.1 + (-0.6) + 8.1]$

The following bar graph shows the daily low temperatures for a week in Sioux Falls, South Dakota.

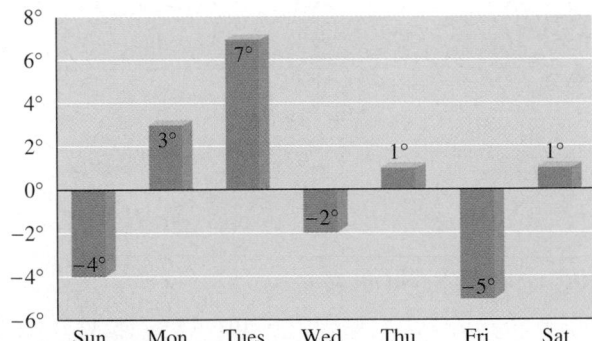

57. On what day of the week was the graphed temperature the highest?

58. On what day of the week was the graphed temperature the lowest?

59. What is the highest temperature shown on the graph?

60. What is the lowest temperature shown on the graph?

61. Find the average daily low temperature for Sunday through Thursday. (*Hint:* To find the average of the five temperatures, find their sum and divide by 5.)

62. Find the average daily low temperature for Tuesday through Thursday.

Solve. See Example 8.

63. The low temperature in Anoka, Minnesota, was $-15°$ last night. During the day it rose only $9°$. Find the high temperature for the day.

64. On January 2, 1943, the temperature was $-4°$ at 7:30 a.m. in Spearfish, South Dakota. Incredibly, it got $49°$ warmer in the next 2 minutes. To what temperature did it rise by 7:32?

65. The lowest elevation on Earth is -1312 feet (that is, 1312 feet below sea level) at the Dead Sea. If you are standing 658 feet above the Dead Sea, what is your elevation? (*Source:* National Geographic Society)

66. The lowest point in Africa is -512 feet at Lake Assal in Djibouti. If you are standing at a point 658 feet above Lake Assal, what is your elevation? (*Source:* Microsoft Encarta)

67. A negative net income results when a company's expenses are more than the money it brings in. Polaroid Corporation had net incomes of $-\$41.1$ million, $-\$126.7$ million, and $-\$51.0$ million in 1996, 1997, and 1998, respectively. What was Polaroid's total net income for these three years? (*Source:* Polaroid Corporation)

68. Apple Computer had net incomes of $-\$816$ million, $-\$1045$ million, and $\$309$ million in 1996, 1997, and 1998, respectively. What was Apple's total net income for these three years? (*Source:* Apple Computer, Inc.)

69. In golf, scores that are under par for the entire round are shown as negative scores, positive scores are shown for scores that are over par, and 0 is par. During the 1999 LPGA Sara Lee Classic, winner Meg Mallon had scores of -6, -7, and -4. What was her overall score? (*Source:* Ladies Professional Golf Association)

70. During the 1999 PGA Masters Tournament, winner Jose Maria Olazabal of Spain had scores of -2, -6, $+1$, and -1. What was his overall score? (*Source:* Professional Golf Association)

Find each additive inverse or opposite. See Example 9.

71. 6   **72.** 4

73. −2   **74.** −8

75. 0   **76.** $-\dfrac{1}{4}$

77. |−6| **78.** |−11|

79. In your own words, explain how to find the opposite of a number.

80. In your own words, explain why 0 is the only number that is its own opposite.

Simplify each of the following. See Example 10.

81. −|−2| **82.** −(−3)

83. −|0| **84.** $\left|-\dfrac{2}{3}\right|$

85. $-\left|-\dfrac{2}{3}\right|$ **86.** −(−7)

87. Explain why adding a negative number to another negative number always gives a negative sum.

88. When a positive and a negative number are added, sometimes the sum is positive, sometimes it is zero, and sometimes it is negative. Explain why this happens.

If a is a positive number and b is a negative number, fill in the blanks with the words positive or negative.

89. −a is a _____.

90. −b is a _____.

91. $a + a$ is a _____.

92. $b + b$ is a _____.

Decide whether the given number is a solution of the given equation.

93. Is −4 a solution of $x + 9 = 5$?

94. Is 10 a solution of $7 = -x + 3$?

95. Is −1 a solution of $y + (-3) = -7$?

96. Is −6 a solution of $1 = y + 7$?

1.6 SUBTRACTING REAL NUMBERS

CD-ROM   SSM

SSG   Video

▶ **OBJECTIVES**

1. Subtract real numbers.
2. Add and subtract real numbers.
3. Evaluate algebraic expressions using real numbers.
4. Solve problems that involve subtraction of real numbers.

1

Now that addition of signed numbers has been discussed, we can explore subtraction. We know that $9 - 7 = 2$. Notice that $9 + (-7) = 2$, also. This means that

$$9 - 7 = 9 + (-7)$$

Notice that the difference of 9 and 7 is the same as the sum of 9 and the opposite of 7. In general, we have the following.

SUBTRACTING TWO REAL NUMBERS

If a and b are real numbers, then $a - b = a + (-b)$.

In other words, to find the difference of two numbers, add the first number to the opposite of the second number.

Example 1   Subtract.

   a. −13 − 4   **b.** 5 − (−6)   **c.** 3 − 6   **d.** −1 − (−7)

Solution

a. $-13 - 4 = -13 + (-4)$ Add -13 to the opposite of $+4$, which is -4.

$= -17$

b. $5 - (-6) = 5 + (6)$ Add 5 to the opposite of -6, which is 6.

$= 11$

c. $3 - 6 = 3 + (-6)$ Add 3 to the opposite of 6, which is -6.

$= -3$

d. $-1 - (-7) = -1 + (7) = 6$

HELPFUL HINT

Study the patterns indicated.

$5 - 11 = 5 + (-11) = -6$

$-3 - 4 = -3 + (-4) = -7$

$7 - (-1) = 7 + (1) = 8$

Example 2   Subtract.

a. $5.3 - (-4.6)$ **b.** $-\dfrac{3}{10} - \dfrac{5}{10}$ **c.** $-\dfrac{2}{3} - \left(-\dfrac{4}{5}\right)$

Solution   **a.** $5.3 - (-4.6) = 5.3 + (4.6) = 9.9$

b. $-\dfrac{3}{10} - \dfrac{5}{10} = -\dfrac{3}{10} + \left(-\dfrac{5}{10}\right) = -\dfrac{8}{10} = -\dfrac{4}{5}$

c. $-\dfrac{2}{3} - \left(-\dfrac{4}{5}\right) = -\dfrac{2}{3} + \left(\dfrac{4}{5}\right) = -\dfrac{10}{15} + \dfrac{12}{15} = \dfrac{2}{15}$ The common denominator is 15.

Example 3   Subtract 8 from -4.

Solution   Be careful when interpreting this: The order of numbers in subtraction is important. 8 is to be subtracted **from** -4.

$$-4 - 8 = -4 + (-8) = -12$$

---

2   If an expression contains additions and subtractions, just write the subtractions as equivalent additions. Then simplify from left to right.

Example 4 Simplify: $-14 - 8 + 10 - (-6)$

Solution $-14 - 8 + 10 - (-6) = -14 + (-8) + 10 + 6 = -6$ ◼

When an expression contains parentheses and brackets, remember the order of operations. Start with the innermost set of parentheses or brackets and work your way outward.

Example 5 Simplify each expression.

a. $-3 + [(-2 - 5) - 2]$ **b.** $2^3 - |10| + [-6 - (-5)]$

Solution **a.** Start with the innermost sets of parentheses. Rewrite $-2 - 5$ as a sum.

$$\begin{aligned} -3 + [(-2 - 5) - 2] &= -3 + [(-2 + (-5)) - 2] \\ &= -3 + [(-7) - 2] &&\text{Add: } -2 + (-5). \\ &= -3 + [-7 + (-2)] &&\text{Write } -7 - 2 \text{ as a sum.} \\ &= -3 + [-9] &&\text{Add.} \\ &= -12 &&\text{Add.} \end{aligned}$$

b. Start simplifying the expression inside the brackets by writing $-6 - (-5)$ as a sum.

$$\begin{aligned} 2^3 - |10| + [-6 - (-5)] &= 2^3 - |10| + [-6 + 5] \\ &= 2^3 - |10| + [-1] &&\text{Add.} \\ &= 8 - 10 + (-1) &&\text{Evaluate } 2^3 \text{ and } |10|. \\ &= 8 + (-10) + (-1) &&\text{Write } 8 - 10 \text{ as a sum.} \\ &= -2 + (-1) &&\text{Add.} \\ &= -3 &&\text{Add.} \end{aligned}$$ ◼

3 Knowing how to evaluate expressions for given replacement values is helpful when checking solutions of equations and when solving problems whose unknowns satisfy given expressions. The next example illustrates this.

Example 6 Find the value of each expression when $x = 2$ and $y = -5$.

a. $\dfrac{x - y}{12 + x}$ **b.** $x^2 - y$

Solution **a.** Replace x with 2 and y with -5. Be sure to put parentheses around -5 to separate signs. Then simplify the resulting expression.

$$\frac{x - y}{12 + x} = \frac{2 - (-5)}{12 + 2} = \frac{2 + 5}{14} = \frac{7}{14} = \frac{1}{2}$$

b. Replace the x with 2 and y with -5 and simplify.

$$x^2 - y = 2^2 - (-5) = 4 - (-5) = 4 + 5 = 9$$ ◼

4 One use of positive and negative numbers is in recording altitudes above and below sea level, as shown in the next example.

Example 7 **FINDING THE VARIATION IN ELEVATION**

The lowest point in North America is in Death Valley, at an elevation of 282 feet below sea level. Nearby, Mount Whitney reaches 14,494 feet, the highest point in the United States outside Alaska. How much of a variation in elevation is there between these two extremes?

Solution To find the variation in elevation between the two heights, find the difference of the high point and the low point.

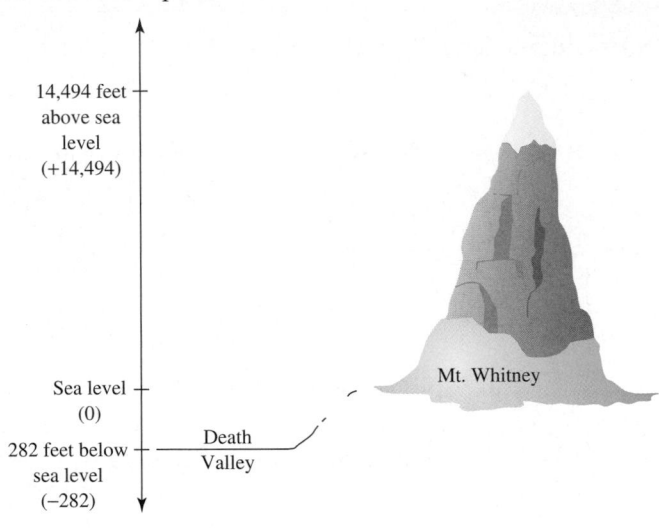

In words:

high point	minus	low point
↓	↓	↓

Translate: 14,494   −   (−282) = 14,494 + 282
                                              = 14,776 feet

Thus, the variation in elevation is 14,776 feet.

SPOTLIGHT ON DECISION MAKING

Suppose you are a dental hygienist. As part of a new patient assessment, you measure the depth of the gum tissue pocket around the patient's teeth with a dental probe and record the results. If these pockets deepen over time, this could indicate a problem with gum health or be an indication of gum disease. Now, a year later, you measure the patient's gum tissue pocket depth again to compare to the initial measurements. Based on these findings, would you alert the dentist to a possible problem with the health of the patient's gums? Explain.

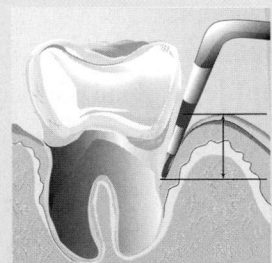

DENTAL CHART

Gum Tissue Pocket Depth (millimeters)					
Tooth:	22	23	24	25	26	27
Initial	2	3	3	2	4	2
Current	2	2	4	5	6	5

A knowledge of geometric concepts is needed by many professionals, such as doctors, carpenters, electronic technicians, gardeners, machinists, and pilots, just to name a few. With this in mind, we review the geometric concepts of **complementary** and **supplementary angles**.

COMPLEMENTARY AND SUPPLEMENTARY ANGLES

Two angles are **complementary** if their sum is 90°.

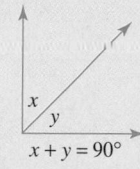

$$x + y = 90°$$

Two angles are **supplementary** if their sum is 180°.

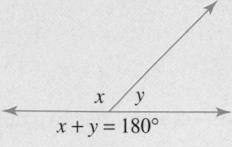

$$x + y = 180°$$

△ **Example 8**   Find each unknown complementary or supplementary angle.

a.

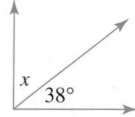

b.

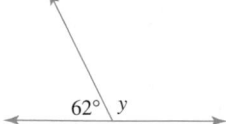

Solution   **a.**   These angles are complementary, so their sum is 90°. This means that x is 90° − 38°.

$$x = 90° - 38° = 52°$$

b.   These angles are supplementary, so their sum is 180°. This means that y is 180° − 62°.

$$y = 180° - 62° = 118°$$

Exercise Set 1.6

Subtract. See Examples 1 through 5.

1. $-6 - 4$

2. $-12 - 8$

3. $4 - 9$

4. $8 - 11$

5. $16 - (-3)$

6. $12 - (-5)$

7. $\dfrac{1}{2} - \dfrac{1}{3}$

8. $\dfrac{3}{4} - \dfrac{7}{8}$

9. $-16 - (-18)$

10. $-20 - (-48)$

11. $-6 - 5$

12. $-8 - 4$

13. $7 - (-4)$

14. $3 - (-6)$

15. $-6 - (-11)$

16. $-4 - (-16)$

17. $16 - (-21)$

18. $15 - (-33)$

19. $9.7 - 16.1$

20. $8.3 - 11.2$

21. $-44 - 27$

22. $-36 - 51$

23. $-21 - (-21)$

24. $-17 - (-17)$

25. $-2.6 - (-6.7)$

26. $-6.1 - (-5.3)$

27. $-\dfrac{3}{11} - \left(-\dfrac{5}{11}\right)$

28. $-\dfrac{4}{7} - \left(-\dfrac{1}{7}\right)$

29. $-\dfrac{1}{6} - \dfrac{3}{4}$

30. $-\dfrac{1}{10} - \dfrac{7}{8}$

31. $8.3 - (-0.62)$

32. $4.3 - (-0.87)$

Perform the operation. See Example 3.

33. Subtract -5 from 8.

34. Subtract 3 from -2.

35. Subtract -1 from -6.

36. Subtract 17 from 1.

37. Subtract 8 from 7.

38. Subtract 9 from -4.

39. Decrease -8 by 15.

40. Decrease 11 by -14.

41. In your own words, explain why $5 - 8$ simplifies to a negative number.

42. Explain why $6 - 11$ is the same as $6 + (-11)$.

Simplify each expression. (Remember the order of operations.) See Examples 4 and 5.

43. $-10 - (-8) + (-4) - 20$

44. $-16 - (-3) + (-11) - 14$

45. $5 - 9 + (-4) - 8 - 8$

46. $7 - 12 + (-5) - 2 + (-2)$

47. $-6 - (2 - 11)$

48. $-9 - (3 - 8)$

49. $3^3 - 8 \cdot 9$

50. $2^3 - 6 \cdot 3$

51. $2 - 3(8 - 6)$

52. $4 - 6(7 - 3)$

53. $(3 - 6) + 4^2$

54. $(2 - 3) + 5^2$

55. $-2 + \left[(8 - 11) - (-2 - 9)\right]$

56. $-5 + \left[(4 - 15) - (-6) - 8\right]$

57. $|-3| + 2^2 + \left[-4 - (-6)\right]$

58. $|-2| + 6^2 + (-3 - 8)$

Evaluate each expression when $x = -5$, $y = 4$, and $t = 10$. See Example 6.

59. $x - y$

60. $y - x$

61. $|x| + 2t - 8y$

62. $|x + t - 7y|$

63. $\dfrac{9 - x}{y + 6}$

64. $\dfrac{15 - x}{y + 2}$

65. $y^2 - x$

66. $t^2 - x$

67. $\dfrac{|x - (-10)|}{2t}$

68. $\dfrac{|5y - x|}{6t}$

The following bar graph shows each month's average daily low temperature in degrees Fahrenheit for Fairbanks, Alaska. Use the graph to answer Exercises 69–74. See Example 7.

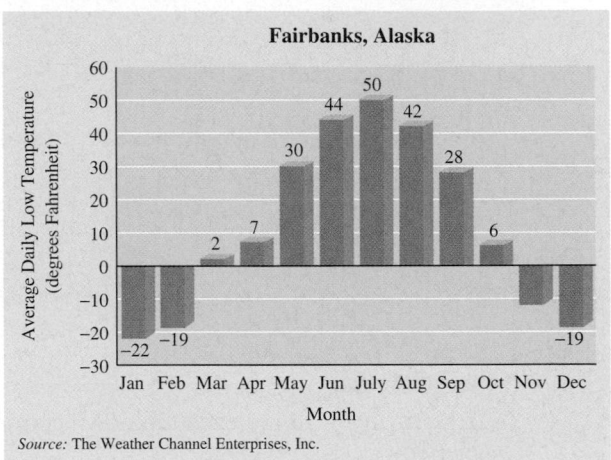

Source: The Weather Channel Enterprises, Inc.

69. In which month is the average daily low temperature the lowest? What is the average daily low temperature for this month?

70. In which month is the average daily low temperature the highest? What is the average daily low temperature for this month?

71. The average daily low temperature for November is 18 degrees less than the average daily low temperature for October. What is the average daily low temperature for November?

72. Which months have the same average daily low temperature?

73. How many degrees warmer is May's average low temperature than February's average low temperature?

74. What is the difference between the average daily low temperatures for July and January?

Solve. See Example 7.

75. Within 24 hours in 1916, the temperature in Browning, Montana, fell from 44 degrees to -56 degrees. How large a drop in temperature was this?

76. Much of New Orleans is just barely above sea level. If George descends 12 feet from an elevation of 5 feet above sea level, what is his new elevation?

77. In a series of plays, the San Francisco 49ers gain 2 yards, lose 5 yards, and then lose another 20 yards. What is their total gain or loss of yardage?

78. In some card games, it is possible to have a negative score. Lavonne Schultz currently has a score of 15 points. She then loses 24 points. What is her new score?

79. Aristotle died in the year −322 (or 322 B.C.). When was he born, if he was 62 years old when he died?

80. Augustus Caesar died in A.D. 14 in his 77th year. When was he born?

81. Tyson Industries stock posted a loss of $1\frac{5}{8}$ points yesterday. If it drops another $\frac{3}{4}$ points today, find its overall change for the two days.

82. A commercial jet liner hits an air pocket and drops 250 feet. After climbing 120 feet, it drops another 178 feet. What is its overall vertical change?

83. The highest point in South America is Mount Aconcagua, Argentina, at an elevation of 22,834 feet. The lowest point is Valdes Peninsula, Argentina, at 131 feet below sea level. How much higher is Mount Aconcagua than Valdes Peninsula? (*Source*: National Geographic Society)

84. The lowest altitude in Antarctica is the Bentley Subglacial Trench at 8327 feet below sea level. The highest altitude is Vinson Massif at an elevation of 16,864 feet above sea level. What is the difference between these altitudes? (*Source*: National Geographic Society)

Find each unknown complementary or supplementary angle. See Example 8.

△ **85.**

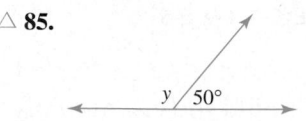

△ **86.**

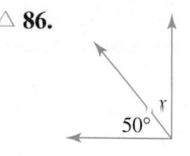

△ **87.**

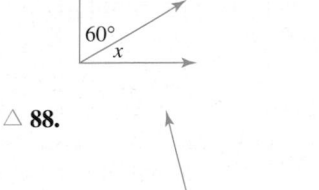

△ **88.**

Decide whether the given number is a solution of the given equation.

89. Is −4 a solution of $x - 9 = 5$?

90. Is 3 a solution of $x - 10 = -7$?

91. Is −2 a solution of $-x + 6 = -x - 1$?

92. Is −10 a solution of $-x - 6 = -x - 1$?

93. Is 2 a solution of $-x - 13 = -15$?

94. Is 5 a solution of $4 = 1 - x$?

If a is a positive number and b is a negative number, determine whether each statement is true or false.

95. $a - b$ is always a positive number.

96. $b - a$ is always a negative number.

97. $|b| - |a|$ is always a positive number.

98. $|b - a|$ is always a positive number.

Without calculating, determine whether each answer is positive or negative. Then use a calculator to find the exact difference.

99. $56{,}875 - 87{,}262$ **100.** $4.362 - 7.0086$

1.7 MULTIPLYING AND DIVIDING REAL NUMBERS

CD-ROM   SSM   ▶ **OBJECTIVES**

SSG   Video

1. Multiply and divide real numbers.
2. Find the values of algebraic expressions.

1   In this section, we discover patterns for multiplying and dividing real numbers. To discover sign rules for multiplication, recall that multiplication is repeated addition. Thus $3 \cdot 2$ means that 2 is an addend 3 times. That is,

$$2 + 2 + 2 = 3 \cdot 2$$

which equals 6. Similarly, $3 \cdot (-2)$ means -2 is an addend 3 times. That is,

$$(-2) + (-2) + (-2) = 3 \cdot (-2)$$

Since $(-2) + (-2) + (-2) = -6$, then $3 \cdot (-2) = -6$. This suggests that the product of a positive number and a negative number is a negative number.

What about the product of two negative numbers? To find out, consider the following pattern.

Factor decreases by 1 each time

$$\left. \begin{array}{l} -3 \cdot 2 = -6 \\ -3 \cdot 1 = -3 \\ -3 \cdot 0 = 0 \end{array} \right\} \text{ Product increases by 3 each time.}$$

This pattern continues as

Factor decreases by 1 each time

$$\left. \begin{array}{l} -3 \cdot -1 = 3 \\ -3 \cdot -2 = 6 \end{array} \right\} \text{ Product increases by 3 each time.}$$

This suggests that the product of two negative numbers is a positive number.

MULTIPLYING REAL NUMBERS

1. The product of two numbers with the *same* sign is a positive number.
2. The product of two numbers with *different* signs is a negative number.

Example 1   Find the product.

a. $(-6)(4)$ b. $2(-1)$ c. $(-5)(-10)$

Solution   a. $(-6)(4) = -24$ b. $2(-1) = -2$ c. $(-5)(-10) = 50$

We know that every whole number multiplied by zero equals zero. This remains true for signed numbers.

ZERO AS A FACTOR

If b is a real number, then $b \cdot 0 = 0$. Also, $0 \cdot b = 0$.

Example 2   Perform the indicated operations.

    a. $(7)(0)(-6)$ **b.** $(-2)(-3)(-4)$ **c.** $(-1)(5)(-9)$ **d.** $(-4)(-11) - (5)(-2)$

Solution   **a.** By the order of operations, we multiply from left to right. Notice that, because one of the factors is 0, the product is 0.

$$(7)(0)(-6) = 0(-6) = 0$$

b. Multiply two factors at a time, from left to right.

$$(-2)(-3)(-4) = (6)(-4) \quad \text{Multiply } (-2)(-3).$$
$$= -24$$

c. Multiply from left to right.

$$(-1)(5)(-9) = (-5)(-9) \quad \text{Multiply } (-1)(5).$$
$$= 45$$

d. Follow the rules for order of operation.

$$(-4)(-11) - (5)(-2) = 44 - (-10) \quad \text{Find each product.}$$
$$= 44 + 10 \quad \text{Add 44 to the opposite of } -10.$$
$$= 54 \quad \text{Add.}$$

Multiplying signed decimals or fractions is carried out exactly the same way as multiplying by integers.

Example 3   Find each product.

    a. $(-1.2)(0.05)$ **b.** $\dfrac{2}{3} \cdot -\dfrac{7}{10}$

Solution   **a.** The product of two numbers with different signs is negative.

$$(-1.2)(0.05) = -\left[(1.2)(0.05)\right]$$
$$= -0.06$$

b. $\dfrac{2}{3} \cdot -\dfrac{7}{10} = -\dfrac{2 \cdot 7}{3 \cdot 10} = -\dfrac{2 \cdot 7}{3 \cdot 2 \cdot 5} = -\dfrac{7}{15}$

Now that we know how to multiply positive and negative numbers, let's see how we find the values of $(-4)^2$ and -4^2, for example. Although these two expressions look similar, the difference between the two is the parentheses. In $(-4)^2$, the parentheses tell us that the base, or repeated factor, is -4. In -4^2, only 4 is the base. Thus,

$$(-4)^2 = (-4)(-4) = 16 \quad \text{The base is } -4.$$
$$-4^2 = -(4 \cdot 4) = -16 \quad \text{The base is 4.}$$

Example 4 Evaluate.

    a. $(-2)^3$ **b.** -2^3 **c.** $(-3)^2$ **d.** -3^2

Solution
a. $(-2)^3 = (-2)(-2)(-2) = -8$ The base is -2.
b. $-2^3 = -(2 \cdot 2 \cdot 2) = -8$ The base is 2.
c. $(-3)^2 = (-3)(-3) = 9$ The base is -3.
d. $-3^2 = -(3 \cdot 3) = -9$ The base is 3.

HELPFUL HINT
Be careful when identifying the base of an exponential expression.

$(-3)^2$ -3^2
Base is -3 Base is 3
$(-3)^2 = (-3)(-3) = 9$ $-3^2 = -(3 \cdot 3) = -9$

Just as every difference of two numbers $a - b$ can be written as the sum $a + (-b)$, so too every quotient of two numbers can be written as a product. For example, the quotient $6 \div 3$ can be written as $6 \cdot \frac{1}{3}$. Recall that the pair of numbers 3 and $\frac{1}{3}$ has a special relationship. Their product is 1 and they are called reciprocals or **multiplicative inverses** of each other.

RECIPROCALS OR MULTIPLICATIVE INVERSES

Two numbers whose product is 1 are called reciprocals or multiplicative inverses of each other.

Notice that **0 has no multiplicative inverse** since 0 multiplied by any number is never 1 but always 0.

Example 5 Find the reciprocal of each number.

    a. 22   **b.** $\frac{3}{16}$ **c.** -10 **d.** $-\frac{9}{13}$

Solution
a. The reciprocal of 22 is $\frac{1}{22}$ since $22 \cdot \frac{1}{22} = 1$.
b. The reciprocal of $\frac{3}{16}$ is $\frac{16}{3}$ since $\frac{3}{16} \cdot \frac{16}{3} = 1$.
c. The reciprocal of -10 is $-\frac{1}{10}$.
d. The reciprocal of $-\frac{9}{13}$ is $-\frac{13}{9}$.

We may now write a quotient as an equivalent product.

QUOTIENT OF TWO REAL NUMBERS

If a and b are real numbers and b is not 0, then

$$\frac{a}{b} = a \cdot \frac{1}{b}$$

In other words, the quotient of two real numbers is the product of the first number and the multiplicative inverse, or reciprocal, of the second number.

Example 6   Use the definition of the quotient of two numbers to find each quotient.

a. $-18 \div 3$

b. $\dfrac{-14}{-2}$

c. $\dfrac{20}{-4}$

Solution   a. $-18 \div 3 = -18 \cdot \dfrac{1}{3} = -6$ b. $\dfrac{-14}{-2} = -14 \cdot -\dfrac{1}{2} = 7$

c. $\dfrac{20}{-4} = 20 \cdot -\dfrac{1}{4} = -5$

Since the quotient $a \div b$ can be written as the product $a \cdot \frac{1}{b}$, it follows that sign patterns for dividing two real numbers are the same as sign patterns for multiplying two real numbers.

MULTIPLYING AND DIVIDING REAL NUMBERS

1. The product or quotient of two numbers with the *same* sign is a positive number.
2. The product or quotient of two numbers with *different* signs is a negative number.

Example 7   Find each quotient.

a. $\dfrac{-24}{-4}$

b. $\dfrac{-36}{3}$

c. $\dfrac{2}{3} \div \left(-\dfrac{5}{4}\right)$

Solution   a. $\dfrac{-24}{-4} = 6$ b. $\dfrac{-36}{3} = -12$ c. $\dfrac{2}{3} \div \left(-\dfrac{5}{4}\right) = \dfrac{2}{3} \cdot \left(-\dfrac{4}{5}\right) = -\dfrac{8}{15}$

The definition of the quotient of two real numbers does not allow for division by 0 because 0 does not have a multiplicative inverse. There is no number we can multiply 0 by to get 1. How then do we interpret $\frac{3}{0}$? We say that division by 0 is not allowed or not defined and that $\frac{3}{0}$ does not represent a real number. The denominator of a fraction can never be 0.

Can the numerator of a fraction be 0? Can we divide 0 by a number? Yes. For example,

$$\frac{0}{3} = 0 \cdot \frac{1}{3} = 0$$

In general, the quotient of 0 and any nonzero number is 0.

ZERO AS A DIVISOR OR DIVIDEND

1. The quotient of any nonzero real number and 0 is undefined. In symbols, if $a \neq 0$, $\dfrac{a}{0}$ is **undefined**.

2. The quotient of 0 and any real number except 0 is 0. In symbols, if $a \neq 0$, $\dfrac{0}{a} = 0$.

Example 8 Perform the indicated operations.

a. $\dfrac{1}{0}$

b. $\dfrac{0}{-3}$

c. $\dfrac{0(-8)}{2}$

Solution a. $\dfrac{1}{0}$ is undefined

b. $\dfrac{0}{-3} = 0$

c. $\dfrac{0(-8)}{2} = \dfrac{0}{2} = 0$

Notice that $\dfrac{12}{-2} = -6, -\dfrac{12}{2} = -6,$ and $\dfrac{-12}{2} = -6.$ This means that

$$\dfrac{12}{-2} = -\dfrac{12}{2} = \dfrac{-12}{2}$$

In words, a single negative sign in a fraction can be written in the denominator, in the numerator, or in front of the fraction without changing the value of the fraction. Thus,

$$\dfrac{1}{-7} = \dfrac{-1}{7} = -\dfrac{1}{7}$$

In general, if a and b are real numbers, $b \neq 0$, $\dfrac{a}{-b} = \dfrac{-a}{b} = -\dfrac{a}{b}$.

Examples combining basic arithmetic operations along with the principles of order of operations help us to review these concepts.

Example 9 Simplify each expression.

a. $\dfrac{(-12)(-3) + 3}{-7 - (-2)}$

b. $\dfrac{2(-3)^2 - 20}{-5 + 4}$

Solution a. First, simplify the numerator and denominator separately, then divide.

$$\dfrac{(-12)(-3) + 3}{-7 - (-2)} = \dfrac{36 + 3}{-7 + 2}$$

$$= \dfrac{39}{-5} \text{ or } -\dfrac{39}{5}$$

b. Simplify the numerator and denominator separately, then divide.

$$\dfrac{2(-3)^2 - 20}{-5 + 4} = \dfrac{2 \cdot 9 - 20}{-5 + 4} = \dfrac{18 - 20}{-5 + 4} = \dfrac{-2}{-1} = 2$$

2 Using what we have learned about multiplying and dividing real numbers, we continue to practice evaluating algebraic expressions.

Example 10 If $x = -2$ and $y = -4$, evaluate each expression.

a. $5x - y$

b. $x^3 - y^2$

c. $\dfrac{3x}{2y}$

Solution   **a.** Replace x with -2 and y with -4 and simplify.

$$5x - y = 5(-2) - (-4) = -10 - (-4) = -10 + 4 = -6$$

b. Replace x with -2 and y with -4.

$$
\begin{aligned}
x^3 - y^2 &= (-2)^3 - (-4)^2 &&\text{Substitute the given values for the variables.}\\
&= -8 - (16) &&\text{Evaluate exponential expressions.}\\
&= -8 + (-16) &&\text{Write as a sum.}\\
&= -24 &&\text{Add.}
\end{aligned}
$$

c. Replace x with -2 and y with -4 and simplify.

$$\frac{3x}{2y} = \frac{3(-2)}{2(-4)} = \frac{-6}{-8} = \frac{3}{4}$$

CALCULATOR EXPLORATIONS

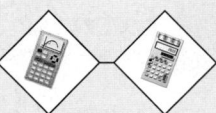

Entering Negative Numbers on a Scientific Calculator

To enter a negative number on a scientific calculator, find a key marked $\boxed{+/-}$. (On some calculators, this key is marked $\boxed{\text{CHS}}$ for "change sign.") To enter -8, for example, press the keys $\boxed{8}$ $\boxed{+/-}$. The display will read $\boxed{-8}$.

Entering Negative Numbers on a Graphing Calculator

To enter a negative number on a graphing calculator, find a key marked $\boxed{(-)}$. Do not confuse this key with the key $\boxed{-}$, which is used for subtraction. To enter -8, for example, press the keys $\boxed{(-)}$ $\boxed{8}$. The display will read $\boxed{-8}$.

Operations with Real Numbers

To evaluate $-2(7 - 9) - 20$ on a calculator, press the keys

$\boxed{2}$ $\boxed{+/-}$ $\boxed{\times}$ $\boxed{(}$ $\boxed{7}$ $\boxed{-}$ $\boxed{9}$ $\boxed{)}$ $\boxed{-}$ $\boxed{2}$ $\boxed{0}$ $\boxed{=}$, or

$\boxed{(-)}$ $\boxed{2}$ $\boxed{(}$ $\boxed{7}$ $\boxed{-}$ $\boxed{9}$ $\boxed{)}$ $\boxed{-}$ $\boxed{2}$ $\boxed{0}$ $\boxed{\text{ENTER}}$.

The display will read $\boxed{-16}$ or $\boxed{\begin{array}{c}-2(7-9)-20\\-16\end{array}}$.

Use a calculator to simplify each expression.

1. $-38(26 - 27)$

2. $-59(-8) + 1726$

3. $134 + 25(68 - 91)$

4. $45(32) - 8(218)$

5. $\dfrac{-50(294)}{175 - 265}$

6. $\dfrac{-444 - 444.8}{-181 - 324}$

7. $9^5 - 4550$

8. $5^8 - 6259$

9. $(-125)^2$ (Be careful.)

10. -125^2 (Be careful.)

MENTAL MATH

Answer the following with positive or negative.

1. The product of two negative numbers is a _____ number.
2. The quotient of two negative numbers is a _____ number.
3. The quotient of a positive number and a negative number is a _____ number.
4. The product of a positive number and a negative number is a _____ number.
5. The reciprocal of a positive number is a _____ number.
6. The opposite of a positive number is a _____ number.

Exercise Set 1.7

Multiply. See Examples 1 through 3.

1. $-6(4)$
2. $-8(5)$
3. $2(-1)$
4. $7(-4)$
5. $-5(-10)$
6. $-6(-11)$
7. $-3 \cdot 4$
8. $-2 \cdot 8$
9. $-6(-7)$
10. $-6(-9)$
11. $2(-9)$
12. $3(-5)$
13. $-\dfrac{1}{2}\left(-\dfrac{3}{5}\right)$
14. $-\dfrac{1}{8}\left(-\dfrac{1}{3}\right)$
15. $-\dfrac{3}{4}\left(-\dfrac{8}{9}\right)$
16. $-\dfrac{5}{6}\left(-\dfrac{3}{10}\right)$
17. $5(-1.4)$
18. $6(-2.5)$
19. $-0.2(-0.7)$
20. $-0.5(-0.3)$
21. $-10(80)$
22. $-20(60)$
23. $4(-7)$
24. $5(-9)$
25. $(-5)(-5)$
26. $(-7)(-7)$
27. $\dfrac{2}{3}\left(-\dfrac{4}{9}\right)$
28. $\dfrac{2}{7}\left(-\dfrac{2}{11}\right)$
29. $-11(11)$
30. $-12(12)$
31. $-\dfrac{20}{25}\left(\dfrac{5}{16}\right)$
32. $-\dfrac{25}{36}\left(\dfrac{6}{15}\right)$
33. $-2.1(-0.4)$
34. $-1.3(-0.6)$
35. $(-1)(2)(-3)(-5)$
36. $(-2)(-3)(-4)(-2)$
37. $(2)(-1)(-3)(5)(3)$
38. $(3)(-5)(-2)(-1)(-2)$

Decide whether each statement is true or false.

39. The product of three negative integers is negative.
40. The product of three positive integers is positive.
41. The product of four negative integers is negative.
42. The product of four positive integers is positive.

Evaluate. See Example 4.

43. $(-2)^4$
44. -2^4
45. -1^5
46. $(-1)^5$
47. $(-5)^2$
48. -5^2
49. -7^2
50. $(-7)^2$

Find each reciprocal or multiplicative inverse. See Example 5.

51. 9
52. 100
53. $\dfrac{2}{3}$
54. $\dfrac{1}{7}$
55. -14
56. -8
57. $-\dfrac{3}{11}$
58. $-\dfrac{6}{13}$
59. 0.2
60. 1.5
61. $\dfrac{1}{-6.3}$
62. $\dfrac{1}{-8.9}$

Divide. See Examples 6 through 8.

63. $\dfrac{18}{-2}$
64. $\dfrac{20}{-10}$
65. $\dfrac{-16}{-4}$
66. $\dfrac{-18}{-6}$
67. $\dfrac{-48}{12}$
68. $\dfrac{-60}{5}$
69. $\dfrac{0}{-4}$
70. $\dfrac{0}{-9}$
71. $-\dfrac{15}{3}$
72. $-\dfrac{24}{8}$

73. $\dfrac{5}{0}$

74. $\dfrac{3}{0}$

75. $\dfrac{-12}{-4}$

76. $\dfrac{-45}{-9}$

77. $\dfrac{30}{-2}$

78. $\dfrac{14}{-2}$

79. $\dfrac{6}{7} \div \left(-\dfrac{1}{3}\right)$

80. $\dfrac{4}{5} \div \left(-\dfrac{1}{2}\right)$

81. $-\dfrac{5}{9} \div \left(-\dfrac{3}{4}\right)$

82. $-\dfrac{1}{10} \div \left(-\dfrac{8}{11}\right)$

83. $-\dfrac{4}{9} \div \dfrac{4}{9}$

84. $-\dfrac{5}{12} \div \dfrac{5}{12}$

Simplify. See Example 9.

85. $\dfrac{-9(-3)}{-6}$

86. $\dfrac{-6(-3)}{-4}$

87. $\dfrac{12}{9-12}$

88. $\dfrac{-15}{1-4}$

89. $\dfrac{-6^2+4}{-2}$

90. $\dfrac{3^2+4}{5}$

91. $\dfrac{8+(-4)^2}{4-12}$

92. $\dfrac{6+(-2)^2}{4-9}$

93. $\dfrac{22+(3)(-2)}{-5-2}$

94. $\dfrac{-20+(-4)(3)}{1-5}$

95. $\dfrac{-3-5^2}{2(-7)}$

96. $\dfrac{-2-4^2}{3(-6)}$

97. $\dfrac{6-2(-3)}{4-3(-2)}$

98. $\dfrac{8-3(-2)}{2-5(-4)}$

99. $\dfrac{-3-2(-9)}{-15-3(-4)}$

100. $\dfrac{-4-8(-2)}{-9-2(-3)}$

101. $\dfrac{|5-9|+|10-15|}{|2(-3)|}$

102. $\dfrac{|-3+6|+|-2+7|}{|-2\cdot2|}$

If $x = -5$ and $y = -3$, evaluate each expression. See Example 10.

103. $3x+2y$

104. $4x+5y$

105. $2x^2-y^2$

106. x^2-2y^2

107. x^3+3y

108. y^3+3x

109. $\dfrac{2x-5}{y-2}$

110. $\dfrac{2y-12}{x-4}$

111. $\dfrac{6-y}{x-4}$

112. $\dfrac{4-2x}{y+3}$

113. Amazon.com is an Internet bookseller. At the end of 1998, Amazon posted a net income of −$124.5 million. If

this continued, what would Amazon's income be after four years? (*Source:* Amazon.com, Inc.)

114. Union Pacific provides rail transportation services. At the end of 1998, Union Pacific posted a net income of −$633 million. If this continued, what would Union Pacific's income be after three years? (*Source:* Union Pacific Corp.)

115. Explain why the product of an even number of negative numbers is a positive number.

116. If a and b are any real numbers, is the statement $a \cdot b = b \cdot a$ always true? Why or why not?

117. Find any real numbers that are their own reciprocal.

118. Explain why 0 has no reciprocal.

If q is a negative number, r is a negative number, and t is a positive number, determine whether each expression simplifies to a positive or negative number If it is not possible to determine, state so.

119. $\dfrac{q}{r \cdot t}$

120. $q^2 \cdot r \cdot t$

121. $q + t$

122. $t + r$

123. $t(q + r)$

124. $r(q - t)$

Write each of the following as an expression and evaluate.

125. The sum of −2 and the quotient of −15 and 3

126. The sum of 1 and the product of −8 and −5

127. Twice the sum of −5 and −3

128. 7 subtracted from the quotient of 0 and 5

Decide whether the given number is a solution of the given equation.

129. Is 7 a solution of $-5x = -35$?

130. Is −4 a solution of $2x = x - 1$?

131. Is −20 a solution of $\dfrac{x}{-10} = 2$?

132. Is −3 a solution of $\dfrac{45}{x} = -15$?

133. Is 5 a solution of $-3x - 5 = -20$?

134. Is −4 a solution of $2x + 4 = x + 8$?

1.8 PROPERTIES OF REAL NUMBERS

CD-ROM   SSM

SSG   Video

▶ **OBJECTIVES**

1. Use the commutative and associative properties.
2. Use the distributive property.
3. Use the identity and inverse properties.

1 In this section we give names to properties of real numbers with which we are already familiar. Throughout this section, the variables a, b, and c represent real numbers.

We know that order does not matter when adding numbers. For example, we know that $7 + 5$ is the same as $5 + 7$. This property is given a special name—the **commutative property of addition**. We also know that order does not matter when multiplying numbers. For example, we know that $-5(6) = 6(-5)$. This property means that multiplication is commutative also and is called the **commutative property of multiplication**.

COMMUTATIVE PROPERTIES

Addition:	$a + b = b + a$
Multiplication:	$a \cdot b = b \cdot a$

These properties state that the *order* in which any two real numbers are added or multiplied does not change their sum or product. For example, if we let $a = 3$ and $b = 5$, then the commutative properties guarantee that

$$3 + 5 = 5 + 3 \quad \text{and} \quad 3 \cdot 5 = 5 \cdot 3$$

> **HELPFUL HINT**
> Is subtraction also commutative? Try an example. Is $3 - 2 = 2 - 3$? **No!** The left side of this statement equals 1; the right side equals −1. There is no commutative property of subtraction. Similarly, there is no commutative property for division. For example, $10 \div 2$ does not equal $2 \div 10$.

Example 1 Use a commutative property to complete each statement.

a. $x + 5 =$ _____   **b.** $3 \cdot x =$ _____

Solution   **a.** $x + 5 = 5 + x$ by the commutative property of addition
              b. $3 \cdot x = x \cdot 3$ by the commutative property of multiplication

Let's now discuss grouping numbers. We know that when we add three numbers, the way in which they are grouped or associated does not change their sum. For example, we know that $2 + (3 + 4) = 2 + 7 = 9$. This result is the same if we group the numbers differently. In other words, $(2 + 3) + 4 = 5 + 4 = 9$, also. Thus, $2 + (3 + 4) = (2 + 3) + 4$. This property is called the **associative property of addition**.

We also know that changing the grouping of numbers when multiplying does not change their product. For example, $2 \cdot (3 \cdot 4) = (2 \cdot 3) \cdot 4$ (check it). This is the **associative property of multiplication**.

ASSOCIATIVE PROPERTIES

Addition:	$(a + b) + c = a + (b + c)$
Multiplication:	$(a \cdot b) \cdot c = a \cdot (b \cdot c)$

These properties state that the way in which three numbers are *grouped* does not change their sum or their product.

Example 2 Use an associative property to complete each statement.

a. $5 + (4 + 6) = $ _____

b. $(-1 \cdot 2) \cdot 5 = $ _____

Solution **a.** $5 + (4 + 6) = (5 + 4) + 6$ by the associative property of addition
b. $(-1 \cdot 2) \cdot 5 = -1 \cdot (2 \cdot 5)$ by the associative property of multiplication

> **HELPFUL HINT**
> Remember the difference between the commutative properties and the associative properties. The commutative properties have to do with the *order* of numbers, and the associative properties have to do with the *grouping* of numbers.

Let's now illustrate how these properties can help us simplify expressions.

Example 3 Simplify each expression.

a. $10 + (x + 12)$

b. $-3(7x)$

Solution **a.** $10 + (x + 12) = 10 + (12 + x)$ by the commutative property of addition
$\qquad\qquad\qquad = (10 + 12) + x$ by the associative property of addition
$\qquad\qquad\qquad = 22 + x$ Add.
b. $-3(7x) = (-3 \cdot 7)x$ by the associative property of multiplication
$\qquad\qquad = -21x$ Multiply.

2 The **distributive property of multiplication over addition** is used repeatedly throughout algebra. It is useful because it allows us to write a product as a sum or a sum as a product.

We know that $7(2 + 4) = 7(6) = 42$. Compare that with $7(2) + 7(4) = 14 + 28 = 42$. Since both original expressions equal 42, they must equal each other, or

$$7(2 + 4) = 7(2) + 7(4)$$

This is an example of the distributive property. The product on the left side of the equal sign is equal to the sum on the right side. We can think of the 7 as being distributed to each number inside the parentheses.

DISTRIBUTIVE PROPERTY OF MULTIPLICATION OVER ADDITION

$$a(b + c) = ab + ac$$

Since multiplication is commutative, this property can also be written as

$$(b + c)a = ba + ca$$

The distributive property can also be extended to more than two numbers inside the parentheses. For example,

$$3(x + y + z) = 3(x) + 3(y) + 3(z)$$
$$= 3x + 3y + 3z$$

Since we define subtraction in terms of addition, the distributive property is also true for subtraction. For example

$$2(x - y) = 2(x) - 2(y)$$
$$= 2x - 2y$$

Example 4 Use the distributive property to write each expression without parentheses. Then simplify the result.

a. $2(x + y)$ **b.** $-5(-3 + 2z)$ **c.** $5(x + 3y - z)$
d. $-1(2 - y)$ **e.** $-(3 + x - w)$ **f.** $4(3x + 7) + 10$

Solution **a.** $2(x + y) = 2 \cdot x + 2 \cdot y$
$$= 2x + 2y$$

b. $-5(-3 + 2z) = -5(-3) + (-5)(2z)$
$$= 15 - 10z$$

c. $5(x + 3y - z) = 5(x) + 5(3y) - 5(z)$
$$= 5x + 15y - 5z$$

d. $-1(2 - y) = (-1)(2) - (-1)(y)$
$$= -2 + y$$

> **HELPFUL HINT**
> Notice in part **e** that
> $-(3 + x - w)$ is first
> rewritten as
> $-1(3 + x - w)$.

e. $-(3 + x - w) = -1(3 + x - w)$
$$= (-1)(3) + (-1)(x) - (-1)(w)$$
$$= -3 - x + w$$

f. $4(3x + 7) + 10 = 4(3x) + 4(7) + 10$ *Apply the distributive property.*
$$= 12x + 28 + 10$$ *Multiply.*
$$= 12x + 38$$ *Add.*

The distributive property can also be used to write a sum as a product.

Example 5 Use the distributive property to write each sum as a product.

a. $8 \cdot 2 + 8 \cdot x$ **b.** $7s + 7t$

Solution **a.** $8 \cdot 2 + 8 \cdot x = 8(2 + x)$ **b.** $7s + 7t = 7(s + t)$

3 Next, we look at the **identity properties**.

The number 0 is called the identity for addition because when 0 is added to any real number, the result is the same real number. In other words, the *identity* of the real number is not changed.

The number 1 is called the identity for multiplication because when a real number is multiplied by 1, the result is the same real number. In other words, the *identity* of the real number is not changed.

IDENTITIES FOR ADDITION AND MULTIPLICATION

0 is the identity element for addition.

$$a + 0 = a \qquad \text{and} \qquad 0 + a = a$$

1 is the identity element for multiplication.

$$a \cdot 1 = a \qquad \text{and} \qquad 1 \cdot a = a$$

Notice that 0 is the *only* number that can be added to any real number with the result that the sum is the same real number. Also, 1 is the *only* number that can be multiplied by any real number with the result that the product is the same real number.

Additive inverses or **opposites** were introduced in Section 1.5. Two numbers are called additive inverses or opposites if their sum is 0. The additive inverse or opposite of 6 is -6 because $6 + (-6) = 0$. The additive inverse or opposite of -5 is 5 because $-5 + 5 = 0$.

Reciprocals or **multiplicative inverses** were introduced in Section 1.3. Two nonzero numbers are called reciprocals or multiplicative inverses if their product is 1. The reciprocal or multiplicative inverse of $\frac{2}{3}$ is $\frac{3}{2}$ because $\frac{2}{3} \cdot \frac{3}{2} = 1$. Likewise, the reciprocal of -5 is $-\frac{1}{5}$ because $-5\left(-\frac{1}{5}\right) = 1$.

ADDITIVE OR MULTIPLICATIVE INVERSES

The numbers a and $-a$ are additive inverses or opposites of each other because their sum is 0; that is,

$$a + (-a) = 0$$

The numbers b and $\frac{1}{b}$ (for $b \neq 0$) are reciprocals or multiplicative inverses of each other because their product is 1; that is,

$$b \cdot \frac{1}{b} = 1$$

Example 6 Name the property illustrated by each true statement.

Solution
a. $3 \cdot y = y \cdot 3$ *Commutative property of multiplication (order changed)*
b. $(x + 7) + 9 = x + (7 + 9)$ *Associative property of addition (grouping changed)*
c. $(b + 0) + 3 = b + 3$ *Identity element for addition*

d. $2 \cdot (z \cdot 5) = 2 \cdot (5 \cdot z)$ — Commutative property of multiplication (order changed)

e. $-2 \cdot \left(-\dfrac{1}{2}\right) = 1$ — Multiplicative inverse property

f. $-2 + 2 = 0$ — Additive inverse property

g. $-6 \cdot (y \cdot 2) = (-6 \cdot 2) \cdot y$ — Commutative and associative properties of multiplication (order and grouping changed)

Exercise Set 1.8

Use a commutative property to complete each statement. See Examples 1 and 3.

1. $x + 16 =$ _____
2. $4 + y =$ _____
3. $-4 \cdot y =$ _____
4. $-2 \cdot x =$ _____
5. $xy =$ _____
6. $ab =$ _____
7. $2x + 13 =$ _____
8. $19 + 3y =$ _____

Use an associative property to complete each statement. See Examples 2 and 3.

9. $(xy) \cdot z =$ _____
10. $3 \cdot (xy) =$ _____
11. $2 + (a + b) =$ _____
12. $(y + 4) + z =$ _____
13. $4 \cdot (ab) =$ _____
14. $(-3y) \cdot z =$ _____
15. $(a + b) + c =$ _____
16. $6 + (r + s) =$ _____

Use the commutative and associative properties to simplify each expression. See Example 3.

17. $8 + (9 + b)$
18. $(r + 3) + 11$
19. $4(6y)$
20. $2(42x)$
21. $\dfrac{1}{5}(5y)$
22. $\dfrac{1}{8}(8z)$
23. $(13 + a) + 13$
24. $7 + (x + 4)$
25. $-9(8x)$
26. $-3(12y)$
27. $\dfrac{3}{4}\left(\dfrac{4}{3}s\right)$
28. $\dfrac{2}{7}\left(\dfrac{7}{2}r\right)$

29. Write an example that shows that division is not commutative.
30. Write an example that shows that subtraction is not commutative.

Use the distributive property to write each expression without parentheses. Then simplify the result. See Example 4.

31. $4(x + y)$
32. $7(a + b)$
33. $9(x - 6)$
34. $11(y - 4)$
35. $2(3x + 5)$
36. $5(7 + 8y)$
37. $7(4x - 3)$
38. $3(8x - 1)$
39. $3(6 + x)$
40. $2(x + 5)$
41. $-2(y - z)$
42. $-3(z - y)$
43. $-7(3y + 5)$
44. $-5(2r + 11)$

45. $5(x + 4m + 2)$
46. $8(3y + z - 6)$
47. $-4(1 - 2m + n)$
48. $-4(4 + 2p + 5)$
49. $-(5x + 2)$
50. $-(9r + 5)$
51. $-(r - 3 - 7p)$
52. $-(q - 2 + 6r)$
53. $\dfrac{1}{2}(6x + 8)$
54. $\dfrac{1}{4}(4x - 2)$
55. $-\dfrac{1}{3}(3x - 9y)$
56. $-\dfrac{1}{5}(10a - 25b)$
57. $3(2r + 5) - 7$
58. $10(4s + 6) - 40$
59. $-9(4x + 8) + 2$
60. $-11(5x + 3) + 10$
61. $-4(4x + 5) - 5$
62. $-6(2x + 1) - 1$

Use the distributive property to write each sum as a product. See Example 5.

63. $4 \cdot 1 + 4 \cdot y$
64. $14 \cdot z + 14 \cdot 5$
65. $11x + 11y$
66. $9a + 9b$
67. $(-1) \cdot 5 + (-1) \cdot x$
68. $(-3)a + (-3)b$
69. $30a + 30b$
70. $25x + 25y$

Name the properties illustrated by each true statement. See Example 6.

71. $3 \cdot 5 = 5 \cdot 3$
72. $4(3 + 8) = 4 \cdot 3 + 4 \cdot 8$
73. $2 + (x + 5) = (2 + x) + 5$
74. $(x + 9) + 3 = (9 + x) + 3$
75. $9(3 + 7) = 9 \cdot 3 + 9 \cdot 7$
76. $1 \cdot 9 = 9$
77. $(4 \cdot y) \cdot 9 = 4 \cdot (y \cdot 9)$
78. $6 \cdot \dfrac{1}{6} = 1$
79. $0 + 6 = 6$
80. $(a + 9) + 6 = a + (9 + 6)$
81. $-4(y + 7) = -4 \cdot y + (-4) \cdot 7$
82. $(11 + r) + 8 = (r + 11) + 8$
83. $-4 \cdot (8 \cdot 3) = (8 \cdot -4) \cdot 3$
84. $r + 0 = r$

Fill in the table with the opposite (additive inverse), and the reciprocal (multiplicative inverse). Assume that the value of each expression is not 0.

	Expression	Opposite	Reciprocal
85.	8		
86.	$-\frac{2}{3}$		
87.	x		
88.	$4y$		
89.	$2x$		
90.	$-7x$		

Determine which pairs of actions are commutative.

91. "taking a test" and "studying for the test"

92. "putting on your shoes" and "putting on your socks"

93. "putting on your left shoe" and "putting on your right shoe"

94. "reading the sports section" and "reading the comics section"

95. Explain why 0 is called the identity element for addition.

96. Explain why 1 is called the identity element for multiplication.

1.9 READING GRAPHS

CD-ROM   SSM

SSG   Video

▶ **O B J E C T I V E S**

1. Read bar graphs.
2. Read line graphs.

In today's world, where the exchange of information must be fast and entertaining, graphs are becoming increasingly popular. They provide a quick way of making comparisons, drawing conclusions, and approximating quantities.

1

    A **bar graph** consists of a series of bars arranged vertically or horizontally. The bar graph in Example 1 shows a comparison of the rates charged by selected electricity companies. The names of the companies are listed horizontally and a bar is shown for each company. Corresponding to the height of the bar for each company is a number along a vertical axis. These vertical numbers are cents charged for each kilowatt-hour of electricity used.

◆ **Example 1**   The following bar graph shows the cents charged per kilowatt-hour for selected electricity companies.

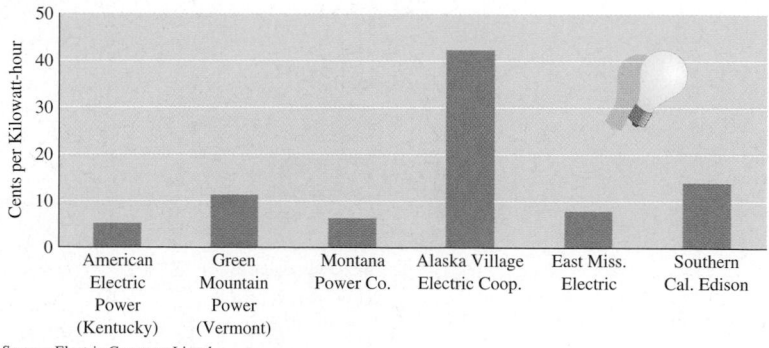

Source: Electric Company Listed

a. Which company charges the highest rate?

b. Which company charges the lowest rate?

c. Approximate the electricity rate charged by the first four companies listed.

d. Approximate the difference in the rates charged by the companies in parts (a) and (b).

Solution   **a.** The tallest bar corresponds to the company that charges the highest rate. Alaska Village Electric Cooperative charges the highest rate.

b. The shortest bar corresponds to the company that charges the lowest rate. American Electric Power in Kentucky charges the lowest rate.

c. To approximate the rate charged by American Electric Power, we go to the top of the bar that corresponds to this company. From the top of the bar, we move horizontally to the left until the vertical axis is reached.

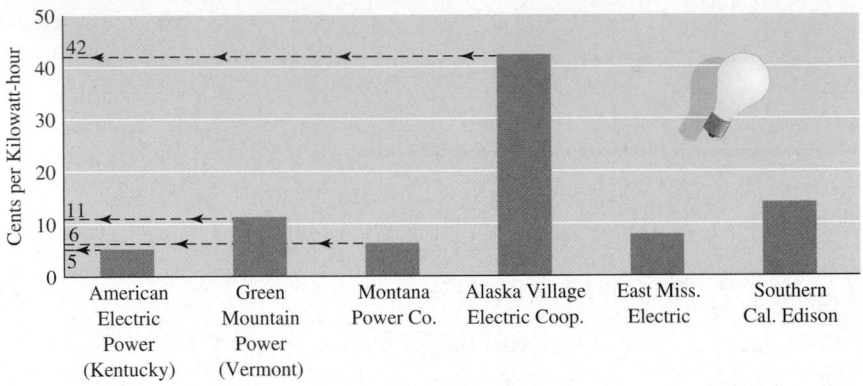

Source: Electric Company Listed

The height of the bar is approximately halfway between the 0 and 10 marks. We therefore conclude that

American Electric Power charges approximately 5¢ per kilowatt-hour.
Green Mountain Power charges approximately 11¢ per kilowatt-hour.
Montana Power Co. charges approximately 6¢ per kilowatt-hour.
Alaska Village Electric charges approximately 42¢ per kilowatt-hour.

d. The difference in rates for Alaska Village Electric Cooperative and American Electric Power is approximately 42¢ − 5¢ or 37¢.

Example 2   The following bar graph shows Disney's top animated films and the amount of money they generated at theaters.

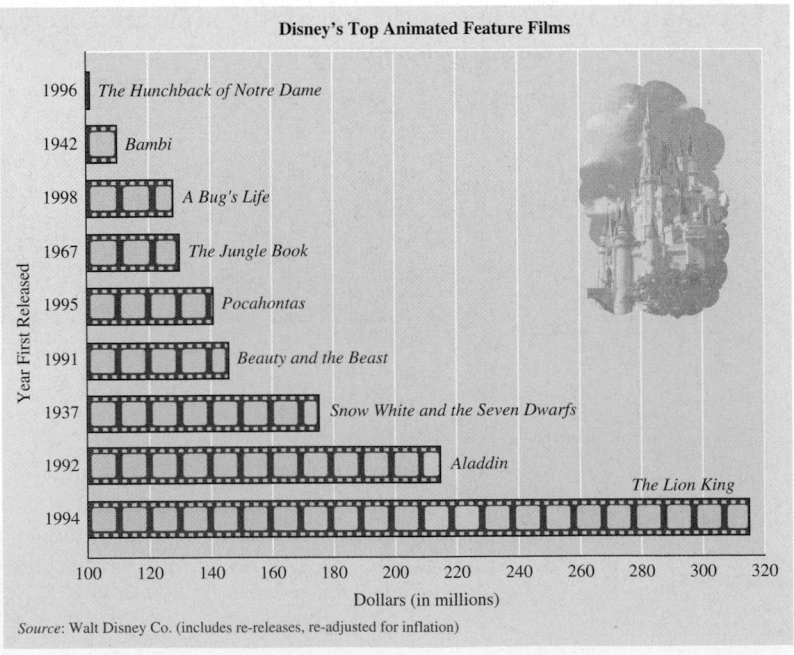

Source: Walt Disney Co. (includes re-releases, re-adjusted for inflation)

a. Find the film shown that generated the most income for Disney and approximate the income.

b. How much more money did the film *Aladdin* make than the film *Beauty and the Beast*?

Solution   **a.** Since these bars are arranged horizontally, we look for the longest bar, which is the bar representing the film *The Lion King*. To approximate the income from this film, we move from the right edge of this bar vertically downward to the dollars axis. This film generated approximately 315 million dollars, or $315,000,000, the most income for Disney.

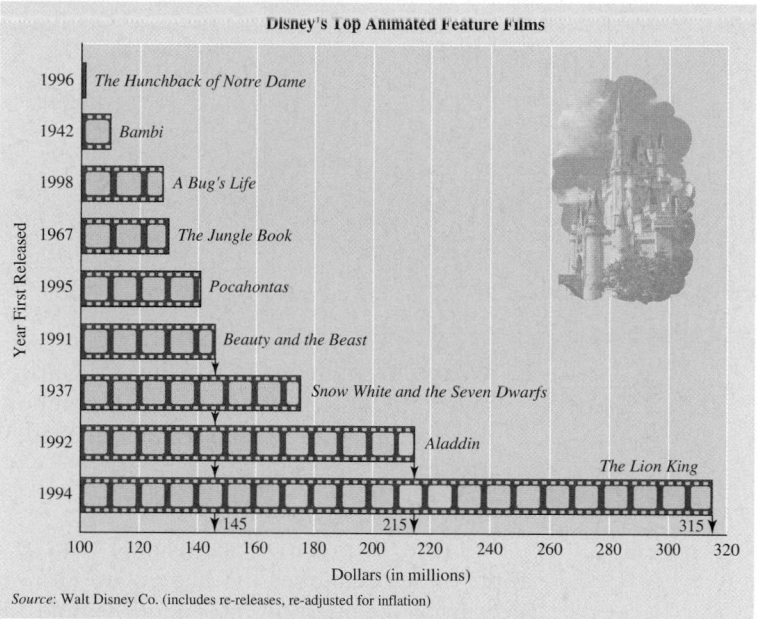

b. *Aladdin* generated approximately 215 million dollars. *Beauty and the Beast* generated approximately 145 million dollars. To find how much more money *Aladdin* generated than *Beauty and the Beast*, we subtract 215 − 145 = 70 million dollars, or $70,000,000.

2   A **line graph** consists of a series of points connected by a line. The graph in Example 3 is a line graph.

Example 3   The line graph below shows the relationship between the distance driven in a 14-foot U-Haul truck in one day and the total cost of renting this truck for that day. Notice that the horizontal axis is labeled Distance and the vertical axis is labeled Total Cost.

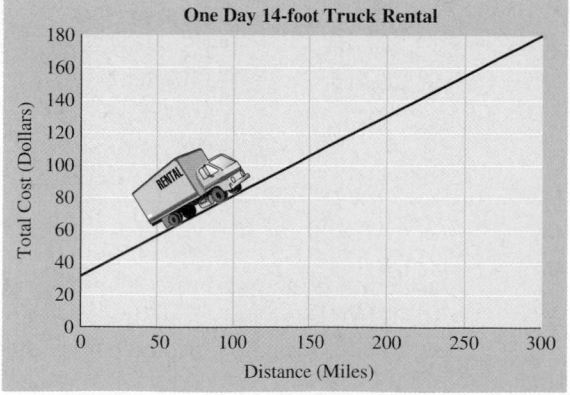

a. Find the total cost of renting the truck if 100 miles are driven.
b. Find the number of miles driven if the total cost of renting is $140.

Solution **a.** Find the number 100 on the horizontal scale and move vertically upward until the line is reached. From this point on the line, we move horizontally to the left until the vertical scale is reached. We find that the total cost of renting the truck if 100 miles are driven is approximately $80.

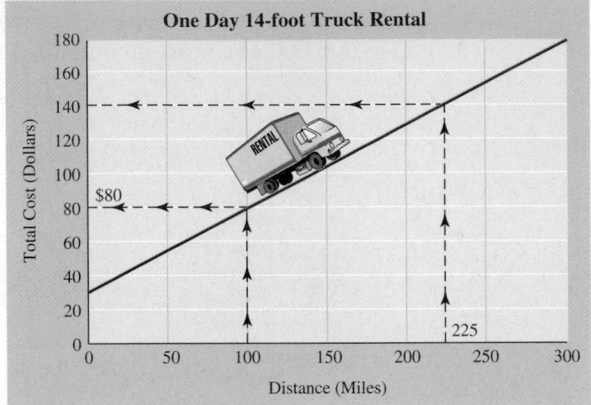

b. We find the number 140 on the vertical scale and move horizontally to the right until the line is reached. From this point on the line, we move vertically downward until the horizontal scale is reached. We find that the truck is driven approximately 225 miles.

From the previous example, we can see that graphing provides a quick way to approximate quantities. In Chapter 6 we show how we can use equations to find exact answers to the questions posed in Example 3. The next graph is another example of a line graph. It is also sometimes called a **broken line graph**.

◆ **Example 4** The line graph shows the relationship between time spent smoking a cigarette and pulse rate. Time is recorded along the horizontal axis in minutes, with 0 minutes being the moment a smoker lights a cigarette. Pulse is recorded along the vertical axis in heartbeats per minute.

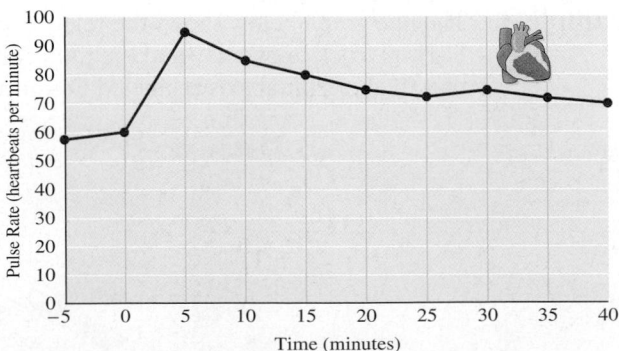

a. What is the pulse rate 15 minutes after lighting a cigarette?
b. When is the pulse rate the lowest?
c. When does the pulse rate show the greatest change?

Solution   **a.** We locate the number 15 along the time axis and move vertically upward until the line is reached. From this point on the line, we move horizontally to the left until the pulse rate axis is reached. Reading the number of beats per minute, we find that the pulse rate is 80 beats per minute 15 minutes after lighting a cigarette.

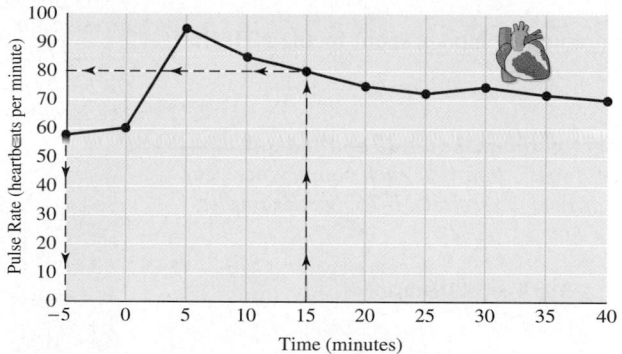

b. We find the lowest point of the line graph, which represents the lowest pulse rate. From this point, we move vertically downward to the time axis. We find that the pulse rate is the lowest at −5 minutes, which means 5 minutes *before* lighting a cigarette.

c. The pulse rate shows the greatest change during the 5 minutes between 0 and 5. Notice that the line graph is *steepest* between 0 and 5 minutes.

Exercise Set 1.9

The following bar graph shows the number of teenagers expected to use the Internet for the years shown. Use this graph to answer Exercises 1–4. See Example 1.

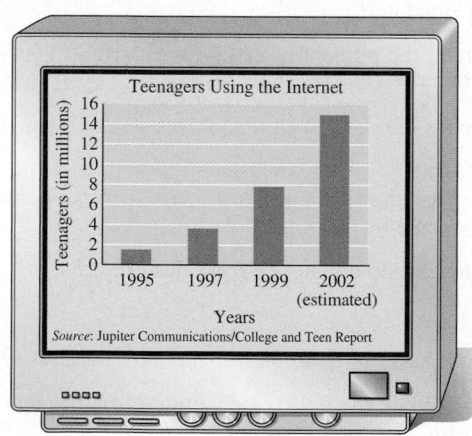

1. Approximate the number of teenagers expected to use the Internet in 1999.

2. Approximate the number of teenagers who use the Internet in 1995.

3. What year shows the greatest *increase* in number of teenagers using the Internet?

4. How many more teenagers are expected to use the Internet in 2002 than in 1999?

The following bar graph shows the amounts of money used by major pro sports for advertising in a recent year. Use this graph to answer Exercises 5–10. See Example 2.

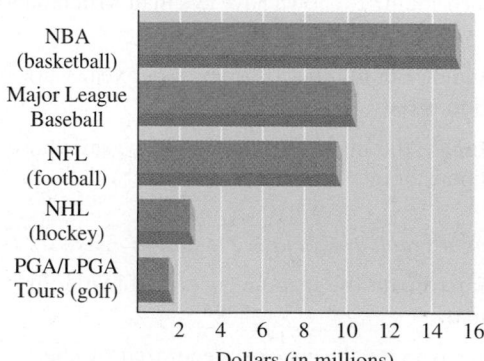

Source: "Competitive Media Reporting," *USA Today,* 6/9/97

5. Which major pro sport used the least amount of money for advertising?

6. Which major pro sport used the greatest amount of money for advertising?

7. Which major pro sports spent over $10,000,000 in advertising?

8. Which major pro sports spent under $5,000,000 in advertising?

9. Estimate the amount of money spent by the NBA for advertising.

10. Estimate the amount of money spent by the NHL for advertising.

The following bar graph shows the top 10 tourist destinations and the number of tourists that visit each country per year. Use this graph to answer Exercises 11–16. See Examples 1 and 2.

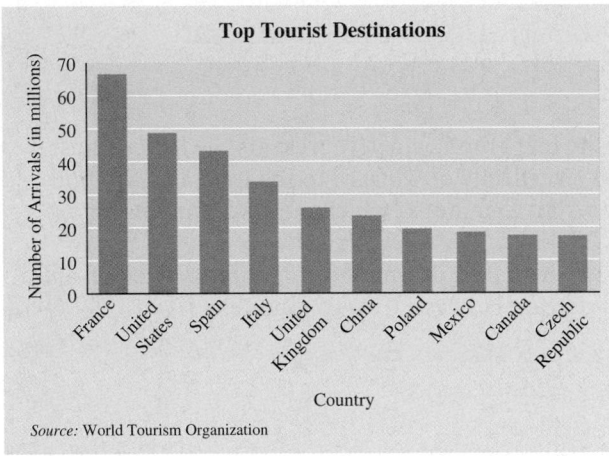

Top Tourist Destinations

Source: World Tourism Organization

11. Which country is the most popular tourist destination?

12. Which country shown is the least popular tourist destination?

13. Which countries have more than 30 million tourists per year?

14. Which countries shown have less than 20 million tourists per year?

15. Estimate the number of tourists per year whose destination is Italy.

16. Estimate the number of tourists per year whose destination is France.

Use the bar graph in Example 2 to answer Exercises 17–22.

17. Approximate the income generated by the film *Pocahontas*.

18. Approximate the income generated by the film *The Hunchback of Notre Dame*.

19. Before 1990, which Disney film generated the most income?

20. After 1990, which Disney film generated the most income?

21. Why do you think that the Disney film *The Little Mermaid* is not shown on this graph?

22. How much less money did the film *The Hunchback of Notre Dame* generate than *Pocahontas*?

Many fires are deliberately set. An increasing number of those arrested for arson are juveniles (age 17 and under). The following line graph shows the percent of deliberately set fires started by juveniles. Use this graph to answer Exercises 23–30. See Examples 3 and 4.

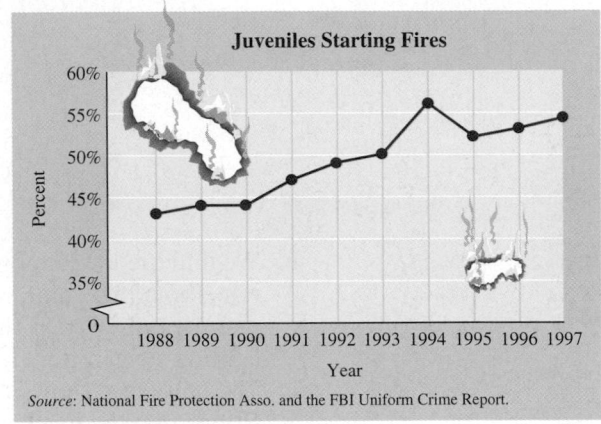

Juveniles Starting Fires

Source: National Fire Protection Asso. and the FBI Uniform Crime Report.

23. What year shows the highest percent of arson fires started by juveniles?

24. What year since 1990 shows a decrease in the percent of fires started by juveniles?

25. Name two consecutive years where the percent appears to remain the same.

26. What year shows the lowest percent of arson fires started by juveniles?

27. Estimate the percent of arson fires started by juveniles in 1997.

28. Estimate the percent of arson fires started by juveniles in 1992.

29. What year shows the greatest increase in the percent of fires started by juveniles?

30. What trend do you notice from this graph?

Use the line graph in Example 4 to answer Exercises 31–34.

31. Approximate the pulse rate 5 minutes before lighting a cigarette.

32. Approximate the pulse rate 10 minutes after lighting a cigarette.

33. Find the difference in pulse rate between 5 minutes before and 10 minutes after lighting a cigarette.

34. When is the pulse rate less than 60 heartbeats per minute?

The line graph below shows the number of students per computer in U.S. public schools. Use this graph for Exercises 35–39. See Examples 3 and 4.

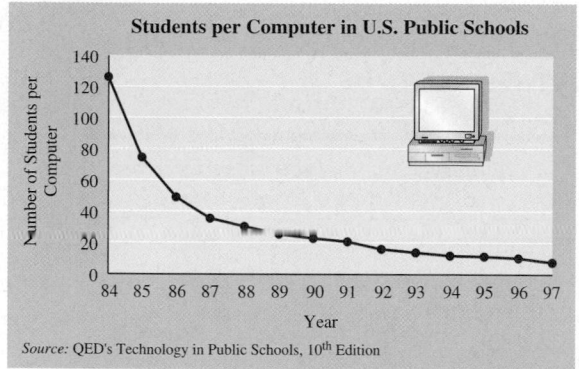

Students per Computer in U.S. Public Schools

Source: QED's Technology in Public Schools, 10ᵗʰ Edition

35. Approximate the number of students per computer in 1991.
36. Approximate the number of students per computer in 1997.
37. During what year was the greatest decrease in number of students per computer?
38. What was the first year that the number of students per computer fell below 20?
39. Discuss any trends shown by this line graph.

The special bar graph shown in the next column is called a double bar graph. This double bar graph is used to compare men and women in the U.S. labor force per year . Use this graph for Exercises 40–48.

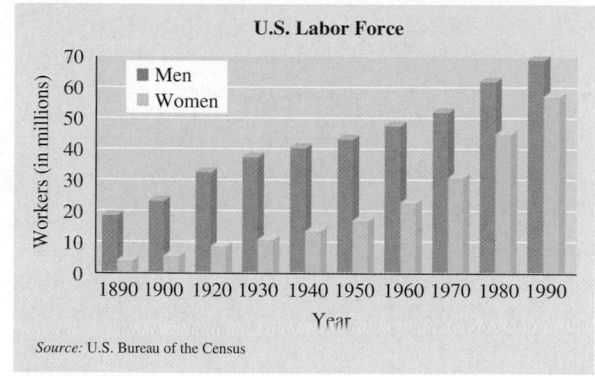

U.S. Labor Force

Source: U.S. Bureau of the Census

40. Estimate the number of men in the workforce in 1890.
41. Estimate the number of women in the workforce in 1890.
42. Estimate the number of women in the workforce in 1990.
43. Estimate the number of men in the workforce in 1990.
44. Give the first year that the number of men in the workforce rose above 20 million.
45. Give the first year that the number of women in the workforce rose above 20 million.
46. Estimate the difference in the number of men and women in the workforce in 1940.
47. Estimate the difference in the number of men and women in the workforce in 1990.
48. Discuss any trends shown by this graph.

Geographic locations can be described by a gridwork of lines called latitudes and longitudes, as shown below. For example, the location of Houston, Texas, can be described by latitude 30° north and longitude 95° west. Use the map shown to answer Exercises 49–52.

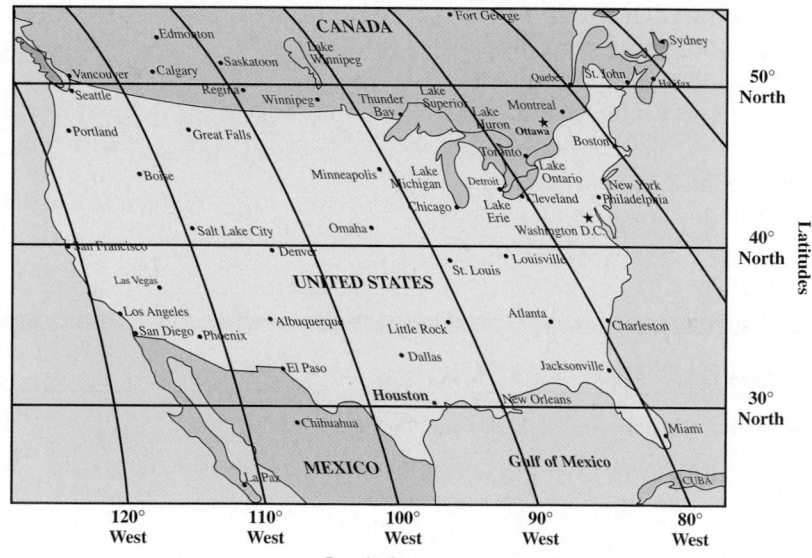

49. Using latitude and longitude, describe the location of New Orleans, Louisiana.
50. Using latitude and longitude, describe the location of Denver, Colorado.
51. Use an atlas and describe the location of your home-town.
52. Give another name for 0° latitude.

For additional Chapter Projects, visit the Real World Activities Website by going to http://www.prenhall.com/martin-gay.

CHAPTER PROJECT

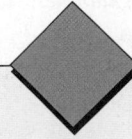

Creating and Interpreting Graphs

Companies often rely on market research to investigate the types of consumers who buy their products and their competitors' products. One way of gathering this type of data is through the use of surveys. The raw data collected from surveys can be difficult to interpret without some organization. Graphs organize data visually. They also allow a user to interpret the data quickly.

In this project, you will conduct a brief survey and create tables and graphs to represent the results. This project may be completed by working in groups or individually.

1. Begin by conducting this survey with fellow students, either in this class or another class.
2. For each survey question, tally the results for each response category. Present the results in a table.
3. For each survey question, find the fraction of the total number of responses that fall in each response category.
4. For each survey question, decide which type of graph would best represent the data: bar graph, line graph, or circle graph. Then create an appropriate graph for each set of responses to the questions.
5. Use the data for the two Internet questions and the two newspaper questions to record the increase or decrease in the number of people who participated in these activities. Complete the table in the next column.
6. Study the tables and graphs. What may you conclude from them? What do they tell you about your survey respondents? Write a paragraph summarizing your findings.

Activity	Increase or Decrease from 3 Years Ago
Using the Internet more than 2 hours per week	
Reading a newspaper at least once per week	

SURVEY

What is your age?
Under 20   20s   30s   40s   50s   60 and older

What is your gender?
Female   Male

Do you currently use the Internet more than 2 hours per week?
Yes   No

Three years ago, did you use the Internet more than 2 hours per week?
Yes   No

Do you currently read a newspaper at least once per week?
Yes   No

Three years ago, did you read a newspaper at least once per week?
Yes   No

CHAPTER 1 VOCABULARY CHECK

Fill in each blank with one of the words or phrases listed below.

set	inequality symbols	opposites	absolute value	numerator
denominator	grouping symbols	exponent	base	reciprocals
variable	equation	solution		

1. The symbols $\neq, <,$ and $>$ are called _____ .
2. A mathematical statement that two expressions are equal is called an _____ .
3. The _____ of a number is the distance between that number and 0 on the number line.
4. A symbol used to represent a number is called a _____ .

5. Two numbers that are the same distance from 0 but lie on opposite sides of 0 are called _____.

6. The number in a fraction above the fraction bar is called the _____.

7. A _____ of an equation is a value for the variable that makes the equation a true statement.

8. Two numbers whose product is 1 are called _____.

9. In 2^3, the 2 is called the _____ and the 3 is called the _____.

10. The number in a fraction below the fraction bar is called the _____.

11. Parentheses and brackets are examples of _____.

12. A _____ is a collection of objects.

CHAPTER 1 HIGHLIGHTS

DEFINITIONS AND CONCEPTS	EXAMPLES

Section 1.2 Symbols and Sets of Numbers

A **set** is a collection of objects, called **elements**, enclosed in braces.

$\{a, c, e\}$

Natural Numbers: $\{1, 2, 3, 4, \dots\}$

Whole Numbers: $\{0, 1, 2, 3, 4, \dots\}$

Integers: $\{\dots, -3, -2, -1, 0, 1, 2, 3, \dots\}$

Rational Numbers: {real numbers that can be expressed as a quotient of integers}

Irrational Numbers: {real numbers that cannot be expressed as a quotient of integers}

Real Numbers: {all numbers that correspond to a point on the number line}

Given the set $\{-3.4, \sqrt{3}, 0, \frac{2}{3}, 5, -4\}$, list the numbers that belong to the set of

Natural numbers 5

Whole numbers 0, 5

Integers $-4, 0, 5$

Rational numbers $-3.4, 0, \frac{2}{3}, 5, -4$

Irrational numbers $\sqrt{3}$

Real numbers $-3.4, \sqrt{3}, 0, \frac{2}{3}, 5, -4$

A line used to picture numbers is called a **number line**.

The **absolute value** of a real number a, denoted by $|a|$, is the distance between a and 0 on the number line.

$|5| = 5$ $|0| = 0$ $|-2| = 2$

Symbols:
$=$ is equal to
$\neq$ is not equal to
$>$ is greater than
$<$ is less than
$\leq$ is less than or equal to
$\geq$ is greater than or equal to

$-7 = -7$

$3 \neq -3$

$4 > 1$

$1 < 4$

$6 \leq 6$

$18 \geq -\dfrac{1}{3}$

Order Property for Real Numbers

For any two real numbers a and b, a is less than b if a is to the left of b on the number line.

$-3 < 0$ $0 > -3$ $0 < 2.5$ $2.5 > 0$

Section 1.3 Fractions

A quotient of two integers is called a **fraction**. The **numerator** of a fraction is the top number. The **denominator** of a fraction is the bottom number.

$\dfrac{13}{17}$ $\leftarrow$ numerator
$\phantom{\dfrac{13}{17}}$ $\leftarrow$ denominator

If $a \cdot b = c$, then a and b are **factors** and c is the **product**.

$$7 \cdot 9 = 63$$
$\downarrow$ $\downarrow$ $\downarrow$
factor   factor   product   *(continued)*

DEFINITIONS AND CONCEPTS	EXAMPLES

Section 1.3 Fractions

A fraction is in **lowest terms** when the numerator and the denominator have no factors in common other than 1.

$\frac{13}{17}$ is in lowest terms.

To write a fraction in lowest terms, factor the numerator and the denominator; then apply the fundamental property.

Write in lowest terms.
$$\frac{6}{14} = \frac{2 \cdot 3}{2 \cdot 7} = \frac{3}{7}$$

Two fractions are **reciprocals** if their product is 1. The reciprocal of $\frac{a}{b}$ is $\frac{b}{a}$.

The reciprocal of
$$\frac{6}{25} \text{ is } \frac{25}{6}$$

To multiply fractions, numerator times numerator is the numerator of the product and denominator times denominator is the denominator of the product.

To divide fractions, multiply the first fraction by the reciprocal of the second fraction.

To add fractions with the same denominator, add the numerators and place the sum over the common denominator.

To subtract fractions with the same denominator, subtract the numerators and place the difference over the common denominator.

Fractions that represent the same quantity are called **equivalent fractions**.

Perform the indicated operations.
$$\frac{2}{5} \cdot \frac{3}{7} = \frac{6}{35}$$
$$\frac{5}{9} \div \frac{2}{7} = \frac{5}{9} \cdot \frac{7}{2} = \frac{35}{18}$$
$$\frac{5}{11} + \frac{3}{11} = \frac{8}{11}$$
$$\frac{13}{15} - \frac{3}{15} = \frac{10}{15} = \frac{2}{3}$$
$$\frac{1}{5} = \frac{1 \cdot 4}{5 \cdot 4} = \frac{4}{20}$$

$\frac{1}{5}$ and $\frac{4}{20}$ are equivalent fractions.

Section 1.4 Introduction to Variable Expressions and Equations

The expression a^n is an **exponential expression**. The number a is called the **base**; it is the repeated factor. The number n is called the **exponent**; it is the number of times that the base is a factor.

$$4^3 = 4 \cdot 4 \cdot 4 = 64$$
$$7^2 = 7 \cdot 7 = 49$$

Order of Operations

Simplify expressions in the following order. If grouping symbols are present, simplify expressions within those first, starting with the innermost set. Also, simplify the numerator and the denominator of a fraction separately.

1. Simplify exponential expressions.
2. Multiply or divide in order from left to right.
3. Add or subtract in order from left to right.

$$\frac{8^2 + 5(7-3)}{3 \cdot 7} = \frac{8^2 + 5(4)}{21}$$
$$= \frac{64 + 5(4)}{21}$$
$$= \frac{64 + 20}{21}$$
$$= \frac{84}{21}$$
$$= 4$$

A symbol used to represent a number is called a **variable**.

Examples of variables are:
$$q, x, z$$

An **algebraic expression** is a collection of numbers, variables, operation symbols, and grouping symbols.

Examples of algebraic expressions are:
$$5x, 2(y-6), \frac{q^2 - 3q + 1}{6}$$

To evaluate an algebraic expression containing a variable, substitute a given number for the variable and simplify.

Evaluate $x^2 - y^2$ if $x = 5$ and $y = 3$.
$$x^2 - y^2 = (5)^2 - 3^2$$
$$= 25 - 9$$
$$= 16 \qquad (continued)$$

DEFINITIONS AND CONCEPTS	EXAMPLES

Section 1.4 Introduction to Variable Expressions and Equations

A mathematical statement that two expressions are equal is called an **equation**.	Equations: $$3x - 9 = 20$$ $$A = \pi r^2$$
A **solution** or **root** of an equation is a value for the variable that makes the equation a true statement.	Determine whether 4 is a solution of $5x + 7 = 27$. $$5x + 7 = 27$$ $$5(4) + 7 = 27$$ $$20 + 7 = 27$$ $$27 = 27 \quad \text{True.}$$ 4 is a solution.

Section 1.5 Adding Real Numbers

To Add Two Numbers with the Same Sign 1. Add their absolute values. 2. Use their common sign as the sign of the sum.	Add. $$10 + 7 = 17$$ $$-3 + (-8) = -11$$
To Add Two Numbers with Different Signs 1. Subtract their absolute values. 2. Use the sign of the number whose absolute value is larger as the sign of the sum.	$$-25 + 5 = -20$$ $$14 + (-9) = 5$$
Two numbers that are the same distance from 0 but lie on opposite sides of 0 are called **opposites** or **additive inverses**. The opposite of a number a is denoted by $-a$.	The opposite of -7 is 7. The opposite of 123 is -123.
The sum of a number a and its opposite, $-a$, is 0. $$a + (-a) = 0$$ If a is a number, then $-(-a) = a$.	$$-4 + 4 = 0$$ $$12 + (-12) = 0$$ $$-(-8) = 8$$ $$-(-14) = 14$$

Section 1.6 Subtracting Real Numbers

To subtract two numbers a and b, add the first number a to the opposite of the second number b. $$a - b = a + (-b)$$	Subtract. $$3 - (-44) = 3 + 44 = 47$$ $$-5 - 22 = -5 + (-22) = -27$$ $$-30 - (-30) = -30 + 30 = 0$$

Section 1.7 Multiplying and Dividing Real Numbers

Quotient of two real numbers $$\frac{a}{b} = a \cdot \frac{1}{b}$$	Multiply or divide. $$\frac{42}{2} = 42 \cdot \frac{1}{2} = 21$$

(continued)

DEFINITIONS AND CONCEPTS	EXAMPLES

Section 1.7 Multiplying and Dividing Real Numbers

Multiplying and Dividing Real Numbers

The product or quotient of two numbers with the same sign is a positive number. The product or quotient of two numbers with different signs is a negative number.

$7 \cdot 8 = 56 \qquad -7 \cdot (-8) = 56$

$-2 \cdot 4 = -8 \qquad 2 \cdot (-4) = -8$

$\dfrac{90}{10} = 9 \qquad \dfrac{-90}{-10} = 9$

Products and Quotients Involving Zero

The product of 0 and any number is 0.

$$b \cdot 0 = 0 \quad \text{and} \quad 0 \cdot b = 0$$

The quotient of a nonzero number and 0 is undefined.

$$\frac{b}{0} \text{ is undefined.}$$

The quotient of 0 and any nonzero number is 0.

$$\frac{0}{b} = 0$$

$\dfrac{42}{-6} = -7 \qquad \dfrac{-42}{6} = -7$

$-4 \cdot 0 = 0 \qquad 0 \cdot \left(-\dfrac{3}{4}\right) = 0$

$\dfrac{-85}{0} \text{ is undefined.}$

$\dfrac{0}{18} = 0 \qquad \dfrac{0}{-47} = 0$

Section 1.8 Properties of Real Numbers

Commutative Properties

Addition: $\quad a + b = b + a$

Multiplication: $\quad a \cdot b = b \cdot a$

$3 + (-7) = -7 + 3$

$-8 \cdot 5 = 5 \cdot (-8)$

Associative Properties

Addition: $\quad (a + b) + c = a + (b + c)$

Multiplication: $\quad (a \cdot b) \cdot c = a \cdot (b \cdot c)$

$(5 + 10) + 20 = 5 + (10 + 20)$

$(-3 \cdot 2) \cdot 11 = -3 \cdot (2 \cdot 11)$

Two numbers whose product is 1 are called **multiplicative inverses** or **reciprocals**. The reciprocal of a nonzero number a is $\dfrac{1}{a}$ because $a \cdot \dfrac{1}{a} = 1$.

The reciprocal of 3 is $\dfrac{1}{3}$.

The reciprocal of $-\dfrac{2}{5}$ is $-\dfrac{5}{2}$.

Distributive Property $\quad a(b + c) = a \cdot b + a \cdot c$

$5(6 + 10) = 5 \cdot 6 + 5 \cdot 10$

$-2(3 + x) = -2 \cdot 3 + (-2)(x)$

Identities $\quad a + 0 = a \qquad 0 + a = a$

$\qquad\qquad\quad a \cdot 1 = a \qquad 1 \cdot a = a$

$5 + 0 = 5 \qquad 0 + (-2) = -2$

$-14 \cdot 1 = -14 \qquad 1 \cdot 27 = 27$

Inverses

Addition or opposite: $\quad a + (-a) = 0$

$7 + (-7) = 0$

Multiplication or reciprocal: $\quad b \cdot \dfrac{1}{b} = 1$

$3 \cdot \dfrac{1}{3} = 1$

Section 1.9 Reading Graphs

To find the value on the vertical axis representing a location on a graph, move horizontally from the location on the graph until the vertical axis is reached. To find the value on the horizontal axis representing a location on a graph, move vertically from the location on the graph until the horizontal axis is reached. The broken line graph to the right shows the average public classroom teachers' salaries for the school year ending in the years shown.

Estimate the average public teacher's salary for the school year ending in 1998.

Find the earliest year that the average salary rose above $37,000.

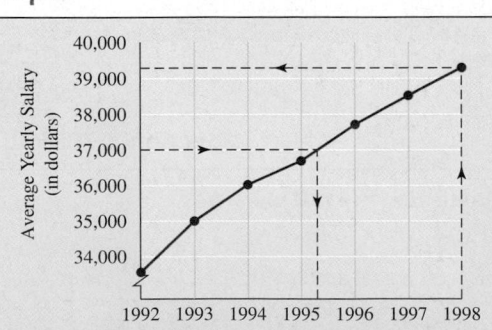

Source: U.S. Bureau of the Census, *Statistical Abstract of the United States: 1999* (114th edition) Washington, D.C., 1999

CHAPTER 1 REVIEW

(1.2) Insert $<$, $>$, or $=$ in the appropriate space to make the following statements true.

1. 8   10

2. 7   2

3. -4 -5

4. $\dfrac{12}{2}$ -8

5. $|-7|$ $|-8|$

6. $|-9|$ -9

7. $-|-1|$ -1

8. $|-14|$ $-(-14)$

9. 1.2   1.02

10. $-\dfrac{3}{2}$ $-\dfrac{3}{4}$

Translate each statement into symbols.

11. Four is greater than or equal to negative three.

12. Six is not equal to five.

13. 0.03 is less than 0.3.

14. Lions and hyenas were featured in the Disney film *The Lion King*. For short distances, lions can run at a rate of 50 miles per hour whereas hyenas can run at a rate of 40 miles per hour. Write an inequality statement comparing the numbers 50 and 40.

Given the following sets of numbers, list the numbers in each set that also belong to the set of:

a. Natural numbers   **b.** Whole numbers

c. Integers   **d.** Rational numbers

e. Irrational numbers   **f.** Real numbers

15. $\left\{-6, 0, 1, 1\dfrac{1}{2}, 3, \pi, 9.62\right\}$

16. $\left\{-3, -1.6, 2, 5, \dfrac{11}{2}, 15.1, \sqrt{5}, 2\pi\right\}$

The following chart shows the gains and losses in dollars of Density Oil and Gas stock for a particular week.

Day	Gain or Loss in Dollars
Monday	$+1$
Tuesday	-2
Wednesday	$+5$
Thursday	$+1$
Friday	-4

17. Which day showed the greatest loss?

18. Which day showed the greatest gain?

(1.3) Write the number as a product of prime factors.

19. 36   **20.** 120

Perform the indicated operations. Write results in lowest terms.

21. $\dfrac{8}{15} \cdot \dfrac{27}{30}$

22. $\dfrac{7}{8} \div \dfrac{21}{32}$

23. $\dfrac{7}{15} + \dfrac{5}{6}$

24. $\dfrac{3}{4} - \dfrac{3}{20}$

25. $2\dfrac{3}{4} + 6\dfrac{5}{8}$

26. $7\dfrac{1}{6} - 2\dfrac{2}{3}$

27. $5 \div \dfrac{1}{3}$

28. $2 \cdot 8\dfrac{3}{4}$

29. Determine the unknown part of the given circle.

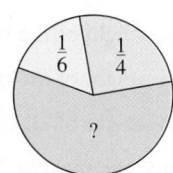

Find the area and the perimeter of each figure.

△ **30.**

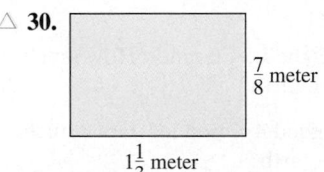

$\dfrac{7}{8}$ meter

$1\dfrac{1}{3}$ meter

△ **31.**

$\dfrac{5}{11}$ in. $\dfrac{3}{11}$ in.

$\dfrac{3}{11}$ in.

$\dfrac{5}{11}$ in.

△ **32.** A trim carpenter needs a piece of quarter round molding $6\dfrac{1}{8}$ feet long for a bathroom. She finds a piece $7\dfrac{1}{2}$ feet long. How long a piece does she need to cut from the $7\dfrac{1}{2}$-foot-long molding in order to use it in the bathroom?

In December 1998, Nkem Chukwu gave birth to the world's first surviving octuplets in Houston, Texas. The following chart gives the octuplets' birthweights. The babies are listed in order of birth.

Baby's Name	Gender	Birthweight (pounds)
Ebuka	girl	$1\frac{1}{2}$
Chidi	girl	$1\frac{11}{16}$
Echerem	girl	$1\frac{3}{4}$
Chima	girl	$1\frac{5}{8}$
Odera	girl	$\frac{11}{16}$
Ikem	boy	$1\frac{1}{8}$
Jioke	boy	$1\frac{13}{16}$
Gorom	girl	$1\frac{1}{8}$

Source: Texas Children's Hospital, Houston, Texas

33. What was the total weight of the boy octuplets?

34. What was the total weight of the girl octuplets?

35. Find the combined weight of all eight octuplets.

36. Which baby weighed the most?

37. Which baby weighed the least?

38. How much more did the heaviest baby weigh than the lightest baby?

39. By March 1999, Chima weighed $5\frac{1}{2}$ pounds. How much weight had she gained since birth?

40. By March 1999, Ikem weighed $4\frac{5}{32}$ pounds. How much weight had he gained since birth?

(1.4) Simplify each expression.

41. 2^4

42. 5^2

43. $\left(\frac{2}{7}\right)^2$

44. $\left(\frac{3}{4}\right)^3$

45. $6 \cdot 3^2 + 2 \cdot 8$

46. $68 - 5 \cdot 2^3$

47. $3(1 + 2 \cdot 5) + 4$

48. $8 + 3(2 \cdot 6 - 1)$

49. $\dfrac{4 + |6 - 2| + 8^2}{4 + 6 \cdot 4}$

50. $5[3(2 + 5) - 5]$

Translate each word statement to symbols.

51. The difference of twenty and twelve is equal to the product of two and four.

52. The quotient of nine and two is greater than negative five.

Evaluate each expression if $x = 6$, $y = 2$, and $z = 8$.

53. $2x + 3y$

54. $x(y + 2z)$

55. $\dfrac{x}{y} + \dfrac{z}{2y}$

56. $x^2 - 3y^2$

△ **57.** The expression $180 - a - b$ represents the measure of the unknown angle of the given triangle. Replace a with 37 and b with 80 to find the measure of the unknown angle.

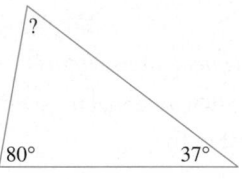

Decide whether the given number is a solution to the given equation.

58. Is $x = 3$ a solution of $7x - 3 = 18$?

59. Is $x = 1$ a solution of $3x^2 + 4 = x - 1$?

(1.5) Find the additive inverse or the opposite.

60. -9

61. $\dfrac{2}{3}$

62. $|-2|$

63. $-|-7|$

Find the following sums.

64. $-15 + 4$

65. $-6 + (-11)$

66. $\dfrac{1}{16} + \left(-\dfrac{1}{4}\right)$

67. $-8 + |-3|$

68. $-4.6 + (-9.3)$

69. $-2.8 + 6.7$

70. The lowest elevation in North America is -282 feet at Death Valley in California. If you are standing at a point 728 feet above Death Valley, what is your elevation? (*Source*: National Geographic Society)

(1.6) Perform the indicated operations.

71. $6 - 20$

72. $-3.1 - 8.4$

73. $-6 - (-11)$

74. $4 - 15$

75. $-21 - 16 + 3(8 - 2)$

76. $\dfrac{11 - (-9) + 6(8 - 2)}{2 + 3 \cdot 4}$

If $x = 3$, $y = -6$, and $z = -9$, evaluate each expression.

77. $2x^2 - y + z$

78. $\dfrac{y - x + 5x}{2x}$

79. At the beginning of the week the price of Density Oil and Gas stock from Exercises 17 and 18 is $50 per share. Find the price of a share of stock at the end of the week.

80. The expression $E - I$ represents a country's merchandise trade balance if the country has exports worth E dollars and imports worth I dollars. Find the merchandise trade balance for a country with exports worth $412 billion and imports worth $536 billion.

(1.7) *Find the multiplicative inverse or reciprocal.*

81. -6

82. $\dfrac{3}{5}$

Simplify each expression.

83. $6(-8)$

84. $(-2)(-14)$

85. $\dfrac{-18}{-6}$

86. $\dfrac{42}{-3}$

87. $-3(-6)(-2)$

88. $(-4)(-3)(0)(-6)$

89. $\dfrac{4(-3) + (-8)}{2 + (-2)}$

90. $\dfrac{3(-2)^2 - 5}{-14}$

91. $\dfrac{-6}{0}$

92. $\dfrac{0}{-2}$

93. During the 1999 LPGA Sara Lee Classic, Michelle Mc-Gann had scores of -9, -7, and $+1$ in three rounds of golf. Find her average score per round. (*Source:* Ladies Professional Golf Association)

94. During the 1999 PGA Masters Tournament, Bob Estes had scores of $-1, 0, -3$, and 0 in four rounds of golf. Find his average score per round. (*Source*: Professional Golf Association)

(1.8) *Name the property illustrated.*

95. $-6 + 5 = 5 + (-6)$

96. $6 \cdot 1 = 6$

97. $3(8 - 5) = 3 \cdot 8 + 3 \cdot (-5)$

98. $4 + (-4) = 0$

99. $2 + (3 + 9) = (2 + 3) + 9$

100. $2 \cdot 8 = 8 \cdot 2$

101. $6(8 + 5) = 6 \cdot 8 + 6 \cdot 5$

102. $(3 \cdot 8) \cdot 4 = 3 \cdot (8 \cdot 4)$

103. $4 \cdot \dfrac{1}{4} = 1$

104. $8 + 0 = 8$

105. $4(8 + 3) = 4(3 + 8)$

(1.9) *Use the following graph to answer Exercises 106–109.*

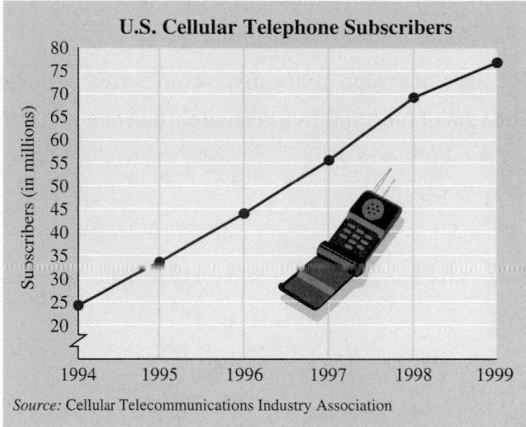

U.S. Cellular Telephone Subscribers

Source: Cellular Telecommunications Industry Association

106. Approximate the number of cellular phone subscribers in 1999.

107. Approximate the increase in cellular phone subscribers in 1999.

108. What year shows the greatest number of subscribers?

109. What trend is shown by this graph?

The following bar graph shows the average annual percent increase in rent from June 1997 to June 1998 in selected major metropolitan areas.

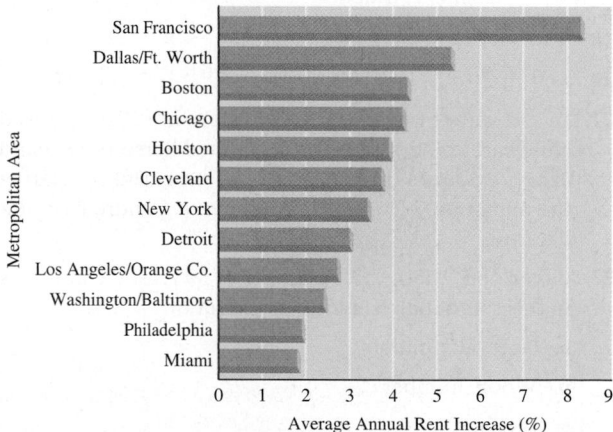

Average Annual Rent Increase (%)

Source: U.S. Department of Labor

110. Which metropolitan area had the greatest increase in annual rent? Approximate the average annual rent increase in this area.

111. Which metropolitan area had the least increase in annual rent? Approximate the average annual rent increase in this area.

112. For this period, the national average increase in rent was 3.2%. Which metropolitan areas had rental increases below the national average?

113. Which metropolitan areas had rental increases above the national average (see Exercise 112)?

CHAPTER 1 TEST

Translate the statement into symbols.

1. The absolute value of negative seven is greater than five.

2. The sum of nine and five is greater than or equal to four.

Simplify the expression.

3. $-13 + 8$

4. $-13 - (-2)$

5. $6 \cdot 3 - 8 \cdot 4$

6. $(13)(-3)$

7. $(-6)(-2)$

8. $\dfrac{|-16|}{-8}$

9. $\dfrac{-8}{0}$

10. $\dfrac{|-6| + 2}{5 - 6}$

11. $\dfrac{1}{2} - \dfrac{5}{6}$

12. $-1\dfrac{1}{8} + 5\dfrac{3}{4}$

13. $-\dfrac{3}{5} + \dfrac{15}{8}$

14. $3(-4)^2 - 80$

15. $6[5 + 2(3 - 8) - 3]$

16. $\dfrac{-12 + 3 \cdot 8}{4}$

17. $\dfrac{(-2)(0)(-3)}{-6}$

Insert <, >, or = in the appropriate space to make each of the following statements true.

18. -3 -7

19. 4 -8

20. $|-3|$ 2

21. $|-2|$ $-1 - (-3)$

22. In the state of Massachusetts, there are 2221 licensed child care centers and 10,993 licensed home-based child care providers. Write an inequality statement comparing the numbers 2221 and 10,993. (*Source*: Children's Foundation)

23. Given $\left\{-5, -1, 0, \frac{1}{4}, 1, 7, 11.6, \sqrt{7}, 3\pi\right\}$, list the numbers in this set that also belong to the set of:

   a. Natural numbers

   b. Whole numbers

   c. Integers

   d. Rational numbers

   e. Irrational numbers

   f. Real numbers

If x = 6, y = −2, and z = −3, evaluate each expression.

24. $x^2 + y^2$

25. $x + yz$

26. $2 + 3x - y$

27. $\dfrac{y + z - 1}{x}$

Identify the property illustrated by each expression.

28. $8 + (9 + 3) = (8 + 9) + 3$

29. $6 \cdot 8 = 8 \cdot 6$

30. $-6(2 + 4) = -6 \cdot 2 + (-6) \cdot 4$

31. $\dfrac{1}{6}(6) = 1$

32. Find the opposite of -9.

33. Find the reciprocal of $-\dfrac{1}{3}$.

The New Orleans Saints were 22 yards from the goal when the following series of gains and losses occurred.

Gains and Losses in Yards	
First Down	5
Second Down	−10
Third Down	−2
Fourth Down	29

34. During which down did the greatest loss of yardage occur?

35. Was a touchdown scored?

36. The temperature at the Winter Olympics was a frigid 14 degrees below zero in the morning, but by noon it had risen 31 degrees. What was the temperature at noon?

37. United HealthCare is a health insurance provider. It had net incomes of $356 million, $460 million, and −$166 million in 1996, 1997, and 1998, respectively. What was United HealthCare's total net income for these three years? (*Source*: United HealthCare Corp.)

38. Jean Avarez decided to sell 280 shares of stock, which decreased in value by $1.50 per share yesterday. How much money did she lose?

Intel is a semiconductor manufacturer that makes almost one-third of the world's computer chips. (You may have seen the slogan "Intel Inside" in commercials on television.) The line graph below shows Intel's net revenues in billions of dollars. Use this figure to answer the questions below.

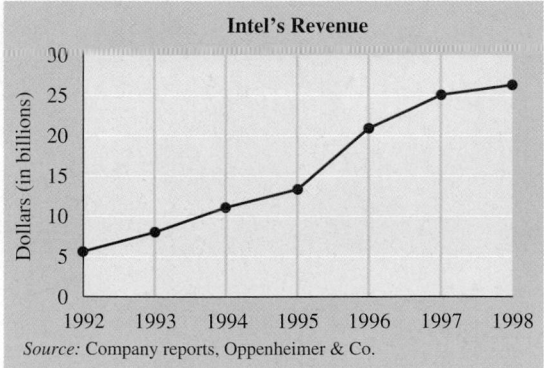

Source: Company reports, Oppenheimer & Co.

39. Estimate Intel's revenue in 1993.
40. Estimate Intel's revenue in 1997.
41. Find the increase in Intel's revenue from 1993 to 1995.
42. What year shows the greatest increase in revenue?

The following bar graph shows the top steel-producing states ranked by total tons of raw steel produced in 1997.

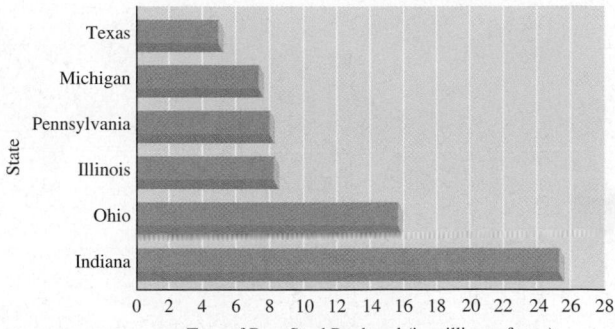

Source: American Iron & Steel Institute

43. Which state was the top steel producer? Approximate the amount of raw steel it produced.
44. Which of the top steel producing states produced the least steel? Approximate the amount of raw steel it produced.
45. Approximate the amount of raw steel produced by Ohio.
46. Approximately how much more steel was produced in Pennsylvania than Texas?

Educational Opportunities

Over 3 million teachers and nearly 1 million teacher aides are at work in this country. While teachers develop and execute lesson plans to help students learn and apply concepts in a variety of subjects, teacher aides provide instructional and clerical support for teachers.

All states require their teachers to have at least a bachelor's degree. Requirements for teacher aides vary from a high school diploma to some college training. Working as a teacher aide provides an excellent opportunity to advance to classroom teaching positions. In fact, according to a survey conducted by the National Education Association, half of all teacher aides aspire to become classroom teachers. Teachers and teacher aides should enjoy working with children, communicate well, and have a solid understanding of topics including science, mathematics, and English. Educators and support staff need good math skills not only for teaching math, but also for figuring grades and other recordkeeping.

 For more information about teachers and teacher aides, visit the National Education Association Website by first going to www.prenhall.com/martin-gay.

In the Spotlight on Decision Making feature on page 93, you will have the opportunity to make a decision involving grades as a teacher aide.

EQUATIONS, INEQUALITIES, AND PROBLEM SOLVING

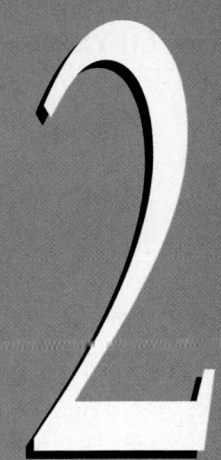

Much of mathematics relates to deciding which statements are true and which are false. When a statement, such as an equation, contains variables, it is usually not possible to decide whether the equation is true or false until the variable has been replaced by a value. For example, the statement $x + 7 = 15$ is an equation stating that the sum $x + 7$ has the same value as 15. Is this statement true or false? It is false for some values of x and true for just one value of x, namely 8. Our purpose in this chapter is to learn ways of deciding which values make an equation or an inequality true.

2.1 SIMPLIFYING ALGEBRAIC EXPRESSIONS

▶ **OBJECTIVES**

CD-ROM SSM

SSG Video

1. Identify terms, like terms, and unlike terms.
2. Combine like terms.
3. Use the distributive property to remove parentheses.
4. Write word phrases as algebraic expressions.

As we explore in this section, an expression such as $3x + 2x$ is not as simple as possible, because—even without replacing x by a value—we can perform the indicated addition.

1

Before we practice simplifying expressions, some new language of algebra is presented. A **term** is a number or the product of a number and variables raised to powers.

Terms

$$-y, \quad 2x^3, \quad -5, \quad 3xz^2, \quad \frac{2}{y}, \quad 0.8z$$

The **numerical coefficient** of a term is the numerical factor. The numerical coefficient of $3x$ is 3. Recall that $3x$ means $3 \cdot x$.

Term	Numerical Coefficient	
$3x$	3	
$\dfrac{y^3}{5}$	$\dfrac{1}{5}$	since $\dfrac{y^3}{5}$ means $\dfrac{1}{5} \cdot y^3$
$0.7ab^3c^5$	0.7	
z	1	
$-y$	-1	
-5	-5	

HELPFUL HINT

The term $-y$ means $-1y$ and thus has a numerical coefficient of -1. The term z means $1z$ and thus has a numerical coefficient of 1.

Example 1 Identify the numerical coefficient in each term.

a. $-3y$ **b.** $22z^4$ **c.** y **d.** $-x$ **e.** $\dfrac{x}{7}$

Solution **a.** The numerical coefficient of $-3y$ is -3.
b. The numerical coefficient of $22z^4$ is 22.
c. The numerical coefficient of y is 1, since y is $1y$.
d. The numerical coefficient of $-x$ is -1, since $-x$ is $-1x$.

e. The numerical coefficient of $\frac{x}{7}$ is $\frac{1}{7}$, since $\frac{x}{7}$ is $\frac{1}{7} \cdot x$.

Terms with the same variables raised to exactly the same powers are called **like terms**. Terms that aren't like terms are called **unlike terms**.

Like Terms	Unlike Terms	
$3x, 2x$	$5x, 5x^2$	Why? Same variable x, but different powers x and x^2.
$-6x^2y, 2x^2y, 4x^2y$	$7y, 3z, 8x^2$	Why? Different variables
$2ab^2c^3, ac^3b^2$	$6abc^3, 6ab^2$	Why? Different variables and different powers

> ▼ **HELPFUL HINT**
> In like terms, each variable and its exponent must match exactly, but these factors don't need to be in the same order.
>
> $$2x^2y \text{ and } 3yx^2 \text{ are like terms.}$$

Example 2   Determine whether the terms are like or unlike.

a. $2x, 3x^2$ **b.** $4x^2y, x^2y, -2x^2y$ **c.** $-2yz, -3zy$ **d.** $-x^4, x^4$

Solution   **a.** Unlike terms, since the exponents on x are not the same.
b. Like terms, since each variable and its exponent match.
c. Like terms, since $zy = yz$ by the commutative property.
d. Like terms.

2   An algebraic expression containing the sum or difference of like terms can be simplified by applying the distributive property. For example, by the distributive property, we rewrite the sum of the like terms $3x + 2x$ as

$$3x + 2x = (3 + 2)x = 5x$$

Also,

$$-y^2 + 5y^2 = (-1 + 5)y^2 = 4y^2$$

Simplifying the sum or difference of like terms is called **combining like terms**.

◆ **Example 3**   Simplify each expression by combining like terms.

a. $7x - 3x$ **b.** $10y^2 + y^2$ **c.** $8x^2 + 2x - 3x$

Solution   **a.** $7x - 3x = (7 - 3)x = 4x$
b. $10y^2 + y^2 = (10 + 1)y^2 = 11y^2$
c. $8x^2 + 2x - 3x = 8x^2 + (2 - 3)x = 8x^2 - x$

Example 4   Simplify each expression by combining like terms.

    a. $2x + 3x + 5 + 2$ **b.** $-5a - 3 + a + 2$

    c. $4y - 3y^2$ **d.** $2.3x + 5x - 6$

Solution   Use the distributive property to combine the numerical coefficients of like terms.

    a. $2x + 3x + 5 + 2 = (2 + 3)x + (5 + 2)$

                                   $= 5x + 7$

    b. $-5a - 3 + a + 2 = -5a + 1a + (-3 + 2)$

                                 $= (-5 + 1)a + (-3 + 2)$

                                 $= -4a - 1$

    c. $4y - 3y^2$ These two terms cannot be combined because they are unlike terms.

    d. $2.3x + 5x - 6 = (2.3 + 5)x - 6$

                            $= 7.3x - 6$

The examples above suggest the following:

COMBINING LIKE TERMS

To **combine like terms**, add the numerical coefficients and multiply the result by the common variable factors.

---

3   Simplifying expressions makes frequent use of the distributive property to remove parentheses.

Example 5   Find each product by using the distributive property to remove parentheses.

    a. $5(x + 2)$ **b.** $-2(y + 0.3z - 1)$ **c.** $-(x + y - 2z + 6)$

Solution   **a.** $5(x + 2) = 5 \cdot x + 5 \cdot 2$ *Apply the distributive property.*

                  $= 5x + 10$ *Multiply.*

    b. $-2(y + 0.3z - 1) = -2(y) + (-2)(0.3z) + (-2)(-1)$ *Apply the distributive property.*

                        $= -2y - 0.6z + 2$ *Multiply.*

    c. $-(x + y - 2z + 6) = -1(x + y - 2z + 6)$ *Distribute −1 over each term.*

                          $= -1(x) - 1(y) - 1(-2z) - 1(6)$

                          $= -x - y + 2z - 6$

HELPFUL HINT

If a "−" sign precedes parentheses, the sign of each term inside the parentheses is changed when the distributive property is applied to remove parentheses.

Examples:

    $-(2x + 1) = -2x - 1$ $-(-5x + y - z) = 5x - y + z$

    $-(x - 2y) = -x + 2y$ $-(-3x - 4y - 1) = 3x + 4y + 1$

When simplifying an expression containing parentheses, we often use the distributive property first to remove parentheses and then again to combine any like terms.

Example 6   Simplify the following expressions.

a. $3(2x - 5) + 1$ **b.** $8 - (7x + 2) + 3x$ **c.** $-2(4x + 7) - (3x - 1)$

Solution   **a.** $3(2x - 5) + 1 = 6x - 15 + 1$ *Apply the distributive property.*
$$= 6x - 14$$ *Combine like terms.*

b. $8 - (7x + 2) + 3x = 8 - 7x - 2 + 3x$ *Apply the distributive property.*
$$= -7x + 3x + 8 - 2$$
$$= -4x + 6$$ *Combine like terms.*

c. $-2(4x + 7) - (3x - 1) = -8x - 14 - 3x + 1$ *Apply the distributive property.*
$$= -11x - 13$$ *Combine like terms.*   ▬

Example 7   Subtract $4x - 2$ from $2x - 3$.

Solution   "Subtract $4x - 2$ **from** $2x - 3$" translates to $(2x - 3) - (4x - 2)$. Next, simplify the algebraic expression.

$$(2x - 3) - (4x - 2) = 2x - 3 - 4x + 2$$ *Apply the distributive property.*
$$= -2x - 1$$ *Combine like terms.*   ▬

---

4   Next, we practice writing word phrases as algebraic expressions.

Example 8   Write the following phrases as algebraic expressions and simplify if possible. Let x represent the unknown number.

a. Twice a number, added to 6
b. The difference of a number and 4, divided by 7
c. Five added to 3 times the sum of a number and 1
d. The sum of twice a number, 3 times the number, and 5 times the number

Solution   **a.** In words:

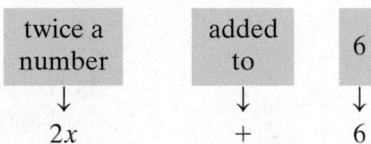

Translate: $2x$ $+$ 6

b. In words:

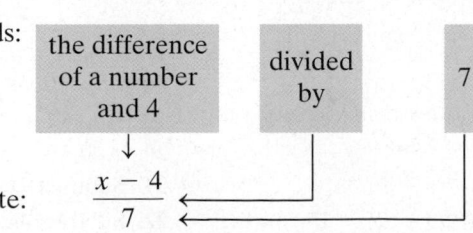

Translate: $\dfrac{x - 4}{7}$

c. In words:

five	added to	3 times	the sum of a number and 1

Translate: 5 $+$ $3\cdot$ $(x + 1)$

Next, we simplify this expression.

$$5 + 3(x + 1) = 5 + 3x + 3 \qquad \text{Use the distributive property.}$$
$$= 8 + 3x \qquad \text{Combine like terms.}$$

d. The phrase "the sum of" means that we add.

In words:

twice a number	added to	3 times the number	added to	5 times the number

Translate: $2x$ $+$ $3x$ $+$ $5x$

Now let's simplify.

$$2x + 3x + 5x = 10x \qquad \text{Combine like terms.}$$

MENTAL MATH

Identify the numerical coefficient of each term. See Example 1.

1. $-7y$

2. $3x$

3. x

4. $-y$

5. $17x^2y$

6. $1.2xyz$

Indicate whether the following lists of terms are like or unlike. See Example 2.

7. $5y, -y$

8. $-2x^2y, 6xy$

9. $2z, 3z^2$

10. $ab^2, -7ab^2$

11. $8wz, \frac{1}{7}zw$

12. $7.4p^3q^2, 6.2p^3q^2r$

Exercise Set 2.1

Simplify each expression by combining any like terms. See Examples 3 and 4.

1. $7y + 8y$

2. $3x + 2x$

3. $8w - w + 6w$

4. $c - 7c + 2c$

5. $3b - 5 - 10b - 4$

6. $6g + 5 - 3g - 7$

7. $m - 4m + 2m - 6$

8. $a + 3a - 2 - 7a$

Simplify each expression. First use the distributive property to remove any parentheses. See Examples 5 and 6.

9. $5(y - 4)$

10. $7(r - 3)$

11. $7(d - 3) + 10$

12. $9(z + 7) - 15$

13. $-(3x - 2y + 1)$

14. $-(y + 5z - 7)$

15. $5(x + 2) - (3x - 4)$

16. $4(2x - 3) - 2(x + 1)$

17. In your own words, explain how to combine like terms.

18. Do like terms contain the same numerical coefficients? Explain your answer.

Write each of the following as an algebraic expression. Simplify if possible. See Example 7.

19. Add $6x + 7$ to $4x - 10$.

20. Add $3y - 5$ to $y + 16$.

21. Subtract $7x + 1$ from $3x - 8$.

22. Subtract $4x - 7$ from $12 + x$.

Simplify each expression.

23. $7x^2 + 8x^2 - 10x^2$
24. $8x + x - 11x$
25. $6x - 5x + x - 3 + 2x$
26. $8h + 13h - 6 + 7h - h$
27. $-5 + 8(x - 6)$
28. $-6 + 5(r - 10)$
29. $5g - 3 - 5 - 5g$
30. $8p + 4 - 8p - 15$
31. $6.2x - 4 + x - 1.2$
32. $7.9y - 0.7 - y + 0.2$
33. $2k - k - 6$
34. $7c - 8 - c$
35. $0.5(m + 2) + 0.4m$
36. $0.2(k + 8) - 0.1k$
37. $-4(3y - 4)$
38. $-3(2x + 5)$
39. $3(2x - 5) - 5(x - 4)$
40. $2(6x - 1) - (x - 7)$
41. $3.4m - 4 - 3.4m - 7$
42. $2.8w - 0.9 - 0.5 - 2.8w$
43. $6x + 0.5 - 4.3x - 0.4x + 3$
44. $0.4y - 6.7 + y - 0.3 - 2.6y$
45. $-2(3x - 4) + 7x - 6$
46. $8y - 2 - 3(y + 4)$
47. $-9x + 4x + 18 - 10x$
48. $5y - 14 + 7y - 20y$
49. $5k - (3k - 10)$
50. $-11c - (4 - 2c)$
51. $(3x + 4) - (6x - 1)$
52. $(8 - 5y) - (4 + 3y)$

Write each of the following phrases as an algebraic expression and simplify if possible. Let x represent the unknown number See Example 8.

53. Twice a number, decreased by four
54. The difference of a number and two, divided by five
55. Three-fourths of a number, increased by twelve
56. Eight more than triple a number
57. The sum of 5 times a number and -2, added to 7 times a number
58. The sum of 3 times a number and 10, **subtracted from** 9 times a number
59. Subtract $5m - 6$ from $m - 9$.
60. Subtract $m - 3$ from $2m - 6$.
61. Eight times the sum of a number and six
62. Five, subtracted from four times a number
63. Double a number, minus the sum of the number and ten
64. Half a number, minus the product of the number and eight
65. Seven, multiplied by the quotient of a number and six
66. The product of a number and ten, less twenty
67. The sum of 2, three times a number, -9, and four times a number
68. The sum of twice a number, -1, five times a number, and -12
△ **69.** Recall that the perimeter of a figure is the total distance around the figure. Given the following rectangle, express the perimeter as an algebraic expression containing the variable x.

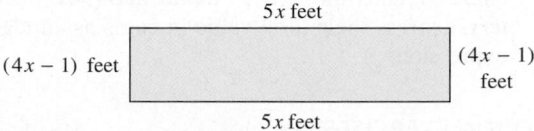

△ **70.** Given the following triangle, express its perimeter as an algebraic expression containing the variable x.

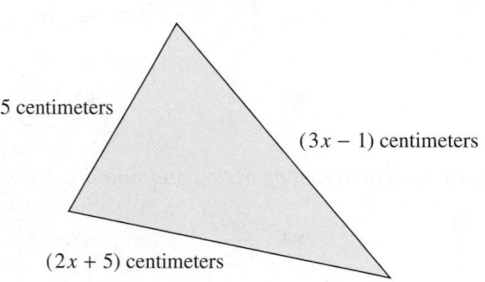

Given the following, determine whether each scale is balanced or not.

1 cone balances 1 cube

1 cylinder balances 2 cubes

71.

72.

73.

74.

75. To convert from feet to inches, we multiply by 12. For example, the number of inches in 2 feet is $12 \cdot 2$ inches. If one board has a length of $(x + 2)$ *feet* and a second board has a length of $(3x - 1)$ *inches*, express their total length in inches as an algebraic expression.

76. The value of 7 nickels is $5 \cdot 7$ cents. Likewise, the value of x nickels is $5x$ cents. If the money box in a drink machine contains x *nickels*, $3x$ *dimes*, and $(30x - 1)$ *quarters*, express their total value in cents as an algebraic expression.

REVIEW EXERCISES

Find the reciprocal or multiplicative inverse of each. See Section 1.7.

77. $\dfrac{5}{8}$ **78.** $\dfrac{7}{6}$ **79.** 2

80. 5   **81.** $-\dfrac{1}{9}$ **82.** $-\dfrac{3}{5}$

Perform each indicated operation and simplify. See Section 1.7.

83. $\dfrac{3x}{3}$ **84.** $\dfrac{-2y}{-2}$ **85.** $-5\left(-\dfrac{1}{5}y\right)$

86. $7\left(\dfrac{1}{7}r\right)$ **87.** $\dfrac{3}{5}\left(\dfrac{5}{3}x\right)$ **88.** $\dfrac{9}{2}\left(\dfrac{2}{9}x\right)$

A Look Ahead

Example
Simplify $-3xy + 2x^2y - (2xy - 1)$.

Solution:
$-3xy + 2x^2y - (2xy - 1)$
$= -3xy + 2x^2y - 2xy + 1 = -5xy + 2x^2y + 1$

Simplify each expression.

89. $5b^2c^3 + 8b^3c^2 - 7b^3c^2$

90. $4m^4p^2 + m^4p^2 - 5m^2p^4$

91. $3x - (2x^2 - 6x) + 7x^2$

92. $9y^2 - (6xy^2 - 5y^2) - 8xy^2$

93. $-(2x^2y + 3z) + 3z - 5x^2y$

94. $-(7c^3d - 8c) - 5c - 4c^3d$

2.2   THE ADDITION AND MULTIPLICATION PROPERTIES OF EQUALITY

CD-ROM   SSM

SSG   Video

▶ **OBJECTIVES**

1. Define linear equation in one variable and equivalent equations.
2. Use the addition property of equality to solve linear equations.
3. Use the multiplication property of equality to solve linear equations.
4. Use both properties of equality to solve linear equations.
5. Write word phrases as algebraic expressions.

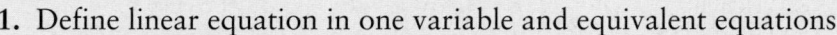
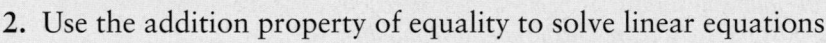

1

Recall from Section 1.4 that an equation is a statement that two expressions have the same value. Also, a value of the variable that makes an equation a true statement is called a solution or root of the equation. The process of finding the solution of an equation is called **solving** the equation for the variable. In this section we concentrate on solving **linear equations** in one variable.

> **LINEAR EQUATION IN ONE VARIABLE**
>
> A **linear equation in one variable** can be written in the form
>
> $$ax + b = c$$
>
> where a, b, and c are real numbers and $a \neq 0$.

Evaluating a linear equation for a given value of the variable, as we did in Section 1.4, can tell us whether that value is a solution, but we can't rely on evaluating an equation as our method of solving it.

Instead, to solve a linear equation in x, we write a series of simpler equations, all *equivalent* to the original equation, so that the final equation has the form

$$x = \textbf{number} \qquad \textbf{or} \qquad \textbf{number} = x$$

Equivalent equations are equations that have the same solution. This means that the "number" above is the solution to the original equation.

2   The first property of equality that helps us write simpler equivalent equations is the **addition property of equality**.

ADDITION PROPERTY OF EQUALITY

If a, b, and c are real numbers, then

$$a = b \quad \text{and} \quad a + c = b + c$$

are equivalent equations.

This property guarantees that adding the same number to both sides of an equation does not change the solution of the equation. Since subtraction is defined in terms of addition, we may also **subtract the same number from both sides** without changing the solution.

A good way to picture a true equation is as a balanced scale. Since it is balanced, each side of the scale weighs the same amount.

$$x - 2 \qquad\qquad 5$$

If the same weight is added to or subtracted from each side, the scale remains balanced.

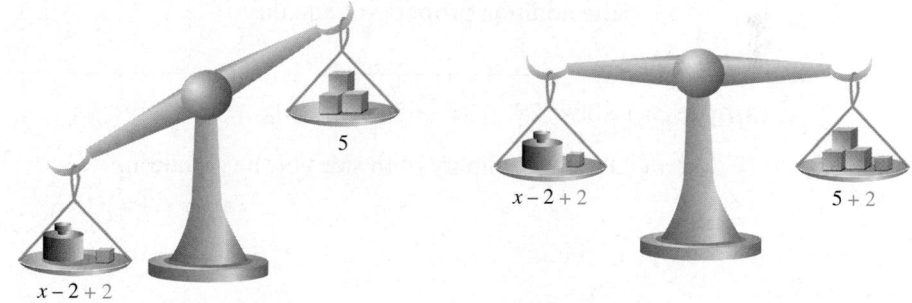

$$5$$
$$x - 2 + 2$$
$$x - 2 + 2 \qquad\qquad 5 + 2$$

We use the addition property of equality to write equivalent equations until the variable is by itself on one side of the equation, and the equation looks like "x = number" or "number = x."

Example 1   Solve $x - 7 = 10$ for x.

Solution   To solve for x, we want x alone on one side of the equation. To do this, we add 7 to both sides of the equation.

$$x - 7 = 10$$
$$x - 7 + 7 = 10 + 7 \qquad \text{Add 7 to both sides.}$$
$$x = 17 \qquad\qquad \text{Simplify.}$$

The solution of the equation $x = 17$ is obviously 17. Since we are writing equivalent equations, the solution of the equation $x - 7 = 10$ is also 17.

To check, replace x with 17 in the original equation.

Check

$$x - 7 = 10$$
$$17 - 7 \overset{?}{=} 10 \qquad \textit{Replace x with 17 in the original equation.}$$
$$10 = 10 \qquad \textit{True.}$$

Since the statement is true, 17 is the solution. ▬

Example 2 Solve $y + 0.6 = -1.0$ for y.

Solution To get y alone on one side of the equation, subtract 0.6 from both sides of the equation.

$$y + 0.6 = -1.0$$
$$y + 0.6 - 0.6 = -1.0 - 0.6 \qquad \textit{Subtract 0.6 from both sides.}$$
$$y = -1.6 \qquad \textit{Combine like terms.}$$

To check the proposed solution, -1.6, replace y with -1.6 in the original equation.

Check

$$y + 0.6 = -1.0$$
$$-1.6 + 0.6 \overset{?}{=} -1.0 \qquad \textit{Replace y with -1.6 in the original equation.}$$
$$-1.0 = -1.0 \qquad \textit{True.}$$

The solution is -1.6. ▬

Many times, it is best to simplify one or both sides of an equation before applying the addition property of equality.

Example 3 Solve: $2x + 3x - 5 + 7 = 10x + 3 - 6x - 4$

Solution First we simplify both sides of the equation.

$$2x + 3x - 5 + 7 = 10x + 3 - 6x - 4$$
$$5x + 2 = 4x - 1 \qquad \textit{Combine like terms on each side of the equation.}$$

Next, we want all terms with a variable on one side of the equation and all numbers on the other side.

$$5x + 2 - 4x = 4x - 1 - 4x \qquad \textit{Subtract 4x from both sides.}$$
$$x + 2 = -1 \qquad \textit{Combine like terms.}$$
$$x + 2 - 2 = -1 - 2 \qquad \textit{Subtract 2 from both sides to get x alone.}$$
$$x = -3 \qquad \textit{Combine like terms.}$$

Check

$$2x + 3x - 5 + 7 = 10x + 3 - 6x - 4 \qquad \textit{Original equation.}$$
$$2(-3) + 3(-3) - 5 + 7 \overset{?}{=} 10(-3) + 3 - 6(-3) - 4 \qquad \textit{Replace x with -3.}$$
$$-6 - 9 - 5 + 7 \overset{?}{=} -30 + 3 + 18 - 4 \qquad \textit{Multiply.}$$
$$-13 = -13 \qquad \textit{True.}$$

The solution is -3. ▬

If an equation contains parentheses, use the distributive property to remove them.

Example 4   Solve $7 = -5(2a - 1) - (-11a + 6)$ for a.

Solution

$$7 = -5(2a - 1) - (-11a + 6)$$
$$7 = -10a + 5 + 11a - 6 \qquad \text{Apply the distributive property.}$$
$$7 = a - 1 \qquad \text{Combine like terms.}$$
$$7 + 1 = a - 1 + 1 \qquad \text{Add 1 to both sides to get } a \text{ alone.}$$
$$8 = a \qquad \text{Combine like terms.}$$

Check to see that 8 is the solution.

> **HELPFUL HINT**
> We may solve an equation so that the variable is alone on either side of the equation. For example, $8 = a$ is equivalent to $a = 8$.

When solving equations, we may sometimes encounter an equation such as

$$-x = 5$$

This equation is not solved for x because x is not isolated. One way to solve this equation for x is to recall that

"$-$" can be read as "the opposite of"

We can read the equation $-x = 5$ then as "the opposite of $x = 5$." If the opposite of x is 5, this means that x is the opposite of 5 or -5.

In summary,

$$-x = 5 \quad \text{and} \quad x = -5$$

are equivalent equations and $x = -5$ is solved for x.

3   As useful as the addition property of equality is, it cannot help us solve every type of linear equation in one variable. For example, adding or subtracting a value on both sides of the equation does not help solve

$$\frac{5}{2}x = 15.$$

Instead, we apply another important property of equality, the **multiplication property of equality**.

MULTIPLICATION PROPERTY OF EQUALITY

If a, b, and c are real numbers and $c \neq 0$, then

$$a = b \qquad \text{and} \qquad ac = bc$$

are equivalent equations.

This property guarantees that multiplying both sides of an equation by the same nonzero number does not change the solution of the equation. Since division is defined in terms of multiplication, we may also **divide both sides of the equation by the same nonzero number** without changing the solution.

Example 5 Solve for x: $\dfrac{5}{2}x = 15$.

Solution To get x alone, multiply both sides of the equation by the reciprocal of $\dfrac{5}{2}$, which is $\dfrac{2}{5}$.

$$\frac{5}{2}x = 15$$

$$\frac{2}{5} \cdot \frac{5}{2}x = \frac{2}{5} \cdot 15 \qquad \text{Multiply both sides by } \frac{2}{5}.$$

> **HELPFUL HINT**
>
> Don't forget to multiply both sides by $\dfrac{2}{5}$.

$$\left(\frac{2}{5} \cdot \frac{5}{2}\right)x = \frac{2}{5} \cdot 15 \qquad \textit{Apply the associative property.}$$

$$1x = 6 \qquad \textit{Simplify.}$$

or

$$x = 6$$

Check Replace x with 6 in the original equation.

$$\frac{5}{2}x = 15 \qquad \textit{Original equation.}$$

$$\frac{5}{2}(6) \stackrel{?}{=} 15 \qquad \textit{Replace x with 6.}$$

$$15 = 15 \qquad \textit{True.}$$

The solution is 6.

In the equation $\dfrac{5}{2}x = 15$, $\dfrac{5}{2}$ is the coefficient of x. When the coefficient of x is a *fraction*, we will get x alone by multiplying by the reciprocal. When the coefficient of x is an integer or a decimal, it is usually more convenient to divide both sides by the coefficient. (Dividing by a number is, of course, the same as multiplying by the reciprocal of the number.)

Example 6 Solve: $-3x = 33$

Solution Recall that $-3x$ means $-3 \cdot x$. To get x alone, we divide both sides by the coefficient of x, that is, -3.

$$-3x = 33$$

$$\frac{-3x}{-3} = \frac{33}{-3} \qquad \textit{Divide both sides by } -3.$$

$$1x = -11 \qquad \textit{Simplify.}$$

$$x = -11$$

Check

$$-3x = 33 \qquad \textit{Original equation.}$$

$$-3(-11) \stackrel{?}{=} 33 \qquad \textit{Replace x with } -11.$$

$$33 = 33 \qquad \textit{True.}$$

The solution is -11.

Example 7 Solve: $\frac{y}{7} = 20$

Solution Recall that $\frac{y}{7} = \frac{1}{7}y$. To get y alone, we multiply both sides of the equation by 7, the reciprocal of $\frac{1}{7}$.

$$\frac{y}{7} = 20$$

$$\frac{1}{7}y - 20$$

$$7 \cdot \frac{1}{7}y = 7 \cdot 20 \qquad \text{Multiply both sides by 7.}$$

$$1y = 140 \qquad \text{Simplify.}$$

$$y = 140$$

Check

$$\frac{y}{7} = 20 \qquad \text{Original equation.}$$

$$\frac{140}{7} \stackrel{?}{=} 20 \qquad \text{Replace } y \text{ with 140.}$$

$$20 = 20 \qquad \text{True.}$$

The solution is 140.

4 Next, we practice solving equations using both properties.

Example 8 Solve: $12a - 8a = 10 + 2a - 13 - 7$

Solution First, simplify both sides of the equation by combining like terms.

$$12a - 8a = 10 + 2a - 13 - 7$$

$$4a = 2a - 10 \qquad \text{Combine like terms.}$$

To get all terms containing a variable on one side, subtract $2a$ from both sides.

$$4a - 2a = 2a - 10 - 2a \qquad \text{Subtract } 2a \text{ from both sides.}$$

$$2a = -10 \qquad \text{Simplify.}$$

$$\frac{2a}{2} = \frac{-10}{2} \qquad \text{Divide both sides by 2.}$$

$$a = -5 \qquad \text{Simplify.}$$

Check Check by replacing a with -5 in the original equation. The solution is -5.

5 Next, we practice writing algebraic expressions.

Example 9 **a.** The sum of two numbers is 8. If one number is 3, find the other number.
b. The sum of two numbers is 8. If one number is x, write an expression representing the other number.

Solution **a.** If the sum of two numbers is 8 and one number is 3, we find the other number by subtracting 3 from 8. The other number is $8 - 3$ or 5.

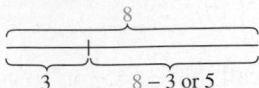

b. If the sum of two numbers is 8 and one number is x, we find the other number by subtracting x from 8. The other number is represented by $8 - x$.

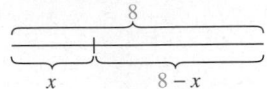

Example 10 If x is the first of three consecutive integers, express the sum of the three integers in terms of x. Simplify if possible.

Solution An example of three consecutive integers is

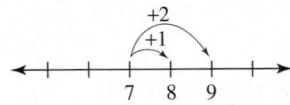

The second consecutive integer is always 1 more than the first, and the third consecutive integer is 2 more than the first. If x is the first of three consecutive integers, the three consecutive integers are

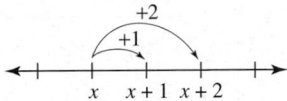

Their sum is

In words:

first integer	+	second integer	+	third integer

Translate: x $+$ $(x + 1)$ $+$ $(x + 2)$

which simplifies to $3x + 3$.

Below are examples of consecutive even and odd integers.

Even integers:

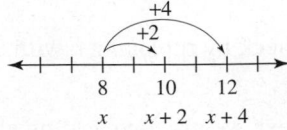

Odd integers:

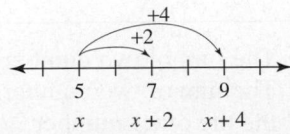

HELPFUL HINT

If x is an odd integer, then $x + 2$ is the next odd integer. This 2 simply means that odd integers are always 2 units from each other.

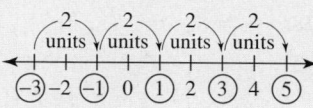

SPOTLIGHT ON DECISION MAKING

Suppose you are a teacher's aide in Mr. Mankato's eighth-grade science class. You are responsible for recording grades and totaling scores. After students took Test 4, Mr. Mankato decided to allow anyone who scored 75 or less on that test the opportunity to do an extra credit project. While preparing a list of students eligible for the extra credit project, you found that one of the entries had gotten smudged and was illegible. Should Demetra Brown be included in the list of students eligible for the extra credit project? Explain.

GRADE BOOK

Teacher: *Mr. Mankato*
Period: *4th*

Student	#1	#2	#3	#4	Total
Brown, Demetra	92	86	94	✹	344
Mankad, Rahul	88	89	91	73	341
Mendoza, Carlos	91	85	83	80	339
Roberge, Ann	93	84	92	74	343

MENTAL MATH

Solve each equation mentally. See Examples 1, 2, and 6.

1. $x + 4 = 6$
2. $x + 7 = 10$
3. $n + 18 = 30$
4. $z + 22 = 40$
5. $b - 11 = 6$
6. $d - 16 = 5$
7. $3a = 27$
8. $9c = 54$
9. $5b = 10$
10. $7t = 14$
11. $6x = -30$
12. $8r = -64$

Exercise Set 2.2

Solve each equation. Check each solution. See Examples 1 and 2.

1. $x + 7 = 10$
2. $x + 14 = 25$
3. $x - 2 = -4$
4. $y - 9 = 1$
5. $3 + x = -11$
6. $8 + z = -8$
7. $r - 8.6 = -8.1$
8. $t - 9.2 = -6.8$
9. $8x = 7x - 3$
10. $2x = x - 5$
11. $5b - 0.7 = 6b$
12. $9x + 5.5 = 10x$

13. $7x - 3 = 6x$
14. $18x - 9 = 19x$

Solve each equation. See Examples 3 and 4.

15. $3x - 6 = 2x + 5$
16. $7y + 2 = 6y + 2$
17. $3t - t - 7 = t - 7$
18. $4c + 8 - c = 8 + 2c$
19. $7x + 2x = 8x - 3$
20. $3n + 2n = 7 + 4n$
21. $-2(x + 1) + 3x = 14$
22. $10 = 8(3y - 4) - 23y + 20$

Solve each equation. See Example 6.

23. $-5x = 20$

24. $-7x = -49$

25. $3x = 0$

26. $-2x = 0$

27. $-x = -12$

28. $-y = 8$

29. $3x + 2x = 50$

30. $-y + 4y = 33$

Solve each equation. See Examples 5 and 7.

31. $\frac{2}{3}x = -8$

32. $\frac{3}{4}n = -15$

33. $\frac{1}{6}d = \frac{1}{2}$

34. $\frac{1}{8}v = \frac{1}{4}$

35. $\frac{a}{-2} = 1$

36. $\frac{d}{15} = 2$

37. $\frac{k}{7} = 0$

38. $\frac{f}{-5} = 0$

39. In your own words, explain the addition property of equality.

40. In your own words, explain the multiplication property of equality.

Solve each equation. Check each solution. See Example 8.

41. $2x - 4 = 16$

42. $3x - 1 = 26$

43. $-x + 2 = 22$

44. $-x + 4 = -24$

45. $6a + 3 = 3$

46. $8t + 5 = 5$

47. $6x + 10 = -20$

48. $-10y + 15 = 5$

49. $5 - 0.3k = 5$

50. $2 + 0.4p = 2$

51. $-2x + \frac{1}{2} = \frac{7}{2}$

52. $-3n - \frac{1}{3} = \frac{8}{3}$

53. $\frac{x}{3} + 2 = -5$

54. $\frac{b}{4} - 1 = -7$

55. $10 = 2x - 1$

56. $12 = 3j - 4$

57. $6z - 8 - z + 3 = 0$

58. $4a + 1 + a - 11 = 0$

59. $10 - 3x - 6 - 9x = 7$

60. $12x + 30 + 8x - 6 = 10$

61. $\frac{5}{6}x = 10$

62. $-\frac{3}{4}x = 9$

63. $1 = 0.4x - 0.6x - 5$

64. $19 = 0.4x - 0.9x - 6$

65. $z - 5z = 7z - 9 - z$

66. $t - 6t = -13 + t - 3t$

67. $0.4x - 0.6x - 5 = 1$

68. $0.4x - 0.9x - 6 = 19$

69. $6 - 2x + 8 = 10$

70. $-5 - 6y + 6 = 19$

71. $-3a + 6 + 5a = 7a - 8a$

72. $4b - 8 - b = 10b - 3b$

73. $20 = -3(2x + 1) + 7x$ **74.** $-3 = -5(4x + 3) + 21x$

See Example 9.

75. Two numbers have a sum of 20. If one number is p, express the other number in terms of p.

76. Two numbers have a sum of 13. If one number is y, express the other number in terms of y.

77. A 10-foot board is cut into two pieces. If one piece is x feet long, express the other length in terms of x.

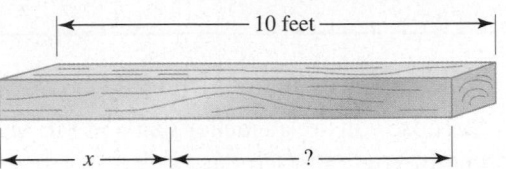

78. A 5-foot piece of string is cut into two pieces. If one piece is x feet long, express the other length in terms of x.

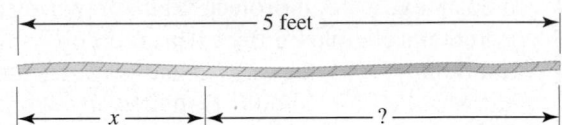

79. Two angles are *supplementary* if their sum is 180°. If one angle measures $x°$, express the measure of its supplement in terms of x.

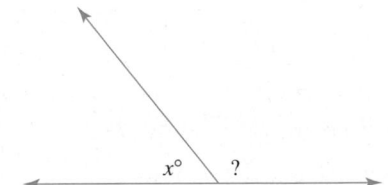

80. Two angles are *complementary* if their sum is 90°. If one angle measures $x°$, express the measure of its complement in terms of x.

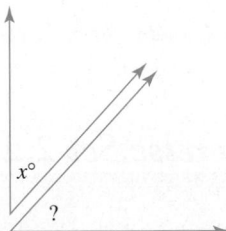

81. In a mayoral election, April Catarella received 284 more votes than Charles Pecot. If Charles received n votes, how many votes did April receive?

82. The length of the top of a computer desk is $1\frac{1}{2}$ feet longer than its width. If its width measures m feet, express its length as an algebraic expression in m.

83. The Verrazano-Narrows Bridge in New York City is the longest suspension bridge in North America. The Golden Gate Bridge in San Francisco is 60 feet shorter than the Verrazano-Narrows Bridge. If the length of the Verrazano-Narrows Bridge is m feet, express the length of the Golden Gate Bridge as an algebraic expression in m. (*Source:* World Almanac, 2000)

84. In a recent U.S. Senate race in Maine, Susan M. Collins received 30,898 more votes than Joseph E. Brennan. If Joseph received n votes, how many did Susan receive? (*Source:* Voter News Service)

△ **85.** The sum of the angles of a triangle is 180°. If one angle of a triangle measures $x°$ and a second angle measures $(2x + 7)°$, express the measure of the third angle in terms of x. Simplify the expression.

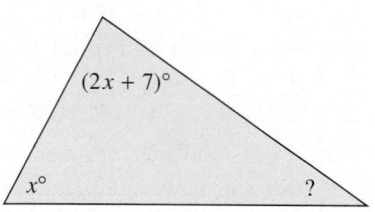

△ **86.** A quadrilateral is a four-sided figure like the one shown below whose angle sum is 360°. If one angle measures $x°$, a second angle measures $3x°$, and a third angle measures $5x°$, express the measure of the fourth angle in terms of x. Simplify the expression.

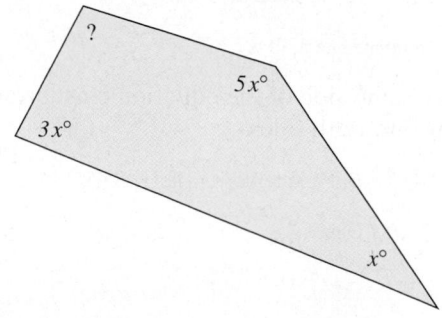

87. A nurse's aide recorded the following fluid intakes for a patient on her night shift: 200 ml, 150 ml, 400 ml. If the patient's doctor requested that a total of 1000 ml of fluid be taken by the patient overnight, how much more fluid must the nurse give the patient? To solve this problem, solve the equation $200 + 150 + 400 + x = 1000$.

88. Let $x = 1$ and then $x = 2$ in the equation $x + 5 = x + 6$. Is either number a solution? How many solutions do you think this equation has? Explain your answer.

89. Let $x = 1$ and then $x = 2$ in the equation $x + 3 = x + 3$. Is either number a solution? How many solutions do you think this equation has? Explain your answer.

Write each algebraic expression described. Simplify if possible. See Example 10.

90. If x represents the first of two consecutive odd integers, express the sum of the two integers in terms of x.

91. If x is the first of four consecutive even integers, write their sum as an algebraic expression in x.

92. If x is the first of three consecutive integers, express the sum of the first integer and the third integer as an algebraic expression containing the variable x.

93. If x is the first of two consecutive integers, express the sum of 20 and the second consecutive integer as an algebraic expression containing the variable x.

94. A licensed practical nurse is instructed to give a patient 2100 milligrams of an antibiotic over a period of 36 hours. If the antibiotic is to be given every 4 hours starting immediately, how much antibiotic should be given in each dose? To answer this question, solve the equation $9x = 2100$.

Solve each equation.

95. $-3.6x = 10.62$

96. $4.95y = -31.185$

97. $7x - 5.06 = -4.92$

98. $0.06y + 2.63 = 2.5562$

REVIEW EXERCISES

Simplify each expression. See Section 2.1.

99. $5x + 2(x - 6)$

100. $-7y + 2y - 3(y + 1)$

101. $6(2z + 4) + 20$

102. $-(3a - 3) + 2a - 6$

103. $-(x - 1) + x$

104. $8(z - 6) + 7z - 1$

2.3 SOLVING LINEAR EQUATIONS

CD-ROM SSM

SSG Video

▶ **O B J E C T I V E S**

1. Apply the general strategy for solving a linear equation.
2. Solve equations containing fractions.
3. Solve equations containing decimals.
4. Recognize identities and equations with no solution.
5. Write sentences as equations and solve.

1

We now present a general strategy for solving linear equations. One new piece of strategy is a suggestion to "clear an equation of fractions" as a first step. Doing so makes the equation more manageable, since operating on integers is more convenient than operating on fractions.

SOLVING LINEAR EQUATIONS IN ONE VARIABLE

Step 1. Multiply on both sides to clear the equation of fractions if they occur.
Step 2. Use the distributive property to remove parentheses if they occur.
Step 3. Simplify each side of the equation by combining like terms.
Step 4. Get all variable terms on one side and all numbers on the other side by using the addition property of equality.
Step 5. Get the variable alone by using the multiplication property of equality.
Step 6. Check the solution by substituting it into the original equation.

Example 1 Solve: $4(2x - 3) + 7 = 3x + 5$

Solution There are no fractions, so we begin with Step 2.

$$4(2x - 3) + 7 = 3x + 5$$

Step 2. $8x - 12 + 7 = 3x + 5$ Apply the distributive property.

Step 3. $8x - 5 = 3x + 5$ Combine like terms.

Step 4. Get all variable terms on the same side of the equation by subtracting $3x$ from both sides, then adding 5 to both sides.

$$8x - 5 - 3x = 3x + 5 - 3x \qquad \text{Subtract } 3x \text{ from both sides.}$$

$$5x - 5 = 5 \qquad \text{Simplify.}$$

$$5x - 5 + 5 = 5 + 5 \qquad \text{Add 5 to both sides.}$$

$$5x = 10 \qquad \text{Simplify.}$$

Step 5.   Use the multiplication property of equality to get x alone.

$$\frac{5x}{5} = \frac{10}{5} \qquad \text{Divide both sides by 5.}$$

$$x = 2 \qquad \text{Simplify.}$$

Step 6.   Check.

$$4(2x - 3) + 7 = 3x + 5 \qquad \text{Original equation}$$
$$4[2(2) - 3] + 7 \stackrel{?}{=} 3(2) + 5 \qquad \text{Replace } x \text{ with 2.}$$
$$4(4 - 3) + 7 \stackrel{?}{=} 6 + 5$$
$$4(1) + 7 \stackrel{?}{=} 11$$
$$4 + 7 \stackrel{?}{=} 11$$
$$11 = 11 \qquad \text{True.}$$

The solution is 2.

> **HELPFUL HINT**
> When checking solutions, use the original written equation.

Example 2   Solve: $8(2 - t) = -5t$

Solution   First, we apply the distributive property.

$$8(2 - t) = -5t$$

Step 2. $16 - 8t = -5t$ Use the distributive property.

Step 4. $16 - 8t + 8t = -5t + 8t$ To get variable terms on one side, add $8t$ to both sides.

$$16 = 3t \qquad \text{Combine like terms.}$$

Step 5. $\dfrac{16}{3} = \dfrac{3t}{3}$ Divide both sides by 3.

$$\frac{16}{3} = t \qquad \text{Simplify.}$$

Step 6.   Check.

$$8(2 - t) = -5t \qquad \text{Original equation}$$
$$8\left(2 - \frac{16}{3}\right) \stackrel{?}{=} -5\left(\frac{16}{3}\right) \qquad \text{Replace } t \text{ with } \frac{16}{3}.$$
$$8\left(\frac{6}{3} - \frac{16}{3}\right) \stackrel{?}{=} -\frac{80}{3} \qquad \text{The LCD is 3.}$$
$$8\left(-\frac{10}{3}\right) \stackrel{?}{=} -\frac{80}{3} \qquad \text{Subtract fractions.}$$
$$-\frac{80}{3} = -\frac{80}{3} \qquad \text{True.}$$

The solution is $\dfrac{16}{3}$.

2   If an equation contains fractions, we can clear the equation of fractions by multiplying both sides by the LCD of all denominators. By doing this, we avoid working with time-consuming fractions.

Example 3   Solve: $\dfrac{x}{2} - 1 = \dfrac{2}{3}x - 3$

Solution   We begin by clearing fractions. To do this, we multiply both sides of the equation by the LCD of 2 and 3, which is 6.

$$\frac{x}{2} - 1 = \frac{2}{3}x - 3$$

Step 1. $6\left(\dfrac{x}{2} - 1\right) = 6\left(\dfrac{2}{3}x - 3\right)$ Multiply both sides by the LCD, 6.

Step 2. $6\left(\dfrac{x}{2}\right) - 6(1) = 6\left(\dfrac{2}{3}x\right) - 6(3)$ Apply the distributive property.

> **HELPFUL HINT**
> Don't forget to multiply *each* term by the LCD.

$$3x - 6 = 4x - 18$$ Simplify.

There are no longer grouping symbols and no like terms on either side of the equation, so we continue with Step 4.

$$3x - 6 = 4x - 18$$

Step 4. $3x - 6 - 3x = 4x - 18 - 3x$ To get variable terms on one side, subtract $3x$ from both sides.
$$-6 = x - 18$$ Simplify.
$$-6 + 18 = x - 18 + 18$$ Add 18 to both sides.
$$12 = x$$ Simplify.

Step 5.   The variable is now alone, so there is no need to apply the multiplication property of equality.

Step 6.   Check.

$$\frac{x}{2} - 1 = \frac{2}{3}x - 3$$ Original equation

$$\frac{12}{2} - 1 \stackrel{?}{=} \frac{2}{3} \cdot 12 - 3$$ Replace x with 12.

$$6 - 1 \stackrel{?}{=} 8 - 3$$ Simplify.

$$5 = 5$$ True.

The solution is 12.

Example 4   Solve: $\dfrac{2(a + 3)}{3} = 6a + 2$

Solution   We clear the equation of fractions first.

$$\frac{2(a + 3)}{3} = 6a + 2$$

Step 1. $3 \cdot \dfrac{2(a + 3)}{3} = 3(6a + 2)$ Clear the fraction by multiplying both sides by the LCD, 3.

Step 2. Next, we use the distributive property and remove parentheses.

$$2a + 6 = 18a + 6$$ Apply the distributive property.

Step 4. $2a + 6 - 6 = 18a + 6 - 6$ Subtract 6 from both sides.

$$2a = 18a$$

$$2a - 18a = 18a - 18a$$ Subtract 18a from both sides.

$$-16a = 0$$

Step 5. $\dfrac{-16a}{-16} = \dfrac{0}{-16}$ Divide both sides by −16.

$$a = 0$$ Write the fraction in simplest form.

Step 6. To check, replace a with 0 in the original equation. The solution is 0.   ▬

3 When solving a problem about money, you may need to solve an equation containing decimals. If you choose, you may multiply to clear the equation of decimals.

◆ **Example 5**   Solve: $0.25x + 0.10(x - 3) = 0.05(22)$

Solution   First we clear this equation of decimals by multiplying both sides of the equation by 100. Recall that multiplying a decimal number by 100 has the effect of moving the decimal point 2 places to the right.

$$0.25x + 0.10(x - 3) = 0.05(22)$$

Step 1. $0.25x + 0.10(x - 3) = 0.05(22)$ Multiply both sides by 100.

$$25x + 10(x - 3) = 5(22)$$

Step 2. $25x + 10x - 30 = 110$ Apply the distributive property.

Step 3. $35x - 30 = 110$ Combine like terms.

Step 4. $35x - 30 + 30 = 110 + 30$ Add 30 to both sides.

$$35x = 140$$ Combine like terms.

Step 5. $\dfrac{35x}{35} = \dfrac{140}{35}$ Divide both sides by 35.

$$x = 4$$

Step 6. To check, replace x with 4 in the original equation. The solution is 4.   ▬

4 So far, each equation that we have solved has had a single solution. However, not every equation in one variable has a single solution. Some equations have no solution, while others have an infinite number of solutions. For example,

$$x + 5 = x + 7$$

has no solution since no matter which **real number** we replace x with, the equation is false.

real number $+ 5 =$ same **real number** $+ 7$ **FALSE**

On the other hand,

$$x + 6 = x + 6$$

has infinitely many solutions since x can be replaced by any real number and the equation is always true.

real number $+ 6 =$ same **real number** $+ 6$ **TRUE**

The equation $x + 6 = x + 6$ is called an **identity**. The next few examples illustrate special equations like these.

Example 6   Solve: $-2(x - 5) + 10 = -3(x + 2) + x$

Solution
$$-2(x - 5) + 10 = -3(x + 2) + x$$
$$-2x + 10 + 10 = -3x - 6 + x \qquad \text{Apply the distributive property on both sides.}$$
$$-2x + 20 = -2x - 6 \qquad \text{Combine like terms.}$$
$$-2x + 20 + 2x = -2x - 6 + 2x \qquad \text{Add } 2x \text{ to both sides.}$$
$$20 = -6 \qquad \text{Combine like terms.}$$

The final equation contains no variable terms, and there is no value for x that makes $20 = -6$ a true equation. We conclude that there is **no solution** to this equation. ∎

Example 7   Solve: $3(x - 4) = 3x - 12$

Solution
$$3(x - 4) = 3x - 12$$
$$3x - 12 = 3x - 12 \qquad \text{Apply the distributive property.}$$

The left side of the equation is now identical to the right side. Every real number may be substituted for x and a true statement will result. We arrive at the same conclusion if we continue.

$$3x - 12 = 3x - 12$$
$$3x - 12 + 12 = 3x - 12 + 12 \qquad \text{Add 12 to both sides.}$$
$$3x = 3x \qquad \text{Combine like terms.}$$
$$3x - 3x = 3x - 3x \qquad \text{Subtract } 3x \text{ from both sides.}$$
$$0 = 0$$

Again, one side of the equation is identical to the other side. Thus, $3(x - 4) = 3x - 12$ is an **identity** and **every real number** is a solution. ∎

---

5   We can apply our equation-solving skills to solving problems written in words. Many times, writing an equation that describes or models a problem involves a direct translation from a word sentence to an equation.

Example 8   **FINDING AN UNKNOWN NUMBER**

Twice a number, added to seven, is the same as three subtracted from the number. Find the number.

Solution   Translate the sentence into an equation and solve.

In words:	twice a number	added to	seven	is the same as	three subtracted from the number
Translate:	$2x$	$+$	7	$=$	$x - 3$

To solve, begin by subtracting x from both sides to isolate the variable term.

$$2x + 7 = x - 3$$
$$2x + 7 - x = x - 3 - x \qquad \text{Subtract } x \text{ from both sides.}$$
$$x + 7 = -3 \qquad \text{Combine like terms.}$$
$$x + 7 - 7 = -3 - 7 \qquad \text{Subtract 7 from both sides.}$$
$$x = -10 \qquad \text{Combine like terms.}$$

Check the solution in the problem as it was originally stated. To do so, replace "number" in the sentence with -10. Twice "-10" added to 7 is the same as 3 subtracted from "-10."

$$2(-10) + 7 = -10 - 3$$
$$-13 = -13$$

The unknown number is -10.

CALCULATOR EXPLORATIONS

Checking Equations

We can use a calculator to check possible solutions of equations. To do this, replace the variable by the possible solution and evaluate both sides of the equation separately.

Equation: $3x - 4 = 2(x + 6)$ Solution: $x = 16$

$$3x - 4 = 2(x + 6) \qquad \text{Original equation}$$
$$3(16) - 4 \overset{?}{=} 2(16 + 6) \qquad \text{Replace } x \text{ with 16.}$$

Now evaluate each side with your calculator.

Evaluate left side: $\boxed{3}$ $\boxed{\times}$ $\boxed{16}$ $\boxed{-}$ $\boxed{4}$ $\boxed{=}$ $\left(\text{or } \boxed{\text{ENTER}}\right)$ Display: $\boxed{44}$

or

$$\boxed{\begin{matrix} 3 * 16 - 4 \\ \qquad\qquad 44 \end{matrix}}$$

Evaluate right side: $\boxed{2}$ $\boxed{(}$ $\boxed{16}$ $\boxed{+}$ $\boxed{6}$ $\boxed{)}$ $\boxed{=}$ $\left(\text{or } \boxed{\text{ENTER}}\right)$ Display: $\boxed{44}$

or

$$\boxed{\begin{matrix} 2(16 + 6) \\ \qquad\qquad 44 \end{matrix}}$$

Since the left side equals the right side, the equation checks.

Use a calculator to check the possible solutions to each equation.

1. $2x = 48 + 6x$; $x = -12$
2. $-3x - 7 = 3x - 1$; $x = -1$
3. $5x - 2.6 = 2(x + 0.8)$; $x = 4.4$
4. $-1.6x - 3.9 = -6.9x - 25.6$; $x = 5$
5. $\dfrac{564x}{4} = 200x - 11(649)$; $x = 121$
6. $20(x - 39) = 5x - 432$; $x = 23.2$

Exercise Set 2.3

Solve each equation. See Examples 1 and 2.

1. $-2(3x - 4) = 2x$ **2.** $-(5x - 1) = 9$

3. $4(2n - 1) = (6n + 4) + 1$

4. $3(4y + 2) = 2(1 + 6y) + 8$

5. $5(2x - 1) - 2(3x) = 1$

6. $3(2 - 5x) + 4(6x) = 12$

7. $6(x - 3) + 10 = -8$ **8.** $-4(2 + n) + 9 = 1$

Solve each equation. See Examples 3 through 5.

9. $\dfrac{3}{4}x - \dfrac{1}{2} = 1$ **10.** $\dfrac{2}{3}x + \dfrac{5}{3} = \dfrac{5}{3}$

11. $x + \dfrac{5}{4} = \dfrac{3}{4}x$ **12.** $\dfrac{7}{8}x + \dfrac{1}{4} = \dfrac{3}{4}x$

13. $\dfrac{x}{2} - 1 = \dfrac{x}{5} + 2$ **14.** $\dfrac{x}{5} - 2 = \dfrac{x}{3}$

15. $\dfrac{6(3 - z)}{5} = -z$ **16.** $\dfrac{4(5 - w)}{3} = -w$

17. $\dfrac{2(x + 1)}{4} = 3x - 2$

18. $\dfrac{3(y + 3)}{5} = 2y + 6$

19. $.50x + .15(70) = .25(142)$

20. $.40x + .06(30) = .20(49)$

21. $.12(y - 6) + .06y = .08y - .07(10)$

22. $.60(z - 300) + .05z = .70z - .41(500)$

Solve each equation. See Examples 6 and 7.

23. $5x - 5 = 2(x + 1) + 3x - 7$

24. $3(2x - 1) + 5 = 6x + 2$

25. $\dfrac{x}{4} + 1 = \dfrac{x}{4}$ **26.** $\dfrac{x}{3} - 2 = \dfrac{x}{3}$

27. $3x - 7 = 3(x + 1)$ **28.** $2(x - 5) = 2x + 10$

29. Explain the difference between simplifying an expression and solving an equation.

30. When solving an equation, if the final equivalent equation is $0 = 5$, what can we conclude? If the final equivalent equation is $-2 = -2$, what can we conclude?

31. On your own, construct an equation for which every real number is a solution.

32. On your own, construct an equation that has no solution.

Solve each equation.

33. $4x + 3 = 2x + 11$ **34.** $6y - 8 = 3y + 7$

35. $-2y - 10 = 5y + 18$ **36.** $7n + 5 = 10n - 10$

37. $.6x - .1 = .5x + .2$ **38.** $.2x - .1 = .6x - 2.1$

39. $2y + 2 = y$ **40.** $7y + 4 = -3$

41. $3(5c - 1) - 2 = 13c + 3$

42. $4(3t + 4) - 20 = 3 + 5t$

43. $x + \dfrac{7}{6} = 2x - \dfrac{7}{6}$ **44.** $\dfrac{5}{2}x - 1 = x + \dfrac{1}{4}$

45. $2(x - 5) = 7 + 2x$ **46.** $-3(1 - 3x) = 9x - 3$

47. $\dfrac{2(z + 3)}{3} = 5 - z$ **48.** $\dfrac{3(w + 2)}{4} = 2w + 3$

49. $\dfrac{4(y - 1)}{5} = -3y$ **50.** $\dfrac{5(1 - x)}{6} = -4x$

51. $8 - 2(a - 1) = 7 + a$

52. $5 - 6(2 + b) = b - 14$

53. $2(x + 3) - 5 = 5x - 3(1 + x)$

54. $4(2 + x) + 1 = 7x - 3(x - 2)$

55. $\dfrac{5x - 7}{3} = x$ **56.** $\dfrac{7n + 3}{5} = -n$

57. $\dfrac{9 + 5v}{2} = 2v - 4$ **58.** $\dfrac{6 - c}{2} = 5c - 8$

59. $-3(t - 5) + 2t = 5t - 4$

60. $-(4a - 7) - 5a = 10 + a$

61. $.02(6t - 3) = .12(t - 2) + .18$

62. $.03(2m + 7) = .06(5 + m) - .09$

63. $.06 - .01(x + 1) = -.02(2 - x)$

64. $-.01(5x + 4) = .04 - .01(x + 4)$

65. $\dfrac{3(x - 5)}{2} = \dfrac{2(x + 5)}{3}$ **66.** $\dfrac{5(x - 1)}{4} = \dfrac{3(x + 1)}{2}$

67. $1000(7x - 10) = 50(412 + 100x)$

68. $10,000(x + 4) = 100(16 + 7x)$

69. $.035x + 5.112 = .010x + 5.107$

70. $.127x - 2.685 = .027x - 2.38$

Write each of the following as equations. Then solve. See Example 8.

71. The sum of twice a number and $\frac{1}{5}$ is equal to the difference between three times the number and $\frac{4}{5}$. Find the number.

72. The sum of four times a number and $\frac{2}{3}$ is equal to the difference of five times the number and $\frac{5}{6}$. Find the number.

73. The sum of twice a number and 7 is equal to the sum of the number and 6. Find the number.

74. The difference of three times a number and 1 is the same as twice the number. Find the number.

75. Three times a number, minus 6, is equal to two times a number, plus 8. Find the number.

76. The sum of 4 times a number and -2 is equal to the sum of 5 times the number and -2. Find the number.

77. One-third of a number is five-sixths. Find the number.

78. Seven-eighths of a number is one-half. Find the number.

79. The difference of a number and four is twice the number. Find the number.

80. The sum of double a number and six is four times the number. Find the number.

81. If the quotient of a number and 4 is added to $\frac{1}{2}$, the result is $\frac{3}{4}$. Find the number.

82. If $\frac{3}{4}$ is added to three times a number, the result is $\frac{1}{2}$ subtracted from twice the number. Find the number.

△ **83.** The perimeter of a geometric figure is the sum of the lengths of its sides. If the perimeter of the following pentagon (five-sided figure) is 28 centimeters, find the length of each side.

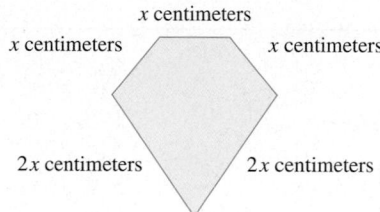

△ **84.** The perimeter of the following triangle is 35 meters. Find the length of each side.

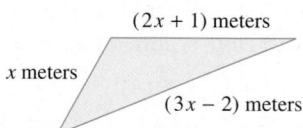

85. Five times a number subtracted from ten is triple the number. Find the number.

86. Nine is equal to ten subtracted from double a number. Find the number.

The graph below is called a three-dimensional bar graph. It shows the five most common names of cities, towns, or villages in the United States.

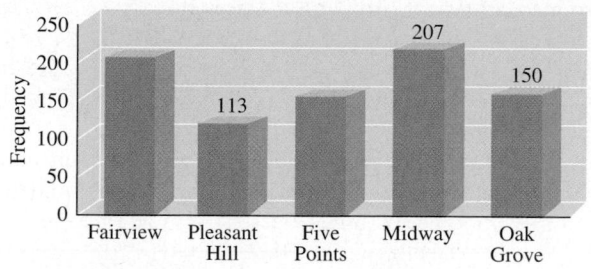

Source: Interior Department

87. What is the most popular name of a city, town, or village in the United States?

88. How many more cities, towns, or villages are named Oak Grove than named Pleasant Hill?

89. Let x represent "the number of towns, cities, or villages named Five Points" and use the information given to determine the unknown number. "The number of towns, cities, or villages named Five Points" added to 55 is equal to twice "the number of towns, cities, or villages named Five Points" minus 90. Check your answer by noticing the height of the bar representing Five Points. Is your answer reasonable?

90. Let x represent "the number of towns, cities, or villages named Fairview" and use the information given to determine the unknown number. Three times "the number of towns, cities, or villages named Fairview" added to 24 is equal to 168 subtracted from 4 times "the number of towns, cities, or villages named Fairview." Check your answer by noticing the height of the bar representing Fairview. Is your answer reasonable?

REVIEW EXERCISES

Evaluate. See Section 1.7.

91. $|2^3 - 3^2| - |5 - 7|$

92. $|5^2 - 2^2| + |9 \div (-3)|$

93. $\dfrac{5}{4 + 3 \cdot 7}$

94. $\dfrac{8}{24 - 8 \cdot 2}$

See Section 2.1.

△ **95.** A plot of land is in the shape of a triangle. If one side is x meters, a second side is $(2x - 3)$ meters and a third side is $(3x - 5)$ meters, express the perimeter of the lot as a simplified expression in x.

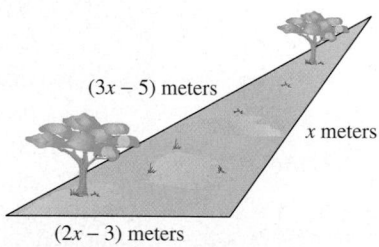

96. A portion of a board has length x feet. The other part has length $(7x - 9)$ feet. Express the total length of the board as a simplified expression in x.

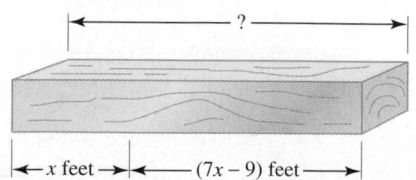

A Look Ahead

Example

Solve $t(t + 4) = t^2 - 2t + 10$.

Solution:
$$t(t + 4) = t^2 - 2t + 10$$
$$t^2 + 4t = t^2 - 2t + 10$$
$$t^2 + 4t - t^2 = t^2 - 2t + 10 - t^2$$
$$4t = -2t + 10$$
$$4t + 2t = -2t + 10 + 2t$$
$$6t = 10$$
$$\frac{6t}{6} = \frac{10}{6}$$
$$t = \frac{5}{3}$$

Solve each equation.

97. $x(x - 3) = x^2 + 5x + 7$

98. $t^2 - 6t = t(8 + t)$

99. $2z(z + 6) = 2z^2 + 12z - 8$

100. $y^2 - 4y + 10 = y(y - 5)$

101. $n(3 + n) = n^2 + 4n$

102. $3c^2 - 8c + 2 = c(3c - 8)$

2.4 AN INTRODUCTION TO PROBLEM SOLVING

CD-ROM

SSM

SSG   Video

▶ **O B J E C T I V E**

1. Apply the steps for problem solving.

1

In previous sections, you practiced writing word phrases and sentences as algebraic expressions and equations to help prepare for problem solving. We now use these translations to help write equations that model a problem. The problem-solving steps given next may be helpful.

GENERAL STRATEGY FOR PROBLEM SOLVING

1. UNDERSTAND the problem. During this step, become comfortable with the problem. Some ways of doing this are:

   Read and reread the problem.

   Choose a variable to represent the unknown.

   Construct a drawing.

   Propose a solution and check. Pay careful attention to how you check your proposed solution. This will help when writing an equation to model the problem.
2. TRANSLATE the problem into an equation.
3. SOLVE the equation.
4. INTERPRET the results: *Check* the proposed solution in the stated problem and *state* your conclusion.

Much of problem solving involves a direct translation from a sentence to an equation.

Example 1   FINDING AN UNKNOWN NUMBER

Twice the sum of a number and 4 is the same as four times the number decreased by 12. Find the number.

Solution   1. UNDERSTAND. Read and reread the problem. If we let

$$x = \text{the unknown number, then}$$

"the sum of a number and 4" translates to "$x + 4$" and "four times the number" translates to "$4x$."

2. TRANSLATE.

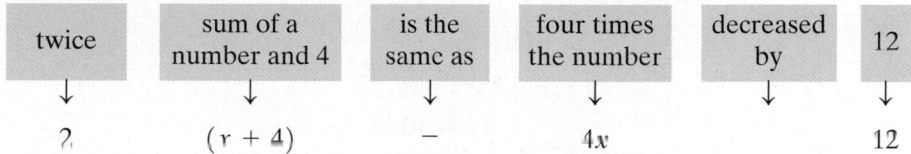

twice	sum of a number and 4	is the same as	four times the number	decreased by	12
↓	↓	↓	↓	↓	↓
2	$(x + 4)$	=	$4x$	−	12

3. SOLVE.

$$2(x + 4) = 4x - 12$$
$$2x + 8 = 4x - 12 \qquad \text{Apply the distributive property.}$$
$$2x + 8 - 4x = 4x - 12 - 4x \qquad \text{Subtract } 4x \text{ from both sides.}$$
$$-2x + 8 = -12$$
$$-2x + 8 - 8 = -12 - 8 \qquad \text{Subtract 8 from both sides.}$$
$$-2x = -20$$
$$\frac{-2x}{-2} = \frac{-20}{-2} \qquad \text{Divide both sides by } -2.$$
$$x = 10$$

4. INTERPRET.

Check: Check this solution in the problem as it was originally stated. To do so, replace "number" with 10. Twice the sum of "10" and 4 is 28, which is the same as 4 times "10" decreased by 12.

State: The number is 10.

Example 2   **FINDING THE LENGTH OF A BOARD**

A 10-foot board is to be cut into two pieces so that the longer piece is 4 times the shorter. Find the length of each piece.

Solution   1. UNDERSTAND the problem. To do so, read and reread the problem. You may also want to propose a solution. For example, if 3 feet represents the length of the shorter piece, then $4(3) = 12$ feet is the length of the longer piece, since it is 4 times the length of the shorter piece. This guess gives a total board length of 3 feet + 12 feet = 15 feet, too long. However, the purpose of proposing a solution is not to guess correctly, but to help better understand the problem and how to model it.

Since the length of the longer piece is given in terms of the length of the shorter piece, let's let

$$x = \text{length of shorter piece, then}$$
$$4x = \text{length of longer piece}$$

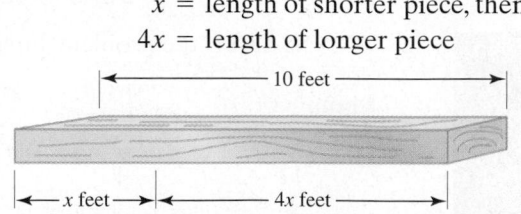

2. TRANSLATE the problem. First, we write the equation in words.

length of shorter piece	added to	length of longer piece	equals	total length of board
↓	↓	↓	↓	↓
x	$+$	$4x$	$=$	10

3. SOLVE.

$$x + 4x = 10$$

$$5x = 10 \qquad \text{Combine like terms.}$$

$$\frac{5x}{5} = \frac{10}{5} \qquad \text{Divide both sides by 5.}$$

$$x = 2$$

4. INTERPRET.

Check: Check the solution in the stated problem. If the shorter piece of board is 2 feet, the longer piece is $4 \cdot (2 \text{ feet}) = 8$ feet and the sum of the two pieces is 2 feet $+$ 8 feet $= 10$ feet.

State: The shorter piece of board is 2 feet and the longer piece of board is 8 feet.   ■

> **HELPFUL HINT**
> Make sure that units are included in your answer, if appropriate.

Example 3   **FINDING THE NUMBER OF REPUBLICAN AND DEMOCRATIC SENATORS**

In a recent year, Congress had 8 more Republican senators than Democratic. If the total number of senators is 100, how many senators of each party were there?

Solution   1. UNDERSTAND the problem. Read and reread the problem. Let's suppose that there are 40 Democratic senators. Since there are 8 more Republicans than Democrats, there must be $40 + 8 = 48$ Republicans. The total number of Democrats and Republicans is then $40 + 48 = 88$. This is incorrect since the total should be 100, but we now have a better understanding of the problem.
    Since the number of Republican senators is given in terms of the number of Democratic senators, let's let

$$x = \text{number of Democrats, then}$$

$$x + 8 = \text{number of Republicans}$$

2. TRANSLATE the problem. First, we write the equation in words.

number of Democrats	added to	number of Republicans	equals	100
↓	↓	↓	↓	↓
x	$+$	$(x + 8)$	$=$	100

3. SOLVE.

$$x + (x + 8) = 100$$

$$2x + 8 = 100 \qquad \text{Combine like terms.}$$

$$2x + 8 - 8 = 100 - 8 \qquad \text{Subtract 8 from both sides.}$$

$$2x = 92$$

$$\frac{2x}{2} = \frac{92}{2} \qquad \text{Divide both sides by 2.}$$

$$x = 46$$

4. INTERPRET.

Check: If there are 46 Democratic senators, then there are $46 + 8 = 54$ Republican senators. The total number of senators is then $46 + 54 = 100$. The results check.

State: There were 46 Democratic and 54 Republican senators.   ▬

Example 4   **CALCULATING CELLULAR PHONE USAGE**

A local cellular phone company charges Mike and Elaine Shubert $50 per month and $0.36 per minute of phone use in their usage category. If they were charged $99.68 for a month's cellular phone use, determine the number of whole minutes of phone use.

Solution   1. UNDERSTAND. Read and reread the problem. Let's propose that the Shuberts use the phone for 70 minutes. Pay careful attention as to how we calculate their bill. For 70 minutes of use, their phone bill will be $50 plus $0.36 per minute of use. This is $50 + 0.36(70) = $75.20, less than $99.68. We now understand the problem and know that the number of minutes is greater than 70.
    If we let

$$x = \text{number of minutes, then}$$
$$0.36x = \text{charge per minute of phone use}$$

2. TRANSLATE.

$50	added to	minute charge	is equal to	$99.68
↓	↓	↓	↓	↓
50	+	0.36x	=	99.68

3. SOLVE.

$$50 + 0.36x = 99.68$$

$$50 + 0.36x - 50 = 99.68 - 50 \qquad \text{Subtract 50 from both sides.}$$

$$0.36x = 49.68 \qquad \text{Simplify.}$$

$$\frac{0.36x}{0.36} = \frac{49.68}{0.36} \qquad \text{Divide both sides by 0.36.}$$

$$x = 138 \qquad \text{Simplify.}$$

4. INTERPRET.

Check: If the Shuberts spend 138 minutes on their cellular phone, their bill is $50 + $0.36(138) = $99.68.

State: The Shuberts spent 138 minutes on their cellular phone this month.   ▬

△ **Example 5** **FINDING ANGLE MEASURES**

If the two walls of the Vietnam Veterans Memorial in Washington D.C. were connected, an isosceles triangle would be formed. The measure of the third angle is 97.5° more than the measure of either of the other two equal angles. Find the measure of the third angle. (*Source:* National Park Service)

Solution 1. UNDERSTAND. Read and reread the problem. We then draw a diagram (recall that an isosceles triangle has two angles with the same measure) and let

$$x = \text{degree measure of one angle}$$
$$x = \text{degree measure of the second equal angle}$$
$$x + 97.5 = \text{degree measure of the third angle}$$

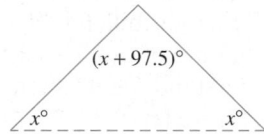

2. TRANSLATE. Recall that the sum of the measures of the angles of a triangle equals 180.

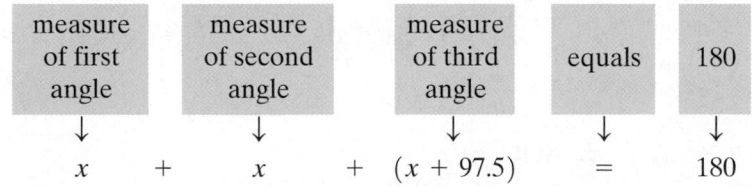

measure of first angle	measure of second angle	measure of third angle	equals	180
↓	↓	↓	↓	↓
x +	x +	$(x + 97.5)$	=	180

3. SOLVE.

$$x + x + (x + 97.5) = 180$$
$$3x + 97.5 = 180 \qquad \text{Combine like terms.}$$
$$3x + 97.5 - 97.5 = 180 - 97.5 \qquad \text{Subtract 97.5 from both sides.}$$
$$3x = 82.5$$
$$\frac{3x}{3} = \frac{82.5}{3} \qquad \text{Divide both sides by 3.}$$
$$x = 27.5$$

4. INTERPRET.

Check: If $x = 27.5$, then the measure of the third angle is $x + 97.5 = 125$. The sum of the angles is then $27.5 + 27.5 + 125 = 180$, the correct sum.

State: The third angle measures 125°.*
(*This is rounded to the nearest whole degree. The two walls actually meet at an angle of 125 degrees 12 minutes.)

Exercise Set 2.4

Solve. See Example 1.

1. The sum of twice a number and $\frac{1}{5}$ is equal to the difference between three times the number and $\frac{4}{5}$. Find the number.

2. The sum of four times a number and $\frac{2}{3}$ is equal to the difference of five times the number and $\frac{5}{6}$. Find the number.

3. Twice the difference of a number and 8 is equal to three times the sum of the number and 3. Find the number.

4. Five times the sum of a number and −1 is the same as 6 times the number. Find the number.

5. The product of twice a number and three is the same as the difference of five times the number and $\frac{3}{4}$. Find the number.

6. If the difference of a number and four is doubled, the result is $\frac{1}{4}$ less than the number. Find the number.

7. If the sum of a number and five is tripled, the result is one less than twice the number. Find the number.

8. Twice the sum of a number and six equals three times the sum of the number and four. Find the number.

Solve. See Examples 2 through 5.

9. The governor of Washington state makes about twice as much money as the governor of Nebraska. If the total of their salaries is $195,000, find the salary of each.

10. In the 1996 Summer Olympics, the United States Team won 24 more gold medals than the German Team. If the total number of gold medals for both is 64, find the number of gold medals that each team won. (*Source: World Almanac*, 2000)

11. A 40-inch board is to be cut into three pieces so that the second piece is twice as long as the first piece and the third piece is 5 times as long as the first piece. If x represents the length of the first piece, find the lengths of all three pieces.

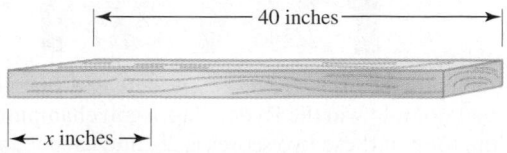

40 inches

← x inches →

12. A 21-foot beam is to be divided so that the longer piece is 1 foot more than 3 times the shorter piece. If x represents the length of the shorter piece, find the lengths of both pieces.

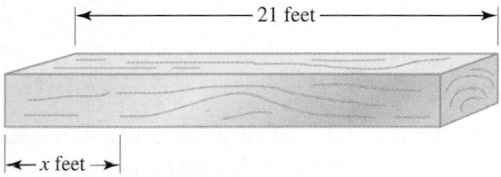

21 feet

← x feet →

13. A car rental agency advertised renting a Buick Century for $24.95 per day and $0.29 per mile. If you rent this car for 2 days, how many whole miles can you drive on a $100 budget?

14. A plumber gave an estimate for the renovation of a kitchen. Her hourly pay is $27 per hour and the plumber's parts will cost $80. If her total estimate is $404, how many hours does she expect this job to take?

△ 15. The flag of Equatorial Guinea contains an isosceles triangle. (Recall that an isosceles triangle contains two angles with the same measure.) If the measure of the third angle of the triangle is 30° more than twice the measure of either of the other two angles, find the measure of each angle of the triangle. (*Hint:* Recall that the sum of the measures of the angles of a triangle is 180°.)

△ 16. The flag of Brazil contains a parallelogram. One angle of the parallelogram is 15° less than twice the measure of the angle next to it. Find the measure of each angle of the parallelogram. (*Hint:* Recall that opposite angles of a parallelogram have the same measure and that the sum of the measures of the angles is 360°.)

17. In a recent election in Florida for a seat in the United States House of Representatives, Corrine Brown received 13,288 more votes than Bill Randall. If the total number of votes was 119,436, find the number of votes for each candidate.

18. In a recent election in Texas for a seat in the United States House of Representatives, Max Sandlin received 25,557 more votes than opponent Dennis Boerner. If the total number of votes was 135,821, find the number of votes for each candidate. (*Source:* Voter News Service)

19. Two angles are supplementary if their sum is 180°. One angle measures three times the measure of the smaller angle. If x represents the measure of the smaller angle and these two angles are supplementary, find the measure of each angle.

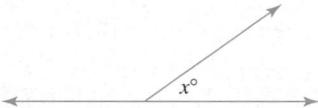

20. Two angles are complementary if their sum is 90°. Given the measures of the complementary angles shown, find the measure of each angle.

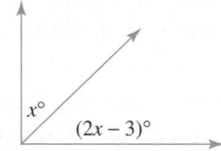

21. To make an international telephone call, you need the code for the country you are calling. The codes for Belgium, France, and Spain are three consecutive integers whose sum is 99. Find the code for each country. (*Source: The World Almanac and Book of Facts,* 1999)

22. To make an international telephone call, you need the code for the country you are calling. The codes for Mali Republic, Côte d'Ivoire, and Niger are three consecutive odd integers whose sum is 675. Find the code for each country.

23. Human fingernails grow much faster than human toenails. Healthy fingernails grow at an average rate of 0.8 inches per year. This is about four times as fast as toenails grow. About how fast do human toenails grow?

24. The human eye blinks once every 5 seconds on average. How many times does the average eye blink in one day? In one year?

25. On December 7, 1995, a probe launched from the robot explorer called Galileo entered the atmosphere of Jupiter at 100,000 miles per hour. The diameter of the probe is 19 inches less than twice its height. If the sum of

the height and the diameter is 83 inches, find each dimension.

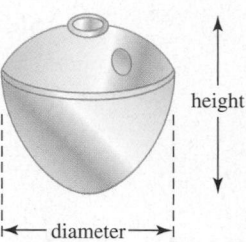

26. Over the past few years the satellite Voyager II has passed by the planets Saturn, Uranus, and Neptune, continually updating information about these planets, including the number of moons for each. Uranus is now believed to have 7 more moons than Neptune. Also, Saturn is now believed to have 2 more than twice the number of moons of Neptune. If the total number of moons for these planets is 41, find the number of moons for each planet. (*Source:* National Space Science Data Center)

27. In 1999, the San Antonio Spurs won the NBA championship against the New York Knicks. The two teams' final scores for the last game of the championship were two consecutive integers whose sum was 155. Find each final score. (*Source: USA Today,* 6/28/99)

28. In September 1999, Team U.S.A. defeated Team Europe by 1 point to win the Ryder Cup, a golf championship. If the total of these two scores is 28, find each team score.

29. Tony Hawk landed the first 900 in skateboard competition history during the X Games "Best Trick" competition in 1999. A 900 means that he rotated 900° in the air with his skateboard. If there are 360° in a single rotation, how many rotations were there?

30. The Pentagon Building in Washington, D.C., is the headquarters for the U.S. Department of Defense. The Pentagon is also the world's largest office building in terms of ground space with a floor area of over 6.5 million square feet. This is three times the floor area of the Empire State Building. About how much floor space does the Empire State Building have? Round to the nearest tenth.

31. After a recent election, there were 14 more Republican governors than Democratic governors in the United States. In addition, 2 of the 50 state governors were neither Republican nor Democrat. How many Democrats and how many Republicans held governor's offices after the 1998 election? (*Source: The World Almanac and Book of Facts, 1999*)

32. In a recent year, the total number of Democrats and Republicans in the U.S. House of Representatives was 434. There were 12 more Republicans than Democrats. Find the number of representatives from each party.

33. A 17-foot piece of string is cut into two pieces so that one piece is 2 feet longer than twice the shorter piece. If the shorter piece is x feet long, find the lengths of both pieces.

34. An 18-foot wire is to be cut so that the longer piece is 5 times longer than the shorter piece. Find the length of each piece.

△ **35.** The measures of the angles of a triangle are 3 consecutive even integers. Find the measure of each angle.

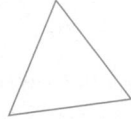

△ **36.** A quadrilateral is a polygon with 4 sides. The sum of the measures of the 4 angles in a quadrilateral is 360°. If the measures of the angles of a quadrilateral are consecutive odd integers, find the measures.

37. Enterprise Car Rental charges a daily rate of $34 plus $0.20 per mile. Suppose that you rent a car for a day and your bill (before taxes) is $104. How many miles did you drive?

38. A woman's $15,000 estate is to be divided so that her husband receives twice as much as her son. If x represents the amount of money that her son receives, find the amount of money that her husband receives and the amount of money that her son receives.

The graph below shows the states with the highest tourism budgets for 1999. Use the graph for Exercises 39–44.

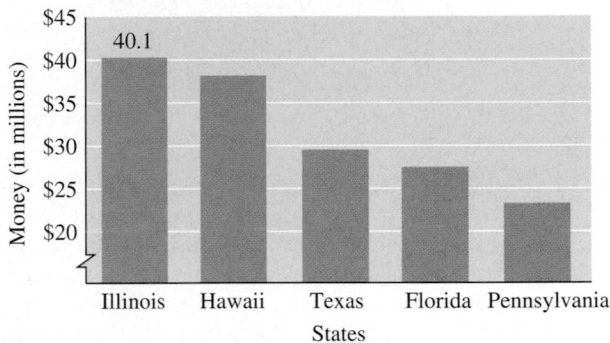

Source: Travel Industry Association of America

39. In your own words, describe what the word *tourism* means.

40. Which state spends the most money on tourism?

41. Which states spend between $25 and $30 million on tourism?

42. The states of Texas and Florida spend a total of $56.6 million for tourism. The state of Texas spends $2.2 million more than the state of Florida. Find the amount that each state spends on tourism.

43. The states of Hawaii and Pennsylvania spend a total of $60.9 million for tourism. The state of Hawaii spends $8.1 million less than twice the amount of money that the state of Pennsylvania spends. Find the amount that each state spends on tourism.

44. Compare the heights of the bars in the graph with your results of Exercises 42 and 43. Are your answers reasonable?

△ **45.** The golden rectangle is a rectangle whose length is approximately 1.6 times its width. The early Greeks thought that a rectangle with these dimensions was the most pleasing to the eye and examples of the golden rectangle are found in many early works of art. For example, the Parthenon in Athens contains many examples of golden rectangles. Mike Hallahan would like to plant a rectangular garden in the shape of a golden rectangle. If he has 78 feet of fencing available, find the dimensions of the garden.

The length-width rectangle approximates the golden rectangle as well as the width-height rectangle.

46. Dr. Dorothy Smith gave the students in her geometry class at the University of New Orleans the following question. Is it possible to construct a triangle such that the second angle of the triangle has a measure that is twice the measure of the first angle and the measure of the third angle is 3 times the measure of the first? If so, find the measure of each angle. (*Hint:* Recall that the sum of the measures of the angles of a triangle is 180°.)

↘ **47.** Give an example of how you recently solved a problem using mathematics.

↘ **48.** In your own words, explain why a solution of a word problem should be checked using the original wording of the problem and not the equation written from the wording.

Recall from Exercise 45 that the golden rectangle is a rectangle whose length is approximately 1.6 times its width.

△ **49.** It is thought that for about 75% of adults, a rectangle in the shape of the golden rectangle is the most pleasing to the eye. Draw 3 rectangles, one in the shape of the golden rectangle, and poll your class. Do the results agree with the percentage given above?

△ **50.** Examples of golden rectangles can be found today in architecture and manufacturing packaging. Find an example of a golden rectangle in your home. A few suggestions: the front face of a book, the floor of a room, the front of a box of food.

△ **51.** Measure the dimensions of each rectangle and decide which one best approximates the shape of the golden rectangle.

a.

b.

c.

REVIEW EXERCISES

Perform the indicated operations. See Sections 1.5 and 1.6.

52. $3 + (-7)$

53. $-2 + (-8)$

54. $4 - 10$

55. $-11 + 2$

56. $-5 - (-1)$

57. $-12 - 3$

Translate each sentence into an equation. See Sections 1.4 and 2.3.

58. Half of the difference of a number and one is thirty-seven.

59. Five times the opposite of a number is the number plus sixty.

60. If three times the sum of a number and 2 is divided by 5, the quotient is 0.

61. If the sum of a number and 9 is subtracted from 50, the result is 0.

2.5 FORMULAS AND PROBLEM SOLVING

CD-ROM SSM

SSG Video

▶ **OBJECTIVES**

1. Use formulas to solve problems.
2. Solve a formula or equation for one of its variables.

1 An equation that describes a known relationship among quantities, such as distance, time, volume, weight, and money is called a **formula**. These quantities are represented by letters and are thus variables of the formula. Here are some common formulas and their meanings.

$A = lw$
Area of a rectangle = length · width

$I = PRT$
Simple Interest = Principal · Rate · Time

$P = a + b + c$
Perimeter of a triangle = side a + side b + side c

$d = rt$
distance = rate · time

$V = lwh$
Volume of a rectangular solid = length · width · height

$F = \left(\dfrac{9}{5}\right)C + 32$

degrees Fahrenheit = $\left(\dfrac{9}{5}\right)$ · degrees Celsius + 32

Formulas are valuable tools because they allow us to calculate measurements as long as we know certain other measurements. For example, if we know we traveled a distance of 100 miles at a rate of 40 miles per hour, we can replace the variables d and r in the formula $d = rt$ and find our time, t.

$$d = rt \qquad \text{Formula}$$
$$100 = 40t \qquad \text{Replace } d \text{ with 100 and } r \text{ with 40.}$$

This is a linear equation in one variable, t. To solve for t, divide both sides of the equation by 40.

$$\frac{100}{40} = \frac{40t}{40} \qquad \text{Divide both sides by 40.}$$
$$\frac{5}{2} = t \qquad \text{Simplify.}$$

The time traveled is $\frac{5}{2}$ hours or $2\frac{1}{2}$ hours.

In this section we solve problems that can be modeled by known formulas. We use the same problem-solving steps that were introduced in the previous section. These steps have been slightly revised to include formulas.

Example 1   FINDING TIME GIVEN RATE AND DISTANCE

A glacier is a giant mass of rocks and ice that flows downhill like a river. Portage Glacier in Alaska is about 6 miles, or 31,680 *feet*, long and moves 400 *feet* per year. Icebergs are created when the front end of the glacier flows into Portage Lake. How long does it take for ice at the head (beginning) of the glacier to reach the lake?

Solution   1.   UNDERSTAND. Read and reread the problem. The appropriate formula needed to solve this problem is the distance formula, $d = rt$. To become familiar with this formula, let's find the distance that ice traveling at a rate of 400 feet per year travels in 100 years. To do so, we let time t be 100 years and rate r be the given 400 feet per year, and substitute these values into the formula $d = rt$. We then have that distance $d = 400(100) = 40,000$ feet. Since we are interested in finding how long it takes ice to travel 31,680 feet, we now know that it is less than 100 years.

Since we are using the formula $d = rt$, we let

t = the time in years for ice to reach the lake

r = rate or speed of ice

d = distance from beginning of glacier to lake

2.   TRANSLATE. To translate to an equation, we use the formula $d = rt$ and let distance $d = 31,680$ feet and rate $r = 400$ feet per year.

$$d = r \cdot t$$

$$31{,}680 = 400 \cdot t \qquad \text{Let } d = 31{,}680 \text{ and } r = 400.$$

3.   SOLVE. Solve the equation for t. To solve for t, divide both sides by 400.

$$\frac{31{,}680}{400} = \frac{400 \cdot t}{400} \qquad \text{Divide both sides by 400.}$$

$$79.2 = t \qquad \text{Simplify.}$$

4.   INTERPRET.

Check: To check, substitute 79.2 for t and 400 for r in the distance formula and check to see that the distance is 31,680 feet.

State: It takes 79.2 years for the ice at the head of Portage Glacier to reach the lake.

> **HELPFUL HINT**
> Don't forget to include units, if appropriate.

△ Example 2   CALCULATING THE LENGTH OF A GARDEN

Charles Pecot can afford enough fencing to enclose a rectangular garden with a perimeter of 140 feet. If the width of his garden must be 30 feet, find the length.

Solution

$w = 30$ feet

1. UNDERSTAND. Read and reread the problem. The formula needed to solve this problem is the formula for the perimeter of a rectangle, $P = 2l + 2w$. Before continuing, let's become familar with this formula.

   l = the length of the rectangular garden

   w = the width of the rectangular garden

   P = perimeter of the garden

2. TRANSLATE. To translate to an equation, we use the formula $P = 2l + 2w$ and let perimeter $P = 140$ feet and width $w = 30$ feet.

$$P = 2l + 2w \qquad \text{Let } P = 140 \text{ and } w = 30.$$
$$\downarrow \qquad\qquad \downarrow$$
$$140 = 2l + 2(30)$$

3. SOLVE.

$$140 = 2l + 2(30)$$
$$140 = 2l + 60 \qquad \text{Multiply } 2(30).$$
$$140 - 60 = 2l + 60 - 60 \qquad \text{Subtract 60 from both sides.}$$
$$80 = 2l \qquad \text{Combine like terms.}$$
$$40 = l \qquad \text{Divide both sides by 2.}$$

4. INTERPRET.

   Check: Substitute 40 for l and 30 for w in the perimeter formula and check to see that the perimeter is 140 feet.

   State: The length of the rectangular garden is 40 feet.   ▬

Example 3   FINDING AN EQUIVALENT TEMPERATURE

The average maximum temperature for January in Algerias, Algeria, is 59° Fahrenheit. Find the equivalent temperature in degrees Celsius.

Solution

1. UNDERSTAND. Read and reread the problem. A formula that can be used to solve this problem is the formula for converting degrees Celsius to degrees Fahrenheit, $F = \frac{9}{5}C + 32$. Before continuing, become familiar with this formula. Using this formula, we let

$$C = \text{temperature in degrees Celsius, and}$$
$$F = \text{temperature in degrees Fahrenheit.}$$

2. TRANSLATE. To translate to an equation, we use the formula $F = \frac{9}{5}C + 32$ and let degrees Fahrenheit $F = 59$.

$$\text{Formula:} \qquad F = \frac{9}{5}C + 32$$

$$\text{Substitute:} \quad 59 = \frac{9}{5}C + 32 \qquad \text{Let } F = 59.$$

3. SOLVE.

$$59 = \frac{9}{5}C + 32$$

$$59 - 32 = \frac{9}{5}C + 32 - 32 \qquad \text{Subtract 32 from both sides.}$$

$$27 = \frac{9}{5}C \qquad \text{Combine like terms.}$$

$$\frac{5}{9} \cdot 27 = \frac{5}{9} \cdot \frac{9}{5}C \qquad \text{Multiply both sides by } \frac{5}{9}.$$

$$15 = C \qquad \text{Simplify.}$$

4. INTERPRET.

Check: To check, replace C with 15 and F with 59 in the formula and see that a true statement results.

State: Thus, 59° Fahrenheit is equivalent to 15° Celsius.

2   We say that the formula $F = \frac{9}{5}C + 32$ is solved for F because F is alone on one side of the equation and the other side of the equation contains no F's. Suppose that we need to convert many Fahrenheit temperatures to equivalent degrees Celsius. In this case, it is easier to perform this task by solving the formula $F = \frac{9}{5}C + 32$ for C. (See Example 7.) For this reason, it is important to be able to solve an equation for any one of its specified variables. For example, the formula $d = rt$ is solved for d in terms of r and t. We can also solve $d = rt$ for t in terms of d and r. To solve for t, divide both sides of the equation by r.

$$d = rt$$

$$\frac{d}{r} = \frac{rt}{r} \qquad \text{Divide both sides by } r.$$

$$\frac{d}{r} = t \qquad \text{Simplify.}$$

To solve a formula or an equation for a specified variable, we use the same steps as for solving a linear equation. These steps are listed next.

SOLVING EQUATIONS FOR A SPECIFIED VARIABLE

Step 1. Multiply on both sides to clear the equation of fractions if they occur.

Step 2. Use the distributive property to remove parentheses if they occur.

Step 3. Simplify each side of the equation by combining like terms.

Step 4. Get all terms containing the specified variable on one side and all other terms on the other side by using the addition property of equality.

Step 5. Get the specified variable alone by using the multiplication property of equality.

⚠ **Example 4**   Solve $V = lwh$ for l.

Solution   This formula is used to find the volume of a box. To solve for l, divide both sides by wh.

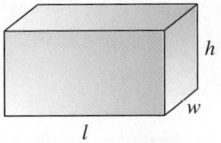

$$V = lwh$$

$$\frac{V}{wh} = \frac{lwh}{wh} \qquad \text{Divide both sides by } wh.$$

$$\frac{V}{wh} = l \qquad \text{Simplify.}$$

Since we have l alone on one side of the equation, we have solved for l in terms of V, w, and h. Remember that it does not matter on which side of the equation we isolate the variable. ▬

Example 5   Solve $y = mx + b$ for x.

Solution   The term containing the variable we are solving for, mx, is on the right side of the equation. Get mx alone by subtracting b from both sides.

$$y = mx + b$$

$$y - b = mx + b - b \qquad \text{Subtract } b \text{ from both sides.}$$

$$y - b = mx \qquad \text{Combine like terms.}$$

Next, solve for x by dividing both sides by m.

$$\frac{y - b}{m} = \frac{mx}{m}$$

$$\frac{y - b}{m} = x \qquad \text{Simplify.} \qquad ▬$$

⚠ **Example 6**   Solve $P = 2l + 2w$ for w.

Solution   This formula relates the perimeter of a rectangle to its length and width. Find the term containing the variable w. To get this term, $2w$, alone subtract $2l$ from both sides.

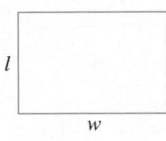

$$P = 2l + 2w$$

$$P - 2l = 2l + 2w - 2l \qquad \text{Subtract } 2l \text{ from both sides.}$$

$$P - 2l = 2w \qquad \text{Combine like terms.}$$

$$\frac{P - 2l}{2} = \frac{2w}{2} \qquad \text{Divide both sides by 2.}$$

$$\frac{P - 2l}{2} = w \qquad \text{Simplify.}$$

The next example has an equation containing a fraction. We will first clear the equation of fractions and then solve for the specified variable. ▬

Example 7   Solve $F = \frac{9}{5}C + 32$ for C.

Solution

$$F = \frac{9}{5}C + 32$$

$$5(F) = 5\left(\frac{9}{5}C + 32\right)$$ Clear the fraction by multiplying both sides by the LCD.

$$5F = 9C + 160$$ Distribute the 5.

$$5F - 160 = 9C + 160 - 160$$ To get the term containing the variable C alone, subtract 160 from both sides.

$$5F - 160 = 9C$$ Combine like terms.

$$\frac{5F - 160}{9} = \frac{9C}{9}$$ Divide both sides by 9.

$$\frac{5F - 160}{9} = C$$ Simplify.

Exercise Set 2.5

Substitute the given values into each given formula and solve for the unknown variable. If necessary, round to one decimal place. See Examples 1 through 3.

1. $A = bh$; $A = 45, b = 15$
   (Area of a parallelogram)

2. $d = rt$; $d = 195, t = 3$
   (Distance formula)

3. $S = 4lw + 2wh$; $S = 102, l = 7, w = 3$
   (Surface area of a special rectangular box)

4. $V = lwh$; $l = 14, w = 8, h = 3$
   (Volume of a rectangular box)

5. $A = \frac{1}{2}h(B + b)$; $A = 180, B = 11, b = 7$
   (Area of a trapezoid)

6. $A = \frac{1}{2}h(B + b)$; $A = 60, B = 7, b = 3$
   (Area of a trapezoid)

7. $P = a + b + c$; $P = 30, a = 8, b = 10$
   (Perimeter of a triangle)

8. $V = \frac{1}{3}Ah$; $V = 45, h = 5$
   (Volume of a pyramid)

9. $C = 2\pi r$; $C = 15.7$ (use the approximation 3.14 or a calculator approximation for π)
   (Circumference of a circle)

10. $A = \pi r^2$; $r = 4.5$ (use the approximation 3.14 or a calculator approximation for π)
    (Area of a circle)

11. $I = PRT$; $I = 3750, P = 25,000, R = 0.05$
    (Simple interest formula)

12. $I = PRT$; $I = 1,056,000, R = 0.055, T = 6$
    (Simple interest formula)

13. $V = \frac{1}{3}\pi r^2 h$; $V = 565.2, r = 6$ (use the calculator approximation for π)
    (Volume of a cone)

14. $V = \frac{4}{3}\pi r^3$; $r = 3$ (use a calculator approximation for π)
    (Volume of a sphere)

Solve each formula for the specified variable. See Examples 4 through 7.

15. $f = 5gh$ for h

16. $C = 2\pi r$ for r

17. $V = LWH$ for W

18. $T = mnr$ for n

19. $3x + y = 7$ for y

20. $-x + y = 13$ for y

21. $A = P + PRT$ for R

22. $A = P + PRT$ for T

23. $V = \frac{1}{3}Ah$ for A

24. $D = \frac{1}{4}fk$ for k

25. $P = a + b + c$ for a

26. $PR = s_1 + s_2 + s_3 + s_4$ for s_3

27. $S = 2\pi rh + 2\pi r^2$ for h

28. $S = 4lw + 2wh$ for h

Solve. See Examples 1 through 3.

29. The world's largest sign for Coca-Cola is located in Arica, Chile. The rectangular sign has a length of 400 feet and has an area of 52,400 square feet. Find the width of the sign. (*Source: Fabulous Facts about Coca-Cola*, Atlanta, GA)

△ **30.** The length of a rectangular garden is 6 meters. If 21 meters of fencing are required to fence the garden, find its width.

6 meters

31. A limousine built in 1968 for the president cost $500,000 and weighed 5.5 tons. This Lincoln Continental Executive could travel at 50 miles per hour with all of its tires shot away. At this rate, how long would it take to travel from Charleston, Virginia, to Washington, D.C., a distance of 375 miles?

32. The SR-71 is a top secret spy plane. It is capable of traveling from Rochester, New York, to San Francisco, California, a distance of approximately 3000 miles, in $1\frac{1}{2}$ hours. Find the rate of the SR-71.

33. Convert Nome, Alaska's 14°F high temperature to Celsius.

34. Convert Paris, France's low temperature of −5°C to Fahrenheit.

△ **35.** Piranha fish require 1.5 cubic feet of water per fish to maintain a healthy environment. Find the maximum number of piranhas you could put in a tank measuring 8 feet by 3 feet by 6 feet.

△ **36.** Find how many goldfish you can put in a cylindrical tank whose diameter is 8 meters and whose height is 3 meters, if each goldfish needs 2 cubic meters of water.

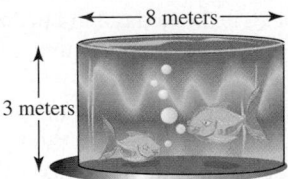

37. The Cat is a high-speed catamaran auto ferry that operates between Bar Harbor, Maine, and Yarmouth, Nova Scotia. The Cat can make the trip in about $2\frac{1}{2}$ hours at a speed of 55 mph. About how far apart are Bar Harbor and Yarmouth? (*Source:* Bay Ferries)

38. A family is planning their vacation to Disney World. They will drive from a small town outside New Orleans, Louisiana, to Orlando, Florida, a distance of 700 miles. They plan to average a rate of 55 miles per hour. How long will this trip take?

△ **39.** A lawn is in the shape of a trapezoid with a height of 60 feet and bases of 70 feet and 130 feet. How many bags of

fertilizer must be purchased to cover the lawn if each bag covers 4000 square feet?

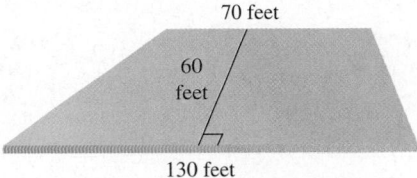

△ **40.** If the area of a right-triangularly shaped sail is 20 square feet and its base is 5 feet, find the height of the sail.

41. The X-30 is a "space plane" that skims the edge of space at 4000 miles per hour. Neglecting altitude, if the circumference of the Earth is approximately 25,000 miles, how long will it take for the X-30 to travel around the Earth?

42. In the United States, a notable hang glider flight was a 303-mile, $8\frac{1}{2}$ hour flight from New Mexico to Kansas. What was the average rate during this flight?

The Dante II is a spider-like robot that is used to map the depths of an active Alaskan volcano.

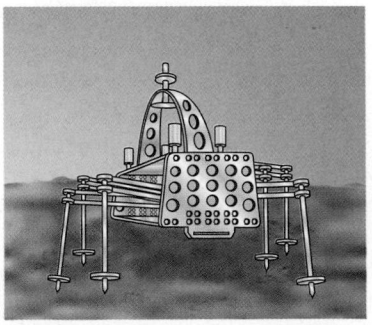

△ **43.** The dimensions of the Dante II are 10 feet long by 8 feet wide by 10 feet high. Find the volume of the smallest box needed to store this robot.

44. The Dante II traveled 600 feet into an active Alaskan volcano in $3\frac{1}{3}$ hours. Find the traveling rate of Dante II in feet per minute. (*Hint:* First convert $3\frac{1}{3}$ hours to minutes.)

△ **45.** Maria's Pizza sells one 16-inch cheese pizza or two 10-inch cheese pizzas for $9.99. Determine which size gives more pizza.

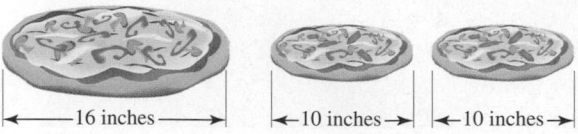

△ **46.** Find how much rope is needed to wrap around the Earth at the equator, if the radius of the Earth is 4000 miles. (*Hint:* Use 3.14 for π and the formula for circumference.)

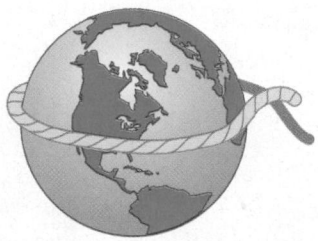

47. A Japanese "bullet" train set a new world record for train speed at 552 kilometers per hour during a manned test run on the Yamanashi Maglev Test Line in April 1999. The Yamanashi Maglev Test Line is 42.8 kilometers long. How many *minutes* would a test run on the Yamanashi Line last at this record-setting speed? Round to the nearest hundredth of a minute. (*Source:* Japan Railways Central Co.)

48. In 1983, the Hawaiian volcano Kilauea began erupting in a series of episodes still occurring at the time of this writing. At times, the lava flows advanced at speeds of up to 0.5 kilometer per hour. In 1983 and 1984, lava flows destroyed 16 homes in the Royal Gardens subdivision, about 6 km away from the eruption site. Roughly how long did it take the lava to reach Royal Gardens? (*Source:* U.S. Geological Survey Hawaiian Volcano Observatory)

49. Find how long it takes Tran Nguyen to drive 135 miles on I-10 if he merges onto I-10 at 10 A.M. and drives nonstop with his cruise control set on 60 mph.

50. Beaumont, Texas, is about 150 miles from Toledo Bend. If Leo Miller leaves Beaumont at 4 A.M. and averages 45 mph, when should he arrive at Toledo Bend?

51. Dry ice is a name given to solidified carbon dioxide. At −78.5° Celsius it changes directly from a solid to a gas. Convert this temperature to Fahrenheit.

52. Lightning bolts can reach a temperature of 50,000° Fahrenheit. Convert this temperature to Celsius.

53. The distance from the sun to the Earth is approximately 93,000,000 miles. If light travels at a rate of 186,000 miles per second, how long does it take light from the sun to reach us?

54. Light travels at a rate of 186,000 miles per second. If our moon is 238,860 miles from the Earth, how long does it take light from the moon to reach us? (Round to the nearest tenth of a second.)

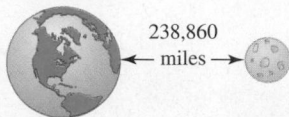

55. On July 16, 1994, the Shoemaker-Levy 9 comet collided with Jupiter. The impact of the largest fragment of the comet, a massive chunk of rock and ice, created a fireball with a radius of 2000 miles. Find the volume of this spherical fireball. (Use a calculator approximation for π. Round to the nearest whole cubic mile.)

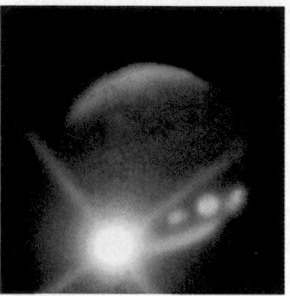

Color image made by the Hubble Space Telescope of the impact site of Fragment G of the comet.

56. The fireball from the largest fragment of the comet (see Exercise 55) immediately collapsed, as it was pulled down by gravity. As it fell, it cooled to approximately −350°F. Convert this temperature to Celsius.

57. Bolts of lightning can travel at 270,000 miles per second. How many times can a lightning bolt travel around the world in one second? (See Exercise 46. Round to the nearest tenth.)

58. A glacier is a giant mass of rocks and ice that flows downhill like a river. Exit Glacier, near Seward, Alaska, moves at a rate of 20 inches a day. Find the distance in feet the glacier moves in a year. (Assume 365 days in a year. Round to 2 decimal places.)

59. Flying fish do not *actually* fly, but glide. They have been known to travel a distance of 1300 feet at a rate of 20 miles per hour. How many seconds did it take to travel this distance? (*Hint:* First convert miles per hour to feet per second. Recall that 1 mile = 5280 feet. Round to the nearest tenth of a second.)

60. Stalactites join stalagmites to form columns. A column found at Natural Bridge Caverns near San Antonio, Texas, rises 15 feet and has a *diameter* of only 2 inches. Find the volume of this column in cubic inches. (*Hint:* Use the formula for volume of a cylinder and use a calculator approximation for π.) Round to the nearest tenth of a cubic inch.

61. Find the temperature at which the Celsius measurement and Fahrenheit measurement are the same number.

62. Normal room temperature is about 78°F. Convert this temperature to Celsius.

63. The formula $V = LWH$ is used to find the volume of a box. If the length of a box is doubled, the width is doubled, and the height is doubled, how does this affect the volume?

64. The formula $A = bh$ is used to find the area of a parallelogram. If the base of a parallelogram is doubled and its height is doubled, how does this affect the area?

REVIEW EXERCISES

Write the following phrases as algebraic expressions. Simplify if possible. See Section 2.1.

65. Nine divided by the sum of a number and 5

66. Half the product of a number and five

67. Three times the sum of a number and four

68. One-third of the quotient of a number and six

69. Double the sum of ten and four times a number

70. Twice a number divided by three times the number

71. Triple the difference of a number and twelve

72. A number minus the sum of the number and six

2.6 PERCENT AND PROBLEM SOLVING

CD-ROM   SSM

SSG   Video

▶ **OBJECTIVES**

1. Write percents as decimals and decimals as percents.
2. Find percents of numbers.
3. Read and interpret graphs containing percents.
4. Solve percent equations.
5. Solve problems containing percents.

1   Much of today's statistics is given in terms of percent: a basketball player's free throw percent, current interest rates, stock market trends, and nutrition labeling, just to name a few. In this section, we explore percent and applications involving percents.

If 37 percent of all households in the United States have a home computer, what does this percent mean? It means that 37 households out of every 100 households have a home computer. In other words,

the word **percent** means **per hundred** so that

$$37 \textbf{ percent} \text{ means } 37 \textbf{ per hundred} \text{ or } \frac{37}{100}.$$

We use the symbol % to denote percent, and we can write a percent as a fraction or a decimal.

$$37\% = \frac{37}{100} = 0.37$$

$$8\% = \frac{8}{100} = 0.08$$

$$100\% = \frac{100}{100} = 1.00$$

This suggests the following:

WRITING A PERCENT AS A DECIMAL

Drop the percent symbol and move the decimal point two places to the left.

Example 1 Write each percent as a decimal.

  a. 35% **b.** 89.5% **c.** 150%

Solution **a.** 35% = 0.35 **b.** 89.5% = 0.895 **c.** 150% = 1.5

To write a decimal as a percent, we reverse the procedure above.

WRITING A DECIMAL AS A PERCENT

Move the decimal point two places to the right and attach the percent symbol, %.

Example 2 Write each number as a percent.

  a. 0.73 **b.** 1.39 **c.** $\frac{1}{4}$

Solution **a.** 0.73 = 73% **b.** 1.39 = 139%

  c. First, write $\frac{1}{4}$ as a decimal.

$$\frac{1}{4} = 0.25 = 25\%.$$

2 To find a percent of a number, recall that the word "of" means multiply.

Example 3 Find 72% of 200.

Solution To find 72% of 200, we multiply.

$$72\% \text{ of } 200 = 72\%(200)$$
$$= 0.72(200) \quad \text{Write 72\% as 0.72.}$$
$$= 144. \quad \text{Multiply.}$$

Thus, 72% of 200 is 144.

3 As mentioned earlier, percents are often used in statistics. Recall that the graph below is called a circle graph or a pie chart. The circle or pie represents a whole, or 100%. Each circle is divided into sectors (shaped like pieces of a pie) which represent various parts of the whole 100%.

Example 4 **PERCENT OF HOME MAINTENANCE**

The circle graph below shows how much money homeowners in the United States spend annually on maintaining their homes. Use this graph to answer the questions below.

Yearly Home Maintenance in the U.S.

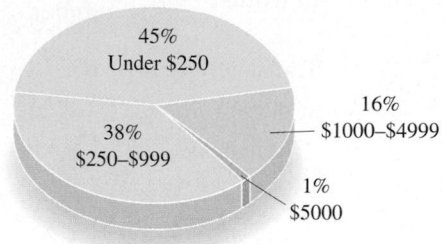

a. What percent of homeowners spend under $250 on yearly home maintenance?
b. What percent of homeowners spend less than $1000 per year on maintenance?
c. How many of the 22,000 homeowners in a town called Fairview might we expect to spend under $250 a year on home maintenance?
d. Find the number of degrees in the 16% sector.

Solution
a. From the circle graph, we see that 45% of homeowners spend under $250 per year on home maintenance.
b. From the circle graph, we know that 45% of homeowners spend under $250 per year and 38% of homeowners spend $250–$999 per year, so that the sum 45% + 38% or 83% of homeowners spend less than $1000 per year.
c. Since 45% of homeowners spend under $250 per year on maintenance, we find 45% of 22,000.

$$45\% \text{ of } 22{,}000 = 0.45(22{,}000)$$
$$= 9900$$

We might then expect that 9900 homeowners in Fairview spend under $250 per year on home maintenance.
d. To find the number of degrees in the 16% sector, recall that the number of degrees around a circle is 360°. Thus, to find the number of degrees in the 16% sector, we find 16% of 360°.

$$16\% \text{ of } 360° = 0.16(360°)$$
$$= 57.6°$$

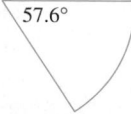

The 16% sector contains 57.6°.

4 Next, we practice writing sentences as percent equations.

Example 5 The number 63 is what percent of 72?

Solution
1. UNDERSTAND. Read and reread the problem. Next, let's guess a solution. Suppose we guess that 63 is 80% of 72. We may check our guess by finding 80% of 72. 80% of 72 = 0.80(72) = 57.6. Close, but not 63. At this point, though, we have a better understanding of the problem, we know the correct answer is close to and greater than 80%, and we know how to check our proposed solution later.

   Let x = the unknown percent.

2. TRANSLATE. Recall that "is" means "equals" and "of" signifies multiplying. Translate the sentence directly.

   In words:

The number 63	is	what percent	of	72	?
Translate: 63	=	x	$\cdot$	72	

3. SOLVE.

$$63 = 72x$$
$$0.875 = x \qquad \text{Divide both sides by 72.}$$
$$87.5\% = x \qquad \text{Write as a percent.}$$

4. INTERPRET.
   Check: Check by verifying that 87.5% of 72 is 63.
   State: The number 63 is 87.5% of 72.

Example 6 The number 120 is 15% of what number?

Solution
1. UNDERSTAND. Read and reread the problem. Guess a solution and check your guess.

   Let x = the unknown number.

2. TRANSLATE.

   In words:

The number 120	is	15%	of	what number	?
Translate: 120	=	15%	$\cdot$	x	

3. SOLVE.

$$120 = 0.15x$$
$$800 = x \qquad \text{Divide both sides by 0.15.}$$

4. INTERPRET.
   Check: Check the proposed solution of 800 by finding 15% of 800 and verifying that the result is 120.
   State: Thus, 120 is 15% of 800.

5 Percent increase or percent decrease is a common way to describe how some measurement has increased or decreased. For example, crime increased by 8%, teachers received a 5.5% increase in salary, or a company decreased its employees by 10%. The next example is a review of percent increase.

Example 7 CALCULATING THE COST OF ATTENDING COLLEGE

The cost of attending a public college rose from $5324 in 1990 to $8086 in 2000. Find the percent increase. (*Source:* U.S. Department of Education. *Note:* These costs include room and board.)

Solution

1. UNDERSTAND. Read and reread the problem. Let's guess that the percent increase is 20%. To see if this is the case, we find 20% of $5324 to find the *increase* in cost. Then we add this increase to $5324 to find the *new cost*. In other words, 20%($5324) = 0.20($5324) = $1064.80, the *increase* in cost. The new cost then would be $5324 + $1064.80 = $6388.80, less than the actual new cost of $8086. We now know that the increase is greater than 20% and we know how to check our proposed solution.

   Let x = the percent increase.

2. TRANSLATE. First, find the **increase**, and then the **percent increase**. The increase in cost is found by:

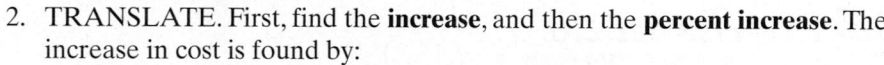

   In words: increase = | new cost | − | old cost | or

   Translate: increase = $8086 − $5324
   = $2762

   Next, find the percent increase. The percent increase or percent decrease is always a percent of the original number or in this case, the old cost.

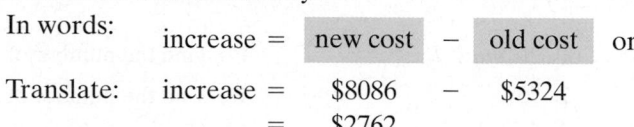

   In words: | increase | is | what percent increase | of | old cost |

   Translate: $2762 = x · $5324

3. SOLVE.

$$2762 = 5324x$$
$$0.519 \approx x$$
$$51.9\% \approx x$$

   Divide both sides by 5324 and round to 3 decimal places.
   Write as a percent.

4. INTERPRET.
   Check: Check the proposed solution.
   State: The percent increase in cost is approximately 51.9%.

SPOTLIGHT ON DECISION MAKING

Suppose you are a personal income tax preparer. Your clients, Jose and Felicia Fernandez, are filing jointly Form 1040 as their individual income tax return. You know that medical expenses may be written off as an itemized deduction if the expenses exceed 7.5% of their adjusted gross income. Furthermore, only the portion of medical expenses that exceed 7.5% of their adjusted gross income can be deducted. Is the Fernandez family eligible to deduct their medical expenses? Explain.

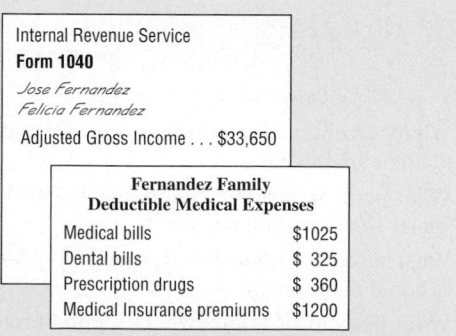

Internal Revenue Service
Form 1040
Jose Fernandez
Felicia Fernandez
Adjusted Gross Income . . . $33,650

Fernandez Family **Deductible Medical Expenses**	
Medical bills	$1025
Dental bills	$ 325
Prescription drugs	$ 360
Medical Insurance premiums	$1200

MENTAL MATH

Tell whether the percent labels in the circle graphs are correct.

1.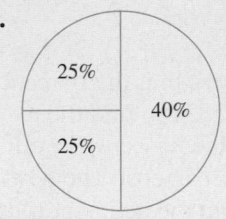

2.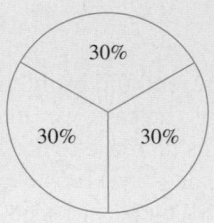

3.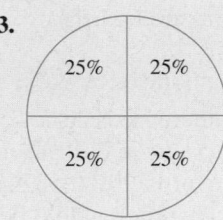

4.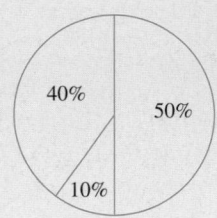

Exercise Set 2.6

Write each percent as a decimal. See Example 1.

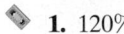 **1.** 120%   **2.** 73%

3. 22.5%   **4.** 4.2%

 5. 0.12%   **6.** 0.86%

Write each number as a percent. See Example 2.

7. 0.75   **8.** 0.3

9. 2   **10.** 5.1

11. $\dfrac{1}{8}$ **12.** $\dfrac{3}{5}$

Use the home maintenance graph to answer Exercises 13 through 20. See Example 4.

Yearly Home Maintenance in the U.S.

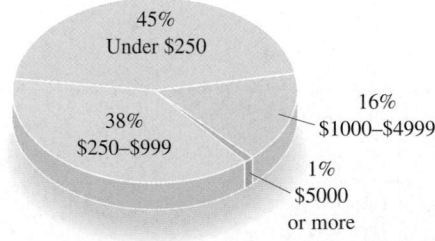

Source: Census Bureau

13. What percent of homeowners spend $250–$999 on yearly home maintenance?

14. What percent of homeowners spend $5000 or more on yearly home maintenance?

15. What percent of homeowners spend $250–$4999 on yearly home maintenance?

16. What percent of homeowners spend $250 or more on yearly home maintenance?

17. Find the number of degrees in the 38% sector.

18. Find the number of degrees in the 1% sector.

19. How many homeowners in your town might you expect to spend $250–$999 on yearly home maintenance?

20. How many homeowners in your town might you expect to spend $5000 or more on yearly home maintenance?

The circle graph below shows the number of minutes that adults spend on their home phone each day. Use this graph for Exercises 21 through 26. See Example 4.

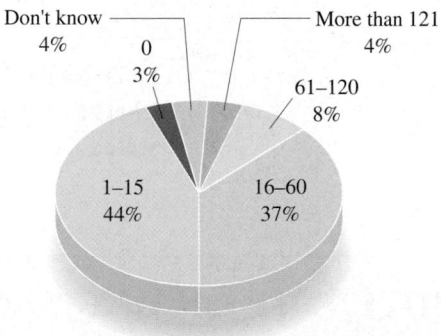

Source: Bruskin/Goldring Research for Sony Electronics

21. What percent of adults spend more than 121 minutes on the phone each day?

22. What percent of adults spend no time on the phone each day?

23. Find the number of degrees in the 4% sector.

24. Find the number of degrees in the 44% sector.

25. Liberty is a town whose adult population is approximately 135,000. How many of these adults might you expect to talk 16–60 minutes on the phone each day?

26. Poll the students in your algebra class. Find what percent of students spend 1–15 minutes on the phone each day. Is this percent close to 44%? Why or why not?

Solve the following. See Examples 3, 5, and 6.

27. What number is 16% of 70?

28. What number is 88% of 1000?

29. The number 28.6 is what percent of 52?

30. The number 87.2 is what percent of 436?

31. The number 45 is 25% of what number?

32. The number 126 is 35% of what number?

33. Find 23% of 20.

34. Find 140% of 86.

35. The number 40 is 80% of what number?

36. The number 56.25 is 45% of what number?

37. The number 144 is what percent of 480?

38. The number 42 is what percent of 35?

Solve. See Example 7. Many applications in this exercise set may be solved more efficiently with the use of a calculator.

39. Dillard's advertised a 25% off sale. If a London Fog coat originally sold for $156, find the decrease and the sale price.

40. Time Saver increased the price of a $0.75 cola by 15%. Find the increase and the new price.

41. At this writing, the women's world record for throwing a discus is held by Gabriele Reinsch of East Germany. Her throw was 252 feet. The men's record is held by Jürgen Schult also of East Germany. His throw was 3.7% shorter than Gabriele's. Find the length of his throw. (Round to the nearest whole foot.) (*Source: Guiness Book of Records*, 1999. *Note:* The women's discus weighs 2 lb 3 oz while the men's weighs 4 lb 8 oz.)

42. Scoville units are used to measure the hotness of a pepper. An alkaloid, capsaicin, is the ingredient that makes a pepper hot and liquid chromatography measures the amount of capsaicin in parts per million. The jalapeno measures around 5000 Scoville units, while the hottest pepper, the habanero, measures around 3000% as hot as the measure of the jalapeno. Find the measure of the habanero pepper.

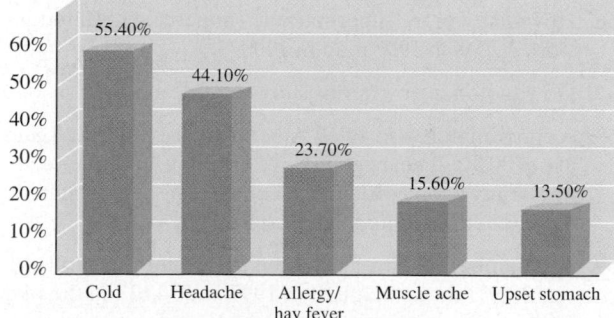

Source: Simmons Market Research Bureau

The graph at the bottom of the previous column shows the percent of people in a survey who have used various over-the-counter drugs in a twelve-month period. Use this graph for Exercises 43–48.

43. What percent of those surveyed used over-the-counter drugs to combat the common cold?

44. What percent of those surveyed used over-the-counter drugs to combat an upset stomach?

45. If 230 people were surveyed, how many of these used over-the-counter drugs for allergies?

46. The city of Chattanooga has a population of approximately 152,000. How many of these people would you expect to have used over-the-counter drugs for relief of a headache?

47. Do the percents shown in the graph have a sum of 100%? Why or why not?

48. Survey your algebra class and find what percent of the class has used over-the-counter drugs for each of the categories listed. Draw a bar graph of the results

49. A recent study showed that 26% of men have dozed off at their place of work. If you currently employ 121 men, how many of these men might you expect to have dozed off at work? (*Source:* Better Sleep Council)

50. A recent study showed that women and girls spend 41% of the household clothes budget. If a family spent $2000 last year on clothing, how much might have been spent on clothing for women and girls? (*Source:* The Interep Radio Store)

Fill in the percent column in each table in Exercises 51 and 52. The first entry has been done for you. See Examples 1 through 3.

51. WOMEN IN THE U.S. ARMED FORCES, 1999

Service	Number of Women	Percent of Total (round to the nearest whole percent)
Army	58,408	$\dfrac{58,408}{186,697} \approx 31\%$
Navy	50,287	
Marines	9696	
Air Force	64,427	
Coast Guard	3879	
Total	186,697	

(*Source:* U.S. Dept. of Defense)

52. U.S. PUBLIC SCHOOLS WITH COMPUTERS, 1999

Type of School	Number of Schools	Percent of Total (round to the nearest whole percent)
Elementary	52,276	$\frac{52{,}276}{84{,}002} \approx 62\%$
Junior High	14,419	
Senior High	17,307	
Total	84,002	

(*Source:* Quality Education Data, Inc.)

53. The hepatitis B vaccine, given as part of the standard regimen of childhood vaccinations, is required for public-school admission in 42 states. Each year about 1800 of the nearly 1,200,000 people who receive the shot have a serious adverse event. What percent of hepatitis B vaccine recipients report a serious adverse event? (*Source:* Vaccine Adverse Event Reporting System)

54. At schools that offer an accredited dental hygiene program, the curriculum requires 1948 clock hours of course and lab work on average. This includes a total of 585 hours of supervised clinical dental hygiene instruction. What percent of dental hygiene students' curriculum is normally devoted to supervised clinical instruction? Round to the nearest whole percent. (*Source:* The American Dental Hygienists' Association)

55. Approximately 10.4 million students are enrolled in community colleges. This represents 44% of all U.S. undergraduates. How many U.S. undergraduates are there in all? Round to the nearest tenth of a million. (*Source:* American Association of Community Colleges)

56. In 1998, a total of 916 million people were solicited by credit card issuers. This represents a 276% increase since 1992. How many people were solicited in 1992? Round to the nearest whole million. (*Source: USA Today,* 5/12/99)

57. Iceberg lettuce is grown and shipped to stores for about 40 cents a head, and consumers purchase it for about 70 cents a head. Find the percent increase.

58. The lettuce consumption per capita in 1968 was about 21.5 pounds, and in 1997 the consumption rose to 24.3 pounds. Find the percent increase. (Round to the nearest tenth of a percent.)

59. During a recent 5-year period, bank fees for bounced checks have risen from $12.62 to $15.65. Find the percent increase. (Round to the nearest whole percent.) If

inflation over the same period has been 16%, do you think the increase in bank fees is fair? (*Source:* Federal Reserve)

60. During a recent 5-year period, the bank fee for depositing a bad check has risen from $5.38 to $6.08. Find the percent increase. (Round to the nearest whole percent.) Given the inflation rate of 16%, do you think that this increase in bank fees is fair?

61. The first Barbie doll was introduced in March 1959 and cost $3. This same 1959 Barbie doll now costs up to $5000. Find the percent increase rounded to the nearest whole percent.

62. The ACT Assessment is a college entrance exam taken by about 60% of college-bound students. The national average score was 20.7 in 1993 and rose to 21.0 in 1998. Find the percent increase. (Round to the nearest hundredth of a percent.)

The double bar graph below shows selected services and products and the percent of supermarkets offering each.

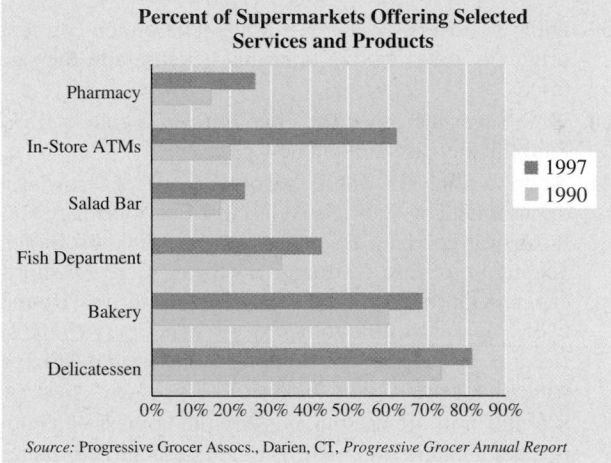

Source: Progressive Grocer Assocs., Darien, CT, *Progressive Grocer Annual Report*

63. What percent of supermarkets had a pharmacy in 1990?

64. What percent of supermarkets had a bakery in 1997?

65. Suppose that there are 50 supermarkets in your city. How many of these 50 supermarkets might have offered customers a salad bar in 1997?

66. How many more supermarkets (in percent) offered in-store ATMs in 1997 than in 1990?

67. Do you notice any trends shown in this graph?

68. Approximately 60% of all Americans are farsighted, and about 30% are nearsighted. In a group of 450 Americans, how many people would you expect to be (a) farsighted? (b) neither farsighted nor nearsighted?

69. The number of crimes reported in Baltimore, Maryland, was 38,831 in the first half of 1997 and 34,611 in the first half of 1998. Find the percent decrease in the number of reported crimes in Baltimore from 1997 to 1998. Round to the nearest whole percent. (*Source:* Federal Bureau of Investigation, *Uniform Crime Reports,* January–June 1998)

Standardized nutrition labels like the ones below have been displayed on food items since 1994. The percent column on the right shows the percent of daily values based on a 2000-calorie diet shown at the bottom of the label. For example, a serving of this food contains 4 grams of total fat, where the recommended daily fat based on a 2000-calorie diet is 65 grams of fat. This means that $\frac{4}{65}$ or approximately 6% (as shown) of your daily recommended fat is taken in by eating a serving of this food.

Nutrition Facts

Serving Size 18 crackers (31g)
Servings Per Container About 9

Amount Per Serving

Calories 130 Calories from Fat 35

	% Daily Value*
Total Fat 4g	**6%**
Saturated Fat 0.5g	**3%**
Polyunsaturated Fat 0g	
Monounsaturated Fat 1.5g	
Cholesterol 0mg	**0%**
Sodium 230mg	x
Total Carbohydrate 23g	y
Dietary Fiber 2g	**8%**
Sugars 3g	
Protein 2g	

Vitamin A 0%	•	Vitamin C 0%
Calcium 2%	•	Iron 6%

*Percent Daily Values are based on a 2,000 calorie diet. Your daily values may be higher or lower depending on your calorie needs.

	Calories:	2,000	2,500
Total Fat	Less than	65g	80g
Sat Fat	Less than	20g	25g
Cholesterol	Less than	300mg	300mg
Sodium	Less than	2400mg	2400mg
Total Carbohydrate		300g	375g
Dietary Fiber		25g	30g

70. Based on a 2000-calorie diet, what percent of daily values of sodium is contained in a serving of this food? In other words, find *x*. (Round to the nearest tenth of a percent.)

71. Based on a 2000-calorie diet, what percent of daily values of total carbohydrate is contained in a serving of this food? In other words, find *y*. (Round to the nearest tenth of a percent.)

72. Notice on the nutrition label that one serving of this food contains 130 calories and 35 of these calories are from fat. Find the percent of calories from fat. (Round to the nearest tenth of a percent.) It is recommended that no

more than 30% of calorie intake come from fat. Does this food satisfy this recommendation?

Below is a nutrition label for a particular food.

NUTRITIONAL INFORMATION PER SERVING

Serving Size: 9.8 oz.   **Servings Per Container: 1**

Calories....................280	Polyunsaturated Fat..........1g		
Protein......................12g	Saturated Fat................3g		
Carbohydrate...............45g	Cholesterol.................20mg		
Fat.........................6g	Sodium.....................520mg		
Percent of Calories from Fat...... ?	Potassium..................220mg		

73. If fat contains approximately 9 calories per gram, find the percent of calories from fat in one serving of this food. (Round to the nearest tenth of a percent.)

74. If protein contains approximately 4 calories per gram, find the percent of calories from protein from one serving of this food. (Round to the nearest tenth of a percent.)

75. Find a food that contains more than 30% of its calories per serving from fat. Analyze the nutrition label and verify that the percents shown are correct.

76. The table shows where lightning strikes. Use this table to draw a circle graph or pie chart of this information.

Fields, Ballparks	Under trees	Bodies of water	Golf courses	Near heavy equipment	Telephone poles	Other
28%	17%	13%	4%	6%	1%	31%

REVIEW EXERCISES

Evaluate the following expressions for the given values. See Section 1.7.

77. $2a + b - c$; $a = 5$, $b = -1$, and $c = 3$

78. $-3a + 2c - b$; $a = -2$, $b = 6$, and $c = -7$

79. $4ab - 3bc$; $a = -5$, $b = -8$, and $c = 2$

80. $ab + 6bc$; $a = 0$, $b = -1$, and $c = 9$

81. $n^2 - m^2$; $n = -3$ and $m = -8$

82. $2n^2 + 3m^2$; $n = -2$ and $m = 7$

2.7 SOLVING LINEAR INEQUALITIES

CD-ROM   SSM   SSG   Video

▶ **OBJECTIVES**

1. Graph solution sets on a number line.
2. Use interval notation.
3. Solve linear inequalities.
4. Solve inequality applications.

1

In Chapter 1, we reviewed these inequality symbols and their meanings:

    < means "is less than"   ≤ means "is less than or equal to"
    > means "is greater than"   ≥ means "is greater than or equal to"

     A linear inequality is similar to a linear equation except that the equality symbol is replaced with an inequality symbol.

Equations	*Inequalities*
$x = 3$	$x \leq 3$
$5n - 6 = 14$	$5n - 6 > 14$
$12 = 7 - 3y$	$12 \leq 7 - 3y$
$\dfrac{x}{4} - 6 = 1$	$\dfrac{x}{4} - 6 > 1$

LINEAR INEQUALITY IN ONE VARIABLE

A **linear inequality in one variable** is an inequality that can be written in the form

$$ax + b < c$$

where a, b, and c are real numbers and a is not 0.

This definition and all other definitions, properties, and steps in this section also hold true for the inequality symbols, $>$, $\geq$, and $\leq$.

     A **solution of an inequality** is a value of the variable that makes the inequality a true statement. The solution set is the set of all solutions. In the inequality $x < 3$, replacing x with any number less than 3, that is, to the left of 3 on the number line, makes the resulting inequality true. This means that any number less than 3 is a solution of the inequality $x < 3$. Since there are infinitely many such numbers, we cannot list all the solutions of the inequality. We *can* use set notation and write

$$\{\ x\ \ |\ \ \ x < 3\ \}. \text{ Recall that this is read}$$

         ↑   ↑

        the   such
       set of   that   *x* is less than 3
       all *x*

     We can also picture the solution set on a number line. To do so, shade the portion of the number line corresponding to numbers less than 3.

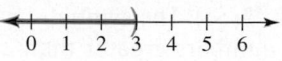

Recall that all the numbers less than 3 lie to the left of 3 on the number line. A parenthesis on the point representing 3 indicates that 3 is not a solution of the inequality: 3 **is not** less than 3. The shaded arrow indicates that the solutions of $x < 3$ continue indefinitely to the left of 3.

Picturing the solutions of an inequality on a number line is called **graphing** the solutions or graphing the inequality, and the picture is called the **graph** of the inequality.

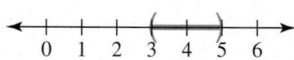

To graph $x \leq 3$, shade the numbers to the left of 3 and place a bracket on the point representing 3. The bracket indicates that 3 is a solution of the inequality $x \leq 3$ and is part of the graph.

Example 1   Graph $x \geq -1$.

Solution   We place a bracket at -1 since the inequality symbol is $\geq$ and -1 is greater than or equal to -1. Then we shade to the right of -1.

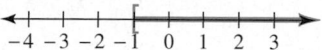

The solution set can be written in set notation as $\{x \mid x \geq -1\}$. Recall that this is read as

$\{x$	$\mid$	$x \geq -1\}$
the set of all x	such that	x is greater than or equal to -1

Inequalities containing one inequality symbol are called **simple inequalities**, whereas inequalities containing two inequality symbols are called **compound inequalities**. A compound inequality is really two simple inequalities in one. The compound inequality

$$3 < x < 5 \quad \text{means} \quad 3 < x \textbf{ and } x < 5$$

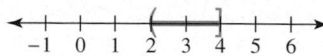

This can be read "x is greater than 3 and less than 5." To graph $3 < x < 5$, place a parenthesis at both 3 and 5 and shade between.

Example 2   Graph $2 < x \leq 4$.

Solution   Graph all numbers greater than 2 and less than or equal to 4. Place a parenthesis at 2, a bracket at 4, and shade between.

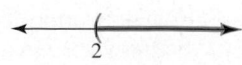

The solution set can be written in set notation as $\{x \mid 2 < x \leq 4\}$.

2   Another way to list solutions of a linear inequality is by interval notation. The graph of $\{x \mid x > 2\}$ looks like and the solutions can be represented in interval notation as $(2, \infty)$. The symbol ∞ is read "infinity" and indicates that the interval includes **all** numbers greater than 2. The left parenthesis indicates that 2 **is not** included in the interval. Using a left bracket would indicate that 2 **is** included in the interval. The following table shows three forms of describing intervals: in set notation, as a graph, and in interval notation.

Set Notation	*Graph*	*Interval Notation*
$\{x \mid x < a\}$	⟵────────)────⟶ a	$(-\infty, a)$
$\{x \mid x > a\}$	⟵────(────────⟶ a	(a, ∞)
$\{x \mid x \le a\}$	⟵────────]────⟶ a	$(-\infty, a]$
$\{x \mid x \ge a\}$	⟵────[────────⟶ a	$[a, \infty)$
$\{x \mid a < x < b\}$	⟵────(──)────⟶ a b	(a, b)
$\{x \mid a \le x \le b\}$	⟵────[──]────⟶ a b	$[a, b]$
$\{x \mid a < x \le b\}$	⟵────(──]────⟶ a b	$(a, b]$
$\{x \mid a \le x < b\}$	⟵────[──)────⟶ a b	$[a, b)$

▼ **HELPFUL HINT**
Notice that a parenthesis is always used to enclose ∞ and $-\infty$.

Example 3 Graph each set on a number line and then write it in interval notation.

    a. $\{x \mid x \ge 2\}$ **b.** $\{x \mid x < -1\}$ **c.** $\{x \mid 0.5 < x \le 3\}$

Solution **a.** ⟵────[────────⟶ $[2, \infty)$
               2

    b. ⟵────)────────⟶ $(-\infty, -1)$
             −1

    c. ⟵────(──]────⟶ $(0.5, 3]$
         0.5 3

3 When solutions to a linear inequality are not immediately obvious, they are found through a process similar to the one used to solve a linear equation. Our goal is to isolate the variable, and we use properties of inequality similar to properties of equality.

ADDITION PROPERTY OF INEQUALITY

If a, b, and c are real numbers, then

$$a < b \quad \text{and} \quad a + c < b + c$$

are equivalent inequalities.

This property also holds true for subtracting values, since subtraction is defined in terms of addition. In other words, adding or subtracting the same quantity from both sides of an inequality does not change the solution of the inequality.

Example 4 Solve $x + 4 \le -6$ for x. Graph the solution set.

Solution To solve for x, subtract 4 from both sides of the inequality.

$$x + 4 \le -6 \qquad \text{Original inequality.}$$
$$x + 4 - 4 \le -6 - 4 \qquad \text{Subtract 4 from both sides.}$$
$$x \le -10 \qquad \text{Simplify.}$$

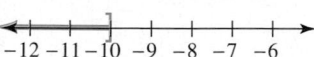

The solution set is $\{x \mid x \le -10\}$, which in interval notation is $(-\infty, -10]$.

> **HELPFUL HINT**
>
> Notice that any number less than or equal to -10 is a solution to $x \le -10$. For example, solutions include
>
> $$-10, \quad -200, \quad -11\frac{1}{2}, \quad -7\pi, \quad -\sqrt{130}, \quad -50.3$$

An important difference between linear equations and linear inequalities is shown when we multiply or divide both sides of an inequality by a nonzero real number. For example, start with the true statement $6 < 8$ and multiply both sides by 2. As we see below, the resulting inequality is also true.

$$6 < 8 \qquad \text{True.}$$
$$2(6) < 2(8) \qquad \text{Multiply both sides by 2.}$$
$$12 < 16 \qquad \text{True.}$$

But if we start with the same true statement $6 < 8$ and multiply both sides by -2, the resulting inequality is not a true statement.

$$6 < 8 \qquad \text{True.}$$
$$-2(6) < -2(8) \qquad \text{Multiply both sides by } -2.$$
$$-12 < -16 \qquad \text{False.}$$

Notice, however, that if we reverse the direction of the inequality symbol, the resulting inequality is true.

$$-12 < -16 \qquad \text{False.}$$
$$-12 > -16 \qquad \text{True.}$$

This demonstrates the multiplication property of inequality.

MULTIPLICATION PROPERTY OF INEQUALITY

1. If a, b, and c are real numbers, and c is **positive**, then

$$a < b \quad \text{and} \quad ac < bc$$

are equivalent inequalities.

2. If a, b, and c are real numbers, and c is **negative**, then

$$a < b \quad \text{and} \quad ac > bc$$

are equivalent inequalities.

Because division is defined in terms of multiplication, this property also holds true when dividing both sides of an inequality by a nonzero number: If we multiply or divide both sides of an inequality by a negative number, **the direction of the inequality sign must be reversed for the inequalities to remain equivalent**.

> **HELPFUL HINT**
> Whenever both sides of an inequality are multiplied or divided by a negative number, the direction of the inequality symbol **must be** reversed to form an equivalent inequality.

Example 5   Solve $-2x \leq -4$. Graph the solution set and write it in interval notation.

Solution   Remember to reverse the direction of the inequality symbol when dividing by a negative number.

> **HELPFUL HINT**
> Don't forget to reverse the direction of the inequality sign.

$$-2x \leq -4$$
$$\frac{-2x}{-2} \geq \frac{-4}{-2} \quad \text{Divide both sides by } -2 \text{ and reverse the inequality sign.}$$
$$x \geq 2 \quad \text{Simplify.}$$

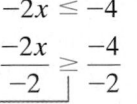

The solution set is $\{x \mid x \geq 2\}$, which in interval notation is $[2, \infty)$. The graph is shown to the left.

Example 6   Solve $2x < -4$. Graph the solution set and write it in interval notation.

Solution

> **HELPFUL HINT**
> Do not reverse the inequality sign.

$$2x < -4$$
$$\frac{2x}{2} < \frac{-4}{2} \quad \text{Divide both sides by 2. Do not reverse the inequality sign.}$$
$$x < -2 \quad \text{Simplify.}$$

The graph of $\{x \mid x < -2\}$ is shown. In interval notation, we have $(-\infty, -2)$.

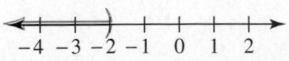

The following steps may be helpful when solving inequalities. Notice that these steps are similar to the ones given in Section 2.3 for solving equations.

SOLVING LINEAR INEQUALITIES IN ONE VARIABLE

Step 1. Clear the inequality of fractions by multiplying both sides of the inequality by the lowest common denominator (LCD) of all fractions in the inequality.

Step 2. Remove grouping symbols such as parentheses by using the distributive property.

Step 3. Simplify each side of the inequality by combining like terms.

Step 4. Write the inequality with variable terms on one side and numbers on the other side by using the addition property of inequality.

Step 5. Get the variable alone by using the multiplication property of inequality.

Don't forget that if both sides of an inequality are multiplied or divided by a negative number, the direction of the inequality sign must be reversed.

Example 7 Solve $-4x + 7 \geq -9$, and graph the solution set.

Solution

$$-4x + 7 \geq -9$$

$$-4x + 7 - 7 \geq -9 - 7 \qquad \text{Subtract 7 from both sides.}$$

$$-4x \geq -16 \qquad \text{Simplify.}$$

$$\frac{-4x}{-4} \leq \frac{-16}{-4} \qquad \text{Divide both sides by } -4 \text{ and reverse the direction of the inequality sign.}$$

$$x \leq 4 \qquad \text{Simplify.}$$

The solution set $\{x \mid x \leq 4\}$ written in interval notation is $(-\infty, 4]$, and its graph is shown.

Example 8 Solve $2x + 7 \leq x - 11$, and graph the solution.

Solution

$$2x + 7 \leq x - 11$$

$$2x + 7 - x \leq x - 11 - x \qquad \text{Subtract } x \text{ from both sides.}$$

$$x + 7 \leq -11 \qquad \text{Simplify.}$$

$$x + 7 - 7 \leq -11 - 7 \qquad \text{Subtract 7 from both sides.}$$

$$x \leq -18 \qquad \text{Simplify.}$$

The solution set $\{x \mid x \leq -18\}$ is graphed. In interval notation, the solution is $(-\infty, -18]$.

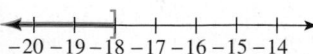

Example 9   Solve for x: $-(x - 3) + 2 < 3(2x - 5) + x$.

Solution

$$-(x - 3) + 2 < 3(2x - 5) + x$$

$$-x + 3 + 2 < 6x - 15 + x \qquad \text{Apply the distributive property.}$$

$$5 - x < 7x - 15 \qquad \text{Combine like terms.}$$

$$5 - x + x < 7x - 15 + x \qquad \text{Add } x \text{ to both sides.}$$

$$5 < 8x - 15 \qquad \text{Combine like terms.}$$

$$5 + 15 < 8x - 15 + 15 \qquad \text{Add 15 to both sides.}$$

$$20 < 8x \qquad \text{Combine like terms.}$$

$$\frac{20}{8} < \frac{8x}{8} \qquad \text{Divide both sides by 8.}$$

$$\frac{5}{2} < x \quad \text{or} \quad x > \frac{5}{2}$$

The solution set written in interval notation is $\left(\dfrac{5}{2}, \infty \right)$ and its graph is

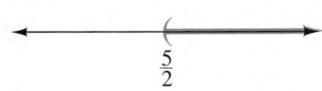

Example 10   Solve for x: $\dfrac{2}{5}(x - 6) \geq x - 1$.

Solution

$$\frac{2}{5}(x - 6) \geq x - 1$$

$$5\left[\frac{2}{5}(x - 6) \right] \geq 5(x - 1) \qquad \text{Multiply both sides by 5 to eliminate fractions.}$$

$$2x - 12 \geq 5x - 5 \qquad \text{Use the distributive property.}$$

$$-3x - 12 \geq -5 \qquad \text{Subtract } 5x \text{ from both sides.}$$

$$-3x \geq 7 \qquad \text{Add 12 to both sides.}$$

$$\frac{-3x}{-3} \leq \frac{7}{-3} \qquad \text{Divide both sides by } -3 \text{ and reverse the inequality symbol.}$$

$$x \leq -\frac{7}{3} \qquad \text{Simplify.}$$

The solution set $\left\{ x \,\middle|\, x \leq -\dfrac{7}{3} \right\}$ is graphed on a number line and is written in interval notation as $\left(-\infty, -\dfrac{7}{3} \right]$.

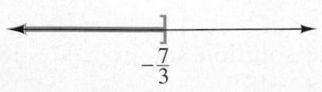

Example 11   Solve for x: $2(x + 3) > 2x + 1$.

Solution

$$2(x + 3) > 2x + 1$$
$$2x + 6 > 2x + 1 \qquad \text{Distribute on left side.}$$
$$2x + 6 - 2x > 2x + 1 - 2x \qquad \text{Subtract } 2x \text{ from both sides.}$$
$$6 > 1 \qquad \text{Simplify.}$$

$6 > 1$ is a true statement for all values of x, so this inequality and the original inequality are true for all numbers. The solution set is $\{x \mid x \text{ is a real number}\}$, or $(-\infty, \infty)$ in interval notation, and its graph is

---

4   Problems containing words such as "at least," "at most," "between," "no more than," and "no less than" usually indicate that an inequality, instead of an equation, be solved. In solving applications involving linear inequalities, we use the same procedure as when we solved applications involving linear equations.

Example 12   **STAYING WITHIN BUDGET**

Marie Chase and Jonathan Edwards are having their wedding reception at the Gallery reception hall. They may spend at most $2000 for the reception. If the reception hall charges a $100 cleanup fee plus $28 per person, find the greatest number of people that they can invite and still stay within their budget.

Solution   1.   UNDERSTAND. Read and reread the problem. Next, guess a solution. If 50 people attend the reception, the cost is $100 + $28(50) = $100 + $1400 = $1500. Let x = the number of people who attend the reception.

2.   TRANSLATE.

In words:

cleanup fee	+	cost per person	must be less than or equal to	$2000
100	+	28x	≤	2000

Translate:

3.   SOLVE.

$$100 + 28x \le 2000$$
$$28x \le 1900 \qquad \text{Subtract 100 from both sides.}$$
$$x \le 67\frac{6}{7} \qquad \text{Divide both sides by 28.}$$

4.   INTERPRET.

Check: Since x represents the number of people, we round down to the nearest whole, or 67. Notice that if 67 people attend, the cost is $100 + $28(67) = $1976. If 68 people attend, the cost is $100 + $28(68) = $2004, which is more than the given $2000.

State: Marie Chase and Jonathan Edwards can invite at most 67 people to the reception.

MENTAL MATH

Solve each of the following inequalities.

1. $5x > 10$ **2.** $4x < 20$ **3.** $2x \geq 16$ **4.** $9x \leq 63$

Exercise Set 2.7

Graph the solution set of each inequality on a number line, and write the solution set in interval notation. See Examples 1 through 3.

1. $\{x \mid x < -3\}$ **2.** $\{x \mid x \geq -7\}$

3. $\{x \mid x \geq 0.3\}$ **4.** $\{x \mid x < -0.2\}$

5. $\{x \mid 5 < x\}$ **6.** $\{x \mid -7 \geq x\}$

7. $\{x \mid -2 < x < 5\}$ **8.** $\{x \mid -5 \leq x \leq -1\}$

9. $\{x \mid 5 > x > -1\}$ **10.** $\{x \mid -3 \geq x \geq -7\}$

Solve each inequality for the variable. Graph the solution set and write the solution set in interval notation. See Examples 4 through 6.

11. $2x < -6$ **12.** $3x > -9$

13. $x - 2 \geq -7$ **14.** $x + 4 \leq 1$

15. $-8x \leq 16$ **16.** $-5x < 20$

Solve each inequality for the variable. Graph the solution set and write it in interval notation. See Examples 7 through 10.

17. $15 + 2x \geq 4x - 7$ **18.** $20 + x < 6x$

19. $\dfrac{3x}{4} \geq 2$ **20.** $\dfrac{5}{6}x \geq -8$

21. $3(x - 5) < 2(2x - 1)$ **22.** $5(x + 4) \leq 4(2x + 3)$

23. $\dfrac{1}{2} + \dfrac{2}{3} \geq \dfrac{x}{6}$ **24.** $\dfrac{3}{4} - \dfrac{2}{3} > \dfrac{x}{6}$

Solve each inequality for the variable. Graph the solution set and write it in interval notation. See Example 11.

25. $4(x - 1) \geq 4x - 8$ **26.** $3x + 1 < 3(x - 2)$

27. $7x < 7(x - 2)$

28. $8(x + 3) \leq 7(x + 5) + x$

29. Explain how solving a linear inequality is similar to solving a linear equation.

30. Explain how solving a linear inequality is different from solving a linear equation.

31. If an inequality in x simplifies to $5 < -2$, what is the solution set and why?

32. If an inequality in x simplifies to $5 > -2$, what is the solution set and why?

Solve each inequality for the variable. Graph the solution set and write it in interval notation.

33. $5x + 3 > 2 + 4x$ **34.** $7x - 1 \geq 6x - 1$

35. $8x - 7 \leq 7x - 5$ **36.** $12x + 14 < 11x - 2$

37. $5x > 10$ **38.** $9x < 45$

39. $-4x \leq 32$ **40.** $-6x \geq 42$

41. $-2x + 7 \geq 9$ **42.** $8 - 5x \leq 23$

43. $4(2x + 1) > 4$ **44.** $6(2 - x) \geq 12$

45. $\dfrac{x + 7}{5} > 1$ **46.** $\dfrac{2x - 4}{3} \leq 2$

47. $-6x + 2 \geq 2(5 - x)$ **48.** $-7x + 4 > 3(4 - x)$

49. $4(3x - 1) \leq 5(2x - 4)$ **50.** $3(5x - 4) \leq 4(3x - 2)$

51. $\dfrac{-5x + 11}{2} \leq 7$ **52.** $\dfrac{4x - 8}{7} < 0$

53. $8x - 16 \le 10x + 2$

54. $18x - 24 < 10x + 64$

55. $2(x - 3) > 70$

56. $3(5x + 6) \ge -12$

57. $-5x + 4 \le -4(x - 1)$

58. $-6x + 2 < -3(x + 4)$

59. $\frac{1}{4}(x - 7) > x + 2$

60. $\frac{3}{5}(x + 1) < x + 1$

61. $\frac{2}{3}(x + 2) < \frac{1}{5}(2x + 7)$

62. $\frac{1}{6}(3x + 10) > \frac{5}{12}(x - 1)$

63. $4(x - 6) + 2x - 4 \ge 3(x - 7) + 10x$

64. $7(2x + 3) + 4x \le 7 + 5(3x - 4)$

65. $\frac{5x + 1}{7} - \frac{2x - 6}{4} \ge -4$

66. $\frac{1 - 2x}{3} + \frac{3x + 7}{7} > 1$

67. $\frac{-x + 2}{2} - \frac{1 - 5x}{8} < -1$

68. $\frac{3 - 4x}{6} - \frac{1 - 2x}{12} \le -2$

Solve the following. See Example 12.

69. High blood cholesterol levels increase the risk of heart disease in adults. Doctors recommend that total blood cholesterol be less than 200 milligrams per deciliter. Total cholesterol levels from 200 up to 240 milligrams per deciliter are considered borderline. Any total cholesterol reading above 240 milligrams per deciliter is considered high.

Letting x represent a patient's total blood cholesterol level, write a series of three inequalities that describe the ranges corresponding to recommended, borderline, and high levels of total blood cholesterol.

70. T. Theodore Fujita created the Fujita Scale (or F-Scale), which uses ratings from 0 to 5 to classify tornadoes. An F-0 tornado has wind speeds between 40 and 72 mph, inclusive. The winds in an F-1 tornado range from 73 to 112 mph. In an F-2 tornado, winds are from 113 to 157 mph. An F-3 tornado has wind speeds ranging from 158 to 206 mph. Wind speeds in an F-4 tornado are clocked at 207 to 260 mph. The most violent tornadoes are ranked at F-5, with wind speeds of at least 261 mph. (*Source:* National Weather Service)

Letting y represent a tornado's wind speed, write a series of six inequalities that describe the wind speed ranges corresponding to each Fujita Scale rank.

71. Six more than twice a number is greater than negative fourteen. Find all numbers that make this statement true.

72. Five times a number increased by one is less than or equal to ten. Find all such numbers.

73. Dennis and Nancy Wood are celebrating their 30th wedding anniversary by having a reception at Tiffany Oaks reception hall. They have budgeted $3000 for their reception. If the reception hall charges a $50.00 cleanup fee plus $34 per person, find the greatest number of people that they may invite and still stay within their budget.

74. A surprise retirement party is being planned for Pratep Puri. A total of $860 has been collected for the event, which is to be held at a local reception hall. This reception hall charges a cleanup fee of $40 and $15 per person for drinks and light snacks. Find the greatest number of people that may be invited and still stay within $860.

△ **75.** The perimeter of a rectangle is to be no greater than 100 centimeters and the width must be 15 centimeters. Find the maximum length of the rectangle.

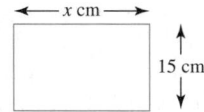

△ **76.** One side of a triangle is four times as long as another side, and the third side is 12 inches long. If the perimeter can be no longer than 87 inches, find the maximum lengths of the other two sides.

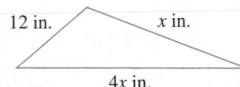

77. A financial planner has a client with $15,000 to invest. If he invests $10,000 in a certificate of deposit paying 11% annual simple interest, at what rate does the remainder of the money need to be invested so that the two investments together yield at least $1600 in yearly interest?

78. Alex earns $600 per month plus 4% of all his sales over $1000. Find the minimum sales that will allow Alex to earn at least $3000 per month.

79. Ben Holladay bowled 146 and 201 in his first two games. What must he bowl in his third game to have an average of at least 180?

80. On an NBA team the two forwards measure 6'8" and 6'6" and the two guards measure 6'0" and 5'9" tall. How tall a center should they hire if they wish to have a starting team average height of at least 6'5"?

REVIEW EXERCISES

Evaluate the following. See Section 1.3.

81. $(2)^3$
82. $(3)^3$
83. $(1)^{12}$
84. 0^5
85. $\left(\dfrac{4}{7}\right)^2$
86. $\left(\dfrac{2}{3}\right)^3$

This broken line graph shows the enrollment of people (members) in a Health Maintenance Organization (HMO). The height of each dot corresponds to the number of members (in millions).

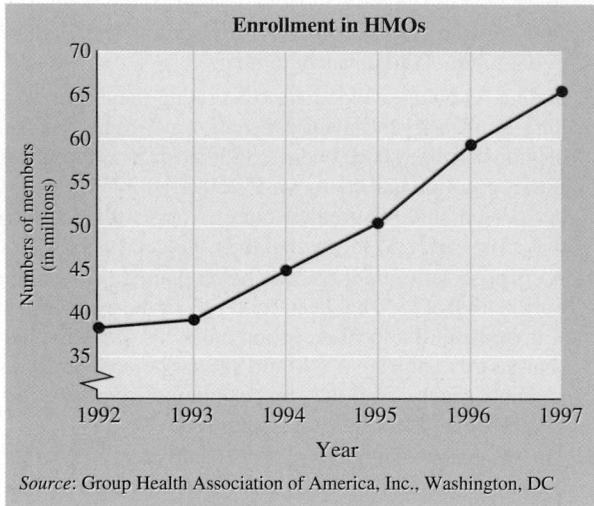

Enrollment in HMOs

Source: Group Health Association of America, Inc., Washington, DC

87. How many people were enrolled in Health Maintenance Organizations in 1995?

88. How many people were enrolled in Health Maintenance Organizations in 1997?

89. Which year shows the greatest increase in number of members?

90. In what years were there over 50,000,000 members of HMOs?

A Look Ahead

Example

Solve $x(x - 6) > x^2 - 5x + 6$, and then graph the solution.

Solution:

$$x(x - 6) > x^2 - 5x + 6$$
$$x^2 - 6x > x^2 - 5x + 6$$
$$x^2 - 6x - x^2 > x^2 - 5x + 6 - x^2$$
$$-6x > -5x + 6$$
$$-x > 6$$
$$\frac{-x}{-1} < \frac{6}{-1}$$
$$x < -6$$

The solution set $\{x \mid x < -6\}$ is graphed as shown.

Solve each inequality and then graph the solution set.

91. $x(x + 4) > x^2 - 2x + 6$

92. $x(x - 3) \geq x^2 - 5x - 8$

93. $x^2 + 6x - 10 < x(x - 10)$

94. $x^2 - 4x + 8 < x(x + 8)$

95. $x(2x - 3) \leq 2x^2 - 5x$

96. $x(4x + 1) < 4x^2 - 3x$

For additional Chapter Projects, visit the Real World Activities Website by going to http://www.prenhall.com/martin-gay.

CHAPTER PROJECT

Developing a Budget

Whether you are rolling in dough or pinching every penny, a budget can help put you in control of your personal finances. Listing your income lets you see where your money comes from, and by tracking your expenses you can analyze your spending practices.

Putting these together in a budget helps you make informed decisions about what you spend.

In this project, you will have the opportunity to develop a personal budget. This project may be completed by working in groups or individually.

1. Keep track of your spending for a week or two. Be sure to write down *everything* that you spend during that time—no matter how small—from a daily newspaper or a can of soda to your tuition bill for the term.

2. Decide on a timeframe for your budget: a month, a term, a school year, or the entire year. How many weeks are in your timeframe?

3. Make a list of all of your income for your timeframe of choice. Be sure to include sources like grants or scholarships, portion of savings earmarked for college expenses for this timeframe, take-home pay from a job, gifts, financial support from family, and interest. Total your income from all sources.

4. Make a list of all of your expected expenses for your timeframe of choice. Be sure to include major expenses such as car payments, auto and health insurance, room and board or rent/mortgage payment, and tuition. Try to estimate more variable expenses like telephone and utilities, books and supplies, groceries and/or restaurant meals, entertainment, transportation or parking, personal care, clothing, dues and/or lab fees, etc.

Don't forget to take into account small or irregular purchases. Use the expense record you made in Question 1 to help you gauge your levels of spending and identify spending categories. Total your expected expenses in all categories.

5. In the equation $I = E + wx$, I represents your income from Question 3, E represents your expected expenses from Question 4, w represents the number of weeks in your timeframe of choice (from Question 2), and x represents the weekly shortage or surplus in your budget. If x is positive, this is the extra amount you have in your budget each week to save or spend as "pocket money." If x is negative, this is the amount by which you fall short each week. To keep your budget balanced, you should try to reduce nonessential expenses by this amount each week.

   Substitute for I, E, and w in the equation and solve for x. Interpret your result.

6. If you have a weekly shortage, which expenses could you try to reduce to balance your budget? If you have a weekly surplus, what would you do with this extra money?

CHAPTER 2 VOCABULARY CHECK

Fill in each blank with one of the words or phrases listed below.

like terms	numerical coefficient	linear equation in one variable
equivalent equations	formula	compound inequalities
linear inequality in one variable	percent	

1. Terms with the same variables raised to exactly the same powers are called _____.
2. A _____ can be written in the form $ax + b = c$.
3. Equations that have the same solution are called _____.
4. Inequalities containing two inequality symbols are called _____.
5. An equation that describes a known relationship among quantities is called a _____.
6. The word _____ means per hundred.
7. A _____ can be written in the form $ax + b < c$, (or $>$, $\leq$, $\geq$).
8. The _____ of a term is its numerical factor.

CHAPTER 2 HIGHLIGHTS

DEFINITIONS AND CONCEPTS	EXAMPLES

Section 2.1 Simplifying Algebraic Expressions

The **numerical coefficient** of a **term** is its numerical factor.

Term	Numerical Coefficient
$-7y$	-7
x	1
$\dfrac{1}{5}a^2b$	$\dfrac{1}{5}$

Terms with the same variables raised to exactly the same powers are **like terms**.

Like Terms	Unlike Terms
$12x, -x$	$3y, 3y^2$
$-2xy, 5yx$	$7a^2b, -2ab^2$

To combine like terms, add the numerical coefficients and multiply the result by the common variable factor.

$$9y + 3y = 12y$$
$$-4z^2 + 5z^2 - 6z^2 = -5z^2$$

To remove parentheses, apply the distributive property.

$$-4(x + 7) + 10(3x - 1)$$
$$= -4x - 28 + 30x - 10$$
$$= 26x - 38$$

Section 2.2 The Addition and Multiplication Properties of Equality

A **linear equation in one variable** can be written in the form $ax + b = c$ where a, b, and c are real numbers and $a \neq 0$.

Linear Equations

$$-3x + 7 = 2$$
$$3(x - 1) = -8(x + 5) + 4$$

Equivalent equations are equations that have the same solution.

$x - 7 = 10$ and $x = 17$
are equivalent equations.

Addition Property of Equality

Adding the same number to or subtracting the same number from both sides of an equation does not change its solution.

$$y + 9 = 3$$
$$y + 9 - 9 = 3 - 9$$
$$y = -6$$

Multiplication Property of Equality

Multiplying both sides or dividing both sides of an equation by the same nonzero number does not change its solution.

$$\frac{2}{3}a = 18$$
$$\frac{3}{2}\left(\frac{2}{3}a\right) = \frac{3}{2}(18)$$
$$a = 27$$

Section 2.3 Solving Linear Equations

To Solve Linear Equations

Solve: $\dfrac{5(-2x + 9)}{6} + 3 = \dfrac{1}{2}$

1. Clear the equation of fractions.

1. $6 \cdot \dfrac{5(-2x + 9)}{6} + 6 \cdot 3 = 6 \cdot \dfrac{1}{2}$

$$5(-2x + 9) + 18 = 3$$

2. Remove any grouping symbols such as parentheses.

2. $-10x + 45 + 18 = 3$ Distributive property.

(continued)

DEFINITIONS AND CONCEPTS	EXAMPLES

Section 2.3 Solving Linear Equations

3. Simplify each side by combining like terms.

3. $-10x + 63 = 3$ *Combine like terms.*

4. Write variable terms on one side and numbers on the other side using the addition property of equality.

4. $-10x + 63 - 63 = 3 - 63$ *Subtract 63.*
$$-10x = -60$$

5. Get the variable alone using the multiplication property of equality.

5. $\dfrac{10x}{-10} = \dfrac{-60}{-10}$ *Divide by −10.*
$$x = 6$$

6. Check by substituting in the original equation.

6. $\dfrac{5(-2x+9)}{6} + 3 = \dfrac{1}{2}$

$$\dfrac{5(-2\cdot6+9)}{6} + 3 \stackrel{?}{=} \dfrac{1}{2}$$

$$\dfrac{5(-3)}{6} + 3 \stackrel{?}{=} \dfrac{1}{2}$$

$$-\dfrac{5}{2} + \dfrac{6}{2} \stackrel{?}{=} \dfrac{1}{2}$$

$$\dfrac{1}{2} = \dfrac{1}{2} \quad \text{True.}$$

Section 2.4 An Introduction to Problem Solving

Problem-Solving Steps

The height of the Hudson volcano in Chili is twice the height of the Kiska volcano in the Aleutian Islands. If the sum of their heights is 12,870 feet, find the height of each.

1. UNDERSTAND the problem.

1. Read and reread the problem. Guess a solution and check your guess.
Let x be the height of the Kiska volcano. Then $2x$ is the height of the Hudson volcano.

x ⊺ $2x$ ⊺

Kiska   Hudson

2. TRANSLATE the problem.

2. In words:

height of Kiska	added to	height of Hudson	is	12,870
Translate: x	$+$	$2x$	$=$	12,870

3. SOLVE.

3. Solve:
$$x + 2x = 12{,}870$$
$$3x = 12{,}870$$
$$x = 4290$$

4. INTERPRET the result.

4. **Check:** If x is 4290 then $2x$ is $2(4290)$ or 8580. Their sum is $4290 + 8580$ or 12,870, the required amount.
State: Kiska volcano is 4290 feet high and Hudson volcano is 8580 feet high.

(continued)

DEFINITIONS AND CONCEPTS	EXAMPLES

Section 2.5 Formulas and Problem Solving

An equation that describes a known relationship among quantities is called a **formula**.	***Formulas*** $A = lw$ (area of a rectangle) $I = PRT$ (simple interest)
To solve a formula for a specified variable, use the same steps as for solving a linear equation. Treat the specified variable as the only variable of the equation.	Solve $P = 2l + 2w$ for l. $$P = 2l + 2w$$ $P - 2w = 2l + 2w - 2w$ Subtract $2w$. $$P - 2w = 2l$$ $$\frac{P - 2w}{2} = \frac{2l}{2}$$ Divide by 2. $$\frac{P - 2w}{2} = l$$ Simplify.
If all values for the variables in a formula are known except for one, this unknown value may be found by substituting in the known values and solving.	If $d = 182$ miles and $r = 52$ miles per hour in the formula $d = r \cdot t$, find t. $$d = r \cdot t$$ $182 = 52 \cdot t$ Let $d = 182$ and $r = 52$. $$3.5 = t$$ The time is 3.5 hours.

Section 2.6 Percent and Problem Solving

The word **percent** means **per hundred**. The symbol % is used to denote percent.	$49\% = \frac{49}{100}, \qquad 1\% = \frac{1}{100}$
To write a percent as a decimal, drop the percent symbol and move the decimal point two places to the left.	$85.\% = 0.85, \qquad 3.5\% = 0.035$
To write a decimal as a percent, move the decimal point two places to the right and attach the percent symbol, %.	$0.35 = 35\%, \quad 10.1 = 1010\%$ $\frac{1}{8} = 0.125 = 12.5\%$
Use the same problem-solving steps to solve a problem containing percents.	32% of what number is 36.8?
1. UNDERSTAND.	1. Read and reread. Guess a solution and check. Let $x =$ the unknown number.
2. TRANSLATE.	2. In words: | 32%	of	what number	is	36.8 | Translate: 32% · x = 36.8
3. SOLVE.	3. Solve: $32\% \cdot x = 36.8$ $$0.32x = 36.8$$ $\frac{0.32x}{0.32} = \frac{36.8}{.32}$ Divide by .32. $x = 115$ Simplify.
4. INTERPRET.	4. **Check:** 32% of 115 is $0.32(115) = 36.8$. **State:** The unknown number is 115.

(continued)

DEFINITIONS AND CONCEPTS	EXAMPLES

Section 2.7 Solving Linear Inequalities

A **linear inequality in one variable** is an inequality that can be written in one of the forms:

$ax + b < c$ $ax + b \leq c$

$ax + b > c$ $ax + b \geq c$

where a, b, and c are real numbers and a is not 0.

Linear Inequalities

$2x + 3 < 6$ $5(x - 6) \geq 10$

$\dfrac{x - 2}{5} > \dfrac{5x + 7}{2}$ $\dfrac{-(x + 8)}{9} \leq \dfrac{-2x}{11}$

Addition Property of Inequality

Adding the same number to or subtracting the same number from both sides of an inequality does not change the solutions.

$y + 4 \leq -1$

$y + 4 - 4 \leq -1 - 4$ Subtract 4.

$y \leq -5$

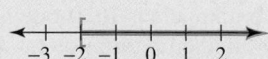

Multiplication Property of Inequality

Multiplying or dividing both sides of an inequality by the same *positive number* does not change its solutions.

$\dfrac{1}{3}x > -2$

$3\left(\dfrac{1}{3}x\right) > 3 \cdot -2$ Multiply by 3.

$x > -6$

Multiplying or dividing both sides of an inequality by the same **negative number and reversing the direction of the inequality sign** does not change its solutions.

$-2x \leq 4$

$\dfrac{-2x}{-2} \geq \dfrac{4}{-2}$ Divide by −2, reverse inequality sign.

$x \geq -2$

To Solve Linear Inequalities

1. Clear the equation of fractions.

2. Remove grouping symbols.

3. Simplify each side by combining like terms.

4. Write variable terms on one side and numbers on the other side using the addition property of inequality.

5. Get the variable alone using the multiplication property of inequality.

Solve: $3(x + 2) \leq -2 + 8$

1. No fractions to clear. $3(x + 2) \leq -2 + 8$

2. $3x + 6 \leq -2 + 8$ Distributive property.

3. $3x + 6 \leq 6$ Combine like terms.

4. $3x + 6 - 6 \leq 6 - 6$ Subtract 6.

$3x \leq 0$

5. $\dfrac{3x}{3} \leq \dfrac{0}{3}$ Divide by 3.

$x \leq 0$

CHAPTER 2 REVIEW

(2.1) Simplify the following expressions.

1. $5x - x + 2x$

2. $0.2z - 4.6x - 7.4z$

3. $\frac{1}{2}x + 3 + \frac{7}{2}x - 5$

4. $\frac{4}{5}y + 1 + \frac{6}{5}y + 2$

5. $2(n - 4) + n - 10$

6. $3(w + 2) - (12 - w)$

7. Subtract $7x - 2$ from $x + 5$.

8. Subtract $1.4y - 3$ from $y - 0.7$.

Write each of the following as algebraic expressions.

9. Three times a number, decreased by 7

10. Twice the sum of a number and 2.8, added to 3 times the number

(2.2) Solve the following.

11. $8x + 4 = 9x$

12. $5y - 3 = 6y$

13. $3x - 5 = 4x + 1$

14. $2x - 6 = x - 6$

15. $4(x + 3) = 3(1 + x)$

16. $6(3 + n) = 5(n - 1)$

Use the addition property to fill in the blank so that the middle equation simplifies to the last equation.

17. $x - 5 = 3$

    $x - 5 + \underline{\quad} = 3 + \underline{\quad}$

         $x = 8$

18. $x + 9 = -2$

    $x + 9 - \underline{\quad} = -2 - \underline{\quad}$

         $x = -11$

Write each as an algebraic expression.

19. The sum of two numbers is 10. If one number is x, express the other number in terms of x.

20. Mandy is 5 inches taller than Melissa. If x inches represents the height of Mandy, express Melissa's height in terms of x.

△ **21.** If one angle measures $(x + 5)°$, express the measure of its supplement in terms of x.

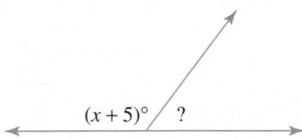

Solve each equation.

22. $\frac{3}{4}x = -9$

23. $\frac{x}{6} = \frac{2}{3}$

24. $-3x + 1 = 19$

25. $5x + 25 = 20$

26. $5x + x = 9 + 4x - 1 + 6$

27. $-y + 4y = 7 - y - 3 - 8$

28. Express the sum of three even consecutive integers as an expression in x. Let x be the first even integer.

(2.3) Solve the following.

29. $\frac{2}{7}x - \frac{5}{7} = 1$

30. $\frac{5}{3}x + 4 = \frac{2}{3}x$

31. $-(5x + 1) = -7x + 3$

32. $-4(2x + 1) = -5x + 5$

33. $-6(2x - 5) = -3(9 + 4x)$

34. $3(8y - 1) = 6(5 + 4y)$

35. $\frac{3(2 - z)}{5} = z$

36. $\frac{4(n + 2)}{5} = -n$

37. $5(2n - 3) - 1 = 4(6 + 2n)$

38. $-2(4y - 3) + 4 = 3(5 - y)$

39. $9z - z + 1 = 6(z - 1) + 7$

40. $5t - 3 - t = 3(t + 4) - 15$

41. $-n + 10 = 2(3n - 5)$

42. $-9 - 5a = 3(6a - 1)$

43. $\frac{5(c + 1)}{6} = 2c - 3$

44. $\frac{2(8 - a)}{3} = 4 - 4a$

45. $200(70x - 3560) = -179(150x - 19{,}300)$

46. $1.72y - .04y = 0.42$

Solve.

47. The quotient of a number and 3 is the same as the difference of the number and two. Find the number.

48. Double the sum of a number and six is the opposite of the number. Find the number.

(2.4) Solve each of the following.

49. The height of the Eiffel Tower is 68 feet more than three times a side of its square base. If the sum of these two dimensions is 1380 feet, find the height of the Eiffel Tower.

50. A 12-foot board is to be divided into two pieces so that one piece is twice as long as the other. If x represents the length of the shorter piece, find the length of each piece.

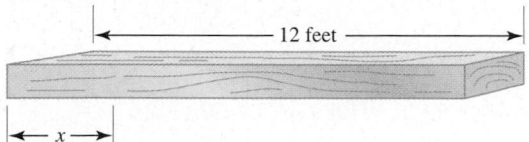

12 feet

x

51. One area code in Ohio is 34 more than three times another area code used in Ohio. If the sum of these area codes is 1262, find the two area codes.

52. Find three consecutive even integers whose sum is negative 114.

(2.5) Substitute the given values into the given formulas and solve for the unknown variable.

△ **53.** $P = 2l + 2w$; $P = 46, l = 14$

△ **54.** $V = lwh$; $V = 192, l = 8, w = 6$

Solve each of the following for the indicated variable.

55. $y = mx + b$ for m **56.** $r = vst - 9$ for s

57. $2y - 5x = 7$ for x **58.** $3x - 6y = -2$ for y

△ **59.** $C = \pi D$ for π △ **60.** $C = 2\pi r$ for π

△ **61.** A swimming pool holds 900 cubic meters of water. If its length is 20 meters and its height is 3 meters, find its width.

?

3 m

20 m

62. The highest temperature on record in Rome, Italy, is 104° Fahrenheit. Convert this temperature to Celsius.

63. A charity 10K race is given annually to benefit a local hospice organization. How long will it take to run/walk a 10K race (10 kilometers or 10,000 meters) if your average pace is 125 **meters** per minute?

(2.6) Solve.

64. Find 12% of 250.   **65.** Find 110% of 85.

66. The number 9 is what percent of 45?

67. The number 59.5 is what percent of 85?

68. The number 137.5 is 125% of what number?

69. The number 768 is 60% of what number?

70. The state of Mississippi has the highest phoneless rate in the United States, 12.6% of households. If a city in Mississippi has 50,000 households, how many of these would you expect to be phoneless?

The graph below shows how business travelers relax when in their hotel rooms.

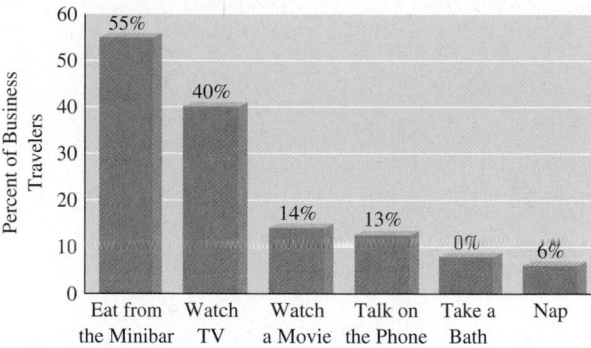

Source: *USA Today,* 1994.

71. What percent of business travelers surveyed relax by taking a nap?

72. What is the most popular way to relax according to the survey?

73. If a hotel in New York currently has 300 business travelers, how many might you expect to relax by watching TV?

74. Do the percents in the graph above have a sum of 100%? Why or why not?

75. The number of employees at Arnold's Box Manufacturers just decreased from 210 to 180. Find the percent decrease. Round to the nearest tenth of a percent.

(2.7) Solve each linear inequality. Graph the solution set and write it in interval notation.

76. $3(x - 5) > -(x + 3)$

77. $-2(x + 7) \geq 3(x + 2)$

78. $4x - (5 + 2x) < 3x - 1$

79. $3(x - 8) < 7x + 2(5 - x)$

80. $24 \geq 6x - 2(3x - 5) + 2x$

81. $48 + x \geq 5(2x + 4) - 2x$

82. $\dfrac{x}{3} + \dfrac{1}{2} > \dfrac{2}{3}$ **83.** $x + \dfrac{3}{4} < \dfrac{-x}{2} + \dfrac{9}{4}$

84. $\dfrac{x - 5}{2} \leq \dfrac{3}{8}(2x + 6)$

85. $\dfrac{3(x - 2)}{5} > \dfrac{-5(x - 2)}{3}$

86. Tina earns \$175 per week plus a 5% commission on all her sales. Find the minimum amount of sales to ensure that she earns at least \$300 per week.

87. Ellen Catarella shot rounds of 76, 82, and 79 golfing. What must she shoot on her next round so that her average will be below 80?

CHAPTER 2 TEST

Simplify each of the following expressions.

1. $2y - 6 - y - 4$

2. $2.7x + 6.1 + 3.2x - 4.9$

3. $4(x - 2) - 3(2x - 6)$

4. $-5(y + 1) + 2(3 - 5y)$

Solve each of the following equations.

5. $-\dfrac{4}{5}x = 4$

6. $4(n - 5) = -(4 - 2n)$

7. $5y - 7 + y = -(y + 3y)$

8. $4z + 1 - z = 1 + z$

9. $\dfrac{2(x + 6)}{3} = x - 5$

10. $\dfrac{4(y - 1)}{5} = 2y + 3$

11. $\dfrac{1}{2} - x + \dfrac{3}{2} = x - 4$

12. $\dfrac{1}{3}(y + 3) = 4y$

13. $-0.3(x - 4) + x = 0.5(3 - x)$

14. $-4(a + 1) - 3a = -7(2a - 3)$

Solve each of the following applications.

15. A number increased by two-thirds of the number is 35. Find the number.

△ **16.** A gallon of water seal covers 200 square feet. How many gallons are needed to paint two coats of water seal on a deck that measures 20 feet by 35 feet?

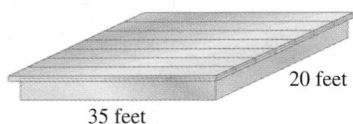

20 feet

35 feet

17. Sedric Angell invested an amount of money in Amoxil stock that earned an annual 10% return, and then he invested twice the original amount in IBM stock that earned an annual 12% return. If his total return from both investments was $2890, find how much he invested in each stock.

18. Two trains leave Los Angeles simultaneously traveling on the same track in opposite directions at speeds of 50 and 64 miles per hour. How long will it take before they are 285 miles apart?

19. Find the value of x if $y = -14$, $m = -2$, and $b = -2$ in the formula $y = mx + b$.

Solve each of the following equations for the indicated variable.

△ **20.** $V = \pi r^2 h$ for h

21. $3x - 4y = 10$ for y

Solve. Graph the solution set and write it in interval notation.

22. $3x - 5 > 7x + 3$

23. $x + 6 > 4x - 6$

24. $\dfrac{2(5x + 1)}{3} > 2$

25. $\dfrac{x}{5} - \dfrac{3}{10} \le \dfrac{1}{2}$

The following graph shows the source of income for charities.

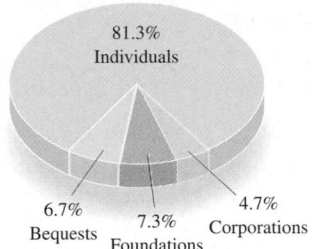

81.3%
Individuals

6.7%
Bequests

7.3%
Foundations

4.7%
Corporations

26. What percent of charity income comes from individuals?

27. If the total annual income for charities is $126.2 billion, find the amount that comes from corporations.

28. Find the number of degrees in the Bequests sector.

Use the following double bar graph for Exercises 29–31.

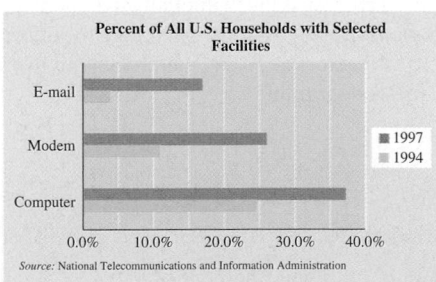

29. What percent of U.S. households had e-mail in 1997?

30. How many more households (in percent) had computers in 1997 than in 1994?

31. In a city with 23,000 households, how many of these would you expect to have had modems in 1997?

CHAPTER 2 CUMULATIVE REVIEW

1. Given the set $\{-2, 0, \frac{1}{4}, 112, -3, 11, \sqrt{2}\}$, list the numbers in this set that belong to the set of:
   a. Natural numbers
   b. Whole numbers
   c. Integers
   d. Rational numbers
   e. Irrational numbers
   f. Real numbers

2. Find the absolute value of each number.
   a. $|4|$ b. $|-5|$ c. $|0|$

3. Write each of the following numbers as a product of primes.
   a. 40   b. 63

4. Write $\frac{2}{5}$ as an equivalent fraction with a denominator of 20.

5. Simplify $3[4(5 + 2) - 10]$.

6. Decide whether 2 is a solution of $3x + 10 = 8x$.

Add.

7. $-1 + (-2)$ 8. $-4 + 6$

9. Simplify each expression.
   a. $-(-10)$ b. $-\left(-\frac{1}{2}\right)$
   c. $-(-2x)$ d. $-|-6|$

10. Subtract.
   a. $5.3 - (-4.6)$ b. $-\frac{3}{10} - \frac{5}{10}$
   c. $-\frac{2}{3} - \left(-\frac{4}{5}\right)$

11. Find each unknown complementary or supplementary angle.
   a.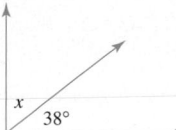
   b.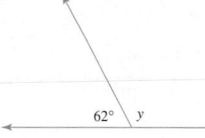

12. Find each product.
   a. $(-1.2)(0.05)$ b. $\frac{2}{3} \cdot -\frac{7}{10}$

13. Find each quotient.
   a. $\frac{-24}{-4}$ b. $\frac{-36}{3}$
   c. $\frac{2}{3} \div \left(-\frac{5}{4}\right)$

14. Use a commutative property to complete each statement.
   a. $x + 5 = $ _____
   b. $3 \cdot x = $ _____

15. Use the distributive property to write each sum as a product.
   a. $8 \cdot 2 + 8 \cdot x$
   b. $7s + 7t$

16. Subtract $4x - 2$ from $2x - 3$.

Solve.

17. $x - 7 = 10$

18. $7 = -5(2a - 1) - (-11a + 6)$

19. $\frac{y}{7} = 20$

20. $4(2x - 3) + 7 = 3x + 5$

21. Twice the sum of a number and 4 is the same as four times the number decreased by 12. Find the number.

22. Solve $V = lwh$ for l.

23. Write each percent as a decimal.
   a. 35% b. 89.5%
   c. 150%

24. The number 63 is what percent of 72?

25. Solve $x + 4 \leq -6$ for x. Graph the solution set and write it in interval notation.

Helping Prepare for Possible Dangerous Situations

Weather affects many aspects of our daily lives. An afternoon thunderstorm can put the damper on picnic plans or yard work. A cold snap can drive up our heating bills or affect the price of orange juice. Heavy rains can cause dangerous driving conditions or catastrophic flooding. High winds can damage roofs or bring down power lines. Accurately predicting the weather can allow people to prepare for possible damage or avoid dangerous weather situations.

Meteorology is the study of the atmosphere, including the science of weather forecasting. Meteorologists must be able to collect and interpret data, read maps and graphs, plot coordinates, make mathematical computations, use mathematical and physical models, and understand basic statistics. They should also have a solid background in chemistry, physics, earth science, and geography, as well as good computer and communication skills. Meteorologists often work as part of a team.

 For more information about careers in meteorology and the atmospheric sciences, visit the National Weather Association Website by first going to www.prenhall.com/martin-gay.

In the Spotlight on Decision Making feature on page 159, you will have the opportunity to make a decision about keeping track of a hurricane as a meteorologist.

GRAPHS AND FUNCTIONS

The linear equations and inequalities we explored in Chapter 2 are statements about a single variable. This chapter examines statements about two variables: linear equations and inequalities in two variables. We focus particularly on graphs of those equations and inequalities which lead to the notion of relation and to the notion of function, perhaps the single most important and useful concept in all of mathematics.

3.1 THE RECTANGULAR COORDINATE SYSTEM

CD-ROM   SSM

SSG   Video

▶ O B J E C T I V E S

1. Define the rectangular coordinate system and plot ordered pairs of numbers.
2. Graph paired data to create a scattergram.
3. Determine whether an ordered pair is a solution of an equation in two variables.
4. Find the missing coordinate of an ordered pair solution, given one coordinate of the pair.

1

In Section 1.9, we learned how to read graphs. Example 4 in Section 1.9 presented the graph below showing the relationship between time spent smoking a cigarette and pulse rate. Notice in this graph that there are two numbers associated with each point of the graph. For example, we discussed earlier that 15 minutes after "lighting up," the pulse rate is 80 beats per minute. If we agree to write the time first and the pulse rate second, we can say there is a point on the graph corresponding to the **ordered pair** of numbers (15, 80). A few more ordered pairs are listed alongside their corresponding points.

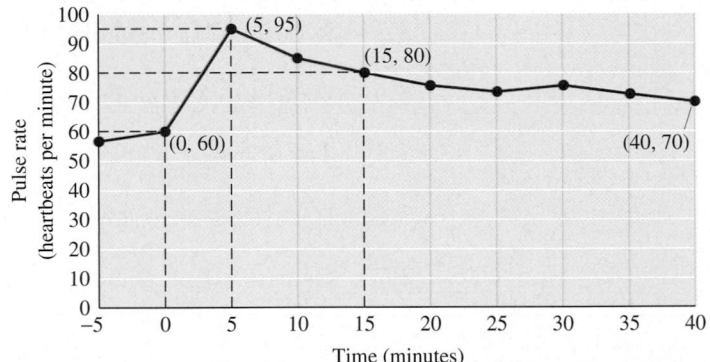

In general, we use this same ordered pair idea to describe the location of a point in a plane (such as a piece of paper). We start with a horizontal and a vertical axis. Each axis is a number line, and for the sake of consistency we construct our axes to intersect at the 0 coordinate of both. This point of intersection is called the **origin**. Notice that these two number lines or axes divide the plane into four regions called **quadrants**. The quadrants are usually numbered with Roman numerals as shown. The axes are not considered to be in any quadrant.

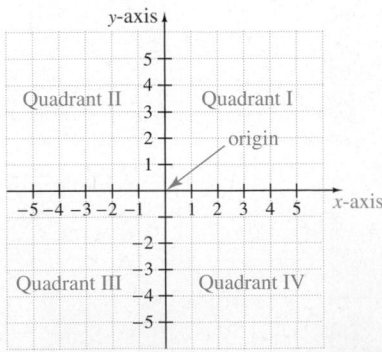

It is helpful to label axes, so we label the horizontal axis the **x-axis** and the vertical axis the **y-axis**. We call the system described on the previous page the **rectangular coordinate system**.

Just as with the pulse rate graph, we can then describe the locations of points by ordered pairs of numbers. We list the horizontal **x-axis** measurement first and the vertical **y-axis** measurement second.

To plot or graph the point corresponding to the ordered pair

$$(a, b)$$

we start at the origin. We then move a units left or right (right if a is positive, left if a is negative). From there, we move b units up or down (up if b is positive, down if b is negative). For example, to plot the point corresponding to the ordered pair $(3, 2)$, we start at the origin, move 3 units right, and from there move 2 units up. (See the figure below.) The x-value, 3, is also called the **x-coordinate** and the y-value, 2, is also called the **y-coordinate**. From now on, we will call the point with coordinates $(3, 2)$ simply the point $(3, 2)$. The point $(-2, 5)$ is graphed below also.

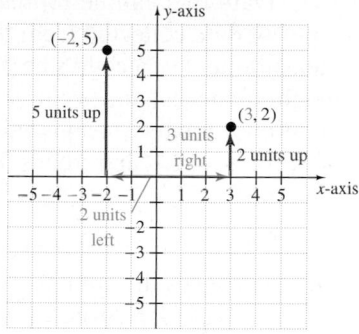

Does the order in which the coordinates are listed matter? Yes! Notice that the point corresponding to the ordered pair $(2, 3)$ is in a different location than the point corresponding to $(3, 2)$. These two ordered pairs of numbers describe two different points of the plane.

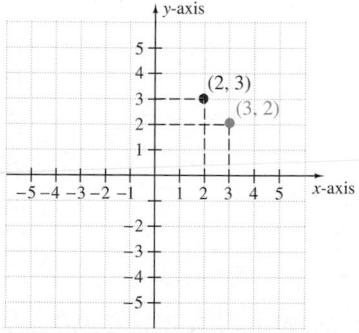

HELPFUL HINT

Don't forget that **each ordered pair corresponds to exactly one point in the plane and that each point in the plane corresponds to exactly one ordered pair.**

Example 1   On a single coordinate system, plot each ordered pair. State in which quadrant, if any, each point lies.

a. $(5, 3)$ **b.** $(-5, 3)$ **c.** $(-2, -4)$ **d.** $(1, -2)$

e. $(0, 0)$ **f.** $(0, 2)$ **g.** $(-5, 0)$ **h.** $\left(0, -5\frac{1}{2}\right)$

Solution   Point $(5, 3)$ lies in quadrant I.
Point $(-5, 3)$ lies in quadrant II.
Point $(-2, -4)$ lies in quadrant III.
Point $(1, -2)$ lies in quadrant IV.

Points $(0, 0)$, $(0, 2)$, $(-5, 0)$, and $\left(0, -5\frac{1}{2}\right)$

lie on axes, so they are not in any quadrant.

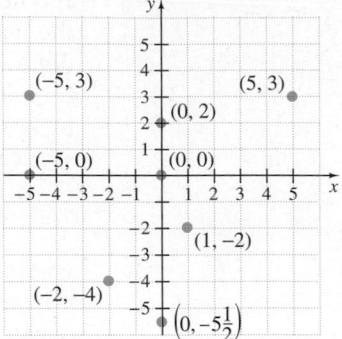

From Example 1, notice that the y-coordinate of any point on the x-axis is 0. For example, the point $(-5, 0)$ lies on the x-axis. Also, the x-coordinate of any point on the y-axis is 0. For example, the point $(0, 2)$ lies on the y-axis.

2   Data that can be represented as an ordered pair is called **paired data**. Many types of data collected from the real world are paired data. For instance, the annual measurement of a child's height can be written as an ordered pair of the form (year, height in inches) and is paired data. The graph of paired data as points in the rectangular coordinate system is called a **scatter diagram**. Scatter diagrams can be used to look for patterns and trends in paired data.

Example 2   The table gives the annual revenues for Wal-Mart Stores for the years shown. (*Source:* Wal-Mart Stores, Inc.)

Year	Wal-Mart Revenue (in billions of dollars)
1993	56
1994	68
1995	83
1996	95
1997	106
1998	121
1999	139

a. Write this paired data as a set of ordered pairs of the form (year, revenue in billions of dollars).
b. Create a scatter diagram of the paired data.
c. What trend in the paired data does the scatter diagram show?

Solution   **a.** The ordered pairs are $(1993, 56)$, $(1994, 68)$, $(1995, 83)$, $(1996, 95)$, $(1997, 106)$, $(1998, 121)$, and $(1999, 139)$.
b. We begin by plotting the ordered pairs. Because the x-coordinate in each ordered pair is a year, we label the x-axis "Year" and mark the horizontal axis with the years given. Then we label the y-axis or vertical axis "Wal-Mart Revenue (in billions of dollars)." It is convenient to mark the vertical axis in multiples of 20, starting with 0.

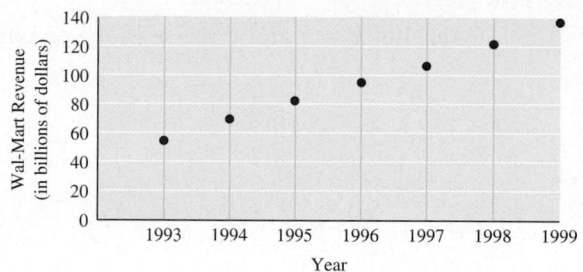

c. The scatter diagram shows that Wal-Mart revenue steadily increased over the years 1993–1999.

---

3   Let's see how we can use ordered pairs to record solutions of equations containing two variables. An equation in one variable such as $x + 1 = 5$ has one solution, which is 4: the number 4 is the value of the variable x that makes the equation true.

   An equation in two variables, such as $2x + y = 8$, has solutions consisting of two values, one for x and one for y. For example, $x = 3$ and $y = 2$ is a solution of $2x + y = 8$ because, if x is replaced with 3 and y with 2, we get a true statement.

$$2x + y = 8$$
$$2(3) + 2 = 8$$
$$8 = 8 \quad \textit{True.}$$

The solution $x = 3$ and $y = 2$ can be written as $(3, 2)$, an **ordered pair** of numbers. The first number, 3, is the x-value and the second number, 2, is the y-value.

   In general, an ordered pair is a **solution** of an equation in two variables if replacing the variables by the values of the ordered pair results in a true statement.

Example 3   Determine whether each ordered pair is a solution of the equation $x - 2y = 6$.

a. $(6, 0)$ **b.** $(0, 3)$ **c.** $\left(1, -\dfrac{5}{2}\right)$

Solution   **a.** Let $x = 6$ and $y = 0$ in the equation $x - 2y = 6$.

$$x - 2y = 6$$
$$6 - 2(0) = 6 \qquad \textit{Replace x with 6 and y with 0.}$$
$$6 - 0 = 6 \qquad \textit{Simplify.}$$
$$6 = 6 \qquad \textit{True.}$$

$(6, 0)$ is a solution, since $6 = 6$ is a true statement.

b. Let $x = 0$ and $y = 3$.

$$x - 2y = 6$$
$$0 - 2(3) = 6 \qquad \textit{Replace x with 0 and y with 3.}$$
$$0 - 6 = 6$$
$$-6 = 6 \qquad \textit{False.}$$

$(0, 3)$ is *not* a solution, since $-6 = 6$ is a false statement.

c. Let $x = 1$ and $y = -\dfrac{5}{2}$ in the equation.

$$x - 2y = 6$$

$$1 - 2\left(-\frac{5}{2}\right) = 6 \qquad \text{Replace } x \text{ with 1 and } y \text{ with } -\frac{5}{2}.$$

$$1 + 5 = 6$$

$$6 = 6 \qquad \text{True.}$$

$\left(1, -\dfrac{5}{2}\right)$ is a solution, since $6 = 6$ is a true statement. ∎

4   If one value of an ordered pair solution of an equation is known, the other value can be determined. To find the unknown value, replace one variable in the equation by its known value. Doing so results in an equation with just one variable that can be solved for the variable using the methods of Chapter 2.

Example 4   Complete the following ordered pair solutions for the equation $3x + y = 12$.

    a. $(0, \ )$ **b.** $(\ , 6)$ **c.** $(-1, \ )$

Solution   **a.** In the ordered pair $(0, \ )$, the x-value is 0. Let $x = 0$ in the equation and solve for y.

$$3x + y = 12$$

$$3(0) + y = 12 \qquad \text{Replace } x \text{ with 0.}$$

$$0 + y = 12$$

$$y = 12$$

The completed ordered pair is $(0, 12)$.

b. In the ordered pair $(\ , 6)$, the y-value is 6. Let $y = 6$ in the equation and solve for x.

$$3x + y = 12$$

$$3x + 6 = 12 \qquad \text{Replace } y \text{ with 6.}$$

$$3x = 6 \qquad \text{Subtract 6 from both sides.}$$

$$x = 2 \qquad \text{Divide both sides by 3.}$$

The ordered pair is $(2, 6)$.

c. In the ordered pair $(-1, \ )$, the x-value is -1. Let $x = -1$ in the equation and solve for y.

$$3x + y = 12$$

$$3(-1) + y = 12 \qquad \text{Replace } x \text{ with } -1.$$

$$-3 + y = 12$$

$$y = 15 \qquad \text{Add 3 to both sides.}$$

The ordered pair is $(-1, 15)$. ∎

Solutions of equations in two variables can also be recorded in a **table of values**, as shown in the next example.

Example 5 Complete the table for the equation $y = 3x$.

x	y
a. −1	
b.	0
c.	−9

Solution **a.** Replace x with −1 in the equation and solve for y.

$$y = 3x$$
$$y = 3(-1) \quad \text{Let } x = -1.$$
$$y = -3$$

The ordered pair is $(-1, -3)$.

b. Replace y with 0 in the equation and solve for x.

$$y = 3x$$
$$0 = 3x \quad \text{Let } y = 0.$$
$$0 = x \quad \text{Divide both sides by 3.}$$

The completed ordered pair is $(0, 0)$.

c. Replace y with −9 in the equation and solve for x.

$$y = 3x$$
$$-9 = 3x \quad \text{Let } y = -9.$$
$$-3 = x \quad \text{Divide both sides by 3.}$$

x	y
−1	−3
0	0
−3	−9

The completed ordered pair is $(-3, -9)$. The completed table is shown to the left.

Example 6 Complete the table for the equation $y = 3$.

x	y
−2	
0	
−5	

Solution The equation $y = 3$ is the same as $0x + y = 3$. Replace x with −2 and we have $0(-2) + y = 3$ or $y = 3$. Notice that no matter what value we replace x by, y always equals 3. The completed table is:

x	y
−2	3
0	3
−5	3

By now, you have noticed that equations in two variables often have more than one solution. We discuss this more in the next section.

A table showing ordered pair solutions may be written vertically or horizontally as shown in the next example.

Example 7   FINDING THE VALUE OF A COMPUTER

A small business purchased a computer for $2000. The business predicts that the computer will be used for 5 years and the value in dollars y of the computer in x years is $y = -300x + 2000$. Complete the table.

x	0	1	2	3	4	5
y						

Solution   To find the value of y when x is 0, replace x with 0 in the equation. We use this same procedure to find y when x is 1 and when x is 2.

When $x = 0$,

$y = -300x + 2000$
$y = -300 \cdot 0 + 2000$
$y = 0 + 2000$
$y = 2000$

When $x = 1$,

$y = -300x + 2000$
$y = -300 \cdot 1 + 2000$
$y = -300 + 2000$
$y = 1700$

When $x = 2$,

$y = -300x + 2000$
$y = -300 \cdot 2 + 2000$
$y = -600 + 2000$
$y = 1400$

We have the ordered pairs $(0, 2000)$, $(1, 1700)$, and $(2, 1400)$. This means that in 0 years the value of the computer is $2000, in 1 year the value of the computer is $1700, and in 2 years the value is $1400. Complete the table of values.

When $x = 3$,

$y = -300x + 2000$
$y = -300 \cdot 3 + 2000$
$y = -900 + 2000$
$y = 1100$

When $x = 4$,

$y = -300x + 2000$
$y = -300 \cdot 4 + 2000$
$y = -1200 + 2000$
$y = 800$

When $x = 5$,

$y = -300x + 2000$
$y = -300 \cdot 5 + 2000$
$y = -1500 + 2000$
$y = 500$

The completed table is

x	0	1	2	3	4	5
y	2000	1700	1400	1100	800	500

The ordered pair solutions recorded in the completed table for the example above are graphed below. Notice that the graph gives a visual picture of the decrease in value of the computer.

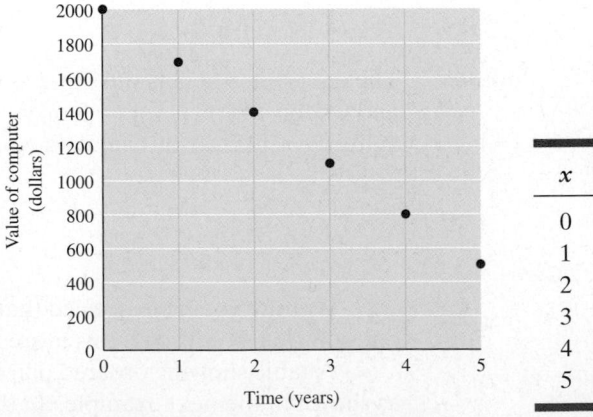

x	y
0	2000
1	1700
2	1400
3	1100
4	800
5	500

SPOTLIGHT ON DECISION MAKING

Suppose you are a meteorologist. You are tracking Hurricane Felix. Hurricane position information is issued every 6 hours. The table lists Felix's most recent positions, given in latitude (vertical scale on the hurricane tracking chart) and longitude (horizontal scale on the hurricane tracking chart). Plot the position of the hurricane on the tracking chart and decide whether Felix is a threat to the United States. If so, what part?

NATIONAL HURRICANE CENTER RECONNAISSANCE REPORT

Felix Coordinates

Date	Time	Latitude	Longitude
9/7	4 PM	23.2	88.0
9/7	10 PM	23.7	88.6
9/8	4 AM	23.4	89.1
9/8	10 AM	22.8	89.9
9/8	4 PM	22.2	91.2
9/8	10 PM	21.7	91.8
9/9	4 AM	21.5	92.4
9/9	10 AM	21.2	93.1

Hurricane Tracking Chart

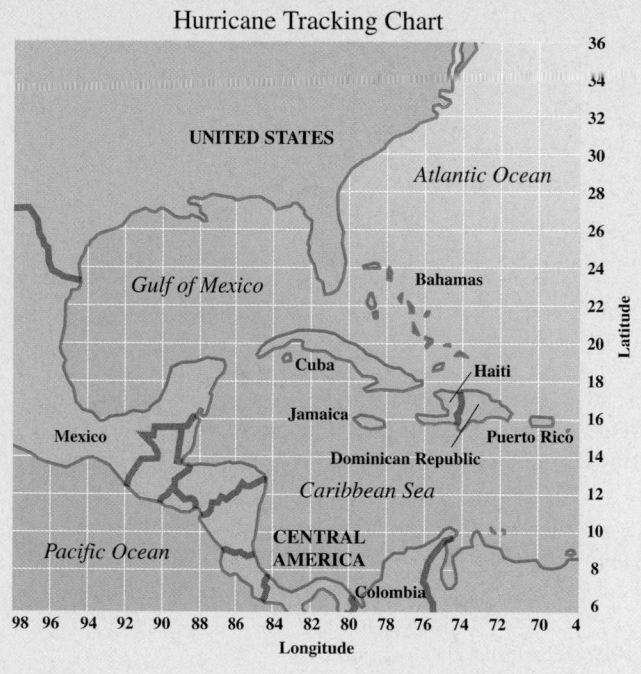

MENTAL MATH

Determine the coordinates of each point on the graph.

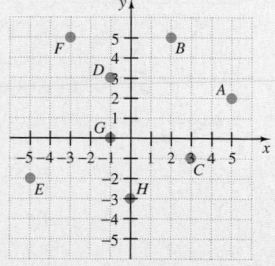

1. Point *A*
2. Point *B*
3. Point *C*
4. Point *D*
5. Point *E*
6. Point *F*
7. Point *G*
8. Point *H*

Exercise Set 3.1

Plot the ordered pairs. State in which quadrant, if any, each point lies. See Example 1.

1. $(1, 5)$
2. $(-5, -2)$
3. $(-3, 0)$
4. $(0, -1)$
5. $(2, -4)$
6. $\left(-1, 4\frac{1}{2}\right)$
7. $\left(4\frac{3}{4}, 0\right)$
8. $\left(0, \frac{7}{8}\right)$
9. $(0, 0)$
10. $(5, 0)$
11. $(0, 4)$
12. $(-3, -3)$

Find the x- and y-coordinates of the following labeled points.

13.

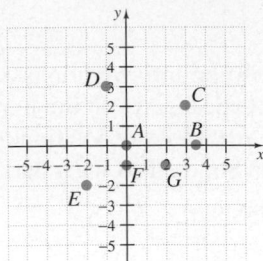

14.

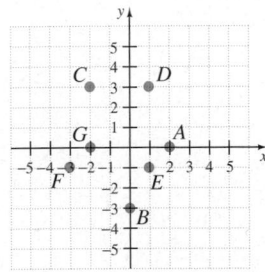

△ **15.** Find the perimeter of the rectangle whose vertices are the points with coordinates $(-1, 5)$, $(3, 5)$, $(3, -4)$, and $(-1, -4)$.

△ **16.** Find the area of the rectangle whose vertices are the points with coordinates $(5, 2)$, $(5, -6)$, $(0, -6)$, and $(0, 2)$.

Solve. See Example 2.

17. The table shows the average price of a gallon of regular unleaded gasoline (in dollars) for the years shown. (*Source:* Energy Information Administration)

Year	Price per Gallon of Unleaded Gasoline (in dollars)
1991	1.14
1992	1.13
1993	1.11
1994	1.11
1995	1.15
1996	1.23
1997	1.23
1998	1.06
1999	1.17

a. Write each paired data as an ordered pair of the form (year, gasoline price).

b. Draw a grid such as the one in Example 2 and create a scatter diagram of the paired data.

18. The table shows the number of regular-season NFL football games won by the winner of the Super Bowl for the years shown. (*Source:* National Football League)

Year	Regular-Season Games Won by Super Bowl Winner
1994	12
1995	13
1996	12
1997	13
1998	12
1999	14
2000	13

a. Write each paired data as an ordered pair of the form (year, games won).

b. Draw a grid such as the one in Example 2 and create a scatter diagram of the paired data.

19. The table shows the average monthly mortgage payment made by Americans during the years shown. (*Source:* National Association of REALTORS®)

Year	Average Monthly Mortgage Payment (in dollars)
1994	578
1995	613
1996	654
1997	675
1998	717

a. Write each paired data as an ordered pair of the form (year, mortgage payment).

b. Draw a grid such as the one in Example 2 and create a scatter diagram of the paired data.

c. What trend in the paired data does the scatter diagram show?

20. The table shows the number of institutions of higher education in the United States for the years shown. (*Source:* U.S. Department of Education)

Year	Number of Institutions of Higher Education
1909	951
1919	1041
1929	1409
1939	1708
1949	1851
1959	2008
1969	2525
1979	3152
1989	3535
1999	3800

a. Write each paired data as an ordered pair of the form (year, number of institutions).

b. Draw a grid such as the one in Example 2 and create a scatter diagram of the paired data.

c. What trend in the paired data does the scatter diagram show?

Determine whether each ordered pair is a solution of the given linear equation. See Example 3.

21. $2x + y = 7$; $(3, 1)$, $(7, 0)$, $(0, 7)$

22. $x - y = 6$; $(5, -1)$, $(7, 1)$, $(0, -6)$

23. $y = -5x$; $(-1, -5)$, $(0, 0)$, $(2, -10)$

24. $x = 2y$; $(0, 0)$, $(2, 1)$, $(-2, -1)$

25. $x = 5$; $(4, 5)$, $(5, 4)$, $(5, 0)$

26. $y = 2$; $(-2, 2)$, $(2, 2)$, $(0, 2)$

27. $x + 2y = 9$; $(5, 2)$, $(0, 9)$

28. $3x + y = 8$; $(2, 3)$, $(0, 8)$

29. $2x - y = 11$; $(3, -4)$, $(9, 8)$

30. $x - 4y = 14$; $(2, -3)$, $(14, 6)$

31. $x = \dfrac{1}{3}y$; $(0, 0)$, $(3, 9)$ **32.** $y = -\dfrac{1}{2}x$; $(0, 0)$, $(4, 2)$

33. $y = -2$; $(-2, -2)$, $(5, -2)$

34. $x = 4$; $(4, 0)$, $(4, 4)$

Complete each ordered pair so that it is a solution of the given linear equation. See Examples 4 through 6.

35. $x - 4y = 4$; $(\ , -2)$, $(4, \ )$

36. $x - 5y = -1$; $(\ , -2)$, $(4, \ )$

37. $3x + y = 9$; $(0, \ )$, $(\ , 0)$

38. $x + 5y = 15$; $(0, \ )$, $(\ , 0)$

39. $y = -7$; $(11, \ )$, $(\ , -7)$

40. $x = \dfrac{1}{2}$; $(\ , 0)$, $\left(\dfrac{1}{2}, \ \right)$

Complete the table of values for each given linear equation; then plot each solution. Use a single coordinate system for each equation. See Examples 4 through 6.

41. $x + 3y = 6$

x	y
0	
	0
	1

42. $2x + y = 4$

x	y
0	
	0
	2

43. $2x - y = 12$

x	y
0	
	-2
-3	

44. $-5x + y = 10$

x	y
	0
	5
2	

45. $2x + 7y = 5$

x	y
0	
	0
	1

46. $x - 6y = 3$

x	y
0	
1	
	-1

47. $x = 3$

x	y
	0
	-0.5
	$\frac{1}{4}$

48. $y = -1$

x	y
-2	
0	
-1	

49. $x = -5y$

x	y
	0
	1
10	

50. $y = -3x$

x	y
0	
-2	
	9

51. Discuss any similarities in the graphs of the ordered pair solutions for Exercises 41–50.

52. Explain why equations in two variables have more than one solution.

Solve. See Example 7.

53. The cost in dollars y of producing x computer desks is given by $y = 80x + 5000$.

   a. Complete the following table and graph the results.

x	100	200	300
y			

   b. Find the number of computer desks that can be produced for $8600. (*Hint:* Find x when $y = 8600$.)

54. The hourly wage y of an employee at a certain production company is given by $y = 0.25x + 9$ where x is the number of units produced in an hour.

   a. Complete the table and graph the results.

x	0	1	5	10
y				

   b. Find the number of units that must be produced each hour to earn an hourly wage of $12.25. (*Hint:* Find x when $y = 12.25$.)

55. The population density y of Minnesota (in people per square mile of land) from 1920 through 1990 is given by $y = 0.364x + 21.939$. In the equation, x represents the number of years after 1900. (*Source:* Based on data from the U.S. Bureau of the Census)

   a. Complete the table.

x	20	65	90
y			

   b. Find the year in which the population density was approximately 50 people per square mile. (*Hint:* Find x when $y = 50$ and round to the nearest whole number.)

56. The percentage y of recorded music sales that were in cassette format from 1991 through 1998 is given by $y = -6.09x + 55.99$. In the equation, x represents the number of years after 1991. (*Source:* Based on data from the Recording Industry Association of America)

   a. Complete the table.

x	1	3	5
y			

   b. Find the year in which approximately 31.6% of recorded music sales were cassettes. (*Hint:* Find x when $y = 31.6$ and round to the nearest whole number.)

The graph below shows the number of Target stores for each year. Use this graph to answer Exercises 57–60.

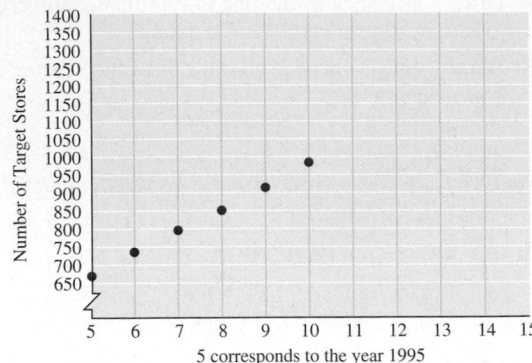

Source: USA Today 1/14/99

57. The ordered pair $(5, 670)$ is a point of the graph. Write a sentence describing the meaning of this ordered pair.

58. The ordered pair $(10, 984)$ is a point of the graph. Write a sentence describing the meaning of this ordered pair.

59. Estimate the increase in Target stores for years 6, 7, and 8.

60. Use a straightedge or ruler and this graph to predict the number of Target stores in the year 2005.

61. When is the graph of the ordered pair (a, b) the same as the graph of the ordered pair (b, a)?

62. In your own words, describe how to plot an ordered pair.

Determine the quadrant or quadrants in which the points described below lie.

63. The first coordinate is positive and the second coordinate is negative.

64. Both coordinates are negative.

65. The first coordinate is negative.

66. The second coordinate is positive.

REVIEW EXERCISES

Solve each equation for y. See Section 2.5.

67. $x + y = 5$

68. $x - y = 3$

69. $2x + 4y = 5$

70. $5x + 2y = 7$

71. $10x = -5y$

72. $4y = -8x$

73. $x - 3y = 6$

74. $2x - 9y = -20$

3.2 GRAPHING EQUATIONS

CD-ROM SSM

SSG Video

▶ **OBJECTIVES**

1. Identify linear equations.
2. Graph a linear equation by finding and plotting ordered pair solutions.
3. Graph a nonlinear equation by finding and plotting ordered pair solutions.

1 In the previous section, we found that equations in two variables may have more than one solution. For example, both $(6, 0)$ and $(2, -2)$ are solutions of the equation $x - 2y = 6$. In fact, this equation has an infinite number of solutions. Other solutions include $(0, -3)$, $(4, -1)$, and $(-2, -4)$. If we graph these solutions, notice that a pattern appears.

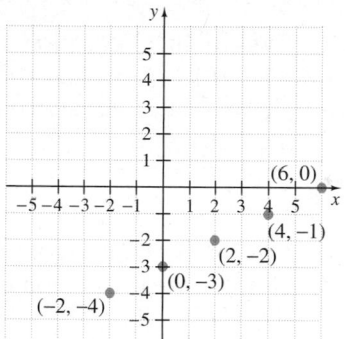

These solutions all appear to lie on the same line, which has been filled in below. It can be shown that every ordered pair solution of the equation corresponds to a point on this line, and every point on this line corresponds to an ordered pair solution. Thus, we say that this line is the graph of the equation $x - 2y = 6$.

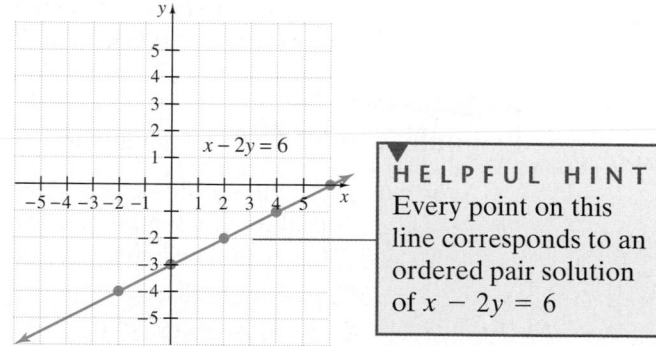

HELPFUL HINT
Every point on this line corresponds to an ordered pair solution of $x - 2y = 6$

The equation $x - 2y = 6$ is called a **linear equation in two variables** and **the graph of every linear equation in two variables is a line.**

LINEAR EQUATION IN TWO VARIABLES

A linear equation in two variables is an equation that can be written in the form

$$Ax + By = C$$

where A, B, and C are real numbers and A and B are not both 0.

The form $Ax + By = C$ is called **standard form**.

> **HELPFUL HINT**
> Notice in the form $Ax + By = C$, the understood exponent on both x and y is 1.

Examples of Linear Equations in Two Variables

$$2x + y = 8 \qquad -2x = 7y \qquad y = \frac{1}{3}x + 2 \qquad y = 7$$

Not all equations in two variables are linear equations. Before we graph linear equations in two variables, let's practice identifying them.

Example 1 Identify the linear equations in two variables.

a. $x - 1.5y = -1.6$ **b.** $y = -2x$ **c.** $x + y^2 = 9$ **d.** $x = 5$

Solution **a.** This is a linear equation in two variables because it is written in the form $Ax + By = C$ with $A = 1$, $B = -1.5$, and $C = -1.6$.

b. This is a linear equation in two variables because it can be written in the form $Ax + By = C$.

$$y = -2x$$
$$2x + y = 0 \qquad \text{Add } 2x \text{ to both sides.}$$

c. This is *not* a linear equation in two variables because y is squared.

d. This is a linear equation in two variables because it can be written in the form $Ax + By = C$.

$$x = 5$$
$$x + 0y = 5 \qquad \text{Add } 0 \cdot y.$$

---

2 From geometry, we know that a straight line is determined by just two points. Graphing a linear equation in two variables, then, requires that we find just two of its infinitely many solutions. Once we do so, we plot the solution points and draw the line connecting the points. Usually, we find a third solution as well, as a check.

Example 2 Graph the linear equation $2x + y = 5$.

Solution Find three ordered pair solutions of $2x + y = 5$. To do this, choose a value for one variable, x or y, and solve for the other variable. For example, let $x = 1$. Then $2x + y = 5$ becomes

$$2x + y = 5$$
$$2(1) + y = 5 \qquad \text{Replace } x \text{ with 1.}$$
$$2 + y = 5 \qquad \text{Multiply.}$$
$$y = \mathbf{3} \qquad \text{Subtract 2 from both sides.}$$

Since $y = 3$ when $x = 1$, the ordered pair $(1, 3)$ is a solution of $2x + y = 5$. Next, let $x = 0$.

$$2x + y = 5$$
$$2(0) + y = 5 \qquad \text{Replace } x \text{ with 0.}$$
$$0 + y = 5$$
$$y = 5$$

The ordered pair $(0, 5)$ is a second solution.

The two solutions found so far allow us to draw the straight line that is the graph of all solutions of $2x + y = 5$. However, we find a third ordered pair as a check. Let $y = -1$.

$$2x + y = 5$$
$$2x + (-1) = 5 \qquad \text{Replace } y \text{ with } -1.$$
$$2x - 1 = 5$$
$$2x = 6 \qquad \text{Add 1 to both sides.}$$
$$x = 3 \qquad \text{Divide both sides by 2.}$$

The third solution is $(3, -1)$. These three ordered pair solutions are listed in table form as shown. The graph of $2x + y = 5$ is the line through the three points.

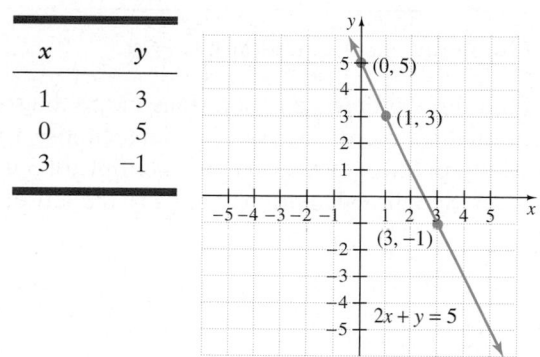

x	y
1	3
0	5
3	-1

HELPFUL HINT

How does a third ordered pair solution above help us check our work? If the three ordered pair solutions do not lie along the same straight line when graphed, either a calculation is wrong or a mistake was made when graphing.

Example 3   Graph the linear equation $y = 3x$.

Solution   To graph this linear equation, we find three ordered pair solutions. Since this equation is solved for y, choose three x values.

x	y
2	6
0	0
-1	-3

If $x = 2$, $y = 3 \cdot 2 = 6$.

If $x = 0$, $y = 3 \cdot 0 = 0$.

If $x = -1$, $y = 3 \cdot -1 = -3$.

Next, graph the ordered pair solutions listed in the table above and draw a line through the plotted points. The line is the graph of $y = 3x$. Every point on the graph represents an ordered pair solution of the equation and every ordered pair solution is a point on this line.

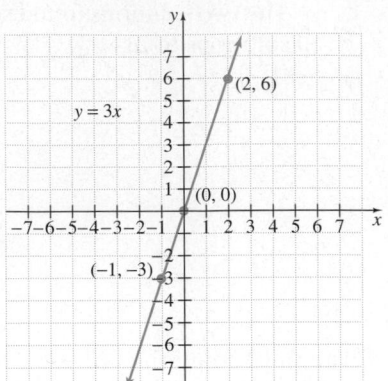

Example 4 Graph the linear equation $y = -\frac{1}{3}x$.

Solution Find three ordered pair solutions, graph the solutions, and draw a line through the plotted solutions. To avoid fractions, choose x values that are multiples of 3 to substitute in the equation. When a multiple of 3 is multiplied by $-\frac{1}{3}$, the result is an integer. See the calculations shown to the left of the table below.

x	y
6	-2
0	0
-3	1

If $x = 6$, then $y = -\dfrac{1}{3} \cdot 6 = -2$.

If $x = 0$, then $y = -\dfrac{1}{3} \cdot 0 = 0$.

If $x = -3$, then $y = -\dfrac{1}{3} \cdot -3 = 1$.

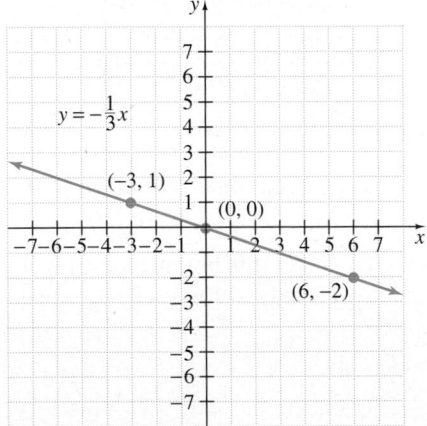

Let's compare the graphs in Examples 3 and 4. The graph of $y = 3x$ tilts upward (as we follow the line from left to right) and the graph of $y = -\frac{1}{3}x$ tilts downward (as we follow the line from left to right). Also notice that both lines go through the origin or that $(0, 0)$ is an ordered pair solution of both equations. In general, the graphs of $y = 3x$ and $y = -\frac{1}{3}x$ are of the form $y = mx$ where m is a constant. The graph of an equation in this form always goes through the origin $(0, 0)$ because when x is 0, $y = mx$ becomes $y = m \cdot 0 = 0$.

Example 5 Graph the linear equation $y = 3x + 6$ and compare this graph with the graph of $y = 3x$ in Example 4.

Solution Find ordered pair solutions, graph the solutions, and draw a line through the plotted solutions. We choose x values and substitute in the equation $y = 3x + 6$.

x	y
-3	-3
0	6
1	9

If $x = -3$, then $y = 3(-3) + 6 = -3$.

If $x = 0$, then $y = 3(0) + 6 = 6$.

If $x = 1$, then $y = 3(1) + 6 = 9$.

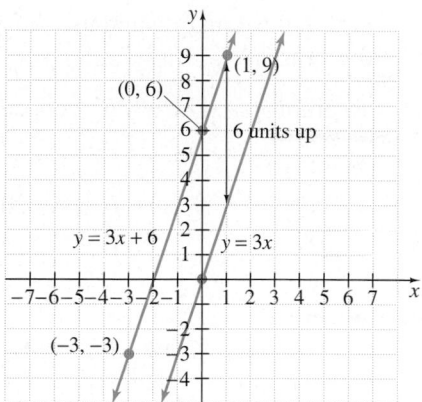

The most startling similarity is that both graphs appear to have the same upward tilt as we move from left to right. Also, the graph of $y = 3x$ crosses the y-axis at the origin, while the graph of $y = 3x + 6$ crosses the y-axis at 6. In fact, the graph of $y = 3x + 6$ is the same as the graph of $y = 3x$ moved vertically upward 6 units. ■

Notice that the graph of $y = 3x + 6$ crosses the y-axis at 6. This happens because when $x = 0$, $y = 3x + 6$ becomes $y = 3 \cdot 0 + 6 = 6$. The graph contains the point $(0, 6)$, which is on the y-axis.

In general, if a linear equation in two variables is solved for y, we say that it is written in the form $y = mx + b$. The graph of this equation contains the point $(0, b)$ because when $x = 0$, $y = mx + b$ is $y = m \cdot 0 + b = b$.

> The graph of $y = mx + b$ crosses the y-axis at b.

Linear equations are often used to model real data as seen in the next example.

Example 6 **ESTIMATING THE NUMBER OF MEDICAL ASSISTANTS**

One of the occupations expected to have the most growth in the next few years is medical assistant. The number of people y (in thousands) employed as medical assistants in the United States can be estimated by the linear equation $y = 11x + 217$, where

x is the number of years after the year 1995. (*Source:* based on data from the Bureau of Labor Statistics)

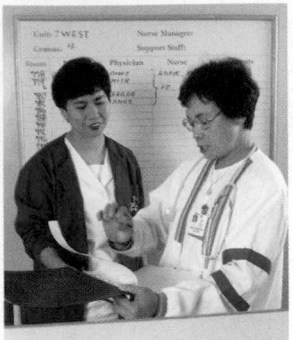

Graph the equation and use the graph to predict the number of medical assistants in the year 2005.

Solution   To graph $y = 11x + 217$, choose x-values and substitute in the equation.

x	y
0	217
2	239
7	294

If $x = 0$, then $y = 11(0) + 217 = 217$.
If $x = 2$, then $y = 11(2) + 217 = 239$.
If $x = 7$, then $y = 11(7) + 217 = 294$.

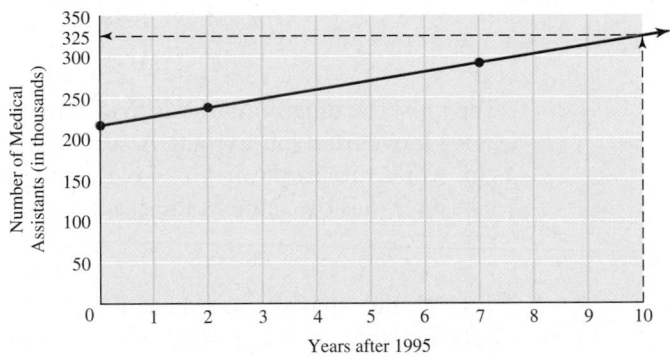

To use the graph to predict the number of medical assistants in the year 2005, we need to find the y-coordinate that corresponds to $x = 10$. (10 years after 1995 is the year 2005.) To do so, find 10 on the x-axis. Move vertically upward to the graphed line and then horizontally to the left. We approximate the number on the y-axis to be 325. Thus in the year 2005, we predict that there will be 325 thousand medical assistants. (The actual value, using 10 for x, is 327.)

3   As we mentioned earlier, not all equations in two variables are linear equations, and not all graphs of equations in two variables are lines.

◆ **Example 7**   Graph $y = x^2$.

Solution   This equation is not linear, and its graph is not a line. We begin by finding ordered pair solutions. Because this graph is solved for y, we choose x-values and find corresponding y-values.

If $x = -3$, then $y = (-3)^2$, or 9.

If $x = -2$, then $y = (-2)^2$, or 4.

If $x = -1$, then $y = (-1)^2$, or 1.

If $x = 0$, then $y = 0^2$, or 0.

If $x = 1$, then $y = 1^2$, or 1.

If $x = 2$, then $y = 2^2$, or 4.

If $x = 3$, then $y = 3^2$, or 9.

x	y
-3	9
-2	4
-1	1
0	0
1	1
2	4
3	9

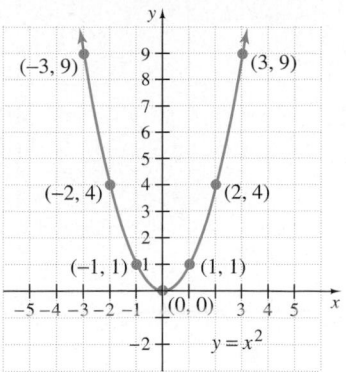

Study the table a moment and look for patterns. Notice that the ordered pair solution $(0, 0)$ contains the smallest y-value because any other x-value squared will give a positive result. This means that the point $(0, 0)$ will be the lowest point on the graph. Also notice that all other y-values correspond to two different x-values. For example, $3^2 = 9$ and also $(-3)^2 = 9$. This means that the graph will be a mirror image of itself across the y-axis. Connect the plotted points with a smooth curve to sketch its graph.

This curve is given a special name, a parabola. We will study more about parabolas in later chapters.

Example 8 Graph the equation $y = |x|$.

Solution This is not a linear equation, and its graph is not a line. Because we do not know the shape of this graph, we find many ordered pair solutions. We will choose x-values and substitute to find corresponding y-values.

If $x = -3$, then $y = |-3|$, or 3.

If $x = -2$, then $y = |-2|$, or 2.

If $x = -1$, then $y = |-1|$, or 1.

If $x = 0$, then $y = |0|$, or 0.

If $x = 1$, then $y = |1|$, or 1.

If $x = 2$, then $y = |2|$, or 2.

If $x = 3$, then $y = |3|$, or 3.

x	y
-3	3
-2	2
-1	1
0	0
1	1
2	2
3	3

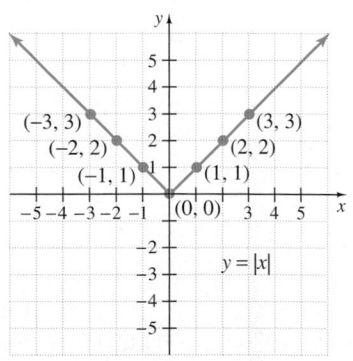

Again, study the table of values for a moment and notice any patterns.

From the plotted ordered pairs, we see that the graph of this absolute value equation is V-shaped. The lowest point of this graph is $(0, 0)$.

Example 9 Graph the equation $y = |x| - 3$.

Solution The graph of $y = |x| - 3$ is simply the graph of $y = |x|$ moved down 3 units. To check, we choose x-values and substitute to find corresponding y-values.

If $x = -4$, then $y = |-4| - 3$, or 1.

If $x = -2$, then $y = |-2| - 3$, or -1.

If $x = 0$, then $y = |0| - 3$, or -3.

If $x = 2$, then $y = |2| - 3$, or -1.

If $x = 4$, then $y = |4| - 3$, or 1.

x	y
-4	1
-2	-1
0	-3
2	-1
4	1

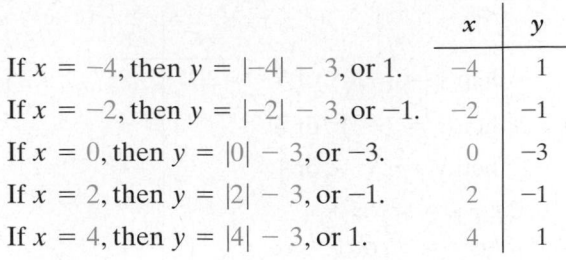

Notice that the graph of $y = |x| - 3$ is the graph of $y = |x|$ moved down 3 units.

GRAPHING CALCULATOR EXPLORATIONS

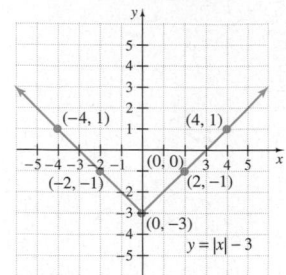

In this section, we begin a study of graphing calculators and graphing software packages for computers. These graphers use the same point plotting technique that we used in this section. The advantage of this graphing technology is, of course, that graphing calculators and computers can find and plot ordered pair solutions much faster than we can. Note, however, that the features described in these boxes may not be available on all graphing calculators.

The rectangular screen where a portion of the rectangular coordinate system is displayed is called a **window**. We call it a **standard window** for graphing when both the x- and y-axes display coordinates between -10 and 10. This information is often displayed in the window menu on a graphing calculator as

Xmin = -10

Xmax = 10

Xscl = 1   The scale on the x-axis is one unit per tick mark.

Ymin = -10

Ymax = 10

Yscl = 1   The scale on the y-axis is one unit per tick mark.

To use a graphing calculator to graph the equation $y = -5x + 4$, press the $\boxed{\text{Y} =}$ key and enter the keystrokes

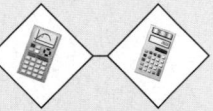

(Check your owner's manual to make sure the "negative" key is pressed here and not the "subtraction" key.)

The top row should now read $Y_1 = -5x + 4$. Next press the $\boxed{\text{GRAPH}}$ key, and the display should look like this:

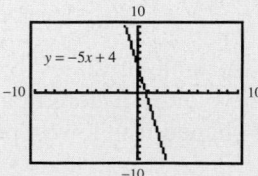

Use a standard window and graph the following equations. (Unless otherwise stated, we will use a standard window when graphing.)

1. $y = -3.2x + 7.9$ **2.** $y = -x + 5.85$ **3.** $y = \dfrac{1}{4}x - \dfrac{2}{3}$ **4.** $y = \dfrac{2}{3}x - \dfrac{1}{5}$

5. $y = |x - 3| + 2$ **6.** $y = |x + 1| - 1$ **7.** $y = x^2 + 3$ **8.** $y = (x + 3)^2$

Exercise Set 3.2

Determine whether each equation is a linear equation in two variables. See Example 1.

1. $-x = 3y + 10$

2. $y = x - 15$

3. $x = y$

4. $x = y^3$

5. $x^2 + 2y = 0$

6. $0.01x - 0.2y = 8.8$

7. $y = -1$

8. $x = 25$

Graph each linear equation. See Examples 2 through 4.

9. $x + y = 4$

10. $x + y = 7$

11. $x - y = -2$

12. $-x + y = 6$

13. $x - 2y = 6$

14. $-x + 5y = 5$

15. $y = 6x + 3$

16. $y = -2x + 7$

17. $x - 2y = -6$

18. $-x + 2y = 5$

19. $y = 6x$

20. $x = -2y$

Graph each pair of linear equations on the same set of axes. Discuss how the graphs are similar and how they are different. See Example 5.

21. $y = 5x$; $y = 5x + 4$

22. $y = 2x$; $y = 2x + 5$

23. $y = -2x$; $y = -2x - 3$

24. $y = x$; $y = x - 7$

25. $y = \dfrac{1}{2}x$; $y = \dfrac{1}{2}x + 2$

26. $y = -\dfrac{1}{4}x$; $y = -\dfrac{1}{4}x + 3$

The graph of $y = 5x$ is below as well as Figures a–d. For Exercises 27 through 30, match each equation with its graph. Hint: Recall that if an equation is written in the form $y = mx + b$, its graph crosses the y-axis at b.

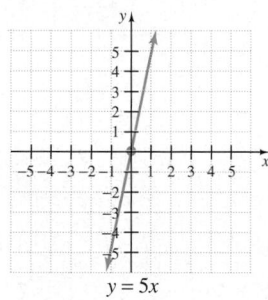
$y = 5x$

a.

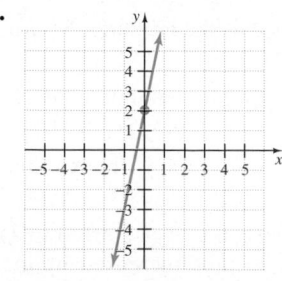

b.

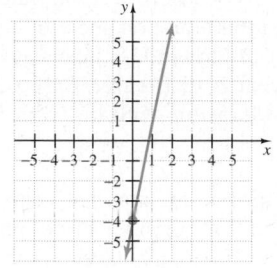

c.

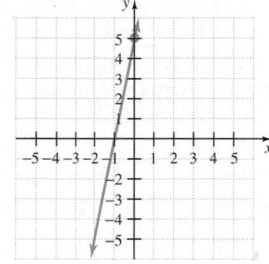

d.

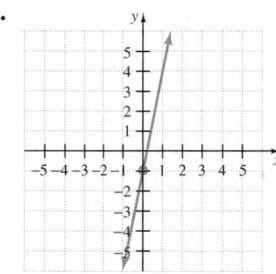

27. $y = 5x + 5$

28. $y = 5x - 4$

29. $y = 5x - 1$

30. $y = 5x + 2$

Solve. See Example 6.

31. The revenue y (in billions of dollars) for Home Depot stores during the years 1996 through 1999 is given by the equation $y = 5x + 15$, where x is the number of years after 1996. Graph this equation and use the graph to predict the revenue for Home Depot Stores in the year 2006. (*Source:* Based on data from The Home Depot Inc.)

32. The number of girls y (in thousands) participating in high school athletic programs in the United States during the years 1993 through 1997 is approximated by the equation $y = 120x + 2004$, where x is the number of years after 1993. Graph this equation and use the graph to predict the number of girls in U.S. high school athletic programs in the year 2010. (*Source:* Based on data from the National Federation of State High School Associations)

33. A fast-growing occupation in the next few years is elementary school teacher. The number of people y (in thousands) employed as elementary teachers in the United States can be estimated by the linear equation $y = 20x + 1539$, where x is the number of years after the year 2000. Graph this equation and use the graph to predict the number of elementary teachers in the year 2008. (*Source:* Based on data from the Bureau of Labor Statistics)

34. The number y of franchised new car dealerships in the United States during the years 1994 through 1997 is given by $y = -50x + 22,850$, where x is the number of years after 1994. Graph this equation and use the graph to predict the number of new car dealerships in the year 2010. (*Source:* Based on data from National Automobile Dealers Association)

Determine whether each equation is linear or not. Then graph the equation. See Examples 2 through 9.

35. $x + y = 3$

36. $y - x = 8$

37. $y = 4x$

38. $y = 6x$

39. $y = 4x - 2$

40. $y = 6x - 5$

41. $y = |x| + 3$

42. $y = |x| + 2$

43. $2x - y = 5$

44. $4x - y = 7$

45. $y = 2x^2$

46. $y = 3x^2$

47. $y = x^2 - 3$

48. $y = x^2 + 3$

49. $y = -2x$

50. $y = -3x$

51. $y = -2x + 3$

52. $y = -3x + 2$

53. $y = |x + 2|$

54. $y = |x - 1|$

55. $y = x^3$

Hint: Let $x = -3, -2, -1, 0, 1, 2.$

56. $y = x^3 - 2$

Hint: Let $x = -3, -2, -1, 0, 1, 2.$

57. $y = -|x|$

58. $y = -x^2$

59. $y = \dfrac{1}{3}x - 1$

60. $y = \dfrac{1}{2}x - 3$

61. $y = -\dfrac{3}{2}x + 1$

62. $y = -\dfrac{2}{3}x + 1$

△ **63.** The perimeter y of a rectangle whose width is a constant 3 inches and whose length is x inches is given by the equation

$$y = 2x + 6$$

  a. Draw a graph of this equation.

  b. Read from the graph the perimeter y of a rectangle whose length x is 4 inches.

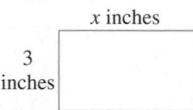

64. The distance y traveled in a train moving at a constant speed of 50 miles per hour is given by the equation

$$y = 50x$$

where x is the time in hours traveled.

  a. Draw a graph of this equation.

  b. Read from the graph the distance y traveled after 6 hours.

This graph shows hourly minimum wages and the years it increased. Use this graph for Exercises 65 through 68.

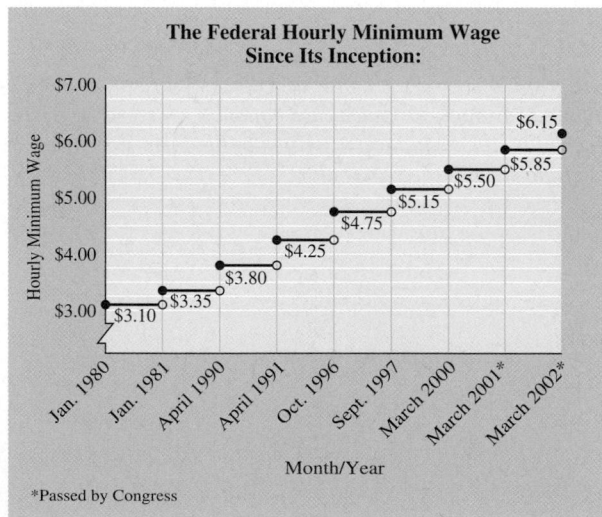

65. What was the first year that the minimum hourly wage rose above $4.00?

66. What was the first year that the minimum hourly wage rose above $5.00?

67. Why do you think that this graph is shaped the way it is?

68. The federal hourly minimum wage started in 1938 at $0.25. How much will it have increased by in 2002?

For exercises 69 through 72, match each description with the graph that best illustrates it.

69. Moe worked 40 hours per week until the fall semester started. He quit and didn't work again until he worked 60 hours a week during the Christmas break.

70. Kawana worked 40 hours a week for her father during the summer. She slowly cut back her hours to not working at all during the fall semester. During the Christmas break, she started working again and increased her hours to 60 hours per week.

71. Wendy worked from July through February, never quitting. She worked between 10 and 30 hours per week.

72. Bartholomew worked from July through February, never quitting. He worked between 10 and 30 hours per week except during the Christmas break. At that time, he worked 40 hours per week.

A 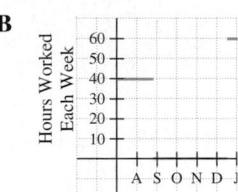   **B**

C 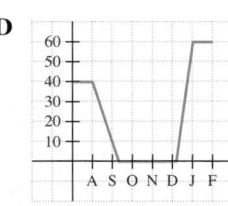   **D**

73. Explain how to find ordered pair solutions of linear equations in two variables.

74. Explain why we generally use three points to graph a line, when only two points are needed.

Write each statement as an equation in two variables. Then graph each equation.

75. The y-value is 5 more than three times the x-value.

76. The y-value is -3 decreased by twice the x-value.

77. The y-value is 2 more than the square of the x-value.

78. The y-value is 5 decreased by the square of the x-value.

Use a graphing calculator to verify the graphs of the following exercises.

79. Exercise 47   **80.** Exercise 48

81. Exercise 55   **82.** Exercise 56

REVIEW EXERCISES

△ **83.** The coordinates of three vertices of a rectangle are $(-2, 5)$, $(4, 5)$, and $(-2, -1)$. Find the coordinates of the fourth vertex. See Section 3.1.

△ **84.** The coordinates of two vertices of a square are $(-3, -1)$ and $(2, -1)$. Find the coordinates of two pairs of points possible for the third and fourth vertices. See Section 3.1.

Solve the following equations. See Section 2.3.

85. $3(x - 2) + 5x = 6x - 16$

86. $5 + 7(x + 1) = 12 + 10x$

87. $3x + \dfrac{2}{5} = \dfrac{1}{10}$ **88.** $\dfrac{1}{6} + 2x = \dfrac{2}{3}$

3.3 INTRODUCTION TO FUNCTIONS

CD-ROM   SSM

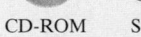

SSG   Video

▶ **OBJECTIVES**

1. Define relation, domain, and range.
2. Identify functions.
3. Use the vertical line test for functions.
4. Find the domain and range of a function.
5. Use function notation.

1 Equations in two variables, such as $y = 2x + 1$, describe **relations** between x-values and y-values. For example, if $x = 1$, then this equation describes how to find the y-value related to $x = 1$. In words, the equation $y = 2x + 1$ says that twice the x-value increased by 1 gives the corresponding y-value. The x-value of 1 corresponds to the y-value of $2(1) + 1 = 3$ for this equation, and we have the ordered pair $(1, 3)$.

There are other ways of describing relations or correspondences between two numbers or, in general, a first set (sometimes called the set of *inputs*) and a second set (sometimes called the set of *outputs*). For example.

First Set: Input	***Correspondence***	***Second Set: Output***
People in a certain city	Each person's age	The set of nonnegative integers

A few examples of ordered pairs from this relation might be (Ana, 4); (Bob, 36); (Trey, 21); and so on.

Below are just a few other ways of describing relations between two sets and the ordered pairs that they generate.

First Set:	***Second Set:***
Input	***Output***

Correspondence

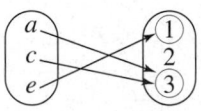

Ordered Pairs
$(a, 3), (c, 3), (e, 1)$

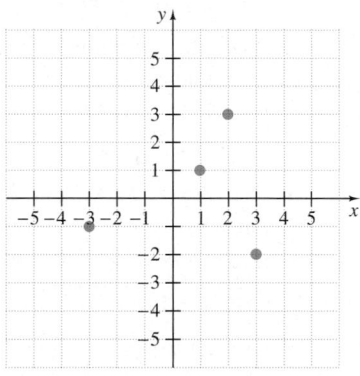

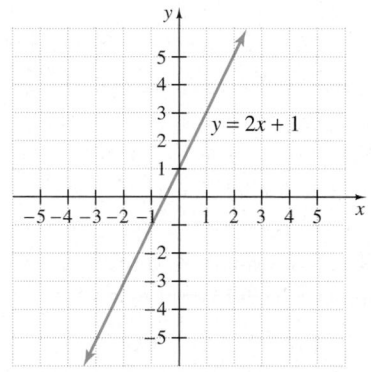

Ordered Pairs	***Some Ordered Pairs***
$(-3, -1), (1, 1), (2, 3), (3, -2)$	$(1, 3), (0, 1)$, and so on

RELATION, DOMAIN, AND RANGE

A **relation** is a set of ordered pairs.
The **domain** of the relation is the set of all first components of the ordered pairs.
The **range** of the relation is the set of all second components of the ordered pairs.

For example, the domain for our first relation above is $\{a, c, e\}$ and the range is $\{1, 3\}$. Notice that the range does not include the element 2 of the second set. This is because no element of the first set is assigned to this element. If a relation is defined in terms of x- and y-values, we will agree that the domain corresponds to x-values and that the range corresponds to y-values.

Example 1   Determine the domain and range of each relation.

a. $\{(2, 3), (2, 4), (0, -1), (3, -1)\}$

b.

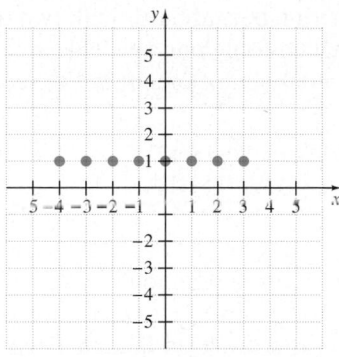

c.

Input:	Output:
Cities	Population (in thousands)

Erie
Miami
Escondido
Waco
Gary

109 200
359
117 52
182 104

Solution   **a.** The domain is the set of all first coordinates of the ordered pairs, $\{2, 0, 3\}$.
The range is the set of all second coordinates, $\{3, 4, -1\}$.

b. Ordered pairs are not listed here, but are given in graph form. The relation is
$\{(-4, 1), (-3, 1), (-2, 1), (-1, 1), (0, 1), (1, 1), (2, 1), (3, 1)\}$. The domain is
$\{-4, -3, -2, -1, 0, 1, 2, 3\}$.
The range is $\{1\}$.

c. The domain is the first set, $\{$Erie, Escondido, Gary, Miami, Waco$\}$.
The range is the numbers in the second set that correspond to elements in the
first set $\{104, 109, 117, 359\}$.

2   Now we consider a special kind of relation called a function.

FUNCTION

A **function** is a relation in which each first component in the ordered pairs
corresponds to *exactly* one second component.

HELPFUL HINT
A function is a special type of relation, so all functions are relations, but not all
relations are functions.

Example 2   Which of the following relations are also functions?

a. $\{(-2, 5), (2, 7), (-3, 5), (9, 9)\}$

b.

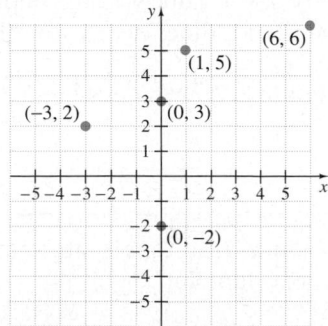

c.

Input	Correspondence	Output
People in a certain city	Each person's age	The set of nonnegative integers

Solution   **a.** Although the ordered pairs $(-2, 5)$ and $(-3, 5)$ have the same y-value, each x-value is assigned to only one y-value, so this set of ordered pairs is a function.
b. The x-value 0 is assigned to two y-values, -2 and 3, in this graph so this relation does not define a function.
c. This relation is a function because although two different people may have the same age, each person has only one age. This means that each element in the first set is assigned to only one element in the second set.   ▬

   We will call an equation such as $y = 2x + 1$ a **relation** since this equation defines a set of ordered pair solutions.

Example 3   Is the relation $y = 2x + 1$ also a function?

Solution   The relation $y = 2x + 1$ is a function if each x-value corresponds to just one y-value. For each x-value substituted in the equation $y = 2x + 1$, the multiplication and addition performed on each gives a single result, so only one y-value will be associated with each x-value. Thus, $y = 2x + 1$ is a function.   ▬

◈ Example 4   Is the relation $x = y^2$ also a function?

Solution   In $x = y^2$, if $y = 3$, then $x = 9$. Also, if $y = -3$, then $x = 9$. In other words, the x-value 9 corresponds to two y-values, 3 and -3. Thus, $x = y^2$ is not a function.   ▬

3   As we have seen so far, not all relations are functions. Consider the graphs of $y = 2x + 1$ and $x = y^2$ shown next. On the graph of $y = 2x + 1$, notice that each x-value corresponds to only one y-value. Recall from Example 3 that $y = 2x + 1$ is a function.

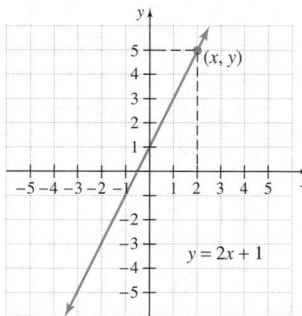   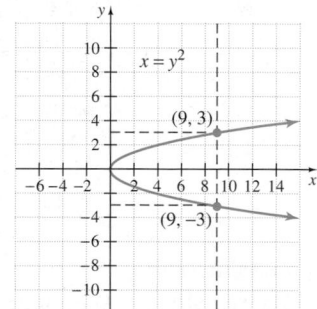

   On the graph of $x = y^2$ the x-value 9, for example, corresponds to two y-values, 3 and -3, as shown by the vertical line. Recall from Example 4 that $x = y^2$ is not a function.
   Graphs can be used to help determine whether a relation is also a function by the following vertical line test.

VERTICAL LINE TEST

If no vertical line can be drawn so that it intersects a graph more than once, the graph is the graph of a function.

Example 5 Which of the following graphs are graphs of functions?

a.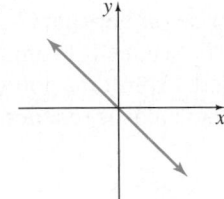

b.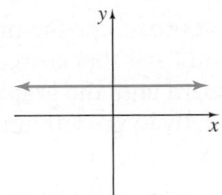

c.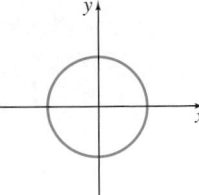

d.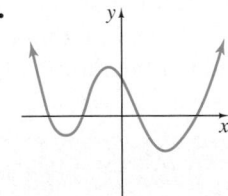

e.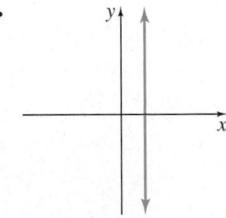

Solution
a. This graph is the graph of a function since no vertical line will intersect this graph more than once.
b. This graph is also the graph of a function.
c. This graph is not the graph of a function. Note that vertical lines can be drawn that intersect the graph in two points.
d. This graph is the graph of a function.
e. This graph is not the graph of a function. A vertical line can be drawn that intersects this line at every point.

Recall that the graph of a linear equation in two variables is a line, and a line that is not vertical will pass the vertical line test. Thus, **all linear equations are functions except those whose graph is a vertical line.**

Examples of functions can often be found in magazines, newspapers, books, and other printed material in the form of tables or graphs such as that in Example 6.

Example 6 The graph shows the sunrise time for Indianapolis, Indiana, for the year. Use this graph to answer the questions.

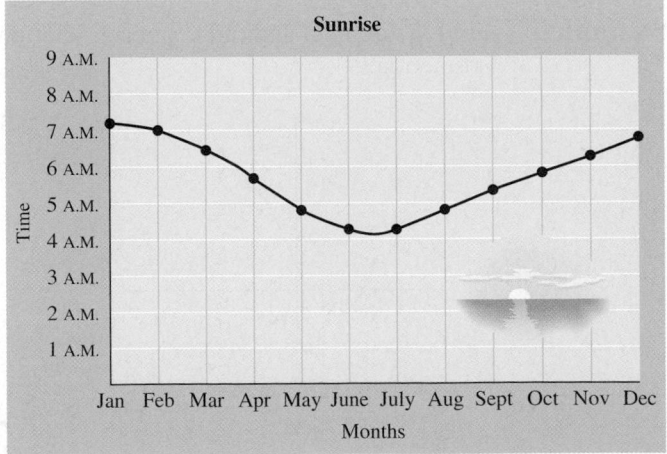

a. Approximate the time of sunrise on February 1.
b. Approximately when does the sun rise at 5 A.M.?
c. Is this the graph of a function?

Solution   **a.** To approximate the time of sunrise on February 1, we find the mark on the horizontal axis that corresponds to February 1. From this mark, we move vertically upward until the graph is reached. From that point on the graph, we move horizontally to the left until the vertical axis is reached. The vertical axis there reads 7 A.M.

b. To approximate when the sun rises at 5 A.M., we find 5 A.M. on the time axis and move horizontally to the right. Notice that we will reach the graph twice, corresponding to two dates for which the sun rises at 5 A.M. We follow both points on the graph vertically downward until the horizontal axis is reached. The sun rises at 5 A.M. at approximately the end of the month of April and the middle of the month of August.

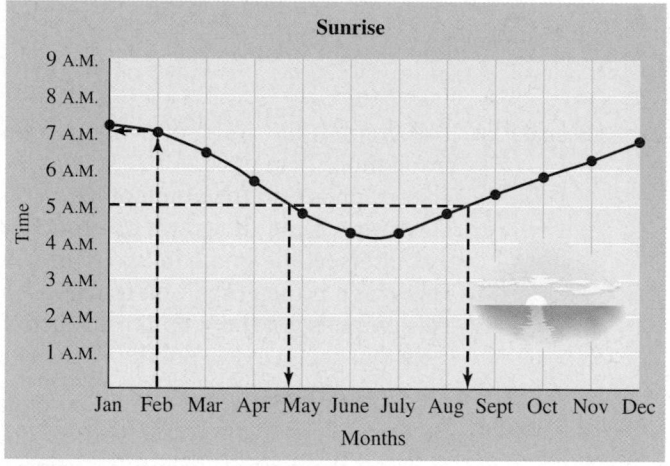

c. The graph is the graph of a function since it passes the vertical line test. In other words, for every day of the year in Indianapolis, there is exactly one sunrise time.

4   Next, we practice finding the domain and range of a relation from its graph.

Example 7   Find the domain and range of each relation. Determine whether the relation is also a function.

a.

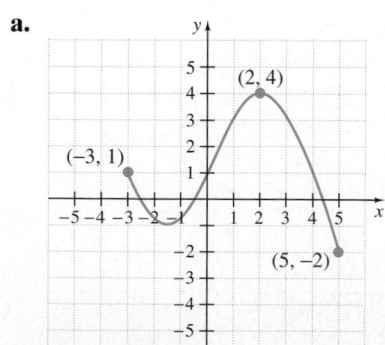

b.

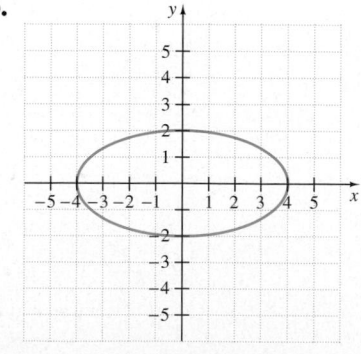

c. 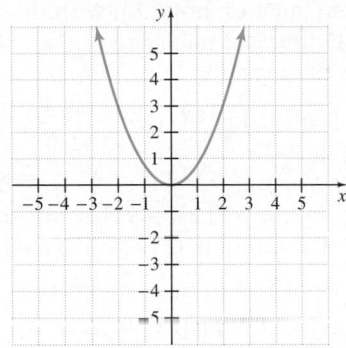   **d.**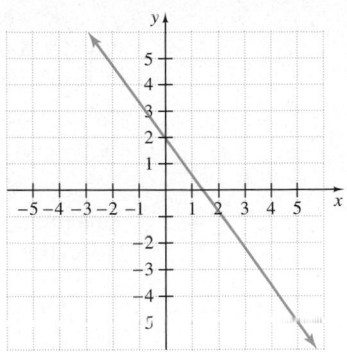

> ▼
> **H E L P F U L   H I N T**
> In Example 7, Part **a**, notice that the graph contains the endpoints $(-3, 1)$ and $(5, -2)$ whereas the graphs in Parts **c** and **d** contain arrows that indicate that they continue forever.

Solution   By the vertical line test, graphs **a**, **c**, and **d** are graphs of functions. The domain is the set of values of x and the range is the set of values of y. We read these values from each graph.

a. 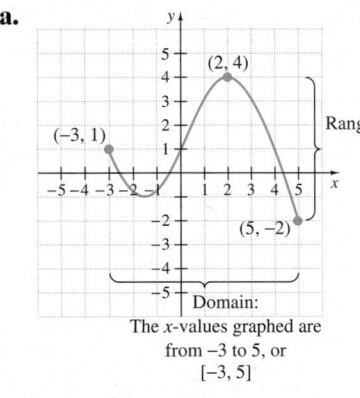   **b.**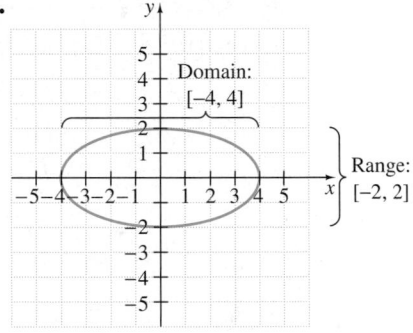

c. 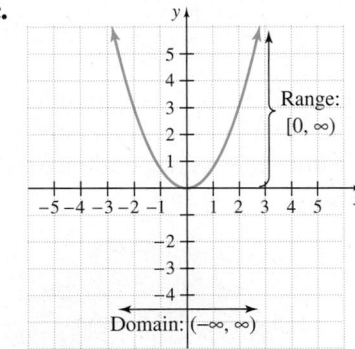   **d.**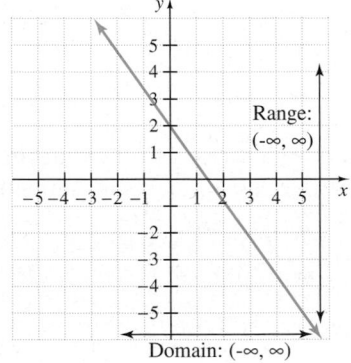

5   Many times letters such as f, g, and h are used to name functions. To denote that y is a function of x, we can write

$$y = f(x)$$

This means that **y is a function of x** or that y *depends on* x. For this reason, y is called the **dependent variable** and x the **independent variable**. The notation $f(x)$ is read "f of x" and is called **function notation**.

For example, to use function notation with the function $y = 4x + 3$, we write $f(x) = 4x + 3$. The notation $f(1)$ means to replace x with 1 and find the resulting y or function value. Since

$$f(x) = 4x + 3$$

then

$$f(1) = 4(1) + 3 = 7$$

This means that when $x = 1$, y or $f(x) = 7$. The corresponding ordered pair is $(1, 7)$. Here, the input is 1 and the output is $f(1)$ or 7. Now let's find $f(2)$, $f(0)$, and $f(-1)$.

$f(x) = 4x + 3$	$f(x) = 4x + 3$	$f(x) = 4x + 3$
$f(2) = 4(2) + 3$	$f(0) = 4(0) + 3$	$f(-1) = 4(-1) + 3$
$= 8 + 3$	$= 0 + 3$	$= -4 + 3$
$= 11$	$= 3$	$= -1$

Ordered Pairs:

$(2, 11)$	$(0, 3)$	$(-1, -1)$

▼
HELPFUL HINT
Note that $f(x)$ is a special symbol in mathematics used to denote a function. The symbol $f(x)$ is read "f of x." It does *not* mean $f \cdot x$ (f times x).

Example 8   If $f(x) = 7x^2 - 3x + 1$ and $g(x) = 3x - 2$, find the following.

a. $f(1)$ **b.** $g(1)$ **c.** $f(-2)$ **d.** $g(0)$

Solution   **a.** Substitute 1 for x in $f(x) = 7x^2 - 3x + 1$ and simplify.

$$f(x) = 7x^2 - 3x + 1$$
$$f(1) = 7(1)^2 - 3(1) + 1 = 5$$

b. $g(x) = 3x - 2$
$$g(1) = 3(1) - 2 = 1$$

c. $f(x) = 7x^2 - 3x + 1$
$$f(-2) = 7(-2)^2 - 3(-2) + 1 = 35$$

d. $g(x) = 3x - 2$
$$g(0) = 3(0) - 2 = -2$$

If it helps, think of a function, f, as a machine that has been programmed with a certain correspondence or rule. An input value (a member of the domain) is then fed into the machine, the machine does the correspondence or rule and the result is the output (a member of the range).

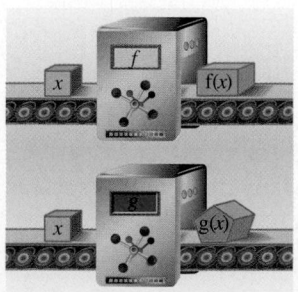

Many types of real-world paired data form functions. The broken-line graph on the next page shows the research and development spending by the Pharmaceutical Manufacturers Association.

Example 9   The following graph shows the research and development expenditures by the Pharmaceutical Manufacturers Association as a function of time.

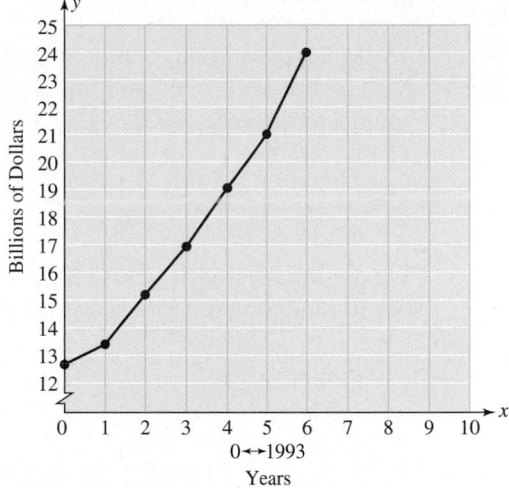

Source: Pharmaceutical Manufacturers Association
*The notation 0 ↔ 1993 means that year 0 corresponds to the year 1993, 1 corresponds to 1994, and so on.

a. Approximate the money spent on research and development in 1997.
b. In 1958, research and development expenditures were \$200 million. Find the increase in expenditures from 1958 to 1999.

Solution   **a.** On the graph, since 0 corresponds to the year 1993, then 4 (0 + 4) corresponds to the year 1997 (1993 + 4). In 1997, approximately \$19 billion was spent.
b. In 1999, approximately \$24 billion, or \$24,000 million, was spent. The increase in spending from 1958 to 1999 is \$24,000 − \$200 = \$23,800 million. ▬

Notice that the graph in Example 9 is the graph of a function since for each year there is only one total amount of money spent by the Pharmaceutical Manufacturers Association on research and development. Also notice that the graph resembles the graph of a line. Often, businesses depend on equations that "closely fit" data-defined functions like this one in order to model the data and predict future trends. For example, by a method called **least squares**, the function $f(x) = 1.882x + 11.79$ approximates the data shown. Its graph and the actual data function are shown below.

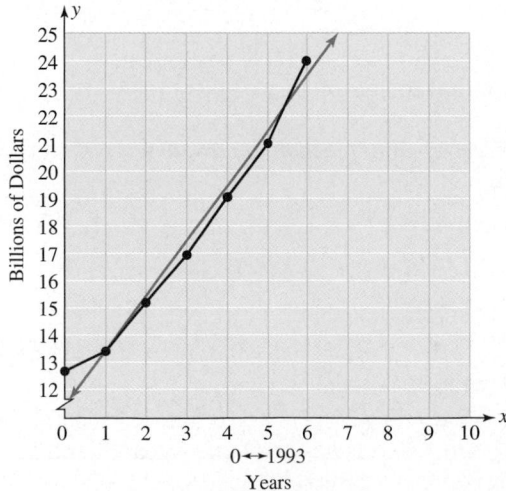

Example 10   Use the function $f(x) = 1.882x + 11.79$ to predict the amount of money that will be spent by the Pharmaceutical Manufacturers Association on research and development in 2006.

Solution   To predict the amount of money that will be spent in the year 2006 we use $f(x) = 1.882x + 11.79$ and find $f(13)$. (Notice that year 0 on the graph corresponds to the year 1993, so year 13 corresponds to the year 2006.)

$$f(x) = 1.882x + 11.79$$
$$f(13) = 1.882(13) + 11.79$$
$$= 36.256$$

We predict that in the year 2006, $36.256 billion dollars will be spent on research and development by the Pharmaceutical Manufacturers Association.

GRAPHING CALCULATOR EXPLORATIONS

It is possible to use a graphing calculator to sketch the graph of more than one equation on the same set of axes. For example, graph the functions $f(x) = x^2$ and $g(x) = x^2 + 4$ on the same set of axes.

To graph on the same set of axes, press the $\boxed{Y =}$ key and enter the equations on the first two lines.

$$Y_1 = x^2$$

$$Y_2 = x^2 + 4$$

Then press the $\boxed{\text{GRAPH}}$ key as usual. The screen should look like this.

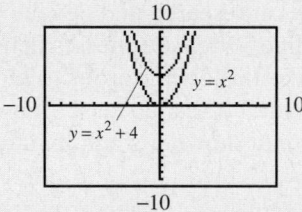

Notice that the graph of y or $g(x) = x^2 + 4$ is the graph of $y = x^2$ moved 4 units upward.

Graph each pair of functions on the same set of axes. Describe the similarities and differences in their graphs.

1. $f(x) = |x|$
    $g(x) = |x| + 1$

2. $f(x) = x^2$
    $h(x) = x^2 - 5$

3. $f(x) = x$
    $H(x) = x - 6$

4. $f(x) = |x|$
    $G(x) = |x| + 3$

5. $f(x) = -x^2$
    $F(x) = -x^2 + 7$

6. $f(x) = x$
    $F(x) = x + 2$

Exercise Set 3.3

Find the domain and the range of each relation. Also determine whether the relation is a function. See Examples 1 and 2.

1. $\{(-1, 7), (0, 6), (-2, 2), (5, 6)\}$

2. $\{(4, 9), (-4, 9), (2, 3), (10, -5)\}$

3. $\{(-2, 4), (6, 4), (-2, -3), (-7, -8)\}$

4. $\{(6, 6), (5, 6), (5, 2), (7, 6)\}$

5. $\{(1, 1), (1, 2), (1, 3), (1, 4)\}$

6. $\{(1, 1), (2, 1), (3, 1), (4, 1)\}$

7. $\left\{\left(\frac{3}{2}, \frac{1}{2}\right), \left(1\frac{1}{2}, -7\right), \left(0, \frac{4}{5}\right)\right\}$

8. $\{(\pi, 0), (0, \pi), (-2, 4), (4, -2)\}$

9. $\{(-3, -3), (0, 0), (3, 3)\}$

10. $\left\{\left(\frac{1}{2}, \frac{1}{4}\right), \left(0, \frac{7}{8}\right), (0.5, \pi)\right\}$

11.
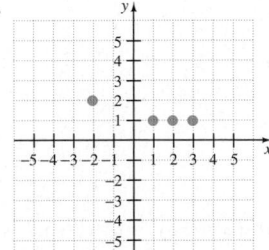

12.
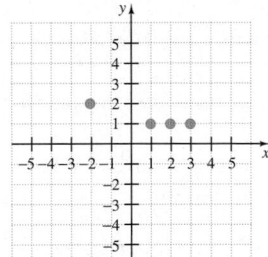

13.
Input:	Output:
State	Number of Congressional Representatives

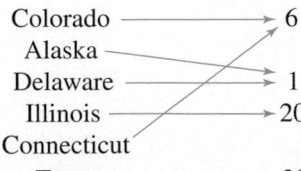

14.
Input:	Output:
Animal	Average Life Span (in years)

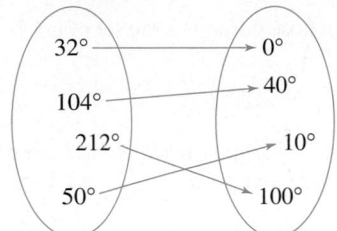

15.
Input:	Output:
Degrees Fahrenheit	Degrees Celsius

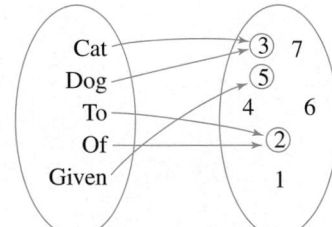

16.
Input:	Output:
Words	Number of Letters

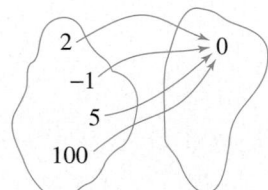

17.
Input	Output

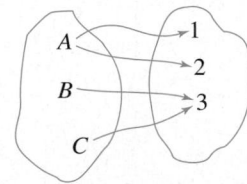

18.
Input	Output

In Exercises 19 and 20, determine whether the relation is a function.

19.

First set: Input	Correspondence	Second set: Output
Class of algebra students	Grade average	Set of nonnegative numbers

20.

First set: Input	Correspondence	Second set: Output
People in New Orleans (population 500,000)	Birthdate	Days of the year

21. Describe a function whose domain is the set of people in your hometown.

22. Describe a function whose domain is the set of people in your algebra class.

Use the vertical line test to determine whether each graph is the graph of a function. See Example 5.

23.

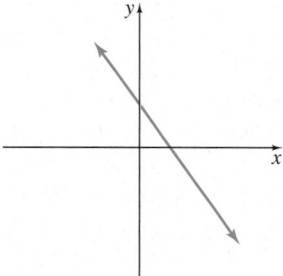

24.

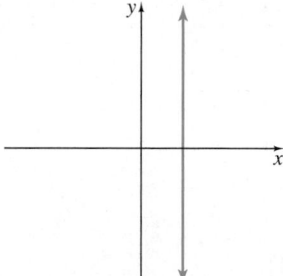

25.

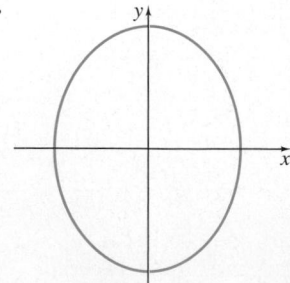

26.

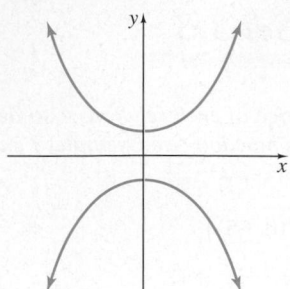

27.

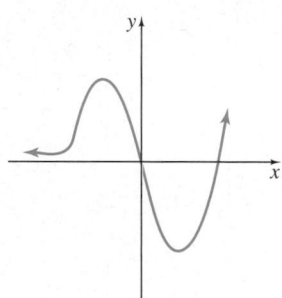

28.

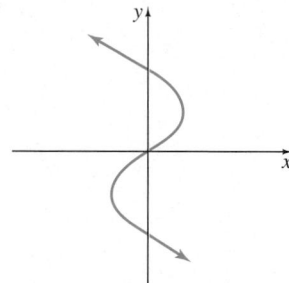

Use the graph in Example 6 to answer Exercises 29–32.

29. Approximate the time of sunrise on September 1 in Indianapolis.

30. Approximate the date(s) when the sun rises in Indianapolis at 7 A.M.

31. Describe the change in sunrise time over the year for Indianapolis.

32. When, in Indianapolis, is the earliest sunrise? What point on the graph does this correspond to?

Find the domain and the range of each relation. Use the vertical line test to determine whether each graph is the graph of a function. See Example 7.

33.

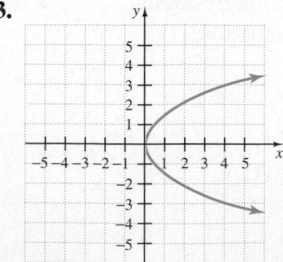

34.

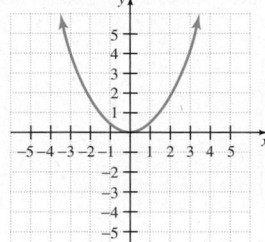

35.

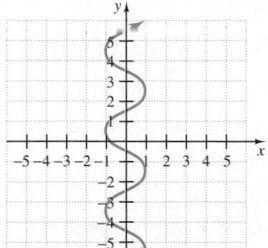

36.

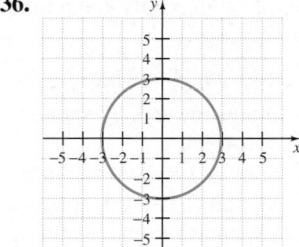

37.

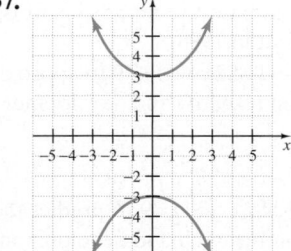

38.

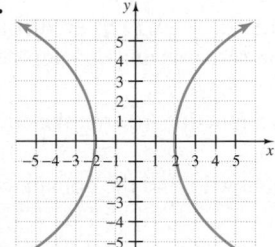

39.

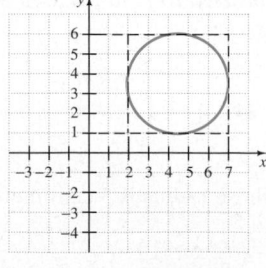

40.

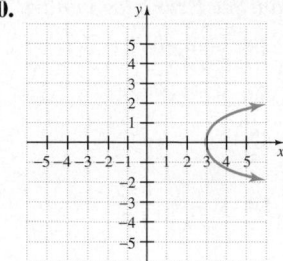

41.

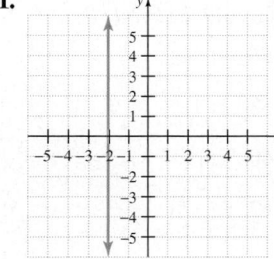

42.

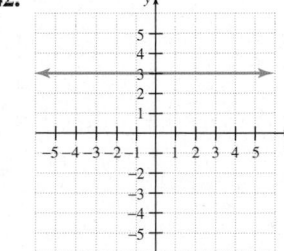

43.

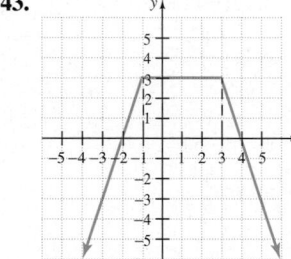

44.

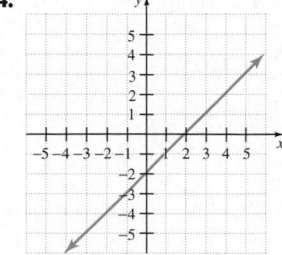

45. In your own words define **(a)** function; **(b)** domain; **(c)** range.

46. Explain the vertical line test and how it is used.

Decide whether each is a function. See Examples 3 and 4.

 47. $y = x + 1$

48. $y = x - 1$

49. $x = 2y^2$

50. $y = x^2$

51. $y - x = 7$

52. $2x - 3y = 9$

53. $y = \dfrac{1}{x}$

54. $y = \dfrac{1}{x - 3}$

If $f(x) = 3x + 3$, $g(x) = 4x^2 - 6x + 3$, and $h(x) = 5x^2 - 7$, find the following. See Example 8.

55. $f(4)$

56. $f(-1)$

57. $h(-3)$

58. $h(0)$

59. $g(2)$

60. $g(1)$

61. $g(0)$

62. $h(-2)$

Given the following functions, find the indicated values. See Example 8.

63. $f(x) = \dfrac{1}{2}x$;

    a. $f(0)$

    b. $f(2)$

    c. $f(-2)$

64. $g(x) = -\dfrac{1}{3}x$;

    a. $g(0)$

    b. $g(-1)$

    c. $g(3)$

65. $g(x) = 2x^2 + 4$;

    a. $g(-11)$

    b. $g(-1)$

    c. $g\left(\dfrac{1}{2}\right)$

66. $h(x) = -x^2$;

    a. $h(-5)$

    b. $h\left(-\dfrac{1}{3}\right)$

    c. $h\left(\dfrac{1}{3}\right)$

67. $f(x) = -5$;

    a. $f(2)$

    b. $f(0)$

    c. $f(606)$

68. $h(x) = 7$;

    a. $h(7)$

    b. $h(542)$

    c. $h\left(-\dfrac{3}{4}\right)$

69. $f(x) = 1.3x^2 - 2.6x + 5.1$ **a.** $f(2)$

    b. $f(-2)$

    c. $f(3.1)$

70. $g(x) = 2.7x^2 + 6.8x - 10.2$

    a. $g(1)$

    b. $g(-5)$

    c. $g(7.2)$

Use the graph of the function below to answer Exercises 71 through 78.

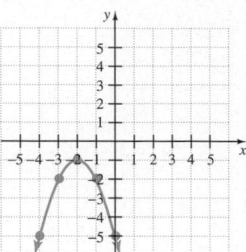

71. If $f(1) = -10$ write the corresponding ordered pair.

72. If $f(-5) = -10$, write the corresponding ordered pair.

73. Find $f(-1)$.   **74.** Find $f(-2)$.

75. Find all values of x such that $f(x) = -5$.

76. Find all values of x such that $f(x) = -2$.

77. What is the greatest number of x-intercepts that a function may have? Explain your answer.

78. What is the greatest number of y-intercepts that a function may have? Explain your answer.

Use the graph in Example 9 to answer the following.

79. a. Use the graph to approximate the money spent on research and development in 1994.

    b. Recall that the function $f(x) = 1.882x + 11.79$ approximates the graph of Example 9. Use this equation to approximate the money spent on research and development in 1994. [*Hint:* Find $f(1)$.]

80. a. Use the graph to approximate the money spent on research and development in 1998.

    b. Use the function $f(x) = 1.882x + 11.79$ to approximate the money spent on research and development in 1998. [*Hint:* Find $f(5)$.]

Solve. See Example 10.

81. Use the function $f(x) = 1.882x + 11.79$ to predict the money that will be spent on research and development in 2005.

82. Use the function $f(x) = 1.882x + 11.79$ to predict the money that will be spent on research and development in 2010.

83. Since $y = x + 7$ describes a function, rewrite the equation using function notation.

84. In your own words, explain how to find the domain of a function given its graph.

The function $A(r) = \pi r^2$ may be used to find the area of a circle if we are given its radius.

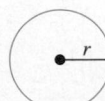

△ **85.** Find the area of a circle whose radius is 5 centimeters. (Do not approximate π.)

△ **86.** Find the area of a circular garden whose radius is 8 feet. (Do not approximate π.)

The function $V(x) = x^3$ may be used to find the volume of a cube if we are given the length x of a side.

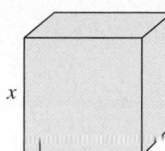

87. Find the volume of a cube whose side is 14 inches.

88. Find the volume of a die whose side is 1.7 centimeters.

Forensic scientists use the following functions to find the height of a woman if they are given the length of her femur bone f or her tibia bone t in centimeters.

$$H(f) = 2.59f + 47.24$$
$$H(t) = 2.72t + 61.28$$

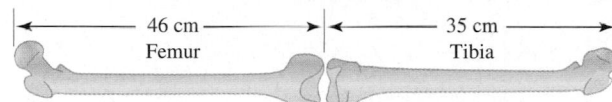

89. Find the height of a woman whose femur measures 46 centimeters.

90. Find the height of a woman whose tibia measures 35 centimeters.

The dosage in milligrams D of Ivermectin, a heartworm preventive, for a dog who weighs x pounds is given by

$$D(x) = \frac{136}{25}x;$$

91. Find the proper dosage for a dog that weighs 30 pounds.

92. Find the proper dosage for a dog that weighs 50 pounds.

93. The per capita consumption (in pounds) of all poultry in the United States is given by the function $C(x) = 1.7x + 88$, where x is the number of years since 1995. (*Source:* Based on actual and estimated data from the Economic Research Service, U.S. Department of Agriculture, 1995–1999)

   a. Find and interpret $C(2)$.

   b. Predict the per capita consumption of all poultry in the United States in 2006.

94. The number of passengers (in millions) aboard airline flights in the United States is given by the function $P(x) = 25.4x + 448.4$, where x is the number of years since 1991. (*Source:* Based on data from the Air Transport Association of America, 1991–1997)

   a. Find and interpret $P(4)$.

   b. Predict the number of airline passengers in 2005.

REVIEW EXERCISES

Complete the given table and use the table to graph the linear equation. See Section 3.2.

95. $x - y = -5$

x	0		1
y		0	

96. $2x + 3y = 10$

x	0		
y		0	2

97. $7x + 4y = 8$

x	0		
y		0	-1

98. $5y - x = -15$

x	0		-2
y		0	

99. $y = 6x$

x	0		-1
y		0	

100. $y = -2x$

x	0		-2
y		0	

△ **101.** Is it possible to find the perimeter of the following geometric figure? If so, find the perimeter.

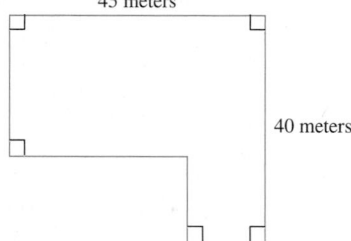

A Look Ahead

Example
If $f(x) = x^2 + 2x + 1$, find the following.
   a. $f(\pi)$ **b.** $f(c)$

Solution

   a. $f(x) = x^2 + 2x + 1$
   $f(\pi) = \pi^2 + 2\pi + 1$
   b. $f(x) = x^2 + 2x + 1$
   $f(c) = (c)^2 + 2(c) + 1$
   $= c^2 + 2c + 1$

Given the following functions, find the indicated values. See the previous example.

102. $f(x) = 2x + 7$;
   a. $f(2)$ **b.** $f(a)$

103. $g(x) = -3x + 12$;
   a. $g(s)$ **b.** $g(r)$

104. $h(x) = x^2 + 7$;
   a. $h(3)$ **b.** $h(a)$

105. $f(x) = x^2 - 12$;
   a. $f(12)$ **b.** $f(a)$

3.4 GRAPHING LINEAR FUNCTIONS

CD-ROM   SSM

SSG   Video

▶ **OBJECTIVES**

1. Graph linear functions.
2. Graph linear functions by finding intercepts.
3. Graph vertical and horizontal lines.

1   In this section, we identify and graph linear functions. By the vertical line test, we know that all linear equations except those whose graphs are vertical lines are functions. For example, we know from Section 3.3 that $y = 2x$ is a linear equation in two variables. Its graph is shown.

x	$y = 2x$
1	2
0	0
-1	-2

Because this graph passes the vertical line test, we know that $y = 2x$ is a function. If we want to emphasize that this equation describes a function, we may write $y = 2x$ as $f(x) = 2x$.

Example 1   Graph $g(x) = 2x + 1$. Compare this graph with the graph of $f(x) = 2x$.

Solution   To graph $g(x) = 2x + 1$, find three ordered pair solutions.

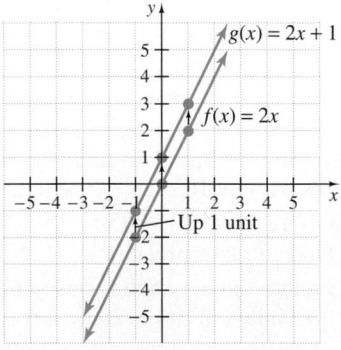

x	$f(x) = 2x$	$g(x) = 2x + 1$
0	0	1
-1	-2	-1
1	2	3

Notice that y-values for the graph of $g(x) = 2x + 1$ are obtained by adding 1 to each y-value of each corresponding point of the graph of $f(x) = 2x$. The graph of $g(x) = 2x + 1$ is the same as the graph of $f(x) = 2x$ shifted upward 1 unit. ▬

In general, a **linear function** is a function that can be written in the form $f(x) = mx + b$. For example, $g(x) = 2x + 1$ is in this form, with $m = 2$ and $b = 1$.

Example 2   Graph the linear functions $f(x) = -3x$ and $g(x) = -3x - 6$ on the same set of axes.

Solution   To graph $f(x)$ and $g(x)$, find ordered pair solutions.

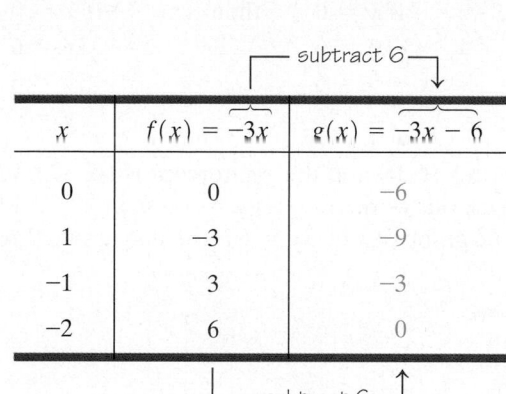

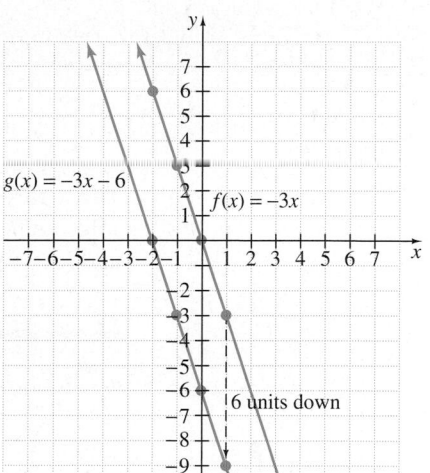

Each y-value for the graph of $g(x) = -3x - 6$ is obtained by subtracting 6 from the y-value of the corresponding point of the graph of $f(x) = -3x$. The graph of $g(x) = -3x - 6$ is the same as the graph of $f(x) = -3x$ shifted down 6 units. ▬

2   Notice that the y-intercept of the graph of $g(x) = -3x - 6$ in the preceding figure is $(0, -6)$. In general, if *a linear function is written in the form $f(x) = mx + b$ or $y = mx + b$, the y-intercept is $(0, b)$.* This is because if x is 0, then $f(x) = mx + b$ becomes $f(0) = m \cdot 0 + b = b$, and we have the ordered pair solution $(0, b)$. We will study this form more in the next section.

Example 3   Find the y-intercept of the graph of each equation.

a. $f(x) = \dfrac{1}{2}x + \dfrac{3}{7}$ 　　　　　　　　　　 **b.** $y = -2.5x - 3.2$

Solution   **a.** The y-intercept of $f(x) = \dfrac{1}{2}x + \dfrac{3}{7}$ is $\left(0, \dfrac{3}{7}\right)$.

b. The y-intercept of $y = -2.5x - 3.2$ is $(0, -3.2)$. ▬

In general, to find the y-intercept of the graph of an equation not in the form $y = mx + b$, let $x = 0$ since any point on the y-axis has an x-coordinate of 0. To find the x-intercept of a line, let $y = 0$ or $f(x) = 0$ since any point on the x-axis has a y-coordinate of 0.

FINDING x- AND y-INTERCEPTS

To find an x-intercept, let $y = 0$ or $f(x) = 0$ and solve for x.
To find a y-intercept, let $x = 0$ and solve for y.

Intercepts are usually easy to find and plot since one coordinate is 0.

Example 4 Graph $x - 3y = 6$ by plotting intercepts.

Solution Let $y = 0$ to find the x-intercept and $x = 0$ to find the y-intercept.

$$
\begin{array}{ll}
\text{If } y = 0 \quad \text{then} & \text{If } x = 0 \quad \text{then} \\
x - 3(0) = 6 & 0 - 3y = 6 \\
x - 0 = 6 & -3y = 6 \\
x = 6 & y = -2
\end{array}
$$

The x-intercept is $(6, 0)$ and the y-intercept is $(0, -2)$. We find a third ordered pair solution to check our work. If we let $y = -1$, then $x = 3$. Plot the points $(6, 0)$, $(0, -2)$, and $(3, -1)$. The graph of $x - 3y = 6$ is the line drawn through these points, as shown.

x	y
6	0
0	-2
3	-1

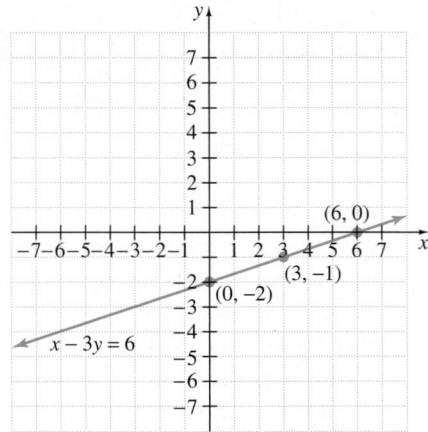

Notice that the equation $x - 3y = 6$ describes a linear function—"linear" because its graph is a line and "function" because the graph passes the vertical line test.

If we want to emphasize that the equation $x - 3y = 6$ from Example 4 describes a function, first solve the equation for y.

$$
\begin{aligned}
x - 3y &= 6 \\
-3y &= -x + 6 \qquad &&\text{Subtract } x \text{ from both sides.} \\
\frac{-3y}{-3} &= \frac{-x}{-3} + \frac{6}{-3} \qquad &&\text{Divide both sides by } -3. \\
y &= \frac{1}{3}x - 2 \qquad &&\text{Simplify.}
\end{aligned}
$$

Next, let $y = f(x)$.

$$
f(x) = \frac{1}{3}x - 2
$$

HELPFUL HINT
Any linear equation that describes a function can be written using function notation. To do so, solve the equation for y and then replace y with $f(x)$, as we did above.

Example 5   Graph $x = -2y$ by plotting intercepts.

Solution   Let $y = 0$ to find the x-intercept and $x = 0$ to find the y-intercept.

If $y = 0$ then	If $x = 0$ then
$x = -2(0)$	$0 = -2y$
$x = 0$	$0 = y$

Ordered pairs $(0, 0)$ $(0, 0)$

Both the x-intercept and y-intercept are $(0, 0)$. This happens when the graph passes through the origin. Since two points are needed to determine a line, we must find at least one more ordered pair that satisfies $x = -2y$. Let $y = -1$ to find a second ordered pair solution and let $y = 1$ as a check point.

If $y = -1$ then	If $y = 1$ then
$x = -2(-1)$	$x = -2(1)$
$x = 2$	$x = -2$

The ordered pairs are $(0, 0)$, $(2, -1)$, and $(-2, 1)$. Plot these points to graph $x = -2y$.

x	y
0	0
2	-1
-2	1

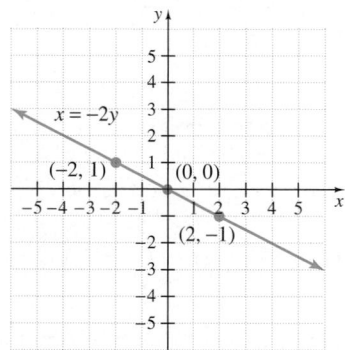

3   The equations $x = c$ and $y = c$, where c is a real number constant, are both linear equations in two variables. Why? Because $x = c$ can be written as $x + 0y = c$ and $y = c$ can be written as $0x + y = c$. We graph these two special linear equations below.

Example 6   Graph $x = 2$.

Solution   The equation $x = 2$ can be written as $x + 0y = 2$. For any y-value chosen, notice that x is 2. No other value for x satisfies $x + 0y = 2$. Any ordered pair whose x-coordinate is 2 is a solution to $x + 0y = 2$ because 2 added to 0 times any value of y is $2 + 0$, or 2. We will use the ordered pairs $(2, 3)$, $(2, 0)$ and $(2, -3)$ to graph $x = 2$.

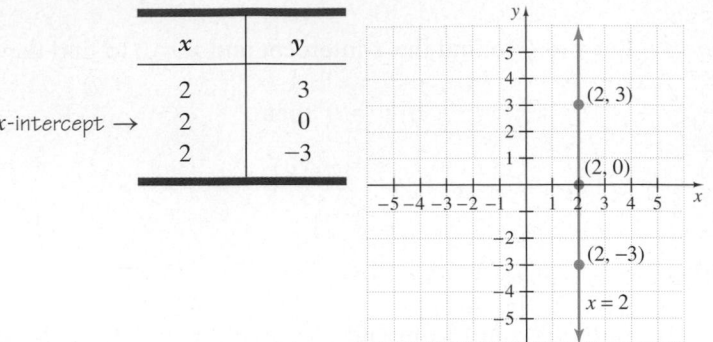

x	y
2	3
x-intercept → 2	0
2	-3

The graph is a vertical line with x-intercept $(2, 0)$. Notice that this graph is not the graph of a function, and it has no y-intercept because x is never 0. ∎

Example 7 Graph $y = -3$.

Solution The equation $y = -3$ can be written as $0x + y = -3$. For any x-value chosen, y is -3. If we choose $4, 0$, and -2 as x-values, the ordered pair solutions are $(4, -3)$, $(0, -3)$, and $(-2, -3)$. We will use these ordered pairs to graph $y = -3$.

x	y
4	-3
0	-3 ← y-intercept
-2	-3

The graph is a horizontal line with y-intercept $(0, -3)$ and no x-intercept. Notice that this graph is the graph of a function. ∎

From Examples 6 and 7, we have the following generalization.

GRAPHING VERTICAL AND HORIZONTAL LINES

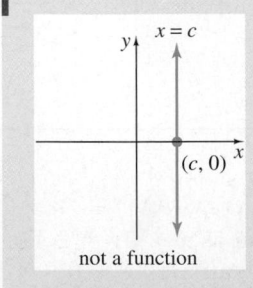

not a function

The graph of $x = c$, where c is a real number, is a vertical line with x-intercept $(c, 0)$.
The graph of $y = c$, where c is a real number, is a horizontal line with y-intercept $(0, c)$.

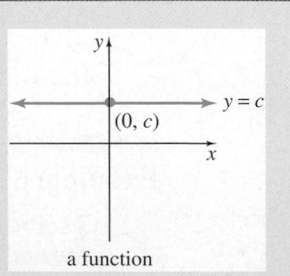

a function

GRAPHING CALCULATOR EXPLORATIONS

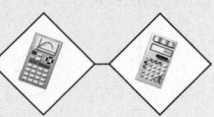

You may have noticed by now that to use the $\boxed{Y=}$ key on a graphing calculator to graph an equation, the equation must be solved for y.

Graph each function by first solving the function for y.

1. $x = 3.5y$ **2.** $-2.7y = x$

3. $5.78x + 2.31y = 10.98$ **4.** $-7.22x + 3.89y = 12.57$

5. $y - |x| = 3.78$ **6.** $3y - 5x^2 = 6x - 4$

7. $y - 5.6x^2 = 7.7x + 1.5$ **8.** $y + 2.6|x| = -3.2$

Exercise Set 3.4

Graph each linear function. See Examples 1 and 2.

1. $f(x) = -2x$ **2.** $f(x) = 2x$

3. $f(x) = -2x + 3$ **4.** $f(x) = 2x + 6$

5. $f(x) = \dfrac{1}{2}x$ **6.** $f(x) = \dfrac{1}{3}x$

7. $f(x) = \dfrac{1}{2}x - 4$ **8.** $f(x) = \dfrac{1}{3}x - 2$

The graph of $f(x) = 5x$ follows. Use this graph to match each linear function with its graph. See Examples 1 through 3.

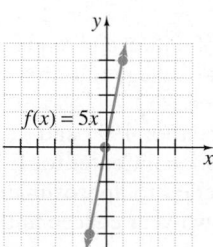

A
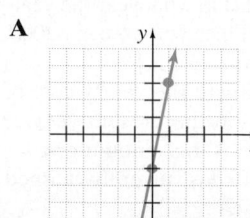

B
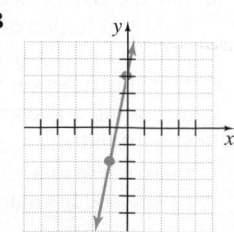

C
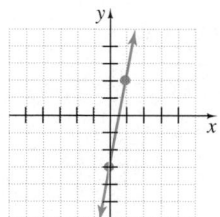

D
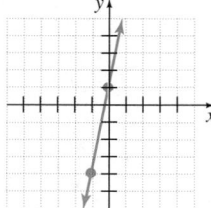

9. $f(x) = 5x - 3$ **10.** $f(x) = 5x - 2$

11. $f(x) = 5x + 1$ **12.** $f(x) = 5x + 3$

Graph each linear function by finding x- and y-intercepts. See Examples 4 and 5.

13. $x - y = 3$ **14.** $x - y = -4$

15. $x = 5y$ **16.** $2x = y$

17. $-x + 2y = 6$ **18.** $x - 2y = -8$

19. $2x - 4y = 8$ **20.** $2x + 3y = 6$

21. In your own words, explain how to find x- and y-intercepts.

22. Explain why it is a good idea to use three points to graph a linear equation.

Graph each linear equation. See Examples 6 and 7.

23. $x = -1$ **24.** $y = 5$

25. $y = 0$ **26.** $x = 0$

27. $y + 7 = 0$ **28.** $x - 3 = 0$

Match each equation with its graph.

A

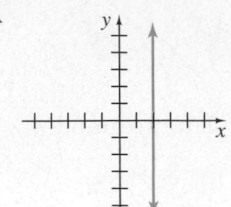

B

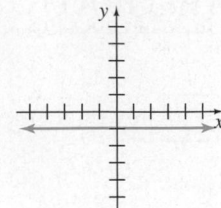

C

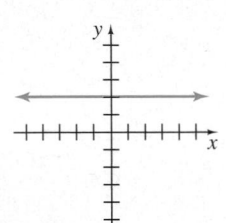

D

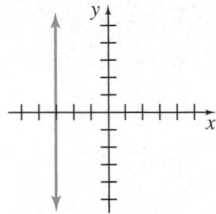

29. $y = 2$

30. $x = -3$

31. $x - 2 = 0$

32. $y + 1 = 0$

33. Discuss whether a vertical line ever has a y-intercept.

34. Discuss whether a horizontal line ever has an x-intercept.

Graph each linear equation.

35. $x + 2y = 8$

36. $x - 3y = 3$

37. $f(x) = \frac{3}{4}x + 2$

38. $f(x) = \frac{4}{3}x + 2$

39. $x = -3$

40. $f(x) = 3$

41. $3x + 5y = 7$

42. $3x - 2y = 5$

43. $f(x) = x$

44. $f(x) = -x$

45. $x + 8y = 8$

46. $x - 3y = 9$

47. $5 = 6x - y$

48. $4 = x - 3y$

49. $-x + 10y = 11$

50. $-x + 9 = -y$

51. $y = 1$

52. $x = 1$

53. $f(x) = \frac{1}{2}x$

54. $f(x) = -2x$

55. $x + 3 = 0$

56. $y - 6 = 0$

57. $f(x) = 4x - \frac{1}{3}$

58. $f(x) = -3x + \frac{3}{4}$

59. $2x + 3y = 6$

60. $4x + y = 5$

Solve.

61. Broyhill Furniture found that it takes 2 hours to manufacture each table for one of its special dining room sets. Each chair takes 3 hours to manufacture. A total of 1500 hours is available to produce tables and chairs of this style. The linear equation that models this situation is $2x + 3y = 1500$, where x represents the number of tables produced and y the number of chairs produced.

    a. Complete the ordered pair solution $(0, \ )$ of this equation. Describe the manufacturing situation this solution corresponds to.

    b. Complete the ordered pair solution $(\ , 0)$ for this equation. Describe the manufacturing situation this solution corresponds to.

    c. If 50 tables are produced, find the greatest number of chairs the company can make.

62. While manufacturing two different camera models, Kodak found that the basic model costs $55 to produce, whereas the deluxe model costs $75. The weekly budget for these two models is limited to $33,000 in production costs. The linear equation that models this situation is $55x + 75y = 33,000$, where x represents the number of basic models and y the number of deluxe models.

    a. Complete the ordered pair solution $(0, \ )$ of this equation. Describe the manufacturing situation this solution corresponds to.

    b. Complete the ordered pair solution $(\ , 0)$ of this equation. Describe the manufacturing situation this solution corresponds to.

    c. If 350 deluxe models are produced, find the greatest number of basic models that can be made in one week.

63. The cost of renting a car for a day is given by the linear function $C(x) = 0.2x + 24$, where $C(x)$ is in dollars and x is the number of miles driven.

    a. Find the cost of driving the car 200 miles.

    b. Graph $C(x) = 0.2x + 24$.

    c. How can you tell from the graph of $C(x)$ that as the number of miles driven increases, the total cost increases also?

64. The cost of renting a piece of machinery is given by the linear function $C(x) = 4x + 10$, where $C(x)$ is in dollars and x is given in hours.

    a. Find the cost of renting the piece of machinery for 8 hours.

    b. Graph $C(x) = 4x + 10$.

    c. How can you tell from the graph of $C(x)$ that as the number of hours increases, the total cost increases also?

65. The yearly cost of tuition and required fees for attending a public two-year college full-time can be estimated by the linear function $f(x) = 72.9x + 785.2$, where x is the number of years after 1990 and $f(x)$ is the total cost. (*Source:* U.S. National Center for Education Statistics)

    a. Use this function to approximate the yearly cost of attending a two-year college in the year 2010. [*Hint:* Find $f(20)$.]

    b. Use the given function to predict in what year the yearly cost of tuition and required fees will exceed $2000. [*Hint:* Let $f(x) = 2000$ and solve for x.]

    c. Use this function to approximate the yearly cost of attending a two-year college in the present year. If you attend a two-year college, is this amount greater than or less than the amount that is currently charged by the college that you attend?

66. The yearly cost of tuition and required fees for attending a public four-year college full-time can be estimated by the linear function $f(x) = 186.1x + 2030$, where x is the number of years after 1990 and $f(x)$ is the total cost in dollars. (*Source:* U.S. National Center for Education Statistics)

   a. Use this function to approximate the yearly cost of attending a four-year college in the year 2010. [*Hint:* Find $f(20)$.]

   b. Use the given function to predict in what year the yearly cost of tuition and required fees will exceed $5000. [*Hint:* Let $f(x) = 5000$ and solve for x.]

   c. Use this function to approximate the yearly cost of attending a four-year college in the present year. If you attend a four-year college, is this amount greater than or less than the amount that is currently charged by the college that you attend?

67. U.S. farm expenses y for livestock feed (in billions of dollars) can be modeled by the linear equation $y = 1.2x + 23.6$, where x represents the number of years after 1995. (*Source:* Based on data from the National Agricultural Statistics Service)

   a. Find the y-intercept of this equation.

   b. What does this y-intercept mean?

68. The number of music cassettes y (in millions) shipped to retailers in the United States can be modeled by the equation $y = -22.5x + 467.0$, where x represents the number of years after 1987. (*Source:* Based on data from the Recording Industry Association of America)

   a. Find the x-intercept of this equation. (Round to the nearest tenth.)

   b. What does this x-intercept mean?

Recall that two lines in the same plane that do not intersect are called **parallel lines.**

△ **69.** Draw a line parallel to the line $x = 5$ that intersects the x-axis at $(1, 0)$. What is the equation of this line?

△ **70.** Draw a line parallel to the line $y = -1$ that intersects the y-axis at $(0, -4)$. What is the equation of this line?

Use a graphing calculator to verify the results of each exercise.

71. Exercise 9   **72.** Exercise 10

73. Exercise 17   **74.** Exercise 18

REVIEW EXERCISES

Simplify. See Sections 1.6 and 1.7.

75. $\dfrac{-6 - 3}{2 - 8}$ **76.** $\dfrac{4 - 5}{-1 - 0}$

77. $\dfrac{-8 - (-2)}{-3 - (-2)}$ **78.** $\dfrac{12 - 3}{10 - 9}$

79. $\dfrac{0 - 6}{5 - 0}$ **80.** $\dfrac{2 - 2}{3 - 5}$

3.5 THE SLOPE OF A LINE

CD-ROM

SSM

SSG

Video

▶ **OBJECTIVES**

1. Find the slope of a line given two points on the line.
2. Solve applications of slope.
3. Find the slope of a line given the equation of a line.
4. Interpret the slope–intercept form in an application.
5. Find the slopes of horizontal and vertical lines.
6. Compare the slopes of parallel and perpendicular lines.

1

You may have noticed by now that different lines often tilt differently. It is very important in many fields to be able to measure and compare the tilt, or **slope**, of lines. For example, a wheelchair ramp with a slope of $\frac{1}{12}$ means that the ramp rises 1 foot for every 12 horizontal feet. A road with a slope or grade of 11% (or $\frac{11}{100}$) means that the road rises 11 feet for every 100 horizontal feet.

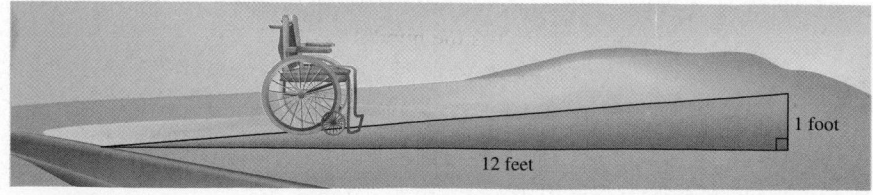
1 foot
12 feet

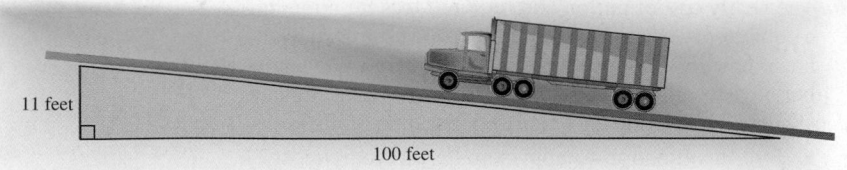

We measure the slope of a line as a ratio of **vertical change** to **horizontal change**. Slope is usually designated by the letter m.

Suppose that we want to measure the slope of the following line.

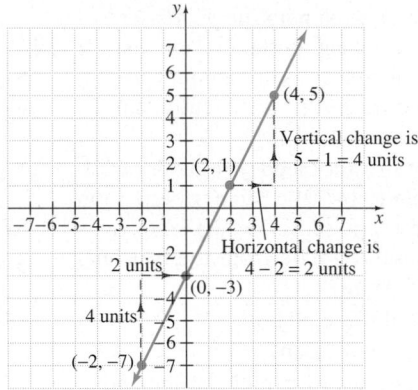

The vertical change between both pairs of points on the line is 4 units per horizontal change of 2 units. Then

$$\text{slope } m = \frac{\text{change in } y \text{ (vertical change)}}{\text{change in } x \text{ (horizontal change)}} = \frac{4}{2} = 2$$

Notice that slope is a rate of change between points. A slope of 2 or $\frac{2}{1}$ means that between pairs of points on the line, the rate of change is a vertical change of 2 units per horizontal change of 1 unit.

Consider the line below, which passes through the points (x_1, y_1) and (x_2, y_2). (The notation x_1 is read "x-sub-one.") The vertical change, or *rise*, between these points is the difference in the y-coordinates: $y_2 - y_1$. The horizontal change, or *run*, between the points is the difference of the x-coordinates: $x_2 - x_1$.

SLOPE OF A LINE

Given a line passing through points (x_1, y_1) and (x_2, y_2) the **slope** m of the line is

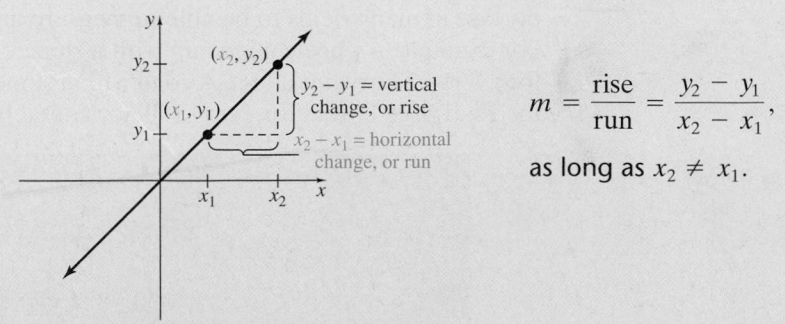

$$m = \frac{\text{rise}}{\text{run}} = \frac{y_2 - y_1}{x_2 - x_1},$$

as long as $x_2 \neq x_1$.

Example 1   Find the slope of the line containing the points $(0, 3)$ and $(2, 5)$. Graph the line.

Solution   We use the slope formula. It does not matter which point we call (x_1, y_1) and which point we call (x_2, y_2). We'll let $(x_1, y_1) = (0, 3)$ and $(x_2, y_2) = (2, 5)$.

$$m = \frac{y_2 - y_1}{x_2 - x_1}$$

$$= \frac{5 - 3}{2 - 0} = \frac{2}{2} = 1$$

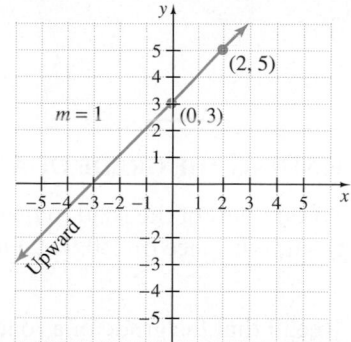

Notice in this example that the slope is positive and that the graph of the line containing $(0, 3)$ and $(2, 5)$ moves upward, or increases, as we go from left to right. ▬

> **HELPFUL HINT**
> When we are trying to find the slope of a line through two given points, it makes no difference which given point is called (x_1, y_1) and which is called (x_2, y_2). Once an x-coordinate is called x_1, however, make sure its corresponding y-coordinate is called y_1.

Example 2   Find the slope of the line containing the points $(5, -4)$ and $(-3, 3)$. Graph the line.

Solution   We use the slope formula, and let $(x_1, y_1) = (5, -4)$ and $(x_2, y_2) = (-3, 3)$.

$$m = \frac{y_2 - y_1}{x_2 - x_1}$$

$$= \frac{3 - (-4)}{-3 - 5} = \frac{7}{-8} = -\frac{7}{8}$$

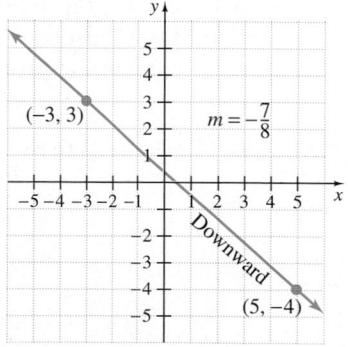

Notice in this example that the slope is negative and that the graph of the line through $(5, -4)$ and $(-3, 3)$ moves downward, or decreases, as we go from left to right. ▬

2   There are many real-world applications of slope. For example, the pitch of a roof, used by builders and architects, is its slope. The pitch of the roof on the next page left is $\frac{7}{10}$ $\left(\frac{\text{rise}}{\text{run}}\right)$. This means that the roof rises vertically 7 feet for every horizontal 10 feet.

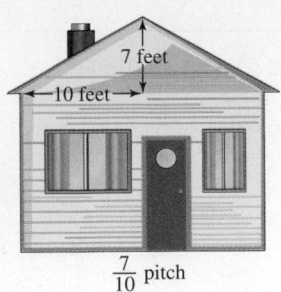

$\frac{7}{10}$ pitch

The grade of a road is its slope written as a percent. A 7% grade, as shown below, means that the road rises (or falls) 7 feet for every horizontal 100 feet. (Recall that $7\% = \frac{7}{100}$.)

$\frac{7}{100} = 7\%$ grade   7 feet   100 feet

Example 3   FINDING THE GRADE OF A ROAD

At one part of the road to the summit of Pikes Peak, the road rises 15 feet for a horizontal distance of 250 feet. Find the grade of the road.

Solution   Recall that the grade of a road is its slope written as a percent.

$$\text{grade} = \frac{\text{rise}}{\text{run}} = \frac{15}{250} = 0.06 = 6\%$$

15 feet
250 feet

The grade is 6%.

Slope can also be interpreted as a rate of change. In other words, slope tells us how fast y is changing with respect to x.

Example 4   FINDING THE SLOPE OF A LINE

The following graph shows the cost y (in cents) of an in-state long-distance telephone call in Massachusetts where x is the length of the call in minutes. Find the slope of the line and attach the proper units for the rate of change.

Solution   Use $(2, 48)$ and $(5, 81)$ to calculate slope.

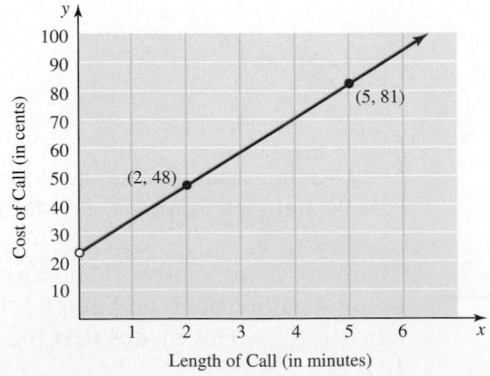

$$m = \frac{81 - 48}{5 - 2} = \frac{33}{3} = \frac{11}{1} \frac{\text{cents}}{\text{minute}}$$

This means that the rate of change of a phone call is 11 cents per 1 minute or the cost of the phone call increases 11 cents per minute. ■

3 As we have seen, the slope of a line is defined by two points on the line. Thus, if we know the equation of a line, we can find its slope.

Example 5 Find the slope of the line whose equation is $f(x) = \frac{2}{3}x + 4$.

Solution Two points are needed on the line defined by $f(x) = \frac{2}{3}x + 4$ or $y = \frac{2}{3}x + 4$ to find its slope. We will use intercepts as our two points.

If $x = 0$, then   If $y = 0$, then

$$y = \frac{2}{3} \cdot 0 + 4 \qquad 0 = \frac{2}{3}x + 4$$

$$y = 4 \qquad\qquad -4 = \frac{2}{3}x \qquad \text{Subtract 4 from both sides.}$$

$$\frac{3}{2}(-4) = \frac{3}{2} \cdot \frac{2}{3}x \qquad \text{Multiply both sides by } \frac{3}{2}.$$

$$-6 = x$$

Use the points $(0, 4)$ and $(-6, 0)$ to find the slope. Let (x_1, y_1) be $(0, 4)$ and (x_2, y_2) be $(-6, 0)$. Then

$$m = \frac{y_2 - y_1}{x_2 - x_1} = \frac{0 - 4}{-6 - 0} = \frac{-4}{-6} = \frac{2}{3}$$ ■

Analyzing the results of Example 5, you may notice a striking pattern:

The slope of $y = \frac{2}{3}x + 4$ is $\frac{2}{3}$, the same as the coefficient of x.

Also, the y-intercept is $(0, 4)$, as expected.
When a linear equation is written in the form $f(x) = mx + b$ or $y = mx + b$, m is the slope of the line and $(0, b)$ is its y-intercept. The form $y = mx + b$ is appropriately called the **slope–intercept form**.

SLOPE–INTERCEPT FORM

When a linear equation in two variables is written in slope–intercept form,

slope y-intercept is $(0, b)$
↓   ↓
$$y = mx + b$$

then m is the slope of the line and $(0, b)$ is the y-intercept of the line.

Example 6   Find the slope and the y-intercept of the line $3x - 4y = 4$.

Solution   We write the equation in slope–intercept form by solving for y.

$$3x - 4y = 4$$

$$-4y = -3x + 4 \qquad \text{Subtract } 3x \text{ from both sides.}$$

$$\frac{-4y}{-4} = \frac{-3x}{-4} + \frac{4}{-4} \qquad \text{Divide both sides by } -4.$$

$$y = \frac{3}{4}x - 1 \qquad \text{Simplify.}$$

The coefficient of x, $\frac{3}{4}$, is the slope, and the y-intercept is $(0, -1)$.

4   Below is the graph of one-day ticket prices at Disney World for the years shown.
Notice that the graph resembles the graph of a line. Recall that businesses often depend on equations that "closely fit" graphs like this one to model the data and predict future trends. By the **least squares** method, the linear function $f(x) = 1.505x + 32.56$ approximates the data shown, where x is the number of years since 1990 and y is the ticket price for that year.

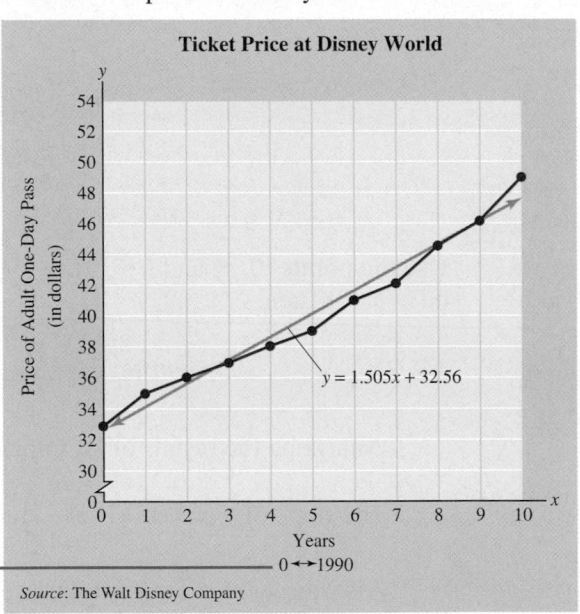

Ticket Price at Disney World

$y = 1.505x + 32.56$

Price of Adult One-Day Pass (in dollars)

Years
$0 \leftrightarrow 1990$

Source: The Walt Disney Company

HELPFUL HINT
The notation $0 \leftrightarrow 1990$ means that the number 0 corresponds to the year 1990, 1 corresponds to the year 1991, and so on.

Example 7   **PREDICTING FUTURE PRICES**

The adult one-day pass price $f(x)$ for Disney World is given by

$$f(x) = 1.505x + 32.56$$

where x is the number of years since 1990

a. Use this equation to predict the ticket price for the year 2004.
b. What does the slope of this equation mean?
c. What does the y-intercept of this equation mean?

Solution **a.** To predict the price of a pass in 2004, we need to find $f(14)$. (Since year 1990 corresponds to $x = 0$, year 2004 corresponds to $x = 14$.)

$$f(x) = 1.505x + 32.56$$
$$f(14) = 1.505(14) + 32.56 \qquad \text{Let } x = 14.$$
$$= 53.63$$

We predict that in the year 2004 the price of an adult one-day pass to Disney World will be about $53.63.

b. The slope of $f(x) = 1.505x + 32.56$ is 1.505. We can think of this number as $\dfrac{\text{rise}}{\text{run}}$ or $\dfrac{1.505}{1}$. This means that the ticket price increases on the average by $1.505 every 1 year.

c. The y-intercept of $y = 1.505 + 32.56$ is $(0, 32.56)$.

↑   ↖
year price

This means that at year $x = 0$ or 1990, the ticket price was about $32.56. ■

---

5   Next we find the slopes of two special types of lines: vertical lines and horizontal lines.

Example 8   Find the slope of the line $x = -5$.

Solution   Recall that the graph of $x = -5$ is a vertical line with x-intercept $(-5, 0)$. To find the slope, we find two ordered pair solutions of $x = -5$. Of course, solutions of $x = -5$ must have an x-value of -5. We will let $(x_1, y_1) = (-5, 0)$ and $(x_2, y_2) = (-5, 4)$. Then

$$m = \frac{y_2 - y_1}{x_2 - x_1}$$

$$= \frac{4 - 0}{-5 - (-5)}$$

$$= \frac{4}{0}$$

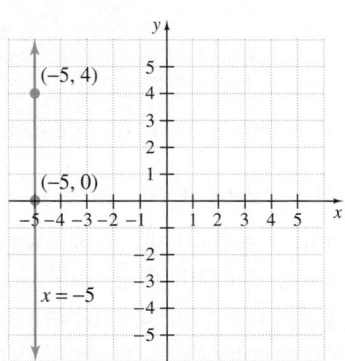

Since $\dfrac{4}{0}$ is undefined, we say that the slope of the vertical line $x = -5$ is undefined. ■

Example 9   Find the slope of the line $y = 2$.

Solution   Recall that the graph of $y = 2$ is a horizontal line with y-intercept $(0, 2)$. To find the slope, we find two points on the line, such as $(0, 2)$ and $(1, 2)$, and use these points to find the slope.

$$m = \frac{2-2}{1-0}$$

$$= \frac{0}{1}$$

$$= 0$$

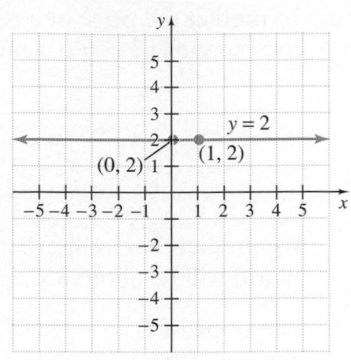

The slope of the horizontal line $y = 2$ is 0.

From the previous two examples, we have the following generalization.

The slope of any vertical line is undefined.
The slope of any horizontal line is 0.

HELPFUL HINT
Slope of 0 and undefined slope are not the same. Vertical lines have undefined slope, whereas horizontal lines have slope of 0.

The following four graphs summarize the overall appearance of lines with positive, negative, zero, or undefined slopes.

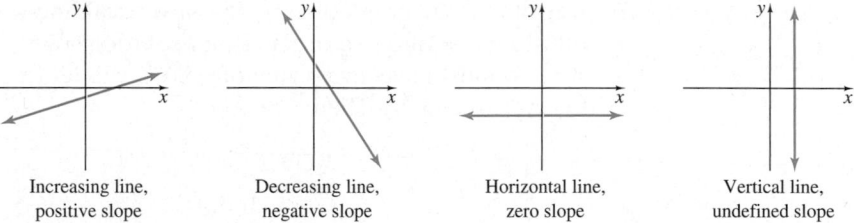

Increasing line,
positive slope

Decreasing line,
negative slope

Horizontal line,
zero slope

Vertical line,
undefined slope

The appearance of a line can give us further information about its slope.

The graphs of $y = \frac{1}{2}x + 1$
and $y = 5x + 1$ are shown to the right. Recall that the graph of $y = \frac{1}{2}x + 1$ has a slope of $\frac{1}{2}$ and that the graph of $y = 5x + 1$ has a slope of 5.

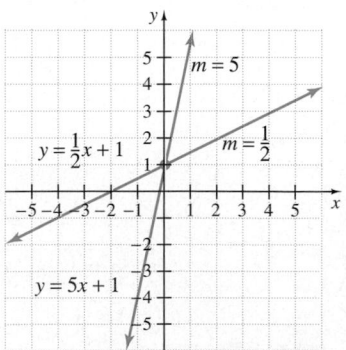

Notice that the line with the slope of 5 is steeper than the line with the slope of $\frac{1}{2}$. This is true in general for positive slopes.

For a line with positive slope m, as m increases, the line becomes steeper.

6   Slopes of lines can help us determine whether lines are parallel. Parallel lines are distinct lines with the same steepness, so it follows that they have the same slope.

PARALLEL LINES

Two nonvertical lines are parallel if they have the same slope and different y-intercepts.

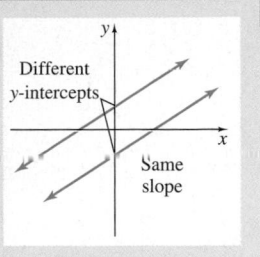

How do the slopes of perpendicular lines compare? (Two lines intersecting at right angles are called **perpendicular lines**.) Suppose that a line has a slope of $\frac{a}{b}$. If the line is rotated 90°, the rise and run are now switched, except that the run is now negative. This means that the new slope is $-\frac{b}{a}$. Notice that

$$\left(\frac{a}{b}\right) \cdot \left(-\frac{b}{a}\right) = -1$$

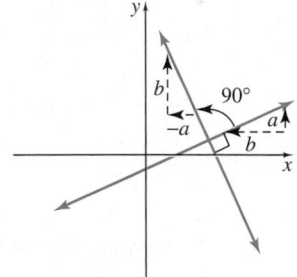

This is how we tell whether two lines are perpendicular.

PERPENDICULAR LINES

Two nonvertical lines are perpendicular if the product of their slopes is −1.

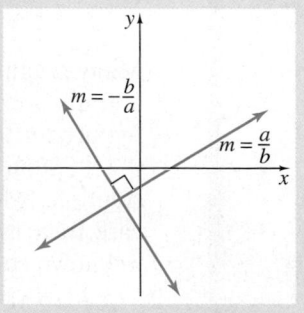

In other words, two nonvertical lines are perpendicular if the slope of one is the negative reciprocal of the slope of the other.

△ **Example 10**   Are the following pairs of lines parallel, perpendicular, or neither?

a. $3x + 7y = 4$
    $6x + 14y = 7$

b. $-x + 3y = 2$
    $2x + 6y = 5$

Solution Find the slope of each line by solving each equation for y.

a. $3x + 7y = 4$ $\qquad\qquad$ $6x + 14y = 7$

$\qquad\quad 7y = -3x + 4$ $\qquad\qquad 14y = -6x + 7$

$\qquad\quad \dfrac{7y}{7} = \dfrac{-3x}{7} + \dfrac{4}{7}$ $\qquad\qquad \dfrac{14y}{14} = \dfrac{-6x}{14} + \dfrac{7}{14}$

$\qquad\quad y = -\dfrac{3}{7}x + \dfrac{4}{7}$ $\qquad\qquad y = -\dfrac{3}{7}x + \dfrac{1}{2}$

$\qquad\qquad\qquad \uparrow \qquad\nwarrow$ $\qquad\qquad\qquad\qquad \uparrow \qquad \uparrow$

$\qquad\qquad$ slope $\quad$ y-intercept $\qquad\qquad$ slope $\quad$ y-intercept

$\qquad\qquad\qquad\qquad\quad \left(0, \dfrac{4}{7}\right)$ $\qquad\qquad\qquad\qquad\quad \left(0, \dfrac{1}{2}\right)$

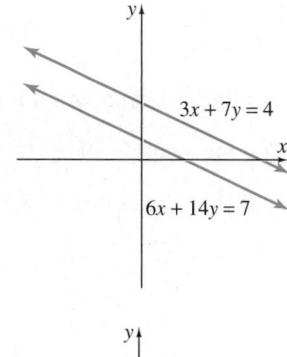

The slopes of both lines are $-\dfrac{3}{7}$.

The y-intercepts are different.

Therefore, the lines are parallel.

b. $-x + 3y = 2$ $\qquad\qquad$ $2x + 6y = 5$

$\qquad\quad 3y = x + 2$ $\qquad\qquad 6y = -2x + 5$

$\qquad\quad \dfrac{3y}{3} = \dfrac{x}{3} + \dfrac{2}{3}$ $\qquad\qquad \dfrac{6y}{6} = \dfrac{-2x}{6} + \dfrac{5}{6}$

$\qquad\quad y = \dfrac{1}{3}x + \dfrac{2}{3}$ $\qquad\qquad y = -\dfrac{1}{3}x + \dfrac{5}{6}$

$\qquad\qquad \uparrow \qquad\qquad \nwarrow$ $\qquad\qquad\quad \uparrow \qquad\qquad \nwarrow$

$\qquad\quad$ slope $\qquad$ y-intercept $\qquad\quad$ slope $\qquad$ y-intercept

$\qquad\qquad\qquad\qquad \left(0, \dfrac{2}{3}\right)$ $\qquad\qquad\qquad\qquad \left(0, \dfrac{5}{6}\right)$

The slopes are not the same and their product is not -1. $\left[\left(\dfrac{1}{3}\right) \cdot \left(-\dfrac{1}{3}\right) = -\dfrac{1}{9}\right]$

Therefore, the lines are neither parallel nor perpendicular.

GRAPHING CALCULATOR EXPLORATIONS

Many graphing calculators have a TRACE feature. This feature allows you to trace along a graph and see the corresponding x- and y-coordinates appear on the screen. Use this feature for the following exercises.

Graph each function and then use the TRACE feature to complete each ordered pair solution. (Many times the tracer will not show an exact x- or y-value asked for. In each case, trace as closely as you can to the given x- or y-coordinate and approximate the other, unknown coordinate to one decimal place.)

1. $y = 2.3x + 6.7$ $\qquad\qquad\qquad$ **2.** $y = -4.8x + 2.9$
$\quad x = 5.1, y = ?$ $\qquad\qquad\qquad\qquad\; x = -1.8, y = ?$

3. $y = -5.9x - 1.6$ $\qquad\qquad\quad\;$ **4.** $y = 0.4x - 8.6$
$\quad x = ?, y = 7.2$ $\qquad\qquad\qquad\qquad\; x = ?, y = -4.4$

5. $y = x^2 + 5.2x - 3.3$ $\qquad\qquad$ **6.** $y = 5x^2 - 6.2x - 8.3$
$\quad x = 2.3, y = ?$ $\qquad\qquad\qquad\qquad\; x = 3.2, y = ?$
$\quad x = ?, y = 36$ $\qquad\qquad\qquad\qquad\;\; x = ?, y = 12$
(There will be two answers here.) $\qquad$ (There will be two answers here.)

SPOTLIGHT ON DECISION MAKING

Suppose you are the manager of an apartment complex. You have just notified residents of a rent increase. Some residents think that the increase may be unjustified and out of line with recent increases. A group of concerned residents asks you to hold an open meeting to answer questions about the increase. You are preparing a set of overheads to use during the meeting to show the history of rent increases at the apartment complex. Which overhead would you use and why?

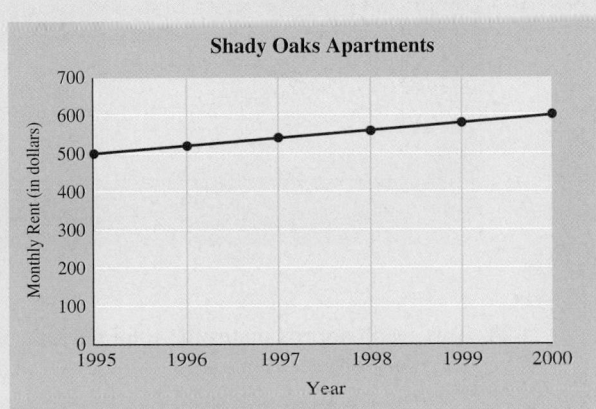

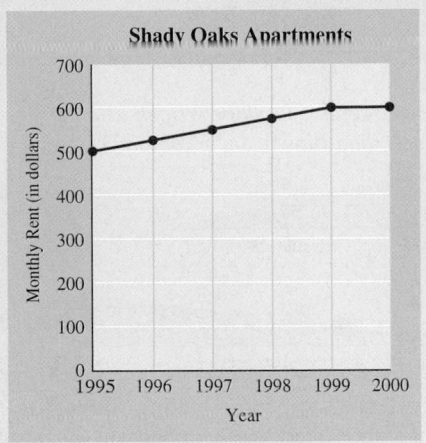

MENTAL MATH

Decide whether a line with the given slope slants upward, downward, horizontally, or vertically from left to right.

1. $m = \dfrac{7}{6}$

2. $m = -3$

3. $m = 0$

4. m is undefined

Exercise Set 3.5

Find the slope of the line that goes through the given points. See Examples 1 and 2.

1. $(3, 2), (8, 11)$

2. $(1, 6), (7, 11)$

3. $(3, 1), (1, 8)$

4. $(2, 9), (6, 4)$

5. $(-2, 8), (4, 3)$

6. $(3, 7), (-2, 11)$

7. $(-2, -6), (4, -4)$

8. $(-3, -4), (-1, 6)$

9. $(-3, -1), (-12, 11)$

10. $(3, -1), (-6, 5)$

11. $(-2, 5), (3, 5)$

12. $(4, 2), (4, 0)$

13. $(-1, 1), (-1, -5)$

14. $(-2, -5), (3, -5)$

15. $(0, 6), (-3, 0)$

16. $(5, 2), (0, 5)$

17. $(-1, 2), (-3, 4)$

18. $(3, -2), (-1, -6)$

The pitch of a roof is its slope. Find the pitch of each roof shown. See Example 3.

19.

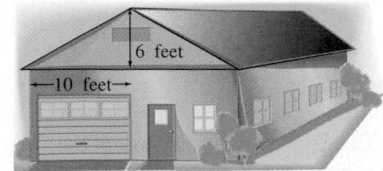

6 feet

←10 feet→

20.

The grade of a road is its slope written as a percent. Find the grade of each road shown. See Example 3.

21.

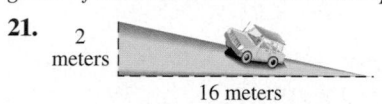

2 meters

16 meters

22.

16 feet

100 feet

23. One of Japan's superconducting "bullet" trains is researched and tested at the Yamanashi Maglev Test Line near Otsuki City. The steepest section of the track has a rise of 2580 for a horizontal distance of 6450 meters. What is the grade for this section of track? (*Source:* Japan Railways Central Co.)

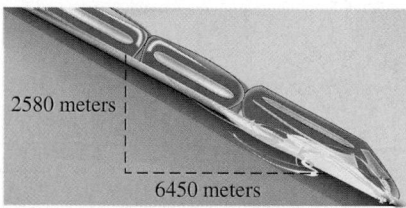

2580 meters

6450 meters

24. The steepest street is Baldwin Street in Dunedin, New Zealand. It has a maximum rise of 10 meters for a horizontal distance of 12.66 meters. Find the grade for this section of road. Round to the nearest whole percent. (*Source: The Guinness Book of Records*)

25. Professional plumbers suggest that a sewer pipe should rise 0.25 inch for every horizontal foot. Find the recommended slope for a sewer pipe. Round to the nearest hundredth.

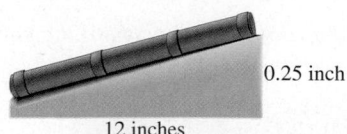

0.25 inch

12 inches

26. According to federal regulations, a wheel chair ramp should rise no more than 1 foot for a horizontal distance of 12 feet. Write the slope as a grade. Round to the nearest tenth of a percent.

Find the slope of each line and write the slope as a rate of change. Don't forget to attach the proper units. See Example 4.

27. This graph approximates the number of U.S. internet users *y* (in millions) for year *x*.

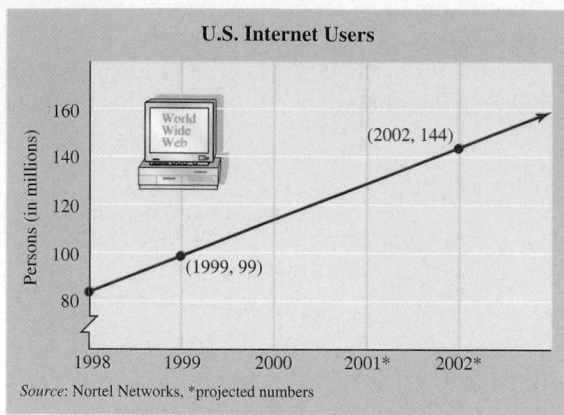

U.S. Internet Users

Source: Nortel Networks, *projected numbers

28. This graph approximates the total number of cosmetic surgeons for year *x*.

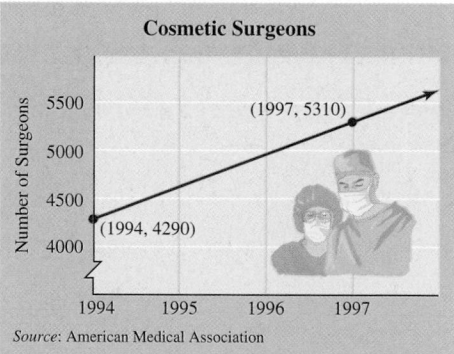

Cosmetic Surgeons

Source: American Medical Association

29. The graph below shows the total cost *y* (in dollars) of owning and operating a compact car where *x* is the number of miles driven.

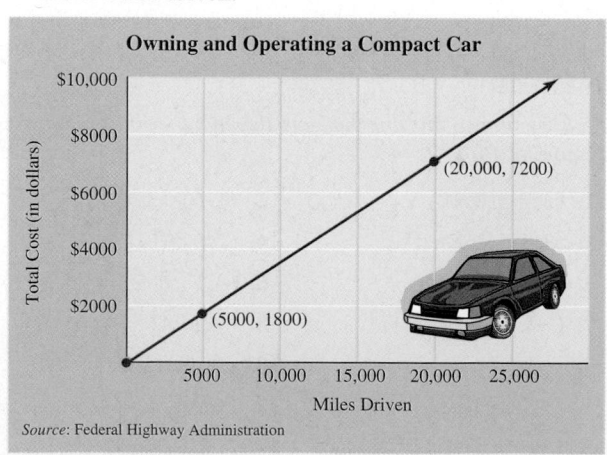

Owning and Operating a Compact Car

Source: Federal Highway Administration

30. The graph below shows the total cost y (in dollars) of owning and operating a full-size pickup truck, where x is the number of miles driven.

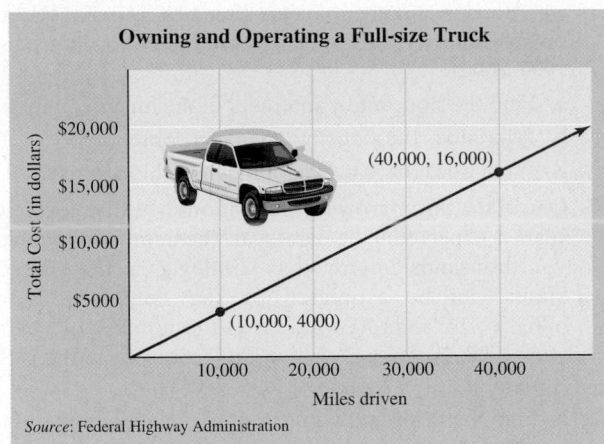

Owning and Operating a Full-size Truck

Source: Federal Highway Administration

Two lines are graphed on each set of axes. Decide whether l_1 or l_2 has the greater slope.

31.

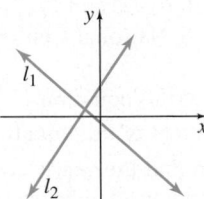

32.

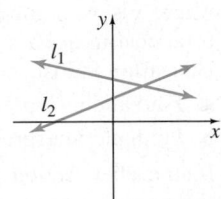

33.

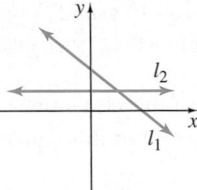

34.

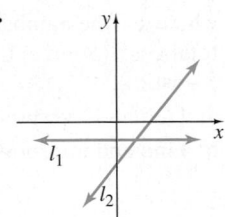

35.

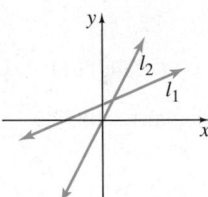

36.

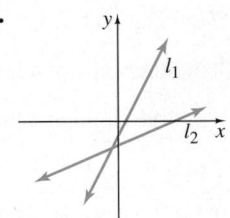

37. Each line in the graph in the next column has negative slope.

a. Find the slope of each line.

b. Use the result of Part **a** to fill in the blank. For lines with negative slopes, the steeper line has the ____ (greater/lesser) slope.

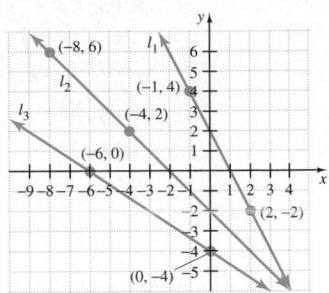

Find the slope and the y-intercept of each line. See Examples 5 and 6.

38. $f(x) = 5x - 2$

39. $f(x) = -2x + 6$

40. $2x + y = 7$

41. $-5x + y = 10$

42. $2x - 3y = 10$

43. $-3x - 4y = 6$

44. $f(x) = \dfrac{1}{2}x$

45. $f(x) = -\dfrac{1}{4}x$

Match each graph with its equation.

A

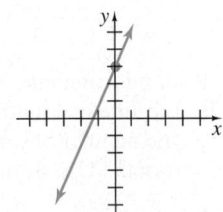

B

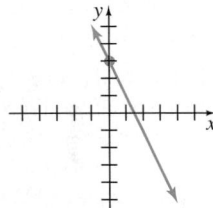

C

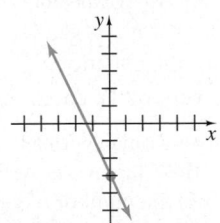

D

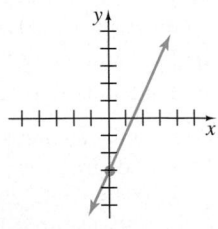

46. $f(x) = 2x + 3$

47. $f(x) = 2x - 3$

48. $f(x) = -2x + 3$

49. $f(x) = -2x - 3$

Find the slope of each line. See Examples 8 and 9.

50. $x = 1$

51. $y = -2$

52. $y = -3$

53. $x = 4$

54. $x + 2 = 0$

55. $y - 7 = 0$

56. Explain how merely looking at a line can tell us whether its slope is negative, positive, undefined, or zero.

57. Explain why the graph of $y = b$ is a horizontal line.

Find the slope and the y-intercept of each line.

58. $f(x) = -x + 5$

59. $f(x) = x + 2$

60. $-6x + 5y = 30$

61. $4x - 7y = 28$

62. $3x + 9 = y$

63. $2y - 7 = x$

64. $y = 4$

65. $x = 7$

66. $f(x) = 7x$

67. $f(x) = \dfrac{1}{7}x$

68. $6 + y = 0$

69. $x - 7 = 0$

70. $2 - x = 3$

71. $2y + 4 = -7$

Determine whether the lines are parallel, perpendicular, or neither. See Example 10.

72. $f(x) = -3x + 6$
$g(x) = 3x + 5$

73. $f(x) = 5x - 6$
$g(x) = 5x + 2$

74. $-4x + 2y = 5$
$2x - y = 7$

75. $2x - y = -10$
$2x + 4y = 2$

76. $-2x + 3y = 1$
$3x + 2y = 12$

77. $x + 4y = 7$
$2x - 5y = 0$

78. Explain whether two lines, both with positive slopes, can be perpendicular.

79. Explain why it is reasonable that nonvertical parallel lines have the same slope.

Solve. See Example 7.

80. The annual average income y of an American man with an associate's degree is given by the linear equation $y = 1431.5x + 31,775.2$, where x is the number of years after 1992. (*Source:* Based on data from the U.S. Bureau of the Census, 1992–1996)

a. Find the average income of an American man with an associate's degree in 1996.

b. Find and interpret the slope of the equation.

c. Find and interpret the y-intercept of the equation.

81. The annual income of an American woman with a bachelor's degree is given by the linear equation $y = 1054.7x + 23,285.9$, where x is the number of years after 1991. (*Source:* Based on data from the U.S. Bureau of the Census, 1991–1996)

a. Find the average income of an American woman with a bachelor's degree in 1996.

b. Find and interpret the slope of the equation.

c. Find and interpret the y-intercept of the equation.

82. One of the top ten occupations in terms of job growth in the next few years is expected to be home health aide. The number of people y in thousands employed as home

health aides in the United States can be estimated by the linear equation

$$378x - 10y = -4950,$$

where x is the number of years after 1996. (*Source:* Based on projections from the U.S. Bureau of Labor Statistics, 1996–2006)

a. Find the slope and y-intercept of the linear equation.

b. What does the slope mean in this context?

c. What does the y-intercept mean in this context?

83. One of the faster growing occupations over the next few years is expected to be paralegal. The number of people y in thousands employed as paralegals in the United States can be estimated by the linear equation $-76x + 10y = 1130$, where x is the number of years after 1996. (*Source:* Based on projections from the U.S. Bureau of Labor Statistics, 1996–2006)

a. Find the slope and y-intercept of the linear equation.

b. What does the slope mean in this context?

c. What does the y-intercept mean in this context?

84. In an earlier section, it was given that the yearly cost of tuition and required fees for attending a public four-year college full-time can be estimated by the linear function

$$f(x) = 186.1x + 2030$$

where x is the number of years after 1990 and $f(x)$ is the total cost in dollars. (*Source:* U.S. National Center for Education Statistics)

a. Find and interpret the slope of this equation.

b. Find and interpret the y-intercept of this equation.

85. If an earlier section, it was given that the yearly cost of tuition and required fees for attending a public two-year college full-time can be estimated by the linear function

$$f(x) = 72.9x + 785.2$$

where x is the number of years after 1990 and $f(x)$ is the total cost. (*Source:* U.S. National Center for Education Statistics)

a. Find and interpret the slope of this equation.

b. Find and interpret the y-intercept of this equation.

Solve.

86. Find the slope of a line parallel to the line
$$f(x) = -\frac{7}{2}x - 6.$$

87. Find the slope of a line parallel to the line $f(x) = x$.

88. Find the slope of a line perpendicular to the line
$$f(x) = -\frac{7}{2}x - 6.$$

89. Find the slope of a line perpendicular to the line $f(x) = x$.

△ **90.** Find the slope of a line parallel to the line $5x - 2y = 6$.

△ **91.** Find the slope of a line parallel to the line
$-3x + 4y = 10$.

△ **92.** Find the slope of a line perpendicular to the line
$5x - 2y = 6$.

93. The following graph shows the altitude of a seagull in flight over a time period of 30 seconds.

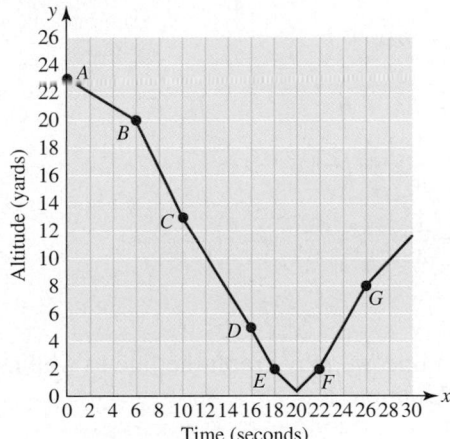

a. Find the coordinates of point B.

b. Find the coordinates of point C.

c. Find the rate of change of altitude between points B and C. (Recall that the rate of change between points is the slope between points. This rate of change will be in yards per second.)

d. Find the rate of change of altitude (in yards per second) between points F and G.

94. Professional plumbers suggest that a sewer pipe should rise 0.25 inch for every foot. Find the recommended slope as a fraction for a sewer pipe. (*Source: Rules of Thumb* by Tom Parker, 1983, Houghton Mifflin Company)

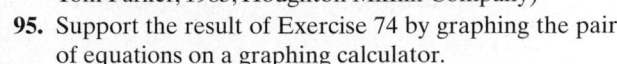 **95.** Support the result of Exercise 74 by graphing the pair of equations on a graphing calculator.

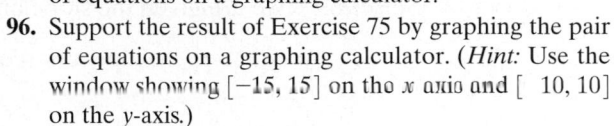 **96.** Support the result of Exercise 75 by graphing the pair of equations on a graphing calculator. (*Hint:* Use the window showing $[-15, 15]$ on the x axis and $[-10, 10]$ on the y-axis.)

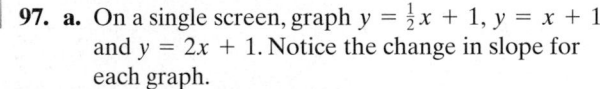 **97. a.** On a single screen, graph $y = \frac{1}{2}x + 1$, $y = x + 1$ and $y = 2x + 1$. Notice the change in slope for each graph.

b. On a single screen, graph $y = -\frac{1}{2}x + 1$, $y = -x + 1$ and $y = -2x + 1$. Notice the change in slope for each graph.

c. Determine whether the following statement is true or false for slope m of a given line. As $|m|$ becomes greater, the line becomes steeper.

REVIEW EXERCISES

Simplify and solve for y. See Section 2.5.

98. $y - 2 = 5(x + 6)$ **99.** $y - 0 = -3[x - (-10)]$

100. $y - (-1) = 2(x - 0)$ **101.** $y - 9 = -8[x - (-4)]$

3.6 EQUATIONS OF LINES

CD-ROM   SSM

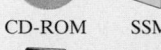

SSG   Video

▶ **OBJECTIVES**

1. Use the slope–intercept form to write the equation of a line.
2. Graph a line using its slope and y-intercept.
3. Use the point–slope form to write the equation of a line.
4. Write equations of vertical and horizontal lines.
5. Find equations of parallel and perpendicular lines.

1

In the last section, we learned that the slope–intercept form of a linear equation is $y = mx + b$. When an equation is written in this form, the slope of the line is the same as the coefficient m of x. Also, the y-intercept of the line is the same as the constant term b. For example, the slope of the line defined by $y = 2x + 3$ is, 2, and its y-intercept is 3.

We may also use the slope–intercept form to write the equation of a line given its slope and y-intercept.

Example 1   Write an equation of the line with y-intercept $(0, -3)$ and slope of $\frac{1}{4}$.

Solution   We are given the slope and the y-intercept. Let $m = \frac{1}{4}$ and $b = -3$, and write the equation in slope–intercept form, $y = mx + b$.

$$y = mx + b$$
$$y = \frac{1}{4}x + (-3) \quad \text{Let } m = \frac{1}{4} \text{ and } b = -3.$$
$$y = \frac{1}{4}x - 3 \quad \text{Simplify.}$$

2   Given the slope and y-intercept of a line, we may graph the line as well as write its equation. Let's graph the line from Example 1.

Example 2   Graph $y = \frac{1}{4}x - 3$.

Solution   Recall that the slope of the graph of $y = \frac{1}{4}x - 3$ is $\frac{1}{4}$ and the y-intercept is $(0, -3)$. To graph the line, we first plot the y-intercept $(0, -3)$. To find another point on the line, we recall that slope is $\frac{\text{rise}}{\text{run}} = \frac{1}{4}$. Another point may then be plotted by starting at $(0, -3)$, rising 1 unit up, and then running 4 units to the right. We are now at the point $(4, -2)$. The graph is the line through these two points.

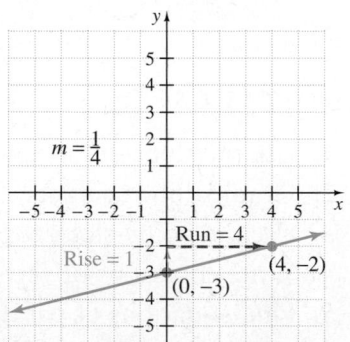

Example 3   Graph $2x + 3y = 12$.

Solution   First, we solve the equation for y to write it in slope–intercept form. In slope–intercept form, the equation is $y = -\frac{2}{3}x + 4$. Next we plot the y-intercept $(0, 4)$. To find another point on the line, we use the slope $-\dfrac{2}{3}$, which can be written as $\dfrac{\text{rise}}{\text{run}} = \dfrac{-2}{3}$. We start at $(0, 4)$ and move down 2 units since the numerator of the slope is -2; then we move 3 units to the right since the denominator of the slope is 3. We arrive at the point $(3, 2)$. The line through these points is the graph.

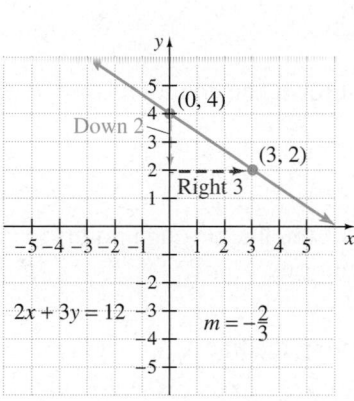
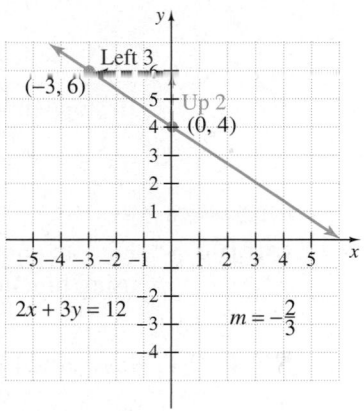

The slope $\dfrac{-2}{3}$ can also be written as $\dfrac{2}{-3}$, so to find another point in Example 3 we could start at $(0, 4)$ and move up 2 units and then 3 units to the left. We would stop at the point $(-3, 6)$. The line through $(-3, 6)$ and $(0, 4)$ is the same line as shown previously through $(3, 2)$ and $(0, 4)$. ∎

3   When the slope of a line and a point on the line are known, the equation of the line can also be found. To do this, use the slope formula to write the slope of a line that passes through points (x_1, y_1) and (x, y). We have

$$m = \frac{y - y_1}{x - x_1}$$

Multiply both sides of this equation by $x - x_1$ to obtain

$$y - y_1 = m(x - x_1)$$

This form is called the **point–slope form** of the equation of a line.

POINT–SLOPE FORM OF THE EQUATION OF A LINE

The point–slope form of the equation of a line is $y - y_1 = m(x - x_1)$, where m is the slope of the line and (x_1, y_1) is a point on the line.

Example 4   Find an equation of the line with slope -3 containing the point $(1, -5)$. Write the equation in slope–intercept form $y = mx + b$.

Solution Because we know the slope and a point of the line, we use the point–slope form with $m = -3$ and $(x_1, y_1) = (1, -5)$.

$$y - y_1 = m(x - x_1) \qquad \text{Point–slope form.}$$

$$y - (-5) = -3(x - 1) \qquad \text{Let } m = -3 \text{ and } (x_1, y_1) = (1, -5).$$

$$y + 5 = -3x + 3 \qquad \text{Apply the distributive property.}$$

$$y = -3x - 2 \qquad \text{Write in slope–intercept form.}$$

In slope–intercept form, the equation is $y = -3x - 2$.

Example 5 Find an equation of the line through points $(4, 0)$ and $(-4, -5)$. Write the equation using function notation.

Solution First, find the slope of the line.

$$m = \frac{-5 - 0}{-4 - 4} = \frac{-5}{-8} = \frac{5}{8}$$

Next, make use of the point–slope form. Replace (x_1, y_1) by either $(4, 0)$ or $(-4, -5)$ in the point–slope equation. We will choose the point $(4, 0)$. The line through $(4, 0)$ with slope $\frac{5}{8}$ is

$$y - y_1 = m(x - x_1) \qquad \text{Point–slope form.}$$

$$y - 0 = \frac{5}{8}(x - 4) \qquad \text{Let } m = \frac{5}{8} \text{ and } (x_1, y_1) = (4, 0).$$

$$8y = 5(x - 4) \qquad \text{Multiply both sides by 8.}$$

$$8y = 5x - 20 \qquad \text{Apply the distributive property.}$$

To write the equation using function notation, we solve for y.

$$8y = 5x - 20$$

$$y = \frac{5}{8}x - \frac{20}{8} \qquad \text{Divide both sides by 8.}$$

$$f(x) = \frac{5}{8}x - \frac{5}{2} \qquad \text{Write using function notation.}$$

The point–slope form of an equation is very useful for solving real-world problems.

Example 6   **PREDICTING SALES**

Southern Star Realty is an established real estate company that has enjoyed constant growth in sales since 1990. In 1992 the company sold 200 houses, and in 1997 the company sold 275 houses. Use these figures to predict the number of houses this company will sell in the year 2006.

Solution   1. UNDERSTAND. Read and reread the problem. Then let

x = the number of years after 1990 and

y = the number of houses sold in the year corresponding to x.

The information provided then gives the ordered pairs $(2, 200)$ and $(7, 275)$. To better visualize the sales of Southern Star Realty, we graph the linear equation that passes through the points $(2, 200)$ and $(7, 275)$.

 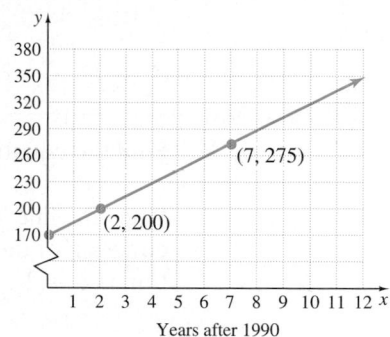

2. TRANSLATE. We write a linear equation that passes through the points $(2, 200)$ and $(7, 275)$. To do so, we first find the slope of the line.

$$m = \frac{275 - 200}{7 - 2} = \frac{75}{5} = 15$$

Then, using the point–slope form to write the equation, we have

$$y - y_1 = m(x - x_1)$$

$y - 200 = 15(x - 2)$ Let $m = 15$ and $(x_1, y_1) = (2, 200)$.

$y - 200 = 15x - 30$ Multiply.

$y = 15x + 170$ Add 200 to both sides.

3. SOLVE. To predict the number of houses sold in the year 2006, we use $y = 15x + 170$ and complete the ordered pair $(16, \quad)$, since $2006 - 1990 = 16$.

$y = 15(16) + 170$ Let $x = 16$.

$y = 410$

4. INTERPRET.

Check: Verify that the point $(16, 410)$ is a point on the line graphed in step 1.

State: Southern Star Realty should expect to sell 410 houses in the year 2006.

4   A few special types of linear equations are linear equations whose graphs are vertical and horizontal lines.

Example 7   Find the equation of the horizontal line containing the point $(2, 3)$.

Solution   Recall that a horizontal line has an equation of the form $y = c$. Since the line contains the point $(2, 3)$, the equation is $y = 3$.

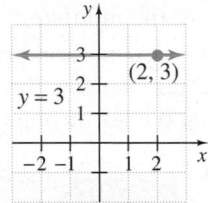

Example 8   Find the equation of the line containing the point $(2, 3)$ with undefined slope.

Solution   Since the line has undefined slope, the line must be vertical. A vertical line has an equation of the form $x = c$, and since the line contains the point $(2, 3)$, the equation is $x = 2$.

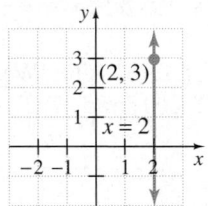

5   Next, we find equations of parallel and perpendicular lines.

△ **Example 9**   Find an equation of the line containing the point $(4, 4)$ and parallel to the line $2x + 3y = -6$. Write the equation in standard form.

Solution   Because the line we want to find is *parallel* to the line $2x + 3y = -6$, the two lines must have equal slopes. Find the slope of $2x + 3y = -6$ by writing it in the form $y = mx + b$.

$$2x + 3y = -6$$

$$3y = -2x - 6 \qquad \text{Subtract } 2x \text{ from both sides.}$$

$$y = \frac{-2x}{3} - \frac{6}{3} \qquad \text{Divide by 3.}$$

$$y = -\frac{2}{3}x - 2 \qquad \text{Write in slope–intercept form.}$$

The slope of this line is $-\dfrac{2}{3}$. Thus, a line parallel to this line will also have a slope of $-\dfrac{2}{3}$. The equation we are asked to find describes a line containing the point $(4, 4)$ with a slope of $-\dfrac{2}{3}$. We use the point–slope form.

$$y - y_1 = m(x - x_1)$$

$$y - 4 = -\frac{2}{3}(x - 4) \qquad \text{Let } m = -\frac{2}{3}, x_1 = 4, \text{ and } y_1 = 4.$$

$$3(y - 4) = -2(x - 4) \qquad \text{Multiply both sides by 3.}$$

$$3y - 12 = -2x + 8 \qquad \text{Apply the distributive property.}$$

$$2x + 3y = 20 \qquad \text{Write in standard form.} \quad \blacksquare$$

HELPFUL HINT
Multiply both sides of the equation $2x + 3y = 20$ by -1, and it becomes $-2x - 3y = -20$. Both equations are in standard form, and their graphs are the same line.

△**Example 10**   Write a function that describes the line containing the point $(4, 4)$ and is perpendicular to the line $2x + 3y = -6$.

Solution   In the previous example, we found that the slope of the line $2x + 3y = -6$ is $-\dfrac{2}{3}$. A line perpendicular to this line will have a slope that is the negative reciprocal of $-\dfrac{2}{3}$, or $\dfrac{3}{2}$. From the point–slope equation, we have

$$y - y_1 = m(x - x_1)$$

$$y - 4 = \frac{3}{2}(x - 4) \qquad \text{Let } x_1 = 4, y_1 = 4 \text{ and } m = \frac{3}{2}.$$

$$2(y - 4) = 3(x - 4) \qquad \text{Multiply both sides by 2.}$$

$$2y - 8 = 3x - 12 \qquad \text{Apply the distributive property.}$$

$$2y = 3x - 4 \qquad \text{Add 8 to both sides.}$$

$$y = \frac{3}{2}x - 2 \qquad \text{Divide both sides by 2.}$$

$$f(x) = \frac{3}{2}x - 2 \qquad \text{Write using function notation.} \quad \blacksquare$$

FORMS OF LINEAR EQUATIONS

$Ax + By = C$	**Standard form** of a linear equation
	A and B are not both 0.
$y = mx + b$	**Slope–intercept form** of a linear equation
	The slope is m, and the y-intercept is $(0, b)$.
$y - y_1 = m(x - x_1)$	**Point–slope form** of a linear equation
	The slope is m, and (x_1, y_1) is a point on the line.
$y = c$	**Horizontal line**
	The slope is 0, and the y-intercept is $(0, c)$.
$x = c$	**Vertical line**
	The slope is undefined and the x-intercept is $(c, 0)$.

PARALLEL AND PERPENDICULAR LINES

Nonvertical parallel lines have the same slope. The product of the slopes of two nonvertical perpendicular lines is -1.

SPOTLIGHT ON DECISION MAKING

Suppose you are a public health official. In 1993, the International Task Force for Disease Eradication (ITFDE) identified mumps as one of six infectious diseases that could probably be eradicated worldwide with current technology. The ITFDE defined "eradication" as reducing the incidence of a disease to zero. Does the graph of reported mumps cases in the United States support the possibility of U.S. mumps eradication? Explain.

   Suppose U.S. officials would like to see mumps eradicated by 2010. If this goal does not currently seem possible, your department will increase eradication efforts with the launch of a new public awareness campaign. Will the new public awareness campaign be necessary? (*Hint:* Use the data for the years 1996 and 1997 to help you decide.)

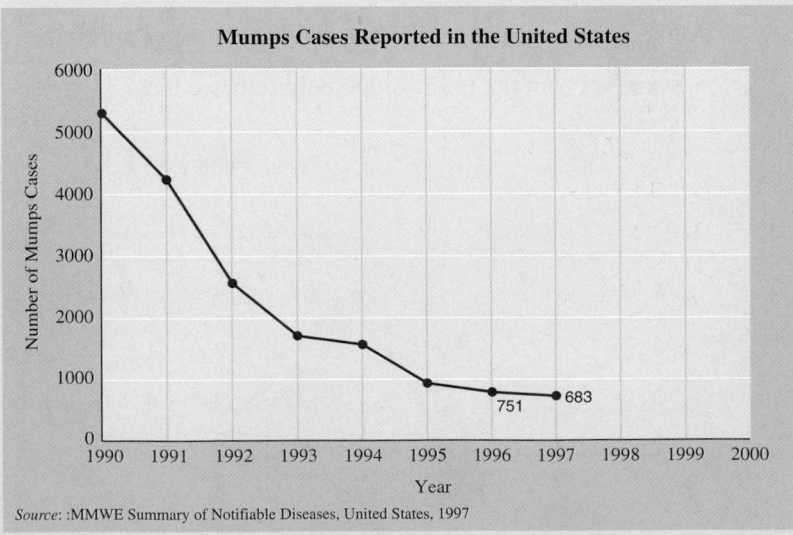

Mumps Cases Reported in the United States

Source: :MMWE Summary of Notifiable Diseases, United States, 1997

MENTAL MATH

State the slope and the y-intercept of each line with the given equation.

1. $y = -4x + 12$

2. $y = \frac{2}{3}x - \frac{7}{2}$

3. $y = 5x$

4. $y = -x$

5. $y = \frac{1}{2}x + 6$

6. $y = -\frac{2}{3}x + 5$

Decide whether the lines are parallel, perpendicular, or neither.

7. $y = 12x + 6$
$y = 12x - 2$

8. $y = -5x + 8$
$y = -5x - 8$

9. $y = -9x + 3$
$y = \frac{3}{2}x - 7$

10. $y = 2x - 12$
$y = \frac{1}{2}x - 6$

Exercise Set 3.6

Use the slope–intercept form of the linear equation to write the equation of each line with the given slope and y-intercept. See Example 1.

1. Slope -1; y-intercept $(0, 1)$

2. Slope $\frac{1}{2}$; y-intercept $(0, -6)$

3. Slope 2; y-intercept $\left(0, \frac{3}{4}\right)$

4. Slope -3; y-intercept $\left(0, -\frac{1}{5}\right)$

5. Slope $\frac{2}{7}$; y-intercept $(0, 0)$

6. Slope $-\frac{4}{5}$; y-intercept $(0, 0)$

Graph each linear equation. See Examples 2 and 3.

7. $y = 5x$

8. $y = 2x + 12$

9. $x + y = 7$

10. $3x + y = 9$

11. $-3x + 2y = 3$

12. $-2x + 5y = -16$

Find an equation of the line with the given slope and containing the given point. Write the equation in slope–intercept form. See Example 4.

13. Slope 3; through $(1, 2)$

14. Slope 4; through $(5, 1)$

15. Slope -2; through $(1, -3)$

16. Slope -4; through $(2, -4)$

17. Slope $\frac{1}{2}$; through $(-6, 2)$

18. Slope $\frac{2}{3}$; through $(-9, 4)$

19. Slope $-\frac{9}{10}$; through $(-3, 0)$

20. Slope $-\frac{1}{5}$; through $(4, -6)$

Find an equation of each line graphed. Write the equation in standard form.

21.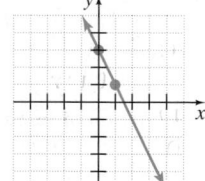

22.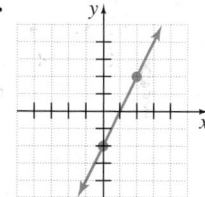

23.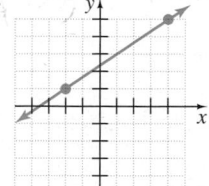

24.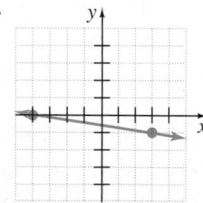

Find an equation of the line passing through the given points. Use function notation to write the equation. See Example 5.

25. $(2, 0), (4, 6)$

26. $(3, 0), (7, 8)$

27. $(-2, 5), (-6, 13)$

28. $(7, -4), (2, 6)$

29. $(-2, -4), (-4, -3)$

30. $(-9, -2), (-3, 10)$

31. $(-3, -8), (-6, -9)$

32. $(8, -3), (4, -8)$

33. Describe how to check to see if the graph of $2x - 4y = 7$ passes through the points $(1.4, -1.05)$ and $(0, -1.75)$. Then follow your directions and check these points.

Use the graph of the following function $f(x)$ to find each value.

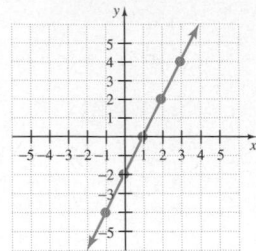

34. $f(1)$ **35.** $f(0)$
36. $f(-1)$ **37.** $f(2)$
38. Find x such that $f(x) = 4$.
39. Find x such that $f(x) = -6$.

Write an equation of each line. See Examples 7 and 8.
40. Vertical; through $(2, 6)$
41. Slope 0; through $(-2, -4)$
42. Horizontal; through $(-3, 1)$
43. Vertical; through $(4, 7)$
44. Undefined slope; through $(0, 5)$
45. Horizontal; through $(0, 5)$
△ **46.** Answer the following true or false. A vertical line is always perpendicular to a horizontal line.

Find an equation of each line. Write the equation using function notation. See Examples 9 and 10.
△ **47.** Through $(3, 8)$; parallel to $f(x) = 4x - 2$
△ **48.** Through $(1, 5)$; parallel to $f(x) = 3x - 4$
49. Through $(2, -5)$; perpendicular to $3y = x - 6$
△ **50.** Through $(-4, 8)$; perpendicular to $2x - 3y = 1$
△ **51.** Through $(-2, -3)$; parallel to $3x + 2y = 5$
△ **52.** Through $(-2, -3)$; perpendicular to $3x + 2y = 5$

Find the equation of each line. Write the equation in standard form unless indicated otherwise.
53. Slope 2; through $(-2, 3)$
54. Slope 3; through $(-4, 2)$
55. Through $(1, 6)$ and $(5, 2)$; use function notation.
56. Through $(2, 9)$ and $(8, 6)$
57. With slope $-\dfrac{1}{2}$; y-intercept 11
58. With slope -4; y-intercept $\dfrac{2}{9}$; use function notation.
59. Through $(-7, -4)$ and $(0, -6)$
60. Through $(2, -8)$ and $(-4, -3)$
61. Slope $-\dfrac{4}{3}$; through $(-5, 0)$
62. Slope $-\dfrac{3}{5}$; through $(4, -1)$
63. Vertical line; through $(-2, -10)$
64. Horizontal line; through $(1, 0)$
△ **65.** Through $(6, -2)$; parallel to the line $2x + 4y = 9$
△ **66.** Through $(8, -3)$; parallel to the line $6x + 2y = 5$
67. Slope 0; through $(-9, 12)$
68. Undefined slope; through $(10, -8)$
△ **69.** Through $(6, 1)$; parallel to the line $8x - y = 9$
△ **70.** Through $(3, 5)$; perpendicular to the line $2x - y = 8$
△ **71.** Through $(5, -6)$; perpendicular to $y = 9$

△ **72.** Through $(-3, -5)$; parallel to $y = 9$
73. Through $(2, -8)$ and $(-6, -5)$; use function notation.
74. Through $(-4, -2)$ and $(-6, 5)$; use function notation.

Solve. Assume that each exercise describes a linear relationship. See Example 6.
75. In 1998 there were 4760 electric-powered vehicles in use in the United States. In 1996 only 3280 electric vehicles were being used. (*Source:* U.S. Energy Information Administration)
   a. Write an equation describing the relationship between time and number of electric-powered vehicles. Use ordered pairs of the form (years past 1996, number of vehicles).
   b. Use this equation to predict the number of electric-powered vehicles in use in 2005.

76. In 1990 there were 403 thousand eating establishments in the United States. In 1996, there were 457 thousand eating establishments.
   a. Write an equation describing the relationship between time and number of eating establishments. Use ordered pairs of the form (years past 1990, number of eating establishments in thousands).
   b. Use this equation to predict the number of eating establishments in 2006.

77. Del Monte Fruit Company recently released a new applesauce. By the end of its first year, profits on this product amounted to $30,000. The anticipated profit for the end of the fourth year is $66,000. The ratio of change in time to change in profit is constant. Let x be years and P be profit.
   a. Write a linear function $P(x)$ that expresses profit as a function of time.
   b. Use this function to predict the company's profit at the end of the seventh year.
   c. Predict when the profit should reach $126,000.
78. The value of a computer bought in 1996 depreciates, or decreases, as time passes. Two years after the computer was bought, it was worth $2600; 4 years after it was bought, it was worth $1000.
   a. If this relationship between number of years past 1996 and value of computer is linear, write an equation describing this relationship. [Use ordered pairs of the form (years past 1996, value of computer).]
   b. Use this equation to estimate the value of the computer in the year 2001.

79. In 1994, the median price of an existing home in the United States was $109,900. In 1998, the median price of an existing home was $128,400. Let y be the median price of an existing home in the year x, where $x = 0$ represents 1994. (*Source:* National Association of REALTORS®)

a. Write a linear equation that models the median existing home price in terms of the year x. [*Hint:* The line must pass through the points $(0, 109{,}900)$ and $(4, 128{,}400)$]

b. Use this equation to predict the median existing home price for the year 2008.

80. The number of births (in thousands) in the United States in 1997 was 3895. The number of births (in thousands) in the United States in 1991 was 4111. Let y be the number of births (in thousands) in the year x, where $x = 0$ represents 1991. (*Source:* National Center for Health Statistics)

a. Write a linear equation that models the number of births (in thousands) in terms of the year x. (See hint for Exercise 79. a.)

b. Use this equation to predict the number of births in the United States for the year 2010.

81. The number of people employed in the United States as medical assistants was 225 thousand in 1996. By the year 2006, this number is expected to rise to 391 thousand. Let y be the number of medical assistants (in thousands) employed in the United States in the year x, where $x = 0$ represents 1996. (*Source:* Bureau of Labor Statistics)

a. Write a linear equation that models the number of people (in thousands) employed as medical assistants in the year x. (See hint for Exercise 79. a.)

b. Use this equation to estimate the number of people who will be employed as medical assistants in the year 2004.

82. The number of people employed in the United States as systems analysts was 506 thousand in 1996. By the year 2006, this number is expected to rise to 1025 thousand. Let y be the number of systems analysts (in thousands) employed in the United States in the year x, where $x = 0$ represents 1996. (*Source:* Bureau of Labor Statistics)

a. Write a linear equation that models the number of people (in thousands) employed as systems analysts in the year x. (See hint for Exercise 79. a.)

b. Use this equation to estimate the number of people who will be employed as systems analysts in the year 2002.

Use a graphing calculator with a TRACE feature to see the results of each exercise.

83. Exercise 55; graph the function and verify that it passes through $(1, 6)$ and $(5, 2)$.

84. Exercise 56; graph the equation and verify that it passes through $(2, 9)$ and $(8, 6)$.

85. Exercise 61; graph the equation. See that it has a negative slope and passes through $(-5, 0)$.

86. Exercise 62; graph the equation. See that it has a negative slope and passes through $(4, -1)$.

REVIEW EXERCISES

Solve and graph the solution. See Section 2.7.

87. $2x - 7 \le 21$

88. $-3x + 1 > 0$

89. $5(x - 2) > 3(r - 1)$

90. $2(x + 1) \le -x + 10$

91. $\frac{x}{2} + \frac{1}{4} < \frac{1}{8}$

92. $\frac{x}{5} - \frac{3}{10} \ge \frac{x}{2} - 1$

A Look Ahead

Example
Find an equation of the perpendicular bisector of the line segment whose endpoints are $(2, 6)$ and $(0, -2)$.

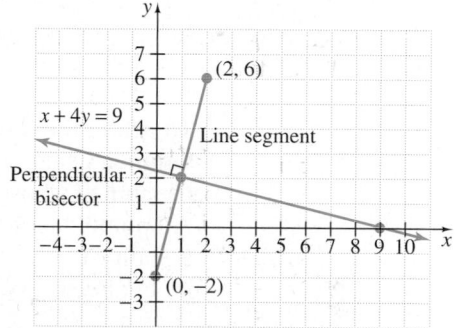

Solution
A perpendicular bisector is a line that contains the midpoint of the given segment and is perpendicular to the segment.

Step 1: The midpoint of the segment with endpoints $(2, 6)$ and $(0, -2)$ is $(1, 2)$.

Step 2: The slope of the segment containing points $(2, 6)$ and $(0, -2)$ is 4.

Step 3: A line perpendicular to this line segment will have slope of $-\frac{1}{4}$.

Step 4: The equation of the line through the midpoint $(1, 2)$ with a slope of $-\frac{1}{4}$ will be the equation of the perpendicular bisector. This equation in standard form is $x + 4y = 9$.

Find an equation of the perpendicular bisector of the line segment whose endpoints are given. See the previous example.

93. $(3, -1); (-5, 1)$

94. $(-6, -3); (-8, -1)$

95. $(-2, 6); (-22, -4)$

96. $(5, 8); (7, 2)$

97. $(2, 3); (-4, 7)$

98. $(-6, 8); (-4, -2)$

3.7 GRAPHING LINEAR INEQUALITIES

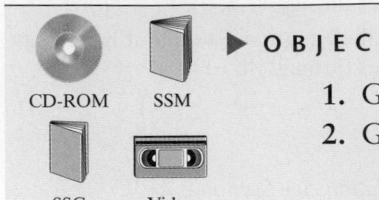

▶ **O B J E C T I V E S**

1. Graph linear inequalities.
2. Graph the intersection or union of two linear inequalities.

CD-ROM SSM

SSG Video

1

Recall that the graph of a linear equation in two variables is the graph of all ordered pairs that satisfy the equation, and we determined that the graph is a line. Here we graph **linear inequalities** in two variables; that is, we graph all the ordered pairs that satisfy the inequality.

If the equal sign in a linear equation in two variables is replaced with an inequality symbol, the result is a linear inequality in two variables.

Examples of Linear Inequalities in Two Variables

$$3x + 5y \geq 6 \qquad 2x - 4y < -3$$

$$4x > 2 \qquad\qquad y \leq 5$$

To graph the linear inequality $x + y < 3$, for example, we first graph the related **boundary** equation $x + y = 3$. The resulting boundary line contains all ordered pairs the sum of whose coordinates is 3. This line separates the plane into two **half-planes**. All points "above" the boundary line $x + y = 3$ have coordinates that satisfy the inequality $x + y > 3$, and all points "below" the line have coordinates that satisfy the inequality $x + y < 3$.

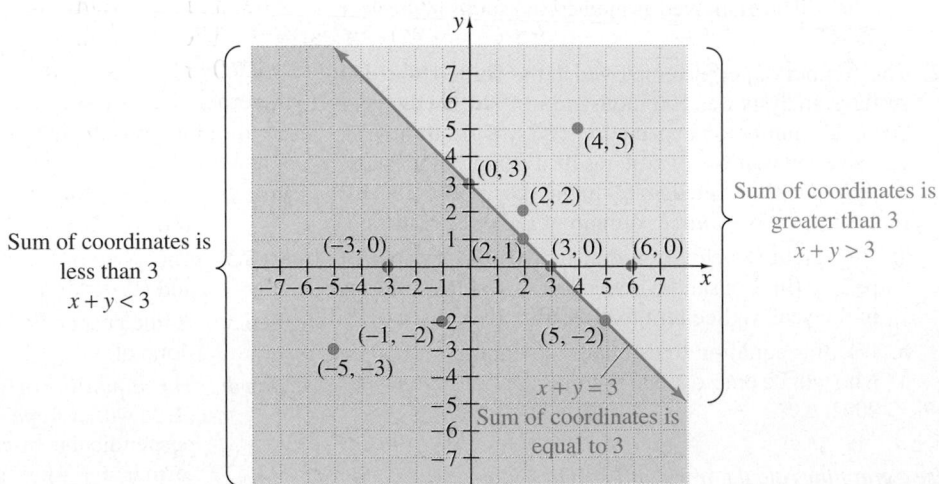

The graph, or **solution region**, for $x + y < 3$, then, is the half-plane below the boundary line and is shown shaded on the next page. The boundary line is shown dashed since it is not a part of the solution region. These ordered pairs on this line satisfy $x + y = 3$ and not $x + y < 3$.

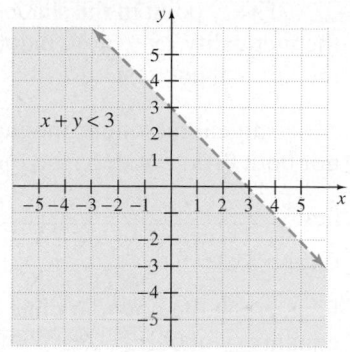

The following steps may be used to graph linear inequalities in two variables.

GRAPHING A LINEAR INEQUALITY IN TWO VARIABLES

Step 1: Graph the boundary line found by replacing the inequality sign with an equal sign. If the inequality sign is $<$ or $>$, graph a dashed line indicating that points on the line are not solutions of the inequality. If the inequality sign is $\leq$ or $\geq$, graph a solid line indicating that points on the line are solutions of the inequality.

Step 2: Choose a **test point not on the boundary line** and substitute the coordinates of this test point into the **original inequality.**

Step 3: If a true statement is obtained in Step 2, shade the half-plane that contains the test point. If a false statement is obtained, shade the half-plane that does not contain the test point.

Example 1 Graph $2x - y < 6$.

Solution First, the boundary line for this inequality is the graph of $2x - y = 6$. Graph a dashed boundary line because the inequality symbol is $<$. Next, choose a test point on either side of the boundary line. The point $(0, 0)$ is not on the boundary line, so we use this point. Replacing x with 0 and y with 0 in the *original inequality* $2x - y < 6$ leads to the following:

$$2x - y < 6$$
$$2(0) - 0 < 6 \qquad \text{Let } x = 0 \text{ and } y = 0.$$
$$0 < 6 \qquad \text{True.}$$

Because $(0, 0)$ satisfies the inequality, so does every point on the same side of the boundary line as $(0, 0)$. Shade the half-plane that contains $(0, 0)$. The half-plane graph of the inequality is shown.

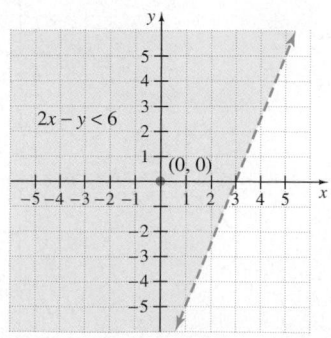

Every point in the shaded half-plane satisfies the original inequality. Notice that the inequality $2x - y < 6$ does not describe a function since its graph does not pass the vertical line test.

In general, linear inequalities of the form $Ax + By \le C$, when A and B are not both 0, do not describe functions.

Example 2 Graph $3x \ge y$.

Solution First, graph the boundary line $3x = y$. Graph a solid boundary line because the inequality symbol is $\ge$. Test a point not on the boundary line to determine which half-plane contains points that satisfy the inequality. We choose $(0, 1)$ as our test point.

$$3x \ge y$$
$$3(0) \ge 1 \quad \text{Let } x = 0 \text{ and } y = 1.$$
$$0 \ge 1 \quad \text{False.}$$

This point does not satisfy the inequality, so the correct half-plane is on the opposite side of the boundary line from $(0, 1)$. The graph of $3x \ge y$ is the boundary line together with the shaded region shown.

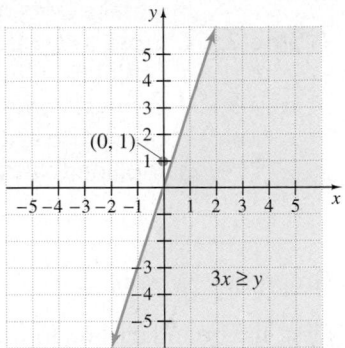

2 The intersection and the union of linear inequalities can also be graphed, as shown in the next two examples.

Example 3 Graph the intersection of $x \ge 1$ and $y \ge 2x - 1$.

Solution Graph each inequality. The intersection of the two graphs is all points common to both regions, as shown by the heaviest shading in the third graph.

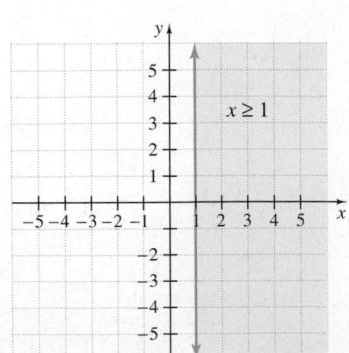
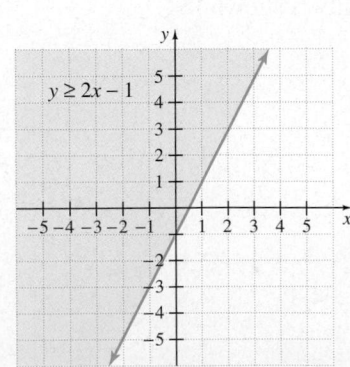
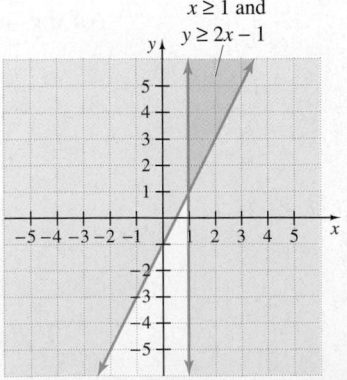

Example 4 Graph the union of $x + \dfrac{1}{2}y \geq -4$ or $y \leq -2$.

Solution Graph each inequality. The union of the two inequalities is both shaded regions, including the solid boundary lines shown in the third graph.

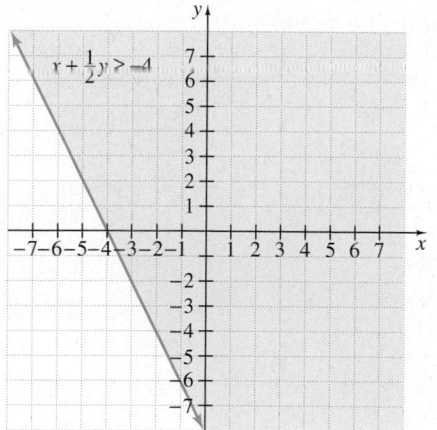
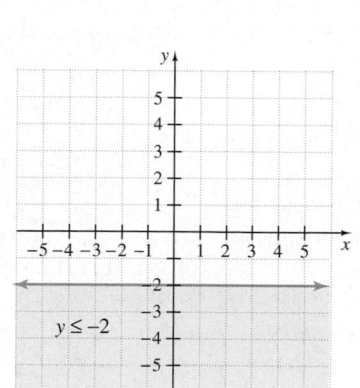
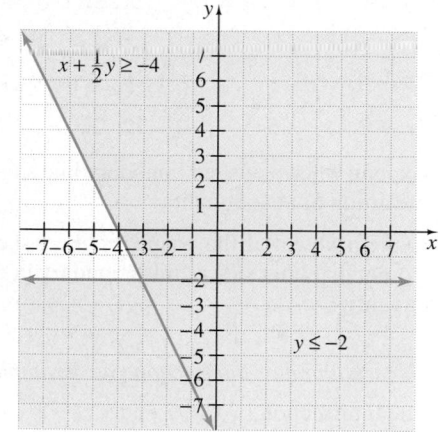

SPOTLIGHT ON DECISION MAKING

Suppose you are a customer service representative for a mail-order medical supply company that sells support stockings. A customer, whose weight is 160 pounds and whose height is 5 feet 9 inches, places an order for support stockings and has asked your assistance in selecting the correct size. What size would you recommend that this customer order? Explain.

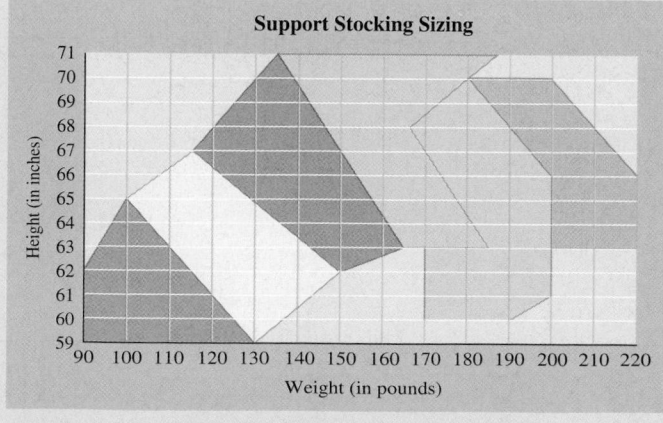

Support Stocking Sizes:

- Small
- Medium
- Tall
- Extra Tall
- Large
- Extra Large

If weight and height fall on a boundary line, order next larger size.

Exercise Set 3.7

Graph each inequality. See Examples 1 and 2.

1. $x < 2$ **2.** $x > -3$

3. $x - y \geq 7$ **4.** $3x + y \leq 1$

 5. $3x + y > 6$ **6.** $2x + y > 2$

7. $y \leq -2x$ **8.** $y \leq 3x$

9. $2x + 4y \geq 8$ **10.** $2x + 6y \leq 12$

11. $5x + 3y > -15$ **12.** $2x + 5y < -20$

13. Explain when a dashed boundary line should be used in the graph of an inequality.

14. Explain why, after the boundary line is sketched, we test a point on either side of this boundary in the original inequality.

Graph each union or intersection. See Examples 3 and 4.

15. The intersection of $x \geq 3$ and $y \leq -2$

16. The union of $x \geq 3$ or $y \leq -2$

17. The union of $x \leq -2$ or $y \geq 4$

18. The intersection of $x \leq -2$ and $y \geq 4$

19. The intersection of $x - y < 3$ and $x > 4$

20. The intersection of $2x > y$ and $y > x + 2$

21. The union of $x + y \leq 3$ or $x - y \geq 5$

22. The union of $x - y \leq 3$ or $x + y > -1$

Graph each inequality.

23. $y \geq -2$ **24.** $y \leq 4$

25. $x - 6y < 12$ **26.** $x - 4y < 8$

27. $x > 5$ **28.** $y \geq -2$

29. $-2x + y \leq 4$ **30.** $-3x + y \leq 9$

31. $x - 3y < 0$ **32.** $x + 2y > 0$

33. $3x - 2y \leq 12$ **34.** $2x - 3y \leq 9$

35. The union of $x - y \geq 2$ or $y < 5$

36. The union of $x - y < 3$ or $x > 4$

37. The intersection of $x + y \leq 1$ and $y \leq -1$

38. The intersection of $y \geq x$ and $2x - 4y \geq 6$

39. The union of $2x + y > 4$ or $x \geq 1$

40. The union of $3x + y < 9$ or $y \leq 2$

41. The intersection of $x \geq -2$ and $x \leq 1$

42. The intersection of $x \geq -4$ and $x \leq 3$

43. The union of $x + y \leq 0$ or $3x - 6y \geq 12$

44. The intersection of $x + y \leq 0$ and $3x - 6y \geq 12$

45. The intersection of $2x - y > 3$ and $x \geq 0$

46. The union of $2x - y > 3$ or $x \geq 0$

Match each inequality with its graph.

A 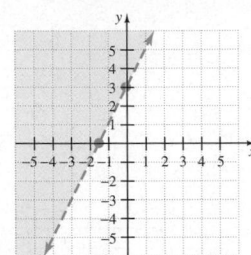   B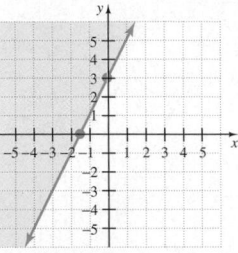

C 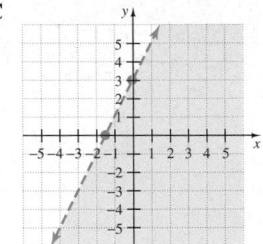   D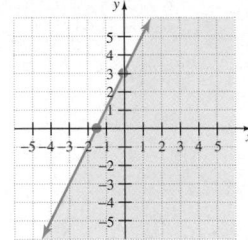

47. $y \leq 2x + 3$ **48.** $y < 2x + 3$

49. $y > 2x + 3$ **50.** $y \geq 2x + 3$

Write the inequality whose graph is given.

51. 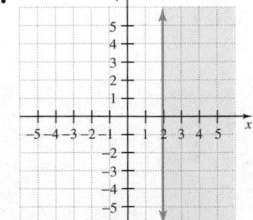   **52.**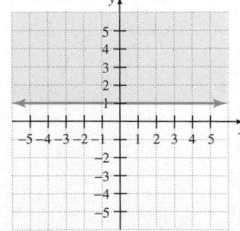

53. 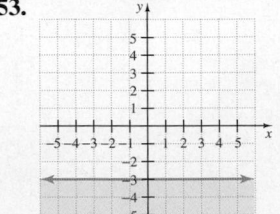   **54.**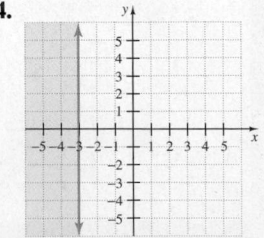

55.

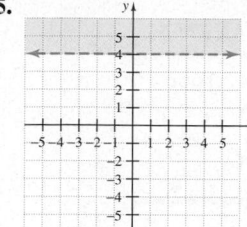

56.

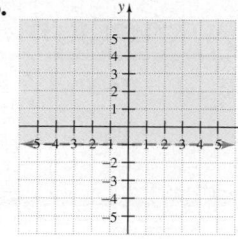

57.

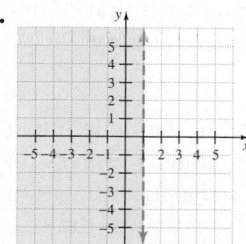

58.

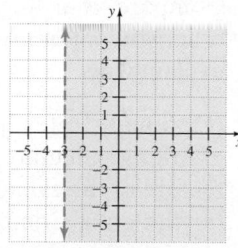

Solve.

59. Rheem Abo-Zahrah decides that she will study at most 20 hours every week and that she must work at least 10 hours every week. Let x represent the hours studying and y represent the hours working. Write two inequalities that model this situation and graph their intersection.

60. The movie and TV critic for the *New York Times* spends between 2 and 6 hours daily reviewing movies and fewer than 5 hours reviewing TV shows. Let x represent the hours watching movies and y represent the time spent watching TV. Write two inequalities that model this situation and graph their intersection.

61. Chris-Craft manufactures boats out of Fiberglas and wood. Fiberglas hulls require 2 hours work, whereas wood hulls require 4 hours work. Employees work at most 40 hours a week. The following inequalities model these restrictions, where x represents the number of Fiberglas hulls produced and y represents the number of wood hulls produced.

$$\begin{cases} x \geq 0 \\ y \geq 0 \\ 2x + 4y \leq 40 \end{cases}$$

Graph the intersection of these inequalities.

REVIEW EXERCISES

Evaluate each expression. See Sections 1.4 and 1.7.

62. 2^3

63. 3^2

64. -5^2

65. $(-5)^2$

66. $(-2)^4$

67. -2^4

68. $\left(\dfrac{3}{5}\right)^3$

69. $\left(\dfrac{2}{7}\right)^2$

Find the domain and the range of each relation. Determine whether the relation is also a function. See Section 3.3.

70.

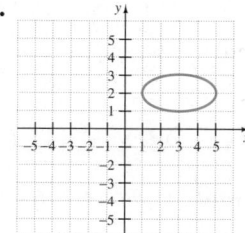

71.

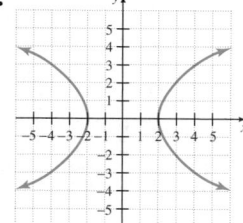

   For additional Chapter Projects, visit the Real World Activities Website by going to http://www.prenhall.com/martin-gay.

3

CHAPTER PROJECT

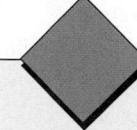

Modeling Real Data

The number of children who live with only one parent has been steadily increasing in the United States since the 1960s. According to the U.S. Bureau of the Census, the percent of children living with one parent varies widely by race/ethnic background. The trend since the 1960s also varies widely, with the data for white families being the most linear. The following table shows the percent of white children (under age 18) living with *both* parents during selected years from 1970 to 1998. In this project, you will have the opportunity to use the data in the table to find a linear function $f(x)$ that represents the data, reflecting the change in living arrangements for children. This project may be completed by working in groups or individually.

PERCENT OF U.S. CHILDREN (WHITE) WHO LIVE WITH BOTH PARENTS

Year	1970	1980	1990	1995	1996	1997	1998
x	0	10	20	25	26	27	28
PERCENT, y	90	83	79	76	75	75	74

Source: U.S. Bureau of the Census

1. Plot the data given in the table as ordered pairs.
2. Use a straight edge to draw on your graph what appears to be the line that "best fits" the data you plotted.
3. Estimate the coordinates of two points that fall on your best-fitting line. Use these points to find a linear function $f(x)$ for the line.
4. What is the slope of your line? Interpret its meaning. Does it make sense in the context of this situation?
5. Find the value of $f(50)$. Write a sentence interpreting its meaning in context.
6. Compare your linear function with that of another student or group. Are they different? If so, explain why.

(Optional) Enter the data from the table into a graphing calculator. Use the linear regression feature of the calculator to find a linear function for the data. Compare this function to the one you found in Question 3. How are they alike or different? Find the value of $f(50)$ using the model you found with the graphing calculator. Compare it to the value of $f(50)$ you found in Question 5.

CHAPTER 3 VOCABULARY CHECK

Fill in each blank with one of the words or phrases listed below.

relation	line	function	standard	slope	domain
slope–intercept	x	y	range	parallel	linear function
point–slope	perpendicular	linear inequality			

1. A _____ is a set of ordered pairs.
2. The graph of every linear equation in two variables is a _____ .
3. The statement $-x + 2y > 0$ is called a _____ in two variables.
4. _____ form of linear equation in two variables is $Ax + By = C$.
5. The _____ of a relation is the set of all second components of the ordered pairs of the relation.
6. _____ lines have the same slope and different y-intercepts.
7. _____ form of a linear equation in two variables is $y = mx + b$.
8. A _____ is a relation in which each first component in the ordered pairs corresponds to exactly one second component.
9. In the equation $y = 4x - 2$, the coefficient of x is the _____ of its corresponding graph.
10. Two lines are _____ if the product of their slopes is -1.
11. To find the x-intercept of a linear equation, let ___ = 0 and solve for the other variable.
12. The _____ of a relation is the set of all first components of the ordered pairs of the relation.
13. A _____ is a function that can be written in the form $f(x) = mx + b$.
14. To find the y-intercept of a linear equation, let ___ = 0 and solve for the other variable.
15. The equation $y - 8 = -5(x + 1)$ is written in _____ form.

CHAPTER 3 HIGHLIGHTS

DEFINITIONS AND CONCEPTS	EXAMPLES

Section 3.1 The Rectangular Coordinate System

The **rectangular coordinate system,** or **Cartesian coordinate system,** consists of a vertical and a horizontal number line intersecting at their 0 coordinates. The vertical number line is called the **y-axis**, and the horizontal number line is called the **x-axis**. The point of intersection of the axes is called the **origin.** The axes divide the plane into four regions called **quadrants.**

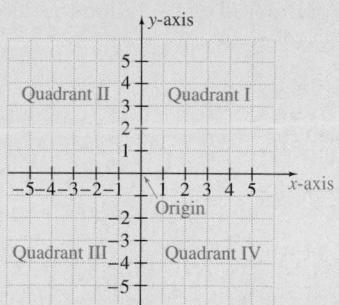

To **plot** or **graph** an ordered pair means to find its corresponding point on a rectangular coordinate system.

To plot or graph the ordered pair $(-2, 5)$, start at the origin. Move 2 units to the left along the x-axis, then 5 units upward parallel to the y-axis.

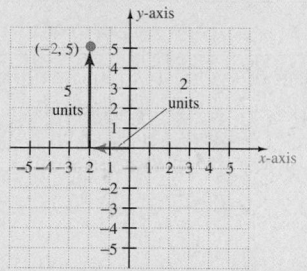

An ordered pair is a **solution** of an equation in two variables if replacing the variables by the corresponding coordinates results in a true statement.

Determine whether $(-2, 3)$ is a solution of
$3x + 2y = 0$

$$3(-2) + 2(3) = 0$$
$$-6 + 6 = 0$$
$$0 = 0 \quad \text{True.}$$

$(-2, 3)$ is a solution.

Section 3.2 Graphing Equations

A **linear equation in two variables** is an equation that can be written in the form $Ax + By = C$, where A, B, and C are real numbers and A and B are not both 0. The form $Ax + By = C$ is called **standard form.**

Linear Equations in Two Variables

$$y = -2x + 5, \quad x = 7$$
$$y - 3 = 0, \quad 6x - 4y = 10$$

$6x - 4y = 10$ is in standard form.

The graph of a linear equation in two variables is a line. To graph a linear equation in two variables, find three ordered pair solutions. Plot the solution points, and draw the line connecting the points.

Graph $3x + y = -6$.

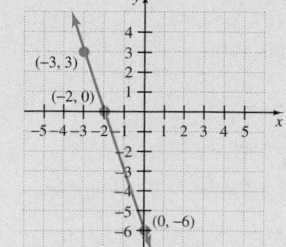

x	y
0	-6
-2	0
-3	3

(continued)

DEFINITIONS AND CONCEPTS	EXAMPLES

Section 3.2 Graphing Equations

To graph an equation that is not linear, find a sufficient number of ordered pair solutions so that a pattern may be discovered.

Graph $y = x^3 + 2$.

x	y
-2	-6
-1	1
0	2
1	3
2	10

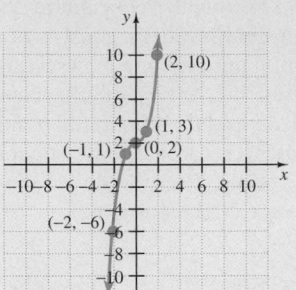

Section 3.3 Introduction to Functions

A **relation** is a set of ordered pairs. The **domain** of the relation is the set of all first components of the ordered pairs. The **range** of the relation is the set of all second components of the ordered pairs.

Relation

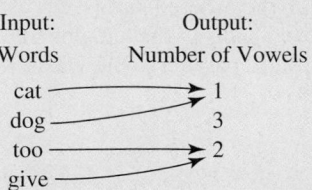

Domain: {cat, dog, too, give}
Range: {1, 2}

A **function** is a relation in which each element of the first set corresponds to exactly one element of the second set.

The previous relation is a function. Each word contains exactly one number of vowels.

Vertical Line Test

If no vertical line can be drawn so that it intersects a graph more than once, the graph is the graph of a function.

Find the domain and the range of the relation. Also determine whether the relation is a function.

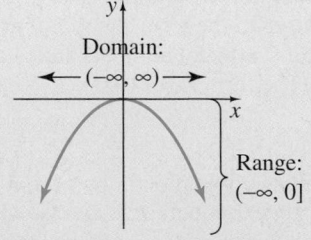

By the vertical line test, this graph is the graph of a function.

The symbol $f(x)$ means **function of x** and is called **function notation.**

If $f(x) = 2x^2 - 5$, find $f(-3)$.

$$f(-3) = 2(-3)^2 - 5 = 2(9) - 5 = 13$$

(continued)

DEFINITIONS AND CONCEPTS	EXAMPLES

Section 3.4 Graphing Linear Functions

A **linear function** is a function that can be written in the form $f(x) = mx + b$.

Linear Functions

$$f(x) = -3, \quad g(x) = 5x, \quad h(x) = -\frac{1}{3}x - 7$$

To graph a linear function, find three ordered pair solutions. Graph the solutions and draw a line through the plotted points.

Graph $f(x) = -2x$.

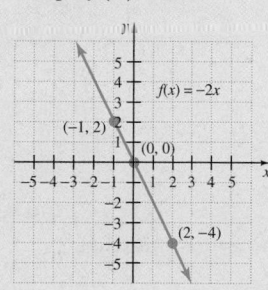

x	y or $f(x)$
-1	2
0	0
2	-4

For any function $f(x)$, the graph of $y = f(x) + K$ is the same as the graph of $y = f(x)$ shifted K units up if K is positive and $|K|$ units down if K is negative.

Graph $g(x) = -2x + 3$.
This is the same as the graph of $f(x) = -2x$ shifted 3 units up.

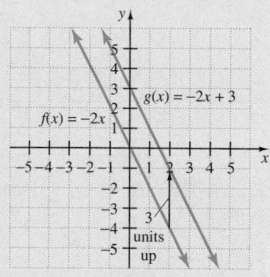

The x-coordinate of a point where a graph crosses the x-axis is called an ***x-intercept***. The y-coordinate of a point where a graph crosses the y-axis is called a ***y-intercept.***

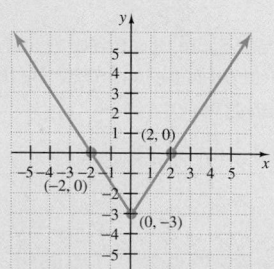

The x-intercepts of the graph are $(-2, 0)$ and $(2, 0)$.
The y-intercept is $(0, -3)$.

To find an x-intercept, let $y = 0$ or $f(x) = 0$ and solve for x.
To find a y-intercept, let $x = 0$ and solve for y.

Graph $5x - y = -5$ by finding intercepts.

$$\begin{array}{ll}
\text{If } x = 0, \text{ then} & \text{If } y = 0, \text{ then} \\
5x - y = -5 & 5x - y = -5 \\
5 \cdot 0 - y = -5 & 5x - 0 = -5 \\
-y = -5 & 5x = -5 \\
y = 5 & x = -1 \\
(0, 5) & (-1, 0)
\end{array}$$

(continued)

DEFINITIONS AND CONCEPTS	EXAMPLES

Section 3.4 Graphing Linear Functions

Ordered pairs are $(0, 5)$ and $(-1, 0)$.

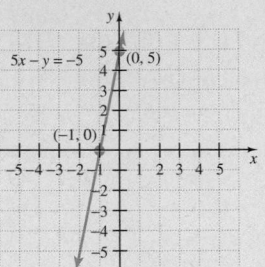

The graph of $x = c$ is a vertical line with x-intercept $(c, 0)$.

The graph of $y = c$ is a horizontal line with y-intercept $(0, c)$.

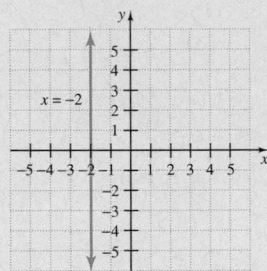 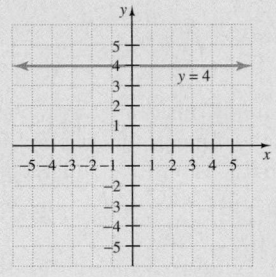

Section 3.5 The Slope of a Line

The **slope** m of the line through (x_1, y_1) and (x_2, y_2) is given by

$$m = \frac{y_2 - y_1}{x_2 - x_1} \text{ as long } x_2 \neq x_1$$

Find the slope of the line through $(-1, 7)$ and $(-2, -3)$.

$$m = \frac{y_2 - y_1}{x_2 - x_1} = \frac{-3 - 7}{-2 - (-1)} = \frac{-10}{-1} = 10$$

The **slope–intercept form** of a linear equation is $y = mx + b$, where m is the slope of the line and $(0, b)$ is the y-intercept.

Find the slope and y-intercept of $-3x + 2y = -8$.

$$2y = 3x - 8$$

$$\frac{2y}{2} = \frac{3x}{2} - \frac{8}{2}$$

$$y = \frac{3}{2}x - 4$$

The slope the line is $\dfrac{3}{2}$, and the y-intercept is $(0, -4)$.

Nonvertical parallel lines have the same slope.

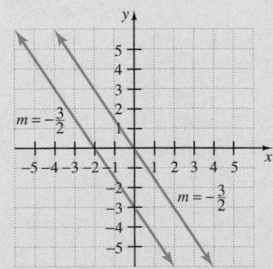

(continued)

DEFINITIONS AND CONCEPTS	EXAMPLES

Section 3.5 The Slope of a Line

If the product of the slopes of two lines is -1, then the lines are perpendicular.

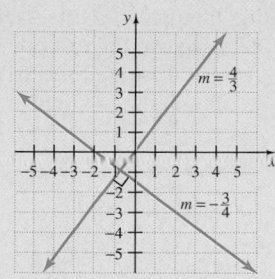

The slope of a horizontal line is 0.
The slope of a vertical line is undefined.

The slope of $y = -2$ is 0.
The slope of $x = 5$ is undefined.

Section 3.6 Equations of Lines

We can use the slope–intercept form to write an equation of a line given its slope and y-intercept.

Write an equation of the line with y-intercept $(0, -1)$ and slope $\dfrac{2}{3}$.

$$y = mx + b$$

$$y = \frac{2}{3}x - 1$$

The point–slope form of the equation of a line is $y - y_1 = m(x - x_1)$, where m is the slope of the line and (x_1, y_1) is a point on the line.

Find an equation of the line with slope 2 containing the point $(1, -4)$. Write the equation in standard form: $Ax + By = C$.

$$y - y_1 = m(x - x_1)$$

$$y - (-4) = 2(x - 1)$$

$$y + 4 = 2x - 2$$

$$-2x + y = -6 \qquad \text{Standard form.}$$

Section 3.7 Graphing Linear Inequalities

If the equal sign in a linear equation in two variables is replaced with an inequality symbol, the result is a **linear inequality in two variables.**

Linear Inequalities in Two Variables

$$x \le -5 \qquad y \ge 2$$

$$3x - 2y > 7 \qquad x < -5$$

To Graph a Linear Inequality

1. Graph the boundary line by graphing the related equation. Draw the line solid if the inequality symbol is $\le$ or $\ge$. Draw the line dashed if the inequality symbol is $<$ or $>$.
2. Choose a test point not on the line. Substitute its coordinates into the original inequality.

Graph $2x - 4y > 4$.

1. Graph $2x - 4y = 4$. Draw a dashed line because the inequality symbol is $>$.

2. Check the test point $(0, 0)$ in the inequality $2x - 4y > 4$.

$$2 \cdot 0 - 4 \cdot 0 > 4 \qquad \text{Let } x = 0 \text{ and } y = 0.$$

$$0 > 4 \qquad \text{False.}$$

(continued)

DEFINITIONS AND CONCEPTS	EXAMPLES

Section 3.7 Graphing Linear Inequalities

3. If the resulting inequality is true, shade the **half-plane** that contains the test point. If the inequality is not true, shade the half-plane that does not contain the test point.

3. The inequality is false, so we shade the half-plane that does not contain $(0, 0)$.

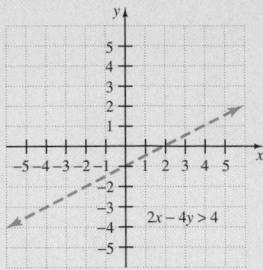

CHAPTER 3 REVIEW

(3.1) *Plot the points and name the quadrant or axis in which each point lies.*

1. $A(2, -1), B(-2, 1), C(0, 3), D(-3, -5)$

2. $A(-3, 4), B(4, -3), C(-2, 0), D(-4, 1)$

Determine whether each ordered pair is a solution to the given equation.

3. $7x - 8y = 56; (0, 56), (8, 0)$

4. $-2x + 5y = 10; (-5, 0), (1, 1)$

5. A local lumberyard uses quantity pricing. The table shows the price per board for different amounts of lumber purchased.

Price per Board (in dollars)	Number of Boards Purchased
8.00	1
7.50	10
6.50	25
5.00	50
2.00	100

a. Write each paired data as an ordered pair of the form (price per board, number of boards purchased).

b. Create a scatter diagram of the paired data. Be sure to label the axes appropriately.

6. The table shows the average cost in cents per mile of owning and operating a vehicle for the years shown.

a. Write each paired data as an ordered pair of the form (year, number of cents per mile).

b. Create a scatter diagram of the paired data. Be sure to label the axes appropriately.

Year	Average Cost of Operating a Vehicle (cents per mile)
1992	46
1993	45
1994	47
1995	49
1996	51
1997	53

(*Source:* American Automobile Manufacturers Association, Inc.)

(3.2) *Determine whether each equation is linear or not. Then graph each equation.*

7. $y = 3x$ **8.** $y = 5x$

9. $3x - y = 4$ **10.** $x - 3y = 2$

11. $y = |x| + 4$ **12.** $y = x^2 + 4$

13. $y = -\dfrac{1}{2}x + 2$ **14.** $y = -x + 5$

15. $y = 2x - 1$ **16.** $y = \dfrac{1}{3}x + 1$

17. $y = -1.36x$ **18.** $y = 2.1x + 5.9$

(3.3) *Find the domain and range of each relation. Also determine whether the relation is a function.*

19. $\left\{ \left(-\dfrac{1}{2}, \dfrac{3}{4} \right), (6, 0.75), (0, -12), (25, 25) \right\}$

20. $\left\{ \left(\dfrac{3}{4}, -\dfrac{1}{2} \right), (0.75, 6), (-12, 0), (25, 25) \right\}$

21.

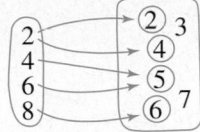

22.

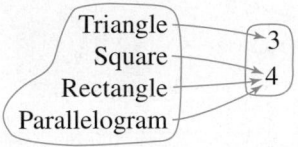

23.

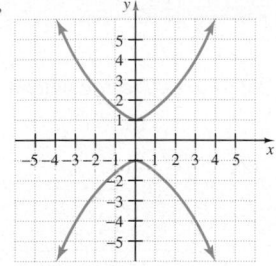

24.

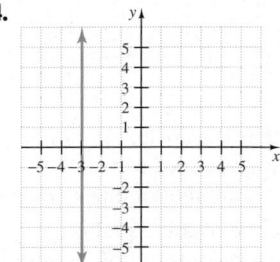

25.

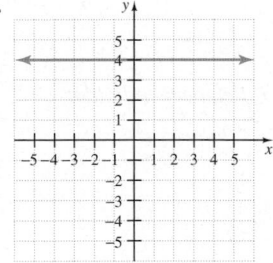

26.

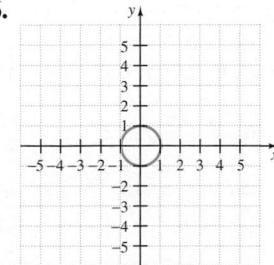

If $f(x) = x - 5$, $g(x) = -3x$, and $h(x) = 2x^2 - 6x + 1$, find the following.

27. $f(2)$ **28.** $g(0)$

29. $g(-6)$ **30.** $h(-1)$

31. $h(1)$ **32.** $f(5)$

The function $J(x) = 2.54x$ may be used to calculate the weight of an object on Jupiter J given its weight on Earth x.

33. If a person weighs 150 pounds on Earth, find the equivalent weight on Jupiter.

34. A 2000-pound probe on Earth weighs how many pounds on Jupiter?

Use the graph of the function below to answer Exercises 35 through 38.

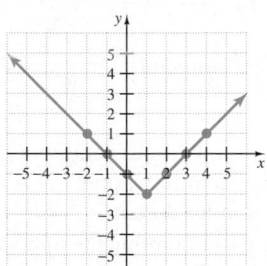

35. Find $f(-1)$.   **36.** Find $f(1)$.

37. Find all values of x such that $f(x) = 1$.

38. Find all values of x such that $f(x) = -1$.

(3.4) *Graph each linear function.*

39. $f(x) = x$ **40.** $f(x) = -\dfrac{1}{3}x$

41. $g(x) = 4x - 1$

The graph of $f(x) = 3x$ is sketched below. Use this graph to match each linear function in Exercises 42–45 with its graph.

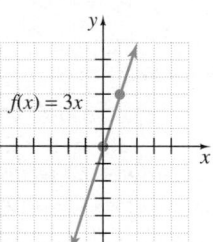

A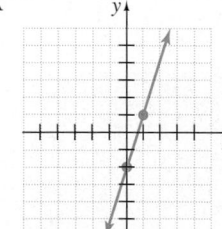

B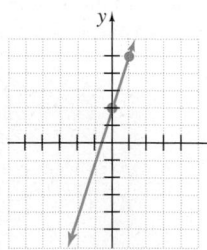

C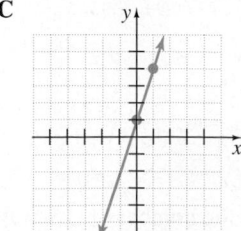

D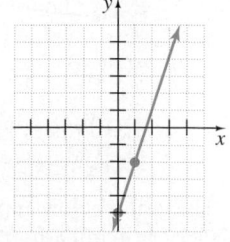

42. $f(x) = 3x + 1$ **43.** $f(x) = 3x - 2$

44. $f(x) = 3x + 2$ **45.** $f(x) = 3x - 5$

Graph each linear equation by finding intercepts if possible.

46. $4x + 5y = 20$ **47.** $3x - 2y = -9$

48. $4x - y = 3$ **49.** $2x + 6y = 9$

50. $y = 5$ **51.** $x = -2$

Graph each linear equation.

52. $x - 2 = 0$ **53.** $y + 3 = 0$

54. The cost C, in dollars, of renting a minivan for a day is given by the linear function $C(x) = 0.3x + 42$, where x is number of miles driven.
   a. Find the cost of renting the minivan for a day and driving it 150 miles.
   b. Graph $C(x) = 0.3x + 42$.

(3.5) *Find the slope of the line through each pair of points.*

55. $(2, 8)$ and $(6, -4)$ **56.** $(-3, 9)$ and $(5, 13)$

57. $(-7, -4)$ and $(-3, 6)$ **58.** $(7, -2)$ and $(-5, 7)$

Find the slope and y-intercept of each line.

59. $6x - 15y = 20$ **60.** $4x + 14y = 21$

Find the slope of each line.

61. $y - 3 = 0$ **62.** $x = -5$

Find the slope of each line and write the slope as a rate of change. Don't forget to attach the proper units.

63. The graph below approximates the number of U.S. persons y (in millions) who have a bachelors degree or higher per year x.

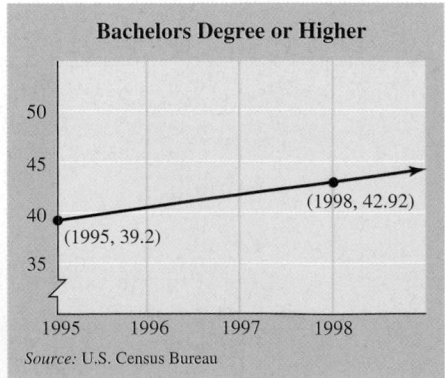

Bachelors Degree or Higher

(1995, 39.2)

(1998, 42.92)

Source: U.S. Census Bureau

64. The graph below approximates the number of U.S. travelers y (in millions) that are vacationing per year x.

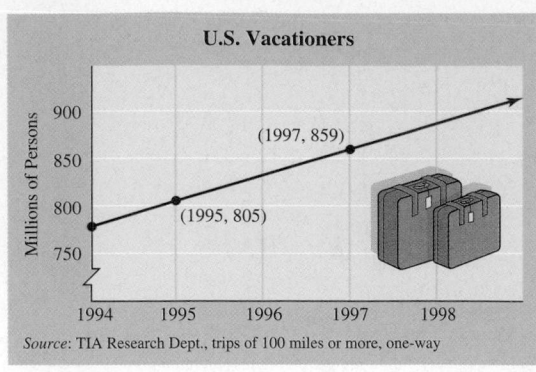

U.S. Vacationers

Millions of Persons

(1997, 859)

(1995, 805)

Source: TIA Research Dept., trips of 100 miles or more, one-way

Two lines are graphed on each set of axes. Decide whether l_1 or l_2 has the greater slope.

65.

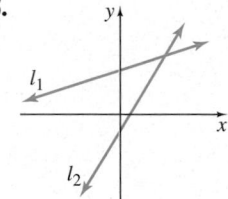

66.

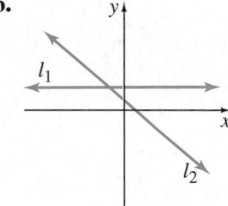

67. Recall from Exercise 54, that the cost C, in dollars, of renting a minivan for a day is given by the linear equation $C(x) = 0.3x + 42$, where x is number of miles driven.
   a. Find and interpret the slope of this equation.
   b. Find and interpret the y-intercept of this equation.

Decide whether the lines are parallel, perpendicular, or neither.

△ **68.** $f(x) = -2x + 6$
        $g(x) = 2x - 1$

△ **69.** $-x + 3y = 2$
        $6x - 18y = 3$

(3.6) *Graph each linear equation using the slope and y-intercept.*

70. $y = -x + 1$ **71.** $y = 4x - 3$

72. $3x - y = 6$ **73.** $y = -5x$

Find an equation of the line satisfying the conditions given.

74. Horizontal; through $(3, -1)$

75. Vertical; through $(-2, -4)$

△ **76.** Parallel to the line $x = 6$; through $(-4, -3)$

77. Slope 0; through $(2, 5)$

Find the standard form equation of each line satisfying the conditions given.

78. Through $(-3, 5)$; slope 3

79. Slope 2; through $(5, -2)$

80. Through $(-6, -1)$ and $(-4, -2)$

81. Through $(-5, 3)$ and $(-4, -8)$

△ **82.** Through $(-2, 3)$; perpendicular to $x = 4$

∧ **83.** Through $(-2, -5)$; parallel to $y = 8$

Find the equation of each line satisfying the given conditions. Write each equation using function notation.

84. Slope $-\dfrac{2}{3}$; y-intercept $(0, 4)$

85. Slope -1; y-intercept $(0, -2)$

△ **86.** Through $(2, -6)$; parallel to $6x + 3y = 5$

△ **87.** Through $(-4, -2)$; parallel to $3x + 2y = 8$

△ **88.** Through $(-6, -1)$; perpendicular to $4x + 3y = 5$

△ **89.** Through $(-4, 5)$; perpendicular to $2x - 3y = 6$

90. In 1996, the number of U.S. paging subscribers (in millions) was 42. The number of subscribers in 1999 (in millions) was 58. Let y be the number of subscribers (in millions) in the year x, where $x = 0$ represents 1996. (*Source:* Strategis Group for Personal Communications Asso.)

a. Write a linear equation that models the number of U.S. paging subscribers (in millions) in terms of the year x. [*Hint:* Write 2 ordered pairs of the form (years past 1996, number of subscribers).]

b. Use this equation to predict the number of U.S. paging subscribers in the year 2007. (Round to the nearest million.)

91. In 1998, the number of people (in millions) reporting arthritis was 43. The number of people (in millions) predicted to be reporting arthritis in 2020 is 60. Let y be the number of people (in millions) reporting arthritis in the year x, where $x = 0$ represents 1998. (*Source:* Arthritis Foundation)

a. Write a linear equation that models the number of people (in millions) reporting arthritis in terms of the year x (See the hint for Exercise 90.)

b. Use this equation to predict the number of people reporting arthritis in 2010. (Round to the nearest million.)

(3.7) *Graph each linear inequality.*

92. $3x + y > 4$

93. $\frac{1}{2}x - y < 2$

94. $5x - 2y \leq 9$

95. $3y \geq x$

96. $y < 1$

97. $x > -2$

98. Graph the union of $y > 2x + 3$ or $x \leq -3$.

99. Graph the intersection of $2x < 3y + 8$ and $y \geq -2$.

CHAPTER 3 TEST

1. Plot the points, and name the quadrant in which each is located: $A(6, -2)$, $B(4, 0)$, $C(-1, 6)$.

2. Complete the ordered pair solution $(-6, \quad)$ of the equation $2y - 3x = 12$.

Graph each line.

3. $2x - 3y = -6$

4. $4x + 6y = 7$

5. $f(x) = \dfrac{2}{3}x$

6. $y = -3$

7. Find the slope of the line that passes through $(5, -8)$ and $(-7, 10)$.

8. Find the slope and the y-intercept of the line $3x + 12y = 8$.

Graph each nonlinear function. Suggested x-values have been given for ordered pair solutions.

9. $f(x) = (x - 1)^2$
   Let $x = -2, -1, 0, 1, 2, 3, 4$

10. $g(x) = |x| + 2$
   Let $x = -3, -2, -1, 0, 1, 2, 3$

Find an equation of each line satisfying the conditions given. Write Exercises 11–15 in standard form. Write Exercises 16–18 using function notation.

11. Horizontal; through $(2, -8)$

12. Vertical; through $(-4, -3)$

△ **13.** Perpendicular to $x = 5$; through $(3, -2)$

14. Through $(4, -1)$; slope -3

15. Through $(0, -2)$; slope 5

16. Through $(4, -2)$ and $(6, -3)$

△ **17.** Through $(-1, 2)$; perpendicular to $3x - y = 4$

△ **18.** Parallel to $2y + x = 3$; through $(3, -2)$

△ **19.** Line L_1 has the equation $2x - 5y = 8$. Line L_2 passes through the points $(1, 4)$ and $(-1, -1)$. Determine whether these lines are parallel, perpendicular, or neither.

Graph each inequality.

20. $x \leq -4$

21. $y > -2$

22. $2x - y > 5$

23. The intersection of $2x + 4y < 6$ and $y \leq -4$

Find the domain and range of each relation. Also determine whether the relation is a function.

24.

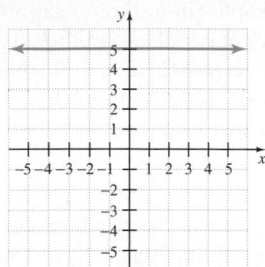

25.

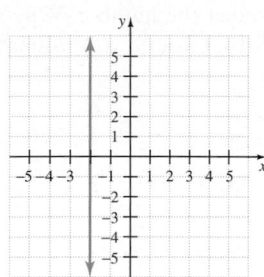

26.

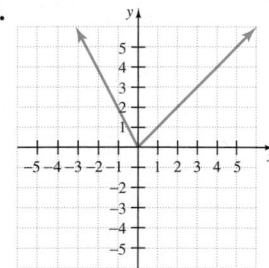

27.

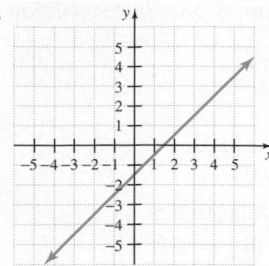

28. The average yearly earnings for high school graduates age 18 and older is given by the linear function
$$f(x) = 732x + 21,428$$
where x is the number of years since 1996 that a person graduated. (*Source:* U.S. Census Bureau)
   a. Find the average earnings in 1998 for high school graduates.
   b. Predict the average earnings for high school graduates in the year 2005.
   c. Predict the first year that the average earnings for high school graduates will be greater than $30,000.
   d. Find and interpret the slope of this equation.
   e. Find and interpret the y-intercept of this equation.

29. This graph approximates the movie ticket sales y (in millions) for the year x. Find the slope of the line and write the slope as a rate of change. Don't forget to attach the proper units.

CHAPTER 3 CUMULATIVE REVIEW

1. Insert $<, >,$ or $=$ in the space between the paired numbers to make each statement true.
   a. 2   3
   b. 7   4
   c. 72   27

2. Find the product of $\dfrac{2}{15}$ and $\dfrac{5}{13}$. Write the product in lowest terms.

3. Simplify $\dfrac{3 + |4 - 3| + 2^2}{6 - 3}$.

4. Add.
   a. $-8 + (-11)$
   b. $-5 + 35$
   c. $0.6 + (-1.1)$
   d. $-\dfrac{7}{10} + \left(-\dfrac{1}{10}\right)$
   e. $11.4 + (-4.7)$
   f. $-\dfrac{3}{8} + \dfrac{2}{5}$

5. Simplify: $-14 - 8 + 10 - (-6)$

6. If $x = -2$ and $y = -4$, evaluate each expression.

   a. $5x - y$ **b.** $x^3 - y^2$

   c. $\dfrac{3x}{2y}$

7. Simplify each expression.

   a. $10 + (x + 12)$ **b.** $-3(7x)$

8. Identify the numerical coefficient in each term.

   a. $-3y$ **b.** $22z^4$

   c. y **d.** $-x$

   e. $\dfrac{x}{7}$

9. Solve $y + 0.6 = -1.0$ for y.

10. Solve for x: $\dfrac{5}{2}x = 15$.

11. If x is the first of three consecutive integers, express the sum of the three integers in terms of x. Simplify if possible.

12. Solve: $\dfrac{2(a + 3)}{3} = 6a + 2$

13. In a recent year, Congress had 8 more Republican senators than Democratic. If the total number of senators is 100, how many senators of each party were there?

14. Charles Pecot can afford enough fencing to enclose a rectangular garden with a perimeter of 140 feet. If the width of his garden must be 30 feet, find the length.

15. Solve $y = mx + b$ for x.

16. Write each number as a percent.

   a. 0.73 **b.** 1.39

   c. $\dfrac{1}{4}$

17. The number 120 is 15% of what number?

18. Graph $x \geq -1$.

19. Determine whether each ordered pair is a solution of the equation $x - 2y = 6$.

   a. $(6, 0)$ **b.** $(0, 3)$

   c. $\left(1, -\dfrac{5}{2}\right)$

20. Identify the linear equations in two variables.

   a. $x - 1.5y = -1.6$ **b.** $y = -2x$

   c. $x + y^2 = 0$ **d.** $x = 5$

21. Is the relation $y = 2x + 1$ also a function?

22. Find the y-intercept of the graph of each equation.

   a. $f(x) = \dfrac{1}{2}x + \dfrac{3}{7}$

   b. $y = -2.5x - 3.2$

23. Find the slope of the line whose equation is $f(x) = \dfrac{2}{3}x + 4$.

24. Write an equation of the line with y-intercept $(0, -3)$ and slope of $\dfrac{1}{4}$.

25. Graph: $2x - y < 6$.

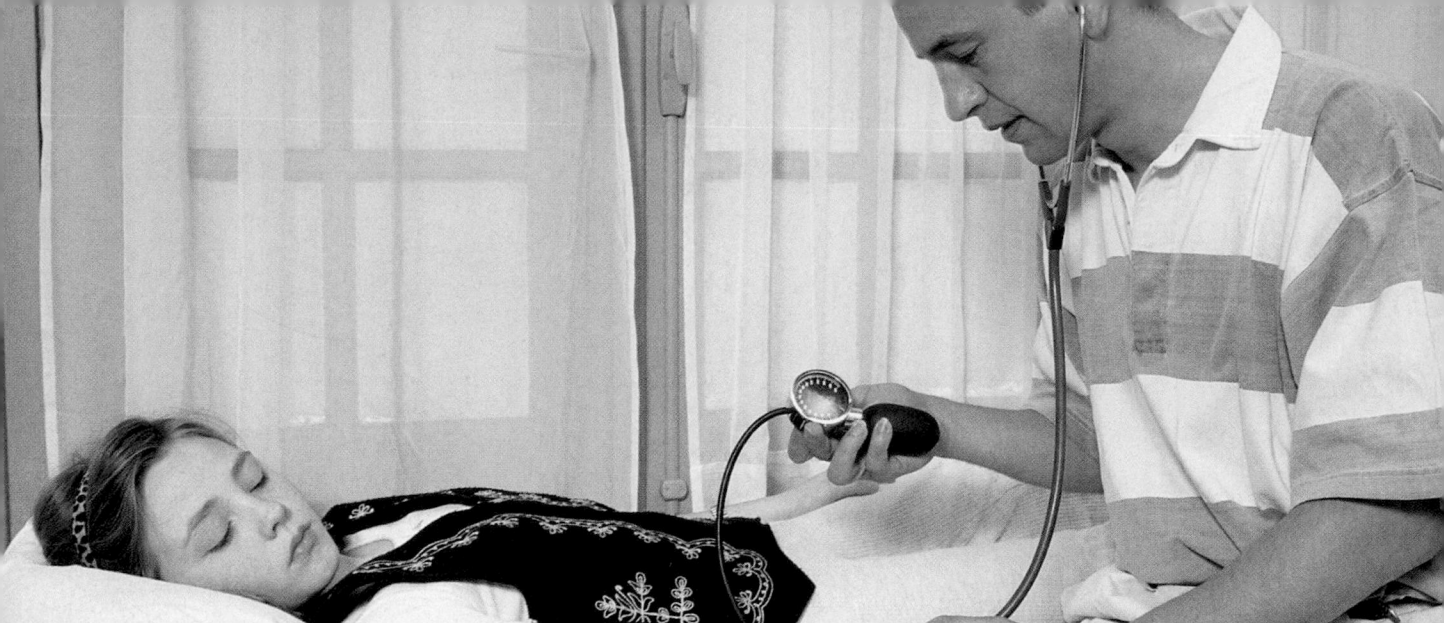

Largest Health Care Occupation in the U.S.

Did you know that nurses make up the largest health care occupation in the United States? Over 2.5 million people work as nurses in settings as varied as hospitals, private homes, corporate offices, nursing homes, overnight camps, doctors' offices, and community health centers. The U.S. Bureau of Labor Statistics predicts that the demand for nurses will continue to grow rapidly as the American population ages, requiring more long-term health care and home health care.

Registered nurses must be licensed in the state in which they work. About 25% of registered nurses hold a diploma from a hospital program, 35% hold an associate's degree, 30% hold a bachelor's degree, and 10% hold a higher degree. Although the focus of study in a nursing degree program is on areas such as anatomy, chemistry, microbiology, nutrition, and clinical experience, it isn't hard to see how good math skills would be useful. In fact, nurses use math skills nearly every day in taking and comparing vital signs, administering medications, and tracking fluid intake and output.

 For more information about a nursing career, visit the National League for Nursing Website by first going to www.prenhall.com/martin-gay.

In the Spotlight on Decision Making feature on page 266, you will have the opportunity, as a registered nurse, to make a decision concerning a patient's blood pressure.

SOLVING SYSTEMS OF LINEAR EQUATIONS

4.1 SOLVING SYSTEMS OF LINEAR EQUATIONS BY GRAPHING

4.2 SOLVING SYSTEMS OF LINEAR EQUATIONS BY SUBSTITUTION

4.3 SOLVING SYSTEMS OF LINEAR EQUATIONS BY ADDITION

4.4 SYSTEMS OF LINEAR EQUATIONS AND PROBLEM SOLVING

In Chapter 3, we graphed equations containing two variables. Equations like these are often needed to represent relationships between two different values. For example, an economist attempts to predict what effects a price change will have on the sales prospects of calculators. There are many real-life opportunities to compare and contrast two such equations, called a system of equations. This chapter presents linear systems and ways we solve these systems and apply them to real-life situations.

4.1 SOLVING SYSTEMS OF LINEAR EQUATIONS BY GRAPHING

CD-ROM SSM

SSG Video

▶ **OBJECTIVES**

1. Determine if an ordered pair is a solution of a system of equations in two variables.
2. Solve a system of linear equations by graphing.
3. Without graphing, determine the number of solutions of a system.

1

A **system of linear equations** consists of two or more linear equations. In this section, we focus on solving systems of linear equations containing two equations in two variables. Examples of such linear systems are

$$\begin{cases} 3x - 3y = 0 \\ x = 2y \end{cases} \qquad \begin{cases} x - y = 0 \\ 2x + y = 10 \end{cases} \qquad \begin{cases} y = 7x - 1 \\ y = 4 \end{cases}$$

A **solution** of a system of two equations in two variables is an ordered pair of numbers that is a solution of both equations in the system.

Example 1 Which of the following ordered pairs is a solution of the given system?

$$\begin{cases} 2x - 3y = 6 & \text{First equation} \\ x = 2y & \text{Second equation} \end{cases}$$

a. $(12, 6)$ **b.** $(0, -2)$

Solution If an ordered pair is a solution of both equations, it is a solution of the system.
a. Replace x with 12 and y with 6 in both equations.

$2x - 3y = 6$ First equation	$x = 2y$ Second equation
$2(12) - 3(6) \stackrel{?}{=} 6$ Let $x = 12$ and $y = 6$.	$12 \stackrel{?}{=} 2(6)$ Let $x = 12$ and $y = 6$.
$24 - 18 \stackrel{?}{=} 6$ Simplify.	$12 = 12$ True
$6 = 6$ True	

Since $(12, 6)$ is a solution of both equations, it is a solution of the system.
b. Start by replacing x with 0 and y with -2 in both equations.

$2x - 3y = 6$ First equation	$x = 2y$ Second equation
$2(0) - 3(-2) \stackrel{?}{=} 6$ Let $x = 0$ and $y = -2$.	$0 \stackrel{?}{=} 2(-2)$ Let $x = 0$ and $y = -2$.
$0 + 6 \stackrel{?}{=} 6$ Simplify.	$0 = -4$ False
$6 = 6$ True	

While $(0, -2)$ is a solution of the first equation, it is not a solution of the second equation, so it is **not** a solution of the system. ∎

2

Since a solution of a system of two equations in two variables is a solution common to both equations, it is also a point common to the graphs of both equations. Let's practice finding solutions of both equations in a system—that is, solutions of a system—by graphing and identifying points of intersection.

Example 2 Solve the system of equations by graphing.

$$\begin{cases} -x + 3y = 10 \\ x + y = 2 \end{cases}$$

Solution On a single set of axes, graph each linear equation.

$-x + 3y = 10$

x	y
0	$\dfrac{10}{3}$
-4	2
2	4

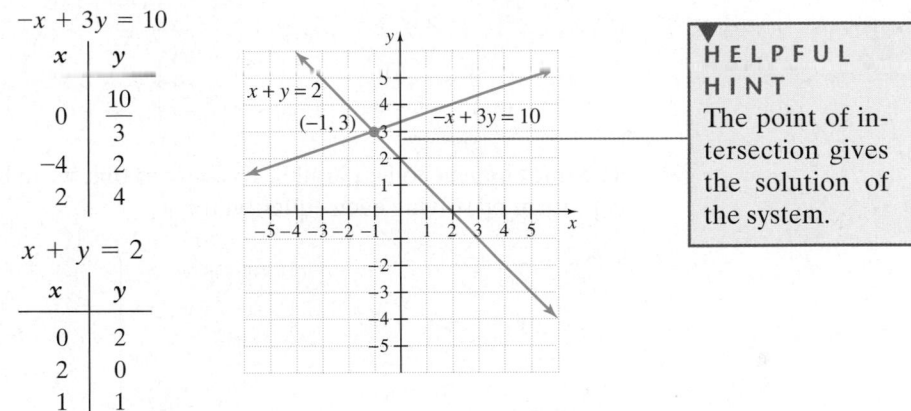

$x + y = 2$

x	y
0	2
2	0
1	1

> **HELPFUL HINT**
> The point of intersection gives the solution of the system.

The two lines appear to intersect at the point $(-1, 3)$. To check, we replace x with -1 and y with 3 in both equations.

$-x + 3y = 10$	First equation	$x + y = 2$	Second equation
$-(-1) + 3(3) \stackrel{?}{=} 10$	Let $x = -1$ and $y = 3$.	$-1 + 3 \stackrel{?}{=} 2$	Let $x = -1$ and $y = 3$.
$1 + 9 \stackrel{?}{=} 10$	Simplify.	$2 = 2$	True
$10 = 10$	True		

$(-1, 3)$ checks, so it is the solution of the system.

> **HELPFUL HINT**
> Neatly drawn graphs can help when you are estimating the solution of a system of linear equations by graphing.

A system of equations that has at least one solution as in Example 2 is said to be a **consistent system**. A system that has no solution is said to be an **inconsistent system**.

Example 3 Solve the following system of equations by graphing.

$$\begin{cases} 2x + y = 7 \\ 2y = -4x \end{cases}$$

Solution Graph each of the two lines in the system.

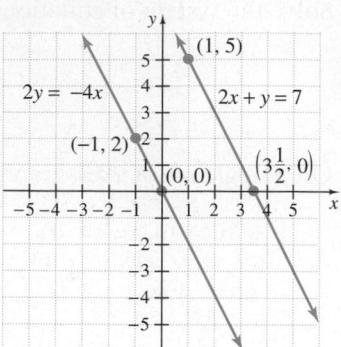

The lines **appear** to be parallel. To confirm this, write both equations in slope-intercept form by solving each equation for y.

$$2x + y = 7 \qquad \text{First equation}$$
$$y = -2x + 7 \qquad \text{Subtract } 2x \text{ from both sides.}$$

$$2y = -4x \qquad \text{Second equation}$$
$$\frac{2y}{2} = \frac{-4x}{2} \qquad \text{Divide both sides by 2.}$$
$$y = -2x$$

Recall that when an equation is written in slope–intercept form, the coefficient of x is the slope. Since both equations have the same slope, -2, but different y-intercepts, the lines are parallel and have no points in common. Thus, there is no solution of the system and the system is inconsistent. ▪

In Examples 2 and 3, the graphs of the two linear equations of each system are different. When this happens, we call these equations **independent equations**. If the graphs of the two equations in a system are identical, we call the equations **dependent equations**.

◆ **Example 4** Solve the system of equations by graphing.

$$\begin{cases} x - y = 3 \\ -x + y = -3 \end{cases}$$

Solution Graph each line.

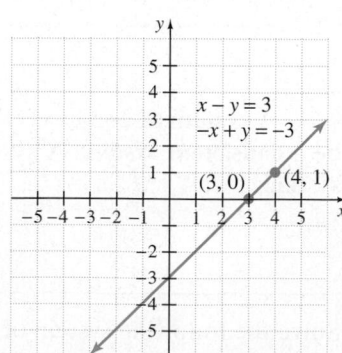

These graphs **appear** to be identical. To confirm this, write each equation in slope–intercept form.

$$x - y = 3 \qquad \text{First equation} \qquad \Big| \qquad -x + y = -3 \qquad \text{Second equation}$$

$$-y = -x + 3 \qquad \text{Subtract } x \text{ from both sides.} \qquad \Big| \qquad y = x - 3 \qquad \text{Add } x \text{ to both sides.}$$

$$\frac{-y}{-1} = \frac{-x}{-1} + \frac{3}{-1} \qquad \text{Divide both sides by } -1.$$

$$y = x - 3$$

The equations are identical and so must be their graphs. The lines have an infinite number of points in common. Thus, there is an infinite number of solutions of the system and this is a consistent system. The equations are dependent equations. ▬

As we have seen, three different situations can occur when graphing the two lines associated with the equations in a linear system:

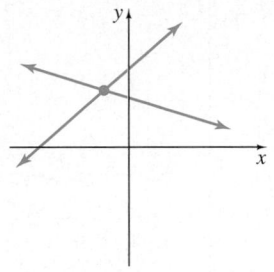

One point of intersection: one solution

Consistent system
(at least one solution)
Independent equations
(graphs of equations differ)

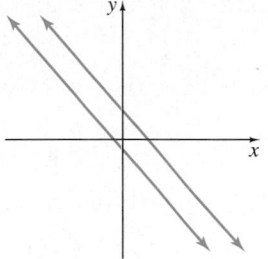

Parallel lines: no solution

Inconsistent system
(no solution)
Independent equations
(graphs of equations differ)

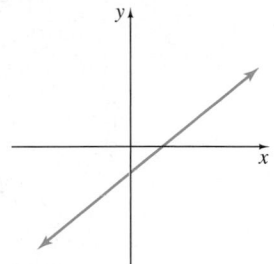

Same line: infinite number of solutions

Consistent system
(at least one solution)
Dependent equations
(graphs of equations identical)

3 You may have suspected by now that graphing alone is not an accurate way to solve a system of linear equations. For example, a solution of $\left(\frac{1}{2}, \frac{2}{9}\right)$ is unlikely to be read correctly from a graph. The next two sections present two accurate methods of solving these systems. In the meantime, we can decide how many solutions a system has by writing each equation in the slope–intercept form.

Example 5 Without graphing, determine the number of solutions of the system.

$$\begin{cases} \dfrac{1}{2}x - y = 2 \\[2mm] x = 2y + 5 \end{cases}$$

Solution First write each equation in slope–intercept form.

$$\frac{1}{2}x - y = 2 \qquad \text{First equation} \qquad \Big| \qquad x = 2y + 5 \qquad \text{Second equation}$$

$$\frac{1}{2}x = y + 2 \qquad \text{Add } y \text{ to both sides.} \qquad \Big| \qquad x - 5 = 2y \qquad \text{Subtract 5 from both sides.}$$

$$\Big| \qquad \frac{x}{2} - \frac{5}{2} = \frac{2y}{2} \qquad \text{Divide both sides by 2.}$$

$$\frac{1}{2}x - 2 = y \qquad \text{Subtract 2 from both sides.} \qquad \Big| \qquad \frac{1}{2}x - \frac{5}{2} = y \qquad \text{Simplify.}$$

The slope of each line is $\frac{1}{2}$, but they have different *y*-intercepts. This tells us that the lines representing these equations are parallel. Since the lines are parallel, the system has no solution and is inconsistent.

Example 6   Determine the number of solutions of the system.

$$\begin{cases} 3x - y = 4 \\ x + 2y = 8 \end{cases}$$

Solution   Once again, the slope–intercept form helps determine how many solutions this system has.

$3x - y = 4$	First equation	$x + 2y = 8$	Second equation
$3x = y + 4$	Add *y* to both sides.	$x = -2y + 8$	Subtract 2*y* from both sides.
$3x - 4 = y$	Subtract 4 from both sides.	$x - 8 = -2y$	Subtract 8 from both sides.
		$\dfrac{x}{-2} - \dfrac{8}{-2} = \dfrac{-2y}{-2}$	Divide both sides by −2.
		$-\dfrac{1}{2}x + 4 = y$	Simplify.

The slope of the second line is $-\frac{1}{2}$, whereas the slope of the first line is 3. Since the slopes are not equal, the two lines are neither parallel, nor identical, and must intersect. Therefore, this system has one solution and is consistent.

GRAPHING CALCULATOR EXPLORATIONS

A graphing calculator may be used to approximate solutions of systems of equations. For example, to approximate the solution of the system

$$\begin{cases} y = -3.14x - 1.35 \\ y = 4.88x + 5.25, \end{cases}$$

first graph each equation on the same set of axes. Then use the intersect feature of your calculator to approximate the point of intersection.

The approximate point of intersection is $(-0.82, 1.23)$.

Solve each system of equations. Approximate the solutions to two decimal places.

1. $\begin{cases} y = -2.68x + 1.21 \\ y = 5.22x - 1.68 \end{cases}$ **2.** $\begin{cases} y = 4.25x + 3.89 \\ y = -1.88x + 3.21 \end{cases}$

3. $\begin{cases} 4.3x - 2.9y = 5.6 \\ 8.1x + 7.6y = -14.1 \end{cases}$ **4.** $\begin{cases} -3.6x - 8.6y = 10 \\ -4.5x + 9.6y = -7.7 \end{cases}$

MENTAL MATH

Each rectangular coordinate system shows the graph of the equations in a system of equations. Use each graph to determine the number of solutions for each associated system. If the system has only one solution, give its coordinates.

1.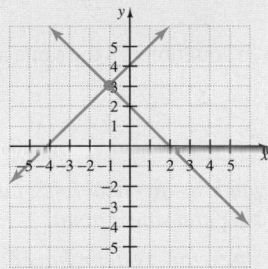

2.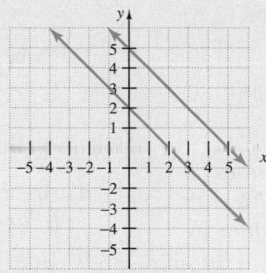

3.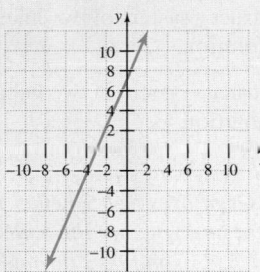

4.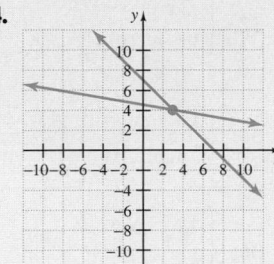

5.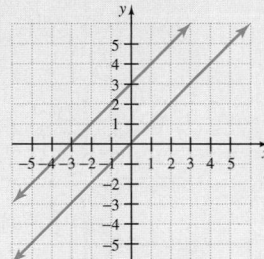

6.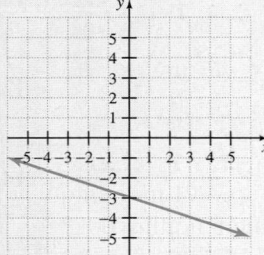

7.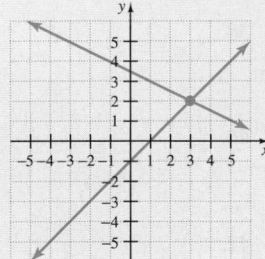

8.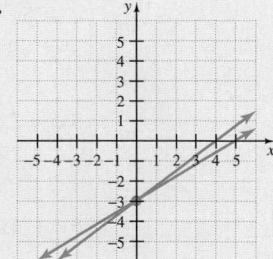

Exercise Set 4.1

Determine whether any ordered pairs satisfy the system of the linear equations. See Example 1.

1. $\begin{cases} x + y = 8 \\ 3x + 2y = 21 \end{cases}$
   a. $(2, 4)$
   b. $(5, 3)$
   c. $(1, 9)$

2. $\begin{cases} 2x + y = 5 \\ x + 3y = 5 \end{cases}$
   a. $(5, 0)$
   b. $(1, 2)$
   c. $(2, 1)$

3. $\begin{cases} 3x - y = 5 \\ x + 2y = 11 \end{cases}$
   a. $(2, -1)$
   b. $(3, 4)$
   c. $(0, -5)$

4. $\begin{cases} 2x - 3y = 8 \\ x - 2y = 6 \end{cases}$
   a. $(4, 0)$
   b. $(-2, -4)$
   c. $(7, 2)$

5. $\begin{cases} 2y = 4x \\ 2x - y = 0 \end{cases}$

  a. $(-3, -6)$
  b. $(0, 0)$
  c. $(1, 2)$

6. $\begin{cases} 4x = 1 - y \\ x - 3y = -8 \end{cases}$

  a. $(0, 1)$
  b. $(1, -3)$
  c. $(-2, 2)$

7. Construct a system of two linear equations that has $(2, 5)$ as a solution.

8. Construct a system of two linear equations that has $(0, 1)$ as a solution.

Solve each system of equations by graphing the equations on the same set of axes. Tell whether the system is consistent or inconsistent and whether the equations are dependent or independent. See Examples 2 through 4.

9. $\begin{cases} y = x + 1 \\ y = 2x - 1 \end{cases}$

10. $\begin{cases} y = 3x - 4 \\ y = x + 2 \end{cases}$

11. $\begin{cases} 2x + y = 0 \\ 3x + y = 1 \end{cases}$

12. $\begin{cases} 2x + y = 1 \\ 3x + y = 0 \end{cases}$

13. $\begin{cases} y = -x - 1 \\ y = 2x + 5 \end{cases}$

14. $\begin{cases} y = x - 1 \\ y = -3x - 5 \end{cases}$

15. $\begin{cases} 2x - y = 6 \\ y = 2 \end{cases}$

16. $\begin{cases} x + y = 5 \\ x = 4 \end{cases}$

17. $\begin{cases} x + y = 5 \\ x + y = 6 \end{cases}$

18. $\begin{cases} 2x + y = 4 \\ x + y = 2 \end{cases}$

19. $\begin{cases} y - 3x = -2 \\ 6x - 2y = 4 \end{cases}$

20. $\begin{cases} y + 2x = 3 \\ 4x = 2 - 2y \end{cases}$

21. $\begin{cases} x - 2y = 2 \\ 3x + 2y = -2 \end{cases}$

22. $\begin{cases} x + 3y = 7 \\ 2x - 3y = -4 \end{cases}$

23. $\begin{cases} \frac{1}{2}x + y = -1 \\ x = 4 \end{cases}$

24. $\begin{cases} x + \frac{3}{4}y = 2 \\ x = -1 \end{cases}$

25. $\begin{cases} y = x - 2 \\ y = 2x + 3 \end{cases}$

26. $\begin{cases} y = x + 5 \\ y = -2x - 4 \end{cases}$

27. $\begin{cases} x + y = 7 \\ x - y = 3 \end{cases}$

28. $\begin{cases} x + y = -4 \\ x - y = 2 \end{cases}$

29. Explain how to use a graph to determine the number of solutions of a system.

30. The ordered pair $(-2, 3)$ is a solution of all three independent equations:

$$x + y = 1$$
$$2x - y = -7$$
$$x + 3y = 7$$

Describe the graph of all three equations on the same axes.

Without graphing, decide.

  a. Are the graphs of the equations identical lines, parallel lines, or lines intersecting at a single point?

  b. How many solutions does the system have? See Examples 5 and 6.

31. $\begin{cases} 4x + y = 24 \\ x + 2y = 2 \end{cases}$

32. $\begin{cases} 3x + y = 1 \\ 3x + 2y = 6 \end{cases}$

33. $\begin{cases} 2x + y = 0 \\ 2y = 6 - 4x \end{cases}$

34. $\begin{cases} 3x + y = 0 \\ 2y = -6x \end{cases}$

35. $\begin{cases} 6x - y = 4 \\ \frac{1}{2}y = -2 + 3x \end{cases}$

36. $\begin{cases} 3x - y = 2 \\ \frac{1}{3}y = -2 + 3x \end{cases}$

37. $\begin{cases} x = 5 \\ y = -2 \end{cases}$

38. $\begin{cases} y = 3 \\ x = -4 \end{cases}$

39. $\begin{cases} 3y - 2x = 3 \\ x + 2y = 9 \end{cases}$

40. $\begin{cases} 2y = x + 2 \\ y + 2x = 3 \end{cases}$

41. $\begin{cases} 6y + 4x = 6 \\ 3y - 3 = -2x \end{cases}$

42. $\begin{cases} 8y + 6x = 4 \\ 4y - 2 = 3x \end{cases}$

43. $\begin{cases} x + y = 4 \\ x + y = 3 \end{cases}$

44. $\begin{cases} 2x + y = 0 \\ y = -2x + 1 \end{cases}$

45. Explain how writing each equation in a linear system in the slope–intercept form helps determine the number of solutions of a system.

46. Is it possible for a system of two linear equations in two variables to be inconsistent, but with dependent equations? Why or why not?

The double line graph below shows the number of pounds of fishery products from U.S. domestic catch and from imports. Use this graph for Exercises 47 and 48.

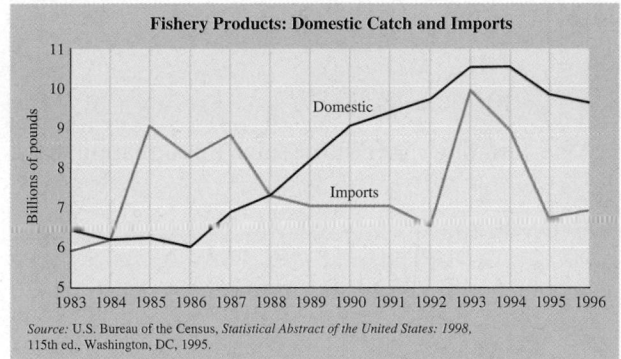

Fishery Products: Domestic Catch and Imports

Source: U.S. Bureau of the Census, *Statistical Abstract of the United States: 1998,* 115th ed., Washington, DC, 1995.

47. In what year(s) is the number of pounds of fishery products imported equal to the number of pounds of domestic catch?

48. In what year(s) is the number of pounds of fishery products imported greater than the number of pounds of domestic catch?

49. Below are tables of values for two linear equations. Using the tables,
  a. find a solution of the corresponding system.
  b. graph several ordered pairs from each table and sketch the two lines.
Does your graph confirm the solution from part **a**?

x	y	x	y
1	3	1	6
2	5	2	7
3	7	3	8
4	9	4	9
5	11	5	10

REVIEW EXERCISES

Solve each equation. See Section 2.3.

50. $5(x - 3) + 3x = 1$

51. $-2x + 3(x + 6) = 17$

52. $4\left(\dfrac{y + 1}{2}\right) + 3y = 0$

53. $-y + 12\left(\dfrac{y - 1}{4}\right) = 3$

54. $8a - 2(3a - 1) = 6$

55. $3z - (4z - 2) = 9$

4.2 SOLVING SYSTEMS OF LINEAR EQUATIONS BY SUBSTITUTION

CD-ROM   SSM

SSG   Video

▶ **OBJECTIVE**

1. Use the substitution method to solve a system of linear equations.

1

As we stated in the preceding section, graphing alone is not an accurate way to solve a system of linear equations. In this section, we discuss a second, more accurate method for solving systems of equations. This method is called the **substitution method** and is introduced in the next example.

Example 1 Solve the system:

$$\begin{cases} 2x + y = 10 & \text{First equation} \\ x = y + 2 & \text{Second equation} \end{cases}$$

Solution   The second equation in this system is $x = y + 2$. This tells us that x and $y + 2$ have the same value. This means that we may substitute $y + 2$ for x in the first equation.

$$2x + y = 10 \qquad \text{First equation}$$

$$2(y + 2) + y = 10 \qquad \text{Substitute } y + 2 \text{ for } x \text{ since } x = y + 2.$$

Notice that this equation now has one variable, y. Let's now solve this equation for y.

$$2(y + 2) + y = 10$$

> **HELPFUL HINT**
> Don't forget the distributive property.

$$2y + 4 + y = 10 \qquad \text{Use the distributive property.}$$
$$3y + 4 = 10 \qquad \text{Combine like terms.}$$
$$3y = 6 \qquad \text{Subtract 4 from both sides.}$$
$$y = 2 \qquad \text{Divide both sides by 3.}$$

Now we know that the y-value of the ordered pair solution of the system is 2. To find the corresponding x-value, we replace y with 2 in the equation $x = y + 2$ and solve for x.

$$x = y + 2$$
$$x = 2 + 2 \qquad \text{Let } y = 2.$$
$$x = 4$$

The solution of the system is the ordered pair $(4, 2)$. Since an ordered pair solution must satisfy both linear equations in the system, we could have chosen the equation $2x + y = 10$ to find the corresponding x-value. The resulting x-value is the same.

Check   We check to see that $(4, 2)$ satisfies both equations of the original system.

First Equation	**Second Equation**
$2x + y = 10$	$x = y + 2$
$2(4) + 2 \stackrel{?}{=} 10$	$4 \stackrel{?}{=} 2 + 2 \qquad \text{Let } x = 4 \text{ and } y = 2.$
$10 = 10 \qquad \text{True.}$	$4 = 4 \qquad \text{True.}$

The solution of the system is $(4, 2)$.

A graph of the two equations shows the two lines intersecting at the point $(4, 2)$.

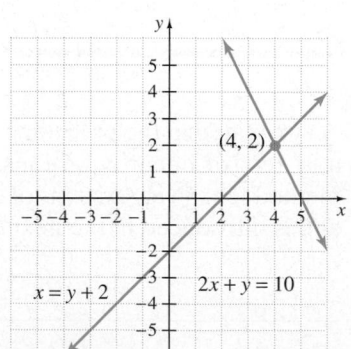

To solve a system of equations by substitution, we first need an equation solved for one of its variables.

Example 2   Solve the system:

$$\begin{cases} x + 2y = 7 \\ 2x + 2y = 13 \end{cases}$$

Solution   We choose one of the equations and solve for x or y. We will solve the first equation for x by subtracting $2y$ from both sides.

$$x + 2y = 7 \qquad \text{First equation}$$
$$x = 7 - 2y \qquad \text{Subtract } 2y \text{ from both sides.}$$

Since $x = 7 - 2y$, we now substitute $7 - 2y$ for x in the second equation and solve for y.

$$2x + 2y = 13 \qquad \text{Second equation}$$
$$2(7 - 2y) + 2y = 13 \qquad \text{Let } x = 7 - 2y.$$
$$14 - 4y + 2y = 13 \qquad \text{Use the distributive property.}$$
$$14 - 2y = 13 \qquad \text{Simplify.}$$
$$-2y = -1 \qquad \text{Subtract 14 from both sides.}$$
$$y = \frac{1}{2} \qquad \text{Divide both sides by } -2.$$

> **HELPFUL HINT**
> Don't forget to insert parentheses when substituting $7 - 2y$ for x.

To find x, we let $y = \dfrac{1}{2}$ in the equation $x = 7 - 2y$.

$$x = 7 - 2y$$
$$x = 7 - 2\left(\frac{1}{2}\right) \qquad \text{Let } y = \frac{1}{2}.$$
$$x = 7 - 1$$
$$x = 6$$

The solution is $\left(6, \dfrac{1}{2}\right)$. Check the solution in both equations of the original system.

The following steps may be used to solve a system of equations by the substitution method.

SOLVING A SYSTEM OF LINEAR EQUATIONS BY THE SUBSTITUTION METHOD

Step 1. Solve one of the equations for one of its variables.
Step 2. Substitute the expression for the variable found in Step 1 into the other equation.
Step 3. Solve the equation from Step 2 to find the value of one variable.
Step 4. Substitute the value found in Step 3 in any equation containing both variables to find the value of the other variable.
Step 5. Check the proposed solution in the original system.

Example 3 Solve the system: $\begin{cases} 7x - 3y = -14 \\ -3x + y = 6 \end{cases}$

Solution To avoid introducing fractions, we will solve the second equation for y.

$$-3x + y = 6 \qquad \textit{Second equation}$$
$$y = 3x + 6$$

Next, substitute $3x + 6$ for y in the first equation.

$$7x - 3y = -14 \qquad \textit{First equation}$$
$$7x - 3(3x + 6) = -14$$
$$7x - 9x - 18 = -14$$
$$-2x - 18 = -14$$
$$-2x = 4$$
$$\frac{-2x}{-2} = \frac{4}{-2}$$
$$x = -2$$

To find the corresponding y-value, substitute -2 for x in the equation $y = 3x + 6$. Then $y = 3(-2) + 6$ or $y = 0$. The solution of the system is $(-2, 0)$. Check this solution in both equations of the system.

> ▼ **HELPFUL HINT**
> When solving a system of equations by the substitution method, begin by solving an equation for one of its variables. If possible, solve for a variable that has a coefficient of 1 or -1. This way, we avoid working with time-consuming fractions.

Example 4 Solve the system: $\begin{cases} \dfrac{1}{2}x - y = 3 \\ x = 6 + 2y \end{cases}$

Solution The second equation is already solved for x in terms of y. Thus we substitute $6 + 2y$ for x in the first equation and solve for y.

$$\frac{1}{2}x - y = 3 \qquad \textit{First equation}$$
$$\frac{1}{2}(6 + 2y) - y = 3 \qquad \textit{Let } x = 6 + 2y.$$
$$3 + y - y = 3$$
$$3 = 3$$

Arriving at a true statement such as $3 = 3$ indicates that the two linear equations in the original system are equivalent. This means that their graphs are identical and there is an infinite number of solutions of the system. Any solution of one equation is also a solution of the other.

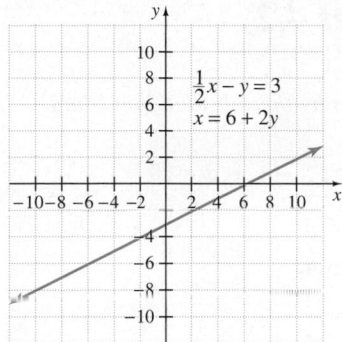

Example 5   Use substitution to solve the system.

$$\begin{cases} 6x + 12y = 5 \\ -4x - 8y = 0 \end{cases}$$

Solution   Choose the second equation and solve for y.

$$-4x - 8y = 0 \qquad \text{Second equation}$$

$$-8y = 4x \qquad \text{Add } 4x \text{ to both sides.}$$

$$\frac{-8y}{-8} = \frac{4x}{-8} \qquad \text{Divide both sides by } -8.$$

$$y = -\frac{1}{2}x \qquad \text{Simplify.}$$

Now replace y with $-\dfrac{1}{2}x$ in the first equation.

$$6x + 12y = 5 \qquad \text{First equation}$$

$$6x + 12\left(-\frac{1}{2}x\right) = 5 \qquad \text{Let } y = -\frac{1}{2}x.$$

$$6x + (-6x) = 5 \qquad \text{Simplify.}$$

$$0 = 5 \qquad \text{Combine like terms.}$$

The false statement $0 = 5$ indicates that this system has no solution and is inconsistent. The graph of the linear equations in the system shows a pair of parallel lines.

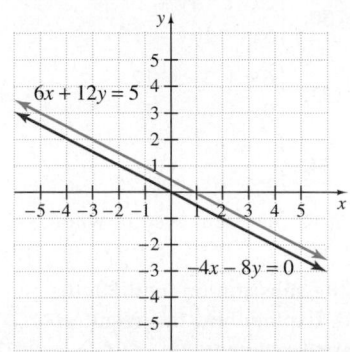

Exercise Set 4.2

Solve each system of equations by the substitution method. See Examples 1 through 5.

1. $\begin{cases} x + y = 3 \\ x = 2y \end{cases}$

2. $\begin{cases} x + y = 20 \\ x = 3y \end{cases}$

3. $\begin{cases} x + y = 6 \\ y = -3x \end{cases}$

4. $\begin{cases} x + y = 6 \\ y = 4x \end{cases}$

5. $\begin{cases} 3x + 2y = 16 \\ x = 3y - 2 \end{cases}$

6. $\begin{cases} 2x + 3y = 18 \\ x = 2y - 5 \end{cases}$

7. $\begin{cases} 3x - 4y = 10 \\ x = 2y \end{cases}$

8. $\begin{cases} 3x - 4y = 10 \\ y = 2x \end{cases}$

9. $\begin{cases} y = 3x + 1 \\ 4y - 8x = 12 \end{cases}$

10. $\begin{cases} y = 2x + 3 \\ 5y - 7x = 18 \end{cases}$

11. $\begin{cases} x + 2y = 6 \\ 2x + 3y = 8 \end{cases}$

12. $\begin{cases} x + 3y = -5 \\ 2x + 2y = 6 \end{cases}$

13. $\begin{cases} 2x - 5y = 1 \\ 3x + y = -7 \end{cases}$

14. $\begin{cases} 4x + 2y = 5 \\ 2x + y = -4 \end{cases}$

15. $\begin{cases} 2y = x + 2 \\ 6x - 12y = 0 \end{cases}$

16. $\begin{cases} 3y = x + 6 \\ 4x + 12y = 0 \end{cases}$

17. $\begin{cases} \dfrac{1}{3}x - y = 2 \\ x - 3y = 6 \end{cases}$

18. $\begin{cases} \dfrac{1}{4}x - 2y = 1 \\ x - 8y = 4 \end{cases}$

19. $\begin{cases} 4x + y = 11 \\ 2x + 5y = 1 \end{cases}$

20. $\begin{cases} 3x + y = -14 \\ 4x + 3y = -22 \end{cases}$

21. $\begin{cases} 2x - 3y = -9 \\ 3x = y + 4 \end{cases}$

22. $\begin{cases} 8x - 3y = -4 \\ 7x = y + 3 \end{cases}$

23. $\begin{cases} 6x - 3y = 5 \\ x + 2y = 0 \end{cases}$

24. $\begin{cases} 10x - 5y = -21 \\ x + 3y = 0 \end{cases}$

25. $\begin{cases} 3x - y = 1 \\ 2x - 3y = 10 \end{cases}$

26. $\begin{cases} 2x - y = -7 \\ 4x - 3y = -11 \end{cases}$

27. $\begin{cases} -x + 2y = 10 \\ -2x + 3y = 18 \end{cases}$

28. $\begin{cases} -x + 3y = 18 \\ -3x + 2y = 19 \end{cases}$

29. $\begin{cases} 5x + 10y = 20 \\ 2x + 6y = 10 \end{cases}$

30. $\begin{cases} 2x + 4y = 6 \\ 5x + 10y = 15 \end{cases}$

31. $\begin{cases} 3x + 6y = 9 \\ 4x + 8y = 16 \end{cases}$

32. $\begin{cases} 6x + 3y = 12 \\ 9x + 6y = 15 \end{cases}$

33. $\begin{cases} y = 2x + 9 \\ y = 7x + 10 \end{cases}$

34. $\begin{cases} y = 5x - 3 \\ y = 8x + 4 \end{cases}$

35. Explain how to identify an inconsistent system when using the substitution method.

36. Occasionally, when using the substitution method, the equation $0 = 0$ is obtained. Explain how this result indicates that the equations are dependent.

Solve each system by the substitution method. First simplify each equation by combining like terms.

37. $-5y + 6y = 3x + 2(x - 5) - 3x + 5$
$4(x + y) - x + y = -12$

38. $5x + 2y - 4x - 2y = 2(2y + 6) - 7$
$3(2x - y) - 4x = 1 + 9$

39. For the years 1960 through 1995, the annual percentage y of U.S. households that used fuel oil to heat their homes is given by the equation $y = -0.65x + 32.02$, where x is the number of years since 1960. For the same period, the annual percentage y of U.S. households that used electricity to heat their homes is given by the equation $y = 0.78x + 1.32$, where x is the number of years since 1960. (*Source:* Based on data from the U.S. Bureau of the Census)

  a. Use the substitution method to solve this system of equations. (Round your final results to the nearest whole numbers.)

  b. Explain the meaning of your answer to part **a**.

  c. Sketch a graph of the system of equations. Write a sentence describing the use of fuel oil and electricity for heating homes between 1960 and 1995.

40. The number y of music CDs (in millions) shipped to retailers in the United States from 1990 through 1998 is given by the equation $y = 74.5x + 289.3$, where x is the number of years since 1990. The number y of music cassettes (in millions) shipped to retailers in the United States from 1990 through 1998 is given by the equation $y = -34.1x + 434.5$, where x is the number of years since 1990. (*Source:* Based on data from the Recording Industry Association of America)

  a. Use the substitution method to solve this system of equations. (Round your final results to the nearest tenth.)

  b. Explain the meaning of your answer to part **a**.

  c. Sketch a graph of the system of equations. Write a sentence describing the trends in the popularity of these two types of music formats.

Use a graphing calculator to solve each system.

41. $\begin{cases} y = 5.1x + 14.56 \\ y = -2x - 3.9 \end{cases}$ **42.** $\begin{cases} y = 3.1x - 16.35 \\ y = -9.7x + 28.45 \end{cases}$

43. $\begin{cases} 3x + 2y = 14.05 \\ 5x + y = 18.5 \end{cases}$ **44.** $\begin{cases} x + y = -15.2 \\ -2x + 5y = -19.3 \end{cases}$

REVIEW EXERCISES

Write equivalent equations by multiplying both sides of the given equation by the given nonzero number. See Section 2.3.

45 $3x + 2y - 6;$ 2   **46.** $-x + y = 10;$ 5

47. $-4x + y = 3;$ 3   **48.** $5a - 7b = -4;$ -4

Add. See Section 2.1.

49. $\begin{array}{r} 3n + 6m \\ \underline{2n - 6m} \end{array}$ **50.** $\begin{array}{r} -2x + 5y \\ \underline{2x + 11y} \end{array}$

51. $\begin{array}{r} -5a - 7b \\ \underline{5a - 8b} \end{array}$ **52.** $\begin{array}{r} 9q + p \\ \underline{-9q - p} \end{array}$

4.3 SOLVING SYSTEMS OF LINEAR EQUATIONS BY ADDITION

CD-ROM

SSM

▶ **OBJECTIVE**

1. Use the addition method to solve a system of linear equations.

SSG

Video

1

We have seen that substitution is an accurate way to solve a linear system. Another method for solving a system of equations accurately is the **addition** or **elimination method**. The addition method is based on the addition property of equality: adding equal quantities to both sides of an equation does not change the solution of the equation. In symbols,

$$\text{if } A = B \text{ and } C = D, \text{ then } A + C = B + D.$$

Example 1 Solve the system: $\begin{cases} x + y = 7 \\ x - y = 5 \end{cases}$

Solution Since the left side of each equation is equal to the right side, we add equal quantities by adding the left sides of the equations together and the right sides of the equations together. If we choose wisely, this adding gives us an equation in one variable, x, which we can solve for x.

$$\begin{array}{ll} x + y = 7 & \text{First equation} \\ \underline{x - y = 5} & \text{Second equation} \\ 2x \quad\;\; = 12 & \text{Add the equations.} \\ x = 6 & \text{Divide both sides by 2.} \end{array}$$

The x-value of the solution is 6. To find the corresponding y-value, let $x = 6$ in either equation of the system. We will use the first equation.

$$\begin{array}{ll} x + y = 7 & \text{First equation} \\ 6 + y = 7 & \text{Let } x = 6. \\ y = 7 - 6 & \text{Solve for } y. \\ y = 1 & \text{Simplify.} \end{array}$$

The solution is $(6, 1)$. Check this in both equations.

First Equation	**Second Equation**	
$x + y = 7$	$x - y = 5$	
$6 + 1 \stackrel{?}{=} 7$	$6 - 1 \stackrel{?}{=} 5$	Let $x = 6$ and $y = 1$.
$7 = 7$ True.	$5 = 5$ True.	

Thus, the solution of the system is $(6, 1)$ and the graphs of the two equations intersect at the point $(6, 1)$ as shown.

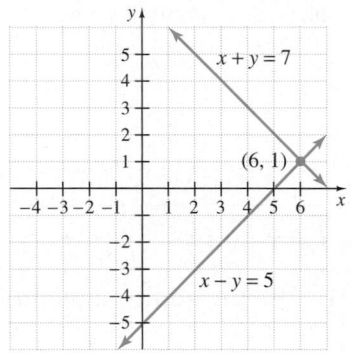

Example 2   Solve the system: $\begin{cases} -2x + y = 2 \\ -x + 3y = -4 \end{cases}$

Solution   If we simply add the two equations, the result is still an equation in two variables. However, our goal is to eliminate one of the variables. Notice what happens if we multiply *both sides* of the first equation by -3, which we are allowed to do by the multiplication property of equality. The system

$$\begin{cases} -3(-2x + y) = -3(2) \\ -x + 3y = -4 \end{cases} \quad \text{simplifies to} \quad \begin{cases} 6x - 3y = -6 \\ -x + 3y = -4 \end{cases}$$

Now add the resulting equations and the y variable is eliminated.

$$
\begin{array}{rl}
6x - 3y = -6 & \\
\underline{-x + 3y = -4} & \\
5x = -10 & \text{Add.} \\
x = -2 & \text{Divide both sides by 5.}
\end{array}
$$

To find the corresponding y-value, let $x = -2$ in any of the preceding equations containing both variables. We use the first equation of the original system.

$$
\begin{array}{rl}
-2x + y = 2 & \text{First equation} \\
-2(-2) + y = 2 & \text{Let } x = -2. \\
4 + y = 2 & \\
y = -2 &
\end{array}
$$

The solution is $(-2, -2)$. Check this ordered pair in both equations of the original system.

In Example 2, the decision to multiply the first equation by -3 was no accident. **To eliminate a variable** when adding two equations, **the coefficient of the variable in one equation must be the opposite of its coefficient in the other equation.**

> ▼ **HELPFUL HINT**
> Be sure to multiply *both sides* of an equation by a chosen number when solving by the addition method. A common mistake is to multiply only the side containing the variables.

Example 3 Solve the system: $\begin{cases} 2x - y = 7 \\ 8x - 4y = 1 \end{cases}$

Solution Multiply both sides of the first equation by -4 and the resulting coefficient of x is -8, the opposite of 8, the coefficient of x in the second equation. The system

▼ **HELPFUL HINT**
Don't forget to multiply both sides by -4.

$$\begin{cases} -4(2x - y) = -4(7) \\ 8x - 4y = 1 \end{cases} \quad \text{simplifies to} \quad \begin{cases} -8x + 4y = -28 \\ 8x - 4y = 1 \end{cases}$$

Now add the resulting equations.

$$\begin{array}{r} -8x + 4y = -28 \\ 8x - 4y = 1 \\ \hline 0 = -27 \quad \text{False.} \end{array}$$

When we add the equations, both variables are eliminated and we have $0 = -27$, a false statement. This means that the system has no solution. The equations, if graphed, are parallel lines. ∎

Example 4 Solve the system: $\begin{cases} 3x - 2y = 2 \\ -9x + 6y = -6 \end{cases}$

Solution First we multiply both sides of the first equation by 3, then we add the resulting equations.

$$\begin{cases} 3(3x - 2y) = 3(2) \\ -9x + 6y = -6 \end{cases} \quad \text{simplifies to} \quad \begin{array}{r} \begin{cases} 9x - 6y = 6 \\ -9x + 6y = -6 \end{cases} \quad \text{Add the equations.} \\ \hline 0 = 0 \end{array}$$

Both variables are eliminated and we have $0 = 0$, a true statement. Whenever you eliminate a variable and get the equation $0 = 0$, the system has an infinite number of solutions. ∎

Example 5 Solve the system: $\begin{cases} 3x + 4y = 13 \\ 5x - 9y = 6 \end{cases}$

Solution We can eliminate the variable y by multiplying the first equation by 9 and the second equation by 4.

$$\begin{cases} 9(3x + 4y) = 9(13) \\ 4(5x - 9y) = 4(6) \end{cases} \quad \text{simplifies to} \quad \begin{array}{r} \begin{cases} 27x + 36y = 117 \\ 20x - 36y = 24 \end{cases} \\ \hline 47x = 141 \\ x = 3 \end{array} \quad \text{Add the equations.}$$

To find the corresponding y-value, we let $x = 3$ in any equation in this example containing two variables. Doing so in any of these equations will give $y = 1$. The solution to this system is $(3, 1)$. Check to see that $(3, 1)$ satisfies each equation in the original system.

If we had decided to eliminate x instead of y in Example 5, the first equation could have been multiplied by 5 and the second by -3. Try solving the original system this way to check that the solution is $(3, 1)$.

The following steps summarize how to solve a system of linear equations by the addition method.

SOLVING A SYSTEM OF TWO LINEAR EQUATIONS BY THE ADDITION METHOD

Step 1. Rewrite each equation in standard form $Ax + By = C$.
Step 2. If necessary, multiply one or both equations by a nonzero number so that the coefficients of a chosen variable in the system are opposites.
Step 3. Add the equations.
Step 4. Find the value of one variable by solving the resulting equation from Step 3.
Step 5. Find the value of the second variable by substituting the value found in Step 4 into either of the original equations.
Step 6. Check the proposed solution in the original system.

Example 6 Solve the system:
$$\begin{cases} -x - \dfrac{y}{2} = \dfrac{5}{2} \\ -\dfrac{x}{2} + \dfrac{y}{4} = 0 \end{cases}$$

Solution We begin by clearing each equation of fractions. To do so, we multiply both sides of the first equation by the LCD 2 and both sides of the second equation by the LCD 4. Then the system

$$\begin{cases} 2\left(-x - \dfrac{y}{2}\right) = 2\left(\dfrac{5}{2}\right) \\ 4\left(-\dfrac{x}{2} + \dfrac{y}{4}\right) = 4(0) \end{cases}$$ simplifies to $$\begin{cases} -2x - y = 5 \\ -2x + y = 0 \end{cases}$$

Now we add the resulting equations in the simplified system.

$$\begin{array}{r} -2x - y = 5 \\ -2x + y = 0 \\ \hline -4x \quad\quad = 5 \end{array}$$ Add.

$$x = -\dfrac{5}{4}$$

To find y, we could replace x with $-\dfrac{5}{4}$ in one of the equations with two variables.

Instead, let's go back to the simplified system and multiply by appropriate factors to eliminate the variable x and solve for y. To do this, we multiply the first equation in the simplified system by -1. Then the system

$$\begin{cases} -1(-2x - y) = -1(5) \\ -2x + y = 0 \end{cases} \quad \text{simplifies to} \quad \begin{cases} 2x + y = -5 \\ \underline{-2x + y = 0} \\ 2y = -5 \quad \text{Add.} \\ y = -\dfrac{5}{2} \end{cases}$$

Check the ordered pair $\left(-\dfrac{5}{4}, -\dfrac{5}{2}\right)$ in both equations of the original system. The solution is $\left(-\dfrac{5}{4}, -\dfrac{5}{2}\right)$. ■

SPOTLIGHT ON DECISION MAKING

Suppose you have been offered two similar positions as a sales associate. In one position, you would be paid a monthly salary of $1500 plus a 2% commission on all sales you make during the month. In the other position, you would be paid a monthly salary of $500 plus a 6% commission on all sales you make during the month. Which position would you choose? Explain your reasoning. Would knowing that the sales positions were at a car dealership affect your choice? What if the positions were at a shoe store?

Exercise Set 4.3

Solve each system of equations by the addition method. See Examples 1 through 5.

1. $\begin{cases} 3x + y = 5 \\ 6x - y = 4 \end{cases}$

2. $\begin{cases} 4x + y = 13 \\ 2x - y = 5 \end{cases}$

3. $\begin{cases} x - 2y = 8 \\ -x + 5y = -17 \end{cases}$

4. $\begin{cases} x - 2y = -11 \\ -x + 5y = 23 \end{cases}$

5. $\begin{cases} x + y = 6 \\ x - y = 6 \end{cases}$

6. $\begin{cases} x - y = 1 \\ -x + 2y = 0 \end{cases}$

7. $\begin{cases} 3x + y = 4 \\ 9x + 3y = 6 \end{cases}$

8. $\begin{cases} 2x + y = 6 \\ 4x + 2y = 12 \end{cases}$

9. $\begin{cases} 3x - 2y = 7 \\ 5x + 4y = 8 \end{cases}$

10. $\begin{cases} 6x - 5y = 25 \\ 4x + 15y = 13 \end{cases}$

11. $\begin{cases} \dfrac{2}{3}x + 4y = -4 \\ 5x + 6y = 18 \end{cases}$

12. $\begin{cases} \dfrac{3}{2}x + 4y = 1 \\ 9x + 24y = 5 \end{cases}$

13. $\begin{cases} 4x - 6y = 8 \\ 6x - 9y = 12 \end{cases}$

14. $\begin{cases} 9x - 3y = 12 \\ 12x - 4y = 18 \end{cases}$

15. $\begin{cases} 3x + y = -11 \\ 6x - 2y = -2 \end{cases}$

16. $\begin{cases} 4x + y = -13 \\ 6x - 3y = -15 \end{cases}$

17. $\begin{cases} 3x + 2y = 11 \\ 5x - 2y = 29 \end{cases}$

18. $\begin{cases} 4x + 2y = 2 \\ 3x - 2y = 12 \end{cases}$

19. $\begin{cases} x + 5y = 18 \\ 3x + 2y = -11 \end{cases}$

20. $\begin{cases} x + 4y = 14 \\ 5x + 3y = 2 \end{cases}$

21. $\begin{cases} 2x - 5y = 4 \\ 3x - 2y = 4 \end{cases}$

22. $\begin{cases} 6x - 5y = 7 \\ 4x - 6y = 7 \end{cases}$

23. $\begin{cases} 2x + 3y = 0 \\ 4x + 6y = 3 \end{cases}$

24. $\begin{cases} -x + 5y = -1 \\ 3x - 15y = 3 \end{cases}$

Solve each system of equations by the addition method. See Example 6.

25. $\begin{cases} \dfrac{x}{3} + \dfrac{y}{6} = 1 \\ \dfrac{x}{2} - \dfrac{y}{4} = 0 \end{cases}$

26. $\begin{cases} \dfrac{x}{2} + \dfrac{y}{8} = 3 \\ x - \dfrac{y}{4} = 0 \end{cases}$

27. $\begin{cases} x - \dfrac{y}{3} = -1 \\ -\dfrac{x}{2} + \dfrac{y}{8} = \dfrac{1}{4} \end{cases}$

28. $\begin{cases} 2x - \dfrac{3y}{4} = -3 \\ x + \dfrac{y}{9} = \dfrac{13}{3} \end{cases}$

29. $\begin{cases} \dfrac{x}{3} - y = 2 \\ -\dfrac{x}{2} + \dfrac{3y}{2} = -3 \end{cases}$

30. $\begin{cases} \dfrac{x}{2} + \dfrac{y}{4} = 1 \\ -\dfrac{x}{4} - \dfrac{y}{8} = 1 \end{cases}$

31. $\begin{cases} 8x = -11y - 16 \\ 2x + 3y = -4 \end{cases}$

32. $\begin{cases} 10x + 3y = -12 \\ 5x = -4y - 16 \end{cases}$

33. When solving a system of equations by the addition method, how do we know when the system has no solution?

34. To solve the system $\begin{cases} 2x - 3y = 5 \\ 5x + 2y = 6 \end{cases}$, explain why the addition method might be preferred rather than the substitution method.

Solve each system by either the addition method or the substitution method.

35. $\begin{cases} 2x - 3y = -11 \\ y = 4x - 3 \end{cases}$

36. $\begin{cases} 4x - 5y = 6 \\ y = 3x - 10 \end{cases}$

37. $\begin{cases} x + 2y = 1 \\ 3x + 4y = -1 \end{cases}$

38. $\begin{cases} x + 3y = 5 \\ 5x + 6y = -2 \end{cases}$

39. $\begin{cases} 2y = x + 6 \\ 3x - 2y = -6 \end{cases}$

40. $\begin{cases} 3y = x + 14 \\ 2x - 3y = -16 \end{cases}$

41. $\begin{cases} y = 2x - 3 \\ y = 5x - 18 \end{cases}$

42. $\begin{cases} y = 6x - 5 \\ y = 4x - 11 \end{cases}$

43. $\begin{cases} x + \dfrac{1}{6}y = \dfrac{1}{2} \\ 3x + 2y = 3 \end{cases}$

44. $\begin{cases} x + \dfrac{1}{3}y = \dfrac{5}{12} \\ 8x + 3y = 4 \end{cases}$

45. $\begin{cases} \dfrac{x + 2}{2} = \dfrac{y + 11}{3} \\ \dfrac{x}{2} = \dfrac{2y + 16}{6} \end{cases}$

46. $\begin{cases} \dfrac{x + 5}{2} = \dfrac{y + 14}{4} \\ \dfrac{x}{3} = \dfrac{2y + 2}{6} \end{cases}$

Solve each system by the addition method.

47. $\begin{cases} 2x + 3y = 14 \\ 3x - 4y = -69.1 \end{cases}$

48. $\begin{cases} 5x - 2y = -19.8 \\ -3x + 5y = -3.7 \end{cases}$

49. Commercial broadcast television stations can be divided into VHF stations (channels 2 through 13) and UHF stations (channels 14 through 83). The number y of VHF stations in the United States from 1985 through 1997 is given by the equation $10x - 2y = -986$, where x is the number of years after 1980. The number y of UHF stations in the United States from 1985 through 1997 is given by the equation $-21x + y = 295$, where x is the number of years after 1980. (*Source:* Based on data from the Television Bureau of Advertising, Inc.)

a. Use the addition method to solve this system of equations. (Round your final results to the nearest whole numbers.)

b. Interpret your solution from part **a**.

c. During which years were there more UHF commercial television stations than VHF stations?

50. In recent years, the number of daily newspapers printed as morning editions has been increasing and the number of daily newspapers printed as evening editions has been decreasing. The number y of daily morning newspapers in existence from 1980 through 1998 is given by the equation $-665x + 36y = 13,800$, where x is the number of years after 1980. The number y of daily evening newspapers in existence from 1980 through 1998 is given by the equation $3239x + 96y = 134,013.6$, where x is the number of years after 1980. (*Source:* Based on data from the *Editor and Publisher International Year Book*, Editor and Publisher Co., New York, NY, annual)

a. Suppose these trends continue in the future. Use the addition method to predict the year in which the number of morning newspapers will equal the number of evening newspapers. (Round to the nearest whole number.)

b. How many of each type of newspaper will be in existence in that year?

51. Use the system of linear equations below to answer the questions.

$$\begin{cases} x + y = 5 \\ 3x + 3y = b \end{cases}$$

a. Find the value of b so that the equations are dependent and the system has an infinite number of solutions.

b. Find a value of b so that the system is inconsistent and there are no solutions to the system.

52. Use the system of linear equations below to answer the questions.

$$\begin{cases} x + y = 4 \\ 2x + by = 8 \end{cases}$$

a. Find the value of b so that the equations are dependent and the system has an infinite number of solutions.

b. Find a value of b so that the system is consistent and the system has a single solution.

53. Suppose you are solving the system

$$\begin{cases} -4x + 7y = 6 \\ x + 2y = 5 \end{cases}$$

by the addition method.

a. What step(s) should you take if you wish to eliminate x when adding the equations?

b. What step(s) should you take if you wish to eliminate y when adding the equations?

54. Suppose you are solving the system

$$\begin{cases} 3x + 8y = -5 \\ 2x - 4y = 3 \end{cases}$$

You decide to use the addition method by multiplying both sides of the second equation by ? In which of the following was the multiplication performed correctly? Explain.

a. $4x - 8y = 3$ b. $4x - 8y = 6$

REVIEW EXERCISES

Rewrite the following sentences using mathematical symbols. Do not solve the equations. See Sections 2.3 and 2.4.

55. Twice a number, added to 6 is 3 less than the number.

56. The sum of three consecutive integers is 66.

57. Three times a number, subtracted from 20 is 2.

58. Twice the sum of 8 and a number is the difference of the number and 20.

59. The product of 4 and the sum of a number and 6 is twice the number.

60. The quotient of twice a number and 7 is subtracted from the reciprocal of the number.

4.4 SYSTEMS OF LINEAR EQUATIONS AND PROBLEM SOLVING

CD-ROM   SSM

SSG   Video

▶ **OBJECTIVE**

1. Use a system of equations to solve problems.

1   Many of the word problems solved earlier using one-variable equations can also be solved using two equations in **two** variables. We use the same problem-solving steps that have been used throughout this text. The only difference is that two variables are assigned to represent the two unknown quantities and that the problem is translated into **two** equations.

PROBLEM-SOLVING STEPS

1. UNDERSTAND the problem. During this step, become comfortable with the problem. Some ways of doing this are to

    Read and reread the problem.
    Choose two variables to represent the two unknowns.
    Construct a drawing.
    Propose a solution and check. Pay careful attention to how you check your proposed solution. This will help when writing equations to model the problem.

2. TRANSLATE the problem into two equations.

3. SOLVE the system of equations.

4. INTERPRET the results: **Check** the proposed solution in the stated problem and **state** your conclusion.

◆**Example 1**   **FINDING UNKNOWN NUMBERS**

Find two numbers whose sum is 37 and whose difference is 21.

Solution   1. UNDERSTAND. Read and reread the problem. Suppose that one number is 20. If their sum is 37, the other number is 17 because $20 + 17 = 37$. Is their difference 21? No; $20 - 17 = 3$. Our proposed solution is incorrect, but we now have a better understanding of the problem.

   Since we are looking for two numbers, we let

   x = first number
   y = second number

2. TRANSLATE. Since we have assigned two variables to this problem, we translate our problem into two equations.

   In words:

two numbers whose sum	is	37
↓	↓	↓

   Translate: $x + y$ $=$ 37

   In words:

two numbers whose difference	is	21
↓	↓	↓

   Translate: $x - y$ $=$ 21

3. SOLVE. Now we solve the system

   $$\begin{cases} x + y = 37 \\ x - y = 21 \end{cases}$$

   Notice that the coefficients of the variable y are opposites. Let's then solve by the addition method and begin by adding the equations.

   $$\begin{array}{r} x + y = 37 \\ \underline{x - y = 21} \\ 2x = 58 \end{array}$$ Add the equations.

   $$x = \frac{58}{2} = 29$$ Divide both sides by 2.

Now we let $x = 29$ in the first equation to find y.

$$x + y = 37 \qquad \text{First equation}$$
$$29 + y = 37$$
$$y = 37 - 29 = 8$$

4. INTERPRET. The solution of the system is $(29, 8)$.

Check: Notice that the sum of 29 and 8 is $29 + 8 = 37$, the required sum. Their difference is $29 - 8 = 21$, the required difference.

State: The numbers are 29 and 8.

Example 2 SOLVING A PROBLEM ABOUT PRICES

A local high school is presenting the play "Grease." Admission for 4 adults and 2 children is $22, while admission for 2 adults and 3 children is $16.

a. What is the price of an adult's ticket?
b. What is the price of a child's ticket?
c. A special rate of $60 is charged for groups of 20 persons. Should a group of 4 adults and 16 children use the group rate? Why or why not?

Solution

1. UNDERSTAND. Read and reread the problem and guess a solution. Let's suppose that the price of an adult's ticket is $5 and the price of a child's ticket is $4. To check our proposed solution, let's see if admission for 4 adults and 2 children is $22. Admission for 4 adults is 4($5) or $20 and admission for 2 children is 2($4) or $8. This gives a total admission of $20 + $8 = $28, not the required $22. Again though, we have accomplished the purpose of this process: We have a better understanding of the problem. To continue, we let

A = the price of an adult's ticket

C = the price of a child's ticket

2. TRANSLATE. We translate the problem into two equations using both variables.

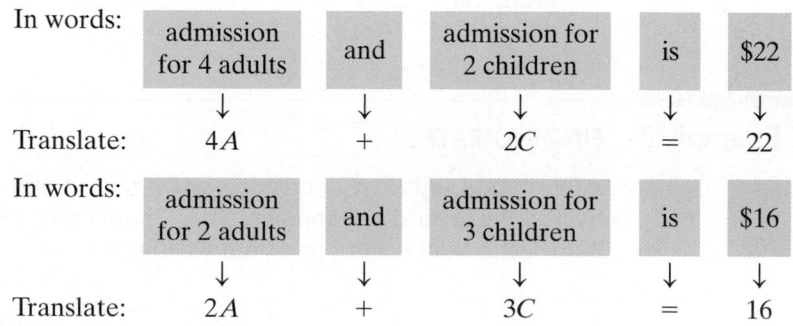

In words:	admission for 4 adults	and	admission for 2 children	is	$22
	↓	↓	↓	↓	↓
Translate:	$4A$	$+$	$2C$	$=$	22

In words:	admission for 2 adults	and	admission for 3 children	is	$16
	↓	↓	↓	↓	↓
Translate:	$2A$	$+$	$3C$	$=$	16

3. SOLVE. We solve the system

$$\begin{cases} 4A + 2C = 22 \\ 2A + 3C = 16 \end{cases}$$

Since both equations are written in standard form, we solve by the addition method. First we multiply the second equation by −2 to eliminate the variable A. Then the system

$$\begin{cases} 4A + 2C = 22 \\ -2(2A + 3C) = -2(16) \end{cases} \quad \text{simplifies to} \quad \begin{cases} 4A + 2C = 22 \\ -4A - 6C = -32 \end{cases} \quad \text{Add the equations.}$$

$$-4C = -10$$

$$C = \frac{-10}{-4} = \frac{5}{2}$$

$$C = \frac{5}{2} = 2.5 \text{ or } \$2.50, \text{ the children's ticket price.}$$

To find A, we replace C with 2.5 in the first equation.

$$4A + 2C = 22 \qquad \text{First equation}$$
$$4A + 2(2.5) = 22 \qquad \text{Let } C = 2.5.$$
$$4A + 5 = 22$$
$$4A = 17$$
$$A = \frac{17}{4} = 4.25 \text{ or } \$4.25, \text{ the adult's ticket price.}$$

4. **INTERPRET.**

   Check: Notice that 4 adults and 2 children will pay $4(\$4.25) + 2(\$2.50) =$ $\$17 + \$5 = \$22$, the required amount. Also, the price for 2 adults and 3 children is $2(\$4.25) + 3(\$2.50) = \$8.50 + \$7.50 = \$16$, the required amount.

   State: Answer the three original questions.

   a. Since $A = 4.25$, the price of an adult's ticket is $4.25.

   b. Since $C = 2.5$, the price of a child's ticket is $2.50.

   c. The regular admission price for 4 adults and 16 children is

   $$4(\$4.25) + 16(\$2.50) = \$17.00 + \$40.00$$
   $$= \$57.00$$

   This is $3 less than the special group rate of $60, so they should *not* request the group rate.

Example 3 FINDING RATES

Albert and Louis live 15 miles away from each other. They decide to meet one day by walking toward one another. After 2 hours they meet. If Louis walks one mile per hour faster than Albert, find both walking speeds.

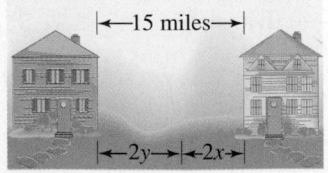

1. **UNDERSTAND.** Read and reread the problem. Let's propose a solution and use the formula $d = r \cdot t$ to check. Suppose that Louis's rate is 4 miles per hour. Since Louis's rate is 1 mile per hour faster, Albert's rate is 3 miles per hour. To check, see if they can walk a total of 15 miles in 2 hours. Louis's distance is rate $\cdot$ time $= 4(2) = 8$ miles and Albert's distance is rate time $= 3(2) = 6$ miles. Their total distance is 8 miles $+$ 6 miles $= 14$ miles, not the required 15 miles. Now that we have a better understanding of the problem, let's model it with a system of equations.

First, we let

$x =$ Albert's rate in miles per hour

$y =$ Louis's rate in miles per hour

Now we use the facts stated in the problem and the formula $d = rt$ to fill in the following chart.

	r	$\cdot$	t	$=$	d
ALBERT	x		2		$2x$
LOUIS	y		2		$2y$

2. **TRANSLATE.** We translate the problem into two equations using both variables.

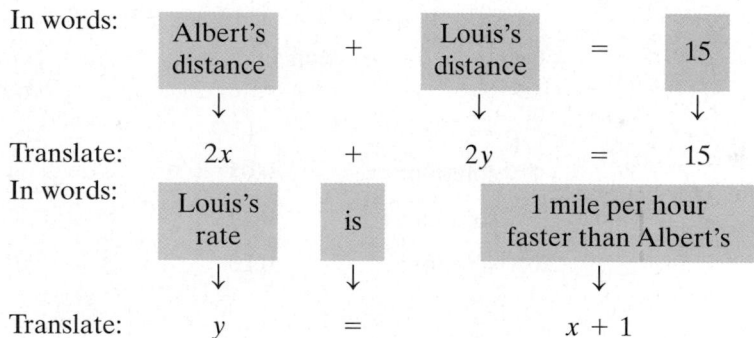

In words:

| Albert's distance | $+$ | Louis's distance | $=$ | 15 |

$\downarrow \qquad\qquad \downarrow \qquad\qquad \downarrow$

Translate: $\qquad 2x \qquad + \qquad 2y \qquad = \qquad 15$

In words:

| Louis's rate | is | 1 mile per hour faster than Albert's |

$\downarrow \qquad \downarrow \qquad\qquad\qquad \downarrow$

Translate: $\qquad y \qquad = \qquad\qquad x + 1$

3. **SOLVE.** The system of equations we are solving is

$$\begin{cases} 2x + 2y = 15 \\ y = x + 1 \end{cases}$$

Let's use substitution to solve the system since the second equation is solved for y.

$$2x + 2y = 15 \qquad \text{First equation}$$

$$2x + 2(x + 1) = 15 \qquad \text{Replace } y \text{ with } x + 1.$$

$$2x + 2x + 2 = 15$$

$$4x = 13$$

$$x = \frac{13}{4} = 3.25$$

$$y = x + 1 = 3.25 + 1 = 4.25$$

4. INTERPRET. Albert's proposed rate is 3.25 miles per hour and Louis's proposed rate is 4.25 miles per hour.

Check: Use the formula $d = rt$ and find that in 2 hours, Albert's distance is $(3.25)(2)$ miles or 6.5 miles. In 2 hours, Louis's distance is $(4.25)(2)$ miles or 8.5 miles. The total distance walked is 6.5 miles + 8.5 miles or 15 miles, the given distance.

State: Albert walks at a rate of 3.25 miles per hour and Louis walks at a rate of 4.25 miles per hour.

Example 4   FINDING AMOUNTS OF SOLUTIONS

Eric Daly, a chemistry teaching assistant, needs 10 liters of a 20% saline solution (salt water) for his 2 p.m. laboratory class. Unfortunately, the only mixtures on hand are a 5% saline solution and a 25% saline solution. How much of each solution should he mix to produce the 20% solution?

Solution   1. **UNDERSTAND.** Read and reread the problem. Suppose that we need 4 liters of the 5% solution. Then we need $10 - 4 = 6$ liters of the 25% solution. To see if this gives us 10 liters of a 20% saline solution, let's find the amount of pure salt in each solution.

	concentration rate	×	amount of solution	=	amount of pure salt
	↓		↓		↓
5% solution:	0.05	×	4 liters	=	0.2 liters
25% solution:	0.25	×	6 liters	=	1.5 liters
20% solution:	0.20	×	10 liters	=	2 liters

Since 0.2 liters + 1.5 liters = 1.7 liters, not 2 liters, our proposed solution is incorrect. But we have gained some insight into how to model and check this problem.

We let

x = number of liters of 5% solution

y = number of liters of 25% solution

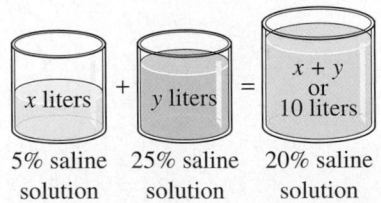

5% saline   25% saline   20% saline
solution   solution   solution

Now we use a table to organize the given data.

	Concentration Rate	Liters of Solution	Liters of Pure Salt
FIRST SOLUTION	5%	x	$0.05x$
SECOND SOLUTION	25%	y	$0.25y$
MIXTURE NEEDED	20%	10	$(0.20)(10)$

2. **TRANSLATE.** We translate into two equations using both variables.

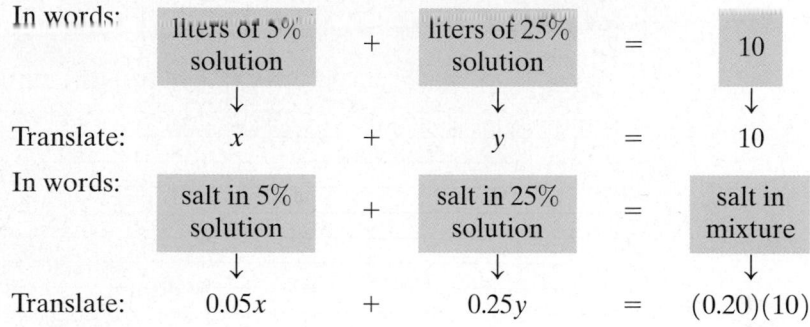

In words:

liters of 5% solution	+	liters of 25% solution	=	10

Translate: x + y =   10

In words:

salt in 5% solution	+	salt in 25% solution	=	salt in mixture

Translate: $0.05x$ + $0.25y$ = $(0.20)(10)$

3. **SOLVE.** Here we solve the system

$$\begin{cases} x + y = 10 \\ 0.05x + 0.25y = 2 \end{cases}$$

To solve by the addition method, we first multiply the first equation by -25 and the second equation by 100. Then the system

$$\begin{cases} -25(x + y) = -25(10) \\ 100(0.05x + 0.25y) = 100(2) \end{cases}$$

simplifies to

$$\begin{cases} -25x - 25y = -250 \\ \underline{5x + 25y = 200} \\ -20x = -50 \quad \text{Add.} \\ x = 2.5 \end{cases}$$

To find y, we let $x = 2.5$ in the first equation of the original system.

$$x + y = 10$$
$$2.5 + y = 10 \quad \text{Let } x = 2.5.$$
$$y = 7.5$$

4. **INTERPRET.** Thus, we propose that Eric needs to mix 2.5 liters of 5% saline solution with 7.5 liters of 25% saline solution.

Check: Notice that $2.5 + 7.5 = 10$, the required number of liters. Also, the sum of the liters of salt in the two solutions equals the liters of salt in the required mixture:

$$0.05(2.5) + 0.25(7.5) = 0.20(10)$$

$$0.125 + 1.875 = 2$$

State: Eric needs 2.5 liters of the 5% saline solution and 7.5 liters of the 25% solution.

SPOTLIGHT ON DECISION MAKING

Suppose you are a registered nurse. Today you are working at a health fair providing free blood pressure screenings. You measure an attendee's blood pressure as 168/82 (read as "168 over 82," where the systolic blood pressure is listed first and the diastolic blood pressure is listed second). What would you recommend that this health fair attendee do? Explain.

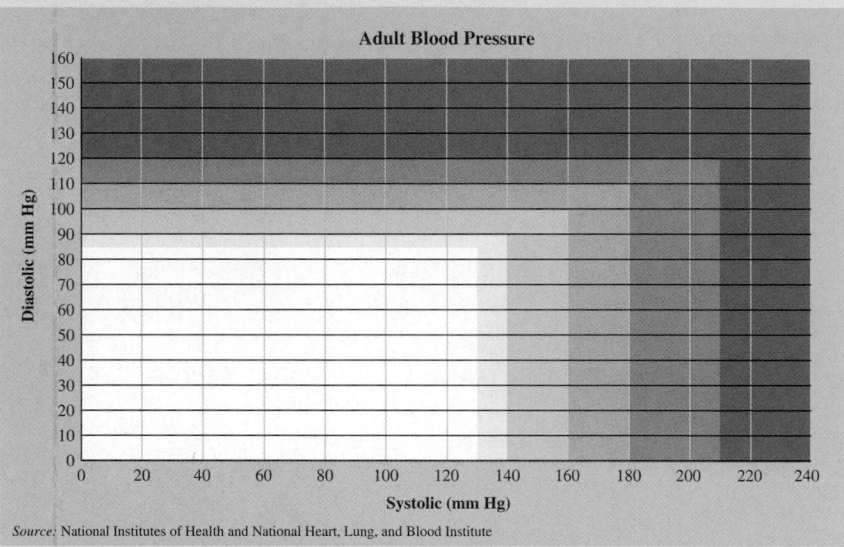

Adult Blood Pressure

Diastolic (mm Hg) / Systolic (mm Hg)

Source: National Institutes of Health and National Heart, Lung, and Blood Institute

Blood Pressure Category and Recommended Follow-up:
- ☐ Normal: recheck in 2 years
- ☐ High normal: recheck in 1 year
- ☐ Mild hypertension: confirm within 2 months
- ☐ Moderate hypertension: see primary care physician within 1 month
- ☐ Severe hypertension: see primary care physician within 1 week
- ☐ Very severe hypertension: see primary care physician immediately

Exercise Set 4.4

Without actually solving each problem, choose each correct solution by deciding which choice satisfies the given conditions.

△ **1.** The length of a rectangle is 3 feet longer than the width. The perimeter is 30 feet. Find the dimensions of the rectangle.
   a. length = 8 feet; width = 5 feet
   b. length = 8 feet; width = 7 feet
   c. length = 9 feet; width = 6 feet

△ **2.** An isosceles triangle, a triangle with two sides of equal length, has a perimeter of 20 inches. Each of the equal sides is one inch longer than the third side. Find the lengths of the three sides.
   a. 6 inches, 6 inches, and 7 inches
   b. 7 inches, 7 inches, and 6 inches
   c. 6 inches, 7 inches, and 8 inches

3. Two computer disks and three notebooks cost $17. However, five computer disks and four notebooks cost $32. Find the price of each.
   a. notebook = $4;   computer disk = $3
   b. notebook = $3;   computer disk = $4
   c. notebook = $5;   computer disk = $2

4. Two music CDs and four music cassette tapes cost a total of $40. However, three music CDs and five cassette tapes cost $55. Find the price of each.
   a. CD = $12; cassette = $4
   b. CD = $15; cassette = $2
   c. CD = $10; cassette = $5

5. Kesha has a total of 100 coins, all of which are either dimes or quarters. The total value of the coins is $13.00. Find the number of each type of coin.
   a. 80 dimes; 20 quarters **b.** 20 dimes; 44 quarters
   c. 60 dimes; 40 quarters

6. Yolanda has 28 gallons of saline solution available in two large containers at her pharmacy. One container holds three times as much as the other container. Find the capacity of each container.
   a. 15 gallons; 5 gallons **b.** 20 gallons; 8 gallons
   c. 21 gallons; 7 gallons

Write a system of equations describing each situation. Do not solve the system. See Example 1.

7. Two numbers add up to 15 and have a difference of 7.

8. The total of two numbers is 16. The first number plus 2 more than 3 times the second equals 18.

9. Keiko has a total of $6500, which she has invested in two accounts. The larger account is $800 greater than the smaller account.

10. Dominique has four times as much money in his savings account as in his checking account. The total amount is $2300.

Solve. See Examples 1 through 4.

11. Two numbers total 83 and have a difference of 17. Find the two numbers.

12. The sum of two numbers is 76 and their difference is 52. Find the two numbers.

13. A first number plus twice a second number is 8. Twice the first number plus the second totals 25. Find the numbers.

14. One number is 4 more than twice the second number. Their total is 25. Find the numbers.

15. The highest scorer during the WNBA 1999 regular season was Cynthia Cooper of the Houston Comets. Over the season, Cooper scored 101 more points than the second-highest scorer, her teammate Sheryl Swoopes. Together, Cooper and Swoopes scored 1271 points during the 1999 regular season. How many points did each player score over the course of the season? (*Source:* Women's National Basketball Association)

16. During the 1998–1999 regular NHL season, Teemu Selanne, of the Mighty Ducks of Anaheim, scored 20 fewer points than the Pittsburgh Penguins' Jaromir Jagr. Together, they scored 234 points during the 1998–1999 regular season. How many points each did Selanne and Jagr score? (*Source:* National Hockey League)

17. Ann Marie Jones has been pricing Amtrak train fares for a group trip to New York. Three adults and four children must pay $159. Two adults and three children must pay $112. Find the price of an adult's ticket, and find the price of a child's ticket.

18. Last month, Jerry Papa purchased five cassettes and two compact discs at Wall-to-Wall Sound for $65. This month he bought three cassettes and four compact discs for $81. Find the price of each cassette, and find the price of each compact disc.

19. Johnston and Betsy Waring have a jar containing 80 coins, all of which are either quarters or nickels. The total value of the coins is $14.60. How many of each type of coin do they have?

20. Art and Bette Meish purchased 40 stamps, a mixture of 32¢ and 19¢ stamps. Find the number of each type of stamp if they spent $12.15.

21. Fred and Staci Whittingham own 50 shares of IBM stock and 40 shares of GA Financial stock. At the close of the markets on March 24, 2000, their stock portfolio was worth $6485.90. The closing price of GA Financial stock was $64.25 more per share than the closing price of IBM stock on that day. What was the closing price of each stock on March 24, 2000? (*Source:* Based on data from Standard & Poor's ComStock)

22. Edie Hall has an investment in Kroger and General Motors stock. On March 24, 2000, Kroger stock closed at $17.875 per share and General Motors stock closed at $85.375 per share. Edie's portfolio was worth $11,317.50 at the end of the day. If Edie owns 60 more shares of General Motors stock than Kroger stock, how many of each type of stock does she own? (*Source:* Based on data from Standard & Poor's ComStock)

23. Pratap Puri rowed 18 miles down the Delaware River in 2 hours, but the return trip took him $4\frac{1}{2}$ hours. Find the rate Pratap could row in still water, and find the rate of the current.

	d	$=$	r	$\cdot$	t
Downstream	18		$x + y$		2
Upstream	18		$x - y$		$4\frac{1}{2}$

24. The Jonathan Schultz family took a canoe 10 miles down the Allegheny River in 1 hour and 15 minutes. After lunch it took them 4 hours to return. Find the rate of the current.

	d	$=$	r	$\cdot$	t
Downstream	10		$x + y$		$1\frac{1}{4}$
Upstream	10		$x - y$		4

25. Dave and Sandy Hartranft are frequent flyers with Delta Airlines. They often fly from Philadelphia to Chicago, a distance of 780 miles. On one particular trip they fly into the wind, and the flight takes 2 hours. The return trip, with the wind behind them, only takes $1\frac{1}{2}$ hours. Find the speed of the wind and find the speed of the plane in still air.

26. With a strong wind behind it, a United Airlines jet flies 2400 miles from Los Angeles to Orlando in 4 hours and 45 minutes. The return trip takes 6 hours, as the plane flies into the wind. Find the speed of the plane in still air, and find the wind speed to the nearest tenth of a mile per hour.

27. Dorren Schmidt is a chemist with Gemco Pharmaceutical. She needs to prepare 12 ounces of a 9% hydrochloric acid solution. Find the amount of 4% and the amount of 12% solution she should mix to get this solution.

28. Elise Everly is preparing 15 liters of a 25% saline solution. Elise has two other saline solutions with strengths of 40% and 10%. Find the amount of 40% solution and the amount of 10% solution she should mix to get 15 liters of a 25% solution.

29. Wayne Osby blends coffee for Maxwell House. He needs to prepare 200 pounds of blended coffee beans selling

for $3.95 per pound. He intends to do this by blending together a high-quality bean costing $4.95 per pound and a cheaper bean costing $2.65 per pound. To the nearest pound, find how much high-quality coffee bean and how much cheaper coffee bean he should blend.

30. Macadamia nuts cost an astounding $16.50 per pound, but research by Planter's Peanuts says that mixed nuts sell better if macadamias are included. The standard mix costs $9.25 per pound. Find how many pounds of macadamias and how many pounds of the standard mix should be combined to produce 40 pounds that will cost $10 per pound. Find the amounts to the nearest tenth of a pound.

△ **31.** Find the measures of two complementary angles if one angle is twice the other. (Recall that two angles are complementary if their sum is 90°.)

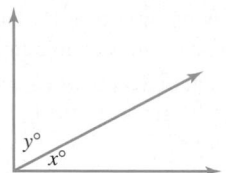

△ **32.** Find the measures of two supplementary angles if one angle is 20° more than four times the other. (Recall that two angles are supplementary if their sum is 180°.)

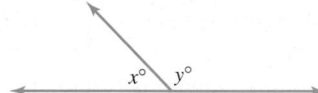

△ **33.** Find the measures of two complementary angles if one angle is 10° more than three times the other.

△ **34.** Find the measures of two supplementary angles if one angle is 18° more than twice the other.

35. Barb Hayes, a pharmacist, needs 50 liters of a 60% alcohol solution. She currently has available a 20% solution and a 70% solution. How many liters of each does she need to make the needed 50 liters of 60% alcohol solution?

36. Two cars are 440 miles apart and traveling toward each other. They meet in 3 hours. If one car's speed is 10 miles per hour faster than the other car's speed, find the speed of each car.

37. Carrie and Raymond McCormick had a pottery stand at the annual Skippack Craft Fair. They sold some of their pottery at the original price of $9.50 each, but later decreased the price of each by $2. If they sold all 90 pieces and took in $721, find how many they sold at the original price and how many they sold at the reduced price.

38. Trinity Church held its annual spaghetti supper and fed a total of 387 people. They charged $6.80 for adults and half-price for children. If they took in $2444.60, find how many adults and how many children attended the supper.

△ **39.** Dale Maxfield has decided to fence off a garden plot behind his house, using his house as the "fence" along one side of the garden. The length (which runs parallel to the house) is 3 feet less than twice the width. Find the dimensions if 33 feet of fencing is used along the three sides requiring it.

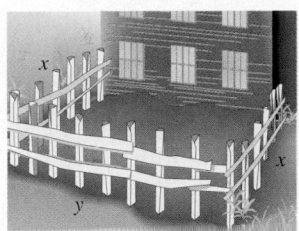

△ **40.** Judy McElroy plans to erect 152 feet of fencing around her rectangular horse pasture. A river bank serves as one side of the rectangle. If each width is 4 feet longer than half the length, find the dimensions.

41. Jim Williamson began a 186-mile bicycle trip to build up stamina for a triathalon competition. Unfortunately, his bicycle chain broke, so he finished the trip walking. The whole trip took 6 hours. If Jim walks at a rate of 4 mph and rides at 40 mph, find the amount of time he spent on the bicycle.

42. Joan Gundersen rented a car from Hertz, which rents its cars for a daily fee plus an additional charge per mile driven. Joan recalls that a car rented for 5 days and driven for 300 miles cost her $178, while a car rented for 4 days and driven for 500 miles cost $197. Find the daily fee, and find the mileage charge.

43. In Canada, eastbound and westbound trains travel along the same track, with sidings to pull onto to avoid accidents. Two trains are now 150 miles apart, with the westbound train traveling twice as fast as the eastbound train. A warning must be issued to pull one train onto a siding or else the trains will crash in $1\frac{1}{4}$ hours. Find the speed of the eastbound train and the speed of the westbound train.

44. Cyril and Anoa Nantambu operate a small construction and supply company. In July they charged the Shaffers $1702.50 for 65 hours of labor and 3 tons of material. In August the Shaffers paid $1349 for 49 hours of labor and $2\frac{1}{2}$ tons of material. Find the cost per hour of labor and the cost per ton of material.

45. Suppose you mix an amount of 30% acid solution with an amount of 50% acid solution. Which of the following acid strengths would be possible for the resulting acid mixture? Explain why.

   a. 22%   **b.** 44%   **c.** 63%

REVIEW EXERCISES

Graph each linear inequality. See Section 3.7.

46. $y < 3 - x$ **47.** $y \geq 4 - 2x$
48. $2x - y \geq 6$ **49.** $3x + 5y < 15$

For additional Chapter Projects, visit the Real World Activities Website by going to http://www.prenhall.com/martin-gay.

CHAPTER PROJECT

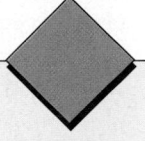

Analyzing the Courses of Ships

From overhead photographs or satellite imagery of ships on the ocean, defense analysts can tell a lot about a ship's immediate course by looking at its wake. Assuming that two ships will maintain their present courses, it is possible to extend their paths, based on the wakes visible in the photograph. This can be used to find possible points of collision for the two ships.

   In this project, you will investigate the courses and possibility of collision of the two ships shown in the figure. This project may be completed by working in groups or individually.

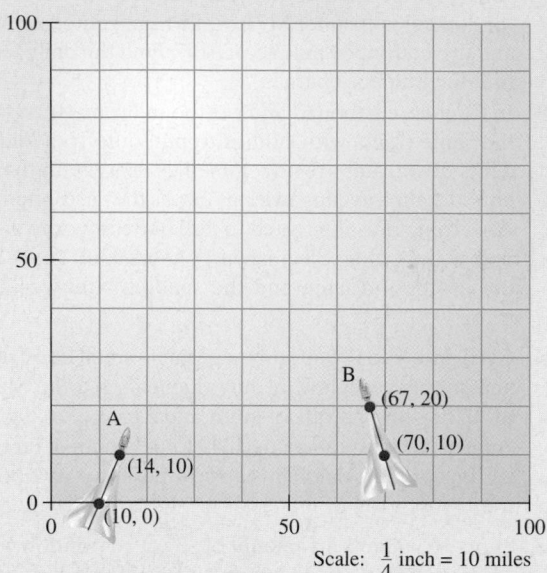

Scale: $\frac{1}{4}$ inch = 10 miles

1. Using each ship's wake as a guide, use a straight-edge to extend the paths of the ships on the figure. Estimate the coordinates of the point of intersection of the ships' courses from the grid. If the ships continue in these courses, they could possibly collide at the point of intersection of their paths.

2. What factor will govern whether or not the ships actually collide at the point found in Question 1?

3. Using the coordinates labeled on each ship's wake, find a linear equation that describes each path.

4. (Optional) Use a graphing calculator to graph both equations in the same window. Use the Intersect or Trace feature to estimate the point of intersection of the two paths. Compare this estimate to your estimate in Question 1.

5. Solve the system of two linear equations using one of the methods in this chapter. The solution is the point of intersection of the two paths. Compare your answer to your estimates from Questions 1 and 4.

6. Plot the point of intersection you found in Question 5 on the figure. Use the figure's scale to find each ship's distance from this point of collision by measuring from the bow (tip) of each ship with a ruler. Suppose that the speed of Ship A is r_1 and the speed of ship B is r_2. Given the present positions and courses of the two ships, find a relationship between their speeds that would ensure their collision.

CHAPTER 4 VOCABULARY CHECK

Fill in each blank with one of the words or phrases listed below.

system of linear equations   solution   consistent   independent
dependent   inconsistent   substitution   addition

1. In a system of linear equations in two variables, if the graphs of the equations are the same, the equations are _____ equations.

2. Two or more linear equations are called a _____.

3. A system of equations that has at least one solution is called a(n) _____ system.

4. A _____ of a system of two equations in two variables is an ordered pair of numbers that is a solution of both equations in the system.

5. Two algebraic methods for solving systems of equations are _____ and _____.

6. A system of equations that has no solution is called a(n) _____ system.

7. In a system of linear equations in two variables, if the graphs of the equations are different, the equations are _____ equations.

CHAPTER 4 HIGHLIGHTS

DEFINITIONS AND CONCEPTS	EXAMPLES

Section 4.1 Solving Systems of Linear Equations by Graphing

A **solution** of a system of two equations in two variables is an ordered pair of numbers that is a solution of both equations in the system.

Determine whether $(-1, 3)$ is a solution of the system:

$$\begin{cases} 2x - y = -5 \\ x = 3y - 10 \end{cases}$$

Replace x with -1 and y with 3 in both equations.

$$2x - y = -5 \qquad\qquad x = 3y - 10$$
$$2(-1) - 3 \stackrel{?}{=} -5 \qquad\qquad -1 \stackrel{?}{=} 3 \cdot 3 - 10$$
$$-5 = -5 \quad \text{True.} \qquad -1 = -1 \qquad \text{True.}$$

$(-1, 3)$ is a solution of the system.

Graphically, a solution of a system is a point common to the graphs of both equations.

Solve by graphing: $\begin{cases} 3x - 2y = -3 \\ x + y = 4 \end{cases}$

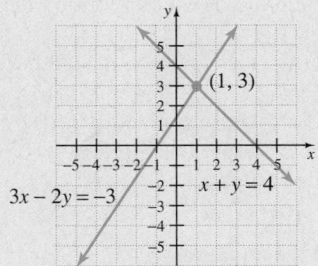

A system of equations with at least one solution is a **consistent system**. A system that has no solution is an **inconsistent system**.

If the graphs of two linear equations are identical, the equations are **dependent**. If their graphs are different, the equations are **independent**.

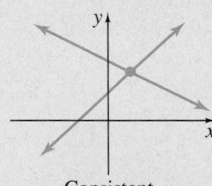

Consistent
and independent

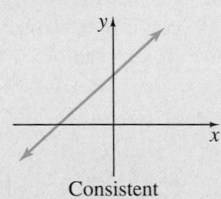

Consistent
and dependent

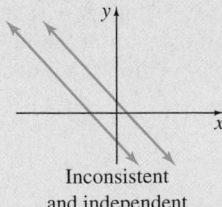

Inconsistent
and independent

(continued)

DEFINITIONS AND CONCEPTS	EXAMPLES

Section 4.2 Solving Systems of Linear Equations by Substitution

To solve a system of linear equations by the substitution method

Step 1. Solve one equation for a variable.

Step 2. Substitute the expression for the variable into the other equation.

Step 3. Solve the equation from Step 2 to find the value of one variable.

Step 4. Substitute the value from Step 3 in either original equation to find the value of the other variable.

Step 5. Check the solution in both equations.

Solve by substitution.

$$\begin{cases} 3x + 2y = 1 \\ x = y - 3 \end{cases}$$

Substitute $y - 3$ for x in the first equation.

$$3x + 2y = 1$$
$$3(y - 3) + 2y = 1$$
$$3y - 9 + 2y = 1$$
$$5y = 10$$
$$y = 2 \qquad \text{Divide by 5.}$$

To find x, substitute 2 for y in $x = y - 3$ so that $x = 2 - 3$ or -1. The solution $(-1, 2)$ checks.

Section 4.3 Solving Systems of Linear Equations by Addition

To solve a system of linear equations by the addition method

Step 1. Rewrite each equation in standard form $Ax + By = C$.

Step 2. Multiply one or both equations by a nonzero number so that the coefficients of a variable are opposites.

Step 3. Add the equations.

Step 4. Find the value of one variable by solving the resulting equation.

Step 5. Substitute the value from Step 4 into either original equation to find the value of the other variable.

Step 6. Check the solution in both equations.

If solving a system of linear equations by substitution or addition yields a true statement such as $-2 = -2$, then the graphs of the equations in the system are identical and there is an infinite number of solutions of the system.

Solve by addition.

$$\begin{cases} x - 2y = 8 \\ 3x + y = -4 \end{cases}$$

Multiply both sides of the first equation by -3.

$$\begin{cases} -3x + 6y = -24 \\ \underline{3x + y = -4} \end{cases}$$
$$7y = -28 \qquad \text{Add.}$$
$$y = -4 \qquad \text{Divide by 7.}$$

To find x, let $y = -4$ in an original equation.

$$x - 2(-4) = 8 \qquad \text{First equation}$$
$$x + 8 = 8$$
$$x = 0$$

The solution $(0, -4)$ checks.

Solve: $\begin{cases} 2x - 6y = -2 \\ x = 3y - 1 \end{cases}$

Substitute $3y - 1$ for x in the first equation.

$$2(3y - 1) - 6y = -2$$
$$6y - 2 - 6y = -2$$
$$-2 = -2 \qquad \text{True.}$$

The system has an infinite number of solutions.

(continued)

DEFINITIONS AND CONCEPTS	EXAMPLES

Section 4.4 Systems of Linear Equations and Problem Solving

Problem-solving steps

1. UNDERSTAND. Read and reread the problem.

Two angles are supplementary if their sum is 180°. The larger of two supplementary angles is three times the smaller, decreased by twelve. Find the measure of each angle. Let

$$x = \text{measure of smaller angle}$$

$$y = \text{measure of larger angle}$$

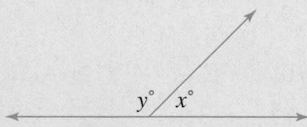

2. TRANSLATE.

In words:

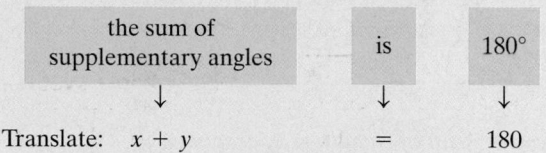

Translate: $x + y$ =   180

In words:

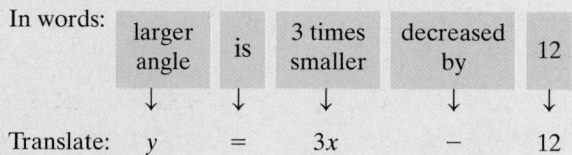

Translate: y = $3x$ −   12

3. SOLVE.

Solve the system:

$$\begin{cases} x + y = 180 \\ y = 3x - 12 \end{cases}$$

Use the substitution method and replace y with $3x - 12$ in the first equation.

$$x + y = 180$$

$$x + (3x - 12) = 180$$

$$4x = 192$$

$$x = 48$$

4. INTERPRET.

Since $y = 3x - 12$, then $y = 3 \cdot 48 - 12$ or 132.

The solution checks. The smaller angle measures 48° and the larger angle measures 132°

CHAPTER 4 REVIEW

(4.1) *Determine whether any of the following ordered pairs satisfy the system of linear equations.*

1. $\begin{cases} 2x - 3y = 12 \\ 3x + 4y = 1 \end{cases}$

2. $\begin{cases} 4x + y = 0 \\ -8x - 5y = 9 \end{cases}$

   a. $(12, 4)$ a. $\left(\dfrac{3}{4}, -3\right)$

   b. $(3, -2)$ b. $(-2, 8)$

   c. $(-3, 6)$ c. $\left(\dfrac{1}{2}, -2\right)$

3. $\begin{cases} 5x - 6y = 18 \\ 2y - x = -4 \end{cases}$

4. $\begin{cases} 2x + 3y = 1 \\ 3y - x = 4 \end{cases}$

   a. $(-6, -8)$ a. $(2, 2)$

   b. $\left(3, \dfrac{5}{2}\right)$ b. $(-1, 1)$

   c. $\left(3, -\dfrac{1}{2}\right)$ c. $(2, -1)$

Solve each system of equations by graphing.

5. $\begin{cases} 2x + y = 5 \\ 3y = -x \end{cases}$

6. $\begin{cases} 3x + y = -2 \\ 2x - y = -3 \end{cases}$

7. $\begin{cases} y - 2x = 4 \\ x + y = -5 \end{cases}$

8. $\begin{cases} y - 3x = 0 \\ 2y - 3 = 6x \end{cases}$

9. $\begin{cases} 3x + y = 2 \\ 3x - 6 = -9y \end{cases}$

10. $\begin{cases} 2y + x = 2 \\ x - y = 5 \end{cases}$

*Without graphing, **(a)** decide whether the graphs of the system are identical lines, parallel lines, or lines intersecting at a single point and **(b)** determine the number of solutions for each system.*

11. $\begin{cases} 2x - y = 3 \\ y = 3x + 1 \end{cases}$

12. $\begin{cases} 3x + y = 4 \\ y = -3x + 1 \end{cases}$

13. $\begin{cases} \dfrac{2}{3}x + \dfrac{1}{6}y = 0 \\ y = -4x \end{cases}$

14. $\begin{cases} \dfrac{1}{4}x + \dfrac{1}{8}y = 0 \\ y = -6x \end{cases}$

(4.2) *Solve the following systems of equations by the substitution method. If there is a single solution, give the ordered pair. If not, state whether the system is inconsistent or whether the equations are dependent.*

15. $\begin{cases} y = 2x + 6 \\ 3x - 2y = -11 \end{cases}$

16. $\begin{cases} y = 3x - 7 \\ 2x - 3y = 7 \end{cases}$

17. $\begin{cases} x + 3y = -3 \\ 2x + y = 4 \end{cases}$

18. $\begin{cases} 3x + y = 11 \\ x + 2y = 12 \end{cases}$

19. $\begin{cases} 4y = 2x - 3 \\ x - 2y = 4 \end{cases}$

20. $\begin{cases} 2x = 3y - 18 \\ x + 4y = 2 \end{cases}$

21. $\begin{cases} 2(3x - y) = 7x - 5 \\ 3(x - y) = 4x - 6 \end{cases}$

22. $\begin{cases} 4(x - 3y) = 3x - 1 \\ 3(4y - 3x) = 1 - 8x \end{cases}$

23. $\begin{cases} \dfrac{3}{4}x + \dfrac{2}{3}y = 2 \\ 3x + y = 18 \end{cases}$

24. $\begin{cases} \dfrac{2}{5}x + \dfrac{3}{4}y = 1 \\ x + 3y = -2 \end{cases}$

(4.3) *Solve the following systems of equations by the addition method. If there is a single solution, give the ordered pair. If not, state whether the system is inconsistent or whether the equations are dependent.*

25. $\begin{cases} 2x + 3y = -6 \\ x - 3y = -12 \end{cases}$

26. $\begin{cases} 4x + y = 15 \\ -4x + 3y = -19 \end{cases}$

27. $\begin{cases} 2x - 3y = -15 \\ x + 4y = 31 \end{cases}$

28. $\begin{cases} x - 5y = -22 \\ 4x + 3y = 4 \end{cases}$

29. $\begin{cases} 2x = 6y - 1 \\ \dfrac{1}{3}x - y = \dfrac{-1}{6} \end{cases}$

30. $\begin{cases} 8x = 3y - 2 \\ \dfrac{4}{7}x - y = \dfrac{-5}{2} \end{cases}$

31. $\begin{cases} 5x = 6y + 25 \\ -2y = 7x - 9 \end{cases}$

32. $\begin{cases} -4x = 8 + 6y \\ -3y = 2x - 3 \end{cases}$

33. $\begin{cases} 3(x - 4) = -2y \\ 2x = 3(y - 19) \end{cases}$

34. $\begin{cases} 4(x + 5) = -3y \\ 3x - 2(y + 18) = 0 \end{cases}$

35. $\begin{cases} \dfrac{2x + 9}{3} = \dfrac{y + 1}{2} \\ \dfrac{x}{3} = \dfrac{y - 7}{6} \end{cases}$

36. $\begin{cases} \dfrac{2 - 5x}{4} = \dfrac{2y - 4}{2} \\ \dfrac{x + 5}{3} = \dfrac{y}{5} \end{cases}$

(4.4) *Solve by writing and solving a system of linear equations.*

37. The sum of two numbers is 16. Three times the larger number decreased by the smaller number is 72. Find the two numbers.

38. The Forrest Theater can seat a total of 360 people. They take in $15,150 when every seat is sold. If orchestra sec-

tion tickets cost $45 and balcony tickets cost $35, find the number of people that can be seated in the orchestra section.

 39. A riverboat can head 340 miles upriver in 19 hours, but the return trip takes only 14 hours. Find the current of the river and find the speed of the ship in still water to the nearest tenth of a mile.

	d	$=$	r	$\cdot$	t
Upriver	340		$x - y$		19
Downriver	340		$x + y$		14

40. Sam Abney invested $9000 one year ago. Part of the money was invested at 6%, the rest at 10%. If the total interest earned in one year was $652.80, find how much was invested at each rate.

△ **41.** Ancient Greeks thought that the most pleasing dimensions for a picture are those where the length is approx-imately 1.6 times longer than the width. This ratio is known as the Golden Ratio. If Sandreka Walker has 6 feet of framing material, find the dimensions of the largest frame she can make that satisfies the Golden Ratio. Find the dimensions to the nearest hundredth of a foot.

42. Find the amount of 6% acid solution and the amount of 14% acid solution Pat should combine to prepare 50 cc (cubic centimeters) of a 12% solution.

43. The Deli charges $3.80 for a breakfast of 3 eggs and 4 strips of bacon. The charge is $2.75 for 2 eggs and 3 strips of bacon. Find the cost of each egg and the cost of each strip of bacon.

44. An exercise enthusiast alternates between jogging and walking. He traveled 15 miles during the past 3 hours. He jogs at a rate of 7.5 miles per hour and walks at a rate of 4 miles per hour. Find how much time, to the nearest hundredth of an hour, he actually spent jogging.

CHAPTER 4 TEST

Answer each question true or false.

1. A system of two linear equations in two variables can have exactly two solutions.

2. Although (1, 4) is not a solution of $x + 2y = 6$, it can still be a solution of the system $\begin{cases} x + 2y = 6 \\ x + y = 5 \end{cases}$.

3. If the two equations in a system of linear equations are added and the result is $3 = 0$, the system has no solution.

4. If the two equations in a system of linear equations are added and the result is $3x = 0$, the system has no solution.

Is the ordered pair a solution of the given linear system?

5. $\begin{cases} 2x - 3y = 5 \\ 6x + y = 1 \end{cases}; (1, -1)$

6. $\begin{cases} 4x - 3y = 24 \\ 4x + 5y = -8 \end{cases}; (3, -4)$

7. Use graphing to find the solutions of the system
$\begin{cases} y - x = 6 \\ y + 2x = -6 \end{cases}$

8. Use the substitution method to solve the system
$\begin{cases} 3x - 2y = -14 \\ x + 3y = -1 \end{cases}$

9. Use the substitution method to solve the system
$\begin{cases} \dfrac{1}{2}x + 2y = -\dfrac{15}{4} \\ 4x = -y \end{cases}$

10. Use the addition method to solve the system
$\begin{cases} 3x + 5y = 2 \\ 2x - 3y = 14 \end{cases}$

11. Use the addition method to solve the system
$\begin{cases} 5x - 6y = 7 \\ 7x - 4y = 12 \end{cases}$

Solve each system using the substitution method or the addition method.

12. $\begin{cases} 3x + y = 7 \\ 4x + 3y = 1 \end{cases}$

13. $\begin{cases} 3(2x + y) = 4x + 20 \\ x - 2y = 3 \end{cases}$

14. $\begin{cases} \dfrac{x-3}{2} = \dfrac{2-y}{4} \\ \dfrac{7-2x}{3} = \dfrac{y}{2} \end{cases}$

15. Lisa has a bundle of money consisting of $1 bills and $5 bills. There are 62 bills in the bundle. The total value of the bundle is $230. Find the number of $1 bills and the number of $5 bills.

16. Don has invested $4000, part at 5% simple annual interest and the rest at 9%. Find how much he invested at each rate if the total interest after 1 year is $311.

17. Although the number of farms in the U.S. is still decreasing, small farms are making a comeback. Texas and

Missouri are the states with the most number of farms. Texas has 116 thousand more farms than Missouri and the total number of farms for these two states is 336 thousand. Find the number of farms for each state.

CHAPTER 4 CUMULATIVE REVIEW

1. Insert $<$, $>$, or $=$ in the space between the paired numbers to make each statement true.
   a. -1 0 **b.** 7 $\frac{14}{2}$
   c. -5 -6

2. Find the value of each expression when $x = 2$ and $y = -5$.
   a. $\dfrac{x-y}{12+x}$ **b.** $x^2 - y$

3. Simplify each expression.
   a. $\dfrac{(-12)(-3)+3}{-7-(-2)}$ **b.** $\dfrac{2(-3)^2-20}{-5+4}$

4. Simplify each expression by combining like terms.
   a. $2x + 3x + 5 + 2$ **b.** $-5a - 3 + a + 2$
   c. $4y - 3y^2$ **d.** $2.3x + 5x - 6$

5. Solve: $2x + 3x - 5 + 7 = 10x + 3 - 6x - 4$

6. Solve: $-3x = 33$

7. Solve: $8(2 - t) = -5t$

8. Solve $P = 2l + 2w$ for w.

9. Complete the table for the equation $y = 3x$.

x	y
-1	
	0
	-9

10. If $f(x) = 7x^2 - 3x + 1$ and $g(x) = 3x - 2$, find the following.
   a. $f(1)$ **b.** $g(1)$
   c. $f(-2)$ **d.** $g(0)$

11. Graph $g(x) = 2x + 1$. Compare this graph with the graph of $f(x) = 2x$.

12. Find the slope and the y-intercept of the line $3x - 4y = 4$.

13. Are the following pairs of lines parallel, perpendicular, or neither?
   a. $3x + 7y = 4$
        $6x + 14y = 7$
   b. $-x + 3y = 2$
        $2x + 6y = 5$

14. Find an equation of the line through points $(4, 0)$ and $(-4, -5)$. Write the equation using function notation.

15. Graph $3x \geq y$.

16. Which of the following ordered pairs is a solution of the given system?
   $$\begin{cases} 2x - 3y = 6 \\ x = 2y \end{cases}$$
   a. $(12, 6)$ **b.** $(0, -2)$

17. Determine the number of solutions of the system.

$$\begin{cases} 3x - y = 4 \\ x + 2y = 8 \end{cases}$$

18. Solve the system:

$$\begin{cases} 2x + y = 10 \\ x = y + 2 \end{cases}$$

Solve each system.

19. $\begin{cases} x + 2y = 7 \\ 2x + 2y = 13 \end{cases}$

20. $\begin{cases} x + y = 7 \\ x - y = 5 \end{cases}$

21. Solve the system:

$$\begin{cases} -x - \dfrac{y}{2} = \dfrac{5}{2} \\ -\dfrac{x}{2} + \dfrac{y}{4} = 0 \end{cases}$$

22. Find two numbers whose sum is 37 and whose difference is 21.

23. Eric Daly, a chemistry teaching assistant, needs 10 liters of a 20% saline solution (salt water) for his 2 p.m. laboratory class. Unfortunately, the only mixtures on hand are a 5% saline solution and a 25% saline solution. How much of each solution should he mix to produce the 20% solution?

A Fast Growing Occupation

According to the U.S. Bureau of Labor Statistics, the paralegal profession is expected to be one of the fastest-growing occupations for the beginning of the 21st century. Paralegals assist lawyers by

- doing legal research
- organizing and tracking case files
- helping draw up legal agreements
- preparing reports to assist attorneys.

Although paralegals provide lawyers with invaluable support in many areas, they are prohibited from arguing cases in court and giving legal advice.

Paralegals should have good communication, research, and problem-solving skills. A well-rounded education is also an asset. Because paralegals collect, organize, and analyze all types of information, including numerical data, a firm grasp of mathematics is helpful.

 For more information about a career as a paralegal, visit the National Federation of Paralegal Associations' Website by first going to www.prenhall.com/martin-gay.

In the Spotlight on Decision Making feature on page 295, you will have the opportunity to make a decision concerning experts' estimates in a legal case as a paralegal.

EXPONENTS AND POLYNOMIALS

Recall from Chapter 1 that an exponent is a shorthand notation for repeated factors. This chapter explores additional concepts about exponents and exponential expressions. An especially useful type of exponential expression is a polynomial. Polynomials model many real-world phenomena. This chapter will focus on operations on polynomials.

5.1 EXPONENTS

CD-ROM   SSM

SSG   Video

▶ **OBJECTIVES**

1. Evaluate exponential expressions.
2. Use the product rule for exponents.
3. Use the power rule for exponents.
4. Use the power rules for products and quotients.
5. Use the quotient rule for exponents, and define a number raised to the 0 power.

1

As we reviewed in Section 1.4, an exponent is a shorthand notation for repeated factors. For example, $2 \cdot 2 \cdot 2 \cdot 2 \cdot 2$ can be written as 2^5. The expression 2^5 is called an **exponential expression**. It is also called the fifth **power** of 2, or we say that 2 is **raised** to the fifth power.

$$5^6 = \underbrace{5 \cdot 5 \cdot 5 \cdot 5 \cdot 5 \cdot 5}_{\text{6 factors; each factor is 5}} \qquad \text{and} \qquad (-3)^4 = \underbrace{(-3) \cdot (-3) \cdot (-3) \cdot (-3)}_{\text{4 factors; each factor is } -3}$$

The **base** of an exponential expression is the repeated factor. The **exponent** is the number of times that the base is used as a factor.

$$5^6 \overset{\text{exponent}}{\underset{\text{base}}{}} \qquad (-3)^4 \overset{\text{exponent}}{\underset{\text{base}}{}}$$

Example 1 Evaluate each expression.

    a. 2^3 **b.** 3^1 **c.** $(-4)^2$ **d.** -4^2 **e.** $\left(\dfrac{1}{2}\right)^4$ **f.** $4 \cdot 3^2$

Solution
    a. $2^3 = 2 \cdot 2 \cdot 2 = 8$
    b. To raise 3 to the first power means to use 3 as a factor only once. Therefore, $3^1 = 3$. Also, when no exponent is shown, the exponent is assumed to be 1.
    c. $(-4)^2 = (-4)(-4) = 16$
    d. $-4^2 = -(4 \cdot 4) = -16$
    e. $\left(\dfrac{1}{2}\right)^4 = \dfrac{1}{2} \cdot \dfrac{1}{2} \cdot \dfrac{1}{2} \cdot \dfrac{1}{2} = \dfrac{1}{16}$
    f. $4 \cdot 3^2 = 4 \cdot 9 = 36$

    Notice how similar -4^2 is to $(-4)^2$ in the example above. The difference between the two is the parentheses. In $(-4)^2$, the parentheses tell us that the base, or repeated factor, is -4. In -4^2, only 4 is the base.

> **HELPFUL HINT**
> Be careful when identifying the base of an exponential expression. Pay close attention to the use of parentheses.
>
> $(-3)^2$ -3^2 $2 \cdot 3^2$
>   The base is -3.   The base is 3.   The base is 3.
> $(-3)^2 = (-3)(-3) = 9$ $-3^2 = -(3 \cdot 3) = -9$ $2 \cdot 3^2 = 2 \cdot 3 \cdot 3 = 18$

2

An exponent has the same meaning whether the base is a number or a variable. If x is a real number and n is a positive integer, then x^n is the product of n factors, each of which is x.

$$x^n = \underbrace{x \cdot x \cdot x \cdot x \cdot x \dots x}_{n \text{ factors of } x}$$

Example 2 Evaluate for the given value of x.

a. $2x^3$; x is 5

b. $\dfrac{9}{x^2}$; x is -3

Solution **a.** If x is 5, $2x^3 = 2 \cdot (5)^3$

$\qquad\qquad\qquad = 2 \cdot (5 \cdot 5 \cdot 5)$

$\qquad\qquad\qquad = 2 \cdot 125$

$\qquad\qquad\qquad = 250$

b. If x is -3, $\dfrac{9}{x^2} = \dfrac{9}{(-3)^2}$

$\qquad\qquad\qquad\qquad = \dfrac{9}{(-3)(-3)}$

$\qquad\qquad\qquad\qquad = \dfrac{9}{9} = 1$

Exponential expressions can be multiplied, divided, added, subtracted, and themselves raised to powers. By our definition of an exponent,

$$5^4 \cdot 5^3 = \underbrace{(5 \cdot 5 \cdot 5 \cdot 5)}_{4 \text{ factors of } 5} \cdot \underbrace{(5 \cdot 5 \cdot 5)}_{3 \text{ factors of } 5}$$

$$= \underbrace{5 \cdot 5 \cdot 5 \cdot 5 \cdot 5 \cdot 5 \cdot 5}_{7 \text{ factors of } 5}$$

$$= 5^7$$

Also,

$$x^2 \cdot x^3 = (x \cdot x) \cdot (x \cdot x \cdot x)$$

$$= x \cdot x \cdot x \cdot x \cdot x$$

$$= x^5$$

In both cases, notice that the result is exactly the same if the exponents are added.

$$5^4 \cdot 5^3 = 5^{4+3} = 5^7 \qquad \text{and} \qquad x^2 \cdot x^3 = x^{2+3} = x^5$$

This suggests the following rule.

PRODUCT RULE FOR EXPONENTS

If m and n are positive integers and a is a real number, then

$$a^m \cdot a^n = a^{m+n}$$

In other words, to multiply two exponential expressions with a **common base**, keep the base and add the exponents.

Example 3 Use the product rule to simplify.

a. $4^2 \cdot 4^5$ **b.** $x^2 \cdot x^5$ **c.** $y^3 \cdot y$ **d.** $y^3 \cdot y^2 \cdot y^7$ **e.** $(-5)^7 \cdot (-5)^8$

Solution
a. $4^2 \cdot 4^5 = 4^{2+5} = 4^7$
b. $x^2 \cdot x^5 = x^{2+5} = x^7$
c. $y^3 \cdot y = y^3 \cdot y^1$
$\quad = y^{3+1}$
$\quad = y^4$

> **HELPFUL HINT**
> Don't forget that if no exponent is written, it is assumed to be 1.

d. $y^3 \cdot y^2 \cdot y^7 = y^{3+2+7} = y^{12}$
e. $(-5)^7 \cdot (-5)^8 = (-5)^{7+8} = (-5)^{15}$

Example 4 Use the product rule to simplify $(2x^2)(-3x^5)$.

Solution Recall that $2x^2$ means $2 \cdot x^2$ and $-3x^5$ means $-3 \cdot x^5$.

$$(2x^2)(-3x^5) = 2 \cdot x^2 \cdot -3 \cdot x^5 \qquad \text{Remove parentheses.}$$
$$= 2 \cdot -3 \cdot x^2 \cdot x^5 \qquad \text{Group factors with common bases.}$$
$$= -6x^7 \qquad \text{Simplify.}$$

> **HELPFUL HINT**
> These examples will remind you of the difference between adding and multiplying terms.
>
> **Addition**
> $$5x^3 + 3x^3 = (5 + 3)x^3 = 8x^3$$
> $$7x + 4x^2 = 7x + 4x^2$$
>
> **Multiplication**
> $$(5x^3)(3x^3) = 5 \cdot 3 \cdot x^3 \cdot x^3 = 15x^{3+3} = 15x^6$$
> $$(7x)(4x^2) = 7 \cdot 4 \cdot x \cdot x^2 = 28x^{1+2} = 28x^3$$

3   Exponential expressions can themselves be raised to powers. Let's try to discover a rule that simplifies an expression like $(x^2)^3$. By definition,

$$(x^2)^3 = \underbrace{(x^2)(x^2)(x^2)}_{3 \text{ factors of } x^2}$$

which can be simplified by the product rule for exponents.

$$(x^2)^3 = (x^2)(x^2)(x^2) = x^{2+2+2} = x^6$$

Notice that the result is exactly the same if we multiply the exponents.

$$(x^2)^3 = x^{2 \cdot 3} = x^6$$

The following property states this result.

POWER RULE FOR EXPONENTS

If m and n are positive integers and a is a real number, then

$$\left(a^m\right)^n = a^{m \cdot n}$$

To raise a power to a power, keep the base and multiply the exponents.

Example 5 Simplify each of the following expressions.

    a. $\left(x^2\right)^5$ **b.** $\left(y^8\right)^2$ **c.** $\left[(-5)^3\right]^7$

Solution   **a.** $\left(x^2\right)^5 = x^{2 \cdot 5} = x^{10}$

         b. $\left(y^8\right)^2 = y^{8 \cdot 2} = y^{16}$

         c. $\left[(-5)^3\right]^7 = (-5)^{21}$

H E L P F U L   H I N T

Take a moment to make sure that you understand when to apply the product rule and when to apply the power rule.

Product Rule → *Add Exponents*	*Power Rule* → *Multiply Exponents*
$x^5 \cdot x^7 = x^{5+7} = x^{12}$ $y^6 \cdot y^2 = y^{6+2} = y^8$	$\left(x^5\right)^7 = x^{5 \cdot 7} = x^{35}$ $\left(y^6\right)^2 = y^{6 \cdot 2} = y^{12}$

---

4   When the base of an exponential expression is a product, the definition of x^n still applies. To simplify $(xy)^3$, for example,

$$
\begin{aligned}
(xy)^3 &= (xy)(xy)(xy) & & (xy)^3 \text{ means 3 factors of } (xy). \\
&= x \cdot x \cdot x \cdot y \cdot y \cdot y & & \text{Group factors with common bases.} \\
&= x^3 y^3 & & \text{Simplify.}
\end{aligned}
$$

Notice that to simplify the expression $(xy)^3$, we raise each factor within the parentheses to a power of 3.

$$(xy)^3 = x^3 y^3$$

In general, we have the following rule.

POWER OF A PRODUCT RULE

If n is a positive integer and a and b are real numbers, then

$$(ab)^n = a^n b^n$$

In other words, to raise a product to a power, we raise each factor to the power.

Example 6 Simplify each expression.

a. $(st)^4$ b. $(2a)^3$ c. $(-5x^2y^3z)^2$

Solution
a. $(st)^4 = s^4 \cdot t^4 = s^4t^4$ Use the power of a product rule.
b. $(2a)^3 = 2^3 \cdot a^3 = 8a^3$ Use the power of a product rule.
c. $(-5x^2y^3z)^2 = (-5)^2 \cdot (x^2)^2 \cdot (y^3)^2 \cdot (z^1)^2$ Use the power of a product rule.
$\qquad\qquad\quad = 25x^4y^6z^2$ Use the power rule for exponents.

Let's see what happens when we raise a quotient to a power. To simplify $\left(\dfrac{x}{y}\right)^3$, for example,

$$\left(\frac{x}{y}\right)^3 = \left(\frac{x}{y}\right)\left(\frac{x}{y}\right)\left(\frac{x}{y}\right) \qquad \left(\frac{x}{y}\right)^3 \text{ means 3 factors of } \left(\frac{x}{y}\right).$$

$$= \frac{x \cdot x \cdot x}{y \cdot y \cdot y} \qquad \text{Multiply fractions.}$$

$$= \frac{x^3}{y^3} \qquad \text{Simplify.}$$

Notice that to simplify the expression $\left(\dfrac{x}{y}\right)^3$, we raise both the numerator and the denominator to a power of 3.

$$\left(\frac{x}{y}\right)^3 = \frac{x^3}{y^3}$$

In general, we have the following.

POWER OF A QUOTIENT RULE

If n is a positive integer and a and c are real numbers, then

$$\left(\frac{a}{c}\right)^n = \frac{a^n}{c^n}, \quad c \neq 0$$

In other words, to raise a quotient to a power, we raise both the numerator and the denominator to the power.

Example 7 Simplify each expression.

a. $\left(\dfrac{m}{n}\right)^7$ b. $\left(\dfrac{x^3}{3y^5}\right)^4$

Solution
a. $\left(\dfrac{m}{n}\right)^7 = \dfrac{m^7}{n^7}, n \neq 0$ Use the power of a quotient rule.

b. $\left(\dfrac{x^3}{3y^5}\right)^4 = \dfrac{(x^3)^4}{3^4 \cdot (y^5)^4}$ Use the power of a product or quotient rule.

$\qquad\qquad\quad = \dfrac{x^{12}}{81y^{20}}, y \neq 0$ Use the power rule for exponents.

<u>**5**</u>   Another pattern for simplifying exponential expressions involves quotients.

To simplify an expression like $\dfrac{x^5}{x^3}$, in which the numerator and the denominator have a common base, we can apply the fundamental principle of fractions and divide the numerator and the denominator by the common base factors. Assume for the remainder of this section that denominators are not 0.

$$\frac{x^5}{x^3} = \frac{x \cdot x \cdot x \cdot x \cdot x}{x \cdot x \cdot x}$$

$$= \frac{x \cdot x \cdot x \cdot x \cdot x}{x \cdot x \cdot x}$$

$$= x \cdot x$$

$$= x^2$$

Notice that the result is exactly the same if we subtract exponents of the common bases.

$$\frac{x^5}{x^3} = x^{5-3} = x^2$$

The quotient rule for exponents states this result in a general way.

QUOTIENT RULE FOR EXPONENTS

If m and n are positive integers and a is a real number, then

$$\frac{a^m}{a^n} = a^{m-n}$$

as long as a is not 0.

In other words, to divide one exponential expression by another with a common base, keep the base and subtract exponents.

Example 8   Simplify each quotient.

a. $\dfrac{x^5}{x^2}$ **b.** $\dfrac{4^7}{4^3}$ **c.** $\dfrac{(-3)^5}{(-3)^2}$ **d.** $\dfrac{2x^5y^2}{xy}$

Solution   **a.** $\dfrac{x^5}{x^2} = x^{5-2} = x^3$ Use the quotient rule.

b. $\dfrac{4^7}{4^3} = 4^{7-3} = 4^4 = 256$ Use the quotient rule.

c. $\dfrac{(-3)^5}{(-3)^2} = (-3)^3 = -27$

d. Begin by grouping common bases.

$$\frac{2x^5y^2}{xy} = 2 \cdot \frac{x^5}{x^1} \cdot \frac{y^2}{y^1}$$

$$= 2 \cdot \left(x^{5-1}\right) \cdot \left(y^{2-1}\right) \qquad \text{Use the quotient rule.}$$

$$= 2x^4y^1 \quad \text{or} \quad 2x^4y$$

Let's now give meaning to an expression such as x^0. To do so, we will simplify $\dfrac{x^3}{x^3}$ in two ways and compare the results.

$$\frac{x^3}{x^3} = x^{3-3} = x^0 \qquad \text{Apply the quotient rule.}$$

$$\frac{x^3}{x^3} = \frac{x \cdot x \cdot x}{x \cdot x \cdot x} = 1 \qquad \text{Apply the fundamental principle for fractions.}$$

Since $\dfrac{x^3}{x^3} = x^0$ and $\dfrac{x^3}{x^3} = 1$, we define that $x^0 = 1$ as long as x is not 0.

ZERO EXPONENT

$a^0 = 1$, as long as a is not 0.

In other words, a base raised to the 0 power is 1, as long as the base is not 0.

Example 9 Simplify the following expressions.

    a. 3^0 **b.** $(ab)^0$ **c.** $(-5)^0$ **d.** -5^0

Solution   **a.** $3^0 = 1$
          b. Assume that neither a nor b is zero.

$$(ab)^0 = a^0 \cdot b^0 = 1 \cdot 1 = 1$$

          c. $(-5)^0 = 1$
          d. $-5^0 = -1 \cdot 5^0 = -1 \cdot 1 = -1$

In the next example, exponential expressions are simplified using two or more of the exponent rules presented in this section.

Example 10 Simplify the following.

    a. $\left(\dfrac{-5x^2}{y^3}\right)^2$ **b.** $\dfrac{(x^3)^4 x}{x^7}$ **c.** $\dfrac{(2x)^5}{x^3}$ **d.** $\dfrac{(a^2b)^3}{a^3b^2}$

Solution   **a.** Use the power of a product or quotient rule; then use the power rule for exponents.

$$\left(\frac{-5x^2}{y^3}\right)^2 = \frac{(-5)^2(x^2)^2}{(y^3)^2} = \frac{25x^4}{y^6}$$

    b. $\dfrac{(x^3)^4 x}{x^7} = \dfrac{x^{12} \cdot x}{x^7} = \dfrac{x^{12+1}}{x^7} = \dfrac{x^{13}}{x^7} = x^{13-7} = x^6$

    c. Use the power of a product rule; then use the quotient rule.

$$\frac{(2x)^5}{x^3} = \frac{2^5 \cdot x^5}{x^3} = 2^5 \cdot x^{5-3} = 32x^2$$

d. Begin by applying the power of a product rule to the numerator.

$$\frac{(a^2b)^3}{a^3b^2} = \frac{(a^2)^3 \cdot b^3}{a^3 \cdot b^2}$$

$$= \frac{a^6b^3}{a^3b^2} \qquad \text{Use the power rule for exponents.}$$

$$= a^{6-3}b^{3-2} \qquad \text{Use the quotient rule.}$$

$$= a^3b^1 \quad \text{or} \quad a^3b$$

SPOTLIGHT ON DECISION MAKING

Suppose you are an auto mechanic and amateur racing enthusiast. You have been modifying the engine in your car and would like to enter a local amateur race. The racing classes depend on the size, or displacement, of the engine.

Engine displacement can be calculated using the formula

$d = \dfrac{\pi}{4} b^2sc$. In the formula,

d = the engine displacement in cubic centimeters (cc)
b = the bore or engine cylinder diameter in centimeters
s = the stroke or distance the piston travels up or down within the cylinder in centimeters
c = the number of cylinders the engine has. You have made the following measurements on your modified engine: 8.4 cm bore, 7.6 cm stroke, and 6 cylinders. In which racing class would you enter your car? Explain.

BRENTWOOD AMATEUR RACING CLUB	
Racing Class	Displacement Limit
A	up to 2000 cc
B	up to 2400 cc
C	up to 2650 cc
D	up to 3000 cc

MENTAL MATH

State the bases and the exponents for each of the following expressions.

1. 3^2

2. 5^4

3. $(-3)^6$

4. -3^7

5. -4^2

6. $(-4)^3$

7. $5 \cdot 3^4$

8. $9 \cdot 7^6$

9. $5x^2$

10. $(5x)^2$

Exercise Set 5.1

Evaluate each expression. See Example 1.

 1. 7^2

2. -3^2

3. $(-5)^1$

4. $(-3)^2$

 5. -2^4

6. -4^3

 7. $(-2)^4$

8. $(-4)^3$

9. $\left(\dfrac{1}{3}\right)^3$

10. $\left(-\dfrac{1}{9}\right)^2$

11. $7 \cdot 2^4$

12. $9 \cdot 1^2$

13. Explain why $(-5)^4 = 625$, while $-5^4 = -625$.

14. Explain why $5 \cdot 4^2 = 80$, while $(5 \cdot 4)^2 = 400$.

Evaluate each expression given the replacement values for x. See Example 2.

15. x^2; $x = -2$

16. x^3; $x = -2$

17. $5x^3$; $x = 3$

18. $4x^2$; $x = -1$

19. $2xy^2$; $x = 3$ and $y = 5$

20. $-4x^2y^3$; $x = 2$ and $y = -1$

21. $\dfrac{2z^4}{5}$; $z = -2$

22. $\dfrac{10}{3y^3}$; $y = 5$

△ **23.** The formula $V = x^3$ can be used to find the volume V of a cube with side length x. Find the volume of a cube with side length 7 meters. (Volume is measured in cubic units.)

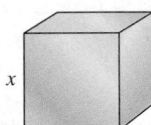

△ **24.** The formula $S = 6x^2$ can be used to find the surface area S of a cube with side length x. Find the surface area of the cube with side length 5 meters. (Surface area is measured in square units.)

△ **25.** To find the amount of water that a swimming pool in the shape of a cube can hold, do we use the formula for volume of the cube or surface area of the cube? (See Exercises 23 and 24.)

△ **26.** To find the amount of material needed to cover an ottoman in the shape of a cube, do we use the formula for volume of the cube or surface area of the cube? (See Exercises 23 and 24.)

Use the product rule to simplify each expression. Write the results using exponents. See Examples 3 and 4.

27. $x^2 \cdot x^5$

28. $y^2 \cdot y$

29. $(-3)^3 \cdot (-3)^9$

30. $(-5)^7 \cdot (-5)^6$

31. $(5y^4)(3y)$

32. $(-2z^3)(-2z^2)$

33. $(4z^{10})(-6z^7)(z^3)$

34. $(12x^5)(-x^6)(x^4)$

Use the power rule and the power of a product or quotient rule to simplify each expression. See Examples 5 through 7.

35. $(pq)^7$

36. $(4s)^3$

37. $\left(\dfrac{m}{n}\right)^9$

38. $\left(\dfrac{xy}{7}\right)^2$

39. $(x^2y^3)^5$

40. $(a^4b)^7$

41. $\left(\dfrac{-2xz}{y^5}\right)^2$

42. $\left(\dfrac{y^4}{-3z^3}\right)^3$

Use the quotient rule and simplify each expression. See Example 8.

43. $\dfrac{x^3}{x}$

44. $\dfrac{y^{10}}{y^9}$

45. $\dfrac{(-2)^5}{(-2)^3}$

46. $\dfrac{(-5)^{14}}{(-5)^{11}}$

47. $\dfrac{p^7q^{20}}{pq^{15}}$

48. $\dfrac{x^8y^6}{y^5}$

49. $\dfrac{7x^2y^6}{14x^2y^3}$

50. $\dfrac{9a^4b^7}{3ab^2}$

Simplify the following. See Example 9.

51. $(2x)^0$

52. $-4x^0$

53. $-2x^0$

54. $(4y)^0$

55. $5^0 + y^0$

56. $-3^0 + 4^0$

Simplify the following. See Example 10.

57. $\left(\dfrac{-3a^2}{b^3}\right)^3$

58. $\left(\dfrac{q^7}{-2p^5}\right)^5$

59. $\dfrac{(x^5)^7 \cdot x^8}{x^4}$

60. $\dfrac{y^{20}}{(y^2)^3 \cdot y^9}$

61. $\dfrac{(z^3)^6}{(5z)^4}$

62. $\dfrac{(3x)^4}{(x^2)^2}$

63. $\dfrac{(6mn)^5}{mn^2}$

64. $\dfrac{(6xy)^2}{9x^2y^2}$

Simplify the following.

65. -5^2

66. $(-5)^2$

67. $\left(\dfrac{1}{4}\right)^3$

68. $\left(\dfrac{2}{3}\right)^3$

69. $(9xy)^2$

70. $(2ab)^5$

71. $(6b)^0$

72. $(5ab)^0$

73. $2^3 + 2^5$

74. $7^2 - 7^0$

75. b^4b^2

76. y^4y^1

77. $a^2a^3a^4$

78. $x^2x^{15}x^9$

79. $(2x^3)(-8x^4)$

80. $(3y^4)(-5y)$

81. $(4a)^3$

82. $(2ab)^4$

83. $(-6xyz^3)^2$

84. $(-3xy^2a^3b)^3$

85. $\left(\dfrac{3y^5}{6x^4}\right)^3$

86. $\left(\dfrac{2ab}{6yz}\right)^4$

87. $\dfrac{x^5}{x^4}$

88. $\dfrac{5x^9}{x^3}$

89. $\dfrac{2x^3y^2z}{xyz}$

90. $\dfrac{x^{12}y^{13}}{x^5y^7}$

91. $\dfrac{(3x^2y^5)^5}{x^3y}$

92. $\dfrac{(4a^2)^4}{a^4b}$

93. In your own words, explain why $5^0 = 1$.

94. In your own words, explain when $(-3)^n$ is positive and when it is negative.

Find the area of each figure.

△ **95.** The following rectangle has width $4x^2$ feet and length $5x^3$ feet. Find its area. (Recall that Area $= l \cdot w$)

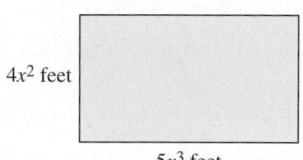

$4x^2$ feet

$5x^3$ feet

△ **96.** The following parallelogram has base length $9y^7$ meters and height $2y^{10}$ meters. Find its area. (Recall that Area $= b \cdot h$)

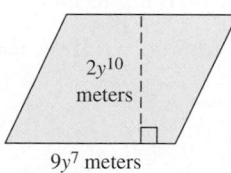

$2y^{10}$ meters

$9y^7$ meters

△ **97.** The following circle has a radius $5y$ centimeters, find its area. Do not approximate π.

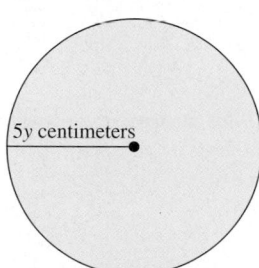

$5y$ centimeters

△ **98.** The square shown has sides of length $8z^5$ decimeters. Find its area.

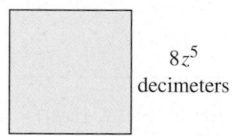

$8z^5$ decimeters

Find the volume of each figure.

△ **99.** The following safe is in the shape of a cube. Each side is $3y^4$ feet, find its volume.

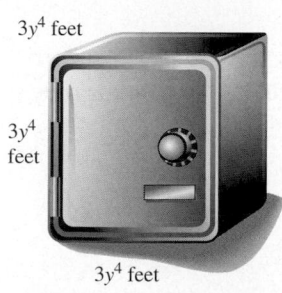

$3y^4$ feet

$3y^4$ feet

$3y^4$ feet

△ **100.** The silo shown is in the shape of a cylinder. Its radius is $4x$ meters and its height is $5x^3$ meters, find its volume. Do not approximate π.

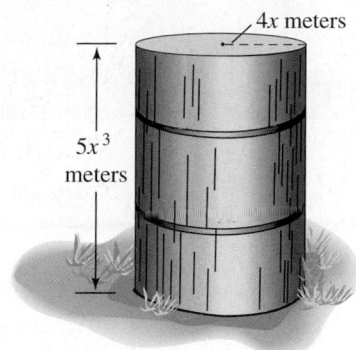

$4x$ meters

$5x^3$ meters

REVIEW EXERCISES

Simplify each expression by combining any like terms. Use the distributive property to remove any parentheses. See Section 2.1.

101. $3x - 5x + 7$

102. $7w + w - 2w$

103. $y - 10 + y$

104. $-6z + 20 - 3z$

105. $7x + 2 - 8x - 6$

106. $10y - 14 - y - 14$

107. $2(x - 5) + 3(5 - x)$

108. $-3(w + 7) + 5(w + 1)$

A Look Ahead

Example
Simplify $x^a \cdot x^{3a}$.

Solution:
Like bases, so add exponents.

$$x^a \cdot x^{3a} = x^{a+3a} = x^{4a}$$

Simplify each expression. Assume that variables represent positive integers. See the example.

109. $x^{5a}x^{4a}$

110. $b^{9a}b^{4a}$

111. $\left(a^b\right)^5$

112. $\left(2a^{4b}\right)^4$

113. $\dfrac{x^{9a}}{x^{4a}}$

114. $\dfrac{y^{15b}}{y^{6b}}$

115. $\left(x^a y^b z^c\right)^{5a}$

116. $\left(9a^2 b^3 c^4 d^5\right)^{ab}$

5.2 NEGATIVE EXPONENTS AND SCIENTIFIC NOTATION

CD-ROM SSM

SSG Video

▶ **OBJECTIVES**

1. Evaluate numbers raised to negative integer powers.
2. Use all the rules and definitions for exponents to simplify exponential expressions.
3. Write numbers in scientific notation.
4. Convert numbers from scientific notation to standard form.

1 Our work with exponential expressions so far has been limited to exponents that are positive integers or 0. Here we expand to give meaning to an expression like x^{-3}.

Suppose that we wish to simplify the expression $\dfrac{x^2}{x^5}$. If we use the quotient rule for exponents, we subtract exponents:

$$\frac{x^2}{x^5} = x^{2-5} = x^{-3}, \quad x \neq 0$$

But what does x^{-3} mean? Let's simplify $\dfrac{x^2}{x^5}$ using the definition of x^n.

$$\frac{x^2}{x^5} = \frac{x \cdot x}{x \cdot x \cdot x \cdot x \cdot x}$$

$$= \frac{x \cdot x}{x \cdot x \cdot x \cdot x \cdot x}$$

$$= \frac{1}{x^3}$$

Divide numerator and denominator by common factors by applying the fundamental principle for fractions.

If the quotient rule is to hold true for negative exponents, then x^{-3} must equal $\dfrac{1}{x^3}$.

From this example, we state the definition for negative exponents.

NEGATIVE EXPONENTS

If a is a real number other than 0 and n is an integer, then

$$a^{-n} = \frac{1}{a^n}$$

In other words, another way to write a^{-n} is to take its reciprocal and change the sign of its exponent.

◆ **Example 1** Simplify by writing each expression with positive exponents only.

a. 3^{-2} **b.** $2x^{-3}$
c. $2^{-1} + 4^{-1}$ **d.** $(-2)^{-4}$

Solution **a.** $3^{-2} = \dfrac{1}{3^2} = \dfrac{1}{9}$ Use the definition of negative exponents.

b. $2x^{-3} = 2 \cdot \dfrac{1}{x^3} = \dfrac{2}{x^3}$ *Use the definition of negative exponents.*

> **HELPFUL HINT**
> Don't forget that since there are no parentheses, only x is the base for the exponent -3.

c. $2^{-1} + 4^{-1} = \dfrac{1}{2} + \dfrac{1}{4} = \dfrac{2}{4} + \dfrac{1}{4} = \dfrac{3}{4}$

d. $(-2)^{-4} = \dfrac{1}{(-2)^4} = \dfrac{1}{(-2)(-2)(-2)(-2)} = \dfrac{1}{16}$

> **HELPFUL HINT**
> A negative exponent *does not affect* the sign of its base.
> Remember: Another way to write a^{-n} is to take its reciprocal and change the sign of its exponent, $a^{-n} = \dfrac{1}{a^n}$. For example,
>
> $$x^{-2} = \dfrac{1}{x^2}, \qquad 2^{-3} = \dfrac{1}{2^3} \text{ or } \dfrac{1}{8}$$
>
> $$\dfrac{1}{y^{-4}} = \dfrac{1}{\dfrac{1}{y^4}} = y^4, \quad \dfrac{1}{5^{-2}} = 5^2 \text{ or } 25$$

Example 2 Simplify each expression. Write results using positive exponents only.

a. $\left(\dfrac{2}{3}\right)^{-3}$ **b.** $\dfrac{1}{x^{-3}}$ **c.** $\dfrac{p^{-4}}{q^{-9}}$

Solution **a.** $\left(\dfrac{2}{3}\right)^{-3} = \dfrac{2^{-3}}{3^{-3}} = \dfrac{3^3}{2^3} = \dfrac{27}{8}$ *Use the definition of negative exponents.*

b. $\dfrac{1}{x^{-3}} = x^3$ **c.** $\dfrac{p^{-4}}{q^{-9}} = \dfrac{q^9}{p^4}$

Example 3 Simplify each expression. Write answers with positive exponents.

a. $\dfrac{y}{y^{-2}}$ **b.** $\dfrac{3}{x^{-4}}$ **c.** $\dfrac{x^{-5}}{x^7}$

Solution **a.** $\dfrac{y}{y^{-2}} = \dfrac{y^1}{y^{-2}} = y^{1-(-2)} = y^3$

b. $\dfrac{3}{x^{-4}} = 3 \cdot \dfrac{1}{x^{-4}} = 3 \cdot x^4 \text{ or } 3x^4$

c. $\dfrac{x^{-5}}{x^7} = x^{-5-7} = x^{-12} = \dfrac{1}{x^{12}}$

2 All the previously stated rules for exponents apply for negative exponents also. Here is a summary of the rules and definitions for exponents.

SUMMARY OF EXPONENT RULES

If m and n are integers and a, b, and c are real numbers, then:

Product rule for exponents: $\quad a^m \cdot a^n = a^{m+n}$

Power rule for exponents: $\quad (a^m)^n = a^{m \cdot n}$

Power of a product: $\quad (ab)^n = a^n b^n$

Power of a quotient: $\quad \left(\dfrac{a}{c}\right)^n = \dfrac{a^n}{c^n}, \quad c \neq 0$

Quotient rule for exponents: $\quad \dfrac{a^m}{a^n} = a^{m-n}, \quad a \neq 0$

Zero exponent: $\quad a^0 = 1, \quad a \neq 0$

Negative exponent: $\quad a^{-n} = \dfrac{1}{a^n}, \quad a \neq 0$

Example 4 Simplify the following expressions. Write each result using positive exponents only.

a. $\dfrac{(x^3)^4 x}{x^7}$

b. $\left(\dfrac{3a^2}{b}\right)^{-3}$

c. $\dfrac{4^{-1}x^{-3}y}{4^{-3}x^2 y^{-6}}$

d. $(y^{-3}z^6)^{-6}$

e. $\left(\dfrac{-2x^3 y}{xy^{-1}}\right)^3$

Solution **a.** $\dfrac{(x^3)^4 x}{x^7} = \dfrac{x^{12} \cdot x}{x^7} = \dfrac{x^{12+1}}{x^7} = \dfrac{x^{13}}{x^7} = x^{13-7} = x^6$ Use the power rule.

b. $\left(\dfrac{3a^2}{b}\right)^{-3} = \dfrac{3^{-3}(a^2)^{-3}}{b^{-3}}$ Raise each factor in the numerator and the denominator to the -3 power.

$= \dfrac{3^{-3}a^{-6}}{b^{-3}}$ Use the power rule.

$= \dfrac{b^3}{3^3 a^6}$ Use the definition of negative exponents.

$= \dfrac{b^3}{27a^6}$ Write 3^3 as 27.

c. $\dfrac{4^{-1}x^{-3}y}{4^{-3}x^2 y^{-6}} = 4^{-1-(-3)}x^{-3-2}y^{1-(-6)} = 4^2 x^{-5} y^7 = \dfrac{4^2 y^7}{x^5} = \dfrac{16y^7}{x^5}$

d. $(y^{-3}z^6)^{-6} = y^{18} \cdot z^{-36} = \dfrac{y^{18}}{z^{36}}$

e. $\left(\dfrac{-2x^3 y}{xy^{-1}}\right)^3 = \dfrac{(-2)^3 x^9 y^3}{x^3 y^{-3}} = \dfrac{-8x^9 y^3}{x^3 y^{-3}} = -8x^{9-3}y^{3-(-3)} = -8x^6 y^6$

3 Both very large and very small numbers frequently occur in many fields of science. For example, the distance between the sun and the planet Pluto is approximately 5,906,000,000 kilometers, and the mass of a proton is approximately 0.00000000000000000000000165 gram. It can be tedious to write these numbers in this standard decimal notation, so **scientific notation** is used as a convenient shorthand for expressing very large and very small numbers.

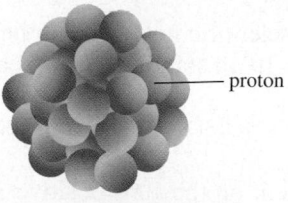

proton

Mass of proton is approximately
0.000 000 000 000 000 000 000 001 65 gram

Sun
5,906,000,000
kilometers
Pluto

SCIENTIFIC NOTATION

A positive number is written in scientific notation if it is written as the product of a number a, where $1 \le a < 10$, and an integer power r of 10:

$$a \times 10^r$$

The numbers below are written in scientific notation. The $\times$ sign for multiplication is used as part of the notation.

2.03×10^2 7.362×10^7 5.906×10^9 (Distance between the sun and Pluto, in kilometers)

1×10^{-3} 8.1×10^{-5} 1.65×10^{-24} (Mass of a proton, in grams)

The following steps are useful when writing numbers in scientific notation.

WRITING A NUMBER IN SCIENTIFIC NOTATION

Step 1. Move the decimal point in the original number so that the new number has a value between 1 and 10.

Step 2. Count the number of decimal places the decimal point is moved in Step 1. If the decimal point is moved to the left, the count is positive. If the decimal point is moved to the right, the count is negative.

Step 3. Multiply the new number in Step 1 by 10 raised to an exponent equal to the count found in Step 2.

Example 5   Write each number in scientific notation.

a. 367,000,000   **b.** 0.000003   **c.** 20,520,000,000   **d.** 0.00085

Solution   **a. Step 1.**   Move the decimal point until the number is between 1 and 10.

367,000,000

8 places

Step 2.   The decimal point is moved to the left 8 places, so the count is positive 8.

Step 3. $367,000,000 = 3.67 \times 10^8$.

b. Step 1.   Move the decimal point until the number is between 1 and 10.

0.000003

6 places

Step 2.   The decimal point is moved 6 places to the right, so the count is -6.

Step 3. $0.000003 = 3.0 \times 10^{-6}$

c. $20,520,000,000 = 2.052 \times 10^{10}$

d. $0.00085 = 8.5 \times 10^{-4}$

4 A number written in scientific notation can be rewritten in standard form. For example, to write 8.63×10^3 in standard form, recall that $10^3 = 1000$.

$$8.63 \times 10^3 = 8.63(1000) = 8630$$

Notice that the exponent on the 10 is positive 3, and we moved the decimal point 3 places to the right.

To write 7.29×10^{-3} in standard form, recall that $10^{-3} = \dfrac{1}{10^3} = \dfrac{1}{1000}$.

$$7.29 \times 10^{-3} = 7.29\left(\dfrac{1}{1000}\right) = \dfrac{7.29}{1000} = 0.00729$$

The exponent on the 10 is negative 3, and we moved the decimal to the left 3 places.

In general, **to write a scientific notation number in standard form**, move the decimal point the same number of places as the exponent on 10. If the exponent is positive, move the decimal point to the right; if the exponent is negative, move the decimal point to the left.

Example 6 Write each number in standard notation, without exponents.

a. 1.02×10^5 **b.** 7.358×10^{-3} **c.** 8.4×10^7 **d.** 3.007×10^{-5}

Solution **a.** Move the decimal point 5 places to the right.

$$1.02 \times 10^5 = 102{,}000.$$

b. Move the decimal point 3 places to the left.

$$7.358 \times 10^{-3} = 0.007358$$

c. $8.4 \times 10^7 = 84{,}000{,}000.$ 7 places to the right

d. $3.007 \times 10^{-5} = 0.00003007$ 5 places to the left

Performing operations on numbers written in scientific notation makes use of the rules and definitions for exponents.

Example 7 Perform each indicated operation. Write each result in standard decimal notation.

a. $(8 \times 10^{-6})(7 \times 10^3)$

b. $\dfrac{12 \times 10^2}{6 \times 10^{-3}}$

Solution   **a.** $\left(8 \times 10^{-6}\right)\left(7 \times 10^{3}\right) = 8 \cdot 7 \cdot 10^{-6} \cdot 10^{3}$
$= 56 \times 10^{-3}$
$= 0.056$

b. $\dfrac{12 \times 10^{2}}{6 \times 10^{-3}} = \dfrac{12}{6} \times 10^{2 - (-3)} = 2 \times 10^{5} = 200{,}000$

CALCULATOR EXPLORATIONS

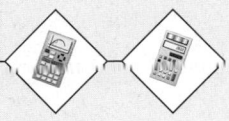

Scientific Notation

To enter a number written in scientific notation on a scientific calculator, locate the scientific notation key, which may be marked $\boxed{\text{EE}}$ or $\boxed{\text{EXP}}$. To enter 3.1×10^{7}, press $\boxed{3.1}\ \boxed{\text{EE}}\ \boxed{7}$. The display should read $\boxed{3.1 \quad 07}$.

Enter each number written in scientific notation on your calculator.

1. 5.31×10^{3}

2. -4.8×10^{14}

3. 6.6×10^{-9}

4. -9.9811×10^{-2}

Multiply each of the following on your calculator. Notice the form of the result.

5. $3{,}000{,}000 \times 5{,}000{,}000$

6. $230{,}000 \times 1000$

Multiply each of the following on your calculator. Write the product in scientific notation.

7. $\left(3.26 \times 10^{6}\right)\left(2.5 \times 10^{13}\right)$

8. $\left(8.76 \times 10^{-4}\right)\left(1.237 \times 10^{9}\right)$

SPOTLIGHT ON DECISION MAKING

Suppose you are a paralegal for a law firm. You are investigating the facts of a mineral rights case. Drilling on property adjacent to the client's has struck a natural gas reserve. The client believes that a portion of this reserve lies within her own property boundaries and she is, therefore, entitled to a portion of the proceeds from selling the natural gas. As part of your investigation, you contact two different experts, who fax you the following estimates for the size of the natural gas reserve:

Expert A	Expert B
Estimate of entire reserve: 4.6×10^{7} cubic feet	Estimate of entire reserve: 6.7×10^{6} cubic feet
Estimate of size of reserve on client's property: 1.84×10^{7} cubic feet	Estimate of size of reserve on client's property: 2.68×10^{6} cubic feet

How, if at all, would you use these estimates in the case? Explain.

MENTAL MATH

State each expression using positive exponents only.

1. $5x^{-2}$

2. $3x^{-3}$

3. $\dfrac{1}{y^{-6}}$

4. $\dfrac{1 \cdot}{x^{-3}}$

5. $\dfrac{4}{y^{-3}}$

6. $\dfrac{16}{y^{-7}}$

Exercise Set 5.2

Simplify each expression. Write each result using positive exponents only. See Examples 1 through 3.

1. 4^{-3}

2. 6^{-2}

3. $7x^{-3}$

4. $(7x)^{-3}$

5. $\left(-\dfrac{1}{4}\right)^{-3}$

6. $\left(-\dfrac{1}{8}\right)^{-2}$

7. $3^{-1} + 2^{-1}$

8. $4^{-1} + 4^{-2}$

9. $\dfrac{1}{p^{-3}}$

10. $\dfrac{1}{q^{-5}}$

11. $\dfrac{p^{-5}}{q^{-4}}$

12. $\dfrac{r^{-5}}{s^{-2}}$

13. $\dfrac{x^{-2}}{x}$

14. $\dfrac{y}{y^{-3}}$

15. $\dfrac{z^{-4}}{z^{-7}}$

16. $\dfrac{x^{-4}}{x^{-1}}$

17. $2^0 + 3^{-1}$

18. $4^{-2} - 4^{-3}$

19. $(-3)^{-2}$

20. $(-2)^{-6}$

21. $\dfrac{-1}{p^{-4}}$

22. $\dfrac{-1}{y^{-6}}$

23. $-2^0 - 3^0$

24. $5^0 + (-5)^0$

Simplify each expression. Write each result using positive exponents only. See Example 4.

25. $\dfrac{x^2 x^5}{x^3}$

26. $\dfrac{y^4 y^5}{y^6}$

27. $\dfrac{p^2 p}{p^{-1}}$

28. $\dfrac{y^3 y}{y^{-2}}$

29. $\dfrac{(m^5)^4 m}{m^{10}}$

30. $\dfrac{(x^2)^8 x}{x^9}$

31. $\dfrac{r}{r^{-3} r^{-2}}$

32. $\dfrac{p}{p^{-3} q^{-5}}$

33. $(x^5 y^3)^{-3}$

34. $(z^5 x^5)^{-3}$

35. $\dfrac{(x^2)^3}{x^{10}}$

36. $\dfrac{(y^4)^2}{y^{12}}$

37. $\dfrac{(a^5)^2}{(a^3)^4}$

38. $\dfrac{(x^2)^5}{(x^4)^3}$

39. $\dfrac{8k^4}{2k}$

40. $\dfrac{27r^4}{3r^6}$

41. $\dfrac{-6m^4}{-2m^3}$

42. $\dfrac{15a^4}{-15a^5}$

43. $\dfrac{-24a^6 b}{6ab^2}$

44. $\dfrac{-5x^4 y^5}{15x^4 y^2}$

45. $\dfrac{6x^2 y^3}{-7xy^5}$

46. $\dfrac{-8xa^2 b}{-5xa^5 b}$

47. $(a^{-5} b^2)^{-6}$

48. $(4^{-1} x^5)^{-2}$

49. $\left(\dfrac{x^{-2} y^4}{x^3 y^7}\right)^2$

50. $\left(\dfrac{a^5 b}{a^7 b^{-2}}\right)^{-3}$

51. $\dfrac{4^2 z^{-3}}{4^3 z^{-5}}$

52. $\dfrac{3^{-1} x^4}{3^3 x^{-7}}$

53. $\dfrac{2^{-3} x^{-4}}{2^2 x}$

54. $\dfrac{5^{-1} z^7}{5^{-2} z^9}$

55. $\dfrac{7ab^{-4}}{7^{-1} a^{-3} b^2}$

56. $\dfrac{6^{-5} x^{-1} y^2}{6^{-2} x^{-4} y^4}$

57. $\left(\dfrac{a^{-5} b}{ab^3}\right)^{-4}$

58. $\left(\dfrac{r^{-2} s^{-3}}{r^{-4} s^{-3}}\right)^{-3}$

59. $\dfrac{(xy^3)^5}{(xy)^{-4}}$

60. $\dfrac{(rs)^{-3}}{(r^2 s^3)^2}$

61. $\dfrac{(-2xy^{-3})^{-3}}{(xy^{-1})^{-1}}$

62. $\dfrac{(-3x^2 y^2)^{-2}}{(xyz)^{-2}}$

△ **63.** Find the volume of the cube.

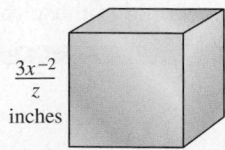

$\dfrac{3x^{-2}}{z}$ inches

△ **64.** Find the area of the triangle.

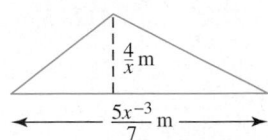

$\frac{4}{x}$ m

$\frac{5x^{-3}}{7}$ m

Write each number in scientific notation. See Example 5.

65. 78,000
66. 9,300,000,000
67. 0.00000167
68. 0.00000017
69. 0.00635
70. 0.00194
71. 1,160,000
72. 700,000

73. The temperature at the interior of the Earth is 20,000,000 degrees Celsius. Write 20,000,000 in scientific notation.

74. The half-life of a carbon isotope is 5000 years. Write 5000 in scientific notation.

75. The distance between the Earth and the sun is 93,000,000 miles. Write 93,000,000 in scientific notation.

76. The population of the world is 6,067,000,000. Write 6,067,000,000 in scientific notation. (*Source*: U.S. Bureau of the Census)

77. On March 23, 1997, Comet Hale-Bopp passed its closest to Earth. It was 120,000,000 miles away. Write 120,000,000 in scientific notation. (*Source: World Almanac and Book of Facts*, 1997)

78. The highest price paid for a diamond is $16,550,000. The diamond is 100.10 carats and pear-shaped and was sold in 1995 in Switzerland. Write this money amount in scientific notation. (*Source: The Guinness Book of Records*, 1999)

Write each number in standard notation. See Example 6.

79. 8.673×10^{-10}
80. 9.056×10^{-4}

81. 3.3×10^{-2}
82. 4.8×10^{-6}
83. 2.032×10^{4}
84. 9.07×10^{10}
85. One coulomb of electricity is 6.25×10^{18} electrons. Write this number in standard notation.
86. The mass of a hydrogen atom is 1.7×10^{-24} grams. Write this number in standard notation.
87. The distance light travels in 1 year is 9.460×10^{12} kilometers. Write this number in standard notation.
88. The population of the United States is 2.68×10^{8}. Write this number in standard notation. (*Source:* U.S. Bureau of the Census)

Evaluate each expression using exponential rules. Write each result in standard notation. See Example 7.

89. $(1.2 \times 10^{-3})(3 \times 10^{-2})$
90. $(2.5 \times 10^{6})(2 \times 10^{-6})$
91. $(4 \times 10^{-10})(7 \times 10^{-9})$
92. $(5 \times 10^{6})(4 \times 10^{-8})$
93. $\dfrac{8 \times 10^{-1}}{16 \times 10^{5}}$
94. $\dfrac{25 \times 10^{-4}}{5 \times 10^{-9}}$
95. $\dfrac{1.4 \times 10^{-2}}{7 \times 10^{-8}}$
96. $\dfrac{0.4 \times 10^{5}}{0.2 \times 10^{11}}$

97. The average amount of water flowing past the mouth of the Amazon River is 4.2×10^{6} cubic feet per second. How much water flows past in an hour? (*Hint:* 1 hour equals 3600 seconds.) Write the result in scientific notation.

98. A beam of light travels 9.460×10^{12} kilometers per year. How far does light travel in 10,000 years? Write the result in scientific notation.

99. Explain why $(a^{-1})^{3}$ has the same value as $(a^{3})^{-1}$.

100. Determine whether each statement is true or false.
    a. $5^{-1} < 5^{-2}$
    b. $\left(\dfrac{1}{5}\right)^{-1} < \left(\dfrac{1}{5}\right)^{-2}$
    c. $a^{-1} < a^{-2}$ for all nonzero numbers.

101. If $a = \dfrac{1}{10}$, then find the value of a^{-2}.

102. It was stated earlier that, for an integer n,

$$x^{-n} = \frac{1}{x^n}, \quad x \neq 0$$

Explain why x may not equal 0.

Simplify. Write results in standard notation.

103. $(2.63 \times 10^{12})(-1.5 \times 10^{-10})$

104. $(6.785 \times 10^{-4})(4.68 \times 10^{10})$

Light travels at a rate of 1.86×10^5 miles per second. Use this information and the distance formula $d = r \cdot t$ to answer Exercises 105 and 106.

105. If the distance from the moon to the Earth is 238,857 miles, find how long it takes the reflected light of the moon to reach the Earth. (Round to the nearest tenth of a second.)

106. If the distance from the sun to the Earth is 93,000,000 miles, find how long it takes the light of the sun to reach the Earth. (Round to the nearest second.)

REVIEW EXERCISES

Simplify each expression. See Section 2.1.

107. $-5y + 4y - 18 - y$ **108.** $12m - 14 - 15m - 1$

109. $-3x - (4x - 2)$ **110.** $-9y - (5 - 6y)$

111. $3(z - 4) - 2(3z + 1)$ **112.** $5(x - 3) - 4(2x - 5)$

A Look Ahead

Example
Simplify the following expressions. Assume that the variable in the exponent represents an integer value.

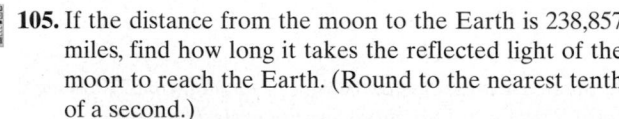

a. $x^{m+1} \cdot x^m$ **b.** $\left(z^{2x+1}\right)^x$ **c.** $\dfrac{y^{6a}}{y^{4a}}$

Solution:

a. $x^{m+1} \cdot x^m = x^{(m+1)+m} = x^{2m+1}$

b. $\left(z^{2x+1}\right)^x = z^{(2x+1)x} = z^{2x^2+x}$

c. $\dfrac{y^{6a}}{y^{4a}} = y^{6a-4a} = y^{2a}$

Simplify each expression. Assume that variables represent positive integers. See the example.

113. $a^{-4m} \cdot a^{5m}$ **114.** $\left(x^{-3s}\right)^3$

115. $\left(3y^{2z}\right)^3$ **116.** $a^{4m+1} \cdot a^4$

117. $\dfrac{y^{4a}}{y^{-a}}$ **118.** $\dfrac{y^{-6a}}{zy^{6a}}$

119. $\left(z^{3a+2}\right)^{-2}$ **120.** $\left(a^{4x-1}\right)^{-1}$

5.3 POLYNOMIAL FUNCTIONS AND ADDING AND SUBTRACTING POLYNOMIALS

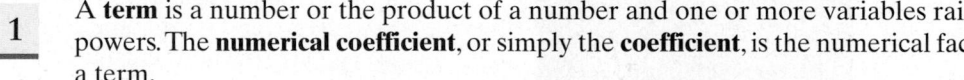

▶ **OBJECTIVES**

CD-ROM   SSM   SSG   Video

1. Identify term, constant, polynomial, monomial, binomial, trinomial, and the degree of a term and of a polynomial.
2. Define polynomial functions.
3. Review combining like terms.
4. Add polynomials.
5. Subtract polynomials.
6. Recognize the graph of a polynomial function from the degree of the polynomial.

1

A **term** is a number or the product of a number and one or more variables raised to powers. The **numerical coefficient**, or simply the **coefficient**, is the numerical factor of a term.

Term	Numerical Coefficient
$-12x^5$	-12
x^3y	1
$-z$	-1
2	2

If a term contains only a number, it is called a **constant term**, or simply a **constant**.

A **polynomial** is a finite sum of terms in which all variables are raised to nonnegative integer powers and no variables appear in any denominator.

Polynomials	Not Polynomials	
$4x^5y + 7xz$	$5x^{-3} + 2x$	negative integer exponent
$-5x^3 + 2x + \dfrac{2}{3}$	$\dfrac{6}{x^2} - 5x + 1$	variable in denominator

A polynomial that contains only one variable is called a **polynomial in one variable**. For example, $3x^2 - 2x + 7$ is a **polynomial in x**. This polynomial in x is written in *descending order* since the terms are listed in descending order of the variable's exponents. (The term 7 can be thought of as $7x^0$.) The following examples are polynomials in one variable written in **descending order**.

$$4x^3 - 7x^2 + 5 \qquad y^2 - 4 \qquad 8a^4 - 7a^2 + 4a$$

A **monomial** is a polynomial consisting of one term. A **binomial** is a polynomial consisting of two terms. A **trinomial** is a polynomial consisting of three terms.

Monomials	Binomials	Trinomials
ax^2	$x + y$	$x^2 + 4xy + y^2$
$-3x$	$6y^2 - 2$	$-x^4 + 3x^3 + 1$
4	$\dfrac{5}{7}z^3 - 2z$	$8y^2 - 2y - 10$

By definition, all monomials, binomials, and trinomials are also polynomials.

Each term of a polynomial has a **degree**.

DEGREE OF A TERM

The **degree of a term** is the sum of the exponents on the *variables* contained in the term.

Example 1 Find the degree of each term.

a. $3x^2$ **b.** -2^3x^5 **c.** y **d.** $12x^2yz^3$ **e.** 5

Solution **a.** The exponent on x is 2, so the degree of the term is 2.
b. The exponent on x is 5, so the degree of the term is 5. (Recall that the degree is the sum of the exponents on only the *variables*.)
c. The degree of y, or y^1, is 1.
d. The degree is the sum of the exponents on the variables, or $2 + 1 + 3 = 6$.
e. The degree of 5, which can be written as $5x^0$, is 0.

From the preceding example, we can say that the degree of a constant is 0. Also, the term 0 has no degree.

Each polynomial also has a degree.

DEGREE OF A POLYNOMIAL

The **degree of a polynomial** is the largest degree of all its terms.

Example 2   Find the degree of each polynomial and also indicate whether the polynomial is a monomial, binomial, or trinomial.

a. $7x^3 - 3x + 2$ **b.** $-xyz$ **c.** $x^4 - 16$

Solution   The degree and classification of each polynomial is summarized in the following table.

	Polynomial	*Degree*	*Classification*
a.	$7x^3 - 3x + 2$	3	Trinomial
b.	$-xyz$	$1 + 1 + 1 = 3$	Monomial
c.	$x^4 - 16$	4	Binomial

Example 3   Find the degree of the polynomial $3xy + x^2y^2 - 5x^2 - 6$.

Solution   The degree of each term is

$$3xy + x^2y^2 - 5x^2 - 6$$
$$\downarrow \quad\quad \downarrow \quad\quad \downarrow \quad\quad \downarrow$$

Degree: 2   4   2   0

The largest degree of any term is 4, so the degree of this polynomial is 4.

2   At times, it is convenient to use function notation to represent polynomials. For example, we may write $P(x)$ to represent the polynomial $3x^2 - 2x - 5$. In symbols, this is

$$P(x) = 3x^2 - 2x - 5$$

This function is called a **polynomial function** because the expression $3x^2 - 2x - 5$ is a polynomial.

> **HELPFUL HINT**
> Recall that the symbol $P(x)$ **does not mean** P times x. It is a special symbol used to denote a function.

Example 4   If $P(x) = 3x^2 - 2x - 5$, find the following.

a. $P(1)$ **b.** $P(-2)$

Solution   **a.** Substitute 1 for x in $P(x) = 3x^2 - 2x - 5$ and simplify.

$$P(x) = 3x^2 - 2x - 5$$
$$P(1) = 3(1)^2 - 2(1) - 5 = -4$$

b. Substitute -2 for x in $P(x) = 3x^2 - 2x - 5$ and simplify.

$$P(x) = 3x^2 - 2x - 5$$
$$P(-2) = 3(-2)^2 - 2(-2) - 5 = 11$$

Many real-world phenomena are modeled by polynomial functions. If the polynomial function model is given, we can often find the solution of a problem by evaluating the function at a certain value.

Example 5 **FINDING THE HEIGHT OF AN OBJECT**

The world's highest bridge, Royal Gorge suspension bridge in Colorado, is 1053 feet above the Arkansas River. An object is dropped from the top of this bridge. Neglecting air resistance, the height of the object at time t seconds is given by the polynomial function $P(t) = -16t^2 + 1053$. Find the height of the object when $t = 1$ second and when $t = 8$ seconds.

Solution To find the height of the object at 1 second, we find $P(1)$.

$$P(t) = -16t^2 + 1053$$
$$P(1) = -16(1)^2 + 1053$$
$$P(1) = 1037$$

When $t = 1$ second, the height of the object is 1037 feet.
To find the height of the object at 8 seconds, we find $P(8)$.

$$P(t) = -16t^2 + 1053$$
$$P(8) = -16(8)^2 + 1053$$
$$P(8) = -1024 + 1053$$
$$P(8) = 29$$

When $t = 8$ seconds, the height of the object is 29 feet. Notice that as time t increases, the height of the object decreases.

3 Before we add polynomials, recall that terms are considered to be **like terms** if they contain exactly the same variables raised to exactly the same powers.

Like Terms	Unlike Terms
$-5x^2, -x^2$	$4x^2, 3x$
$7xy^3z, -2xzy^3$	$12x^2y^3, -2xy^3$

To simplify a polynomial, **combine like terms** by using the distributive property. For example, by the distributive property,

$$5x + 7x = (5 + 7)x = 12x$$

Example 6 Simplify by combining like terms.

a. $-12x^2 + 7x^2 - 6x$ **b.** $3xy - 2x + 5xy - x$

Solution By the distributive property,

a. $-12x^2 + 7x^2 - 6x = (-12 + 7)x^2 - 6x = -5x^2 - 6x$
b. Use the associative and commutative properties to group together like terms; then combine.

$$3xy - 2x + 5xy - x = 3xy + 5xy - 2x - x$$
$$= (3 + 5)xy + (-2 - 1)x$$
$$= 8xy - 3x$$

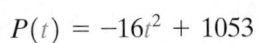

4 Now we have reviewed the necessary skills to add polynomials.

ADDING POLYNOMIALS

Combine all like terms.

Example 7 Add.

a. $(7x^3y - xy^3 + 11) + (6x^3y - 4)$ **b.** $(3a^3 - b + 2a - 5) + (a + b + 5)$

Solution **a.** To add, remove the parentheses and group like terms.

$$(7x^3y - xy^3 + 11) + (6x^3y - 4)$$
$$= 7x^3y - xy^3 + 11 + 6x^3y - 4$$
$$= 7x^3y + 6x^3y - xy^3 + 11 - 4 \quad \text{Group like terms.}$$
$$= 13x^3y - xy^3 + 7 \quad \text{Combine like terms.}$$

b. $(3a^3 - b + 2a - 5) + (a + b + 5)$

$$= 3a^3 - b + 2a - 5 + a + b + 5$$
$$= 3a^3 - b + b + 2a + a - 5 + 5 \quad \text{Group like terms.}$$
$$= 3a^3 + 3a \quad \text{Combine like terms.} \ \blacksquare$$

Example 8 Add $11x^3 - 12x^2 + x - 3$ and $x^3 - 10x + 5$.

Solution $(11x^3 - 12x^2 + x - 3) + (x^3 - 10x + 5)$

$$= 11x^3 + x^3 - 12x^2 + x - 10x - 3 + 5 \quad \text{Group like terms.}$$
$$= 12x^3 - 12x^2 - 9x + 2 \quad \text{Combine like terms.} \ \blacksquare$$

Sometimes it is more convenient to add polynomials vertically. To do this, line up like terms beneath one another and add like terms.

5 The definition of subtraction of real numbers can be extended to apply to polynomials. To subtract a number, we add its opposite.

$$a - b = a + (-b)$$

Likewise, to subtract a polynomial, we add its opposite. In other words, if P and Q are polynomials, then

$$P - Q = P + (-Q)$$

The polynomial $-Q$ is the **opposite**, or **additive inverse**, of the polynomial Q. We can find $-Q$ by writing the opposite of each term of Q.

SUBTRACTING POLYNOMIALS

To subtract a polynomial, add its opposite.

For example,

To subtract, add its opposite (found by writing the opposite of each term).

$$(3x^2 + 4x - 7) - (3x^2 - 2x - 5) = (3x^2 + 4x - 7) + (-3x^2 + 2x + 5)$$

$$= 3x^2 + 4x - 7 - 3x^2 + 2x + 5$$

$$= 6x - 2 \qquad \text{Combine like terms.}$$

Example 9 Subtract $(12z^5 - 12z^3 + z) - (-3z^4 + z^3 + 12z)$.

Solution To subtract, add the opposite of the second polynomial to the first polynomial.

$$(12z^5 - 12z^3 + z) - (-3z^4 + z^3 + 12z)$$

$$= 12z^5 - 12z^3 + z + 3z^4 - z^3 - 12z)$$

$$= 12z^5 + 3z^4 - 12z^3 - z^3 + z - 12z \qquad \text{Group like terms.}$$

$$= 12z^5 + 3z^4 - 13z^3 - 11z \qquad \text{Combine like terms.} \quad \blacksquare$$

Example 10 Subtract $4x^3y^2 - 3x^2y^2 + 2y^2$ from $10x^3y^2 - 7x^2y^2$.

Solution If we subtract 2 from 8, the difference is $8 - 2 = 6$. Notice the order of the numbers, and then write "Subtract $4x^3y^2 - 3x^2y^2 + 2y^2$ from $10x^3y^2 - 7x^2y^2$" as a mathematical expression.

$$(10x^3y^2 - 7x^2y^2) - (4x^3y^2 - 3x^2y^2 + 2y^2)$$

$$= 10x^3y^2 - 7x^2y^2 - 4x^3y^2 + 3x^2y^2 - 2y^2 \qquad \text{Remove parentheses.}$$

$$= 6x^3y^2 - 4x^2y^2 - 2y^2 \qquad \text{Combine like terms.} \quad \blacksquare$$

Polynomials can also be added or subtracted vertically. Just remember to line up like terms. For example, perform the subtraction $(10x^3y^2 - 7x^2y^2) - (4x^3y^2 - 3x^2y^2 + 2y^2)$ vertically.

Add the opposite of the second polynomial.

$$\begin{array}{l} 10x^3y^2 - 7x^2y^2 \\ -(4x^3y^2 - 3x^2y^2 + 2y^2) \end{array} \quad \text{is equivalent to} \quad \begin{array}{l} 10x^3y^2 - 7x^2y^2 \\ \underline{-4x^3y^2 + 3x^2y^2 - 2y^2} \\ 6x^3y^2 - 4x^2y^2 - 2y^2 \end{array}$$

Polynomial functions, like polynomials, can be added, subtracted, multiplied, and divided. For example, if

$$P(x) = x^2 + x + 1$$

then

$$2P(x) = 2(x^2 + x + 1) = 2x^2 + 2x + 2 \qquad \text{Use the distributive property.}$$

Also, if $Q(x) = 5x^2 - 1$, then $P(x) + Q(x) = (x^2 + x + 1) + (5x^2 - 1)$
$$= 6x^2 + x.$$

A useful business and economics application of subtracting polynomial functions is finding the profit function $P(x)$ when given a revenue function $R(x)$ and a cost function $C(x)$. In business, it is true that

$$\text{profit} = \text{revenue} - \text{cost, or}$$
$$P(x) = R(x) - C(x)$$

For example, if the revenue function is $R(x) = 7x$ and the cost function is $C(x) = 2x + 5000$, then the profit function is

$$P(x) = R(x) - C(x)$$

or

$$P(x) = 7x - (2x + 5000) \quad \text{Substitute } R(x) = 7x$$
$$P(x) = 5x - 5000 \quad \text{and } C(x) = 2x + 5000.$$

Problem-solving exercises involving profit are in the exercise set.

6   In this section, we reviewed how to find the degree of a polynomial. Knowing the degree of a polynomial can help us recognize the graph of the related polynomial function. For example, we know from Section 3.2 that the graph of the polynomial function $f(x) = x^2$ is a parabola as shown to the left.

The polynomial x^2 has degree 2. The graphs of all polynomial functions of degree 2 will have this same general shape—opening upward, as shown, or downward. Graphs of polynomial functions of degree 2 or 3 will, in general, resemble one of the graphs shown next.

Degree 2

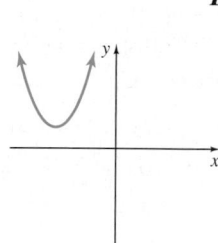

Coefficient of x^2
is a positive number.

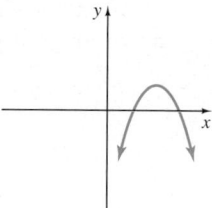

Coefficient of x^2
is a negative number.

Degree 3

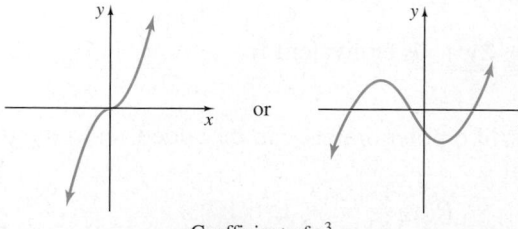

Coefficient of x^3
is a positive number.

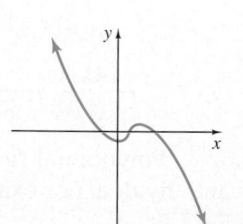

Coefficient of x^3
is a negative number.

General Shapes of Graphs of Polynomial Functions

Example 11   Determine which of the following graphs is the graph of
$$f(x) = 5x^3 - 6x^2 + 2x + 3$$

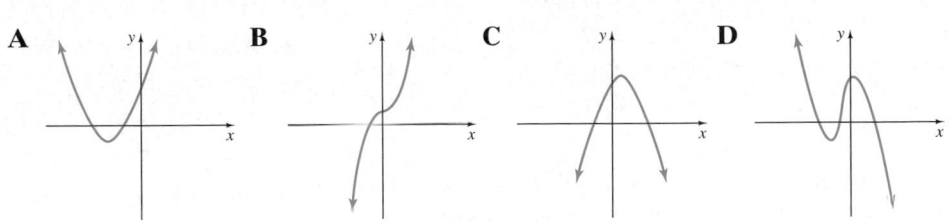

Solution   The degree of $f(x)$ is 3, which means that its graph has the shape of B or D. The coefficient of x^3 is 5, a positive number, so the graph has the shape of B.

GRAPHING CALCULATOR EXPLORATIONS

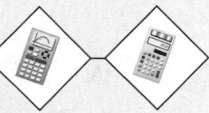

A graphing calculator may be used to visualize addition and subtraction of polynomials in one variable. For example, to visualize the following polynomial subtraction statement
$$(3x^2 - 6x + 9) - (x^2 - 5x + 6) = 2x^2 - x + 3$$
graph both

$$Y_1 = (3x^2 - 6x + 9) - (x^2 - 5x + 6) \qquad \text{\textit{left side of equation}}$$

and

$$Y_2 = 2x^2 - x + 3 \qquad \text{\textit{right side of equation}}$$

on the same screen and see that their graphs coincide. (*Note:* If the graphs do not coincide, we can be sure that a mistake has been made in combining polynomials or in calculator keystrokes. If the graphs appear to coincide, we cannot be sure that our work is correct. This is because it is possible for the graphs to differ so slightly that we do not notice it.)

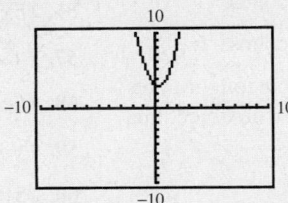

The graphs of Y_1 and Y_2 are shown. The graphs appear to coincide, so the subtraction statement
$$(3x^2 - 6x + 9) - (x^2 - 5x + 6) = 2x^2 - x + 3$$
appears to be correct.

Perform the indicated operations. Then visualize by using the procedure described above.

1. $(2x^2 + 7x + 6) + (x^3 - 6x^2 - 14)$

2. $(-14x^3 - x + 2) + (-x^3 + 3x^2 + 4x)$

3. $(1.8x^2 - 6.8x - 1.7) - (3.9x^2 - 3.6x)$

4. $(-4.8x^2 + 12.5x - 7.8) - (3.1x^2 - 7.8x)$

5. $(1.29x - 5.68) + (7.69x^2 - 2.55x + 10.98)$

6. $(-0.98x^2 - 1.56x + 5.57) + (4.36x - 3.71)$

Exercise Set 5.3

Find the degree of each term. See Example 1.

1. 4

2. 7

3. $5x^2$

4. $-z^3$

5. $-3xy^2$

6. $12x^3z$

Find the degree of each polynomial and indicate whether the polynomial is a monomial, binomial, trinomial, or none of these. See Examples 2 and 3.

7. $6x + 3$

8. $7x - 8$

9. $3x^2 - 2x + 5$

10. $5x^2 - 3x^2y - 2x^3$

11. $-xyz$

12. -9

13. $x^2y - 4xy^2 + 5x + y$

14. $-2x^2y - 3y^2 + 4x + y^5$

15. In your own words, describe how to find the degree of a term.

16. In your own words, describe how to find the degree of a polynomial.

If $P(x) = x^2 + x + 1$ and $Q(x) = 5x^2 - 1$, find the following. See Example 4.

17. $P(7)$

18. $Q(4)$

19. $Q(-10)$

20. $P(-4)$

21. $P(0)$

22. $Q(0)$

Refer to Example 5 for Exercises 23 through 26.

23. Find the height of the object at $t = 2$ seconds.

24. Find the height of the object at $t = 4$ seconds.

25. Find the height of the object at $t = 6$ seconds.

26. Approximate (to the nearest second) how long it takes before the object hits the ground. (*Hint:* The object hits the ground when $P(t) = 0$.)

Simplify by combining like terms. See Example 6.

27. $5y + y$

28. $-x + 3x$

29. $4x + 7x - 3$

30. $-8y + 9y + 4y^2$

31. $4xy + 2x - 3xy - 1$

32. $-8xy^2 + 4x - x + 2xy^2$

Perform the indicated operations. See Examples 7 through 10.

33. $(9y^2 - 8) + (9y^2 - 9)$

34. $(x^2 + 4x - 7) + (8x^2 + 9x - 7)$

35. Add $(x^2 + xy - y^2)$ and $(2x^2 - 4xy + 7y^2)$.

36. Add $(4x^3 - 6x^2 + 5x + 7)$ and $(2x^2 + 6x - 3)$.

37. $\begin{array}{r} x^2 - 6x + 3 \\ + \quad (2x + 5) \\ \hline \end{array}$

38. $\begin{array}{r} -2x^2 + 3x - 9 \\ + \quad (2x - 3) \\ \hline \end{array}$

39. $(9y^2 - 7y + 5) - (8y^2 - 7y + 2)$

40. $(2x^2 + 3x + 12) - (5x - 7)$

41. Subtract $(6x^2 - 3x)$ from $(4x^2 + 2x)$.

42. Subtract $(xy + x - y)$ from $(xy + x - 3)$.

43. $\begin{array}{r} 3x^2 - 4x + 8 \\ - \quad (5x^2 - 7) \\ \hline \end{array}$

44. $\begin{array}{r} -3x^2 - 4x + 8 \\ - \quad (5x + 12) \\ \hline \end{array}$

45. $(5x - 11) + (-x - 2)$

46. $(3x^2 - 2x) + (5x^2 - 9x)$

47. $(7x^2 + x + 1) - (6x^2 + x - 1)$

48. $(4x - 4) - (-x - 4)$

49. $(7x^3 - 4x + 8) + (5x^3 + 4x + 8x)$

50. $(9xyz + 4x - y) + (-9xyz - 3x + y + 2)$

51. $(9x^3 - 2x^2 + 4x - 7) - (2x^3 - 6x^2 - 4x + 3)$

52. $(3x^2 + 6xy + 3y^2) - (8x^2 - 6xy - y^2)$

53. Add $(y^2 + 4yx + 7)$ and $(-19y^2 + 7yx + 7)$.

54. Subtract $(x - 4)$ from $(3x^2 - 4x + 5)$.

55. $(3x^3 - b + 2a - 6) + (-4x^3 + b + 6a - 6)$

56. $(5x^2 - 6) + (2x^2 - 4x + 8)$

57. $(4x^2 - 6x + 2) - (-x^2 + 3x + 5)$

58. $(5x^2 + x + 9) - (2x^2 - 9)$

59. $(-3x + 8) + (-3x^2 + 3x - 5)$

60. $(5y^2 - 2y + 4) + (3y + 7)$

61. $(-3 + 4x^2 + 7xy^2) + (2x^3 - x^2 + xy^2)$

62. $(-3x^2y + 4) - (-7x^2y - 8y)$

63. $\begin{array}{r} 6y^2 - 6y + 4 \\ -(-y^2 - 6y + 7) \\ \hline \end{array}$

64. $\begin{array}{r} -4x^3 + 4x^2 - 4x \\ -(2x^3 - 2x^2 + 3x) \\ \hline \end{array}$

65. $\begin{array}{r} 3x^2 + 15x + 8 \\ +(2x^2 + 7x + 8) \\ \hline \end{array}$

66. $\begin{array}{r} 9x^2 + 9x - 4 \\ +(7x^2 - 3x - 4) \\ \hline \end{array}$

67. Find the sum of $(5q^4 - 2q^2 - 3q)$ and $(-6q^4 + 3q^2 + 5)$.

68. Find the sum of $\left(5y^4 - 7y^2 + x^2 - 3\right)$ and $\left(-3y^4 + 2y^2 + 4\right)$.

69. Subtract $(3x + 7)$ from the sum of $\left(7x^2 + 4x + 9\right)$ and $\left(8x^2 + 7x - 8\right)$.

70. Subtract $(9x + 8)$ from the sum of $\left(3x^2 - 2x - x^3 + 2\right)$ and $\left(5x^2 - 8x - x^3 + 4\right)$.

71. Find the sum of $\left(4x^4 - 7x^2 + 3\right)$ and $\left(2 - 3x^4\right)$.

72. Find the sum of $\left(8x^4 - 14x^2 + 6\right)$ and $\left(-12x^6 - 21x^4 - 9x^2\right)$.

73. $\left(8x^{2y} - 7x^y + 3\right) + \left(-4x^{2y} + 9x^y - 14\right)$

74. $\left(14z^{5x} + 3z^{2x} + z\right) - \left(2z^{5x} - 10z^{2x} + 3z\right)$

△ **75.** Given the following triangle, find its perimeter.

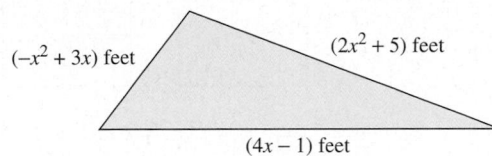

$(-x^2 + 3x)$ feet $(2x^2 + 5)$ feet

$(4x - 1)$ feet

△ **76.** Given the following quadrilateral, find its perimeter.

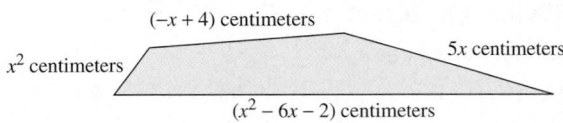

$(-x + 4)$ centimeters

x^2 centimeters $5x$ centimeters

$(x^2 - 6x - 2)$ centimeters

△ **77.** A wooden beam is $\left(4y^2 + 4y + 1\right)$ meters long. If a piece $\left(y^2 - 10\right)$ meters is cut, express the length of the remaining piece of beam as a polynomial in y.

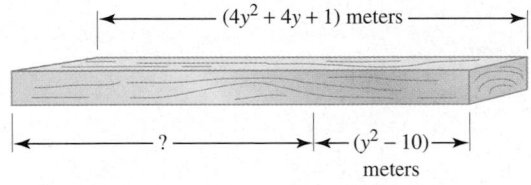

$(4y^2 + 4y + 1)$ meters

? $(y^2 - 10)$→
meters

△ **78.** A piece of quarter-round molding is $(13x - 7)$ inches long. If a piece $(2x + 2)$ inches is removed, express the length of the remaining piece of molding as a polynomial in x.

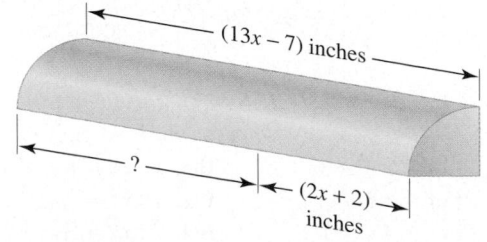

$(13x - 7)$ inches

? $(2x + 2)$→
inches

Solve.

79. The polynomial $P(t) = -32t + 500$ models the relationship between the length of time t in seconds a particle flies through space, beginning at a velocity of 500 feet per second, and its accrued velocity, $P(t)$. Find $P(3)$, the accrued velocity after 3 seconds.

80. The polynomial function $P(x) = 45x - 100{,}000$ models the relationship between the number of lamps x that Sherry's Lamp Shop sells and the profit the shop makes, $P(x)$. Find $P(4000)$, the profit from selling 4000 lamps.

81. An object is thrown upward with an initial velocity of 50 feet per second from the top of the 350-foot high City Hall in Milwaukee, Wisconsin. The height of the object at any time t can be described by the polynomial function $P(t) = -16t^2 + 50t + 350$. Find the height of the projectile when $t = 1$ second, $t = 2$ seconds, and $t = 3$ seconds. (*Source: World Almanac*)

82. An object is thrown upward with an initial velocity of 25 feet per second from the top of the 984-foot-high Eiffel Tower in Paris, France. The height of the object at any time t can be described by the polynomial function $P(t) = -16t^2 + 25t + 984$. Find the height of the projectile when $t = 1$ second, $t = 3$ seconds, and $t = 5$ seconds. (*Source:* Council on Tall Buildings and Urban Habitat, Lehigh University)

A projectile is fired upward from the ground with an initial velocity of 300 feet per second. Neglecting air resistance, the height of the projectile at any time t can be described by the polynomial function

$$P(t) = -16t^2 + 300t$$

Use this in Exercises 83–85.

83. Find the height of the projectile at the given times.

    a. $t = 1$ second

    b. $t = 2$ seconds

    c. $t = 3$ seconds

    d. $t = 4$ seconds

84. Explain why the height increases and then decreases as time passes.

85. Approximate (to the nearest second) how long before the object hits the ground.

86. The total cost (in dollars) for MCD, Inc., Manufacturing Company to produce x blank audiocassette tapes per week is given by the polynomial function $C(x) = 0.8x + 10{,}000$. Find the total cost in producing 20,000 tapes per week.

87. The function $f(x) = 0.43x^2 + 164.6x + 949.3$ can be used to approximate spending for health care in the United States, where x is the number of years since 1980 and $f(x)$ is the amount of money spent per capita. (*Source:* U.S. Health Care Financing Administration)

a. Approximate the amount of money spent on health care per capita in the year 1985.

b. Approximate the amount of money spent on health care per capita in the year 1995.

c. Use the given function to predict the amount of money that will be spent on health care per capita in the year 2010.

d. From parts **a**, **b**, and **c**, is the amount of money spent rising at a steady rate? Why or why not?

88. The function $f(x) = -0.85x^3 + 14.28x^2 - 49.38x + 574.16$ can be used to approximate the number of health maintenance organizations (HMO's) in the United States x years after 1990. (Round each answer to the nearest whole.) (*Source:* Based on data from Interstudy, Minneapolis, MN)

a. Approximate the number of HMO's in 1995.

b. Approximate the number of HMO's in 1998.

c. Use the given model to predict the number of HMO's in the United States in 2003.

Match each equation with its graph. See Example 11.

89. $f(x) = 3x^2 - 2$

90. $h(x) = 5x^3 - 6x + 2$

91. $g(x) = -2x^3 - 3x^2 + 3x - 2$

92. $g(x) = -2x^2 - 6x + 2$

A

B

C

D

Find the area of each figure. Write a polynomial that describes the total area of the rectangles and squares shown in Exercises 93–94. Then simplify the polynomial.

△ **93.**

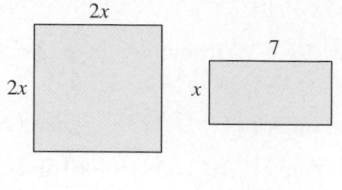
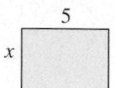

△ **94.**

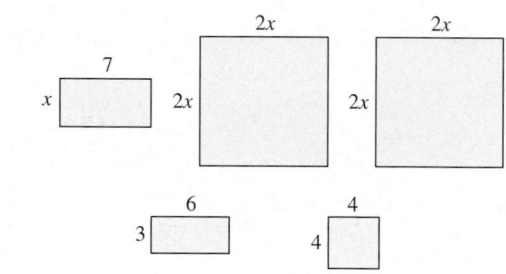

REVIEW EXERCISES

Multiply. See Section 1.4.

95. $5(3x - 2)$

96. $-7(2z - 6y)$

97. $-2(x^2 - 5x + 6)$

98. $5(-3y^2 - 2y + 7)$

A Look Ahead

Example

If $P(x) = -3x + 5$, find the following.

  a. $P(a)$ **b.** $P(-x)$ **c.** $P(x + h)$

Solution:

  a. $P(x) = -3x + 5$

      $P(a) = -3a + 5$

  b. $P(x) = -3x + 5$

     $P(-x) = -3(-x) + 5$

          $= 3x + 5$

  c. $P(x) = -3x + 5$

    $P(x + h) = -3(x + h) + 5$

            $= -3x - 3h + 5$

*If $P(x)$ is the polynomial given, find **a.** $P(a)$, **b.** $P(-x)$, and **c.** $P(x + h)$. See the preceding example.*

99. $P(x) = 2x - 3$ **100.** $P(x) = 8x + 3$

101. $P(x) = 4x$ **102.** $P(x) = -4x$

103. $P(x) = 4x - 1$ **104.** $P(x) = 3x - 2$

5.4 MULTIPLYING POLYNOMIALS

CD-ROM SSM

SSG Video

▶ **O B J E C T I V E S**

1. Use the distributive property to multiply polynomials.
2. Multiply polynomials vertically.

---

1 To multiply polynomials, we apply our knowledge of the rules and definitions of exponents.

To multiply two monomials such as $(-5x^3)$ and $(-2x^4)$, use the associative and commutative properties and regroup. Remember that to multiply exponential expressions with a common base we add exponents.

$$(-5x^3)(-2x^4) = (-5)(-2)(x^3)(x^4) = 10x^7$$

To multiply polynomials that are not monomials, use the distributive property.

Example 1 Use the distributive property to find each product.

a. $5x(2x^3 + 6)$

b. $-3x^2(5x^2 + 6x - 1)$

Solution **a.** $5x(2x^3 + 6) = 5x(2x^3) + 5x(6)$ Use the distributive property.

$$= 10x^4 + 30x$$ Multiply.

b. $-3x^2(5x^2 + 6x - 1)$
$= (-3x^2)(5x^2) + (-3x^2)(6x) + (-3x^2)(-1)$ Use the distributive property.
$= -15x^4 - 18x^3 + 3x^2$ Multiply.   ■

We also use the distributive property to multiply two binomials. To multiply $(x + 3)$ by $(x + 1)$, distribute the factor $(x + 1)$ first.

$(x + 3)(x + 1) = x(x + 1) + 3(x + 1)$ Distribute $(x + 1)$.

$= x(x) + x(1) + 3(x) + 3(1)$ Apply distributive property a second time.

$= x^2 + x + 3x + 3$ Multiply.

$= x^2 + 4x + 3$ Combine like terms.

This idea can be expanded so that we can multiply any two polynomials.

MULTIPLYING POLYNOMIALS

Multiply each term of the first polynomial by each term of the second polynomial, and then combine like terms.

Example 2 Find the product: $(3x + 2)(2x - 5)$.

Solution Multiply each term of the first binomial by each term of the second.

$$(3x + 2)(2x - 5) = 3x(2x) + 3x(-5) + 2(2x) + 2(-5)$$
$$= 6x^2 - 15x + 4x - 10 \qquad \text{Multiply.}$$
$$= 6x^2 - 11x - 10 \qquad \text{Combine like terms.}$$

Example 3 Multiply: $(2x - y)^2$.

Solution Recall that $a^2 = a \cdot a$, so $(2x - y)^2 = (2x - y)(2x - y)$. Multiply each term of the first polynomial by each term of the second.

$$(2x - y)(2x - y) = 2x(2x) + 2x(-y) + (-y)(2x) + (-y)(-y)$$
$$= 4x^2 - 2xy - 2xy + y^2 \qquad \text{Multiply.}$$
$$= 4x^2 - 4xy + y^2 \qquad \text{Combine like terms.}$$

Example 4 Multiply: $(t + 2)$ by $(3t^2 - 4t + 2)$.

Solution Multiply each term of the first polynomial by each term of the second.

$$(t + 2)(3t^2 - 4t + 2) = t(3t^2) + t(-4t) + t(2) + 2(3t^2) + 2(-4t) + 2(2)$$
$$= 3t^3 - 4t^2 + 2t + 6t^2 - 8t + 4$$
$$= 3t^3 + 2t^2 - 6t + 4 \qquad \text{Combine like terms.}$$

Example 5 Multiply: $(3a + b)^3$.

Solution Write $(3a + b)^3$ as $(3a + b)(3a + b)(3a + b)$.

$$(3a + b)(3a + b)(3a + b) = (9a^2 + 3ab + 3ab + b^2)(3a + b)$$
$$= (9a^2 + 6ab + b^2)(3a + b)$$
$$= (9a^2 + 6ab + b^2)3a + (9a^2 + 6ab + b^2)b$$
$$= 27a^3 + 18a^2b + 3ab^2 + 9a^2b + 6ab^2 + b^3$$
$$= 27a^3 + 27a^2b + 9ab^2 + b^3$$

2 Another convenient method for multiplying polynomials is to use a vertical format similar to the format used to multiply real numbers. We demonstrate this method by multiplying $(3y^2 - 4y + 1)$ by $(y + 2)$.

Example 6 Multiply: $(3y^2 - 4y + 1)(y + 2)$. Use a vertical format.

Solution **Step 1.** Multiply 2 by each term of the top polynomial. Write the first **partial product** below the line.

$$
\begin{array}{r}
3y^2 - 4y + 1 \\
\times \qquad y + 2 \\
\hline
6y^2 - 8y + 2 \qquad \text{partial product}
\end{array}
$$

Step 2. Multiply y by each term of the top polynomial. Write this partial product underneath the previous one, being careful to line up like terms.

$$
\begin{array}{r}
3y^2 - 4y + 1 \\
\times \quad\quad y + 2 \\
\hline
6y^2 - 8y + 2 \\
3y^3 - 4y^2 + \ y
\end{array}
$$

partial product
partial product

Step 3. Combine like terms of the partial products.

$$
\begin{array}{r}
3y^2 - 4y + 1 \\
\times \quad\quad y + 2 \\
\hline
6y^2 - 8y + 2 \\
3y^3 - 4y^2 + \ y \\
\hline
3y^3 + 2y^2 - 7y + 2
\end{array}
$$

Combine like terms.

Thus, $(y + 2)(3y^2 - 4y + 1) = 3y^3 + 2y^2 - 7y + 2$.

When multiplying vertically, be careful if a power is missing. You may want to leave space in the partial products and take care that like terms are lined up.

Example 7 Multiply: $(2x^3 - 3x + 4)(x^2 + 1)$. Use a vertical format.

Solution

$$
\begin{array}{r}
2x^3 - 3x \ + 4 \\
\times \quad\quad x^2 + 1 \\
\hline
2x^3 \quad\quad - 3x \ + 4 \\
2x^5 - 3x^3 + 4x^2 \\
\hline
2x^5 - \ x^3 + 4x^2 - 3x \ + 4
\end{array}
$$

Leave space for missing powers of x.

Combine like terms.

MENTAL MATH

Find the following products mentally.

1. $5x(2y)$
2. $7a(4b)$
3. $x^2 \cdot x^5$
4. $z \cdot z^4$
5. $6x(3x^2)$
6. $5a^2(3a^2)$

Exercise Set 5.4

Find the following products. See Example 1.

1. $2a(2a - 4)$
2. $3a(2a + 7)$
3. $7x(x^2 + 2x - 1)$
4. $-5y(y^2 + y - 10)$
5. $3x^2(2x^2 - x)$
6. $-4y^2(5y - 6y^2)$

△ 7. The area of the larger rectangle below is $x(x + 3)$. Find another expression for this area by finding the sum of the areas of the smaller rectangles.

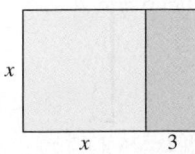

△ **8.** Write an expression for the area of the larger rectangle below in two different ways.

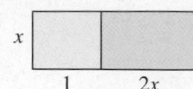

Find the following products. See Examples 2 and 3.

9. $(a + 7)(a - 2)$ **10.** $(y + 5)(y + 7)$

11. $(2y - 4)^2$ **12.** $(6x - 7)^2$

13. $(5x - 9y)(6x - 5y)$ **14.** $(3x - 7y)(7x + 2y)$

15. $(2x^2 - 5)^2$ **16.** $(x^2 - 4)^2$

△ **17.** The area of the figure below is $(x + 2)(x + 3)$. Find another expression for this area by finding the sum of the areas of the smaller rectangles.

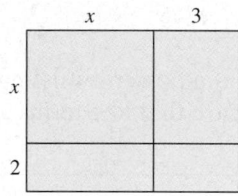

△ **18.** Write an expression for the area of the figure below in two different ways.

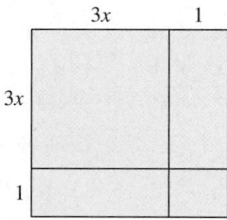

Find the following products. See Example 4.

19. $(x - 2)(x^2 - 3x + 7)$ **20.** $(x + 3)(x^2 + 5x - 8)$

21. $(x + 5)(x^3 - 3x + 4)$ **22.** $(a + 2)(a^3 - 3a^2 + 7)$

23. $(2a - 3)(5a^2 - 6a + 4)$

24. $(3 + b)(2 - 5b - 3b^2)$

Find the following products. See Example 5.

25. $(x + 2)^3$ **26.** $(y - 1)^3$

27. $(2y - 3)^3$ **28.** $(3x + 4)^3$

Find the following products. Use the vertical multiplication method. See Examples 6 and 7.

29. $(x + 3)(2x^2 + 4x - 1)$

30. $(2x - 5)(3x^2 - 4x + 7)$

31. $(x^3 + 5x - 7)(x^2 - 9)$

32. $(3x^4 - x^2 + 2)(2x^3 + 1)$

33. Evaluate each of the following.

     a. $(2 + 3)^2; 2^2 + 3^2$

     b. $(8 + 10)^2; 8^2 + 10^2$

     c. Does $(a + b)^2 = a^2 + b^2$ no matter what the values of a and b are? Why or why not?

34. Perform the indicated operations. Explain the difference between the two expressions.

     a. $(3x + 5) + (3x + 7)$

     b. $(3x + 5)(3x + 7)$

Find the following products.

35. $2a(a + 4)$ **36.** $-3a(2a + 7)$

37. $3x(2x^2 - 3x + 4)$ **38.** $-4x(5x^2 - 6x - 10)$

39. $(5x + 9y)(3x + 2y)$ **40.** $(5x - 5y)(2x - y)$

41. $(x + 2)(x^2 + 5x + 6)$

42. $(x - 7)(x^2 - 15x + 56)$

43. $(7x + 4)^2$ **44.** $(3x - 2)^2$

45. $-2a^2(3a^2 - 2a + 3)$

46. $-4b^2(3b^3 - 12b^2 - 6)$

47. $(x + 3)(x^2 + 7x + 12)$

48. $(n + 1)(n^2 - 7n - 9)$

49. $(a + 1)^3$ **50.** $(x - y)^3$

51. $(x + y)(x + y)$ **52.** $(x + 3)(7x + 1)$

53. $(x - 7)(x - 6)$ **54.** $(4x + 5)(-3x + 2)$

55. $3a(a^2 + 2)$ **56.** $x^3(x + 12)$

57. $-4y(y^2 + 3y - 11)$ **58.** $-2x(5x^2 - 6x + 1)$

59. $(5x + 1)(5x - 1)$ **60.** $(2x + y)(3x - y)$

61. $(5x + 4)(x^2 - x + 4)$ **62.** $(x - 2)(x^2 - x + 3)$

63. $(2x - 5)^3$ **64.** $(3y - 1)^3$

65. $(4x + 5)(8x^2 + 2x - 4)$ **66.** $(x + 7)(x^2 - 7x - 8)$

67. $(7xy - y)^2$ **68.** $(x + 2y)^2$

69. $(5y^2 - y + 3)(y^2 - 3y - 2)$

70. $(2x^2 + x - 1)(x^2 + 3x + 4)$

71. $(3x^2 + 2x - 4)(2x^2 - 4x + 3)$

72. $(a^2 + 3a - 2)(2a^2 - 5a - 1)$

Express each of the following as polynomials.

△ **73.** Find the area of the rectangle.

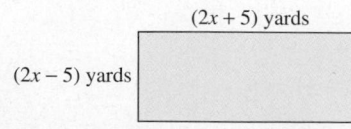

(2x + 5) yards

(2x − 5) yards

△ 74. Find the area of the square field.

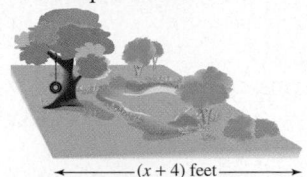

(x + 4) feet

△ 75. Find the area of the triangle.

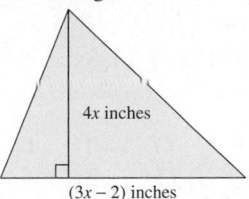

4x inches

(3x − 2) inches

△ 76. Find the volume of the cube-shaped glass block.

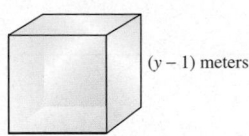

(y − 1) meters

△ 77. Write a polynomial that describes the area of the shaded region.

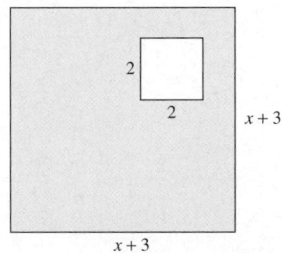

2

2

x + 3

x + 3

△ 78. Write a polynomial that describes the area of the shaded region.

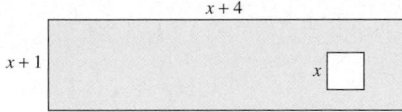

x + 4

x + 1

x

79. Multiply the following polynomials.
   a. $(a + b)(a - b)$
   b. $(2x + 3y)(2x - 3y)$
   c. $(4x + 7)(4x - 7)$
   d. Can you make a general statement about all products of the form $(x + y)(x - y)$?

REVIEW EXERCISES

Perform the indicated operation. See Section 5.1.

80. $(5x)^7$

81. $(4p)^2$

82. $(-3y^3)^2$

83. $(-7m^2)^2$

*For income tax purposes, Rob Calcutta, the owner of Copy Services, uses a method called **straight-line depreciation** to show the depreciated (or decreased) value of a copy machine he recently purchased. Rob assumes that he can use the machine for 7 years. The graph below shows the depreciated value of the machine over the years. Use this graph to answer Exercises 84–89. See Sections 1.9 and 3.1.*

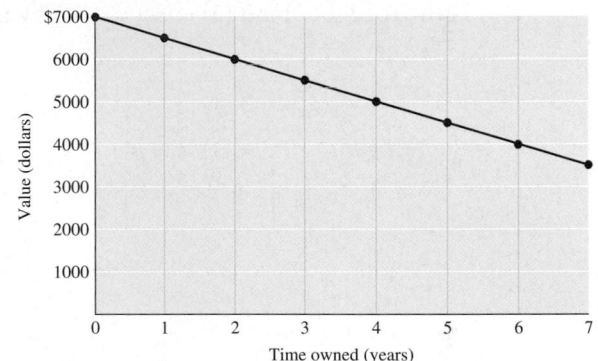

84. What was the purchase price of the copy machine?
85. What is the depreciated value of the machine in 7 years?
86. What loss in value occurred during the first year?
87. What loss in value occurred during the second year?
88. Why do you think this method of depreciating is called straight-line depreciation?
89. Why is the line tilted downward?

5.5   SPECIAL PRODUCTS

CD-ROM   SSM

SSG   Video

▶ **OBJECTIVES**

   1. Multiply two binomials using the FOIL method.
   2. Square a binomial.
   3. Multiply the sum and difference of two terms.
   4. Evaluate polynomial functions.

1   In this section, we multiply binomials using special products. First, a special order for multiplying binomials called the FOIL order or method is introduced. This method is demonstrated by multiplying $(3x + 1)$ by $(2x + 5)$.

F   stands for the product of the **First** terms. $(3x + 1)(2x + 5)$

$$(3x)(2x) = 6x^2 \quad \textbf{F}$$

O   stands for the product of the **Outer** terms. $(3x + 1)(2x + 5)$

$$(3x)(5) = 15x \quad \textbf{O}$$

I   stands for the product of the **Inner** terms. $(3x + 1)(2x + 5)$

$$(1)(2x) = 2x \quad \textbf{I}$$

L   stands for the product of the **Last** terms. $(3x + 1)(3x + 5)$

$$(1)(5) = 5 \quad \textbf{L}$$

$$\overset{\text{F} \quad\quad \text{O} \quad\quad \text{I} \quad\quad \text{L}}{(3x + 1)(2x + 5) = 6x^2 + 15x + 2x + 5}$$
$$= 6x^2 + 17x + 5 \qquad \textit{Combine like terms.}$$

Example 1   Find $(x - 3)(x + 4)$ by the FOIL method.

Solution
$$\overset{\text{F} \quad\quad\quad \text{O} \quad\quad\quad \text{I} \quad\quad\quad \text{L}}{(x - 3)(x + 4) = (x)(x) + (x)(4) + (-3)(x) + (-3)(4)}$$

$$= x^2 + 4x - 3x - 12$$
$$= x^2 + x - 12 \qquad \textit{Combine like terms.} \quad \blacksquare$$

Example 2   Find $(5x - 7)(x - 2)$ by the FOIL method.

Solution
$$\overset{\text{F} \quad\quad\quad \text{O} \quad\quad\quad \text{I} \quad\quad\quad \text{L}}{(5x - 7)(x - 2) = 5x(x) + 5x(-2) + (-7)(x) + (-7)(-2)}$$

$$= 5x^2 - 10x - 7x + 14$$
$$= 5x^2 - 17x + 14 \qquad \textit{Combine like terms.} \quad \blacksquare$$

Example 3   Multiply: $(y + 6)(2y - 1)$.

Solution
$$\overset{\text{F} \quad\quad \text{O} \quad\quad \text{I} \quad\quad \text{L}}{(y + 6)(2y - 1) = 2y^2 - 1y + 12y - 6}$$
$$= 2y^2 + 11y - 6 \qquad\qquad\qquad \blacksquare$$

2   Now, try squaring a binomial using the FOIL method.

Example 4   Multiply: $(3y + 1)^2$.

Solution $(3y + 1)^2 = (3y + 1)(3y + 1)$

$$\qquad\quad\begin{array}{cccc} \text{F} & \text{O} & \text{I} & \text{L} \end{array}$$
$$= (3y)(3y) + (3y)(1) + 1(3y) + 1(1)$$
$$= 9y^2 + 3y + 3y + 1$$
$$= 9y^2 + 6y + 1$$

Notice the pattern that appears in Example 4.

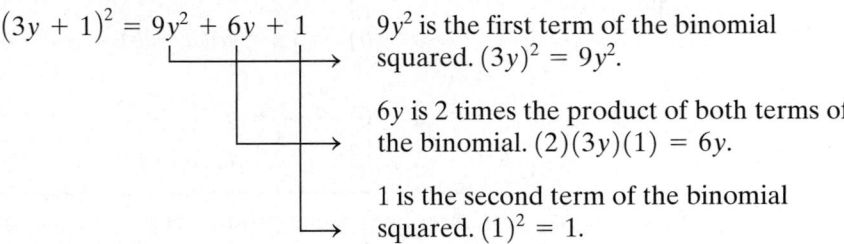

$9y^2$ is the first term of the binomial squared. $(3y)^2 = 9y^2$.

$6y$ is 2 times the product of both terms of the binomial. $(2)(3y)(1) = 6y$.

1 is the second term of the binomial squared. $(1)^2 = 1$.

This pattern leads to the following, which can be used when squaring a binomial. We call these **special products**.

SQUARING A BINOMIAL

A binomial squared is equal to the square of the first term plus or minus twice the product of both terms plus the square of the second term.

$$(a + b)^2 = a^2 + 2ab + b^2$$
$$(a - b)^2 = a^2 - 2ab + b^2$$

This product can be visualized geometrically.

The area of the large square is side · side.

Area $= (a + b)(a + b) = (a + b)^2$

The area of the large square is also the sum of the areas of the smaller rectangles.

Area $= a^2 + ab + ab + b^2 = a^2 + 2ab + b^2$

Thus, $(a + b)^2 = a^2 + 2ab + b^2$.

Example 5   Use a special product to square each binomial.

a. $(t + 2)^2$ **b.** $(p - q)^2$ **c.** $(2x + 5)^2$ **d.** $(x^2 - 7y)^2$

Solution

first term squared	plus or minus	twice the product of the terms	plus	second term squared

a. $(t + 2)^2 = t^2 + 2(t)(2) + 2^2 = t^2 + 4t + 4$

b. $(p - q)^2 = p^2 - 2(p)(q) + q^2 = p^2 - 2pq + q^2$

c. $(2x + 5)^2 = (2x)^2 + 2(2x)(5) + 5^2 = 4x^2 + 20x + 25$

d. $(x^2 - 7y)^2 = (x^2)^2 - 2(x^2)(7y) + (7y^2) = x^4 - 14x^2y + 49y^2$

> **HELPFUL HINT**
>
> Notice that
>
> $$(a + b)^2 \neq a^2 + b^2 \qquad \text{The middle term } 2ab \text{ is missing.}$$
>
> $$(a + b)^2 = (a + b)(a + b) = a^2 + 2ab + b^2$$
>
> Likewise,
>
> $$(a - b)^2 \neq a^2 - b^2$$
>
> $$(a - b)^2 = (a - b)(a - b) = a^2 - 2ab + b^2$$

Another special product is the product of the sum and difference of the same two terms, such as $(x + y)(x - y)$. Finding this product by the FOIL method, we see a pattern emerge.

$$(x + y)(x - y) = x^2 - xy + xy - y^2$$

$$= x^2 - y^2$$

Notice that the middle two terms subtract out. This is because the **O**uter product is the opposite of the **I**nner product. Only the **difference of squares** remains.

3   **MULTIPLYING THE SUM AND DIFFERENCE OF TWO TERMS**

The product of the sum and difference of two terms is the square of the first term minus the square of the second term.

$$(a + b)(a - b) = a^2 - b^2$$

Example 6   Use a special product to multiply.

a. $(x + 4)(x - 4)$ **b.** $(6t + 7)(6t - 7)$ **c.** $\left(x - \dfrac{1}{4}\right)\left(x + \dfrac{1}{4}\right)$

d. $(2p - q)(2p + q)$ **e.** $(3x^2 - 5y)(3x^2 + 5y)$

Solution

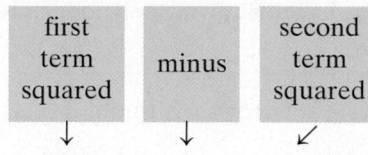

first term squared	minus	second term squared

a. $(x + 4)(x - 4) = x^2 \quad - \quad 4^2 = x^2 - 16$

b. $(6t + 7)(6t - 7) = (6t)^2 \quad - \quad 7^2 = 36t^2 - 49$

c. $\left(x - \dfrac{1}{4}\right)\left(x + \dfrac{1}{4}\right) = x^2 - \left(\dfrac{1}{4}\right)^2 = x^2 - \dfrac{1}{16}$

d. $(2p - q)(2p + q) = (2p)^2 - q^2 = 4p^2 - q^2$

e. $(3x^2 - 5y)(3x^2 + 5y) = (3x^2)^2 - (5y)^2 = 9x^4 - 25y^2$ ◼

Let's now practice multiplying polynomials in general. If possible, use a special product.

Example 7   Multiply.

a. $(x - 5)(3x + 4)$ **b.** $(7x + 4)^2$ **c.** $(y - 0.6)(y + 0.6)$

d. $(a - 3)(a^2 + 2a - 1)$

Solution   **a.** $(x - 5)(3x + 4) = 3x^2 + 4x - 15x - 20$ FOIL.

$\qquad\qquad\qquad\quad = 3x^2 - 11x - 20$

b. $(7x + 4)^2 = (7x)^2 + 2(7x)(4) + 4^2$ Squaring a binomial.

$\qquad\qquad\quad = 49x^2 + 56x + 16$

c. $(y - 0.6)(y + 0.6) = y^2 - (0.6)^2 = y^2 - 0.36$ Multiplying the sum and difference of 2 terms.

d. $(a - 3)(a^2 + 2a - 1) = a(a^2 + 2a - 1) - 3(a^2 + 2a - 1)$ Multiplying each term of the binomial by each term of the trinomial.

$\qquad\qquad\qquad\qquad\quad = a^3 + 2a^2 - a - 3a^2 - 6a + 3$

$\qquad\qquad\qquad\qquad\quad = a^3 - a^2 - 7a + 3$ ◼

Example 8   Multiply $[3 + (2x + y)]^2$.

Solution   Think of 3 as the first term and $(2x + y)$ as the second term, and apply the method for squaring a binomial.

$$[a + b]^2 \quad = a^2 + 2\,(a)\cdot\ \ b\ \ +\ \ b^2$$

$$[3 + (2x + y)]^2 = 3^2 + 2(3)(2x + y) + (2x + y)^2$$

$$= 9 + 6(2x + y) + (2x + y)^2$$

$$= 9 + 12x + 6y + (2x)^2 + 2(2x)(y) + y^2 \qquad \text{Square } (2x + y).$$

$$= 9 + 12x + 6y + 4x^2 + 4xy + y^2$$ ◼

Example 9   Multiply $[(5x - 2y) - 1][(5x - 2y) + 1]$.

Solution   Think of $(5x - 2y)$ as the first term and 1 as the second term, and apply the method for the product of the sum and difference of two terms.

$$(a \quad - b)\ \ (a \quad + b) = \quad a^2 \quad - b^2$$

$$[(5x - 2y) - 1][(5x - 2y) + 1] = (5x - 2y)^2 - 1^2$$

$$= (5x)^2 - 2(5x)(2y) + (2y)^2 - 1 \qquad \begin{array}{l}\text{Square}\\ (5x - 2y).\end{array}$$

$$= 25x^2 - 20xy + 4y^2 - 1$$ ◼

4   Our work in multiplying polynomials is often useful in evaluating polynomial functions.

Example 10   If $f(x) = x^2 + 5x - 2$, find $f(a + 1)$.

Solution   To find $f(a + 1)$, replace x with the expression $a + 1$ in the polynomial function $f(x)$.

$$f(x) = x^2 + 5x - 2$$
$$f(a + 1) = (a + 1)^2 + 5(a + 1) - 2$$
$$= a^2 + 2a + 1 + 5a + 5 - 2$$
$$= a^2 + 7a + 4$$

GRAPHING CALCULATOR EXPLORATIONS

In the previous section, we used a graphing calculator to visualize addition and subtraction of polynomials in one variable. In this section, the same method is used to visualize multiplication of polynomials in one variable. For example, to see that

$$(x - 2)(x + 1) = x^2 - x - 2,$$

graph both $Y_1 = (x - 2)(x + 1)$ and $Y_2 = x^2 - x - 2$ on the same screen and see whether their graphs coincide.

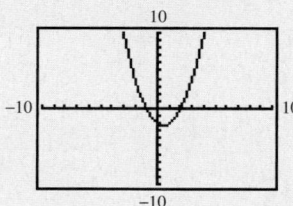

By tracing along both graphs, we see that the graphs of Y_1 and Y_2 appear to coincide, and thus $(x - 2)(x + 1) = x^2 - x - 2$ appears to be correct.

Multiply. Use a graphing calculator to visualize the results.

1. $(x + 4)(x - 4)$ **2.** $(x + 3)(x + 3)$

3. $(3x - 7)^2$ **4.** $(5x - 2)^2$

5. $(5x + 1)(x^2 - 3x - 2)$ **6.** $(7x + 4)(2x^2 + 3x - 5)$

MENTAL MATH

Answer each exercise true or false.

1. $(x + 4)^2 = x^2 + 16$ **2.** For $(x + 6)(2x - 1)$ the product of the first terms is $2x^2$.

3. $(x + 4)(x - 4) = x^2 + 16$ **4.** The product $(x - 1)(x^3 + 3x - 1)$ is a polynomial of degree 5.

Exercise Set 5.5

Find each product using the FOIL method. See Examples 1 through 3.

1. $(x + 3)(x + 4)$ **2.** $(x + 5)(x - 1)$
3. $(x - 5)(x + 10)$ **4.** $(y - 12)(y + 4)$
5. $(5x - 6)(x + 2)$ **6.** $(3y - 5)(2y - 7)$
7. $(y - 6)(4y - 1)$ **8.** $(2x - 9)(x - 11)$
9. $(2x + 5)(3x - 1)$ **10.** $(6x + 2)(x - 2)$

Find each product. See Examples 4 and 5.

11. $(x - 2)^2$ **12.** $(x + 7)^2$
13. $(2x - 1)^2$ **14.** $(7x - 3)^2$
15. $(3a - 5)^2$ **16.** $(5a + 2)^2$
17. $(5x + 9)^2$ **18.** $(6s - 2)^2$

19. Using your own words, explain how to square a binomial such as $(a + b)^2$.

20. Explain how to find the product of two binomials using the FOIL method.

Find each product. See Example 6.

21. $(a - 7)(a + 7)$ **22.** $(b + 3)(b - 3)$
23. $(3x - 1)(3x + 1)$ **24.** $(4x - 5)(4x + 5)$
25. $\left(3x - \dfrac{1}{2}\right)\left(3x + \dfrac{1}{2}\right)$ **26.** $\left(10x + \dfrac{2}{7}\right)\left(10x - \dfrac{2}{7}\right)$
27. $(9x + y)(9x - y)$ **28.** $(2x - y)(2x + y)$

Express each of the following as a polynomial in x.

△ **29.** Find the area of the square rug shown if its side is $(2x + 1)$ feet.

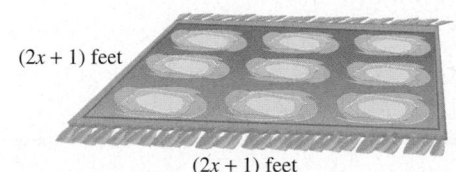

(2x + 1) feet

(2x + 1) feet

△ **30.** Find the area of the rectangular canvas if its length is $(3x - 2)$ inches and its width is $(x - 4)$ inches.

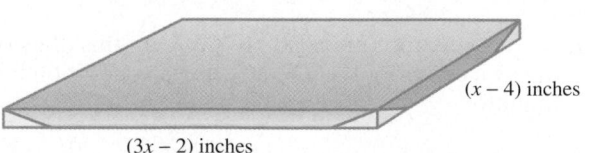

(x − 4) inches

(3x − 2) inches

Find each product. See Example 7.

31. $(a + 5)(a + 4)$ **32.** $(a - 5)(a - 7)$
33. $(a + 7)^2$ **34.** $(b - 2)^2$
35. $(4a + 1)(3a - 1)$ **36.** $(6a + 7)(6a + 5)$
37. $(x + 2)(x - 2)$ **38.** $(x - 10)(x + 10)$
39. $(3a + 1)^2$ **40.** $(4a - 2)^2$
41. $(x + y)(4x - y)$ **42.** $(3x + 2)(4x - 2)$
43. $(x + 3)(x^2 - 6x + 1)$ **44.** $(x - 2)(x^2 - 4x + 2)$
45. $(2a - 3)^2$ **46.** $(5b - 4x)^2$
47. $(5x - 6z)(5x + 6z)$ **48.** $(11x - 7y)(11x + 7y)$
49. $(x - 3)(x - 5)$ **50.** $(a + 5b)(a + 6b)$
51. $\left(x - \dfrac{1}{3}\right)\left(x + \dfrac{1}{3}\right)$ **52.** $\left(3x + \dfrac{1}{5}\right)\left(3x - \dfrac{1}{5}\right)$
53. $(a + 11)(a - 3)$ **54.** $(2x + 5)(x - 8)$
55. $(x - 2)^2$ **56.** $(3b + 7)^2$
57. $(3b + 7)(2b - 5)$ **58.** $(3y - 13)(y - 3)$
59. $(7p - 8)(7p + 8)$ **60.** $(3s - 4)(3s + 4)$
61. $\left(\dfrac{1}{3}a^2 - 7\right)\left(\dfrac{1}{3}a^2 + 7\right)$ **62.** $\left(\dfrac{2}{3}a - b^2\right)\left(\dfrac{2}{3}a - b^2\right)$
63. $5x^2(3x^2 - x + 2)$ **64.** $4x^3(2x^2 + 5x - 1)$
65. $(2r - 3s)(2r + 3s)$ **66.** $(6r - 2x)(6r + 2x)$
67. $(3x - 7y)^2$ **68.** $(4s - 2y)^2$
69. $(4x + 5)(4x - 5)$ **70.** $(3x + 5)(3x - 5)$
71. $(x + 4)(x + 4)$ **72.** $(3x + 2)(3x + 2)$
73. $\left(a - \dfrac{1}{2}y\right)\left(a + \dfrac{1}{2}y\right)$ **74.** $\left(\dfrac{a}{2} + 4y\right)\left(\dfrac{a}{2} - 4y\right)$
75. $\left(\dfrac{1}{5}x - y\right)\left(\dfrac{1}{5}x + y\right)$ **76.** $\left(\dfrac{y}{6} - 8\right)\left(\dfrac{y}{6} + 8\right)$
77. $(a + 1)(3a^2 - a + 1)$ **78.** $(b + 3)(2b^2 + b - 3)$

Find the area of each shaded region.

△ **79.**

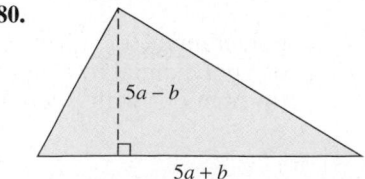

2x + 3

2x − 3

x

x

△ **80.**

5a − b

5a + b

△ **81.**

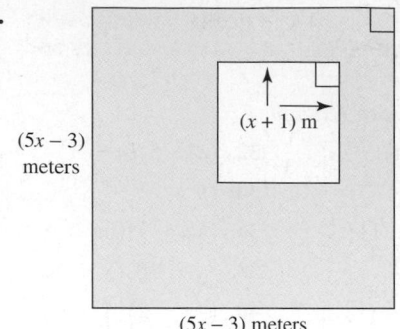

(5x − 3) meters

(x + 1) m

(5x − 3) meters

△ **82.**

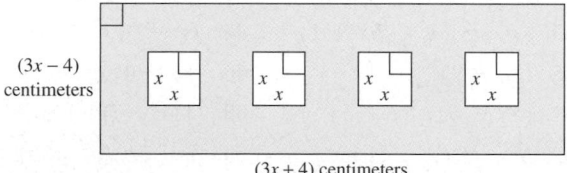

(3x − 4) centimeters

x
x   x
x   x
x   x
x

(3x + 4) centimeters

Multiply, using special product methods. See Examples 8 and 9.

83. $[3 + (4b + 1)]^2$ **84.** $[5 - (3b - 3)]^2$

85. $[(2s - 3) - 1][(2s - 3) + 1]$

86. $[(2y + 5) + 6][(2y + 5) - 6]$

87. $[(x + 4) - y]^2$ **88.** $[(2a^2 + 4a) + 1]^2$

89. $[(x + y) - 3][(x + y) + 3]$
90. $[(a + c) - 5][(a + c) + 5]$
91. $[(a - 3) + b][(a - 3) - b]$
92. $[(x - 2) + y][(x - 2) - y]$

REVIEW EXERCISES

Find the slope of each line. See Section 3.5.

93.

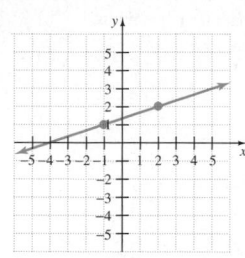

94.

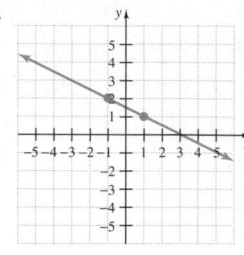

95.

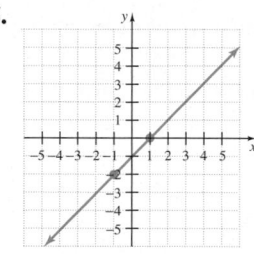

96.

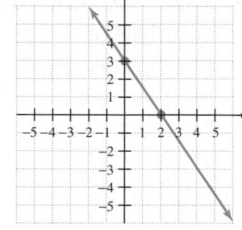

5.6 DIVIDING POLYNOMIALS

     ▶ **OBJECTIVES**

CD-ROM   SSM

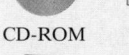

SSG   Video

1. Divide a polynomial by a monomial.
2. Divide by a polynomial.

---

1   Recall that a rational expression is a quotient of polynomials. An equivalent form of a rational expression can be obtained by performing the indicated division. For example, the rational expression $\dfrac{10x^3 - 5x^2 + 20x}{5x}$ can be thought of as the polynomial $10x^3 - 5x^2 + 20x$ divided by the monomial $5x$. To perform this division of a polynomial by a monomial (which we do below) recall the following addition fact for fractions with a common denominator.

$$\frac{a}{c} + \frac{b}{c} = \frac{a + b}{c}$$

If a, b, and c are monomials, we might read this equation from right to left and gain insight into dividing a polynomial by a monomial.

DIVIDING A POLYNOMIAL BY A MONOMIAL

Divide each term in the polynomial by the monomial.

$$\frac{a + b}{c} = \frac{a}{c} + \frac{b}{c}, \text{ where } c \neq 0$$

Example 1   Divide $10x^3 - 5x^2 + 20x$ by $5x$.

Solution   We divide each term of $10x^3 - 5x^2 + 20x$ by $5x$ and simplify.

$$\frac{10x^3 - 5x^2 + 20x}{5x} = \frac{10x^3}{5x} - \frac{5x^2}{5x} + \frac{20x}{5x} = 2x^2 - x + 4$$

Check: To check, see that (quotient)(divisor) = dividend, or

$$\left(2x^2 - x + 4\right)(5x) = 10x^3 - 5x^2 + 20x.$$

Example 2   Divide $\dfrac{3x^5y^2 - 15x^3y - x^2y - 6x}{x^2y}$

Solution   We divide each term in the numerator by x^2y.

$$\frac{3x^5y^2 - 15x^3y - x^2y - 6x}{x^2y} = \frac{3x^5y^2}{x^2y} - \frac{15x^3y}{x^2y} - \frac{x^2y}{x^2y} - \frac{6x}{x^2y}$$

$$= 3x^3y - 15x - 1 - \frac{6}{xy}$$

---

2   To divide a polynomial by a polynomial other than a monomial, we use **long division.** Polynomial long division is similar to long division of real numbers. We review long division of real numbers by dividing 7 into 296.

$$
\begin{array}{r}
42 \\
\text{Divisor:} \quad 7{\overline{\smash{\big)}\,296}} \\
\underline{-28} \\
16 \\
\underline{-14} \\
2
\end{array}
$$

$4(7) = 28$

Subtract and bring down the next digit in the dividend.

$2(7) = 14$

Subtract. The remainder is 2.

The quotient is $42\dfrac{2}{7}$ $\dfrac{\text{(remainder)}}{\text{(divisor)}}$.

Check: To check, notice that

$$42(7) + 2 = 296, \text{ the dividend.}$$

This same division process can be applied to polynomials, as shown next.

Example 3   Divide $2x^2 - x - 10$ by $x + 2$.

Solution $2x^2 - x - 10$ is the dividend, and $x + 2$ is the divisor.

Step 1:   Divide $2x^2$ by x.

$$x + 2 \overline{\smash{)}2x^2 - x - 10} \quad\quad \overset{\displaystyle 2x}{}$$

$\dfrac{2x^2}{x} = 2x$, so $2x$ is the first term of the quotient.

Step 2:   Multiply $2x(x + 2)$.

$$\begin{array}{r} 2x \\ x + 2 \overline{\smash{)}2x^2 -\ x - 10} \\ 2x^2 + 4x \end{array}$$

$2x(x + 2)$

Like terms are lined up vertically.

Step 3:   Subtract $(2x^2 + 4x)$ from $(2x^2 - x - 10)$ by changing the signs of $(2x^2 + 4x)$ and adding.

$$\begin{array}{r} 2x \\ x + 2 \overline{\smash{)}2x^2 -\ x - 10} \\ -2x^2 - 4x \\ \hline -5x \end{array}$$

Step 4:   Bring down the next term, -10, and start the process over.

$$\begin{array}{r} 2x \\ x + 2 \overline{\smash{)}2x^2 -\ x - 10} \\ -2x^2 - 4x \\ \hline -5x - 10 \end{array}$$

Step 5:   Divide $-5x$ by x.

$$\begin{array}{r} 2x -\ 5 \\ x + 2 \overline{\smash{)}2x^2 -\ x - 10} \\ -2x^2 - 4x \\ \hline -5x - 10 \end{array}$$

$\dfrac{-5x}{x} = -5$, so -5 is the second term of the quotient.

Step 6:   Multiply $-5(x + 2)$.

$$\begin{array}{r} 2x -\ 5 \\ x + 2 \overline{\smash{)}2x^2 -\ x - 10} \\ -2x^2 - 4x \\ \hline -5x - 10 \\ -5x - 10 \end{array}$$

$-5(x + 2)$

Like terms are lined up vertically.

Step 7:   Subtract $(-5x - 10)$ from $(-5x - 10)$.

$$\begin{array}{r} 2x -\ 5 \\ x + 2 \overline{\smash{)}2x^2 -\ x - 10} \\ -2x^2 - 4x \\ \hline -5x - 10 \\ +5x + 10 \\ \hline 0 \end{array}$$

Then $\dfrac{2x^2 - x - 10}{x + 2} = 2x - 5$. There is no remainder.

Check: Check this result by multiplying $2x - 5$ by $x + 2$. Their product is

$$(2x - 5)(x + 2) = 2x^2 - x - 10, \text{ the dividend.}$$

Example 4 Divide: $(6x^2 - 19x + 12) \div (3x - 5)$

Solution

$$\begin{array}{r} 2x \\ 3x - 5\overline{)6x^2 - 19x + 12} \\ \underline{6x^2 - 10x} \\ -9x + 12 \end{array}$$

Divide $\dfrac{6x^2}{3x} = 2x$.

Multiply $2x(3x - 5)$.
Subtract by adding the opposite.
Bring down the next term, $+ 12$.

$$\begin{array}{r} 2x - 3 \\ 3x - 5\overline{)6x^2 - 19x + 12} \\ \underline{6x^2 - 10x} \\ -9x + 12 \\ \underline{-9x - 15} \\ -3 \end{array}$$

Divide $\dfrac{-9x}{3x} = -3$.

Multiply $-3(3x - 5)$.
Subtract by adding the opposite.

Check:

| divisor | · | quotient | + | remainder |

$$(3x - 5) \qquad (2x - 3) + (-3) = 6x^2 - 19x + 15 - 3$$

$$= 6x^2 - 19x + 12 \quad \text{the dividend}$$

The division checks, so

$$\frac{6x^2 - 19x + 12}{3x - 5} = 2x - 3 - \frac{3}{3x - 5}$$

HELPFUL HINT
This fraction is the remainder over the divisor.

Example 5 Divide $3x^4 + 2x^3 - 8x + 6$ by $x^2 - 1$.

Solution Before dividing, we represent any "missing powers" by the product of 0 and the variable raised to the missing power. There is no x^2 term in the dividend, so we include $0x^2$ to represent the missing term. Also, there is no x term in the divisor, so we include $0x$ in the divisor.

$$\begin{array}{r} 3x^2 + 2x + 3 \\ x^2 + 0x - 1\overline{)3x^4 + 2x^3 + 0x^2 - 8x + 6} \\ \underline{3x^4 + 0x^3 - 3x^2} \\ 2x^3 + 3x^2 - 8x \\ \underline{2x^3 + 0x^2 - 2x} \\ 3x^2 - 6x + 6 \\ \underline{3x^2 + 0x - 3} \\ -6x + 9 \end{array}$$

$\dfrac{3x^4}{x^2} = 3x^2$

$3x^2(x^2 + 0x - 1)$
Subtract. Bring down $-8x$.
$\dfrac{2x^3}{x^2} = 2x$, a term of the quotient
$2x(x^2 + 0x - 1)$
Subtract. Bring down 6.
$\dfrac{3x^2}{x^2} = 3$, a term of the quotient
$3(x^2 + 0x - 1)$
Subtract.

The division process is finished when the degree of the remainder polynomial is less than the degree of the divisor. Thus,

$$\frac{3x^4 + 2x^3 - 8x + 6}{x^2 - 1} = 3x^2 + 2x + 3 + \frac{-6x + 9}{x^2 - 1}$$

Example 6   Divide $27x^3 + 8$ by $3x + 2$.

Solution   We replace the missing terms in the dividend with $0x^2$ and $0x$.

$$
\begin{array}{r}
9x^2 - 6x + 4 \\
3x + 2 \overline{)27x^3 + 0x^2 + 0x + 8} \\
\underline{27x^3 \mp 18x^2} \\
-18x^2 + 0x \\
\underline{\mp 18x^2 \mp 12x} \\
12x + 8 \\
\underline{12x \mp 8} \\
0
\end{array}
$$

$9x^2(3x + 2)$

Subtract. Bring down $0x$.

$-6x(3x + 2)$

Subtract. Bring down 8.

$4(3x + 2)$

Subtract.

Thus, $\dfrac{27x^3 + 8}{3x + 2} = 9x^2 - 6x + 4.$

MENTAL MATH

Simplify each expression mentally.

1. $\dfrac{a^6}{a^4}$ **2.** $\dfrac{y^2}{y}$ **3.** $\dfrac{a^3}{a}$ **4.** $\dfrac{p^8}{p^3}$ **5.** $\dfrac{k^5}{k^2}$

6. $\dfrac{k^7}{k^5}$ **7.** $\dfrac{p^8}{p^3}$ **8.** $\dfrac{k^5}{k^2}$ **9.** $\dfrac{k^7}{k^5}$

Exercise Set 5.6

Divide. See Examples 1 and 2.

1. Divide $4a^2 + 8a$ by $2a$.

2. Divide $6x^4 - 3x^3$ by $3x^2$.

3. $\dfrac{12a^5b^2 + 16a^4b}{4a^4b}$

4. $\dfrac{4x^3y + 12x^2y^2 - 4xy^3}{4xy}$

5. $\dfrac{4x^2y^2 + 6xy^2 - 4y^2}{2x^2y}$

6. $\dfrac{6x^5 + 74x^4 + 24x^3}{2x^3}$

7. $\dfrac{4x^2 + 8x + 4}{4}$

8. $\dfrac{15x^3 - 5x^2 + 10x}{5x^2}$

9. A board of length $(3x^4 + 6x^2 - 18)$ meters is to be cut into three pieces of the same length. Find the length of each piece.

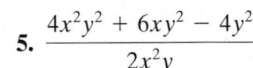
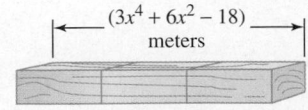

(3x⁴ + 6x² – 18) meters

△ **10.** The perimeter of a regular hexagon is given to be $12x^5 - 48x^3 + 3$ miles. Find the length of each side.

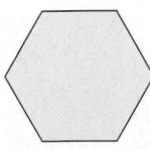

Divide. See Examples 3 through 6

11. $(x^2 + 3x + 2) \div (x + 2)$

12. $(y^2 + 7y + 10) \div (y + 5)$

13. $(2x^2 - 6x - 8) \div (x + 1)$

14. $(3x^2 + 19x + 20) \div (x + 5)$

15. $2x^2 + 3x - 2$ by $2x + 4$

16. $6x^2 - 17x - 3$ by $3x - 9$

17. $(4x^3 + 7x^2 + 8x + 20) \div (2x + 4)$

18. $(18x^3 + x^2 - 90x - 5) \div (9x^2 - 45)$

△ **19.** If the area of the rectangle is $(15x^2 - 29x - 14)$ square inches and its length is $(5x + 2)$ inches, find its width.

(5x + 2) inches

△ **20.** If the area of a parallelogram is $(2x^2 - 17x + 35)$ square centimeters and its base is $(2x - 7)$ centimeters, find its height.

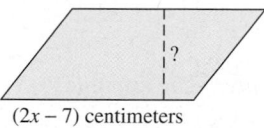

(2x − 7) centimeters

Divide.

21. $25a^2b^{12}$ by $10a^5b^7$ **22.** $12a^2b^3$ by $8a^7b$

23. $(x^6y^6 - x^3y^3) \div x^3y^3$

24. $(25xy^2 + 75xyz + 125x^2yz) \div -5x^2y$

25. $(a^2 + 4a + 3) \div (a + 1)$

26. $(3x^2 - 14x + 16) \div (x - 2)$

27. $(2x^2 + x - 10) \div (x - 2)$

28. $(x^2 - 7x + 12) \div (x - 5)$

29. $-16y^3 + 24y^4$ by $-4y^2$

30. $-20a^2b + 12ab^2$ by $-4ab$

31. $(2x^2 + 13x + 15) \div (x - 5)$

32. $(2x^2 + 13x + 5) \div (2x + 3)$

33. $(20x^2y^3 + 6xy^4 - 12x^3y^5) \div 2xy^3$

34. $(3x^2y + 6x^2y^2 + 3xy) \div 3xy$

35. $(6x^2 + 16x + 8) \div (3x + 2)$

36. $(x^2 - 25) \div (x + 5)$

37. $(2y^2 + 7y - 15) \div (2y - 3)$

38. $(3x^2 - 4x + 6) \div (x - 2)$

39. $4x^2 - 9$ by $2x - 3$

40. $8x^2 + 6x - 27$ by $4x + 9$

41. $2x^3 + 6x - 4$ by $x + 4$

42. $4x^3 - 5x$ by $2x - 1$

43. $3x^2 - 4$ by $x - 1$

44. $x^2 - 9$ by $x + 4$

45. $(-13x^3 + 2x^4 + 16x^2 - 9x + 20) \div (5 - x)$

46. $(5x^2 - 5x + 2x^3 + 20) \div (4 + x)$

47. $3x^5 - x^3 + 4x^2 - 12x - 8$ by $x^2 - 2$

48. $-8x^3 + 2x^4 + 19x^2 - 33x + 15$ by $x^2 - x + 5$

49. $(3x^3 - 5) \div 3x^2$

50. $(14x^3 - 2) \div (7x - 1)$

△ **51.** The perimeter of a square is $(12x^3 + 4x - 16)$ feet. Find the length of its side.

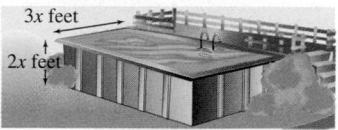

Perimeter is
$(12x^3 + 4x - 16)$
feet

△ **52.** The volume of the swimming pool shown is $(36x^5 - 12x^3 + 6x^2)$ cubic feet. If its height is $2x$ feet and its width is $3x$ feet, find its length.

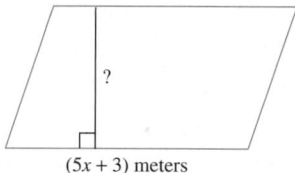

3x feet

2x feet

△ **53.** The area of the following parallelogram is $(10x^2 + 31x + 15)$ square meters. If its base is $(5x + 3)$ meters, find its height.

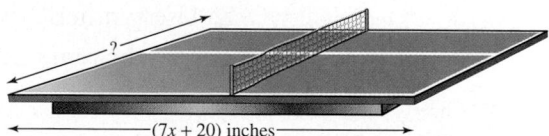

(5x + 3) meters

△ **54.** The area of the top of the Ping-Pong table is $(49x^2 + 70x - 200)$ square inches. If its length is $(7x + 20)$ inches, find its width.

(7x + 20) inches

55. Find $P(1)$ for the polynomial function $P(x) = 3x^3 + 2x^2 - 4x + 3$. Next, divide $3x^3 + 2x^2 - 4x + 3$ by $x - 1$. Compare the remainder with $P(1)$.

56. Find $P(-2)$ for the polynomial function $P(x) = x^3 - 4x^2 - 3x + 5$. Next, divide $x^3 - 4x^2 - 3x + 5$ by $x + 2$. Compare the remainder with $P(-2)$.

57. Find $P(-3)$ for the polynomial $P(x) = 5x^4 - 2x^2 + 3x - 6$. Next, divide $5x^4 - 2x^2 + 3x - 6$ by $x + 3$. Compare the remainder with $P(-3)$.

58. Find $P(2)$ for the polynomial function $P(x) = -4x^4 + 2x^3 - 6x + 3$. Next, divide $-4x^4 + 2x^3 - 6x + 3$ by $x - 2$. Compare the remainder with $P(2)$.

59. Write down any patterns you noticed from Exercises 55–58.

60. Explain how to check polynomial long division.

61. Try performing the following division without changing the order of the terms. Describe why this makes the process more complicated. Then perform the division again after putting the terms in the dividend in descending order of exponents.

$$\frac{4x^2 - 12x - 12 + 3x^3}{x - 2}$$

REVIEW EXERCISES

Insert <, >, or = to make each statement true. See Section 1.4.

62. 3^2 ____ $(-3)^2$

63. $(-5)^2$ ____ 5^2

64. -2^3 ____ $(-2)^3$

65. 3^4 ____ $(-3)^4$

Solve each inequality. See Section 2.7.

66. $x + 5 < 4$

67. $2x + 1 \geq 9$

68. $3x + 5 < 14$

69. $2x - 6 \leq x + 11$

A Look Ahead

Example

$$\left(x^2 - \frac{7}{2}x + 4\right) \div (x + 2)$$

Solution

$$
\begin{array}{r}
x - \frac{11}{2} \\
x + 2 \overline{)\, x^2 - \frac{7}{2}x + 4}\\
\underline{x^2 + 2x }\\
-\frac{11}{2}x + 4\\
\underline{-\frac{11}{2}x - 11}\\
15
\end{array}
$$

The quotient is $x - \frac{11}{2} + \frac{15}{x + 2}$.

Divide. See the preceding example.

70. $\left(x^4 + \frac{2}{3}x^3 + x\right) \div (x - 1)$

71. $\left(2x^3 + \frac{9}{2}x^2 - 4x - 10\right) \div (x + 2)$

72. $\left(3x^4 - x - x^3 + \frac{1}{2}\right) \div (2x - 1)$

73. $\left(2x^4 + \frac{1}{2}x^3 + x^2 + x\right) \div (x - 2)$

74. $(5x^4 - 2x^2 + 10x^3 - 4x) \div (5x + 10)$

75. $(9x^5 + 6x^4 - 6x^2 - 4x) \div (3x + 2)$

5.7 SYNTHETIC DIVISION AND THE REMAINDER THEOREM

CD-ROM   SSM   SSG   Video

▶ **OBJECTIVES**

1. Use synthetic division to divide a polynomial by a binomial.

2. Use the remainder theorem to evaluate polynomials.

1 When a polynomial is to be divided by a binomial of the form $x - c$, a shortcut process called **synthetic division** may be used. On the left is an example of long division, and on the right, the same example showing the coefficients of the variables only.

$$
\begin{array}{r}
2x^2 + 5x + 2\\
x - 3 \overline{)\,2x^3 - x^2 - 13x + 1}\\
\underline{2x^3 - 6x^2}\\
5x^2 - 13x\\
\underline{5x^2 - 15x}\\
2x + 1\\
\underline{2x - 6}\\
7
\end{array}
$$

$$
\begin{array}{r}
2 5 2\\
1 - 3 \overline{)\,2 - 1 - 13 + 1}\\
\underline{2 - 6}\\
5 - 13\\
\underline{5 - 15}\\
2 + 1\\
\underline{2 - 6}\\
7
\end{array}
$$

Notice that as long as we keep coefficients of powers of x in the same column, we can perform division of polynomials by performing algebraic operations on the coefficients only. This shortcut process of dividing with coefficients only in a special format is called synthetic division. To find $(2x^3 - x^2 - 13x + 1) \div (x - 3)$ by synthetic division, follow the next example.

Example 1   Use synthetic division to divide $2x^3 - x^2 - 13x + 1$ by $x - 3$.

Solution   To use synthetic division, the divisor must be in the form $x - c$. Since we are dividing by $x - 3$, c is 3. Write down 3 and the coefficients of the dividend.

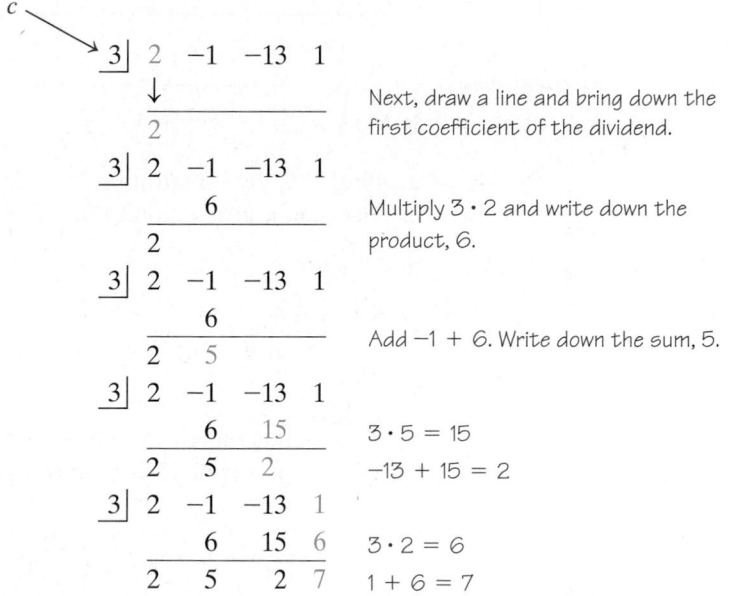

c

$$\begin{array}{r|rrrr} 3 & 2 & -1 & -13 & 1 \\ & & \downarrow & & \\ \hline & 2 & & & \end{array}$$

Next, draw a line and bring down the first coefficient of the dividend.

$$\begin{array}{r|rrrr} 3 & 2 & -1 & -13 & 1 \\ & & 6 & & \\ \hline & 2 & & & \end{array}$$

Multiply $3 \cdot 2$ and write down the product, 6.

$$\begin{array}{r|rrrr} 3 & 2 & -1 & -13 & 1 \\ & & 6 & & \\ \hline & 2 & 5 & & \end{array}$$

Add $-1 + 6$. Write down the sum, 5.

$$\begin{array}{r|rrrr} 3 & 2 & -1 & -13 & 1 \\ & & 6 & 15 & \\ \hline & 2 & 5 & 2 & \end{array}$$

$3 \cdot 5 = 15$
$-13 + 15 = 2$

$$\begin{array}{r|rrrr} 3 & 2 & -1 & -13 & 1 \\ & & 6 & 15 & 6 \\ \hline & 2 & 5 & 2 & 7 \end{array}$$

$3 \cdot 2 = 6$
$1 + 6 = 7$

The quotient is found in the bottom row. The numbers 2, 5, and 2 are the coefficients of the quotient polynomial, and the number 7 is the remainder. The degree of the quotient polynomial is one less than the degree of the dividend. In our example, the degree of the dividend is 3, so the degree of the quotient polynomial is 2. As we found when we performed the long division, the quotient is

$$2x^2 + 5x + 2, \qquad \text{remainder } 7$$

or

$$2x^2 + 5x + 2 + \frac{7}{x - 3}$$

Example 2   Use synthetic division to divide $x^4 - 2x^3 - 11x^2 + 5x + 34$ by $x + 2$.

Solution   The divisor is $x + 2$, which we write in the form $x - c$ as $x - (-2)$. Thus, c is -2. The dividend coefficients are $1, -2, -11, 5,$ and 34.

c

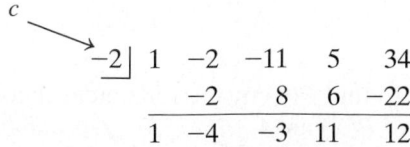

$$\begin{array}{r|rrrrr} -2 & 1 & -2 & -11 & 5 & 34 \\ & & -2 & 8 & 6 & -22 \\ \hline & 1 & -4 & -3 & 11 & 12 \end{array}$$

The dividend is a fourth-degree polynomial, so the quotient polynomial is a third-degree polynomial. The quotient is $x^3 - 4x^2 - 3x + 11$ with a remainder of 12. Thus,

$$\frac{x^4 - 2x^3 - 11x^2 + 5x + 34}{x + 2} = x^3 - 4x^2 - 3x + 11 + \frac{12}{x + 2}$$

HELPFUL HINT

Before dividing by synthetic division, write the dividend in descending order of variable exponents. Any "missing powers" of the variable should be represented by 0 times the variable raised to the missing power.

Example 3 If $P(x) = 2x^3 - 4x^2 + 5$

    a. Find $P(2)$ by substitution.
    b. Use synthetic division to find the remainder when $P(x)$ is divided by $x - 2$.

Solution   **a.** $P(x) = 2x^3 - 4x^2 + 5$
            $P(2) = 2(2)^3 - 4(2)^2 + 5$
                 $= 2(8) - 4(4) + 5 = 16 - 16 + 5 = 5$

Thus, $P(2) = 5$.

    b. The coefficients of $P(x)$ are $2, -4, 0,$ and 5. The number 0 is a coefficient of the missing power of x^1. The divisor is $x - 2$, so c is 2.

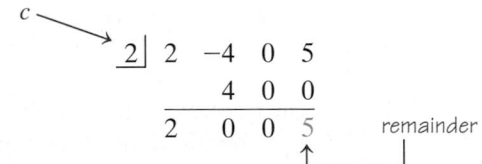

The remainder when $P(x)$ is divided by $x - 2$ is 5.

---

2 Notice in the preceding example that $P(2) = 5$ and that the remainder when $P(x)$ is divided by $x - 2$ is 5. This is no accident. This illustrates the **remainder theorem.**

REMAINDER THEOREM

If a polynomial $P(x)$ is divided by $x - c$, then the remainder is $P(c)$.

Example 4 Use the remainder theorem and synthetic division to find $P(4)$ if

$$P(x) = 4x^6 - 25x^5 + 35x^4 + 17x^2.$$

Solution To find $P(4)$ by the remainder theorem, we divide $P(x)$ by $x - 4$. The coefficients of $P(x)$ are $4, -25, 35, 0, 17, 0,$ and 0. Also, c is 4.

$$c \searrow$$

$$\begin{array}{r|rrrrrr} 4 & 4 & -25 & 35 & 0 & 17 & 0 & 0 \\ & & 16 & -36 & -4 & -16 & 4 & 16 \\ \hline & 4 & -9 & -1 & -4 & 1 & 4 & 16 \end{array} \text{ remainder}$$

Thus, $P(4) = 16$, the remainder.

Exercise Set 5.7

Use synthetic division to divide. See Examples 1 and 2.

1. $(x^2 + 3x - 40) \div (x - 5)$

2. $(x^2 - 14x + 24) \div (x - 2)$

3. $(x^2 + 5x - 6) \div (x + 6)$

4. $(x^2 + 12x + 32) \div (x + 4)$

5. $(x^3 - 7x^2 - 13x + 5) \div (x - 2)$

6. $(x^3 + 6x^2 + 4x - 7) \div (x + 5)$

7. $(4x^2 - 9) \div (x - 2)$

8. $(3x^2 - 4) \div (x - 1)$

For the given polynomial $P(x)$ and the given c, find $P(c)$ by (a) direct substitution and (b) the remainder theorem. See Examples 3 and 4.

9. $P(x) = 3x^2 - 4x - 1; P(2)$

10. $P(x) = x^2 - x + 3; P(5)$

11. $P(x) = 4x^4 + 7x^2 + 9x - 1; P(-2)$

12. $P(x) = 8x^5 + 7x + 4; P(-3)$

13. $P(x) = x^5 + 3x^4 + 3x - 7; P(-1)$

14. $P(x) = 5x^4 - 4x^3 + 2x - 1; P(-1)$

Use synthetic division to divide.

15. $(x^3 - 3x^2 + 2) \div (x - 3)$

16. $(x^2 + 12) \div (x + 2)$

17. $(6x^2 + 13x + 8) \div (x + 1)$

18. $(x^3 - 5x^2 + 7x - 4) \div (x - 3)$

19. $(2x^4 - 13x^3 + 16x^2 - 9x + 20) \div (x - 5)$

20. $(3x^4 + 5x^3 - x^2 + x - 2) \div (x + 2)$

21. $(3x^2 - 15) \div (x + 3)$

22. $(3x^2 + 7x - 6) \div (x + 4)$

23. $(3x^3 - 6x^2 + 4x + 5) \div \left(x - \dfrac{1}{2}\right)$

24. $(8x^3 - 6x^2 - 5x + 3) \div \left(x + \dfrac{3}{4}\right)$

25. $(3x^3 + 2x^2 - 4x + 1) \div \left(x - \dfrac{1}{3}\right)$

26. $(9y^3 + 9y^2 - y + 2) \div \left(y + \dfrac{2}{3}\right)$

27. $(7x^2 - 4x + 12 + 3x^3) \div (x + 1)$

28. $(x^4 + 4x^3 - x^2 - 16x - 4) \div (x - 2)$

29. $(x^3 - 1) \div (x - 1)$ 30. $(y^3 - 8) \div (y - 2)$

31. $(x^2 - 36) \div (x + 6)$

32. $(4x^3 + 12x^2 + x - 12) \div (x + 3)$

For the given polynomial $P(x)$ and the given c, use the remainder theorem to find $P(c)$.

33. $P(x) = x^3 + 3x^2 - 7x + 4; 1$

34. $P(x) = x^3 + 5x^2 - 4x - 6; 2$

35. $P(x) = 3x^3 - 7x^2 - 2x + 5; -3$

36. $P(x) = 4x^3 + 5x^2 - 6x - 4; -2$

37. $P(x) = 4x^4 + x^2 - 2; -1$

38. $P(x) = x^4 - 3x^2 - 2x + 5; -2$

39. $P(x) = 2x^4 - 3x^2 - 2; \dfrac{1}{3}$

40. $P(x) = 4x^4 - 2x^3 + x^2 - x - 4; \dfrac{1}{2}$

41. $P(x) = x^5 + x^4 - x^3 + 3; \dfrac{1}{2}$

42. $P(x) = x^5 - 2x^3 + 4x^2 - 5x + 6; \dfrac{2}{3}$

43. Explain an advantage of using the remainder theorem instead of direct substitution.

44. Explain an advantage of using synthetic division instead of long division.

We say that 2 is a factor of 8 because 2 divides 8 evenly, or with a remainder of 0. In the same manner, the polynomial $x - 2$ is a factor of the polynomial $x^3 - 14x^2 + 24x$ because the remainder is 0 when $x^3 - 14x^2 + 24x$ is divided by $x - 2$. Use this information for Exercises 45 through 47.

45. Use synthetic division to show that $x + 3$ is a factor of $x^3 + 3x^2 + 4x + 12$.

46. Use synthetic division to show that $x - 2$ is a factor of $x^3 - 2x^2 - 3x + 6$.

47. From the remainder theorem, the polynomial $x - c$ is a factor of a polynomial function $P(x)$ if $P(c)$ is what value?

48. If a polynomial is divided by $x - 5$, the quotient is $2x^2 + 5x - 6$ and the remainder is 3. Find the original polynomial.

49. If a polynomial is divided by $x + 3$, the quotient is $x^2 - x + 10$ and the remainder is -2. Find the original polynomial.

△ **50.** If the area of a parallelogram is $(x^4 - 23x^2 + 9x - 5)$ square centimeters and its base is $(x + 5)$ centimeters, find its height.

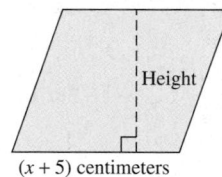

Height

$(x + 5)$ centimeters

△ **51.** If the volume of a box is $(x^4 + 6x^3 - 7x^2)$ cubic meters, its height is x^2 meters, and its length is $(x + 7)$ meters, find its width.

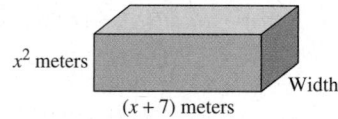

x^2 meters

$(x + 7)$ meters

Width

REVIEW EXERCISES

Multiply each expression. See Section 5.4.

52. $2a(a^2 + 1)$

53. $-4a(3a^2 - 4)$

54. $2x(x^2 + 7x - 5)$

55. $4y(y^2 - 8y - 4)$

56. $-3xy(xy^2 + 7x^2y + 8)$

57. $-9xy(4xyz + 7xy^2z + 2)$

58. $9ab(ab^2c + 4bc - 8)$

59. $-7sr(6s^2r + 9sr^2 + 9rs + 8)$

Use the bar graph below to answer Exercises 60–63. See Section 1.9.

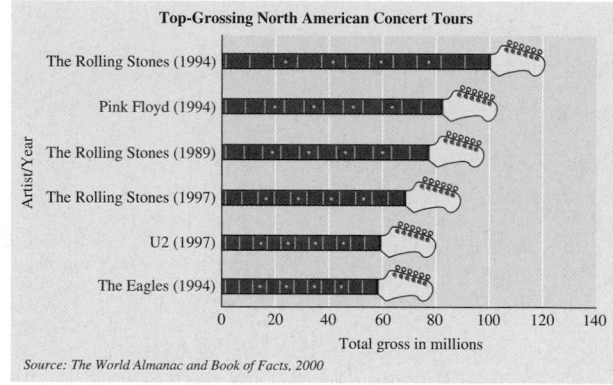

Source: The World Almanac and Book of Facts, 2000

60. Which artist has grossed the most money on tour?

61. Estimate the amount of money made by the concert tour of Pink Floyd.

62. Estimate the amount of money made by the 1997 concert tour of U2.

63. Which artist shown has grossed the least amount of money on a tour?

For additional Chapter Projects, visit the Real World Activities Website by going to http://www.prenhall.com/martin-gay.

CHAPTER PROJECT

Modeling with Polynomials

The polynomial model $1.65x^2 + 14.77x + 364.27$ dollars represents consumer spending per person per year on all U.S. media from 1990 to 1996. This includes spending on subscription TV services, recorded music, newspapers, magazines, books, home video, theater movies, video games, and educational software. The polynomial model $0.81x^2 + 3.41x + 88.98$ dollars represents consumer spending per person per year on subscription TV services alone during this same

period. In both models, x is the number of years after 1990. (*Source:* Based on data from Veronis, Suhler & Associates, New York, NY, *Communications Industry Report*, annual)

In this project, you will have the opportunity to investigate these polynomial models numerically, algebraically, and graphically. This project may be completed by working in groups or individually.

1. Use the polynomials to complete the following table showing the annual consumer spending per person over the period 1990–1996 by evaluating each polynomial at the given values of x. Then subtract each value in the fourth column from the corresponding value in the third column. Record the result in the last column, "Difference." What do you think these values represent? What trends do you notice in the data?

Year	x	Consumer Spending per Person per Year on ALL U.S. Media	Consumer spending per Person per Year on Subscription TV	Difference
1990	0			
1992	2			
1994	4			
1996	6			

2. Use the polynomial models to find a new polynomial model representing the amount of consumer spending per person on U.S. media other than subscription TV services (such as recorded music, newspapers, magazines, books, home video, theater movies, video games, and educational software). Then use this new polynomial model to complete the following table.

Year	x	Consumer Spending per Person per Year on Media Other Than Subscription TV
1990	0	
1992	2	
1994	4	
1996	6	

3. Compare the values in the last column of the table in Question 1 to the values in the last column of the table in Question 2. What do you notice? What can you conclude?

4. Use the polynomial models to estimate consumer spending on (a) all U.S. media, (b) subscription TV, and (c) media other than subscription TV for the year 1998.

5. Use the polynomial models to estimate consumer spending on (a) all U.S. media, (b) subscription TV, and (c) media other than subscription TV for the year 2000.

6. Create a bar graph that represents the data for consumer spending on all U.S. media in the years 1990, 1992, 1994, and 1996 along with your estimates for 1998 and 2000. Study your bar graph. Discuss what the graph implies about the future.

CHAPTER 5 VOCABULARY CHECK

Fill in each blank with one of the words or phrases listed below.

term   coefficient   monomial   binomial   trinomial   synthetic division
polynomials   degree of a term   degree of a polynomial   FOIL

1. A _____ is a number or the product of numbers and variables raised to powers.
2. The _____ method may be used when multiplying two binomials.
3. A polynomial with exactly 3 terms is called a _____.
4. The _____ is the greatest degree of any term of the polynomial.
5. A polynomial with exactly 2 terms is called a _____.
6. The _____ of a term is its numerical factor.
7. The _____ is the sum of the exponents on the variables in the term.
8. A polynomial with exactly 1 term is called a _____.
9. Monomials, binomials, and trinomials are all examples of _____.
10. When a polynomial is to be divided by a binomial of the form $x - c$, a shortcut process called _____ may be used.

CHAPTER 5 HIGHLIGHTS

DEFINITIONS AND CONCEPTS	EXAMPLES

Section 5.1 Exponents

a^n means the product of n factors, each of which is a.

$$3^2 = 3 \cdot 3 = 9$$
$$(-5)^3 = (-5)(-5)(-5) = -125$$
$$\left(\frac{1}{2}\right)^4 = \frac{1}{2} \cdot \frac{1}{2} \cdot \frac{1}{2} \cdot \frac{1}{2} = \frac{1}{16}$$

If m and n are integers and no denominators are 0,

Product Rule: $a^m \cdot a^n = a^{m+n}$

Power Rule: $(a^m)^n = a^{m \cdot n}$

Power of a Product Rule: $(ab)^n = a^n b^n$

Power of a Quotient Rule: $\left(\dfrac{a}{b}\right)^n = \dfrac{a^n}{b^n}$

Quotient Rule: $\dfrac{a^m}{a^n} = a^{m-n}$

Zero Exponent: $a^0 = 1, a \neq 0$.

$$x^2 \cdot x^7 = x^{2+7} = x^9$$
$$(5^3)^8 = 5^{3 \cdot 8} = 5^{24}$$
$$(7y)^4 = 7^4 y^4$$
$$\left(\frac{x}{8}\right)^3 = \frac{x^3}{8^3}$$
$$\frac{x^9}{x^4} = x^{9-4} = x^5$$
$$5^0 = 1, x^0 = 1, x \neq 0$$

Section 5.2 Negative Exponents and Scientific Notation

If $a \neq 0$ and n is an integer,

$$a^{-n} = \frac{1}{a^n}$$

Rules for exponents are true for positive and negative integers.

$$3^{-2} = \frac{1}{3^2} = \frac{1}{9}; 5x^{-2} = \frac{5}{x^2}$$

Simplify: $\left(\dfrac{x^{-2}y}{x^5}\right)^{-2} = \dfrac{x^4 y^{-2}}{x^{-10}}$

$$= x^{4-(-10)} y^{-2}$$
$$= \frac{x^{14}}{y^2}$$

A positive number is written in scientific notation if it is as the product of a number a, $1 \leq a < 10$, and an integer power r of 10.

$$a \times 10^r$$

Write each number in scientific notation.

$$12{,}000 = 1.2 \times 10^4$$

$$0.00000568 = 5.68 \times 10^{-6}$$

Section 5.3 Polynomial Functions and Adding and Subtracting Polynomials

A **polynomial** is a finite sum of terms in which all variables have exponents raised to nonnegative integer powers and no variables appear in the denominator.

Polynomials

$$1.3x^2 \qquad \text{(monomial)}$$
$$-\frac{1}{3}y + 5 \qquad \text{(binomial)}$$
$$6z^2 - 5z + 7 \qquad \text{(trinomial)}$$

A function P is a **polynomial function** if $P(x)$ is a polynomial.

For the polynomial function

$$P(x) = -x^2 + 6x - 12, \text{find } P(-2)$$
$$P(-2) = -(-2)^2 + 6(-2) - 12 = -28.$$

(continued)

DEFINITIONS AND CONCEPTS	EXAMPLES

Section 5.3 Polynomial Functions and Adding and Subtracting Polynomials

To add polynomials, combine all like terms.

Add

$$(3y^2x - 2yx + 11) + (-5y^2x - 7)$$
$$= -2y^2x - 2yx + 4$$

To subtract polynomials, change the signs of the terms of the polynomial being subtracted, then add.

Subtract

$$(-2z^3 - z + 1) - (3z^3 + z - 6)$$
$$= -2z^3 - z + 1 - 3z^3 - z + 6$$
$$= -5z^3 - 2z + 7$$

Section 5.4 Multiplying Polynomials

To multiply two polynomials, multiply each term of one polynomial by each term of the other polynomial, and then combine like terms.

Multiply:

$$(2x + 1)(5x^2 - 6x + 2)$$
$$= 2x(5x^2 - 6x + 2) + 1(5x^2 - 6x + 2)$$
$$= 10x^3 - 12x^2 + 4x + 5x^2 - 6x + 2$$
$$= 10x^3 - 7x^2 - 2x + 2$$

Section 5.5 Special Products

The **FOIL method** may be used when multiplying two binomials.

Multiply: $(5x - 3)(2x + 3)$

$$(5x - 3)(2x + 3) = (5x)(2x) + (5x)(3) + (-3)(2x) + (-3)(3)$$
$$= 10x^2 + 15x - 6x - 9$$
$$= 10x^2 + 9x - 9$$

Squaring a Binomial

$$(a + b)^2 = a^2 + 2ab + b^2$$

$$(a - b)^2 = a^2 - 2ab + b^2$$

Square each binomial.

$$(x + 5)^2 = x^2 + 2(x)(5) + 5^2$$
$$= x^2 + 10x + 25$$

$$(3x - 2y)^2 = (3x)^2 - 2(3x)(2y) + (2y)^2$$
$$= 9x^2 - 12xy + 4y^2$$

Multiplying the Sum and Difference of Two Terms

$$(a + b)(a - b) = a^2 - b^2$$

Multiply.

$$(6y + 5)(6y - 5) = (6y)^2 - 5^2$$
$$= 36y^2 - 25$$

(continued)

DEFINITIONS AND CONCEPTS	EXAMPLES

Section 5.6 Dividing Polynomials

To divide a polynomial by a monomial:
Divide each term in the polynomial by the monomial.

Divide $\dfrac{12a^5b^3 - 6a^2b^2 + ab}{6a^2b^2}$.

$$= \frac{12a^5b^3}{6a^2b^2} - \frac{6a^2b^2}{6a^2b^2} + \frac{ab}{6a^2b^2}$$

$$= 2a^3b - 1 + \frac{1}{6ab}$$

To divide a polynomial by a polynomial, other than a monomial:
Use **long division.**

Divide $2x^3 - x^2 - 8x - 1$ by $x - 2$.

$$
\begin{array}{r}
2x^2 + 3x - 2 \\
x - 2\overline{)2x^3 - x^2 - 8x - 1} \\
\underline{2x^3 - 4x^2} \\
3x^2 - 8x \\
\underline{3x^2 - 6x} \\
-2x - 1 \\
\underline{-2x + 4} \\
-5
\end{array}
$$

The quotient is $2x^2 + 3x - 2 - \dfrac{5}{x-2}$.

Section 5.7 Synthetic Division and the Remainder Theorem

A shortcut method called **synthetic division** may be used to divide a polynomial by a binomial of the form $x - c$.

Use synthetic division to divide $2x^3 - x^2 - 8x - 1$ by $x - 2$.

$$
\begin{array}{c|cccc}
2 & 2 & -1 & -8 & -1 \\
  &   & 4 & 6 & -4 \\
\hline
  & 2 & 3 & -2 & -5
\end{array}
$$

The quotient is $2x^2 + 3x - 2 - \dfrac{5}{x-2}$.

CHAPTER 5 REVIEW

(5.1) *State the base and the exponent for each expression.*

1. 3^2 **2.** $(-5)^4$

3. -5^4

Evaluate each expression.

4. 8^3 **5.** $(-6)^2$

6. -6^2 **7.** $-4^3 - 4^0$

8. $(3b)^0$ **9.** $\dfrac{8b}{8b}$

Simplify each expression.

10. $5b^3b^5a^6$ **11.** $2^3 \cdot x^0$

12. $[(-3)^2]^3$ **13.** $(2x^3)(-5x^2)$

14. $\left(\dfrac{mn}{q}\right)^2 \cdot \left(\dfrac{mn}{q}\right)$ **15.** $\left(\dfrac{3ab^2}{6ab}\right)^4$

16. $\dfrac{x^9}{x^4}$ **17.** $\dfrac{2x^7y^8}{8xy^2}$

18. $\dfrac{12xy^6}{3x^4y^{10}}$ **19.** $5a^7(2a^4)^3$

20. $(2x)^2(9x)$ **21.** $\dfrac{(-4)^2(3^3)}{(4^5)(3^2)}$

22. $\dfrac{(-7)^2(3^5)}{(-7)^3(3^4)}$ **23.** $\dfrac{(2x)^0(-4)^2}{16x}$

24. $\dfrac{(8xy)(3xy)}{18x^2y^2}$ **25.** $m^0 + p^0 + 3q^0$

26. $(-5a)^0 + 7^0 + 8^0$ **27.** $(3xy^2 + 8x + 9)^0$

28. $8x^0 + 9^0$ **29.** $6(a^2b^3)^3$

30. $\dfrac{(x^3z)^a}{x^2z^2}$

(5.2) *Simplify each expression.*

31. 7^{-2} **32.** -7^{-2}

33. $2x^{-4}$

34. $(2x)^{-4}$

35. $\left(\dfrac{1}{5}\right)^{-3}$

36. $\left(\dfrac{-2}{3}\right)^{-2}$

37. $2^0 + 2^{-4}$

38. $6^{-1} - 7^{-1}$

Simplify each expression. Assume that variables in an exponent represent positive integers only. Write each answer using positive exponents.

39. $\dfrac{1}{(2q)^{-3}}$

40. $\dfrac{-1}{(qr)^{-3}}$

41. $\dfrac{r^{-3}}{s^{-4}}$

42. $\dfrac{rs^{-3}}{r^{-4}}$

43. $\dfrac{-6}{8x^{-3}r^4}$

44. $\dfrac{-4s}{16s^{-3}}$

45. $\left(2x^{-5}\right)^{-3}$

46. $\left(3y^{-6}\right)^{-1}$

47. $\left(3a^{-1}b^{-1}c^{-2}\right)^{-2}$

48. $\left(4x^{-2}y^{-3}z\right)^{-3}$

49. $\dfrac{5^{-2}x^8}{5^{-3}x^{11}}$

50. $\dfrac{7^5y^{-2}}{7^7y^{-10}}$

51. $\left(\dfrac{bc^{-2}}{bc^{-3}}\right)^4$

52. $\left(\dfrac{x^{-3}y^{-4}}{x^{-2}y^{-5}}\right)^{-3}$

53. $\dfrac{x^{-4}y^{-6}}{x^2y^7}$

54. $\dfrac{a^5b^{-5}}{a^{-5}b^5}$

55. $-2^0 + 2^{-4}$

56. $-3^{-2} - 3^{-3}$

57. $a^{6m}a^{5m}$

58. $\dfrac{\left(x^{5+h}\right)^3}{x^5}$

59. $\left(3xy^{2z}\right)^3$

60. $a^{m+2}a^{m+3}$

Write each number in scientific notation.

61. 0.00027

62. 0.8868

63. $80{,}800{,}000$

64. $-868{,}000$

65. The population of California is 32,667,000. Write this number in scientific notation. (*Source:* Federal–State Cooperative Program for Population Estimates)

$\triangle$ **66.** The radius of the Earth is approximately 4000 miles. Write 4000 in scientific notation.

4000 miles

Write each number in standard form.

67. 8.67×10^5

68. 3.86×10^{-3}

69. 8.6×10^{-4}

70. 8.936×10^5

71. The number of photons of light emitted by a 100-watt bulb every second is 1×10^{20}. Write 1×10^{20} in standard notation.

72. The real mass of all the galaxies in the constellation of Virgo is 3×10^{-25}. Write 3×10^{-25} in standard notation.

Simplify. Express each result in standard form.

73. $\left(8 \times 10^4\right)\left(2 \times 10^{-7}\right)$

74. $\dfrac{8 \times 10^4}{2 \times 10^{-7}}$

(5.3)

75. a. Complete the table for the polynomial
$3a^2b - 2a^2 + ab - b^2 - 6$.

Term	Numerical Coefficient	Degree of Term
$3a^2b$		
$-2a^2$		
ab		
$-b^2$		
-6		

b. What is the degree of the polynomial?

76. a. Complete the table for the polynomial
$x^2y^2 + 5x^2 - 7y^2 + 11xy - 1$.

Term	Numerical Coefficient	Degree of Term
x^2y^2		
$5x^2$		
$-7y^2$		
$11xy$		
-1		

b. What is the degree of the polynomial?

Simplify by combining like terms.

77. $4x + 8x - 6x^2 - 6x^2y$

78. $-8xy^3 + 4xy^3 - 3x^3y$

Add or subtract as indicated.

79. $(3x + 7y) + (4x^2 - 3x + 7) + (y - 1)$

80. $(4x^2 - 6xy + 9y^2) - (8x^2 - 6xy - y^2)$

81. $(3x^2 - 4b + 28) + (9x^2 - 30) - (4x^2 - 6b + 20)$

82. Add $(9xy + 4x^2 + 18)$ and $(7xy - 4x^3 - 9x)$.

83. Subtract $(x - 7)$ from the sum of $(3x^2y - 7xy - 4)$ and $(9x^2y + x)$.

84. $x^2 - 5x + 7$
$\underline{\quad - (\ x + 4)}$

85. $x^3 \quad + 2xy^2 - y$
$\underline{\quad + (x - 4xy^2 \quad - 7)}$

If $P(x) = 9x^2 - 7x + 8$, find the following.

86. $P(6)$

87. $P(-2)$

88. $P(-3)$

△ **89.** Find the perimeter of the rectangle.

$x^2y + 5$
cm

$2x^2y - 6x + 1$
cm

(5.4) *Multiply each expression.*

90. $9x(x^2y)$

91. $-7(8xz^2)$

92. $(6xa^2)(xya^3)$

93. $(4xy)(-3xa^2y^3)$

94. $6(x + 5)$

95. $9(x - 7)$

96. $4(2a + 7)$

97. $9(6a - 3)$

98. $-7x(x^2 + 5)$

99. $-8y(4y^2 - 6)$

100. $-2(x^3 - 9x^2 + x)$

101. $-3a(a^2b + ab + b^2)$

102. $(3a^3 - 4a + 1)(-2a)$

103. $(6b^3 - 4b + 2)(7b)$

104. $(2x + 2)(x - 7)$

105. $(2x - 5)(3x + 2)$

106. $(4a - 1)(a + 7)$

107. $(6a - 1)(7a + 3)$

108. $(x + 7)(x^3 + 4x - 5)$

109. $(x + 2)(x^5 + x + 1)$

110. $(x^2 + 2x + 4)(x^2 + 2x - 4)$

111. $(x^3 + 4x + 4)(x^3 + 4x - 4)$

(5.5) *Multiply.*

112. $2x(3x^2 - 7x + 1)$

113. $3y(5y^2 - y + 2)$

114. $(6x - 1)(4x + 3)$

115. $(4a - 1)(3a + 7)$

116. $(x + 7)^2$

117. $(x - 5)^2$

118. $(3x - 7)^2$

119. $(4x + 2)^2$

120. $(y + 1)(y^2 - 6y - 5)$

121. $(x - 2)(x^2 - x - 2)$

122. $(5x - 9)^2$

123. $(5x + 1)(5x - 1)$

124. $(7x + 4)(7x - 4)$

125. $(a + 2b)(a - 2b)$

126. $(2x - 6)(2x + 6)$

127. $(4a^2 - 2b)(4a^2 + 2b)$

128. $[4 + (3a - b)][4 - (3a - b)]$

(5.6) *Divide.*

129. Divide $3x^5yb^9$ by $9xy^7$.

130. Divide $-9xb^4z^3$ by $-4axb^2$.

131. $(4xy + 2x^2 - 9) \div 4xy$

132. Divide $12xb^2 + 16xb^4$ by $4xb^3$.

133. $(3x^4 - 25x^2 - 20) \div (x - 3)$

134. $(-x^2 + 2x^4 + 5x - 12) \div (x + 2)$

135. $(3x^4 + 5x^3 + 7x^2 + 3x - 2) \div (x^2 + x + 2)$

136. $(9x^4 - 6x^3 + 3x^2 - 12x - 30) \div (3x^2 - 2x - 5)$

(5.7) *Use synthetic division to find each quotient.*

137. $(3x^3 + 12x - 4) \div (x - 2)$

138. $(3x^3 + 2x^2 - 4x - 1) \div \left(x + \dfrac{3}{2}\right)$

139. $(x^5 - 1) \div (x + 1)$

140. $(x^3 - 81) \div (x - 3)$

141. $(x^3 - x^2 + 3x^4 - 2) \div (x - 4)$

142. $(3x^4 - 2x^2 + 10) \div (x + 2)$

If $P(x) = 3x^5 - 9x + 7$, use the remainder theorem to find the following.

143. $P(4)$

144. $P(-5)$

△ **145.** If the area of the rectangle is $(x^4 - x^3 - 6x^2 - 6x + 18)$ square miles and its width is $(x - 3)$ miles, find the length.

$x^4 - x^3 - 6x^2 - 6x + 18$
square miles

$x - 3$
miles

CHAPTER 5 TEST

Evaluate each expression.

1. 2^5

2. $(-3)^4$

3. -3^4

4. 4^{-3}

Simplify each exponential expression.

5. $(3x^2)(-5x^9)$

6. $\dfrac{y^7}{y^2}$

7. $\dfrac{r^{-8}}{r^{-3}}$

Simplify each expression. Write the result using only positive exponents.

8. $\left(\dfrac{x^2y^3}{x^3y^{-4}}\right)^2$

9. $\dfrac{6^2x^{-4}y^{-1}}{6^3x^{-3}y^7}$

Express each number in scientific notation.

10. 563,000

11. 0.0000863

Write each number in standard form.

12. 1.5×10^{-3}

13. 6.23×10^4

14. Simplify. Write the answer in standard form.
$(1.2 \times 10^5)(3 \times 10^{-7})$

15. **a.** Complete the table for the polynomial $4xy^2 + 7xyz + x^3y - 2$.

Term	Numerical Coefficient	Degree of Term
$4xy^2$		
$7xyz$		
x^3y		
-2		

b. What is the degree of the polynomial?

16. Simplify by combining like terms.
$5x^2 + 4xy - 7x^2 + 11 + 8xy$

Perform each indicated operation.

17. $(8x^3 + 7x^2 + 4x - 7) + (8x^3 - 7x - 6)$

18. $\begin{array}{r} 5x^3 + x^2 + 5x - 2 \\ -(8x^3 - 4x^2 + x - 7) \\ \hline \end{array}$

19. Subtract $(4x + 2)$ from the sum of $(8x^2 + 7x + 5)$ and $(x^3 - 8)$.

Multiply.

20. $(3x + 7)(x^2 + 5x + 2)$

21. $3x^2(2x^2 - 3x + 7)$

22. $(x + 7)(3x - 5)$

23. $(3x - 7)(3x + 7)$

24. $(4x - 2)^2$

25. $(8x + 3)^2$

26. $(r^2 - 9b)(r^2 + 9b)$

27. The height of the Bank of China in Hong Kong is 1001 feet. Neglecting air resistance, the height of an object dropped from this building at time t seconds is given by the polynomial $-16t^2 + 1001$. Find the height of the object at the given times below.

	0 seconds	1 second	3 seconds	5 seconds
t				
$-16t^2 + 1001$				

Divide.

28. $\dfrac{8xy^2}{4x^3y^3z}$

29. $\dfrac{4x^2 + 2xy - 7x}{8xy}$

30. $(x^2 + 7x + 10) \div (x + 5)$

31. $\dfrac{27x^3 - 8}{3x + 2}$

32. The number of bankruptcy cases (in thousands) filed in the United States x years after 1993 is given by the polynomial function $f(x) = 62x^2 - 149x + 922$ thousand cases for 1993 through 1997. Use this model to predict the number of bankruptcy cases in 2000. (Hint: Find $f(7)$.) *Source:* Based on data from the Administrative Office of the U.S. Courts)

33. Use synthetic division to divide $(4x^4 - 3x^3 + 2x^2 - x - 1)$ by $(x + 3)$.

34. If $P(x) = 4x^4 + 7x^2 - 2x - 5$, use the remainder theorem to find $P(-2)$.

△ 35. Write the area of the shaded region as a polynomial.

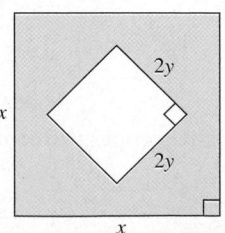

36. A pebble is hurled upward from the top of the Canada Trust Tower, which is 880 feet tall, with an initial velocity of 96 feet per second. Neglecting air resistance, the height $h(t)$ of the pebble after t seconds is given by the polynomial function

$$h(t) = -16t^2 + 96t + 880$$

a. Find the height of the pebble when $t = 1$.

b. Find the height of the pebble when $t = 5.1$.

CHAPTER 5 CUMULATIVE REVIEW

1. Tell whether each statement is true or false.

a. $8 \geq 8$ **b.** $8 \leq 8$

c. $23 \leq 0$ **d.** $23 \geq 0$

2. Find each quotient. Write all answers in lowest terms.

a. $\dfrac{4}{5} \div \dfrac{5}{16}$ **b.** $\dfrac{7}{10} \div 14$

c. $\dfrac{3}{8} \div \dfrac{3}{10}$

3. Evaluate the following:

a. 3^2 **b.** 5^3

c. 2^4 **d.** 7^1

e. $\left(\dfrac{3}{7}\right)^2$

4. Add.

a. $-3 + (-7)$ **b.** $-1 + (-20)$

c. $-2 + (-10)$

5. Subtract 8 from -4.

6. Find the reciprocal of each number.

a. 22 **b.** $\dfrac{3}{16}$

c. -10 **d.** $-\dfrac{9}{13}$

7. Use an associative property to complete each statement.

a. $5 + (4 + 6) = $ _____

b. $(-1 \cdot 2) \cdot 5 = $ _____

8. The following bar graph shows the cents charged per kilowatt-hour for selected electricity companies.

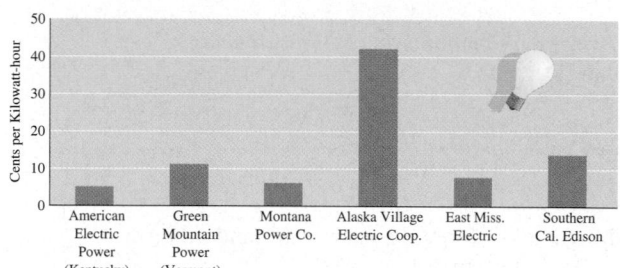

Source: Electric Company Listed

a. Which company charges the highest rate?

b. Which company charges the lowest rate?

c. Approximate the electricity rate charged by the first four companies listed.

d. Approximate the difference in the rates charged by the companies in parts (a) and (b).

9. Find each product by using the distributive property to remove parentheses.

a. $5(x + 2)$

b. $-2(y + 0.3z - 1)$

c. $-(x + y - 2z + 6)$

10. Solve $y + 0.6 = -1.0$ for y.

11. Twice a number, added to seven, is the same as three subtracted from the number. Find the number.

12. Solve $F = \dfrac{9}{5}C + 32$ for C.

13. Find 72% of 200.

14. Graph $2 < x \leq 4$.

15. Complete the following ordered pair solutions for the equation $3x + y = 12$.

a. $(0, \ )$ **b.** $(\ , 6)$

c. $(-1, \ )$

16. Graph the linear equation $2x + y = 5$.

17. Is the relation $x = y^2$ also a function?

18. Graph $x = 2$.

19. Find the slope of the line $x = -5$.

20. Find the equation of the horizontal line containing the point $(2, 3)$.

21. Graph: $2x - y < 6$

22. Use the product rule to simplify $(2x^2)(-3x^5)$.

23. Find the degree of the polynomial $3xy + x^2y^2 - 5x^2 - 6$.

24. Multiply: $(2x - y)^2$

Outdoor Opportunities

Professional landscapers, landscape technicians, and landscape architects work together to design, install, and maintain attractive outdoor planting schemes in parks, around buildings, and in homeowners' yards. There are over 70,000 professional lawn care and landscape companies in the United States, employing more than 623,000 people. In recent years, the demand for lawn care and landscape services has skyrocketed.

People working in the landscape industry enjoy working outdoors. They also are knowledgeable about plants and how best to care for them. Landscapers find math and geometry useful in situations such as estimating and pricing landscaping jobs, mixing fertilizers or pesticides, and designing or laying out areas to be landscaped.

 For more information about a career in the landscaping industry, visit the Associated Landscape Contractors of America Website by first going to www.prenhall.com/martin-gay.

In the Spotlight on Decision Making feature on page 386, you will have the opportunity as a landscaper to make a decision about a flower bed design.

FACTORING POLYNOMIALS

In Chapter 5, you learned how to multiply polynomials. This chapter deals with an operation that is the reverse process of multiplying, called *factoring*. Factoring is an important algebraic skill because this process allows us to write a sum as a product.

At the end of this chapter, we use factoring to help us solve equations other than linear equations, and in Chapter 7 we use factoring to simplify and perform arithmetic operations on rational expressions.

6.1 THE GREATEST COMMON FACTOR AND FACTORING BY GROUPING

CD-ROM SSM

SSG Video

▶ **OBJECTIVES**

1. Find the greatest common factor of a list of integers.
2. Find the greatest common factor of a list of terms.
3. Factor out the greatest common factor from a polynomial.
4. Factor a polynomial by grouping.

When an integer is written as the product of two or more other integers, each of these integers is called a **factor** of the product. This is true for polynomials, also. When a polynomial is written as the product of two or more other polynomials, each of these polynomials is called a factor of the product.

The process of writing a polynomial as a product is called **factoring** the polynomial.

$$\underset{\text{factor}}{2} \cdot \underset{\text{factor}}{3} = \underset{\text{product}}{6} \qquad \underset{\text{factor}}{x^2} \cdot \underset{\text{factor}}{x^3} = \underset{\text{product}}{x^5} \qquad \underset{\text{factor}}{(x+2)}\underset{\text{factor}}{(x+3)} = \underset{\text{product}}{x^2 + 5x + 6}$$

Notice that factoring is the reverse process of multiplying.

$$x^2 + 5x + 6 = (x + 2)(x + 3)$$

factoring

multiplying

The first step in factoring a polynomial is to see whether the terms of the polynomial have a common factor. If there is one, we can write the polynomial as a product by **factoring out** the common factor. We will usually factor out the **greatest common factor (GCF)**.

1 The GCF of a list of integers is the largest integer that is a factor of all the integers in the list. For example, the GCF of 12 and 20 is 4 because 4 is the largest integer that is a factor of both 12 and 20. With large integers, the GCF may not be easily found by inspection. When this happens, use the following steps.

FINDING THE GCF OF A LIST OF INTEGERS

Step 1. Write each number as a product of prime numbers.
Step 2. Identify the common prime factors.
Step 3. The product of all common prime factors found in Step 2 is the greatest common factor. If there are no common prime factors, the greatest common factor is 1.

Recall from Section 1.3 that a prime number is a whole number other than 1, whose only factors are 1 and itself.

Example 1 Find the GCF of each list of numbers.

    a. 28 and 40   **b.** 55 and 21   **c.** 15, 18, and 66

Solution   **a.** Write each number as a product of primes.

$$28 = 2 \cdot 2 \cdot 7 = 2^2 \cdot 7$$
$$40 = 2 \cdot 2 \cdot 2 \cdot 5 = 2^3 \cdot 5$$

There are two common factors, each of which is 2, so the GCF is 4.

$$\text{GCF} = 2 \cdot 2 = 4$$

b. $55 = 5 \cdot 11$
$21 = 3 \cdot 7$
There are no common prime factors; thus, the GCF is 1.

c. $15 = 3 \cdot 5$
$18 = 2 \cdot 3 \cdot 3 = 2 \cdot 3^2$
$66 = 2 \cdot 3 \cdot 11$
The only prime factor common to all three numbers is 3, so the GCF is 3.   ▬

2   The greatest common factor of a list of variables raised to powers is found in a similar way. For example, the GCF of x^2, x^3, and x^5 is x^2 because each term contains a factor of x^2 and no higher power of x is a factor of each term.

$$x^2 = x \cdot x$$
$$x^3 = x \cdot x \cdot x$$
$$x^5 = x \cdot x \cdot x \cdot x \cdot x$$

There are two common factors of x, so the GCF $= x \cdot x$ or x^2. From this example, we see that **the GCF of a list of common variables raised to powers is the variable raised to the smallest exponent in the list**.

◆ **Example 2**   Find the GCF of each list of terms.

a. x^3, x^7, and x^5 **b.** y, y^4, and y^7

Solution   **a.** The GCF is x^3, since 3 is the smallest exponent to which x is raised.
b. The GCF is y^1 or y, since 1 is the smallest exponent on y.   ▬

In general, the **greatest common factor (GCF) of a list of terms** is the product of the GCF of the numerical coefficients and the GCF of the variable factors.

Example 3   Find the greatest common factor of each list of terms.

a. $6x^2$, $10x^3$, and $-8x$ **b.** $8y^2$, y^3, and y^5 **c.** a^3b^2, a^5b, and a^6b^2

Solution   **a.** The GCF of the numerical coefficients 6, 10, and -8 is 2.
The GCF of variable factors x^2, x^3, and x is x.
Thus, the GCF of the terms $6x^2$, $10x^3$, and $-8x$ is $2x$.
b. The GCF of the numerical coefficients 8, 1, and 1 is 1.
The GCF of variable factors y^2, y^3, and y^5 is y^2.
Thus, the GCF of the terms $8y^2$, y^3, and y^5 is $1y^2$ or y^2.
c. The GCF of a^3, a^5, and a^6 is a^3.
The GCF of b^2, b, and b^2 is b. Thus, the GCF of the terms is a^3b.   ▬

3 The first step in factoring a polynomial is to find the GCF of its terms. Once we do so, we can write the polynomial as a product by **factoring out** the GCF.

The polynomial $8x + 14$, for example, contains two terms: $8x$ and 14. The GCF of these terms is 2. We factor out 2 from each term by writing each term as a product of 2 and the term's remaining factors.

$$8x + 14 = 2 \cdot 4x + 2 \cdot 7$$

Using the distributive property, we can write

$$8x + 14 = 2 \cdot 4x + 2 \cdot 7$$
$$= 2(4x + 7)$$

Thus, a factored form of $8x + 14$ is $2(4x + 7)$.

> **HELPFUL HINT**
> A factored form of $8x + 14$ is *not*
>
> $$2 \cdot 4x + 2 \cdot 7$$
>
> Although the *terms* have been factored (written as a product), the *polynomial* $8x + 14$ has not been factored. A factored form of $8x + 14$ is the *product* $2(4x + 7)$.

Example 4 Factor each polynomial by factoring out the GCF.

    a. $6t + 18$ **b.** $y^5 - y^7$

Solution   **a.** The GCF of terms $6t$ and 18 is 6.

$$6t + 18 = 6 \cdot t + 6 \cdot 3$$
$$= 6(t + 3) \qquad \text{Apply the distributive property.}$$

Our work can be checked by multiplying 6 and $(t + 3)$.

$$6(t + 3) = 6 \cdot t + 6 \cdot 3 = 6t + 18, \text{ the original polynomial.}$$

  b. The GCF of y^5 and y^7 is y^5. Thus,

$$y^5 - y^7 = (y^5)1 - (y^5)y^2$$
$$= y^5(1 - y^2)$$

> **HELPFUL HINT**
> Don't forget the 1.

Example 5 Factor $-9a^5 + 18a^2 - 3a$.

Solution $-9a^5 + 18a^2 - 3a = (3a)(-3a^4) + (3a)(6a) + (3a)(-1)$
                            $= 3a(-3a^4 + 6a - 1)$

In Example 5 we could have chosen to factor out a $-3a$ instead of $3a$. If we factor out a $-3a$, we have

$$-9a^5 + 18a^2 - 3a = (-3a)(3a^4) + (-3a)(-6a) + (-3a)(1)$$
$$= -3a(3a^4 - 6a + 1)$$

Example 6   Factor $25x^4z + 15x^3z + 5x^2z$.

Solution   The greatest common factor is $5x^2z$.

$$25x^4z + 15x^3z + 5x^2z = 5x^2z(5x^2 + 3x + 1)$$

> **HELPFUL HINT**
> Be careful when the GCF of the terms is the same as one of the terms in the poly-
> nomial. The greatest common factor of the terms of $8x^2 - 6x^3 + 2x$ is $2x$. When
> factoring out $2x$ from the terms of $8x^2 - 6x^3 + 2x$, don't forget a term of 1.
>
> $$8x^2 - 6x^3 + 2x = 2x(4x) - 2x(3x^2) + 2x(1)$$
> $$= 2x(4x - 3x^2 + 1)$$
>
> Check by multiplying.
>
> $$2x(4x - 3x^2 + 1) = 8x^2 - 6x^3 + 2x$$

Example 7   Factor $5(x + 3) + y(x + 3)$.

Solution   The binomial $(x + 3)$ is the greatest common factor. Use the distributive property
to factor out $(x + 3)$.

$$5(x + 3) + y(x + 3) = (x + 3)(5 + y)$$

Example 8   Factor $3m^2n(a + b) - (a + b)$.

Solution   The greatest common factor is $(a + b)$.

$$3m^2n(a + b) - 1(a + b) = (a + b)(3m^2n - 1)$$

4   Once the GCF is factored out, we can often continue to factor the polynomial, using
a variety of techniques. We discuss here a technique for factoring polynomials called
grouping.

Example 9   Factor $xy + 2x + 3y + 6$ by grouping. Check by multiplying.

Solution   The GCF of the first two terms is x, and the GCF of the last two terms is 3.

$$xy + 2x + 3y + 6 = x(y + 2) + 3(y + 2)$$

> **HELPFUL HINT**
> Notice that this is *not* a factored form of the original polynomial. It is a sum, not
> a product.

Next, factor out the common binomial factor of $(y + 2)$.

$$x(y + 2) + 3(y + 2) = (y + 2)(x + 3)$$

To check, multiply $(y + 2)$ by $(x + 3)$.

$$(y + 2)(x + 3) = xy + 2x + 3y + 6, \text{ the original polynomial.}$$

Thus, the factored form of $xy + 2x + 3y + 6$ is $(y + 2)(x + 3)$.

FACTORING A FOUR-TERM POLYNOMIAL BY GROUPING

Step 1. Arrange the terms so that the first two terms have a common factor and the last two terms have a common factor.

Step 2. For each pair of terms, use the distributive property to factor out the pair's greatest common factor.

Step 3. If there is now a common binomial factor, factor it out.

Step 4. If there is no common binomial factor in Step 3, begin again, rearranging the terms differently. If no rearrangement leads to a common binomial factor, the polynomial cannot be factored.

Example 10   Factor $3x^2 + 4xy - 3x - 4y$ by grouping.

Solution   The first two terms have a common factor of x. Factor -1 from the last two terms so that the common binomial factor of $(3x + 4y)$ appears.

$$3x^2 + 4xy - 3x - 4y = x(3x + 4y) - 1(3x + 4y)$$

Next, factor out the common factor $(3x + 4y)$.

$$= (3x + 4y)(x - 1)$$

> **HELPFUL HINT**
> One more reminder: When **factoring** a polynomial, make sure the polynomial is written as a **product**. For example, it is true that
>
> $$3x^2 + 4xy - 3x - 4y = x(3x + 4y) - 1(3x + 4y)$$
>
> but $x(3x + 4y) - 1(3x + 4y)$ is not a **factored form** of the original polynomial since it is a **sum**, not a **product**. The factored form of $3x^2 + 4xy - 3x - 4y$ is $(3x + 4y)(x - 1)$.

Factoring out a greatest common factor first makes factoring by any method easier, as we see in the next example.

Example 11   Factor $4ax - 4ab - 2bx + 2b^2$.

Solution   First, factor out the common factor 2 from all four terms.

$$4ax - 4ab - 2bx + 2b^2$$
$$= 2(2ax - 2ab - bx + b^2) \qquad \text{Factor out 2 from all four terms.}$$
$$= 2[2a(x - b) - b(x - b)] \qquad \begin{array}{l}\text{Factor out common factors from each} \\ \text{pair of terms.}\end{array}$$
$$= 2(x - b)(2a - b) \qquad \text{Factor out the common binomial.}$$

Notice that we factored out $-b$ instead of b from the second pair of terms so that the binomial factor of each pair is the same.

MENTAL MATH

Find the prime factorization of the following integers.

1. 14   **2.** 15   **3.** 10   **4.** 70

Find the GCF of the following pairs of integers.

5. $6, 15$ **6.** $20, 15$ **7.** $3, 18$ **8.** $14, 35$

Exercise Set 6.1

Find the GCF for each list. See Examples 1 through 3.

1. $32, 36$ **2.** $36, 90$
3. $12, 18, 36$ **4.** $24, 14, 21$
5. y^2, y^4, y^7 **6.** x^3, x^2, x^3
7. $x^{10}y^2, xy^2, x^3y^3$ **8.** p^7q, p^8q^2, p^9q^3
9. $8x, 4$ **10.** $9y, y$
11. $12y^4, 20y^3$ **12.** $32x, 18x^2$
13. $12x^3, 6x^4, 3x^5$ **14.** $15y^2, 5y^7, 20y^3$
15. $18x^2y, 9x^3y^3, 36x^3y$ **16.** $7x, 21x^2y^2, 14xy$

Factor out the GCF from each polynomial. See Examples 4 through 6.

17. $30x - 15$ **18.** $42x - 7$
19. $24cd^3 - 18c^2d$ **20.** $25x^4y^3 - 15x^2y^2$
21. $-24a^4x + 18a^3x$ **22.** $-15a^2x + 9ax$
23. $12x^3 + 16x^2 - 8x$ **24.** $6x^3 - 9x^2 + 12x$
25. $5x^3y - 15x^2y + 10xy$ **26.** $14x^3y + 7x^2y - 7xy$
27. Construct a binomial whose greatest common factor is $5a^3$.
28. Construct a trinomial whose greatest common factor is $2x^2$.

Factor out the GCF from each polynomial. See Examples 7 and 8.

29. $y(x + 2) + 3(x + 2)$ **30.** $z(y + 4) + 3(y + 4)$
31. $x(y - 3) - 4(y - 3)$ **32.** $6(x + 2) - y(x + 2)$
33. $2x(x + y) - (x + y)$ **34.** $xy(y + 1) - (y + 1)$

Factor the following four-term polynomials by grouping. See Examples 9 through 11.

35. $5x + 15 + xy + 3y$ **36.** $xy + y + 2x + 2$
37. $2y - 8 + xy - 4x$ **38.** $6x - 42 + xy - 7y$
39. $3xy - 6x + 8y - 16$ **40.** $xy - 2yz + 5x - 10z$
41. $y^3 + 3y^2 + y + 3$ **42.** $x^3 + 4x + x^2 + 4$

Write an expression for the area of each shaded region. Then write the expression as a factored polynomial.

△ **43.**

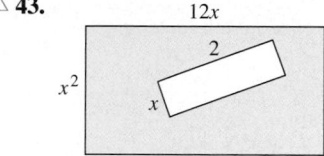

△ **44.**

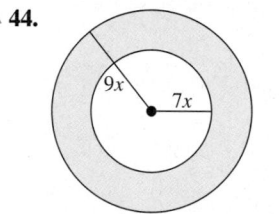

△ **45.**

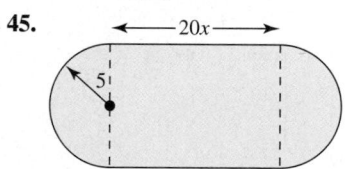

△ **46.**

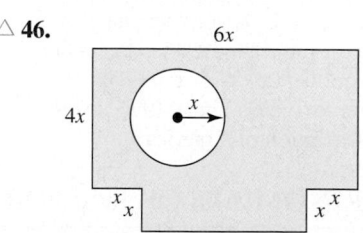

Factor the following polynomials.

47. $3x - 6$ **48.** $4x - 16$
49. $32xy - 18x^2$ **50.** $10xy - 15x^2$
51. $4x - 8y + 4$ **52.** $7x + 21y - 7$
53. $8(x + 2) - y(x + 2)$
54. $x(y^2 + 1) - 3(y^2 + 1)$
55. $-40x^8y^6 - 16x^9y^5$ **56.** $-21x^3y - 49x^2y^2$
57. $-3x + 12$ **58.** $-10x + 20$
59. $18x^3y^3 - 12x^3y^2 + 6x^5y^2$
60. $32x^3y^3 - 24x^2y^3 + 8x^2y^4$

61. $y^2(x - 2) + (x - 2)$ **62.** $x(y + 4) + (y + 4)$

63. $5xy + 15x + 6y + 18$

64. $2x^3 + x^2 + 8x + 4$

65. $4x^2 - 8xy - 3x + 6y$

66. $2x^3 - x^2 - 10x + 5$

67. $126x^3yz + 210y^4z^3$

68. $231x^3y^2z - 143yz^2$

69. $3y - 5x + 15 - xy$

70. $2x - 9y + 18 - xy$

71. $12x^2y - 42x^2 - 4y + 14$

72. $90 + 15y^2 - 18x - 3xy^2$

73. Explain how you can tell whether a polynomial is written in factored form.

74. Construct a 4-term polynomial that can be factored by grouping.

Which of the following expressions is factored?

75. $(a + 6)(b - 2)$ **76.** $(x + 5)(x + y)$

77. $5(2y + z) - b(2y + z)$

78. $3x(a + 2b) + 2(a + 2b)$

Write an expression for the length of each rectangle.

△ **79.**

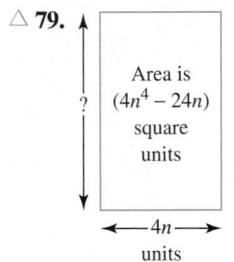

Area is $(4n^4 - 24n)$ square units

? (height)

$4n$ units

△ **80.**

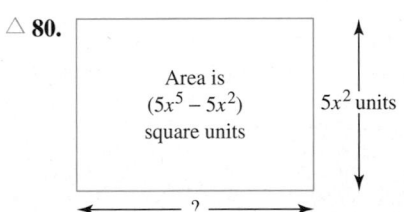

Area is $(5x^5 - 5x^2)$ square units

$5x^2$ units

?

81. The number (in millions) of CDs sold annually in the United States each year during 1996–1998 can be modeled by the polynomial function $f(x) = 60x^2 - 85x + 780$, where x is the number of years since 1996. (*Source:* Recording Industry Association of America)

a. Find the number of CDs sold in 1998. To do so, find $f(2)$.

b. Use this expression to predict the number of CDs sold in 2001.

c. Factor the polynomial $60x^2 - 85x + 780$.

82. The number (in thousands) of students who graduated from U.S. high schools each year during 1993–1997 can be modeled by the polynomial function $f(x) = 8x^2 + 20x + 2488$, where x is the number of years since 1993. (*Source:* U.S. Bureau of the Census)

a. Find the number of students who graduated from U.S. high schools in 1995. To do so, find $f(2)$.

b. Use this expression to predict the number of students who will graduate from U.S. high schools in 2002.

c. Factor the polynomial $8x^2 + 20x + 2488$.

REVIEW EXERCISES

Multiply. See Section 5.5.

83. $(x + 2)(x + 5)$ **84.** $(y + 3)(y + 6)$

85. $(a - 7)(a - 8)$ **86.** $(z - 4)(z - 4)$

A Look Ahead

Fill in the chart by finding two numbers that have the given product and sum. The first row is filled in for you.

	Two Numbers	Their Product	Their Sum
	4, 7	28	11
87.		12	8
88.		20	9
89.		8	-9
90.		16	-10
91.		-10	3
92.		-9	0
93.		-24	-5
94.		-36	-5

6.2 FACTORING TRINOMIALS OF THE FORM $x^2 + bx + c$

 CD-ROM

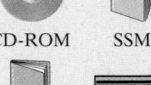 SSM

SSG

Video

▶ **OBJECTIVES**

1. Factor trinomials of the form $x^2 + bx + c$.

2. Factor out the greatest common factor and then factor a trinomial of the form $x^2 + bx + c$.

1

In this section, we factor trinomials of the form $x^2 + bx + c$, such as

$$x^2 + 4x + 3, \quad x^2 - 8x + 15, \quad x^2 + 4x - 12, \quad r^2 - r - 42$$

Notice that for these trinomials, the coefficient of the squared variable is 1.

Recall that factoring means to write as a product and that factoring and multiplying are reverse processes. Using the FOIL method of multiplying binomials, we have that

$$\overset{\text{F} \quad \text{O} \quad \text{I} \quad \text{L}}{(x + 3)(x + 1) = x^2 + 1x + 3x + 3}$$

$$= x^2 + 4x + 3$$

Thus, a factored form of $x^2 + 4x + 3$ is $(x + 3)(x + 1)$.

Notice that the product of the first terms of the binomials is $x \cdot x = x^2$, the first term of the trinomial. Also, the product of the last two terms of the binomials is $3 \cdot 1 = 3$, the third term of the trinomial. The sum of these same terms is $3 + 1 = 4$, the coefficient of the middle term, x, of the trinomial.

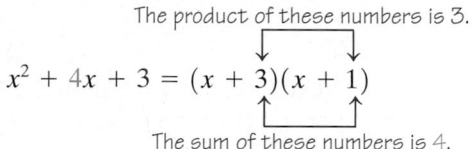

The product of these numbers is 3.

$$x^2 + 4x + 3 = (x + 3)(x + 1)$$

The sum of these numbers is 4.

Many trinomials, such as the one above, factor into two binomials. To factor $x^2 + 7x + 10$, let's assume that it factors into two binomials and begin by writing two pairs of parentheses. The first term of the trinomial is x^2, so we use x and x as the first terms of the binomial factors.

$$x^2 + 7x + 10 = (x + \quad)(x + \quad)$$

To determine the last term of each binomial factor, we look for two integers whose product is 10 and whose sum is 7. Since our numbers must have a positive product and a positive sum, we list pairs of positive integer factors of 10 only.

Positive Factors of 10	*Sum of Factors*
1, 10	$1 + 10 = 11$
2, 5	$2 + 5 = 7$

The correct pair of numbers is 2 and 5 because their product is 10 and their sum is 7. Now we can fill in the last terms of the binomial factors.

$$x^2 + 7x + 10 = (x + 2)(x + 5)$$

To see if we have factored correctly, multiply.

$$(x + 2)(x + 5) = x^2 + 2x + 5x + 10$$

$$= x^2 + 7x + 10 \qquad \text{Combine like terms.}$$

▼
H E L P F U L   H I N T
Since multiplication is commutative, the factored form of $x^2 + 7x + 10$ can be written as either $(x + 2)(x + 5)$ or $(x + 5)(x + 2)$.

FACTORING A TRINOMIAL OF THE FORM $x^2 + bx + c$

To factor a trinomial of the form $x^2 + bx + c$, look for two numbers whose product is c and whose sum is b. The factored form of $x^2 + bx + c$ is

$$\overbrace{(x + \text{one number})(x + \text{other number})}^{\text{product is } c}$$
$$\underbrace{\phantom{(x + \text{one number})(x + \text{other number})}}_{\text{sum is } b}$$

Example 1 Factor $x^2 + 7x + 12$.

Solution Begin by writing the first terms of the binomial factors.

$$(x + \quad)(x + \quad)$$

Next, look for two numbers whose product is 12 and whose sum is 7. Since our numbers must have a positive product and a positive sum, we look at positive pairs of factors of 12 only.

Positive Factors of 12	Sum of Factors
1, 12	$1 + 12 = 13$
2, 6	$2 + 6 = 8$
3, 4	$3 + 4 = 7$

The correct pair of numbers is 3 and 4 because their product is 12 and their sum is 7. Use these numbers as the last terms of the binomial factors.

$$x^2 + 7x + 12 = (x + 3)(x + 4)$$

To check, multiply $(x + 3)$ by $(x + 4)$.

Example 2 Factor $x^2 - 8x + 15$.

Solution Begin by writing the first terms of the binomials.

$$(x + \quad)(x + \quad)$$

Now look for two numbers whose product is 15 and whose sum is -8. Since our numbers must have a positive product and a negative sum, we look at negative factors of 15 only.

Negative Factors of 15	Sum of Factors
$-1, -15$	$-1 + (-15) = -16$
$-3, -5$	$-3 + (-5) = -8$

The correct pair of numbers is -3 and -5 because their product is 15 and their sum is -8. Then

$$x^2 - 8x + 15 = (x - 3)(x - 5)$$

Example 3   Factor $x^2 + 4x - 12$.

Solution $x^2 + 4x - 12 = (x + \quad)(x + \quad)$

Look for two numbers whose product is -12 and whose sum is 4. Since our numbers have a negative product, their signs must be different.

Factors of -12	*Sum of Factors*
$-1, 12$	$-1 + 12 = 11$
$1, -12$	$1 + (-12) = -11$
$-2, 6$	$-2 + 6 = 4$
$2, -6$	$2 + (-6) = -4$
$-3, 4$	$-3 + 4 = 1$
$3, -4$	$3 + (-4) = -1$

The correct pair of numbers is -2 and 6 since their product is -12 and their sum is 4. Hence

$$x^2 + 4x - 12 = (x - 2)(x + 6)$$

Example 4   Factor $r^2 - r - 42$.

Solution   Because the variable in this trinomial is r, the first term of each binomial factor is r.

$$r^2 - r - 42 = (r + \quad)(r + \quad)$$

Find two numbers whose product is -42 and whose sum is -1, the numerical coefficient of r. The numbers are 6 and -7. Therefore,

$$r^2 - r - 42 = (r + 6)(r - 7)$$

Example 5   Factor $a^2 + 2a + 10$.

Solution   Look for two numbers whose product is 10 and whose sum is 2. Neither 1 and 10 nor 2 and 5 give the required sum, 2. We conclude that $a^2 + 2a + 10$ is not factorable with integers. The polynomial $a^2 + 2a + 10$ is called a **prime polynomial**.

Example 6   Factor $x^2 + 5yx + 6y^2$.

Solution $x^2 + 5yx + 6y^2 = (x + \quad)(x + \quad)$

Look for two terms whose product is $6y^2$ and whose sum is $5y$, the coefficient of x in the middle term of the trinomial. The terms are $2y$ and $3y$ because $2y \cdot 3y = 6y^2$ and $2y + 3y = 5y$. Therefore,

$$x^2 + 5yx + 6y^2 = (x + 2y)(x + 3y)$$

The following sign patterns may be useful when factoring trinomials.

HELPFUL HINT—SIGN PATTERNS
A positive constant in a trinomial tells us to look for two numbers with the same sign. The sign of the coefficient of the middle term tells us whether the signs are both positive or both negative.

$$x^2 + 10x + 16 = (x + 2)(x + 8)$$

both positive / same sign

$$x^2 - 10x + 16 = (x - 2)(x - 8)$$

both negative / same sign

A negative constant in a trinomial tells us to look for two numbers with opposite signs.

$$x^2 + 6x - 16 = (x + 8)(x - 2) \quad x^2 - 6x - 16 = (x - 8)(x + 2)$$

opposite signs

---

2 Remember that the first step in factoring any polynomial is to factor out the greatest common factor (if there is one other than 1 or −1).

Example 7 Factor $3m^2 - 24m - 60$.

Solution First factor out the greatest common factor, 3, from each term.

$$3m^2 - 24m - 60 = 3(m^2 - 8m - 20)$$

Next, factor $m^2 - 8m - 20$ by looking for two factors of −20 whose sum is −8. The factors are −10 and 2.

$$3m^2 - 24m - 60 = 3(m + 2)(m - 10)$$

Remember to write the common factor 3 as part of the answer.
Check by multiplying.

$$3(m + 2)(m - 10) = 3(m^2 - 8m - 20)$$
$$= 3m^2 - 24m - 60$$

HELPFUL HINT
When factoring a polynomial, remember that factored out common factors are part of the final factored form. For example,

$$5x^2 - 15x - 50 = 5(x^2 - 3x - 10)$$
$$= 5(x + 2)(x - 5)$$

Thus, $5x^2 - 15x - 50$ **factored completely** is $5(x + 2)(x - 5)$.

MENTAL MATH

Complete the following.

1. $x^2 + 9x + 20 = (x + 4)(x \quad)$ **2.** $x^2 + 12x + 35 = (x + 5)(x \quad)$ **3.** $x^2 - 7x + 12 = (x - 4)(x \quad)$
4. $x^2 - 13x + 22 = (x - 2)(x \quad)$ **5.** $x^2 + 4x + 4 = (x + 2)(x \quad)$ **6.** $x^2 + 10x + 24 = (x + 6)(x \quad)$

Exercise Set 6.2

Factor each trinomial completely. If the polynomial can't be factored, write prime. See Examples 1 through 5.

1. $x^2 + 7x + 6$
2. $x^2 + 6x + 8$
3. $x^2 + 9x + 8$
4. $x^2 + 13x + 30$
5. $x^2 - 8x + 15$
6. $x^2 - 9x + 14$
7. $x^2 - 10x + 9$
8. $x^2 - 6x + 9$
9. $x^2 - 15x + 5$
10. $x^2 - 13x + 30$
11. $x^2 - 3x - 18$
12. $x^2 - x - 30$
13. $x^2 + 5x + 2$
14. $x^2 - 7x + 5$

Factor each trinomial completely. See Example 6.

15. $x^2 + 8xy + 15y^2$
16. $x^2 + 6xy + 8y^2$
17. $x^2 - 2xy + y^2$
18. $x^2 - 11xy + 30y^2$
19. $x^2 - 3xy - 4y^2$
20. $x^2 - 4xy - 77y^2$

Factor each trinomial completely. See Example 7.

21. $2z^2 + 20z + 32$
22. $3x^2 + 30x + 63$
23. $2x^3 - 18x^2 + 40x$
24. $x^3 - x^2 - 56x$
25. $7x^2 + 14xy - 21y^2$
26. $6r^2 - 3rs - 3s^2$
27. To factor $x^2 + 13x + 42$, think of two numbers whose _____ is 42 and whose _____ is 13.
28. Write a polynomial that factors as $(x - 3)(x + 8)$.

Factor each trinomial completely.

29. $x^2 + 15x + 36$
30. $x^2 + 19x + 60$
31. $x^2 - x - 2$
32. $x^2 - 5x - 14$
33. $r^2 - 16r + 48$
34. $r^2 - 10r + 21$
35. $x^2 - 4x - 21$
36. $x^2 - 4x - 32$
37. $x^2 + 7xy + 10y^2$
38. $x^2 - 3xy - 4y^2$
39. $r^2 - 3r + 6$
40. $x^2 + 4x - 10$
41. $2t^2 + 24t + 64$
42. $2t^2 + 20t + 50$
43. $x^3 - 2x^2 - 24x$
44. $x^3 - 3x^2 - 28x$
45. $x^2 - 16x + 63$
46. $x^2 - 19x + 88$

47. $x^2 + xy - 2y^2$
48. $x^2 - xy - 6y^2$
49. $3x^2 - 60x + 108$
50. $2x^2 - 24x + 70$
51. $x^2 - 18x - 144$
52. $x^2 + x - 42$
53. $6x^3 + 54x^2 + 120x$
54. $3x^3 + 3x^2 - 126x$
55. $2t^5 - 14t^4 + 24t^3$
56. $3x^6 + 30x^5 + 72x^4$
57. $5x^3y - 25x^2y^2 - 120xy^3$
58. $3x^2 - 6xy - 72y^2$
59. $4x^2y + 4xy - 12y$
60. $3x^2y - 9xy + 45y$
61. $2a^2b - 20ab^2 + 42b^3$
62. $-1x^2z + 14xz^2 - 28z^3$

Find a positive value of b so that each trinomial is factorable.

63. $x^2 + bx + 15$
64. $y^2 + by + 20$
65. $m^2 + bm - 27$
66. $x^2 + bx - 14$

Find a positive value of c so that each trinomial is factorable.

67. $x^2 + 6x + c$
68. $t^2 + 8t + c$
69. $y^2 - 4y + c$
70. $n^2 - 16n + c$

Complete the following sentences in your own words.

71. If $x^2 + bx + c$ is factorable and c is negative, then the signs of the last term factors of the binomial are opposite because ....

72. If $x^2 + bx + c$ is factorable and c is positive, then the signs of the last term factors of the binomials are the same because ....

REVIEW EXERCISES

Multiply. See Section 5.5.

73. $(2x + 1)(x + 5)$
74. $(3x + 2)(x + 4)$
75. $(5y - 4)(3y - 1)$
76. $(4z - 7)(7z - 1)$
77. $(a + 3)(9a - 4)$
78. $(y - 5)(6y + 5)$

Graph each linear equation. See Section 3.2.

79. $y = -3x$ **80.** $y = 5x$

81. $y = 2x - 7$ **82.** $y = -x + 4$

A Look Ahead

Example

Factor $2t^5y^2 - 22t^4y^2 + 56t^3y^2$.

Solution:

First, factor out the greatest common factor of $2t^3y^2$.

$$2t^5y^2 - 22t^4y^2 + 56t^3y^2 = 2t^3y^2(t^2 - 11t + 28)$$
$$= 2t^3y^2(t - 4)(t - 7)$$

Factor each trinomial completely.

83. $2x^2y + 30xy + 100y$

84. $3x^2z^2 + 9xz^2 + 6z^2$

85. $-12x^2y^3 - 24xy^3 - 36y^3$

86. $-4x^2t^4 + 4xt^4 + 24t^4$

87. $y^2(x + 1) - 2y(x + 1) - 15(x + 1)$

88. $z^2(x + 1) - 3z(x + 1) - 70(x + 1)$

6.3 FACTORING TRINOMIALS OF THE FORM $ax^2 + bx + c$

CD-ROM   SSM

SSG   Video

▶ **OBJECTIVES**

1. Factor trinomials of the form $ax^2 + bx + c$.
2. Factor out a GCF before factoring a trinomial of the form $ax^2 + bx + c$.
3. Factor perfect square trinomials.
4. Factor trinomials of the form $ax^2 + bx + c$ by an alternate method.

1

In this section, we factor trinomials of the form $ax^2 + bx + c$, such as

$$3x^2 + 11x + 6, \qquad 8x^2 - 22x + 5, \qquad 2x^2 + 13x - 7$$

Notice that the coefficient of the squared variable in these trinomials is a number other than 1. We will factor these trinomials using a trial-and-check method based on our work in the last section.

To begin, let's review the relationship between the numerical coefficients of the trinomial and the numerical coefficients of its factored form. For example, since $(2x + 1)(x + 6) = 2x^2 + 13x + 6$, the factored form of $2x^2 + 13x + 6$ is

$$2x^2 + 13x + 6 = (2x + 1)(x + 6)$$

Notice that $2x$ and x are factors of $2x^2$, the first term of the trinomial. Also, 6 and 1 are factors of 6, the last term of the trinomial, as shown:

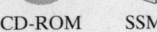

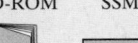

$$2x^2 + 13x + 6 = (2x + 1)(x + 6)$$

Also notice that $13x$, the middle term, is the sum of the following products:

$$2x^2 + 13x + 6 = (2x + 1)(x + 6)$$

	$1x$
$+$	$12x$
	$13x$ middle term

Let's use this pattern to factor $5x^2 + 7x + 2$. First, we find factors of $5x^2$. Since all numerical coefficients in this trinomial are positive, we will use factors with positive numerical coefficients only. Thus, the factors of $5x^2$ are $5x$ and x. Let's try these factors as first terms of the binomials. Thus far, we have

$$5x^2 + 7x + 2 = (5x + \quad)(x + \quad)$$

Next, we need to find positive factors of 2. Positive factors of 2 are 1 and 2. Now we try possible combinations of these factors as second terms of the binomials until we obtain a middle term of $7x$.

$$(5x + \underline{1})(x + 2) = 5x^2 + 11x + 2$$

$$\begin{array}{l} 1x \\ + 10x \\ \hline 11x \longrightarrow \textbf{incorrect} \text{ middle term} \end{array}$$

Let's try switching factors 2 and 1.

$$(5x + \underline{2})(x + 1) = 5x^2 + 7x + 2$$

$$\begin{array}{l} 2x \\ + 5x \\ \hline 7x \longrightarrow \textbf{correct} \text{ middle term} \end{array}$$

Thus the factored form of $5x^2 + 7x + 2$ is $(5x + 2)(x + 1)$. To check, we multiply $(5x + 2)$ and $(x + 1)$. The product is $5x^2 + 7x + 2$.

Example 1   Factor: $3x^2 + 11x + 6$

Solution   Since all numerical coefficients are positive, we use factors with positive numerical coefficients. We first find factors of $3x^2$.

Factors of $3x^2$: $3x^2 = 3x \cdot x$

If factorable, the trinomial will be of the form

$$3x^2 + 11x + 6 = (3x + \quad)(x + \quad)$$

Next we factor 6.

Factors of 6: $6 = 1 \cdot 6,\qquad 6 = 2 \cdot 3$

Now we try combinations of factors of 6 until a middle term of $11x$ is obtained. Let's try 1 and 6 first.

$$(3x + \underline{1})(x + 6) = 3x^2 + 19x + 6$$

$$\begin{array}{l} 1x \\ + 18x \\ \hline 19x \longrightarrow \textbf{incorrect} \text{ middle term} \end{array}$$

Now let's next try 6 and 1.

$$(3x + 6)(x + 1)$$

Before multiplying, notice that the terms of the factor $3x + 6$ have a common factor of 3. The terms of the original trinomial $3x^2 + 11x + 6$ have no common factor other than 1, so the terms of the factored form of $3x^2 + 11x + 6$ can contain no common factor other than 1. This means that $(3x + 6)(x + 1)$ is not a factored form.

Next let's try 2 and 3 as last terms.

$$(3x + 2)(x + 3) = 3x^2 + 11x + 6$$

$$\begin{array}{r} 2x \\ + \quad 9x \\ \hline 11x \end{array} \longrightarrow \textbf{correct} \text{ middle term}$$

Thus the factored form of $3x^2 + 11x + 6$ is $(3x + 2)(x + 3)$.   ■

> **HELPFUL HINT**
> If the terms of a trinomial have no common factor (other than 1), then the terms of neither of its binomial factors will contain a common factor (other than 1).

Example 2   Factor: $8x^2 - 22x + 5$

Solution   Factors of $8x^2$: $8x^2 = 8x \cdot x$, $8x^2 = 4x \cdot 2x$
We'll try $8x$ and x.

$$8x^2 - 22x + 5 = (8x + \quad)(x + \quad)$$

Since the middle term, $-22x$, has a negative numerical coefficient, we factor 5 into negative factors.

$$\text{Factors of 5:}   5 = -1 \cdot -5$$

Let's try -1 and -5.

$$(8x - 1)(x - 5) = 8x^2 - 41x + 5$$

$$\begin{array}{r} -1x \\ + \quad (-40x) \\ \hline -41x \end{array} \longrightarrow \textbf{incorrect} \text{ middle term}$$

Now let's try -5 and -1.

$$(8x - 5)(x - 1) = 8x^2 - 13x + 5$$

$$\begin{array}{r} -5x \\ + \quad (-8x) \\ \hline -13x \end{array} \longrightarrow \textbf{incorrect} \text{ middle term}$$

Don't give up yet! We can still try other factors of $8x^2$. Let's try $4x$ and $2x$ with -1 and -5.

$$(4x - 1)(2x - 5) = 8x^2 - 22x + 5$$

$$\begin{array}{r} -2x \\ + \quad (-20x) \\ \hline -22x \end{array} \longrightarrow \textbf{correct} \text{ middle term}$$

The factored form of $8x^2 - 22x + 5$ is $(4x - 1)(2x - 5)$.   ■

Example 3   Factor: $2x^2 + 13x - 7$

Solution   Factors of $2x^2$: $2x^2 = 2x \cdot x$

Factors of -7: $-7 = -1 \cdot 7$, $-7 = 1 \cdot -7$

We try possible combinations of these factors:

$$(2x + 1)(x - 7) = 2x^2 - 13x - 7 \qquad \textbf{incorrect} \text{ middle term}$$
$$(2x - 1)(x + 7) = 2x^2 + 13x - 7 \qquad \textbf{correct} \text{ middle term}$$

The factored form of $2x^2 + 13x - 7$ is $(2x - 1)(x + 7)$.

Example 4   Factor: $10x^2 - 13xy - 3y^2$

Solution   Factors of $10x^2$: $10x^2 = 10x \cdot x$, $10x^2 = 2x \cdot 5x$

Factors of $-3y^2$: $-3y^2 = -3y \cdot y$, $-3y^2 = 3y \cdot -y$

We try some combinations of these factors:

$$(10x - 3y)(x + y) = 10x^2 + 7xy - 3y^2$$
$$(x + 3y)(10x - y) = 10x^2 + 29xy - 3y^2$$
$$(5x + 3y)(2x - y) = 10x^2 + xy - 3y^2$$
$$(2x - 3y)(5x + y) = 10x^2 - 13xy - 3y^2 \qquad \textbf{correct} \text{ middle term}$$

The factored form of $10x^2 - 13xy - 3y^2$ is $(2x - 3y)(5x + y)$.

2   Don't forget that the best first step in factoring any polynomial is to look for a common factor to factor out.

Example 5   Factor: $24x^4 + 40x^3 + 6x^2$

Solution   Notice that all three terms have a common factor of $2x^2$. First, factor out $2x^2$.

$$24x^4 + 40x^3 + 6x^2 = 2x^2(12x^2 + 20x + 3)$$

Next, factor $12x^2 + 20x + 3$.

Factors of $12x^2$: $12x^2 = 6x \cdot 2x$, $12x^2 = 4x \cdot 3x$, $12x^2 = 12x \cdot x$

Since all terms in the trinomial have positive numerical coefficients, factor 3 using positive factors only.

Factors of 3: $3 = 1 \cdot 3$

We try some combinations of the factors.

$$2x^2(4x + 3)(3x + 1) = 2x^2(12x^2 + 13x + 3)$$
$$2x^2(12x + 1)(x + 3) = 2x^2(12x^2 + 37x + 3)$$
$$2x^2(2x + 3)(6x + 1) = 2x^2(12x^2 + 20x + 3) \quad \textbf{correct} \text{ middle term}$$

The factored form of $24x^4 + 40x^3 + 6x^2$ is $2x^2(2x + 3)(6x + 1)$.

HELPFUL HINT
Don't forget to include the common factor in the factored form.

Example 6   Factor: $4x^2 - 12x + 9$

Solution   Factors of $4x^2$: $4x^2 = 2x \cdot 2x$, $4x^2 = 4x \cdot x$
Since the middle term $-12x$ has a negative numerical coefficient, factor 9 into negative factors only.

Factors of 9: $9 = -3 \cdot -3$, $9 = -1 \cdot -9$

The correct combination is

$$(2x - \underline{3})(2x - 3) = 4x^2 - 12x + 9$$

$$\underbrace{\quad\quad\quad}_{-6x}$$
$$\underline{+ \;(-6x)}$$
$$-12x \longrightarrow \textbf{correct} \text{ middle term}$$

Thus, $4x^2 - 12x + 9 = (2x - 3)(2x - 3)$, which can also be written as $(2x - 3)^2$.

---

3   Notice in Example 6 that $4x^2 - 12x + 9 = (2x - 3)^2$. The trinomial $4x^2 - 12x + 9$ is called a **perfect square trinomial** since it is the square of the binomial $2x - 3$.

In the last chapter, we learned a shortcut special product for squaring a binomial, recognizing that

$$(a + b)^2 = a^2 + 2ab + b^2$$

The trinomial $a^2 + 2ab + b^2$ is a perfect square trinomial, since it is the square of the binomial $a + b$. We can use this pattern to help us factor perfect square trinomials. To use this pattern, we must first be able to recognize a perfect square trinomial. A trinomial is a perfect square when its first term is the square of some expression a, its last term is the square of some expression b, and its middle term is twice the product of the expressions a and b. When a trinomial fits this description, its factored form is $(a + b)^2$.

PERFECT SQUARE TRINOMIALS

$$a^2 + 2ab + b^2 = (a + b)^2$$
$$a^2 - 2ab + b^2 = (a - b)^2$$

Example 7   Factor: $x^2 + 12x + 36$

Solution   This trinomial is a perfect square trinomial since:
1. The first term is the square of x: $x^2 = (x)^2$.
2. The last term is the square of 6: $36 = (6)^2$.
3. The middle term is twice the product of x and 6: $12x = 2 \cdot x \cdot 6$.
Thus, $x^2 + 12x + 36 = (x + 6)^2$.

Example 8   Factor: $25x^2 + 25xy + 4y^2$

Solution   Determine whether or not this trinomial is a perfect square by considering the same three questions.
1. Is the first term a square? Yes, $25x^2 = (5x)^2$.
2. Is the last term a square? Yes, $4y^2 = (2y)^2$.
3. Is the middle term twice the product of $5x$ and $2y$? **No.** $2 \cdot 5x \cdot 2y = 20xy$, not $25xy$.

Therefore, $25x^2 + 25xy + 4y^2$ is not a perfect square trinomial. It is factorable, though. Using earlier techniques, we find that

$$25x^2 + 25xy + 4y^2 = (5x + 4y)(5x + y).$$

> **HELPFUL HINT**
> A perfect square trinomial that is not recognized as such can be factored by other methods.

Example 9   Factor: $4m^2 - 4m + 1$

Solution   This is a perfect square trinomial since $4m^2 = (2m)^2$, $1 = (1)^2$, and $4m = 2 \cdot 2m \cdot 1$.

$$4m^2 - 4m + 1 = (2m - 1)^2$$

4   Grouping can also be used to factor trinomials of the form $ax^2 + bx + c$. To use this method, write the trinomial as a four-term polynomial. For example, to factor $2x^2 + 11x + 12$ using grouping, find two numbers whose product is $2 \cdot 12 = 24$ and whose sum is 11. Since we want a positive product and a positive sum, we consider positive pairs of factors of 24 only.

Factors of 24	*Sum of Factors*
1, 24	$1 + 24 = 25$
2, 12	$2 + 12 = 14$
3, 8	$3 + 8 = 11$

The factors are 3 and 8. Use these factors to write the middle term $11x$ as $3x + 8x$. Replace $11x$ with $3x + 8x$ in the original trinomial and factor by grouping.

$$2x^2 + 11x + 12 = 2x^2 + 3x + 8x + 12$$
$$= (2x^2 + 3x) + (8x + 12)$$
$$= x(2x + 3) + 4(2x + 3)$$
$$= (2x + 3)(x + 4)$$

In general, we have the following:

FACTORING TRINOMIALS OF THE FORM $ax^2 + bx + c$ BY GROUPING

Step 1. Find two numbers whose product is $a \cdot c$ and whose sum is b.
Step 2. Write the middle term, bx, using the factors found in Step 1.
Step 3. Factor by grouping.

Example 10   Factor $8x^2 - 14x + 5$ by grouping.

Solution   This trinomial is of the form $ax^2 + bx + c$ with $a = 8, b = -14$, and $c = 5$.

Step 1.   Find two numbers whose product is $a \cdot c$ or $8 \cdot 5 = 40$, and whose sum is b or -14. The numbers are -4 and -10.

Step 2.   Write $-14x$ as $-4x - 10x$ so that

$$8x^2 - 14x + 5 = 8x^2 - 4x - 10x + 5$$

Step 3.   Factor by grouping.

$$8x^2 - 4x - 10x + 5 = 4x(2x - 1) - 5(2x - 1)$$
$$= (2x - 1)(4x - 5)$$

Example 11   Factor $3x^2 - x - 10$ by grouping.

Solution   In $3x^2 - x - 10$, $a = 3, b = -1$, and $c = -10$.

Step 1.   Find two numbers whose product is $a \cdot c$ or $3(-10) = -30$ and whose sum is b or -1. The numbers are -6 and 5.

Step 2. $3x^2 - x - 10 = 3x^2 - 6x + 5x - 10$

Step 3. $ = 3x(x - 2) + 5(x - 2)$
$$= (x - 2)(3x + 5)$$

MENTAL MATH

State whether or not each trinomial is a perfect trinomial square.

1. $x^2 + 14x + 49$
2. $9x^2 - 12x + 4$
3. $y^2 + 2y + 4$
4. $x^2 - 4x + 2$
5. $9y^2 + 6y + 1$
6. $y^2 - 16y + 64$

Exercise Set 6.3

Factor completely. See Examples 1 through 5. (See Examples 10 and 11 for alternate method.)

1. $2x^2 + 13x + 15$
2. $3x^2 + 8x + 4$
3. $2x^2 - 9x - 5$
4. $3x^2 + 20x - 63$
5. $2y^2 - y - 6$
6. $8y^2 - 17y + 9$
7. $16a^2 - 24a + 9$
8. $25x^2 + 20x + 4$
9. $36r^2 - 5r - 24$
10. $20r^2 + 27r - 8$
11. $10x^2 + 17x + 3$
12. $21x^2 - 41x + 10$
13. $21x^2 - 48x - 45$
14. $12x^2 - 14x - 10$
15. $12x^2 - 14x - 6$
16. $20x^2 - 2x + 6$
17. $4x^3 - 9x^2 - 9x$
18. $6x^3 - 31x^2 + 5x$

Factor the following perfect square trinomials. See Examples 6 through 9.

19. $x^2 + 22x + 121$
20. $x^2 + 18x + 81$
21. $x^2 - 16x + 64$
22. $x^2 - 12x + 36$
23. $16y^2 - 40y + 25$
24. $9y^2 + 48y + 64$
25. $x^2y^2 - 10xy + 25$
26. $4x^2y^2 - 28xy + 49$
27. Describe a perfect square trinomial.
28. Write a perfect square trinomial that factors as $(x + 3y)^2$.

Factor the following completely.

29. $2x^2 - 7x - 99$
30. $2x^2 + 7x - 72$
31. $4x^2 - 8x - 21$
32. $6x^2 - 11x - 10$
33. $30x^2 - 53x + 21$
34. $21x^3 - 6x - 30$
35. $24x^2 - 58x + 9$
36. $36x^2 + 55x - 14$
37. $9x^2 - 24xy + 16y^2$
38. $25x^2 + 60xy + 36y^2$
39. $x^2 - 14xy + 49y^2$
40. $x^2 + 10xy + 25y^2$
41. $2x^2 + 7x + 5$
42. $2x^2 + 7x + 3$
43. $3x^2 - 5x + 1$
44. $3x^2 - 7x + 6$
45. $-2y^2 + y + 10$
46. $-4x^2 - 23x + 6$
47. $16x^2 + 24xy + 9y^2$
48. $4x^2 - 36xy + 81y^2$
49. $8x^2y + 34xy - 84y$
50. $6x^2y^2 - 2xy^2 - 60y^2$
51. $3x^2 + x - 2$
52. $8y^2 + y - 9$

53. $x^2y^2 + 4xy + 4$
54. $x^2y^2 - 6xy + 9$
55. $49y^2 + 42xy + 9x^2$
56. $16x^2 - 8xy + y^2$
57. $3x^2 - 42x + 63$
58. $5x^2 - 75x + 60$
59. $42a^2 - 43a + 6$
60. $54a^2 + 39ab - 8b^2$
61. $18x^2 - 9x - 14$
62. $8x^2 + 6x - 27$
63. $25p^2 - 70pq + 49q^2$
64. $36p^2 - 18pq + 9q^2$
65. $15x^2 - 16x - 15$
66. $12x^2 + 7x - 12$
67. $-27t + 7t^2 - 4$
68. $4t^2 - 7 - 3t$

The area of the largest square in the figure is $(a + b)^2$. Use this figure to answer Exercises 69 and 70.

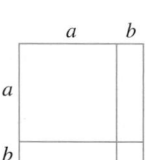

△ 69. Write the area of the largest square as the sum of the areas of the smaller squares and rectangles.

△ 70. What factoring formula from this section is visually represented by this square?

Find a positive value of b so that each trinomial is factorable.

71. $3x^2 + bx - 5$
72. $2y^2 + by + 3$
73. $2z^2 + bz - 7$
74. $5x^2 + bx - 1$

Find a positive value of c so that each trinomial is factorable.

75. $5x^2 + 7x + c$
76. $7x^2 + 22x + c$
77. $3x^2 - 8x + c$
78. $11y^2 - 40y + c$

REVIEW EXERCISES

Multiply the following. See Section 5.4.

79. $(x - 2)(x + 2)$
80. $(y^2 + 3)(y^2 - 3)$
81. $(a + 3)(a^2 - 3a + 9)$
82. $(z - 2)(z^2 + 2z + 4)$
83. $(y - 5)(y^2 + 5y + 25)$
84. $(m + 7)(m^2 - 7m + 49)$

As of 1998, approximately 42% of U.S. households had a personal computer. The following graph shows the percent of selected households having a computer grouped according to household income. See Section 1.9.

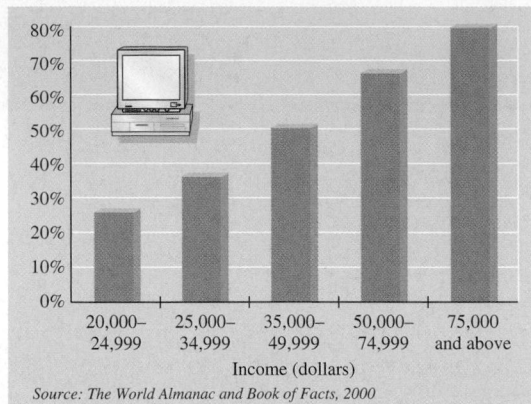

Income (dollars)

Source: The World Almanac and Book of Facts, 2000

85. Which range of household income corresponds to the greatest percent of households having a personal computer?

86. Which range of household income corresponds to the greatest increase in percent of households having a personal computer?

87. Describe any trend you notice from this graph.

88. Why don't the percents shown in the graph add to 42%?

A Look Ahead

Example
Factor $10x^4y - 5x^3y^2 - 15x^2y^3$.

Solution:
Start by factoring out the greatest common factor of $5x^2y$.

$$10x^4y - 5x^3y^2 - 15x^2y^3 = 5x^2y(2x^2 - xy - 3y^2)$$

Next, factor $2x^2 - xy - 3y^2$. Try $2x$ and x as factors of $2x^2$ and $-3y$ and y as factors of $-3y^2$.

$$
\underbrace{(2x - \underbrace{3y})(x + y)}_{\substack{-3xy \\ +2xy}}
$$

$-xy$ \qquad The middle term is correct.

Hence $10x^4y - 5x^3y^2 - 15x^2y^3 = 5x^2y(2x - 3y)(x + y)$.

Factor completely. See the example.

89. $-12x^3y^2 + 3x^2y^2 + 15xy^2$

90. $-12r^3x^2 + 38r^2x^2 + 14rx^2$

91. $-30p^3q + 88p^2q^2 + 6pq^3$

92. $3x^3y^2 + 3x^2y^3 - 18xy^4$

93. $4x^2(y - 1)^2 + 10x(y - 1)^2 + 25(y - 1)^2$

94. $3x^2(a + 3)^3 - 28x(a + 3)^3 + 25(a + 3)^3$

6.4 FACTORING BINOMIALS

CD-ROM   SSM

SSG   Video

▶ **OBJECTIVES**

1. Factor the difference of two squares.
2. Factor the sum or difference of two cubes.

1 When learning to multiply binomials in Chapter 5, we studied a special product, the product of the sum and difference of two terms, *a* and *b*:

$$(a + b)(a - b) = a^2 - b^2$$

For example, the product of $x + 3$ and $x - 3$ is

$$(x + 3)(x - 3) = x^2 - 9$$

The binomial $x^2 - 9$ is called a **difference of squares**. In this section, we use the pattern for the product of a sum and difference to factor the binomial difference of squares.

To use this pattern to help us factor, we must be able to recognize a difference of squares. A binomial is a difference of squares when it is the difference of the square of some expression *a* and the square of some expression *b*.

DIFFERENCE OF TWO SQUARES

$$a^2 - b^2 = (a + b)(a - b)$$

Example 1 Factor $x^2 - 25$.

Solution $x^2 - 25$ is the difference of two squares since $x^2 - 25 = x^2 - 5^2$. Therefore,

$$x^2 - 25 = x^2 - 5^2 = (x + 5)(x - 5)$$

Multiply to check.

Example 2 Factor each difference of squares.

a. $4x^2 - 1$ **b.** $25a^2 - 9b^2$

Solution **a.** $4x^2 - 1 = (2x)^2 - 1^2 = (2x + 1)(2x - 1)$
b. $25a^2 - 9b^2 = (5a)^2 - (3b)^2 = (5a + 3b)(5a - 3b)$

Example 3 Factor $x^4 - y^6$.

Solution Write x^4 as $(x^2)^2$ and y^6 as $(y^3)^2$.

$$x^4 - y^6 = (x^2)^2 - (y^3)^2 = (x^2 + y^3)(x^2 - y^3)$$

Example 4 Factor $9x^2 - 36$.

Solution Remember when factoring always to check first for common factors. If there are common factors, factor out the GCF and then factor the resulting polynomial.

$$9x^2 - 36 = 9(x^2 - 4) \qquad \text{Factor out the GCF 9.}$$
$$= 9(x^2 - 2^2)$$
$$= 9(x + 2)(x - 2)$$

In this example, if we forget to factor out the GCF first, we still have the difference of two squares.

$$9x^2 - 36 = (3x)^2 - (6)^2 = (3x + 6)(3x - 6)$$

This binomial has not been factored completely since both terms of both binomial factors have a common factor of 3.

$$3x + 6 = 3(x + 2) \quad \text{and} \quad 3x - 6 = 3(x - 2)$$

Then

$$9x^2 - 36 = (3x + 6)(3x - 6) = 3(x + 2)3(x - 2) = 9(x + 2)(x - 2)$$

Factoring is easier if the GCF is factored out first before using other methods.

Example 5 Factor $x^2 + 4$.

Solution The binomial $x^2 + 4$ is the **sum** of squares since we can write $x^2 + 4$ as $x^2 + 2^2$. We might try to factor using $(x + 2)(x + 2)$ or $(x - 2)(x - 2)$. But when multiplying to check, neither factoring is correct.

$$(x + 2)(x + 2) = x^2 + 4x + 4$$
$$(x - 2)(x - 2) = x^2 - 4x + 4$$

In both cases, the product is a trinomial, not the required binomial. In fact, $x^2 + 4$ is a prime polynomial.

> **HELPFUL HINT**
> After the greatest common factor has been removed, the *sum* of two squares cannot be factored further using real numbers.

2 Although the sum of two squares usually does not factor, the sum or difference of two cubes can be factored and reveals factoring patterns. The pattern for the sum of cubes is illustrated by multiplying the binomial $x + y$ and the trinomial $x^2 - xy + y^2$.

$$
\begin{array}{r}
x^2 - xy + y^2 \\
x + y \\
\hline
x^2y - xy^2 + y^3 \\
x^3 - x^2y + xy^2 \\
\hline
x^3 \qquad\qquad + y^3
\end{array}
$$

$$(x + y)(x^2 - xy + y^2) = x^3 + y^3 \qquad \text{Sum of cubes.}$$

The pattern for the difference of two cubes is illustrated by multiplying the binomial $x - y$ by the trinomial $x^2 + xy + y^2$. The result is

$$(x - y)(x^2 + xy + y^2) = x^3 - y^3 \qquad \text{Difference of cubes.}$$

SUM OR DIFFERENCE OF TWO CUBES

$$a^3 + b^3 = (a + b)(a^2 - ab + b^2)$$
$$a^3 - b^3 = (a - b)(a^2 + ab + b^2)$$

Example 6 Factor $x^3 + 8$.

Solution First, write the binomial in the form $a^3 + b^3$.

$$x^3 + 8 = x^3 + 2^3 \qquad \text{Write in the form } a^3 + b^3.$$

If we replace a with x and b with 2 in the formula above, we have

$$x^3 + 2^3 = (x + 2)[x^2 - (x)(2) + 2^2]$$
$$= (x + 2)(x^2 - 2x + 4)$$

> **HELPFUL HINT**
> When factoring sums or differences of cubes, notice the sign patterns.
>
> same sign
> $$x^3 + y^3 = (x + y)(x^2 - xy + y^2)$$
> opposite sign   always positive
>
> same sign
> $$x^3 - y^3 = (x - y)(x^2 + xy + y^2)$$
> opposite sign   always positive

Example 7   Factor $y^3 - 27$.

Solution $y^3 - 27 = y^3 - 3^3$ Write in the form $a^3 - b^3$.
$$= (y - 3)[y^2 + (y)(3) + 3^2]$$
$$= (y - 3)(y^2 + 3y + 9)$$

Example 8   Factor $64x^3 + 1$.

Solution $64x^3 + 1 = (4x)^3 + 1^3$
$$= (4x + 1)[(4x)^2 - (4x)(1) + 1^2]$$
$$= (4x + 1)(16x^2 - 4x + 1)$$

◆ **Example 9**   Factor $54a^3 - 16b^3$.

Solution   Remember to factor out common factors first before using other factoring methods.

$$54a^3 - 16b^3 = 2(27a^3 - 8b^3) \qquad \text{Factor out the GCF 2.}$$
$$= 2[(3a)^3 - (2b)^3] \qquad \text{Difference of two cubes.}$$
$$= 2(3a - 2b)[(3a)^2 + (3a)(2b) + (2b)^2]$$
$$= 2(3a - 2b)(9a^2 + 6ab + 4b^2)$$

GRAPHING CALCULATOR EXPLORATIONS

A graphing calculator is a convenient tool for evaluating an expression at a given replacement value. For example, let's evaluate $x^2 - 6x$ when $x = 2$. To do so, store the value 2 in the variable x and then enter and evaluate the algebraic expression.

```
2 → X
                    2
X² − 6X
                  −8
```

The value of $x^2 - 6x$ when $x = 2$ is -8. You may want to use this method for evaluating expressions as you explore the following.

We can use a graphing calculator to explore factoring patterns numerically. Use your calculator to evaluate $x^2 - 2x + 1$, $x^2 - 2x - 1$, and $(x - 1)^2$ for each value of x given in the table. What do you observe?

	$x^2 - 2x + 1$	$x^2 - 2x - 1$	$(x - 1)^2$
$x = 5$			
$x = -3$			
$x = 2.7$			
$x = -12.1$			
$x = 0$			

Notice in each case that $x^2 - 2x - 1 \neq (x - 1)^2$. Because for each x in the table the value of $x^2 - 2x + 1$ and the value of $(x - 1)^2$ are the same, we might guess that $x^2 - 2x + 1 = (x - 1)^2$. We can verify our guess algebraically with multiplication:

$$(x - 1)(x - 1) = x^2 - x - x + 1 = x^2 - 2x + 1$$

MENTAL MATH

State each number as a square.

1. 1   **2.** 25   **3.** 81

4. 64   **5.** 9   **6.** 100

State each number as a cube.

7. 1   **8.** 64   **9.** 8   **10.** 27

Exercise Set 6.4

Factor the difference of two squares. See Examples 1 through 4.

1. $x^2 - 4$

2. $y^2 - 81$

3. $y^2 - 49$

4. $x^2 - 100$

5. $25y^2 - 9$

6. $49a^2 - 16$

7. $121 - 100x^2$

8. $144 - 81x^2$

9. $12x^2 - 27$

10. $36x^2 - 64$

11. $169a^2 - 49b^2$

12. $225a^2 - 81b^2$

13. $x^2y^2 - 1$

14. $16 - a^2b^2$

15. $x^4 - 9$

16. $y^4 - 25$

17. $49a^4 - 16$

18. $49b^4 - 1$

19. $x^4 - y^{10}$

20. $x^{14} - y^4$

21. What binomial multiplied by $(x - 6)$ gives the difference of two squares?

22. What binomial multiplied by $(5 + y)$ gives the difference of two squares?

Factor the sum or difference of two cubes. See Examples 6 through 9.

23. $a^3 + 27$

24. $b^3 - 8$

25. $8a^3 + 1$

26. $64x^3 - 1$

27. $5k^3 + 40$

28. $6r^3 - 162$

29. $x^3y^3 - 64$

30. $8x^3 - y^3$

31. $x^3 + 125$

32. $a^3 - 216$

33. $24x^4 - 81xy^3$

34. $375y^6 - 24y^3$

35. What binomial multiplied by $(4x^2 - 2xy + y^2)$ gives the sum or difference of two cubes?

36. What binomial multiplied by $(1 + 4y + 16y^2)$ gives the sum or difference of two cubes?

Factor the binomials completely.

37. $x^2 - 4$

38. $x^2 - 36$

39. $81 - p^2$

40. $100 - t^2$

41. $4r^2 - 1$

42. $9t^2 - 1$

43. $9x^2 - 16$

44. $36y^2 - 25$

45. $16r^2 + 1$

46. $49y^2 + 1$

47. $27 - t^3$

48. $125 + r^3$

49. $8r^3 - 64$

50. $54r^3 + 2$

51. $t^3 - 343$

52. $s^3 + 216$

53. $x^2 - 169y^2$

54. $x^2 - 225y^2$

55. $x^2y^2 - z^2$

56. $x^3y^3 - z^3$

57. $x^3y^3 + 1$

58. $x^2y^2 + z^2$

59. $s^3 - 64t^3$

60. $8t^3 + s^3$

61. $18r^3 - 8$

62. $32t^4 - 50$

63. $9xy^2 - 4x$

64. $16xy^2 - 64x$

65. $25y^4 - 100y^2$

66. $xy^3 - 9xyz^2$

67. $x^3y - 4xy^3$

68. $12s^3t^3 + 192s^5t$

69. $8s^6t^3 + 100s^3t^6$

70. $25x^5y + 121x^3y$

71. $27x^2y^3 - xy^2$

72. $8x^3y^3 + x^3y$

73. An object is dropped from the top of Pittsburgh's USX Towers, which is 841 feet tall. (*Source: World Almanac* research) The height of the object after t seconds is given by the polynomial function $f(t) = 841 - 16t^2$.

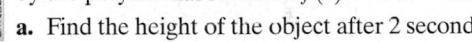
a. Find the height of the object after 2 seconds.

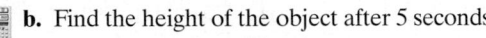
b. Find the height of the object after 5 seconds.

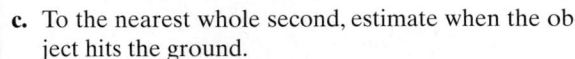
c. To the nearest whole second, estimate when the object hits the ground.

d. Factor $841 - 16t^2$.

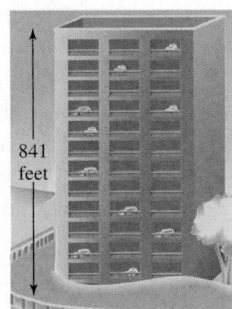

841 feet

74. A worker on the top of the Aetna Life Building in San Francisco accidentally drops a bolt. The Aetna Life Building is 529 feet tall. (*Source: World Almanac* research) The height of the bolt after t seconds is given by the polynomial function $f(t) = 529 - 16t^2$.

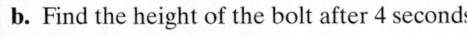
a. Find the height of the bolt after 1 second.

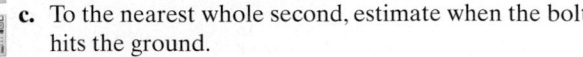
b. Find the height of the bolt after 4 seconds.

c. To the nearest whole second, estimate when the bolt hits the ground.

d. Factor $529 - 16t^2$.

75. In your own words, explain how to tell whether a binomial is a difference of squares. Then explain how to factor a difference of squares.

76. In your own words, explain how to tell whether a binomial is a sum of cubes. Then explain how to factor a sum of cubes.

REVIEW EXERCISES

Divide the following. See Section 5.6.

77. $\dfrac{8x^4 + 4x^3 - 2x + 6}{2x}$

78. $\dfrac{3y^4 + 9y^2 - 6y + 1}{3y^2}$

Use long division to divide the following. See Section 5.6.

79. $\dfrac{2x^2 - 3x - 2}{x - 2}$

80. $\dfrac{4x^2 - 21x + 21}{x - 3}$

81. $\dfrac{3x^2 + 13x + 10}{x + 3}$

82. $\dfrac{5x^2 + 14x + 12}{x + 2}$

A Look Ahead

Example

Factor $(x + y)^2 - (x - y)^2$.

Solution:

Use the method for factoring the difference of squares.

$(x + y)^2 - (x - y)^2 = \big[(x + y) + (x - y)\big]$
$$\big[(x + y) - (x - y)\big]$$
$$= (x + y + x - y)(x + y - x + y)$$
$$= (2x)(2y)$$
$$= 4xy$$

Factor each difference of squares. See the example.

83. $a^2 - (2 + b)^2$

84. $(x + 3)^2 - y^2$

85. $(x^2 - 4)^2 - (x - 2)^2$

86. $(x^2 - 9) - (3 - x)$

6.5 CHOOSING A FACTORING STRATEGY

CD-ROM   SSM

SSG   Video

▶ **O B J E C T I V E**

1. Factor polynomials completely.

1

A polynomial is factored completely when it is written as the product of prime polynomials. This section uses the various methods of factoring polynomials that have been discussed in earlier sections. Since these methods are applied throughout the remainder of this text, as well as in later courses, it is important to master the skills of factoring. The following is a set of guidelines for factoring polynomials.

FACTORING A POLYNOMIAL

Step 1. Are there any common factors? If so, factor out the GCF.

Step 2. How many terms are in the polynomial?

    a. If there are **two** terms, decide if one of the following can be applied.

        i. Difference of two squares: $a^2 - b^2 = (a - b)(a + b)$.

        ii. Difference of two cubes: $a^3 - b^3 = (a - b)(a^2 + ab + b^2)$.

        iii. Sum of two cubes: $a^3 + b^3 = (a + b)(a^2 - ab + b^2)$.

    b. If there are **three** terms, try one of the following.

        i. Perfect square trinomial: $a^2 + 2ab + b^2 = (a + b)^2$.

        ii. If not a perfect square trinomial, factor using the methods presented in Sections 6.2 and 6.3.

    c. If there are **four** or more terms, try factoring by grouping.

Step 3. See if any factors in the factored polynomial can be factored further.

Step 4. Check by multiplying.

Example 1   Factor $10t^2 - 17t + 3$.

Solution   **Step 1.**   The terms of this polynomial have no common factor (other than 1).

Step 2.   There are three terms, so this polynomial is a trinomial. This trinomial is not a perfect square trinomial, so factor using methods from earlier sections.

Factors of $10t^2$: $10t^2 = 2t \cdot 5t$, $10t^2 = t \cdot 10t$

Since the middle term, $-17t$, has a negative numerical coefficient, find negative factors of 3.

Factors of 3: $3 = -1 \cdot -3$

Try different combinations of these factors. The correct combination is

$$(2t - 3)(5t - 1) = 10t^2 - 17t + 3$$
$$\underbrace{-15t}$$
$$\underline{-2t}$$
$$-17t \qquad \qquad \textit{correct middle term}$$

Step 3.   No factor can be factored further, so we have factored completely.

Step 4.   To check, multiply $2t - 3$ and $5t - 1$.

$$(2t - 3)(5t - 1) = 10t^2 - 2t - 15t + 3 = 10t^2 - 17t + 3$$

The factored form of $10t^2 - 17t + 3$ is $(2t - 3)(5t - 1)$. ∎

Example 2   Factor $2x^3 + 3x^2 - 2x - 3$.

Solution   **Step 1.**   There are no factors common to all terms.

Step 2.   Try factoring by grouping since this polynomial has four terms.

$$2x^3 + 3x^2 - 2x - 3 = x^2(2x + 3) - 1(2x + 3) \qquad \textit{Factor out the greatest common factor for each pair of terms.}$$

$$= (2x + 3)(x^2 - 1) \qquad \textit{Factor out } 2x + 3.$$

Step 3.   The binomial $x^2 - 1$ can be factored further. It is the difference of two squares.

$$= (2x + 3)(x + 1)(x - 1) \qquad \textit{Factor } x^2 - 1 \textit{ as a difference of squares.}$$

Step 4.   Check by finding the product of the three binomials.
The polynomial factored completely is $(2x + 3)(x + 1)(x - 1)$. ∎

Example 3   Factor $12m^2 - 3n^2$.

Solution   **Step 1.**   The terms of this binomial contain a greatest common factor of 3.

$$12m^2 - 3n^2 = 3(4m^2 - n^2) \qquad \textit{Factor out the greatest common factor.}$$

CHAPTER 6 FACTORING POLYNOMIALS

Step 2. The binomial $4m^2 - n^2$ is a difference of squares.

$$= 3(2m + n)(2m - n) \quad \text{\textit{Factor the difference of squares.}}$$

Step 3. No factor can be factored further.

Step 4. We check by multiplying.

$$3(2m + n)(2m - n) = 3(4m^2 - n^2) = 12m^2 - 3n^2$$

The factored form of $12m^2 - 3n^2$ is $3(2m + n)(2m - n)$.

Example 4   Factor $x^3 + 27y^3$.

Solution   **Step 1.** The terms of this binomial contain no common factor (other than 1).

Step 2. This binomial is the sum of two cubes.

$$x^3 + 27y^3 = (x)^3 + (3y)^3$$
$$= (x + 3y)[x^2 - 3xy + (3y)^2]$$
$$= (x + 3y)(x^2 - 3xy + 9y^2)$$

Step 3. No factor can be factored further.

Step 4. We check by multiplying.

$$(x + 3y)(x^2 - 3xy + 9y^2) = x(x^2 - 3xy + 9y^2)$$
$$+ 3y(x^2 - 3xy + 9y^2)$$
$$= x^3 - 3x^2y + 9xy^2 + 3x^2y$$
$$- 9xy^2 + 27y^3$$
$$= x^3 + 27y^3$$

Thus, $x^3 + 27y^3$ factored completely is $(x + 3y)(x^2 - 3xy + 9y^2)$.

Example 5   Factor $30a^2b^3 + 55a^2b^2 - 35a^2b$.

Solution   **Step 1.** $30a^2b^3 + 55a^2b^2 - 35a^2b = 5a^2b(6b^2 + 11b - 7)$ \text{\textit{Factor out the GCF.}}

Step 2. $= 5a^2b(2b - 1)(3b + 7)$ \text{\textit{Factor the resulting trinomial.}}

Step 3. No factor can be factored further.

Step 4. Check by multiplying.

The trinomial factored completely is $5a^2b(2b - 1)(3b + 7)$.

SPOTLIGHT ON DECISION MAKING

Suppose you are shopping for a credit card and have received the following credit card offers in the mail.

Credit Card Offer #1	Credit Card Offer #2	Credit Card Offer #3
We offer a low 9.8% APR interest rate, coupled with a generous 20-day grace period and low $35 annual fee. We can offer you a $2000 credit limit. In addition, you'll earn one frequent flier mile for every dollar charged on your card. And every time you use your card to pay for airline tickets, we'll even give you $250,000 in flight insurance–absolutely free!	With our card, you'll get a 14.5% APR interest rate, absolutely *no* annual fee, a 25-day grace period, and a $2500 credit limit. And, only with our card, you can receive a year-end cash bonus of up to 1% of the total value of purchases made to your card during the year! That's our way of saying *"Thank you"* for choosing our card.	*Just say "yes" to this offer, and you can get our lowest 17% APR interest rate, along with a 30-day grace period and easy-to-swallow $20 annual fee. You also qualify for a $3000 credit limit.* *That's not all: you earn gift certificates for a local mall with every purchase! Plus, you'll receive $100,000 of flight insurance whenever you charge airline tickets.*

You construct a decision grid to help make your choice. In the decision grid, give each of the decision criteria a rank reflecting its importance to you, with 1 being not important to 10 being very important. Then for each card offer, decide how well the criteria are supported, assigning a rating of 1 for poor support to a rating of 10 for excellent support. For Credit Card Offer #1, fill in the Score column by multiplying rank by rating for each criteria. Repeat for each credit card offer. Finally, total the scores for each credit card offer. The offer with the highest score is likely to be the best choice for you.

Based on your decision-grid analysis, which credit card would you choose? Explain.

Criteria	Rank	CREDIT CARD OFFER #1		CREDIT CARD OFFER #2		CREDIT CARD OFFER #3	
		Rating	*Score*	*Rating*	*Score*	*Rating*	*Score*
Interest rate							
Annual fee							
Grace period							
Credit limit							
Automatic flight insurance							
Rebates/incentives/bonuses							
TOTAL							

Exercise Set 6.5

Factor the following completely. See Examples 1 through 5.

1. $a^2 + 2ab + b^2$
2. $a^2 - 2ab + b^2$
3. $a^2 + a - 12$
4. $a^2 - 7a + 10$
5. $a^2 - a - 6$
6. $a^2 + 2a + 1$
7. $x^2 + 2x + 1$
8. $x^2 + x - 2$
9. $x^2 + 4x + 3$
10. $x^2 + x - 6$
11. $x^2 + 7x + 12$
12. $x^2 + x - 12$
13. $x^2 + 3x - 4$
14. $x^2 - 7x + 10$
15. $x^2 + 2x - 15$
16. $x^2 + 11x + 30$
17. $x^2 - x - 30$
18. $x^2 + 11x + 24$
19. $2x^2 - 98$
20. $3x^2 - 75$
21. $x^2 + 3x + xy + 3y$
22. $3y - 21 + xy - 7x$

23. $x^2 + 6x - 16$

24. $x^2 - 3x - 28$

25. $4x^3 + 20x^2 - 56x$

26. $6x^3 - 6x^2 - 120x$

27. $12x^2 + 34x + 24$

28. $8a^2 + 6ab - 5b^2$

29. $4a^2 - b^2$

30. $28 - 13x - 6x^2$

31. $20 - 3x - 2x^2$

32. $x^2 - 2x + 4$

33. $a^2 + a - 3$

34. $6y^2 + y - 15$

35. $4x^2 - x - 5$

36. $x^2y - y^3$

37. $4t^2 + 36$

38. $x^2 + x + xy + y$

39. $ax + 2x + a + 2$

40. $18x^3 - 63x^2 + 9x$

41. $12a^3 - 24a^2 + 4a$

42. $x^2 + 14x - 32$

43. $x^2 - 14x - 48$

44. $16a^2 - 56ab + 49b^2$

45. $25p^2 - 70pq + 49q^2$

46. $7x^2 + 24xy + 9y^2$

47. $125 - 8y^3$

48. $64x^3 + 27$

49. $-x^2 - x + 30$

50. $-x^2 + 6x - 8$

51. $14 + 5x - x^2$

52. $3 - 2x - x^2$

53. $3x^4y + 6x^3y - 72x^2y$

54. $2x^3y + 8x^2y^2 - 10xy^3$

55. $5x^3y^2 - 40x^2y^3 + 35xy^4$

56. $4x^4y - 8x^3y - 60x^2y$

57. $12x^3y + 243xy$

58. $6x^3y^2 + 8xy^2$

59. $(x - y)^2 - z^2$

60. $(x + 2y)^2 - 9$

61. $3rs - s + 12r - 4$

62. $x^3 - 2x^2 + 3x - 6$

63. $4x^2 - 8xy - 3x + 6y$

64. $4x^2 - 2xy - 7yz + 14xz$

65. $6x^2 + 18xy + 12y^2$

66. $12x^2 + 46xy - 8y^2$

67. $xy^2 - 4x + 3y^2 - 12$

68. $x^2y^2 - 9x^2 + 3y^2 - 27$

69. $5(x + y) + x(x + y)$

70. $7(x - y) + y(x - y)$

71. $14t^2 - 9t + 1$

72. $3t^2 - 5t + 1$

73. $3x^2 + 2x - 5$

74. $7x^2 + 19x - 6$

75. $x^2 + 9xy - 36y^2$

76. $3x^2 + 10xy - 8y^2$

77. $1 - 8ab - 20a^2b^2$

78. $1 - 7ab - 60a^2b^2$

79. $x^4 - 10x^2 + 9$

80. $x^4 - 13x^2 + 36$

81. $x^4 - 14x^2 - 32$

82. $x^4 - 22x^2 - 75$

83. $x^2 - 23x + 120$

84. $y^2 + 22y + 96$

85. $6x^3 - 28x^2 + 16x$

86. $6y^3 - 8y^2 - 30y$

87. $27x^3 - 125y^3$

88. $216y^3 - z^3$

89. $x^3y^3 + 8z^3$

90. $27a^3b^3 + 8$

91. $2xy - 72x^3y$

92. $2x^3 - 18x$

93. $x^3 + 6x^2 - 4x - 24$

94. $x^3 - 2x^2 - 36x + 72$

95. $6a^3 + 10a^2$

96. $4n^2 - 6n$

97. $a^2(a + 2) + 2(a + 2)$

98. $a - b + x(a - b)$

99. $x^3 - 28 + 7x^2 - 4x$

100. $a^3 - 45 - 9a + 5a^2$

101. Explain why it makes good sense to factor out the GCF first, before using other methods of factoring.

102. The sum of two squares usually does not factor. Is the sum of two squares $9x^2 + 81y^2$ factorable?

REVIEW EXERCISES

Solve each equation. See Section 2.2.

103. $x - 6 = 0$

104. $y + 5 = 0$

105. $2m + 4 = 0$

106. $3x - 9 = 0$

107. $5z - 1 = 0$

108. $4a + 2 = 0$

Solve the following. See Section 2.5.

△ **109.** A suitcase has a volume of 960 cubic inches. Find x.

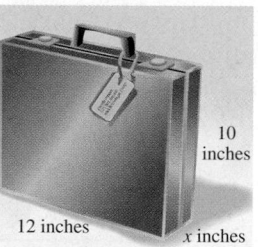

10 inches

12 inches

x inches

△ **110.** The sail shown has an area of 25 square feet. Find its height, x.

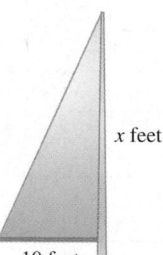

x feet

10 feet

List the x- and y-intercepts for each graph. See Section 3.4.

111.

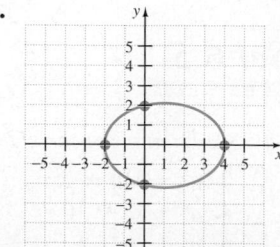

112.

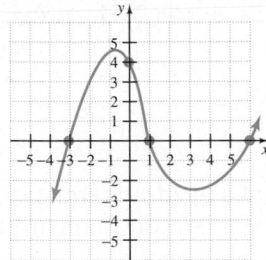

114.

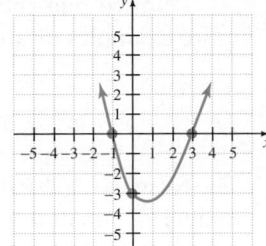

113.

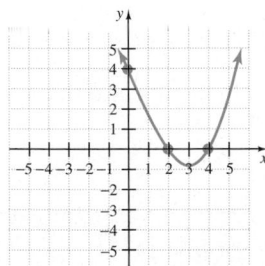

6.6 SOLVING QUADRATIC EQUATIONS BY FACTORING

CD-ROM

SSM

SSG   Video

▶ **OBJECTIVES**

1. Define quadratic equation.
2. Solve quadratic equations by factoring.
3. Solve equations with degree greater than 2 by factoring.

1

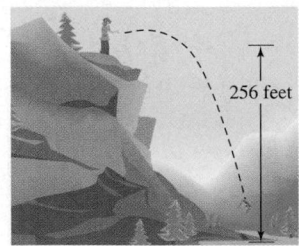

256 feet

Linear equations, while versatile, are not versatile enough to model many real-life phenomena. For example, let's suppose an object is dropped from the top of a 256-foot cliff and we want to know how long before the object strikes the ground. The answer to this question is found by solving the equation $-16t^2 + 256 = 0$. (See Example 1 in the next section.) This equation is called a **quadratic equation** because it contains a variable with an exponent of 2 and no other variable in the equation contains an exponent greater than 2. In this section, we solve quadratic equations by factoring.

QUADRATIC EQUATION

A quadratic equation is one that can be written in the form

$$ax^2 + bx + c = 0,$$

where a, b, and c are real numbers, and $a \neq 0$.

Notice that the degree of the polynomial $ax^2 + bx + c$ is 2. Here are a few more examples of quadratic equations.

Quadratic Equations

$$3x^2 + 5x + 6 = 0 \qquad x^2 = 9 \qquad y^2 + y = 1$$

The form $ax^2 + bx + c = 0$ is called the **standard form** of a quadratic equation. The quadratic equations $3x^2 + 5x + 6 = 0$ and $-16t^2 + 256 = 0$ are in standard form. One side of the equation is 0 and the other side is a polynomial of degree 2 written in descending powers of the variable.

2   Some quadratic equations can be solved by making use of factoring and the **zero factor theorem**.

> ### ZERO FACTOR THEOREM
>
> If a and b are real numbers and if $ab = 0$, then $a = 0$ or $b = 0$.

This theorem states that if the product of two numbers is 0 then at least one of the numbers must be 0.

Example 1   Solve $(x - 3)(x + 1) = 0$.

Solution   If this equation is to be a true statement, then either the factor $x - 3$ must be 0 or the factor $x + 1$ must be 0. In other words, either

$$x - 3 = 0 \qquad \text{or} \qquad x + 1 = 0$$

If we solve these two linear equations, we have

$$x = 3 \qquad \text{or} \qquad x = -1$$

Thus, 3 and -1 are both solutions of the equation $(x - 3)(x + 1) = 0$. To check, we replace x with 3 in the original equation. Then we replace x with -1 in the original equation.

Check

$(x - 3)(x + 1) = 0$	$(x - 3)(x + 1) = 0$
$(3 - 3)(3 + 1) \stackrel{?}{=} 0$ Replace x with 3.	$(-1 - 3)(-1 + 1) \stackrel{?}{=} 0$ Replace x with -1.
$0(4) = 0$ True.	$(-4)(0) = 0$ True.

The solutions are 3 and -1.

> ▼ **HELPFUL HINT**
>
> The zero factor property says that *if a product is 0, then a factor is 0.*
>
> If $a \cdot b = 0$, then $a = 0$ or $b = 0$.
>
> If $x(x + 5) = 0$, then $x = 0$ or $x + 5 = 0$.
>
> If $(x + 7)(2x - 3) = 0$, then $x + 7 = 0$ or $2x - 3 = 0$.
>
> Use this property only when the product is 0. For example, if $a \cdot b = 8$, we do not know the value of a or b. The values may be $a = 2, b = 4$ or $a = 8, b = 1$, or any other two numbers whose product is 8.

Example 2   Solve $x^2 - 9x = -20$.

Solution   First, write the equation in standard form; then factor.

$$x^2 - 9x = -20$$
$$x^2 - 9x + 20 = 0 \qquad \text{Write in standard form by adding 20 to both sides.}$$
$$(x - 4)(x - 5) = 0 \qquad \text{Factor.}$$

Next, use the zero factor theorem and set each factor equal to 0.

$$x - 4 = 0 \quad \text{or} \quad x - 5 = 0 \qquad \text{Set each factor equal to 0.}$$
$$x = 4 \quad \text{or} \quad x = 5 \qquad \text{Solve.}$$

Check the solutions by replacing x with each value in the original equation. The solutions are 4 and 5.

The following steps may be used to solve a quadratic equation by factoring.

SOLVING QUADRATIC EQUATIONS BY FACTORING

Step 1. Write the equation in standard form: $ax^2 + bx + c = 0$.
Step 2. Factor the quadratic completely.
Step 3. Set each factor containing a variable equal to 0.
Step 4. Solve the resulting equations.
Step 5. Check each solution in the original equation.

Since it is not always possible to factor a quadratic polynomial, not all quadratic equations can be solved by factoring. Other methods of solving quadratic equations are presented in Chapter 10.

Example 3   Solve $x(2x - 7) = 4$.

Solution   First, write the equation in standard form; then factor.

$$x(2x - 7) = 4$$
$$2x^2 - 7x = 4 \qquad \text{Multiply.}$$
$$2x^2 - 7x - 4 = 0 \qquad \text{Write in standard form.}$$
$$(2x + 1)(x - 4) = 0 \qquad \text{Factor.}$$
$$2x + 1 = 0 \quad \text{or} \quad x - 4 = 0 \qquad \text{Set each factor equal to zero.}$$
$$2x = -1 \quad \text{or} \quad x = 4 \qquad \text{Solve.}$$
$$x = -\frac{1}{2}$$

Check both solutions $-\frac{1}{2}$ and 4.

HELPFUL HINT
One more reminder: To apply the zero factor theorem, one side of the equation must be 0 and the other side of the equation must be factored. To solve the equation $x(2x - 7) = 4$, for example, you may **not** set each factor equal to 4.

Example 4 Solve $-2x^2 - 4x + 30 = 0$.

Solution The equation is in standard form so we begin by factoring out a common factor of -2.

$$-2x^2 - 4x + 30 = 0$$
$$-2(x^2 + 2x - 15) = 0 \qquad \text{Factor out } -2.$$
$$-2(x + 5)(x - 3) = 0 \qquad \text{Factor the quadratic.}$$

Next, set each factor **containing a variable** equal to 0.

$$x + 5 = 0 \qquad \text{or} \qquad x - 3 = 0 \qquad \text{Set each factor containing a}$$
$$x = -5 \qquad \text{or} \qquad x = 3 \qquad \text{variable equal to 0.}$$
$$\text{Solve.}$$

Note that the factor -2 is a constant term containing no variables and can never equal 0. The solutions are -5 and 3. ∎

3 Some equations involving polynomials of degree higher than 2 may also be solved by factoring and then applying the zero factor theorem.

Example 5 Solve $3x^3 - 12x = 0$.

Solution Factor the left side of the equation. Begin by factoring out the common factor of $3x$.

$$3x^3 - 12x = 0$$
$$3x(x^2 - 4) = 0 \qquad \text{Factor out the GCF } 3x.$$
$$3x(x + 2)(x - 2) = 0 \qquad \text{Factor } x^2 - 4, \text{ a difference}$$
$$\text{of squares.}$$

$$3x = 0 \quad \text{or} \quad x + 2 = 0 \quad \text{or} \quad x - 2 = 0 \qquad \text{Set each factor equal to 0.}$$
$$x = 0 \quad \text{or} \quad x = -2 \quad \text{or} \quad x = 2 \qquad \text{Solve.}$$

Thus, the equation $3x^3 - 12x = 0$ has three solutions: $0, -2,$ and 2. To check, replace x with each solution in the original equation.

Let $x = 0$.	**Let $x = -2$.**	**Let $x = 2$.**
$3(0)^3 - 12(0) \stackrel{?}{=} 0$	$3(-2)^3 - 12(-2) \stackrel{?}{=} 0$	$3(2)^3 - 12(2) \stackrel{?}{=} 0$
$0 = 0$	$3(-8) + 24 \stackrel{?}{=} 0$	$3(8) - 24 \stackrel{?}{=} 0$
	$0 = 0$	$0 = 0$

Substituting $0, -2,$ or 2 into the original equation results each time in a true equation. The solutions are $0, -2,$ and 2. ∎

Example 6 Solve $(5x - 1)(2x^2 + 15x + 18) = 0$.

Solution
$$(5x - 1)(2x^2 + 15x + 18) = 0$$
$$(5x - 1)(2x + 3)(x + 6) = 0 \qquad \text{Factor the trinomial.}$$

$$5x - 1 = 0 \quad \text{or} \quad 2x + 3 = 0 \quad \text{or} \quad x + 6 = 0 \qquad \text{Set each factor equal to 0.}$$
$$5x = 1 \quad \text{or} \quad 2x = -3 \quad \text{or} \quad x = -6 \qquad \text{Solve.}$$
$$x = \frac{1}{5} \quad \text{or} \quad x = -\frac{3}{2}$$

The solutions are $\frac{1}{5}, -\frac{3}{2},$ and -6. Check by replacing x with each solution in the original equation. The solutions are $-6, -\frac{3}{2},$ and $\frac{1}{5}$. ∎

Example 7   Solve $2x^3 - 4x^2 - 30x = 0$.

Solution   Begin by factoring out the GCF $2x$.

$$2x^3 - 4x^2 - 30x = 0$$
$$2x(x^2 - 2x - 15) = 0 \qquad \text{Factor out the GCF } 2x.$$
$$2x(x - 5)(x + 3) = 0 \qquad \text{Factor the quadratic.}$$

$2x = 0$	or	$x - 5 = 0$	or	$x + 3 = 0$	Set each factor containing a variable equal to 0.
$x = 0$	or	$x = 5$	or	$x = -3$	Solve.

Check by replacing x with each solution in the cubic equation. The solutions are -3, 0, and 5.

In Chapter 3, we graphed linear equations in two variables, such as $y = 5x - 6$. Recall that to find the x-intercept of the graph of a linear equation, let $y = 0$ and solve for x. This is also how to find the x-intercepts of the graph of a **quadratic equation in two variables**, such as $y = x^2 - 5x + 4$.

Example 8   Find the x-intercepts of the graph of $y = x^2 - 5x + 4$.

Solution   Let $y = 0$ and solve for x.

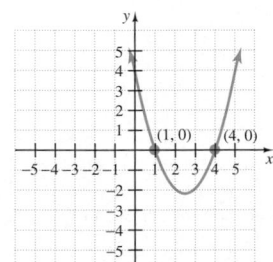

$$y = x^2 - 5x + 4$$
$$0 = x^2 - 5x + 4 \qquad \text{Let } y = 0.$$
$$0 = (x - 1)(x - 4) \qquad \text{Factor.}$$

$x - 1 = 0$	or	$x - 4 = 0$	Set each factor equal to 0.
$x = 1$	or	$x = 4$	Solve.

The x-intercepts of the graph of $y = x^2 - 5x + 4$ are $(1, 0)$ and $(4, 0)$.
The graph of $y = x^2 - 5x + 4$ is shown in the margin.

In general, a quadratic equation in two variables is one that can be written in the form $y = ax^2 + bx + c$ where $a \neq 0$. The graph of such an equation is called a **parabola** and will open up or down depending on the value of a.

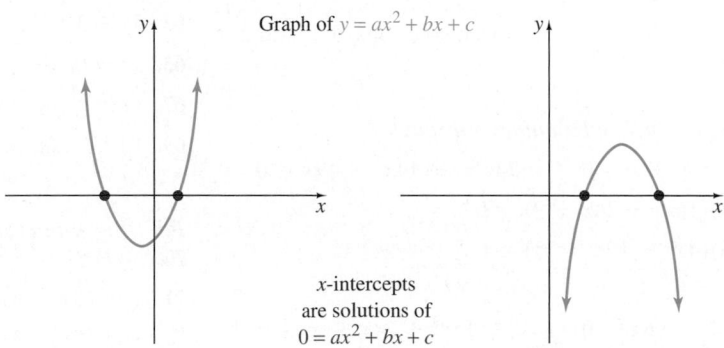

Graph of $y = ax^2 + bx + c$

x-intercepts
are solutions of
$0 = ax^2 + bx + c$

Notice that the x-intercepts of the graph of $y = ax^2 + bx + c$ are the real number solutions of $0 = ax^2 + bx + c$. Also, the real number solutions of $0 = ax^2 + bx + c$ are the x-intercepts of the graph of $y = ax^2 + bx + c$. We study more about graphs of quadratic equations in two variables in Chapter 10.

M E N T A L   M A T H

Solve each equation by inspection.

1. $(a - 3)(a - 7) = 0$

2. $(a - 5)(a - 2) = 0$

3. $(x + 8)(x + 6) = 0$

4. $(x + 2)(x + 3) = 0$

5. $(x + 1)(x - 3) = 0$

6. $(x - 1)(x + 2) = 0$

Exercise Set 6.6

Solve each equation. See Example 1.

1. $(x - 2)(x + 1) = 0$

2. $(x + 3)(x + 2) = 0$

3. $x(x + 6) = 0$

4. $2x(x - 7) = 0$

5. $(2x + 3)(4x - 5) = 0$

6. $(3x - 2)(5x + 1) = 0$

7. $(2x - 7)(7x + 2) = 0$

8. $(9x + 1)(4x - 3) = 0$

9. Write a quadratic equation that has two solutions, 6 and −1. Leave the polynomial in the equation in factored form.

10. Write a quadratic equation that has two solutions, 0 and −2. Leave the polynomial in the equation in factored form.

Solve each equation. See Examples 2 through 4.

11. $x^2 - 13x + 36 = 0$

12. $x^2 + 2x - 63 = 0$

13. $x^2 + 2x - 8 = 0$

14. $x^2 - 5x + 6 = 0$

15. $x^2 - 4x = 32$

16. $x^2 - 5x = 24$

17. $x(3x - 1) = 14$

18. $x(4x - 11) = 3$

19. $3x^2 + 19x - 72 = 0$

20. $36x^2 + x - 21 = 0$

21. Write a quadratic equation in standard form that has two solutions, 5 and 7.

22. Write an equation that has three solutions, 0, 1, and 2.

Solve each equation. See Examples 5 through 7.

23. $x^3 - 12x^2 + 32x = 0$

24. $x^3 - 14x^2 + 49x = 0$

25. $(4x - 3)(16x^2 - 24x + 9) = 0$

26. $(2x + 5)(4x^2 - 10x + 25) = 0$

27. $4x^3 - x = 0$

28. $4y^3 - 36y = 0$

29. $32x^3 - 4x^2 - 6x = 0$

30. $15x^3 + 24x^2 - 63x = 0$

Solve each equation. Be careful. Some of the equations are quadratic and higher degree and some are linear.

31. $x(x + 7) = 0$

32. $y(6 - y) = 0$

33. $(x + 5)(x - 4) = 0$

34. $(x - 8)(x - 1) = 0$

35. $x^2 - x = 30$

36. $x^2 + 13x = -36$

37. $6y^2 - 22y - 40 = 0$

38. $3x^2 - 6x - 9 = 0$

39. $(2x + 3)(2x^2 - 5x - 3) = 0$

40. $(2x - 9)(x^2 + 5x - 36) = 0$

41. $x^2 - 15 = -2x$

42. $x^2 - 26 = -11x$

43. $x^2 - 16x = 0$

44. $x^2 + 5x = 0$

45. $-18y^2 - 33y + 216 = 0$

46. $-20y^2 + 145y - 35 = 0$

47. $12x^2 - 59x + 55 = 0$

48. $30x^2 - 97x + 60 = 0$

49. $18x^2 + 9x - 2 = 0$

50. $28x^2 - 27x - 10 = 0$

51. $x(6x + 7) = 5$

52. $4x(8x + 9) = 5$

53. $4(x - 7) = 6$

54. $5(3 - 4x) = 9$

55. $5x^2 - 6x - 8 = 0$

56. $9x^2 + 6x + 2 = 0$

57. $(y - 2)(y + 3) = 6$

58. $(y - 5)(y - 2) = 28$

59. $4y^2 - 1 = 0$

60. $4y^2 - 81 = 0$

61. $t^2 + 13t + 22 = 0$

62. $x^2 - 9x + 18 = 0$

63. $5t - 3 = 12$

64. $9 - t = -1$

65. $x^2 + 6x - 17 = -26$

66. $x^2 - 8x - 4 = -20$

67. $12x^2 + 7x - 12 = 0$

68. $30x^2 - 11x - 30 = 0$

69. $10t^3 - 25t - 15t^2 = 0$

70. $36t^3 - 48t - 12t^2 = 0$

Find the x-intercepts of the graph of each equation. See Example 8.

71. $y = (3x + 4)(x - 1)$

72. $y = (5x - 3)(x - 4)$

73. $y = x^2 - 3x - 10$

74. $y = x^2 + 7x + 6$

75. $y = 2x^2 + 11x - 6$

76. $y = 4x^2 + 11x + 6$

For Exercises 77 through 82, match each equation with its graph.

A
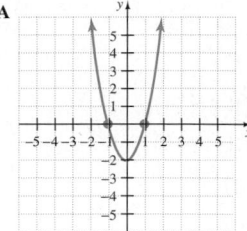

B
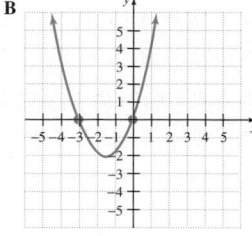

C
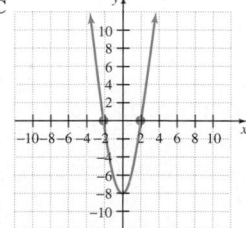

D
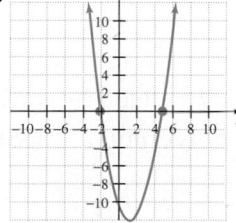

E
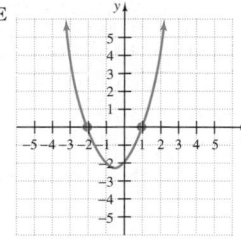

F
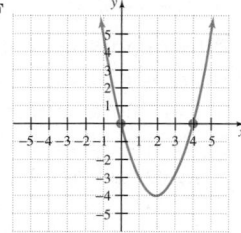

77. $y = (x + 2)(x - 1)$ **78.** $y = (x - 5)(x + 2)$

79. $y = x(x + 3)$ **80.** $y = x(x - 4)$

81. $y = 2x^2 - 8$ **82.** $y = 2x^2 - 2$

 83. A compass is accidentally thrown upward and out of an air balloon at a height of 300 feet. The height y of the compass at time x in seconds is given by the equation
$$y = -16x^2 + 20x + 300.$$

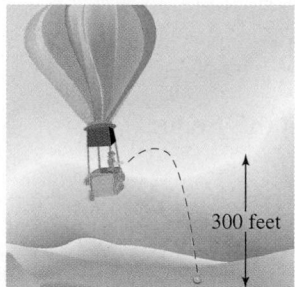

300 feet

a. Find the height of the compass at the given times by filling in the table below.

time x	0	1	2	3	4	5	6
height y							

b. Use the table to determine when the compass strikes the ground.

c. Use the table to approximate the maximum height of the compass.

d. Plot the points (x, y) on a rectangular coordinate system and connect them with a smooth curve. Explain your results.

 84. A rocket is fired upward from the ground with an initial velocity of 100 feet per second. The height y of the rocket at any time x is given by the equation
$$y = -16x^2 + 100x.$$

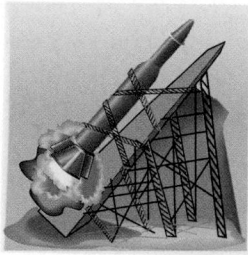

a. Find the height of the rocket at the given times by filling in the table below.

time x	0	1	2	3	4	5	6	7
height y								

b. Use the table to approximate when the rocket strikes the ground to the nearest second.

c. Use the table to approximate the maximum height of the rocket.

d. Plot the points (x, y) on a rectangular coordinate system and connect them with a smooth curve. Explain your results.

REVIEW EXERCISES

Perform the following operations. Write all results in lowest terms. See Section 1.3.

85. $\dfrac{3}{5} + \dfrac{4}{9}$ **86.** $\dfrac{2}{3} + \dfrac{3}{7}$

87. $\dfrac{7}{10} - \dfrac{5}{12}$ **88.** $\dfrac{5}{9} - \dfrac{5}{12}$

89. $\dfrac{7}{8} \div \dfrac{7}{15}$ **90.** $\dfrac{5}{12} - \dfrac{3}{10}$

91. $\dfrac{4}{5} \cdot \dfrac{7}{8}$ **92.** $\dfrac{3}{7} \cdot \dfrac{12}{17}$

A Look Ahead

Example

Solve $(x - 6)(2x - 3) = (x + 2)(x + 9)$.

Solution:
$$(x - 6)(2x - 3) = (x + 2)(x + 9)$$
$$2x^2 - 15x + 18 = x^2 + 11x + 18$$
$$x^2 - 26x = 0$$
$$x(x - 26) = 0$$
$$x = 0 \quad \text{or} \quad x - 26 = 0$$
$$x = 26$$

Solve each equation. See the example.

93. $(x - 3)(3x + 4) = (x + 2)(x - 6)$

94. $(2x - 3)(x + 6) = (x - 9)(x + 2)$

95. $(2x - 3)(x + 8) = (x - 6)(x + 4)$

96. $(x + 6)(x - 6) = (2x - 9)(x + 4)$

97. $(4x - 1)(x - 8) = (x + 2)(x + 4)$

98. $(5x - 2)(x + 3) = (2x - 3)(x + 2)$

6.7 QUADRATIC EQUATIONS AND PROBLEM SOLVING

CD-ROM   SSM

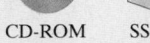
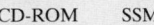

SSG   Video

▶ **OBJECTIVES**

1. Solve problems that can be modeled by quadratic equations.
2. Find the x-intercepts of a polynomial function.

---

1 Some problems may be modeled by quadratic equations. To solve these problems, we use the same problem-solving steps that were introduced in Section 2.4. When solving these problems, keep in mind that a solution of an equation that models a problem may not be a solution to the problem. For example, a person's age or the length of a rectangle is always a positive number. Discard solutions that do not make sense as solutions of the problem.

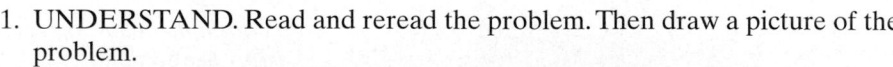

Example 1   **FINDING THE LENGTH OF TIME**

For a TV commercial, a piece of luggage is dropped from a cliff 256 feet above the ground to show the durability of the luggage. Neglecting air resistance, the height $h(t)$ in feet of the luggage above the ground after t seconds is given by the quadratic equation

$$h(t) = -16t^2 + 256$$

Find how long it takes for the luggage to hit the ground.

Solution

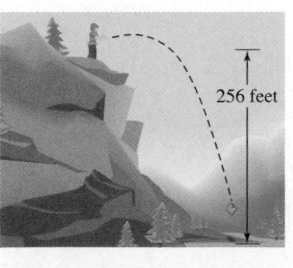

256 feet

1. UNDERSTAND. Read and reread the problem. Then draw a picture of the problem.

The function $h(t) = -16t^2 + 256$ models the height of the falling luggage at time t. Familiarize yourself with this function by finding a few function values.

When $t = 1$ second, the height of the suitcase is
$$h(1) = -16(1)^2 + 256 = 240 \text{ feet.}$$

When $t = 2$ seconds, the height of the suitcase is
$$h(2) = -16(2)^2 + 256 = 192 \text{ feet.}$$

2. TRANSLATE. To find how long it takes the luggage to hit the ground, we want to know the value of t for which the height $h(t) = 0$.
$$0 = -16t^2 + 256$$

3. SOLVE. We solve the quadratic equation by factoring.

$$0 = -16t^2 + 256$$
$$0 = -16(t^2 - 16)$$
$$0 = -16(t - 4)(t + 4)$$
$$t - 4 = 0 \quad \text{or} \quad t + 4 = 0$$
$$t = 4 \qquad\qquad t = -4$$

4. INTERPRET. Since the time t cannot be negative, the proposed solution is 4 seconds.

    Check: Verify that the height of the luggage when t is 4 seconds, $h(4)$, is 0.
    $$h(4) = -16(4)^2 + 256 = -256 + 256 = 0 \text{ feet.}$$

    State: The solution checks and the luggage hits the ground 4 seconds after it is dropped.

Example 2   **FINDING THE LENGTH AND WIDTH OF A DEN**

In May 1995, 24 inches of rain fell on Slidell, Louisiana, in just two days, causing approximately one-third of the houses in that community to flood. When a home floods, all flooring in the home must be removed and replaced. The Callacs' home flooded and the soiled rug in their den was removed and now needs to be replaced. The length of their den is 4 feet more than the width. If the area of the floor is 117 square feet, find its length and width.

Solution

1. UNDERSTAND. Read and reread the problem. Propose and check a solution. Let

    $$x = \text{ the width of the floor; then}$$

    $$x + 4 = \text{ the length of the floor since it is 4 feet longer.}$$

2. TRANSLATE. Here, we use the formula for the area of a rectangle.

    In words:

width	·	length	=	area

    Translate: x · $(x + 4)$ = 117

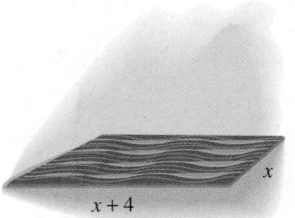

3. SOLVE. $$x(x + 4) = 117$$

    $$x^2 + 4x = 117 \qquad \text{Multiply.}$$
    $$x^2 + 4x - 117 = 0 \qquad \text{Write in standard form.}$$
    $$(x + 13)(x - 9) = 0 \qquad \text{Factor.}$$

    Next, set each factor equal to 0.

    $$x + 13 = 0 \qquad \text{or} \qquad x - 9 = 0$$
    $$x = -13 \qquad \text{or} \qquad x = 9 \qquad \text{Solve.}$$

4. INTERPRET. The solutions are -13 and 9. Since x represents the width of the room, the solution -13 must be discarded. The proposed width is 9 feet and the proposed length is $x + 4$ or $9 + 4$ or 13 feet.

    Check: The area of a 9-foot by 13-foot room is (9 feet)(13 feet) = 117 square feet. The proposed solution checks.

    State: The floor is 9 feet by 13 feet.

△**Example 3** **FINDING THE BASE AND HEIGHT OF A SAIL**

The height of a triangular sail is 2 meters less than twice the length of the base. If the sail has an area of 30 square meters, find the length of its base and the height.

Solution

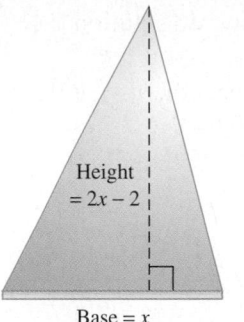

Height
= $2x - 2$

Base = x

1. UNDERSTAND. Read and reread the problem. Since we are finding the length of the base and the height, we let

$$x = \text{the length of the base}$$

and since the height is 2 meters less than twice the base,

$$2x - 2 = \text{the height}$$

An illustration is shown to the left.

2. TRANSLATE. We are given that the area of the triangle is 30 square meters, so we use the formula for area of a triangle.

area of triangle	=	$\frac{1}{2}$	·	base	·	height
↓		↓		↓		↓
30	=	$\frac{1}{2}$	·	x	·	$(2x - 2)$

3. SOLVE. Now we solve the quadratic equation.

$$30 = \frac{1}{2}x(2x - 2)$$

$$30 = x^2 - x \qquad \text{Multiply.}$$

$$x^2 - x - 30 = 0 \qquad \text{Write in standard form.}$$

$$(x - 6)(x + 5) = 0 \qquad \text{Factor.}$$

$$x - 6 = 0 \quad \text{or} \quad x + 5 = 0 \qquad \text{Set each factor equal to 0.}$$

$$x = 6 \qquad\qquad x = -5$$

4. INTERPRET. Since x represents the length of the base, we discard the solution −5. The base of a triangle cannot be negative. The base is then 6 feet and the height is $2(6) - 2 = 10$ feet.

Check: To check this problem, we recall that $\frac{1}{2}$ base · height = area, or

$$\frac{1}{2}(6)(10) = 30 \qquad \text{the required area}$$

State: The base of the triangular sail is 6 meters and the height is 10 meters.

The next example makes use of the **Pythagorean theorem** and consecutive integers. Before we review this theorem, recall that a **right triangle** is a triangle that contains a 90° or right angle. The **hypotenuse** of a right triangle is the side opposite the right angle and is the longest side of the triangle. The **legs** of a right triangle are the other sides of the triangle.

PYTHAGOREAN THEOREM

In a right triangle, the sum of the squares of the lengths of the two legs is equal to the square of the length of the hypotenuse.

$$(\text{leg})^2 + (\text{leg})^2 = (\text{hypotenuse})^2 \qquad \text{or} \qquad a^2 + b^2 = c^2$$

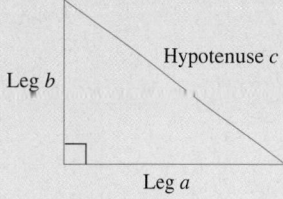

Hypotenuse c

Leg b

Leg a

> **HELPFUL HINT**
> If you use this formula, don't forget that c represents the length of the hypotenuse.

Study the following diagrams for a review of consecutive integers.

Consecutive integers: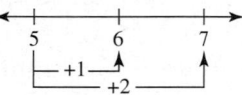

If x is the first integer: $x, x + 1, x + 2$

Consecutive even integers: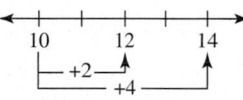

If x is the first even integer: $x, x + 2, x + 4$

Consecutive odd integers: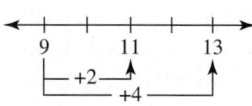

If x is the first odd integer: $x, x + 2, x + 4$

△ **Example 4**   **FINDING THE DIMENSIONS OF A TRIANGLE**

Find the lengths of the sides of a right triangle if the lengths can be expressed as three consecutive even integers.

Solution   1. UNDERSTAND. Read and reread the problem. Let's suppose that the length of one leg of the right triangle is 4 units. Then the other leg is the next even integer, or 6 units, and the hypotenuse of the triangle is the next even integer, or 8 units. Remember that the hypotenuse is the longest side. Let's see if a triangle with sides of these lengths forms a right triangle. To do this, we check to see whether the Pythagorean theorem holds true.

$$4^2 + 6^2 \overset{?}{=} 8^2$$
$$16 + 36 \overset{?}{=} 64$$
$$52 = 64 \qquad \text{False.}$$

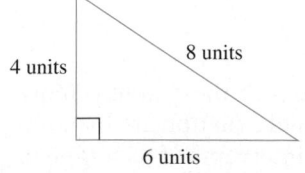

Our proposed numbers do not check, but we now have a better understanding of the problem.

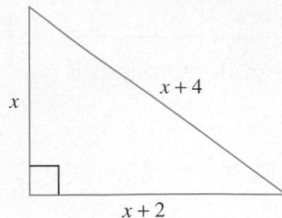

We let x, $x + 2$, and $x + 4$ be three consecutive even integers. Since these integers represent lengths of the sides of a right triangle, we have

$$x = \text{one leg}$$
$$x + 2 = \text{other leg}$$
$$x + 4 = \text{hypotenuse (longest side)}$$

2. TRANSLATE. By the Pythagorean theorem, we have that

$$(\text{hypotenuse})^2 = (\text{leg})^2 + (\text{leg})^2$$
$$(x + 4)^2 = (x)^2 + (x + 2)^2$$

3. SOLVE. Now we solve the equation.

$$(x + 4)^2 = x^2 + (x + 2)^2$$

$x^2 + 8x + 16 = x^2 + x^2 + 4x + 4$	Multiply.
$x^2 + 8x + 16 = 2x^2 + 4x + 4$	Combine like terms.
$x^2 - 4x - 12 = 0$	Write in standard form.
$(x - 6)(x + 2) = 0$	Factor.
$x - 6 = 0 \qquad \text{or} \qquad x + 2 = 0$	Set each factor equal to 0.
$x = 6 \qquad\qquad\qquad x = -2$	

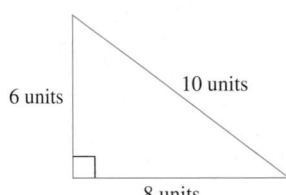

4. INTERPRET. We discard $x = -2$ since length cannot be negative. If $x = 6$, then $x + 2 = 8$ and $x + 4 = 10$.

Check: Verify that $(\text{hypotenuse})^2 = (\text{leg})^2 + (\text{leg})^2$, or $10^2 = 6^2 + 8^2$, or $100 = 36 + 64$.

State: The sides of the right triangle have lengths 6 units, 8 units, and 10 units. ▬

2

Recall that to find the x-intercepts of the graph of a function, let $f(x) = 0$, or $y = 0$, and solve for x. This fact gives us a visual interpretation of the results of this section.

From Section 6.6, we know that the solutions of the equation $(x + 2)(x - 6) = 0$ are -2 and 6. These solutions give us important information about the related polynomial function $p(x) = (x + 2)(x - 6)$. We know that when x is -2 or when x is 6, the value of $p(x)$ is 0.

$$p(x) = (x + 2)(x - 6)$$
$$p(-2) = (-2 + 2)(-2 - 6) = (0)(-8) = 0$$
$$p(6) = (6 + 2)(6 - 6) = (8)(0) = 0$$

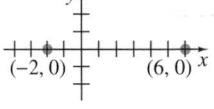

Thus, we know that $(-2, 0)$ and $(6, 0)$ are the x-intercepts of the graph of $p(x)$.

We also know that the graph of $p(x)$ does not cross the x-axis at any other point. For this reason, and the fact that $p(x) = (x + 2)(x - 6) = x^2 - 4x - 12$ has degree 2, we conclude that the graph of p must look something like one of these two graphs:

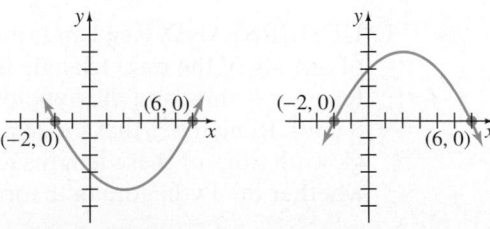

In the following section and in a later chapter, we explore these graphs more fully. For the moment, know that the solutions of a polynomial equation are the x-intercepts of the graph of the related function and that the x-intercepts of the graph of a polynomial function are the solutions of the related polynomial equation. These values are also called **roots**, or **zeros**, of a polynomial function.

Example 5   Match each function with its graph.

$$f(x) = (x - 3)(x + 2) \qquad g(x) = x(x + 2)(x - 2) \qquad h(x) = (x - 2)(x + 2)(x - 1)$$

A 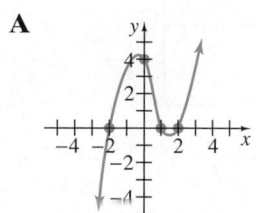   **B** 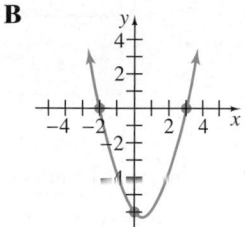   **C**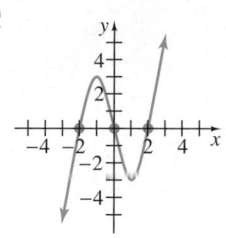

Solution   The graph of the function $f(x) = (x - 3)(x + 2)$ has two x-intercepts, $(3, 0)$ and $(-2, 0)$, because the equation $0 = (x - 3)(x + 2)$ has two solutions, 3 and -2.

The graph of $f(x)$ is graph B.

The graph of the function $g(x) = x(x + 2)(x - 2)$ has three x-intercepts $(0, 0)$, $(-2, 0)$, and $(2, 0)$, because the equation $0 = x(x + 2)(x - 2)$ has three solutions, $0, -2,$ and 2.

The graph of $g(x)$ is graph C.

The graph of the function $h(x) = (x - 2)(x + 2)(x - 1)$ has three x-intercepts, $(-2, 0)$, $(1, 0)$, and $(2, 0)$, because the equation $0 = (x - 2)(x + 2)(x - 1)$ has three solutions, $-2, 1,$ and 2.

The graph of $h(x)$ is graph A.

GRAPHING CALCULATOR EXPLORATIONS

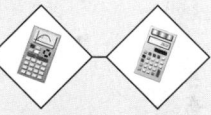

We can use a graphing calculator to approximate real number solutions of any quadratic equation in standard form, whether the associated polynomial is factorable or not. For example, let's solve the quadratic equation $x^2 - 2x - 4 = 0$. The solutions of this equation will be the x-intercepts of the graph of the function $f(x) = x^2 - 2x - 4$. (Recall that to find x-intercepts, we let $f(x) = 0$, or $y = 0$.) When we use a standard window, the graph of this function looks like this.

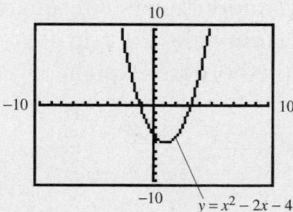

$y = x^2 - 2x - 4$

The graph appears to have one x-intercept between -2 and -1 and one between 3 and 4. To find the x-intercept between 3 and 4 to the nearest hundredth, we can use a zero feature, a Zoom feature, which magnifies a portion of the graph around the cursor, or we can redefine our window. If we redefine our window to

$$\text{Xmin} = 2 \qquad \qquad \text{Ymin} = -1$$
$$\text{Xmax} = 5 \qquad \qquad \text{Ymax} = 1$$
$$\text{Xscl} = 1 \qquad \qquad \text{Yscl} = 1 \qquad \qquad \textit{(continued)}$$

the resulting screen is

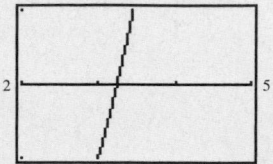

By using the Trace feature, we can now see that one of the intercepts is between 3.21 and 3.25. To approximate to the nearest hundredth, Zoom again or redefine the window to

$$\text{Xmin} = 3.2 \qquad \text{Ymin} = -0.1$$
$$\text{Xmax} = 3.3 \qquad \text{Ymax} = 0.1$$
$$\text{Xscl} = 1 \qquad \text{Yscl} = 1$$

If we use the Trace feature again, we see that, to the nearest hundredth, the x-intercept is 3.24. By repeating this process, we can approximate the other x-intercept to be -1.24.

To check, find $f(3.24)$ and $f(-1.24)$. Both of these values should be close to 0. (They will not be exactly 0 since we approximated these solutions.)

$$f(3.24) = 0.0176 \quad \text{and} \quad f(-1.24) = 0.0176$$

Solve each of these quadratic equations by graphing a related function and approximating the x-intercepts to the nearest thousandth.

1. $x^2 + 3x - 2 = 0$ **2.** $5x^2 - 7x + 1 = 0$

3. $2.3x^2 - 4.4x - 5.6 = 0$ **4.** $0.2x^2 + 6.2x + 2.1 = 0$

5. $0.09x^2 - 0.13x - 0.08 = 0$ **6.** $x^2 + 0.08x - 0.01 = 0$

△ SPOTLIGHT ON DECISION MAKING

Suppose you are a landscaper. You are landscaping a public park and have just put in a flower bed measuring 8 feet by 12 feet. You would also like to surround the bed with a decorative floral border consisting of low-growing, spreading plants. Each plant will cover approximately 1 square foot when mature, and you have 224 plants to use. How wide of a strip of ground should you prepare around the flower bed for the border? Explain.

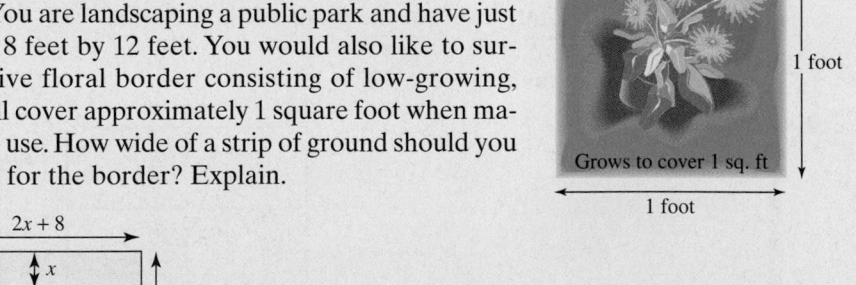

Grows to cover 1 sq. ft

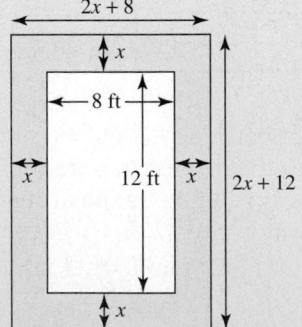

Exercise Set 6.7

Represent each given condition using a single variable, x. See Examples 1 through 4.

△ **1.** The length and width of a rectangle whose length is 4 centimeters more than its width

△ **2.** The length and width of a rectangle whose length is twice its width

3. Two consecutive odd integers

4. Two consecutive even integers

△ **5.** The base and height of a triangle whose height is one more than four times its base

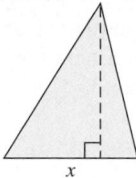

△ **6.** The base and height of a trapezoid whose base is three less than five times its height

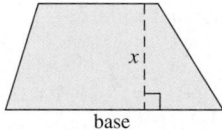

Use the information given to find the dimensions of each figure.

△ **7.** The *area* of the square is 121 square units. Find the length of its sides.

△ **8.** The *area* of the rectangle is 84 square inches. Find its length and width.

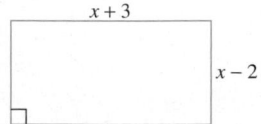

△ **9.** The *perimeter* of the quadrilateral is 120 centimeters. Find the lengths of the sides.

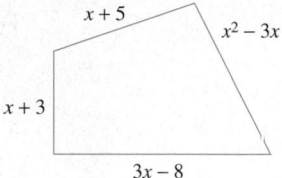

△ **10.** The *perimeter* of the triangle is 85 feet. Find the lengths of its sides.

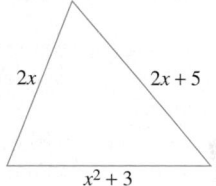

△ **11.** The *area* of the parallelogram is 96 square miles. Find its base and height.

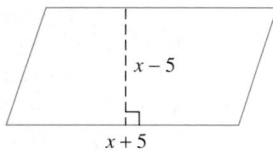

△ **12.** The *area* of the circle is 25π square kilometers. Find its radius.

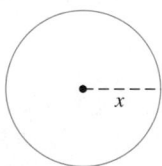

Solve. See Examples 1 through 4.

13. An object is thrown upward from the top of an 80-foot building with an initial velocity of 64 feet per second. The height $h(t)$ of the object after t seconds is given by the quadratic equation $h(t) = -16t^2 + 64t + 80$. When will the object hit the ground?

14. A hang glider pilot accidentally drops her compass from the top of a 400-foot cliff. The height $h(t)$ of the compass after t seconds is given by the quadratic equation $h(t) = -16t^2 + 400$. When will the compass hit the ground?

△ **15.** The length of a rectangle is 7 centimeters less than twice its width. Its area is 30 square centimeters. Find the dimensions of the rectangle.

△ **16.** The length of a rectangle is 9 inches more than its width. Its area is 112 square inches. Find the dimensions of the rectangle.

The equation $D = \frac{1}{2}n(n-3)$ gives the number of diagonals D for a polygon with n sides. For example, a polygon with 6 sides has $D = \frac{1}{2} \cdot 6(6-3)$ or $D = 9$ diagonals. (See if you can count all 9 diagonals. Some are shown in the figure.) Use this equation, $D = \frac{1}{2}n(n-3)$, for Exercises 17–20.

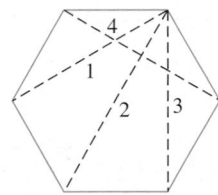

△ **17.** Find the number of diagonals for a polygon that has 12 sides.

△ **18.** Find the number of diagonals for a polygon that has 15 sides.

△ **19.** Find the number of sides n for a polygon that has 35 diagonals.

△ **20.** Find the number of sides n for a polygon that has 14 diagonals.

Solve.

21. The sum of a number and its square is 132. Find the number.

22. The sum of a number and its square is 182. Find the number.

△ **23.** Two boats travel at a right angle to each other after leaving the same dock at the same time. One hour later the boats are 17 miles apart. If one boat travels 7 miles

per hour faster than the other boat, find the rate of each boat.

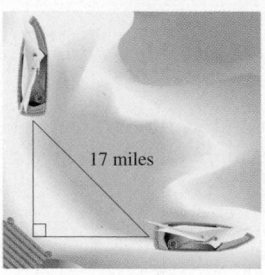

△ **24.** The side of a square equals the width of a rectangle. The length of the rectangle is 6 meters longer than its width. The sum of the areas of the square and the rectangle is 176 square meters. Find the side of the square.

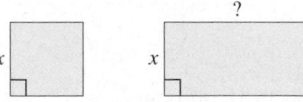

25. The sum of two numbers is 20, and the sum of their squares is 218. Find the numbers.

26. The sum of two numbers is 25, and the sum of their squares is 325. Find the numbers.

△ **27.** If the sides of a square are increased by 3 inches, the area becomes 64 square inches. Find the length of the sides of the original square.

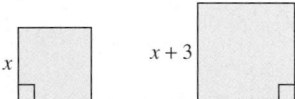

△ **28.** If the sides of a square are increased by 5 meters, the area becomes 100 square meters. Find the length of the sides of the original square.

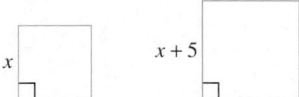

△ **29.** One leg of a right triangle is 4 millimeters longer than the smaller leg and the hypotenuse is 8 millimeters longer than the smaller leg. Find the lengths of the sides of the triangle.

△ **30.** One leg of a right triangle is 9 centimeters longer than the other leg and the hypotenuse is 45 centimeters. Find the lengths of the legs of the triangle.

△ **31.** The length of the base of a triangle is twice its height. If the area of the triangle is 100 square kilometers, find the height.

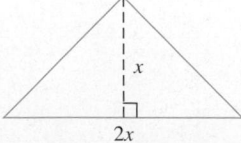

△ **32.** The height of a triangle is 2 millimeters less than the base. If the area is 60 square millimeters, find the base.

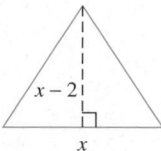

△ **33.** Find the length of the shorter leg of a right triangle if the longer leg is 12 feet more than the shorter leg and the hypotenuse is 12 feet less than twice the shorter leg

△ **34.** Find the length of the shorter leg of a right triangle if the longer leg is 10 miles more than the shorter leg and the hypotenuse is 10 miles less than twice the shorter leg.

35. An object is dropped from the top of the 625-foot-tall Waldorf-Astoria Hotel on Park Avenue in New York City. (*Source: World Almanac* research) The height $h(t)$ of the object after t seconds is given by the equation $h(t) = -16t^2 + 625$. Find how many seconds pass before the object reaches the ground.

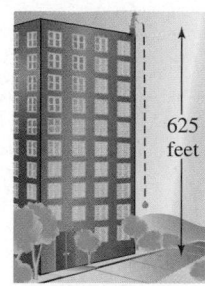

36. A 6-foot-tall person drops an object from the top of the Westin Peachtree Plaza in Atlanta, Georgia. The Westin building is 723 feet tall. (*Source: World Almanac* research) The height $h(t)$ of the object after t seconds is given by the equation $h(t) = -16t^2 + 729$. Find how many seconds pass before the object reaches the ground.

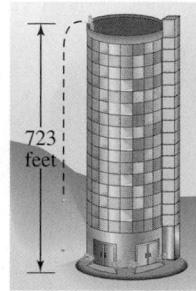

37. While hovering near the top of Ribbon Falls in Yosemite National Park at 1600 feet, a helicopter pilot accidentally drops his sunglasses. The height $h(t)$ of the sunglasses after t seconds is given by the polynomial function
$$h(t) = -16t^2 + 1600$$

When will the sunglasses hit the ground?

38. After t seconds, the height $h(t)$ of a model rocket launched from the ground into the air is given by the function
$$h(t) = -16t^2 + 80t$$
Find how long it takes the rocket to reach a height of 96 feet.

△ **39.** The floor of a shed has an area of 91 square feet. The floor is in the shape of a rectangle whose length is 6 feet more than the width. Find the length and the width of the floor of the shed.

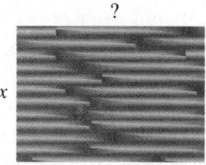

△ **40** A vegetable garden with an area of 143 square feet is to be fertilized. If the width of the garden is 2 feet less than the length, find the dimensions of the garden.

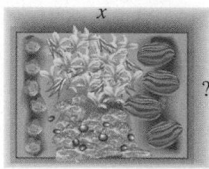

41. The function $W(x) = 0.5x^2$ gives the number of servings of wedding cake that can be obtained from a two-layer x-inch square wedding cake tier. What size square wedding cake tier is needed to serve 50 people? (*Source: Based on data from the Wilton 2000 Yearbook of Cake Decorating*)

42. Use the function in Exercise 41 to determine what size wedding cake tier is needed to serve 200 people.

43. At the end of 2 years, P dollars invested at an interest rate r compounded annually increases to an amount, A dollars, given by
$$A = P(1 + r)^2$$
Find the interest rate if $100 increased to $144 in 2 years.

44. At the end of 2 years, P dollars invested at an interest rate r compounded annually increases to an amount, A dollars, given by

$$A = P(1 + r)^2$$

Find the interest rate if $2000 increased to $2420 in 2 years.

△ **45.** Find the dimensions of a rectangle whose width is 7 miles less than its length and whose area is 120 square miles.

△ **46.** Find the dimensions of a rectangle whose width is 2 inches less than half its length and whose area is 160 square inches.

47. If the cost, C, for manufacturing x units of a certain product is given by $C = x^2 - 15x + 50$, find the number of units manufactured at a cost of $9500.

48. If a switchboard handles n telephones, the number C of telephone connections it can make simultaneously is given by the equation $C = \dfrac{n(n-1)}{2}$. Find how many telephones are handled by a switchboard making 120 telephone connections simultaneously.

△ **49.** According to the International America's Cup Class (IACC) rule, a sailboat competing in the America's Cup match must have a 110-foot-tall mast and a combined mainsail and jib sail area of 3000 square feet. *(Source: America's Cup Organizing Committee)* A design for an IACC-class sailboat calls for the mainsail to be 60% of the combined sail area. If the height of the triangular mainsail is 28 feet more than twice the length of the boom, find the length of the boom and the height of the mainsail.

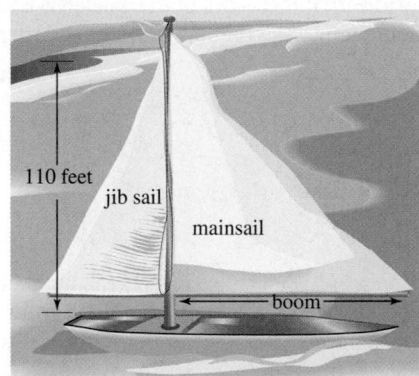

Match each polynomial function with its graph (A–F). See Example 5.

50. $f(x) = (x - 2)(x + 5)$

51. $g(x) = (x + 1)(x - 6)$

52. $h(x) = x(x + 3)(x - 3)$

53. $F(x) = (x + 1)(x - 2)(x + 5)$

54. $G(x) = 2x^2 + 9x + 4$

55. $H(x) = 2x^2 - 7x - 4$

A
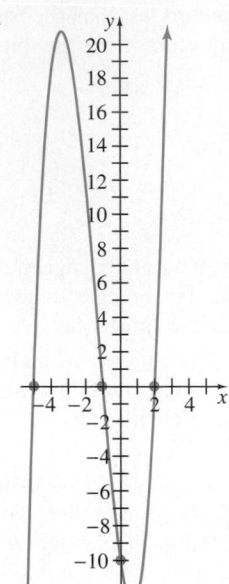

B
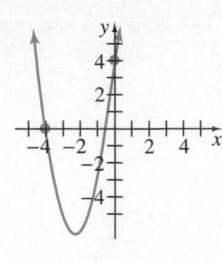

C
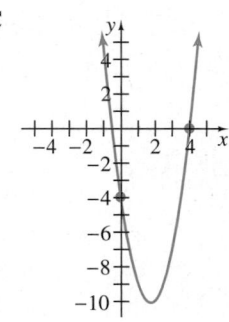

D
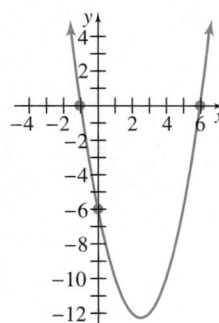

E
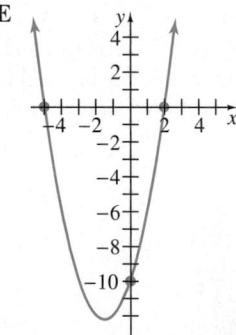

F
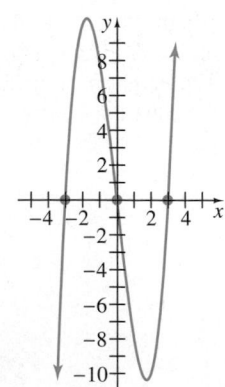

Write a quadratic function that has the given numbers as roots or zeros. (Hint: See page 384).

56. 5, 3

57. 6, 7

58. −1, 2

59. 4, −3

REVIEW EXERCISES

Write the x- and y-intercepts for each graph and determine whether the graph is the graph of a function. See Sections 3.3 and 3.4.

60.

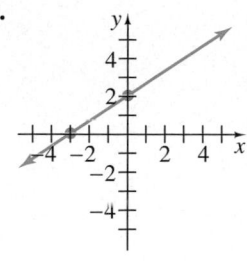

61.

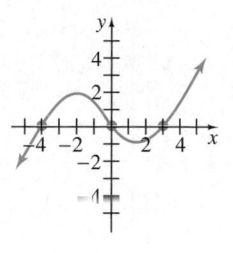

62.

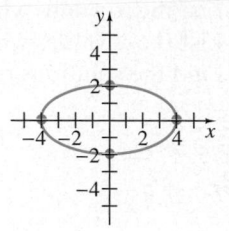

63.

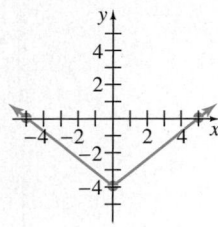

64. Draw a function with intercepts $(-3, 0), (5, 0),$ and $(0, 4).$

65. Draw a function with intercepts $(-7, 0), \left(-\dfrac{1}{2}, 0\right), (4, 0),$ and $(0, -1).$

6.8 AN INTRODUCTION TO GRAPHING POLYNOMIAL FUNCTIONS

CD-ROM   SSM

SSG   Video

▶ **OBJECTIVES**

1. Analyze the graph of a polynomial function.
2. Graph quadratic functions.
3. Find the vertex of a parabola by using the vertex formula.
4. Graph cubic functions.

1 We discussed linear functions of the form $f(x) = mx + b$ in Chapter 3. In Chapter 5, we briefly discussed polynomial functions. In this section, we further discuss polynomial functions. As mentioned earlier, some polynomial functions are given special names according to their degree. For example,

$f(x) = 2x - 6$ is called a **linear function**; its **degree is one**.

$f(x) = 5x^2 - x + 3$ is called a **quadratic function**; its **degree is two**.

$f(x) = 7x^3 + 3x^2 - 1$ is called a **cubic function**; its **degree is three**.

$f(x) = -8x^4 - 3x^3 + 2x^2 + 20$ is called a **quartic function**; its **degree is four**.

All the above functions are also polynomial functions.

Before we practice graphing polynomial functions, let's analyze the graph of a polynomial function.

Example 1 Given the graph of the function $g(x),$

a. Find the domain and the range of the function.
b. List the x- and y-intercepts
c. Find the coordinates of the point with the greatest y-value.
d. Find the coordinates of the point with the least y-value.

  e. List the x-values whose y-values are equal to 0.
  f. List the x-values whose y-values are greater than 0.
  g. Find the solutions of $g(x) = 0$.

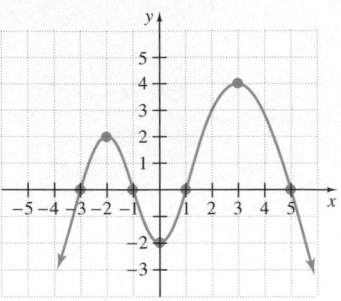

Solution   **a.** The domain is the set of all real numbers, or in interval notation, $(-\infty, \infty)$. The range is $(-\infty, 4]$.

  b. The x-intercepts are $(-3, 0)$, $(-1, 0)$, $(1, 0)$, and $(5, 0)$. The y-intercept is $(0, -2)$.

  c. The point with the greatest y-value corresponds to the "highest" point. This is the point with coordinates $(3, 4)$. (This means that for all real number values for x, the greatest y-value, or $f(x)$ value, is 4.)

  d. The point with the least y-value corresponds to the "lowest" point. This graph contains no "lowest" point, so there is no point with the least y-value.

  e. The y-values are equal to 0 when the graph lies on the x-axis. The x-values when this occurs are the x-intercepts $(-3, 0)$, $(-1, 0)$, $(1, 0)$, and $(5, 0)$. Notice that this tells us that $g(-3) = 0$, $g(-1) = 0$, $g(1) = 0$, and $g(5) = 0$.

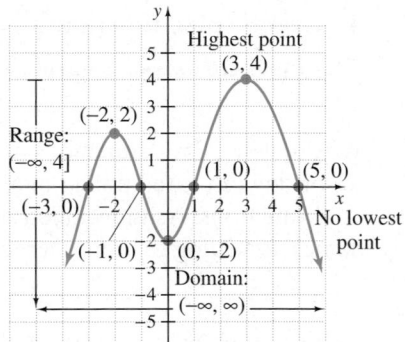

  f. The y-values are greater than 0 when the graph lies above the x-axis. The x-values when this occurs are between $x = -3$ and $x = -1$ and between $x = 1$ and $x = 5$.

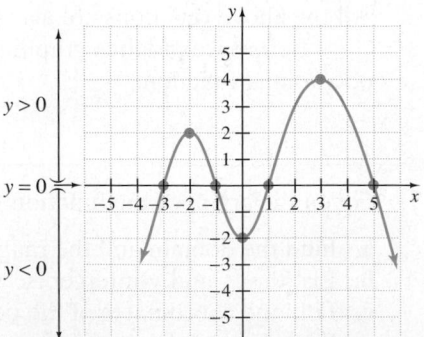

g. The solutions of $g(x) = 0$ are the x-intercepts of the graph. The x-intercepts are $(-3, 0)$, $(-1, 0)$, $(1, 0)$, and $(5, 0)$. This means that when $x = -3, -1, 1,$ or $5, y$ or $g(x) = 0$. The solutions are $-3, -1, 1,$ and 5.   ■

The graph of any polynomial function (linear, quadratic, cubic, and so on) can be sketched by plotting a sufficient number of ordered pairs that satisfy the function and connecting them to form a smooth curve. The graph of all polynomial functions will pass the vertical line test since they are graphs of functions. To graph a linear function defined by $f(x) = mx + b$, recall that two ordered pair solutions will suffice since its graph is a line. To graph other polynomial functions, we need to find and plot more ordered pair solutions to ensure a reasonable picture of its graph.

2   Since we know how to graph linear functions (see Chapter 3), we will now graph quadratic functions and discuss special characteristics of their graphs.

QUADRATIC FUNCTION

A quadratic function is a function that can be written in the form

$$f(x) = ax^2 + bx + c$$

where a, b, and c are real numbers and $a \neq 0$.

We know that an equation of the form $f(x) = ax^2 + bx + c$ may be written as $y = ax^2 + bx + c$. Thus, both $f(x) = ax^2 + bx + c$ and $y = ax^2 + bx + c$ define quadratic functions as long as a is not 0.

Recall the graph of the quadratic function defined by $f(x) = x^2$ by plotting points. Choose $-3, -2, -1, 0, 1, 2,$ and 3 as x-values, and find corresponding $f(x)$ or y-values.

x	$y = f(x)$
-3	9
-2	4
-1	1
0	0
1	1
2	4
3	9

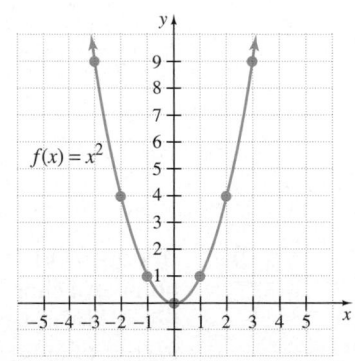

Notice that the graph passes the vertical line test, as it should since it is a function. Recall that this curve is called a **parabola.** The highest point on a parabola that opens downward or the lowest point on a parabola that opens upward is called the **vertex** of the parabola. The vertex of this parabola is $(0, 0)$, the lowest point on the graph. If we fold the graph along the y-axis, we can see that the two sides of the graph coincide. This means that this curve is symmetric about the y-axis, and the y-axis, or the line $x = 0$, is called the **axis of symmetry.** The graph of every quadratic function is a parabola and has an axis of symmetry: the vertical line that passes through the vertex of the parabola.

Example 2   Graph the quadratic function $f(x) = -x^2 + 2x - 3$ by plotting points.

Solution   To graph, choose values for x and find corresponding $f(x)$ or y-values.

x	$y = f(x)$
-2	-11
-1	-6
0	-3
1	-2
2	-3
3	-6

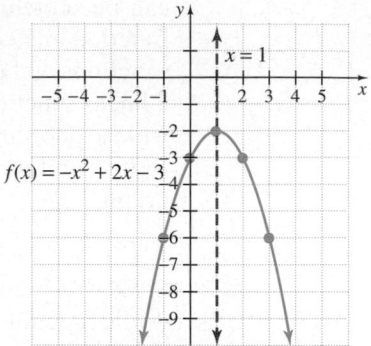

The vertex of this parabola is $(1, -2)$, the highest point on the graph. The vertical line $x = 1$ is the axis of symmetry. Recall that to find the x-intercepts of a graph, let $f(x)$ or $y = 0$. Since this graph has no x-intercepts, it means that $0 = -x^2 + 2x - 3$ has no real number solutions.

Notice that the parabola $f(x) = -x^2 + 2x - 3$ opens downward, whereas $f(x) = x^2$ opens upward. When the equation of a quadratic function is written in the form $f(x) = ax^2 + bx + c$, recall that the coefficient of the squared variable a determines whether the parabola opens downward or upward. If $a > 0$, the parabola opens upward, and if $a < 0$, the parabola opens downward.

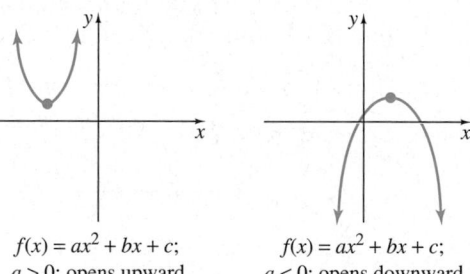

$f(x) = ax^2 + bx + c$;
$a > 0$; opens upward

$f(x) = ax^2 + bx + c$;
$a < 0$; opens downward

3   In both $f(x) = x^2$ and $f(x) = -x^2 + 2x - 3$, the vertex happens to be one of the points we chose to plot. Since this is not always the case, and since plotting the vertex allows us to draw the graph quickly, we need a consistent method for finding the vertex. One method is to use the following formula, which we shall derive in Chapter 10.

VERTEX FORMULA

The graph of $f(x) = ax^2 + bx + c, a \neq 0$, is a parabola with vertex

$$\left(\frac{-b}{2a}, f\left(\frac{-b}{2a}\right)\right)$$

We can also find the x- and y-intercepts of a parabola to aid in graphing. Recall that x-intercepts of the graph of any equation may be found by letting $y = 0$ in the equation and solving for x. Also, y-intercepts may be found by letting $x = 0$ in the equation and solving for y or $f(x)$.

Example 3   Graph $f(x) = x^2 + 2x - 3$. Find the vertex and any intercepts.

Solution   To find the vertex, use the vertex formula. For the function $f(x) = x^2 + 2x - 3$, $a = 1$ and $b = 2$. Thus,

$$x = \frac{-b}{2a} = \frac{-2}{2(1)} = -1$$

Next find $f(-1)$.

$$f(-1) = (-1)^2 + 2(-1) - 3$$
$$= 1 - 2 - 3$$
$$= -4$$

The vertex is $(-1, -4)$, and since $a = 1$ is greater than 0, this parabola opens upward. This parabola will have two x-intercepts because its vertex lies below the x-axis and it opens upward. To find the x-intercepts, let y or $f(x) = 0$ and solve for x.

$$f(x) = x^2 + 2x - 3$$
$$0 = x^2 + 2x - 3 \qquad \text{Let } f(x) = 0.$$
$$0 = (x + 3)(x - 1) \qquad \text{Factor.}$$
$$x + 3 = 0 \quad \text{or} \quad x - 1 = 0 \qquad \text{Set each factor equal to 0.}$$
$$x = -3 \quad \text{or} \qquad x = 1 \qquad \text{Solve.}$$

The x-intercepts are $(-3, 0)$ and $(1, 0)$.
   To find the y-intercept, let $x = 0$.

$$f(x) = x^2 + 2x - 3$$
$$f(0) = 0^2 + 2(0) - 3$$
$$f(0) = -3$$

The y-intercept is $(0, -3)$
   Now plot these points and connect them with a smooth curve.

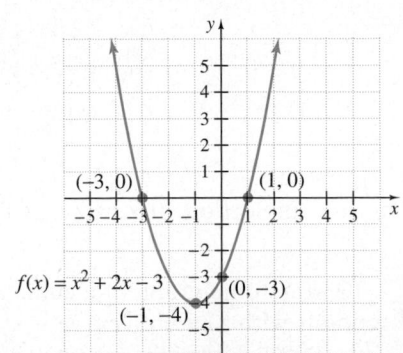

▼
HELPFUL HINT
Not all graphs of parabolas have x-intercepts. To see this, first plot the vertex of the parabola and decide whether the parabola opens upward or downward. Then use this information to decide whether the graph of the parabola has x-intercepts.

Example 4   Graph $f(x) = 3x^2 - 12x + 13$. Find the vertex and any intercepts.

Solution   To find the vertex, use the vertex formula. For the function $y = 3x^2 - 12x + 13$, $a = 3$ and $b = -12$. Thus,

$$x = \frac{-b}{2a} = \frac{-(-12)}{2(3)} = \frac{12}{6} = 2$$

Next find $f(2)$.

$$f(2) = 3(2)^2 - 12(2) + 13$$
$$= 3(4) - 24 + 13$$
$$= 1$$

The vertex is $(2, 1)$. Also, this parabola opens upward, since $a = 3$, which is greater than 0. Notice that this parabola has no x-intercepts: Its vertex lies above the x-axis, and it opens upward.
   To find the y-intercept, let $x = 0$.

$$f(0) = 3(0)^2 - 12(0) + 13$$
$$= 0 - 0 + 13$$
$$= 13$$

The y-intercept is $(0, 13)$. Use this information along with symmetry of a parabola to sketch the graph of $f(x) = 3x^2 - 12x + 13$.

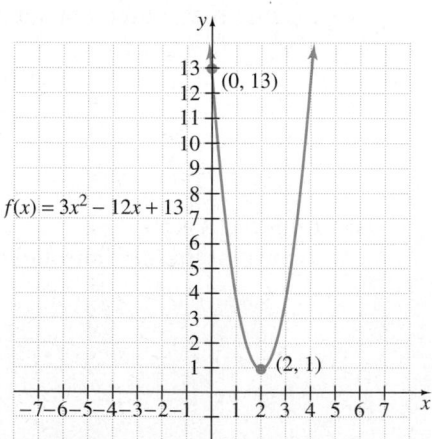

In Section 10.6, we study the graphing of quadratic functions further.

4   To sketch the graph of a cubic function, we again plot points and then connect the points with a smooth curve.
   When graphing cubic functions, keep in mind their general shape from Section 5.3.

Graph of a Polynomial Function of Degree 3

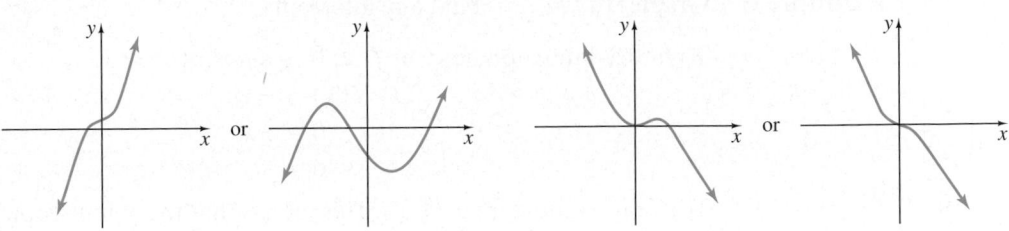

Coefficient of x^3
is a positive number.

Coefficient of x^3
is a negative number.

Example 5   Graph $f(x) = x^3 - 4x$. Find any intercepts.

Solution   To find x-intercepts, let y or $f(x) = 0$ and solve for x.

$$f(x) = x^3 - 4x$$
$$0 = x^3 - 4x \qquad \text{Let } f(x) = 0.$$
$$0 = x(x^2 - 4)$$
$$0 = x(x + 2)(x - 2) \qquad \text{Factor.}$$

$x = 0$ or $x + 2 = 0$ or $x - 2 = 0$ Set each factor equal to 0.
$x = 0$ or $x = -2$ or $x = 2$ Solve.

This graph has three x-intercepts. They are $(0, 0)$, $(-2, 0)$ and $(2, 0)$.
To find the y-intercept, let $x = 0$.

$$f(0) = 0^3 - 4(0) = 0$$

The y-intercept is $(0, 0)$. Next select some x-values and find their corresponding $f(x)$ or y-values.

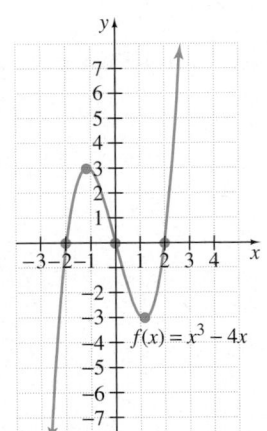

$f(x) = x^3 - 4x$

$f(-3) = (-3)^3 - 4(-3) = -27 + 12 = -15$

$f(-1) = (-1)^3 - 4(-1) = -1 + 4 = 3$

$f(1) = 1^3 - 4(1) = 1 - 4 = -3$

$f(3) = 3^3 - 4(3) = 27 - 12 = 15$

x	$f(x)$
-3	-15
-1	3
1	-3
3	15

Plot the intercepts and points and connect them with a smooth curve.

HELPFUL HINT
When a graph has an x-intercept of $(0, 0)$, notice that the y-intercept will also be $(0, 0)$.

HELPFUL HINT
If unsure about the graph of a function, plot more points.

Example 6   Graph $f(x) = -x^3$. Find any intercepts.

Solution   To find x-intercepts, let y or $f(x) = 0$ and solve for x.

$$f(x) = -x^3$$
$$0 = -x^3$$
$$0 = x$$

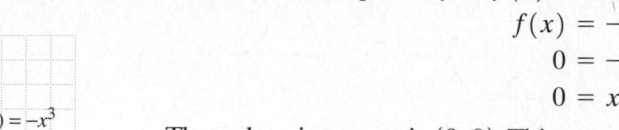

The only x-intercept is $(0, 0)$. This means that the y-intercept is $(0, 0)$ also. Next choose some x-values and find corresponding y-values.

$$f(x) = -x^3$$
$$f(-2) = -(-2)^3 = 8$$
$$f(-1) = -(-1)^3 = 1$$
$$f(1) = -(1)^3 = -1$$
$$f(2) = -2^3 = -8$$

x	$f(x)$
-2	8
-1	1
1	-1
2	-8

Plot the points and sketch the graph of $f(x) = -x^3$.

MENTAL MATH

State whether the graph of each quadratic function, a parabola, opens upward or downward.

1. $f(x) = 2x^2 + 7x + 10$ **2.** $f(x) = -3x^2 - 5x$

3. $f(x) = -x^2 + 5$ **4.** $f(x) = x^2 + 3x + 7$

Exercise Set 6.8

For the graph of each function $f(x)$, answer the following. See Example 1.

  a. Find the domain and the range of the function.
  b. List the x- and y-intercepts.
  c. Find the coordinates of the point with the greatest y-value.
  d. Find the coordinates of the point with the least y-value.
  e. List the x-values whose y-values are equal to 0.
  f. List the x-values whose y-values are greater than 0.
  g. Find the solutions of $f(x) = 0$.

1.

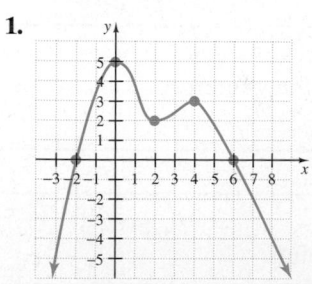

2.

3. The graph in Example 3 of this section.
4. The graph in Example 4 of this section.
5. The graph in Example 5 of this section.
6. The graph in Example 6 of this section.

Graph each quadratic function by plotting points. See Example 2.

7. $f(x) = 2x^2$ **8.** $f(x) = -3x^2$

9. $f(x) = x^2 + 1$ **10.** $f(x) = x^2 - 2$

11. $f(x) = -x^2$ **12.** $f(x) = \dfrac{1}{2}x^2$

Find the vertex of the graph of each function. See Examples 3 and 4.

13. $f(x) = x^2 + 8x + 7$

14. $f(x) = x^2 + 6x + 5$

15. $f(x) = 3x^2 + 6x + 4$

16. $f(x) = -2x^2 + 2x + 1$

17. $f(x) = -x^2 + 10x + 5$

18. $f(x) = -x^2 - 8x + 2$

19. If the vertex of a parabola lies below the x-axis and the parabola opens upward, how many x-intercepts will the graph have?

20. If the vertex of a parabola lies below the x-axis and the parabola opens downward, how many x-intercepts will the graph have?

21. If the vertex of a parabola lies above the x-axis and the parabola opens upward, how many x-intercepts will the graph have?

22. If the vertex of a parabola lies above the x-axis and the parabola opens downward, how many x-intercepts will the graph have?

23. If the vertex of a parabola is the origin, how many x-intercepts and how many y-intercepts will the graph have?

Graph each quadratic function. Find and label the vertex and intercepts. See Examples 3 and 4.

24. $f(x) = x^2 + 8x + 7$ ◆ **25.** $f(x) = x^2 + 6x + 5$

26. $f(x) = x^2 - 2x - 24$ **27.** $f(x) = x^2 - 12x + 35$

28. $f(x) = 2x^2 - 6x$ **29.** $f(x) = -3x^2 + 6x$

Graph each cubic function. Find any intercepts. See Examples 5 and 6.

30. $f(x) = 4x^3 - 9x$ ◆ **31.** $f(x) = 2x^3 - 5x^2 - 3x$

32. $f(x) = x^3 + 3x^2 - x - 3$

33. $f(x) = x^3 + x^2 - 4x - 4$

34. Can the graph of a function ever have more than one y-intercept point? Why?

35. In general, is there a limit to the number of x-intercepts for the graph of a function?

Graph each function. Find intercepts. If the function is a quadratic function, find the vertex.

36. $f(x) = x^2 + 4x - 5$ **37.** $f(x) = x^2 + 2x - 3$

38. $f(x) = (x - 2)(x + 2)(x + 1)$

39. $f(x) = x^3 - 4x^2 + 3x$

40. $f(x) = x^2 + 1$

41. $f(x) = x^2 + 4$

42. $f(x) = -5x^2 + 5x$

43. $f(x) = 3x^2 - 12x$

44. $f(x) = x^3 - 9x$

45. $f(x) = x^3 + x^2 - 12x$

46. $f(x) = -x^3 - x^2 + 2x$

47. $f(x) = x^3 + x^2 - 9x - 9$

48. $f(x) = x^2 - 4x + 4$

49. $f(x) = x^2 - 2x + 1$

50. $f(x) = -x^3 + x$

51. $f(x) = x^2 + 6x$

52. $f(x) = 2x^2 - x - 3$

53. $f(x) = (x + 2)(x - 2)$

54. $f(x) = -x^3 + 3x^2 + x - 3$

55. $f(x) = -x^3 + 25x$

56. $f(x) = x^2 - 10x + 26$

57. $f(x) = x^2 + 2x + 4$

58. $f(x) = x(x - 4)(x + 2)$

59. $f(x) = 3x(x - 3)(x + 5)$

60. $g(x) = x(x - 2)(x + 3)(x + 5)$

61. $h(x) = (x - 4)(x - 2)(2x + 1)(x + 3)$

Use a graphing calculator to verify the graph in each exercise.

62. Exercise 36   **63.** Exercise 37

64. Exercise 54   **65.** Exercise 55

Use a graphing calculator to approximate all x-intercepts to the nearest tenth.

66. $F(x) = -x^4 + 2.1x^2 + 5.6$

67. $G(x) = x^4 - 6.2x^2 - 6.2$

REVIEW EXERCISES

Simplify each fraction. See Sections 5.1 and 5.2.

68. $-\dfrac{8}{10}$

69. $-\dfrac{45}{100}$

70. $\dfrac{x^7 y^{10}}{x^3 y^{15}}$

71. $\dfrac{a^{14} b^2}{ab^4}$

72. $\dfrac{7n^{-9}m^{-2}}{14nm^{-5}}$

73. $\dfrac{20x^{-3}y^5}{25y^{-2}x}$

 For additional Chapter Projects, visit the Real World Activities
Website by going to http://www.prenhall.com/martin-gay.

CHAPTER PROJECT

Choosing Among Building Options

Whether putting in a new floor, hanging new wallpaper, or retiling a bathroom, it may be necessary to choose among several different materials with different pricing schemes. If a fixed amount of money is available for projects like these, it can be helpful to compare the choices by calculating how much area can be covered by a fixed dollar-value of material.

In this project, you will have the opportunity to choose among three different choices of materials for building a patio around a swimming pool. This project may be completed by working in groups or individually.

Option	Material	Price
A	Poured cement	$5 per square foot
B	Brick	$7.50 per square foot plus a $30 flat fee for delivering the bricks
C	Outdoor carpeting	$4.50 per square foot plus $10.86 per foot of the pool's perimeter to install an edging

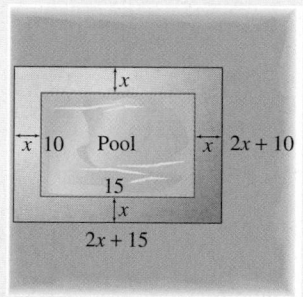

Situation: Suppose you have just had a 10-foot-by-15-foot inground swimming pool installed in your backyard. You have $3000 left from the building project that you would like to spend on surrounding the pool with a patio, equally wide on all sides (see figure). You have talked to several local suppliers about options for building this patio and must choose among the following.

△ 1. Find the area of the swimming pool.

△ 2. Write an algebraic expression for the total area of the region containing both the pool and the patio.

△ 3. Use subtraction to find an algebraic expression for the area of just the patio (not including the area of the pool).

△ 4. Find the perimeter of the swimming pool alone.

5. For each patio material option, write an algebraic expression for the total cost of installing the patio based on its area and the given price information.

6. If you plan to spend the entire $3000 on the patio, how wide would the patio in option A be?

7. If you plan to spend the entire $3000 on the patio, how wide would the patio in option B be?

8. If you plan to spend the entire $3000 on the patio, how wide would the patio in option C be?

9. Which option would you choose? Why? Discuss the pros and cons of each option.

CHAPTER 6 VOCABULARY CHECK

Fill in each blank with one of the words or phrases listed below.

factoring   quadratic equation   perfect square trinomial
greatest common factor

1. An equation that can be written in the form $ax^2 + bx + c = 0$ (with a not 0) is called a

   _____ .

2. _____ is the process of writing an expression as a product.

3. The _____ of a list of terms is the product of all common factors.

4. A trinomial that is the square of some binomial is called a _____ .

CHAPTER 6 HIGHLIGHTS

DEFINITIONS AND CONCEPTS	EXAMPLES

Section 6.1 The Greatest Common Factor and Factoring by Grouping

Factoring is the process of writing an expression as a product.

Factor: $6 = 2 \cdot 3$

$$x^2 + 5x + 6 = (x + 2)(x + 3)$$

To Find the GCF of a List of Integers

Step 1. Write each number as a product of primes.

Step 2. Identify the common prime factors.

Step 3. The product of all common factors is the greatest common factor. If there are no common prime factors, the GCF is 1.

Find the GCF of 12, 36, and 48.

$12 = 2 \cdot 2 \cdot 3$

$36 = 2 \cdot 2 \cdot 3 \cdot 3$

$48 = 2 \cdot 2 \cdot 2 \cdot 2 \cdot 3$

$GCF = 2 \cdot 2 \cdot 3 = 12.$

The GCF of a list of common variables raised to powers is the variable raised to the smallest exponent in the list.

The GCF of z^5, z^3, and z^{10} is z^3.

The GCF of a list of terms is the product of all common factors.

Find the GCF of $8x^2y$, $10x^3y^2$, and $26x^2y^3$.
The GCF of 8, 10, and 26 is 2.
The GCF of x^2, x^3, and x^2 is x^2.
The GCF of y, y^2, and y^3 is y.
The GCF of the terms is $2x^2y$.

To Factor by Grouping

Step 1. Arrange the terms so that the first two terms have a common factor and the last two have a common factor.

Step 2. For each pair of terms, factor out the pair's GCF.

Step 3. If there is now a common binomial factor, factor it out.

Step 4. If there is no common binomial factor, begin again, rearranging the terms differently. If no rearrangement leads to a common binomial factor, the polynomial cannot be factored.

Factor $10ax + 15a - 6xy - 9y$.

Step 1. $10ax + 15a - 6xy - 9y$

Step 2. $5a(2x + 3) - 3y(2x + 3)$

Step 3. $(2x + 3)(5a - 3y)$

Section 6.2 Factoring Trinomials of the Form $x^2 + bx + c$

To Factor a Trinomial of the Form $x^2 + bx + c$

Look for two numbers whose product is c and whose sum is b. The factored form is

$$(x + \text{one number})(x + \text{other number})$$

Factor: $x^2 + 7x + 12$

$3 + 4 = 7 \qquad 3 \cdot 4 = 12$

$(x + 3)(x + 4)$

(continued)

DEFINITIONS AND CONCEPTS	EXAMPLES

Section 6.3 Factoring Trinomials of the Form $ax^2 + bx + c$

Method 1: To factor $ax^2 + bx + c$, try various combinations of factors of ax^2 and c until a middle term of bx is obtained when checking.	Factor: $3x^2 + 14x - 5$ Factors of $3x^2$: $3x, x$ Factors of -5: $-1, 5$ and $1, -5$. $(3x - 1)(x + 5)$ $-1x$ $15x$ $14x$ **correct** middle term
Method 2: Factor $ax^2 + bx + c$ by grouping. **Step 1.** Find two numbers whose product is $a \cdot c$ and whose sum is b. **Step 2.** Rewrite bx, using the factors found in Step 1. **Step 3.** Factor by grouping.	Factor: $3x^2 + 14x - 5$ **Step 1.** Find two numbers whose product is $3 \cdot (-5)$ or -15 and whose sum is 14. They are 15 and -1. **Step 2.** $3x^2 + 14x - 5$ $= 3x^2 + 15x - 1x - 5$ **Step 3.** $= 3x(x + 5) - 1(x + 5)$ $= (x + 5)(3x - 1)$
A **perfect square trinomial** is a trinomial that is the square of some binomial.	Perfect square trinomial = square of binomial $x^2 + 4x + 4 \quad = (x + 2)^2$ $25x^2 - 10x + 1 \quad = (5x - 1)^2$
To Factor Perfect Square Trinomials $a^2 + 2ab + b^2 = (a + b)^2$ $a^2 - 2ab + b^2 = (a - b)^2$	Factor: $x^2 + 6x + 9 = x^2 + 2 \cdot x \cdot 3 + 3^2 = (x + 3)^2$ $4x^2 - 12x + 9 = (2x)^2 - 2 \cdot 2x \cdot 3 + 3^2 = (2x - 3)^2$

Section 6.4 Factoring Binomials

Difference of Squares $a^2 - b^2 = (a + b)(a - b)$ *Sum or Difference of Cubes* $a^3 + b^3 = (a + b)(a^2 - ab + b^2)$ $a^3 - b^3 = (a - b)(a^2 + ab + b^2)$	Factor: $x^2 - 9 = x^2 - 3^2 = (x + 3)(x - 3)$ $y^3 + 8 = y^3 + 2^3 = (y + 2)(y^2 - 2y + 4)$ $125z^3 - 1 = (5z)^3 - 1^3 = (5z - 1)(25z^2 + 5z + 1)$

Section 6.5 Choosing a Factoring Strategy

To Factor a Polynomial
 Step 1. Factor out the GCF.
 Step 2. **a.** If two terms,
      i. $a^2 - b^2 = (a - b)(a + b)$
     ii. $a^3 - b^3 = (a - b)(a^2 + ab + b^2)$
    iii. $a^3 + b^3 = (a + b)(a^2 - ab + b^2)$
	Factor: $2x^4 - 6x^2 - 8$
	 Step 1. $2x^4 - 6x^2 - 8 = 2(x^4 - 3x^2 - 4)$
	
	 (continued)

DEFINITIONS AND CONCEPTS	EXAMPLES

Section 6.5 Choosing a Factoring Strategy

b. If three terms,

    i. $a^2 + 2ab + b^2 = (a + b)^2$

    ii. Methods in Sections 6.2 and 6.3

c. If four or more terms, try factoring by grouping.

Step 3. See if any factors can be factored further.

Step 4. Check by multiplying.

Step 2. b. ii. $= 2(x^2 + 1)(x^2 - 4)$

Step 3. $= 2(x^2 + 1)(x + 2)(x - 2)$

Step 4.   Check by multiplying.

$$2(x^2 + 1)(x + 2)(x - 2) = 2(x^2 + 1)(x^2 - 4)$$
$$= 2(x^4 - 3x^2 - 4)$$
$$= 2x^4 - 6x^2 - 8$$

Section 6.6 Solving Quadratic Equations by Factoring

A **quadratic equation** is an equation that can be written in the form $ax^2 + bx + c = 0$ with a not 0.

The form $ax^2 + bx + c = 0$ is called the **standard form** of a quadratic equation.

Quadratic Equation	Standard Form
$x^2 = 16$	$x^2 - 16 = 0$
$y = -2y^2 + 5$	$2y^2 + y - 5 = 0$

Zero Factor Theorem

If a and b are real numbers and if $ab = 0$, then $a = 0$ or $b = 0$.

If $(x + 3)(x - 1) = 0$, then $x + 3 = 0$ or $x - 1 = 0$

To solve quadratic equations by factoring,

Solve: $3x^2 = 13x - 4$

Step 1. Write the equation in standard form: $ax^2 + bx + c = 0$.

Step 1. $3x^2 - 13x + 4 = 0$

Step 2. Factor the quadratic.

Step 2. $(3x - 1)(x - 4) = 0$

Step 3. Set each factor containing a variable equal to 0.

Step 3. $3x - 1 = 0$ or $x - 4 = 0$

Step 4. Solve the equations.

Step 4. $3x = 1$ or $x = 4$

$$x = \frac{1}{3}$$

Step 5. Check in the original equation.

Step 5. Check both $\frac{1}{3}$ and 4 in the original equation.

Section 6.7 Quadratic Equations and Problem Solving

Problem-Solving Steps

1. UNDERSTAND the problem.

A garden is in the shape of a rectangle whose length is two feet more than its width. If the area of the garden is 35 square feet, find its dimensions.

  1. Read and reread the problem. Guess a solution and check your guess.

     Let x be the width of the rectangular garden. Then $x + 2$ is the length.

(continued)

DEFINITIONS AND CONCEPTS	EXAMPLES

Section 6.7 Quadratic Equations and Problem Solving

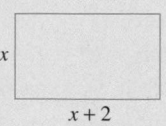

$x + 2$

2. TRANSLATE.

2. In words:

length	·	width	=	area

Translate: $(x + 2)$ · x = 35

3. SOLVE.

3.
$$(x + 2)x = 35$$
$$x^2 + 2x - 35 = 0$$
$$(x - 5)(x + 7) = 0$$
$$x - 5 = 0 \quad \text{or} \quad x + 7 = 0$$
$$x = 5 \quad \text{or} \quad x = -7$$

4. INTERPRET.

4. Discard the solution of -7 since x represents width. *Check:* If x is 5 feet then $x + 2 = 5 + 2 = 7$ feet. The area of a rectangle whose width is 5 feet and whose length is 7 feet is (5 feet)(7 feet) or 35 square feet. *State:* The garden is 5 feet by 7 feet.

Section 6.8 An Introduction to Graphing Polynomial Functions

To Graph a Polynomial Function

Find and plot x- and y-intercepts and a sufficient number of ordered pair solutions. Then connect the plotted points with a smooth curve.

Graph $f(x) = x^3 + 2x^2 - 3x$.
$$0 = x^3 + 2x^2 - 3x$$
$$0 = x(x - 1)(x + 3)$$
$$x = 0 \text{ or } x = 1 \text{ or } x = -3$$

The x-intercepts are $(0, 0)$, $(1, 0)$, and $(-3, 0)$.
$$f(0) = 0^3 + 2 \cdot 0^2 - 3 \cdot 0 = 0.$$
The y-intercept is $(0, 0)$

x	$f(x)$
-4	-20
-2	6
-1	4
$\frac{1}{2}$	$-\frac{7}{8}$
2	10

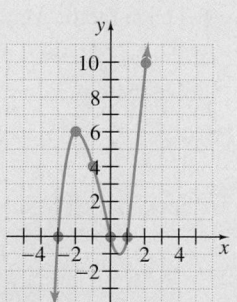

A quadratic function is a function that can be written in the form
$$f(x) = ax^2 + bx + c, a \neq 0$$

The graph of this quadratic function is a parabola with vertex $\left(\dfrac{-b}{2a}, f\left(\dfrac{-b}{2a}\right)\right)$.

Find the vertex of the graph of the quadratic function
$$f(x) = 2x^2 - 8x + 1$$
Here $a = 2$ and $b = -8$.
$$\frac{-b}{2a} = \frac{-(-8)}{2 \cdot 2} = 2$$
$$f(2) = 2 \cdot 2^2 - 8 \cdot (2) + 1 = -7$$
The vertex has coordinates $(2, -7)$.

CHAPTER 6 REVIEW

(6.1) *Complete the factoring.*

1. $6x^2 - 15x = 3x(\quad)$

2. $2x^3y - 6x^2y^2 - 8xy^3 = 2xy(\quad)$

Factor the GCF from each polynomial.

3. $20x^2 + 12x$

4. $6x^2y^2 - 3xy^3$

5. $-8x^4y + 6x^2y^2$

6. $3x(2x + 3) - 5(2x + 3)$

7. $5x(x + 1) - (x + 1)$

Factor.

8. $3x^2 - 3x + 2x - 2$

9. $6x^2 + 10x - 3x - 5$

10. $3a^2 + 9ab + 3b^2 + ab$

(6.2) *Factor each trinomial.*

11. $x^2 + 6x + 8$

12. $x^2 - 11x + 24$

13. $x^2 + x + 2$

14. $x^2 - 5x - 6$

15. $x^2 + 2x - 8$

16. $x^2 + 4xy - 12y^2$

17. $x^2 + 8xy + 15y^2$

18. $3x^2y + 6xy^2 + 3y^3$

19. $72 - 18x - 2x^2$

20. $32 + 12x - 4x^2$

(6.3) *Factor each trinomial.*

21. $2x^2 + 11x - 6$

22. $4x^2 - 7x + 4$

23. $4x^2 + 4x - 3$

24. $6x^2 + 5xy - 4y^2$

25. $6x^2 - 25xy + 4y^2$

26. $18x^2 - 60x + 50$

27. $2x^2 - 23xy - 39y^2$

28. $4x^2 - 28xy + 49y^2$

29. $18x^2 - 9xy - 20y^2$

30. $36x^3y + 24x^2y^2 - 45xy^3$

(6.4) *Factor each binomial.*

31. $4x^2 - 9$

32. $9t^2 - 25s^2$

33. $16x^2 + y^2$

34. $x^3 - 8y^3$

35. $8x^3 + 27$

36. $2x^3 + 8x$

37. $54 - 2x^3y^3$

38. $9x^2 - 4y^2$

39. $16x^4 - 1$

40. $x^4 + 16$

(6.5) *Factor.*

41. $2x^2 + 5x - 12$

42. $3x^2 - 12$

43. $x(x - 1) + 3(x - 1)$

44. $x^2 + xy - 3x - 3y$

45. $4x^2y - 6xy^2$

46. $8x^2 - 15x - x^3$

47. $125x^3 + 27$

48. $24x^2 - 3x - 18$

49. $(x + 7)^2 - y^2$

50. $x^2(x + 3) - 4(x + 3)$

(6.6) *Solve the following equations.*

51. $(x + 6)(x - 2) = 0$

52. $3x(x + 1)(7x - 2) = 0$

53. $4(5x + 1)(x + 3) = 0$

54. $x^2 + 8x + 7 = 0$

55. $x^2 - 2x - 24 = 0$

56. $x^2 + 10x = -25$

57. $x(x - 10) = -16$

58. $(3x - 1)(9x^2 + 3x + 1) = 0$

59. $56x^2 - 5x - 6 = 0$

60. $20x^2 - 7x - 6 = 0$

61. $5(3x + 2) = 4$

62. $6x^2 - 3x + 8 = 0$

63. $12 - 5t = -3$

64. $5x^3 + 20x^2 + 20x = 0$

65. $4t^3 - 5t^2 - 21t = 0$

(6.7) *Solve the following problems.*

△ 66. A flag for a local organization is in the shape of a rectangle whose length is 15 inches less than twice its width. If the area of the flag is 500 square inches, find its dimensions.

△ 67. The base of a triangular sail is four times its height. If the area of the triangle is 162 square yards, find the base.

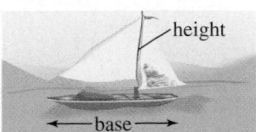

68. Find two consecutive positive integers whose product is 380.

69. A rocket is fired from the ground with an initial velocity of 440 feet per second. Its height $h(t)$ after t seconds is given by the function

$$h(t) = -16t^2 + 440t$$

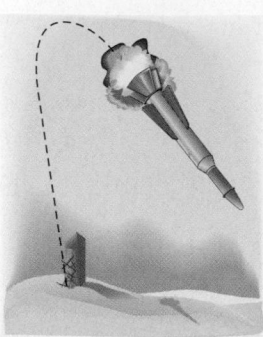

  ✎ **a.** Find how many seconds pass before the rocket reaches a height of 2800 feet. Explain why two answers are obtained.
   b. Find how many seconds pass before the rocket reaches the ground again.

△ **70.** An architect's squaring instrument is in the shape of a right triangle. Find the length of the long leg of the right triangle if the hypotenuse is 8 centimeters longer than the long leg and the short leg is 8 centimeters shorter than the long leg.

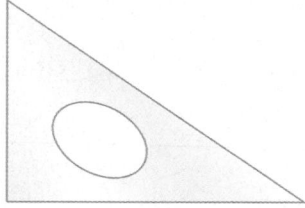

(6.8) *Exercises 71–74 refer to the following graph.*

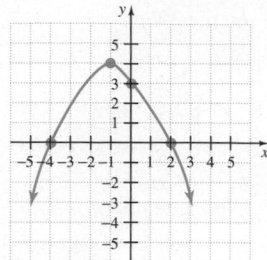

71. Find the domain and the range of the function.

72. List the x- and y-intercepts.

73. Find the coordinates of the point with the greatest y-value.

74. List the x-values for which the y-values are greater than 0.

Graph each polynomial function defined by the equation. Find all intercepts. If the function is a quadratic function, find the vertex.

75. $f(x) = x^2 + 6x + 9$

76. $f(x) = x^2 - 5x + 4$

77. $f(x) = (x - 1)(x^2 - 2x - 3)$

78. $f(x) = (x + 3)(x^2 - 4x + 3)$

79. $f(x) = 2x^2 - 4x + 5$

80. $f(x) = x^2 - 2x + 3$

81. $f(x) = x^3 - 16x$

82. $f(x) = x^3 + 5x^2 + 6x$

CHAPTER 6 TEST

Factor each polynomial completely. If a polynomial cannot be factored, write "prime."

1. $9x^3 + 39x^2 + 12x$
2. $x^2 + x - 10$
3. $x^2 + 4$
4. $y^2 - 8y - 48$
5. $3a^2 + 3ab - 7a - 7b$
6. $3x^2 - 5x + 2$
7. $x^2 + 20x + 90$
8. $x^2 + 14xy + 24y^2$
9. $26x^6 - x^4$
10. $50x^3 + 10x^2 - 35x$
11. $180 - 5x^2$
12. $64x^3 - 1$
13. $6t^2 - t - 5$
14. $xy^2 - 7y^2 - 4x + 28$
15. $x - x^5$
16. $-xy^3 - x^3y$

Solve each equation.

17. $x^2 + 5x = 14$
18. $(x + 3)^2 = 16$

19. $3x(2x - 3)(3x + 4) = 0$

20. $5t^3 - 45t = 0$

21. $3x^2 = -12x$
22. $t^2 - 2t - 15 = 0$

23. $7x^2 = 168 + 35x$
24. $6x^2 = 15x$

Solve each problem.

△ **25.** Find the dimensions of a rectangular garden whose length is 5 feet longer than its width and whose area is 66 square feet.

△ **26.** A deck for a home is in the shape of a triangle. The length of the base of the triangle is 9 feet longer than its altitude. If the area of the triangle is 68 square feet, find the length of the base.

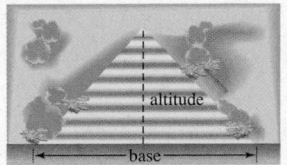

27. The sum of two numbers is 17, and the sum of their squares is 145. Find the numbers.

△ **28.** Write the area of the shaded region as a factored polynomial.

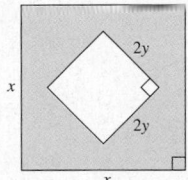

29. A pebble is hurled upward from the top of the Canada Trust Tower, which is 880 feet tall, with an initial velocity of 96 feet per second. Neglecting air resistance, the height $h(t)$ of the pebble after t seconds is given by the polynomial function

$$h(t) = -16t^2 + 96t + 880$$

   a. Find the height of the pebble when $t = 1$.
   b. Find the height of the pebble when $t = 5.1$.
   c. When will the pebble hit the ground?

Graph. Find and label x- and y-intercepts. If the graph is a parabola, find its vertex.

30. $f(x) = x^2 - 4x - 5$ **31.** $f(x) = x^3 - 1$

CHAPTER 6 CUMULATIVE REVIEW

1. Translate each sentence into a mathematical statement.
   a. Nine is less than or equal to eleven.
   b. Eight is greater than one.
   c. Three is not equal to four.

2. Write each fraction in lowest terms.
   a. $\dfrac{42}{49}$ **b.** $\dfrac{11}{27}$
   c. $\dfrac{88}{20}$

3. Simplify: $\dfrac{8 + 2 \cdot 3}{2^2 - 1}$

4. Add.
   a. $3 + (-7) + (-8)$
   b. $\left[7 + (-10)\right] + \left[-2 + (-4)\right]$

5. Find the product.
   a. $(-6)(4)$ **b.** $2(-1)$
   c. $(-5)(-10)$

6. Simplify each expression by combining like terms.
   a. $7x - 3x$ **b.** $10y^2 + y^2$
   c. $8x^2 + 2x - 3x$

Solve.

7. A 10-foot board is to be cut into two pieces so that the longer piece is 4 times the shorter. Find the length of each piece.

8. Graph the linear equation $y = -\dfrac{1}{3}x$.

9. Is the relation $x = y^2$ also a function?

10. Graph $x = 2$.

11. Find the slope of the line $y = 2$.

12. Find the equation of the horizontal line containing the point $(2, 3)$.

13. Graph the union of $x + \dfrac{1}{2}y \geq -4$ or $y \leq -2$.

14. Evaluate for the given value of x.
   a. $2x^3$; x is 5
   b. $\dfrac{9}{x^2}$; x is -3

15. Simplify by writing each expression with positive exponents only.
   a. 3^{-2} **b.** $2x^{-3}$ —
   c. $2^{-1} + 4^{-1}$ **d.** $(-2)^{-4}$

16. Write each number in scientific notation.
   a. 367,000,000
   b. 0.000003
   c. 20,520,000,000
   d. 0.00085

17. Find the degree of the polynomial $3xy + x^2y^2 - 5x^2 - 6$.

18. Find the product: $(3x + 2)(2x - 5)$

19. Multiply: $(3y + 1)^2$

20. Find the GCF of each list of terms.
   a. x^3, x^7, x^5
   b. y, y^4, and y^7

Factor.

21. $x^2 + 7x + 12$

22. $8x^2 - 22x + 5$

23. $25a^2 - 9b^2$

24. Solve: $(x - 3)(x + 1) = 0$

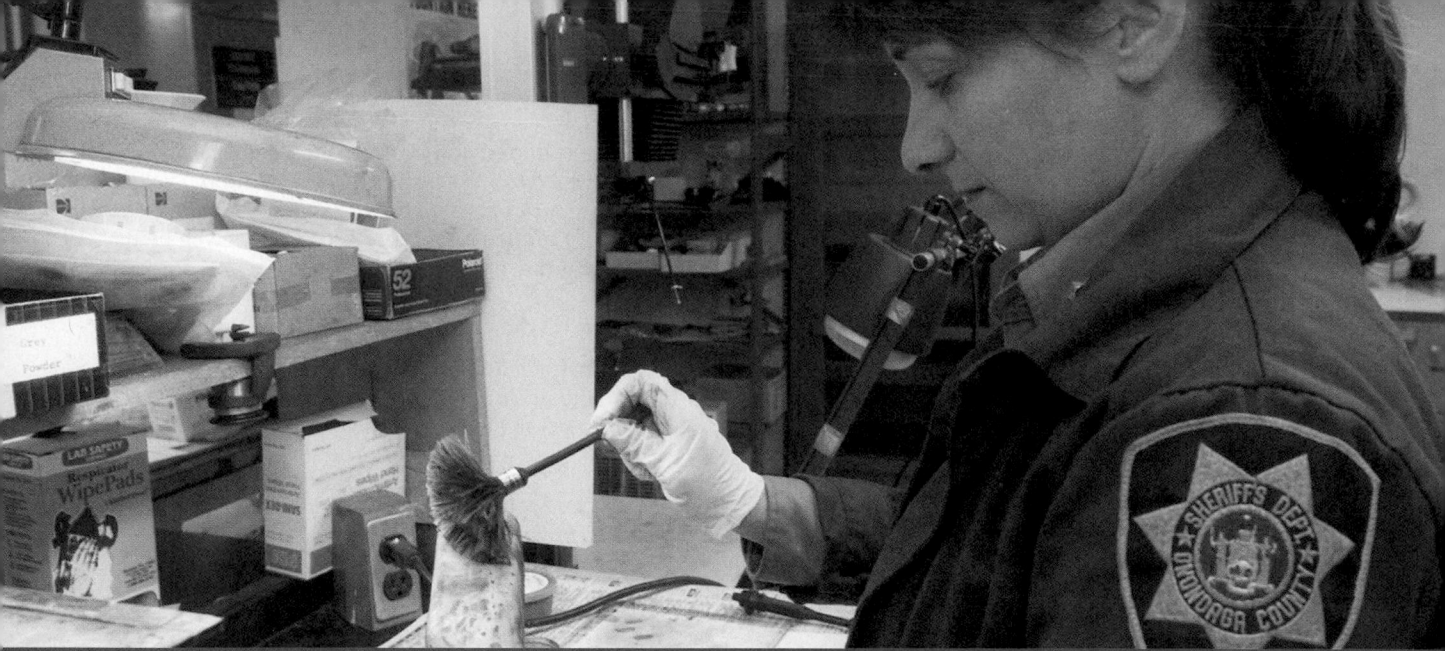

A Widely Known Aspect of Forensic Science

Forensic science is the application of science and technology to the res-
olution of criminal and civil issues. Areas of forensic science include forensic engineering, forensic pathology, forensic dentistry, forensic psychiatry,
physical anthropology, and forensic toxicology. One of the most widely
known aspects of forensic science is criminalistics—the analysis of crime
scenes and the examination of physical evidence. Criminalistic investiga-
tions are often carried out in what is popularly known as a crime
laboratory.

Crime lab workers might be expected to analyze and identify chem-
ical compounds, human hair and tissues, or materials such as metals, glass,
or wood. They might also make measurements on items of physical evidence and compare footprints, firearms, or bullets. Anyone wishing to
work in criminalistics, or the forensic sciences in general, should have a
firm grasp of scientific principles. They also need excellent problem-solving and communication skills. Because mathematics goes hand-in-hand
with the sciences, forensic scientists should also have a good mathematical background.

 *For more information about a career in the forensic sciences, visit
the American Academy of Forensic Sciences Website by first going to
www.prenhall.com/martin-gay.*

*In the Spotlight on Decision Making feature on page 414, you will
have the opportunity, as a forensic lab technician, to decide what material a piece of physical evidence is made of.*

RATIONAL EXPRESSIONS

7

In this chapter, we expand our knowledge of algebraic expressions to include another category called rational expressions, such as $\dfrac{x+1}{x}$. We explore the operations of addition, subtraction, multiplication, and division for these algebraic fractions, using principles similar to the principles for number fractions. Thus, the material in this chapter will make full use of your knowledge of number fractions.

7.1 RATIONAL FUNCTIONS AND SIMPLIFYING RATIONAL EXPRESSIONS

CD-ROM   SSM

SSG   Video

▶ **OBJECTIVES**

1. Define a rational expression and a rational function.
2. Find the domain of a rational function.
3. Simplify rational expressions.

1 Recall that a *rational number*, or *fraction*, is a number that can be written as the quotient $\dfrac{p}{q}$ of two integers p and q as long as q is not 0. A **rational expression** is an expression that can be written as the quotient $\dfrac{P}{Q}$ of two polynomials P and Q as long as Q is not 0.

Examples of Rational Expressions

$$\frac{3x + 7}{2} \qquad \frac{5x^2 - 3}{x - 1} \qquad \frac{7x - 2}{2x^2 + 7x + 6}$$

Rational expressions are sometimes used to describe functions. For example, we call the function $f(x) = \dfrac{x^2 + 2}{x - 3}$ a **rational function** since $\dfrac{x^2 + 2}{x - 3}$ is a rational expression.

Example 1 **COST FOR PRESSING COMPACT DISCS**

For the ICL Production Company, the rational function $C(x) = \dfrac{2.6x + 10,000}{x}$ describes the company's cost per disc of pressing x compact discs. Find the cost per disc for pressing

a. 100 compact discs
b. 1000 compact discs

Solution **a.** $C(100) = \dfrac{2.6(100) + 10,000}{100} = \dfrac{10,260}{100} = 102.6$

The cost per disc for pressing 100 compact discs is \$102.60.

b. $C(1000) = \dfrac{2.6(1000) + 10,000}{1000} = \dfrac{12,600}{1000} = 12.6$

The cost per disc for pressing 1000 compact discs is \$12.60. Notice that as more compact discs are produced, the cost per disc decreases. ■

2 As with fractions, a rational expression is **undefined** if the denominator is 0. If a variable in a rational expression is replaced with a number that makes the denominator 0, we say that the rational expression is **undefined** for this value of the variable. For example, the rational expression $\dfrac{x^2 + 2}{x - 3}$ is undefined when x is 3, because replacing

x with 3 results in a denominator of 0. For this reason, we must exclude 3 from the domain of the function defined by $f(x) = \dfrac{x^2 + 2}{x - 3}$.

The domain of f is then

$$\{x \mid x \text{ is a real number and } x \neq 3\}$$

"The set of all x such that x is a real number and x is not equal to 3."

Unless told otherwise, we assume that the domain of a function described by an equation is the set of all real numbers for which the equation is defined.

Example 2 Find the domain of each rational function.

a. $f(x) = \dfrac{8x^3 + 7x^2 + 20}{2}$ **b.** $g(x) = \dfrac{5x^2 - 3}{x - 1}$ **c.** $f(x) = \dfrac{7x - 2}{x^2 - 2x - 15}$

Solution The domain of each function will contain all real numbers except those values that make the denominator 0.

a. No matter what the value of x, the denominator of $f(x) = \dfrac{8x^3 + 7x^2 + 20}{2}$ is never 0, so the domain of f is $\{x \mid x \text{ is a real number}\}$.

b. To find the values of x that make the denominator of $g(x)$ equal to 0, we solve the equation "denominator $= 0$":

$$x - 1 = 0, \quad \text{or} \quad x = 1$$

The domain of $g(x)$ must exclude 1 since the rational expression is undefined when x is 1. The domain of g is $\{x \mid x \text{ is a real number and } x \neq 1\}$.

c. We find the domain by setting the denominator equal to 0.

$$x^2 - 2x - 15 = 0 \qquad \textit{Set the denominator equal to 0 and solve.}$$
$$(x - 5)(x + 3) = 0$$
$$x - 5 = 0 \quad \text{or} \quad x + 3 = 0$$
$$x = 5 \quad \text{or} \quad x = -3$$

If x is replaced with 5 or with -3, the rational expression is undefined. The domain of f is $\{x \mid x \text{ is a real number and } x \neq 5 \text{ and } x \neq -3\}$. ∎

---

3 Recall that a fraction is in lowest terms or simplest form if the numerator and denominator have no common factors other than 1 (or -1). For example, $\dfrac{3}{13}$ is in lowest terms since 3 and 13 have no common factors other than 1 (or -1).

To **simplify** a rational expression, or to write it in lowest terms, we use the fundamental principle of rational expressions.

FUNDAMENTAL PRINCIPLE OF RATIONAL EXPRESSIONS

For any rational expression $\dfrac{P}{Q}$ and any polynomial R, where $R \neq 0$,

$$\frac{PR}{QR} = \frac{P}{Q}$$

Thus, the fundamental principle says that multiplying or dividing the numerator and denominator of a rational expression by the same nonzero polynomial yields an equivalent rational expression.

To simplify a rational expression such as $\dfrac{(x+2)^2}{x^2-4}$, factor the numerator and the denominator and then use the fundamental principle of rational expressions to divide out common factors.

$$\frac{(x+2)^2}{x^2-4} = \frac{(x+2)(x+2)}{(x+2)(x-2)} = \frac{x+2}{x-2}$$

This means that the rational expression $\dfrac{(x+2)^2}{x^2-4}$ has the same value as the rational expression $\dfrac{x+2}{x-2}$ for all values of x except 2 and −2. (Remember that when x is 2, the denominators of both rational expressions are 0 and that when x is −2, the original rational expression has a denominator of 0.)

As we simplify rational expressions, we will assume that the simplified rational expression is equivalent to the original rational expression for all real numbers except those for which either denominator is 0.

In general, the following steps may be used to simplify rational expressions or to write a rational expression in lowest terms.

SIMPLIFYING OR WRITING A RATIONAL EXPRESSION IN LOWEST TERMS

Step 1: Completely factor the numerator and denominator of the rational expression.

Step 2: Apply the fundamental principle of rational expressions to divide out factors common to both the numerator and denominator.

For now, we assume that variables in a rational expression do not represent values that make the denominator 0.

Example 3 Write each rational expression in lowest terms.

a. $\dfrac{24x^6y^5}{8x^7y}$

b. $\dfrac{2x^2}{10x^3 - 2x^2}$

Solution **a.** The GCF of the numerator and denominator is $8x^6y$.

$$\frac{24x^6y^5}{8x^7y} = \frac{(8x^6y)\,3y^4}{(8x^6y)\,x} = \frac{3y^4}{x}$$

b. Factor out $2x^2$ from the denominator. Then divide numerator and denominator by their GCF, $2x^2$.

$$\frac{2x^2}{10x^3 - 2x^2} = \frac{2x^2}{2x^2\,(5x-1)} = \frac{1}{5x-1}$$

Example 4   Write $\dfrac{x^2 + 8x + 7}{x^2 - 4x - 5}$ in simplest form.

Solution   Factor the numerator and denominator and apply the fundamental principle.

$$\frac{x^2 + 8x + 7}{x^2 - 4x - 5} = \frac{(x + 7)(x + 1)}{(x - 5)(x + 1)} = \frac{x + 7}{x - 5}$$

> ▼ **HELPFUL HINT**
> When simplifying a rational expression, the fundamental principle applies to
> common *factors*, **not common *terms*.**
>
> $$\frac{x \cdot (x + 2)}{x \cdot x} = \frac{x + 2}{x}$$
>
> Common factors. These can be
> divided out.
>
> $$\frac{x + 2}{x}$$
>
> Common terms. Fundamental principle
> does not apply. This is in simplest form.

Example 5   Write each rational expression in lowest terms.

a. $\dfrac{2 + x}{x + 2}$ **b.** $\dfrac{2 - x}{x - 2}$ **c.** $\dfrac{18 - 2x^2}{x^2 - 2x - 3}$

Solution   **a.**   By the commutative property of addition, $2 + x = x + 2$, so

$$\frac{2 + x}{x + 2} = \frac{x + 2}{x + 2} = 1$$

b.   The terms in the numerator of $\dfrac{2 - x}{x - 2}$ differ by sign from the terms of the

denominator, so the polynomials are opposites of each other and the expression
simplifies to -1. To see this, factor out -1 from the numerator or the denominator. If -1 is factored from the numerator, then

$$\frac{2 - x}{x - 2} = \frac{-1(-2 + x)}{x - 2} = \frac{-1(x - 2)}{x - 2} = -1$$

> ▼ **HELPFUL HINT**
> When the numerator and the denominator of a rational expression are oppo-
> sites of each other, the expression simplifies to -1.

---

c. $\dfrac{18 - 2x^2}{x^2 - 2x - 3} = \dfrac{2(9 - x^2)}{(x + 1)(x - 3)}$

$$= \frac{2(3 + x)(3 - x)}{(x + 1)(x - 3)} \qquad \text{Factor.}$$

Notice the opposites $3 - x$ and $x - 3$. We write $3 - x$ as $-1(x - 3)$ and simplify.

$$\frac{2(3 + x)(3 - x)}{(x + 1)(x - 3)} = \frac{2(3 + x) \cdot -1(x - 3)}{(x + 1)(x - 3)} = -\frac{2(3 + x)}{x + 1}$$

HELPFUL HINT

Recall from Section 1.7 that, for a fraction $\frac{a}{b}$,

$$\frac{a}{-b} = \frac{-a}{b} = -\frac{a}{b}$$

For example,

$$\frac{-(x + 1)}{(x + 2)} = \frac{(x + 1)}{-(x + 2)} = -\frac{x + 1}{x + 2}$$

Example 6 Write each rational expression in lowest terms.

a. $\dfrac{x^3 + 8}{2 + x}$

b. $\dfrac{2y^2 + 2}{y^3 - 5y^2 + y - 5}$

Solution a. $\dfrac{x^3 + 8}{2 + x} = \dfrac{(x + 2)(x^2 - 2x + 4)}{x + 2}$ Factor the sum of two cubes.

$= x^2 - 2x + 4$ Divide out common factors.

b. First factor the denominator by grouping.

$$y^3 - 5y^2 + y - 5 = (y^3 - 5y^2) + (y - 5)$$
$$= y^2(y - 5) + (y - 5)$$
$$= (y - 5)(y^2 + 1)$$

Then

$$\frac{2y^2 + 2}{y^3 - 5y^2 + y - 5} = \frac{2(y^2 + 1)}{(y - 5)(y^2 + 1)} = \frac{2}{y - 5}$$

SPOTLIGHT ON DECISION MAKING

Suppose you are a forensic lab technician. You have been asked to try to identify a piece of metal found at a crime scene. You know that one way to analyze the piece of metal is to find its density (mass per unit volume), using the formula *density* $= \dfrac{mass}{volume}$. After weighing the metal, you find its mass as 36.2 grams. You have also determined the volume of the metal to be 4.5 milliliters. Use this information to decide which type of metal this piece is most likely made of, and explain your reasoning. What other characteristics might help the identification?

Densities	
Metal	*Density (g/ml)*
Aluminum	2.7
Iron	7.8
Lead	11.5
Silver	10.5

GRAPHING CALCULATOR EXPLORATIONS

Recall that since the rational expression $\dfrac{7x - 2}{(x - 2)(x + 5)}$ is not defined when $x = 2$ or when $x = -5$, we say that the domain of the rational function $f(x) = \dfrac{7x - 2}{(x - 2)(x + 5)}$ is all real numbers except 2 and −5. This domain can be written as $\{x \mid x$ is a real number and $x \neq 2$, $x \neq -5\}$. This means that the graph of $f(x)$ should not cross the vertical lines $x = 2$ and $x = -5$. The graph of $f(x)$ in *connected* mode follows. In connected mode the graphing calculator tries to connect all dots of the graph so that the result is a smooth curve. This is what has happened in the graph. Notice that the graph appears to contain vertical lines at $x = 2$ and at $x = -5$. We know that this cannot happen because the function is not defined at $x = 2$ and at $x = -5$. We also know that this cannot happen because the graph of this function would not pass the vertical line test.

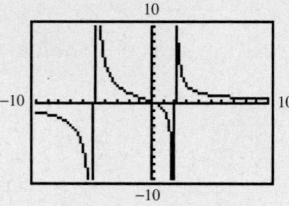

If we graph $f(x)$ in *dot* mode, the graph appears as follows. In dot mode the graphing calculator will not connect dots with a smooth curve. Notice that the vertical lines have disappeared, and we have a better picture of the graph. The graph, however, actually appears more like the hand-drawn graph to its right. By using a Table feature, a Calculate Value feature, or by tracing, we can see that the function is not defined at $x = 2$ and at $x = -5$.

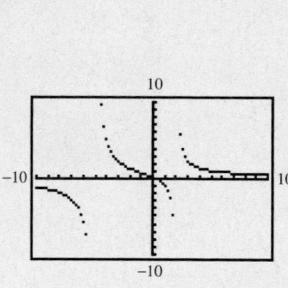
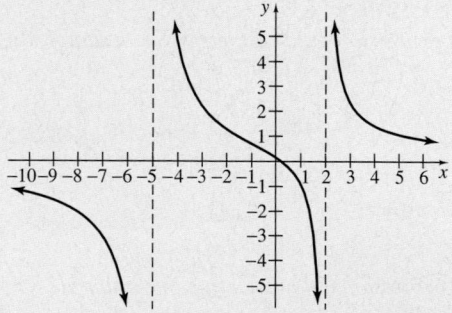

Find the domain of each rational function. Then graph each rational function and use the graph to confirm the domain.

1. $f(x) = \dfrac{x + 1}{x^2 - 4}$

2. $g(x) = \dfrac{5x}{x^2 - 9}$

3. $h(x) = \dfrac{x^2}{2x^2 + 7x - 4}$

4. $f(x) = \dfrac{3x + 2}{4x^2 - 19x - 5}$

Exercise Set 7.1

Find each function value. See Example 1.

1. $f(x) = \dfrac{x + 8}{2x - 1}; f(2), f(0), f(-1)$

2. $f(y) = \dfrac{y - 2}{-5 + y}; f(-5), f(0), f(10)$

3. $g(x) = \dfrac{x^2 + 8}{x^3 - 25x}; g(3), g(-2), g(1)$

4. $s(t) = \dfrac{t^3 + 1}{t^2 + 1}; s(-1), s(1), s(2)$

Find the domain of each rational function. See Example 2.

5. $f(x) = \dfrac{5x - 7}{4}$

6. $g(x) = \dfrac{4 - 3x}{2}$

7. $s(t) = \dfrac{t^2 + 1}{2t}$

8. $v(t) = -\dfrac{5t + t^2}{3t}$

9. $f(x) = \dfrac{3x}{7 - x}$

10. $f(x) = \dfrac{-4x}{-2 + x}$

11. $R(x) = \dfrac{3 + 2x}{x^3 + x^2 - 2x}$

12. $h(x) = \dfrac{5 - 3x}{2x^2 - 14x + 20}$

13. $C(x) = \dfrac{x + 3}{x^2 - 4}$

14. $R(x) = \dfrac{5}{x^2 - 7x}$

15. In your own words, explain how to find the domain of a rational function.

16. In your own words, explain how to simplify a rational expression or to write it in lowest terms.

Write each rational expression in lowest terms. See Examples 3 through 6.

17. $\dfrac{10x^3}{18x}$

18. $\dfrac{48a^7}{16a^{10}}$

19. $\dfrac{9x^6y^3}{18x^2y^5}$

20. $\dfrac{10ab^5}{15a^3b^5}$

Write each rational expression in lowest terms. See Example 5.

21. $\dfrac{x + 5}{5 + x}$

22. $\dfrac{x - 5}{5 - x}$

23. $\dfrac{x - 1}{1 - x^2}$

24. $\dfrac{10 + 5x}{x^2 + 2x}$

25. $\dfrac{4x - 8}{3x - 6}$

26. $\dfrac{12 - 6x}{30 - 15x}$

27. $\dfrac{2x - 14}{7 - x}$

28. $\dfrac{9 - x}{5x - 45}$

29. $\dfrac{x^2 - 2x - 3}{x^2 - 6x + 9}$

30. $\dfrac{x^2 + 10x + 25}{x^2 + 8x + 15}$

31. $\dfrac{2x^2 + 12x + 18}{x^2 - 9}$

32. $\dfrac{x^2 - 4}{2x^2 + 8x + 8}$

33. $\dfrac{3x + 6}{x^2 + 2x}$

34. $\dfrac{3x + 4}{9x^2 + 4}$

35. $\dfrac{2x^2 - x - 3}{2x^3 - 3x^2 + 2x - 3}$

36. $\dfrac{3x^2 - 5x - 2}{6x^3 + 2x^2 + 3x + 1}$

37. $\dfrac{8q^2}{16q^3 - 16q^2}$

38. $\dfrac{3y}{6y^2 - 30y}$

39. $\dfrac{x^2 + 6x - 40}{10 + x}$

40. $\dfrac{x^2 - 8x + 16}{4 - x}$

41. $\dfrac{x^3 - 125}{5 - x}$

42. $\dfrac{4x + 4}{2x^3 + 2}$

43. $\dfrac{8x^3 - 27}{4x - 6}$

44. $\dfrac{9x^2 - 15x + 25}{27x^3 + 125}$

Solve.

45. The dose of medicine prescribed for a child depends on the child's age A in years and the adult dose D for the medication. Young's Rule is a formula used by pediatricians that gives a child's dose C as

$$C = \dfrac{DA}{A + 12}$$

Suppose that an 8-year-old child needs medication, and the normal adult dose is 1000 mg. What size dose should the child receive?

46. During a storm, water treatment engineers monitor how quickly rain is falling. If too much rain comes too fast, there is a danger of sewers backing up. A formula that gives the rainfall intensity i in millimeters per hour for a certain strength storm in eastern Virginia is

$$i = \dfrac{5840}{t + 29}$$

where t is the duration of the storm in minutes. What rainfall intensity should engineers expect for a storm of this strength in eastern Virginia that lasts for 80 minutes? Round your answer to one decimal place.

47. Calculating body-mass index is a way to gauge whether a person should lose weight. Doctors recommend that body-mass index values fall between 19 and 25. The formula for body-mass index B is

$$B = \frac{705w}{h^2}$$

where w is weight in pounds and h is height in inches. Should a 148-pound person who is 5 feet 6 inches tall lose weight?

48. Anthropologists and forensic scientists use a measure called the cephalic index to help classify skulls. The cephalic index of a skull with width W and length L from front to back is given by the formula

$$C = \frac{100W}{L}$$

A long skull has an index value less than 75, a medium skull has an index value between 75 and 85, and a broad skull has an index value over 85. Find the cephalic index of a skull that is 5 inches wide and 6.4 inches long. Classify the skull.

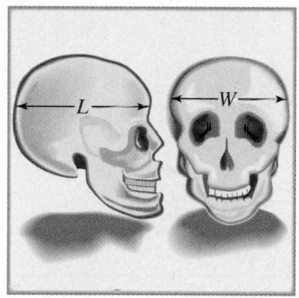

49. The total revenue $R(x)$ from the sale of a popular music compact disc is approximately given by the function

$$R(x) = \frac{150x^2}{x^2 + 3}$$

where x is the number of years since the CD has been released and revenue $R(x)$ is in millions of dollars.
   a. Find the total revenue generated by the end of the first year.
   b. Find the total revenue generated by the end of the second year.
   c. Find the total revenue generated in the second year only.

50. For a certain model fax machine, the manufacturing cost $C(x)$ per machine is given by the function

$$C(x) = \frac{250x + 10,000}{x}$$

where x is the number of fax machines manufactured and cost $C(x)$ is in dollars per machine.
   a. Find the cost per fax machine when manufacturing 100 fax machines.
   b. Find the cost per fax machine when manufacturing 1000 fax machines.
   c. Does the cost per machine decrease or increase when more machines are manufactured? Explain why this is so.

51. Recall that the fundamental principle applies to common *factors* only. Which of the following are *not* true? Explain why.
   a. $\dfrac{3 - 1}{3 + 5} = -\dfrac{1}{5}$
   b. $\dfrac{2x + 10}{2} = x + 5$
   c. $\dfrac{37}{72} = \dfrac{3}{2}$
   d. $\dfrac{2x + 6}{2} = x + 3$

52. Graph a portion of the function $f(x) = \dfrac{20x}{100 - x}$. To do so, complete the given table, plot the points, and then connect the plotted points with a smooth curve.

x	0	10	30	50	70	90	95	99
y or $f(x)$								

53. The domain of the function $f(x) = \dfrac{1}{x}$ is all real numbers except 0. This means that the graph of this function will be in two pieces: one piece corresponding to x values less than 0 and one piece corresponding to x values greater than 0. Graph the function by completing the following tables, separately plotting the points, and connecting each set of plotted points with a smooth curve.

x	$\frac{1}{4}$	$\frac{1}{2}$	1	2	4
y or $f(x)$					

x	-4	-2	-1	$-\frac{1}{2}$	$-\frac{1}{4}$
y or $f(x)$					

54. The function $f(x) = \dfrac{100,000x}{100 - x}$ models the cost in dollars for removing x percent of the pollutants from a bayou in which a nearby company dumped creosote.
   a. What is the domain of $f(x)$?
   b. Find the cost of removing 30% of the pollutants from the bayou. (*Hint:* Find $f(30)$.)
   c. Find the cost of removing 60% of the pollutants and then 80% of the pollutants.
   d. Find $f(90)$, then $f(95)$, and then $f(99)$. What happens to the cost as x approaches 100%?

55. The total revenue from the sale of a popular book is approximated by the rational function $R(x) = \dfrac{1000x^2}{x^2 + 4}$

where x is the number of years since publication and $R(x)$ is the total revenue in millions of dollars.

   a. Find the total revenue at the end of the first year.
   b. Find the total revenue at the end of the second year.
   c. Find the revenue during the second year only.

How does the graph of $y = \dfrac{x^2 - 9}{x - 3}$ *compare to the graph of*
$y = x + 3$?

Recall that $\dfrac{x^2 - 9}{x - 3} = \dfrac{(x + 3)(x - 3)}{x - 3} = x + 3$ *as long as x*

is not 3. This means that the graph of $y = \dfrac{x^2 - 9}{x - 3}$ *is the same*

as the graph of $y = x + 3$ *with* $x \neq 3$. *To graph* $y = \dfrac{x^2 - 9}{x - 3}$,
then, graph the linear equation $y = x + 3$ *and place an open
dot on the graph at 3. This open dot or interruption of the line
at 3 means* $x \neq 3$.

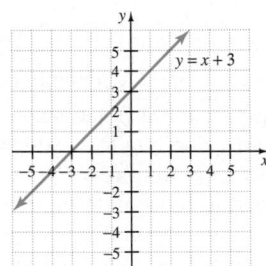

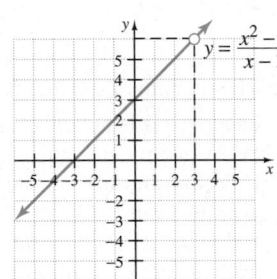

56. Graph $y = \dfrac{x^2 - 25}{x + 5}$.

57. Graph $y = \dfrac{x^2 - 16}{x - 4}$.

58. Graph $y = \dfrac{x^2 + x - 12}{x + 4}$.

59. Graph $y = \dfrac{x^2 - 6x + 8}{x - 2}$.

REVIEW EXERCISES

Perform the indicated operations. See Section 1.3.

60. $\dfrac{1}{3} \cdot \dfrac{9}{11}$

61. $\dfrac{5}{27} \cdot \dfrac{2}{5}$

62. $\dfrac{1}{3} \div \dfrac{1}{4}$

63. $\dfrac{7}{8} \div \dfrac{1}{2}$

64. $\dfrac{5}{6} \cdot \dfrac{10}{11} \cdot \dfrac{2}{3}$

65. $\dfrac{4}{3} \cdot \dfrac{1}{7} \cdot \dfrac{10}{13}$

66. $\dfrac{13}{20} \div \dfrac{2}{9}$

67. $\dfrac{8}{15} \div \dfrac{5}{8}$

7.2 MULTIPLYING AND DIVIDING RATIONAL EXPRESSIONS

CD-ROM   SSM

SSG   Video

▶ **OBJECTIVES**

1. Multiply rational expressions.
2. Divide rational expressions.

1 Just as simplifying rational expressions is similar to simplifying number fractions, multiplying and dividing rational expressions is similar to multiplying and dividing number fractions. To find the product of fractions and rational expressions, multiply the numerators and multiply the denominators.

$$\frac{3}{5} \cdot \frac{1}{4} = \frac{3 \cdot 1}{5 \cdot 4} = \frac{3}{20} \quad \text{and} \quad \frac{x}{y + 1} \cdot \frac{x + 3}{y - 1} = \frac{x(x + 3)}{(y + 1)(y - 1)}$$

MULTIPLYING RATIONAL EXPRESSIONS

Let P, Q, R, and S be polynomials. Then

$$\frac{P}{Q} \cdot \frac{R}{S} = \frac{PR}{QS}$$

as long as $Q \neq 0$ and $S \neq 0$.

Example 1   Multiply.

a. $\dfrac{25x}{2} \cdot \dfrac{1}{y^3}$

b. $\dfrac{-7x^2}{5y} \cdot \dfrac{3y^5}{14x^2}$

Solution   To multiply rational expressions, multiply the numerators and then multiply the denominators of both expressions. Then simplify if possible.

a. $\dfrac{25x}{2} \cdot \dfrac{1}{y^3} = \dfrac{25x \cdot 1}{2 \cdot y^3} = \dfrac{25x}{2y^3}$

The expression $\dfrac{25x}{2y^3}$ is in simplest form.

b. $\dfrac{-7x^2}{5y} \cdot \dfrac{3y^5}{14x^2} = \dfrac{-7x^2 \cdot 3y^5}{5y \cdot 14x^2}$ Multiply.

The expression $\dfrac{-7x^2 \cdot 3y^5}{5y \cdot 14x^2}$ is not in simplest form, so we factor the numerator and the denominator and apply the fundamental principle.

$$= \dfrac{-1 \cdot 7 \cdot 3 \cdot x^2 \cdot y \cdot y^4}{5 \cdot 2 \cdot 7 \cdot x^2 \cdot y}$$

$$= -\dfrac{3y^4}{10}$$

When multiplying rational expressions, it is usually best to factor each numerator and denominator. This will help us when we apply the fundamental principle to write the product in lowest terms.

Example 2   Multiply: $\dfrac{x^2 + x}{3x} \cdot \dfrac{6}{5x + 5}$.

Solution $\dfrac{x^2 + x}{3x} \cdot \dfrac{6}{5x + 5} = \dfrac{x(x + 1)}{3x} \cdot \dfrac{2 \cdot 3}{5(x + 1)}$ Factor numerators and denominators.

$$= \dfrac{x(x + 1) \cdot 2 \cdot 3}{3x \cdot 5(x + 1)}$$ Multiply.

$$= \dfrac{2}{5}$$ Simplify.

The following steps may be used to multiply rational expressions.

MULTIPLYING RATIONAL EXPRESSIONS

Step 1.   Completely factor numerators and denominators.

Step 2.   Multiply numerators and multiply denominators.

Step 3.   Simplify or write the product in lowest terms by applying the fundamental principle to all common factors.

Example 3   Multiply: $\dfrac{3x + 3}{5x - 5x^2} \cdot \dfrac{2x^2 + x - 3}{4x^2 - 9}$.

Solution $\dfrac{3x + 3}{5x - 5x^2} \cdot \dfrac{2x^2 + x - 3}{4x^2 - 9} = \dfrac{3(x + 1)}{5x(1 - x)} \cdot \dfrac{(2x + 3)(x - 1)}{(2x - 3)(2x + 3)}$ Factor.

$= \dfrac{3(x + 1)(2x + 3)(x - 1)}{5x(1 - x)(2x - 3)(2x + 3)}$ Multiply.

$= \dfrac{3(x + 1)(x - 1)}{5x(1 - x)(2x - 3)}$ Apply the fundamental principle.

Next, recall that $x - 1$ and $1 - x$ are opposites so that $x - 1 = -1(1 - x)$.

$= \dfrac{3(x + 1)(-1)(1 - x)}{5x(1 - x)(2x - 3)}$ Write $x - 1$ as $-1(1 - x)$.

$= \dfrac{-3(x + 1)}{5x(2x - 3)}$ or $-\dfrac{3(x + 1)}{5x(2x - 3)}$ Apply the fundamental principle.

2   We can divide by a rational expression in the same way we divide by a fraction. To divide by a fraction, multiply by its reciprocal.

HELPFUL HINT

Don't forget how to find reciprocals. The reciprocal of $\dfrac{a}{b}$ is $\dfrac{b}{a}$, $a \neq 0, b \neq 0$.

For example, to divide $\frac{3}{2}$ by $\frac{7}{8}$, multiply $\frac{3}{2}$ by $\frac{8}{7}$.

$$\frac{3}{2} \div \frac{7}{8} = \frac{3}{2} \cdot \frac{8}{7} = \frac{3 \cdot 4 \cdot 2}{2 \cdot 7} = \frac{12}{7}$$

DIVIDING RATIONAL EXPRESSIONS

Let $P, Q, R,$ and S be polynomials. Then,

$$\frac{P}{Q} \div \frac{R}{S} = \frac{P}{Q} \cdot \frac{S}{R} = \frac{PS}{QR}$$

as long as $Q \neq 0, S \neq 0,$ and $R \neq 0$.

Example 4   Divide: $\dfrac{3x^3y^7}{40} \div \dfrac{4x^3}{y^2}$.

Solution $\dfrac{3x^3y^7}{40} \div \dfrac{4x^3}{y^2} = \dfrac{3x^3y^7}{40} \cdot \dfrac{y^2}{4x^3}$ Multiply by the reciprocal of $\dfrac{4x^3}{y^2}$.

$= \dfrac{3x^3y^9}{160x^3}$

$= \dfrac{3y^9}{160}$ Simplify.

Example 5   Divide $\dfrac{(x-1)(x+2)}{10}$ by $\dfrac{2x+4}{5}$.

Solution $\dfrac{(x-1)(x+2)}{10} \div \dfrac{2x+4}{5} = \dfrac{(x-1)(x+2)}{10} \cdot \dfrac{5}{2x+4}$ Multiply by the reciprocal of $\dfrac{2x+4}{5}$.

$= \dfrac{(x-1)(x+2) \cdot 5}{5 \cdot 2 \cdot 2 \cdot (x+2)}$ Factor and multiply.

$= \dfrac{x-1}{4}$ Simplify.

The following may be used to divide by a rational expression.

DIVIDING BY A RATIONAL EXPRESSION

Multiply by its reciprocal.

Example 6   Divide: $\dfrac{6x+2}{x^2-1} \div \dfrac{3x^2+x}{x-1}$.

Solution $\dfrac{6x+2}{x^2-1} \div \dfrac{3x^2+x}{x-1} = \dfrac{6x+2}{x^2-1} \cdot \dfrac{x-1}{3x^2+x}$ Multiply by the reciprocal.

$= \dfrac{2(3x+1)(x-1)}{(x+1)(x-1) \cdot x(3x+1)}$ Factor and multiply.

$= \dfrac{2}{x(x+1)}$ Simplify.

Example 7   Divide: $\dfrac{2x^2-11x+5}{5x-25} \div \dfrac{4x-2}{10}$.

Solution $\dfrac{2x^2 - 11x + 5}{5x - 25} \div \dfrac{4x - 2}{10} = \dfrac{2x^2 - 11x + 5}{5x - 25} \cdot \dfrac{10}{4x - 2}$ Multiply by the reciprocal.

$$= \dfrac{(2x - 1)(x - 5) \cdot 2 \cdot 5}{5(x - 5) \cdot 2(2x - 1)}$$ Factor and multiply.

$$= \dfrac{1}{1} \quad \text{or} \quad 1$$ Simplify.

Now that we know how to multiply fractions and rational expressions, we can use this knowledge to help us convert between units of measure. To do so, we will use **unit fractions**. A unit fraction is a fraction that equals 1. For example, since 12 in. = 1 ft, we have the unit fractions

$$\dfrac{12 \text{ in.}}{1 \text{ ft}} = 1 \quad \text{and} \quad \dfrac{1 \text{ ft}}{12 \text{ in.}} = 1$$

Example 8   CONVERTING FROM SQUARE YARDS TO SQUARE FEET

The largest casino in the world is the Foxwoods Resort Casino in Ledyard, CT. The gaming area for this casino is 21,444 *square yards*. Find the size of the gaming area in *square feet*. (*Source: The Guinness Book of Records*, 1996)

Solution   There are 9 square feet in 1 square yard.

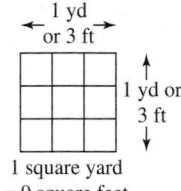

1 yd
or 3 ft

1 yd or
3 ft

1 square yard
= 9 square feet

unit fraction

$$21{,}444 \text{ square yards} = 21{,}444 \text{ sq. yd} \cdot \dfrac{9 \text{ sq. ft}}{1 \text{ sq. yd}}$$

$$= 192{,}996 \text{ square feet}$$

> **HELPFUL HINT**
> When converting a unit of measurement, if possible, write the unit fraction so that **the numerator contains units converting to** and **the denominator contains original units.**
>
> unit fraction
>
> $$48 \text{ in.} = \dfrac{48 \text{ in.}}{1} \cdot \dfrac{1 \text{ ft}}{12 \text{ in.}} \quad \begin{array}{l} \leftarrow \text{ units converting to} \\ \leftarrow \text{ original units} \end{array}$$
>
> $$= \dfrac{48}{12} \text{ ft} = 4 \text{ ft}$$

MENTAL MATH

Find the following products. See Example 1.

1. $\dfrac{2}{y} \cdot \dfrac{x}{3}$

2. $\dfrac{3x}{4} \cdot \dfrac{1}{y}$

3. $\dfrac{5}{7} \cdot \dfrac{y^2}{x^2}$

4. $\dfrac{x^5}{11} \cdot \dfrac{4}{z^3}$

5. $\dfrac{9}{x} \cdot \dfrac{x}{5}$

6. $\dfrac{y}{7} \cdot \dfrac{3}{y}$

Exercise Set 7.2

Multiply. Simplify if possible. See Examples 1 through 3.

1. $\dfrac{3x}{y^2} \cdot \dfrac{7y}{4x}$

2. $\dfrac{9x^2}{y} \cdot \dfrac{4y}{3x^2}$

 **3.** $\dfrac{8x}{2} \cdot \dfrac{x^5}{4x^2}$

4. $\dfrac{6x^2}{10x^3} \cdot \dfrac{5x}{12}$

5. $-\dfrac{5a^2b}{30a^2b^2} \cdot b^3$

6. $-\dfrac{9x^3y^2}{18xy^5} \cdot y^3$

7. $\dfrac{x}{2x-14} \cdot \dfrac{x^2-7x}{5}$

8. $\dfrac{4x-24}{20x} \cdot \dfrac{5}{x-6}$

9. $\dfrac{6x+6}{5} \cdot \dfrac{10}{36x+36}$

10. $\dfrac{x^2+x}{8} \cdot \dfrac{16}{x+1}$

11. $\dfrac{m^2-n^2}{m+n} \cdot \dfrac{m}{m^2-mn}$

12. $\dfrac{(m-n)^2}{m+n} \cdot \dfrac{m}{m^2-mn}$

13. $\dfrac{x^2-25}{x^2-3x-10} \cdot \dfrac{x+2}{x}$

14. $\dfrac{a^2+6a+9}{a^2-4} \cdot \dfrac{a+3}{a-2}$

 **15.** Find the area of the following rectangle.

△

$\dfrac{2x}{x^2-25}$ feet

$\dfrac{x+5}{9x^3}$ feet

△ **16.** Find the area of the following square.

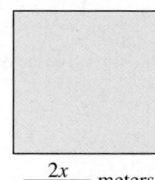

$\dfrac{2x}{5x^2+3x}$ meters

Divide. Simplify if possible. See Examples 4 through 7.

17. $\dfrac{5x^7}{2x^5} \div \dfrac{10x}{4x^3}$

18. $\dfrac{9y^4}{6y} \div \dfrac{y^2}{3}$

19. $\dfrac{8x^2}{y^3} \div \dfrac{4x^2y^3}{6}$

20. $\dfrac{7a^2b}{3ab^2} \div \dfrac{21a^2b^2}{14ab}$

21. $\dfrac{(x-6)(x+4)}{4x} \div \dfrac{2x-12}{8x^2}$

22. $\dfrac{(x+3)^2}{5} \div \dfrac{5x+15}{25}$

23. $\dfrac{3x^2}{x^2-1} \div \dfrac{x^5}{(x+1)^2}$

24. $\dfrac{(x+1)}{(x+1)(2x+3)} \div \dfrac{20}{2x+3}$

25. $\dfrac{m^2-n^2}{m+n} \div \dfrac{m}{m^2+nm}$

26. $\dfrac{(m-n)^2}{m+n} \div \dfrac{m^2-mn}{m}$

 **27.** $\dfrac{x+2}{7-x} \div \dfrac{x^2-5x+6}{x^2-9x+14}$

28. $(x-3) \div \dfrac{x^2+3x-18}{x}$

29. $\dfrac{x^2+7x+10}{1-x} \div \dfrac{x^2+2x-15}{x-1}$

30. $\dfrac{a^2-b^2}{9} \div \dfrac{3b-3a}{27x^2}$

✎ **31.** Explain how to multiply rational expressions.

✎ **32.** Explain how to divide rational expressions.

Perform the indicated operations.

33. $\dfrac{5a^2b}{30a^2b^2} \cdot \dfrac{1}{b^3}$

34. $\dfrac{9x^2y^2}{42xy^5} \cdot \dfrac{6}{x^5}$

35. $\dfrac{12x^3y}{8xy^7} \div \dfrac{7x^5y}{6x}$

36. $\dfrac{4y^2z}{3y^7z^7} \div \dfrac{12y}{6z}$

37. $\dfrac{5x-10}{12} \div \dfrac{4x-8}{8}$

38. $\dfrac{6x+6}{5} \div \dfrac{3x+3}{10}$

39. $\dfrac{x^2+5x}{8} \cdot \dfrac{9}{3x+15}$

40. $\dfrac{3x^2+12x}{6} \cdot \dfrac{9}{2x+8}$

41. $\dfrac{7}{6p^2+q} \div \dfrac{14}{18p^2+3q}$

42. $\dfrac{5x-10}{12} \div \dfrac{4x-8}{8}$

43. $\dfrac{3x + 4y}{x^2 + 4xy + 4y^2} \cdot \dfrac{x + 2y}{2}$ 44. $\dfrac{2a + 2b}{3} \div \dfrac{a^2 - b^2}{a - b}$

45. $\dfrac{x^2 - 9}{x^2 + 8} \div \dfrac{3 - x}{2x^2 + 16}$

46. $\dfrac{x^2 - y^2}{3x^2 + 3xy} \cdot \dfrac{3x^2 + 6x}{3x^2 - 2xy - y^2}$

47. $\dfrac{(x + 2)^2}{x - 2} \div \dfrac{x^2 - 4}{2x - 4}$ 48. $\dfrac{x^2 - 4}{2y} \div \dfrac{2 - x}{6xy}$

49. $\dfrac{a^2 + 7a + 12}{a^2 + 5a + 6} \cdot \dfrac{a^2 + 8a + 15}{a^2 + 5a + 4}$

50. $\dfrac{b^2 + 2b - 3}{b^2 + b - 2} \cdot \dfrac{b^2 - 4}{b^2 + 6b + 8}$

51. $\dfrac{1}{-x - 4} \div \dfrac{x^2 - 7x}{x^2 - 3x - 28}$

52. $\dfrac{x^2 - 10x + 21}{7 - x} \div (x + 3)$

53. $\dfrac{x^2 - 5x - 24}{2x^2 - 2x - 24} \cdot \dfrac{4x^2 + 4x - 24}{x^2 - 10x + 16}$

54. $\dfrac{a^2 - b^2}{a} \cdot \dfrac{a + b}{a^2 + ab}$ 55. $(x - 5) \div \dfrac{5 - x}{x^2 + 2}$

56. $\dfrac{2x^2 + 3xy + y^2}{x^2 - y^2} \div \dfrac{1}{2x + 2y}$

57. $\dfrac{x^2 - y^2}{x^2 - 2xy + y^2} \cdot \dfrac{y - x}{x + y}$ 58. $\dfrac{x + 3}{x^2 - 9} \cdot \dfrac{x^2 - 8x + 15}{5x}$

59. Find the quotient of $\dfrac{x^2 - 9}{2x}$ and $\dfrac{x + 3}{8x^4}$.

60. Find the quotient of $\dfrac{4x^2 + 4x + 1}{4x + 2}$ and $\dfrac{4x + 2}{16}$.

Convert as indicated. See Example 8.

61. 10 square feet = _____ square inches.

62. 1008 square inches = _____ square feet.

63. The Pentagon, headquarters for the Department of Defense, contains 3,707,745 square feet of office space. Convert this to square yards. (Round to the nearest square yard.) (*Source: World Almanac,* 2000)

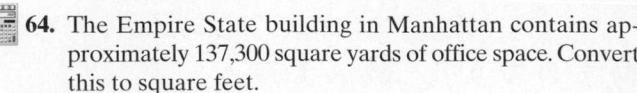
64. The Empire State building in Manhattan contains approximately 137,300 square yards of office space. Convert this to square feet.

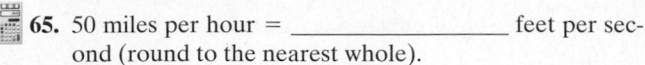
65. 50 miles per hour = _____ feet per second (round to the nearest whole).

66. 10 feet per second = _____ miles per hour (round to the nearest tenth).

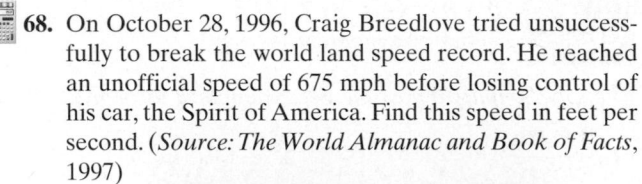
67. The speed of sound is 5023 feet per second in ocean water whose temperature is 77°F. Convert this speed of sound to miles per hour. Round to the nearest tenth. (*Source: CRC Handbook of Chemistry and Physics,* 65th edition)

68. On October 28, 1996, Craig Breedlove tried unsuccessfully to break the world land speed record. He reached an unofficial speed of 675 mph before losing control of his car, the Spirit of America. Find this speed in feet per second. (*Source: The World Almanac and Book of Facts,* 1997)

Simplify. These expressions contain 4-term polynomials and sums and differences of cubes.

69. $\dfrac{a^2 + ac + ba + bc}{a - b} \div \dfrac{a + c}{a + b}$

70. $\dfrac{x^2 + 2x - xy - 2y}{x^2 - y^2} \div \dfrac{2x + 4}{x + y}$

71. $\dfrac{3x^2 + 8x + 5}{x^2 + 8x + 7} \cdot \dfrac{x + 7}{x^2 + 4}$

72. $\dfrac{16x^2 + 2x}{16x^2 + 10x + 1} \cdot \dfrac{1}{4x^2 + 2x}$

73. $\dfrac{x^3 + 8}{x^2 - 2x + 4} \cdot \dfrac{4}{x^2 - 4}$

74. $\dfrac{9y}{3y - 3} \cdot \dfrac{y^3 - 1}{y^3 + y^2 + y}$

75. $\dfrac{a^2 - ab}{6a^2 + 6ab} \div \dfrac{a^3 - b^3}{a^2 - b^2}$

76. $\dfrac{x^3 + 27y^3}{6x} \div \dfrac{x^2 - 9y^2}{x^2 - 3xy}$

REVIEW EXERCISES

Perform each operation. See Section 1.3.

77. $\dfrac{1}{5} + \dfrac{4}{5}$ **78.** $\dfrac{3}{15} + \dfrac{6}{15}$

79. $\dfrac{9}{9} - \dfrac{19}{9}$ **80.** $\dfrac{4}{3} - \dfrac{8}{3}$

81. $\dfrac{6}{5} + \left(\dfrac{1}{5} - \dfrac{8}{5}\right)$ **82.** $-\dfrac{3}{2} + \left(\dfrac{1}{2} - \dfrac{3}{2}\right)$

See Section 3.2.

83. Graph the linear equation $x - 2y = 6$.

84. Graph the linear equation $5x + y = 10$.

A Look Ahead

Example

Perform the indicated operations.

$$\dfrac{15x^2 - x - 6}{12x^3} \cdot \dfrac{4x}{9 - 25x^2} \div \dfrac{x}{3x - 2}$$

Solution:

$$\dfrac{15x^2 - x - 6}{12x^3} \cdot \dfrac{4x}{9 - 25x^2} \div \dfrac{x}{3x - 2}$$

$$= \left(\dfrac{15x^2 - x - 6}{12x^3} \cdot \dfrac{4x}{9 - 25x^2}\right) \cdot \dfrac{3x - 2}{x}$$

$$= \dfrac{(3x - 2)(5x + 3) \cdot 4x(3x - 2)}{12x^3(3 - 5x)(3 + 5x) \cdot x}$$

$$= \dfrac{(3x - 2)^2}{3x^3(3 - 5x)}$$

Perform the following operations.

85. $\left(\dfrac{x^2 - y^2}{x^2 + y^2} \div \dfrac{x^2 - y^2}{3x}\right) \cdot \dfrac{x^2 + y^2}{6}$

86. $\left(\dfrac{x^2 - 9}{x^2 - 1} \cdot \dfrac{x^2 + 2x + 1}{2x^2 + 9x + 9}\right) \div \dfrac{2x + 3}{1 - x}$

87. $\left(\dfrac{2a + b}{b^2} \cdot \dfrac{3a^2 - 2ab}{ab + 2b^2}\right) \div \dfrac{a^2 - 3ab + 2b^2}{5ab - 10b^2}$

88. $\left(\dfrac{x^2y^2 - xy}{4x - 4y} \div \dfrac{3y - 3x}{8x - 8y}\right) \cdot \dfrac{y - x}{8}$

7.3 ADDING AND SUBTRACTING RATIONAL EXPRESSIONS WITH COMMON DENOMINATORS AND LEAST COMMON DENOMINATOR

CD-ROM   SSM

SSG   Video

▶ **OBJECTIVES**

1. Add and subtract rational expressions with the same denominator.
2. Find the least common denominator of a list of rational expressions.
3. Write a rational expression as an equivalent expression whose denominator is given.

1 Like multiplication and division, addition and subtraction of rational expressions is similar to addition and subtraction of rational numbers. In this section, we add and subtract rational expressions with a common (or the same) denominator.

Add: $\dfrac{6}{5} + \dfrac{2}{5}$ Add: $\dfrac{9}{x + 2} + \dfrac{3}{x + 2}$

Add the numerators and place the sum over the common denominator.

$\dfrac{6}{5} + \dfrac{2}{5} = \dfrac{6 + 2}{5}$ $\dfrac{9}{x + 2} + \dfrac{3}{x + 2} = \dfrac{9 + 3}{x + 2}$

$= \dfrac{8}{5}$ Simplify. $= \dfrac{12}{x + 2}$ Simplify.

ADDING AND SUBTRACTING RATIONAL EXPRESSIONS WITH COMMON DENOMINATORS

If $\dfrac{P}{R}$ and $\dfrac{Q}{R}$ are rational expressions, then

$$\frac{P}{R} + \frac{Q}{R} = \frac{P + Q}{R} \quad \text{and} \quad \frac{P}{R} - \frac{Q}{R} = \frac{P - Q}{R}$$

To add or subtract rational expressions, add or subtract numerators and place the sum or difference over the common denominator.

Example 1   Add: $\dfrac{5m}{2n} + \dfrac{m}{2n}$

Solution

$$\frac{5m}{2n} + \frac{m}{2n} = \frac{5m + m}{2n} \qquad \text{Add the numerators.}$$

$$= \frac{6m}{2n} \qquad \text{Simplify the numerator by combining like terms.}$$

$$= \frac{3m}{n} \qquad \text{Simplify by applying the fundamental principle.}$$

Example 2   Subtract: $\dfrac{2y}{2y - 7} - \dfrac{7}{2y - 7}$

Solution

$$\frac{2y}{2y - 7} - \frac{7}{2y - 7} = \boxed{\frac{2y - 7}{2y - 7}} \qquad \text{Subtract the numerators.}$$

$$= \frac{1}{1} \quad \text{or} \quad 1 \qquad \text{Simplify.}$$

Example 3   Subtract: $\dfrac{3x^2 + 2x}{x - 1} - \dfrac{10x - 5}{x - 1}$

Solution

$$\frac{3x^2 + 2x}{x - 1} - \frac{10x - 5}{x - 1} = \frac{3x^2 + 2x - (10x - 5)}{x - 1} \qquad \begin{array}{l}\text{Subtract the numerators.}\\ \text{Notice the parentheses.}\end{array}$$

$$= \frac{3x^2 + 2x - 10x + 5}{x - 1} \qquad \text{Use the distributive property.}$$

$$= \frac{3x^2 - 8x + 5}{x - 1} \qquad \text{Combine like terms.}$$

$$= \frac{(x - 1)(3x - 5)}{x - 1} \qquad \text{Factor.}$$

$$= 3x - 5 \qquad \text{Simplify.}$$

> **HELPFUL HINT**
> Notice how the numerator $10x - 5$ has been subtracted in Example 3.
>
> This $-$ sign applies to the   So parentheses are inserted
> entire numerator of $10x - 5$.   here to indicate this.
>
> $$\frac{3x^2 + 2x}{x - 1} - \frac{10x - 5}{x - 1} = \frac{3x^2 + 2x - (10x - 5)}{x - 1}$$

2

To add and subtract fractions with **unlike** denominators, first find a least common denominator (LCD), and then write all fractions as equivalent fractions with the LCD.

For example, suppose we add $\frac{8}{3}$ and $\frac{2}{5}$. The LCD of denominators 3 and 5 is 15, since 15 is the smallest number that both 3 and 5 divide into evenly. Rewrite each fraction so that its denominator is 15. (Notice how we apply the fundamental principle.)

$$\frac{8}{3} + \frac{2}{5} = \frac{8(5)}{3(5)} + \frac{2(3)}{5(3)} = \frac{40}{15} + \frac{6}{15} = \frac{40 + 6}{15} = \frac{46}{15}$$

To add or subtract rational expressions with unlike denominators, we also first find an LCD and then write all rational expressions as equivalent expressions with the LCD. The **least common denominator (LCD) of a list of rational expressions** is a polynomial of least degree whose factors include all the factors of the denominators in the list.

FINDING THE LEAST COMMON DENOMINATOR (LCD)

Step 1. Factor each denominator completely.
Step 2. The least common denominator (LCD) is the product of all unique factors found in Step 1, each raised to a power equal to the greatest number of times that the factor appears in any one factored denominator.

Example 4 Find the LCD for each pair.

a. $\dfrac{1}{8}, \dfrac{3}{22}$ **b.** $\dfrac{7}{5x}, \dfrac{6}{15x^2}$

Solution **a.** Start by finding the prime factorization of each denominator.

$$8 = 2 \cdot 2 \cdot 2 = 2^3 \qquad \text{and} \qquad 22 = 2 \cdot 11$$

Next, write the product of all the unique factors, each raised to a power equal to the greatest number of times that the factor appears.

The greatest number of times that the factor 2 appears is 3.
The greatest number of times that the factor 11 appears is 1.

$$\text{LCD} = 2^3 \cdot 11^1 = 8 \cdot 11 = 88$$

b. Factor each denominator.

$$5x = 5 \cdot x \qquad \text{and} \qquad 15x^2 = 3 \cdot 5 \cdot x^2$$

The greatest number of times that the factor 5 appears is 1.
The greatest number of times that the factor 3 appears is 1.
The greatest number of times that the factor x appears is 2.

$$\text{LCD} = 3^1 \cdot 5^1 \cdot x^2 = 15x^2$$

Example 5 Find the LCD of $\dfrac{7x}{x + 2}$ and $\dfrac{5x^2}{x - 2}$.

Solution The denominators $x + 2$ and $x - 2$ are completely factored already. The factor $x + 2$ appears once and the factor $x - 2$ appears once.

$$\text{LCD} = (x + 2)(x - 2)$$

Example 6 Find the LCD of $\dfrac{6m^2}{3m + 15}$ and $\dfrac{2}{(m + 5)^2}$.

Solution Factor each denominator.

$$3m + 15 = 3(m + 5)$$
$$(m + 5)^2 \text{ is already factored.}$$

The greatest number of times that the factor 3 appears is 1.
The greatest number of times that the factor $m + 5$ appears *in any one denominator* is 2.

$$\text{LCD} = 3(m + 5)^2$$

Example 7 Find the LCD of $\dfrac{t - 10}{t^2 - t - 6}$ and $\dfrac{t + 5}{t^2 + 3t + 2}$.

Solution Start by factoring each denominator.

$$t^2 - t - 6 = (t - 3)(t + 2)$$
$$t^2 + 3t + 2 = (t + 1)(t + 2)$$
$$\text{LCD} = (t - 3)(t + 2)(t + 1)$$

Example 8 Find the LCD of $\dfrac{2}{x - 2}$ and $\dfrac{10}{2 - x}$.

Solution The denominators $x - 2$ and $2 - x$ are opposites. That is, $2 - x = -1(x - 2)$. Use $x - 2$ or $2 - x$ as the LCD.

$$\text{LCD} = x - 2 \qquad \text{or} \qquad \text{LCD} = 2 - x$$

3 Next we practice writing a rational expression as an equivalent rational expression with a given denominator. To do this, we apply the fundamental principle, which says that $\dfrac{PR}{QR} = \dfrac{P}{Q}$, or equivalently that $\dfrac{P}{Q} = \dfrac{PR}{QR}$. This can be seen by recalling that multiplying an expression by 1 produces an equivalent expression. In other words,

$$\frac{P}{Q} = \frac{P}{Q} \cdot 1 = \frac{P}{Q} \cdot \frac{R}{R} = \frac{PR}{QR}.$$

Example 9 Write $\dfrac{4b}{9a}$ as an equivalent fraction with the given denominator.

$$\frac{4b}{9a} = \frac{}{27a^2b}$$

Solution Ask yourself: "What do we multiply $9a$ by to get $27a^2b$?" The answer is $3ab$, since $9a(3ab) = 27a^2b$. Multiply the numerator and denominator by $3ab$.

$$\frac{4b}{9a} = \frac{4b(3ab)}{9a(3ab)} = \frac{12ab^2}{27a^2b}$$

Example 10 Write the rational expression as an equivalent rational expression with the given denominator.

$$\frac{5}{x^2 - 4} = \frac{}{(x - 2)(x + 2)(x - 4)}$$

Solution First, factor the denominator $x^2 - 4$ as $(x - 2)(x + 2)$.
If we multiply the original denominator $(x - 2)(x + 2)$ by $x - 4$, the result is the new denominator $(x - 2)(x + 2)(x - 4)$. Thus, multiply the numerator and the denominator by $x - 4$.

$$\frac{5}{\underbrace{x^2 - 4}_{\substack{\text{factored} \\ \text{denominator}}}} = \frac{5}{(x - 2)(x + 2)} = \frac{5(x - 4)}{(x - 2)(x + 2)(x - 4)}$$

$$= \frac{5x - 20}{(x - 2)(x + 2)(x - 4)}$$

MENTAL MATH

Perform the indicated operations.

1. $\dfrac{2}{3} + \dfrac{1}{3}$

2. $\dfrac{5}{11} + \dfrac{1}{11}$

3. $\dfrac{3x}{9} + \dfrac{4x}{9}$

4. $\dfrac{3y}{8} + \dfrac{2y}{8}$

5. $\dfrac{8}{9} - \dfrac{7}{9}$

6. $-\dfrac{4}{12} - \dfrac{3}{12}$

7. $\dfrac{7}{5} - \dfrac{10y}{5}$

8. $\dfrac{12x}{7} - \dfrac{4x}{7}$

Exercise Set 7.3

Add or subtract as indicated. Simplify the result if possible. See Examples 1 through 3.

1. $\dfrac{a}{13} + \dfrac{9}{13}$

2. $\dfrac{x+1}{7} + \dfrac{6}{7}$

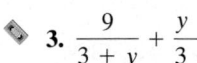 **3.** $\dfrac{9}{3+y} + \dfrac{y+1}{3+y}$

4. $\dfrac{9}{y+9} + \dfrac{y}{y+9}$

5. $\dfrac{4m}{3n} + \dfrac{5m}{3n}$

6. $\dfrac{3p}{2} + \dfrac{11p}{2}$

7. $\dfrac{2x+1}{x-3} + \dfrac{3x+6}{x-3}$

8. $\dfrac{4p-3}{2p+7} + \dfrac{3p+8}{2p+7}$

9. $\dfrac{7}{8} - \dfrac{3}{8}$

10. $\dfrac{4}{5} - \dfrac{13}{5}$

11. $\dfrac{4m}{m-6} - \dfrac{24}{m-6}$

12. $\dfrac{8y}{y-2} - \dfrac{16}{y-2}$

13. $\dfrac{2x^2}{x-5} - \dfrac{25+x^2}{x-5}$

14. $\dfrac{6x^2}{2x-5} - \dfrac{25+2x^2}{2x-5}$

15. $\dfrac{-3x^2-4}{x-4} - \dfrac{12-4x^2}{x-4}$

16. $\dfrac{7x^2-9}{2x-5} - \dfrac{16+3x^2}{2x-5}$

17. $\dfrac{2x+3}{x+1} - \dfrac{x+2}{x+1}$

18. $\dfrac{1}{x^2-2x-15} - \dfrac{4-x}{x^2-2x-15}$

19. $\dfrac{3}{x^3} + \dfrac{9}{x^3}$

20. $\dfrac{5}{xy} + \dfrac{8}{xy}$

21. $\dfrac{5}{x+4} - \dfrac{10}{x+4}$

22. $\dfrac{4}{2x+1} - \dfrac{8}{2x+1}$

23. $\dfrac{x}{x+y} - \dfrac{2}{x+y}$

24. $\dfrac{y+1}{y+2} - \dfrac{3}{y+2}$

25. $\dfrac{8x}{2x+5} + \dfrac{20}{2x+5}$

26. $\dfrac{12y-5}{3y-1} + \dfrac{1}{3y-1}$

27. $\dfrac{5x+4}{x-1} - \dfrac{2x+7}{x-1}$

28. $\dfrac{x^2+9x}{x+7} - \dfrac{4x+14}{x+7}$

29. $\dfrac{a}{a^2+2a-15} - \dfrac{3}{a^2+2a-15}$

30. $\dfrac{3y}{y^2+3y-10} - \dfrac{6}{y^2+3y-10}$

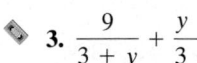 **31.** $\dfrac{2x+3}{x^2-x-30} - \dfrac{x-2}{x^2-x-30}$

32. $\dfrac{3x-1}{x^2+5x-6} - \dfrac{2x-7}{x^2+5x-6}$

△ **33.** A square-shaped pasture has a side of length $\dfrac{5}{x-2}$ meters. Express its perimeter as a rational expression.

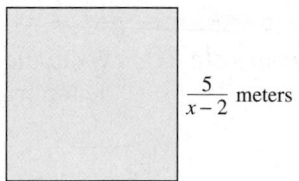

$\dfrac{5}{x-2}$ meters

△ **34.** The following trapezoid has sides of indicated length. Find its perimeter.

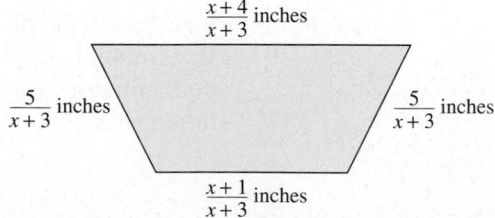

$\dfrac{x+4}{x+3}$ inches

$\dfrac{5}{x+3}$ inches

$\dfrac{5}{x+3}$ inches

$\dfrac{x+1}{x+3}$ inches

35. Describe the process for adding and subtracting two rational expressions with the same denominators.

36. Explain the similarities between subtracting $\dfrac{3}{8}$ from $\dfrac{7}{8}$ and subtracting $\dfrac{6}{x+3}$ from $\dfrac{9}{x+3}$.

Find the LCD for the following lists of rational expressions. See Examples 4 through 8.

37. $\dfrac{2}{3}, \dfrac{4}{33}$

38. $\dfrac{8}{20}, \dfrac{4}{15}$

39. $\dfrac{19}{2x}, \dfrac{5}{4x^3}$

40. $\dfrac{17x}{4y^5}, \dfrac{2}{8y}$

41. $\dfrac{9}{8x}, \dfrac{3}{2x + 4}$

42. $\dfrac{1}{6y}, \dfrac{3x}{4y + 12}$

43. $\dfrac{1}{3x + 3}, \dfrac{8}{2x^2 + 4x + 2}$

44. $\dfrac{19x + 5}{4x - 12}, \dfrac{3}{2x^2 - 12x + 18}$

45. $\dfrac{5}{x - 8}, \dfrac{3}{8 - x}$

46. $\dfrac{2x + 5}{3x - 7}, \dfrac{5}{7 - 3x}$

47. $\dfrac{4 + x}{8x^2(x - 1)^2}, \dfrac{17}{10x^3(x - 1)}$

48. $\dfrac{2x + 3}{9x(x + 2)}, \dfrac{9x + 5}{12(x + 2)^2}$

49. $\dfrac{9x + 1}{2x + 1}, \dfrac{3x - 5}{2x - 1}$

50. $\dfrac{5}{4x - 2}, \dfrac{7}{4x + 2}$

51. $\dfrac{5x + 1}{2x^2 + 7x - 4}, \dfrac{3x}{2x^2 + 5x - 3}$

52. $\dfrac{4}{x^2 + 4x + 3}, \dfrac{4x - 2}{x^2 + 10x + 21}$

53. Write some instructions to help a friend who is having difficulty finding the LCD of two rational expressions.

54. Explain why the LCD of rational expressions $\dfrac{7}{x + 1}$ and $\dfrac{9x}{(x + 1)^2}$ is $(x + 1)^2$ and not $(x + 1)^3$.

Rewrite each rational expression as an equivalent rational expression whose denominator is the given polynomial. See Examples 9 and 10.

55. $\dfrac{3}{2x} = \dfrac{}{4x^2}$

56. $\dfrac{3}{9y^5} = \dfrac{}{72y^9}$

57. $\dfrac{6}{3a} = \dfrac{}{12ab^2}$

58. $\dfrac{17a}{4y^2x} = \dfrac{}{32y^3x^2z}$

59. $\dfrac{9}{x + 3} = \dfrac{}{2(x + 3)}$

60. $\dfrac{4x + 1}{3x + 6} = \dfrac{}{3y(x + 2)}$

61. $\dfrac{9a + 2}{5a + 10} = \dfrac{}{5b(a + 2)}$

62. $\dfrac{5 + y}{2x^2 + 10} = \dfrac{}{4(x^2 + 5)}$

63. $\dfrac{x}{x^2 + 6x + 8} = \dfrac{}{(x + 4)(x + 2)(x + 1)}$

64. $\dfrac{5x}{x^2 + 2x - 3} = \dfrac{}{(x - 1)(x - 5)(x + 3)}$

65. $\dfrac{9y - 1}{15x^2 - 30} = \dfrac{}{30x^2 - 60}$

66. $\dfrac{8x + 3}{7y^2 - 21} = \dfrac{}{21y^2 - 63}$

67. $\dfrac{5}{2x^2 - 9x - 5} = \dfrac{}{3x(2x + 1)(x - 7)(x - 5)}$

68. $\dfrac{x - 9}{3x^2 + 10x + 3} = \dfrac{}{x(x + 3)(x + 5)(3x + 1)}$

Write each rational expression as an equivalent expression with a denominator of $x - 2$.

69. $\dfrac{5}{2 - x}$

70. $\dfrac{8y}{2 - x}$

71. $-\dfrac{7 + x}{2 - x}$

72. $\dfrac{x - 3}{-(x - 2)}$

73. The planet Mercury revolves around the sun in 88 Earth days. It takes Jupiter 4332 Earth days to make one revolution around the sun. (*Source:* National Space Science Data Center) If the two planets are aligned as shown in the figure, how long will it take for them to align again?

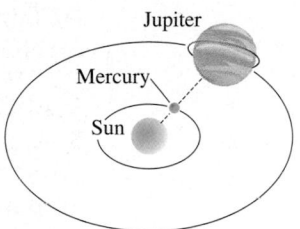

74. You are throwing a barbecue and you want to make sure that you purchase the same number of hot dogs as hot dog buns. Hot dogs come 8 to a package and hot dog buns come 12 to a package. What is the least number of each type of package you should buy?

75. An algebra student approaches you with a problem. He's tried to subtract two rational expressions, but his result does not match the book's. Check to see if the student has made an error. If so, correct his work shown below.

$$\dfrac{2x - 6}{x - 5} - \dfrac{x + 4}{x - 5}$$
$$= \dfrac{2x - 6 - x + 4}{x - 5}$$
$$= \dfrac{x - 2}{x - 5}$$

REVIEW EXERCISES

Solve the following quadratic equations by factoring. See Section 6.6.

76. $x(x - 3) = 0$

77. $2x(x + 5) = 0$

78. $x^2 + 6x + 5 = 0$

79. $x^2 - 6x + 5 = 0$

Perform each operation. See Section 1.3.

80. $\dfrac{2}{3} + \dfrac{5}{7}$

81. $\dfrac{9}{10} - \dfrac{3}{5}$

82. $\dfrac{2}{6} - \dfrac{3}{4}$

83. $\dfrac{11}{15} + \dfrac{5}{9}$

7.4 ADDING AND SUBTRACTING RATIONAL EXPRESSIONS WITH UNLIKE DENOMINATORS

CD-ROM   SSM

SSG   Video

▶ O B J E C T I V E

1. Add and subtract rational expressions with unlike denominators.

1   In the previous section, we practiced all the skills we need to add and subtract rational expressions with unlike or different denominators. The steps are as follows:

ADDING OR SUBTRACTING RATIONAL EXPRESSIONS WITH UNLIKE DENOMINATORS

Step 1. Find the LCD of the rational expressions.
Step 2. Rewrite each rational expression as an equivalent expression whose denominator is the LCD found in Step 1.
Step 3. Add or subtract numerators and write the sum or difference over the common denominator.
Step 4. Simplify or write the rational expression in simplest form.

Example 1   Perform each indicated operation.

a. $\dfrac{a}{4} - \dfrac{2a}{8}$

b. $\dfrac{3}{10x^2} + \dfrac{7}{25x}$

Solution   **a.** First, we must find the LCD. Since $4 = 2^2$ and $8 = 2^3$, the LCD $= 2^3 = 8$. Next we write each fraction as an equivalent fraction with the denominator 8, then we subtract.

$$\frac{a}{4} - \frac{2a}{8} = \frac{a(2)}{4(2)} - \frac{2a}{8} = \frac{2a}{8} - \frac{2a}{8} = \frac{2a - 2a}{8} = \frac{0}{8} = 0$$

b. Since $10x^2 = 2 \cdot 5 \cdot x \cdot x$ and $25x = 5 \cdot 5 \cdot x$, the LCD $= 2 \cdot 5^2 \cdot x^2 = 50x^2$. We write each fraction as an equivalent fraction with a denominator of $50x^2$.

$$\frac{3}{10x^2} + \frac{7}{25x} = \frac{3(5)}{10x^2(5)} + \frac{7(2x)}{25x(2x)}$$

$$= \frac{15}{50x^2} + \frac{14x}{50x^2}$$

$$= \frac{15 + 14x}{50x^2} \qquad \text{Add numerators. Write the sum over the common denominator.}$$

Example 2 Subtract: $\dfrac{6x}{x^2 - 4} - \dfrac{3}{x + 2}$

Solution Since $x^2 - 4 = (x + 2)(x - 2)$, the LCD $= (x + 2)(x - 2)$. We write equivalent expressions with the LCD as denominators.

$$\frac{6x}{x^2 - 4} - \frac{3}{x + 2} = \frac{6x}{(x + 2)(x - 2)} - \frac{3(x - 2)}{(x + 2)(x - 2)}$$

$$= \frac{6x - 3(x - 2)}{(x + 2)(x - 2)} \qquad \text{Subtract numerators. Write the difference over the common denominator.}$$

$$= \frac{6x - 3x + 6}{(x + 2)(x - 2)} \qquad \text{Apply the distributive property in the numerator.}$$

$$= \frac{3x + 6}{(x + 2)(x - 2)} \qquad \text{Combine like terms in the numerator.}$$

Next we factor the numerator to see if this rational expression can be simplified.

$$= \frac{3(x + 2)}{(x + 2)(x - 2)} \qquad \text{Factor.}$$

$$= \frac{3}{x - 2} \qquad \text{Apply the fundamental principle to simplify.}$$

Example 3 Add: $\dfrac{2}{3t} + \dfrac{5}{t + 1}$

Solution The LCD is $3t(t + 1)$. We write each rational expression as an equivalent rational expression with a denominator of $3t(t + 1)$.

$$\frac{2}{3t} + \frac{5}{t + 1} = \frac{2(t + 1)}{3t(t + 1)} + \frac{5(3t)}{(t + 1)(3t)}$$

$$= \frac{2(t + 1) + 5(3t)}{3t(t + 1)} \qquad \text{Add numerators. Write the sum over the common denominator.}$$

$$= \frac{2t + 2 + 15t}{3t(t + 1)} \qquad \text{Apply the distributive property in the numerator.}$$

$$= \frac{17t + 2}{3t(t + 1)} \qquad \text{Combine like terms in the numerator.}$$

Example 4 Subtract: $\dfrac{7}{x - 3} - \dfrac{9}{3 - x}$

Solution To find a common denominator, we notice that $x - 3$ and $3 - x$ are opposites. That is, $3 - x = -(x - 3)$. We write the denominator $3 - x$ as $-(x - 3)$ and simplify.

$$\frac{7}{x - 3} - \frac{9}{3 - x} = \frac{7}{x - 3} - \frac{9}{-(x - 3)}$$

$$= \frac{7}{x - 3} - \frac{-9}{x - 3} \qquad \text{Apply } \frac{a}{-b} = \frac{-a}{b}.$$

$$= \frac{7 - (-9)}{x - 3} \qquad \begin{array}{l}\text{Subtract numerators. Write the difference}\\ \text{over the common denominator.}\end{array}$$

$$= \frac{16}{x - 3}$$

Example 5 Add: $1 + \dfrac{m}{m + 1}$

Solution Recall that 1 is the same as $\dfrac{1}{1}$. The LCD of and $\dfrac{1}{1}$ and $\dfrac{m}{m + 1}$ is $m + 1$.

$$1 + \frac{m}{m + 1} = \frac{1}{1} + \frac{m}{m + 1} \qquad \text{Write 1 as } \tfrac{1}{1}.$$

$$= \frac{1(m + 1)}{1(m + 1)} + \frac{m}{m + 1} \qquad \begin{array}{l}\text{Multiply both the numerator and the}\\ \text{denominator of } \tfrac{1}{1} \text{ by } m + 1.\end{array}$$

$$= \frac{m + 1 + m}{m + 1} \qquad \begin{array}{l}\text{Add numerators. Write the sum over the}\\ \text{common denominator.}\end{array}$$

$$= \frac{2m + 1}{m + 1} \qquad \text{Combine like terms in the numerator.}$$

Example 6 Subtract: $\dfrac{3}{2x^2 + x} - \dfrac{2x}{6x + 3}$

Solution First, we factor the denominators.

$$\frac{3}{2x^2 + x} - \frac{2x}{6x + 3} = \frac{3}{x(2x + 1)} - \frac{2x}{3(2x + 1)}$$

The LCD is $3x(2x + 1)$. We write equivalent expressions with denominators of $3x(2x + 1)$.

$$= \frac{3(3)}{x(2x + 1)(3)} - \frac{2x(x)}{3(2x + 1)(x)}$$

$$= \frac{9 - 2x^2}{3x(2x + 1)} \qquad \begin{array}{l}\text{Subtract numerators. Write the difference}\\ \text{over the common denominator.}\end{array}$$

Example 7   Add: $\dfrac{2x}{x^2 + 2x + 1} + \dfrac{x}{x^2 - 1}$

Solution   First we factor the denominators.

$$\frac{2x}{x^2 + 2x + 1} + \frac{x}{x^2 - 1} = \frac{2x}{(x + 1)(x + 1)} + \frac{x}{(x + 1)(x - 1)}$$

Now we write the rational expressions as equivalent expressions with denominators of $(x + 1)(x + 1)(x - 1)$, the LCD.

$$= \frac{2x(x - 1)}{(x + 1)(x + 1)(x - 1)} + \frac{x(x + 1)}{(x + 1)(x - 1)(x + 1)}$$

$$= \frac{2x(x - 1) + x(x + 1)}{(x + 1)^2(x - 1)} \qquad \text{Add numerators. Write the sum over the common denominator.}$$

$$= \frac{2x^2 - 2x + x^2 + x}{(x + 1)^2(x - 1)} \qquad \text{Apply the distributive property in the numerator.}$$

$$= \frac{3x^2 - x}{(x + 1)^2(x - 1)} \quad \text{or} \quad \frac{x(3x - 1)}{(x + 1)^2(x - 1)}$$

The numerator was factored as a last step to see if the rational expression could be simplified further. Since there are no factors common to the numerator and the denominator, we can't simplify further. ■

GRAPHING CALCULATOR EXPLORATIONS

A graphing calculator can be used to support the results of operations on rational expressions. For example, to verify the result of Example 2, graph

$$Y_1 = \frac{6x}{x^2 - 4} - \frac{3}{x + 2} \quad \text{and} \quad Y_2 = \frac{3}{x - 2}$$

on the same set of axes. The graphs should be the same. Use a Table feature or a Trace feature to see that this is true.

MENTAL MATH

Match each exercise with the first step needed to perform the operation. Do not actually perform the operation.

1. $\dfrac{3}{4} - \dfrac{y}{4}$ 　　　 **2.** $\dfrac{2}{a} \cdot \dfrac{3}{(a + 6)}$ 　　　 **3.** $\dfrac{x + 1}{x} \div \dfrac{x - 1}{x}$ 　　　 **4.** $\dfrac{9}{x - 2} - \dfrac{x}{x + 2}$

A. Multiply the first rational expression by the reciprocal of the second rational expression.

B. Find the LCD. Write each expression as an equivalent expression with the LCD as denominator.

C. Multiply numerators, then multiply denominators.

D. Subtract numerators. Place the difference over a common denominator.

Exercise Set 7.4

Perform the indicated operations. See Example 1.

1. $\dfrac{4}{2x} + \dfrac{9}{3x}$

2. $\dfrac{15}{7a} + \dfrac{8}{6a}$

3. $\dfrac{15a}{b} + \dfrac{6b}{5}$

4. $\dfrac{4c}{d} + \dfrac{8x}{5}$

5. $\dfrac{3}{x} + \dfrac{5}{2x^2}$

6. $\dfrac{14}{3x^2} + \dfrac{6}{x}$

Perform the indicated operations. See Examples 2 and 3.

7. $\dfrac{6}{x+1} + \dfrac{9}{2x+2}$

8. $\dfrac{8}{x+4} - \dfrac{3}{3x+12}$

9. $\dfrac{15}{2x-4} + \dfrac{x}{x^2-4}$

10. $\dfrac{3}{x+2} - \dfrac{1}{x^2-4}$

11. $\dfrac{3}{4x} + \dfrac{8}{x-2}$

12. $\dfrac{x}{x+1} + \dfrac{3}{x-1}$

13. $\dfrac{5}{y^2} - \dfrac{y}{2y+1}$

14. $\dfrac{x}{4x-3} - \dfrac{3}{8x-6}$

15. In your own words, explain how to add two rational expressions with unlike denominators.

16. In your own words, explain how to subtract two rational expressions with unlike denominators.

Add or subtract as indicated. See Example 4.

17. $\dfrac{6}{x-3} + \dfrac{8}{3-x}$

18. $\dfrac{9}{x-3} + \dfrac{9}{3-x}$

19. $\dfrac{-8}{x^2-1} - \dfrac{7}{1-x^2}$

20. $\dfrac{-9}{25x^2-1} + \dfrac{7}{1-25x^2}$

21. $\dfrac{x}{x^2-4} - \dfrac{2}{4-x^2}$

22. $\dfrac{5}{2x-6} - \dfrac{3}{6-2x}$

Add or subtract as indicated. See Example 5.

23. $\dfrac{5}{x} + 2$

24. $\dfrac{7}{x^2} - 5x$

25. $\dfrac{5}{x-2} + 6$

26. $\dfrac{6y}{y+5} + 1$

27. $\dfrac{y+2}{y+3} - 2$

28. $\dfrac{7}{2x-3} - 3$

△ **29.** Two angles are said to be complementary if their sum is 90°. If one angle measures $\frac{40}{x}$ degrees, find the measure of its complement.

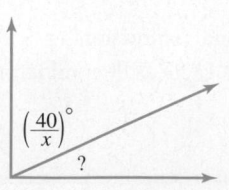

△ **30.** Two angles are said to be supplementary if their sum is 180°. If one angle measures $\dfrac{x+2}{x}$ degrees, find the measure of its supplement.

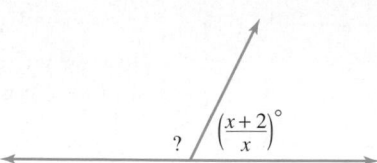

Perform the indicated operations. See Examples 1 through 7.

31. $\dfrac{5x}{x+2} - \dfrac{3x-4}{x+2}$

32. $\dfrac{7x}{x-3} - \dfrac{4x+9}{x-3}$

33. $\dfrac{3x^4}{x} - \dfrac{4x^2}{x^2}$

34. $\dfrac{5x}{6} + \dfrac{15x^2}{2}$

35. $\dfrac{1}{x+3} - \dfrac{1}{(x+3)^2}$

36. $\dfrac{5x}{(x-2)^2} - \dfrac{3}{x-2}$

37. $\dfrac{4}{5b} + \dfrac{1}{b-1}$

38. $\dfrac{1}{y+5} + \dfrac{2}{3y}$

39. $\dfrac{2}{m} + 1$

40. $\dfrac{6}{x} - 1$

41. $\dfrac{6}{1-2x} - \dfrac{4}{2x-1}$

42. $\dfrac{10}{3n-4} - \dfrac{5}{4-3n}$

43. $\dfrac{7}{(x+1)(x-1)} + \dfrac{8}{(x+1)^2}$

44. $\dfrac{5x+2}{(x+1)(x+5)} - \dfrac{2}{x+5}$

45. $\dfrac{x}{x^2-1} - \dfrac{2}{x^2-2x+1}$

46. $\dfrac{x}{x^2-4} - \dfrac{5}{x^2-4x+4}$

47. $\dfrac{3a}{2a+6} - \dfrac{a-1}{a+3}$

48. $\dfrac{1}{x+y} - \dfrac{y}{x^2-y^2}$

49. $\dfrac{5}{2-x} + \dfrac{x}{2x-4}$

50. $\dfrac{-1}{a-2} + \dfrac{4}{4-2a}$

51. $\dfrac{-7}{y^2-3y+2} - \dfrac{2}{y-1}$

52. $\dfrac{2}{x^2+4x+4} + \dfrac{1}{x+2}$

53. $\dfrac{13}{x^2-5x+6} - \dfrac{5}{x-3}$

54. $\dfrac{27}{y^2-81} + \dfrac{3}{2(y+9)}$

55. $\dfrac{8}{(x+2)(x-2)} + \dfrac{4}{(x+2)(x-3)}$

56. $\dfrac{5}{6x^2(x+2)} + \dfrac{4x}{x(x+2)^2}$

57. $\dfrac{5}{9x^2-4} + \dfrac{2}{3x-2}$

58. $\dfrac{4}{x^2 - x - 6} + \dfrac{x}{x^2 + 5x + 6}$

59. 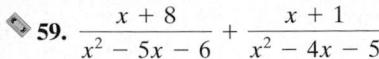 $\dfrac{x + 8}{x^2 - 5x - 6} + \dfrac{x + 1}{x^2 - 4x - 5}$

60. $\dfrac{x}{x^2 + 12x + 20} - \dfrac{1}{x^2 + 8x - 20}$

61. A board of length $\dfrac{3}{x + 4}$ inches was cut into two pieces. If one piece is $\dfrac{1}{x - 4}$ inches, express the length of the other board as a rational expression.

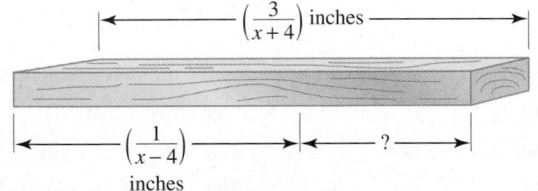

$\left(\dfrac{3}{x+4}\right)$ inches

$\left(\dfrac{1}{x-4}\right)$ inches

?

△ **62.** The length of the rectangle is $\dfrac{3}{y - 5}$ feet, while its width is $\dfrac{2}{y}$ feet. Find its perimeter and then find its area.

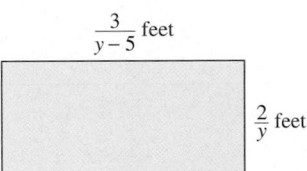

$\dfrac{3}{y-5}$ feet

$\dfrac{2}{y}$ feet

73. $\dfrac{5}{x^2 - 3x + 2} + \dfrac{1}{x - 2}$

74. $\dfrac{4}{2x^2 + 5x - 3} + \dfrac{2}{x + 3}$

75. $\left(\dfrac{2}{3} - \dfrac{1}{x}\right) \cdot \left(\dfrac{3}{x} + \dfrac{1}{2}\right)$

76. $\left(\dfrac{2}{3} - \dfrac{1}{x}\right) \div \left(\dfrac{3}{x} + \dfrac{1}{2}\right)$

77. $\left(\dfrac{1}{x} + \dfrac{2}{3}\right) - \left(\dfrac{1}{x} - \dfrac{2}{3}\right)$

78. $\left(\dfrac{1}{2} + \dfrac{2}{x}\right) - \left(\dfrac{1}{2} - \dfrac{1}{x}\right)$

79. $\left(\dfrac{2a}{3}\right)^2 \div \left(\dfrac{a^2}{a + 1} - \dfrac{1}{a + 1}\right)$

80. $\left(\dfrac{x + 2}{2x} - \dfrac{x - 2}{2x}\right) \cdot \left(\dfrac{5x}{4}\right)^2$

81. $\left(\dfrac{2x}{3}\right)^2 \div \left(\dfrac{x}{3}\right)^2$

82. $\left(\dfrac{2x}{3}\right)^2 \cdot \left(\dfrac{3}{x}\right)^2$

Use a graphing calculator to support the results of each exercise.

83. Exercise 5   **84.** Exercise 6

85. Exercise 31   **86.** Exercise 32

87. Explain when the LCD is the product of the denominators.

88. Explain when the LCD is the same as one of the denominators of a rational expression to be added or subtracted.

REVIEW EXERCISES

Simplify. Follow the order shown.

89. $\dfrac{\left.\dfrac{3}{4} + \dfrac{1}{4}\right\}}{\left.\dfrac{3}{8} + \dfrac{13}{8}\right\}}$ ← ① Add. ③ Divide. ② Add.

90. $\dfrac{\left.\dfrac{9}{5} + \dfrac{6}{5}\right\}}{\left.\dfrac{17}{6} + \dfrac{7}{6}\right\}}$ ← ① Add. ③ Divide. ② Add.

91. $\dfrac{\left.\dfrac{2}{5} + \dfrac{1}{5}\right\}}{\left.\dfrac{7}{10} + \dfrac{7}{10}\right\}}$ ← ① Add. ③ Divide. ② Add.

92. $\dfrac{\left.\dfrac{1}{4} + \dfrac{5}{4}\right\}}{\left.\dfrac{3}{8} + \dfrac{7}{8}\right\}}$ ← ① Add. ③ Divide. ② Add.

Multiple choice. Select the correct result.

63. $\dfrac{3}{x} + \dfrac{y}{x} =$ **A.** $\dfrac{3 + y}{x^2}$ **B.** $\dfrac{3 + y}{2x}$ **C.** $\dfrac{3 + y}{x}$

64. $\dfrac{3}{x} - \dfrac{y}{x} =$ **A.** $\dfrac{3 - y}{x^2}$ **B.** $\dfrac{3 - y}{2x}$ **C.** $\dfrac{3 - y}{x}$

65. $\dfrac{3}{x} \cdot \dfrac{y}{x} =$ **A.** $\dfrac{3y}{x}$ **B.** $\dfrac{3y}{x^2}$ **C.** $3y$

66. $\dfrac{3}{x} \div \dfrac{y}{x} =$ **A.** $\dfrac{3}{y}$ **B.** $\dfrac{y}{3}$ **C.** $\dfrac{3}{x^2 y}$

Perform the indicated operations. Addition, subtraction, multiplication, and division of rational expressions are included here.

67. $\dfrac{15x}{x + 8} \cdot \dfrac{2x + 16}{3x}$

68. $\dfrac{9z + 5}{15} \cdot \dfrac{5z}{81z^2 - 25}$

69. $\dfrac{8x + 7}{3x + 5} - \dfrac{2x - 3}{3x + 5}$

70. $\dfrac{2z^2}{4z - 1} - \dfrac{z - 2z^2}{4z - 1}$

71. $\dfrac{5a + 10}{18} \div \dfrac{a^2 - 4}{10a}$

72. $\dfrac{9}{x^2 - 1} \div \dfrac{12}{3x + 3}$

Find the slope of each line. See Section 3.5.

93.
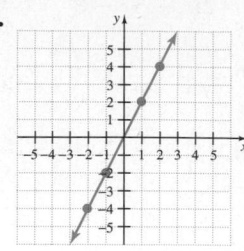

94.
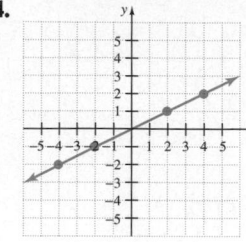

95.
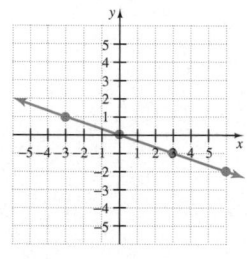

96.
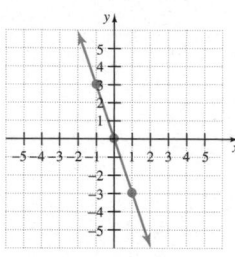

A Look Ahead

Example
Add $x^{-1} + 3x^{-2}$.

Solution

$$x^{-1} + 3x^{-2} = \frac{1}{x} + \frac{3}{x^2}$$

$$= \frac{1 \cdot x}{x \cdot x} + \frac{3}{x^2}$$

$$= \frac{x}{x^2} + \frac{3}{x^2}$$

$$= \frac{x + 3}{x^2}$$

Perform the indicated operation. See the preceding example.

97. $x^{-1} + (2x)^{-1}$ **98.** $3y^{-1} + (4y)^{-1}$

99. $4x^{-2} - 3x^{-1}$ **100.** $(4x)^{-2} - (3x)^{-1}$

101. $x^{-3}(2x + 1) - 5x^{-2}$ **102.** $4x^{-3} + x^{-4}(5x + 7)$

7.5 SIMPLIFYING COMPLEX FRACTIONS

CD-ROM   SSM   SSG   Video

▶ **OBJECTIVES**

1. Simplify complex fractions by simplifying the numerator and denominator and then dividing.
2. Simplify complex fractions by multiplying by a common denominator.
3. Simplify expressions with negative exponents.

1 A rational expression whose numerator, denominator, or both contain one or more rational expressions is called a **complex rational expression** or a **complex fraction**.

Complex Fractions

$$\frac{\frac{1}{a}}{\frac{b}{2}} \qquad \frac{\frac{x}{2y^2}}{\frac{6x - 2}{9y}} \qquad \frac{x + \frac{1}{y}}{y + 1}$$

The parts of a complex fraction are

$$\frac{\frac{x}{y + 2}}{7 + \frac{1}{y}}$$

← numerator of complex fraction
← main fraction bar
← denominator of complex fraction

Our goal in this section is to simplify complex fractions. A complex fraction is simplified when it is in the form $\dfrac{P}{Q}$, where P and Q are polynomials that have no common factors. Two methods of simplifying complex fractions are introduced. The first method evolves from the definition of a fraction as a quotient.

SIMPLIFYING A COMPLEX FRACTION: METHOD I

Step 1: Simplify the numerator and the denominator of the complex fraction so that each is a single fraction.

Step 2: Perform the indicated division by multiplying the numerator of the complex fraction by the reciprocal of the denominator of the complex fraction.

Step 3: Simplify if possible.

Example 1 Simplify each complex fraction.

a. $\dfrac{\dfrac{2x}{27y^2}}{\dfrac{6x^2}{9}}$

b. $\dfrac{\dfrac{5x}{x+2}}{\dfrac{10}{x-2}}$

c. $\dfrac{\dfrac{x}{y^2}+\dfrac{1}{y}}{\dfrac{y}{x^2}+\dfrac{1}{x}}$

Solution **a.** The numerator of the complex fraction is already a single fraction, and so is the denominator. Perform the indicated division by multiplying the numerator, $\dfrac{2x}{27y^2}$, by the reciprocal of the denominator, $\dfrac{6x^2}{9}$. Then simplify.

$$\dfrac{\dfrac{2x}{27y^2}}{\dfrac{6x^2}{9}} = \dfrac{2x}{27y^2} \div \dfrac{6x^2}{9}$$

$$= \dfrac{2x}{27y^2} \cdot \dfrac{9}{6x^2} \qquad \text{Multiply by the reciprocal of } \dfrac{6x^2}{9}.$$

$$= \dfrac{2x \cdot 9}{27y^2 \cdot 6x^2}$$

$$= \dfrac{1}{9xy^2}$$

HELPFUL HINT
Both the numerator and denominator are single fractions, so we perform the indicated division.

b. $\dfrac{\left\{\dfrac{5x}{x+2}\right.}{\left\{\dfrac{10}{x-2}\right.} = \dfrac{5x}{x+2} \cdot \dfrac{x-2}{10}$ Multiply by the reciprocal of $\dfrac{10}{x-2}$.

$= \dfrac{5x(x-2)}{2 \cdot 5(x+2)}$

$= \dfrac{x(x-2)}{2(x+2)}$ Simplify.

c. First simplify the numerator and the denominator of the complex fraction separately so that each is a single fraction. Then perform the indicated division.

$\dfrac{\dfrac{x}{y^2} + \dfrac{1}{y}}{\dfrac{y}{x^2} + \dfrac{1}{x}} = \dfrac{\dfrac{x}{y^2} + \dfrac{1 \cdot y}{y \cdot y}}{\dfrac{y}{x^2} + \dfrac{1 \cdot x}{x \cdot x}}$ The LCD is y^2.

   The LCD is x^2.

$= \dfrac{\dfrac{x+y}{y^2}}{\dfrac{y+x}{x^2}}$ Add.

   Add.

$= \dfrac{x+y}{y^2} \cdot \dfrac{x^2}{y+x}$ Multiply by the reciprocal of $\dfrac{y+x}{x^2}$.

$= \dfrac{x^2(x+y)}{y^2(y+x)}$

$= \dfrac{x^2}{y^2}$ Simplify.

2 Next we look at another method of simplifying complex fractions. With this method we multiply the numerator and the denominator of the complex fraction by the LCD of all fractions in the complex fraction.

SIMPLIFYING A COMPLEX FRACTION: METHOD II

Step 1: Multiply the numerator and the denominator of the complex fraction by the LCD of the fractions in both the numerator and the denominator.

Step 2: Simplify.

Example 2   Simplify each complex fraction.

a. $\dfrac{\dfrac{5x}{x+2}}{\dfrac{10}{x-2}}$

b. $\dfrac{\dfrac{x}{y^2} + \dfrac{1}{y}}{\dfrac{y}{x^2} + \dfrac{1}{x}}$

Solution **a.** The least common denominator of $\dfrac{5x}{x+2}$ and $\dfrac{10}{x-2}$ is $(x+2)(x-2)$.

Multiply both the numerator, $\dfrac{5x}{x+2}$, and the denominator, $\dfrac{10}{x-2}$, by the LCD.

$$\frac{\dfrac{5x}{x+2}}{\dfrac{10}{x-2}} = \frac{\left(\dfrac{5x}{x+2}\right) \cdot (x+2)(x-2)}{\left(\dfrac{10}{x-2}\right) \cdot (x+2)(x-2)} \qquad \text{Multiply numerator and denominator by the LCD.}$$

$$= \frac{5x \cdot (x-2)}{2 \cdot 5 \cdot (x+2)} \qquad \text{Simplify.}$$

$$= \frac{x(x-2)}{2(x+2)} \qquad \text{Simplify.}$$

b. The least common denominator of $\dfrac{x}{y^2}, \dfrac{1}{y}, \dfrac{y}{x^2}$, and $\dfrac{1}{x}$ is x^2y^2.

$$\frac{\dfrac{x}{y^2}+\dfrac{1}{y}}{\dfrac{y}{x^2}+\dfrac{1}{x}} = \frac{\left(\dfrac{x}{y^2}+\dfrac{1}{y}\right) \cdot x^2y^2}{\left(\dfrac{y}{x^2}+\dfrac{1}{x}\right) \cdot x^2y^2} \qquad \text{Multiply the numerator and denominator by the LCD.}$$

$$= \frac{\dfrac{x}{y^2} \cdot x^2y^2 + \dfrac{1}{y} \cdot x^2y^2}{\dfrac{y}{x^2} \cdot x^2y^2 + \dfrac{1}{x} \cdot x^2y^2} \qquad \text{Use the distributive property.}$$

$$= \frac{x^3 + x^2y}{y^3 + xy^2} \qquad \text{Simplify.}$$

$$= \frac{x^2(x+y)}{y^2(y+x)} \qquad \text{Factor.}$$

$$= \frac{x^2}{y^2} \qquad \text{Simplify.}$$

---

3 If an expression contains negative exponents, write the expression as an equivalent expression with positive exponents.

Example 3 Simplify.

$$\frac{x^{-1} + 2xy^{-1}}{x^{-2} - x^{-2}y^{-1}}$$

Solution This fraction does not appear to be a complex fraction. If we write it by using only positive exponents, however, we see that it is a complex fraction.

$$\frac{x^{-1} + 2xy^{-1}}{x^{-2} - x^{-2}y^{-1}} = \frac{\dfrac{1}{x}+\dfrac{2x}{y}}{\dfrac{1}{x^2}-\dfrac{1}{x^2y}}$$

The LCD of $\dfrac{1}{x}$, $\dfrac{2x}{y}$, $\dfrac{1}{x^2}$, and $\dfrac{1}{x^2y}$ is x^2y. Multiply both the numerator and denominator by x^2y.

$$= \dfrac{\left(\dfrac{1}{x} + \dfrac{2x}{y}\right) \cdot x^2y}{\left(\dfrac{1}{x^2} - \dfrac{1}{x^2y}\right) \cdot x^2y}$$

$$= \dfrac{\dfrac{1}{x} \cdot x^2y + \dfrac{2x}{y} \cdot x^2y}{\dfrac{1}{x^2} \cdot x^2y - \dfrac{1}{x^2y} \cdot x^2y}$$ Apply the distributive property.

$$= \dfrac{xy + 2x^3}{y - 1}$$ Simplify.

Exercise Set 7.5

Simplify each complex fraction. See Examples 1 and 2.

1. $\dfrac{\dfrac{1}{3}}{\dfrac{2}{5}}$

2. $\dfrac{\dfrac{3}{5}}{\dfrac{4}{5}}$

3. $\dfrac{\dfrac{4}{x}}{\dfrac{5}{2x}}$

4. $\dfrac{\dfrac{5}{2x}}{\dfrac{4}{x}}$

5. $\dfrac{\dfrac{10}{3x}}{\dfrac{5}{6x}}$

6. $\dfrac{\dfrac{15}{2x}}{\dfrac{5}{6x}}$

7. $\dfrac{1 + \dfrac{2}{5}}{2 + \dfrac{3}{5}}$

8. $\dfrac{2 + \dfrac{1}{7}}{3 - \dfrac{4}{7}}$

9. $\dfrac{\dfrac{4}{x-1}}{\dfrac{x}{x-1}}$

10. $\dfrac{\dfrac{x}{x+2}}{\dfrac{2}{x+2}}$

11. $\dfrac{1 - \dfrac{2}{x}}{x - \dfrac{4}{9x}}$

12. $\dfrac{5 - \dfrac{3}{x}}{x + \dfrac{2}{3x}}$

13. $\dfrac{\dfrac{1}{x+1} - 1}{\dfrac{1}{x-1} + 1}$

14. $\dfrac{1 + \dfrac{1}{x-1}}{1 - \dfrac{1}{x+1}}$

Simplify. See Example 3.

15. $\dfrac{x^{-1}}{x^{-2} + y^{-2}}$

16. $\dfrac{a^{-3} + b^{-1}}{a^{-2}}$

17. $\dfrac{2a^{-1} + 3b^{-2}}{a^{-1} - b^{-1}}$

18. $\dfrac{x^{-1} + y^{-1}}{3x^{-2} + 5y^{-2}}$

19. $\dfrac{1}{x - x^{-1}}$

20. $\dfrac{x^{-2}}{x + 3x^{-1}}$

Simplify.

21. $\dfrac{\dfrac{x+1}{7}}{\dfrac{x+2}{7}}$

22. $\dfrac{\dfrac{y}{10}}{\dfrac{x+1}{10}}$

23. $\dfrac{\dfrac{1}{2} - \dfrac{1}{3}}{\dfrac{3}{4} + \dfrac{2}{5}}$

24. $\dfrac{\dfrac{5}{6} - \dfrac{1}{2}}{\dfrac{1}{3} + \dfrac{1}{8}}$

25. $\dfrac{\dfrac{x+1}{3}}{\dfrac{2x-1}{6}}$

26. $\dfrac{\dfrac{x+3}{12}}{\dfrac{4x-5}{15}}$

27. $\dfrac{\dfrac{x}{3}}{\dfrac{2}{x+1}}$

28. $\dfrac{\dfrac{x-1}{5}}{\dfrac{3}{x}}$

29. $\dfrac{\dfrac{2}{x} + 3}{\dfrac{4}{x^2} - 9}$

30. $\dfrac{2 + \dfrac{1}{x}}{4x - \dfrac{1}{x}}$

31. $\dfrac{1 - \dfrac{x}{y}}{\dfrac{x^2}{y^2} - 1}$

32. $\dfrac{1 - \dfrac{2}{x}}{x - \dfrac{4}{x}}$

33. $\dfrac{\dfrac{-2x}{x - y}}{\dfrac{y}{x^2}}$

34. $\dfrac{\dfrac{7y}{x^2 + xy}}{\dfrac{y^2}{x^2}}$

35. $\dfrac{\dfrac{2}{x} + \dfrac{1}{x^2}}{\dfrac{y}{x^2}}$

36. $\dfrac{\dfrac{3}{x^2} - \dfrac{2}{x}}{\dfrac{1}{x} + 2}$

37. $\dfrac{\dfrac{x}{9} - \dfrac{1}{x}}{1 + \dfrac{3}{x}}$

38. $\dfrac{\dfrac{x}{4} - \dfrac{4}{x}}{1 - \dfrac{4}{x}}$

39. $\dfrac{\dfrac{x - 1}{x^2 - 4}}{1 + \dfrac{1}{x - 2}}$

40. $\dfrac{\dfrac{2}{x + 5} + \dfrac{4}{x + 3}}{\dfrac{3x + 13}{x^2 + 8x + 15}}$

41. $\dfrac{\dfrac{4}{5 - x} + \dfrac{5}{x - 5}}{\dfrac{2}{x} + \dfrac{3}{x - 5}}$

42. $\dfrac{\dfrac{3}{x - 4} - \dfrac{2}{4 - x}}{\dfrac{2}{x - 4} - \dfrac{2}{x}}$

43. $\dfrac{\dfrac{x + 2}{x} - \dfrac{2}{x - 1}}{\dfrac{x + 1}{x} + \dfrac{x + 1}{x - 1}}$

44. $\dfrac{\dfrac{5}{a + 2} - \dfrac{1}{a - 2}}{\dfrac{3}{2 + a} + \dfrac{6}{2 - a}}$

45. $\dfrac{\dfrac{x - 2}{x + 2} + \dfrac{x + 2}{x - 2}}{\dfrac{x - 2}{x + 2} - \dfrac{x + 2}{x - 2}}$

46. $\dfrac{\dfrac{x - 1}{x + 1} - \dfrac{x + 1}{x - 1}}{\dfrac{x - 1}{x + 1} + \dfrac{x + 1}{x - 1}}$

47. $\dfrac{\dfrac{2}{y^2} - \dfrac{5}{xy} - \dfrac{3}{x^2}}{\dfrac{2}{y^2} + \dfrac{7}{xy} + \dfrac{3}{x^2}}$

48. $\dfrac{\dfrac{2}{x^2} - \dfrac{1}{xy} - \dfrac{1}{y^2}}{\dfrac{1}{x^2} - \dfrac{3}{xy} + \dfrac{2}{y^2}}$

49. $\dfrac{a^{-1} + 1}{a^{-1} - 1}$

50. $\dfrac{a^{-1} - 4}{4 + a^{-1}}$

51. $\dfrac{3x^{-1} + (2y)^{-1}}{x^{-2}}$

52. $\dfrac{5x^{-2} - 3y^{-1}}{x^{-1} + y^{-1}}$

53. $\dfrac{2a^{-1} + (2a)^{-1}}{a^{-1} + 2a^{-2}}$

54. $\dfrac{a^{-1} + 2a^{-2}}{2a^{-1} + (2a)^{-1}}$

55. $\dfrac{5x^{-1} + 2y^{-1}}{x^{-2}y^{-2}}$

56. $\dfrac{x^{-2}y^{-2}}{5x^{-1} + 2y^{-1}}$

57. $\dfrac{5x^{-1} - 2y^{-1}}{25x^{-2} - 4y^{-2}}$

58. $\dfrac{3x^{-1} + 3y^{-1}}{4x^{-2} - 9y^{-2}}$

59. $\left(x^{-1} + y^{-1}\right)^{-1}$

60. $\dfrac{xy}{x^{-1} + y^{-1}}$

61. $\dfrac{x}{1 - \dfrac{1}{1 + \dfrac{1}{x}}}$

62. $\dfrac{1}{1 - \dfrac{1}{1 - \dfrac{1}{x}}}$

63. When the source of a sound is traveling toward a listener, the pitch that the listener hears due to the Doppler effect is given by the complex rational compression

$$\dfrac{a}{1 - \dfrac{s}{770}}, \text{ where } a \text{ is the actual pitch of the sound and } s$$

is the speed of the sound source. Simplify this expression.

64. Which of the following are equivalent to $\dfrac{\dfrac{1}{x}}{\dfrac{3}{y}}$?

a. $\dfrac{1}{x} \div \dfrac{3}{y}$

b. $\dfrac{1}{x} \cdot \dfrac{y}{3}$

c. $\dfrac{1}{x} \div \dfrac{y}{3}$

To find the average of two numbers, we find their sum and divide by 2. For example, the average of 65 and 81 is found by simplifying $\dfrac{65 + 81}{2}$. *This simplifies to* $\dfrac{146}{2} = 73$.

65. Find the average of $\dfrac{1}{3}$ and $\dfrac{3}{4}$.

66. Write the average of $\dfrac{3}{n}$ and $\dfrac{5}{n^2}$ as a simplified rational expression.

Solve.

67. In electronics, when two resistors R_1 (read R sub 1) and R_2 (read R sub 2) are connected in parallel, the total resistance is given by the complex fraction $\dfrac{1}{\dfrac{1}{R_1} + \dfrac{1}{R_2}}$. Simplify this expression.

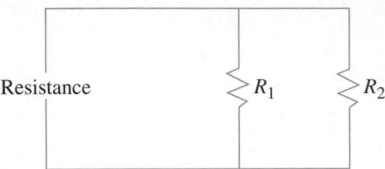

Resistance R_1 R_2

68. Astronomers occasionally need to know the day of the week a particular date fell on. The complex fraction

$$\frac{J + \dfrac{3}{2}}{7},$$

where J is the *Julian day number*, is used to make this calculation. Simplify this expression.

69. If the distance formula $d = r \cdot t$ is solved for t, then $t = \dfrac{d}{r}$. Use this formula to find t if distance d is $\dfrac{20x}{3}$ miles and rate r is $\dfrac{5x}{9}$ miles per hour. Write t in simplified form.

△ **70.** If the formula for area of a rectangle, $A = l \cdot w$, is solved for w, then $w = \dfrac{A}{l}$. Use this formula to find w if area A is $\dfrac{4x - 2}{3}$ square meters and length l is $\dfrac{6x - 3}{5}$ meters. Write w in simplified form.

*In the study of calculus, the difference quotient $\dfrac{f(a + h) - f(a)}{h}$ is often found and simplified. Find and simplify this quotient for each function $f(x)$ by following steps **a** through **d**.*

  a. *Find $f(a + h)$.*

  b. *Find $f(a)$.*

  c. *Use steps **a** and **b** to find $\dfrac{f(a + h) - f(a)}{h}$*

  d. *Simplify the result of step **c**.*

71. $f(x) = \dfrac{1}{x}$ **72.** $f(x) = \dfrac{5}{x}$

73. $\dfrac{3}{x + 1}$ **74.** $\dfrac{2}{x^2}$

REVIEW EXERCISES

Simplify. See Sections 5.1 and 5.2.

75. $\dfrac{3x^3 y^2}{12x}$ **76.** $\dfrac{-36xb^3}{9xb^2}$

77. $\dfrac{144x^5 y^5}{-16x^2 y}$ **78.** $\dfrac{48x^3 y^2}{-4xy}$

A Look Ahead

Example

Simplify $\dfrac{2(a + b)^{-1} - 5(a - b)^{-1}}{4(a^2 - b^2)^{-1}}$

Solution

$$\frac{2(a + b)^{-1} - 5(a - b)^{-1}}{4(a^2 - b^2)^{-1}} = \frac{\dfrac{2}{a + b} - \dfrac{5}{a - b}}{\dfrac{4}{a^2 - b^2}}$$

$$= \frac{\left(\dfrac{2}{a + b} - \dfrac{5}{a - b}\right) \cdot (a + b)(a - b)}{\left[\dfrac{4}{(a + b)(a - b)}\right] \cdot (a + b)(a - b)}$$

$$= \frac{\dfrac{2}{a + b} \cdot (a + b)(a - b) - \dfrac{5}{a - b} \cdot (a + b)(a - b)}{\dfrac{4(a + b)(a - b)}{(a + b)(a - b)}}$$

$$= \frac{2(a - b) - 5(a + b)}{4}$$

$$= \frac{-3a - 7b}{4}, \text{ or } -\frac{3a + 7b}{4}$$

Simplify. See the preceding example.

79. $\dfrac{1}{1 - (1 - x)^{-1}}$

80. $\dfrac{1}{1 + (1 + x)^{-1}}$

81. $\dfrac{(x + 2)^{-1} + (x - 2)^{-1}}{(x^2 - 4)^{-1}}$

82. $\dfrac{(y - 1)^{-1} - (y + 4)^{-1}}{(y^2 + 3y - 4)^{-1}}$

83. $\dfrac{3(a + 1)^{-1} + 4a^{-2}}{(a^3 + a^2)^{-1}}$

84. $\dfrac{9x^{-1} - 5(x - y)^{-1}}{4(x - y)^{-1}}$

7.6 SOLVING EQUATIONS CONTAINING RATIONAL EXPRESSIONS

CD-ROM   SSM

SSG   Video

▶ **OBJECTIVES**

1. Solve equations containing rational expressions.
2. Understand the difference between solving an equation and performing arithmetic operations on rational expressions.
3. Solve equations containing rational expressions for a specified variable.

1 In Chapter 2, we solved equations containing fractions. In this section, we continue the work we began in Chapter 2 by solving equations containing rational expressions.

Examples of Equations Containing Rational Expressions

$$\frac{x}{5} + \frac{x+2}{9} = 8 \quad \text{and} \quad \frac{x+1}{9x-5} = \frac{2}{3x}$$

To solve equations such as these, use the multiplication property of equality to clear the equation of fractions by multiplying both sides of the equation by the LCD.

Example 1 Solve: $\dfrac{x}{2} + \dfrac{8}{3} = \dfrac{1}{6}$

Solution The LCD of denominators 2, 3, and 6 is 6, so we multiply both sides of the equation by 6.

$$6\left(\frac{x}{2} + \frac{8}{3}\right) = 6\left(\frac{1}{6}\right)$$

$$6\left(\frac{x}{2}\right) + 6\left(\frac{8}{3}\right) = 6\left(\frac{1}{6}\right) \qquad \text{Use the distributive property.}$$

> **HELPFUL HINT**
> Make sure that *each* term is multiplied by the LCD.

$$3 \cdot x + 16 = 1 \qquad \text{Multiply and simplify.}$$

$$3x = -15 \qquad \text{Subtract 16 from both sides.}$$

$$x = -5 \qquad \text{Divide both sides by 3.}$$

Check To check, we replace x with -5 in the original equation.

$$\frac{-5}{2} + \frac{8}{3} \stackrel{?}{=} \frac{1}{6} \qquad \text{Replace } x \text{ with } -5.$$

$$\frac{1}{6} = \frac{1}{6} \qquad \text{True.}$$

This number checks, so the solution is -5.

Example 2   Solve: $\dfrac{t-4}{2} - \dfrac{t-3}{9} = \dfrac{5}{18}$

Solution   The LCD of denominators 2, 9, and 18 is 18, so we multiply both sides of the equation by 18.

$$18\left(\frac{t-4}{2} - \frac{t-3}{9}\right) = 18\left(\frac{5}{18}\right)$$

$$18\left(\frac{t-4}{2}\right) - 18\left(\frac{t-3}{9}\right) = 18\left(\frac{5}{18}\right) \qquad \text{Use the distributive property.}$$

> **HELPFUL HINT**
> Multiply *each* term by 18.

$$9(t-4) - 2(t-3) = 5 \qquad \text{Simplify.}$$

$$9t - 36 - 2t + 6 = 5 \qquad \text{Use the distributive property.}$$

$$7t - 30 = 5 \qquad \text{Combine like terms.}$$

$$7t = 35 \qquad \text{Add 30 to both sides.}$$

$$t = 5 \qquad \text{Solve for } t.$$

Check

$$\frac{t-4}{2} - \frac{t-3}{9} = \frac{5}{18}$$

$$\frac{5-4}{2} - \frac{5-3}{9} \stackrel{?}{=} \frac{5}{18} \qquad \text{Replace } t \text{ with 5.}$$

$$\frac{1}{2} - \frac{2}{9} \stackrel{?}{=} \frac{5}{18} \qquad \text{Simplify.}$$

$$\frac{5}{18} = \frac{5}{18} \qquad \text{True.}$$

The solution is 5.

Recall from Section 7.1 that a rational expression is defined for all real numbers except those that make the denominator of the expression 0. This means that if an equation contains *rational expressions with variables in the denominator*, we must be certain that the proposed solution does not make the denominator 0. If replacing the variable with the proposed solution makes the denominator 0, the rational expression is undefined and this proposed solution must be rejected.

Example 3   Solve: $3 - \dfrac{6}{x} = x + 8$

Solution In this equation, 0 cannot be a solution because if x is 0, the rational expression $\frac{6}{x}$ is undefined. The LCD is x, so we multiply both sides of the equation by x.

$$x\left(3 - \frac{6}{x}\right) = x(x + 8)$$

> **HELPFUL HINT**
> Multiply *each* term by x.

$$x(3) - x\left(\frac{6}{x}\right) = x \cdot x + x \cdot 8 \qquad \textit{Use the distributive property.}$$

$$3x - 6 = x^2 + 8x \qquad \textit{Simplify.}$$

Now we write the quadratic equation in standard form and solve for x.

$$0 = x^2 + 5x + 6$$

$$0 = (x + 3)(x + 2) \qquad \textit{Factor.}$$

$$x + 3 = 0 \qquad \text{or} \qquad x + 2 = 0 \qquad \textit{Set each factor equal to 0 and solve.}$$

$$x = -3 \qquad\qquad x = -2$$

Notice that neither -3 nor -2 makes the denominator in the original equation equal to 0.

Check To check these solutions, we replace x in the original equation by -3, and then by -2.

If $x = -3$:

$$3 - \frac{6}{x} = x + 8$$

$$3 - \frac{6}{-3} \stackrel{?}{=} -3 + 8$$

$$3 - (-2) \stackrel{?}{=} 5$$

$$5 = 5 \qquad \textit{True.}$$

If $x = -2$:

$$3 - \frac{6}{x} = x + 8$$

$$3 - \frac{6}{-2} \stackrel{?}{=} -2 + 8$$

$$3 - (-3) \stackrel{?}{=} 6$$

$$6 = 6 \qquad \textit{True.}$$

Both -3 and -2 are solutions. $\blacksquare$

The following steps may be used to solve an equation containing rational expressions.

SOLVING AN EQUATION CONTAINING RATIONAL EXPRESSIONS

Step 1. Multiply both sides of the equation by the LCD of all rational expressions in the equation.

Step 2. Remove any grouping symbols and solve the resulting equation.

Step 3. Check the solution in the original equation.

Example 4 Solve: $\dfrac{4x}{x^2 - 25} + \dfrac{2}{x - 5} = \dfrac{1}{x + 5}$

Solution The denominator $x^2 - 25$ factors as $(x + 5)(x - 5)$. The LCD is then $(x + 5)(x - 5)$, so we multiply both sides of the equation by this LCD.

$$(x + 5)(x - 5)\left(\dfrac{4x}{(x + 5)(x - 5)} + \dfrac{2}{x - 5}\right) = (x + 5)(x - 5)\left(\dfrac{1}{x + 5}\right)$$

Multiply by the LCD. Notice that −5 and 5 cannot be solutions.

$$(x + 5)(x - 5) \cdot \dfrac{4x}{x^2 - 25} + (x + 5)(x - 5) \cdot \dfrac{2}{x - 5}$$

Use the distributive property.

$$= (x + 5)(x - 5) \cdot \dfrac{1}{x + 5}$$

$$4x + 2(x + 5) = x - 5 \qquad \text{Simplify.}$$

$$4x + 2x + 10 = x - 5 \qquad \text{Use the distributive property.}$$

$$6x + 10 = x - 5 \qquad \text{Combine like terms.}$$

$$5x = -15$$

$$x = -3 \qquad \text{Divide both sides by 5.}$$

Check: Check by replacing x with −3 in the original equation. The solution is −3. ∎

Example 5 Solve: $\dfrac{2x}{x - 4} = \dfrac{8}{x - 4} + 1$

Solution Multiply both sides by the LCD, $x - 4$.

$$(x - 4)\left(\dfrac{2x}{x - 4}\right) = (x - 4)\left(\dfrac{8}{x - 4} + 1\right)$$

Multiply by the LCD. Notice that 4 cannot be a solution.

$$(x - 4) \cdot \dfrac{2x}{x - 4} = (x - 4) \cdot \dfrac{8}{x - 4} + (x - 4) \cdot 1$$

Use the distributive property.

$$2x = 8 + (x - 4) \qquad \text{Simplify.}$$

$$2x = 4 + x$$

$$x = 4$$

Notice that 4 makes the denominator 0 in the original equation. Therefore, 4 is *not* a solution and this equation has *no solution*. ∎

HELPFUL HINT
As we can see from Example 5, it is important to check the proposed solution(s) in the *original* equation.

Example 6   Solve: $x + \dfrac{14}{x-2} = \dfrac{7x}{x-2} + 1$

Solution   Notice the denominators in this equation. We can see that 2 can't be a solution. The LCD is $x - 2$, so we multiply both sides of the equation by $x - 2$.

$$(x-2)\left(x + \dfrac{14}{x-2}\right) = (x-2)\left(\dfrac{7x}{x-2} + 1\right)$$

$$(x-2)(x) + (x-2)\left(\dfrac{14}{x-2}\right) = (x-2)\left(\dfrac{7x}{x-2}\right) + (x-2)(1)$$

$$x^2 - 2x + 14 = 7x + x - 2 \qquad \text{Simplify.}$$

$$x^2 - 2x + 14 = 8x - 2 \qquad \text{Combine like terms.}$$

$$x^2 - 10x + 16 = 0 \qquad \text{Write the quadratic equation in standard form.}$$

$$(x-8)(x-2) = 0 \qquad \text{Factor.}$$

$$x - 8 = 0 \quad \text{or} \quad x - 2 = 0 \qquad \text{Set each factor equal to 0.}$$

$$x = 8 \qquad\qquad x = 2 \qquad \text{Solve.}$$

As we have already noted, 2 can't be a solution of the original equation. So we need only replace x with 8 in the original equation. We find that 8 is a solution; the only solution is 8.   ▬

2   At this point, let's make sure you understand the difference between solving an equation containing rational expressions and performing operations on rational expressions.

Example 7   **a.** Solve for x: $\dfrac{x}{4} + 2x = 9$.   **b.** Add: $\dfrac{x}{4} + 2x$.

Solution   **a.** This is an equation to solve for x. Begin by multiplying both sides by the LCD, 4.

$$4\left(\dfrac{x}{4} + 2x\right) = 4(9)$$

$$4\left(\dfrac{x}{4}\right) + 4(2x) = 4(9) \qquad \text{Apply the distributive property.}$$

$$x + 8x = 36 \qquad \text{Simplify.}$$

$$9x = 36 \qquad \text{Combine like terms.}$$

$$x = 4 \qquad \text{Solve.}$$

Check to see that 4 is the solution.

b. This example is **not an equation** to solve; it is an addition to perform. To add these rational expressions, find the LCD and write each rational expression as an equivalent expression whose denominator is the LCD. The LCD is 4.

$$\frac{x}{4} + 2x = \frac{x}{4} + \frac{2x(4)}{4}$$

$$= \frac{x + 8x}{4} \qquad \text{Add.}$$

$$= \frac{9x}{4} \qquad \text{Combine like terms in the numerator.} \quad ▬$$

3

The last example in this section is an equation containing several variables. We are directed to solve for one of them. The steps used in the preceeding examples can be applied to solve equations for a specified variable as well.

Example 8   Solve $\dfrac{1}{a} + \dfrac{1}{b} = \dfrac{1}{x}$ for x.

Solution   (This type of equation often models a work problem, as we shall see in Section 7.8.) The LCD is abx, so we multiply both sides by abx.

$$abx\left(\frac{1}{a} + \frac{1}{b}\right) = abx\left(\frac{1}{x}\right)$$

$$abx\left(\frac{1}{a}\right) + abx\left(\frac{1}{b}\right) = abx \cdot \frac{1}{x}$$

$$bx + ax = ab \qquad \text{Simplify.}$$

$$x(b + a) = ab \qquad \text{Factor out } x \text{ from each term on the left side.}$$

$$\frac{x(b + a)}{b + a} = \frac{ab}{b + a} \qquad \text{Divide both sides by } b + a.$$

$$x = \frac{ab}{b + a} \qquad \text{Simplify.}$$

This equation is now solved for x.   ▬

GRAPHING CALCULATOR EXPLORATIONS

A graphing calculator may be used to check solutions of equations containing rational expressions. For example, to check the solution of Example 1, $\dfrac{x}{2} + \dfrac{8}{3} = \dfrac{1}{6}$, graph $y_1 = x/2 + 8/3$ and $y_2 = 1/6$.

Use TRACE and ZOOM, or use INTERSECT, to find the point of intersection. The point of intersection has an x-value of -5, so the solution of the equation is -5.

Use a graphing calculator to check the examples of this section.

1. Example 2   **2.** Example 7a   **3.** Example 3

4. Example 4   **5.** Example 5   **6.** Example 6

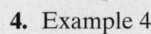

MENTAL MATH

Solve each equation for the variable.

1. $\dfrac{x}{5} = 2$ **2.** $\dfrac{x}{8} = 4$ **3.** $\dfrac{z}{6} = 6$ **4.** $\dfrac{y}{7} = 8$

Exercise Set 7.6

Solve each equation. See Examples 1 and 2.

1. $\dfrac{x}{5} + 3 = 9$ **2.** $\dfrac{x}{5} - 2 = 9$

3. $\dfrac{x}{2} + \dfrac{5x}{4} = \dfrac{x}{12}$ **4.** $\dfrac{x}{6} + \dfrac{4x}{3} = \dfrac{x}{18}$

5. $2 + \dfrac{10}{x} = x + 5$ **6.** $6 + \dfrac{5}{y} = y - \dfrac{2}{y}$

7. $\dfrac{a}{5} = \dfrac{a - 3}{2}$ **8.** $\dfrac{2b}{5} = \dfrac{b + 2}{6}$

9. $\dfrac{x - 3}{5} + \dfrac{x - 2}{2} = \dfrac{1}{2}$ **10.** $\dfrac{a + 5}{4} + \dfrac{a + 5}{2} = \dfrac{a}{8}$

Recall that two angles are supplementary if the sum of their measures is 180°. Find the measures of the following supplementary angles.

△ **11.**

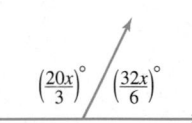

△ **12.**

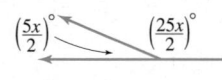

Recall that two angles are complementary if the sum of their measures is 90°. Find the measures of the following complementary angles.

△ **13.**

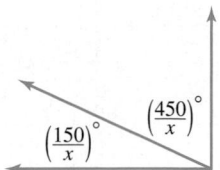

△ **14.**

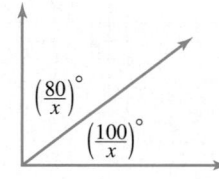

Solve each equation. See Examples 3 through 6.

15. $\dfrac{9}{2a - 5} = -2$ **16.** $\dfrac{6}{4 - 3x} = 3$

17. $\dfrac{y}{y + 4} + \dfrac{4}{y + 4} = 3$ **18.** $\dfrac{5y}{y + 1} - \dfrac{3}{y + 1} = 4$

19. $\dfrac{2x}{x + 2} - 2 = \dfrac{x - 8}{x - 2}$ **20.** $\dfrac{4y}{y - 3} - 3 = \dfrac{3y - 1}{y + 3}$

21. $\dfrac{4y}{y - 4} + 5 = \dfrac{5y}{y - 4}$ **22.** $\dfrac{2a}{a + 2} - 5 = \dfrac{7a}{a + 2}$

23. $\dfrac{7}{x - 2} + 1 = \dfrac{x}{x + 2}$ **24.** $1 + \dfrac{3}{x + 1} = \dfrac{x}{x - 1}$

25. $\dfrac{x + 1}{x + 3} = \dfrac{2x^2 - 15x}{x^2 + x - 6} - \dfrac{x - 3}{x - 2}$

26. $\dfrac{3}{x + 3} = \dfrac{12x + 19}{x^2 + 7x + 12} - \dfrac{5}{x + 4}$

27. $\dfrac{y}{2y + 2} + \dfrac{2y - 16}{4y + 4} = \dfrac{2y - 3}{y + 1}$

28. $\dfrac{1}{x + 2} = \dfrac{4}{x^2 - 4} - \dfrac{1}{x - 2}$

Determine whether each of the following is an equation or an expression. If it is an equation, then solve it for its variable. If it is an expression, perform the indicated operation. See Example 7.

29. $\dfrac{1}{x} + \dfrac{2}{3}$ **30.** $\dfrac{3}{a} + \dfrac{5}{6}$

31. $\dfrac{1}{x} + \dfrac{2}{3} = \dfrac{3}{x}$ **32.** $\dfrac{3}{a} + \dfrac{5}{6} = 1$

33. $\dfrac{2}{x + 1} - \dfrac{1}{x}$ **34.** $\dfrac{4}{x - 3} - \dfrac{1}{x}$

35. $\dfrac{2}{x + 1} - \dfrac{1}{x} = 1$

36. $\dfrac{4}{x - 3} - \dfrac{1}{x} = \dfrac{6}{x(x - 3)}$

37. Explain the difference between solving an equation such as $\dfrac{x}{2} + \dfrac{3}{4} = \dfrac{x}{4}$ for x and performing an operation such as adding $\dfrac{x}{2} + \dfrac{3}{4}$.

38. When solving an equation such as $\dfrac{y}{4} = \dfrac{y}{2} - \dfrac{1}{4}$, we may multiply all terms by 4. When subtracting two rational expressions such as $\dfrac{y}{2} - \dfrac{1}{4}$, we may not. Explain why.

Solve each equation.

39. $\dfrac{2x}{7} - 5x = 9$

40. $\dfrac{4x}{8} - 5x = 10$

41. $\dfrac{2}{y} + \dfrac{1}{2} = \dfrac{5}{2y}$

42. $\dfrac{6}{3y} + \dfrac{3}{y} = 1$

43. $\dfrac{4x + 10}{7} = \dfrac{8}{2}$

44. $\dfrac{1}{2} = \dfrac{x + 1}{8}$

45. $2 + \dfrac{3}{a - 3} = \dfrac{a}{a - 3}$

46. $\dfrac{2y}{y - 2} - \dfrac{4}{y - 2} = 4$

47. $\dfrac{5}{x} + \dfrac{2}{3} = \dfrac{7}{2x}$

48. $\dfrac{5}{3} - \dfrac{3}{2x} = \dfrac{5}{4}$

49. $\dfrac{2a}{a + 4} = \dfrac{3}{a - 1}$

50. $\dfrac{5}{3x - 8} = \dfrac{x}{x - 2}$

51. $\dfrac{x + 1}{3} - \dfrac{x - 1}{6} = \dfrac{1}{6}$

52. $\dfrac{3x}{5} - \dfrac{x - 6}{3} = \dfrac{1}{5}$

53. $\dfrac{4r - 1}{r^2 + 5r - 14} + \dfrac{2}{r + 7} = \dfrac{1}{r - 2}$

54. $\dfrac{2t + 3}{t - 1} - \dfrac{2}{t + 3} = \dfrac{5 - 6t}{t^2 + 2t - 3}$

55. $\dfrac{t}{t - 4} = \dfrac{t + 4}{6}$

56. $\dfrac{15}{x + 4} = \dfrac{x - 4}{x}$

57. $\dfrac{x}{2x + 6} + \dfrac{x + 1}{3x + 9} = \dfrac{2}{4x + 12}$

58. $\dfrac{a}{5a - 5} - \dfrac{a - 2}{2a - 2} = \dfrac{5}{4a - 4}$

Solve each equation for the indicated variable. See Example 8.

59. $\dfrac{D}{R} = T$; for R

60. $\dfrac{A}{W} = L$; for W

61. $\dfrac{3}{x} = \dfrac{5y}{x + 2}$; for y

62. $\dfrac{7x - 1}{2x} = \dfrac{5}{y}$; for y

63. $\dfrac{3a + 2}{3b - 2} = -\dfrac{4}{2a}$; for b

64. $\dfrac{6x + y}{7x} = \dfrac{3x}{h}$; for h

65. $\dfrac{A}{BH} = \dfrac{1}{2}$; for B

66. $\dfrac{V}{\pi r^2 h} = 1$; for h

67. $\dfrac{C}{\pi r} = 2$; for r

68. $\dfrac{3V}{A} = H$; for V

69. $\dfrac{1}{a} = \dfrac{1}{b} + \dfrac{1}{c}$; for a

70. $\dfrac{1}{2} - \dfrac{1}{x} = \dfrac{1}{y}$; for x

71. $\dfrac{m^2}{6} - \dfrac{n}{3} = \dfrac{p}{2}$; for n

72. $\dfrac{x^2}{r} + \dfrac{y^2}{t} = 1$; for r

Solve each equation.

73. $\dfrac{5}{a^2 + 4a + 3} + \dfrac{2}{a^2 + a - 6} - \dfrac{3}{a^2 - a - 2} = 0$

74. $-\dfrac{2}{a^2 + 2a - 8} + \dfrac{1}{a^2 + 9a + 20} = \dfrac{-4}{a^2 + 3a - 10}$

Solve each equation. Begin by writing each equation with positive exponents only.

75. $x^{-2} - 19x^{-1} + 48 = 0$

76. $x^{-2} - 5x^{-1} - 36 = 0$

77. $p^{-2} + 4p^{-1} - 5 = 0$

78. $6p^{-2} - 5p^{-1} + 1 = 0$

Solve each equation. Round solutions to two decimal places.

79. $\dfrac{1.4}{x - 2.6} = \dfrac{-3.5}{x + 7.1}$

80. $\dfrac{-8.5}{x + 1.9} = \dfrac{5.7}{x - 3.6}$

81. $\dfrac{10.6}{y} - 14.7 = \dfrac{9.92}{3.2} + 7.6$

82. $\dfrac{12.2}{x} + 17.3 = \dfrac{9.6}{x} - 14.7$

REVIEW EXERCISES

Identify the x- and y-intercepts. See Section 3.4.

83.

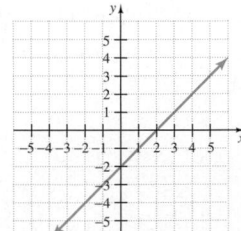

84.

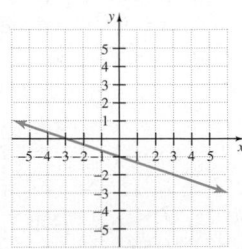

85.

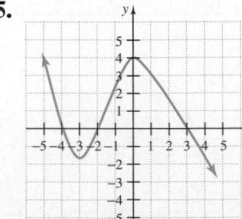

86.

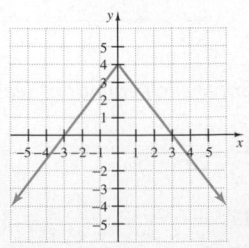

7.7 RATIO AND PROPORTION

CD-ROM   SSM

SSG   Video

▶ OBJECTIVES

1. Use fractional notation to express ratios.
2. Identify and solve proportions.
3. Use proportions to solve problems.
4. Determine unit pricing.

1   A **ratio** is the quotient of two numbers or two quantities.

RATIO

If a and b are two numbers and $b \neq 0$, the **ratio of a to b** is the quotient of a and b. The ratio of a to b can also be written as

$$a:b \text{ or as } \frac{a}{b}$$

Example 1   Write a ratio for each phrase. Use fractional notation.

a. The ratio of 2 parts salt to 5 parts water
b. The ratio of 18 inches to 2 feet

Solution   **a.** The ratio of 2 parts salt to 5 parts water is $\frac{2}{5}$.

b. When comparing measurements, use the same unit of measurement in the numerator as in the denominator. Here, we write 2 feet as $2 \cdot 12$ inches, or 24 inches. The ratio of 18 inches to 2 feet is then $\frac{18}{24} = \frac{3}{4}$ in simplest form.

2   If two ratios are equal, we say the ratios are **in proportion** to each other. A **proportion** is a mathematical statement that two ratios are equal.

For example, the equation $\frac{1}{2} = \frac{4}{8}$ is a proportion, as is $\frac{x}{5} = \frac{8}{10}$, because both sides of the equations are ratios. When we want to emphasize the equation as a proportion, we

read the proportion $\frac{1}{2} = \frac{4}{8}$ as "one is to two as four is to eight"

In a proportion, cross products are equal. To understand cross products, let's start with the proportion

$$\frac{a}{b} = \frac{c}{d}$$

and multiply both sides by the LCD, bd.

$$bd\left(\frac{a}{b}\right) = bd\left(\frac{c}{d}\right) \qquad \text{Multiply both sides by the LCD, } bd.$$

$$\underbrace{ad}_{\nearrow} = \underbrace{bc}_{\nwarrow} \qquad \text{Simplify.}$$

$$\text{cross product} \qquad \text{cross product}$$

Notice why ad and bc are called cross products.

$$\frac{a}{b} = \frac{c}{d}$$

$$bc$$
$$ad$$

> ## CROSS PRODUCTS
>
> If $\dfrac{a}{b} = \dfrac{c}{d}$, then $ad = bc$.

For example, since

$$\frac{1}{2} = \frac{4}{8}, \quad \text{then} \quad 1 \cdot 8 = 2 \cdot 4 \quad \text{or}$$
$$8 = 8$$

Example 2   Solve for x: $\dfrac{45}{x} = \dfrac{5}{7}$

Solution   To solve, we set cross products equal.

$$\frac{45}{x} = \frac{5}{7}$$

$$45 \cdot 7 = x \cdot 5 \qquad \text{Set cross products equal.}$$
$$315 = 5x \qquad \text{Multiply.}$$
$$\frac{315}{5} = \frac{5x}{5} \qquad \text{Divide both sides by 5.}$$
$$63 = x \qquad \text{Simplify.}$$

Check: To check, substitute 63 for x in the original proportion. The solution is 63. ■

Example 3   Solve for x: $\dfrac{x-5}{3} = \dfrac{x+2}{5}$

Solution

$$\frac{x-5}{3} = \frac{x+2}{5}$$

$5(x - 5) = 3(x + 2)$ Set cross products equal.

$5x - 25 = 3x + 6$ Multiply.

$5x = 3x + 31$ Add 25 to both sides.

$2x = 31$ Subtract $3x$ from both sides.

$\dfrac{2x}{2} = \dfrac{31}{2}$ Divide both sides by 2.

$x = \dfrac{31}{2}$

Check: Verify that $\dfrac{31}{2}$ is the solution.   ▬

3 Proportions can be used to model and solve many real-life problems. When using proportions in this way, it is important to judge whether the solution is reasonable. Doing so helps us to decide if the proportion has been formed correctly. We use the same problem-solving steps that were introduced in Section 2.4.

Example 4   **CALCULATING THE COST OF RECORDABLE COMPACT DISCS**

Three boxes of CD-Rs (recordable compact discs) cost $37.47. How much should 5 boxes cost?

Solution

1. UNDERSTAND. Read and reread the problem. We know that the cost of 5 boxes is more than the cost of 3 boxes, or $37.47, and less than the cost of 6 boxes, which is double the cost of 3 boxes, or 2($37.47) = $74.94. Let's suppose that 5 boxes cost $60.00. To check, we see if 3 boxes is to 5 boxes as the *price* of 3 boxes is to the *price* of 5 boxes. In other words, we see if

$$\frac{3 \text{ boxes}}{5 \text{ boxes}} = \frac{\text{price of 3 boxes}}{\text{price of 5 boxes}}$$

or

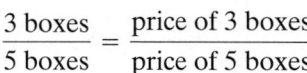

$$\frac{3}{5} = \frac{37.47}{60.00}$$

or

$3(60.00) = 5(37.47)$ Set cross products equal.

$180.00 = 187.35$ Not a true statement.

Thus, $60 is not correct, but we now have a better understanding of the problem.

Let x = price of 5 boxes of CD-Rs.

2. TRANSLATE.

$$\frac{3 \text{ boxes}}{5 \text{ boxes}} = \frac{\text{price of 3 boxes}}{\text{price of 5 boxes}}$$

$$\frac{3}{5} = \frac{37.47}{x}$$

3. SOLVE.

$$\frac{3}{5} = \frac{37.47}{x}$$

$3x = 5(37.47)$ *Set cross products equal.*

$3x = 187.35$

$x = 62.45$ *Divide both sides by 3.*

4. INTERPRET.

Check: Verify that 3 boxes is to 5 boxes as $37.47 is to $62.45. Also, notice that our solution is a reasonable one as discussed in Step 1.

State: Five boxes of CD-Rs cost $62.45.

HELPFUL HINT

The proportion $\dfrac{5 \text{ boxes}}{3 \text{ boxes}} = \dfrac{\text{price of 5 boxes}}{\text{price of 3 boxes}}$ could also have been used to solve the problem above. Notice that the cross products are the same.

---

4 When shopping for an item offered in many different sizes, it is important to be able to determine the best buy, or the best price per unit. To find the unit price of an item, divide the total price of the item by the total number of units.

$$\text{unit price} = \frac{\text{total price}}{\text{number of units}}$$

For example, if a 16-ounce can of green beans is priced at $0.88, its unit price is

$$\text{unit price} = \frac{\$0.88}{16} = \$0.055$$

Example 5 **COMPARING CEREAL COSTS**

A supermarket offers a 14-ounce box of cereal for $3.79 and an 18-ounce box of the same brand of cereal for $4.99. Which is the better buy?

Solution To find the better buy, we compare unit prices. The following unit prices were rounded to three decimal places.

Size	Price	Unit Price
14-ounce	$3.79	$\dfrac{\$3.79}{14} \approx \$0.271$
18-ounce	$4.99	$\dfrac{\$4.99}{18} \approx \$0.277$

The 14-ounce box of cereal has the lower unit price so it is the better buy. ▬

SPOTLIGHT ON DECISION MAKING

Suppose you must select a child care center for your two-year-old daughter. You have compiled information on two possible choices in the table shown at the right. Which child care center would you choose? Why?

	Center A	Center B
Weekly cost	$130	$110
Number of 2-year-olds	7	13
Number of adults for 2-year-olds	2	3
Distance from home	6 miles	11 miles

Exercise Set 7.7

Write each ratio in fractional notation in simplest form. See Example 1.

1. 2 megabytes to 15 megabytes
2. 18 disks to 41 disks
△ 3. 10 inches to 12 inches
4. 15 miles to 40 miles
5. 5 quarts to 3 gallons
6. 8 inches to 3 feet
◆ 7. 4 nickels to 2 dollars
8. 12 quarters to 2 dollars
9. 175 centimeters to 5 meters
10. 90 centimeters to 4 meters
11. 190 minutes to 3 hours
12. 60 hours to 2 days
13. Suppose someone tells you that the ratio of 11 inches to 2 feet is $\frac{11}{2}$. How do you correct that person and explain the error?
14. Write a ratio that can be written in fractional notation as $\frac{3}{2}$.

Solve each proportion. See Examples 2 and 3.

15. $\dfrac{2}{3} = \dfrac{x}{6}$

16. $\dfrac{x}{2} = \dfrac{16}{6}$

◆ 17. $\dfrac{x}{10} = \dfrac{5}{9}$

18. $\dfrac{9}{4x} = \dfrac{6}{2}$

19. $\dfrac{4x}{6} = \dfrac{7}{2}$

20. $\dfrac{a}{5} = \dfrac{3}{2}$

21. $\dfrac{a}{25} = \dfrac{12}{10}$

22. $\dfrac{n}{10} = 9$

23. $\dfrac{x-3}{x} = \dfrac{4}{7}$

24. $\dfrac{y}{y-16} = \dfrac{5}{3}$

25. $\dfrac{5x+1}{x} = \dfrac{6}{3}$

26. $\dfrac{3x-2}{5} = \dfrac{4x}{1}$

◆ 27. $\dfrac{x+1}{2x+3} = \dfrac{2}{3}$

28. $\dfrac{x+1}{x+2} = \dfrac{5}{3}$

29. $\dfrac{9}{5} = \dfrac{12}{3x+2}$

30. $\dfrac{6}{11} = \dfrac{27}{3x-2}$

31. $\dfrac{3}{x+1} = \dfrac{5}{2x}$

32. $\dfrac{7}{x-3} = \dfrac{8}{2x}$

Solve. See Example 4.

33. The ratio of the weight of an object on Earth to the weight of the same object on Pluto is 100 to 3. If an elephant weighs 4100 pounds on Earth, find the elephant's weight on Pluto.

34. If a 170-pound person weighs approximately 65 pounds on Mars, how much does a 9000-pound satellite weigh?

◆ 35. There are 110 calories per 28.4 grams of Crispy Rice cereal. Find how many calories are in 42.6 grams of this cereal.

36. On an architect's blueprint, 1 inch corresponds to 4 feet. Find the length of a wall represented by a line that is $3\frac{7}{8}$ inches long on the blueprint.

CHAPTER 7 RATIONAL EXPRESSIONS

37. A recent headline read, "Women Earn Bigger Checks in 1 of Every 6 Couples." If there are 23,000 couples in a nearby metropolitan area, how many women would you expect to earn bigger paychecks?

38. A human factors expert recommends that there be at least 9 square feet of floor space in a college classroom for every student in the class. Find the minimum floor space that 40 students need.

39. To mix weed killer with water correctly, it is necessary to mix 8 teaspoons of weed killer with 2 gallons of water. Find how many gallons of water are needed to mix with the entire box if it contains 36 teaspoons of weed killer.

40. Ken Hall, a tailback, holds the high school sports record for total yards rushed in a season. In 1953, he rushed for 4045 total yards in 12 games. Find his average rushing yards per game.

41. To estimate the number of people in Jackson, population 50,000, who have no health insurance, 250 people were polled. Of those polled, 39 had no insurance. How many people in the city might we expect to be uninsured?

42. The manufacturers of cans of salted mixed nuts state that the ratio of peanuts to other nuts is 3 to 2. If 324 peanuts are in a can, find how many other nuts should also be in the can.

43. There are 1280 calories in a 14-ounce portion of Eagle Brand Milk. Find how many calories are in 2 ounces of Eagle Brand Milk.

44. Due to space problems at a local university, a 20-foot by 12-foot conference room is converted into a classroom. Find the maximum number of students the room can accommodate. (See Exercise 38.)

Given the following prices charged for various sizes of an item, find the best buy. See Example 5.

45. Laundry detergent
110 ounces for $5.79
240 ounces for $13.99

46. Jelly
10 ounces for $1.14
15 ounces for $1.69

47. Tuna (in cans)
6 ounces for $0.69
8 ounces for $0.90
16 ounces for $1.89

48. Picante sauce
10 ounces for $0.99
16 ounces for $1.69
30 ounces for $3.29

49. Blank video cassettes
4-pack for $8.99
6-pack for $13.99

50. Recordable compact disc
10-pack for $14.99
100-pack for $99.99

51. Milk
1 quart for $1.57
half gallon (2 qt) $2.10
gallon (4 qt) for $3.99

52. Pens
3-pack for $1.99
6-pack for $3.69
dozen for $6.99

The following bar graph shows the capacity of the United States to generate electricity from the wind in the years shown. Use this graph for Exercises 53 and 54.

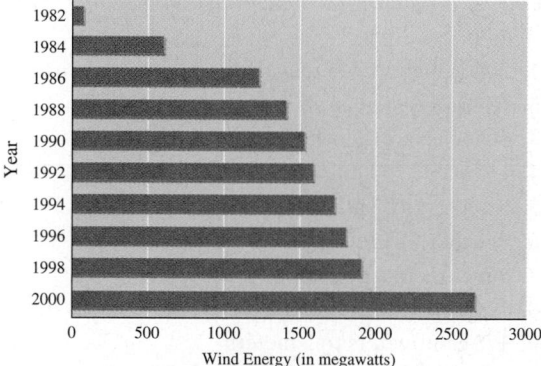

53. Find the approximate increase in megawatt capacity during the 2-year period from 1982 to 1984.

54. Find the approximate increase in megawatt capacity during the 2-year period from 1998 to 2000.

In general, 1000 megawatts will serve the average electricity needs of 560,000 people. Use this fact and the preceding graph to answer Exercises 55 and 56.

55. In 2000, the number of megawatts that can be generated from wind will serve the electricity needs of how many people?

56. How many megawatts of electricity are needed to serve the city or town in which you live?

57. If x is 10, is $\frac{2}{x}$ in proportion to $\frac{x}{50}$? Explain why or why not.

58. For what value of x is $\dfrac{x}{x-1}$ in proportion to $\dfrac{x+1}{x}$? Explain your result.

59. For which of the following equations can we immediately use cross products to solve for x?

a. $\dfrac{2-x}{5} = \dfrac{1+x}{3}$ **b.** $\dfrac{2}{5} - x = \dfrac{1+x}{3}$

Find the slope of the line through each pair of points. Use the slope to determine whether the line is vertical, horizontal, or moves upward or downward from left to right. See Section 3.5.

60. $(-2, 5), (4, -3)$ **61.** $(0, 4)\ (2, 10)$

62. $(-3, -6)(1, 5)$ **63.** $(-2, 7)(3, -2)$

64. $(3, 7)\ (3, -2)$ **65.** $(0, -4)\ (2, -4)$

7.8   RATIONAL EQUATIONS AND PROBLEM SOLVING

 CD-ROM

 SSM

SSG

Video

▶ **O B J E C T I V E S**

1. Translate sentences to equations containing rational expressions.
2. Solve problems involving work.
3. Solve problems involving distance.
4. Solve problems involving similar triangles.

1   In this section, we solve problems that can be modeled by equations containing rational expressions. To solve these problems, we use the same problem-solving steps that were first introduced in Section 2.4. In our first example, our goal is to find an unknown number.

Example 1   **FINDING AN UNKNOWN NUMBER**

The quotient of a number and 6 minus $\dfrac{5}{3}$ is the quotient of the number and 2. Find the number.

Solution   1.   UNDERSTAND. Read and reread the problem. Suppose that the unknown number is 2, then we see if the quotient of 2 and 6, or $\dfrac{2}{6}$, minus $\dfrac{5}{3}$ is equal to the quotient of 2 and 2, or $\dfrac{2}{2}$.

$$\frac{2}{6} - \frac{5}{3} = \frac{1}{3} - \frac{5}{3} = -\frac{4}{3}, \text{ not } \frac{2}{2}$$

Don't forget that the purpose of a proposed solution is to better understand the problem.

Let x = the unknown number.

2.   TRANSLATE.

In words:	the quotient of x and 6	minus	$\dfrac{5}{3}$	is	the quotient of x and 2
	↓	↓	↓	↓	↓
Translate:	$\dfrac{x}{6}$	$-$	$\dfrac{5}{3}$	$=$	$\dfrac{x}{2}$

3. SOLVE. Here, we solve the equation $\dfrac{x}{6} - \dfrac{5}{3} = \dfrac{x}{2}$. We begin by multiplying both sides of the equation by the LCD, 6.

$$6\left(\dfrac{x}{6} - \dfrac{5}{3}\right) = 6\left(\dfrac{x}{2}\right)$$

$$6\left(\dfrac{x}{6}\right) - 6\left(\dfrac{5}{3}\right) = 6\left(\dfrac{x}{2}\right) \qquad \text{Apply the distributive property.}$$

$$x - 10 = 3x \qquad \text{Simplify.}$$

$$-10 = 2x \qquad \text{Subtract } x \text{ from both sides.}$$

$$-\dfrac{10}{2} = \dfrac{2x}{2} \qquad \text{Divide both sides by 2.}$$

$$-5 = x \qquad \text{Simplify.}$$

4. INTERPRET.

Check: To check, we verify that "the quotient of -5 and 6 minus $\dfrac{5}{3}$ is the quotient of -5 and 2," or $-\dfrac{5}{6} - \dfrac{5}{3} = -\dfrac{5}{2}$.

State: The unknown number is -5.

2

The next example is often called a work problem. Work problems usually involve people or machines doing a certain task.

Example 2   **FINDING WORK RATES**

Sam Waterton and Frank Schaffer work in a plant that manufactures automobiles. Sam can complete a quality control check of the plant in 3 hours while his assistant, Frank, needs 7 hours to complete the same job. The regional manager is coming to inspect the plant facilities, so both Sam and Frank are directed to complete a quality control check together. How long will this take?

Solution   1. UNDERSTAND. Read and reread the problem. The key idea here is the relationship between the **time** (hours) it takes to complete the job and the **part of the job** completed in 1 unit of time (hour). For example, if the **time** it takes Sam to complete the job is 3 hours, the **part of the job** he can complete in 1 hour is $\dfrac{1}{3}$. Similarly, Frank can complete $\dfrac{1}{7}$ of the job in 1 hour.

Let $x =$ the **time** in hours it takes Sam and Frank to complete the job together. Then $\dfrac{1}{x} =$ the **part of the job** they complete in 1 hour.

	Hours to Complete Total Job	Part of Job Completed in 1 Hour
Sam	3	$\dfrac{1}{3}$
Frank	7	$\dfrac{1}{7}$
Together	x	$\dfrac{1}{x}$

2. TRANSLATE.

In words:

part of job Sam completed in 1 hour	added to	part of job Frank completed in 1 hour	is equal to	part of job they completed together in 1 hour
↓	↓	↓	↓	↓

Translate:

$$\frac{1}{3} \qquad + \qquad \frac{1}{7} \qquad = \qquad \frac{1}{x}$$

3. SOLVE. Here, we solve the equation $\frac{1}{3} + \frac{1}{7} = \frac{1}{x}$. We begin by multiplying both sides of the equation by the LCD, $21x$.

$$21x\left(\frac{1}{3}\right) + 21x\left(\frac{1}{7}\right) = 21x\left(\frac{1}{x}\right)$$

$$7x + 3x = 21 \qquad\qquad \text{Simplify.}$$

$$10x = 21$$

$$x = \frac{21}{10} \quad \text{or} \quad 2\frac{1}{10} \text{ hours}$$

4. INTERPRET.

Check: Our proposed solution is $2\frac{1}{10}$ hours. This proposed solution is reasonable since $2\frac{1}{10}$ hours is more than half of Sam's time and less than half of Frank's time. Check this solution in the originally *stated* problem.

State: Sam and Frank can complete the quality control check in $2\frac{1}{10}$ hours. ▄

---

3   Next we look at a problem solved by the distance formula.

Example 3   **FINDING SPEEDS OF VEHICLES**

A car travels 180 miles in the same time that a truck travels 120 miles. If the car's speed is 20 miles per hour faster than the truck's, find the car's speed and the truck's speed.

Solution   1. UNDERSTAND. Read and reread the problem. Suppose that the truck's speed is 45 miles per hour. Then the car's speed is 20 miles per hour more, or 65 miles per hour.

We are given that the car travels 180 miles in the same time that the truck travels 120 miles. To find the time it takes the car to travel 180 miles, remember that since $d = rt$, we know that $\frac{d}{r} = t$.

Car's Time   ***Truck's Time***

$$t = \frac{d}{r} = \frac{180}{65} = 2\frac{50}{65} = 2\frac{10}{13} \text{ hours} \qquad t = \frac{d}{r} = \frac{120}{45} = 2\frac{30}{45} = 2\frac{2}{3} \text{ hours}$$

Since the times are not the same, our proposed solution is not correct. But we have a better understanding of the problem.

Let $x =$ the speed of the truck.
Since the car's speed is 20 miles per hour faster than the truck's, then

$$x + 20 = \text{the speed of the car}$$

Use the formula $d = r \cdot t$ or **distance** = **rate** · **time**. Prepare a chart to organize the information in the problem.

	Distance	=	Rate	·	Time
TRUCK	120		x		$\dfrac{120}{x}$ ← distance ← rate
CAR	180		$x + 20$		$\dfrac{180}{x + 20}$ ← distance ← rate

HELPFUL HINT

If $d = r \cdot t$,

then $t = \dfrac{d}{r}$

or $time = \dfrac{distance}{rate}$.

2. **TRANSLATE.** Since the car and the truck traveled the same amount of time, we have that

In words: car's time   =   truck's time
                    ↓   ↓

Translate: $\dfrac{180}{x + 20}$ = $\dfrac{120}{x}$

3. **SOLVE.** We begin by multiplying both sides of the equation by the LCD, $x(x + 20)$, or cross multiplying.

$$\frac{180}{x + 20} = \frac{120}{x}$$

$$180x = 120(x + 20)$$
$$180x = 120x + 2400 \qquad \text{Use the distributive property.}$$
$$60x = 2400 \qquad \text{Subtract } 120x \text{ from both sides.}$$
$$x = 40 \qquad \text{Divide both sides by 60.}$$

4. **INTERPRET.** The speed of the truck is 40 miles per hour. The speed of the car must then be $x + 20$ or 60 miles per hour.

    Check: Find the time it takes the car to travel 180 miles and the time it takes the truck to travel 120 miles.

    Car's Time   *Truck's Time*

    $t = \dfrac{d}{r} = \dfrac{180}{60} = 3$ hours $t = \dfrac{d}{r} = \dfrac{120}{40} = 3$ hours

    Since both travel the same amount of time, the proposed solution is correct.
    State: The car's speed is 60 miles per hour and the truck's speed is 40 miles per hour. ∎

4   **Similar triangles** have the same shape but not necessarily the same size. In similar triangles, the measures of corresponding angles are equal, and corresponding sides are in proportion.

If triangle ABC and triangle XYZ shown are similar, then we know that the measure of angle $A =$ the measure of angle X, the measure of angle $B =$ the measure of angle Y, and the measure of angle $C =$ the measure of angle Z. We also know that corresponding sides are in proportion: $\dfrac{a}{x} = \dfrac{b}{y} = \dfrac{c}{z}$.

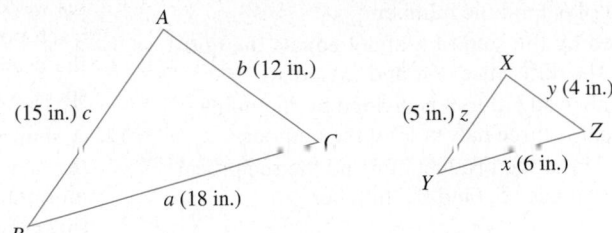

In this section, we will position similar triangles so that they have the same orientation.

To show that corresponding sides are in proportion for the triangles above, we write the ratios of the corresponding sides.

$$\frac{a}{x} = \frac{18}{6} = 3 \qquad \frac{b}{y} = \frac{12}{4} = 3 \qquad \frac{c}{z} = \frac{15}{5} = 3$$

△ **Example 4**   **FINDING THE LENGTH OF A SIDE OF A TRIANGLE**

If the following two triangles are similar, find the missing length x.

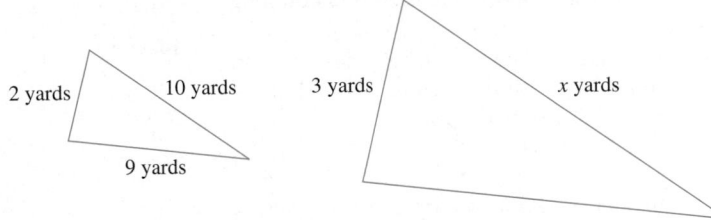

Solution   Since the triangles are similar, their corresponding sides are in proportion and we have

$$\frac{2}{3} = \frac{10}{x}$$

To solve, we multiply both sides by the LCD, $3x$, or cross multiply.

$$2x = 30$$
$$x = 15 \qquad \text{Divide both sides by 2.}$$

The missing length is 15 yards.

SPOTLIGHT ON DECISION MAKING

Suppose you coach your company's softball team. Two new employees are interested in playing on the company team, and you have only one open position. Employee A reports that last season he had 32 hits in 122 times at bat. Employee B reports that last season she had 19 hits in 56 times at bat. Which would you try to recruit first? Why? What other factors would you want to consider?

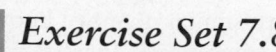

Exercise Set 7.8

Solve the following. See Example 1.

1. Three times the reciprocal of a number equals 9 times the reciprocal of 6. Find the number.

2. Twelve divided by the sum of x and 2 equals the quotient of 4 and the difference of x and 2. Find x.

3. If twice a number added to 3 is divided by the number plus 1, the result is three halves. Find the number.

4. A number added to the product of 6 and the reciprocal of the number equals -5. Find the number.

See Example 2.

5. Smith Engineering found that an experienced surveyor surveys a roadbed in 4 hours. An apprentice surveyor needs 5 hours to survey the same stretch of road. If the two work together, find how long it takes them to complete the job.

6. An experienced bricklayer constructs a small wall in 3 hours. The apprentice completes the job in 6 hours. Find how long it takes if they work together.

7. In 2 minutes, a conveyor belt moves 300 pounds of recyclable aluminum from the delivery truck to a storage area. A smaller belt moves the same quantity of cans the same distance in 6 minutes. If both belts are used, find how long it takes to move the cans to the storage area.

8. Find how long it takes the conveyor belts described in Exercise 7 to move 1200 pounds of cans. (*Hint:* Think of 1200 pounds as four 300-pound jobs.)

See Example 3.

9. A jogger begins her workout by jogging to the park, a distance of 12 miles. She then jogs home at the same speed but along a different route. This return trip is 18 miles and her time is one hour longer. Find her jogging speed. Complete the accompanying chart and use it to find her jogging speed.

	distance	=	rate	·	time
Trip to park	12				x
Return trip	18				$x + 1$

10. A boat can travel 9 miles upstream in the same amount of time it takes to travel 11 miles downstream. If the current of the river is 3 miles per hour, complete the chart below and use it to find the speed of the boat in still water.

	distance	=	rate	·	time
Upstream	9		$r - 3$		
Downstream	11		$r + 3$		

11. A cyclist rode the first 20-mile portion of his workout at a constant speed. For the 16-mile cooldown portion of his workout, he reduced his speed by 2 miles per hour. Each portion of the workout took the same time. Find the cyclist's speed during the first portion and find his speed during the cooldown portion.

12. A semi truck travels 300 miles through the flatland in the same amount of time that it travels 180 miles through mountains. The rate of the truck is 20 miles per hour slower in the mountains than in the flatland. Find both the flatland rate and mountain rate.

Given that the following pairs of triangles are similar, find the missing length, x. See Example 4.

13.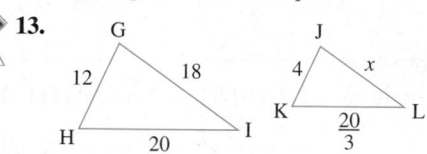

14.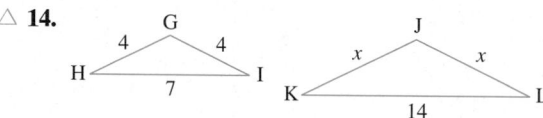

15.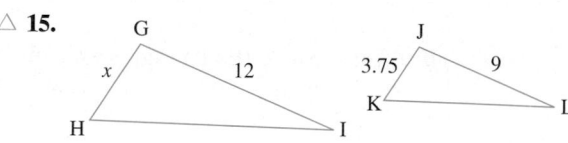

16.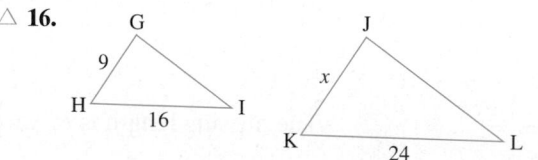

Solve the following.

17. One-fourth equals the quotient of a number and 8. Find the number.

18. Four times a number added to 5 is divided by 6. The result is $\frac{7}{2}$. Find the number.

19. Marcus and Tony work for Lombardo's Pipe and Concrete. Mr. Lombardo is preparing an estimate for a customer. He knows that Marcus lays a slab of concrete in

6 hours. Tony lays the same size slab in 4 hours. If both work on the job and the cost of labor is $45.00 per hour, decide what the labor estimate should be.

20. Mr. Dodson can paint his house by himself in 4 days. His son needs an additional day to complete the job if he works by himself. If they work together, find how long it takes to paint the house.

21. While road testing a new make of car, the editor of a consumer magazine finds that he can go 10 miles into a 3-mile-per-hour wind in the same amount of time he can go 11 miles with a 3-mile-per-hour wind behind him. Find the speed of the car in still air.

22. A fisherman on Pearl River rows 9 miles downstream in the same amount of time he rows 3 miles upstream. If the current is 6 miles per hour, find how long it takes him to cover the 12 miles.

Find the unknown length y in the following pairs of similar triangles.

△ 23.

△ 24.

△ 25.

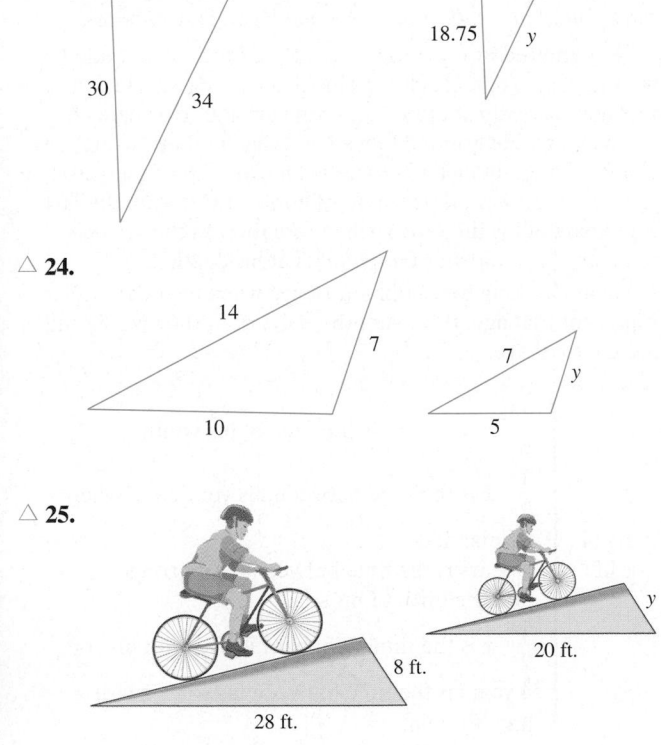

△ 26.

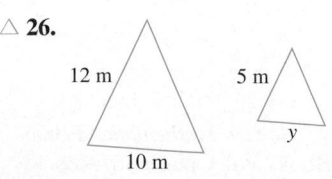

△ 27.

△ 28.

Solve the following.

29. Two divided by the difference of a number and 3 minus 4 divided by a number plus 3, equals 8 times the reciprocal of the difference of the number squared and 9. What is the number?

30. If 15 times the reciprocal of a number is added to the ratio of 9 times a number minus 7 and the number plus 2, the result is 9. What is the number?

31. A pilot flies 630 miles with a tail wind of 35 miles per hour. Against the wind, he flies only 455 miles in the same amount of time. Find the rate of the plane in still air.

32. A marketing manager travels 1080 miles in a corporate jet and then an additional 240 miles by car. If the car ride takes one hour longer than the jet ride takes, and if the rate of the jet is 6 times the rate of the car, find the time the manager travels by jet and find the time the manager travels by car.

33. A cyclist rides 16 miles per hour on level ground on a still day. He finds that he rides 48 miles with the wind behind him in the same amount of time that he rides 16 miles into the wind. Find the rate of the wind.

34. The current on a portion of the Mississippi River is 3 miles per hour. A barge can go 6 miles upstream in the same amount of time it takes to go 10 miles downstream. Find the speed of the boat in still water.

35. One custodian cleans a suite of offices in 3 hours. When a second worker is asked to join the regular custodian, the job takes only $1\frac{1}{2}$ hours. How long does it take the second worker to do the same job alone?

36. One person proofreads a copy for a small newspaper in 4 hours. If a second proofreader is also employed, the job can be done in $2\frac{1}{2}$ hours. How long does it take for the second proofreader to do the same job alone?

37. One pipe fills a storage pool in 20 hours. A second pipe fills the same pool in 15 hours. When a third pipe is added and all three are used to fill the pool, it takes only 6 hours. Find how long it takes the third pipe to do the job.

38. One pump fills a tank 2 times as fast as another pump. If the pumps work together, they fill the tank in 18 minutes. How long does it take for each pump to fill the tank?

△ **39.** An architech is completing the plans for a triangular deck. Use the diagram below to find the missing dimension.

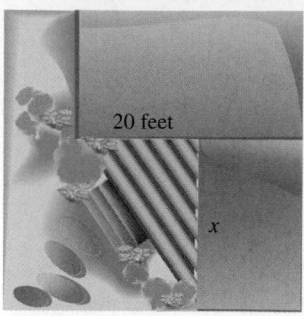

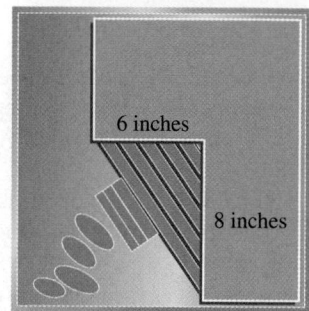

△ **40.** A student wishes to make a small model of a triangular mainsail in order to study the effects of wind on the sail. The smaller model will be the same shape as a regular-size sailboat's mainsail. Use the following diagram to find the missing dimensions.

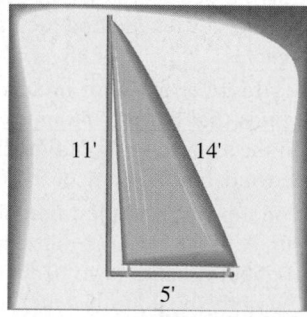

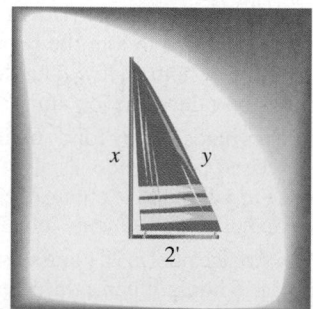

41. Andrew and Timothy Larson volunteer at a local recycling plant. Andrew can sort a batch of recyclables in 2 hours alone while his brother Timothy needs 3 hours to complete the same job. If they work together, how long will it take them to sort one batch?

42. A car travels 280 miles in the same time that a motorcycle travels 240 miles. If the car's speed is 10 miles per hour more than the motorcycle's, find the speed of the car and the speed of the motorcycle.

43. In 6 hours, an experienced cook prepares enough pies to supply a local restaurant's daily order. Another cook prepares the same number of pies in 7 hours. Together with a third cook, they prepare the pies in 2 hours. Find the work rate of the third cook.

44. Mrs. Smith balances the company books in 8 hours. It takes her assistant half again as long to do the same job. If they work together, find how long it takes them to balance the books.

45. One pump fills a tank 3 times as fast as another pump. If the pumps work together, they fill the tank in 21 minutes. How long does it take for each pump to fill the tank?

One of the great algebraists of ancient times was a man named Diophantus. Little is known of his life other than that he lived and worked in Alexandria. Some historians believe he lived during the first century of the Christian era, about the time of Nero. The only clue to his personal life is the following epigram found in a collection called the Palatine Anthology.

God granted him youth for a sixth of his life and added a twelfth part to this. He clothed his cheeks in down. He lit him the light of wedlock after a seventh part and five years after his marriage, He granted him a son. Alas, lateborn wretched child. After attaining the measure of half his father's life, cruel fate overtook him, thus leaving Diophantus during the last four years of his life only such consolation as the science of numbers. How old was Diophantus at his death?*

We are looking for Diophantus' age when he died, so let x represent that age. If we sum the parts of his life, we should get the total age.

Parts of his life
$\begin{cases} \dfrac{1}{6} \cdot x + \dfrac{1}{12} \cdot x \text{ is the time of his youth.} \\[2mm] \dfrac{1}{7} \cdot x \text{ is the time between his youth and when he married.} \\[2mm] \text{5 years is the time between his marriage and the birth of his son.} \\[2mm] \dfrac{1}{2} \cdot x \text{ is the time Diophantus had with his son.} \\[2mm] \text{4 years is the time between his son's death and his own.} \end{cases}$

The sum of these parts should equal Diophantus' age when he died.

$$\frac{1}{6} \cdot x + \frac{1}{12} \cdot x + \frac{1}{7} \cdot x + 5 + \frac{1}{2} \cdot x + 4 = x$$

*From *The Nature and Growth of Modern Mathematics*, Edna Kramer, 1970, Fawcett Premier Books, Vol. 1, pages 107–108.

46. Solve the epigram.

47. How old was Diophantus when his son was born? How old was the son when he died?

48. Solve the following epigram:
I was four when my mother packed my lunch and sent me off to school. Half my life was spent in school and another sixth was spent on a farm. Alas, hard times befell me. My crops and cattle fared poorly and my land was sold. I returned to school for 3 years and have spent one tenth of my life teaching. How old am I?

49. Write an epigram describing your life. Be sure that none of the time periods in your epigram overlap.

50. During the 2000 Grand Prix of Miami auto race, Juan Montoya posted the fastest lap speed but Max Papis won the race. The track is 1.5 miles long. When traveling at their fastest lap speeds, Papis drove 1.48 miles in the same time that it took Montoya to complete an entire lap. Montoya's fastest lap speed was 2.61 mph faster than Papis's fastest lap speed. Find each driver's fastest lap speed. (*Source:* Based on data from Championship Auto Racing Teams, Inc.)

51. A hyena spots a giraffe 0.5 mile away and begins running toward it. The giraffe starts running away from the hyena just as the hyena begins running toward it. A hyena can

run at a speed of 40 mph and a giraffe can run at 32 mph. How long will it take for the hyena to overtake the giraffe? (*Source:* Based on data from *Natural History*, March 1974)

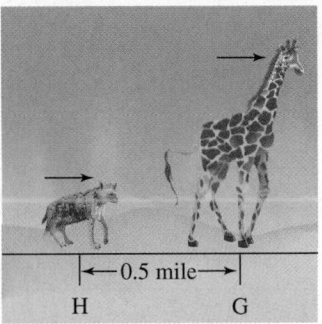

0.5 mile
H   G

REVIEW EXERCISES

Graph each linear equation by finding intercepts. See Section 3.4.

52. $5x + y = 10$ **53.** $-x + 3y = 6$

54. $x = -3y$ **55.** $y = 2x$

56. $x - y = -2$ **57.** $y - x = -5$

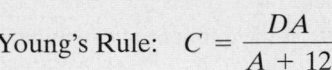

For additional Chapter Projects, visit the Real World Activities Website by going to http://www.prenhall.com/martin-gay.

CHAPTER PROJECT

Comparing Dosage Formulas

In this project, you will have the opportunity to investigate two well-known formulas for predicting the correct doses of medication for children. This project may be completed by working in groups or individually.

Young's Rule and Cowling's Rule are dose formulas for prescribing medicines to children. Unlike formulas for, say area or distance, these dose formulas describe only an approximate relationship. The formulas relate a child's age A in years and an adult dose D of medication to the proper child's dose C. The formulas are most accurate when applied to chil-

dren between the ages of 2 and 13.

$$\text{Young's Rule:} \quad C = \frac{DA}{A + 12}$$

$$\text{Cowling's Rule:} \quad C = \frac{D(A + 1)}{24}$$

1. Let the adult dose $D = 1000$ mg. Complete the Young's Rule and Cowling's Rule columns of the following table comparing the doses predicted by both formulas for ages 2 through 13.

Age A	Young's Rule	Cowling's Rule	Difference
2			
3			
4			
5			
6			
7			
8			
9			
10			
11			
12			
13			

2. Use the data from the table in Question 1 to form sets of ordered pairs of the form (age, child's dose) for each formula. Graph the ordered pairs for each formula on the same graph. Describe the shapes of the graphed data.

3. Use your table, graph, or both, to decide whether either formula will consistently predict a larger dose than the other. If so, which one? If not, is there an age at which the doses predicted by one becomes greater than the doses predicted by the other? If so, estimate that age.

4. Use your graph to estimate for what age the difference in the two predicted doses is greatest.

5. Return to the table in Question 1 and complete the last column, titled "Difference," by finding the absolute value of the difference between the Young's dose and the Cowling's dose for each age. Use this column in the table to verify your graphical estimate found in Question 4.

6. Does Cowling's Rule ever predict exactly the adult dose? If so, at what age? Explain. Does Young's Rule ever predict exactly the adult dose? If so, at what age? Explain.

7. Many doctors prefer to use formulas that relate doses to factors other than a child's age. Why is age not necessarily the most important factor when predicting a child's dose? What other factors might be used?

CHAPTER 7 VOCABULARY CHECK

Fill in each blank with one of the words or phrases listed below.

rational expression   complex fraction   ratio   proportion
cross products

1. A _____ is the quotient of two numbers. It can be written as a fraction, using a colon, or using the word *to*.

2. $\dfrac{x}{2} = \dfrac{7}{16}$ is an example of a _____ .

3. If $\dfrac{a}{b} = \dfrac{c}{d}$, then ad and bc are called _____ .

4. A _____ is an expression that can be written in the form P/Q, where P and Q are polynomials and Q is not 0.

5. In a _____ , the numerator or denominator or both may contain fractions.

CHAPTER 7 HIGHLIGHTS

DEFINITIONS AND CONCEPTS	EXAMPLES

Section 7.1 Rational Functions and Simplifying Rational Expressions

A **rational expression** is an expression that can be written in the form $\frac{P}{Q}$, where P and Q are polynomials and Q does not equal 0.

Rational Expressions
$$\frac{7y^3}{4}, \frac{x^2 + 6x + 1}{x - 3}, \frac{-5}{s^3 + 8}$$

Fundamental Principle of Rational Expressions
If P and Q are polynomials, and Q and R are not 0, then
$$\frac{PR}{QR} = \frac{P}{Q}$$

By the fundamental principle,
$$\frac{(x - 3)(x + 1)}{x(x + 1)} = \frac{x - 3}{x}$$
as long as $x \neq 0$ and $x \neq -1$.

To Simplify a Rational Expression
Step 1. Factor the numerator and denominator.
Step 2. Apply the fundamental principle to divide out common factors.

Simplify: $\frac{4x + 20}{x^2 - 25}$
$$\frac{4x + 20}{x^2 - 25} = \frac{4(x + 5)}{(x + 5)(x - 5)} = \frac{4}{x - 5}$$

Section 7.2 Multiplying and Dividing Rational Expressions

To Multiply Rational Expressions
Step 1. Factor numerators and denominators.
Step 2. Multiply numerators and multiply denominators.
Step 3. Write the product in simplest form.
$$\frac{P}{Q} \cdot \frac{R}{S} = \frac{PR}{QS}$$
$$Q \neq 0, S \neq 0$$

Multiply: $\frac{4x + 4}{2x - 3} \cdot \frac{2x^2 + x - 6}{x^2 - 1}$
$$\frac{4x + 4}{2x - 3} \cdot \frac{2x^2 + x - 6}{x^2 - 1} = \frac{4(x + 1)}{2x - 3} \cdot \frac{(2x - 3)(x + 2)}{(x + 1)(x - 1)}$$
$$= \frac{4(x + 1)(2x - 3)(x + 2)}{(2x - 3)(x + 1)(x - 1)}$$
$$= \frac{4(x + 2)}{x - 1}$$

To Divide by a Rational Expression
Multiply by the reciprocal.
$$\frac{P}{Q} \div \frac{R}{S} = \frac{P}{Q} \cdot \frac{S}{R} = \frac{PS}{QR}$$
$$Q \neq 0, R \neq 0, S \neq 0$$

Divide: $\frac{15x + 5}{3x^2 - 14x - 5} \div \frac{15}{3x - 12}$
$$\frac{15x + 5}{3x^2 - 14x - 5} \div \frac{15}{3x - 12} = \frac{5(3x + 1)}{(3x + 1)(x - 5)} \cdot \frac{3(x - 4)}{3 \cdot 5}$$
$$= \frac{x - 4}{x - 5}$$

Section 7.3 Adding and Subtracting Rational Expressions with Common Denominators and Least Common Denominator

To Add or Subtract Rational Expressions with the Same Denominator
Add or subtract numerators, and place the sum or difference over the common denominator.
$$\frac{P}{R} + \frac{Q}{R} = \frac{P + Q}{R}$$
$$\frac{P}{R} - \frac{Q}{R} = \frac{P - Q}{R}$$

Perform indicated operations.
$$\frac{5}{x + 1} + \frac{x}{x + 1} = \frac{5 + x}{x + 1}$$
$$\frac{2y + 7}{y^2 - 9} - \frac{y + 4}{y^2 - 9} = \frac{(2y + 7) - (y + 4)}{y^2 - 9}$$
$$= \frac{2y + 7 - y - 4}{y^2 - 9}$$
$$= \frac{y + 3}{(y + 3)(y - 3)}$$
$$= \frac{1}{y - 3}$$

(continued)

DEFINITIONS AND CONCEPTS	EXAMPLES

Section 7.3 Adding and Subtracting Rational Expressions with Common Denominators and Least Common Denominator

To Find the Least Common Denominator (LCD)

Step 1. Factor the denominators.

Step 2. The LCD is the product of all unique factors, each raised to a power equal to the greatest number of times that it appears in any one factored denominator.

Find the LCD for

$$\frac{7x}{x^2 + 10x + 25} \text{ and } \frac{11}{3x^2 + 15x}$$

$$x^2 + 10x + 25 = (x + 5)(x + 5)$$

$$3x^2 + 15x = 3x(x + 5)$$

$$\text{LCD is } 3x(x + 5)(x + 5) \text{ or } 3x(x + 5)^2$$

Section 7.4 Adding and Subtracting Rational Expressions with Unlike Denominators

To Add or Subtract Rational Expressions with Unlike Denominators

Step 1. Find the LCD.

Step 2. Rewrite each rational expression as an equivalent expression whose denominator is the LCD.

Step 3. Add or subtract numerators and place the sum or difference over the common denominator.

Step 4. Write the result in simplest form.

Perform the indicated operation.

$$\frac{9x + 3}{x^2 - 9} - \frac{5}{x - 3}$$

$$= \frac{9x + 3}{(x + 3)(x - 3)} - \frac{5}{x - 3}$$

LCD is $(x + 3)(x - 3)$.

$$= \frac{9x + 3}{(x + 3)(x - 3)} - \frac{5(x + 3)}{(x - 3)(x + 3)}$$

$$= \frac{9x + 3 - 5(x + 3)}{(x + 3)(x - 3)}$$

$$= \frac{9x + 3 - 5x - 15}{(x + 3)(x - 3)}$$

$$= \frac{4x - 12}{(x + 3)(x - 3)}$$

$$= \frac{4(x - 3)}{(x + 3)(x - 3)} = \frac{4}{x + 3}$$

Section 7.5 Simplifying Complex Fractions

Method 1: Simplify the numerator and the denominator so that each is a single fraction. Then perform the indicated division and simplify if possible.

Simplify $\dfrac{\dfrac{x + 2}{x}}{x - \dfrac{4}{x}}$.

Method 1: $\dfrac{\dfrac{x + 2}{x}}{\dfrac{x \cdot x}{1 \cdot x} - \dfrac{4}{x}} = \dfrac{\dfrac{x + 2}{x}}{\dfrac{x^2 - 4}{x}}$

$$= \frac{x + 2}{x} \cdot \frac{x}{(x + 2)(x - 2)} = \frac{1}{x - 2}$$

(continued)

DEFINITIONS AND CONCEPTS	EXAMPLES

Section 7.5 Simplifying Complex Fractions

Method 2: Multiply the numerator and the denominator of the complex fraction by the LCD of the fractions in both the numerator and the denominator. Then simplify if possible.

Method 2: $\dfrac{\left(\dfrac{x+2}{x}\right)\cdot x}{\left(x-\dfrac{4}{x}\right)\cdot x} = \dfrac{x+2}{x\cdot x - \dfrac{4}{x}\cdot x}$

$= \dfrac{x+2}{x^2-4} = \dfrac{x+2}{(x+2)(x-2)} = \dfrac{1}{x-2}$

Section 7.6 Solving Equations Containing Rational Expressions

To Solve an Equation Containing Rational Expressions

Step 1. Multiply both sides of the equation by the LCD of all rational expressions in the equation.

Step 2. Remove any grouping symbols and solve the resulting equation.

Step 3. Check the solution in the original equation.

Solve: $\dfrac{5x}{x+2} + 3 = \dfrac{4x-6}{x+2}$

$(x+2)\left(\dfrac{5x}{x+2}+3\right) = (x+2)\left(\dfrac{4x-6}{x+2}\right)$

$(x+2)\left(\dfrac{5x}{x+2}\right) + (x+2)(3) = (x+2)\left(\dfrac{4x-6}{x+2}\right)$

$5x + 3x + 6 = 4x - 6$

$4x = -12$

$x = -3$

The solution checks and the solution is -3.

Section 7.7 Ratio and Proportion

A **ratio** is the quotient of two numbers or two quantities.

The ratio of a to b can be written as

Fractional Notation   *Colon Notation*

$\dfrac{a}{b}$ $a:b$

A **proportion** is a mathematical statement that two ratios are equal.

In the proportion $\dfrac{a}{b} = \dfrac{c}{d}$, the products ad and bc are called **cross products.**

Cross products:

If $\dfrac{a}{b} = \dfrac{c}{d}$, then $ad = bc$.

Write the ratio of 5 hours to 1 day using fractional notation.

$\dfrac{5\text{ hours}}{1\text{ day}} = \dfrac{5\text{ hours}}{24\text{ hours}} = \dfrac{5}{24}$

Proportions

$\dfrac{2}{3} = \dfrac{8}{12}$ $\dfrac{x}{7} = \dfrac{15}{35}$

Cross Products

$\dfrac{2}{3} = \dfrac{8}{12}$ → $3\cdot 8$ or 24 / $2\cdot 12$ or 24

Solve: $\dfrac{3}{4} = \dfrac{x}{x-1}$

$\dfrac{3}{4} = \dfrac{x}{x-1}$

$3(x-1) = 4x$ Set cross products equal.

$3x - 3 = 4x$

$-3 = x$

(continued)

DEFINITIONS AND CONCEPTS	EXAMPLES

Section 7.8 Rational Equations and Problem Solving

Problem-Solving Steps

    1. UNDERSTAND. Read and reread the problem.

A small plane and a car leave Kansas City, Missouri, and head for Minneapolis, Minnesota, a distance of 450 miles. The speed of the plane is 3 times the speed of the car, and the plane arrives 6 hours ahead of the car. Find the speed of the car.

Let x = the speed of the car.
Then $3x$ = the speed of the plane.

	Distance	=	Rate	·	Time
Car	450		x		$\dfrac{450}{x}\left(\dfrac{\text{distance}}{\text{rate}}\right)$
Plane	450		$3x$		$\dfrac{450}{3x}\left(\dfrac{\text{distance}}{\text{rate}}\right)$

In words:

plane's time	+	6 hours	=	car's time
↓	↓	↓		↓

    2. TRANSLATE.

Translate: $\dfrac{450}{3x} \quad + \quad 6 \quad = \quad \dfrac{450}{x}$

    3. SOLVE.

$$\frac{450}{3x} + 6 = \frac{450}{x}$$

$$3x\left(\frac{450}{3x}\right) + 3x(6) = 3x\left(\frac{450}{x}\right)$$

$$450 + 18x = 1350$$

$$18x = 900$$

$$x = 50$$

    4. INTERPRET.

Check the solution by replacing x with 50 in the original equation.
State the conclusion: The speed of the car is 50 miles per hour.

CHAPTER 7 REVIEW

(7.1) *Find the domain for each rational function.*

1. $F(x) = \dfrac{-3x^2}{x - 5}$

2. $h(x) = \dfrac{4x}{3x - 12}$

3. $f(x) = \dfrac{x^3 + 2}{x^2 + 8x}$

4. $G(x) = \dfrac{20}{3x^2 - 48}$

Write each rational expression in lowest terms.

5. $\dfrac{15x^4}{45x^2}$

6. $\dfrac{x + 2}{2 + x}$

7. $\dfrac{18m^6 p^2}{10m^4 p}$

8. $\dfrac{x - 12}{12 - x}$

9. $\dfrac{5x - 15}{25x - 75}$

10. $\dfrac{22x + 8}{11x + 4}$

11. $\dfrac{2x}{2x^2 - 2x}$

12. $\dfrac{x + 7}{x^2 - 49}$

13. $\dfrac{2x^2 + 4x - 30}{x^2 + x - 20}$

14. $\dfrac{xy - 3x + 2y - 6}{x^2 + 4x + 4}$

15. The average cost of manufacturing x bookcases is given by the rational function

$$C(x) = \frac{35x + 4200}{x}$$

  a. Find the average cost per bookcase of manufacturing 50 bookcases.

  b. Find the average cost per bookcase of manufacturing 100 bookcases.

  c. As the number of bookcases increases, does the average cost per bookcase increase or decrease? (See parts **a** and **b**.)

(7.2) *Perform the indicated operations and simplify.*

16. $\dfrac{15x^3y^2}{z} \cdot \dfrac{z}{5xy^3}$

17. $\dfrac{-y^3}{8} \cdot \dfrac{9x^2}{y^3}$ **18.** $\dfrac{x^2 - 9}{x^2 - 4} \cdot \dfrac{x - 2}{x + 3}$

19. $\dfrac{2x + 5}{x - 6} \cdot \dfrac{2x}{-x + 6}$

20. $\dfrac{x^2 - 5x - 24}{x^2 - x - 12} \div \dfrac{x^2 - 10x + 16}{x^2 + x - 6}$

21. $\dfrac{4x + 4y}{xy^2} \div \dfrac{3x + 3y}{x^2y}$

22. $\dfrac{x^2 + x - 42}{x - 3} \cdot \dfrac{(x - 3)^2}{x + 7}$

23. $\dfrac{2a + 2b}{3} \cdot \dfrac{a - b}{a^2 - b^2}$

24. $\dfrac{x^2 - 9x + 14}{x^2 - 5x + 6} \cdot \dfrac{x + 2}{x^2 - 5x - 14}$

25. $(x - 3) \cdot \dfrac{x}{x^2 + 3x - 18}$

26. $\dfrac{2x^2 - 9x + 9}{8x - 12} \div \dfrac{x^2 - 3x}{2x}$

27. $\dfrac{x^2 - y^2}{x^2 + xy} \div \dfrac{3x^2 - 2xy - y^2}{3x^2 + 6x}$

28. $\dfrac{x^2 - y^2}{8x^2 - 16xy + 8y^2} \div \dfrac{x + y}{4x - y}$

29. $\dfrac{x - y}{4} \div \dfrac{y^2 - 2y - xy + 2x}{16x + 24}$

30. $\dfrac{y - 3}{4x + 3} \div \dfrac{9 - y^2}{4x^2 - x - 3}$

(7.3) *Perform the indicated operations and simplify.*

31. $\dfrac{5x - 4}{3x - 1} + \dfrac{6}{3x - 1}$ **32.** $\dfrac{4x - 5}{3x^2} - \dfrac{2x + 5}{3x^2}$

33. $\dfrac{9x + 7}{6x^2} - \dfrac{3x + 4}{6x^2}$

Find the LCD of each pair of rational expressions.

34. $\dfrac{x + 4}{2x}, \dfrac{3}{7x}$

35. $\dfrac{x - 2}{x^2 - 5x - 24}, \dfrac{3}{x^2 + 11x + 24}$

Rewrite the following rational expressions as equivalent expressions whose denominator is the given polynomial.

36. $\dfrac{x + 2}{x^2 + 11x + 18}, (x + 2)(x - 5)(x + 9)$

37. $\dfrac{3x - 5}{x^2 + 4x + 4}, (x + 2)^2(x + 3)$

(7.4) *Perform the indicated operations and simplify.*

38. $\dfrac{4}{5x^2} - \dfrac{6}{y}$ **39.** $\dfrac{2}{x - 3} - \dfrac{4}{x - 1}$

40. $\dfrac{x + 7}{x + 3} - \dfrac{x - 3}{x + 7}$

41. $\dfrac{4}{x + 3} - 2$

42. $\dfrac{3}{x^2 + 2x - 8} + \dfrac{2}{x^2 - 3x + 2}$

43. $\dfrac{2x - 5}{6x + 9} - \dfrac{4}{2x^2 + 3x}$

44. $\dfrac{x - 1}{x^2 - 2x + 1} - \dfrac{x + 1}{x - 1}$

45. $\dfrac{x - 1}{x^2 + 4x + 4} + \dfrac{x - 1}{x + 2}$

Find the perimeter and the area of each figure.

△ **46.**

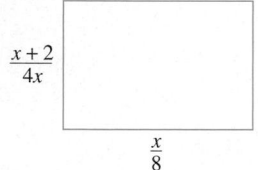

$\dfrac{x + 2}{4x}$ $\dfrac{x}{8}$

△ **47.**

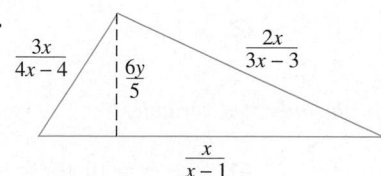

$\dfrac{3x}{4x - 4}$ $\dfrac{6y}{5}$ $\dfrac{2x}{3x - 3}$ $\dfrac{x}{x - 1}$

(7.5) *Simplify each complex fraction.*

48. $\dfrac{\dfrac{5x}{27}}{-\dfrac{10xy}{21}}$ **49.** $\dfrac{\dfrac{8x}{x^2 - 9}}{\dfrac{4}{x + 3}}$

50. $\dfrac{\dfrac{3}{5} + \dfrac{2}{7}}{\dfrac{1}{5} + \dfrac{5}{6}}$

51. $\dfrac{\dfrac{2}{a} + \dfrac{1}{2a}}{a + \dfrac{a}{2}}$

52. $\dfrac{3 - y^{-1}}{2 - y^{-1}}$

53. $\dfrac{2 + x^{-2}}{x^{-1} + 2x^{-2}}$

54. $\dfrac{\dfrac{6}{x + 2} + 4}{\dfrac{8}{x + 2} - 4}$

55. If $f(x) = \dfrac{3}{x}$, find each of the following:

  a. $f(a + h)$ **b.** $f(a)$

  c. Use parts **a** and **b** to find $\dfrac{f(a + h) - f(a)}{h}$.

  d. Simplify the results of part **c**.

(7.6) Solve each equation for the variable or perform the indicated operation.

56. $\dfrac{x + 4}{9} = \dfrac{5}{9}$

57. $\dfrac{n}{10} = 9 - \dfrac{n}{5}$

58. $\dfrac{5y - 3}{7} = \dfrac{15y - 2}{28}$

59. $\dfrac{2}{x + 1} - \dfrac{1}{x - 2} = -\dfrac{1}{2}$

60. $\dfrac{1}{a + 3} + \dfrac{1}{a - 3} = -\dfrac{5}{a^2 - 9}$

61. $\dfrac{y}{2y + 2} + \dfrac{2y - 16}{4y + 4} = \dfrac{y - 3}{y + 1}$

62. $\dfrac{4}{x + 3} + \dfrac{8}{x^2 - 9} = 0$

63. $\dfrac{2}{x - 3} - \dfrac{4}{x + 3} = \dfrac{8}{x^2 - 9}$

64. $\dfrac{x - 3}{x + 1} - \dfrac{x - 6}{x + 5} = 0$ **65.** $x + 5 = \dfrac{6}{x}$

Solve the equation for the indicated variable.

66. $\dfrac{4A}{5b} = x^2$, for b **67.** $\dfrac{x}{7} + \dfrac{y}{8} = 10$, for y

(7.7) Write each phrase as a ratio in fractional notation.

68. 20 cents to 1 dollar

69. four parts red to six parts white

Solve each proportion.

70. $\dfrac{x}{2} = \dfrac{12}{4}$

71. $\dfrac{20}{1} = \dfrac{x}{25}$

72. $\dfrac{32}{100} = \dfrac{100}{x}$

73. $\dfrac{20}{2} = \dfrac{c}{5}$

74. $\dfrac{2}{x - 1} = \dfrac{3}{x + 3}$

75. $\dfrac{4}{y - 3} = \dfrac{2}{y - 3}$

76. $\dfrac{y + 2}{y} = \dfrac{5}{3}$

77. $\dfrac{x - 3}{3x + 2} = \dfrac{2}{6}$

Given the following prices charged for various sizes of an item, find the best buy.

 78. Shampoo

      10 ounces for $1.29

      16 ounces for $2.15

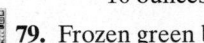 **79.** Frozen green beans

      8 ounces for $0.89

      15 ounces for $1.63

      20 ounces for $2.36

Solve.

80. A machine can process 300 parts in 20 minutes. Find how many parts can be processed in 45 minutes.

81. As his consulting fee, Mr. Visconti charges $90.00 per day. Find how much he charges for 3 hours of consulting. Assume an 8-hour work day.

82. One fund raiser can address 100 letters in 35 minutes. Find how many he can address in 55 minutes.

(7.8) Solve each problem.

83. Five times the reciprocal of a number equals the sum of $\frac{3}{2}$ times the reciprocal of the number and $\frac{7}{6}$. What is the number?

84. The reciprocal of a number equals the reciprocal of the difference of 4 and the number. Find the number.

85. A car travels 90 miles in the same time that a car traveling 10 miles per hour slower travels 60 miles. Find the speed of each car.

86. The speed of a bayou near Lafayette, Louisiana, is 4 miles per hour. A paddle boat travels 48 miles upstream in the same amount of time it takes to travel 72 miles downstream. Find the speed of the boat in still water.

87. When Mark and Maria manicure Mr. Stergeon's lawn, it takes them 5 hours. If Mark works alone, it takes 7 hours. Find how long it takes Maria alone.

88. It takes pipe A 20 days to fill a fish pond. Pipe B takes 15 days. Find how long it takes both pipes together to fill the pond.

Given that the pairs of triangles are similar, find each missing length x.

△ **89.**

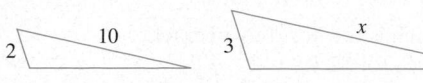

△ **90.**

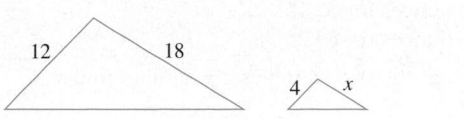

△ **91.**

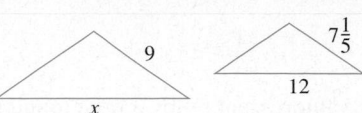

△ **92.**

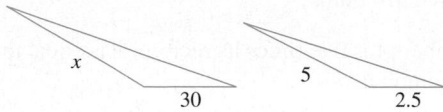

CHAPTER 7 TEST

1. Find the domain of
$$f(x) = \frac{x + 5}{x^2 - 6x}$$

2. For a certain computer desk, the manufacturing cost $C(x)$ per desk (in dollars) is
$$C(x) = \frac{100x + 3000}{x}$$

where x is the number of desks manufactured.

a. Find the average cost per desk when manufacturing 200 computer desks.

b. Find the average cost per desk when manufacturing 1000 computer desks.

Simplify each rational expression.

3. $\dfrac{3x - 6}{5x - 10}$

4. $\dfrac{x + 10}{x^2 - 100}$

5. $\dfrac{x + 6}{x^2 + 12x + 36}$

6. $\dfrac{x + 3}{x^3 + 27}$

7. $\dfrac{2m^3 - 2m^2 - 12m}{m^2 - 5m + 6}$

8. $\dfrac{ay + 3a + 2y + 6}{ay + 3a + 5y + 15}$

9. $\dfrac{y - x}{x^2 - y^2}$

Perform the indicated operation and simplify if possible.

10. $\dfrac{x^2 - 13x + 42}{x^2 + 10x + 21} \div \dfrac{x^2 - 4}{x^2 + x - 6}$

11. $\dfrac{3}{x - 1} \cdot (5x - 5)$

12. $\dfrac{y^2 - 5y + 6}{2y + 4} \cdot \dfrac{y + 2}{2y - 6}$

13. $\dfrac{5}{2x + 5} - \dfrac{6}{2x + 5}$

14. $\dfrac{5a}{a^2 - a - 6} - \dfrac{2}{a - 3}$

15. $\dfrac{6}{x^2 - 1} + \dfrac{3}{x + 1}$

16. $\dfrac{x^2 - 9}{x^2 - 3x} \div \dfrac{xy + 5x + 3y + 15}{2x + 10}$

17. $\dfrac{x + 2}{x^2 + 11x + 18} + \dfrac{5}{x^2 - 3x - 10}$

18. $\dfrac{4y}{y^2 + 6y + 5} - \dfrac{3}{y^2 + 5y + 4}$

Solve each equation.

19. $\dfrac{4}{y} - \dfrac{5}{3} = \dfrac{-1}{5}$

20. $\dfrac{5}{y + 1} = \dfrac{4}{y + 2}$

21. $\dfrac{a}{a - 3} = \dfrac{3}{a - 3} - \dfrac{3}{2}$

22. $\dfrac{10}{x^2 - 25} = \dfrac{3}{x + 5} + \dfrac{1}{x - 5}$

Simplify each complex fraction.

23. $\dfrac{\dfrac{5x^2}{yz^2}}{\dfrac{10x}{z^3}}$

24. $\dfrac{\dfrac{b}{a} - \dfrac{a}{b}}{\dfrac{b}{a} + \dfrac{b}{a}}$

25. $\dfrac{5 - y^{-2}}{y^{-1} + 2y^{-2}}$

26. In a sample of 85 fluorescent bulbs, 3 were found to be defective. At this rate, how many defective bulbs should be found in 510 bulbs?

27. One number plus five times its reciprocal is equal to six. Find the number.

28. A pleasure boat traveling down the Red River takes the same time to go 14 miles upstream as it takes to go 16 miles downstream. If the current of the river is 2 miles per hour, find the speed of the boat in still water.

29. An inlet pipe can fill a tank in 12 hours. A second pipe can fill the tank in 15 hours. If both pipes are used, find how long it takes to fill the tank.

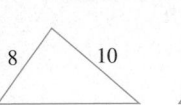 **30.** Decide which is the best buy in crackers.
6 ounces for $1.19
10 ounces for $2.15
16 ounces for $3.25

△ **31.** Given that the two triangles are similar, find x.

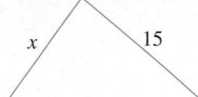

CHAPTER 7 CUMULATIVE REVIEW

1. Write each sentence as an equation. Let x represent the unknown number.
  a. The quotient of 15 and a number is 4.
  b. Three subtracted from 12 is a number.
  c. Four times a number added to 17 is 21.

2. Which of the following relations are also functions?
  a. $\{(-2, 5), (2, 7), (-3, 5), (9, 9)\}$
  b.

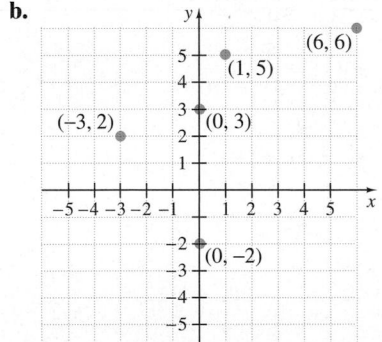

  c.

Input	Correspondence	Output
People in a certain city	Each person's age	The set of nonnegative integers

3. Graph $x - 3y = 6$ by plotting intercept.

4. Find an equation of the line with slope -3 containing the point $(1, -5)$. Write the equation in slope–intercept form $y = mx + b$.

5. Graph the intersection of $x \geq 1$ and $y \geq 2x - 1$.

6. Use the product rule to simplify.
  a. $4^2 \cdot 4^5$ **b.** $x^2 \cdot x^5$
  c. $y^3 \cdot y$ **d.** $y^3 \cdot y^2 \cdot y^7$
  e. $(-5)^7 \cdot (-5)^8$

7. Simplify each expression. Write results using positive exponents only.
  a. $\left(\dfrac{2}{3}\right)^{-3}$ **b.** $\dfrac{1}{x^{-3}}$
  c. $\dfrac{p^{-4}}{q^{-9}}$

8. Multiply: $(3a + b)^3$

9. Use a special product to square each binomial.
  a. $(t + 2)^2$
  b. $(p - q)^2$
  c. $(2x + 5)^2$
  d. $(x^2 - 7y)^2$

10. Find the GCF of each list of numbers.
  a. 28 and 40   **b.** 55 and 21
  c. 15, 18, and 66

Factor.

11. $-9a^5 + 18a^2 - 3a$

12. $3m^2 - 24m - 60$

13. $3x^2 + 11x + 6$

14. $x^2 + 12x + 36$

15. $x^2 + 4$

16. $x^3 + 8$

17. $2x^3 + 3x^2 - 2x - 3$

18. $12m^2 - 3n^2$

19. Solve: $x(2x - 7) = 4$

20. Find the x-intercepts of the graph of $y = x^2 - 5x + 4$.

21. The height of a triangular sail is 2 meters less than twice the length of the base. If the sail has an area of 30 square meters, find the length of its base and the height.

22. Divide: $\dfrac{6x + 2}{x^2 - 1} \div \dfrac{3x^2 + x}{x - 1}$

23. A supermarket offers a 14-ounce box of cereal for $3.79 and an 18-ounce box of the same brand of cereal for $4.99. Which is the better buy?

Keeping Us Informed

The field of journalism includes newspaper, radio, and television reporters, as well as photographers, graphic artists, copy editors, and other communications specialists. In the United States, the news media are responsible for keeping us informed about current national and international events.

Excellent communication skills are the top requirement for journalists. However, journalists also must be informed about a wide variety of subjects so they can report effectively on their news assignments. A liberal-arts education, including courses in economics, statistics, and sciences, can be a huge asset for journalists. Journalists must be able to do research, conduct interviews, study documents, and interpret numerical data and statistics.

 For more information about careers in journalism, visit the American Society of Newspaper Editors Website by first going to www.prenhall.com/martin-gay.

In the Spotlight on Decision Making feature on page 491, you will have the opportunity to use percents to make a decision about how to approach a story as a reporter.

TRANSITIONS TO INTERMEDIATE ALGEBRA

8

Mathematics is a tool for solving problems in such diverse fields as transportation, engineering, economics, medicine, business, and biology. By now, you have seen how we solve problems using mathematics by modeling real-world phenomena with mathematical equations or inequalities. Our ability to solve problems using mathematics, then, depends in part on our ability to solve various types of equations and inequalities. In this chapter, we review solving linear and quadratic equations and introduce solving other types of equations as well as systems of linear equations in three variables.

8.1 REVIEW SOLVING EQUATIONS

CD-ROM   SSM

SSG   Video

▶ **OBJECTIVE**

1. Solve linear and quadratic equations.

1

Recall that equations model many real-life problems. For example, we can use a simple linear equation to calculate the increase in digital camera sales.

With the help of your computer, digital cameras allow you to see your pictures and make copies immediately, send them in e-mail or use them on a Web page. Current sales and projected sales of these cameras are shown in the graph below.

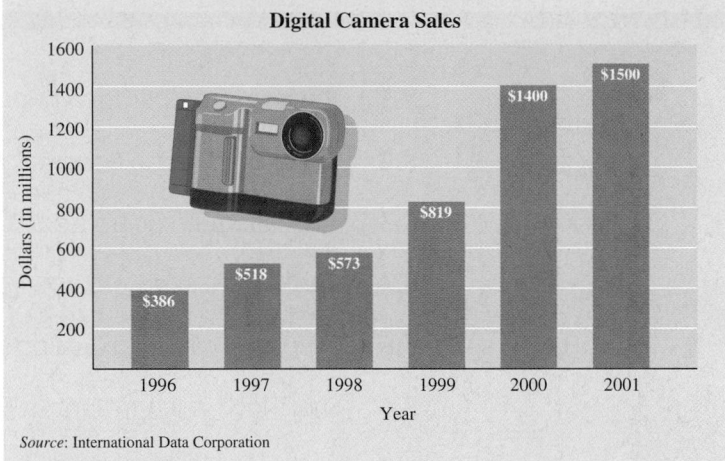

Digital Camera Sales

Source: International Data Corporation

To find the increase in sales from 1999 to 2000, for example, we can use the equation below.

In words:

Increase in sales	is	sales in 2000	minus	sales in 1999
Translate: x	$=$	1400	$-$	819

Since our variable x (increase in sales) is by itself on one side of the equation, we can find the value of x by simplifying the right side.

$$x = 581$$

The increase in sales of digital cameras from 1999 to 2000 is $581 million.

The equation, $x = 1400 - 819$ is a linear equation in one variable. In this section, we review solving linear equations and quadratic equations that can be solved by factoring. We will learn other methods for solving quadratic equations in Chapter 10.

Example 1   Solve: $2(x - 3) = 5x - 9$.

Solution   First, use the distributive property.

$$2(x - 3) = 5x - 9$$
$$2x - 6 = 5x - 9 \qquad \text{Use the distributive property.}$$

This is a linear equation so next, get variable terms on the same side of the equation by subtracting $5x$ from both sides.

$$2x - 6 - 5x = 5x - 9 - 5x \quad \text{Subtract } 5x \text{ from both sides.}$$
$$-3x - 6 = -9 \quad \text{Simplify.}$$
$$-3x - 6 + 6 = -9 + 6 \quad \text{Add 6 to both sides.}$$
$$-3x = -3 \quad \text{Simplify.}$$
$$\frac{-3x}{-3} = \frac{-3}{-3} \quad \text{Divide both sides by } -3.$$
$$x = 1$$

Let $x = 1$ in the original equation to see that 1 is the solution.

Example 2 Solve for y: $\frac{y}{3} - \frac{y}{4} = \frac{1}{6}$.

Solution First, clear the equation of fractions by multiplying both sides of the equation by 12, the LCD of denominators 3, 4, and 6.

$$\frac{y}{3} - \frac{y}{4} = \frac{1}{6}$$

$$12\left(\frac{y}{3} - \frac{y}{4}\right) = 12\left(\frac{1}{6}\right) \quad \text{Multiply both sides by the LCD 12.}$$

$$12\left(\frac{y}{3}\right) - 12\left(\frac{y}{4}\right) = 2 \quad \text{Apply the distributive property.}$$

$$4y - 3y = 2 \quad \text{Simplify.}$$

This is a linear equation, so we simply combine like terms on the left side to solve.

$$y = 2 \quad \text{Simplify.}$$

Check: To check, let $y = 2$ in the original equation.

$$\frac{y}{3} - \frac{y}{4} = \frac{1}{6} \quad \text{Original equation}$$

$$\frac{2}{3} - \frac{2}{4} \stackrel{?}{=} \frac{1}{6} \quad \text{Let } y = 2.$$

$$\frac{8}{12} - \frac{6}{12} \stackrel{?}{=} \frac{1}{6} \quad \text{Write fractions with the LCD.}$$

$$\frac{2}{12} \stackrel{?}{=} \frac{1}{6} \quad \text{Subtract.}$$

$$\frac{1}{6} = \frac{1}{6} \quad \text{Simplify.}$$

This is a true statement, so the solution is 2.

Example 3 Solve $3(x^2 + 4) + 5 = -6(x^2 + 2x) + 13$.

Solution First, use the distributive property.

$$3(x^2 + 4) + 5 = -6(x^2 + 2x) + 13$$
$$3x^2 + 12 + 5 = -6x^2 - 12x + 13 \quad \text{Apply the distributive property.}$$

This is a quadratic equation so we write it in standard form. In other words, we write the equation so that one side is 0.

$$9x^2 + 12x + 4 = 0$$ Rewrite the equation so that one side is 0.

$$(3x + 2)(3x + 2) = 0$$ Factor.

$$3x + 2 = 0 \quad \text{or} \quad 3x + 2 = 0$$ Set each factor equal to 0.

$$3x = -2 \quad \text{or} \quad 3x = -2$$

$$x = -\frac{2}{3} \quad \text{or} \quad x = -\frac{2}{3}$$ Solve each equation.

The solution is $-\frac{2}{3}$. Check by substituting $-\frac{2}{3}$ into the original equation.

Example 4   Solve for x: $\dfrac{x + 5}{2} + \dfrac{1}{2} = 2x - \dfrac{x - 3}{8}$.

Solution   Multiply both sides of the equation by 8, the LCD of 2 and 8.

$$8\left(\frac{x + 5}{2} + \frac{1}{2}\right) = 8\left(2x - \frac{x - 3}{8}\right)$$ Multiply both sides by 8.

$$8\left(\frac{x + 5}{2}\right) + 8 \cdot \frac{1}{2} = 8 \cdot 2x - 8\left(\frac{x - 3}{8}\right)$$ Apply the distributive property.

$$4(x + 5) + 4 = 16x - (x - 3)$$ Simplify.

$$4x + 20 + 4 = 16x - x + 3$$

$$4x + 24 = 15x + 3$$ Combine like terms.

This equation is linear, so we get variable terms on one side and numbers on the other side.

$$-11x + 24 = 3$$ Subtract 15x from both sides.

$$-11x = -21$$ Subtract 24 from both sides.

$$\frac{-11x}{-11} = \frac{-21}{-11}$$ Divide both sides by −11.

$$x = \frac{21}{11}$$ Simplify.

To check, verify that replacing x with $\dfrac{21}{11}$ makes the original equation true. The solution is $\dfrac{21}{11}$.

Example 5   Solve $2x^2 = \dfrac{17}{3}x + 1$.

Solution

$$2x^2 = \frac{17}{3}x + 1$$

$$3(2x^2) = 3\left(\frac{17}{3}x + 1\right)$$ Clear the equation of fractions.

$$6x^2 = 17x + 3$$ Apply the distributive property.

This equation is quadratic, so we write the equation so that one side is 0.

$$6x^2 - 17x - 3 = 0$$ Rewrite the equation in standard form.

$$(6x + 1)(x - 3) = 0$$ Factor.

$$6x + 1 = 0 \quad \text{or} \quad x - 3 = 0$$ Set each factor equal to zero.

$$6x = -1$$

$$x = -\frac{1}{6} \quad \text{or} \quad x = 3$$ Solve each equation.

The solutions are $-\dfrac{1}{6}$ and 3.

Exercise Set 8.1

Solve each equation. See Examples 1 through 5.

1. $x^2 + 11x + 24 = 0$ **2.** $y^2 - 10y + 24 = 0$

3. $3x - 4 - 5x = x + 4 + x$

4. $13x - 15x + 8 = 4x + 2 - 24$

5. $12x^2 + 5x - 2 = 0$ **6.** $3y^2 - y - 14 = 0$

7. $z^2 + 9 = 10z$ **8.** $n^2 + n = 72$

9. $5(y + 4) = 4(y + 5)$ **10.** $6(y - 4) = 3(y - 8)$

11. $0.6x - 10 = 1.4x - 14$ **12.** $0.3x + 2.4 = 0.1x + 4$

13. $x(5x + 2) = 3$ **14.** $n(2n - 3) = 2$

15. $6x - 2(x - 3) = 4(x + 1) + 4$

16. $10x - 2(x + 4) = 8(x - 2) + 6$

17. $\dfrac{3}{8} + \dfrac{b}{3} = \dfrac{5}{12}$ **18.** $\dfrac{a}{2} + \dfrac{7}{4} = 5$

19. $x^2 - 6x = x(8 + x)$ **20.** $n(3 + n) = n^2 + 4n$

21. $\dfrac{z^2}{6} - \dfrac{z}{2} - 3 = 0$ **22.** $\dfrac{c^2}{20} - \dfrac{c}{4} + \dfrac{1}{5} = 0$

23. $z + 3(2 + 4z) = 6(z + 1) + 5z$

24. $4(m - 6) - m = 8(m - 3) - 5m$

25. $\dfrac{x^2}{2} + \dfrac{x}{20} = \dfrac{1}{10}$ **26.** $\dfrac{y^2}{30} = \dfrac{y}{15} + \dfrac{1}{2}$

27. $\dfrac{4t^2}{5} = \dfrac{t}{5} + \dfrac{3}{10}$ **28.** $\dfrac{5x^2}{6} - \dfrac{7x}{2} + \dfrac{2}{3} = 0$

29. $\dfrac{3t + 1}{8} = \dfrac{5 + 2t}{7} + 2$ **30.** $4 - \dfrac{2z + 7}{9} = \dfrac{7 - z}{12}$

31. $\dfrac{m - 4}{3} - \dfrac{3m - 1}{5} = 1$ **32.** $\dfrac{n + 1}{8} - \dfrac{2 - n}{3} = \dfrac{5}{6}$

33. $3x^2 = -x$ **34.** $y^2 = -5y$

35. $x(x - 3) = x^2 + 5x + 7$

36. $z^2 - 4z + 10 = z(z - 5)$

37. $3(t - 8) + 2t = 7 + t$

38. $7c - 2(3c + 1) = 5(4 - 2c)$

39. $-3(x - 4) + x = 5(3 - x)$

40. $-4(a + 1) - 3a = -7(2a - 3)$

41. $(x - 1)(x + 4) = 24$

42. $(2x - 1)(x + 2) = -3$

43. $\dfrac{x^2}{4} - \dfrac{5}{2}x + 6 = 0$ **44.** $\dfrac{x^2}{18} + \dfrac{x}{2} + 1 = 0$

45. $y^2 + \dfrac{1}{4} = -y$ **46.** $\dfrac{x^2}{10} + \dfrac{5}{2} = x$

47. Which solution strategies are incorrect? Why?
   a. Solve $(y - 2)(y + 2) = 4$ by setting each factor equal to 4.
   b. Solve $(x + 1)(x + 3) = 0$ by setting each factor equal to 0.
   c. Solve $z^2 + 5z + 6 = 0$ by factoring $z^2 + 5z + 6$ and setting each factor equal to 0.
   d. Solve $x^2 + 6x + 8 = 10$ by factoring $x^2 + 6x + 8$ and setting each factor equal to 0.

48. Describe two ways a linear equation differs from a quadratic equation.

Find the value of K such that the equations are equivalent.

49. $3.2x + 4 = 5.4x - 7$
$3.2x = 5.4x + K$

50. $-7.6y - 10 = -1.1y + 12$
$-7.6y = -1.1y + K$

51. $\dfrac{x}{6} + 4 = \dfrac{x}{3}$
$x + K = 2x$

52. $\dfrac{5x}{4} + \dfrac{1}{2} = \dfrac{x}{2}$

$5x + K = 2x$

 Solve and check.

53. $2.569x = -12.48534$

54. $-9.112y = -47.537304$

55. $2.86z - 8.1258 = -3.75$

56. $1.25x - 20.175 = -8.15$

REVIEW EXERCISES

Translate each phrase into an expression. Use the variable x to represent each unknown number. See Section 1.4.

57. the quotient of 8 and a number

58. the sum of 8 and a number

59. the product of 8 and a number

60. the difference of 8 and a number

61. 2 more than three times a number

62. 5 subtracted from twice a number

8.2 FURTHER PROBLEM SOLVING

CD-ROM   SSM

SSG   Video

▶ **OBJECTIVES**

1. Write algebraic expressions that can be simplified.
2. Apply the steps for problem solving.

1 In order to review problem solving, we practice writing algebraic expressions that can be simplified.

Our first example involves consecutive integers and perimeter. Recall that *consecutive integers* are integers that follow one another in order. Study the examples of consecutive, even, and odd integers and their representations.

Consecutive Integers:

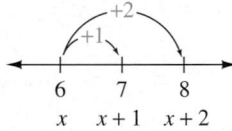

Consecutive Even Integers:

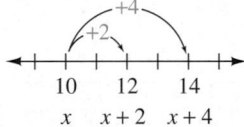

Consecutive Odd Integers:

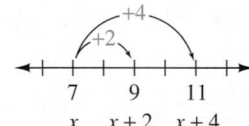

Example 1   Write the following as algebraic expressions. Then simplify.

a. The sum of two consecutive integers, if x is the first consecutive integer.

△ **b.** The perimeter of the triangle with sides of length x, $5x$, and $6x - 3$.

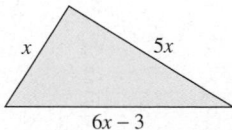

Solution   **a.** Recall that if x is the first integer, then the next consecutive integer is 1 more, or $x + 1$.

In words: | first integer | plus | next consecutive integer |

Translate: x $+$ $(x + 1)$

Then $x + (x + 1) = x + x + 1$

$= 2x + 1$ Simplify by combining like terms.

b. The perimeter of a triangle is the sum of the lengths of the sides.

In words: side + side + side

Translate: x + $5x$ + $(6x - 3)$

Then $x + 5x + (6x - 3) = x + 5x + 6x - 3$

$\qquad\qquad\qquad\qquad\qquad = 12x - 3$ Simplify.

Example 2   The three busiest airports in the United States are in the cities of Chicago, Atlanta, and Dallas/Ft. Worth. The airport in Atlanta has 7.7 million more arrivals and departures than the Dallas/Ft. Worth airport. The Chicago airport has 9.8 million more arrivals and departures than the Dallas/Ft Worth airport. Write the sum of the arrivals and departures from these three cities as a simplified algebraic expression. Let x be the number of arrivals and departures at the Dallas/Ft. Worth airport.

Solution   If x = millions of arrivals and departures at the Dallas/Ft. Worth airport, then
$x + 7.7$ = millions of arrivals and departures at the Atlanta airport and
$x + 9.8$ = millions of arrivals and departures at the Chicago airport
Since we want their sum, we have

In words:

arrivals and departures at Dallas/Ft. Worth	+	arrivals and departures at Atlanta	+	arrivals and departures at Chicago.

Translate: x + $(x + 7.7)$ + $(x + 9.8)$

Then $x + (x + 7.7) + (x + 9.8) = x + x + 7.7 + x + 9.8$

$\qquad\qquad\qquad\qquad\qquad\qquad = 3x + 17.5$ Combine like terms.

In Exercise 21, we will find the actual number of arrivals and departures at these airports.

2   Recall that our main purpose for studying algebra is to solve problems. Below is a review of the general strategy for problem solving that was introduced in Section 2.4.

GENERAL STRATEGY FOR PROBLEM SOLVING

1. UNDERSTAND the problem. During this step, become comfortable with the problem. Some ways of doing this are:

   Read and reread the problem.

   Choose a variable to represent the unknown.

   Construct a drawing.

   Propose a solution and check. Pay careful attention to how you check your proposed solution. This will help when writing an equation to model the problem.

2. TRANSLATE the problem into an equation.

3. SOLVE the equation.

4. INTERPRET the results: *Check* the proposed solution in the stated problem and *state* your conclusion.

Let's review this strategy as we solve the problem below.

Example 3 **FINDING THE ORIGINAL PRICE OF A COMPUTER**

Suppose that The Digital Store just announced an 8% decrease in the price of their Compaq Presario computers. If one particular computer model sells for $2162 after the decrease, find the original price of this computer.

Solution 1. UNDERSTAND. Read and reread the problem. Recall that a percent decrease means a percent of the original price. Let's guess that the original price of the computer is $2500. The amount of decrease is then 8% of $2500, or $(0.08)($2500) = 200. This means that the new price of the computer is the original price minus the decrease, or $2500 - $200 = 2300. Our guess is incorrect, but we now have an idea of how to model this problem. In our model, we will let $x =$ the original price of the computer.

2. TRANSLATE.

In words:

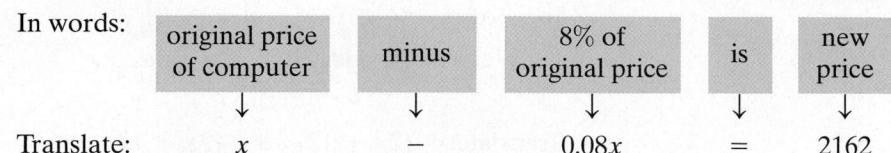

original price of computer	minus	8% of original price	is	new price
↓	↓	↓	↓	↓

Translate: x $-$ $0.08x$ $=$ 2162

3. SOLVE the equation.

$$x - 0.08x = 2162$$
$$0.92x = 2162 \qquad \text{Combine like terms.}$$
$$x = \frac{2162}{0.92} = 2350 \qquad \text{Divide both sides by 0.92.}$$

4. INTERPRET.

Check: If the original price of the computer was $2350, the new price is

$$\$2350 - (0.08)(\$2350) = \$2350 - \$188$$
$$= \$2162 \qquad \text{The given new price.}$$

State: The original price of the computer was $2350.

⬛

△ **Example 4**   **FINDING THE LENGTHS OF A TRIANGLE'S SIDES**

A pennant in the shape of an isosceles triangle is to be constructed for the Slidell High School Athletic Club and sold at a fund-raiser. The company manufacturing the pennant charges according to perimeter, and the athletic club has determined that a perimeter of 149 centimeters should make a nice profit. If each equal side of the triangle is twice the length of the third side, increased by 12 centimeters, find the lengths of the sides of the triangular pennant.

Solution   1. UNDERSTAND. Read and reread the problem. Recall that the perimeter of a triangle is the distance around. Let's guess that the third side of the triangular pennant is 20 centimeters. This means that each equal side is twice 20 centimeters, increased by 12 centimeters, or $2(20) + 12 = 52$ centimeters.

This gives a perimeter of $20 + 52 + 52 = 124$ centimeters. Our guess is incorrect, but we now have a better understanding of how to model this problem.

Now we let the third side of the triangle $= x$

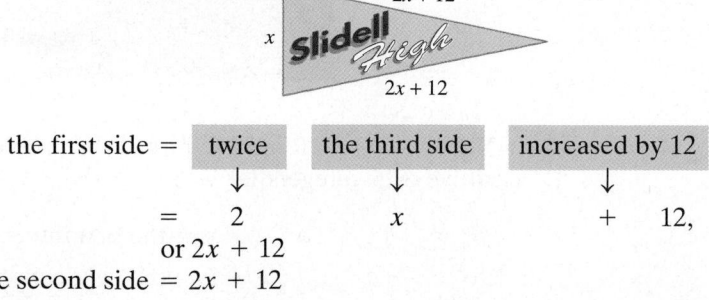

the first side =	twice	the third side	increased by 12
	↓	↓	↓
=	2	x	+ 12,

or $2x + 12$

the second side $= 2x + 12$

2. TRANSLATE.

In words:

first side	+	second side	+	third side	=	149
↓		↓		↓		↓

Translate: $(2x + 12)$ + $(2x + 12)$ + x =   149

3. SOLVE the equation.

$$(2x + 12) + (2x + 12) + x = 149$$

$2x + 12 + 2x + 12 + x = 149$ Remove parentheses.

$5x + 24 = 149$ Combine like terms.

$5x = 125$ Subtract 24 from both sides.

$x = 25$ Divide both sides by 5.

4. INTERPRET. If the third side is 25 centimeters, then the first side is $2(25) + 12 = 62$ centimeters and the second side is 62 centimeters also.

Check: The first and second sides are each twice 25 centimeters increased by 12 centimeters or 62 centimeters. Also, the perimeter is $25 + 62 + 62 = 149$ centimeters, the required perimeter.

State: The lengths of the sides of the triangle are 25 centimeters, 62 centimeters, and 62 centimeters.

Example 5   Kelsey Ohleger was helping her friend Benji Burnstine study for an algebra exam. Kelsey told Benji that her two latest art history quiz scores are two consecutive even integers whose sum is 174. Help Benji find the scores.

Solution   1. UNDERSTAND. Read and reread the problem. Since we are looking for consecutive even integers, let

x = the first integer. Then

$x + 2$ = the next consecutive even integer.

2. TRANSLATE.

In words:

first integer	+	next even integer	=	174
↓		↓		↓

Translate: x $+$ $(x + 2)$ $=$ 174

3. SOLVE.

$$x + (x + 2) = 174$$

$$2x + 2 = 174 \quad \text{Combine like terms}$$

$$2x = 172 \quad \text{Subtract 2 from both sides.}$$

$$x = 86 \quad \text{Divide both sides by 2.}$$

4. INTERPRET. If $x = 86$, then $x + 2 = 86 + 2$ or 88.

Check: The numbers 86 and 88 are two consecutive even integers. Their sum is 174, the required sum.

State: Kelsey's art history quiz scores are 86 and 88.

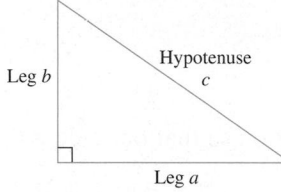

Leg b Hypotenuse c Leg a

Some of the exercises at the end of this section make use of the **Pythagorean theorem.** Recall that a **right triangle** is a triangle that contains a 90° angle, or right angle. The **hypotenuse** of a right triangle is the side opposite the right angle and is the longest side of the triangle. The **legs** of a right triangle are the other sides of the triangle.

PYTHAGOREAN THEOREM

In a right triangle, the sum of the squares of the lengths of the two legs is equal to the square of the length of the hypotenuse.

$$(\text{leg})^2 + (\text{leg})^2 = (\text{hypotenuse})^2 \quad \text{or} \quad a^2 + b^2 = c^2$$

△ **Example 6**   **USING THE PYTHAGOREAN THEOREM**

While framing an addition to an existing home, Kim Menzies, a carpenter, used the Pythagorean theorem to determine whether a wall was "square"—that is, whether the wall formed a right angle with the floor. He used a triangle whose sides are three consecutive integers. Find a right triangle whose sides are three consecutive integers.

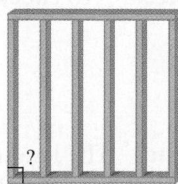

Solution   1. UNDERSTAND. Read and reread the problem.

Let x, $x + 1$, and $x + 2$ be three consecutive integers. Since these integers represent lengths of the sides of a right triangle, we have

$$x = \text{one leg}$$
$$x + 1 = \text{other leg}$$
$$x + 2 = \text{hypotenuse (longest side)}$$

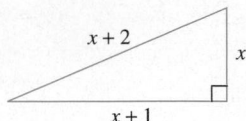

2. TRANSLATE. By the Pythagorean theorem, we have

In words:

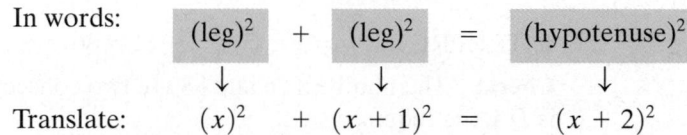

Translate: $(x)^2$ + $(x + 1)^2$ = $(x + 2)^2$

3. SOLVE the equation.

$$x^2 + (x + 1)^2 = (x + 2)^2$$
$$x^2 + x^2 + 2x + 1 = x^2 + 4x + 4 \qquad \text{Multiply.}$$
$$2x^2 + 2x + 1 = x^2 + 4x + 4$$

Notice this equation is quadratic. To solve, we write the equation so that one side is 0. Then we solve by factoring.

$$x^2 - 2x - 3 = 0 \qquad \text{Write in standard form.}$$
$$(x - 3)(x + 1) = 0$$
$$x - 3 = 0 \quad \text{or} \quad x + 1 = 0$$
$$x = 3 \qquad\qquad x = -1$$

4. INTERPRET. Discard $x = -1$ since length cannot be negative. If $x = 3$, then $x + 1 = 4$ and $x + 2 = 5$.

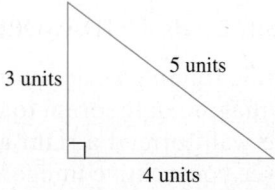

Check: To check, see that $(\text{leg})^2 + (\text{leg})^2 = (\text{hypotenuse})^2$

$$3^2 + 4^2 = 5^2$$
$$9 + 16 = 25 \qquad \text{True.}$$

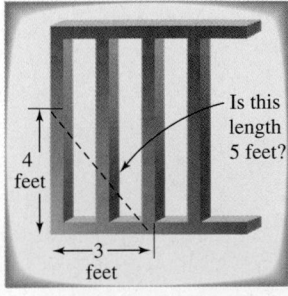

State: The lengths of the sides of the right triangle are 3, 4, and 5 units. Kim used this information, for example, by marking off lengths of 3 and 4 feet on the floor and framing respectively. If the diagonal length between these marks was 5 feet, the wall was "square." If not, adjustments were made.

SPOTLIGHT ON DECISION MAKING

Suppose you are a reporter for the *Marston Gazette*, a daily newspaper serving the medium-sized industrial city of Marston. Your editor has assigned you to a feature story about food pantries, soup kitchens, and other efforts to alleviate hunger among the city's homeless, poor, and working poor.

While researching your story assignment, you find that, according to the U.S. Department of Agriculture, 10.2% of all households in the United States do not have access to enough food to meet their basic needs. A survey conducted by Marston Social Services reveals that approximately 4950 of the estimated 39,400 Marston households must cope with hunger.

Armed with these basic facts, you now need to decide what angle to take with your story. Which of the following approaches would you choose? Why? What other information would you want to consider?

a. Marston lags behind nation in fight against hunger.

b. Marston mirrors national hunger picture.

c. Marston makes progress against hunger.

Exercise Set 8.2

Write the following as algebraic expressions. Then simplify. See Examples 1 and 2.

 **1.** The perimeter of the square with side length y.

2. The perimeter of the rectangle with length x and width $x - 5$.

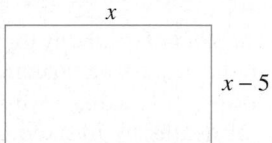

3. The sum of three consecutive integers if the first integer is z.

4. The sum of three consecutive odd integers if the first integer is x.

5. The total amount of money (in cents) in x nickels and $(x + 3)$ dimes. (*Hint:* the value of a nickel is 5 cents and the value of a dime is 10 cents.)

6. The total amount of money (in cents) in y quarters and $(2y - 1)$ nickels. (Use the hint for Exercise 5.)

△ **7.** A piece of land along Bayou Liberty is to be fenced and subdivided as shown so that each rectangle has the same dimensions. Express the total amount of fencing needed as an algebraic expression in x.

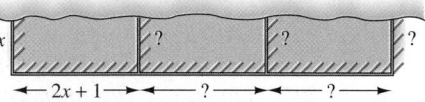

△ **8.** Write the perimeter of the floor plan shown as an algebraic expression in x.

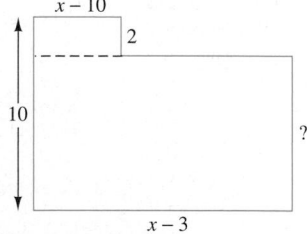

Solve. See Examples 3 through 5.

9. Four times the difference of a number and 2 is the same as 6 times the number, increased by 2. Find the number.

10. Twice the sum of a number and 3 is the same as 1 subtracted from the number. Find the number.

11. One number is 5 times another number. If the sum of the two numbers is 270, find the numbers.

12. One number is 6 less than another number. If the sum of the two numbers is 150, find the numbers.

13. The United States consists of 2271 million acres of land. Approximately 29% of this land is federally owned. Find the number of acres that are federally owned. (*Source:* U.S. General Services Administration)

14. The state of Nevada contains the most federally owned acres of land in the United States. If 90% of the state's 70 million acres of land is federally owned, find the number of federally owned acres. (*Source:* U.S. General Services Administration)

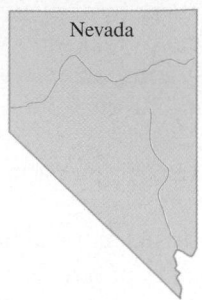

Nevada

15. Recently, 47% of homes in the United States contained computers. If Charlotte, North Carolina, contains 110,000 homes, how many of these homes would you expect to have computers? (*Source:* Telecommunication Research survey)

16. Recently, 26% of homes in the United States contained online services. If Abilene, Texas, contains 40,000 homes, how many of these homes would you expect to have online services? (*Source:* Telecommunication Research survey)

The following graph is called a circle graph or a pie chart. The circle represents a whole, or in this case, 100%. This particular graph shows the kind of loans that customers get from credit unions. Use this graph to answer Exercises 17–20.

Credit Union Loans

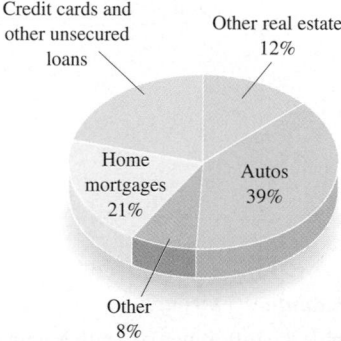

Credit cards and other unsecured loans

Other real estate 12%

Home mortgages 21%

Autos 39%

Other 8%

Source: National Credit Union Administration.

17. What percent of credit union loans are for credit cards and other unsecured loans?

18. What types of loans make up most credit union loans?

19. If the University of New Orleans Credit Union processed 300 loans last year, how many of these might we expect to be automobile loans?

20. If Homestead's Credit Union processed 537 loans last year, how many of these do you expect to be either home mortgages or other real estate? (Round to the nearest whole.)

21. The airports in Chicago, Atlanta, and Dallas/Ft. Worth have a total of 199 million annual arrivals and departures. Use this information and Example 2 in this section to find the number from each individual airport.

22. The perimeter of the triangle in Example 1b is 483 feet. Find the length of each side.

23. The B767-300ER aircraft has 104 more seats than the B737-200 aircraft. If their total number of seats is 328, find the number of seats for each aircraft. (*Source:* Air Transport Association of America)

24. The governor of Connecticut makes $29,000 less per year than the governor of Delaware. If the total of these salaries is $185,000, find the salary of each governor. (*Source: 2000 World Almanac*)

25. A new fax machine was recently purchased for an office in Hopedale for $464.40 including tax. If the tax rate in Hopedale is 8%, find the price of the fax machine before taxes.

26. A premedical student at a local university was complaining that she had just paid $86.11 for her human anatomy book, including tax. Find the price of the book before taxes if the tax rate at this university is 9%.

27. According to government statistics, the number of telephone company operators in the United States is expected to decrease to 26,000 by the year 2006. This represents a decrease of 47% from the number of telephone operators in 1996. (*Source:* U.S. Bureau of Labor Statistics)

  a. Find the number of telephone company operators in 1996. Round to the nearest whole number.

  b. In your own words, explain why you think that the need for telephone company operators is decreasing.

28. The number of deaths by tornadoes from the 1940s to the 1980s has decreased by 70.86%. There were 521 deaths from tornadoes in the 1980s. (*Source:* National Weather Service)

  a. Find the number of deaths by tornadoes in the 1940s. Round to the nearest whole number.

  b. In your own words, explain why you think that the number of deaths by tornadoes has decreased so much since the 1940s.

29. Manufacturers claim that a CD-ROM disc will last 20 years. Recently, statements made by the U.S. National Archives and Records Administration suggest that 20 years decreased by 75% is a more realistic lifespan because the aluminum substratum on which the data is recorded can be affected by oxidation. Find the lifespan of a CD-ROM according to the U.S. National Archives and Records Administration.

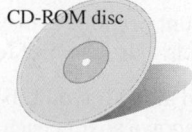

CD-ROM disc

30. In one year, 2.7% of India's forest was lost to deforestation. This percent represents 10,000 square kilometers of forest. Find the total square kilometers of forest in India before this decrease. (Round to the nearest whole square kilometer.)

31. Americans used computers to electronically file 21.1 million federal income tax returns in 1999. This represented a 24% increase over the previous year. How many federal income tax returns were filed electronically in 1998? Round to the nearest million. (*Source: Associated Press, April 1999*)

32. HDPE (high-density polyethylene) plastics are used to make milk and water jugs, as well as bottles for juices and laundry detergents. When these types of bottles are recycled, they can be used to make bags, recycling bins, motor oil bottles, and agricultural pipe. HDPE bottle recycling increased 7% to 704 million pounds from 1996 to 1997. How many pounds of HDPE bottles were recycled in 1996? Round to the nearest million. (*Source:* American Plastics Council)

33. Two frames are needed with the same outside perimeter: one frame in the shape of a square and one in the shape of an equilateral triangle. Each side of the triangle is 6 centimeters longer than each side of the square. Find the dimensions of each frame.

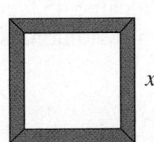 

x

34. The length of a rectangular sign is 2 feet less than three times its width. Find the dimensions if the perimeter is 28 feet.

x feet

35. In a blueprint of a rectangular room, the length is to be 2 centimeters greater than twice its width. Find the dimensions if the perimeter is to be 40 centimeters.

36. A plant food solution contains 5 cups of water for every 1 cup of concentrate. If the solution contains 78 cups of these two ingredients, find the number of cups of concentrate in the solution.

37. The external tank of a NASA space shuttle contains the propellants used for the first 8.5 minutes after launch. Its height is 5 times the sum of its width and 1. If the sum of the height and width is 55.4 meters, find the dimensions of this tank.

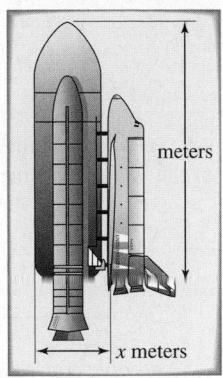

meters

x meters

38. The blue whale is the largest of the whales. Its average weight is 3 times the difference of the average weight of a humpback whale and 5 tons. If the total of the average weights is 117 tons, find the average weight of each type of whale.

Recall that the sum of the angle measures of a triangle is 180°.

39. Find the measures of the angles of a triangle if the measure of one angle is twice the measure of a second angle and the third angle measures 3 times the second angle decreased by 12.

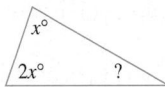

$x°$

$2x°$?

40. Find the angles of a triangle whose two base angles are equal and whose third angle is 10° less than three times a base angle.

x x

Recall that two angles are complements of each other if their sum is 90°. Two angles are supplements of each other if their sum is 180°. Find the measure of each angle.

41.

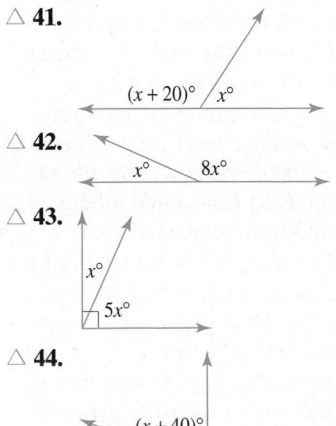

$(x + 20)°$ $x°$

42.

$x°$ $8x°$

43.

$x°$

$5x°$

44.

$(x + 40)°$

$x°$

△ **45.** One angle is three times its supplement increased by 20°. Find the measures of the two supplementary angles.

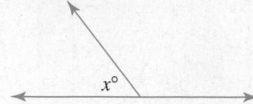

△ **46.** One angle is twice its complement increased by 30°. Find the measures of the two complementary angles.

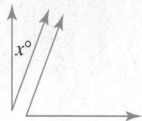

Solve. Exercises 47–52 involve consecutive integers.

47. The sum of three consecutive integers is 228. Find the integers.

48. The sum of three consecutive odd integers is 327. Find the integers.

49. The zip codes of three Nevada locations—Fallon, Fernley, and Gardnerville Ranchos—are three consecutive even integers. If twice the first integer added to the third is 268,222, find each zip code.

50. During a recent year, the average SAT scores in math for the states of Alabama, Louisiana, and Michigan were 3 consecutive integers. If the sum of the first integer, second integer, and three times the third integer is 2637, find each score.

51. Determine whether there are three consecutive integers such that their sum is three times the second integer.

52. Determine whether there are two consecutive odd integers such that 7 times the first exceeds 5 times the second by 54.

To break even in a manufacturing business, income or revenue R must equal the cost of production C. Use this information for Exercises 53 through 58.

53. The cost C to produce x number of skateboards is $C = 100 + 20x$. The skateboards are sold wholesale for $24 each, so revenue R is given by $R = 24x$. Find how many skateboards the manufacturer needs to produce and sell to break even. (*Hint:* Set the cost expression equal to the revenue expression and solve for x.)

54. The revenue R from selling x number of computer boards is given by $R = 60x$, and the cost C of producing them is given by $C = 50x + 5000$. Find how many boards must be sold to break even. Find how much money is needed to produce the break-even number of boards.

55. The cost C of producing x number of paperback books is given by $C = 4.50x + 2400$. Income R from these books is given by $R = 7.50x$. Find how many books should be produced and sold to break even.

56. Find the break-even quantity for a company that makes x number of computer monitors at a cost C given by $C = 875 + 70x$ and receives revenue R given by $R = 105x$.

✎ **57.** In your own words, explain what happens if a company makes and sells fewer products than the break-even point.

✎ **58.** In your own words, explain what happens if more products than the break-even point are made and sold.

△ **59.** Newsprint is either discarded or recycled. Americans recycle about 27% of all newsprint, but an amount of newsprint equivalent to 30 million trees is discarded every year. About how many trees' worth of newsprint is *recycled* in the United States each year? (*Source:* The Earth Works Group)

△ **60.** Find an angle such that its supplement is equal to twice its complement increased by 50°.

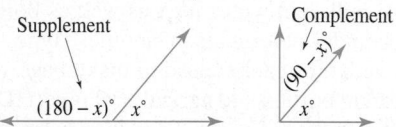

61. One number exceeds another by five, and their product is 66. Find the numbers.

62. If the sum of two numbers is 4 and their product is $\frac{15}{4}$, find the numbers.

Solve. See Example 6.

◈ **63.** An electrician needs to run a cable from the top of a 60-foot tower to a transmitter box located 45 feet away from the base of the tower. Find how long he should make the cable.

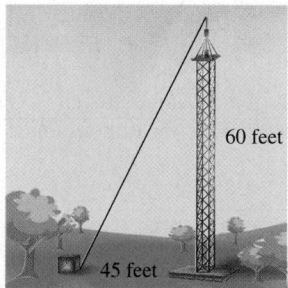

△ **64.** A stereo system installer needs to run speaker wire along the two diagonals of a rectangular room whose dimensions are 40 feet by 75 feet. Find how much speaker wire she needs.

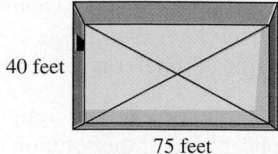

65. If the cost, $C(x)$, for manufacturing x units of a certain product is given by $C(x) = x^2 - 15x + 50$, find the number of units manufactured at a cost of $9500.

△ **66.** Determine whether any three consecutive integers represent the lengths of the sides of a right triangle.

△ **67.** The shorter leg of a right triangle is 3 centimeters less than the other leg. Find the length of the two legs if the hypotenuse is 15 centimeters.

△ **68.** The longer leg of a right triangle is 4 feet longer than the other leg. Find the length of the two legs if the hypotenuse is 20 feet.

△ **69.** Marie Mulroney has a rectangular board 12 inches by 16 inches around which she wants to put a uniform border of shells. If she has enough shells for a border whose area is 128 square inches, determine the width of the border.

△ **70.** A gardener has a rose garden that measures 30 feet by 20 feet. He wants to put a uniform border of pine bark around the outside of the garden. Find how wide the border should be if he has enough pine bark to cover 336 square feet.

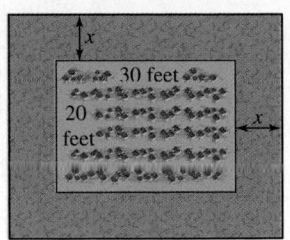

REVIEW EXERCISES

Graph each set on a number line and write it in interval notation. See Section 2.7.

71. $\{x \mid 0 \le x \le 5\}$

72. $\{x \mid -7 < x \le 1\}$

73. $\left\{x \mid -\dfrac{1}{2} < x < \dfrac{3}{2}\right\}$

74. $\{x \mid -2.5 \le x < 5.3\}$

8.3 COMPOUND INEQUALITIES

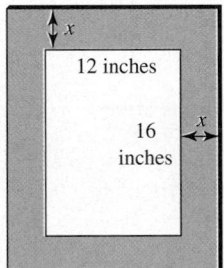

CD-ROM   SSM

SSG   Video

▶ **OBJECTIVES**

1. Find the intersection of two sets.
2. Solve compound inequalities containing **and**.
3. Find the union of two sets.
4. Solve compound inequalities containing **or**.

Two inequalities joined by the words **and** or **or** are called **compound inequalities.**

Compound Inequalities

$$x + 3 < 8 \text{ and } x > 2$$

$$\frac{2x}{3} \ge 5 \text{ or } -x + 10 < 7$$

1 The solution set of a compound inequality formed by the word **and** is the **intersection** of the solution sets of the two inequalities.

INTERSECTION OF TWO SETS

The intersection of two sets, A and B, is the set of all elements common to both sets. A intersect B is denoted by

$$A \cap B$$

Example 1   Find the intersection: $\{2, 4, 6, 8\} \cap \{3, 4, 5, 6\}$

Solution   The numbers 4 and 6 are in both sets. The intersection is $\{4, 6\}$.

2   A value is a solution of a compound inequality formed by the word **and** if it is a solution of *both* inequalities. For example, the solution set of the compound inequality $x \leq 5$ and $x \geq 3$ contains all values of x that make the inequality $x \leq 5$ a true statement **and** the inequality $x \geq 3$ a true statement. The first graph shown below is the graph of $x \leq 5$, the second graph is the graph of $x \geq 3$, and the third graph shows the intersection of the two graphs. The third graph is the graph of $x \leq 5$ **and** $x \geq 3$.

$\{x \mid x \leq 5\}$ $(-\infty, 5]$

$\{x \mid x \geq 3\}$ $[3, \infty)$

$\{x \mid x \leq 5 \text{ and } x \geq 3\}$ $[3, 5]$

The compound inequality $x \leq 5$ and $x \geq 3$ can be written in a more compact form as $3 \leq x \leq 5$. The solution set $\{x \mid 3 \leq x \leq 5\}$ includes all numbers that are less than or equal to 5 and at the same time greater than or equal to 3. In interval notation, the solution set is $[3, 5]$.

Example 2   Solve $x - 7 < 2$ and $2x + 1 < 9$.

Solution   First we solve each inequality separately.

$$
\begin{array}{ccc}
x - 7 < 2 & \text{and} & 2x + 1 < 9 \\
x < 9 & \text{and} & 2x < 8 \\
x < 9 & \text{and} & x < 4
\end{array}
$$

Now we can graph the two intervals on two number lines and find their intersection. Their intersection is shown on the third number line.

$\{x \mid x < 9\}$ $(-\infty, 9)$

$\{x \mid x < 4\}$ $(-\infty, 4)$

$\{x \mid x < 9 \text{ and } x < 4\}$ $(-\infty, 4)$

$\phantom{\{x \mid x < 9 \text{ and } x < 4\}} = \{x \mid x < 4\}$

The solution set is $(-\infty, 4)$.

Example 3   Solve $2x \geq 0$ and $4x - 1 \leq -9$.

Solution   First we solve each inequality separately.

$$
\begin{array}{ccc}
2x \geq 0 & \text{and} & 4x - 1 \leq -9 \\
x \geq 0 & \text{and} & 4x \leq -8 \\
x \geq 0 & \text{and} & x \leq -2
\end{array}
$$

Now we can graph the two intervals and find their intersection.

$\{x \mid x \geq 0\}$

$[0, \infty)$

$\{x \mid x \leq -2\}$

$(-\infty, -2]$

$\{x \mid x \geq 0 \text{ and } x \leq -2\} = \varnothing$

$\varnothing$

There is no number that is greater than or equal to 0 *and* less than or equal to -2. The solution set is $\varnothing$.

> ▼ HELPFUL HINT
> Example 3 shows that some compound inequalities have no solution. Also, some have all real numbers as solutions.

To solve a compound inequality written in a compact form, such as $2 < 4 - x < 7$, we get x alone in the "middle part." Since a compound inequality is really two inequalities in one statement, we must perform the same operations on all three parts of the inequality.

Example 4 Solve $2 < 4 - x < 7$.

Solution To get x alone, we first subtract 4 from all three parts.

$$2 < 4 - x < 7$$

$$2 - 4 < 4 - x - 4 < 7 - 4 \qquad \text{Subtract 4 from all three parts.}$$

$$-2 < -x < 3 \qquad \text{Simplify.}$$

$$\frac{-2}{-1} > \frac{-x}{-1} > \frac{3}{-1} \qquad \begin{array}{l}\text{Divide all three parts by } -1 \text{ and}\\ \text{reverse the inequality symbols.}\end{array}$$

$$2 > x > -3$$

> ▼ HELPFUL HINT
> Don't forget to reverse both inequality symbols.

This is equivalent to $-3 < x < 2$.
The solution set in interval notation is $(-3, 2)$, and its graph is shown.

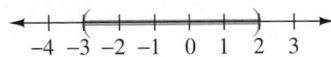

Example 5 Solve $-1 \leq \dfrac{2x}{3} + 5 \leq 2$.

Solution First, clear the inequality of fractions by multiplying all three parts by the LCD of 3.

$$-1 \leq \frac{2x}{3} + 5 \leq 2$$

$$3(-1) \leq 3\left(\frac{2x}{3} + 5\right) \leq 3(2) \qquad \text{Multiply all three parts by the LCD of 3.}$$

$$-3 \leq 2x + 15 \leq 6 \qquad \text{Use the distributive property and multiply.}$$

$$-3 - 15 \leq 2x + 15 - 15 \leq 6 - 15 \qquad \text{Subtract 15 from all three parts.}$$

$$-18 \leq 2x \leq -9 \qquad \text{Simplify.}$$

$$\frac{-18}{2} \leq \frac{2x}{2} \leq \frac{-9}{2} \qquad \text{Divide all three parts by 2.}$$

$$-9 \leq x \leq -\frac{9}{2} \qquad \text{Simplify.}$$

The graph of the solution is shown.

$$-\frac{9}{2}$$

-10 -9 -8 -7 -6 -5 -4 -3

The solution set in interval notation is $\left[-9, -\frac{9}{2}\right]$.

3 The solution set of a compound inequality formed by the word **or** is the **union** of the solution sets of the two inequalities.

UNION OF TWO SETS

The union of two sets, A and B, is the set of elements that belong to *either* of the sets. A union B is denoted by

$$A \cup B$$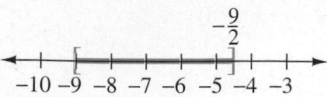

Example 6 Find the union: $\{2, 4, 6, 8\} \cup \{3, 4, 5, 6\}$

Solution The numbers that are in either set or both sets are $\{2, 3, 4, 5, 6, 8\}$. This set is the union.

4 A value is a solution of a compound inequality formed by the word **or** if it is a solution of **either** inequality. For example, the solution set of the compound inequality $x \leq 1$ **or** $x \geq 3$ contains all numbers that make the inequality $x \leq 1$ a true statement **or** the inequality $x \geq 3$ a true statement.

$\{x \mid x \leq 1\}$

-1 0 1 2 3 4 5 6

$(-\infty, 1]$

$\{x \mid x \geq 3\}$

-1 0 1 2 3 4 5 6

$[3, \infty)$

$\{x \mid x \leq 1 \text{ or } x \geq 3\}$

-1 0 1 2 3 4 5 6

$(-\infty, 1] \cup [3, \infty)$

In interval notation, the set $\{x \mid x \leq 1 \text{ or } x \geq 3\}$ is written as $(-\infty, 1] \cup [3, \infty)$.

Example 7 Solve $5x - 3 \leq 10$ or $x + 1 \geq 5$.

Solution First we solve each inequality separately.

$$5x - 3 \leq 10 \quad \text{or} \quad x + 1 \geq 5$$
$$5x \leq 13 \quad \text{or} \quad x \geq 4$$
$$x \leq \frac{13}{5} \quad \text{or} \quad x \geq 4$$

Now we can graph each interval and find their union.

$\left\{ x \mid x \leq \dfrac{13}{5} \right\}$
$\left(-\infty, \dfrac{13}{5} \right]$

$\{ x \mid x \geq 4 \}$
$[4, \infty)$

$\left\{ x \mid x \leq \dfrac{13}{5} \text{ or } x \geq 4 \right\}$
$\left(-\infty, \dfrac{13}{5} \right] \cup [4, \infty)$

The solution set is $\left(-\infty, \dfrac{13}{5} \right] \cup [4, \infty)$.

Example 8

Solve: $-2x - 5 < -3$ or $6x < 0$.

Solution   First we solve each inequality separately.

$$
\begin{array}{ccc}
-2x - 5 < -3 & \text{or} & 6x < 0 \\
-2x < 2 & \text{or} & x < 0 \\
x > -1 & \text{or} & x < 0
\end{array}
$$

Now we can graph each interval and find their union.

$\{ x \mid x > -1 \}$
$(-1, \infty)$

$\{ x \mid x < 0 \}$
$(-\infty, 0)$

$\{ x \mid x > -1 \text{ or } x < 0 \}$
$= $ all real numbers
$(-\infty, \infty)$

The solution set is $(-\infty, \infty)$.

Exercise Set 8.3

If $A = \{ x \mid x \text{ is an even integer} \}$, $B = \{ x \mid x \text{ is an odd integer} \}$, $C = \{ 2, 3, 4, 5 \}$, and $D = \{ 4, 5, 6, 7 \}$, list the elements of each set. See Examples 1 and 6.

1. $C \cup D$

2. $C \cap D$

3. $A \cap D$

4. $A \cup D$

5. $A \cup B$

6. $A \cap B$

7. $B \cap D$

8. $B \cup D$

9. $B \cup C$

10. $B \cap C$

11. $A \cap C$

12. $A \cup C$

Solve each compound inequality. Graph the solution set and write it in interval notation. See Examples 2 and 3.

13. $x < 5$ and $x > -2$

14. $x \leq 7$ and $x \leq 1$

 15. $x + 1 \geq 7$ and $3x - 1 \geq 5$

16. $-2x < -8$ and $x - 5 < 5$

17. $4x + 2 \leq -10$ and $2x \leq 0$

18. $x + 4 > 0$ and $4x > 0$

Solve each compound inequality. Graph the solution set and write it in interval notation. See Examples 4 and 5.

19. $5 < x - 6 < 11$

20. $-2 \le x + 3 \le 0$

21. $-2 \le 3x - 5 \le 7$

22. $1 < 4 + 2x < 7$

23. $1 \le \dfrac{2}{3}x + 3 \le 4$

24. $-2 < \dfrac{1}{2}x - 5 < 1$

25. $-5 \le \dfrac{x + 1}{4} \le -2$

26. $-4 \le \dfrac{2x + 5}{3} \le 1$

Solve each compound inequality. Graph the solution set and write it in interval notation. See Examples 7 and 8.

27. $x < -1$ or $x > 0$

28. $x \le 1$ or $x \le -3$

29. $-2x \le -4$ or $5x - 20 \ge 5$

30. $x + 4 < 0$ or $6x > -12$

31. $3(x - 1) < 12$ or $x + 7 > 10$

32. $5(x - 1) \ge -5$ or $5 - x \le 11$

33. Explain how solving an and-compound inequality is similar to finding the intersection of two sets.

34. Explain how solving an or-compound inequality is similar to finding the union of two sets.

Solve each compound inequality. Graph the solution set and write it in interval notation.

35. $x < 2$ and $x > -1$

36. $x < 5$ and $x < 1$

37. $x < 2$ or $x > -1$

38. $x < 5$ or $x < 1$

39. $x \ge -5$ and $x \ge -1$

40. $x \le 0$ or $x \ge -3$

41. $x \ge -5$ or $x \ge -1$

42. $x \le 0$ and $x \ge -3$

43. $0 \le 2x - 3 \le 9$

44. $3 < 5x + 1 < 11$

45. $\dfrac{1}{2} < x - \dfrac{3}{4} < 2$

46. $\dfrac{2}{3} < x + \dfrac{1}{2} < 4$

47. $x + 3 \ge 3$ and $x + 3 \le 2$

48. $2x - 1 \ge 3$ and $-x > 2$

49. $3x \ge 5$ or $-x - 6 < 1$

50. $\dfrac{3}{8}x + 1 \le 0$ or $-2x < -4$

51. $0 < \dfrac{5 - 2x}{3} < 5$

52. $-2 < \dfrac{-2x - 1}{3} < 2$

53. $-6 < 3(x - 2) \le 8$

54. $-5 < 2(x + 4) < 8$

55. $-x + 5 > 6$ and $1 + 2x \le -5$

56. $5x \le 0$ and $-x + 5 < 8$

◆ **57.** $3x + 2 \leq 5$ or $7x > 29$

58. $-x < 7$ or $3x + 1 < -20$

59. $5 - x > 7$ and $2x + 3 \geq 13$

60. $-2x < -6$ or $1 - x > -2$

61. $-\frac{1}{2} \leq \frac{4x - 1}{6} < \frac{5}{6}$

62. $-\frac{1}{2} \leq \frac{3x - 1}{10} < \frac{1}{2}$

63. $\frac{1}{15} < \frac{8 - 3x}{15} < \frac{4}{5}$

64. $-\frac{1}{4} < \frac{6 - x}{12} < -\frac{1}{6}$

65. $0.3 < 0.2x - 0.9 < 1.5$

66. $-0.7 \leq 0.4x + 0.8 < 0.5$

The formula for converting Fahrenheit temperatures to Celsius temperatures is $C = \frac{5}{9}(F - 32)$. Use this formula for Exercises 67 and 68.

67. During a recent year, the temperatures in Chicago ranged from $-29°$ to $35°$C. Use a compound inequality to convert these temperatures to Fahrenheit temperatures.

68. In Oslo, the average temperature ranges from $-10°$ to $18°$ Celsius. Use a compound inequality to convert these temperatures to the Fahrenheit scale.

Solve.

69. Christian D'Angelo has scores of 68, 65, 75, and 78 on his algebra tests. Use a compound inequality to find the scores he can make on his final exam to receive a C in the course. The final exam counts as two tests, and a C is received if the final course average is from 70 to 79.

70. Wendy Wood has scores of 80, 90, 82, and 75 on her chemistry tests. Use a compound inequality to find the range of scores she can make on her final exam to receive a B in the course. The final exam counts as two tests, and a B is received if the final course average is from 80 to 89.

Use the graph to answer Exercises 71 and 72.

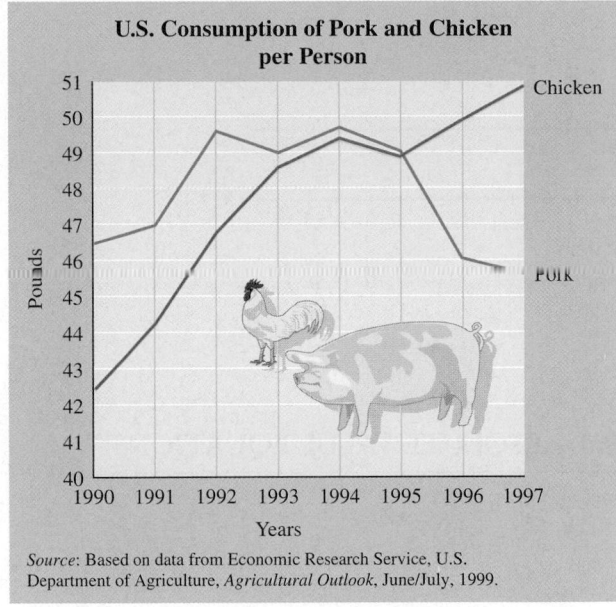

U.S. Consumption of Pork and Chicken per Person

Source: Based on data from Economic Research Service, U.S. Department of Agriculture, *Agricultural Outlook*, June/July, 1999.

71. For what years was the consumption of pork greater than 48 pounds per person *and* the consumption of chicken greater than 48 pounds per person?

72. For what years was the consumption of pork less than 48 pounds per person *or* the consumption of chicken greater than 49 pounds per person?

REVIEW EXERCISES

Evaluate the following. See Sections 1.5 and 1.6.

73. $|-7| - |19|$

74. $|-7 - 19|$

75. $-(-6) - |-10|$

76. $|-4| - (-4) + |-20|$

Find by inspection all values for x that make each equation true See Section 1.2.

77. $|x| = 7$

78. $|x| = 5$

79. $|x| = 0$

80. $|x| = -2$

A Look Ahead

Example
Solve $x - 6 < 3x < 2x + 5$.

Solution:
Notice that this inequality contains a variable not only in the middle, but also on the left and the right. When this occurs, we solve by rewriting the inequality using the word ***and***.

$x - 6 < 3x$ and $3x < 2x + 5$

$-6 < 2x$ and $x < 5$

$-3 < x$

$x > -3$ and $x < 5$

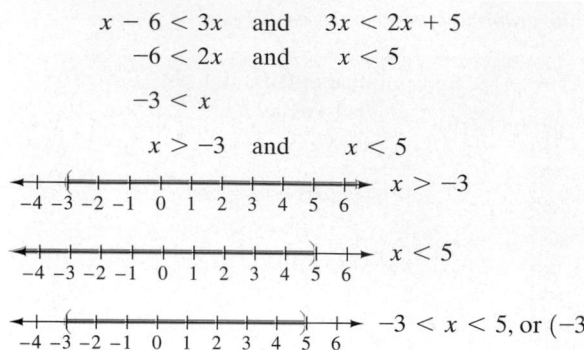

$x > -3$

$x < 5$

$-3 < x < 5$, or $(-3, 5)$

Solve each compound inequality for x. See the example.

81. $2x - 3 < 3x + 1 < 4x - 5$

82. $x + 3 < 2x + 1 < 4x + 6$

83. $-3(x - 2) \leq 3 - 2x \leq 10 - 3x$

84. $7x - 1 \leq 7 + 5x \leq 3(1 + 2x)$

85. $5x - 8 < 2(2 + x) < -2(1 + 2x)$

86. $1 + 2x < 3(2 + x) < 1 + 4x$

8.4 ABSOLUTE VALUE EQUATIONS

CD-ROM   SSM

SSG   Video

▶ **OBJECTIVE**

1. Solve absolute value equations.

1 In Chapter 1, we defined the absolute value of a number as its distance from 0 on a number line.

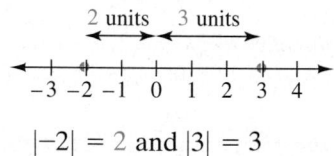

$|-2| = 2$ and $|3| = 3$

In this section, we concentrate on solving equations containing the absolute value of a variable or a variable expression. Examples of absolute value equations are

$$|x| = 3 \qquad -5 = |2y + 7| \qquad |z - 6.7| = |3z + 1.2|$$

Since distance and absolute value are so closely related, absolute value equations and inequalities (see Section 8.5) are extremely useful in solving distance-type problems, such as calculating the possible error in a measurement.

For the absolute value equation $|x| = 3$, its solution set will contain all numbers whose distance from 0 is 3 units. Two numbers are 3 units away from 0 on the number line: 3 and -3.

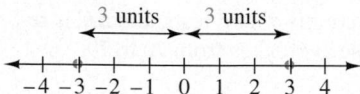

Thus, the solution set of the equation $|x| = 3$ is $\{3, -3\}$. This suggests the following:

SOLVING EQUATIONS OF THE FORM $|x| = a$

If a is a positive number, then $|x| = a$ is equivalent to $x = a$ or $x = -a$.

Example 1   Solve $|p| = 2$.

Solution   Since 2 is positive, $|p| = 2$ is equivalent to $p = 2$ or $p = -2$.
To check, let $p = 2$ and then $p = -2$ in the original equation.

$|p| = 2$	*Original equation*	$|p| = 2$	*Original equation*
$|2| = 2$	*Let $p = 2$.*	$|-2| = 2$	*Let $p = -2$.*
$2 = 2$	*True.*	$2 = 2$	*True.*

The solutions are 2 and -2 or the solution set is $\{2, -2\}$.

If the expression inside the absolute value bars is more complicated than a single variable x, we can still apply the absolute value property.

Example 2   Solve $|5w + 3| = 7$.

Solution   Here the expression inside the absolute value bars is $5w + 3$. If we think of the expression $5w + 3$ as x in the absolute value property, we see that $|x| = 7$ is equivalent to

$$x = 7 \quad \text{or} \quad x = -7$$

Then substitute $5w + 3$ for x, and we have

$$5w + 3 = 7 \quad \text{or} \quad 5w + 3 = -7$$

Solve these two equations for w.

$$5w + 3 = 7 \quad \text{or} \quad 5w + 3 = -7$$
$$5w = 4 \quad \text{or} \quad 5w = -10$$
$$w = \frac{4}{5} \quad \text{or} \quad w = -2$$

Check   To check, let $w = -2$ and then $w = \frac{4}{5}$ in the original equation.

Let $w = -2$	Let $w = \frac{4}{5}$
$|5(-2) + 3| = 7$	$\left|5\left(\frac{4}{5}\right) + 3\right| = 7$
$|-10 + 3| = 7$	$|4 + 3| = 7$
$|-7| = 7$	$|7| = 7$
$7 = 7$ *True.*	$7 = 7$ *True.*

Both solutions check, and the solutions are -2 and $\frac{4}{5}$.

Example 3   Solve $\left|\dfrac{x}{2} - 1\right| = 11$.

Solution $\left|\dfrac{x}{2} - 1\right| = 11$ is equivalent to

$$\frac{x}{2} - 1 = 11 \qquad \text{or} \qquad \frac{x}{2} - 1 = -11$$

$$2\left(\frac{x}{2} - 1\right) = 2(11) \quad \text{or} \quad 2\left(\frac{x}{2} - 1\right) = 2(-11) \qquad \text{Clear fractions.}$$

$$x - 2 = 22 \qquad \text{or} \qquad x - 2 = -22 \qquad \text{Apply the distributive property.}$$

$$x = 24 \qquad \text{or} \qquad x = -20$$

The solutions are 24 and −20.

To apply the absolute value property, first make sure that the absolute value expression is isolated.

> **HELPFUL HINT**
> If the equation has a single absolute value expression containing variables, isolate the absolute value expression first.

Example 4   Solve $|2x| + 5 = 7$.

Solution   We want the absolute value expression alone on one side of the equation, so begin by subtracting 5 from both sides. Then apply the absolute value property.

$$|2x| + 5 = 7$$
$$|2x| = 2 \qquad \text{Subtract 5 from both sides.}$$
$$2x = 2 \quad \text{or} \quad 2x = -2$$
$$x = 1 \quad \text{or} \quad x = -1$$

The solutions are −1 and 1.

Example 5   Solve $|y| = 0$.

Solution   We are looking for all numbers whose distance from 0 is zero units. The only number is 0. The solution is 0.

The next two examples illustrate a special case for absolute value equations. This special case occurs when an isolated absolute value is equal to a negative number.

Example 6   Solve $2|x| + 25 = 23$.

Solution   First, isolate the absolute value.

$$2|x| + 25 = 23$$
$$2|x| = -2 \qquad \text{Subtract 25 from both sides.}$$
$$|x| = -1 \qquad \text{Divide both sides by 2.}$$

The absolute value of a number is never negative, so this equation has no solution.

Example 7   Solve $\left| \dfrac{3x + 1}{2} \right| = -2$.

Solution   Again, the absolute value of any expression is never negative, so no solution exists.

Given two absolute value expressions, we might ask, when are the absolute values of two expressions equal? To see the answer, notice that

$$|2| = |2|, \quad |-2| = |-2|, \quad |-2| = |2|, \quad \text{and} \quad |2| = |-2|$$

$$\nwarrow \; \nearrow \qquad \nwarrow \; \nearrow \qquad \nwarrow \; \nearrow \qquad\qquad \nwarrow \; \nearrow$$

$$\text{same} \qquad\quad \text{same} \qquad\quad \text{opposites} \qquad\quad \text{opposites}$$

Two absolute value expressions are equal when the expressions inside the absolute value bars are equal to or are opposites of each other.

Example 8   Solve $|3x + 2| = |5x - 8|$.

Solution   This equation is true if the expressions inside the absolute value bars are equal to or are opposites of each other.

$$3x + 2 = 5x - 8 \quad \text{or} \quad 3x + 2 = -(5x - 8)$$

Next, solve each equation.

$$
\begin{aligned}
3x + 2 &= 5x - 8 &\quad \text{or} \quad& 3x + 2 = -5x + 8 \\
-2x + 2 &= -8 &\quad \text{or} \quad& 8x + 2 = 8 \\
-2x &= -10 &\quad \text{or} \quad& 8x = 6 \\
x &= 5 &\quad \text{or} \quad& x = \frac{3}{4}
\end{aligned}
$$

The solutions are $\dfrac{3}{4}$ and 5.

Example 9   Solve $|x - 3| = |5 - x|$.

Solution

$$
\begin{aligned}
x - 3 &= 5 - x &\quad \text{or} \quad& x - 3 = -(5 - x) \\
2x - 3 &= 5 &\quad \text{or} \quad& x - 3 = -5 + x \\
2x &= 8 &\quad \text{or} \quad& x - 3 - x = -5 + x - x \\
x &= 4 &\quad \text{or} \quad& -3 = -5 \qquad \text{False.}
\end{aligned}
$$

Recall from Section 2.3 that when an equation simplifies to a false statement, the equation has no solution. Thus, the only solution for the original absolute value equation is 4.

The following box summarizes the methods shown for solving absolute value equations.

ABSOLUTE VALUE EQUATIONS

$|x| = a$
- If a is positive, then solve $x = a$ or $x = -a$.
- If a is 0, solve $x = 0$.
- If a is negative, the equation $|x| = a$ has no solution.

$|x| = |y|$ Solve $x = y$ or $x = -y$.

MENTAL MATH

Simplify each expression.

1. $|-7|$ **2.** $|-8|$ **3.** $-|5|$

4. $-|10|$ **5.** $-|-6|$ **6.** $-|-3|$

7. $|-3| + |-2| + |-7|$ **8.** $|-1| + |-6| + |-8|$

Exercise Set 8.4

Solve each absolute value equation. See Examples 1–7.

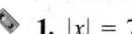

1. $|x| = 7$ **2.** $|y| = 15$

3. $|3x| = 12.6$ **4.** $|6n| = 12.6$

5. $|2x - 5| = 9$ **6.** $|6 + 2n| = 4$

7. $\left|\dfrac{x}{2} - 3\right| = 1$ **8.** $\left|\dfrac{n}{3} + 2\right| = 4$

9. $|z| + 4 = 9$ **10.** $|x| + 1 = 3$

11. $|3x| + 5 = 14$ **12.** $|2x| - 6 = 4$

13. $|2x| = 0$ **14.** $|7z| = 0$

15. $|4n + 1| + 10 = 4$ **16.** $|3z - 2| + 8 = 1$

17. $|5x - 1| = 0$ **18.** $|3y + 2| = 0$

19. Write an absolute value equation representing all numbers x whose distance from 0 is 5 units.

20. Write an absolute value equation representing all numbers x whose distance from 0 is 2 units.

Solve. See Examples 8 and 9.

21. $|5x - 7| = |3x + 11|$ **22.** $|9y + 1| = |6y + 4|$

23. $|z + 8| = |z - 3|$ **24.** $|2x - 5| = |2x + 5|$

25. Describe how solving an absolute value equation such as $|2x - 1| = 3$ is similar to solving an absolute value equation such as $|2x - 1| = |x - 5|$.

26. Describe how solving an absolute value equation such as $|2x - 1| = 3$ is different from solving an absolute value equation such as $|2x - 1| = |x - 5|$.

Solve each absolute value equation.

27. $|x| = 4$ **28.** $|x| = 1$

29. $|y| = 0$ **30.** $|y| = 8$

31. $|z| = -2$ **32.** $|y| = -9$

33. $|7 - 3x| = 7$ **34.** $|4m + 5| = 5$

35. $|6x| - 1 = 11$ **36.** $|7z| + 1 = 22$

37. $|4p| = -8$ **38.** $|5m| = -10$

39. $|x - 3| + 3 = 7$ **40.** $|x + 4| - 4 = 1$

41. $\left|\dfrac{z}{4} + 5\right| = -7$ **42.** $\left|\dfrac{c}{5} - 1\right| = -2$

43. $|9v - 3| = -8$ **44.** $|1 - 3b| = -7$

45. $|8n + 1| = 0$ **46.** $|5x - 2| = 0$

47. $|1 + 6c| - 7 = -3$ **48.** $|2 + 3m| - 9 = -7$

49. $|5x + 1| = 11$ **50.** $|8 - 6c| = 1$

51. $|4x - 2| = |-10|$ **52.** $|3x + 5| = |-4|$

53. $|5x + 1| = |4x - 7|$

54. $|3 + 6n| = |4n + 11|$

55. $|6 + 2x| = -|-7|$

56. $|4 - 5y| = -|-3|$

57. $|2x - 6| = |10 - 2x|$

58. $|4n + 5| = |4n + 3|$

59. $\left|\dfrac{2x - 5}{3}\right| = 7$

60. $\left|\dfrac{1 + 3n}{4}\right| = 4$

61. $2 + |5n| = 17$

62. $8 + |4m| = 24$

63. $\left|\dfrac{2x - 1}{3}\right| = |-5|$

64. $\left|\dfrac{5x + 2}{2}\right| = |-6|$

65. $|2y - 3| = |9 - 4y|$

66. $|5z - 1| = |7 - z|$

67. $\left|\dfrac{3n + 2}{8}\right| = |-1|$

68. $\left|\dfrac{2r - 6}{5}\right| = |-2|$

69. $|x + 4| = |7 - x|$

70. $|8 - y| = |y + 2|$

71. $\left|\dfrac{8c - 7}{3}\right| = -|-5|$

72. $\left|\dfrac{5d + 1}{6}\right| = -|-9|$

73. Explain why some absolute value equations have two solutions.

74. Explain why some absolute value equations have one solution.

REVIEW EXERCISES

The circle graph shows the sources of Walt Disney Company's operating income for 1999. Use this graph to answer Exercises 75–77. See Section 2.6.

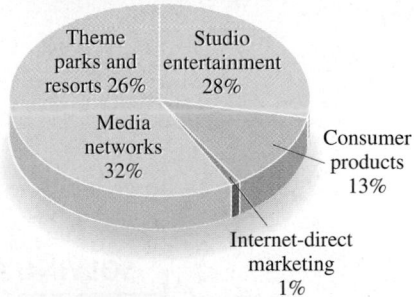

Walt Disney Company Operating Income 1999

Theme parks and resorts 26%
Studio entertainment 28%
Media networks 32%
Consumer products 13%
Internet-direct marketing 1%

Source: Walt Disney Company.

75. What percent of Disney's operating income came from the consumer products?

△ **76.** A circle contains 360°. Find the number of degrees found in the 26% sector for theme parks and resorts.

77. If Disney's operating income for all of 1999 was $3.4 billion, find the amount of income expected from the media networks segment.

List five integer solutions of each inequality.

78. $|x| \le 3$

79. $|x| \ge -2$

80. $|y| > -10$

81. $|y| < 0$

8.5 ABSOLUTE VALUE INEQUALITIES

CD-ROM SSM

SSG Video

▶ **OBJECTIVES**

1. Solve absolute value inequalities of the form $|x| < a$.
2. Solve absolute value inequalities of the form $|x| > a$.

1 The solution set of an absolute value inequality such as $|x| < 2$ contains all numbers whose distance from 0 is less than 2 units, as shown below.

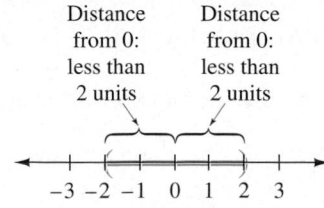

Distance from 0: less than 2 units   Distance from 0: less than 2 units

−3 −2 −1 0 1 2 3

The solution set is $\{x | -2 < x < 2\}$, or $(-2, 2)$ in interval notation.

Example 1 Solve $|x| \le 3$.

Solution The solution set of this inequality contains all numbers whose distance from 0 is less than or equal to 3. Thus 3, −3, and all numbers between 3 and −3 are in the solution set.

$$\xleftarrow{\quad} \overset{-4\ -3\ -2\ -1\ \ 0\ \ 1\ \ 2\ \ 3\ \ 4\ \ 5}{\quad} \xrightarrow{\quad}$$

The solution set is $[-3, 3]$.

In general, we have the following.

SOLVING ABSOLUTE VALUE INEQUALITIES OF THE FORM $|x| < a$

If a is a positive number, then $|x| < a$ is equivalent to $-a < x < a$.

This property also holds true for the inequality symbol $\le$.

Example 2 Solve for m: $|m - 6| < 2$.

Solution Replace x with $m - 6$ and a with 2 in the preceding property, and we see that

$$|m - 6| < 2 \quad \text{is equivalent to} \quad -2 < m - 6 < 2$$

Solve this compound inequality for m by adding 6 to all three parts.

$$-2 < m - 6 < 2$$
$$-2 + 6 < m - 6 + 6 < 2 + 6 \qquad \text{Add 6 to all three parts.}$$
$$4 < m < 8 \qquad \text{Simplify.}$$

The solution set is $(4, 8)$, and its graph is shown.

$$\xleftarrow{\quad} \overset{3\ \ 4\ \ 5\ \ 6\ \ 7\ \ 8\ \ 9}{\quad} \xrightarrow{\quad}$$

> **HELPFUL HINT**
> Before using an absolute value inequality property, isolate the absolute value expression on one side of the inequality.

Example 3 Solve for x: $|5x + 1| + 1 \le 10$.

Solution First, isolate the absolute value expression by subtracting 1 from both sides.

$$|5x + 1| + 1 \le 10$$
$$|5x + 1| \le 10 - 1 \qquad \text{Subtract 1 from both sides.}$$
$$|5x + 1| \le 9 \qquad \text{Simplify.}$$

Since 9 is positive, we apply the absolute value property for $|x| \leq a$.

$$-9 \leq 5x + 1 \leq 9$$

$$-9 - 1 \leq 5x + 1 - 1 \leq 9 - 1 \qquad \text{Subtract 1 from all three parts.}$$

$$-10 \leq 5x \leq 8 \qquad \text{Simplify.}$$

$$-2 \leq x \leq \frac{8}{5} \qquad \text{Divide all three parts by 5.}$$

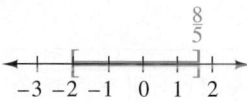

The solution set is $\left[-2, \dfrac{8}{5}\right]$, and the graph is shown above.

Example 4   Solve for x: $\left|2x - \dfrac{1}{10}\right| < -13$.

Solution   The absolute value of a number is always nonnegative and can never be less than -13. Thus this absolute value inequality has no solution. The solution set is $\{\ \}$ or $\varnothing$.

2   Let us now solve an absolute value inequality of the form $|x| > a$, such as $|x| \geq 3$. The solution set contains all numbers whose distance from 0 is 3 or more units. Thus the graph of the solution set contains 3 and all points to the right of 3 on the number line or -3 and all points to the left of -3 on the number line.

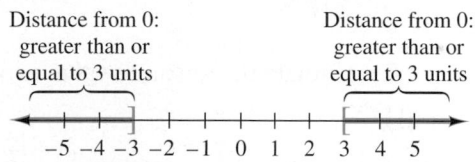

This solution set is written as $\{x \mid x \leq -3 \text{ or } x \geq 3\}$. In interval notation, the solution is $(-\infty, -3] \cup [3, \infty)$, since "or" means "union." In general, we have the following.

SOLVING ABSOLUTE VALUE INEQUALITIES OF THE FORM $|x| > a$

If a is a positive number, then $|x| > a$ is equivalent to $x < -a$ or $x > a$.

This property also holds true for the inequality symbol $\geq$.

Example 5   Solve for y: $|y - 3| > 7$.

Solution   Since 7 is positive, we apply the property for $|x| > a$.

$$|y - 3| > 7 \text{ is equivalent to } y - 3 < -7 \text{ or } y - 3 > 7$$

Next, solve the compound inequality.

$$
\begin{array}{llll}
y - 3 < -7 & \text{or} & y - 3 > 7 & \\
y - 3 + 3 < -7 + 3 & \text{or} & y - 3 + 3 > 7 + 3 & \text{Add 3 to both sides.} \\
y < -4 & \text{or} & y > 10 & \text{Simplify.}
\end{array}
$$

The solution set is $(-\infty, -4) \cup (10, \infty)$, and its graph is shown.

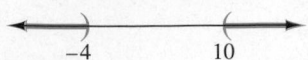

Examples 6 and 8 illustrate special cases of absolute value inequalities. These special cases occur when an isolated absolute value expression is less than, less than or equal to, greater than, or greater than or equal to a negative number or 0.

Example 6   Solve $|2x + 9| + 5 > 3$.

Solution   First isolate the absolute value expression by subtracting 5 from both sides.

$$|2x + 9| + 5 > 3$$

$$|2x + 9| + 5 - 5 > 3 - 5 \qquad \text{Subtract 5 from both sides.}$$

$$|2x + 9| > -2 \qquad \text{Simplify.}$$

The absolute value of any number is always nonnegative and thus is always greater than -2. This inequality and the original inequality are true for all values of x. The solution set is $\{x \mid x \text{ is a real number}\}$ or $(-\infty, \infty)$ and its graph is shown.

Example 7   Solve $\left|\dfrac{x}{3} - 1\right| - 7 \geq -5$.

Solution   First, isolate the absolute value expression by adding 7 to both sides.

$$\left|\dfrac{x}{3} - 1\right| - 7 \geq -5$$

$$\left|\dfrac{x}{3} - 1\right| - 7 + 7 \geq -5 + 7 \qquad \text{Add 7 to both sides.}$$

$$\left|\dfrac{x}{3} - 1\right| \geq 2 \qquad \text{Simplify.}$$

Next, write the absolute value inequality as an equivalent compound inequality and solve.

$$\dfrac{x}{3} - 1 \leq -2 \qquad \text{or} \qquad \dfrac{x}{3} - 1 \geq 2$$

$$3\left(\dfrac{x}{3} - 1\right) \leq 3(-2) \quad \text{or} \quad 3\left(\dfrac{x}{3} - 1\right) \geq 3(2) \qquad \text{Clear the inequalities of fractions.}$$

$$x - 3 \leq -6 \qquad \text{or} \qquad x - 3 \geq 6 \qquad \text{Apply the distributive property.}$$

$$x \leq -3 \qquad \text{or} \qquad x \geq 9 \qquad \text{Add 3 to both sides.}$$

The solution set is $(-\infty, -3] \cup [9, \infty)$, and its graph is shown.

Example 8   Solve for x: $\left|\dfrac{2(x + 1)}{3}\right| \le 0$

Solution   Recall that "$\le$" means "less than or equal to." The absolute value of any expression will never be less than 0, but it may be equal to 0. Thus, to solve $\left|\dfrac{2(x + 1)}{3}\right| \le 0$ we solve $\left|\dfrac{2(x + 1)}{3}\right| = 0$

$$\frac{2(x + 1)}{3} = 0$$

$$3\left[\frac{2(x + 1)}{3}\right] = 3(0) \qquad \text{Clear the equation of fractions.}$$

$$2x + 2 = 0 \qquad \text{Apply the distributive property.}$$

$$2x = -2 \qquad \text{Subtract 2 from both sides.}$$

$$x = -1 \qquad \text{Divide both sides by 2.}$$

The solution set is $\{-1\}$.

The following box summarizes the types of absolute value equations and inequalities.

SOLVING ABSOLUTE VALUE EQUATIONS AND INEQUALITIES WITH $a > 0$

Algebraic Solution	*Solution Graph*
$\mid x \mid = a$ is equivalent to $x = a$ or $x = -a$.	
$\mid x \mid < a$ is equivalent to $-a < x < a$.	
$\mid x \mid > a$ is equivalent to $x < -a$ or $x > a$.	

Exercise Set 8.5

Solve each inequality. Then graph the solution set. See Examples 1 through 4.

1. $|x| \le 4$

2. $|x| < 6$

3. $|x - 3| < 2$

4. $|y| \le 5$

5. $|x + 3| < 2$

6. $|x + 4| < 6$

7. $|2x + 7| \le 13$

8. $|5x - 3| \le 18$

9. $|x| + 7 \le 12$

10. $|x| + 6 \le 7$

11. $|3x - 1| < -5$

12. $|8x - 3| < -2$

13. $|x - 6| - 7 \le -1$

14. $|z + 2| - 7 < -3$

Solve each inequality. Graph the solution set. See Examples 5 through 7.

15. $|x| > 3$

16. $|y| \ge 4$

17. $|x + 10| \ge 14$

18. $|x - 9| \ge 2$

19. $|x| + 2 > 6$

20. $|x| - 1 > 3$

21. $|5x| > -4$

22. $|4x - 11| > -1$

23. $|6x - 8| + 3 > 7$

24. $|10 + 3x| + 1 > 2$

Solve each inequality. Graph the solution set. See Example 8.

25. $|x| \le 0$

26. $|x| \ge 0$

27. $|8x + 3| > 0$ **28.** $|5x - 6| < 0$

29. Write an absolute value inequality representing all numbers x whose distance from 0 is less than 7 units.

30. Write an absolute value inequality representing all numbers x whose distance from 0 is greater than 4 units.

31. Write $-5 \leq x \leq 5$ as an equivalent inequality containing an absolute value.

32. Write $x > 1$ or $x < -1$ as an equivalent inequality containing an absolute value.

Solve each inequality. Graph the solution set.

33. $|x| \leq 2$ **34.** $|z| < 6$

35. $|y| > 1$ **36.** $|x| \geq 10$

37. $|x - 3| < 8$ **38.** $|-3 + x| \leq 10$

39. $|0.6x - 3| > 0.6$ **40.** $|1 + 0.3x| \geq 0.1$

41. $5 + |x| \leq 2$

42. $8 + |x| < 1$

43. $|x| > -4$

44. $|x| \leq -7$

45. $|2x - 7| \leq 11$

46. $|5x + 2| < 8$

47. $|x + 5| + 2 \geq 8$

48. $|-1 + x| - 6 > 2$

49. $|x| > 0$

50. $|x| < 0$

51. $9 + |x| > 7$

52. $5 + |x| \geq 4$

53. $6 + |4x - 1| \leq 9$

54. $-3 + |5x - 2| \leq 4$

55. $\left|\dfrac{2}{3}x + 1\right| > 1$

56. $|5x - 1| \geq 2$

57. $|5x + 3| < -6$

58. $|4 + 9x| \geq -6$

59. $|8x + 3| \geq 0$

60. $|5x - 6| \leq 0$

61. $|1 + 3x| + 4 < 5$

62. $|7x - 3| - 1 \leq 10$

63. $|x| - 3 \geq -3$

64. $|x| + 6 < 6$

65. $|8x| - 10 > -2$

66. $|6x| - 13 \geq -7$

67. $\left|\dfrac{x + 6}{3}\right| > 2$

68. $\left|\dfrac{7 + x}{2}\right| \geq 4$

69. $|2(3 + x)| > 6$

70. $|5(x - 3)| \geq 10$

71. $\left|\dfrac{5(x + 2)}{3}\right| < 7$

72. $\left|\dfrac{6(3 + x)}{5}\right| \leq 4$

73. $-15 + |2x - 7| \leq -6$

74. $-9 + |3 + 4x| < -4$

75. $\left|2x + \dfrac{3}{4}\right| - 7 \leq -2$

76. $\left|\dfrac{3}{5} + 4x\right| - 6 < -1$

Solve each equation or inequality for x.

77. $|2x - 3| < 7$ **78.** $|2x - 3| > 7$

79. $|2x - 3| = 7$ **80.** $|5 - 6x| = 29$

81. $|x - 5| \geq 12$ **82.** $|x + 4| \geq 20$

83. $|9 + 4x| = 0$

84. $|9 + 4x| \geq 0$

85. $|2x + 1| + 4 < 7$

86. $8 + |5x - 3| \geq 11$

87. $|3x - 5| + 4 = 5$

88. $|8x| = -5$

89. $|x + 11| = -1$

90. $|4x - 4| = -3$

91. $\left|\dfrac{2x - 1}{3}\right| = 6$

92. $\left|\dfrac{6 - x}{4}\right| = 5$

93. $\left|\dfrac{3x - 5}{6}\right| > 5$

94. $\left|\dfrac{4x - 7}{5}\right| < 2$

95. Describe how solving $|x - 3| = 5$ is different from solving $|x - 3| < 5$.

96. Describe how solving $|x + 4| = 0$ is similar to solving $|x + 4| \leq 0$.

The expression $|x_T - x|$ is defined to be the absolute error in x, where x_T is the true value of a quantity and x is the measured value or value as stored in a computer.

97. If the true value of a quantity is 3.5 and the absolute error must be less than 0.05, find the acceptable measured values.

98. If the true value of a quantity is 0.2 and the approximate value stored in a computer is $\dfrac{51}{256}$, find the absolute error.

REVIEW EXERCISES

Consider the equation $3x - 4y = 12$. For each value of x or y given, find the corresponding value of the other variable that makes the statement true. See Section 2.3.

99. If $x = 2$, find y

100. If $y = -1$, find x.

101. If $y = -3$, find x

102. If $x = 4$, find y

8.6 SOLVING SYSTEMS OF LINEAR EQUATIONS IN THREE VARIABLES

CD-ROM SSM

SSG Video

▶ **OBJECTIVES**

1. Solve a system of three linear equations in three variables.
2. Solve problems that can be modeled by a system of three linear equations.

In this section, the algebraic methods of solving systems of two linear equations in two variables are extended to systems of three linear equations in three variables. We call the equation $3x - y + z = -15$, for example, a **linear equation in three variables** since there are three variables and each variable is raised only to the power 1. A solution of this equation is an **ordered triple (x, y, z)** that makes the equation a true statement. For example, the ordered triple $(2, 0, -21)$ is a solution of $3x - y + z = -15$ since replacing x with 2, y with 0, and z with -21 yields the true statement $3(2) - 0 + (-21) = -15$. The graph of this equation is a plane in three-dimensional space, just as the graph of a linear equation in two variables is a line in two-dimensional space.

Although we will not discuss the techniques for graphing equations in three variables, visualizing the possible patterns of intersecting planes gives us insight into the possible patterns of solutions of a system of three three-variable linear equations. There are four possible patterns.

1. Three planes have a single point in common. This point represents the single solution of the system. This system is **consistent**.

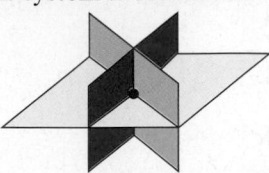

2. Three planes intersect at no point common to all three. This system has no solution. A few ways that this can occur are shown. This system is **inconsistent**.

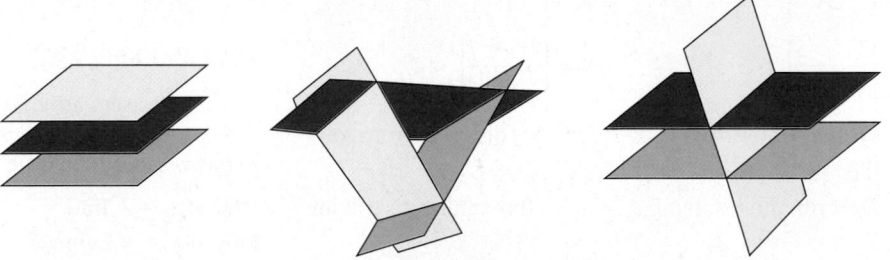

3. Three planes intersect at all the points of a single line. The system has infinitely many solutions. This system is **consistent**.

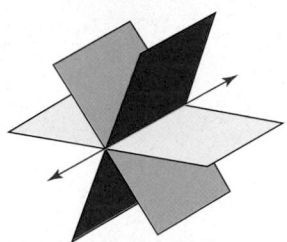

4. Three planes coincide at all points on the plane. The system is consistent, and the equations are **dependent**.

1

To use the elimination method to solve a system in three variables, we eliminate a variable and obtain a system in two variables.

Example 1   Solve the system.

$$\begin{cases} 3x - y + z = -15 & \text{Equation (1)} \\ x + 2y - z = 1 & \text{Equation (2)} \\ 2x + 3y - 2z = 0 & \text{Equation (3)} \end{cases}$$

Solution   Add equations (1) and (2) to eliminate z.

$$\begin{array}{r} 3x - y + z = -15 \\ x + 2y - z = 1 \\ \hline 4x + y = -14 \end{array} \qquad \text{Equation (4)}$$

HELPFUL HINT
Don't forget to add two other equations besides equations (1) and (2) *and* to eliminate the same variable.

Next, add two *other* equations and *eliminate z again.* To do so, multiply both sides of equation (1) by 2 and add this resulting equation to equation (3). Then

$$\begin{cases} 2(3x - y + z) = 2(-15) \\ 2x + 3y - 2z = 0 \end{cases} \quad \begin{array}{c} \text{simplifies} \\ \text{to} \end{array} \quad \begin{cases} 6x - 2y + 2z = -30 \\ \underline{2x + 3y - 2z = 0} \\ 8x + y = -30 \quad \text{Equation (5)} \end{cases}$$

Now solve equations (4) and (5) for x and y. To solve by elimination, multiply both sides of equation (4) by -1 and add this resulting equation to equation (5). Then

$$\begin{cases} -1(4x + y) = -1(-14) \\ 8x + y = -30 \end{cases} \quad \begin{array}{c} \text{simplifies} \\ \text{to} \end{array} \quad \begin{cases} -4x - y = 14 \\ \underline{8x + y = -30} \\ 4x = -16 \quad \text{Add the equations.} \\ x = -4 \quad \text{Solve for } x. \end{cases}$$

Replace x with -4 in equation (4) or (5).

$$\begin{aligned} 4x + y &= -14 &&\text{Equation (4)} \\ 4(-4) + y &= -14 &&\text{Let } x = -4. \\ y &= 2 &&\text{Solve for } y. \end{aligned}$$

Finally, replace x with -4 and y with 2 in equation (1), (2), or (3).

$$\begin{aligned} x + 2y - z &= 1 &&\text{Equation (2)} \\ -4 + 2(2) - z &= 1 &&\text{Let } x = -4 \text{ and } y = 2. \\ -4 + 4 - z &= 1 \\ -z &= 1 \\ z &= -1 \end{aligned}$$

The solution is $(-4, 2, -1)$. To check, let $x = -4$, $y = 2$, and $z = -1$ in all three original equations of the system.

Equation (1)	*Equation (2)*	*Equation (3)*
$3x - y + z = -15$	$x + 2y - z = 1$	$2x + 3y - 2z = 0$
$3(-4) - 2 + (-1) = -15$	$-4 + 2(2) - (-1) = 1$	$2(-4) + 3(2) - 2(-1) = 0$
$-12 - 2 - 1 = -15$	$-4 + 4 + 1 = 1$	$-8 + 6 + 2 = 0$
$-15 = -15$	$1 = 1$	$0 = 0$
True.	True.	True.

All three statements are true, so the solution is $(-4, 2, -1)$.

Example 2   Solve the system.

$$\begin{cases} 2x - 4y + 8z = 2 & (1) \\ -x - 3y + z = 11 & (2) \\ x - 2y + 4z = 0 & (3) \end{cases}$$

Solution   Add equations (2) and (3) to eliminate x, and the new equation is

$$-5y + 5z = 11 \quad (4)$$

To eliminate x again, multiply both sides of equation (2) by 2, and add the resulting equation to equation (1). Then

$$\begin{cases} 2x - 4y + 8z = 2 \\ 2(-x - 3y + z) = 2(11) \end{cases} \quad \begin{array}{c} \text{simplifies} \\ \text{to} \end{array} \quad \begin{cases} 2x - 4y + 8z = 2 \\ \underline{-2x - 6y + 2z = 22} \\ {-10y + 10z = 24} \quad (5) \end{cases}$$

Next, solve for y and z using equations (4) and (5). Multiply both sides of equation (4) by -2, and add the resulting equation to equation (5).

$$\begin{cases} -2(-5y + 5z) = -2(11) \\ -10y + 10z = 24 \end{cases} \quad \begin{array}{c} \text{simplifies} \\ \text{to} \end{array} \quad \begin{cases} 10y - 10z = -22 \\ \underline{-10y + 10z = 24} \\ {0 = 2} \quad \text{False.} \end{cases}$$

Since the statement is false, this system is inconsistent and has no solution. ■

The elimination method is summarized next.

SOLVING A SYSTEM OF THREE LINEAR EQUATIONS BY THE ELIMINATION METHOD

Step 1: Write each equation in standard form $Ax + By + Cz = D$.

Step 2: Choose a pair of equations and use the equations to eliminate a variable.

Step 3: Choose any other pair of equations and eliminate the **same variable** as in Step 2.

Step 4: Two equations in two variables should be obtained from Step 2 and Step 3. Use methods from Section 4.3 to solve this system for both variables.

Step 5: To solve for the third variable, substitute the values of the variables found in Step 4 into any of the original equations containing the third variable.

Example 3 Solve the system.

$$\begin{cases} 2x + 4y = 1 & (1) \\ 4x - 4z = -1 & (2) \\ y - 4z = -3 & (3) \end{cases}$$

Solution Notice that equation (2) has no term containing the variable y. Let us eliminate y using equations (1) and (3). Multiply both sides of equation (3) by -4, and add the resulting equation to equation (1). Then

$$\begin{cases} 2x + 4y = 1 \\ -4(y - 4z) = -4(-3) \end{cases} \quad \begin{array}{c} \text{simplifies} \\ \text{to} \end{array} \quad \begin{cases} 2x + 4y = 1 \\ - 4y + 16z = 12 \\ \underline{2x + 16z = 13} \quad (4) \end{cases}$$

Next, solve for z using equations (4) and (2). Multiply both sides of equation (4) by -2 and add the resulting equation to equation (2).

$$\begin{cases} -2(2x + 16z) = -2(13) \\ 4x - 4z = -1 \end{cases} \quad \begin{array}{c} \text{simplifies} \\ \text{to} \end{array} \quad \begin{cases} -4x - 32z = -26 \\ \underline{4x - 4z = -1} \\ -36z = -27 \\ z = \dfrac{3}{4} \end{cases}$$

Replace z with $\dfrac{3}{4}$ in equation (3) and solve for y.

$$y - 4\left(\dfrac{3}{4}\right) = -3 \qquad \text{Let } z = \dfrac{3}{4} \text{ in equation (3).}$$
$$y - 3 = -3$$
$$y = 0$$

Replace y with 0 in equation (1) and solve for x.

$$2x + 4(0) = 1$$
$$2x = 1$$
$$x = \dfrac{1}{2}$$

The solution is $\left(\dfrac{1}{2}, 0, \dfrac{3}{4}\right)$. Check to see that this solution satisfies all three equations of the system.

Example 4 Solve the system.

$$\begin{cases} x - 5y - 2z = 6 & (1) \\ -2x + 10y + 4z = -12 & (2) \\ \dfrac{1}{2}x - \dfrac{5}{2}y - z = 3 & (3) \end{cases}$$

Solution Multiply both sides of equation (3) by 2 to eliminate fractions, and multiply both sides of equation (2) by $-\dfrac{1}{2}$ so that the coefficient of x is 1. The resulting system is then

$$\begin{cases} x - 5y - 2z = 6 & (1) \\ x - 5y - 2z = 6 & \text{Multiply (2) by } -\dfrac{1}{2}. \\ x - 5y - 2z = 6 & \text{Multiply (3) by 2.} \end{cases}$$

All three equations are identical, and therefore equations (1), (2), and (3) are all equivalent. There are infinitely many solutions of this system. The equations are dependent. The solution set can be written as $\{(x, y, z) \mid x - 5y - 2z = 6\}$.

2 To introduce problem solving by writing a system of three linear equations in three variables, we solve a problem about triangles.

◆ **Example 5** **FINDING ANGLE MEASURES**

The measure of the largest angle of a triangle is 80° more than the measure of the smallest angle, and the measure of the remaining angle is 10° more than the measure of the smallest angle. Find the measure of each angle.

Solution 1. UNDERSTAND. Read and reread the problem. Recall that the sum of the measures of the angles of a triangle is 180°. Then guess a solution. If the smallest angle measures 20°, the measure of the largest angle is 80° more, or $20° + 80° = 100°$. The measure of the remaining angle is 10° more than the measure of the smallest angle, or $20° + 10° = 30°$. The sum of these three angles is $20° + 100° + 30° = 150°$, not the required 180°. We now know that the measure of the smallest angle is greater than 20°.

To model this problem we will let

$$x = \text{degree measure of the smallest angle}$$
$$y = \text{degree measure of the largest angle}$$
$$z = \text{degree measure of the remaining angle}$$

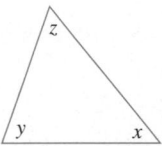

2. TRANSLATE. We translate the given information into three equations.

In words:

the sum of the measures	=	180

Translate: $x + y + z = 180$

In words:

the largest angle	is	80 more than the smallest angle

Translate: $y = x + 80$

In words:

the remaining angle	is	10 more than the smallest angle

Translate: $z = x + 10$

3. SOLVE. We solve the system

$$\begin{cases} x + y + z = 180 \\ \qquad\quad y = x + 80 \\ \qquad\quad z = x + 10 \end{cases}$$

Since y and z are both expressed in terms of x, we will solve using the subsitution method. We substitute $y = x + 80$ and $z = x + 10$ in the first equation. Then

$$x + y + z = 180 \qquad \text{First equation}$$

$$x + (x + 80) + (x + 10) = 180 \qquad \text{Let } y = x + 80 \text{ and } z = x + 10.$$

$$3x + 90 = 180$$

$$3x = 90$$

$$x = 30$$

Then $y = x + 80 = 30 + 80 = 110$, and $z = x + 10 = 30 + 10 = 40$. The ordered triple solution is $(30, 110, 40)$.

4. **INTERPRET.**

Check: Notice that $30° + 40° + 110° = 180°$. Also, the measure of the largest angle, $110°$, is $80°$ more than the measure of the smallest angle, $30°$. The measure of the remaining angle, $40°$, is $10°$ more than the measure of the smallest angle, $30°$.

State: The angles measure $30°$, $110°$, and $40°$.

Exercise Set 8.6

Solve each system. See Examples 1 and 3.

1. $\begin{cases} x + y = 3 \\ 2y = 10 \\ 3x + 2y - 3z = 1 \end{cases}$

2. $\begin{cases} 5x = 5 \\ 2x + y = 4 \\ 3x + y - 4z = -15 \end{cases}$

3. $\begin{cases} 2x + 2y + z = 1 \\ -x + y + 2z = 3 \\ x + 2y + 4z = 0 \end{cases}$

4. $\begin{cases} 2x - 3y + z = 5 \\ x + y + z = 0 \\ 4x + 2y + 4z = 4 \end{cases}$

Solve each system. See Examples 2 and 4.

5. $\begin{cases} x - 2y + z = -5 \\ -3x + 6y - 3z = 15 \\ 2x - 4y + 2z = -10 \end{cases}$

6. $\begin{cases} 3x + y - 2z = 2 \\ -6x - 2y + 4z = -2 \\ 9x + 3y - 6z = 6 \end{cases}$

7. $\begin{cases} 4x - y + 2z = 5 \\ 2y + z = 4 \\ 4x + y + 3z = 10 \end{cases}$

8. $\begin{cases} 5y - 7z = 14 \\ 2x + y + 4z = 10 \\ 2x + 6y - 3z = 30 \end{cases}$

9. Write a system of linear equations in three variables that has $(-1, 2, -4)$ as a solution. (There are many possibilities.)

10. Write a system of three linear equations in three variables that has $(2, 1, 5)$ as a solution. (There are many possibilities.)

Solve each system.

11. $\begin{cases} x + 5z = 0 \\ 5x + y = 0 \\ y - 3z = 0 \end{cases}$

12. $\begin{cases} x - 5y = 0 \\ x - z = 0 \\ -x + 5z = 0 \end{cases}$

13. $\begin{cases} 6x - 5z = 17 \\ 5x - y + 3z = -1 \\ 2x + y = -41 \end{cases}$

14. $\begin{cases} x + 2y = 6 \\ 7x + 3y + z = -33 \\ x - z = 16 \end{cases}$

15. $\begin{cases} x + y + z = 8 \\ 2x - y - z = 10 \\ x - 2y - 3z = 22 \end{cases}$

16. $\begin{cases} 5x + y + 3z = 1 \\ x - y + 3z = -7 \\ -x + y = 1 \end{cases}$

17. $\begin{cases} x + 2y - z = 5 \\ 6x + y + z = 7 \\ 2x + 4y - 2z = 5 \end{cases}$

18. $\begin{cases} 4x - y + 3z = 10 \\ x + y - z = 5 \\ 8x - 2y + 6z = 10 \end{cases}$

19. $\begin{cases} 2x - 3y + z = 2 \\ x - 5y + 5z = 3 \\ 3x + y - 3z = 5 \end{cases}$

20. $\begin{cases} 4x + y - z = 8 \\ x - y + 2z = 3 \\ 3x - y + z = 6 \end{cases}$

21. $\begin{cases} -2x - 4y + 6z = -8 \\ x + 2y - 3z = 4 \\ 4x + 8y - 12z = 16 \end{cases}$

22. $\begin{cases} -6x + 12y + 3z = -6 \\ 2x - 4y - z = 2 \\ -x + 2y + \dfrac{z}{2} = -1 \end{cases}$

23. $\begin{cases} 2x + 2y - 3z = 1 \\ y + 2z = -14 \\ 3x - 2y = -1 \end{cases}$

24. $\begin{cases} 7x + 4y = 10 \\ x - 4y + 2z = 6 \\ y - 2z = -1 \end{cases}$

25. $\begin{cases} \dfrac{3}{4}x - \dfrac{1}{3}y + \dfrac{1}{2}z = 9 \\ \dfrac{1}{6}x + \dfrac{1}{3}y - \dfrac{1}{2}z = 2 \\ \dfrac{1}{2}x - y + \dfrac{1}{2}z = 2 \end{cases}$

26. $\begin{cases} \dfrac{1}{3}x - \dfrac{1}{4}y + z = -9 \\ \dfrac{1}{2}x - \dfrac{1}{3}y - \dfrac{1}{4}z = -6 \\ x - \dfrac{1}{2}y - z = -8 \end{cases}$

27. The fraction $\dfrac{1}{24}$ can be written as the following sum:

$$\frac{1}{24} = \frac{x}{8} + \frac{y}{4} + \frac{z}{3}$$

where the numbers x, y, and z are solutions of

$$\begin{cases} x + y + z = 1 \\ 2x - y + z = 0 \\ -x + 2y + 2z = -1 \end{cases}$$

Solve the system and see that the sum of the fractions is $\dfrac{1}{24}$.

28. The fraction $\dfrac{1}{18}$ can be written as the following sum.

$$\frac{1}{18} = \frac{x}{2} + \frac{y}{3} + \frac{z}{9}$$

where the numbers x, y, and z are solutions of

$$\begin{cases} x + 3y + z = -3 \\ -x + y + 2z = -14 \\ 3x + 2y - z = 12 \end{cases}$$

Solve the system and see that the sum of the fractions is $\dfrac{1}{18}$.

Solve. See Example 5.

29. Rabbits in a lab are to be kept on a strict daily diet to include 30 grams of protein, 16 grams of fat, and 24 grams of carbohydrates. The scientist has only three food mixes available with the following grams of nutrients per unit. Find how many units of each mix are needed daily to meet each rabbit's dietary needs.

	Protein	*Fat*	*Carbohydrate*
Mix A	4	6	3
Mix B	6	1	2
Mix C	4	1	12

30. Gary Gundersen mixes different solutions with concentrations of 25%, 40%, and 50% to get 200 liters of a 32% solution. If he uses twice as much of the 25% solution as of the 40% solution, find how many liters of each kind he uses.

31. In 1999 the WNBA's top scorer was Cynthia Cooper of the Houston Comets. She scored a total of 686 points during the regular season. The number of two-point field goals Cooper made was 20 less than three times the number of three-point field goals she made. She also made 50 more free throws (each worth one point) than two-point field goals. Find how many free throws, two-point field goals, and three-point field goals Cynthia Cooper made during the 1999 season. (*Source:* Women's National Basketball Association)

32. During the 2000 NBA All-Star Game, the top-scoring player was Allen Iverson of the Philadelphia 76ers. Iverson, playing for the Eastern Conference All-Star Team, scored a total of 26 points during the All-Star Game. The number of free throws (each worth 1 point) he made was 2 more than the number of three-point field goals he made. Iverson also made 4 more two-point field goals than free throws. How many free throws, two-point field goals, and three-point field goals did Allen Iverson make during the 2000 NBA All-Star Game? (*Source:* National Basketball Association)

△ 33. The measure of the largest angle of a triangle is 90° more than the measure of the smallest angle, and the measure of the remaining angle is 30° more than the measure of the smallest angle. Find the measure of each angle.

34. Suppose you mix an amount of 25% acid solution with an amount of 60% acid solution. You then calculate the acid strength of the resulting acid mixture. For which of the following results should you suspect an error in your calculation? Why?

  a. 14%   **b.** 32%   **c.** 55%

35. Find the values of a, b, and c such that the equation $y = ax^2 + bx + c$ has ordered pair solutions $(1, 6)$, $(-1, -2)$, and $(0, -1)$. To do so, substitute each ordered pair solution into the equation. Each time, the result is an equation in three unknowns: a, b, and c. Then solve the resulting system of three linear equations in three unknowns, a, b, and c.

36. Find the values of a, b, and c such that the equation $y = ax^2 + bx + c$ has ordered pair solutions $(1, 2)$, $(2, 3)$ and $(-1, 6)$. (*Hint:* See Exercise 35.)

△ 37. Find the values of x, y, and z in the following triangle.

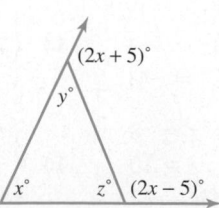

△ **38.** The sum of the measures of the angles of a quadrilateral is 360°. Find the values of x, y, and z in the following quadrilateral.

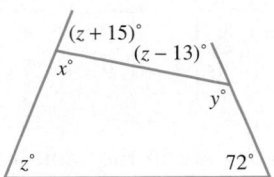

39. Data (x, y) for the total number y (in thousands) of college-bound students who took the ACT assessment in the year x are $(5, 945)$, $(6, 925)$, and $(9, 1019)$, where $x = 5$ represents 1995 and $x = 9$ represents 1999. Find the values of a, b, and c such that the equation $y = ax^2 + bx + c$ models this data. According to your model, how many students will take the ACT in 2005? (*Source:* ACT, Inc.)

40. Monthly normal rainfall data (x, y) for Portland, Oregon, are $(4, 2.47)$, $(7, 0.6)$, $(8, 1.1)$, where x represents time in months (with $x = 1$ representing January) and y represents rainfall in inches. Find the values of a, b, and c rounded to 2 decimal places such that the equation $y = ax^2 + bx + c$ models this data. According to your model, how much rain should Portland expect during September? (*Source:* National Climatic Data Center)

Multiply both sides of equation (1) by 2, and add the resulting equation to equation (2).

41. $3x - y + z = 2$ (1)
$-x + 2y + 3z = 6$ (2)

42. $2x + y + 3z = 7$ (1)
$-4x + y + 2z = 4$ (2)

Multiply both sides of equation (1) by -3, and add the resulting equation to equation (2).

43. $x + 2y - z = 0$ (1)
$3x + y - z = 2$ (2)

44. $2x - 3y + 2z = 5$ (1)
$x - 9y + z = -1$ (2)

A Look Ahead

Solve each system.

45. $\begin{cases} x + y \qquad - w = 0 \\ \qquad y + 2z + w = 3 \\ x \qquad - z \qquad = 1 \\ 2x - y \qquad - w = -1 \end{cases}$

46. $\begin{cases} 5x + 4y \qquad = 29 \\ \qquad y + z - w = -2 \\ 5x \qquad + z \qquad = 23 \\ \qquad y - z + w = 4 \end{cases}$

47. $\begin{cases} x + y + z + w = 5 \\ 2x + y + z + w = 6 \\ x + y + z \qquad = 2 \\ x + y \qquad = 0 \end{cases}$

48. $\begin{cases} 2x \qquad - z \qquad = -1 \\ \qquad y + z + w = 9 \\ \qquad y \qquad - 2w = -6 \\ x + y \qquad = 3 \end{cases}$

8.7   SOLVING SYSTEMS OF EQUATIONS BY MATRICES

CD-ROM

SSM

SSG

Video

▶ **OBJECTIVES**

1. Use matrices to solve a system of two equations in two variables.

2. Use matrices to solve a system of three equations in three variables.

By now, you may have noticed that the solution of a system of equations depends on the coefficients of the equations in the system and not on the variables. In this section, we introduce solving a system of equations by a **matrix.**

1 A matrix (plural: **matrices**) is a rectangular array of numbers. The following are examples of matrices.

$$\begin{bmatrix} 1 & 0 \\ 0 & 1 \end{bmatrix} \quad \begin{bmatrix} 2 & 1 & 3 & -1 \\ 0 & -1 & 4 & 5 \\ -6 & 2 & 1 & 0 \end{bmatrix} \quad \begin{bmatrix} a & b & c \\ d & e & f \end{bmatrix}$$

The numbers aligned horizontally in a matrix are in the same **row**. The numbers aligned vertically are in the same **column**.

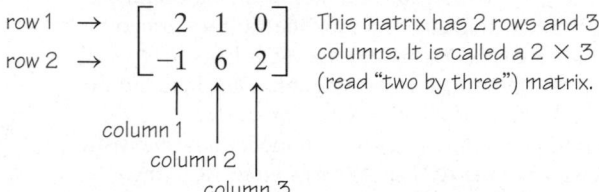

To see the relationship between systems of equations and matrices, study the example below.

System of Equations		Corresponding Matrix	

$$\begin{cases} 2x - 3y = 6 & \text{Equation 1} \\ x + y = 0 & \text{Equation 2} \end{cases} \qquad \begin{bmatrix} 2 & -3 & \vdots & 6 \\ 1 & 1 & \vdots & 0 \end{bmatrix} \begin{array}{l} \text{Row 1} \\ \text{Row 2} \end{array}$$

Notice that the rows of the matrix correspond to the equations in the system. The coefficients of each variable are placed to the left of a vertical dashed line. The constants are placed to the right. Each of these numbers in the matrix is called an **element**.

The method of solving systems by matrices is to write this matrix as an equivalent matrix from which we easily identify the solution. Two matrices are equivalent if they represent systems that have the same solution set. The following **row operations** can be performed on matrices, and the result is an equivalent matrix.

ELEMENTARY ROW OPERATIONS

1. Any two rows in a matrix may be interchanged.
2. The elements of any row may be multiplied (or divided) by the same nonzero number.
3. The elements of any row may be multiplied (or divided) by a nonzero number and added to their corresponding elements in any other row.

> **HELPFUL HINT**
> Notice that these *row* operations are the same operations that we can perform on *equations* in a system.

Example 1 Use matrices to solve the system.

$$\begin{cases} x + 3y = 5 \\ 2x - y = -4 \end{cases}$$

Solution The corresponding matrix is $\begin{bmatrix} 1 & 3 & \vdots & 5 \\ 2 & -1 & \vdots & -4 \end{bmatrix}$. We use elementary row operations to write an equivalent matrix that looks like $\begin{bmatrix} 1 & a & \vdots & b \\ 0 & 1 & \vdots & c \end{bmatrix}$.

For the matrix given, the element in the first row, first column is already 1, as desired. Next we write an equivalent matrix with a 0 below the 1. To do this, we multiply row 1 by -2 and add to row 2. *We will change only row 2.*

$$\begin{bmatrix} 1 & 3 & \vdots & 5 \\ -2(1) + 2 & -2(3) + (-1) & \vdots & -2(5) + (-4) \end{bmatrix} \quad \text{simplifies to} \quad \begin{bmatrix} 1 & 3 & \vdots & 5 \\ 0 & -7 & \vdots & -14 \end{bmatrix}$$

↑ ↑ ↑ ↑ ↑ ↑
row 1 row 2 row 1 row 2 row 1 row 2
element element element element element element

Now we change the -7 to a 1 by use of an elementary row operation. We divide row 2 by -7, then

$$\begin{bmatrix} 1 & 3 & \vdots & 5 \\ \dfrac{0}{-7} & \dfrac{-7}{-7} & \vdots & \dfrac{-14}{-7} \end{bmatrix} \quad \text{simplifies to} \quad \begin{bmatrix} 1 & 3 & \vdots & 5 \\ 0 & 1 & \vdots & 2 \end{bmatrix}$$

This last matrix corresponds to the system

$$\begin{cases} x + 3y = 5 \\ y = 2 \end{cases}$$

To find x, we let $y = 2$ in the first equation, $x + 3y = 5$.

$$x + 3y = 5 \qquad \text{First equation}$$
$$x + 3(2) = 5 \qquad \text{Let } y = 2.$$
$$x = -1$$

The ordered pair solution is $(-1, 2)$. Check to see that this ordered pair satisfies both equations. ∎

Example 2 Use matrices to solve the system.

$$\begin{cases} 2x - y = 3 \\ 4x - 2y = 5 \end{cases}$$

Solution The corresponding matrix is $\begin{bmatrix} 2 & -1 & \vdots & 3 \\ 4 & -2 & \vdots & 5 \end{bmatrix}$. To get 1 in the row 1, column 1 position, we divide the elements of row 1 by 2.

$$\begin{bmatrix} \dfrac{2}{2} & -\dfrac{1}{2} & \vdots & \dfrac{3}{2} \\ 4 & -2 & \vdots & 5 \end{bmatrix} \quad \text{simplifies to} \quad \begin{bmatrix} 1 & -\dfrac{1}{2} & \vdots & \dfrac{3}{2} \\ 4 & -2 & \vdots & 5 \end{bmatrix}$$

To get 0 under the 1, we multiply the elements of row 1 by -4 and add the new elements to the elements of row 2.

$$\begin{bmatrix} 1 & -\dfrac{1}{2} & \vdots & \dfrac{3}{2} \\ -4(1)+4 & -4\left(-\dfrac{1}{2}\right)-2 & \vdots & -4\left(\dfrac{3}{2}\right)+5 \end{bmatrix} \quad \text{simplifies to} \quad \begin{bmatrix} 1 & -\dfrac{1}{2} & \vdots & \dfrac{3}{2} \\ 0 & 0 & \vdots & -1 \end{bmatrix}$$

The corresponding system is $\begin{cases} x - \dfrac{1}{2}y = \dfrac{3}{2} \\ \quad\quad 0 = -1 \end{cases}$. The equation $0 = -1$ is false for all y

or x values; hence the system is inconsistent and has no solution. ▬

2 To solve a system of three equations in three variables using matrices, we will write the corresponding matrix in the form

$$\begin{bmatrix} 1 & a & b & \vdots & d \\ 0 & 1 & c & \vdots & e \\ 0 & 0 & 1 & \vdots & f \end{bmatrix}$$

Example 3 Use matrices to solve the system.

$$\begin{cases} x + 2y + z = 2 \\ -2x - y + 2z = 5 \\ x + 3y - 2z = -8 \end{cases}$$

Solution The corresponding matrix is $\begin{bmatrix} 1 & 2 & 1 & \vdots & 2 \\ -2 & -1 & 2 & \vdots & 5 \\ 1 & 3 & -2 & \vdots & -8 \end{bmatrix}$. Our goal is to write an

equivalent matrix with 1's along the diagonal (see the numbers in red) and 0's below the 1's. The element in row 1, column 1 is already 1. Next we get 0's for each element in the rest of column 1. To do this, first we multiply the elements of row 1 by 2 and add the new elements to row 2. Also, we multiply the elements of row 1 by -1 and add the new elements to the elements of row 3. We *do not change row 1*. Then

$$\begin{bmatrix} 1 & 2 & 1 & \vdots & 2 \\ 2(1)-2 & 2(2)-1 & 2(1)+2 & \vdots & 2(2)+5 \\ -1(1)+1 & -1(2)+3 & -1(1)-2 & \vdots & -1(2)-8 \end{bmatrix} \quad \text{simplifies to} \quad \begin{bmatrix} 1 & 2 & 1 & \vdots & 2 \\ 0 & 3 & 4 & \vdots & 9 \\ 0 & 1 & -3 & \vdots & -10 \end{bmatrix}$$

We continue down the diagonal and use elementary row operations to get 1 where the element 3 is now. To do this, we interchange rows 2 and 3.

$$\begin{bmatrix} 1 & 2 & 1 & \vdots & 2 \\ 0 & 3 & 4 & \vdots & 9 \\ 0 & 1 & -3 & \vdots & -10 \end{bmatrix} \quad \text{is equivalent to} \quad \begin{bmatrix} 1 & 2 & 1 & \vdots & 2 \\ 0 & 1 & -3 & \vdots & -10 \\ 0 & 3 & 4 & \vdots & 9 \end{bmatrix}$$

Next we want the new row 3, column 2 element to be 0. We multiply the elements of row 2 by -3 and add the result to the elements of row 3.

$$\begin{bmatrix} 1 & 2 & 1 & \vdots & 2 \\ 0 & 1 & -3 & \vdots & -10 \\ -3(0)+0 & -3(1)+3 & -3(-3)+4 & \vdots & -3(-10)+9 \end{bmatrix}$$ simplifies to $$\begin{bmatrix} 1 & 2 & 1 & \vdots & 2 \\ 0 & 1 & -3 & \vdots & -10 \\ 0 & 0 & 13 & \vdots & 39 \end{bmatrix}$$

Finally, we divide the elements of row 3 by 13 so that the final diagonal element is 1.

$$\begin{bmatrix} 1 & 2 & 1 & \vdots & 2 \\ 0 & 1 & -3 & \vdots & -10 \\ \frac{0}{13} & \frac{0}{13} & \frac{13}{13} & \vdots & \frac{39}{13} \end{bmatrix}$$ simplifies to $$\begin{bmatrix} 1 & 2 & 1 & \vdots & 2 \\ 0 & 1 & -3 & \vdots & -10 \\ 0 & 0 & 1 & \vdots & 3 \end{bmatrix}$$

This matrix corresponds to the system

$$\begin{cases} x + 2y + z = 2 \\ y - 3z = -10 \\ z = 3 \end{cases}$$

We identify the z-coordinate of the solution as 3. Next we replace z with 3 in the second equation and solve for y.

$$y - 3z = -10 \qquad \textit{Second equation}$$
$$y - 3(3) = -10 \qquad \textit{Let } z = 3.$$
$$y = -1$$

To find x, we let $z = 3$ and $y = -1$ in the first equation.

$$x + 2y + z = 2 \qquad \textit{First equation}$$
$$x + 2(-1) + 3 = 2 \qquad \textit{Let } z = 3 \textit{ and } y = -1.$$
$$x = 1$$

The ordered triple solution is $(1, -1, 3)$. Check to see that it satisfies all three equations in the original system.

SPOTLIGHT ON DECISION MAKING

Suppose you are an urban planner working for the public transportation authority of a large city. Currently, more commuters travel into the downtown area during the morning rush hour than travel out of it. However, a strong economy is creating new jobs in the suburbs. As suburban job growth continues, more city dwellers will travel out of the downtown area during the morning rush hour. You have been assigned to study the situation, focusing on how the trend will affect existing bus routes and schedules and how soon.

After a detailed study of public transportation utilization, you find that the number of commuters into downtown each morning can be described by the equation $y = 40,000 + 200x$, where x is the number of months from now. The number of commuters out of downtown each morning can be described by the equation $y = 20,000 + 1000x$, where x is the number of months from now. In the short term, as the numbers of commuters in each direction increase, additional buses can be put on the existing routes to handle the load. But once the number of commuters leaving downtown exceeds the number of commuters traveling into downtown, the bus routes must be totally revamped.

It usually takes the public transportation authority $1\frac{1}{2}$ years to plan major changes to bus routes. If changes are needed more quickly than that, a consulting firm can be hired to speed up the process. Decide whether a consulting firm will be needed to help revamp the bus routes.

Exercise Set 8.7

Solve each system of linear equations using matrices. See Example 1.

1. $\begin{cases} x + y = 1 \\ x - 2y = 4 \end{cases}$

2. $\begin{cases} 2x - y = 8 \\ x + 3y = 11 \end{cases}$

3. $\begin{cases} x + 3y = 2 \\ x + 2y = 0 \end{cases}$

4. $\begin{cases} 4x - y = 5 \\ 3x + 3y = 0 \end{cases}$

Solve each system of linear equations using matrices. See Example 2.

5. $\begin{cases} x - 2y = 4 \\ 2x - 4y = 4 \end{cases}$

6. $\begin{cases} -x + 3y = 6 \\ 3x - 9y = 9 \end{cases}$

7. $\begin{cases} 3x - 3y = 9 \\ 2x - 2y = 6 \end{cases}$

8. $\begin{cases} 9x - 3y = 6 \\ -18x + 6y = -12 \end{cases}$

Solve each system of linear equations using matrices. See Example 3.

9. $\begin{cases} x + y = 3 \\ 2y = 10 \\ 3x + 2y - 4z = 12 \end{cases}$

10. $\begin{cases} 5x = 5 \\ 2x + y = 4 \\ 3x + y - 5z = -15 \end{cases}$

11. $\begin{cases} 2y - z = -7 \\ x + 4y + z = -4 \\ 5x - y + 2z = 13 \end{cases}$

12. $\begin{cases} 4y + 3z = -2 \\ 5x - 4y = 1 \\ -5x + 4y + z = -3 \end{cases}$

Solve each system of linear equations using matrices.

13. $\begin{cases} x - 4 = 0 \\ x + y = 1 \end{cases}$

14. $\begin{cases} 3y = 6 \\ x + y = 7 \end{cases}$

15. $\begin{cases} x + y + z = 2 \\ 2x - z = 5 \\ 3y + z = 2 \end{cases}$

16. $\begin{cases} x + 2y + z = 5 \\ x - y - z = 3 \\ y + z = 2 \end{cases}$

17. $\begin{cases} 5x - 2y = 27 \\ -3x + 5y = 18 \end{cases}$

18. $\begin{cases} 4x - y = 9 \\ 2x + 3y = -27 \end{cases}$

19. $\begin{cases} 4x - 7y = 7 \\ 12x - 21y = 24 \end{cases}$

20. $\begin{cases} 2x - 5y = 12 \\ -4x + 10y = 20 \end{cases}$

21. $\begin{cases} 4x - y + 2z = 5 \\ 2y + z = 4 \\ 4x + y + 3z = 10 \end{cases}$

22. $\begin{cases} 5y - 7z = 14 \\ 2x + y + 4z = 10 \\ 2x + 6y - 3z = 30 \end{cases}$

23. $\begin{cases} 4x + y + z = 3 \\ -x + y - 2z = -11 \\ x + 2y + 2z = -1 \end{cases}$

24. $\begin{cases} x + y + z = 9 \\ 3x - y + z = -1 \\ -2x + 2y - 3z = -2 \end{cases}$

25. Consider the system
$$\begin{cases} 2x - 3y = 8 \\ x + 5y = -3 \end{cases}$$

What is wrong with its corresponding matrix shown below?

$$\left[\begin{array}{cc|c} 2 & 3 & 8 \\ 0 & 5 & 3 \end{array}\right]$$

26. The percent y of U.S. households that owned a black-and-white television set between the years 1980 and 1993 can be modeled by the linear equation $2.3x + y = 52$, where x represents the number of years after 1980. Similarly, the percent y of U.S. households that owned a microwave oven during this same period can be modeled by the linear equation $-5.4x + y = 14$. (*Source:* Based on data from the Energy Information Administration, U.S. Department of Energy)

a. The data used to form these two models was incomplete. It is impossible to tell from the data the year in which the percent of households owning black-and-white television sets was the same as the percent of households owning microwave ovens. Use matrix methods to estimate the year in which this occurred.

b. Did more households own black-and-white television sets or microwave ovens in 1980? In 1993? What trends do these models show? Does this seem to make sense? Why or why not?

c. According to the models, when will the percent of households owning black-and-white television sets reach 0%?

REVIEW EXERCISES

Determine whether each graph is the graph of a function. See Section 3.3.

27.

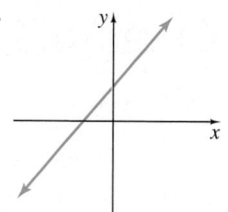

28.

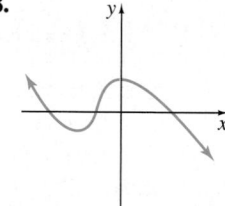

29.

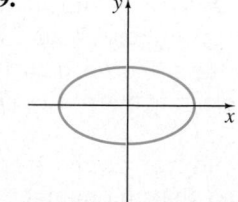

30.

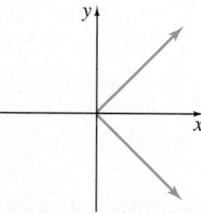

Evaluate. See Section 1.7.

31. $(-1)(-5) - (6)(3)$

32. $(2)(-8) - (-4)(1)$

33. $(4)(-10) - (2)(-2)$

34. $(-7)(3) - (-2)(-6)$

35. $(-3)(-3) - (-1)(-9)$

36. $(5)(6) - (10)(10)$

8.8 SOLVING SYSTEMS OF EQUATIONS BY DETERMINANTS

CD-ROM   SSM

SSG   Video

▶ **O B J E C T I V E S**

1. Define and evaluate a 2 × 2 determinant.
2. Use Cramer's rule to solve a system of two linear equations in two variables.
3. Define and evaluate a 3 × 3 determinant.
4. Use Cramer's rule to solve a linear system of three equations in three variables.

1   We have solved systems of two linear equations in two variables in four different ways: graphically, by substitution, by elimination, and by matrices. Now we analyze another method called **Cramer's rule**.

Recall that a matrix is a rectangular array of numbers. If a matrix has the same number of rows and columns, it is called a **square matrix**. Examples of square matrices are

$$\begin{bmatrix} 1 & 6 \\ 5 & 2 \end{bmatrix} \qquad \begin{bmatrix} 2 & 4 & 1 \\ 0 & 5 & 2 \\ 3 & 6 & 9 \end{bmatrix}$$

A **determinant** is a real number associated with a square matrix. The determinant of a square matrix is denoted by placing vertical bars about the array of numbers. Thus,

The determinant of the square matrix $\begin{bmatrix} 1 & 6 \\ 5 & 2 \end{bmatrix}$ is $\begin{vmatrix} 1 & 6 \\ 5 & 2 \end{vmatrix}$.

The determinant of the square matrix $\begin{bmatrix} 2 & 4 & 1 \\ 0 & 5 & 2 \\ 3 & 6 & 9 \end{bmatrix}$ is $\begin{vmatrix} 2 & 4 & 1 \\ 0 & 5 & 2 \\ 3 & 6 & 9 \end{vmatrix}$.

We define the determinant of a 2 × 2 matrix first. (Recall that 2 × 2 is read "two by two." It means that the matrix has 2 rows and 2 columns.)

DETERMINANT OF A 2 × 2 MATRIX

$$\begin{vmatrix} a & b \\ c & d \end{vmatrix} = ad - bc$$

Example 1   Evaluate each determinant.

a. $\begin{vmatrix} -1 & 2 \\ 3 & -4 \end{vmatrix}$ **b.** $\begin{vmatrix} 2 & 0 \\ 7 & -5 \end{vmatrix}$

Solution   First we identify the values of a, b, c, and d. Then we perform the evaluation.

a. Here $a = -1$, $b = 2$, $c = 3$, and $d = -4$.

$$\begin{vmatrix} -1 & 2 \\ 3 & -4 \end{vmatrix} = ad - bc = (-1)(-4) - (2)(3) = -2$$

b. In this example, $a = 2, b = 0, c = 7$, and $d = -5$.

$$\begin{vmatrix} 2 & 0 \\ 7 & -5 \end{vmatrix} = ad - bc = 2(-5) - (0)(7) = -10$$

2   To develop Cramer's rule, we solve the system $\begin{cases} ax + by = h \\ cx + dy = k \end{cases}$ using elimination.

First, we eliminate y by multiplying both sides of the first equation by d and both sides of the second equation by $-b$ so that the coefficients of y are opposites. The result is that

$$\begin{cases} d(ax + by) = d \cdot h \\ -b(cx + dy) = -b \cdot k \end{cases} \quad \text{simplifies to} \quad \begin{cases} adx + bdy = hd \\ -bcx - bdy = -kb \end{cases}$$

We now add the two equations and solve for x.

$$\begin{array}{rl} adx + bdy &= hd \\ -bcx - bdy &= -kb \\ \hline adx - bcx &= hd - kb \qquad \text{Add the equations.} \\ (ad - bc)x &= hd - kb \\ x &= \dfrac{hd - kb}{ad - bc} \qquad \text{Solve for } x. \end{array}$$

When we replace x with $\dfrac{hd - kb}{ad - bc}$ in the equation $ax + by = h$ and solve for y, we find that $y = \dfrac{ak - ch}{ad - bc}$.

Notice that the numerator of the value of x is the determinant of

$$\begin{vmatrix} h & b \\ k & d \end{vmatrix} = hd - kb$$

Also, the numerator of the value of y is the determinant of

$$\begin{vmatrix} a & h \\ c & k \end{vmatrix} = ak - hc$$

Finally, the denominators of the values of x and y are the same and are the determinant of

$$\begin{vmatrix} a & b \\ c & d \end{vmatrix} = ad - bc$$

This means that the values of x and y can be written in determinant notation.

$$x = \frac{\begin{vmatrix} h & b \\ k & d \end{vmatrix}}{\begin{vmatrix} a & b \\ c & d \end{vmatrix}} \quad \text{and} \quad y = \frac{\begin{vmatrix} a & h \\ c & k \end{vmatrix}}{\begin{vmatrix} a & b \\ c & d \end{vmatrix}}$$

For convenience, we label the determinants D, D_x, and D_y.

$$\underbrace{\begin{vmatrix} a & b \\ c & d \end{vmatrix}}_{} = D \qquad \begin{vmatrix} h & b \\ k & d \end{vmatrix} = D_x \qquad \begin{vmatrix} a & h \\ c & k \end{vmatrix} = D_y$$

x-coefficients → ↘ y-coefficients ↴

↑ x-column replaced by constants   ↑ y-column replaced by constants

These determinant formulas for the coordinates of the solution of a system are known as **Cramer's rule**.

CRAMER'S RULE FOR TWO LINEAR EQUATIONS IN TWO VARIABLES

The solution of the system $\begin{cases} ax + by = h \\ cx + dy = k \end{cases}$ is given by

$$x = \frac{\begin{vmatrix} h & b \\ k & d \end{vmatrix}}{\begin{vmatrix} a & b \\ c & d \end{vmatrix}} = \frac{D_x}{D} \qquad y = \frac{\begin{vmatrix} a & h \\ c & k \end{vmatrix}}{\begin{vmatrix} a & b \\ c & d \end{vmatrix}} = \frac{D_y}{D}$$

as long as $D = ad - bc$ is not 0.

When $D = 0$, the system is either inconsistent or the equations are dependent. When this happens, we need to use another method to see which is the case.

Example 2 Use Cramer's rule to solve each system.

a. $\begin{cases} 3x + 4y = -7 \\ x - 2y = -9 \end{cases}$ **b.** $\begin{cases} 5x + y = 5 \\ -7x - 2y = -7 \end{cases}$

Solution **a.** First we find D, D_x, and D_y.

$$\begin{matrix} a & b & h \\ \downarrow & \downarrow & \downarrow \end{matrix}$$
$$\begin{cases} 3x + 4y = -7 \\ x - 2y = -9 \end{cases}$$
$$\begin{matrix} \uparrow & \uparrow & \uparrow \\ c & d & k \end{matrix}$$

$$D = \begin{vmatrix} a & b \\ c & d \end{vmatrix} = \begin{vmatrix} 3 & 4 \\ 1 & -2 \end{vmatrix} = 3(-2) - 4(1) = -10$$

$$D_x = \begin{vmatrix} h & b \\ k & d \end{vmatrix} = \begin{vmatrix} -7 & 4 \\ -9 & -2 \end{vmatrix} = (-7)(-2) - 4(-9) = 50$$

$$D_y = \begin{vmatrix} a & h \\ c & k \end{vmatrix} = \begin{vmatrix} 3 & -7 \\ 1 & -9 \end{vmatrix} = 3(-9) - (-7)(1) = -20$$

Then $x = \dfrac{D_x}{D} = \dfrac{50}{-10} = -5$ and $y = \dfrac{D_y}{D} = \dfrac{-20}{-10} = 2$. The ordered pair solution is $(-5, 2)$.

As always, check the solution in both original equations.

b. Find $D, D_x,$ and D_y for $\begin{cases} 5x + y = 5 \\ -7x - 2y = -7 \end{cases}$.

$$D = \begin{vmatrix} 5 & 1 \\ -7 & -2 \end{vmatrix} = 5(-2) - (1)(-7) = -3$$

$$D_x = \begin{vmatrix} 5 & 1 \\ -7 & -2 \end{vmatrix} = 5(-2) - (1)(-7) = -3$$

$$D_y = \begin{vmatrix} 5 & 5 \\ -7 & -7 \end{vmatrix} = 5(-7) - 5(-7) = 0$$

Then $x = \dfrac{D_x}{D} = \dfrac{-3}{-3} = 1$ and $y = \dfrac{D_y}{D} = \dfrac{0}{-3} = 0$.

The ordered pair solution is $(1, 0)$. Check this solution in both original equations.

3

A 3×3 determinant can be used to solve a system of three equations in three variables. The determinant of a 3×3 matrix, however, is considerably more complex than a 2×2 one.

DETERMINANT OF A 3×3 MATRIX

$$\begin{vmatrix} a_1 & b_1 & c_1 \\ a_2 & b_2 & c_2 \\ a_3 & b_3 & c_3 \end{vmatrix} = a_1 \cdot \begin{vmatrix} b_2 & c_2 \\ b_3 & c_3 \end{vmatrix} - a_2 \cdot \begin{vmatrix} b_1 & c_1 \\ b_3 & c_3 \end{vmatrix} + a_3 \cdot \begin{vmatrix} b_1 & c_1 \\ b_2 & c_2 \end{vmatrix}$$

Notice that the determinant of a 3×3 matrix is related to the determinants of three 2×2 matrices. Each determinant of these 2×2 matrices is called a **minor**, and every element of a 3×3 matrix has a minor associated with it. For example, the minor of c_2 is the determinant of the 2×2 matrix found by deleting the row and column containing c_2.

The minor of c_2 is $\begin{vmatrix} a_1 & b_1 \\ a_3 & b_3 \end{vmatrix}$

Also, the minor of element a_1 is the determinant of the 2×2 matrix that has no row or column containing a_1.

The minor of a_1 is $\begin{vmatrix} b_2 & c_2 \\ b_3 & c_3 \end{vmatrix}$

So the determinant of a 3×3 matrix can be written as

$$a_1 \cdot (\text{minor of } a_1) - a_2 \cdot (\text{minor of } a_2) + a_3 \cdot (\text{minor of } a_3)$$

Finding the determinant by using minors of elements in the first column is called **expanding** by the minors of the first column. *The value of a determinant can be found by expanding by the minors of any row or column.* The following **array of signs** is helpful in determining whether to add or subtract the product of an element and its minor.

$$+ \quad - \quad +$$
$$- \quad + \quad -$$
$$+ \quad - \quad +$$

If an element is in a position marked $+$, we add. If marked $-$, we subtract.

Example 3 Evaluate by expanding by the minors of the given row or column.

$$\begin{vmatrix} 0 & 5 & 1 \\ 1 & 3 & -1 \\ -2 & 2 & 4 \end{vmatrix}$$

a. First column   **b.** Second row

Solution **a.** The elements of the first column are $0, 1$, and -2. The first column of the array of signs is $+, -, +$.

$$\begin{vmatrix} 0 & 5 & 1 \\ 1 & 3 & -1 \\ -2 & 2 & 4 \end{vmatrix} = 0 \cdot \begin{vmatrix} 3 & -1 \\ 2 & 4 \end{vmatrix} - 1 \cdot \begin{vmatrix} 5 & 1 \\ 2 & 4 \end{vmatrix} + (-2) \cdot \begin{vmatrix} 5 & 1 \\ 3 & -1 \end{vmatrix}$$

$$= 0(12 - (-2)) - 1(20 - 2) + (-2)(-5 - 3)$$
$$= 0 - 18 + 16 = -2$$

b. The elements of the second row are $1, 3$, and -1. This time, the signs begin with $-$ and again alternate.

$$\begin{vmatrix} 0 & 5 & 1 \\ 1 & 3 & -1 \\ -2 & 2 & 4 \end{vmatrix} = -1 \cdot \begin{vmatrix} 5 & 1 \\ 2 & 4 \end{vmatrix} + 3 \cdot \begin{vmatrix} 0 & 1 \\ -2 & 4 \end{vmatrix} - (-1) \cdot \begin{vmatrix} 0 & 5 \\ -2 & 2 \end{vmatrix}$$

$$= -1(20 - 2) + 3(0 - (-2)) - (-1)(0 - (-10))$$
$$= -18 + 6 + 10 = -2$$

Notice that the determinant of the 3×3 matrix is the same regardless of the row or column you select to expand by.

4 A system of three equations in three variables may be solved with Cramer's rule also. Using the elimination process to solve a system with unknown constants as coefficients leads to the following.

CRAMER'S RULE FOR THREE EQUATIONS IN THREE VARIABLES

The solution of the system $\begin{cases} a_1 x + b_1 y + c_1 z = k_1 \\ a_2 x + b_2 y + c_2 z = k_2 \\ a_3 x + b_3 y + c_3 z = k_3 \end{cases}$ is given by

$$x = \frac{D_x}{D} \qquad y = \frac{D_y}{D} \qquad \text{and} \qquad z = \frac{D_z}{D}$$

(continued)

where

$$D = \begin{vmatrix} a_1 & b_1 & c_1 \\ a_2 & b_2 & c_2 \\ a_3 & b_3 & c_3 \end{vmatrix} \qquad D_x = \begin{vmatrix} k_1 & b_1 & c_1 \\ k_2 & b_2 & c_2 \\ k_3 & b_3 & c_3 \end{vmatrix}$$

$$D_y = \begin{vmatrix} a_1 & k_1 & c_1 \\ a_2 & k_2 & c_2 \\ a_3 & k_3 & c_3 \end{vmatrix} \qquad D_z = \begin{vmatrix} a_1 & b_1 & k_1 \\ a_2 & b_2 & k_2 \\ a_3 & b_3 & k_3 \end{vmatrix}$$

as long as D is not 0.

Example 4   Use Cramer's rule to solve the system.

$$\begin{cases} x - 2y + z = 4 \\ 3x + y - 2z = 3 \\ 5x + 5y + 3z = -8 \end{cases}$$

Solution   First we find $D, D_x, D_y,$ and D_z. Beginning with D, we expand by the minors of the first column.

$$D = \begin{vmatrix} 1 & -2 & 1 \\ 3 & 1 & -2 \\ 5 & 5 & 3 \end{vmatrix} = 1 \cdot \begin{vmatrix} 1 & -2 \\ 5 & 3 \end{vmatrix} - 3 \cdot \begin{vmatrix} -2 & 1 \\ 5 & 3 \end{vmatrix} + 5 \cdot \begin{vmatrix} -2 & 1 \\ 1 & -2 \end{vmatrix}$$

$$= 1(3 - (-10)) - 3(-6 - 5) + 5(4 - 1)$$

$$= 13 + 33 + 15 = 61$$

$$D_x = \begin{vmatrix} 4 & -2 & 1 \\ 3 & 1 & -2 \\ -8 & 5 & 3 \end{vmatrix} = 4 \cdot \begin{vmatrix} 1 & -2 \\ 5 & 3 \end{vmatrix} - 3 \cdot \begin{vmatrix} -2 & 1 \\ 5 & 3 \end{vmatrix} + (-8) \cdot \begin{vmatrix} -2 & 1 \\ 1 & -2 \end{vmatrix}$$

$$= 4(3 - (-10)) - 3(-6 - 5) + (-8)(4 - 1)$$

$$= 52 + 33 - 24 = 61$$

$$D_y = \begin{vmatrix} 1 & 4 & 1 \\ 3 & 3 & -2 \\ 5 & -8 & 3 \end{vmatrix} = 1 \cdot \begin{vmatrix} 3 & -2 \\ -8 & 3 \end{vmatrix} - 3 \cdot \begin{vmatrix} 4 & 1 \\ -8 & 3 \end{vmatrix} + 5 \cdot \begin{vmatrix} 4 & 1 \\ 3 & -2 \end{vmatrix}$$

$$= 1(9 - 16) - 3(12 - (-8)) + 5(-8 - 3)$$

$$= -7 - 60 - 55 = -122$$

$$D_z = \begin{vmatrix} 1 & -2 & 4 \\ 3 & 1 & 3 \\ 5 & 5 & -8 \end{vmatrix} = 1 \cdot \begin{vmatrix} 1 & 3 \\ 5 & -8 \end{vmatrix} - 3 \cdot \begin{vmatrix} -2 & 4 \\ 5 & -8 \end{vmatrix} + 5 \cdot \begin{vmatrix} -2 & 4 \\ 1 & 3 \end{vmatrix}$$

$$= 1(-8 - 15) - 3(16 - 20) + 5(-6 - 4)$$

$$= -23 + 12 - 50 = -61$$

From these determinants, we calculate the solution.

$$x = \frac{D_x}{D} = \frac{61}{61} = 1 \qquad y = \frac{D_y}{D} = \frac{-122}{61} = -2 \qquad z = \frac{D_z}{D} = \frac{-61}{61} = -1$$

The ordered triple solution is $(1, -2, -1)$. Check this solution by verifying that it satisfies each equation of the system.

Exercise Set 8.8

Evaluate. See Example 1.

1. $\begin{vmatrix} 3 & 5 \\ -1 & 7 \end{vmatrix}$

2. $\begin{vmatrix} -5 & 1 \\ 0 & -4 \end{vmatrix}$

3. $\begin{vmatrix} 9 & -2 \\ 4 & -3 \end{vmatrix}$

4. $\begin{vmatrix} 4 & 0 \\ 9 & 8 \end{vmatrix}$

5. $\begin{vmatrix} -2 & 9 \\ 4 & -18 \end{vmatrix}$

6. $\begin{vmatrix} -40 & 8 \\ 70 & -14 \end{vmatrix}$

Use Cramer's rule, if possible, to solve each system of linear equations. See Example 2.

7. $\begin{cases} 2y - 4 = 0 \\ x + 2y = 5 \end{cases}$

8. $\begin{cases} 4x - y = 5 \\ 3x - 3 = 0 \end{cases}$

9. $\begin{cases} 3x + y = 1 \\ \quad 2y = 2 - 6x \end{cases}$

10. $\begin{cases} y = 2x - 5 \\ 8x - 4y = 20 \end{cases}$

11. $\begin{cases} 5x - 2y = 27 \\ -3x + 5y = 18 \end{cases}$

12. $\begin{cases} 4x - y = 9 \\ 2x + 3y = -27 \end{cases}$

Evaluate. See Example 3.

13. $\begin{vmatrix} 2 & 1 & 0 \\ 0 & 5 & -3 \\ 4 & 0 & 2 \end{vmatrix}$

14. $\begin{vmatrix} -6 & 4 & 2 \\ 1 & 0 & 5 \\ 0 & 3 & 1 \end{vmatrix}$

15. $\begin{vmatrix} 4 & -6 & 0 \\ -2 & 3 & 0 \\ 4 & -6 & 1 \end{vmatrix}$

16. $\begin{vmatrix} 5 & 2 & 1 \\ 3 & -6 & 0 \\ -2 & 8 & 0 \end{vmatrix}$

17. $\begin{vmatrix} 3 & 6 & -3 \\ -1 & -2 & 3 \\ 4 & -1 & 6 \end{vmatrix}$

18. $\begin{vmatrix} 2 & -2 & 1 \\ 4 & 1 & 3 \\ 3 & 1 & 2 \end{vmatrix}$

Use Cramer's rule, if possible, to solve each system of linear equations. See Example 4.

19. $\begin{cases} 3x + z = -1 \\ -x - 3y + z = 7 \\ 3y + z = 5 \end{cases}$

20. $\begin{cases} 4y - 3z = -2 \\ 8x - 4y = 4 \\ -8x + 4y + z = -2 \end{cases}$

21. $\begin{cases} x + y + z = 8 \\ 2x - y - z = 10 \\ x - 2y + 3z = 22 \end{cases}$

22. $\begin{cases} 5x + y + 3z = 1 \\ x - y - 3z = -7 \\ -x + y = 1 \end{cases}$

Evaluate.

23. $\begin{vmatrix} 10 & -1 \\ -4 & 2 \end{vmatrix}$

24. $\begin{vmatrix} -6 & 2 \\ 5 & -1 \end{vmatrix}$

25. $\begin{vmatrix} 1 & 0 & 4 \\ 1 & -1 & 2 \\ 3 & 2 & 1 \end{vmatrix}$

26. $\begin{vmatrix} 0 & 1 & 2 \\ 3 & -1 & 2 \\ 3 & 2 & -2 \end{vmatrix}$

27. $\begin{vmatrix} \frac{3}{4} & \frac{5}{2} \\ -\frac{1}{6} & \frac{7}{3} \end{vmatrix}$

28. $\begin{vmatrix} \frac{5}{7} & \frac{1}{3} \\ \frac{6}{7} & \frac{2}{3} \end{vmatrix}$

29. $\begin{vmatrix} 4 & -2 & 2 \\ 6 & -1 & 3 \\ 2 & 1 & 1 \end{vmatrix}$

30. $\begin{vmatrix} 1 & 5 & 0 \\ 7 & 9 & -4 \\ 3 & 2 & -2 \end{vmatrix}$

31. $\begin{vmatrix} -2 & 5 & 4 \\ 5 & -1 & 3 \\ 4 & 1 & 2 \end{vmatrix}$

32. $\begin{vmatrix} 5 & -2 & 4 \\ -1 & 5 & 3 \\ 1 & 4 & 2 \end{vmatrix}$

33. If all the elements in a single row of a determinant are zero, to what does the determinant evaluate? Explain your answer.

34. If all the elements in a single column of a determinant are 0, to what does the determinant evaluate? Explain your answer.

Find the value of x such that each is a true statement.

35. $\begin{vmatrix} 1 & x \\ 2 & 7 \end{vmatrix} = -3$

36. $\begin{vmatrix} 6 & 1 \\ -2 & x \end{vmatrix} = 26$

Use Cramer's rule, if possible, to solve each system of linear equations.

37. $\begin{cases} 2x - 5y = 4 \\ x + 2y = -7 \end{cases}$

38. $\begin{cases} 3x - y = 2 \\ -5x + 2y = 0 \end{cases}$

39. $\begin{cases} 4x + 2y = 5 \\ 2x + y = -1 \end{cases}$

40. $\begin{cases} 3x + 6y = 15 \\ 2x + 4y = 3 \end{cases}$

41. $\begin{cases} 2x + 2y + z = 1 \\ -x + y + 2z = 3 \\ x + 2y + 4z = 0 \end{cases}$

42. $\begin{cases} 2x - 3y + z = 5 \\ x + y + z = 0 \\ 4x + 2y + 4z = 4 \end{cases}$

43. $\begin{cases} \dfrac{2}{3}x - \dfrac{3}{4}y = -1 \\ -\dfrac{1}{6}x + \dfrac{3}{4}y = \dfrac{5}{2} \end{cases}$

44. $\begin{cases} \dfrac{1}{2}x - \dfrac{1}{3}y = -3 \\ \dfrac{1}{8}x + \dfrac{1}{6}y = 0 \end{cases}$

45. $\begin{cases} 0.7x - 0.2y = -1.6 \\ 0.2x - y = -1.4 \end{cases}$

46. $\begin{cases} -0.7x + 0.6y = 1.3 \\ 0.5x - 0.3y = -0.8 \end{cases}$

47. $\begin{cases} -2x + 4y - 2z = 6 \\ x - 2y + z = -3 \\ 3x - 6y + 3z = -9 \end{cases}$

48. $\begin{cases} -x - y + 3z = 2 \\ 4x + 4y - 12z = -8 \\ -3x - 3y + 9z = 6 \end{cases}$

49. $\begin{cases} x - 2y + z = -5 \\ 3y + 2z = 4 \\ 3x - y = -2 \end{cases}$

50. $\begin{cases} 4x + 5y = 10 \\ 3y + 2z = -6 \\ x + y + z = 3 \end{cases}$

51. Suppose you are interested in finding the determinant of a 4 × 4 matrix. Study the pattern shown in the array of signs for a 3 × 3 matrix. Use the pattern to expand the array of signs for use with a 4 × 4 matrix.

52. Why would expanding by minors of the second row be a good choice for the determinant $\begin{vmatrix} 3 & 4 & -2 \\ 5 & 0 & 0 \\ 6 & -3 & 7 \end{vmatrix}$?

REVIEW EXERCISES

Simplify each expression. See Section 2.1.

53. $5x - 6 + x - 12$ **54.** $4y + 3 - 15y - 1$

55. $2(3x - 6) + 3(x - 1)$

56. $-3(2y - 7) - 1(11 + 12y)$

Graph each function. See Section 3.4.

57. $f(x) = 5x - 6$ **58.** $g(x) = -x + 1$

59. $h(x) = 3$ **60.** $f(x) = -3$

A Look Ahead

Example

Evaluate the determinant.

$$\begin{vmatrix} 2 & 0 & -1 & 3 \\ 0 & 5 & -2 & -1 \\ 3 & 1 & 0 & 1 \\ 4 & 2 & -2 & 0 \end{vmatrix}$$

Solution

To evaluate a 4 × 4 determinant, select any row or column and expand by the minors. The array of signs for a 4 × 4 determinant is the same as for a 3 × 3 determinant except expanded. We expand using the fourth row.

$$\rightarrow \begin{vmatrix} 2 & 0 & -1 & 3 \\ 0 & 5 & -2 & -1 \\ 3 & 1 & 0 & 1 \\ 4 & 2 & -2 & 0 \end{vmatrix}$$

$$= -4 \cdot \begin{vmatrix} 0 & -1 & 3 \\ 5 & -2 & -1 \\ 1 & 0 & 1 \end{vmatrix} + 2 \cdot \begin{vmatrix} 2 & -1 & 3 \\ 0 & -2 & -1 \\ 3 & 0 & 1 \end{vmatrix}$$

$$-(-2) \cdot \begin{vmatrix} 2 & 0 & 3 \\ 0 & 5 & -1 \\ 3 & 1 & 1 \end{vmatrix} + 0 \cdot \begin{vmatrix} 2 & 0 & -1 \\ 0 & 5 & -2 \\ 3 & 1 & 0 \end{vmatrix}$$

Now find the value of each 3 × 3 determinant. The value of the 4 × 4 determinant is

$$-4(12) + 2(17) + 2(-33) + 0 = -80$$

Find the value of each determinant. See the preceding example.

61. $\begin{vmatrix} 5 & 0 & 0 & 0 \\ 0 & 4 & 2 & -1 \\ 1 & 3 & -2 & 0 \\ 0 & -3 & 1 & 2 \end{vmatrix}$ **62.** $\begin{vmatrix} 1 & 7 & 0 & -1 \\ 1 & 3 & -2 & 0 \\ 1 & 0 & -1 & 2 \\ 0 & -6 & 2 & 4 \end{vmatrix}$

63. $\begin{vmatrix} 4 & 0 & 2 & 5 \\ 0 & 3 & -1 & 1 \\ 0 & 0 & 2 & 0 \\ 0 & 0 & 0 & 1 \end{vmatrix}$ **64.** $\begin{vmatrix} 2 & 0 & -1 & 4 \\ 6 & 0 & 4 & 1 \\ 2 & 4 & 3 & -1 \\ 4 & 0 & 5 & -4 \end{vmatrix}$

For additional Chapter Projects, visit the Real World Activities Website by going to http://www.prenhall.com/martin-gay.

CHAPTER PROJECT

Analyzing Municipal Budgets

Nearly all cities, towns, and villages operate with an annual budget. Budget items might include expenses for fire and police protection as well as for street maintenance and parks. No matter how big or small the budget, city officials need to know if municipal spending is over or under budget. In this project, you will have the opportunity to analyze a municipal budget and make budgetary recommendations. This project may be completed by working in groups or individually.

Suppose that each year your town creates a municipal budget. The next year's annual municipal budget is submitted for approval by the town's citizens at the annual town meeting. This year's budget was printed in the town newspaper earlier in the year.

You have joined a group of citizens who are concerned about your town's budgeting and spending processes. Your group plans to analyze this year's budget along with what was actually spent by the town this year. You hope to present your findings at

the annual town meeting and make some budgetary recommendations for next year's budget. The municipal budget contains many different areas of spending. To help focus your group's analysis, you have decided to research spending habits only for categories in which the actual expenses differ from the budgeted amount by more than 12% of the budgeted amount.

1. For each category in the budget, write a specific absolute value inequality that describes the condition that must be met before your group will research spending habits for that category. In each case, let the variable x represent the actual expense for a budget category.

2. For each category in the budget, write an equivalent compound inequality for the condition described in Question 1. Again, let the variable x represent the actual expense for a budget category.

3. Below is a listing of the actual expenditures made this year for each budget category. Use the inequalities from either Question 1 or Question 2 to complete the Budget Worksheet given at the end of this project. (The first category has been filled in.) From the Budget Worksheet, decide which categories must be researched.

	Department/Program	Actual Expenditure
I.	**Board of Health**	
	Immunization Programs	$14,800
	Inspections	$41,900
II.	**Fire Department**	
	Equipment	$375,000
	Salaries	$268,500
III.	**Libraries**	
	Book/Periodical Purchases	$107,300
	Equipment	$29,000
	Salaries	$118,400
IV.	**Parks and Recreation**	
	Maintenance	$82,500
	Playground Equipment	$45,000
	Salaries	$118,000
	Summer Programs	$96,200
V.	**Police Department**	
	Equipment	$328,000
	Salaries	$405,000
VI.	**Public Works**	
	Recycling	$48,100
	Sewage	$92,500
	Snow Removal & Road Salt	$268,300
	Street Maintenance	$284,000
	Water Treatment	$94,100
	TOTAL	$2,816,600

THE TOWN CRIER
Annual Budget Set at Town Meeting
ANYTOWN, USA (MG)—This year's annual budget is as follows:

	Amount Budgeted
BOARD OF HEALTH	
Immunization Programs	$15,000
Inspections	$50,000
FIRE DEPARTMENT	
Equipment	$450,000
Salaries	$275,000
LIBRARIES	
Book/Periodical Purchases	$90,000
Equipment	$30,000
Salaries	$120,000
PARKS AND RECREATION	
Maintenance	$70,000
Playground Equipment	$50,000
Salaries	$140,000
Summer Programs	$80,000
POLICE DEPARTMENT	
Equipment	$300,000
Salaries	$400,000
PUBLIC WORKS	
Recycling	$50,000
Sewage	$100,000
Snow Removal & Road Salt	$200,000
Street Maintenance	$250,000
Water Treatment	$100,000
TOTAL	**$2,770,000**

4. Can you think of possible reasons why spending in the categories that must be researched were over or under budget?

5. Based on this year's municipal budget and actual expenses, what recommendations would you make for next year's budget? Explain your reasoning.

6. (Optional) Research the annual budget used by your own town or your college or university. Conduct a similar analysis of the budget with respect to actual expenses. What can you conclude?

BUDGET WORKSHEET

Budget category	Budgeted amount	Minimum allowed	Actual expense	Maximum allowed	Within budget?	Amt over/ under budget
Immunization Programs	$15,000	$13,200	$14,800	$16,800	Yes	Under $200

CHAPTER 8 VOCABULARY CHECK

Fill in each blank with one of the words or phrases listed below.

consecutive integers   matrix   determinant   square
absolute value   compound inequality   intersection   union

1. The statement "$x < 5$ or $x > 7$" is called a(n) _____ .

2. The _____ of two sets is the set of all elements common to both sets.

3. The _____ of two sets is the set of all elements that belong to either of the sets.

4. A number's distance from 0 is called its _____ .

5. The integers 17, 18, 19 are examples of _____ .

6. If a matrix has the same number of rows and columns, it is called a _____ matrix.

7. A real number associated with a square matrix is called its _____ .

8. A _____ is a rectangular array of numbers.

CHAPTER 8 HIGHLIGHTS

DEFINITIONS AND CONCEPTS	EXAMPLES

Section 8.1 Review Solving Equations

Solving Linear and Quadratic Equations in One Variable

Step 1: Multiply on both sides to clear the equation of fractions if they occur.

Step 2: Use the distributive property to remove parentheses if they occur.

Step 3: Simplify each side of the equation by combining like terms.

Step 4: Decide whether the equation is linear or quadratic.

If linear ($ax + b = c$),

Step 5: Get all variable terms on one side and all numbers on the other side by using the addition property of equality.

Step 6: Get the variable alone by using the multiplication property of equality.

If quadratic ($ax^2 + bx + c = 0$),

Step 5: Write the equation in standard form: $ax^2 + bx + c = 0$.

Step 6: Factor completely.

Step 7: Set each factor containing a variable equal to 0.

Step 8: Solve.

Final Step Check each solution in the original equation.

Solve for x:

$$x - \frac{x - 2}{6} = \frac{x - 7}{3} + \frac{2}{3}$$

(continued)

DEFINITIONS AND CONCEPTS	EXAMPLES

Section 8.1 Review Solving Equations

1. Clear the equation of fractions.

1. $6\left(x - \dfrac{x-2}{6}\right) = 6\left(\dfrac{x-7}{3} + \dfrac{2}{3}\right)$ Multiply both sides by 6.

$6x - (x-2) = 2(x-7) + 2(2)$

2. Remove grouping symbols such as parentheses.

2. $6x - x + 2 = 2x - 14 + 4$ Remove grouping symbols

3. Simplify by combining like terms.

4. This equation is linear.

4. $5x + 2 = 2x - 10$

5. Write variable terms on one side and numbers on the other side using the addition property of equality.

5. $5x + 2 - 2 = 2x - 10 - 2$ Subtract 2.

$5x = 2x - 12$

$5x - 2x = 2x - 12 - 2x$ Subtract $2x$.

$3x = -12$

6. Get the variable alone using the multiplication property of equality.

6. $\dfrac{3x}{3} = \dfrac{-12}{3}$ Divide by 3.

$x = -4$

Final Step   Check the proposed solution in the original equation.

$-4 - \dfrac{-4-2}{6} \overset{?}{=} \dfrac{-4-7}{3} + \dfrac{2}{3}$ Replace x with -4 in the original equation.

$-4 - \dfrac{-6}{6} \overset{?}{=} \dfrac{-11}{3} + \dfrac{2}{3}$

$-4 - (-1) \overset{?}{=} \dfrac{-9}{3}$

$-3 = -3$ True.

Section 8.2 Further Problem Solving

Problem-Solving Strategy

Colorado is shaped like a rectangle whose length is about 1.3 times its width. If the perimeter of Colorado is 2070 kilometers, find its dimensions.

1. UNDERSTAND the problem.

1. Read and reread the problem. Guess a solution and check your guess.

Let x = width of Colorado in kilometers. Then $1.3x$ = length of Colorado in kilometers

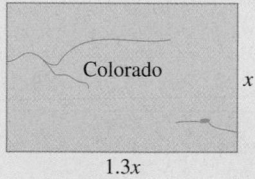

Colorado x

$1.3x$

2. TRANSLATE the problem.

2. In words:

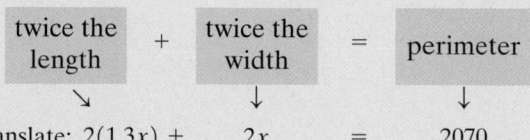

twice the length	+	twice the width	=	perimeter
↘		↓		↓

Translate: $2(1.3x) +$ $2x$ =   2070

(continued)

DEFINITIONS AND CONCEPTS	EXAMPLES

Section 8.2 Further Problem Solving

3. SOLVE the equation.

4. INTERPRET the results.

3. $2.6x + 2x = 2070$
$$4.6x = 2070$$
$$x = 450$$

4. If $x = 450$ kilometers, then $1.3x = 1.3(450) = 585$ kilometers. *Check:* The perimeter of a rectangle whose width is 450 kilometers and length is 585 kilometers is $2(450) + 2(585) = 2070$ kilometers, the required perimeter. *State:* The dimensions of Colorado are 450 kilometers by 585 kilometers

Section 8.3 Compound Inequalities

Two inequalities joined by the words **and** or **or** are called **compound inequalities.**

Compound inequalities:
$$x - 7 \leq 4 \quad \text{and} \quad x \geq -21$$
$$2x + 7 > x - 3 \quad \text{or} \quad 5x + 2 > -3$$

The solution set of a compound inequality formed by the word **and** is the **intersection,** $\cap$, of the solution sets of the two inequalities.

Solve for x:
$$x < 5 \text{ and } x < 3$$

$\{x \mid x < 5\}$ $(-\infty, 5)$

$\{x \mid x < 3\}$ $(-\infty, 3)$

$\{x \mid x < 3$ and $x < 5\}$ $(-\infty, 3)$

The solution set of a compound inequality formed by the word **or** is the **union,** $\cup$, of the solution sets of the two inequalities.

Solve for x:
$$x - 2 \geq -3 \quad \text{or} \quad 2x \leq -4$$
$$x \geq -1 \quad \text{or} \quad x \leq -2$$

$\{x \mid x \geq -1\}$ $[-1, \infty)$

$\{x \mid x \leq -2\}$ $(-\infty, -2]$

$\{x \mid x \leq -2$ or $x \geq -1\}$ $(-\infty, -2]$ $\cup [-1, \infty)$

Section 8.4 Absolute Value Equations

If a is a positive number, then $|x| = a$ is equivalent to $x = a$ or $x = -a$.

Solve for y:
$$|5y - 1| - 7 = 4$$
$$|5y - 1| = 11 \qquad \text{Add 7.}$$
$$5y - 1 = 11 \quad \text{or} \quad 5y - 1 = -11$$
$$5y = 12 \quad \text{or} \quad 5y = -10 \qquad \text{Add 1.}$$
$$y = \frac{12}{5} \quad \text{or} \quad y = -2 \qquad \text{Divide by 5.}$$

The solutions are -2 and $\frac{12}{5}$.

(continued)

DEFINITIONS AND CONCEPTS	EXAMPLES

Section 8.4 Absolute Value Equations

If a is negative, then $|x| = a$ has no solution.

Solve for x:

$$\left|\frac{x}{2} - 7\right| = -1$$

There is no solution.

If an absolute value equation is of the form $|x| = |y|$, solve the equations $x = y$ and $x = -y$.

Solve for x:

$$|x - 7| = |2x + 1|$$

$$x - 7 = 2x + 1 \quad \text{or} \quad x - 7 = -(2x + 1)$$

$$x = 2x + 8 \qquad\qquad x - 7 = -2x - 1$$

$$-x = 8 \qquad\qquad\qquad x = -2x + 6$$

$$x = -8 \qquad \text{or} \qquad 3x = 6$$

$$x = 2$$

The solutions are -8 and 2.

Section 8.5 Absolute Value Inequalities

If a is a positive number, then $|x| < a$ is equivalent to the compound inequality $-a < x < a$.

Solve for y:

$$|y - 5| \le 3$$

$$-3 \le y - 5 \le 3$$

$$-3 + 5 \le y - 5 + 5 \le 3 + 5 \qquad \text{Add 5.}$$

$$2 \le y \le 8$$

The solution set is $[2, 8]$.

If a is a positive number, then $|x| > a$ is equivalent to the compound inequality $x < -a$ or $x > a$.

Solve for x:

$$\left|\frac{x}{2} - 3\right| > 7$$

$$\frac{x}{2} - 3 < -7 \quad \text{or} \quad \frac{x}{2} - 3 > 7$$

$$x - 6 < -14 \quad \text{or} \quad x - 6 > 14 \qquad \text{Multiply by 2.}$$

$$x < -8 \quad \text{or} \quad x > 20 \qquad \text{Add 6.}$$

The solution set is $(-\infty, -8) \cup (20, \infty)$.

Section 8.6 Solving Systems of Linear Equations in Three Variables

A **solution** of an equation in three variables x, y, and z is an **ordered triple** (x, y, z) that makes the equation a true statement.

Verify that $(-2, 1, 3)$ is a solution of $2x + 3y - 2z = -7$.

Replace x with -2, y with 1, and z with 3.

$$2(-2) + 3(1) - 2(3) = -7$$

$$-4 + 3 - 6 = -7$$

$$-7 = -7 \qquad \text{True.}$$

$(-2, 1, 3)$ is a solution.

(continued)

DEFINITIONS AND CONCEPTS	EXAMPLES

Section 8.6 Solving Systems of Linear Equations in Three Variables

To solve a system of three linear equations by the elimination method:

Step 1: Write each equation in standard form, $Ax + By + Cz = D$.

Step 2: Choose a pair of equations and use the equations to eliminate a variable.

Step 3: Choose any other pair of equations and eliminate the same variable.

Step 4: Solve the system of two equations in two variables from Steps 2 and 3.

Step 5: Solve for the third variable by substituting the values of the variables from Step 4 into any of the original equations.

Solve

$$\begin{cases} 2x + y - z = 0 \ (1) \\ x - y - 2z = -6 \ (2) \\ -3x - 2y + 3z = -22 \ (3) \end{cases}$$

1. Each equation is written in standard form.

2.
$$\begin{array}{l} 2x + y - z = 0 \ (1) \\ \underline{x - y - 2z = -6 \ (2)} \\ 3x - 3z = -6 \ (4) \qquad \text{Add.} \end{array}$$

3. Eliminate y from equations (1) and (3) also.

$$\begin{array}{ll} 4x + 2y - 2z = 0 & \text{Multiply equation} \\ \underline{-3x - 2y + 3z = -22} \ (3) & \text{(1) by 2.} \\ x + z = -22 \ (5) & \text{Add.} \end{array}$$

4. Solve
$$\begin{cases} 3x - 3z = -6 \ (4) \\ x + z = -22 \ (5) \end{cases}$$

$$\begin{array}{ll} x - z = -2 & \text{Divide equation} \\ \underline{x + z = -22} \ (5) & \text{(4) by 3.} \\ 2x = -24 & \\ x = -12 & \end{array}$$

To find z, use equation (5).
$$x + z = -22$$
$$-12 + z = -22$$
$$z = -10$$

5. To find y, use equation (1).
$$2x + y - z = 0$$
$$2(-12) + y - (-10) = 0$$
$$-24 + y + 10 = 0$$
$$y = 14$$

The solution is $(-12, 14, -10)$.

Section 8.7 Solving Systems of Equations by Matrices

A **matrix** is a rectangular array of numbers.

$$\begin{bmatrix} -7 & 0 & 3 \\ 1 & 2 & 4 \end{bmatrix} \qquad \begin{bmatrix} a & b & c \\ d & e & f \\ g & h & i \end{bmatrix}$$

The **corresponding matrix of the system** is obtained by writing a matrix composed of the coefficients of the variables and the constants of the system.

The corresponding matrix of the system

$$\begin{cases} x - y = 1 \\ 2x + y = 11 \end{cases} \quad \text{is} \quad \begin{bmatrix} 1 & -1 & | & 1 \\ 2 & 1 & | & 11 \end{bmatrix}$$

(continued)

| **DEFINITIONS AND CONCEPTS** | **EXAMPLES** |

Section 8.7   Solving Systems of Equations by Matrices

The following **row operations** can be performed on matrices, and the result is an equivalent matrix.

Elementary row operations

1. Interchange any two rows.
2. Multiply (or divide) the elements of one row by the same nonzero number.
3. Multiply (or divide) the elements of one row by the same nonzero number and add to its corresponding elements in any other row.

Use matrices to solve: $\begin{cases} x - y = 1 \\ 2x + y = 11 \end{cases}$

The corresponding matrix is

$$\left[\begin{array}{cc|c} 1 & -1 & 1 \\ 2 & 1 & 11 \end{array}\right]$$

Use row operations to write an equivalent matrix with 1's along the diagonal and 0's below each 1 in the diagonal. Multiply row 1 by -2 and add to row 2. Change row 2 only.

$$\left[\begin{array}{cc|c} 1 & -1 & 1 \\ -2(1) + 2 & -2(-1) + 1 & -2(1) + 11 \end{array}\right]$$

simplifies to $\left[\begin{array}{cc|c} 1 & -1 & 1 \\ 0 & 3 & 9 \end{array}\right]$

Divide row 2 by 3.

$$\left[\begin{array}{cc|c} 1 & -1 & 1 \\ \dfrac{0}{3} & \dfrac{3}{3} & \dfrac{9}{3} \end{array}\right] \quad \text{simplifies to} \quad \left[\begin{array}{cc|c} 1 & -1 & 1 \\ 0 & 1 & 3 \end{array}\right]$$

This matrix corresponds to the system

$$\begin{cases} x - y = 1 \\ y = 3 \end{cases}$$

Let $y = 3$ in the first equation.

$$x - 3 = 1$$
$$x = 4$$

The ordered pair solution is $(4, 3)$.

Section 8.8 Solving Systems of Equations by Determinants

A **square matrix** is a matrix with the same number of rows and columns.

$$\left[\begin{array}{cc} -2 & 1 \\ 6 & 8 \end{array}\right] \qquad \left[\begin{array}{ccc} 4 & -1 & 6 \\ 0 & 2 & 5 \\ 1 & 1 & 2 \end{array}\right]$$

A **determinant** is a real number associated with a square matrix. To denote the determinant, place vertical bars about the array of numbers.

The determinant of $\left[\begin{array}{cc} -2 & 1 \\ 6 & 8 \end{array}\right]$ is $\left|\begin{array}{cc} -2 & 1 \\ 6 & 8 \end{array}\right|$.

The determinant of a 2×2 matrix is

$$\left|\begin{array}{cc} a & b \\ c & d \end{array}\right| = ad - bc$$

$$\left|\begin{array}{cc} -2 & 1 \\ 6 & 8 \end{array}\right| = -2 \cdot 8 - 1 \cdot 6 = -22$$

(continued)

DEFINITIONS AND CONCEPTS	EXAMPLES

Section 8.8 Solving Systems of Equations by Determinants

Cramer's Rule for Two Linear Equations in Two Variables

The solution of the system $\begin{cases} ax + by = h \\ cx + dy = k \end{cases}$ is given by

$$x = \dfrac{\begin{vmatrix} h & b \\ k & d \end{vmatrix}}{\begin{vmatrix} a & b \\ c & d \end{vmatrix}} = \dfrac{D_x}{D} \qquad y = \dfrac{\begin{vmatrix} a & h \\ c & k \end{vmatrix}}{\begin{vmatrix} a & b \\ c & d \end{vmatrix}} = \dfrac{D_y}{D}$$

as long as $D = ad - bc$ is not 0.

Determinant of a 3 × 3 Matrix

$$\begin{vmatrix} a_1 & b_1 & c_1 \\ a_2 & b_2 & c_2 \\ a_3 & b_3 & c_3 \end{vmatrix} = a_1 \cdot \begin{vmatrix} b_2 & c_2 \\ b_3 & c_3 \end{vmatrix}$$

$$- a_2 \cdot \begin{vmatrix} b_1 & c_1 \\ b_3 & c_3 \end{vmatrix} + a_3 \cdot \begin{vmatrix} b_1 & c_1 \\ b_2 & c_2 \end{vmatrix}$$

Each 2 × 2 matrix above is called a **minor**.

Cramer's Rule for Three Equations in Three Variables

The solution of the system $\begin{cases} a_1x + b_1y + c_1z = k_1 \\ a_2x + b_2y + c_2z = k_2 \\ a_3x + b_3y + c_3z = k_3 \end{cases}$ is given by

$$x = \dfrac{D_x}{D}, \qquad y = \dfrac{D_y}{D}, \qquad \text{and} \qquad z = \dfrac{D_z}{D}$$

where

$$D = \begin{vmatrix} a_1 & b_1 & c_1 \\ a_2 & b_2 & c_2 \\ a_3 & b_3 & c_3 \end{vmatrix} \qquad D_x = \begin{vmatrix} k_1 & b_1 & c_1 \\ k_2 & b_2 & c_2 \\ k_3 & b_3 & c_3 \end{vmatrix}$$

$$D_y = \begin{vmatrix} a_1 & k_1 & c_1 \\ a_2 & k_2 & c_2 \\ a_3 & k_3 & c_3 \end{vmatrix} \qquad D_z = \begin{vmatrix} a_1 & b_1 & k_1 \\ a_2 & b_2 & k_2 \\ a_3 & b_3 & k_3 \end{vmatrix}$$

as long as D is not 0.

Use Cramer's rule to solve

$$\begin{cases} 3x + 2y = 8 \\ 2x - y = -11 \end{cases}$$

$$D = \begin{vmatrix} 3 & 2 \\ 2 & -1 \end{vmatrix} = 3(-1) - 2(2) = -7$$

$$D_x = \begin{vmatrix} 8 & 2 \\ -11 & -1 \end{vmatrix} = 8(-1) - 2(-11) = 14$$

$$D_y = \begin{vmatrix} 3 & 8 \\ 2 & -11 \end{vmatrix} = 3(-11) - 8(2) = -49$$

$$x = \frac{D_x}{D} = \frac{14}{-7} = -2 \qquad y = \frac{D_y}{D} = \frac{-49}{-7} = 7$$

The ordered pair solution is $(-2, 7)$.

$$\begin{vmatrix} 0 & 2 & -1 \\ 5 & 3 & 0 \\ 2 & -2 & 4 \end{vmatrix} = 0\begin{vmatrix} 3 & 0 \\ -2 & 4 \end{vmatrix} - 2\begin{vmatrix} 5 & 0 \\ 2 & 4 \end{vmatrix} + (-1)\begin{vmatrix} 5 & 3 \\ 2 & -2 \end{vmatrix}$$

$$= 0(12 - 0) - 2(20 - 0) - 1(-10 - 6)$$

$$= 0 - 40 + 16 = -24$$

Use Cramer's rule to solve

$$\begin{cases} 3y + 2z = 8 \\ x + y + z = 3 \\ 2x - y + z = 2 \end{cases}$$

$$D = \begin{vmatrix} 0 & 3 & 2 \\ 1 & 1 & 1 \\ 2 & -1 & 1 \end{vmatrix} = -3$$

$$D_x = \begin{vmatrix} 8 & 3 & 2 \\ 3 & 1 & 1 \\ 2 & -1 & 1 \end{vmatrix} = 3$$

$$D_y = \begin{vmatrix} 0 & 8 & 2 \\ 1 & 3 & 1 \\ 2 & 2 & 1 \end{vmatrix} = 0$$

$$D_z = \begin{vmatrix} 0 & 3 & 8 \\ 1 & 1 & 3 \\ 2 & -1 & 2 \end{vmatrix} = -12$$

$$x = \frac{D_x}{D} = \frac{3}{-3} = -1 \qquad y = \frac{D_y}{D} = \frac{0}{-3} = 0$$

$$z = \frac{D_z}{D} = \frac{-12}{-3} = 4$$

The ordered-triple solution is $(-1, 0, 4)$.

CHAPTER 8 REVIEW

(8.1) *Solve each equation.*

1. $4(x - 5) = 2x - 14$

2. $x + 7 = -2(x + 8)$

3. $3(2y - 1) = -8(6 + y)$

4. $-(z + 12) = 5(2z - 1)$

5. $w^2 - 5w = 36$

6. $x^2 + 32 = 12x$

7. $0.3(x - 2) - 1.2$

8. $1.5 = 0.2(c - 0.3)$

9. $-4(2 - 3h) = 2(3h - 4) + 6h$

10. $6(m - 1) + 3(2 - m) = 0$

11. $6 - 3(2g + 4) - 4g = 5(1 - 2g)$

12. $20 - 5(p + 1) + 3p = -(2p - 15)$

13. $x^2 - 2x - 15 = 0$

14. $x^2 + 6x - 7 = 0$

15. $12x^2 + 2x - 2 = 0$

16. $8x^2 + 13x + 5 = 0$

17. $\dfrac{y}{4} - \dfrac{y}{2} = -8$

18. $\dfrac{2x}{3} - \dfrac{8}{3} = x$

19. $\dfrac{b - 2}{3} = \dfrac{b + 2}{5}$

20. $\dfrac{2t - 1}{3} = \dfrac{3t + 2}{15}$

21. $x^2 + (x + 1)^2 = 61$

22. $y^2 + (y + 2)^2 = 34$

23. $\dfrac{x - 2}{5} + \dfrac{x + 2}{2} = \dfrac{x + 4}{3}$

24. $\dfrac{2z - 3}{4} - \dfrac{4 - z}{2} = \dfrac{z + 1}{3}$

(8.2) *Solve.*

25. Twice the difference of a number and 3 is the same as 1 added to three times the number. Find the number.

26. One number is 5 more than another number. If the sum of the numbers is 285, find the numbers.

27. In 1998, the average annual earnings for a worker with an associate's degree was $29,872. This represents a 30.47% increase over the average annual earnings for a high school graduate in 1998. Find the average annual earnings for a high school graduate in 1998. Round to the nearest whole dollar. (*Source:* U.S. Bureau of the Census)

28. Find four consecutive integers such that twice the first subtracted from the sum of the other three integers is sixteen.

29. Determine whether there are two consecutive odd integers such that 5 times the first exceeds 3 times the second by 54.

△ **30.** The length of a rectangular playing field is 5 meters less than twice its width. If 230 meters of fencing goes around the field, find the dimensions of the field.

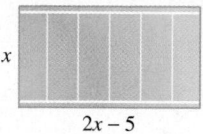

x

$2x - 5$

31. A car rental company charges $29.95 per day for a compact car plus 15 cents per mile for every mile over 100 miles driven per day. If Mr. Woo's bill for 2 days' use is $83.60 before taxes, find how many miles to the nearest whole mile he drove.

32. The cost C of producing x number of scientific calculators is given by $C = 4.50x + 3000$, and the revenue R from selling them is given by $R = 16.50x$. Find the number of calculators that must be sold to break even. (To break even, revenue = cost.)

33. The sum of a number and twice its square is 105. Find the number.

△ **34.** The length of a rectangular piece of carpet is 2 meters less than 5 times its width. Find the dimensions of the carpet if its area is 16 square meters.

35. A scene from an adventure film calls for a stunt dummy to be dropped from above the second-story platform of the Eiffel Tower, a distance of 400 feet. Its height $h(t)$ at time t seconds is given by

$$h(t) = -16t^2 + 400$$

Determine when the stunt dummy will reach the ground.

400 feet

(8.3) *Solve each inequality.*

36. $1 \leq 4x - 7 \leq 3$

37. $-2 \leq 8 + 5x < -1$

38. $-3 < 4(2x - 1) < 12$

39. $-6 < x - (3 - 4x) < -3$

40. $\dfrac{1}{6} < \dfrac{4x - 3}{3} \le \dfrac{4}{5}$

41. $0 \le \dfrac{2(3x + 4)}{5} \le 3$

42. $x \le 2$ and $x > -5$

43. $x \le 2$ or $x > -5$

44. $3x - 5 > 6$ or $-x < -5$

45. $-2x \le 6$ and $-2x + 3 < -7$

(8.4) Solve each absolute value equation.

46. $|x - 7| = 9$

47. $|8 - x| = 3$

48. $|2x + 9| = 9$

49. $|-3x + 4| = 7$

50. $|3x - 2| + 6 = 10$

51. $5 + |6x + 1| = 5$

52. $-5 = |4x - 3|$

53. $|5 - 6x| + 8 = 3$

54. $|7x| - 26 = -5$

55. $-8 = |x - 3| - 10$

56. $\left|\dfrac{3x - 7}{4}\right| = 2$

57. $\left|\dfrac{9 - 2x}{5}\right| = -3$

58. $|6x + 1| = |15 + 4x|$

59. $|x - 3| = |7 + 2x|$

(8.5) Solve each absolute value inequality. Graph the solution set and write in interval notation.

60. $|5x - 1| < 9$

61. $|6 + 4x| \ge 10$

62. $|3x| - 8 > 1$

63. $9 + |5x| < 24$

64. $|6x - 5| \le -1$

65. $|6x - 5| \ge -1$

66. $\left|3x + \dfrac{2}{5}\right| \ge 4$

67. $\left|\dfrac{4x - 3}{5}\right| < 1$

68. $\left|\dfrac{x}{3} + 6\right| - 8 > -5$

69. $\left|\dfrac{4(x - 1)}{7}\right| + 10 < 2$

(8.6) Solve each system of equations in three variables.

70. $\begin{cases} x \quad\ + z = 4 \\ 2x - y \quad\ = 4 \\ x + y - z = 0 \end{cases}$

71. $\begin{cases} 2x + 5y \quad\quad = 4 \\ x - 5y + z = -1 \\ 4x \quad\quad - z = 11 \end{cases}$

72. $\begin{cases} \quad\ 4y + 2z = 5 \\ 2x + 8y \quad\quad = 5 \\ 6x \quad\quad + 4z = 1 \end{cases}$

73. $\begin{cases} 5x + 7y \quad\quad = 9 \\ \quad\ 14y - z = 28 \\ 4x \quad\quad + 2z = -4 \end{cases}$

74. $\begin{cases} 3x - 2y + 2z = 5 \\ -x + 6y + z = 4 \\ 3x + 14y + 7z = 20 \end{cases}$

75. $\begin{cases} x + 2y + 3z = 11 \\ \quad\ y + 2z = 3 \\ 2x \quad\quad + 2z = 10 \end{cases}$

76. $\begin{cases} 7x - 3y + 2z = 0 \\ 4x - 4y - z = 2 \\ 5x + 2y + 3z = 1 \end{cases}$

77. $\begin{cases} x - 3y - 5z = -5 \\ 4x - 2y + 3z = 13 \\ 5x + 3y + 4z = 22 \end{cases}$

Use systems of equations to solve the following applications.

78. The sum of three numbers is 98. The sum of the first and second is two more than the third number, and the second is four times the first. Find the numbers.

79. An employee at a See's Candy Store needs a special mixture of candy. She has creme-filled chocolates that sell for $3.00 per pound, chocolate-covered nuts that sell for $2.70 per pound, and chocolate-covered raisins that sell for $2.25 per pound. She wants to have twice as many raisins as nuts in the mixture. Find how many pounds of each she should use to make 45 pounds worth $2.80 per pound.

80. Chris Kringler has $2.77 in his coin jar—all in pennies, nickels, and dimes. If he has 53 coins in all and four more nickels than dimes, find how many of each type of coin he has.

△ **81.** The perimeter of an isosceles (two sides equal) triangle is 73 centimeters. If two sides are of equal length and the third side is 7 centimeters longer than the others, find the lengths of the three sides.

82. The sum of three numbers is 295. One number is five more than a second and twice the third. Find the numbers.

(8.7) Use matrices to solve each system.

83. $\begin{cases} 3x + 10y = 1 \\ x + 2y = -1 \end{cases}$

84. $\begin{cases} 3x - 6y = 12 \\ 2y = x - 4 \end{cases}$

85. $\begin{cases} 3x - 2y = -8 \\ 6x + 5y = 11 \end{cases}$

86. $\begin{cases} 6x - 6y = -5 \\ 10x - 2y = 1 \end{cases}$

87. $\begin{cases} 3x - 6y = 0 \\ 2x + 4y = 5 \end{cases}$

88. $\begin{cases} 5x - 3y = 10 \\ -2x + y = -1 \end{cases}$

89. $\begin{cases} 0.2x - 0.3y = -0.7 \\ 0.5x + 0.3y = 1.4 \end{cases}$

90. $\begin{cases} 3x + 2y = 8 \\ 3x - y = 5 \end{cases}$

91. $\begin{cases} x + z = 4 \\ 2x - y = 0 \\ x + y - z = 0 \end{cases}$

92. $\begin{cases} 2x + 5y = 4 \\ x - 5y + z = -1 \\ 4x - z = 11 \end{cases}$

93. $\begin{cases} 3x - y = 11 \\ x + 2z = 13 \\ y - z = -7 \end{cases}$

94. $\begin{cases} 5x + 7y + 3z = 9 \\ 14y - z = 28 \\ 4x + 2z = -4 \end{cases}$

95. $\begin{cases} 7x - 3y + 2z = 0 \\ 4x - 4y - z = 2 \\ 5x + 2y + 3z = 1 \end{cases}$

96. $\begin{cases} x + 2y + 3z = 14 \\ y + 2z = 3 \\ 2x - 2z = 10 \end{cases}$

(8.8) *Evaluate.*

97. $\begin{vmatrix} -1 & 3 \\ 5 & 2 \end{vmatrix}$

98. $\begin{vmatrix} 3 & -1 \\ 2 & 5 \end{vmatrix}$

99. $\begin{vmatrix} 2 & -1 & -3 \\ 1 & 2 & 0 \\ 3 & -2 & 2 \end{vmatrix}$

100. $\begin{vmatrix} -2 & 3 & 1 \\ 4 & 4 & 0 \\ 1 & -2 & 3 \end{vmatrix}$

Use Cramer's rule, if possible, to solve each system of equations.

101. $\begin{cases} 3x - 2y = -8 \\ 6x + 5y = 11 \end{cases}$

102. $\begin{cases} 6x - 6y = -5 \\ 10x - 2y = 1 \end{cases}$

103. $\begin{cases} 3x + 10y = 1 \\ x + 2y = -1 \end{cases}$

104. $\begin{cases} y = \dfrac{1}{2}x + \dfrac{2}{3} \\ 4x + 6y = 4 \end{cases}$

105. $\begin{cases} 2x - 4y = 22 \\ 5x - 10y = 16 \end{cases}$

106. $\begin{cases} 3x - 6y = 12 \\ 2y = x - 4 \end{cases}$

107. $\begin{cases} x + z = 4 \\ 2x - y = 0 \\ x + y - z = 0 \end{cases}$

108. $\begin{cases} 2x + 5y = 4 \\ x - 5y + z = -1 \\ 4x - z = 11 \end{cases}$

109. $\begin{cases} x + 3y - z = 5 \\ 2x - y - 2z = 3 \\ x + 2y + 3z = 4 \end{cases}$

110. $\begin{cases} 2x - z = 1 \\ 3x - y + 2z = 3 \\ x + y + 3z = -2 \end{cases}$

111. $\begin{cases} x + 2y + 3z = 14 \\ y + 2z = 3 \\ 2x - 2z = 10 \end{cases}$

112. $\begin{cases} 5x + 7y = 9 \\ 14y - z = 28 \\ 4x + 2z = -4 \end{cases}$

CHAPTER 8 TEST

Solve each equation.

1. $8x + 14 = 5x + 44$

2. $3(x + 2) = 11 - 2(2 - x)$

3. $\dfrac{x^2}{5} - \dfrac{2x}{5} = 3$

4. $\dfrac{x^2}{7} + \dfrac{6x}{7} = 1$

5. $\dfrac{z}{2} + \dfrac{z}{3} = 10$

6. $\dfrac{7w}{4} + 5 = \dfrac{3w}{10} + 1$

7. $|6x - 5| = 1$

8. $|8 - 2t| = -6$

Solve each inequality.

9. $-3 < 2(x - 3) \le 4$

10. $|3x + 1| > 5$

11. $x \ge 5$ and $x \ge 4$

12. $x \ge 5$ or $x \ge 4$

13. $-x > 1$ and $3x + 3 \ge x - 3$

14. $6x + 1 > 5x + 4$ or $1 - x > -4$

Evaluate each determinant.

15. $\begin{vmatrix} 4 & -7 \\ 2 & 5 \end{vmatrix}$

16. $\begin{vmatrix} 4 & 0 & 2 \\ 1 & -3 & 5 \\ 0 & -1 & 2 \end{vmatrix}$

Solve each system.

17. $\begin{cases} 2x - 3y = 4 \\ 3y + 2z = 2 \\ x - z = -5 \end{cases}$

18. $\begin{cases} 3x - 2y - z = -1 \\ 2x - 2y = 4 \\ 2x - 2z = -12 \end{cases}$

19. $\begin{cases} \dfrac{x}{2} + \dfrac{y}{4} = -\dfrac{3}{4} \\ x + \dfrac{3}{4}y = -4 \end{cases}$

Use Cramer's rule, if possible, to solve each system.

20. $\begin{cases} 3x - y = 7 \\ 2x + 5y = -1 \end{cases}$

21. $\begin{cases} 4x - 3y = -6 \\ -2x + y = 0 \end{cases}$

22. $\begin{cases} x + y + z = 4 \\ 2x + 5y \quad\;\; = 1 \\ x - y - 2z = 0 \end{cases}$ **23.** $\begin{cases} 3x + 2y + 3z = 3 \\ x \qquad\;\; - z = 9 \\ 4y + z = -4 \end{cases}$

Use matrices to solve each system.

24. $\begin{cases} x - y = -2 \\ 3x - 3y = -6 \end{cases}$ **25.** $\begin{cases} x + 2y = -1 \\ 2x + 5y = -5 \end{cases}$

26. $\begin{cases} x - y - z = 0 \\ 3x - y - 5z = -2 \\ 2x + 3y \quad\;\; = -5 \end{cases}$ **27.** $\begin{cases} 2x - y + 3z = 4 \\ 3x \qquad - 3z = -2 \\ -5x + y \qquad = 0 \end{cases}$

Solve.

28. A motel in New Orleans charges $90 per day for double occupancy and $80 per day for single occupancy. If 80 rooms are occupied for a total of $6930, how many rooms of each kind are there?

29. A company that manufactures boxes recently purchased $2000 worth of new equipment to offer gift boxes to its customers. The cost of producing a package of gift boxes is $1.50 and it is sold for $4.00. Find the number of packages that must be sold for the company to break even.

30. In 2006, the number of people employed as database administrators, computer support specialists, and all other computer scientists is expected to be 461,000 in the United States. This represents a 118% increase over the number of people employed in these occupations in 1996. Find the number of database administrators, computer support specialists, and all other computer scientists employed in 1996. (*Source:* U.S. Bureau of Labor Statistics)

△ **31.** A circular dog pen has a circumference of 78.5 feet. Approximate π by 3.14 and estimate how many hunting dogs could be safely kept in the pen if each dog needs at least 60 square feet of room.

32. The company that makes Photoray sunglasses figures that the cost C to make x number of sunglasses weekly is given by $C = 3910 + 2.8x$, and the weekly revenue R is given by $R = 7.4x$. Use an inequality to find the number of sunglasses that must be made and sold to make a profit. (Revenue must exceed cost in order to make a profit.)

33. The sum of two numbers is 5, and the sum of their squares is 73. Find the numbers.

CHAPTER 8 CUMULATIVE REVIEW

1. If $f(x) = 7x^2 - 3x + 1$ and $g(x) = 3x - 2$, find the following.
   a. $f(1)$ **b.** $g(1)$
   c. $f(-2)$ **d.** $g(0)$

2. Graph $g(x) = 2x + 1$. Compare this graph with the graph of $f(x) = 2x$.

3. Find the slope and the y-intercept of the line $3x - 4y = 4$.

4. Are the following pairs of lines parallel, perpendicular, or neither?
   a. $3x + 7y = 4$
       $6x + 14y = 7$
   b. $-x + 3y = 2$
       $2x + 6y = 5$

5. Find an equation of the line through points $(4, 0)$ and $(-4, -5)$. Write the equation using function notation.

6. Graph $3x \ge y$.

7. Solve: $2(x - 3) = 5x - 9$

8. Solve: $2x^2 = \dfrac{17}{3}x + 1$

9. Find the intersection: $\{2, 4, 6, 8\} \cap \{3, 4, 5, 6\}$

10. Solve: $x - 7 < 2$ and $2x + 1 < 9$

11. Solve: $2 < 4 - x < 7$

12. Find the union: $\{2, 4, 6, 8\} \cup \{3, 4, 5, 6\}$

13. Solve: $-2x - 5 < -3$ or $6x < 0$

Solve:

14. $|p| = 2$

15. $\left| \dfrac{x}{2} - 1 \right| = 11$

16. Solve: $|2x| + 5 = 7$

17. $|x - 3| = |5 - x|$

18. $|x| \leq 3$

19. Solve for m: $|m - 6| < 2$

20. $|2x + 9| + 5 > 3$

21. A pennant in the shape of an isosceles triangle is to be constructed for the Slidell High School Athletic Club and sold at a fund-raiser. The company manufacturing the pennant charges according to perimeter, and the athletic club has determined that a perimeter of 149 centimeters should make a nice profit. If each equal side of the triangle is twice the length of the third side, increased by 12 centimeters, find the lengths of the sides of the triangular pennant.

22. Solve the system.
$$\begin{cases} 3x - y + z = -15 \\ x + 2y - z = 1 \\ 2x + 3y - 2z = 0 \end{cases}$$

23. Use matrices to solve the system.
$$\begin{cases} x + 3y = 5 \\ 2x - y = -4 \end{cases}$$

24. Evaluate each determinant.

a. $\begin{vmatrix} -1 & 2 \\ 3 & -4 \end{vmatrix}$

b. $\begin{vmatrix} 2 & 0 \\ 7 & -5 \end{vmatrix}$

Studying Human Behavior

Psychology is the scientific study of the human mind and behavior. Over 150,000 psychologists practice in the United States in such diverse areas as experimental psychology, clinical psychology, industrial psychology, educational psychology, counseling psychology, psychotherapy, military psychology, consumer psychology, family psychology, and sports psychology.

Although psychology careers in teaching, research, and counseling frequently require advanced degrees, there are many career paths in which a two- or four-year degree is useful. Employment counselors, child protection workers, corrections officers, social service directors, day-care-center supervisors, and hospital patient service representatives are all examples of positions for which advanced psychology degrees are not necessarily required. No matter which educational path is chosen, psychologists must have good communication, interpersonal, research, and analytical skills, including the ability to reason numerically, interpret statistics, read tables and graphs, and solve problems.

 For more information about psychology careers, visit the American Psychological Association Website by first going to <u>www.prenhall.com/martin-gay</u>.

In the Spotlight on Decision Making feature on page 585, you will have the opportunity to make a decision about assigning subjects to the appropriate test groups for a psychology experiment.

RADICALS, RATIONAL EXPONENTS, AND COMPLEX NUMBERS

In this chapter, radical notation is reviewed, and then rational exponents are introduced. As the name implies, rational exponents are exponents that are rational numbers. We present an interpretation of rational exponents that is consistent with the meaning and rules already established for integer exponents, and we present two forms of notation for roots: radical and exponent. We conclude this chapter with complex numbers, a natural extension of the real number system.

9.1 RADICALS AND RADICAL FUNCTIONS

CD-ROM   SSM

SSG   Video

▶ **O B J E C T I V E S**

1. Find square roots.
2. Approximate roots using a calculator.
3. Find cube roots.
4. Find nth roots.
5. Find $\sqrt[n]{a^n}$ where a is a real number.
6. Graph square and cube root functions.

1

To find a **square root** of a number a, we find a number that was squared to get a. Thus, because

$$5^2 = 25 \qquad \text{and} \qquad (-5)^2 = 25, \text{then}$$

both 5 and −5 are square roots of 25.

Recall that we denote the **nonnegative**, or **principal**, **square root** with the **radical sign**.

$$\sqrt{25} = 5$$

We denote the **negative square root** with the **negative radical sign**.

$$-\sqrt{25} = -5$$

An expression containing a radical sign is called a **radical expression**. An expression within, or "under," a radical sign is called a **radicand**.

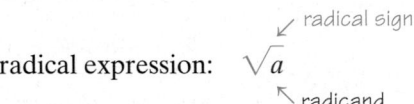

radical expression: $\sqrt{a}$

PRINCIPAL AND NEGATIVE SQUARE ROOTS

The **principal square root** of a nonnegative number a is its nonnegative square root. The principal square root is written as $\sqrt{a}$. The **negative square root** of a is written as $-\sqrt{a}$.

◆ **Example 1**   Simplify. Assume that all variables represent positive numbers.

    a. $\sqrt{36}$ **b.** $\sqrt{0}$ **c.** $\sqrt{\dfrac{4}{49}}$ **d.** $\sqrt{0.25}$ **e.** $\sqrt{x^6}$ **f.** $\sqrt{9x^{10}}$ **g.** $-\sqrt{81}$

Solution   **a.** $\sqrt{36} = 6$ because $6^2 = 36$ and 6 is not negative.

    b. $\sqrt{0} = 0$ because $0^2 = 0$ and 0 is not negative.

    c. $\sqrt{\dfrac{4}{49}} = \dfrac{2}{7}$ because $\left(\dfrac{2}{7}\right)^2 = \dfrac{4}{49}$ and $\dfrac{2}{7}$ is not negative.

    d. $\sqrt{0.25} = 0.5$ because $(0.5)^2 = 0.25$.

    e. $\sqrt{x^6} = x^3$ because $\left(x^3\right)^2 = x^6$.

f. $\sqrt{9x^{10}} = 3x^5$ because $(3x^5)^2 = 9x^{10}$.

g. $-\sqrt{81} = -9$. The negative in front of the radical indicates the negative square root of 81.

Can we find the square root of a negative number, say $\sqrt{-4}$? That is, can we find a real number whose square is -4? No, there is no real number whose square is -4, and we say that $\sqrt{-4}$ is not a real number. In general:

The square root of a negative number is not a real number.

> **HELPFUL HINT**
> Don't forget, the square root of a negative number, such as $\sqrt{-9}$, is not a real number. In Section 9.7, we will see what kind of a number $\sqrt{-9}$ is.

2 Recall that numbers such as 1, 4, 9, and 25 are called **perfect squares**, since $1 = 1^2$, $4 = 2^2$, $9 = 3^2$, and $25 = 5^2$. Square roots of perfect square radicands simplify to rational numbers. What happens when we try to simplify a root such as $\sqrt{3}$? Since 3 is not a perfect square, $\sqrt{3}$ is not a rational number. It is called an **irrational number**, and we can find a decimal **approximation** of it. To find decimal approximations, use a calculator. For example, an approximation for $\sqrt{3}$ is

$$\sqrt{3} \approx 1.732$$
$$\uparrow$$
approximation symbol

To see if the approximation is reasonable, notice that since

$$1 < 3 < 4, \text{ then}$$
$$\sqrt{1} < \sqrt{3} < \sqrt{4}, \text{ or}$$
$$1 < \sqrt{3} < 2.$$

We found $\sqrt{3} \approx 1.732$, a number between 1 and 2, so our result is reasonable.

Example 2 Use a calculator to approximate $\sqrt{20}$. Round the approximation to 3 decimal places and check to see that your approximation is reasonable.

Solution $$\sqrt{20} \approx 4.472$$

Is this reasonable? Since $16 < 20 < 25$, then $\sqrt{16} < \sqrt{20} < \sqrt{25}$, or $4 < \sqrt{20} < 5$. The approximation is between 4 and 5 and thus is reasonable.

3 Finding roots can be extended to other roots such as cube roots. For example, since $2^3 = 8$, we call 2 the **cube root** of 8. In symbols, we write

$$\sqrt[3]{8} = 2$$

CUBE ROOT

The **cube root** of a real number a is written as $\sqrt[3]{a}$, and

$$\sqrt[3]{a} = b \text{ only if } b^3 = a$$

From this definition, we have

$$\sqrt[3]{64} = 4 \text{ since } 4^3 = 64$$
$$\sqrt[3]{-27} = -3 \text{ since } (-3)^3 = -27$$
$$\sqrt[3]{x^3} = x \text{ since } x^3 = x^3$$

Notice that, unlike with square roots, *it is possible to have a negative radicand when finding a cube root.* This is so because the *cube* of a negative number is a negative number. Therefore, the *cube root* of a negative number is a negative number.

Example 3   Find the cube roots.

a. $\sqrt[3]{1}$ **b.** $\sqrt[3]{-64}$ **c.** $\sqrt[3]{\dfrac{8}{125}}$ **d.** $\sqrt[3]{x^6}$ **e.** $\sqrt[3]{-8x^9}$

Solution   **a.** $\sqrt[3]{1} = 1$ because $1^3 = 1$.

b. $\sqrt[3]{-64} = -4$ because $(-4)^3 = -64$.

c. $\sqrt[3]{\dfrac{8}{125}} = \dfrac{2}{5}$ because $\left(\dfrac{2}{5}\right)^3 = \dfrac{8}{125}$.

d. $\sqrt[3]{x^6} = x^2$ because $(x^2)^3 = x^6$.

e. $\sqrt[3]{-8x^9} = -2x^3$ because $(-2x^3)^3 = -8x^9$.

4   Just as we can raise a real number to powers other than 2 or 3, we can find roots other than square roots and cube roots. In fact, we can find the **nth root** of a number, where n is any natural number. In symbols, the *n*th root of a is written as $\sqrt[n]{a}$, where n is called the **index**. The index 2 is usually omitted for square roots.

HELPFUL HINT
If the index is even, such as $\sqrt{\ }$, $\sqrt[4]{\ }$, $\sqrt[6]{\ }$, and so on, the radicand must be nonnegative for the root to be a real number. For example,

$$\sqrt[4]{16} = 2, \text{ but } \sqrt[4]{-16} \text{ is not a real number.}$$
$$\sqrt[6]{64} = 2, \text{ but } \sqrt[6]{-64} \text{ is not a real number.}$$

If the index is odd, such as $\sqrt[3]{\ }$, $\sqrt[5]{\ }$, and so on, the radicand may be any real number. For example,

$$\sqrt[3]{64} = 4 \text{ and } \sqrt[3]{-64} = -4$$
$$\sqrt[5]{32} = 2 \text{ and } \sqrt[5]{-32} = -2$$

Example 4   Simplify the following expressions.

a. $\sqrt[4]{81}$ **b.** $\sqrt[5]{-243}$ **c.** $-\sqrt{25}$ **d.** $\sqrt[4]{-81}$ **e.** $\sqrt[3]{64x^3}$

Solution   **a.** $\sqrt[4]{81} = 3$ because $3^4 = 81$ and 3 is positive.

b. $\sqrt[5]{-243} = -3$ because $(-3)^5 = -243$.

c. $-\sqrt{25} = -5$ because -5 is the opposite of $\sqrt{25}$.

d. $\sqrt[4]{-81}$ is not a real number. There is no real number that, when raised to the fourth power, is -81.

e. $\sqrt[3]{64x^3} = 4x$ because $(4x)^3 = 64x^3$. ∎

---

5   Recall that the notation $\sqrt{a^2}$ indicates the positive square root of a^2 only. For example,

$$\sqrt{(-5)^2} = \sqrt{25} = 5$$

When variables are present in the radicand and it is unclear whether the variable represents a positive number or a negative number, absolute value bars are sometimes needed to ensure that the result is a positive number. For example,

$$\sqrt{x^2} = |x|$$

This ensures that the result is positive. This same situation may occur when the index is any *even* positive integer. When the index is any *odd* positive integer, absolute value bars are not necessary.

FINDING $\sqrt[n]{a^n}$

If n is an *even* positive integer, then $\sqrt[n]{a^n} = |a|$.

If n is an *odd* positive integer, then $\sqrt[n]{a^n} = a$.

Example 5   Simplify.

a. $\sqrt{(-3)^2}$ **b.** $\sqrt{x^2}$ **c.** $\sqrt[4]{(x-2)^4}$ **d.** $\sqrt[3]{(-5)^3}$ **e.** $\sqrt[5]{(2x-7)^5}$

Solution   **a.** $\sqrt{(-3)^2} = |-3| = 3$ When the index is even, the absolute value bars ensure us that our result is not negative.

b. $\sqrt{x^2} = |x|$

c. $\sqrt[4]{(x-2)^4} = |x-2|$

d. $\sqrt[3]{(-5)^3} = -5$ Absolute value bars are not needed when the index is odd.

e. $\sqrt[5]{(2x-7)^5} = 2x - 7$ ∎

---

6   Recall that an equation in x and y describes a function if each x-value is paired with exactly one y-value. With this in mind, does the equation

$$y = \sqrt{x}$$

describe a function? First, notice that replacement values for x must be nonnegative real numbers, since $\sqrt{x}$ is not a real number if $x < 0$. The notation $\sqrt{x}$ denotes the principal square root of x, so for every nonnegative number x, there is exactly one number, $\sqrt{x}$. Therefore, $y = \sqrt{x}$ describes a function, and we may write it as

$$f(x) = \sqrt{x}$$

Recall that the domain of a function in x is the set of all possible replacement values for x. This means that the domain of this function is the set of all nonnegative numbers, or $\{x \mid x \geq 0\}$.

We find function values for $f(x)$ as usual. For example,

$$f(0) = \sqrt{0} = 0$$
$$f(1) = \sqrt{1} = 1$$
$$f(4) = \sqrt{4} = 2$$
$$f(9) = \sqrt{9} = 3$$

Choosing perfect squares for x ensures us that $f(x)$ is a rational number, but it is important to stress that $f(x) = \sqrt{x}$ is defined for all nonnegative real numbers. For example,

$$f(3) = \sqrt{3} \approx 1.732$$

Example 6   Graph the square root function $f(x) = \sqrt{x}$.

Solution   To graph, we identify the domain, evaluate the function for several values of x, plot the resulting points, and connect the points with a smooth curve. The domain of this function is the set of all nonnegative numbers or $\{x \mid x \geq 0\}$. The table comes from the function values obtained earlier.

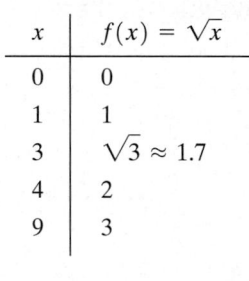

x	$f(x) = \sqrt{x}$
0	0
1	1
3	$\sqrt{3} \approx 1.7$
4	2
9	3

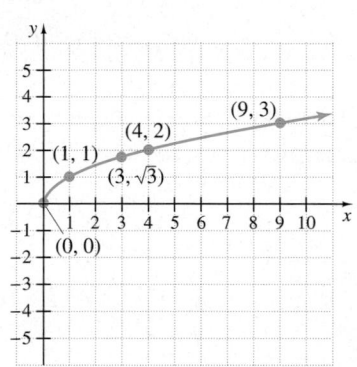

Notice that the graph of this function passes the vertical line test, as expected.

The equation $f(x) = \sqrt[3]{x}$ also describes a function. Here x may be any real number, so the domain of this function is the set of all real numbers. A few function values are given next.

$$f(0) = \sqrt[3]{0} = 0$$
$$f(1) = \sqrt[3]{1} = 1$$
$$f(-1) = \sqrt[3]{-1} = -1$$
$$f(6) = \sqrt[3]{6}$$
$$f(-6) = \sqrt[3]{-6}$$
$$f(8) = \sqrt[3]{8} = 2$$
$$f(-8) = \sqrt[3]{-8} = -2$$

Here, the radicands are not perfect cubes. The radicals do not simplify to rational numbers.

Example 7   Graph the function $f(x) = \sqrt[3]{x}$.

Solution   To graph, we identify the domain, plot points, and connect the points with a smooth curve. The domain of this function is the set of all real numbers. The table comes from the function values obtained earlier. We have approximated $\sqrt[3]{6}$ and $\sqrt[3]{-6}$ for graphing purposes.

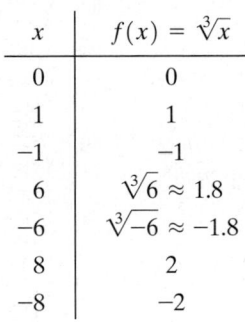

x	$f(x) = \sqrt[3]{x}$
0	0
1	1
−1	−1
6	$\sqrt[3]{6} \approx 1.8$
−6	$\sqrt[3]{-6} \approx -1.8$
8	2
−8	−2

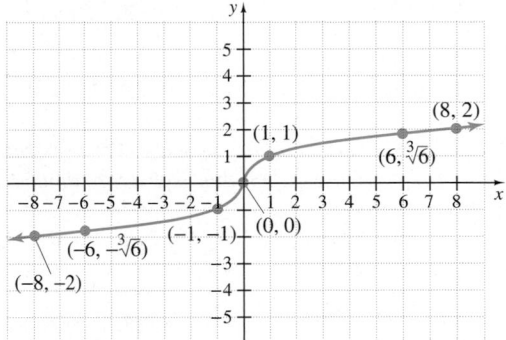

The graph of this function passes the vertical line test, as expected.

SPOTLIGHT ON DECISION MAKING

Suppose you are a scientist working for NASA. A new moon, S/2001U1, has been discovered orbiting the planet Uranus in our outer solar system. You have been asked to check whether it is possible for this moon to have an oxygen atmosphere. You can do so by comparing the average speed of an oxygen molecule (480 meters per second) to the moon's **escape velocity**, the speed an object must travel to permanently leave the moon's gravitational pull. If the moon's escape velocity is greater than the average speed of oxygen molecules, then it is possible for the moon to retain oxygen in its atmosphere—that is, if oxygen exists on the moon at all.

Data about the new moon are listed in the table. Use that along with the escape velocity formula given below to decide whether it is possible for S/2001U1 to have an oxygen atmosphere.

$$v = \sqrt{\frac{2GM}{r}}, \text{ where}$$

v is the escape velocity (in meters per second, m/s),

M is the mass of the moon (in kilograms, kg),

r is the radius of the moon (in meters, m), and

G is the universal constant of gravitation where

$$\left(G = 6.67 \times 10^{-11} \frac{m^3}{kg \cdot s^2}\right).$$

S/2001U1 Parameters	
Mass	9.07×10^{20} kg
Radius	620,000 m
Visual geometric albedo	0.07
Orbital period	7.2 days

Exercise Set 9.1

Simplify. Assume that variables represent positive real numbers. See Example 1.

1. $\sqrt{100}$

2. $\sqrt{400}$

3. $\sqrt{\dfrac{1}{4}}$

4. $\sqrt{\dfrac{9}{25}}$

5. $\sqrt{0.0001}$

6. $\sqrt{0.04}$

7. $-\sqrt{36}$

8. $-\sqrt{9}$

9. $\sqrt{x^{10}}$

10. $\sqrt{x^{16}}$

11. $\sqrt{16y^6}$

12. $\sqrt{64y^{20}}$

Use a calculator to approximate each square root to 3 decimal places. Check to see that each approximation is reasonable. See Example 2.

13. $\sqrt{7}$

14. $\sqrt{11}$

15. $\sqrt{38}$

16. $\sqrt{56}$

17. $\sqrt{200}$

18. $\sqrt{300}$

Find each cube root. See Example 3.

19. $\sqrt[3]{64}$

20. $\sqrt[3]{27}$

21. $\sqrt[3]{\dfrac{1}{8}}$

22. $\sqrt[3]{\dfrac{27}{64}}$

23. $\sqrt[3]{-1}$

24. $\sqrt[3]{-125}$

25. $\sqrt[3]{x^{12}}$

26. $\sqrt[3]{x^{15}}$

27. $\sqrt[3]{-27x^9}$

28. $\sqrt[3]{-64x^6}$

Find each root. Assume that all variables represent nonnegative real numbers. See Example 4.

29. $-\sqrt[4]{16}$

30. $\sqrt[5]{-243}$

31. $\sqrt[4]{-16}$

32. $\sqrt{-16}$

33. $\sqrt[5]{-32}$

34. $\sqrt[5]{-1}$

35. $\sqrt[5]{x^{20}}$

36. $\sqrt[4]{x^{20}}$

37. $\sqrt[6]{64x^{12}}$

38. $\sqrt[5]{-32x^{15}}$

39. $\sqrt{81x^4}$

40. $\sqrt[4]{81x^4}$

41. $\sqrt[4]{256x^8}$

42. $\sqrt{256x^8}$

Simplify. Assume that the variables represent any real number. See Example 5.

43. $\sqrt{(-8)^2}$

44. $\sqrt{(-7)^2}$

45. $\sqrt[3]{(-8)^3}$

46. $\sqrt[5]{(-7)^5}$

47. $\sqrt{4x^2}$

48. $\sqrt[4]{16x^4}$

49. $\sqrt[3]{x^3}$

50. $\sqrt[5]{x^5}$

51. $\sqrt{(x-5)^2}$

52. $\sqrt{(y-6)^2}$

53. $\sqrt{x^2 + 4x + 4}$
(*Hint:* Factor the polynomial first.)

54. $\sqrt{x^2 - 8x + 16}$
(*Hint:* Factor the polynomial first.)

Simplify each radical. Assume that all variables represent positive real numbers.

55. $-\sqrt{121}$

56. $-\sqrt[3]{125}$

57. $\sqrt[3]{8x^3}$

58. $\sqrt{16x^8}$

59. $\sqrt{y^{12}}$

60. $\sqrt[3]{y^{12}}$

61. $\sqrt{25a^2b^{20}}$

62. $\sqrt{9x^4y^6}$

63. $\sqrt[3]{-27x^{12}y^9}$

64. $\sqrt[3]{-8a^{21}b^6}$

65. $\sqrt[4]{a^{16}b^4}$

66. $\sqrt[4]{x^8y^{12}}$

67. $\sqrt[5]{-32x^{10}y^5}$

68. $\sqrt[5]{-243z^{15}}$

69. $\sqrt{\dfrac{25}{49}}$

70. $\sqrt{\dfrac{4}{81}}$

71. $\sqrt{\dfrac{x^2}{4y^2}}$

72. $\sqrt{\dfrac{y^{10}}{9x^6}}$

73. $-\sqrt[3]{\dfrac{z^{21}}{27x^3}}$

74. $-\sqrt[3]{\dfrac{64a^3}{b^9}}$

75. $\sqrt[4]{\dfrac{x^4}{16}}$

76. $\sqrt[4]{\dfrac{y^4}{81x^4}}$

If $f(x) = \sqrt{2x + 3}$ and $g(x) = \sqrt[3]{x - 8}$, find the following function values. See Examples 6 and 7.

77. $f(0)$

78. $g(0)$

79. $g(7)$

80. $f(-1)$

81. $g(-19)$

82. $f(3)$

83. $f(2)$

84. $g(1)$

Identify the domain and then graph each function. See Example 6.

85. $f(x) = \sqrt{x} + 2$

86. $f(x) = \sqrt{x} - 2$

87. $f(x) = \sqrt{x - 3}$; use the following table.

x	$f(x)$
3	
4	
7	
12	

88. $f(x) = \sqrt{x + 1}$; use the following table.

x	$f(x)$
-1	
0	
3	
8	

Identify the domain and then graph each function. See Example 7.

89. $f(x) = \sqrt[3]{x} + 1$

90. $f(x) = \sqrt[3]{x} - 2$

91. $g(x) = \sqrt[3]{x - 1}$; use the following table.

x	$g(x)$
1	
2	
0	
9	
−7	

92. $g(x) = \sqrt[3]{x + 1}$; use the following table.

x	$g(x)$
−1	
0	
−2	
7	
−9	

93. Suppose that a friend tells you that $\sqrt{13} \approx 5.7$. Without a calculator, how can you convince your friend that he must have made an error?

 94. Escape velocity is the minimum speed that an object must reach to escape a planet's pull of gravity. Escape velocity v is given by the equation $v = \sqrt{\dfrac{2GM}{r}}$, where M is the mass of the planet, r is its radius, and G is the universal gravitational constant, which has a value of $G = 6.67 \times 10^{-11} \ \text{m}^3/\text{kg} \cdot \text{s}^2$. The mass of Earth is 5.97×10^{24} kg and its radius is 6.37×10^6 m. Use this information to find the escape velocity for Earth. Round to the nearest whole number. (*Source:* National Space Science Data Center)

Use a graphing calculator to verify the domain of each function and its graph.

95. Exercise 85 **96.** Exercise 86

97. Exercise 89 **98.** Exercise 90

REVIEW EXERCISES

Simplify each exponential expression. See Sections 5.1 and 5.2.

99. $\left(-2x^3y^2\right)^5$ **100.** $\left(4y^6z^7\right)^3$

101. $\left(-3x^2y^3z^5\right)\left(20x^5y^7\right)$ **102.** $\left(-14a^5bc^2\right)\left(2abc^4\right)$

103. $\dfrac{7x^{-1}y}{14\left(x^5y^2\right)^{-2}}$ **104.** $\dfrac{\left(2a^{-1}b^2\right)^3}{\left(8a^2b\right)^{-2}}$

9.2 RATIONAL EXPONENTS

CD-ROM SSM

SSG Video

▶ **OBJECTIVES**

1. Understand the meaning of $a^{1/n}$.
2. Understand the meaning of $a^{m/n}$.
3. Understand the meaning of $a^{-m/n}$.
4. Use rules for exponents to simplify expressions that contain rational exponents.
5. Use rational exponents to simplify radical expressions.

1

So far in this text, we have not defined expressions with rational exponents such as $3^{1/2}$, $x^{2/3}$, and $-9^{-1/4}$. We will define these expressions so that the rules for exponents will apply to these rational exponents as well.

Suppose that $x = 5^{1/3}$. Then

$$x^3 = \left(5^{1/3}\right)^3 = 5^{1/3 \cdot 3} = 5^1 \text{ or } 5$$
$$\llcorner \text{ using rules } \uparrow$$
$$\text{for exponents}$$

Since $x^3 = 5$, then x is the number whose cube is 5, or $x = \sqrt[3]{5}$. Notice that we also know that $x = 5^{1/3}$. This means

$$5^{1/3} = \sqrt[3]{5}$$

DEFINITION OF $a^{1/n}$

If n is a positive integer greater than 1 and $\sqrt[n]{a}$ is a real number, then

$$a^{1/n} = \sqrt[n]{a}$$

Notice that the denominator of the rational exponent corresponds to the index of the radical.

Example 1 Use radical notation to write the following. Simplify if possible.

a. $4^{1/2}$ **b.** $64^{1/3}$ **c.** $x^{1/4}$ **d.** $0^{1/6}$ **e.** $-9^{1/2}$ **f.** $(81x^8)^{1/4}$ **g.** $(5y)^{1/3}$

Solution   **a.** $4^{1/2} = \sqrt{4} = 2$ **b.** $64^{1/3} = \sqrt[3]{64} = 4$
c. $x^{1/4} = \sqrt[4]{x}$ **d.** $0^{1/6} = \sqrt[6]{0} = 0$
e. $-9^{1/2} = -\sqrt{9} = -3$ **f.** $(81x^8)^{1/4} = \sqrt[4]{81x^8} = 3x^2$
g. $(5y)^{1/3} = \sqrt[3]{5y}$

2 As we expand our use of exponents to include $\frac{m}{n}$, we define their meaning so that rules for exponents still hold true. For example, by properties of exponents,

$$8^{2/3} = (8^{1/3})^2 = (\sqrt[3]{8})^2 \quad \text{or}$$
$$8^{2/3} = (8^2)^{1/3} = \sqrt[3]{8^2}$$

DEFINITION OF $a^{m/n}$

If m and n are positive integers greater than 1 with $\frac{m}{n}$ in lowest terms, then

$$a^{m/n} = \sqrt[n]{a^m} = (\sqrt[n]{a})^m$$

as long as $\sqrt[n]{a}$ is a real number.

Notice that the denominator n of the rational exponent corresponds to the index of the radical. The numerator m of the rational exponent indicates that the base is to be raised to the mth power. This means

$$8^{2/3} = \sqrt[3]{8^2} = \sqrt[3]{64} = 4 \quad \text{or}$$
$$8^{2/3} = (\sqrt[3]{8})^2 = 2^2 = 4$$

HELPFUL HINT
Most of the time, $(\sqrt[n]{a})^m$ will be easier to calculate than $\sqrt[n]{a^m}$.

Example 2 Use radical notation to write the following. Then simplify if possible.

a. $4^{3/2}$ **b.** $-16^{3/4}$ **c.** $(-27)^{2/3}$ **d.** $\left(\dfrac{1}{9}\right)^{3/2}$ **e.** $(4x-1)^{3/5}$

Solution **a.** $4^{3/2} = \left(\sqrt{4}\right)^3 = 2^3 = 8$

b. $-16^{3/4} = -\left(\sqrt[4]{16}\right)^3 = -(2)^3 = -8$

c. $(-27)^{2/3} = \left(\sqrt[3]{-27}\right)^2 = (-3)^2 = 9$

d. $\left(\dfrac{1}{9}\right)^{3/2} = \left(\sqrt{\dfrac{1}{9}}\right)^3 = \left(\dfrac{1}{3}\right)^3 = \dfrac{1}{27}$

e. $(4x-1)^{3/5} = \sqrt[5]{(4x-1)^3}$

> **HELPFUL HINT**
> The *denominator* of a rational exponent is the index of the corresponding radical.
> For example, $x^{1/5} = \sqrt[5]{x}$ and $z^{2/3} = \sqrt[3]{z^2}$, or $z^{2/3} = \left(\sqrt[3]{z}\right)^2$.

3 The rational exponents we have given meaning to exclude negative rational numbers. To complete the set of definitions, we define $a^{-m/n}$.

DEFINITION OF $a^{-m/n}$

$$a^{-m/n} = \dfrac{1}{a^{m/n}}$$

as long as $a^{m/n}$ is a nonzero real number.

Example 3 Write each expression with a positive exponent, and then simplify.

a. $16^{-3/4}$ **b.** $(-27)^{-2/3}$

Solution **a.** $16^{-3/4} = \dfrac{1}{16^{3/4}} = \dfrac{1}{\left(\sqrt[4]{16}\right)^3} = \dfrac{1}{2^3} = \dfrac{1}{8}$

b. $(-27)^{-2/3} = \dfrac{1}{(-27)^{2/3}} = \dfrac{1}{\left(\sqrt[3]{-27}\right)^2} = \dfrac{1}{(-3)^2} = \dfrac{1}{9}$

HELPFUL HINT

If an expression contains a negative rational exponent, such as $9^{-3/2}$, you may want to first write the expression with a positive exponent and then interpret the rational exponent. Notice that the sign of the base is not affected by the sign of its exponent. For example,

$$9^{-3/2} = \frac{1}{9^{3/2}} = \frac{1}{(\sqrt{9})^3} = \frac{1}{27}$$

Also,

$$(-27)^{-1/3} = \frac{1}{(-27)^{1/3}} = -\frac{1}{3}$$

---

4 It can be shown that the properties of integer exponents hold for rational exponents. By using these properties and definitions, we can now simplify expressions that contain rational exponents.

These rules are repeated here for review.

SUMMARY OF EXPONENT RULES

If m and n are rational numbers, and $a, b,$ and c are numbers for which the expressions below exist, then

Product rule for exponents: $\qquad\qquad a^m \cdot a^n = a^{m+n}$

Power rule for exponents: $\qquad\qquad (a^m)^n = a^{m \cdot n}$

Power rules for products and quotients: $\qquad (ab)^n = a^n b^n$ and

$$\left(\frac{a}{c}\right)^n = \frac{a^n}{c^n}, c \neq 0$$

Quotient rule for exponents: $\qquad\qquad \dfrac{a^m}{a^n} = a^{m-n}, a \neq 0$

Zero exponent: $\qquad\qquad\qquad\qquad a^0 = 1, a \neq 0$

Negative exponent: $\qquad\qquad\qquad a^{-n} = \dfrac{1}{a^n}, a \neq 0$

Example 4 Use properties of exponents to simplify. Write results with only positive exponents.

a. $x^{1/2}x^{1/3}$

b. $\dfrac{7^{1/3}}{7^{4/3}}$

c. $\dfrac{(2x^{2/5}y^{-1/3})^5}{x^2 y}$

Solution **a.** $x^{1/2}x^{1/3} = x^{1/2+1/3} = x^{3/6+2/6} = x^{5/6}$

b. $\dfrac{7^{1/3}}{7^{4/3}} = 7^{1/3-4/3} = 7^{-3/3} = 7^{-1} = \dfrac{1}{7}$

c. We begin by using the power rule $(ab)^m = a^m b^m$ to simplify the numerator.

$$\frac{(2x^{2/5}y^{-1/3})^5}{x^2 y} = \frac{2^5 (x^{2/5})^5 (y^{-1/3})^5}{x^2 y} = \frac{32 x^2 y^{-5/3}}{x^2 y}$$

$$= 32 x^{2-2} y^{-5/3 - 3/3} \qquad \text{Apply the quotient rule.}$$

$$= 32 x^0 y^{-8/3}$$

$$= \frac{32}{y^{8/3}}$$

Example 5 Multiply.

     a. $z^{2/3}(z^{1/3} - z^5)$ **b.** $(x^{1/3} - 5)(x^{1/3} + 2)$

Solution **a.** $z^{2/3}(z^{1/3} - z^5) = z^{2/3} z^{1/3} - z^{2/3} z^5$ Apply the distributive property.

$$= z^{2/3 + 1/3} - z^{2/3 + 5} \qquad \text{Use the product rule.}$$

$$= z^{3/3} - z^{2/3 + 15/3}$$

$$= z - z^{17/3}$$

b. $(x^{1/3} - 5)(x^{1/3} + 2) = x^{2/3} + 2x^{1/3} - 5x^{1/3} - 10$ Think of $(x^{1/3} - 5)$ and $(x^{1/3} + 2)$ as 2 binomials, and use FOIL.

$$= x^{2/3} - 3x^{1/3} - 10$$

Example 6 Factor $x^{-1/2}$ from the expression $3x^{-1/2} - 7x^{5/2}$. Assume that all variables represent positive numbers.

Solution $3x^{-1/2} - 7x^{5/2} = (x^{-1/2})(3) - (x^{-1/2})(7x^{6/2})$

$$= x^{-1/2}(3 - 7x^3)$$

To check, multiply $x^{-1/2}(3 - 7x^3)$ to see that the product is $3x^{-1/2} - 7x^{5/2}$.

5 Some radical expressions are easier to simplify when we first write them with rational exponents. We can simplify some radical expressions by first writing the expression with rational exponents. Use properties of exponents to simplify, and then convert back to radical notation.

Example 7 Use rational exponents to simplify. Assume that variables represent positive numbers.

     a. $\sqrt[6]{25}$ **b.** $\sqrt[8]{x^4}$ **c.** $\sqrt[4]{r^2 s^6}$

Solution **a.** $\sqrt[6]{25} = 25^{1/6} = (5^2)^{1/6} = 5^{2/6} = 5^{1/3} = \sqrt[3]{5}$

b. $\sqrt[8]{x^4} = x^{4/8} = x^{1/2} = \sqrt{x}$

c. $\sqrt[4]{r^2 s^6} = (r^2 s^6)^{1/4} = r^{2/4} s^{6/4} = r^{1/2} s^{3/2} = (rs^3)^{1/2} = \sqrt{rs^3}$

Example 8   Use rational exponents to write as a single radical.

a. $\sqrt{x} \cdot \sqrt[4]{x}$

b. $\dfrac{\sqrt{x}}{\sqrt[3]{x}}$

c. $\sqrt[3]{3} \cdot \sqrt{2}$

Solution   **a.** $\sqrt{x} \cdot \sqrt[4]{x} = x^{1/2} \cdot x^{1/4} = x^{1/2+1/4}$

$$= x^{3/4} = \sqrt[4]{x^3}$$

b. $\dfrac{\sqrt{x}}{\sqrt[3]{x}} = \dfrac{x^{1/2}}{x^{1/3}} = x^{1/2-1/3} = x^{3/6-2/6}$

$$= x^{1/6} = \sqrt[6]{x}$$

c. $\sqrt[3]{3} \cdot \sqrt{2} = 3^{1/3} \cdot 2^{1/2}$ Write with rational exponents.

$$= 3^{2/6} \cdot 2^{3/6}$$ Write the exponents so that they have the same denominator.

$$= (3^2 \cdot 2^3)^{1/6}$$ Use $a^n b^n = (ab)^n$.

$$= \sqrt[6]{3^2 \cdot 2^3}$$ Write with radical notation.

$$= \sqrt[6]{72}$$ Multiply $3^2 \cdot 2^3$.

SPOTLIGHT ON DECISION MAKING

Suppose you are a telecommunications industry analyst. A colleague has just formulated a mathematical model for the number of cellular telephone subscriptions in the United States from 1985 to 1998. The model is $y = 341.8x^{21/5}$, where y is the number of cellular telephone subscriptions x years after 1980. The actual data from 1985 to 1998 are listed in the table.

Your colleague has asked for your help in evaluating whether this model represents the actual data well. By comparing the numbers of subscriptions given by the model to the actual data given in the table, decide whether this mathematical model is acceptable. Explain your reasoning.

U.S. CELLULAR TELEPHONE SUBSCRIPTIONS, 1985–1998

Year	Subscriptions
1985	340,213
1986	681,825
1987	1,230,855
1988	2,069,441
1989	3,508,944
1990	5,283,055
1991	7,557,148
1992	11,032,753
1993	16,009,461
1994	24,134,421
1995	33,785,661
1996	44,042,992
1997	55,312,293
1998	69,209,321

(*Source:* The CTIA Semi-Annual Wireless Survey)

Exercise Set 9.2

Use radical notation to write each expression. Simplify if possible. See Example 1.

1. $49^{1/2}$

2. $64^{1/3}$

3. $27^{1/3}$

4. $8^{1/3}$

5. $\left(\dfrac{1}{16}\right)^{1/4}$

6. $\left(\dfrac{1}{64}\right)^{1/2}$

7. $169^{1/2}$

8. $81^{1/4}$

9. $2m^{1/3}$

10. $(2m)^{1/3}$

11. $(9x^4)^{1/2}$

12. $(16x^8)^{1/2}$

13. $(-27)^{1/3}$

14. $-64^{1/2}$

15. $-16^{1/4}$

16. $(-32)^{1/5}$

Use radical notation to write each expression. Simplify if possible. See Example 2.

17. $16^{3/4}$

18. $4^{5/2}$

19. $(-64)^{2/3}$

20. $(-8)^{4/3}$

21. $(-16)^{3/4}$

22. $(-9)^{3/2}$

23. $(2x)^{3/5}$

24. $2x^{3/5}$

25. $(7x+2)^{2/3}$

26. $(x-4)^{3/4}$

27. $\left(\dfrac{16}{9}\right)^{3/2}$

28. $\left(\dfrac{49}{25}\right)^{3/2}$

Write with positive exponents. Simplify if possible. See Example 3.

29. $8^{-4/3}$

30. $64^{-2/3}$

31. $(-64)^{-2/3}$

32. $(-8)^{-4/3}$

33. $(-4)^{-3/2}$

34. $(-16)^{-5/4}$

35. $x^{-1/4}$

36. $y^{-1/6}$

37. $\dfrac{1}{a^{-2/3}}$

38. $\dfrac{1}{n^{-8/9}}$

39. $\dfrac{5}{7x^{-3/4}}$

40. $\dfrac{2}{3y^{-5/7}}$

41. Explain how writing x^{-7} with positive exponents is similar to writing $x^{-1/4}$ with positive exponents.

42. Explain how writing $2x^{-5}$ with positive exponents is similar to writing $2x^{-3/4}$ with positive exponents.

Use the properties of exponents to simplify each expression. Write with positive exponents. See Example 4.

43. $a^{2/3}a^{5/3}$

44. $b^{9/5}b^{8/5}$

45. $x^{-2/5}\cdot x^{7/5}$

46. $y^{4/3}\cdot y^{-1/3}$

47. $3^{1/4}\cdot 3^{3/8}$

48. $5^{1/2}\cdot 5^{1/6}$

49. $\dfrac{y^{1/3}}{y^{1/6}}$

50. $\dfrac{x^{3/4}}{x^{1/8}}$

51. $(4u^2)^{3/2}$

52. $(32^{1/5}x^{2/3})^3$

53. $\dfrac{b^{1/2}b^{3/4}}{-b^{1/4}}$

54. $\dfrac{a^{1/4}a^{-1/2}}{a^{2/3}}$

55. $\dfrac{(3x^{1/4})^3}{x^{1/12}}$

56. $\dfrac{(2x^{1/5})^4}{x^{3/10}}$

Multiply. See Example 5.

57. $v^{1/2}(v^{1/2}-v^{2/3})$

58. $r^{1/2}(r^{1/2}+r^{3/2})$

59. $x^{2/3}(2x-2)$

60. $3x^{1/2}(x+y)$

61. $(2x^{1/3}+3)(2x^{1/3}-3)$

62. $(y^{1/2}+5)(y^{1/2}+5)$

Factor the common factor from the given expression. See Example 6.

63. $x^{8/3}$; $x^{8/3}+x^{10/3}$

64. $x^{3/2}$; $x^{5/2}-x^{3/2}$

65. $x^{1/5}$; $x^{2/5}-3x^{1/5}$

66. $x^{2/7}$; $x^{3/7}-2x^{2/7}$

67. $x^{-1/3}$; $5x^{-1/3}+x^{2/3}$

68. $x^{-3/4}$; $x^{-3/4}+3x^{1/4}$

Use rational exponents to simplify each radical. Assume that all variables represent positive numbers. See Example 7.

69. $\sqrt[6]{x^3}$

70. $\sqrt[9]{a^3}$

71. $\sqrt[6]{4}$

72. $\sqrt[4]{36}$

73. $\sqrt[4]{16x^2}$

74. $\sqrt[8]{4y^2}$

75. $\sqrt[8]{x^4y^4}$

76. $\sqrt[9]{y^6z^3}$

Use rational expressions to write as a single radical expression. See Example 8.

77. $\sqrt[3]{y}\cdot\sqrt[5]{y^2}$

78. $\sqrt[3]{y^2}\cdot\sqrt[6]{y}$

79. $\dfrac{\sqrt[3]{b^2}}{\sqrt[4]{b}}$

80. $\dfrac{\sqrt[4]{a}}{\sqrt[5]{a}}$

81. $\dfrac{\sqrt[3]{a^2}}{\sqrt[6]{a}}$

82. $\dfrac{\sqrt[5]{b^2}}{\sqrt[10]{b^3}}$

83. $\sqrt{3}\cdot\sqrt[3]{4}$

84. $\sqrt[3]{5}\cdot\sqrt{2}$

85. $\sqrt[5]{7}\cdot\sqrt[3]{y}$

86. $\sqrt[4]{5}\cdot\sqrt[3]{x}$

87. In physics, the speed of a wave traveling over a stretched string with tension t and density u is given by the expression $\dfrac{\sqrt{t}}{\sqrt{u}}$. Write this expression with rational exponents.

88. In electronics, the angular frequency of oscillations in a certain type of circuit is given by the expression $(LC)^{-1/2}$. Use radical notation to write this expression.

Basal metabolic rate (BMR) is the number of calories per day a person needs to maintain life. A person's basal metabolic rate $B(w)$ in calories per day can be estimated with the function $B(w) = 70w^{3/4}$, where w is the person's weight in kilograms.

89. Estimate the BMR for a person who weighs 50 kilograms. Round to the nearest calorie. (Note: 50 kilograms is approximately 110 pounds.)

90. Estimate the BMR for a person who weighs 85 kilograms. Round to the nearest calorie. (Note: 85 kilograms is approximately 187 pounds.)

91. Hewlett-Packard (HP) is a global leader in computing and imaging products. HP's annual net revenue can be modeled by the function $f(x) = 6550x^{43/50}$, where $f(x)$ is net revenue in millions of dollars in the year x, and $x = 0$ represents the year 1990. (*Source:* Hewlett-Packard Company, 1995–1999)

   a. Use this model to find HP's net revenue in 1999.
   b. Predict HP's net revenue in 2004.

Fill in the box with the correct expression.

92. $\boxed{} \cdot a^{2/3} = a^{3/3}$, or a **93.** $\boxed{} \cdot x^{1/8} = x^{4/8}$, or $x^{1/2}$

94. $\dfrac{\boxed{}}{x^{-2/5}} = x^{3/5}$ **95.** $\dfrac{\boxed{}}{y^{-3/4}} = y^{4/4}$, or y

Use a calculator to write a four-decimal-place approximation of each.

96. $8^{1/4}$ **97.** $20^{1/5}$

98. $18^{3/5}$ **99.** $76^{5/7}$

REVIEW EXERCISES

Write each integer as a product of two integers such that one of the factors is a perfect square. For example, write 18 as 9 · 2, because 9 is a perfect square.

100. 75   **101.** 20

102. 48   **103.** 45

Write each integer as a product of two integers such that one of the factors is a perfect cube. For example, write 24 as 8 · 3, because 8 is a perfect cube.

104. 16   **105.** 56

106. 54   **107.** 80

9.3 SIMPLIFYING RADICAL EXPRESSIONS

CD-ROM

SSM

SSG

Video

▶ **OBJECTIVES**

1. Use the product rule for radicals.
2. Use the quotient rule for radicals.
3. Simplify radicals.

1 It is possible to simplify some radicals that do not evaluate to rational numbers. To do so, we use a product rule and a quotient rule for radicals. To discover the product rule, notice the following pattern.

$$\sqrt{9} \cdot \sqrt{4} = 3 \cdot 2 = 6$$
$$\sqrt{9 \cdot 4} = \sqrt{36} = 6$$

Since both expressions simplify to 6, it is true that

$$\sqrt{9} \cdot \sqrt{4} = \sqrt{9 \cdot 4}$$

This pattern suggests the following product rule for radicals.

PRODUCT RULE FOR RADICALS

If $\sqrt[n]{a}$ and $\sqrt[n]{b}$ are real numbers, then

$$\sqrt[n]{a} \cdot \sqrt[n]{b} = \sqrt[n]{ab}$$

Notice that the product rule is the relationship $a^{1/n} \cdot b^{1/n} = (ab)^{1/n}$ stated in radical notation.

Example 1 Multiply.

 a. $\sqrt{3} \cdot \sqrt{5}$ b. $\sqrt{21} \cdot \sqrt{x}$ c. $\sqrt[3]{4} \cdot \sqrt[3]{2}$ d. $\sqrt[4]{5y^2} \cdot \sqrt[4]{2x^3}$ e. $\sqrt{\dfrac{2}{a}} \cdot \sqrt{\dfrac{b}{3}}$

Solution a. $\sqrt{3} \cdot \sqrt{5} = \sqrt{3 \cdot 5} = \sqrt{15}$

 b. $\sqrt{21} \cdot \sqrt{x} = \sqrt{21x}$

 c. $\sqrt[3]{4} \cdot \sqrt[3]{2} = \sqrt[3]{4 \cdot 2} = \sqrt[3]{8} = 2$

 d. $\sqrt[4]{5y^2} \cdot \sqrt[4]{2x^3} = \sqrt[4]{5y^2 \cdot 2x^3} = \sqrt[4]{10y^2x^3}$

 e. $\sqrt{\dfrac{2}{a}} \cdot \sqrt{\dfrac{b}{3}} = \sqrt{\dfrac{2}{a} \cdot \dfrac{b}{3}} = \sqrt{\dfrac{2b}{3a}}$

2 To discover a quotient rule for radicals, notice the following pattern.

$$\sqrt{\dfrac{4}{9}} = \dfrac{2}{3}$$

$$\dfrac{\sqrt{4}}{\sqrt{9}} = \dfrac{2}{3}$$

Since both expressions simplify to $\dfrac{2}{3}$, it is true that

$$\sqrt{\dfrac{4}{9}} = \dfrac{\sqrt{4}}{\sqrt{9}}$$

This pattern suggests the following quotient rule for radicals.

QUOTIENT RULE FOR RADICALS

If $\sqrt[n]{a}$ and $\sqrt[n]{b}$ are real numbers and $\sqrt[n]{b}$ is not zero, then

$$\sqrt[n]{\dfrac{a}{b}} = \dfrac{\sqrt[n]{a}}{\sqrt[n]{b}}$$

Notice that the quotient rule is the relationship $\left(\dfrac{a}{b}\right)^{1/n} = \dfrac{a^{1/n}}{b^{1/n}}$ stated in radical notation. We can use the quotient rule to simplify radical expressions by reading the rule from left to right, or to divide radicals by reading the rule from right to left.

For example,

$$\sqrt{\dfrac{x}{16}} = \dfrac{\sqrt{x}}{\sqrt{16}} = \dfrac{\sqrt{x}}{4} \qquad \text{using } \sqrt[n]{\dfrac{a}{b}} = \dfrac{\sqrt[n]{a}}{\sqrt[n]{b}}$$

$$\dfrac{\sqrt{75}}{\sqrt{3}} = \sqrt{\dfrac{75}{3}} = \sqrt{25} = 5 \qquad \text{using } \dfrac{\sqrt[n]{a}}{\sqrt[n]{b}} = \sqrt[n]{\dfrac{a}{b}}$$

Note: *For the remainder of this chapter, we will assume that variables represent positive real numbers. Since this is so, we need not insert absolute value bars when we simplify even roots.*

Example 2 Use the quotient rule to simplify.

a. $\sqrt{\dfrac{25}{49}}$ **b.** $\sqrt{\dfrac{x}{9}}$ **c.** $\sqrt[3]{\dfrac{8}{27}}$ **d.** $\sqrt[4]{\dfrac{3}{16y^4}}$

Solution   **a.** $\sqrt{\dfrac{25}{49}} = \dfrac{\sqrt{25}}{\sqrt{49}} = \dfrac{5}{7}$

b. $\sqrt{\dfrac{x}{9}} = \dfrac{\sqrt{x}}{\sqrt{9}} = \dfrac{\sqrt{x}}{3}$

c. $\sqrt[3]{\dfrac{8}{27}} = \dfrac{\sqrt[3]{8}}{\sqrt[3]{27}} = \dfrac{2}{3}$

d. $\sqrt[4]{\dfrac{3}{16y^4}} = \dfrac{\sqrt[4]{3}}{\sqrt[4]{16y^4}} = \dfrac{\sqrt[4]{3}}{2y}$

3   Both the product and quotient rules can be used to simplify a radical. If the product rule is read from right to left, we have that $\sqrt[n]{ab} = \sqrt[n]{a} \cdot \sqrt[n]{b}$. This is used to simplify the following radicals.

Example 3 Simplify the following.

a. $\sqrt{50}$ **b.** $\sqrt[3]{24}$ **c.** $\sqrt{26}$ **d.** $\sqrt[4]{32}$

Solution   **a.** Factor 50 such that one factor is the largest perfect square that divides 50. The largest perfect square factor of 50 is 25, so we write 50 as $25 \cdot 2$ and use the product rule for radicals to simplify.

$$\sqrt{50} = \sqrt{25 \cdot 2} = \sqrt{25} \cdot \sqrt{2} = 5\sqrt{2}$$

The largest perfect square factor of 50.

> **HELPFUL HINT**
> Don't forget that, for example, $5\sqrt{2}$ means $5 \cdot \sqrt{2}$.

b. $\sqrt[3]{24} = \sqrt[3]{8 \cdot 3} = \sqrt[3]{8} \cdot \sqrt[3]{3} = 2\sqrt[3]{3}$

The largest perfect cube factor of 24.

c. $\sqrt{26}$ The largest perfect square factor of 26 is 1, so $\sqrt{26}$ cannot be simplified further.

d. $\sqrt[4]{32} = \sqrt[4]{16 \cdot 2} = \sqrt[4]{16} \cdot \sqrt[4]{2} = 2\sqrt[4]{2}$

The largest fourth power factor of 32.

After simplifying a radical such as a square root, always check the radicand to see that it contains no other perfect square factors. It may, if the largest perfect square factor of the radicand was not originally recognized. For example,

$$\sqrt{200} = \sqrt{4 \cdot 50} = \sqrt{4} \cdot \sqrt{50} = 2\sqrt{50}$$

Notice that the radicand 50 still contains the perfect square factor 25. This is because 4 is not the largest perfect square factor of 200. We continue as follows.

$$2\sqrt{50} = 2\sqrt{25 \cdot 2} = 2 \cdot \sqrt{25} \cdot \sqrt{2} = 2 \cdot 5 \cdot \sqrt{2} = 10\sqrt{2}$$

The radical is now simplified since 2 contains no perfect square factors (other than 1).

HELPFUL HINT

To help you recognize largest perfect power factors of a radicand, it will help if you are familiar with some perfect powers. A few are listed below.

Perfect Squares   1,   4,   9,   16,   25,   36,   49,   64,   81,   100,   121,   144
                   1^2 2^2 3^2 4^2 5^2 6^2 7^2 8^2 9^2 10^2 11^2 12^2

Perfect Cubes   1,   8,   27,   64,   125
                 1^3 2^3 3^3 4^3 5^3

Perfect Fourth
Powers   1,   16,   81,   256
                 1^4 2^4 3^4 4^4

In general, we say that a radicand of the form $\sqrt[n]{a}$ is simplified when a contains no factors that are perfect nth powers (other than 1 or -1).

Example 4   Use the product rule to simplify.

a. $\sqrt{25x^3}$ **b.** $\sqrt[3]{54x^6y^8}$ **c.** $\sqrt[4]{81z^{11}}$

Solution   **a.** $\sqrt{25x^3} = \sqrt{25x^2 \cdot x}$ Find the largest perfect square factor.

$= \sqrt{25x^2} \cdot \sqrt{x}$ Apply the product rule.

$= 5x\sqrt{x}$ Simplify.

b. $\sqrt[3]{54x^6y^8} = \sqrt[3]{27 \cdot 2 \cdot x^6 \cdot y^6 \cdot y^2}$ Factor the radicand and identify perfect cube factors.

$= \sqrt[3]{27x^6y^6 \cdot 2y^2}$

$= \sqrt[3]{27x^6y^6} \cdot \sqrt[3]{2y^2}$ Apply the product rule.

$= 3x^2y^2\sqrt[3]{2y^2}$ Simplify.

c. $\sqrt[4]{81z^{11}} = \sqrt[4]{81 \cdot z^8 \cdot z^3}$ Factor the radicand and identify perfect fourth power factors.

$= \sqrt[4]{81z^8} \cdot \sqrt[4]{z^3}$ Apply the product rule.

$= 3z^2\sqrt[4]{z^3}$ Simplify.

Example 5   Use the quotient rule to divide, and simplify if possible.

a. $\dfrac{\sqrt{20}}{\sqrt{5}}$ b. $\dfrac{\sqrt{50x}}{2\sqrt{2}}$ c. $\dfrac{7\sqrt[3]{48x^4y^8}}{\sqrt[3]{6y^2}}$

Solution   a. $\dfrac{\sqrt{20}}{\sqrt{5}} = \sqrt{\dfrac{20}{5}}$ Apply the quotient rule.

$= \sqrt{4}$ Simplify.

$= 2$ Simplify.

b. $\dfrac{\sqrt{50x}}{2\sqrt{2}} = \dfrac{1}{2} \cdot \sqrt{\dfrac{50x}{2}}$ Apply the quotient rule.

$= \dfrac{1}{2} \cdot \sqrt{25x}$ Simplify.

$= \dfrac{1}{2} \cdot \sqrt{25} \cdot \sqrt{x}$ Factor 25x.

$= \dfrac{1}{2} \cdot 5 \cdot \sqrt{x}$ Simplify.

$= \dfrac{5}{2} \sqrt{x}$

c. $\dfrac{7\sqrt[3]{48x^4y^8}}{\sqrt[3]{6y^2}} = 7 \cdot \sqrt[3]{\dfrac{48x^4y^8}{6y^2}}$ Apply the quotient rule.

$= 7 \cdot \sqrt[3]{8x^4y^6}$ Simplify.

$= 7\sqrt[3]{8x^3y^6 \cdot x}$ Factor.

$= 7 \cdot \sqrt[3]{8x^3y^6} \cdot \sqrt[3]{x}$ Apply the product rule.

$= 7 \cdot 2xy^2 \cdot \sqrt[3]{x}$ Simplify.

$= 14xy^2\sqrt[3]{x}$

Exercise Set 9.3

Use the product rule to multiply. See Example 1.

1. $\sqrt{7} \cdot \sqrt{2}$ 2. $\sqrt{11} \cdot \sqrt{10}$

3. $\sqrt[4]{8} \cdot \sqrt[4]{2}$ 4. $\sqrt[4]{27} \cdot \sqrt[4]{3}$

5. $\sqrt[3]{4} \cdot \sqrt[3]{9}$ 6. $\sqrt[3]{10} \cdot \sqrt[3]{5}$

7. $\sqrt{2} \cdot \sqrt{3x}$ 8. $\sqrt{3y} \cdot \sqrt{5x}$

9. $\sqrt{\dfrac{7}{x}} \cdot \sqrt{\dfrac{2}{y}}$ 10. $\sqrt{\dfrac{6}{m}} \cdot \sqrt{\dfrac{n}{5}}$

11. $\sqrt[4]{4x^3} \cdot \sqrt[4]{5}$ 12. $\sqrt[4]{ab^2} \cdot \sqrt[4]{27ab}$

Use the quotient rule to simplify. See Examples 2 and 3.

13. $\sqrt{\dfrac{6}{49}}$ 14. $\sqrt{\dfrac{8}{81}}$

15. $\sqrt{\dfrac{2}{49}}$ 16. $\sqrt{\dfrac{5}{121}}$

17. $\sqrt[4]{\dfrac{x^3}{16}}$ 18. $\sqrt[4]{\dfrac{y}{81x^4}}$

19. $\sqrt[3]{\dfrac{4}{27}}$ 20. $\sqrt[3]{\dfrac{3}{64}}$

21. $\sqrt[4]{\dfrac{8}{x^8}}$

22. $\sqrt[4]{\dfrac{a^3}{81}}$

23. $\sqrt[3]{\dfrac{2x}{81y^{12}}}$

24. $\sqrt[3]{\dfrac{3}{8x^6}}$

25. $\sqrt{\dfrac{x^2y}{100}}$

26. $\sqrt{\dfrac{y^2z}{36}}$

27. $\sqrt{\dfrac{5x^2}{4y^2}}$

28. $\sqrt{\dfrac{y^{10}}{9x^6}}$

29. $-\sqrt[3]{\dfrac{z^7}{27x^3}}$

30. $-\sqrt[3]{\dfrac{64a}{b^9}}$

Simplify. See Examples 3 and 4.

31. $\sqrt{32}$

32. $\sqrt{27}$

33. $\sqrt[3]{192}$

34. $\sqrt[3]{108}$

35. $5\sqrt{75}$

36. $3\sqrt{8}$

37. $\sqrt{24}$

38. $\sqrt{20}$

39. $\sqrt{100x^5}$

40. $\sqrt{64y^9}$

41. $\sqrt[3]{16y^7}$

42. $\sqrt[3]{64y^9}$

43. $\sqrt[4]{a^8b^7}$

44. $\sqrt[5]{32z^{12}}$

45. $\sqrt{y^5}$

46. $\sqrt[3]{y^5}$

47. $\sqrt{25a^2b^3}$

48. $\sqrt{9x^5y^7}$

49. $\sqrt[5]{-32x^{10}y}$

50. $\sqrt[5]{-243z^9}$

51. $\sqrt[3]{50x^{14}}$

52. $\sqrt[3]{40y^{10}}$

53. $-\sqrt{32a^8b^7}$

54. $-\sqrt{20ab^6}$

55. $\sqrt{9x^7y^9}$

56. $\sqrt{12r^9s^{12}}$

57. $\sqrt[3]{125r^9s^{12}}$

58. $\sqrt[3]{8a^6b^9}$

Use the quotient rule to divide. Then simplify if possible. See Example 5.

59. $\dfrac{\sqrt{14}}{\sqrt{7}}$

60. $\dfrac{\sqrt{45}}{\sqrt{9}}$

61. $\dfrac{\sqrt[3]{24}}{\sqrt[3]{3}}$

62. $\dfrac{\sqrt[3]{10}}{\sqrt[3]{2}}$

63. $\dfrac{5\sqrt[4]{48}}{\sqrt[4]{3}}$

64. $\dfrac{7\sqrt[4]{162}}{\sqrt[4]{2}}$

65. $\dfrac{\sqrt{x^5y^3}}{\sqrt{xy}}$

66. $\dfrac{\sqrt{a^7b^6}}{\sqrt{a^3b^2}}$

67. $\dfrac{8\sqrt[3]{54m^7}}{\sqrt[3]{2m}}$

68. $\dfrac{\sqrt[3]{128x^3}}{-3\sqrt[3]{2x}}$

69. $\dfrac{3\sqrt{100x^2}}{2\sqrt{2x^{-1}}}$

70. $\dfrac{\sqrt{270y^2}}{5\sqrt{3y^{-4}}}$

71. $\dfrac{\sqrt[4]{96a^{10}b^3}}{\sqrt[4]{3a^2b^3}}$

72. $\dfrac{\sqrt[5]{64x^{10}y^3}}{\sqrt[5]{2x^3y^{-7}}}$

73. The formula for the surface area A of a cone with height h and radius r is given by

$$A = \pi r\sqrt{r^2 + h^2}$$

   a. Find the surface area of a cone whose height is 3 centimeters and whose radius is 4 centimeters.

   b. Approximate to two decimal places the surface area of a cone whose height is 7.2 feet and whose radius is 6.8 feet.

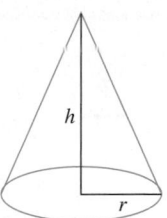

74. Before Mount Vesuvius, a volcano in Italy, erupted violently in 79 A.D., its height was 4190 feet. Vesuvius was roughly cone-shaped, and its base had a radius of approximately 25,200 feet. Use the formula for the surface area of a cone, given in Exercise 73, to approximate the surface area this volcano had before it erupted. (*Source: Global Volcanism Network*)

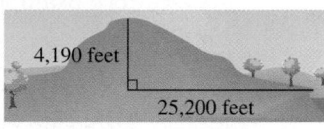

75. The owner of Knightime Video has determined that the demand equation for renting older releases is given by the equation $F(x) = 0.6\sqrt{49 - x^2}$, where x is the price in dollars per two-day rental and $F(x)$ is the number of times the video is demanded per week.

   a. Approximate to one decimal place the demand per week of an older release if the rental price is $3 per two-day rental.

   b. Approximate to one decimal place the demand per week of an older release if the rental price is $5 per two-day rental.

   c. Explain how the owner of the video store can use this equation to predict the number of copies of each tape that should be in stock.

REVIEW EXERCISES

Perform each indicated operation. See Sections 5.3 and 5.4.

76. $6x + 8x$

77. $(6x)(8x)$

78. $(2x + 3)(x - 5)$

79. $(2x + 3) + (x - 5)$

80. $9y^2 - 8y^2$

81. $(9y^2)(-8y^2)$

82. $-3(x + 5)$

83. $-3 + x + 5$

84. $(x - 4)^2$

85. $(2x + 1)^2$

9.4 ADDING, SUBTRACTING, AND MULTIPLYING RADICAL EXPRESSIONS

CD-ROM   SSM

SSG   Video

▶ **OBJECTIVES**

1. Add or subtract radical expressions.
2. Multiply radical expressions.

1

We have learned that sums or differences of like terms can be simplified. To simplify these sums or differences, we use the distributive property. For example,

$$2x + 3x = (2 + 3)x = 5x \quad \text{and} \quad 7x^2y - 4x^2y = (7 - 4)x^2y = 3x^2y$$

The distributive property can also be used to add **like radicals.**

LIKE RADICALS

Radicals with the same index and the same radicand are like radicals.

For example, $2\sqrt{7} + 3\sqrt{7} = (2 + 3)\sqrt{7} = 5\sqrt{7}$. Also,

$$5\sqrt{3x} - 7\sqrt{3x} = (5 - 7)\sqrt{3x} = -2\sqrt{3x}$$

The expression $2\sqrt{7} + 2\sqrt[3]{7}$ cannot be simplified further since $2\sqrt{7}$ and $2\sqrt[3]{7}$ are not like radicals.

Example 1 Add or subtract. Assume that variables represent positive real numbers.

   a. $\sqrt{20} + 2\sqrt{45}$ **b.** $\sqrt[3]{54} - 5\sqrt[3]{16} + \sqrt[3]{2}$ **c.** $\sqrt{27x} - 2\sqrt{9x} + \sqrt{72x}$

   d. $\sqrt[3]{98} + \sqrt{98}$ **e.** $\sqrt[3]{48y^4} + \sqrt[3]{6y^4}$

Solution First, simplify each radical. Then add or subtract any like radicals.

   a. $\sqrt{20} + 2\sqrt{45} = \sqrt{4 \cdot 5} + 2\sqrt{9 \cdot 5}$ Factor 20 and 45.

$$= \sqrt{4} \cdot \sqrt{5} + 2 \cdot \sqrt{9} \cdot \sqrt{5} \qquad \text{Use the product rule.}$$

$$= 2 \cdot \sqrt{5} + 2 \cdot 3 \cdot \sqrt{5} \qquad \text{Simplify } \sqrt{4} \text{ and } \sqrt{9}.$$

$$= 2\sqrt{5} + 6\sqrt{5}$$

$$= 8\sqrt{5} \qquad \text{Add like radicals.}$$

   b. $\sqrt[3]{54} - 5\sqrt[3]{16} + \sqrt[3]{2}$

$$= \sqrt[3]{27} \cdot \sqrt[3]{2} - 5 \cdot \sqrt[3]{8} \cdot \sqrt[3]{2} + \sqrt[3]{2} \qquad \text{Factor and use the product rule.}$$

$$= 3 \cdot \sqrt[3]{2} - 5 \cdot 2 \cdot \sqrt[3]{2} + \sqrt[3]{2} \qquad \text{Simplify } \sqrt[3]{27} \text{ and } \sqrt[3]{8}.$$

$$= 3\sqrt[3]{2} - 10\sqrt[3]{2} + \sqrt[3]{2} \qquad \text{Write } 5 \cdot 2 \text{ as } 10.$$

$$= -6\sqrt[3]{2} \qquad \text{Combine like radicals.}$$

c. $\sqrt{27x} - 2\sqrt{9x} + \sqrt{72x}$

$\quad = \sqrt{9} \cdot \sqrt{3x} - 2 \cdot \sqrt{9} \cdot \sqrt{x} + \sqrt{36} \cdot \sqrt{2x}$ Factor and use the product rule.

$\quad = 3 \cdot \sqrt{3x} - 2 \cdot 3 \cdot \sqrt{x} + 6 \cdot \sqrt{2x}$ Simplify $\sqrt{9}$ and $\sqrt{36}$.

$\quad = 3\sqrt{3x} - 6\sqrt{x} + 6\sqrt{2x}$ Write $2 \cdot 3$ as 6.

HELPFUL HINT

None of these terms contain like radicals. We can simplify no further.

---

d. $\sqrt[3]{98} + \sqrt{98} = \sqrt[3]{98} + \sqrt{49} \cdot \sqrt{2}$ Factor and use the product rule.

$\quad\quad\quad\quad\quad = \sqrt[3]{98} + 7\sqrt{2}$ No further simplification is possible.

e. $\sqrt[3]{48y^4} + \sqrt[3]{6y^4} = \sqrt[3]{8y^3} \cdot \sqrt[3]{6y} + \sqrt[3]{y^3} \cdot \sqrt[3]{6y}$ Factor and use the product rule.

$\quad\quad\quad\quad\quad\quad = 2y\sqrt[3]{6y} + y\sqrt[3]{6y}$ Simplify $\sqrt[3]{8y^3}$ and $\sqrt[3]{y^3}$.

$\quad\quad\quad\quad\quad\quad = 3y\sqrt[3]{6y}$ Combine like radicals.   ■

Example 2   Add or subtract as indicated.

a. $\dfrac{\sqrt{45}}{4} - \dfrac{\sqrt{5}}{3}$ **b.** $\sqrt[3]{\dfrac{7x}{8}} + 2\sqrt[3]{7x}$

Solution   **a.** $\dfrac{\sqrt{45}}{4} - \dfrac{\sqrt{5}}{3} = \dfrac{3\sqrt{5}}{4} - \dfrac{\sqrt{5}}{3}$ To subtract, notice that the LCD is 12.

$\quad\quad\quad\quad\quad = \dfrac{3\sqrt{5} \cdot 3}{4 \cdot 3} - \dfrac{\sqrt{5} \cdot 4}{3 \cdot 4}$ Write each expression as an equivalent expression with a denominator of 12.

$\quad\quad\quad\quad\quad = \dfrac{9\sqrt{5}}{12} - \dfrac{4\sqrt{5}}{12}$ Multiply factors in the numerator and the denominator.

$\quad\quad\quad\quad\quad = \dfrac{5\sqrt{5}}{12}$ Subtract.

b. $\sqrt[3]{\dfrac{7x}{8}} + 2\sqrt[3]{7x} = \dfrac{\sqrt[3]{7x}}{\sqrt[3]{8}} + 2\sqrt[3]{7x}$ Apply the quotient rule for radicals.

$\quad\quad\quad\quad\quad = \dfrac{\sqrt[3]{7x}}{2} + 2\sqrt[3]{7x}$ Simplify.

$\quad\quad\quad\quad\quad = \dfrac{\sqrt[3]{7x}}{2} + \dfrac{2\sqrt[3]{7x} \cdot 2}{2}$ Write each expression as an equivalent expression with a denominator of 2.

$\quad\quad\quad\quad\quad = \dfrac{\sqrt[3]{7x}}{2} + \dfrac{4\sqrt[3]{7x}}{2}$

$\quad\quad\quad\quad\quad = \dfrac{5\sqrt[3]{7x}}{2}$ Add.   ■

2   We can multiply radical expressions by using many of the same properties used to multiply polynomial expressions. For instance, to multiply $\sqrt{2}(\sqrt{6} - 3\sqrt{2})$, we use the distributive property and multiply $\sqrt{2}$ by each term inside the parentheses.

$$\sqrt{2}(\sqrt{6} - 3\sqrt{2}) = \sqrt{2}(\sqrt{6}) - \sqrt{2}(3\sqrt{2}) \qquad \text{Use the distributive property.}$$

$$= \sqrt{2 \cdot 6} - 3\sqrt{2 \cdot 2}$$

$$= \sqrt{2 \cdot 2 \cdot 3} - 3 \cdot 2 \qquad \text{Use the product rule for radicals.}$$

$$= 2\sqrt{3} - 6$$

Example 3   Multiply.

a. $\sqrt{3}(5 + \sqrt{30})$ **b.** $(\sqrt{5} - \sqrt{6})(\sqrt{7} + 1)$ **c.** $(7\sqrt{x} + 5)(3\sqrt{x} - \sqrt{5})$

d. $(4\sqrt{3} - 1)^2$ **e.** $(\sqrt{2x} - 5)(\sqrt{2x} + 5)$

Solution   **a.** $\sqrt{3}(5 + \sqrt{30}) = \sqrt{3}(5) + \sqrt{3}(\sqrt{30})$

$$= 5\sqrt{3} + \sqrt{3 \cdot 30}$$

$$= 5\sqrt{3} + \sqrt{3 \cdot 3 \cdot 10}$$

$$= 5\sqrt{3} + 3\sqrt{10}$$

b.   To multiply, we can use the FOIL method.

$$\overset{\text{First}}{}\quad\overset{\text{Outer}}{}\quad\overset{\text{Inner}}{}\quad\overset{\text{Last}}{}$$

$$(\sqrt{5} - \sqrt{6})(\sqrt{7} + 1) = \sqrt{5} \cdot \sqrt{7} + \sqrt{5} \cdot 1 - \sqrt{6} \cdot \sqrt{7} - \sqrt{6} \cdot 1$$

$$= \sqrt{35} + \sqrt{5} - \sqrt{42} - \sqrt{6}$$

c. $(7\sqrt{x} + 5)(3\sqrt{x} - \sqrt{5}) = 7\sqrt{x}(3\sqrt{x}) - 7\sqrt{x}(\sqrt{5}) + 5(3\sqrt{x}) - 5(\sqrt{5})$

$$= 21x - 7\sqrt{5x} + 15\sqrt{x} - 5\sqrt{5}$$

d. $(4\sqrt{3} - 1)^2 = (4\sqrt{3} - 1)(4\sqrt{3} - 1)$

$$= 4\sqrt{3}(4\sqrt{3}) - 4\sqrt{3}(1) - 1(4\sqrt{3}) - 1(-1)$$

$$= 16 \cdot 3 - 4\sqrt{3} - 4\sqrt{3} + 1$$

$$= 48 - 8\sqrt{3} + 1$$

$$= 49 - 8\sqrt{3}$$

e. $(\sqrt{2x} - 5)(\sqrt{2x} + 5) = \sqrt{2x} \cdot \sqrt{2x} + 5\sqrt{2x} - 5\sqrt{2x} - 5 \cdot 5$

$$= 2x - 25$$

MENTAL MATH

Simplify. Assume that all variables represent positive real numbers.

1. $2\sqrt{3} + 4\sqrt{3}$ **2.** $5\sqrt{7} + 3\sqrt{7}$ **3.** $8\sqrt{x} - 5\sqrt{x}$

4. $3\sqrt{y} + 10\sqrt{y}$ **5.** $7\sqrt[3]{x} + 5\sqrt[3]{x}$ **6.** $8\sqrt[3]{z} - 2\sqrt[3]{z}$

Exercise Set 9.4

Add or subtract. See Examples 1 and 2.

1. $\sqrt{8} - \sqrt{32}$ **2.** $\sqrt{27} - \sqrt{75}$

3. $2\sqrt{2x^3} + 4x\sqrt{8x}$ **4.** $3\sqrt{45x^3} + x\sqrt{5x}$

5. $2\sqrt{50} - 3\sqrt{125} + \sqrt{98}$

6. $4\sqrt{32} - \sqrt{18} + 2\sqrt{128}$

7. $\sqrt[3]{16x} - \sqrt[3]{54x}$ **8.** $2\sqrt[3]{3a^4} - 3a\sqrt[3]{81a}$

9. $\sqrt{9b^3} - \sqrt{25b^3} + \sqrt{49b^3}$

10. $\sqrt{4x^7} + 9x^2\sqrt{x^3} - 5x\sqrt{x^5}$

11. $\dfrac{5\sqrt{2}}{3} + \dfrac{2\sqrt{2}}{5}$ **12.** $\dfrac{\sqrt{3}}{2} + \dfrac{4\sqrt{3}}{3}$

13. $\sqrt[3]{\dfrac{11}{8}} - \dfrac{\sqrt[3]{11}}{6}$ **14.** $\dfrac{2\sqrt[3]{4}}{7} - \dfrac{\sqrt[3]{4}}{14}$

15. $\dfrac{\sqrt{20x}}{9} + \sqrt{\dfrac{5x}{9}}$ **16.** $\dfrac{3x\sqrt{7}}{5} + \sqrt{\dfrac{7x^2}{100}}$

17. $7\sqrt{9} - 7 + \sqrt{3}$ **18.** $\sqrt{16} - 5\sqrt{10} + 7$

19. $2 + 3\sqrt{y^2} - 6\sqrt{y^2} + 5$

20. $3\sqrt{7} - \sqrt[3]{x} + 4\sqrt{7} - 3\sqrt[3]{x}$

21. $3\sqrt{108} - 2\sqrt{18} - 3\sqrt{48}$

22. $-\sqrt{75} + \sqrt{12} - 3\sqrt{3}$

23. $-5\sqrt[3]{625} + \sqrt[3]{40}$ **24.** $-2\sqrt[3]{108} - \sqrt[3]{32}$

25. $\sqrt{9b^3} - \sqrt{25b^3} + \sqrt{16b^3}$

26. $\sqrt{4x^7y^5} + 9x^2\sqrt{x^3y^5} - 5xy\sqrt{x^5y^3}$

27. $5y\sqrt{8y} + 2\sqrt{50y^3}$ **28.** $3\sqrt{8x^2y^3} - 2x\sqrt{32y^3}$

29. $\sqrt[3]{54xy^3} - 5\sqrt[3]{2xy^3} + y\sqrt[3]{128x}$

30. $2\sqrt[3]{24x^3y^4} + 4x\sqrt[3]{81y^4}$

31. $6\sqrt[3]{11} + 8\sqrt{11} - 12\sqrt{11}$ **32.** $3\sqrt[3]{5} + 4\sqrt{5}$

33. $-2\sqrt[4]{x^7} + 3\sqrt[4]{16x^7}$

34. $6\sqrt[3]{24x^3} - 2\sqrt[3]{81x^3} - x\sqrt[3]{3}$

35. $\dfrac{4\sqrt{3}}{3} - \dfrac{\sqrt{12}}{3}$ **36.** $\dfrac{\sqrt{45}}{10} + \dfrac{7\sqrt{5}}{10}$

37. $\dfrac{\sqrt[3]{8x^4}}{7} + \dfrac{3x\sqrt[3]{x}}{7}$ **38.** $\dfrac{\sqrt[4]{48}}{5x} - \dfrac{2\sqrt[4]{3}}{10x}$

39. $\sqrt{\dfrac{28}{x^2}} + \sqrt{\dfrac{7}{4x^2}}$ **40.** $\dfrac{\sqrt{99}}{5x} - \sqrt{\dfrac{44}{x^2}}$

41. $\sqrt[3]{\dfrac{16}{27}} - \dfrac{\sqrt[3]{54}}{6}$ **42.** $\dfrac{\sqrt[3]{3}}{10} + \sqrt[3]{\dfrac{24}{125}}$

43. $-\dfrac{\sqrt[3]{2x^4}}{9} + \sqrt[3]{\dfrac{250x^4}{27}}$ **44.** $\dfrac{\sqrt[3]{y^5}}{8} + \dfrac{5y\sqrt[3]{y^2}}{4}$

△ **45.** Find the perimeter of the trapezoid.

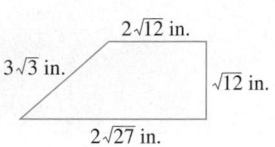

△ **46.** Find the perimeter of the triangle.

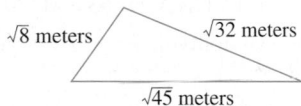

Multiply, and then simplify if possible. See Example 3.

47. $\sqrt{7}(\sqrt{5} + \sqrt{3})$ **48.** $\sqrt{5}(\sqrt{15} - \sqrt{35})$

49. $(\sqrt{5} - \sqrt{2})^2$ **50.** $(3x - \sqrt{2})(3x - \sqrt{2})$

51. $\sqrt{3x}(\sqrt{3} - \sqrt{x})$ **52.** $\sqrt{5y}(\sqrt{y} + \sqrt{5})$

53. $(2\sqrt{x} - 5)(3\sqrt{x} + 1)$ **54.** $(8\sqrt{y} + z)(4\sqrt{y} - 1)$

55. $(\sqrt[3]{a} - 4)(\sqrt[3]{a} + 5)$ **56.** $(\sqrt[3]{a} + 2)(\sqrt[3]{a} + 7)$

57. $6(\sqrt{2} - 2)$ **58.** $\sqrt{5}(6 - \sqrt{5})$

59. $\sqrt{2}(\sqrt{2} + x\sqrt{6})$ **60.** $\sqrt{3}(\sqrt{3} - 2\sqrt{5x})$

61. $(2\sqrt{7} + 3\sqrt{5})(\sqrt{7} - 2\sqrt{5})$

62. $(\sqrt{6} - 4\sqrt{2})(3\sqrt{6} + 1)$

63. $(\sqrt{x} - y)(\sqrt{x} + y)$ **64.** $(3\sqrt{x} + 2)(\sqrt{3x} - 2)$

65. $(\sqrt{3} + x)^2$ **66.** $(\sqrt{y} - 3x)^2$

67. $(\sqrt{5x} - 3\sqrt{2})(\sqrt{5x} - 3\sqrt{3})$

68. $(5\sqrt{3x} - \sqrt{y})(4\sqrt{x} + 1)$

69. $(\sqrt[3]{4} + 2)(\sqrt[3]{2} - 1)$

70. $(\sqrt[3]{3} + \sqrt[3]{2})(\sqrt[3]{9} - \sqrt[3]{4})$

71. $(\sqrt[3]{x} + 1)(\sqrt[3]{x} - 4\sqrt{x} + 7)$

72. $(\sqrt[3]{3x} + 3)(\sqrt[3]{2x} - 3x - 1)$

△ **73.** Baseboard needs to be installed around the perimeter of a rectangular room.

    a. Find how much baseboard should be ordered by finding the perimeter of the room.

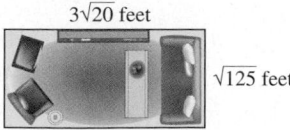

    b. Find the area of the room.

△ **74.** A border of wallpaper is to be used around the perimeter of the odd-shaped room shown.

    a. Find how much wallpaper border is needed by finding the perimeter of the room.

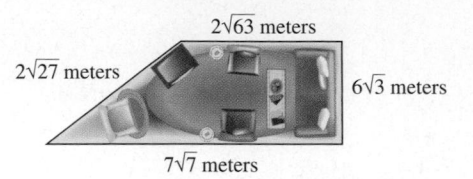

2√63 meters

2√27 meters

6√3 meters

7√7 meters

    b. Find the area of the room. (*Hint:* The area of a trapezoid is the product of half the height $6\sqrt{3}$ meters and the sum of the bases $2\sqrt{63}$ and $7\sqrt{7}$ meters.)

 75. Explain how simplifying $2x + 3x$ is similar to simplifying $2\sqrt{x} + 3\sqrt{x}$.

 76. Explain how multiplying $(x - 2)(x + 3)$ is similar to multiplying $(\sqrt{x} - \sqrt{2})(\sqrt{x} + 3)$.

REVIEW EXERCISES

Factor each numerator and denominator. Then simplify if possible. See Section 7.1.

77. $\dfrac{2x - 14}{2}$

78. $\dfrac{8x - 24y}{4}$

79. $\dfrac{7x - 7y}{x^2 - y^2}$

80. $\dfrac{x^3 - 8}{4x - 8}$

81. $\dfrac{6a^2b - 9ab}{3ab}$

82. $\dfrac{14r - 28r^2s^2}{7rs}$

83. $\dfrac{-4 + 2\sqrt{3}}{6}$

84. $\dfrac{-5 + 10\sqrt{7}}{5}$

9.5 RATIONALIZING DENOMINATORS AND NUMERATORS OF RADICAL EXPRESSIONS

▶ **OBJECTIVES**

CD-ROM   SSM

SSG   Video

1. Rationalize denominators.
2. Rationalize numerators.
3. Rationalize denominators or numerators having two terms.

1   Often in mathematics, it is helpful to write a radical expression such as $\dfrac{\sqrt{3}}{\sqrt{2}}$ either without a radical in the denominator or without a radical in the numerator. The process of writing this expression as an equivalent expression but without a radical in the denominator is called **rationalizing the denominator**. To rationalize the denominator of $\dfrac{\sqrt{3}}{\sqrt{2}}$, we use the fundamental principle of fractions and multiply the numerator and the denominator by $\sqrt{2}$. Recall that this is the same as multiplying by $\dfrac{\sqrt{2}}{\sqrt{2}}$, which simplifies to 1.

$$\frac{\sqrt{3}}{\sqrt{2}} = \frac{\sqrt{3} \cdot \sqrt{2}}{\sqrt{2} \cdot \sqrt{2}} = \frac{\sqrt{6}}{\sqrt{4}} = \frac{\sqrt{6}}{2}$$

Example 1   Rationalize the denominator of each expression.

    a. $\dfrac{2}{\sqrt{5}}$ **b.** $\dfrac{2\sqrt{16}}{\sqrt{9x}}$ **c.** $\sqrt[3]{\dfrac{1}{2}}$

Solution   **a.** To rationalize the denominator, we multiply the numerator and denominator by a factor that makes the radicand in the denominator a perfect square.

$$\frac{2}{\sqrt{5}} = \frac{2 \cdot \sqrt{5}}{\sqrt{5} \cdot \sqrt{5}} = \frac{2\sqrt{5}}{5} \qquad \text{\small The denominator is now rationalized.}$$

b. First, we simplify the radicals and then rationalize the denominator.

$$\frac{2\sqrt{16}}{\sqrt{9x}} = \frac{2(4)}{3\sqrt{x}} = \frac{8}{3\sqrt{x}}$$

To rationalize the denominator, multiply the numerator and denominator by $\sqrt{x}$. Then

$$\frac{8}{3\sqrt{x}} = \frac{8 \cdot \sqrt{x}}{3\sqrt{x} \cdot \sqrt{x}} = \frac{8\sqrt{x}}{3x}$$

c. $\sqrt[3]{\dfrac{1}{2}} = \dfrac{\sqrt[3]{1}}{\sqrt[3]{2}} = \dfrac{1}{\sqrt[3]{2}}$. Now we rationalize the denominator. Since $\sqrt[3]{2}$ is a cube root, we want to multiply by a value that will make the radicand 2 a perfect cube. If we multiply by $\sqrt[3]{2^2}$, we get $\sqrt[3]{2^3} = \sqrt[3]{8} = 2$.

$$\frac{1 \cdot \sqrt[3]{2^2}}{\sqrt[3]{2} \cdot \sqrt[3]{2^2}} = \frac{\sqrt[3]{4}}{\sqrt[3]{2^3}} = \frac{\sqrt[3]{4}}{2} \qquad \text{Multiply numerator and denominator by } \sqrt[3]{2^2} \text{ and then simplify.}$$

Example 2   Rationalize the denominator of $\sqrt{\dfrac{7x}{3y}}$.

Solution $\sqrt{\dfrac{7x}{3y}} = \dfrac{\sqrt{7x}}{\sqrt{3y}}$ Use the quotient rule. No radical may be simplified further.

$$= \frac{\sqrt{7x} \cdot \sqrt{3y}}{\sqrt{3y} \cdot \sqrt{3y}} \qquad \text{Multiply numerator and denominator by } \sqrt{3y} \text{ so that the radicand in the denominator is a perfect square.}$$

$$= \frac{\sqrt{21xy}}{3y} \qquad \text{Use the product rule in the numerator and denominator. Remember that } \sqrt{3y} \cdot \sqrt{3y} = 3y.$$

Example 3   Rationalize the denominator of $\dfrac{\sqrt[4]{x}}{\sqrt[4]{81y^5}}$.

Solution   First, simplify each radical if possible.

$$\frac{\sqrt[4]{x}}{\sqrt[4]{81y^5}} = \frac{\sqrt[4]{x}}{\sqrt[4]{81y^4} \cdot \sqrt[4]{y}} \qquad \text{Use the product rule in the denominator.}$$

$$= \frac{\sqrt[4]{x}}{3y\sqrt[4]{y}} \qquad \text{Write } \sqrt[4]{81y^4} \text{ as } 3y.$$

$$= \frac{\sqrt[4]{x} \cdot \sqrt[4]{y^3}}{3y\sqrt[4]{y} \cdot \sqrt[4]{y^3}} \qquad \text{Multiply numerator and denominator by } \sqrt[4]{y^3} \text{ so that the radicand in the denominator is a perfect fourth power.}$$

$$= \frac{\sqrt[4]{xy^3}}{3y\sqrt[4]{y^4}} \qquad \text{Use the product rule in the numerator and denominator.}$$

$$= \frac{\sqrt[4]{xy^3}}{3y^2} \qquad \text{In the denominator, } \sqrt[4]{y^4} = y \text{ and } 3y \cdot y = 3y^2.$$

2   As mentioned earlier, it is also often helpful to write an expression such as $\dfrac{\sqrt{3}}{\sqrt{2}}$ as an equivalent expression without a radical in the numerator. This process is called **rationalizing the numerator**. To rationalize the numerator of $\dfrac{\sqrt{3}}{\sqrt{2}}$, we multiply the numerator and the denominator by $\sqrt{3}$.

$$\frac{\sqrt{3}}{\sqrt{2}} = \frac{\sqrt{3} \cdot \sqrt{3}}{\sqrt{2} \cdot \sqrt{3}} = \frac{\sqrt{9}}{\sqrt{6}} = \frac{3}{\sqrt{6}}$$

Example 4   Rationalize the numerator of $\dfrac{\sqrt{7}}{\sqrt{45}}$.

Solution   First we simplify $\sqrt{45}$.

$$\frac{\sqrt{7}}{\sqrt{45}} = \frac{\sqrt{7}}{\sqrt{9 \cdot 5}} = \frac{\sqrt{7}}{3\sqrt{5}}$$

Next we rationalize the numerator by multiplying the numerator and the denominator by $\sqrt{7}$.

$$\frac{\sqrt{7}}{3\sqrt{5}} = \frac{\sqrt{7} \cdot \sqrt{7}}{3\sqrt{5} \cdot \sqrt{7}} = \frac{7}{3\sqrt{5 \cdot 7}} = \frac{7}{3\sqrt{35}}$$

Example 5   Rationalize the numerator of $\dfrac{\sqrt[3]{2x^2}}{\sqrt[3]{5y}}$.

Solution   The numerator and the denominator of this expression are already simplified. To rationalize the numerator, $\sqrt[3]{2x^2}$, we multiply the numerator and denominator by a factor that will make the radicand a perfect cube. If we multiply $\sqrt[3]{2x^2}$ by $\sqrt[3]{4x}$, we get $\sqrt[3]{8x^3} = 2x$.

$$\frac{\sqrt[3]{2x^2}}{\sqrt[3]{5y}} = \frac{\sqrt[3]{2x^2} \cdot \sqrt[3]{4x}}{\sqrt[3]{5y} \cdot \sqrt[3]{4x}} = \frac{\sqrt[3]{8x^3}}{\sqrt[3]{20xy}} = \frac{2x}{\sqrt[3]{20xy}}$$

3   Remember the product of the sum and difference of two terms?

$$(a + b)(a - b) = a^2 - b^2$$

These two expressions are called conjugates of each other.

To rationalize a numerator or denominator that is a sum or difference of two terms, we use conjugates. To see how and why this works, let's rationalize the denominator of the expression $\dfrac{5}{\sqrt{3} - 2}$. To do so, we multiply both the numerator and the denominator by $\sqrt{3} + 2$, the **conjugate** of the denominator $\sqrt{3} - 2$, and see what happens.

$$\frac{5}{\sqrt{3} - 2} = \frac{5(\sqrt{3} + 2)}{(\sqrt{3} - 2)(\sqrt{3} + 2)}$$

$$= \frac{5(\sqrt{3} + 2)}{(\sqrt{3})^2 - 2^2} \qquad \text{Multiply the sum and difference of two terms: } (a + b)(a - b) = a^2 - b^2.$$

$$= \frac{5(\sqrt{3} + 2)}{3 - 4}$$

$$= \frac{5(\sqrt{3} + 2)}{-1}$$

$$= -5(\sqrt{3} + 2) \quad \text{or} \quad -5\sqrt{3} - 10$$

Notice in the denominator that the product of $(\sqrt{3} - 2)$ and its conjugate, $(\sqrt{3} + 2)$, is -1. In general, the product of an expression and its conjugate will contain no radical terms. This is why, when rationalizing a denominator or a numerator containing two terms, we multiply by its conjugate. Examples of conjugates are

$$\sqrt{a} - \sqrt{b} \qquad \text{and} \qquad \sqrt{a} + \sqrt{b}$$

$$x + \sqrt{y} \qquad \text{and} \qquad x - \sqrt{y}$$

Example 6   Rationalize each denominator.

a. $\dfrac{2}{3\sqrt{2} + 4}$ **b.** $\dfrac{\sqrt{6} + 2}{\sqrt{5} - \sqrt{3}}$ **c.** $\dfrac{2\sqrt{m}}{3\sqrt{x} + \sqrt{m}}$

Solution   **a.** Multiply the numerator and denominator by the conjugate of the denominator, $3\sqrt{2} + 4$.

$$\frac{2}{3\sqrt{2} + 4} = \frac{2(3\sqrt{2} - 4)}{(3\sqrt{2} + 4)(3\sqrt{2} - 4)}$$

$$= \frac{2(3\sqrt{2} - 4)}{(3\sqrt{2})^2 - 4^2}$$

$$= \frac{2(3\sqrt{2} - 4)}{18 - 16}$$

$$= \frac{2(3\sqrt{2} - 4)}{2}, \quad \text{or} \quad 3\sqrt{2} - 4$$

It is often helpful to leave a numerator in factored form to help determine whether the expression can be simplified.

b. Multiply the numerator and denominator by the conjugate of $\sqrt{5} - \sqrt{3}$.

$$\frac{\sqrt{6} + 2}{\sqrt{5} - \sqrt{3}} = \frac{(\sqrt{6} + 2)(\sqrt{5} + \sqrt{3})}{(\sqrt{5} - \sqrt{3})(\sqrt{5} + \sqrt{3})}$$

$$= \frac{\sqrt{6}\sqrt{5} + \sqrt{6}\sqrt{3} + 2\sqrt{5} + 2\sqrt{3}}{(\sqrt{5})^2 - (\sqrt{3})^2}$$

$$= \frac{\sqrt{30} + \sqrt{18} + 2\sqrt{5} + 2\sqrt{3}}{5 - 3}$$

$$= \frac{\sqrt{30} + 3\sqrt{2} + 2\sqrt{5} + 2\sqrt{3}}{2}$$

c. Multiply by the conjugate of $3\sqrt{x} + \sqrt{m}$ to eliminate the radicals from the denominator.

$$\frac{2\sqrt{m}}{3\sqrt{x} + \sqrt{m}} = \frac{2\sqrt{m}(3\sqrt{x} - \sqrt{m})}{(3\sqrt{x} + \sqrt{m})(3\sqrt{x} - \sqrt{m})} = \frac{6\sqrt{mx} - 2m}{(3\sqrt{x})^2 - (\sqrt{m})^2}$$

$$= \frac{6\sqrt{mx} - 2m}{9x - m}$$

Example 7   Rationalize the numerator of $\dfrac{\sqrt{x} + 2}{5}$.

Solution   We multiply the numerator and the denominator by the conjugate of the numerator, $\sqrt{x} + 2$.

$$\frac{\sqrt{x} + 2}{5} = \frac{(\sqrt{x} + 2)(\sqrt{x} - 2)}{5(\sqrt{x} - 2)} \qquad \text{Multiply by } \sqrt{x} - 2, \text{ the conjugate of } \sqrt{x} + 2.$$

$$= \frac{(\sqrt{x})^2 - 2^2}{5(\sqrt{x} - 2)} \qquad (a + b)(a - b) = a^2 - b^2$$

$$= \frac{x - 4}{5(\sqrt{x} - 2)}$$

MENTAL MATH

Find the conjugate of each expression.

1. $\sqrt{2} + x$ **2.** $\sqrt{3} + y$ **3.** $5 - \sqrt{a}$

4. $6 - \sqrt{b}$ **5.** $7\sqrt{5} + 8\sqrt{x}$ **6.** $9\sqrt{2} - 6\sqrt{y}$

Exercise Set 9.5

Rationalize each denominator. See Examples 1 through 3.

1. $\dfrac{\sqrt{2}}{\sqrt{7}}$ **2.** $\dfrac{\sqrt{3}}{\sqrt{2}}$ **3.** $\sqrt{\dfrac{1}{5}}$ **4.** $\sqrt{\dfrac{1}{2}}$

5. $\sqrt[3]{\dfrac{3}{4}}$

6. $\sqrt[3]{\dfrac{2}{9}}$

7. $\dfrac{4}{\sqrt[3]{3}}$

8. $\dfrac{6}{\sqrt[3]{9}}$

9. $\dfrac{3}{\sqrt{8x}}$

10. $\dfrac{5}{\sqrt{27a}}$

11. $\dfrac{3}{\sqrt[3]{4x^2}}$

12. $\dfrac{5}{\sqrt[3]{3y}}$

13. $\sqrt{\dfrac{4}{x}}$

14. $\sqrt{\dfrac{25}{y}}$

15. $\dfrac{9}{\sqrt{3a}}$

16. $\dfrac{x}{\sqrt{5}}$

17. $\dfrac{3}{\sqrt[3]{2}}$

18. $\dfrac{5}{\sqrt[3]{9}}$

19. $\dfrac{2\sqrt{3}}{\sqrt{7}}$

20. $\dfrac{-5\sqrt{2}}{\sqrt{11}}$

21. $\sqrt{\dfrac{2x}{5y}}$

22. $\sqrt{\dfrac{13a}{2b}}$

23. $\sqrt[4]{\dfrac{81}{8}}$

24. $\sqrt[4]{\dfrac{1}{9}}$

25. $\sqrt[4]{\dfrac{16}{9x^7}}$

26. $\sqrt[5]{\dfrac{32}{m^6n^{13}}}$

27. $\dfrac{5a}{\sqrt[5]{8a^9b^{11}}}$

28. $\dfrac{9y}{\sqrt[4]{4y^9}}$

45. $\sqrt{\dfrac{18x^4y^6}{3z}}$

46. $\sqrt{\dfrac{8x^5y}{2z}}$

47. When rationalizing the denominator of $\dfrac{\sqrt{5}}{\sqrt{7}}$, explain why both the numerator and the denominator must be multiplied by $\sqrt{7}$.

48. When rationalizing the numerator of $\dfrac{\sqrt{5}}{\sqrt{7}}$, explain why both the numerator and the denominator must be multiplied by $\sqrt{5}$.

Rationalize each denominator. See Example 6.

49. $\dfrac{6}{2 - \sqrt{7}}$

50. $\dfrac{3}{\sqrt{7} - 4}$

51. $\dfrac{-7}{\sqrt{x} - 3}$

52. $\dfrac{-8}{\sqrt{y} + 4}$

53. $\dfrac{\sqrt{2} - \sqrt{3}}{\sqrt{2} + \sqrt{3}}$

54. $\dfrac{\sqrt{3} + \sqrt{4}}{\sqrt{2} + \sqrt{3}}$

55. $\dfrac{\sqrt{a} + 1}{2\sqrt{a} - \sqrt{b}}$

56. $\dfrac{2\sqrt{a} - 3}{2\sqrt{a} - \sqrt{b}}$

57. $\dfrac{8}{1 + \sqrt{10}}$

58. $\dfrac{-3}{\sqrt{6} - 2}$

59. $\dfrac{\sqrt{x}}{\sqrt{x} + \sqrt{y}}$

60. $\dfrac{2\sqrt{a}}{2\sqrt{x} - \sqrt{y}}$

61. $\dfrac{2\sqrt{3} + \sqrt{6}}{4\sqrt{3} - \sqrt{6}}$

62. $\dfrac{4\sqrt{5} + \sqrt{2}}{2\sqrt{5} - \sqrt{2}}$

Rationalize each numerator. See Examples 4 and 5.

29. $\sqrt{\dfrac{5}{3}}$

30. $\sqrt{\dfrac{3}{2}}$

31. $\sqrt{\dfrac{18}{5}}$

32. $\sqrt{\dfrac{12}{7}}$

33. $\dfrac{\sqrt{4x}}{7}$ $\quad \sqrt{x}$

34. $\dfrac{\sqrt{3x^5}}{6}$

35. $\dfrac{\sqrt[3]{5y^2}}{\sqrt[3]{4x}}$

36. $\dfrac{\sqrt[3]{4x}}{\sqrt[3]{z^4}}$

37. $\sqrt{\dfrac{2}{5}}$

38. $\sqrt{\dfrac{3}{7}}$

39. $\dfrac{\sqrt{2x}}{11}$

40. $\dfrac{\sqrt{y}}{7}$

41. $\sqrt[3]{\dfrac{7}{8}}$

42. $\sqrt[3]{\dfrac{25}{2}}$

43. $\dfrac{\sqrt[3]{3x^5}}{10}$

44. $\sqrt[3]{\dfrac{9y}{7}}$

Rationalize each numerator. See Example 7.

63. $\dfrac{2 - \sqrt{11}}{6}$

64. $\dfrac{\sqrt{15} + 1}{2}$

65. $\dfrac{2 - \sqrt{7}}{-5}$

66. $\dfrac{\sqrt{5} + 2}{\sqrt{2}}$

67. $\dfrac{\sqrt{x} + 3}{\sqrt{x}}$

68. $\dfrac{5 + \sqrt{2}}{\sqrt{2x}}$

69. $\dfrac{\sqrt{2} - 1}{\sqrt{2} + 1}$

70. $\dfrac{\sqrt{8} - \sqrt{3}}{\sqrt{2} + \sqrt{3}}$

71. $\dfrac{\sqrt{x} + 1}{\sqrt{x} - 1}$

72. $\dfrac{\sqrt{x} + \sqrt{y}}{\sqrt{x} - \sqrt{y}}$

△ 73. The formula of the radius of a sphere r with surface area A is given by the formula

$$r = \sqrt{\dfrac{A}{4\pi}}$$

Rationalize the denominator of the radical expression in this formula.

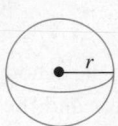

△ **74.** The formula for the radius of a cone r with height 7 centimeters and volume V is given by the formula

$$r = \sqrt{\frac{3V}{7\pi}}$$

Rationalize the numerator of the radical expression in this formula.

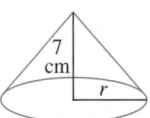

75. Explain why rationalizing the denominator does not change the value of the original expression.

76. Explain why rationalizing the numerator does not change the value of the original expression.

REVIEW EXERCISES

Solve each equation. See Sections 2.2, 6.6, and 6.7.

77. $2x - 7 = 3(x - 4)$ **78.** $9x - 4 = 7(x - 2)$

79. $(x - 6)(2x + 1) = 0$ **80.** $(y + 2)(5y + 4) = 0$

81. $x^2 - 8x = -12$ **82.** $x^3 = x$

9.6 RADICAL EQUATIONS AND PROBLEM SOLVING

CD-ROM SSM

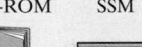

SSG Video

▶ **OBJECTIVES**

1. Solve equations that contain radical expressions.
2. Use the Pythagorean theorem to model problems.

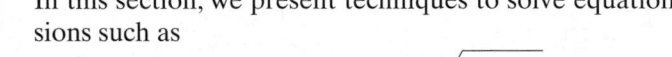

1 In this section, we present techniques to solve equations containing radical expressions such as

$$\sqrt{2x - 3} = 9$$

We use the power rule to help us solve these radical equations.

> **POWER RULE**
>
> If both sides of an equation are raised to the same power, **all** solutions of the original equation are **among** the solutions of the new equation.

This property *does not* say that raising both sides of an equation to a power yields an equivalent equation. A solution of the new equation *may or may not* be a solution of the original equation. Thus, *each solution of the new equation must be checked* to make sure it is a solution of the original equation. Recall that a proposed solution that is not a solution of the original equation is called an **extraneous solution**.

Example 1 Solve $\sqrt{2x - 3} = 9$.

Solution We use the power rule to square both sides of the equation to eliminate the radical.

$$\sqrt{2x - 3} = 9$$
$$(\sqrt{2x - 3})^2 = 9^2$$
$$2x - 3 = 81$$
$$2x = 84$$
$$x = 42$$

Now we, check the solution in the original equation.

Check:
$$\sqrt{2x - 3} = 9$$
$$\sqrt{2(42) - 3} \stackrel{?}{=} 9 \qquad \text{Let } x = 42.$$
$$\sqrt{84 - 3} \stackrel{?}{=} 9$$
$$\sqrt{81} \stackrel{?}{=} 9$$
$$9 = 9 \qquad \text{True.}$$

The solution checks, so we conclude that the solution is 42.

To solve a radical equation, first isolate a radical on one side of the equation.

Example 2   Solve $\sqrt{-10x - 1} + 3x = 0$.

Solution   First, isolate the radical on one side of the equation. To do this, we subtract $3x$ from both sides.

$$\sqrt{-10x - 1} + 3x = 0$$
$$\sqrt{-10x - 1} + 3x - 3x = 0 - 3x$$
$$\sqrt{-10x - 1} = -3x$$

Next we use the power rule to eliminate the radical.

$$\left(\sqrt{-10x - 1}\right)^2 = (-3x)^2$$
$$-10x - 1 = 9x^2$$

Since this is a quadratic equation, we can set the equation equal to 0 and try to solve by factoring.

$$9x^2 + 10x + 1 = 0$$
$$(9x + 1)(x + 1) = 0 \qquad \text{Factor.}$$
$$9x + 1 = 0 \quad \text{or} \quad x + 1 = 0 \qquad \text{Set each factor equal to 0.}$$
$$x = -\frac{1}{9} \qquad\qquad x = -1$$

Check: Let $x = -\frac{1}{9}$. $\qquad\qquad\qquad$ Let $x = -1$.

$$\sqrt{-10x - 1} + 3x = 0 \qquad\qquad \sqrt{-10x - 1} + 3x = 0$$
$$\sqrt{-10\left(-\frac{1}{9}\right) - 1} + 3\left(-\frac{1}{9}\right) \stackrel{?}{=} 0 \qquad \sqrt{-10(-1) - 1} + 3(-1) \stackrel{?}{=} 0$$
$$\sqrt{\frac{10}{9} - \frac{9}{9}} - \frac{3}{9} \stackrel{?}{=} 0 \qquad\qquad \sqrt{10 - 1} - 3 \stackrel{?}{=} 0$$
$$\sqrt{\frac{1}{9}} - \frac{1}{3} \stackrel{?}{=} 0 \qquad\qquad\qquad \sqrt{9} - 3 \stackrel{?}{=} 0$$
$$\frac{1}{3} - \frac{1}{3} = 0 \quad \text{True.} \qquad\qquad 3 - 3 = 0 \quad \text{True.}$$

Both solutions check. The solutions are $-\frac{1}{9}$ and -1.

The following steps may be used to solve a radical equation.

SOLVING A RADICAL EQUATION

Step 1: Isolate one radical on one side of the equation.
Step 2: Raise each side of the equation to a power equal to the index of the radical and simplify.
Step 3: If the equation still contains a radical term, repeat Steps 1 and 2. If not, solve the equation.
Step 4: Check all proposed solutions in the original equation.

Example 3 Solve $\sqrt[3]{x + 1} + 5 = 3$.

Solution First we isolate the radical by subtracting 5 from both sides of the equation.

$$\sqrt[3]{x + 1} + 5 = 3$$
$$\sqrt[3]{x + 1} = -2$$

Next we raise both sides of the equation to the third power to eliminate the radical.

$$\left(\sqrt[3]{x + 1}\right)^3 = (-2)^3$$
$$x + 1 = -8$$
$$x = -9$$

The solution checks in the original equation, so the solution is -9.

Example 4 Solve $\sqrt{4 - x} = x - 2$.

Solution

$$\sqrt{4 - x} = x - 2$$
$$\left(\sqrt{4 - x}\right)^2 = (x - 2)^2$$
$$4 - x = x^2 - 4x + 4$$
$$x^2 - 3x = 0 \qquad \text{Write the quadratic equation in standard form.}$$
$$x(x - 3) = 0 \qquad \text{Factor.}$$
$$x = 0 \quad \text{or} \quad x - 3 = 0 \qquad \text{Set each factor equal to 0.}$$
$$x = 3$$

Check:

$$\sqrt{4 - x} = x - 2 \qquad\qquad \sqrt{4 - x} = x - 2$$
$$\sqrt{4 - 0} \stackrel{?}{=} 0 - 2 \quad \text{Let } x = 0. \qquad \sqrt{4 - 3} \stackrel{?}{=} 3 - 2 \quad \text{Let } x = 3.$$
$$2 = -2 \qquad \text{False.} \qquad\qquad 1 = 1 \qquad \text{True.}$$

The proposed solution 3 checks, but 0 does not. Since 0 is an extraneous solution, the only solution is 3.

HELPFUL HINT
In Example 4, notice that $(x - 2)^2 = x^2 - 4x + 4$. Make sure binomials are squared correctly.

Example 5 Solve $\sqrt{2x + 5} + \sqrt{2x} = 3$.

Solution We get one radical alone by subtracting $\sqrt{2x}$ from both sides.

$$\sqrt{2x + 5} + \sqrt{2x} = 3$$
$$\sqrt{2x + 5} = 3 - \sqrt{2x}$$

Now we use the power rule to begin eliminating the radicals. First we square both sides.

$$\left(\sqrt{2x + 5}\right)^2 = \left(3 - \sqrt{2x}\right)^2$$

$$2x + 5 = 9 - 6\sqrt{2x} + 2x \qquad \text{Multiply} \\ \left(3 - \sqrt{2x}\right)\left(3 - \sqrt{2x}\right).$$

There is still a radical in the equation, so we get the radical alone again. Then we square both sides.

$$2x + 5 = 9 - 6\sqrt{2x} + 2x \qquad \text{Get the radical alone.}$$

$$6\sqrt{2x} = 4$$

$$36(2x) = 16 \qquad \text{Square both sides of the equation to eliminate the radical.}$$

$$72x = 16 \qquad \text{Multiply.}$$

$$x = \frac{16}{72} \qquad \text{Solve.}$$

$$x = \frac{2}{9} \qquad \text{Simplify.}$$

The proposed solution, $\frac{2}{9}$, checks in the original equation. The solution is $\frac{2}{9}$.

HELPFUL HINT
Make sure expressions are squared correctly. In Example 5, we squared $\left(3 - \sqrt{2x}\right)$ as

$$\left(3 - \sqrt{2x}\right)^2 = \left(3 - \sqrt{2x}\right)\left(3 - \sqrt{2x}\right)$$
$$= 3 \cdot 3 - 3\sqrt{2x} - 3\sqrt{2x} + \sqrt{2x} \cdot \sqrt{2x}$$
$$= 9 - 6\sqrt{2x} + 2x$$

---

2 Recall that the Pythagorean theorem states that in a right triangle, the length of the hypotenuse squared equals the sum of the lengths of each of the legs squared.

PYTHAGOREAN THEOREM

If a and b are the lengths of the legs of a right triangle and c is the length of the hypotenuse, then $a^2 + b^2 = c^2$.

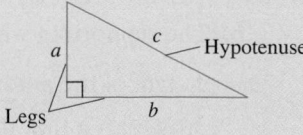

△ **Example 6** Find the length of the unknown leg of the right triangle.

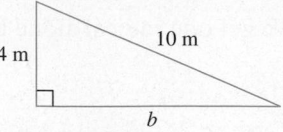

Solution In the formula $a^2 + b^2 = c^2$, c is the length of the hypotenuse. Here, $c = 10$, the length of the hypotenuse, and $a = 4$. We solve for b. Then $a^2 + b^2 = c^2$ becomes

$$4^2 + b^2 = 10^2$$
$$16 + b^2 = 100$$
$$b^2 = 84 \qquad \text{Subtract 16 from both sides.}$$

Since b is a length and thus is positive, we have that

$$b = \sqrt{84} = \sqrt{4 \cdot 21} = 2\sqrt{21}$$

The unknown leg of the triangle is $2\sqrt{21}$ meters long. ▬

△ **Example 7** **CALCULATING PLACEMENT OF A WIRE**

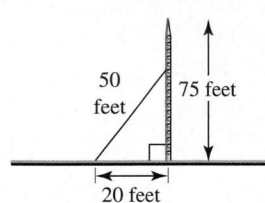

A 50-foot supporting wire is to be attached to a 75-foot antenna. Because of surrounding buildings, sidewalks, and roadways, the wire must be anchored exactly 20 feet from the base of the antenna.

a. How high from the base of the antenna is the wire attached?
b. Local regulations require that a supporting wire be attached at a height no less than $\frac{3}{5}$ of the total height of the antenna. From part **a**, have local regulations been met?

Solution 1. UNDERSTAND. Read and reread the problem. From the diagram we notice that a right triangle is formed with hypotenuse 50 feet and one leg 20 feet. Let x be the height from the base of the antenna to the attached wire.

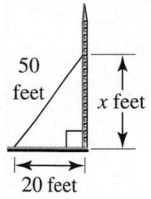

2. TRANSLATE. Use the Pythagorean theorem.

$$a^2 + b^2 = c^2$$
$$20^2 + x^2 = 50^2 \qquad a = 20, c = 50$$

3. SOLVE. $20^2 + x^2 = 50^2$

$$400 + x^2 = 2500$$
$$x^2 = 2100 \qquad \text{Subtract 400 from both sides.}$$
$$x = \sqrt{2100}$$
$$= 10\sqrt{21}$$

4. INTERPRET. *Check* the work and *state* the solution.

a. The wire is attached exactly $10\sqrt{21}$ feet from the base of the pole, or approximately 45.8 feet.

b. The supporting wire must be attached at a height no less than $\frac{3}{5}$ of the total height of the antenna. This height is $\frac{3}{5}$ (75 feet), or 45 feet. Since we know from part **a** that the wire is to be attached at a height of approximately 45.8 feet, local regulations have been met. ▬

GRAPHING CALCULATOR EXPLORATIONS

We can use a graphing calculator to solve radical equations. For example, to use a graphing calculator to approximate the solutions of the equation solved in Example 4, we graph the following.

$$Y_1 = \sqrt{4 - x} \quad \text{and} \quad Y_2 = x - 2$$

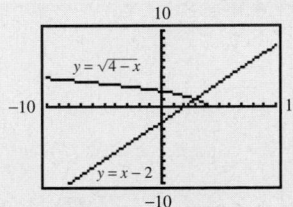

The x-value of the point of intersection is the solution. Use the Intersect feature or the Zoom and Trace features of your graphing calculator to see that the solution is 3.

Use a graphing calculator to solve each radical equation. Round all solutions to the nearest hundredth.

1. $\sqrt{x + 7} = x$
2. $\sqrt{3x + 5} = 2x$
3. $\sqrt{2x + 1} = \sqrt{2x + 2}$
4. $\sqrt{10x - 1} = \sqrt{-10x + 10} - 1$
5. $1.2x = \sqrt{3.1x + 5}$
6. $\sqrt{1.9x^2 - 2.2} = -0.8x + 3$

SPOTLIGHT ON DECISION MAKING

Suppose you are a psychologist studying a person's ability to recognize patterns. You theorize that IQ is linked to pattern-recognition ability and design a research study using human subjects to test your theory. As part of your study, you have decided to form three groups of test subjects based roughly on IQ. Group A will consist of subjects having IQ's under 90, Group B will consist of subjects having IQ's from 90 to 105, and Group C will consist of subjects having IQ's over 105.

While preparing for your research study, you came across the findings of another psychologist suggesting that the number S of nonsense syllables that a person can repeat consecutively depends on his or her IQ score I according to the equation $S = 2\sqrt{I} - 9$. Because administering IQ tests can be time-consuming and because your groupings by IQ need only be approximate, you decide to use this equation as a quick way to assign test subjects

to Groups A, B, and C. Each subject is individually screened by listening to a string of 20 random nonsense syllables and then repeating as many as possible. The results for the first 5 test subjects are listed in the table. For each subject, decide to which group—A, B, or C—the subject should be assigned.

TEST SUBJECT SCREENING

Subject	S—the number of Nonsense Syllables Successfully Repeated
1	11
2	13
3	9
4	12
5	10

Exercise Set 9.6

Solve. See Examples 1 and 2.

1. $\sqrt{2x} = 4$

2. $\sqrt{3x} = 3$

3. $\sqrt{x - 3} = 2$

4. $\sqrt{x + 1} = 5$

5. $\sqrt{2x} = -4$

6. $\sqrt{5x} = -5$

7. $\sqrt{4x - 3} - 5 = 0$

8. $\sqrt{x - 3} - 1 = 0$

9. $\sqrt{2x - 3} - 2 = 1$

10. $\sqrt{3x + 3} - 4 = 8$

Solve. See Example 3.

11. $\sqrt[3]{6x} = -3$

12. $\sqrt[3]{4x} = -2$

13. $\sqrt[3]{x - 2} - 3 = 0$

14. $\sqrt[3]{2x - 6} - 4 = 0$

Solve. See Examples 4 and 5.

15. $\sqrt{13 - x} = x - 1$

16. $\sqrt{2x - 3} = 3 - x$

17. $x - \sqrt{4 - 3x} = -8$

18. $2x + \sqrt{x + 1} = 8$

19. $\sqrt{y + 5} = 2 - \sqrt{y - 4}$

20. $\sqrt{x + 3} + \sqrt{x - 5} = 3$

21. $\sqrt{x - 3} + \sqrt{x + 2} = 5$

22. $\sqrt{2x - 4} - \sqrt{3x + 4} = -2$

Solve. See Examples 1 through 5.

23. $\sqrt{3x - 2} = 5$

24. $\sqrt{5x - 4} = 9$

25. $-\sqrt{2x} + 4 = -6$

26. $-\sqrt{3x + 9} = -12$

27. $\sqrt{3x + 1} + 2 = 0$

28. $\sqrt{3x + 1} - 2 = 0$

29. $\sqrt[4]{4x + 1} - 2 = 0$

30. $\sqrt[4]{2x - 9} - 3 = 0$

31. $\sqrt{4x - 3} = 7$

32. $\sqrt{3x + 9} = 6$

33. $\sqrt[3]{6x - 3} - 3 = 0$

34. $\sqrt[3]{3x} + 4 = 7$

35. $\sqrt[3]{2x - 3} - 2 = -5$

36. $\sqrt[3]{x - 4} - 5 = -7$

37. $\sqrt{x + 4} = \sqrt{2x - 5}$

38. $\sqrt{3y + 6} = \sqrt{7y - 6}$

39. $x - \sqrt{1 - x} = -5$

40. $x - \sqrt{x - 2} = 4$

41. $\sqrt[3]{-6x - 1} = \sqrt[3]{-2x - 5}$

42. $x + \sqrt{x + 5} = 7$

43. $\sqrt{5x - 1} - \sqrt{x + 2} = 3$

44. $\sqrt{2x - 1} - 4 = -\sqrt{x - 4}$

45. $\sqrt{2x - 1} = \sqrt{1 - 2x}$

46. $\sqrt{7x - 4} = \sqrt{4 - 7x}$

47. $\sqrt{3x + 4} - 1 = \sqrt{2x + 1}$

48. $\sqrt{x - 2} + 3 = \sqrt{4x + 1}$

49. $\sqrt{y + 3} - \sqrt{y - 3} = 1$

50. $\sqrt{x + 1} - \sqrt{x - 1} = 2$

51. What is wrong with the following steps?

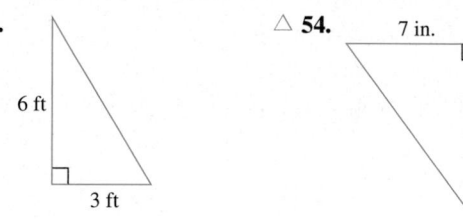

$$\sqrt{2x + 5} + \sqrt{4 - x} = 8$$
$$\left(\sqrt{2x + 5} + \sqrt{4 - x}\right)^2 = 8^2$$
$$(2x + 5) + (4 - x) = 64$$
$$x + 9 = 64$$
$$x = 55$$

52. How can you immediately tell that the equation $\sqrt{2y + 3} = -4$ has no real solution?

Find the length of the unknown side of each triangle. See Example 6.

53.

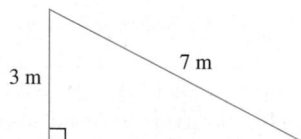

6 ft   3 ft

54.   7 in.   8 in.

55.

3 m   7 m

56.

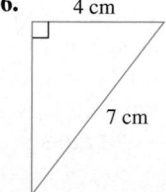

4 cm   7 cm

Find the length of the unknown side of each triangle. Give the exact length and a one-decimal-place approximation. See Example 6.

57.

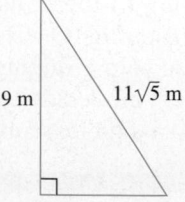

9 m $11\sqrt{5}$ m

△ **58.**
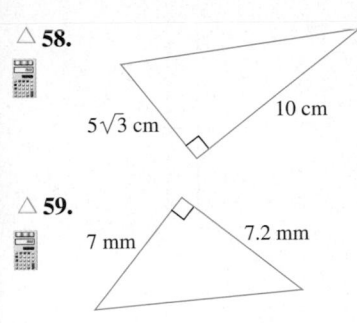
5√3 cm

10 cm

△ **59.**
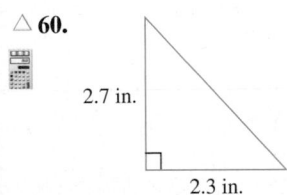
7 mm   7.2 mm

△ **60.**
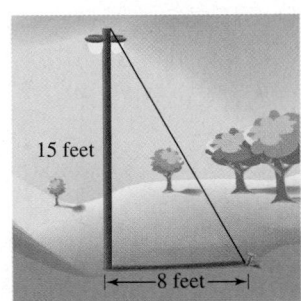
2.7 in.

2.3 in.

Solve. See Example 7. Give exact answers and two-decimal-place approximations where appropriate.

△ **61.** A wire is needed to support a vertical pole 15 feet high. The cable will be anchored to a stake 8 feet from the base of the pole. How much cable is needed?

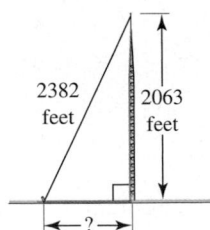
15 feet

|←———8 feet———→|

△ **62.** The tallest structure in the United States is a TV tower in Blanchard, North Dakota. Its height is 2063 feet. A 2382-foot length of wire is to be used as a guy wire attached to the top of the tower. Approximate to the nearest foot how far from the base of the tower the guy wire must be anchored. (*Source:* U.S. Geological Survey)

2382 feet   2063 feet

|←—?—→|

△ **63.** A spotlight is mounted on the eaves of a house 12 feet above the ground. A flower bed runs between the house and the sidewalk, so the closest the ladder can be placed to the house is 5 feet. How long a ladder is needed so

that an electrician can reach the place where the light is mounted?

12 feet

|← 5 →|
feet

△ **64.** A wire is to be attached to support a telephone pole. Because of surrounding buildings, sidewalks, and roadways, the wire must be anchored exactly 15 feet from the base of the pole. Telephone company workers have only 30 feet of cable, and 2 feet of that must be used to attach the cable to the pole and to the stake on the ground. How high from the base of the pole can the wire be attached?

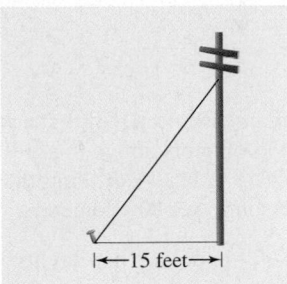
|←—15 feet—→|

△ **65.** The radius of the Moon is 1080 miles. Use the formula for the radius r of a sphere given its surface area A,

$$r = \sqrt{\frac{A}{4\pi}}$$

to find the surface area of the Moon. Round to the nearest square mile. (*Source:* National Space Science Data Center)

66. Police departments find it very useful to be able to approximate the speed of a car when they are given the distance that the car skidded before it came to a stop. If the road surface is wet concrete, the function $S(x) = \sqrt{10.5x}$ is used, where $S(x)$ is the speed of the car in miles per hour and x is the distance skidded in feet. Find how fast a car was moving if it skidded 280 feet on wet concrete.

67. The formula $v = \sqrt{2gh}$ gives the velocity v, in feet per second, of an object when it falls h feet accelerated by gravity g, in feet per second squared. If g is approximately 32 feet per second squared, find how far an object has fallen if its velocity is 80 feet per second.

68. Two tractors are pulling a tree stump from a field. If two forces A and B pull at right angles (90°) to each other, the size of the resulting force R is given by the formula $R = \sqrt{A^2 + B^2}$. If tractor A is exerting 600 pounds of force and the resulting force is 850 pounds, find how much force tractor B is exerting.

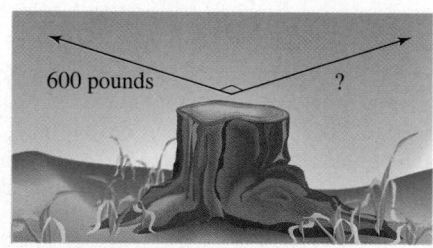
600 pounds ?

69. Solve: $\sqrt{\sqrt{x + 3} + \sqrt{x}} = \sqrt{3}$

70. The maximum distance $D(h)$ that a person can see from a height h kilometers above the ground is given by the function $D(h) = 111.7\sqrt{h}$. Find the height that would allow a person to see 80 kilometers.

71. The cost $C(x)$ in dollars per day to operate a small delivery service is given by $C(x) = 80\sqrt[3]{x} + 500$, where x is the number of deliveries per day. In July, the manager decides that it is necessary to keep delivery costs below \$1620. Find the greatest number of deliveries this company can make per day and still keep overhead below \$1620.

72. Explain why proposed solutions of radical equations must be checked.

73. Consider the equations $\sqrt{2x} = 4$ and $\sqrt[3]{2x} = 4$.
  a. Explain the difference in solving these equations.
  b. Explain the similarity in solving these equations.

REVIEW EXERCISES

Use the vertical line test to determine whether each graph represents the graph of a function. See Section 3.3.

74.
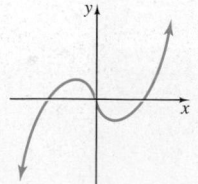

75.
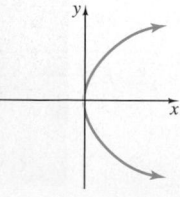

76.
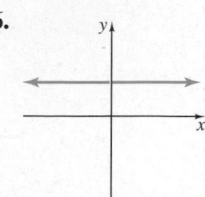

77.
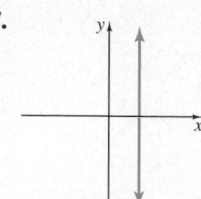

78.
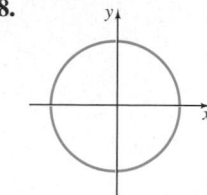

79.
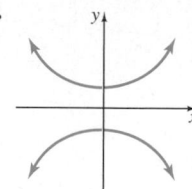

Simplify. See Section 7.5.

80. $\dfrac{\dfrac{x}{6}}{\dfrac{2x}{3} + \dfrac{1}{2}}$

81. $\dfrac{\dfrac{1}{y} + \dfrac{4}{5}}{\dfrac{-3}{20}}$

82. $\dfrac{\dfrac{z}{5} + \dfrac{1}{10}}{\dfrac{z}{20} - \dfrac{z}{5}}$

83. $\dfrac{\dfrac{1}{y} + \dfrac{1}{x}}{\dfrac{1}{y} - \dfrac{1}{x}}$

A Look Ahead

Example
Solve $(t^2 - 3t) - 2\sqrt{t^2 - 3t} = 0$.

Solution
Substitution can be used to make this problem somewhat simpler. Since $t^2 - 3t$ occurs more than once, let $x = t^2 - 3t$.

$$(t^2 - 3t) - 2\sqrt{t^2 - 3t} = 0$$
$$x - 2\sqrt{x} = 0 \qquad \text{Let } x = t^2 - 3t.$$
$$x = 2\sqrt{x}$$
$$x^2 = \left(2\sqrt{x}\right)^2$$
$$x^2 = 4x$$
$$x^2 - 4x = 0$$
$$x(x - 4) = 0$$
$$x = 0 \quad \text{or} \quad x - 4 = 0$$
$$x = 4$$

Now we "undo" the substitution.

$$x = 0 \qquad \text{Replace } x \text{ with } t^2 - 3t.$$
$$t^2 - 3t = 0$$
$$t(t - 3) = 0$$
$$t = 0 \quad \text{or} \quad t - 3 = 0$$
$$t = 3$$

$x = 4$ Replace x with $t^2 - 3t$.
$$t^2 - 3t = 4$$
$$t^2 - 3t - 4 = 0$$
$$(t - 4)(t + 1) = 0$$
$$t - 4 = 0 \quad \text{or} \quad t + 1 = 0$$
$$t = 4 \qquad\qquad t = -1$$

In this problem, we have four possible solutions: $0, 3, 4,$ and -1. All four solutions check in the original equation, so the solutions are $-1, 0, 3, 4$.

Solve. See the preceding example.

84. $3\sqrt{x^2 - 8x} = x^2 - 8x$

85. $\sqrt{(x^2 - x) + 7} = 2(x^2 - x) - 1$

86. $7 - (x^2 - 3x) = \sqrt{(x^2 - 3x) + 5}$

87. $x^2 + 6x = 4\sqrt{x^2 + 6x}$

9.7 COMPLEX NUMBERS

CD-ROM   SSM

SSG   Video

▶ **O B J E C T I V E S**

1. Define imaginary and complex numbers.
2. Add or subtract complex numbers.
3. Multiply complex numbers.
4. Divide complex numbers.
5. Raise i to powers.

---

1 Our work with radical expressions has excluded expressions such as $\sqrt{-16}$ because $\sqrt{-16}$ is not a real number; there is no real number whose square is -16. In this section, we discuss a number system that includes roots of negative numbers. This number system is the **complex number system**, and it includes the set of real numbers as a subset. The complex number system allows us to solve equations such as $x^2 + 1 = 0$ that have no real number solutions. The set of complex numbers includes the **imaginary unit**.

IMAGINARY UNIT

The imaginary unit, written i, is the number whose square is -1. That is,
$$i^2 = -1 \quad \text{and} \quad i = \sqrt{-1}$$

To write the square root of a negative number in terms of i, use the property that if a is a positive number, then
$$\sqrt{-a} = \sqrt{-1} \cdot \sqrt{a}$$
$$= i \cdot \sqrt{a}$$

Using i, we can write $\sqrt{-16}$ as
$$\sqrt{-16} = \sqrt{-1 \cdot 16} = \sqrt{-1} \cdot \sqrt{16} = i \cdot 4, \text{ or } 4i$$

Example 1   Write with i notation.

a. $\sqrt{-36}$ **b.** $\sqrt{-5}$ **c.** $-\sqrt{-20}$

Solution   **a.** $\sqrt{-36} = \sqrt{-1 \cdot 36} = \sqrt{-1} \cdot \sqrt{36} = i \cdot 6$, or $6i$
b. $\sqrt{-5} = \sqrt{-1(5)} = \sqrt{-1} \cdot \sqrt{5} = i\sqrt{5}$. Since $\sqrt{5}i$ can easily be confused with $\sqrt{5i}$, we write $\sqrt{5}i$ as $i\sqrt{5}$.
c. $-\sqrt{-20} = -\sqrt{-1 \cdot 20} = -\sqrt{-1} \cdot \sqrt{4 \cdot 5} = -i \cdot 2\sqrt{5} = -2i\sqrt{5}$

The product rule for radicals does not necessarily hold true for imaginary numbers. *To multiply square roots of negative numbers, first we write each number in terms of the imaginary unit i.* For example, to multiply $\sqrt{-4}$ and $\sqrt{-9}$, we first write each number in the form bi.

$$\sqrt{-4}\,\sqrt{-9} = 2i(3i) = 6i^2 = 6(-1) = -6$$

We will also use this method to simplify quotients of square roots of negative numbers.

Example 2 Multiply or divide as indicated.

    a. $\sqrt{-3}\cdot\sqrt{-5}$ **b.** $\sqrt{-36}\cdot\sqrt{-1}$ **c.** $\sqrt{8}\cdot\sqrt{-2}$ **d.** $\dfrac{\sqrt{-125}}{\sqrt{5}}$

Solution   **a.** $\sqrt{-3}\cdot\sqrt{-5} = i\sqrt{3}(i\sqrt{5}) = i^2\sqrt{15} = -1\sqrt{15} = -\sqrt{15}$

    b. $\sqrt{-36}\cdot\sqrt{-1} = 6i(i) = 6i^2 = 6(-1) = -6$

    c. $\sqrt{8}\cdot\sqrt{-2} = 2\sqrt{2}\,(i\sqrt{2}) = 2i(\sqrt{2}\,\sqrt{2}) = 2i(2) = 4i$

    d. $\dfrac{\sqrt{-125}}{\sqrt{5}} = \dfrac{i\sqrt{125}}{\sqrt{5}} = i\sqrt{25} = 5i$

Now that we have practiced working with the imaginary unit, we define complex numbers.

COMPLEX NUMBERS

A **complex number** is a number that can be written in the form $a + bi$, where a and b are real numbers.

Notice that the set of real numbers is a subset of the complex numbers since any real number can be written in the form of a complex number. For example,

$$16 = 16 + 0i$$

In general, a complex number $a + bi$ is a real number if $b = 0$. Also, a complex number is called an **imaginary number** if $a = 0$. For example,

$$3i = 0 + 3i \qquad \text{and} \qquad i\sqrt{7} = 0 + i\sqrt{7}$$

are imaginary numbers.

The following diagram shows the relationship between complex numbers and their subsets.

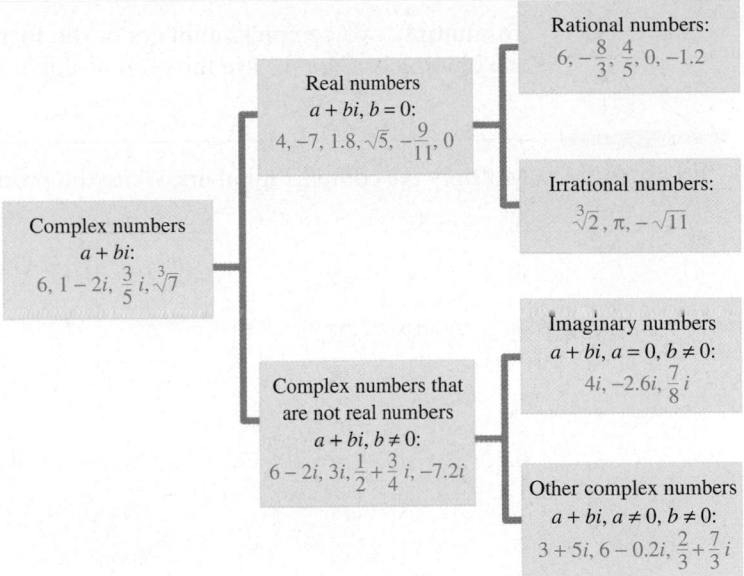

Complex numbers
$a + bi$:
$6, 1 - 2i, \frac{3}{5} i, \sqrt[3]{7}$

Real numbers
$a + bi, b = 0$:
$4, -7, 1.8, \sqrt{5}, -\frac{9}{11}, 0$

Rational numbers:
$6, -\frac{8}{3}, \frac{4}{5}, 0, -1.2$

Irrational numbers:
$\sqrt[3]{2}, \pi, -\sqrt{11}$

Complex numbers that
are not real numbers
$a + bi, b \neq 0$:
$6 - 2i, 3i, \frac{1}{2} + \frac{3}{4} i, -7.2i$

Imaginary numbers
$a + bi, a = 0, b \neq 0$:
$4i, -2.6i, \frac{7}{8} i$

Other complex numbers
$a + bi, a \neq 0, b \neq 0$:
$3 + 5i, 6 - 0.2i, \frac{2}{3} + \frac{7}{3} i$

2 Two complex numbers $a + bi$ and $c + di$ are equal if and only if $a = c$ and $b = d$. Complex numbers can be added or subtracted by adding or subtracting their real parts and then adding or subtracting their imaginary parts.

> ### SUM OR DIFFERENCE OF COMPLEX NUMBERS
>
> If $a + bi$ and $c + di$ are complex numbers, then their sum is
> $$(a + bi) + (c + di) = (a + c) + (b + d)i$$
> Their difference is
> $$(a + bi) - (c + di) = a + bi - c - di = (a - c) + (b - d)i$$

Example 3 Add or subtract the complex numbers. Write the sum or difference in the form $a + bi$.

a. $(2 + 3i) + (-3 + 2i)$ **b.** $(5i) - (1 - i)$ **c.** $(-3 - 7i) - (-6)$

Solution **a.** $(2 + 3i) + (-3 + 2i) = (2 - 3) + (3 + 2)i = -1 + 5i$

b. $5i - (1 - i) = 5i - 1 + i$

$$= -1 + (5 + 1)i$$

$$= -1 + 6i$$

c. $(-3 - 7i) - (-6) = -3 - 7i + 6$

$$= (-3 + 6) - 7i$$

$$= 3 - 7i$$

3 To multiply two complex numbers of the form $a + bi$, we multiply as though they are binomials. Then we use the relationship $i^2 = -1$ to simplify.

Example 4 Multiply the complex numbers. Write the product in the form $a + bi$.

a. $-7i \cdot 3i$ **b.** $3i(2 - i)$ **c.** $(2 - 5i)(4 + i)$

d. $(2 - i)^2$ **e.** $(7 + 3i)(7 - 3i)$

Solution **a.** $-7i \cdot 3i = -21i^2$

$= -21(-1)$ Replace i^2 with -1.

$= 21$

b. $3i(2 - i) = 3i \cdot 2 - 3i \cdot i$ Use the distributive property.

$= 6i - 3i^2$ Multiply.

$= 6i - 3(-1)$ Replace i^2 with -1.

$= 6i + 3$

$= 3 + 6i$

Use the FOIL method. (First, Outer, Inner, Last)

c. $(2 - 5i)(4 + i) = 2(4) + 2(i) - 5i(4) - 5i(i)$

$\qquad\qquad\qquad\quad$ F $\quad$ O $\quad$ I $\quad\quad$ L

$= 8 + 2i - 20i - 5i^2$

$= 8 - 18i - 5(-1)$ $i^2 = -1$

$= 8 - 18i + 5$

$= 13 - 18i$

d. $(2 - i)^2 = (2 - i)(2 - i)$

$= 2(2) - 2(i) - 2(i) + i^2$

$= 4 - 4i + (-1)$ $i^2 = -1$

$= 3 - 4i$

e. $(7 + 3i)(7 - 3i) = 7(7) - 7(3i) + 3i(7) - 3i(3i)$

$= 49 - 21i + 21i - 9i^2$

$= 49 - 9(-1)$ $i^2 = -1$

$= 49 + 9$

$= 58$

Notice that if you add, subtract, or multiply two complex numbers, just like real numbers, the result is a complex number.

4 From Example 4e, notice that the product of $7 + 3i$ and $7 - 3i$ is a real number. These two complex numbers are called **complex conjugates** of one another. In general, we have the following definition.

COMPLEX CONJUGATES

The complex numbers $(a + bi)$ and $(a - bi)$ are called **complex conjugates** of each other, and $(a + bi)(a - bi) = a^2 + b^2$.

To see that the product of a complex number $a + bi$ and its conjugate $a - bi$ is the real number $a^2 + b^2$ we multiply.

$$(a + bi)(a - bi) = a^2 - abi + abi - b^2i^2$$
$$= a^2 - b^2(-1)$$
$$= a^2 + b^2$$

We use complex conjugates to divide by a complex number.

Example 5 Find each quotient. Write in the form $a + bi$.

a. $\dfrac{2 + i}{1 - i}$ **b.** $\dfrac{7}{3i}$

Solution **a.** Multiply the numerator and denominator by the complex conjugate of $1 - i$ to eliminate the imaginary number in the denominator.

$$\frac{2 + i}{1 - i} = \frac{(2 + i)(1 + i)}{(1 - i)(1 + i)}$$

$$= \frac{2(1) + 2(i) + 1(i) + i^2}{1^2 - i^2}$$

$$= \frac{2 + 3i - 1}{1 + 1}$$

$$= \frac{1 + 3i}{2} \quad \text{or} \quad \frac{1}{2} + \frac{3}{2}i$$

HELPFUL HINT

Recall that division can be checked by multiplication.

To check that $\dfrac{2 + i}{1 - i} = \dfrac{1}{2} + \dfrac{3}{2}i$, in Example 5a, multiply $\left(\dfrac{1}{2} + \dfrac{3}{2}i\right)(1 - i)$ to verify that the product is $2 + i$.

---

b. Multiply the numerator and denominator by the conjugate of $3i$. Note that $3i = 0 + 3i$, so its conjugate is $0 - 3i$ or $-3i$.

$$\frac{7}{3i} = \frac{7(-3i)}{(3i)(-3i)} = \frac{-21i}{-9i^2} = \frac{-21i}{-9(-1)} = \frac{-21i}{9} = \frac{-7i}{3} \quad \text{or} \quad 0 - \frac{7}{3}i$$

5

We can use the fact that $i^2 = -1$ to find higher powers of i. To find i^3, we rewrite it as the product of i^2 and i.

$$i^3 = i^2 \cdot i = (-1)i = -i$$
$$i^4 = i^2 \cdot i^2 = (-1) \cdot (-1) = 1$$

We continue this process and use the fact that $i^4 = 1$ and $i^2 = -1$ to simplify i^5 and i^6.

$$i^5 = i^4 \cdot i = 1 \cdot i = i$$
$$i^6 = i^4 \cdot i^2 = 1 \cdot (-1) = -1$$

If we continue finding powers of i, we generate the following pattern. Notice that the values i, -1, $-i$, and 1 repeat as i is raised to higher and higher powers.

$i^1 = i$	$i^5 = i$	$i^9 = i$
$i^2 = -1$	$i^6 = -1$	$i^{10} = -1$
$i^3 = -i$	$i^7 = -i$	$i^{11} = -i$
$i^4 = 1$	$i^8 = 1$	$i^{12} = 1$

This pattern allows us to find other powers of i. To do so, we will use the fact that $i^4 = 1$ and rewrite a power of i in terms of i^4.

For example, $i^{22} = i^{20} \cdot i^2 = \left(i^4\right)^5 \cdot i^2 = 1^5 \cdot (-1) = 1 \cdot (-1) = -1$.

Example 6 Find the following powers of i.

    a. i^7 **b.** i^{20} **c.** i^{46} **d.** i^{-12}

Solution   **a.** $i^7 = i^4 \cdot i^3 = 1(-i) = -i$

        b. $i^{20} = \left(i^4\right)^5 = 1^5 = 1$

        c. $i^{46} = i^{44} \cdot i^2 = \left(i^4\right)^{11} \cdot i^2 = 1^{11}(-1) = -1$

        d. $i^{-12} = \dfrac{1}{i^{12}} = \dfrac{1}{\left(i^4\right)^3} = \dfrac{1}{(1)^3} = \dfrac{1}{1} = 1$

MENTAL MATH

Simplify. See Example 1.

1. $\sqrt{-81}$ **2.** $\sqrt{-49}$ **3.** $\sqrt{-7}$ **4.** $\sqrt{-3}$
5. $-\sqrt{16}$ **6.** $-\sqrt{4}$ **7.** $\sqrt{-64}$ **8.** $\sqrt{-100}$

Exercise Set 9.7

Write in terms of i. See Example 1.

1. $\sqrt{-24}$ **2.** $\sqrt{-32}$

3. $-\sqrt{-36}$ **4.** $-\sqrt{-121}$

5. $8\sqrt{-63}$ **6.** $4\sqrt{-20}$

7. $-\sqrt{54}$ **8.** $\sqrt{-63}$

Multiply or divide. See Example 2.

9. $\sqrt{-2} \cdot \sqrt{-7}$ **10.** $\sqrt{-11} \cdot \sqrt{-3}$

11. $\sqrt{-5} \cdot \sqrt{-10}$ **12.** $\sqrt{-2} \cdot \sqrt{-6}$

13. $\sqrt{16} \cdot \sqrt{-1}$ **14.** $\sqrt{3} \cdot \sqrt{-27}$

15. $\dfrac{\sqrt{-9}}{\sqrt{3}}$ **16.** $\dfrac{\sqrt{49}}{\sqrt{-10}}$

17. $\dfrac{\sqrt{-80}}{\sqrt{-10}}$

18. $\dfrac{\sqrt{-40}}{\sqrt{-8}}$

Add or subtract. Write the sum or difference in the form $a + bi$. See Example 3.

19. $(4 - 7i) + (2 + 3i)$

20. $(2 - 4i) - (2 - i)$

21. $(6 + 5i) - (8 - i)$

22. $(8 - 3i) + (-8 + 3i)$

23. $6 - (8 + 4i)$

24. $(9 - 4i) - 9$

Multiply. Write the product in the form $a + bi$. See Example 4.

25. $6i(2 - 3i)$

26. $5i(4 - 7i)$

27. $(\sqrt{3} + 2i)(\sqrt{3} - 2i)$

28. $(\sqrt{5} - 5i)(\sqrt{5} + 5i)$

29. $(4 - 2i)^2$

30. $(6 - 3i)^2$

Write each quotient in the form $a + bi$. See Example 5.

31. $\dfrac{4}{i}$

32. $\dfrac{5}{6i}$

33. $\dfrac{7}{4 + 3i}$

34. $\dfrac{9}{1 - 2i}$

35. $\dfrac{3 + 5i}{1 + i}$

36. $\dfrac{6 + 2i}{4 - 3i}$

37. $\dfrac{5 - i}{3 - 2i}$

38. $\dfrac{6 - i}{2 + i}$

Perform the indicated operation. Write the result in the form $a + bi$.

39. $(7i)(-9i)$

40. $(-6i)(-4i)$

41. $(6 - 3i) - (4 - 2i)$

42. $(-2 - 4i) - (6 - 8i)$

43. $(6 - 2i)(3 + i)$

44. $(2 - 4i)(2 - i)$

45. $(8 - 3i) + (2 + 3i)$

46. $(7 + 4i) + (4 - 4i)$

47. $(1 - i)(1 + i)$

48. $(6 + 2i)(6 - 2i)$

49. $\dfrac{16 + 15i}{-3i}$

50. $\dfrac{2 - 3i}{-7i}$

51. $(9 + 8i)^2$

52. $(4 - 7i)^2$

53. $\dfrac{2}{3 + i}$

54. $\dfrac{5}{3 - 2i}$

55. $(5 - 6i) - 4i$

56. $(6 - 2i) + 7i$

57. $\dfrac{2 - 3i}{2 + i}$

58. $\dfrac{6 + 5i}{6 - 5i}$

59. $(2 + 4i) + (6 - 5i)$

60. $(5 - 3i) + (7 - 8i)$

Find each power of i. See Example 6.

61. i^8

62. i^{10}

63. i^{21}

64. i^{15}

65. i^{11}

66. i^{40}

67. i^{-6}

68. i^{-9}

Write in the form $a + bi$.

69. $i^3 + i^4$

70. $i^8 - i^7$

71. $i^6 + i^8$

72. $i^4 + i^{12}$

73. $2 + \sqrt{-9}$

74. $5 - \sqrt{-16}$

75. $\dfrac{6 + \sqrt{-18}}{3}$

76. $\dfrac{4 - \sqrt{-8}}{2}$

77. $\dfrac{5 - \sqrt{-75}}{10}$

78. Describe how to find the conjugate of a complex number.

79. Explain why the product of a complex number and its complex conjugate is a real number.

Simplify.

80. $(8 - \sqrt{-3}) - (2 + \sqrt{-12})$

81. $(8 - \sqrt{-4}) - (2 + \sqrt{-16})$

82. Determine whether $2i$ is a solution of $x^2 + 4 = 0$.

83. Determine whether $-1 + i$ is a solution of $x^2 + 2x = -2$.

REVIEW EXERCISES

Recall that the sum of the measures of the angles of a triangle is $180°$. Find the unknown angle in each triangle. See Section 2.4.

△ **84.**

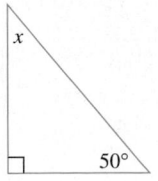

△ **85.**

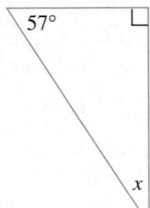

Use synthetic division to divide the following. See Section 5.7.

86. $(x^3 - 6x^2 + 3x - 4) \div (x - 1)$

87. $(5x^4 - 3x^2 + 2) \div (x + 2)$

Thirty people were recently polled about their average month-ly balance in their checking accounts. The results of this poll are shown in the following histogram. Use this graph to answer Exercises 88 through 93. See Section 1.9.

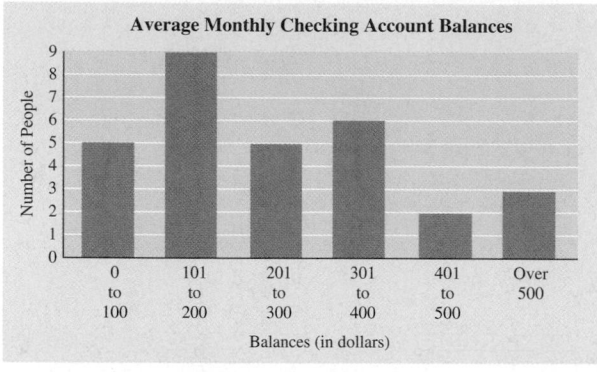

Average Monthly Checking Account Balances

88. How many people polled reported an average checking balance of $201 to $300?

89. How many people polled reported an average checking balance of $0 to $100?

90. How many people polled reported an average checking balance of $200 or less?

91. How many people polled reported an average checking balance of $301 or more?

92. What percent of people polled reported an average checking balance of $201 to $300?

93. What percent of people polled reported an average checking balance of $0 to $100?

For additional Chapter Projects, visit the Real World Activities Website by going to http://www.prenhall.com/martin-gay.

CHAPTER PROJECT

Calculating the Length and Period of a Pendulum

A simple pendulum of a given length, like the kind found in a clock, has a unique property. The time required to complete one full back-and-forth swing (called the **period**) is the same regardless of the mass of the pendulum or the distance it travels. The time to complete one full swing *does*, however, depend on the pendulum's length. In this project, you will have the opportunity to investigate the relationship between the length of a pendulum and its period. You will need at least 1 meter of string, a weight of some sort, a meter stick, a stopwatch, and a calculator. This project may be completed by working in groups or individually.

Make a simple pendulum by securely tying the string to the weight.

The formula relating a pendulum's period T (in seconds) to its length l (in centimeters) is

$$T = 2\pi\sqrt{\frac{l}{980}}$$

The period of a pendulum is defined as the time it takes the pendulum to complete one full back-and-forth swing. In this project, you will be measuring your simple pendulum's period with a stopwatch. Because the periods will be only a few seconds long, it will be more accurate for you to time a total of 5 complete

swings and then find the average time of one complete swing.

1. For each of the pendulum (string) lengths l given in Table 1, measure the time required for 5 complete swings and record it in the appropriate column. Next, divide this value by 5 to find the measured period of the pendulum for the given length and record it in the Measured Period T_m column in the table. Use the given formula to calculate the theoretical period T for the same pendulum length and record it in the appropriate column. (Round to two decimal places.) Find and record in the last column the difference between the measured period and the theoretical period.

2. For each of the periods T given in Table 2, use the given formula to calculate the theoretical pendulum length l required to yield the given period. Record l in the appropriate column; round to one decimal place. Next, using this length l, measure and record the time for 5 complete swings. Divide this time by 5 to find the measured period T_m, and record it. Then find and record in the last column the difference between the theoretical period and the measured period.

3. Use the general trends you find in the tables to describe the relationship between a pendulum's period and its length.

4. Discuss the differences you found between the values of the theoretical period and the measured period. What factors contributed to these differences?

TABLE 1

Length l (in centimeters)	Time for 5 Swings (in seconds)	Measured Period T_m (in seconds)	Theoretical Period T (in seconds)	Difference $\lvert T - T_m \rvert$
30				
55				
70				

TABLE 2

Period T (in seconds)	Theoretical Length l (in centimeters)	Time for 5 swings (in seconds)	Measured Period T_m	Difference $\lvert T - T_m \rvert$
1				
1.25				
2				

CHAPTER 9 VOCABULARY CHECK

Fill in each blank with one of the words or phrases listed below.

index   rationalizing   conjugate   principal square root   cube root
complex number   like radicals   radicand   imaginary unit

1. The _____ of $\sqrt{3} + 2$ is $\sqrt{3} - 2$.

2. The _____ of a nonnegative number a is written as $\sqrt{a}$.

3. The process of writing a radical expression as an equivalent expression but without a radical in the denominator is called _____ the denominator.

4. The _____, written i, is the number whose square is -1.

5. The _____ of a number is written as $\sqrt[3]{a}$.

6. In the notation $\sqrt[n]{a}$, n is called the _____ and a is called the _____.

7. Radicals with the same index and the same radicand are called _____.

8. A _____ is a number that can be written in the form $a + bi$ where a and b are real numbers.

CHAPTER 9 HIGHLIGHTS

DEFINITIONS AND CONCEPTS	EXAMPLES

Section 9.1 Radicals and Radical Functions

The **positive**, or **principal**, **square root** of a nonnegative number a is written as $\sqrt{a}$.

$$\sqrt{a} = b \text{ only if } b^2 = a \text{ and } b \geq 0$$

The **negative square root** of a is written as $-\sqrt{a}$.

The **cube root** of a real number a is written as $\sqrt[3]{a}$.

$$\sqrt[3]{a} = b \text{ only if } b^3 = a$$

If n is an even positive integer, then $\sqrt[n]{a^n} = |a|$.

If n is an odd positive integer, then $\sqrt[n]{a^n} = a$.

A **radical function** in x is a function defined by an expression containing a root of x.

$$\sqrt{36} = 6 \qquad \sqrt{\frac{9}{100}} = \frac{3}{10}$$

$$-\sqrt{36} = -6 \qquad \sqrt{0.04} = 0.2$$

$$\sqrt[3]{27} = 3 \qquad \sqrt[3]{-\frac{1}{8}} = -\frac{1}{2}$$

$$\sqrt[3]{y^6} = y^2 \qquad \sqrt[3]{64x^9} = 4x^3$$

$$\sqrt{(-3)^2} = |-3| = 3$$

$$\sqrt[3]{(-7)^3} = -7$$

If $f(x) = \sqrt{x} + 2$,

$$f(1) = \sqrt{1} + 2 = 1 + 2 = 3$$

$$f(3) = \sqrt{3} + 2 \approx 3.73$$

Section 9.2 Rational Exponents

$a^{1/n} = \sqrt[n]{a}$ if $\sqrt[n]{a}$ is a real number.

If m and n are positive integers greater than 1 with $\dfrac{m}{n}$ in lowest terms and $\sqrt[n]{a}$ is a real number, then

$$a^{m/n} = \left(a^{1/n}\right)^m = \left(a^m\right)^{1/n} = \sqrt[n]{a^m} = \left(\sqrt[n]{a}\right)^m$$

$a^{-m/n} = \dfrac{1}{a^{m/n}}$ as long as $a^{m/n}$ is a nonzero number.

Exponent rules are true for rational exponents.

$$81^{1/2} = \sqrt{81} = 9$$

$$\left(-8x^3\right)^{1/3} = \sqrt[3]{-8x^3} = -2x$$

$$4^{5/2} = \left(\sqrt{4}\right)^5 = 2^5 = 32$$

$$27^{2/3} = \left(\sqrt[3]{27}\right)^2 = 3^2 = 9$$

$$16^{-3/4} = \frac{1}{16^{3/4}} = \frac{1}{\left(\sqrt[4]{16}\right)^3} = \frac{1}{2^3} = \frac{1}{8}$$

$$x^{2/3} \cdot x^{-5/6} = x^{2/3 - 5/6} = x^{-1/6} = \frac{1}{x^{1/6}}$$

$$\left(8^{14}\right)^{1/7} = 8^2 = 64$$

$$\frac{a^{4/5}}{a^{-2/5}} = a^{4/5 - (-2/5)} = a^{6/5}$$

Section 9.3 Simplifying Radical Expressions

Product and Quotient Rules

If $\sqrt[n]{a}$ and $\sqrt[n]{b}$ are real numbers,

$$\sqrt[n]{a} \cdot \sqrt[n]{b} = \sqrt[n]{a \cdot b}$$

$$\frac{\sqrt[n]{a}}{\sqrt[n]{b}} = \sqrt[n]{\frac{a}{b}}, \text{ provided } \sqrt[n]{b} \neq 0$$

A radical of the form $\sqrt[n]{a}$ is **simplified** when a contains no factors that are perfect nth powers.

Multiply or divide as indicated:

$$\sqrt{11} \cdot \sqrt{3} = \sqrt{33}$$

$$\frac{\sqrt[3]{40x}}{\sqrt[3]{5x}} = \sqrt[3]{8} = 2$$

$$\sqrt{40} = \sqrt{4 \cdot 10} = 2\sqrt{10}$$

$$\sqrt{36x^5} = \sqrt{36x^4 \cdot x} = 6x^2\sqrt{x}$$

$$\sqrt[3]{24x^7y^3} = \sqrt[3]{8x^6y^3 \cdot 3x} = 2x^2y\sqrt[3]{3x}$$

(continued)

DEFINITIONS AND CONCEPTS	EXAMPLES

Section 9.4 Adding, Subtracting, and Multiplying Radical Expressions

Radicals with the same index and the same radicand are **like radicals**.

The distributive property can be used to add like radicals.

Radical expressions are multiplied by using many of the same properties used to multiply polynomials.

$$5\sqrt{6} + 2\sqrt{6} = (5 + 2)\sqrt{6} = 7\sqrt{6}$$
$$-\sqrt[3]{3x} - 10\sqrt[3]{3x} + 3\sqrt[3]{10x}$$
$$= (-1 - 10)\sqrt[3]{3x} + 3\sqrt[3]{10x}$$
$$= -11\sqrt[3]{3x} + 3\sqrt[3]{10x}$$

Multiply.
$$(\sqrt{5} - \sqrt{2x})(\sqrt{2} + \sqrt{2x})$$
$$= \sqrt{10} + \sqrt{10x} - \sqrt{4x} - 2x$$
$$= \sqrt{10} + \sqrt{10x} - 2\sqrt{x} - 2x$$
$$(2\sqrt{3} - \sqrt{8x})(2\sqrt{3} + \sqrt{8x})$$
$$= 4(3) - 8x = 12 - 8x$$

Section 9.5 Rationalizing Denominators and Numerators of Radical Expressions

The **conjugate** of $a + b$ is $a - b$.

The process of writing the denominator of a radical expression without a radical is called **rationalizing the denominator.**

The conjugate of $\sqrt{7} + \sqrt{3}$ is $\sqrt{7} - \sqrt{3}$
Rationalize each denominator.
$$\frac{\sqrt{5}}{\sqrt{3}} = \frac{\sqrt{5} \cdot \sqrt{3}}{\sqrt{3} \cdot \sqrt{3}} = \frac{\sqrt{15}}{3}$$
$$\frac{6}{\sqrt{7} + \sqrt{3}} = \frac{6(\sqrt{7} - \sqrt{3})}{(\sqrt{7} + \sqrt{3})(\sqrt{7} - \sqrt{3})}$$
$$= \frac{6(\sqrt{7} - \sqrt{3})}{7 - 3}$$
$$= \frac{6(\sqrt{7} - \sqrt{3})}{4} = \frac{3(\sqrt{7} - \sqrt{3})}{2}$$

The process of writing the numerator of a radical expression without a radical is called **rationalizing the numerator.**

Rationalize each numerator.
$$\frac{\sqrt[3]{9}}{\sqrt[3]{5}} = \frac{\sqrt[3]{9} \cdot \sqrt[3]{3}}{\sqrt[3]{5} \cdot \sqrt[3]{3}} = \frac{\sqrt[3]{27}}{\sqrt[3]{15}} = \frac{3}{\sqrt[3]{15}}$$
$$\frac{\sqrt{9} + \sqrt{3x}}{12} = \frac{(\sqrt{9} + \sqrt{3x})(\sqrt{9} - \sqrt{3x})}{12(\sqrt{9} - \sqrt{3x})}$$
$$= \frac{9 - 3x}{12(\sqrt{9} - \sqrt{3x})}$$
$$= \frac{3(3 - x)}{3 \cdot 4(3 - \sqrt{3x})} = \frac{3 - x}{4(3 - \sqrt{3x})}$$

Section 9.6 Radical Equations and Problem Solving

To Solve a Radical Equation

Step 1: Write the equation so that one radical is by itself on one side of the equation.

Step 2: Raise each side of the equation to a power equal to the index of the radical and simplify.

Step 3: If the equation still contains a radical, repeat Steps 1 and 2. If not, solve the equation.

Step 4: Check all proposed solutions in the original equation.

Solve $x = \sqrt{4x + 9} + 3$.
1. $x - 3 = \sqrt{4x + 9}$

2. $(x - 3)^2 = (\sqrt{4x + 9})^2$
$x^2 - 6x + 9 = 4x + 9$
3. $x^2 - 10x = 0$
$x(x - 10) = 0$
$x = 0 \quad \text{or} \quad x = 10$
4. The proposed solution 10 checks, but 0 does not.
The solution is 10. *(continued)*

DEFINITIONS AND CONCEPTS	EXAMPLES

Section 9.7 Complex Numbers

$i^2 = -1$ and $i = \sqrt{-1}$

Simplify $\sqrt{-9}$.

$$\sqrt{-9} = \sqrt{-1 \cdot 9} = \sqrt{-1} \cdot \sqrt{9} = i \cdot 3 \text{ or } 3i$$

A **complex number** is a number that can be written in the form $a + bi$, where a and b are real numbers.

Complex Numbers	*Written in form $a + bi$*
12	$12 + 0i$
$-5i$	$0 + (-5)i$
$-2 - 3i$	$-2 + (-3)i$

Multiply.

$$\sqrt{-3} \cdot \sqrt{-7} = i\sqrt{3} \cdot i\sqrt{7}$$
$$= i^2\sqrt{21}$$
$$= -\sqrt{21}$$

To add or subtract complex numbers, add or subtract their real parts and then add or subtract their imaginary parts.

Perform each indicated operation.

$$(-3 + 2i) - (7 - 4i) = -3 + 2i - 7 + 4i$$
$$= -10 + 6i$$

To multiply complex numbers, multiply as though they are binomials.

$$(-7 - 2i)(6 + i) = -42 - 7i - 12i - 2i^2$$
$$= -42 - 19i - 2(-1)$$
$$= -42 - 19i + 2$$
$$= -40 - 19i$$

The complex numbers $(a + bi)$ and $(a - bi)$ are called **complex conjugates.**

The complex conjugate of

$$(3 + 6i) \text{ is } (3 - 6i).$$

Their product is a real number.

$$(3 - 6i)(3 + 6i) = 9 - 36i^2$$
$$= 9 - 36(-1) = 9 + 36 = 45$$

To divide complex numbers, multiply the numerator and the denominator by the conjugate of the denominator.

Divide.

$$\frac{4}{2 - i} = \frac{4(2 + i)}{(2 - i)(2 + i)}$$
$$= \frac{4(2 + i)}{4 - i^2}$$
$$= \frac{4(2 + i)}{5}$$
$$= \frac{8 + 4i}{5} = \frac{8}{5} + \frac{4}{5}i$$

CHAPTER 9 REVIEW

(9.1) *Find the root. Assume that all variables represent positive numbers.*

1. $\sqrt{81}$

2. $\sqrt[4]{81}$

3. $\sqrt[3]{-8}$

4. $\sqrt[4]{-16}$

5. $-\sqrt{\dfrac{1}{49}}$

6. $\sqrt{x^{64}}$

7. $-\sqrt{36}$

8. $\sqrt[3]{64}$

9. $\sqrt[3]{-a^6b^9}$

10. $\sqrt{16a^4b^{12}}$

11. $\sqrt[5]{32a^5b^{10}}$

12. $\sqrt[5]{-32x^{15}y^{20}}$

13. $\sqrt{\dfrac{x^{12}}{36y^2}}$

14. $\sqrt[3]{\dfrac{27y^3}{z^{12}}}$

Simplify. Use absolute value bars when necessary.

15. $\sqrt{(-x)^2}$

16. $\sqrt[4]{(x^2-4)^4}$

17. $\sqrt[3]{(-27)^3}$

18. $\sqrt[5]{(-5)^5}$

19. $-\sqrt[5]{x^5}$

20. $\sqrt[4]{16(2y+z)^{12}}$

21. $\sqrt{25(x-y)^{10}}$

22. $\sqrt[5]{-y^5}$

23. $\sqrt[9]{-x^9}$

Identify the domain and then graph each function.

24. $f(x) = \sqrt{x} + 3$

25. $g(x) = \sqrt[3]{x} - 3$; use the accompanying table.

x	-5	2	3	4	11
$g(x)$					

(9.2) *Evaluate the following.*

26. $\left(\dfrac{1}{81}\right)^{1/4}$

27. $\left(-\dfrac{1}{27}\right)^{1/3}$

28. $(-27)^{-1/3}$

29. $(-64)^{-1/3}$

30. $-9^{3/2}$

31. $64^{-1/3}$

32. $(-25)^{5/2}$

33. $\left(\dfrac{25}{49}\right)^{-3/2}$

34. $\left(\dfrac{8}{27}\right)^{-2/3}$

35. $\left(-\dfrac{1}{36}\right)^{-1/4}$

Write with rational exponents.

36. $\sqrt[3]{x^2}$

37. $\sqrt[5]{5x^2y^3}$

Write with radical notation.

38. $y^{4/5}$

39. $5(xy^2z^5)^{1/3}$

40. $(x+2y)^{-1/2}$

Simplify each expression. Assume that all variables represent positive numbers. Write with only positive exponents.

41. $a^{1/3}a^{4/3}a^{1/2}$

42. $\dfrac{b^{1/3}}{b^{4/3}}$

43. $(a^{1/2}a^{-2})^3$

44. $(x^{-3}y^6)^{1/3}$

45. $\left(\dfrac{b^{3/4}}{a^{-1/2}}\right)^8$

46. $\dfrac{x^{1/4}x^{-1/2}}{x^{2/3}}$

47. $\left(\dfrac{49c^{5/3}}{a^{-1/4}b^{5/6}}\right)^{-1}$

48. $a^{-1/4}(a^{5/4}-a^{9/4})$

Use a calculator and write a three-decimal-place approximation.

49. $\sqrt{20}$

50. $\sqrt[3]{-39}$

51. $\sqrt[4]{726}$

52. $56^{1/3}$

53. $-78^{3/4}$

54. $105^{-2/3}$

Use rational exponents to write each radical with the same index. Then multiply.

55. $\sqrt[3]{2} \cdot \sqrt{7}$

56. $\sqrt[3]{3} \cdot \sqrt[4]{x}$

(9.3) *Perform the indicated operations and then simplify if possible. For the remainder of this review, assume that variables represent positive numbers only.*

57. $\sqrt{3} \cdot \sqrt{8}$

58. $\sqrt[3]{7y} \cdot \sqrt[3]{x^2z}$

59. $\dfrac{\sqrt{44x^3}}{\sqrt{11x}}$

60. $\dfrac{\sqrt[4]{a^6b^{13}}}{\sqrt[4]{a^2b}}$

Simplify.

61. $\sqrt{60}$

62. $-\sqrt{75}$

63. $\sqrt[3]{162}$

64. $\sqrt[3]{-32}$

65. $\sqrt{36x^7}$

66. $\sqrt[3]{24a^5b^7}$

67. $\sqrt{\dfrac{p^{17}}{121}}$

68. $\sqrt[3]{\dfrac{y^5}{27x^6}}$

69. $\sqrt[4]{\dfrac{xy^6}{81}}$

70. $\sqrt{\dfrac{2x^3}{49y^4}}$

△ **71.** The formula for the radius r of a circle of area A is

$$r = \sqrt{\dfrac{A}{\pi}}$$

a. Find the exact radius of a circle whose area is 25 square meters.

b. Approximate to two decimal places the radius of a circle whose area is 104 square inches.

(9.4) *Perform the indicated operation.*

72. $x\sqrt{75xy} - \sqrt{27x^3y}$

73. $2\sqrt{32x^2y^3} - xy\sqrt{98y}$

74. $\sqrt[3]{128} + \sqrt[3]{250}$

75. $3\sqrt[4]{32a^5} - a\sqrt[4]{162a}$

76. $\dfrac{5}{\sqrt{4}} + \dfrac{\sqrt{3}}{3}$

77. $\sqrt{\dfrac{8}{x^2}} - \sqrt{\dfrac{50}{16x^2}}$

78. $2\sqrt{50} - 3\sqrt{125} + \sqrt{98}$

79. $2a\sqrt[4]{32b^5} - 3b\sqrt[4]{162a^4b} + \sqrt[4]{2a^4b^5}$

Multiply and then simplify if possible.

80. $\sqrt{3}(\sqrt{27} - \sqrt{3})$

81. $(\sqrt{x} - 3)^2$

82. $(\sqrt{5} - 5)(2\sqrt{5} + 2)$

83. $(2\sqrt{x} - 3\sqrt{y})(2\sqrt{x} + 3\sqrt{y})$

84. $(\sqrt{a} + 3)(\sqrt{a} - 3)$

85. $(\sqrt[3]{a} + 2)^2$

86. $(\sqrt[3]{5x} + 9)(\sqrt[3]{5x} - 9)$

87. $(\sqrt[3]{a} + 4)(\sqrt[3]{a^2} - 4\sqrt[3]{a} + 16)$

(9.5) *Rationalize each denominator.*

88. $\dfrac{3}{\sqrt{7}}$

89. $\sqrt{\dfrac{x}{12}}$

90. $\dfrac{5}{\sqrt[3]{4}}$

91. $\sqrt{\dfrac{24x^5}{3y^2}}$

92. $\sqrt[3]{\dfrac{15x^6y^7}{z^2}}$

93. $\dfrac{5}{2 - \sqrt{7}}$

94. $\dfrac{3}{\sqrt{y} - 2}$

95. $\dfrac{\sqrt{2} - \sqrt{3}}{\sqrt{2} + \sqrt{3}}$

Rationalize each numerator.

96. $\dfrac{\sqrt{11}}{3}$

97. $\sqrt{\dfrac{18}{y}}$

98. $\dfrac{\sqrt[3]{9}}{7}$

99. $\sqrt{\dfrac{24x^5}{3y^2}}$

100. $\sqrt[3]{\dfrac{xy^2}{10z}}$

101. $\dfrac{\sqrt{x} + 5}{-3}$

(9.6) *Solve each equation for the variable.*

102. $\sqrt{y - 7} = 5$

103. $\sqrt{2x + 10} = 4$

104. $\sqrt[3]{2x - 6} = 4$

105. $\sqrt{x + 6} = \sqrt{x + 2}$

106. $2x - 5\sqrt{x} = 3$

107. $\sqrt{x + 9} = 2 + \sqrt{x - 7}$

Find each unknown length.

△ **108.**

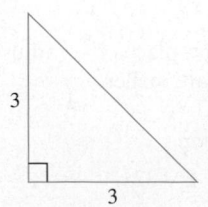

△ **109.**

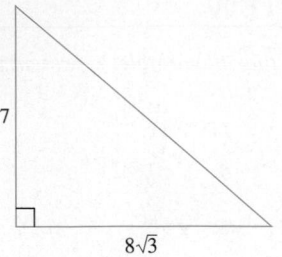

△ **110.** Beverly Hillis wants to determine the distance x across a pond on her property. She is able to measure the distances shown on the following diagram. Find how wide the lake is at the crossing point, indicated by the triangle, to the nearest tenth of a foot.

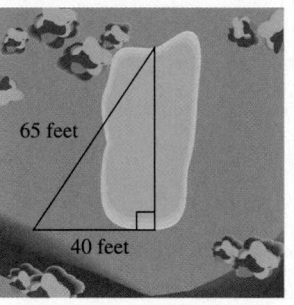

△ **111.** A pipe fitter needs to connect two underground pipelines that are offset by 3 feet, as pictured in the diagram. Neglecting the joints needed to join the pipes, find the length of the shortest possible connecting pipe rounded to the nearest hundredth of a foot.

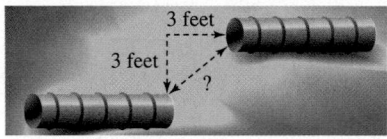

(9.7) *Perform the indicated operation and simplify. Write the result in the form $a + bi$.*

112. $\sqrt{-8}$

113. $-\sqrt{-6}$

114. $\sqrt{-4} + \sqrt{-16}$

115. $\sqrt{-2} \cdot \sqrt{-5}$

116. $(12 - 6i) + (3 + 2i)$

117. $(-8 - 7i) - (5 - 4i)$

118. $(\sqrt{3} + \sqrt{2}) + (3\sqrt{2} - \sqrt{-8})$

119. $2i(2 - 5i)$

120. $-3i(6 - 4i)$

121. $(3 + 2i)(1 + i)$

122. $(2 - 3i)^2$

123. $(\sqrt{6} - 9i)(\sqrt{6} + 9i)$

124. $\dfrac{2 + 3i}{2i}$

125. $\dfrac{1 + i}{-3i}$

CHAPTER 9 TEST

Raise to the power or find the root. Assume that all variables represent positive numbers. Write with only positive exponents.

1. $\sqrt{216}$

2. $-\sqrt[4]{x^{64}}$

3. $\left(\dfrac{1}{125}\right)^{1/3}$

4. $\left(\dfrac{1}{125}\right)^{-1/3}$

5. $\left(\dfrac{8x^3}{27}\right)^{2/3}$

6. $\sqrt[3]{-a^{18}b^9}$

7. $\left(\dfrac{64c^{4/3}}{a^{-2/3}b^{5/6}}\right)^{1/2}$

8. $a^{-2/3}\left(a^{5/4} - a^3\right)$

Find the root. Use absolute value bars when necessary.

9. $\sqrt[4]{(4xy)^4}$

10. $\sqrt[3]{(-27)^3}$

Rationalize the denominator. Assume that all variables represent positive numbers.

11. $\sqrt{\dfrac{9}{y}}$

12. $\dfrac{4 - \sqrt{x}}{4 + 2\sqrt{x}}$

13. $\dfrac{\sqrt[3]{ab}}{\sqrt[3]{ab^2}}$

14. Rationalize the numerator of $\dfrac{\sqrt{6} + x}{8}$ and simplify.

Perform the indicated operations. Assume that all variables represent positive numbers.

15. $\sqrt{125x^3} - 3\sqrt{20x^3}$

16. $\sqrt{3}(\sqrt{16} - \sqrt{2})$

17. $(\sqrt{x} + 1)^2$

18. $(\sqrt{2} - 4)(\sqrt{3} + 1)$

19. $(\sqrt{5} + 5)(\sqrt{5} - 5)$

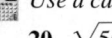 *Use a calculator to approximate each to three decimal places.*

20. $\sqrt{561}$

21. $386^{-2/3}$

Solve.

22. $x = \sqrt{x - 2} + 2$

23. $\sqrt{x^2 - 7} + 3 = 0$

24. $\sqrt{x + 5} = \sqrt{2x - 1}$

Perform the indicated operation and simplify. Write the result in the form $a + bi$.

25. $\sqrt{-2}$

26. $-\sqrt{-8}$

27. $(12 - 6i) - (12 - 3i)$

28. $(6 - 2i)(6 + 2i)$

29. $(4 + 3i)^2$

30. $\dfrac{1 + 4i}{1 - i}$

△ **31.** Find x.

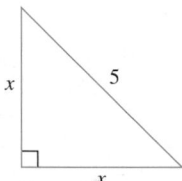

32. Identify the domain of $g(x)$. Then complete the accompanying table and graph $g(x)$.

$$g(x) = \sqrt{x + 2}$$

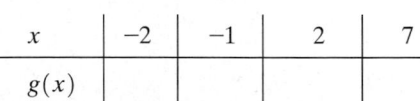

x	-2	-1	2	7
$g(x)$				

 Solve.

33. The function $V(r) = \sqrt{2.5r}$ can be used to estimate the maximum safe velocity V in miles per hour at which a car can travel if it is driven along a curved road with a *radius of curvature r* in feet. To the nearest whole number, find the maximum safe speed if a cloverleaf exit on an expressway has a radius of curvature of 300 feet.

34. Use the formula from Exercise 33 to find the radius of curvature if the safe velocity is 30 mph.

CHAPTER 9 CUMULATIVE REVIEW

1. Simplify each expression.

a. $\dfrac{(-12)(-3) + 3}{-7 - (-2)}$ **b.** $\dfrac{2(-3)^2 - 20}{-5 + 4}$

2. Complete the table for the equation $y = 3x$.

x	y
-1	
	0
	-9

3. Determine the domain and range of each relation.

a. $\{(2, 3), (2, 4), (0, -1), (3, -1)\}$

b.

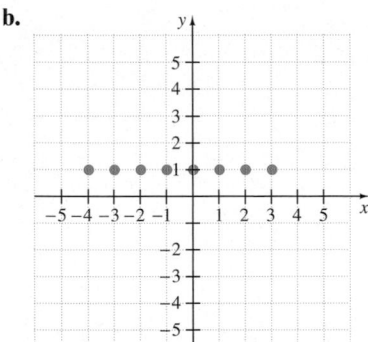

c.

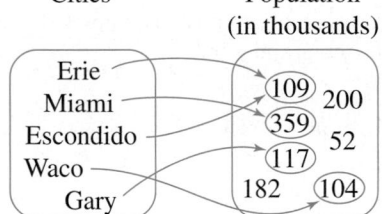

4. Graph $y = -3$.

5. Find the slope of the line whose equation is $f(x) = \frac{2}{3}x + 4$.

6. Determine the number of solutions of the system:

$$\begin{cases} 3x - y = 4 \\ x + 2y = 8 \end{cases}$$

7. Solve the system:

$$\begin{cases} x + 2y = 7 \\ 2x + 2y = 13 \end{cases}$$

8. Solve the system:

$$\begin{cases} 2x - y = 7 \\ 8x - 4y = 1 \end{cases}$$

9. Eric Daly, a chemistry teaching assistant, needs 10 liters of a 20% saline solution (salt water) for his 2 p.m. laboratory class. Unfortunately, the only mixtures on hand are a 5% saline solution and a 25% saline solution. How much of each solution should he mix to produce the 20% solution?

10. If $P(x) = 3x^2 - 2x - 5$, find the following.

a. $P(1)$

b. $P(-2)$

11. Divide $10x^3 - 5x^2 + 20x$ by $5x$.

12. Use synthetic division to divide $2x^3 - x^2 - 13x + 1$ by $x - 3$.

13. Factor $x^2 + 5yx + 6y^2$.

14. Write each rational expression in lowest terms.

a. $\dfrac{x^3 + 8}{2 + x}$

b. $\dfrac{2y^2 + 2}{y^3 - 5y^2 + y - 5}$

15. Divide: $\dfrac{3x^3y^7}{40} \div \dfrac{4x^3}{y^2}$.

16. Subtract: $\dfrac{2y}{2y - 7} - \dfrac{7}{2y - 7}$.

17. Add: $\dfrac{2x}{x^2 + 2x + 1} + \dfrac{x}{x^2 - 1}$.

18. Simplify each complex fraction.

a. $\dfrac{\dfrac{5x}{x + 2}}{\dfrac{10}{x - 2}}$

b. $\dfrac{\dfrac{x}{y^2} + \dfrac{1}{y}}{\dfrac{y}{x^2} + \dfrac{1}{x}}$

19. Solve: $\dfrac{x}{2} + \dfrac{8}{3} = \dfrac{1}{6}$

20. If the following two triangles are similar, find the missing length x.

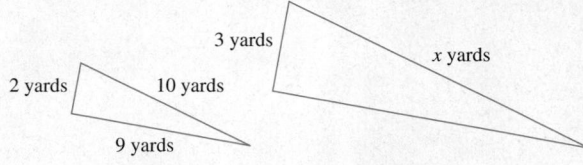

21. Write each expression with a positive exponent, and then simplify.

    a. $16^{-3/4}$ **b.** $(-27)^{-2/3}$

22. Rationalize the numerator of $\dfrac{\sqrt{x}+2}{5}$.

Allies of Good Nutrition

Diet and nutrition play a major role in good health. Diets low in saturated fats and dietary cholesterol tend to lower the risk of cardiovascular disease. High sodium intakes have been associated with high blood pressure and stroke. Antioxidants, such as beta carotene and vitamin C, can help protect against heart disease and some cancers. It is no wonder, then, that public interest in healthy eating habits is soaring.

Registered dieticians, food and nutrition experts, are valuable allies in an effort to eat right. They plan nutrition programs, supervise food preparation, and educate about the health benefits of good nutrition. Registered dieticians work in diverse environments: hospitals, schools, day-care centers, nursing homes, government or university laboratories, private practice, corporate wellness programs, food-producing companies, pharmaceutical companies, restaurant management, and community health settings. Dieticians use math and problem-solving skills in tasks such as analyzing the nutritional content of a recipe or food product, assessing a client's diet, and determining an individual's nutritional requirements.

 For more information about careers in dietetics, visit the American Dietetic Association Web site by first going to www.prenhall.com/martin-gay.

In the Spotlight on Decision Making feature on page 624, you will have the opportunity to make a decision about the adequacy of vitamin A intake as a registered dietician.

QUADRATIC EQUATIONS AND FUNCTIONS

An important part of the study of algebra is learning to model and solve problems. Often, the model of a problem is a quadratic equation or a function containing a second-degree polynomial. In this chapter, we continue the work begun in Chapter 6, when we solved polynomial equations in one variable by factoring. Two additional methods of solving quadratic equations are analyzed, as well as methods of solving nonlinear inequalities in one variable.

10.1 SOLVING QUADRATIC EQUATIONS BY COMPLETING THE SQUARE

▶ **OBJECTIVES**

1. Use the square root property to solve quadratic equations.
2. Solve quadratic equations by completing the square.
3. Use quadratic equations to solve problems.

1   In Chapter 6, we solved quadratic equations by factoring. Recall that a **quadratic, or second-degree, equation** is an equation that can be written in the form $ax^2 + bx + c = 0$, where a, b, and c are real numbers and a is not 0. To solve a quadratic equation such as $x^2 = 9$ by factoring, we use the zero-factor theorem. To use the zero-factor theorem, the equation must first be written in standard form, $ax^2 + bx + c = 0$.

$$x^2 = 9$$

$$x^2 - 9 = 0 \qquad \text{Subtract 9 from both sides.}$$

$$(x + 3)(x - 3) = 0 \qquad \text{Factor.}$$

$$x + 3 = 0 \quad \text{or} \quad x - 3 = 0 \qquad \text{Set each factor equal to 0.}$$

$$x = -3 \qquad\qquad x = 3 \qquad \text{Solve.}$$

The solutions are -3 and 3, the positive and negative square roots of 9. Not all quadratic equations can be solved by factoring, so we need to explore other methods. Notice that the solutions of the equation $x^2 = 9$ are two numbers whose square is 9.

$$3^2 = 9 \qquad \text{and} \qquad (-3)^2 = 9$$

Thus, we can solve the equation $x^2 = 9$ by taking the square root of both sides. Be sure to include both $\sqrt{9}$ and $-\sqrt{9}$ as solutions since both $\sqrt{9}$ and $-\sqrt{9}$ are numbers whose square is 9.

$$x^2 = 9$$

$$\sqrt{x^2} = \pm\sqrt{9} \qquad \text{The notation } \pm\sqrt{9} \text{ (read as "plus or minus } \sqrt{9}\text{")}$$

$$x = \pm 3 \qquad \text{indicates the pair of numbers } +\sqrt{9} \text{ and } -\sqrt{9}.$$

This illustrates the square root property.

HELPFUL HINT
The notation ± 3, for example, is read as "plus or minus 3." It is a shorthand notation for the pair of numbers $+3$ and -3.

SQUARE ROOT PROPERTY

If b is a real number and if $a^2 = b$, then $a = \pm\sqrt{b}$.

Example 1   Use the square root property to solve $x^2 = 50$.

Solution $x^2 = 50$

$\qquad x = \pm\sqrt{50}$ Use the square root property.

$\qquad x = \pm 5\sqrt{2}$ Simplify the radical.

Check   Let $x = 5\sqrt{2}$.   Let $x = -5\sqrt{2}$.

$\qquad\qquad x^2 = 50$ $x^2 = 50$

$\qquad (5\sqrt{2})^2 \overset{?}{=} 50$ $(-5\sqrt{2})^2 \overset{?}{=} 50$

$\qquad\qquad 25\cdot 2 \overset{?}{=} 50$ $25\cdot 2 \overset{?}{=} 50$

$\qquad\qquad 50 = 50$ True. $50 = 50$ True.

The solutions are $5\sqrt{2}$ and $-5\sqrt{2}$.

Example 2   Use the square root property to solve $2x^2 = 14$.

Solution   First we get the squared variable alone on one side of the equation.

$$2x^2 = 14$$

$$x^2 = 7 \qquad \text{Divide both sides by 2.}$$

$$x = \pm\sqrt{7} \qquad \text{Use the square root property.}$$

Check to see that the solutions are $\sqrt{7}$ and $-\sqrt{7}$.

Example 3   Use the square root property to solve $(x + 1)^2 = 12$.

Solution $(x + 1)^2 = 12$

$\qquad x + 1 = \pm\sqrt{12}$ Use the square root property.

$\qquad x + 1 = \pm 2\sqrt{3}$ Simplify the radical.

$\qquad\qquad x = -1 \pm 2\sqrt{3}$ Subtract 1 from both sides.

Check   Below is a check for $-1 + 2\sqrt{3}$. The check for $-1 - 2\sqrt{3}$ is almost the same and is left for you to do on your own.

$$(x + 1)^2 = 12$$

$$(-1 + 2\sqrt{3} + 1)^2 \overset{?}{=} 12$$

$$(2\sqrt{3})^2 \overset{?}{=} 12$$

$$4\cdot 3 \overset{?}{=} 12$$

$$12 = 12 \qquad \text{True.}$$

The solutions are $-1 + 2\sqrt{3}$, $-1 - 2\sqrt{3}$.

Example 4 Use the square root property to solve $(2x - 5)^2 = -16$.

Solution $(2x - 5)^2 = -16$

$2x - 5 = \pm\sqrt{-16}$ Use the square root property.

$2x - 5 = \pm 4i$ Simplify the radical.

$2x = 5 \pm 4i$ Add 5 to both sides.

$x = \dfrac{5 \pm 4i}{2}$ Divide both sides by 2.

The solutions are $\dfrac{5 + 4i}{2}$ and $\dfrac{5 - 4i}{2}$.

Notice from Examples 3 and 4 that, if we write a quadratic equation so that one side is the square of a binomial, we can solve by using the square root property. To write the square of a binomial, we write perfect square trinomials. Recall that a perfect square trinomial is a trinomial that can be factored into two identical binomial factors.

Perfect Square Trinomials	*Factored Form*
$x^2 + 8x + 16$	$(x + 4)^2$
$x^2 - 6x + 9$	$(x - 3)^2$
$x^2 + 3x + \dfrac{9}{4}$	$\left(x + \dfrac{3}{2}\right)^2$

Notice that for each perfect square trinomial, **the constant term of the trinomial is the square of half the coefficient of the x-term.** For example,

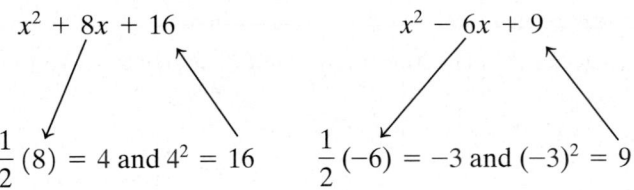

$x^2 + 8x + 16$ $x^2 - 6x + 9$

$\dfrac{1}{2}(8) = 4$ and $4^2 = 16$ $\dfrac{1}{2}(-6) = -3$ and $(-3)^2 = 9$

2   The process of writing a quadratic equation so that one side is a perfect square trinomial is called **completing the square.**

Example 5 Solve $p^2 + 2p = 4$ by completing the square.

Solution First, add the square of half the coefficient of p to both sides so that the resulting trinomial will be a perfect square trinomial. The coefficient of p is 2.

$$\frac{1}{2}(2) = 1 \quad \text{and} \quad 1^2 = 1$$

Add 1 to both sides of the original equation.

$$p^2 + 2p = 4$$

$$p^2 + 2p + 1 = 4 + 1 \qquad \text{Add 1 to both sides.}$$

$$(p + 1)^2 = 5 \qquad \text{Factor the trinomial; simplify the right side.}$$

We may now use the square root property and solve for p.

$$p + 1 = \pm\sqrt{5} \qquad \text{Use the square root property.}$$

$$p = -1 \pm \sqrt{5} \qquad \text{Subtract 1 from both sides.}$$

Notice that there are two solutions: $-1 + \sqrt{5}$ and $-1 - \sqrt{5}$. ▬

Example 6   Solve $m^2 - 7m - 1 = 0$ for m by completing the square.

Solution   First, add 1 to both sides of the equation so that the left side has no constant term.

$$m^2 - 7m - 1 = 0$$

$$m^2 - 7m = 1$$

Now find the constant term that makes the left side a perfect square trinomial by squaring half the coefficient of m. Add this constant to both sides of the equation.

$$\frac{1}{2}(-7) = -\frac{7}{2} \quad \text{and} \quad \left(-\frac{7}{2}\right)^2 = \frac{49}{4}$$

$$m^2 - 7m + \frac{49}{4} = 1 + \frac{49}{4} \qquad \text{Add } \frac{49}{4} \text{ to both sides of the equation.}$$

$$\left(m - \frac{7}{2}\right)^2 = \frac{53}{4} \qquad \text{Factor the perfect square trinomial and simplify the right side.}$$

$$m - \frac{7}{2} = \pm\sqrt{\frac{53}{4}} \qquad \text{Apply the square root property.}$$

$$m = \frac{7}{2} \pm \frac{\sqrt{53}}{2} \qquad \text{Add } \frac{7}{2} \text{ to both sides and simplify } \sqrt{\frac{53}{4}}.$$

$$m = \frac{7 \pm \sqrt{53}}{2} \qquad \text{Simplify.}$$

The solutions are $\dfrac{7 + \sqrt{53}}{2}$ and $\dfrac{7 - \sqrt{53}}{2}$. ▬

Example 7   Solve $2x^2 - 8x + 3 = 0$.

Solution   Our procedure for finding the constant term to complete the square works only if the coefficient of the squared variable term is 1. Therefore, to solve this equation, the first step is to divide both sides by 2, the coefficient of x^2.

$$2x^2 - 8x + 3 = 0$$

$$x^2 - 4x + \frac{3}{2} = 0 \qquad \text{Divide both sides by 2.}$$

$$x^2 - 4x = -\frac{3}{2} \qquad \text{Subtract } \frac{3}{2} \text{ from both sides.}$$

Next find the square of half of -4.

$$\frac{1}{2}(-4) = -2 \quad \text{and} \quad (-2)^2 = 4$$

Add 4 to both sides of the equation to complete the square.

$$x^2 - 4x + 4 = -\frac{3}{2} + 4$$

$$(x - 2)^2 = \frac{5}{2} \qquad \text{Factor the perfect square and simplify the right side.}$$

$$x - 2 = \pm\sqrt{\frac{5}{2}} \qquad \text{Apply the square root property.}$$

$$x - 2 = \pm\frac{\sqrt{10}}{2} \qquad \text{Rationalize the denominator.}$$

$$x = 2 \pm \frac{\sqrt{10}}{2} \qquad \text{Add 2 to both sides.}$$

$$= \frac{4}{2} \pm \frac{\sqrt{10}}{2} \qquad \text{Find the common denominator.}$$

$$= \frac{4 \pm \sqrt{10}}{2} \qquad \text{Simplify.}$$

The solutions are $\dfrac{4 + \sqrt{10}}{2}$ and $\dfrac{4 - \sqrt{10}}{2}$.   ■

The following steps may be used to solve a quadratic equation such as $ax^2 + bx + c = 0$ by completing the square. This method may be used whether or not the polynomial $ax^2 + bx + c$ is factorable.

SOLVING A QUADRATIC EQUATION IN x BY COMPLETING THE SQUARE

Step 1:   If the coefficient of x^2 is 1, go to Step 2. Otherwise, divide both sides of the equation by the coefficient of x^2.

Step 2:   Isolate all variable terms on one side of the equation.

Step 3:   Complete the square for the resulting binomial by adding the square of half of the coefficient of x to both sides of the equation.

Step 4:   Factor the resulting perfect square trinomial and write it as the square of a binomial.

Step 5:   Use the square root property to solve for x.

Example 8   Solve $3x^2 - 9x + 8 = 0$ by completing the square.

Solution $3x^2 - 9x + 8 = 0$

Step 1: $x^2 - 3x + \dfrac{8}{3} = 0$ Divide both sides of the equation by 3.

Step 2: $x^2 - 3x = -\dfrac{8}{3}$ Subtract $\dfrac{8}{3}$ from both sides.

Since $\dfrac{1}{2}(-3) = -\dfrac{3}{2}$ and $\left(-\dfrac{3}{2}\right)^2 = \dfrac{9}{4}$, we add $\dfrac{9}{4}$ to both sides of the equation.

Step 3: $x^2 - 3x + \dfrac{9}{4} = -\dfrac{8}{3} + \dfrac{9}{4}$

Step 4: $\left(x - \dfrac{3}{2}\right)^2 = -\dfrac{5}{12}$ Factor the perfect square trinomial.

Step 5: $x - \dfrac{3}{2} = \pm\sqrt{-\dfrac{5}{12}}$ Apply the square root property.

$x - \dfrac{3}{2} = \pm\dfrac{i\sqrt{5}}{2\sqrt{3}}$ Simplify the radical.

$x - \dfrac{3}{2} = \pm\dfrac{i\sqrt{15}}{6}$ Rationalize the denominator.

$x = \dfrac{3}{2} \pm \dfrac{i\sqrt{15}}{6}$ Add $\dfrac{3}{2}$ to both sides.

$= \dfrac{9}{6} \pm \dfrac{i\sqrt{15}}{6}$ Find a common denominator.

$= \dfrac{9 \pm i\sqrt{15}}{6}$ Simplify.

The solutions are $\dfrac{9 + i\sqrt{15}}{6}$ and $\dfrac{9 - i\sqrt{15}}{6}$.

3

Recall the **simple interest** formula $I = Prt$, where I is the interest earned, P is the principal, r is the rate of interest, and t is time. If \$100 is invested at a simple interest rate of 5% annually, at the end of 3 years the total interest I earned is

$$I = P \cdot r \cdot t$$

or

$$I = 100 \cdot 0.05 \cdot 3 = \$15$$

and the new principal is

$$\$100 + \$15 = \$115$$

Most of the time, the interest computed on money borrowed or money deposited is **compound interest.** Compound interest, unlike simple interest, is computed on original principal *and* on interest already earned. To see the difference between simple interest and compound interest, suppose that \$100 is invested at a rate of 5% compounded annually. To find the total amount of money at the end of 3 years, we calculate as follows.

$$I = P \cdot r \cdot t$$

First year: Interest $= \$100 \cdot 0.05 \cdot 1 = \$5.00$
New principal $= \$100.00 + \$5.00 = \$105.00$

Second year: Interest $= \$105.00 \cdot 0.05 \cdot 1 = \$5.25$
New principal $= \$105.00 + \$5.25 = \$110.25$

Third year: Interest $= \$110.25 \cdot 0.05 \cdot 1 \approx \$5.51$
New principal $= \$110.25 + \$5.51 = \$115.76$

At the end of the third year, the total compound interest earned is \$15.76, whereas the total simple interest earned is \$15.

It is tedious to calculate compound interest as we did above, so we use a compound interest formula. The formula for calculating the total amount of money when interest is compounded annually is

$$A = P(1 + r)^t$$

where P is the original investment, r is the interest rate per compounding period, and t is the number of periods. For example, the amount of money A at the end of 3 years if \$100 is invested at 5% compounded annually is

$$A = \$100(1 + 0.05)^3 \approx \$100(1.1576) = \$115.76$$

as we previously calculated.

Example 9   FINDING INTEREST RATES

Find the interest rate r if \$2000 compounded annually grows to \$2420 in 2 years.

Solution   1. UNDERSTAND the problem. Since the \$2000 is compounded annually, we use the compound interest formula. For this example, make sure that you understand the formula for compounding interest annually.

2. TRANSLATE. We substitute the given values into the formula.

$$A = P(1 + r)^t$$

$$2420 = 2000(1 + r)^2 \qquad \text{Let } A = 2420, P = 2000, \text{ and } t = 2.}$$

3. **SOLVE.** Solve the equation for r.

$$2420 = 2000(1 + r)^2$$

$$\frac{2420}{2000} = (1 + r)^2 \qquad \text{Divide both sides by 2000.}$$

$$\frac{121}{100} = (1 + r)^2 \qquad \text{Simplify the fraction.}$$

$$\pm\sqrt{\frac{121}{100}} = 1 + r \qquad \text{Use the square root property.}$$

$$\pm\frac{11}{10} = 1 + r \qquad \text{Simplify.}$$

$$-1 \pm \frac{11}{10} = r$$

$$-\frac{10}{10} \pm \frac{11}{10} = r$$

$$\frac{1}{10} = r \quad \text{or} \quad -\frac{21}{10} = r$$

4. **INTERPRET.** The rate cannot be negative, so we reject $-\dfrac{21}{10}$.

Check: $\dfrac{1}{10} = 0.10 = 10\%$ per year. If we invest \$2000 at 10% compounded annually, in 2 years the amount in the account would be $2000(1 + 0.10)^2 = 2420$ dollars, the desired amount.

State: The interest rate is 10% compounded annually.

GRAPHING CALCULATOR EXPLORATIONS

In Section 6.7, we showed how we can use a graphing calculator to approximate real number solutions of a quadratic equation written in standard form. We can also use a graphing calculator to solve a quadratic equation when it is not written in standard form. For example, to solve $(x + 1)^2 = 12$, the quadratic equation in Example 3, we graph the following on the same set of axes. Use Xmin $= -10$, Xmax $= 10$, Ymin $= -13$, and Ymax $= 13$.

$$Y_1 = (x + 1)^2 \quad \text{and} \quad Y_2 = 12$$

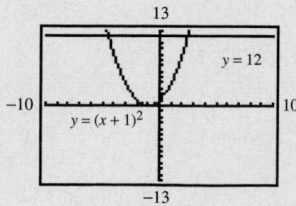

Use the Intersect feature or the Zoom and Trace features to locate the points of intersection of the graphs. The x-values of these points are the solutions of $(x + 1)^2 = 12$. The solutions, rounded to two decimal places, are 2.46 and −4.46.

Check to see that these numbers are approximations of the exact solutions $-1 \pm 2\sqrt{3}$.

Use a graphing calculator to solve each quadratic equation. Round all solutions to the nearest hundredth.

1. $x(x - 5) = 8$

2. $x(x + 2) = 5$

3. $x^2 + 0.5x = 0.3x + 1$

4. $x^2 - 2.6x = -2.2x + 3$

5. Use a graphing calculator and solve $(2x - 5)^2 = -16$, Example 4 in this section, using the window

$$\text{Xmin} = -20$$
$$\text{Xmax} = 20$$
$$\text{Xscl} = 1$$
$$\text{Ymin} = -20$$
$$\text{Ymax} = 20$$
$$\text{Yscl} = 1$$

Explain the results. Compare your results with the solution found in Example 4.

6. What are the advantages and disadvantages of using a graphing calculator to solve quadratic equations?

Exercise Set 10.1

Use the square root property to solve each equation. These equations have real-number solutions. See Examples 1 through 3.

1. $x^2 = 16$

2. $x^2 = 49$

3. $x^2 - 7 = 0$

4. $x^2 - 11 = 0$

5. $x^2 = 18$

6. $y^2 = 20$

7. $3z^2 - 30 = 0$

8. $2x^2 = 4$

9. $(x + 5)^2 = 9$

10. $(y - 3)^2 = 4$

11. $(z - 6)^2 = 18$

12. $(y + 4)^2 = 27$

13. $(2x - 3)^2 = 8$

14. $(4x + 9)^2 = 6$

Use the square root property to solve each equation. See Examples 1 through 4.

15. $x^2 + 9 = 0$

16. $x^2 + 4 = 0$

17. $x^2 - 6 = 0$

18. $y^2 - 10 = 0$

19. $2z^2 + 16 = 0$

20. $3p^2 + 36 = 0$

21. $(x - 1)^2 = -16$

22. $(y + 2)^2 = -25$

23. $(z + 7)^2 = 5$

24. $(x + 10)^2 = 11$

25. $(x + 3)^2 = -8$

26. $(y - 4)^2 = -18$

Add the proper constant to each binomial so that the resulting trinomial is a perfect square trinomial. Then factor the trinomial.

27. $x^2 + 16x$

28. $y^2 + 2y$

29. $z^2 - 12z$

30. $x^2 - 8x$

31. $p^2 + 9p$

32. $n^2 + 5n$

33. $x^2 + x$

34. $y^2 - y$

Find two possible missing terms so that each is a perfect square trinomial.

35. $x^2 + \ \blacksquare\ + 16$

36. $y^2 + \ \blacksquare\ + 9$

37. $z^2 + \ \blacksquare\ + \dfrac{25}{4}$

38. $x^2 + \ \blacksquare\ + \dfrac{1}{4}$

Solve each equation by completing the square. These equations have real number solutions. See Examples 5 through 7.

39. $x^2 + 8x = -15$

40. $y^2 + 6y = -8$

41. $x^2 + 6x + 2 = 0$

42. $x^2 - 2x - 2 = 0$

43. $x^2 + x - 1 = 0$

44. $x^2 + 3x - 2 = 0$

45. $x^2 + 2x - 5 = 0$

46. $y^2 + y - 7 = 0$

47. $3p^2 - 12p + 2 = 0$

48. $2x^2 + 14x - 1 = 0$

49. $4y^2 - 12y - 2 = 0$

50. $6x^2 - 3 = 6x$

51. $2x^2 + 7x = 4$

52. $3x^2 - 4x = 4$

53. $x^2 - 4x - 5 = 0$

54. $y^2 + 6y - 8 = 0$

55. $x^2 + 8x + 1 = 0$

56. $x^2 - 10x + 2 = 0$

57. $3y^2 + 6y - 4 = 0$

58. $2y^2 + 12y + 3 = 0$

59. $2x^2 - 3x - 5 = 0$

60. $5x^2 + 3x - 2 = 0$

Solve each equation by completing the square. See Examples 5 through 8.

61. $y^2 + 2y + 2 = 0$

62. $x^2 + 4x + 6 = 0$

63. $x^2 - 6x + 3 = 0$

64. $x^2 - 7x - 1 = 0$

65. $2a^2 + 8a = -12$

66. $3x^2 + 12x = -14$

67. $5x^2 + 15x - 1 = 0$

68. $16y^2 + 16y - 1 = 0$

69. $2x^2 - x + 6 = 0$

70. $4x^2 - 2x + 5 = 0$

71. $x^2 + 10x + 28 = 0$

72. $y^2 + 8y + 18 = 0$

73. $z^2 + 3z - 4 = 0$

74. $y^2 + y - 2 = 0$

75. $2x^2 - 4x + 3 = 0$

76. $9x^2 - 36x = -40$

77. $3x^2 + 3x = 5$

78. $5y^2 - 15y = 1$

Use the formula $A = P(1 + r)^t$ to solve Exercises 79–82. See Example 9.

79. Find the rate r at which \$3000 grows to \$4320 in 2 years.

80. Find the rate r at which \$800 grows to \$882 in 2 years.

81. Find the rate at which \$810 grows to \$1000 in 2 years.

82. Find the rate at which \$2000 grows to \$2880 in 2 years.

83. In your own words, what is the difference between simple interest and compound interest?

84. If you are depositing money in an account that pays 4%, would you prefer the interest to be simple or compound? Explain why.

85. If you are borrowing money at a rate of 10%, would you prefer the interest to be simple or compound? Explain why.

Neglecting air resistance, the distance s(t) in feet traveled by a freely falling object is given by the function $s(t) = 16t^2$, where t is time in seconds. Use this formula to solve Exercises 86 through 89. Round answers to two decimal places.

86. The Petronas Towers in Kuala Lumpur, built in 1997, are the tallest buildings in Malaysia. Each tower is 1483 feet tall. How long would it take an object to fall to the ground from the top of one of the towers? (*Source:* Council on Tall Buildings and Urban Habitat, Lehigh University)

87. The height of the Chicago Beach Tower Hotel, built in 1998 in Dubai, United Arab Emirates, is 1053 feet. How long would it take an object to fall to the ground from the top of the building? (*Source:* Council on Tall Buildings and Urban Habitat, Lehigh University)

88. The height of the Nurek Dam in Tajikistan (part of the former USSR that borders Afghanistan) is 984 feet. How long would it take an object to fall from the top to the base of the dam? (*Source:* U.S. Committee on Large Dams of the International Commission on Large Dams)

89. The Hoover Dam, located on the Colorado River on the border of Nevada and Arizona near Las Vegas, is 725 feet tall. How long would it take an object to fall from the top to the base of the dam? (*Source:* U.S. Committee on Large Dams of the International Commission on Large Dams)

Solve.

△ **90.** The area of a square room is 225 square feet. Find the dimensions of the room.

△ **91.** The area of a circle is 36π square inches. Find the radius of the circle.

△ **92.** An isosceles right triangle has legs of equal length. If the hypotenuse is 20 centimeters long, find the length of each leg.

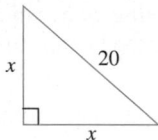

△ **93.** A 27-inch TV is advertised in the *Daily Sentry* newspaper. If 27 inches is the measure of the diagonal of the picture tube, find the measure of the side of the picture tube.

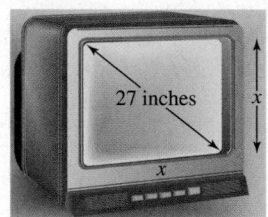

 *A common equation used in business is a demand equation. It expresses the relationship between the unit price of some commodity and the quantity demanded. For Exercises 94 and 95, p represents the unit price and x represents the quantity demanded in thousands.*

94. A manufacturing company has found that the demand equation for a certain type of scissors is given by the equation $p = -x^2 + 47$. Find the demand for the scissors if the price is $11 per pair.

95. Acme, Inc., sells desk lamps and has found that the demand equation for a certain style of desk lamp is given by the equation $p = -x^2 + 15$. Find the demand for the desk lamp if the price is $7 per lamp.

REVIEW EXERCISES

Simplify each expression. See Sections 9.1 and 9.4.

96. $\dfrac{3}{4} - \sqrt{\dfrac{25}{16}}$

97. $\dfrac{3}{5} + \sqrt{\dfrac{16}{25}}$

98. $\dfrac{1}{2} - \sqrt{\dfrac{9}{4}}$

99. $\dfrac{9}{10} - \sqrt{\dfrac{49}{100}}$

Simplify each expression. See Section 9.4.

100. $\dfrac{6 + 4\sqrt{5}}{2}$

101. $\dfrac{10 - 20\sqrt{3}}{2}$

102. $\dfrac{3 - 9\sqrt{5}}{6}$

103. $\dfrac{12 - 8\sqrt{7}}{16}$

Evaluate $\sqrt{b^2 - 4ac}$ for each set of values. See Section 9.3.

104. $a = 2, b = 4, c = -1$

105. $a = 1, b = 6, c = 2$

106. $a = 3, b = -1, c = -2$

107. $a = 1, b = -3, c = -1$

10.2 SOLVING QUADRATIC EQUATIONS BY THE QUADRATIC FORMULA

CD-ROM   SSM

SSG   Video

▶ **OBJECTIVES**

1. Solve quadratic equations by using the quadratic formula.
2. Determine the number and type of solutions of a quadratic equation by using the discriminant.
3. Solve geometric problems modeled by quadratic equations.

1

Any quadratic equation can be solved by completing the square. Since the same sequence of steps is repeated each time we complete the square, let's complete the square for a general quadratic equation, $ax^2 + bx + c = 0$. By doing so, we find a pattern for the solutions of a quadratic equation known as the **quadratic formula.**

Recall that to complete the square for an equation such as $ax^2 + bx + c = 0$, we first divide both sides by the coefficient of x^2.

$$ax^2 + bx + c = 0$$

$$x^2 + \frac{b}{a}x + \frac{c}{a} = 0 \qquad \text{Divide both sides by } a, \text{ the coefficient of } x^2.$$

$$x^2 + \frac{b}{a}x = -\frac{c}{a} \qquad \text{Subtract the constant } \frac{c}{a} \text{ from both sides.}$$

Next, find the square of half $\frac{b}{a}$, the coefficient of x.

$$\frac{1}{2}\left(\frac{b}{a}\right) = \frac{b}{2a} \quad \text{and} \quad \left(\frac{b}{2a}\right)^2 = \frac{b^2}{4a^2}$$

Add this result to both sides of the equation.

$$x^2 + \frac{b}{a}x + \frac{b^2}{4a^2} = -\frac{c}{a} + \frac{b^2}{4a^2} \qquad \text{Add } \frac{b^2}{4a^2} \text{ to both sides.}$$

$$x^2 + \frac{b}{a}x + \frac{b^2}{4a^2} = \frac{-c \cdot 4a}{a \cdot 4a} + \frac{b^2}{4a^2} \qquad \text{Find a common denominator on the right side.}$$

$$x^2 + \frac{b}{a}x + \frac{b^2}{4a^2} = \frac{b^2 - 4ac}{4a^2} \qquad \text{Simplify the right side.}$$

$$\left(x + \frac{b}{2a}\right)^2 = \frac{b^2 - 4ac}{4a^2} \qquad \text{Factor the perfect square trinomial on the left side.}$$

$$x + \frac{b}{2a} = \pm\sqrt{\frac{b^2 - 4ac}{4a^2}} \qquad \text{Apply the square root property.}$$

$$x + \frac{b}{2a} = \pm\frac{\sqrt{b^2 - 4ac}}{2a} \qquad \text{Simplify the radical.}$$

$$x = -\frac{b}{2a} \pm \frac{\sqrt{b^2 - 4ac}}{2a} \qquad \text{Subtract } \frac{b}{2a} \text{ from both sides.}$$

$$x = \frac{-b \pm \sqrt{b^2 - 4ac}}{2a} \qquad \text{Simplify.}$$

This equation identifies the solutions of the general quadratic equation in standard form and is called the quadratic formula. It can be used to solve any equation written in standard form $ax^2 + bx + c = 0$ as long as a is not 0.

QUADRATIC FORMULA

A quadratic equation written in the form $ax^2 + bx + c = 0$ has the solutions

$$x = \frac{-b \pm \sqrt{b^2 - 4ac}}{2a}$$

Example 1 Solve $3x^2 + 16x + 5 = 0$ for x.

Solution This equation is in standard form, so $a = 3, b = 16$, and $c = 5$. Substitute these values into the quadratic formula.

$$x = \frac{-b \pm \sqrt{b^2 - 4ac}}{2a} \qquad \text{Quadratic formula}$$

$$= \frac{-16 \pm \sqrt{16^2 - 4(3)(5)}}{2 \cdot 3} \qquad \text{Use } a = 3, b = 16, \text{ and } c = 5.$$

$$= \frac{-16 \pm \sqrt{256 - 60}}{6}$$

$$= \frac{-16 \pm \sqrt{196}}{6} = \frac{-16 \pm 14}{6}$$

$$x = \frac{-16 + 14}{6} = -\frac{1}{3} \quad \text{or} \quad x = \frac{-16 - 14}{6} = -\frac{30}{6} = -5$$

The solutions are $-\dfrac{1}{3}$ and -5.

Example 2 Solve $2x^2 - 4x = 3$.

Solution First write the equation in standard form by subtracting 3 from both sides.

$$2x^2 - 4x - 3 = 0$$

Now $a = 2, b = -4$, and $c = -3$. Substitute these values into the quadratic formula.

> **HELPFUL HINT**
> To replace a, b, and c correctly in the quadratic formula, write the quadratic equation in standard form $ax^2 + bx + c = 0$.

$$x = \frac{-b \pm \sqrt{b^2 - 4ac}}{2a}$$

$$= \frac{-(-4) \pm \sqrt{(-4)^2 - 4(2)(-3)}}{2 \cdot 2}$$

$$= \frac{4 \pm \sqrt{16 + 24}}{4}$$

$$= \frac{4 \pm \sqrt{40}}{4} = \frac{4 \pm 2\sqrt{10}}{4}$$

$$= \frac{2(2 \pm \sqrt{10})}{2 \cdot 2} = \frac{2 \pm \sqrt{10}}{2}$$

The solutions are $\dfrac{2 + \sqrt{10}}{2}$ and $\dfrac{2 - \sqrt{10}}{2}$.

> **HELPFUL HINT**
> To simplify the expression $\dfrac{4 \pm 2\sqrt{10}}{4}$ in the preceding example, note that 2 is factored out of both terms of the numerator *before* simplifying.
>
> $$\frac{4 \pm 2\sqrt{10}}{4} = \frac{2(2 \pm \sqrt{10})}{2 \cdot 2} = \frac{2 \pm \sqrt{10}}{2}$$

Example 3   Solve $\frac{1}{4}m^2 - m + \frac{1}{2} = 0$.

Solution   We could use the quadratic formula with $a = \frac{1}{4}$, $b = -1$, and $c = \frac{1}{2}$. Instead, we find a simpler, equivalent standard form equation whose coefficients are not fractions. Multiply both sides of the equation by 4 to clear fractions.

$$4\left(\frac{1}{4}m^2 - m + \frac{1}{2}\right) = 4 \cdot 0$$

$$m^2 - 4m + 2 = 0 \qquad \text{Simplify.}$$

Substitute $a = 1$, $b = -4$, and $c = 2$ into the quadratic formula and simplify.

$$m = \frac{-(-4) \pm \sqrt{(-4)^2 - 4(1)(2)}}{2 \cdot 1} = \frac{4 \pm \sqrt{16 - 8}}{2}$$

$$= \frac{4 \pm \sqrt{8}}{2} = \frac{4 \pm 2\sqrt{2}}{2} = \frac{2(2 \pm \sqrt{2})}{2}$$

$$= 2 \pm \sqrt{2}$$

The solutions are $2 + \sqrt{2}$ and $2 - \sqrt{2}$.

Example 4   Solve $x = -3x^2 - 3$.

Solution   The equation in standard form is $3x^2 + x + 3 = 0$. Thus, let $a = 3$, $b = 1$, and $c = 3$ in the quadratic formula.

$$x = \frac{-1 \pm \sqrt{1^2 - 4(3)(3)}}{2 \cdot 3} = \frac{-1 \pm \sqrt{1 - 36}}{6} = \frac{-1 \pm \sqrt{-35}}{6} = \frac{-1 \pm i\sqrt{35}}{6}$$

The solutions are $\dfrac{-1 + i\sqrt{35}}{6}$ and $\dfrac{-1 - i\sqrt{35}}{6}$.

In Example 1, the equation $3x^2 + 16x + 5 = 0$ had 2 real roots, $-\frac{1}{3}$ and -5. In Example 4, the equation $3x^2 + x + 3 = 0$ (written in standard form) had no real roots. How do their related graphs compare? Recall that the x-intercepts of $f(x) = 3x^2 + 16x + 5$ occur where $f(x) = 0$ or where $3x^2 + 16x + 5 = 0$. Since this equation has 2 real roots, the graph has 2 x-intercepts. Similarly, since the equation $3x^2 + x + 3 = 0$ has no real roots, the graph of $f(x) = 3x^2 + x + 3$ has no x-intercepts.

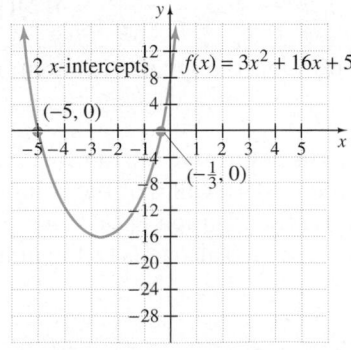

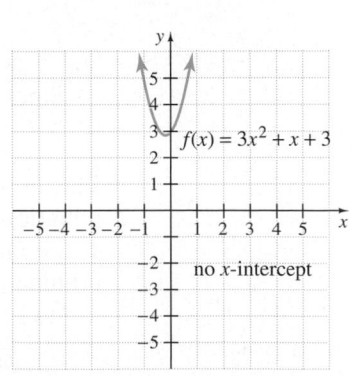

<u>2</u> In the quadratic formula, $x = \dfrac{-b \pm \sqrt{b^2 - 4ac}}{2a}$, the radicand $b^2 - 4ac$ is called the **discriminant** because, by knowing its value, we can **discriminate** among the possible number and type of solutions of a quadratic equation. Possible values of the discriminant and their meanings are summarized next.

DISCRIMINANT

The following table corresponds the discriminant $b^2 - 4ac$ of a quadratic equation of the form $ax^2 + bx + c = 0$ with the number and type of solutions of the equation.

$b^2 - 4ac$	*Number and Type of Solutions*
Positive	Two real solutions
Zero	One real solution
Negative	Two complex but not real solutions

Example 5 Use the discriminant to determine the number and type of solutions of each quadratic equation.

a. $x^2 + 2x + 1 = 0$ **b.** $3x^2 + 2 = 0$ **c.** $2x^2 - 7x - 4 = 0$

Solution **a.** In $x^2 + 2x + 1 = 0$, $a = 1$, $b = 2$, and $c = 1$. Thus,

$$b^2 - 4ac = 2^2 - 4(1)(1) = 0$$

Since $b^2 - 4ac = 0$, this quadratic equation has one real solution.

b. In this equation, $a = 3$, $b = 0$, $c = 2$. Then $b^2 - 4ac = 0 - 4(3)(2) = -24$. Since $b^2 - 4ac$ is negative, the quadratic equation has two complex but not real solutions.

c. In this equation, $a = 2$, $b = -7$, and $c = -4$. Then

$$b^2 - 4ac = (-7)^2 - 4(2)(-4) = 81$$

Since $b^2 - 4ac$ is positive, the quadratic equation has two real solutions.   ▬

The discriminant helps us determine the number and type of solutions of a quadratic equation, $ax^2 + bx + c = 0$. Recall that the solutions of this equation are the same as the x-intercepts of its related graph $f(x) = ax^2 + bx + c$. This means that the discriminant of $ax^2 + bx + c = 0$ also tells us the number of x-intercepts for the graph of $f(x) = ax^2 + bx + c$.

GRAPH OF $f(x) = ax^2 + bx + c$

$b^2 - 4ac > 0$,
$f(x)$ has 2 x-intercepts

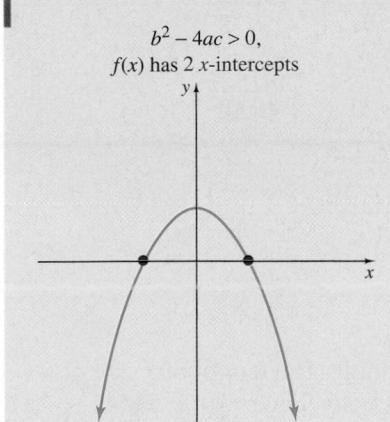

$b^2 - 4ac = 0$,
$f(x)$ has 1 x-intercept

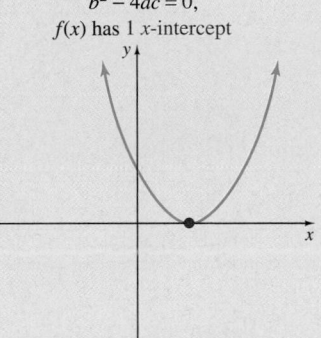

$b^2 - 4ac < 0$,
$f(x)$ has no x-intercepts

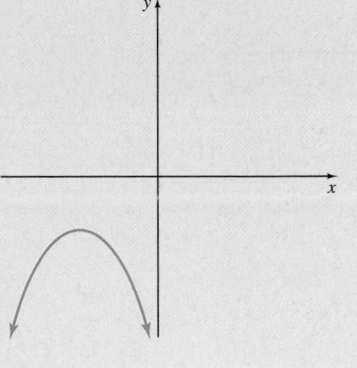

3   The quadratic formula is useful in solving problems that are modeled by quadratic equations.

△ **Example 6**   **CALCULATING DISTANCE SAVED**

At a local university, students often leave the sidewalk and cut across the lawn to save walking distance. Given the diagram below of a favorite place to cut across the lawn, approximate how many feet of walking distance a student saves by cutting across the lawn instead of walking on the sidewalk.

Solution   1. UNDERSTAND. Read and reread the problem. In the diagram, notice that a triangle is formed. Since the corner of the block forms a right angle, we use the Pythagorean theorem for right triangles. You may want to review this theorem.

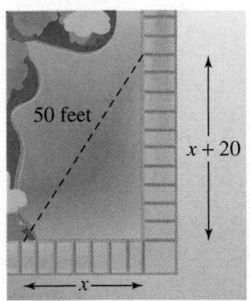

2. TRANSLATE. By the Pythagorean theorem, we have

In words: $(\text{leg})^2 + (\text{leg})^2 = (\text{hypotenuse})^2$

Translate: $x^2 + (x + 20)^2 = 50^2$

3. SOLVE. Use the quadratic formula to solve.

$$x^2 + x^2 + 40x + 400 = 2500 \qquad \text{Square } (x+20) \text{ and } 50.$$

$$2x^2 + 40x - 2100 = 0 \qquad \text{Set the equation equal to 0.}$$

$$x^2 + 20x - 1050 = 0 \qquad \text{Divide by 2.}$$

Here, $a = 1, b = 20, c = -1050$. By the quadratic formula,

$$x = \frac{-20 \pm \sqrt{20^2 - 4(1)(-1050)}}{2 \cdot 1}$$

$$= \frac{-20 \pm \sqrt{400 + 4200}}{2} = \frac{-20 \pm \sqrt{4600}}{2}$$

$$= \frac{-20 \pm \sqrt{100 \cdot 46}}{2} = \frac{-20 \pm 10\sqrt{46}}{2}$$

$$= -10 \pm 5\sqrt{46} \qquad \text{Simplify.}$$

4. INTERPRET.

Check: Your calculations in the quadratic formula. The length of a side of a triangle can't be negative, so we reject $-10 - 5\sqrt{46}$. Since $-10 + 5\sqrt{46} \approx 24$ feet, the walking distance along the sidewalk is

$$x + (x + 20) \approx 24 + (24 + 20) = 68 \text{ feet.}$$

State: A student saves $68 - 50$ or 18 feet of walking distance by cutting across the lawn.

SPOTLIGHT ON DECISION MAKING

Suppose you are a registered dietician. Recently, you read an article in a nutrition journal that described a relationship between weight and the Recommended Dietary Allowance (RDA) for vitamin A in children up to age 10. The relationship is $y = 0.149x^2 - 4.475x + 406.478$, where y is the RDA for vitamin A in micrograms for a child whose weight is x pounds. (*Source:* Food and Nutrition Board, National Academy of Sciences—Institute of Medicine, 1989)

You are working with a 4-year-old patient who weighs 40 pounds. After analyzing her diet, you are able to determine that she is currently getting an average of 400 micrograms of vitamin A daily. Decide whether her current vitamin A intake is adequate. If not, how much more is needed each day? In either case, determine how much weight she will need to gain before a daily intake of 500 micrograms of vitamin A is appropriate.

Exercise Set 10.2

Use the quadratic formula to solve each equation. These equations have real number solutions. See Examples 1 through 3.

1. $m^2 + 5m - 6 = 0$

2. $p^2 + 11p - 12 = 0$

3. $2y = 5y^2 - 3$

4. $5x^2 - 3 = 14x$

5. $x^2 - 6x + 9 = 0$

6. $y^2 + 10y + 25 = 0$

7. $x^2 + 7x + 4 = 0$

8. $y^2 + 5y + 3 = 0$

9. $8m^2 - 2m = 7$

10. $11n^2 - 9n = 1$

11. $3m^2 - 7m = 3$

12. $x^2 - 13 = 5x$

13. $\frac{1}{2}x^2 - x - 1 = 0$

14. $\frac{1}{6}x^2 + x + \frac{1}{3} = 0$

15. $\frac{2}{5}y^2 + \frac{1}{5}y = \frac{3}{5}$

16. $\frac{1}{8}x^2 + x = \frac{5}{2}$

17. $\frac{1}{3}y^2 - y - \frac{1}{6} = 0$

18. $\frac{1}{2}y^2 = y + \frac{1}{2}$

19. Solve Exercise 1 by factoring. Explain the result.

20. Solve Exercise 2 by factoring. Explain the result.

Use the quadratic formula to solve each equation. See Example 4.

21. $6 = -4x^2 + 3x$

22. $9x^2 + x + 2 = 0$

23. $(x + 5)(x - 1) = 2$

24. $x(x + 6) = 2$

25. $10y^2 + 10y + 3 = 0$

26. $3y^2 + 6y + 5 = 0$

The solutions of the quadratic equation $ax^2 + bx + c = 0$ are

$$\frac{-b + \sqrt{b^2 - 4ac}}{2a} \quad and \quad \frac{-b - \sqrt{b^2 - 4ac}}{2a}$$

27. Show that the sum of these solutions is $\dfrac{-b}{a}$.

28. Show that the product of these solutions is $\dfrac{c}{a}$.

Use the discriminant to determine the number and type of solutions of each equation. See Example 5.

29. $9x - 2x^2 + 5 = 0$

30. $5 - 4x + 12x^2 = 0$

31. $4x^2 + 12x = -9$

32. $9x^2 + 1 = 6x$

33. $3x = -2x^2 + 7$

34. $3x^2 = 5 - 7x$

35. $6 = 4x - 5x^2$

36. $8x = 3 - 9x^2$

Use the quadratic formula to solve each equation. These equations have real number solutions.

37. $x^2 + 5x = -2$

38. $y^2 - 8 = 4y$

39. $(m + 2)(2m - 6) = 5(m - 1) - 12$

40. $7p(p - 2) + 2(p + 4) = 3$

41. $\dfrac{x^2}{3} - x = \dfrac{5}{3}$

42. $\dfrac{x^2}{2} - 3 = -\dfrac{9}{2}x$

43. $x(6x + 2) - 3 = 0$

44. $x(7x + 1) = 2$

Use the quadratic formula to solve each equation.

45. $x^2 + 6x + 13 = 0$

46. $x^2 + 2x + 2 = 0$

47. $\dfrac{2}{5}y^2 + \dfrac{1}{5}y + \dfrac{3}{5} = 0$

48. $\dfrac{1}{8}x^2 + x + \dfrac{5}{2} = 0$

49. $\dfrac{1}{2}y^2 = y - \dfrac{1}{2}$

50. $\dfrac{2}{3}x^2 - \dfrac{20}{3}x = -\dfrac{100}{6}$

51. $(n - 2)^2 = 15n$

52. $\left(p - \dfrac{1}{2}\right)^2 = \dfrac{p}{2}$

Solve. See Example 6.

53. Nancy, Thelma, and John Varner live on a corner lot. Often, neighborhood children cut across their lot to save walking distance. Given the diagram below, approximate to the nearest foot how many feet of walking distance is saved by cutting across their property instead of walking around the lot.

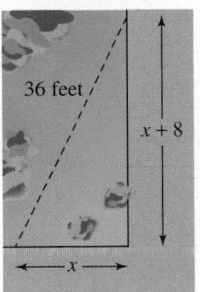

54. Given the diagram below, approximate to the nearest foot how many feet of walking distance a person saves by cutting across the lawn instead of walking on the sidewalk.

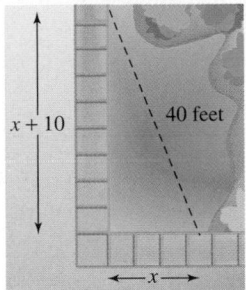

55. The hypotenuse of an isosceles right triangle is 2 centimeters longer than either of its legs. Find the exact length of each side. (*Hint:* An isosceles right triangle is a right triangle whose legs are the same length.)

56. The hypotenuse of an isosceles right triangle is one meter longer than either of its legs. Find the length of each side.

57. Uri Chechov's rectangular dog pen for his Irish setter must have an area of 400 square feet. Also, the length must be 10 feet longer than the width. Find the dimensions of the pen.

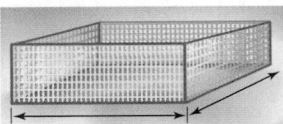

58. An entry in the Peach Festival Poster Contest must be rectangular and have an area of 1200 square inches. Furthermore, its length must be 20 inches longer than its width. Find the dimensions each entry must have.

59. A holding pen for cattle must be square and have a diagonal length of 100 meters.

a. Find the length of a side of the pen.

b. Find the area of the pen.

60. A rectangle is three times longer than it is wide. It has a diagonal of length 50 centimeters.

a. Find the dimensions of the rectangle.

b. Find the perimeter of the rectangle.

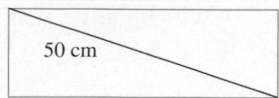

61. If a point B divides a line segment such that the smaller portion is to the larger portion as the larger is to the whole, the whole is the length of the *golden ratio*.

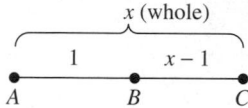

The golden ratio was thought by the Greeks to be the most pleasing to the eye, and many of their buildings contained numerous examples of the golden ratio. The value of the golden ratio is the positive solution of

$$\underset{\text{(larger)}}{\overset{\text{(smaller)}}{\frac{x-1}{1}}} = \underset{\text{(whole)}}{\overset{\text{(larger)}}{\frac{1}{x}}}$$

Find this value.

△ **62.** The base of a triangle is four more than twice its height. If the area of the triangle is 42 square centimeters, find its base and height.

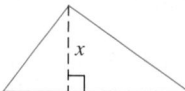

The Wollomombi Falls in Australia have a height of 1100 feet. A pebble is thrown upward from the top of the falls with an initial velocity of 20 feet per second. The height of the pebble h after t seconds is given by the equation $h = -16t^2 + 20t + 1100$. Use this equation for Exercises 63 and 64.

63. How long after the pebble is thrown will it hit the ground? Round to the nearest tenth of a second.

64. How long after the pebble is thrown will it be 550 feet from the ground? Round to the nearest tenth of a second.

A ball is thrown downward from the top of a 180-foot building with an initial velocity of 20 feet per second. The height of the ball h after t seconds is given by the equation $h = -16t^2 - 20t + 180$. Use this equation to answer Exercises 65 and 66.

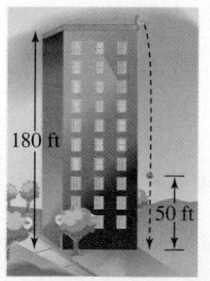

65. How long after the ball is thrown will it strike the ground? Round the result to the nearest tenth of a second.

66. How long after the ball is thrown will it be 50 feet from the ground? Round the result to the nearest tenth of a second.

The accompanying graph shows the daily low temperatures for one week in New Orleans, Louisiana.

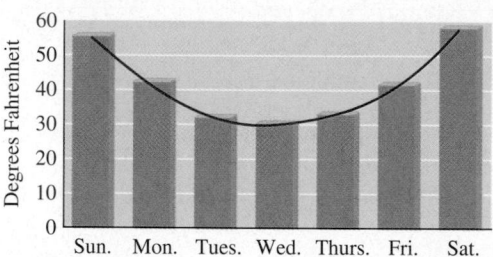

67. Which day of the week shows the greatest decrease in low temperature?

68. Which day of the week shows the greatest increase in low temperature?

69. Which day of the week had the lowest low temperature?

70. Use the graph to estimate the low temperature on Thursday.

Notice that the shape of the temperature graph is similar to a parabola (see Section 6.8). In fact, this graph can be modeled by the quadratic function $f(x) = 3x^2 - 18x + 56$, where $f(x)$ is the temperature in degrees Fahrenheit and x is the number of days from Sunday. (This graph is shown in red above.) Use this function to answer Exercises 71 and 72.

71. Use the quadratic function given to approximate the temperature on Thursday. Does your answer agree with the graph above?

72. Use the function given and the quadratic formula to find when the temperature was 35°F. [*Hint:* Let $f(x) = 35$ and solve for x.] Round your answer to one decimal place and interpret your result.

73. Wal-Mart Stores' net income can be modeled by the quadratic function $f(x) = 128.5x^2 - 69.5x + 2681$, where $f(x)$ is net income in millions of dollars and x is the number of years after 1995. (*Source:* Based on data from Wal-Mart Stores, Inc.)

a. Find Wal-Mart's net income in 1997.

b. If the trend described by the model continues, predict the year after 1995 in which Wal-Mart's net income will be $15,000 million. Round to the nearest whole year.

74. The number of inmates in custody in U.S. prisons and jails can be modeled by the quadratic function

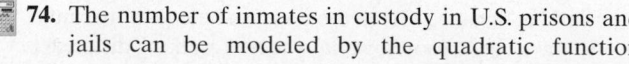

$p(x) = -716.2x^2 + 87{,}453.7x + 1{,}148{,}702$ where $p(x)$ is the number of inmates and x is the number of years after 1990. (*Source:* Based on data from the Bureau of Justice Statistics, U.S. Department of Justice, 1990–1998) Round **a** and **b** to the nearest ten thousand.

   a. Find the number of prison inmates in the United States in 1992.

   b. Find the number of prison inmates in the United States in 1998.

75. Use a graphing calculator to solve Exercises 63 and 65.

76. Use a graphing calculator to solve Exercises 64 and 66.

Recall that the discriminant also tells us the number of x-intercepts of the related function.

77. Check the results of Exercise 29 by graphing $y = 9x - 2x^2 + 5$.

78. Check the results of Exercise 30 by graphing $y = 5 - 4x + 12x^2$.

REVIEW EXERCISES

Solve each equation. See Sections 7.6 and 9.6.

79. $\sqrt{5x - 2} = 3$

80. $\sqrt{y + 2} + 7 = 12$

81. $\dfrac{1}{x} + \dfrac{2}{5} = \dfrac{7}{x}$

82. $\dfrac{10}{z} = \dfrac{5}{z} - \dfrac{1}{3}$

Factor. See Section 6.5.

83. $x^4 + x^2 - 20$

84. $2y^4 + 11y^2 - 6$

85. $z^4 - 13z^2 + 36$

86. $x^4 - 1$

A Look Ahead

Example

Solve $x^2 - 3\sqrt{2}x + 2 = 0$.

Solution

In this equation, $a = 1$, $b = -3\sqrt{2}$, and $c = 2$. By the quadratic formula, we have

$$x = \frac{-b \pm \sqrt{b^2 - 4ac}}{2a}$$

$$= \frac{3\sqrt{2} \pm \sqrt{(-3\sqrt{2})^2 - 4(1)(2)}}{2(1)}$$

$$= \frac{3\sqrt{2} \pm \sqrt{18 - 8}}{2} = \frac{3\sqrt{2} \pm \sqrt{10}}{2}$$

The solutions are $\dfrac{3\sqrt{2} + \sqrt{10}}{2}$ and $\dfrac{3\sqrt{2} - \sqrt{10}}{2}$.

Use the quadratic formula to solve each quadratic equation. See the preceding example.

87. $3x^2 - \sqrt{12}x + 1 = 0$

88. $5x^2 + \sqrt{20}x + 1 = 0$

89. $x^2 + \sqrt{2}x + 1 = 0$

90. $x^2 - \sqrt{2}x + 1 = 0$

91. $2x^2 - \sqrt{3}x - 1 = 0$

92. $7x^2 + \sqrt{7}x - 2 = 0$

10.3 SOLVING EQUATIONS BY USING QUADRATIC METHODS

CD-ROM   SSM

SSG   Video

▶ **OBJECTIVES**

   1. Solve various equations that are quadratic in form.

   2. Solve problems that lead to quadratic equations.

1   In this section, we discuss various types of equations that can be solved in part by using the methods for solving quadratic equations.

   Once each equation is simplified, you may want to use these steps when deciding what method to use to solve the quadratic equation.

SOLVING A QUADRATIC EQUATION

Step 1: If the equation is in the form $(ax + b)^2 = c$, use the square root property and solve. If not, go to Step 2.

Step 2: Write the equation in standard form: $ax^2 + bx + c = 0$.

Step 3: Try to solve the equation by the factoring method. If not possible, go to Step 4.

Step 4: Solve the equation by the quadratic formula.

The first example is a radical equation that becomes a quadratic equation once we square both sides.

Example 1 Solve $x - \sqrt{x} - 6 = 0$.

Solution Recall that to solve a radical equation, first get the radical alone on one side of the equation. Then square both sides.

$$x - 6 = \sqrt{x} \qquad \text{Add } \sqrt{x} \text{ to both sides.}$$

$$(x - 6)^2 = (\sqrt{x})^2 \qquad \text{Square both sides.}$$

$$x^2 - 12x + 36 = x$$

$$x^2 - 13x + 36 = 0 \qquad \text{Set the equation equal to 0.}$$

$$(x - 9)(x - 4) = 0$$

$$x - 9 = 0 \quad \text{or} \quad x - 4 = 0$$

$$x = 9 \qquad\qquad x = 4$$

Check

$$\text{Let } x = 9 \qquad\qquad\qquad \text{Let } x = 4$$

$$x - \sqrt{x} - 6 = 0 \qquad\qquad x - \sqrt{x} - 6 = 0$$

$$9 - \sqrt{9} - 6 \overset{?}{=} 0 \qquad\qquad 4 - \sqrt{4} - 6 \overset{?}{=} 0$$

$$9 - 3 - 6 \overset{?}{=} 0 \qquad\qquad 4 - 2 - 6 \overset{?}{=} 0$$

$$0 = 0 \quad \text{True.} \qquad\qquad -4 = 0 \quad \text{False.}$$

The solution is 9.

Example 2 Solve $\dfrac{3x}{x - 2} - \dfrac{x + 1}{x} = \dfrac{6}{x(x - 2)}$.

Solution In this equation, x cannot be either 2 or 0, because these values cause denominators to equal zero. To solve for x, we first multiply both sides of the equation by $x(x - 2)$ to clear the fractions. By the distributive property, this means that we multiply each term by $x(x - 2)$.

$$x(x - 2)\left(\frac{3x}{x - 2}\right) - x(x - 2)\left(\frac{x + 1}{x}\right) = x(x - 2)\left[\frac{6}{x(x - 2)}\right]$$

$$3x^2 - (x - 2)(x + 1) = 6 \quad \text{Simplify.}$$

$$3x^2 - (x^2 - x - 2) = 6 \quad \text{Multiply.}$$

$$3x^2 - x^2 + x + 2 = 6$$

$$2x^2 + x - 4 = 0 \quad \text{Simplify.}$$

This equation cannot be factored using integers, so we solve by the quadratic formula.

$$x = \frac{-1 \pm \sqrt{1^2 - 4(2)(-4)}}{2 \cdot 2} \quad \begin{array}{l}\text{Use } a = 2, b = 1, \text{and } c = -4 \text{ in}\\ \text{the quadratic formula.}\end{array}$$

$$= \frac{-1 \pm \sqrt{1 + 32}}{4} \quad \text{Simplify.}$$

$$= \frac{-1 \pm \sqrt{33}}{4}$$

Neither proposed solution will make the denominators 0.

The solutions are $\dfrac{-1 + \sqrt{33}}{4}$ and $\dfrac{-1 - \sqrt{33}}{4}$.   ▪

Example 3   Solve $p^4 - 3p^2 - 4 = 0$.

Solution   First we factor the trinomial.

$$p^4 - 3p^2 - 4 = 0$$

$$(p^2 - 4)(p^2 + 1) = 0 \qquad\qquad \text{Factor.}$$

$$(p - 2)(p + 2)(p^2 + 1) = 0 \qquad\qquad \begin{array}{l}\text{Factor further.}\\ \text{Set each factor}\end{array}$$

$$p - 2 = 0 \quad \text{or} \quad p + 2 = 0 \quad \text{or} \quad p^2 + 1 = 0 \qquad \begin{array}{l}\text{equal to 0 and solve.}\end{array}$$

$$p = 2 \qquad\qquad p = -2 \qquad\qquad p^2 = -1$$

$$p = \pm\sqrt{-1} = \pm i$$

The solutions are $2, -2, i$ and $-i$.   ▪

> **HELPFUL HINT**
> Example 3 can be solved using substitution also. Think of $p^4 - 3p^2 - 4 = 0$ as
>
> $(p^2)^2 - 3p^2 - 4 = 0$ Then let $x = p^2$, and solve and substitute back. The
> solutions will be the same.
> $x^2 - 3x - 4 = 0$

Example 4   Solve $(x - 3)^2 - 3(x - 3) - 4 = 0$.

Solution   Notice that the quantity $(x - 3)$ is repeated in this equation. Sometimes it is helpful to substitute a variable (in this case other than x) for the repeated quantity. We will let $y = x - 3$. Then

$$(x - 3)^2 - 3(x - 3) - 4 = 0$$

becomes

$$y^2 - 3y - 4 = 0 \qquad \text{Let } x - 3 = y.$$

$$(y - 4)(y + 1) = 0 \qquad \text{Factor.}$$

To solve, we use the zero factor property.

$$y - 4 = 0 \quad \text{or} \quad y + 1 = 0 \qquad \text{Set each factor equal to 0.}$$

$$y = 4 \qquad\qquad y = -1 \qquad \text{Solve.}$$

To find values of x, we substitute back. That is, we substitute $x - 3$ for y.

$$x - 3 = 4 \quad \text{or} \quad x - 3 = -1$$

$$x = 7 \qquad\qquad x = 2$$

Both 2 and 7 check. The solutions are 2 and 7.

> **HELPFUL HINT**
> When using substitution, don't forget to substitute back to the original variable.

Example 5 Solve $x^{2/3} - 5x^{1/3} + 6 = 0$.

Solution The key to solving this equation is recognizing that $x^{2/3} = (x^{1/3})^2$. We replace $x^{1/3}$ with m so that

$$(x^{1/3})^2 - 5x^{1/3} + 6 = 0$$

becomes

$$m^2 - 5m + 6 = 0$$

Now we solve by factoring.

$$m^2 - 5m + 6 = 0$$

$$(m - 3)(m - 2) = 0 \qquad \text{Factor.}$$

$$m - 3 = 0 \quad \text{or} \quad m - 2 = 0 \qquad \text{Set each factor equal to 0.}$$

$$m = 3 \qquad\qquad m = 2$$

Since $m = x^{1/3}$, we have

$$x^{1/3} = 3 \qquad \text{or} \quad x^{1/3} = 2$$

$$x = 3^3 = 27 \quad \text{or} \qquad x = 2^3 = 8$$

Both 8 and 27 check. The solutions are 8 and 27.

2   The next example is a work problem. This problem is modeled by a rational equation that simplifies to a quadratic equation.

Example 6   **FINDING WORK TIME**

Together, an experienced typist and an apprentice typist can process a document in 6 hours. Alone, the experienced typist can process the document 2 hours faster than the apprentice typist can. Find the time in which each person can process the document alone.

Solution   1. UNDERSTAND. Read and reread the problem. The key idea here is the relationship between the *time* (hours) it takes to complete the job and the *part of the job* completed in one unit of time (hour). For example, because they can complete the job together in 6 hours, the *part of the job* they can complete in 1 hour is $\frac{1}{6}$. Let

x = the *time* in hours it takes the apprentice typist to complete the job alone

$x - 2$ = the *time* in hours it takes the experienced typist to complete the job alone

We can summarize in a chart the information discussed

	Total Hours to Complete Job	*Part of Job Completed in 1 Hour*
APPRENTICE TYPIST	x	$\frac{1}{x}$
EXPERIENCED TYPIST	$x - 2$	$\frac{1}{x - 2}$
TOGETHER	6	$\frac{1}{6}$

2. TRANSLATE.

In words:

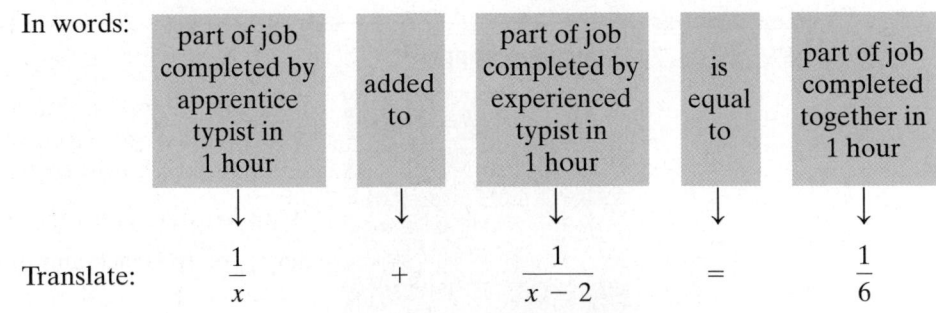

part of job completed by apprentice typist in 1 hour	added to	part of job completed by experienced typist in 1 hour	is equal to	part of job completed together in 1 hour
↓	↓	↓	↓	↓

Translate: $\dfrac{1}{x}$ $+$ $\dfrac{1}{x - 2}$ $=$ $\dfrac{1}{6}$

3. SOLVE.

$$\frac{1}{x} + \frac{1}{x-2} = \frac{1}{6}$$

$$6x(x-2)\left(\frac{1}{x} + \frac{1}{x-2}\right) = 6x(x-2) \cdot \frac{1}{6}$$ Multiply both sides by the LCD, $6x(x-2)$.

$$6x(x-2) \cdot \frac{1}{x} + 6x(x-2) \cdot \frac{1}{x-2} = 6x(x-2) \cdot \frac{1}{6}$$ Use the distributive property.

$$6(x-2) + 6x = x(x-2)$$

$$6x - 12 + 6x = x^2 - 2x$$

$$0 = x^2 - 14x + 12$$

Now we can substitute $a = 1$, $b = -14$, and $c = 12$ into the quadratic formula and simplify.

$$x = \frac{-(-14) \pm \sqrt{(-14)^2 - 4(1)(12)}}{2 \cdot 1} = \frac{14 \pm \sqrt{148}}{2}$$

Using a calculator or a square root table, we see that $\sqrt{148} \approx 12.2$ rounded to one decimal place. Thus,

$$x \approx \frac{14 \pm 12.2}{2}$$

$$x \approx \frac{14 + 12.2}{2} = 13.1 \quad \text{or} \quad x \approx \frac{14 - 12.2}{2} = 0.9$$

4. INTERPRET.

Check: If the apprentice typist completes the job alone in 0.9 hours, the experienced typist completes the job alone in $x - 2 = 0.9 - 2 = -1.1$ hours. Since this is not possible, we reject the solution of 0.9. The approximate solution thus is 13.1 hours.

State: The apprentice typist can complete the job alone in approximately 13.1 hours, and the experienced typist can complete the job alone in approximately

$$x - 2 = 13.1 - 2 = 11.1 \text{ hours.}$$ ▬

Example 7   FINDING SPEED

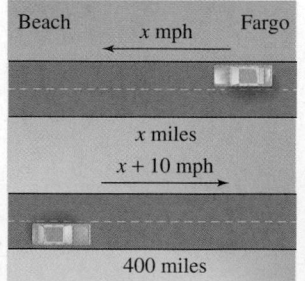

Beach and Fargo are about 400 miles apart. A salesperson travels from Fargo to Beach one day at a certain speed. She returns to Fargo the next day and drives 10 mph faster. Her total travel time was $14\frac{2}{3}$ hours. Find her speed to Beach and the return speed to Fargo.

Solution   1. UNDERSTAND. Read and reread the problem. Let

$$x = \text{the speed to Beach, so}$$

$$x + 10 = \text{the return speed to Fargo.}$$

Then organize the given information in a table.

	distance	=	rate	·	time	
TO BEACH	400		x		$\dfrac{400}{x}$	←distance ←rate
RETURN TO FARGO	400		$x + 10$		$\dfrac{400}{x + 10}$	←distance ←rate

HELPFUL HINT
Since $d = rt$, then
$t = \dfrac{d}{r}$. The time
column was
completed using $\dfrac{d}{r}$.

2. TRANSLATE.

In words:

time to Beach	+	return time to Fargo	=

Translate: $\dfrac{400}{x}$ $+$ $\dfrac{400}{x + 10}$ $=$ $\dfrac{44}{3}$

3. SOLVE.

$$\frac{400}{x} + \frac{400}{x + 10} = \frac{44}{3}$$

$$\frac{100}{x} + \frac{100}{x + 10} = \frac{11}{3} \qquad \text{Divide both sides by 4.}$$

$$3x(x + 10)\left(\frac{100}{x} + \frac{100}{x + 10}\right) = 3x(x + 10) \cdot \frac{11}{3} \qquad \text{Multiply both sides by the LCD } 3x(x + 10).$$

$$3x(x + 10) \cdot \frac{100}{x} + 3x(x + 10) \cdot \frac{100}{x + 10} = 3x(x + 10) \cdot \frac{11}{3} \qquad \text{Use the distributive property.}$$

$$3(x + 10) \cdot 100 + 3x \cdot 100 = x(x + 10) \cdot 11$$

$$300x + 3000 + 300x = 11x^2 + 110x$$

$$0 = 11x^2 - 490x - 3000 \qquad \text{Set equation equal to 0.}$$

$$0 = (11x + 60)(x - 50) \qquad \text{Factor.}$$

$$11x + 60 = 0 \qquad \text{or} \quad x - 50 = 0 \qquad \text{Set each factor equal to 0.}$$

$$x = -\frac{60}{11} = -5\frac{5}{11} \qquad \text{or} \qquad x = 50$$

4. INTERPRET.

Check: The speed is not negative, so it's not $-5\dfrac{5}{11}$. The number 50 does check.

State: The speed to Beach was 50 mph and her return speed to Fargo was 60 mph.

Exercise Set 10.3

Solve. See Example 1.

1. $2x = \sqrt{10 + 3x}$

2. $3x = \sqrt{8x + 1}$

3. $x - 2\sqrt{x} = 8$

4. $x - \sqrt{2x} = 4$

5. $\sqrt{9x} = x + 2$

6. $\sqrt{16x} = x + 3$

Solve. See Example 2.

7. $\dfrac{2}{x} + \dfrac{3}{x - 1} = 1$

8. $\dfrac{6}{x^2} = \dfrac{3}{x + 1}$

9. $\dfrac{3}{x} + \dfrac{4}{x + 2} = 2$

10. $\dfrac{5}{x - 2} + \dfrac{4}{x + 2} = 1$

11. $\dfrac{7}{x^2 - 5x + 6} = \dfrac{2x}{x - 3} - \dfrac{x}{x - 2}$

12. $\dfrac{11}{2x^2 + x - 15} = \dfrac{5}{2x - 5} - \dfrac{x}{x + 3}$

Solve. See Example 3.

13. $p^4 - 16 = 0$

14. $x^4 + 2x^2 - 3 = 0$

15. $4x^4 + 11x^2 = 3$

16. $z^4 = 81$

17. $z^4 - 13z^2 + 36 = 0$

18. $9x^4 + 5x^2 - 4 = 0$

Solve. See Examples 4 and 5.

19. $x^{2/3} - 3x^{1/3} - 10 = 0$

20. $x^{2/3} + 2x^{1/3} + 1 = 0$

21. $(5n + 1)^2 + 2(5n + 1) - 3 = 0$

22. $(m - 6)^2 + 5(m - 6) + 4 = 0$

23. $2x^{2/3} - 5x^{1/3} = 3$

24. $3x^{2/3} + 11x^{1/3} = 4$

25. $1 + \dfrac{2}{3t - 2} = \dfrac{8}{(3t - 2)^2}$

26. $2 - \dfrac{7}{x + 6} = \dfrac{15}{(x + 6)^2}$

27. $20x^{2/3} - 6x^{1/3} - 2 = 0$

28. $4x^{2/3} + 16x^{1/3} = -15$

Solve. See Examples 1 through 5.

29. $a^4 - 5a^2 + 6 = 0$

30. $x^4 - 12x^2 + 11 = 0$

31. $\dfrac{2x}{x - 2} + \dfrac{x}{x + 3} = -\dfrac{5}{x + 3}$

32. $\dfrac{5}{x - 3} + \dfrac{x}{x + 3} = \dfrac{19}{x^2 - 9}$

33. $(p + 2)^2 = 9(p + 2) - 20$

34. $2(4m - 3)^2 - 9(4m - 3) = 5$

35. $2x = \sqrt{11x + 3}$

36. $4x = \sqrt{2x + 3}$

37. $x^{2/3} - 8x^{1/3} + 15 = 0$

38. $x^{2/3} - 2x^{1/3} - 8 = 0$

39. $y^3 + 9y - y^2 - 9 = 0$

40. $x^3 + x - 3x^2 - 3 = 0$

41. $2x^{2/3} + 3x^{1/3} - 2 = 0$

42. $6x^{2/3} - 25x^{1/3} - 25 = 0$

43. $x^{-2} - x^{-1} - 6 = 0$

44. $y^{-2} - 8y^{-1} + 7 = 0$

45. $x - \sqrt{x} = 2$

46. $x - \sqrt{3x} = 6$

47. $\dfrac{x}{x - 1} + \dfrac{1}{x + 1} = \dfrac{2}{x^2 - 1}$

48. $\dfrac{x}{x - 5} + \dfrac{5}{x + 5} = -\dfrac{1}{x^2 - 25}$

49. $p^4 - p^2 - 20 = 0$

50. $x^4 - 10x^2 + 9 = 0$

51. $2x^3 = -54$

52. $y^3 - 216 = 0$

53. $1 = \dfrac{4}{x - 7} + \dfrac{5}{(x - 7)^2}$

54. $3 + \dfrac{1}{2p + 4} = \dfrac{10}{(2p + 4)^2}$

55. $27y^4 + 15y^2 = 2$

56. $8z^4 + 14z^2 = -5$

Solve. See Examples 6 and 7.

57. A jogger ran 3 miles, decreased her speed by 1 mile per hour, and then ran another 4 miles. If her total time jogging was $1\dfrac{3}{5}$ hours, find her speed for each part of her run.

58. Mark Keaton's workout consists of jogging for 3 miles, and then riding his bike for 5 miles at a speed 4 miles per hour faster than he jogs. If his total workout time is 1 hour, find his jogging speed and his biking speed.

59. A Chinese restaurant in Mandeville, Louisiana, has a large goldfish pond around the restaurant. Suppose that an inlet pipe and a hose together can fill the pond in 8 hours. The inlet pipe alone can complete the job in one hour less time than the hose alone. Find the time that the hose can complete the job alone and the time that the inlet pipe can complete the job alone. Round each to the nearest tenth of an hour.

60. A water tank on a farm in Flatonia, Texas, can be filled with a large inlet pipe and a small inlet pipe in 3 hours. The large inlet pipe alone can fill the tank in 2 hours less time than the small inlet pipe alone. Find the time to the nearest tenth of an hour each pipe can fill the tank alone.

61. Roma Sherry drove 330 miles from her hometown to Tucson. During her return trip, she was able to increase her speed by 11 mph. If her return trip took 1 hour less time, find her original speed and her speed returning home.

62. A salesperson drove to Portland, a distance of 300 miles. During the last 80 miles of his trip, heavy rainfall forced him to decrease his speed by 15 mph. If his total driving time was 6 hours, find his original speed and his speed during the rainfall.

63. Bill Shaughnessy and his son Billy can clean the house together in 4 hours. When the son works alone, it takes him an hour longer to clean than it takes his dad alone. Find how long to the nearest tenth of an hour it takes the son to clean alone.

64. Together, Noodles and Freckles eat a 50-pound bag of dog food in 30 days. Noodles by himself eats a 50-pound

bag in 2 weeks less time than Freckles does by himself. How many days to the nearest whole day would a 50-pound bag of dog food last Freckles?

65. The product of a number and 4 less than the number is 96. Find the number.

66. A whole number increased by its square is two more than twice itself. Find the number.

△ **67.** Suppose that an open box is to be made from a square sheet of cardboard by cutting out squares from each corner as shown and then folding along the dotted lines. If the box is to have a volume of 300 cubic centimeters, find the original dimensions of the sheet of cardboard.

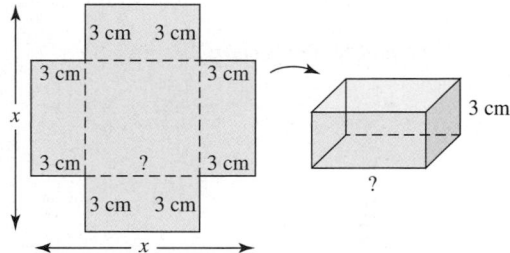

a. The ? in the drawing to the left will be the length (and also the width) of the box as shown in the drawing above. Represent this length in terms of x.

b. Use the formula for volume of a box, $V = l \cdot w \cdot h$, to write an equation in x.

c. Solve the equation for x and give the dimensions of the sheet of cardboard. Check your solution.

△ **68.** Suppose that an open box is to be made from a square sheet of cardboard by cutting out squares from each corner as shown and then folding along the dotted lines. If the box is to have a volume of 128 cubic inches, find the original dimensions of the sheet of cardboard. (*Hint:* Use Exercise 67 Parts **a, b,** and **c** to help you.)

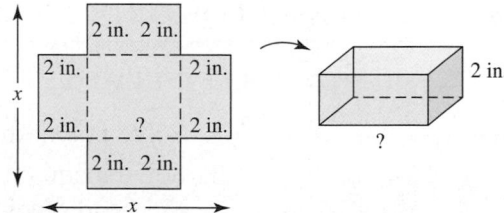

69. During the 2000 Grand Prix of Miami auto race, Juan Montoya posted the fastest lap speed but Max Papis won the race. The track is 7920 feet (1.5 miles) long. Montoya's fastest lap speed was 3.8 feet per second faster than Papis' fastest lap speed. Traveling at these fastest speeds, Papis would have taken 0.376 seconds longer than Montoya to complete a lap. (*Source:* Championship Auto Racing Teams, Inc.)

a. Find Max Papis' fastest lap speed during the race. Round to one decimal place.

b. Find Juan Montoya's fastest lap speed during the race. Round to one decimal place.

c. Convert each speed to miles per hour. Round to one decimal place.

 70. Use a graphing calculator to solve Exercise 29. Compare the solution with the solution from Exercise 29. Explain any differences.

71. Write a polynomial equation that has three solutions: 2, 5, and −7.

72. Write a polynomial equation that has three solutions: 0, 2i, and −2i.

REVIEW EXERCISES

Solve each inequality. See Section 2.7.

73. $\dfrac{5x}{3} + 2 \le 7$

74. $\dfrac{2x}{3} + \dfrac{1}{6} \ge 2$

75. $\dfrac{y-1}{15} > -\dfrac{2}{5}$

76. $\dfrac{z-2}{12} < \dfrac{1}{4}$

Find the domain and range of each graphed relation. Decide which relations are also functions. See Section 3.3.

77.

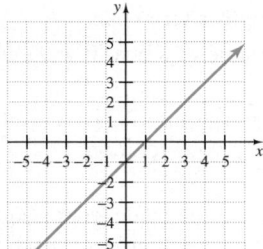

78.

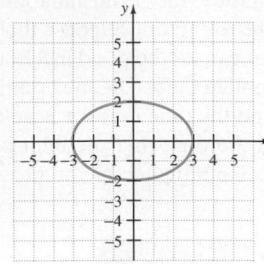

79.

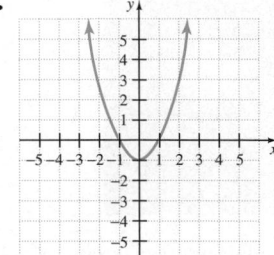

80.

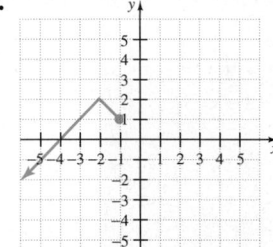

10.4 NONLINEAR INEQUALITIES IN ONE VARIABLE

CD-ROM

SSM

SSG

Video

▶ **OBJECTIVES**

1. Solve polynomial inequalities of degree 2 or greater.
2. Solve inequalities that contain rational expressions with variables in the denominator.

1

Just as we can solve linear inequalities in one variable, so can we also solve quadratic inequalities in one variable. A **quadratic inequality** is an inequality that can be written so that one side is a quadratic expression and the other side is 0. Here are examples of quadratic inequalities in one variable. Each is written in **standard form.**

$$x^2 - 10x + 7 \le 0 \qquad 3x^2 + 2x - 6 > 0$$

$$2x^2 + 9x - 2 < 0 \qquad x^2 - 3x + 11 \ge 0$$

A solution of a quadratic inequality in one variable is a value of the variable that makes the inequality a true statement.

The value of an expression such as $x^2 - 3x - 10$ can sometimes be positive, sometimes negative, and sometimes 0, depending on the value substituted for x. To solve the inequality $x^2 - 3x - 10 < 0$, we are looking for all values of x that make the expression $x^2 - 3x - 10$ **less than 0,** or **negative.** To understand how we find these values, we'll study the graph of the quadratic function $y = x^2 - 3x - 10$.

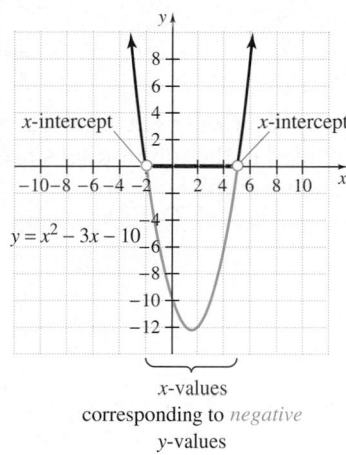

x-values
corresponding to *negative*
y-values

Notice that the x-values for which y is positive are separated from the x-values for which y is negative by the x-intercepts. (Recall that the x-intercepts correspond to values of x for which $y = 0$.) Thus, the solution set of $x^2 - 3x - 10 < 0$ consists of all real numbers from -2 to 5, or in interval notation, $(-2, 5)$.

It is not necessary to graph $y = x^2 - 3x - 10$ to solve the related inequality $x^2 - 3x - 10 < 0$. Instead, we can draw a number line representing the x-axis and keep the following in mind: *A region on the number line for which the value of* $x^2 - 3x - 10$ *is positive is separated from a region on the number line for which the value of* $x^2 - 3x - 10$ *is negative by a value for which the expression is* 0.

Let's find these values for which the expression is 0 by solving the related equation:

$$x^2 - 3x - 10 = 0$$

$$(x - 5)(x + 2) = 0 \qquad \text{Factor.}$$

$$x - 5 = 0 \quad \text{or} \quad x + 2 = 0 \qquad \text{Set each factor equal to 0.}$$

$$x = 5 \quad \text{or} \qquad x = -2 \qquad \text{Solve.}$$

These two numbers -2 and 5, divide the number line into three regions. We will call the regions A, B, and C. These regions are important because, if the value of $x^2 - 3x - 10$ is negative when a number from a region is substituted for x, then $x^2 - 3x - 10$ is negative when any number in that region is substituted for x. The same is true if the value of $x^2 - 3x - 10$ is positive for a particular value of x in a region.

To see whether the inequality $x^2 - 3x - 10 < 0$ is true or false in each region, we choose a test point from each region and substitute its value for x in the inequality

$x^2 - 3x - 10 < 0$. If the resulting inequality is true, the region containing the test point is a solution region.

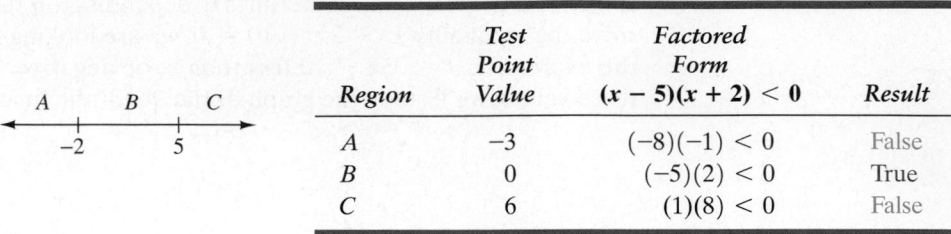

Region	Test Point Value	Factored Form $(x - 5)(x + 2) < 0$	Result
A	−3	$(-8)(-1) < 0$	False
B	0	$(-5)(2) < 0$	True
C	6	$(1)(8) < 0$	False

The values in region B satisfy the inequality. The numbers −2 and 5 are not included in the solution set since the inequality symbol is $<$. The solution set is $(-2, 5)$, and its graph is shown.

$$\overset{A \qquad B \qquad C}{\underset{F \ -2 \quad T \quad 5 \ F}{\longleftarrow \!\!\!\! \longrightarrow}}$$

Example 1 Solve $(x + 3)(x - 3) > 0$.

Solution First we solve the related equation $(x + 3)(x - 3) = 0$.

$$(x + 3)(x - 3) = 0$$

$$x + 3 = 0 \quad \text{or} \quad x - 3 = 0$$

$$x = -3 \quad \text{or} \quad x = 3$$

The two numbers −3 and 3 separate the number line into three regions, A, B, and C.

Now we substitute the value of a test point from each region. If the test value satisfies the inequality, every value in the region containing the test value is a solution.

Region	Test Point Value	$(x + 3)(x - 3) > 0$	Result
A	−4	$(-1)(-7) > 0$	True
B	0	$(3)(-3) > 0$	False
C	4	$(7)(1) > 0$	True

The points in regions A and C satisfy the inequality. The numbers −3 and 3 are not included in the solution since the inequality symbol is $>$. The solution set is $(-\infty, -3) \cup (3, \infty)$, and its graph is shown.

$$\overset{A \qquad B \qquad C}{\underset{T \ -3 \quad F \quad 3 \ T}{\longleftarrow \!\!\!\! \longrightarrow}}$$

The following steps may be used to solve a polynomial inequality.

SOLVING A POLYNOMIAL INEQUALITY

Step 1: Write the inequality in standard form and then solve the related equation.

Step 2: Separate the number line into regions with the solutions from Step 1.

Step 3: For each region, choose a test point and determine whether its value satisfies the *original inequality*.

Step 4: The solution set includes the regions whose test point value is a solution. If the inequality symbol is $\leq$ or $\geq$, the values from Step 1 are solutions; if $<$ or $>$, they are not.

Example 2 Solve $x^2 - 4x \leq 0$.

Solution First we solve the related equation $x^2 - 4x = 0$.

$$x^2 - 4x = 0$$
$$x(x - 4) = 0$$
$$x = 0 \quad \text{or} \quad x = 4$$

The numbers 0 and 4 separate the number line into three regions, A, B, and C.

Check a test value in each region in the original inequality. Values in region B satisfy the inequality. The numbers 0 and 4 are included in the solution since the inequality symbol is $\leq$. The solution set is $[0, 4]$, and its graph is shown.

Example 3 Solve $(x + 2)(x - 1)(x - 5) \leq 0$.

Solution First we solve $(x + 2)(x - 1)(x - 5) = 0$. By inspection, we see that the solutions are -2, 1, and 5. They separate the number line into four regions, A, B, C, and D. Next we check test points from each region.

Region	Test Point Value	$(x + 2)(x - 1)$ $(x - 5) \leq 0$	Result
A	-3	$(-1)(-4)(-8) \leq 0$	True
B	0	$(2)(-1)(-5) \leq 0$	False
C	2	$(4)(1)(-3) \leq 0$	True
D	6	$(8)(5)(1) \leq 0$	False

The solution set is $(-\infty, -2] \cup [1, 5]$, and its graph is shown. We include the numbers $-2, 1,$ and 5 because the inequality symbol is $\leq$.

2 Inequalities containing rational expressions with variables in the denominator are solved by using a similar procedure.

Example 4 Solve $\dfrac{x + 2}{x - 3} \leq 0$.

Solution First we find all values that make the denominator equal to 0. To do this, we solve $x - 3 = 0$ and find that $x = 3$.

Next, we solve the related equation $\dfrac{x + 2}{x - 3} = 0$.

$$\frac{x + 2}{x - 3} = 0$$

$$x + 2 = 0 \qquad \text{Multiply both sides by the LCD, } x - 3.$$

$$x = -2$$

Now we place these numbers on a number line and proceed as before, checking test point values in the original inequality.

Choose -3 from region A.

$$\frac{x + 2}{x - 3} \leq 0$$

$$\frac{-3 + 2}{-3 - 3} \leq 0$$

$$\frac{-1}{-6} \leq 0$$

$$\frac{1}{6} \leq 0 \qquad \text{False.}$$

Choose 0 from region B.

$$\frac{x + 2}{x - 3} \leq 0$$

$$\frac{0 + 2}{0 - 3} \leq 0$$

$$-\frac{2}{3} \leq 0 \qquad \text{True.}$$

Choose 4 from region C.

$$\frac{x + 2}{x - 3} \leq 0$$

$$\frac{4 + 2}{4 - 3} \leq 0$$

$$6 \leq 0 \qquad \text{False.}$$

The solution set is $[-2, 3)$. This interval includes -2 because -2 satisfies the original inequality. This interval does not include 3, because 3 would make the denominator 0.

The following steps may be used to solve a rational inequality with variables in the denominator.

SOLVING A RATIONAL INEQUALITY

Step 1: Solve for values that make any denominator 0.
Step 2: Solve the related equation.
Step 3: Separate the number line into regions with the solutions from Steps 1 and 2.
Step 4: For each region, choose a test point and determine whether its value satisfies the *original inequality.*
Step 5: The solution set includes the regions whose test point value is a solution. Check whether to include values from Step 2. Be sure *not* to include values that make any denominator 0.

Example 5   Solve $\dfrac{5}{x+1} < -2$.

Solution   First we find values for x that make the denominator equal to 0.

$$x + 1 = 0$$
$$x = -1$$

Next we solve $\dfrac{5}{x+1} = -2$.

$$(x+1) \cdot \frac{5}{x+1} = (x+1) \cdot -2 \qquad \text{Multiply both sides by the LCD, } x+1.$$

$$5 = -2x - 2 \qquad \text{Simplify.}$$

$$7 = -2x$$

$$-\frac{7}{2} = x$$

We use these two solutions to divide a number line into three regions and choose test points. Only a test point value from region B satisfies the *original inequality.* The solution set is $\left(-\dfrac{7}{2}, -1\right)$, and its graph is shown.

Exercise Set 10.4

Solve each quadratic inequality. Graph the solution set and write the solution set in interval notation. See Examples 1 through 3.

1. $(x+1)(x+5) > 0$
2. $(x+1)(x+5) \le 0$
3. $(x-3)(x+4) \le 0$
4. $(x+4)(x-1) > 0$
5. $x^2 - 7x + 10 \le 0$
6. $x^2 + 8x + 15 \ge 0$
7. $3x^2 + 16x < -5$
8. $2x^2 - 5x < 7$

9. $(x-6)(x-4)(x-2) > 0$
10. $(x-6)(x-4)(x-2) \le 0$
11. $x(x-1)(x+4) \le 0$
12. $x(x-6)(x+2) > 0$
13. $(x^2 - 9)(x^2 - 4) > 0$
14. $(x^2 - 16)(x^2 - 1) \le 0$

Solve each inequality. Graph the solution set and write the solution set in interval notation. See Example 4.

15. $\dfrac{x+7}{x-2} < 0$ **16.** $\dfrac{x-5}{x-6} > 0$ **17.** $\dfrac{5}{x+1} > 0$

18. $\dfrac{3}{y-5} < 0$ **19.** $\dfrac{x+1}{x-4} \ge 0$ **20.** $\dfrac{x+1}{x-4} \le 0$

21. Explain why $\dfrac{x+2}{x-3} > 0$ and $(x+2)(x-3) > 0$ have the same solutions.

22. Explain why $\dfrac{x+2}{x-3} \ge 0$ and $(x+2)(x-3) \ge 0$ do not have the same solutions.

Solve each inequality. Graph the solution set and write the solution set in interval notation. See Example 5.

23. $\dfrac{3}{x-2} < 4$ **24.** $\dfrac{-2}{y+3} > 2$ **25.** $\dfrac{x^2+6}{5x} \ge 1$

26. $\dfrac{y^2+15}{8y} \le 1$

Solve each inequality. Graph the solution set and write the solution set in interval notation.

27. $(x-8)(x+7) > 0$ **28.** $(x-5)(x+1) < 0$

29. $(2x-3)(4x+5) \le 0$ **30.** $(6x+7)(7x-12) > 0$

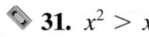 **31.** $x^2 > x$

32. $x^2 < 25$

33. $(2x-8)(x+4)(x-6) \le 0$

34. $(3x-12)(x+5)(2x-3) \ge 0$

35. $6x^2 - 5x \ge 6$

36. $12x^2 + 11x \le 15$

37. $4x^3 + 16x^2 - 9x - 36 > 0$

38. $x^3 + 2x^2 - 4x - 8 < 0$

39. $x^4 - 26x^2 + 25 \ge 0$

40. $16x^4 - 40x^2 + 9 \le 0$

41. $(2x-7)(3x+5) > 0$

42. $(4x-9)(2x+5) < 0$

43. $\dfrac{x}{x-10} < 0$ **44.** $\dfrac{x+10}{x-10} > 0$

45. $\dfrac{x-5}{x+4} \ge 0$ **46.** $\dfrac{x-3}{x+2} \le 0$

47. $\dfrac{x(x+6)}{(x-7)(x+1)} \ge 0$

48. $\dfrac{(x-2)(x+2)}{(x+1)(x-4)} \le 0$

49. $\dfrac{-1}{x-1} > -1$ **50.** $\dfrac{4}{y+2} < -2$

51. $\dfrac{x}{x+4} \le 2$ **52.** $\dfrac{4x}{x-3} \ge 5$

53. $\dfrac{z}{z-5} \ge 2z$ **54.** $\dfrac{p}{p+4} \le 3p$

55. $\dfrac{(x+1)^2}{5x} > 0$ **56.** $\dfrac{(2x-3)^2}{x} < 0$

Find all numbers that satisfy each of the following.

57. A number minus its reciprocal is less than zero. Find the numbers.

58. Twice a number added to its reciprocal is nonnegative. Find the numbers.

59. The total profit function $P(x)$ for a company producing x thousand units is given by
$$P(x) = -2x^2 + 26x - 44$$
Find the values of x for which the company makes a profit. [*Hint:* The company makes a profit when $P(x) > 0$.]

60. A projectile is fired straight up from the ground with an initial velocity of 80 feet per second. Its height $s(t)$ in feet at any time t is given by the function
$$s(t) = -16t^2 + 80t$$
Find the interval of time for which the height of the projectile is greater than 96 feet.

Use a graphing calculator to check each exercise.

61. Exercise 27   **62.** Exercise 28

63. Exercise 39   **64.** Exercise 40

REVIEW EXERCISES

Recall that the graph of $f(x) + K$ is the same as the graph of $f(x)$ shifted K units upward if $K > 0$ and $|K|$ units downward if $K < 0$. Use the graph of $f(x) = |x|$ below to sketch the graph of each function. (See Sections 3.2 and 3.3.)

65. $g(x) = |x| + 2$ **66.** $H(x) = |x| - 2$

67. $F(x) = |x| - 1$ **68.** $h(x) = |x| + 5$

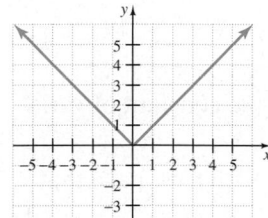

Use the graph of $f(x) = x^2$ below to sketch the graph of each function.

69. $F(x) = x^2 - 3$ **70.** $h(x) = x^2 - 4$

71. $H(x) = x^2 + 1$ **72.** $g(x) = x^2 + 3$

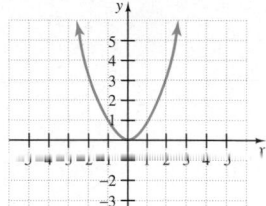

10.5   VARIATION AND PROBLEM SOLVING

CD-ROM   SSM

SSG   Video

▶ **OBJECTIVES**

1. Solve problems involving direct variation.
2. Solve problems involving inverse variation.
3. Solve problems involving joint variation.
4. Solve problems involving combined variation.

1   A very familiar example of direct variation is the relationship of the circumference C of a circle to its radius r. The formula $C = 2\pi r$ expresses that the circumference is always 2π times the radius. In other words, C is always a constant multiple (2π) of r. Because it is, we say that **C varies directly as r**, that **C varies directly with r**, or that **C is directly proportional to r.**

> **DIRECT VARIATION**
>
> **y varies directly as x,** or **y is directly proportional to x,** if there is a nonzero constant k such that
>
> $$y = kx$$
>
> The number k is called the **constant of variation** or the **constant of proportionality.**

     In the above definition, the relationship described between x and y is a linear one. In other words, the graph of $y = kx$ is a line. The slope of the line is k, and the line passes through the origin.

     For example, the graph of the direct variation equation $C = 2\pi r$ is shown. The horizontal axis represents the radius r, and the vertical axis is the circumference C. From the graph we can read that when the radius is 6 units, the circumference is approximately 38 units. Also, when the circumference is 45 units, the radius is between 7 and 8 units. Notice that as the radius increases, the circumference increases.

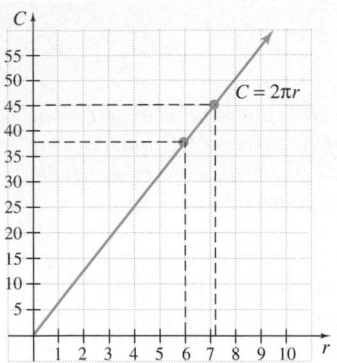

Example 1 Suppose that y varies directly as x. If y is 5 when x is 30, find the constant of variation and the direct variation equation.

Solution Since y varies directly as x, we write $y = kx$. If $y = 5$ when $x = 30$, we have that

$$y = kx$$

$$5 = k(30) \qquad \text{Replace } y \text{ with 5 and } x \text{ with 30.}$$

$$\frac{1}{6} = k \qquad \text{Solve for } k.$$

The constant of variation is $\frac{1}{6}$.

After finding the constant of variation k, the direct variation equation can be written as $y = \frac{1}{6}x$.

Example 2 **USING DIRECT VARIATION AND HOOKE'S LAW**

Hooke's law states that the distance a spring stretches is directly proportional to the weight attached to the spring. If a 40-pound weight attached to the spring stretches the spring 5 inches, find the distance that a 65-pound weight attached to the spring stretches the spring.

Solution 1. UNDERSTAND. Read and reread the problem. Notice that we are given that the distance a spring stretches is **directly proportional** to the weight attached. We let

$$d = \text{the distance stretched}$$
$$w = \text{the weight attached}$$

The constant of variation is represented by k.

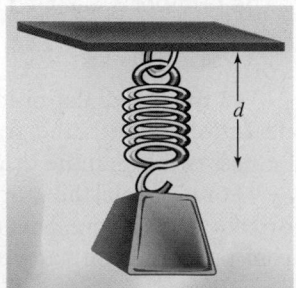

2. TRANSLATE. Because d is directly proportional to w, we write

$$d = kw$$

3. SOLVE. When a weight of 40 pounds is attached, the spring stretches 5 inches. That is, when $w = 40, d = 5$.

$$d = kw$$
$$5 = k(40) \qquad \text{Replace } d \text{ with 5 and } w \text{ with 40.}$$
$$\frac{1}{8} = k \qquad \text{Solve for } k.$$

Now when we replace k with $\frac{1}{8}$ in the equation

$$d = kw, \text{ we have}$$
$$d = \frac{1}{8} w$$

To find the stretch when a weight of 65 pounds is attached, we replace w with 65 to find d.

$$d = \frac{1}{8}(65)$$
$$= \frac{65}{8} = 8\frac{1}{8} \quad \text{or} \quad 8.125$$

4. INTERPRET.

Check: Check the proposed solution of 8.125 inches in the original problem.

State: The spring stetches 8.125 inches when a 65-pound weight is attached.   ▬

2

When y is proportional to the **reciprocal** of another variable x, we say that **y varies inversely as x**, or that **y is inversely proportional to x**. An example of the inverse variation relationship is the relationship between the pressure that a gas exerts and the volume of its container. As the volume of a container decreases, the pressure of the gas it contains increases.

INVERSE VARIATION

y varies inversely as x, or **y is inversely proportional to x**, if there is a nonzero constant k such that

$$y = \frac{k}{x}$$

The number k is called the **constant of variation** or the **constant of proportionality**.

Notice that $y = \frac{k}{x}$ is a rational equation. Its graph for $k > 0$ and $x > 0$ is shown.

From the graph, we can see that as x increases, y decreases.

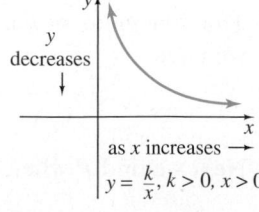

$y = \frac{k}{x}, k > 0, x > 0$

Example 3   Suppose that u varies inversely as w. If u is 3 when w is 5, find the constant of variation and the inverse variation equation.

Solution   Since u varies inversely as w, we have $u = \dfrac{k}{w}$. We let $u = 3$ and $w = 5$, and we solve for k.

$$u = \frac{k}{w}$$

$$3 = \frac{k}{5} \qquad \text{Let } u = 3 \text{ and } w = 5.$$

$$15 = k \qquad \text{Multiply both sides by 5.}$$

The constant of variation k is 15. This gives the inverse variation equation

$$u = \frac{15}{w}$$

Example 4   **USING INVERSE VARIATION AND BOYLE'S LAW**

Boyle's law says that if the temperature stays the same, the pressure P of a gas is inversely proportional to the volume V. If a cylinder in a steam engine has a pressure of 960 kilopascals when the volume is 1.4 cubic meters, find the pressure when the volume increases to 2.5 cubic meters.

Solution   1. UNDERSTAND. Read and reread the problem. Notice that we are given that the pressure of a gas is *inversely proportional* to the volume. We will let $P =$ the pressure and $V =$ the volume. The constant of variation is represented by k.

2. TRANSLATE. Because P is inversely proportional to V, we write

$$P = \frac{k}{V}$$

When $P = 960$ kilopascals, the volume $V = 1.4$ cubic meters. We use this information to find k.

$$960 = \frac{k}{1.4} \qquad \text{Let } P = 960 \text{ and } V = 1.4.$$

$$1344 = k \qquad \text{Multiply both sides by 1.4.}$$

Thus, the value of k is 1344. Replacing k with 1344 in the variation equation, we have

$$P = \frac{1344}{V}$$

Next we find P when V is 2.5 cubic meters.

3. SOLVE.

$$P = \frac{1344}{2.5} \qquad \text{Let } V = 2.5.$$

$$= 537.6$$

4. INTERPRET.

Check: Check the proposed solution in the original problem.

State: When the volume is 2.5 cubic meters, the pressure is 537.6 kilopascals. ■

---

3 Sometimes the ratio of a variable to the product of many other variables is constant. For example, the ratio of distance traveled to the product of speed and time traveled is always 1.

$$\frac{d}{rt} = 1 \qquad \text{or} \qquad d = rt$$

Such a relationship is called **joint variation.**

> **JOINT VARIATION**
>
> If the ratio of a variable y to the product of two or more variables is constant, then y **varies jointly as,** or **is jointly proportional to,** the other variables. If
>
> $$y = kxz$$
>
> then the number k is the **constant of variation** or the **constant of proportionality.**

△ **Example 5** **EXPRESSING SURFACE AREA**

The surface area of a cylinder varies jointly as its radius and height. Express surface area S in terms of radius r and height h.

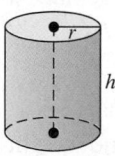

Solution Because the surface area varies jointly as the radius r and the height h, we equate S to a constant multiple of r and h.

$$S = krh$$

In the equation, $S = krh$, it can be determined that the constant k is 2π, and we then have the formula $S = 2\pi rh$. (This formula does not include the areas of the two circular bases.) ■

---

4 Some examples of variation involve combinations of direct, inverse, and joint variation. We will call these variations **combined variation.**

△ **Example 6**   **FINDING COLUMN WEIGHT**

The maximum weight that a circular column can support is directly proportional to the fourth power of its diameter and is inversely proportional to the square of its height. A 2-meter-diameter column that is 8 meters in height can support 1 ton. Find the weight that a 1-meter-diameter column that is 4 meters in height can support.

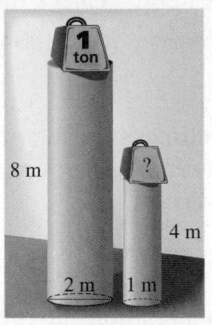

Solution   1. UNDERSTAND. Read and reread the problem. Let w = weight, d = diameter, h = height, and k = the constant of variation.

2. TRANSLATE. Since w is directly proportional to d^4 and inversely proportional to h^2, we have

$$w = \frac{kd^4}{h^2}$$

3. SOLVE. To find k, we are given that a 2-meter-diameter column that is 8 meters in height can support 1 ton. That is, $w = 1$ when $d = 2$ and $h = 8$, or

$$1 = \frac{k \cdot 2^4}{8^2} \qquad \text{Let } w = 1, d = 2, \text{ and } h = 8.$$

$$1 = \frac{k \cdot 16}{64}$$

$$4 = k \qquad \text{Solve for } k.$$

Now replace k with 4 in the equation $w = \dfrac{kd^4}{h^2}$ and we have

$$w = \frac{4d^4}{h^2}$$

To find weight w for a 1-meter-diameter column that is 4 meters in height, let $d = 1$ and $h = 4$.

$$w = \frac{4 \cdot 1^4}{4^2}$$

$$w = \frac{4}{16} = \frac{1}{4}$$

4. INTERPRET.

Check: Check the proposed solution in the original problem.

State: The 1-meter-diameter column that is 4 meters in height can hold $\frac{1}{4}$ ton of weight.

SPOTLIGHT ON DECISION MAKING

Suppose you are painting the ceilings of your one-story home, whose layout is shown in the figure. The amount of paint you need is directly proportional to the area of what is to be painted. A clerk at the paint store says that 450 square feet can be painted with 4 quarts of paint. Quarts of paint cost $5.95 each and gallons of paint cost $21.50 each. You have brought only $50 with you to the store.

Can you get all the paint you need for the project with the money you have brought? If so, explain how. If not, how much more money will you need?

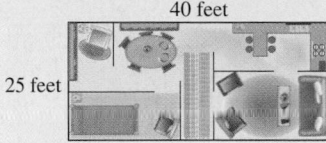

40 feet

25 feet

Exercise Set 10.5

If y varies directly as x, find the constant of variation k and the direct variation equation for each situation. See Example 1.

1. $y = 4$ when $x = 20$

2. $y = 5$ when $x = 30$

3. $y = 6$ when $x = 4$

4. $y = 12$ when $x = 8$

5. $y = 7$ when $x = \dfrac{1}{2}$

6. $y = 11$ when $x = \dfrac{1}{3}$

7. $y = 0.2$ when $x = 0.8$

8. $y = 0.4$ when $x = 2.5$

Solve. See Example 2.

9. The weight of a synthetic ball varies directly with the cube of its radius. A ball with a radius of 2 inches weighs 1.20 pounds. Find the weight of a ball of the same material with a 3-inch radius.

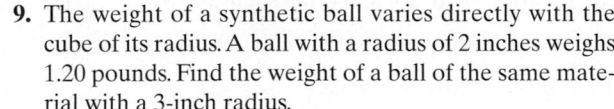

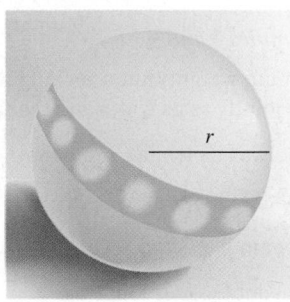

r

10. At sea, the distance to the horizon is directly proportional to the square root of the elevation of the observer. If a person who is 36 feet above the water can see 7.4 miles, find how far a person 64 feet above the water can see. Round answer to one decimal place.

11. The amount P of pollution varies directly with the population N of people. Kansas City has a population of 450,000 and produces 260,000 tons of pollutants. Find how many tons of pollution we should expect St. Louis to produce, if we know that its population is 980,000. Round answer to the nearest whole ton.

12. Charles' law states that if the pressure P stays the same, the volume V of a gas is directly proportional to its temperature T. If a balloon is filled with 20 cubic meters of a gas at a temperature of 300 K, find the new volume if the temperature rises to 360 K while the pressure stays the same.

If y varies inversely as x, find the constant of variation k and the inverse variation equation for each situation. See Example 3.

13. $y = 6$ when $x = 5$

14. $y = 20$ when $x = 9$

15. $y = 100$ when $x = 7$

16. $y = 63$ when $x = 3$

17. $y = \dfrac{1}{8}$ when $x = 16$

18. $y = \dfrac{1}{10}$ when $x = 40$

19. $y = 0.2$ when $x = 0.7$

20. $y = 0.6$ when $x = 0.3$

Solve. See Example 4.

21. Pairs of markings a set distance apart are made on highways so that police can detect drivers exceeding the speed limit. Over a fixed distance, the speed R varies inversely with the time T. In one particular pair of markings, R is 45 mph when T is 6 seconds. Find the speed of a car that travels the given distance in 5 seconds.

22. The weight of an object on or above the surface of Earth varies inversely as the square of the distance between the object and Earth's center. If a person weighs 160 pounds on Earth's surface, find the individual's weight if he moves 200 miles above Earth. Round answer to the nearest pound. (Assume that Earth's radius is 4000 miles.)

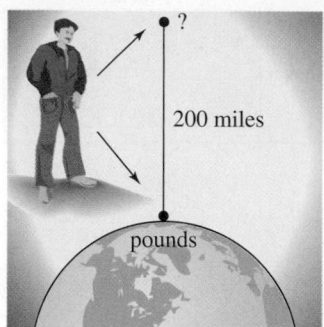

23. If the voltage V in an electric circuit is held constant, the current I is inversely proportional to the resistance R. If the current is 40 amperes when the resistance is 270 ohms, find the current when the resistance is 150 ohms.

24. Because it is more efficient to produce larger numbers of items, the cost of producing Dysan computer disks is inversely proportional to the number produced. If 4000 can be produced at a cost of $1.20 each, find the cost per disk when 6000 are produced.

25. The intensity I of light varies inversely as the square of the distance d from the light source. If the distance from the light source is doubled (see the figure at the bottom of this column and the figure at the top of the next column), determine what happens to the intensity of light at the new location.

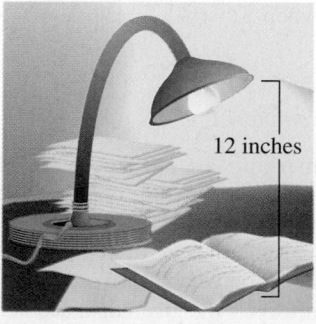

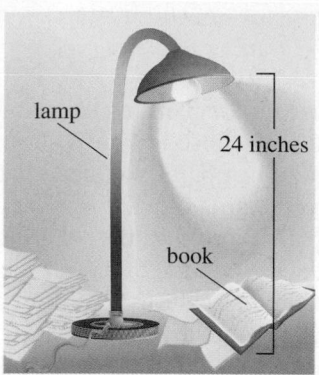

△ **26.** The maximum weight that a circular column can hold is inversely proportional to the square of its height. If an 8-foot column can hold 2 tons, find how much weight a 10-foot column can hold.

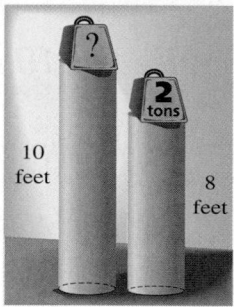

Write each statement as an equation. See Example 5.

27. x varies jointly as y and z.

28. P varies jointly as R and the square of S.

29. r varies jointly as s and the cube of t.

30. a varies jointly as b and c.

Solve. See Examples 5 and 6.

△ **31.** The maximum weight that a rectangular beam can support varies jointly as its width and the square of its height and inversely as its length. If a beam $\frac{1}{2}$ foot wide, $\frac{1}{3}$ foot high, and 10 feet long can support 12 tons, find how much a similar beam can support if the beam is $\frac{2}{3}$ foot wide, $\frac{1}{2}$ foot high, and 16 feet long.

32. The number of cars manufactured on an assembly line at a General Motors plant varies jointly as the number of workers and the time they work. If 200 workers can produce 60 cars in 2 hours, find how many cars 240 workers should be able to make in 3 hours.

△ **33.** The volume of a cone varies jointly as the square of its radius and its height. If the volume of a cone is 32π cubic inches when the radius is 4 inches and the height is 6 inches, find the volume of a cone when the radius is 3 inches and the height is 5 inches.

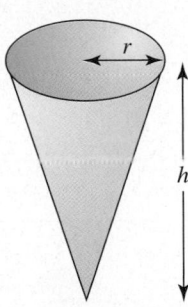

34. When a wind blows perpendicularly against a flat surface, its force is jointly proportional to the surface area and the speed of the wind. A sail whose surface area is 12 square feet experiences a 20-pound force when the wind speed is 10 miles per hour. Find the force on an 8-square-foot sail if the wind speed is 12 miles per hour.

35. The horsepower that can be safely transmitted to a shaft varies jointly as the shaft's angular speed of rotation (in revolutions per minute) and the cube of its diameter. A 2-inch shaft making 120 revolutions per minute safely transmits 40 horsepower. Find how much horsepower can be safely transmitted by a 3-inch shaft making 80 revolutions per minute.

△ **36.** The maximum weight that a rectangular beam can support varies jointly as its width and the square of its height and inversely as its length. If a beam $\frac{1}{3}$ foot wide, 1 foot high, and 10 feet long can support 3 tons, find how much weight a similar beam can support if it is 1 foot wide, $\frac{1}{3}$ foot high, and 9 feet long.

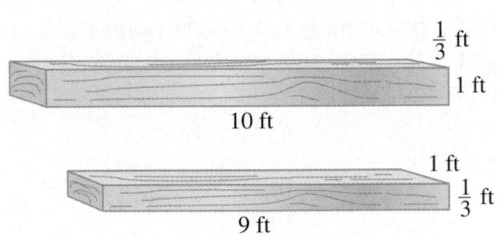

37. The atmospheric pressure y (in millibars) is inversely proportional to the altitude x (in kilometers). If the atmospheric pressure is 400 millibars at an altitude of 8 kilometers, find the atmospheric pressure at an altitude of 4 kilometers.

38. The horsepower to drive a boat varies directly as the cube of the speed of the boat. If the speed of the boat is to double, determine the corresponding increase in horsepower required.

△ **39.** The volume of a cylinder varies jointly as the height and the square of the radius. If the height is halved and the radius is doubled, determine what happens to the volume.

40. Suppose that y varies directly as x. If x is doubled, what is the effect on y?

41. Suppose that y varies directly as x^2. If x is doubled, what is the effect on y?

Complete the following table for the inverse variation $y = \dfrac{k}{x}$ over each given value of k. Plot the points on a rectangular coordinate system.

x	$\frac{1}{4}$	$\frac{1}{2}$	1	2	4
$y = \dfrac{k}{x}$					

42. $k = 1$ **43.** $k = 3$

44. $k = 5$ **45.** $k = \dfrac{1}{2}$

REVIEW EXERCISES

Find the exact circumference and area of each circle. See the inside cover for a list of geometric formulas.

△ **46.**

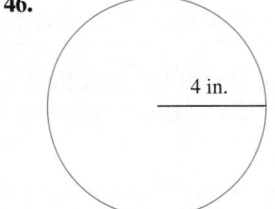

4 in.

△ **47.**

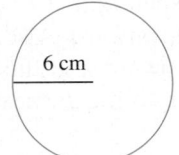

6 cm

△ **48.**

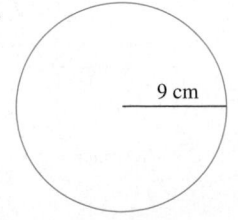

9 cm

△ **49.**

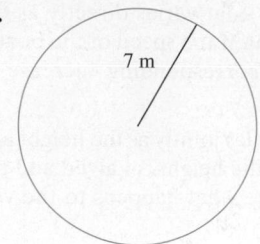

7 m

Find each square root. See Section 9.1.

50. $\sqrt{81}$ **51.** $\sqrt{36}$

52. $\sqrt{1}$ **53.** $\sqrt{4}$

54. $\sqrt{\dfrac{1}{4}}$ **55.** $\sqrt{\dfrac{1}{25}}$

56. $\sqrt{\dfrac{4}{9}}$ **57.** $\sqrt{\dfrac{25}{121}}$

10.6 QUADRATIC FUNCTIONS AND THEIR GRAPHS

CD-ROM

SSM

SSG   Video

▶ **OBJECTIVES**

1. Graph quadratic functions of the form $f(x) = x^2 + k$.
2. Graph quadratic functions of the form $f(x) = (x - h)^2$.
3. Graph quadratic functions of the form $f(x) = (x - h)^2 + k$.
4. Graph quadratic functions of the form $f(x) = ax^2$.
5. Graph quadratic functions of the form $f(x) = a(x - h)^2 + k$.

1

We first graphed the quadratic equation $y = x^2$ in Section 3.2. In Section 3.3, we learned that this graph defines a function, and we wrote $y = x^2$ as $f(x) = x^2$. Quadratic functions and their graphs were studied further in Section 6.8. In those sections, we discovered that the graph of a quadratic function is a parabola opening upward or downward. In this section, we continue our study of quadratic functions and their graphs.

First, let's recall the definition of a quadratic function.

QUADRATIC FUNCTION

A quadratic function is a function that can be written in the form $f(x) = ax^2 + bx + c$, where a, b, and c are real numbers and $a \neq 0$.

Notice that equations of the form $y = ax^2 + bx + c$, where $a \neq 0$, define quadratic functions, since y is a function of x or $y = f(x)$.

Recall that if $a > 0$, the parabola opens upward and if $a < 0$, the parabola opens downward. Also, the vertex of a parabola is the lowest point if the parabola opens upward and the highest point if the parabola opens downward. The axis of symmetry is the vertical line that passes through the vertex.

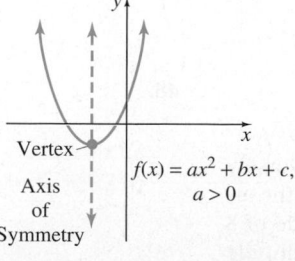

Vertex
Axis
of
Symmetry
$f(x) = ax^2 + bx + c,$
$a > 0$

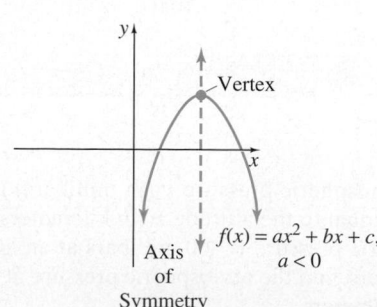

Vertex
$f(x) = ax^2 + bx + c,$
$a < 0$
Axis
of
Symmetry

Example 1 Graph $f(x) = x^2$ and $g(x) = x^2 + 3$ on the same set of axes.

Solution   First we construct a table of values for $f(x)$ and plot the points. Notice that for each x-value, the corresponding value of $g(x)$ must be 3 more than the corresponding value of $f(x)$ since $f(x) = x^2$ and $g(x) = x^2 + 3$. In other words, the graph of $g(x) = x^2 + 3$ is the same as the graph of $f(x) = x^2$ shifted upward 3 units. The axis of symmetry for both graphs is the y-axis.

x	$f(x) = x^2$	$g(x) = x^2 + 3$
-2	4	7
-1	1	4
0	0	3
1	1	4
2	4	7

Each y-value ↑ is increased by 3.

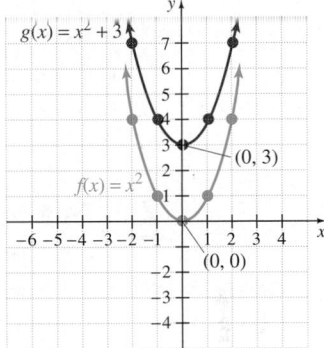

In general, we have the following properties.

GRAPHING THE PARABOLA DEFINED BY $f(x) = x^2 + k$

If k is positive, the graph of $f(x) = x^2 + k$ is the graph of $y = x^2$ shifted upward k units.

If k is negative, the graph of $f(x) = x^2 + k$ is the graph of $y = x^2$ shifted downward $|k|$ units.

The vertex is $(0, k)$, and the axis of symmetry is the y-axis.

Example 2 Graph each function.

    a. $F(x) = x^2 + 2$ **b.** $g(x) = x^2 - 3$

Solution   **a.** $F(x) = x^2 + 2$

The graph of $F(x) = x^2 + 2$ is obtained by shifting the graph of $y = x^2$ upward 2 units.

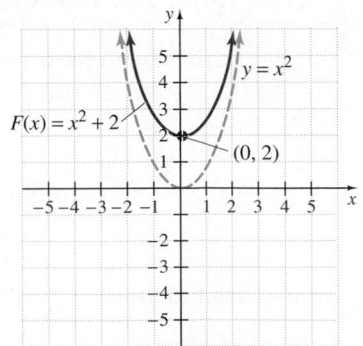

b. $g(x) = x^2 - 3$

The graph of $g(x) = x^2 - 3$ is obtained by shifting the graph of $y = x^2$ downward 3 units.

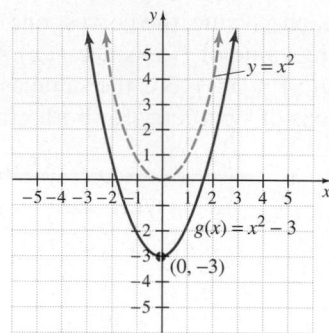

2 Now we will graph functions of the form $f(x) = (x - h)^2$.

Example 3 Graph $f(x) = x^2$ and $g(x) = (x - 2)^2$ on the same set of axes.

Solution By plotting points, we see that for each x-value, the corresponding value of $g(x)$ is the same as the value of $f(x)$ when the x-value is increased by 2. Thus, the graph of $g(x) = (x - 2)^2$ is the graph of $f(x) = x^2$ shifted to the right 2 units. The axis of symmetry for the graph of $g(x) = (x - 2)^2$ is also shifted 2 units to the right and is the line $x = 2$.

x	$f(x) = x^2$	x	$g(x) = (x - 2)^2$
-2	4	0	4
-1	1	1	1
0	0	2	0
1	1	3	1
2	4	4	4

Each x-value ↑
increased by 2
corresponds to
same y-value.

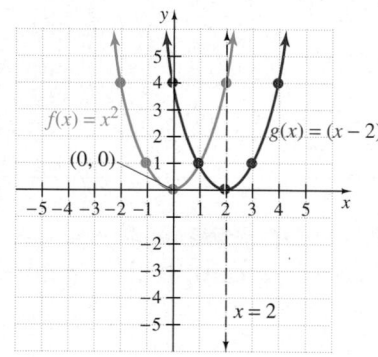

In general, we have the following properties.

> **GRAPHING THE PARABOLA DEFINED BY $f(x) = (x - h)^2$**
>
> If h is positive, the graph of $f(x) = (x - h)^2$ is the graph of $y = x^2$ shifted to the right h units.
>
> If h is negative, the graph of $f(x) = (x - h)^2$ is the graph of $y = x^2$ shifted to the left $|h|$ units.
>
> The vertex is $(h, 0)$, and the axis of symmetry is the vertical line $x = h$.

Example 4 Graph each function.

a. $G(x) = (x - 3)^2$ **b.** $F(x) = (x + 1)^2$

Solution   **a.** The graph of $G(x) = (x - 3)^2$ is obtained by shifting the graph of $y = x^2$ to the right 3 units. The graph of $G(x)$ is below on the left.

b. The equation $F(x) = (x + 1)^2$ can be written as $F(x) = [x - (-1)]^2$. The graph of $F(x) = [x - (-1)]^2$ is obtained by shifting the graph of $y = x^2$ to the left 1 unit. The graph of $F(x)$ is below on the right.

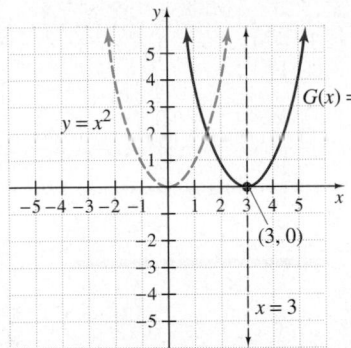 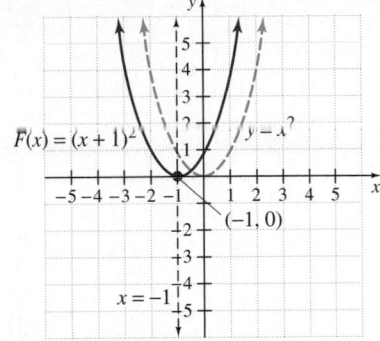

3   As we will see in graphing functions of the form $f(x) = (x - h)^2 + k$, it is possible to combine vertical and horizontal shifts.

GRAPHING THE PARABOLA DEFINED BY $f(x) = (x - h)^2 + k$

The parabola has the same shape as $y = x^2$.
The vertex is (h, k), and the axis of symmetry is the vertical line $x = h$.

Example 5   Graph $F(x) = (x - 3)^2 + 1$.

Solution   The graph of $F(x) = (x - 3)^2 + 1$ is the graph of $y = x^2$ shifted 3 units to the right and 1 unit up. The vertex is then $(3, 1)$, and the axis of symmetry is $x = 3$. A few ordered pair solutions are plotted to aid in graphing.

x	$F(x) = (x - 3)^2 + 1$
1	5
2	2
4	2
5	5

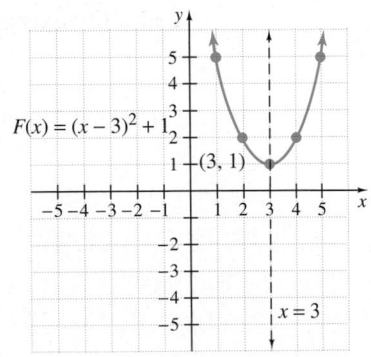

4   Next, we discover the change in the shape of the graph when the coefficient of x^2 is not 1.

Example 6   Graph $f(x) = x^2$, $g(x) = 3x^2$, and $h(x) = \dfrac{1}{2}x^2$ on the same set of axes.

Solution Comparing the tables of values, we see that for each x-value, the corresponding value of $g(x)$ is triple the corresponding value of $f(x)$. Similarly, the value of $h(x)$ is half the value of $f(x)$.

x	$f(x) = x^2$
-2	4
-1	1
0	0
1	1
2	4

x	$g(x) = 3x^2$
-2	12
-1	3
0	0
1	3
2	12

x	$h(x) = \frac{1}{2}x^2$
-2	2
-1	$\frac{1}{2}$
0	0
1	$\frac{1}{2}$
2	2

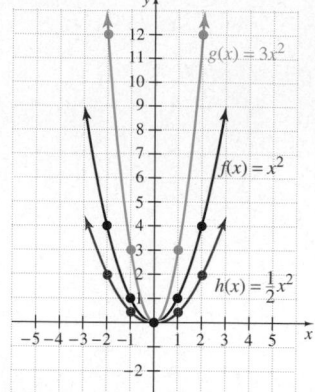

The result is that the graph of $g(x) = 3x^2$ is narrower than the graph of $f(x) = x^2$ and the graph of $h(x) = \frac{1}{2}x^2$ is wider. The vertex for each graph is $(0, 0)$, and the axis of symmetry is the y-axis.

GRAPHING THE PARABOLA DEFINED BY $f(x) = ax^2$

If a is positive, the parabola opens upward, and if a is negative, the parabola opens downward.

If $|a| > 1$, the graph of the parabola is narrower than the graph of $y = x^2$.

If $|a| < 1$, the graph of the parabola is wider than the graph of $y = x^2$.

Example 7 Graph $f(x) = -2x^2$.

Solution Because $a = -2$, a negative value, this parabola opens downward. Since $|-2| = 2$ and $2 > 1$, the parabola is narrower than the graph of $y = x^2$. The vertex is $(0, 0)$, and the axis of symmetry is the y-axis. We verify this by plotting a few points.

x	$f(x) = -2x^2$
-2	-8
-1	-2
0	0
1	-2
2	-8

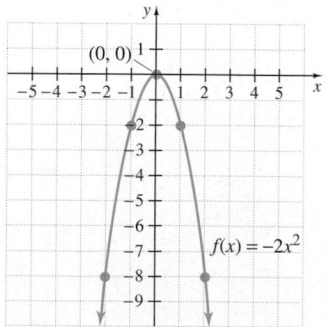

5 Now we will see the shape of the graph of a quadratic function of the form $f(x) = a(x - h)^2 + k$.

Example 8 Graph $g(x) = \frac{1}{2}(x + 2)^2 + 5$. Find the vertex and the axis of symmetry.

Solution   The function $g(x) = \frac{1}{2}(x + 2)^2 + 5$ may be written as $g(x) = \frac{1}{2}[x - (-2)]^2 + 5$. Thus, this graph is the same as the graph of $y = x^2$ shifted 2 units to the left and 5 units up, and it is wider because a is $\frac{1}{2}$. The vertex is $(-2, 5)$, and the axis of symmetry is $x = -2$. We plot a few points to verify.

x	$g(x) = \frac{1}{2}(x + 2)^2 + 5$
-4	7
-3	$5\frac{1}{2}$
-2	5
-1	$5\frac{1}{2}$
0	7

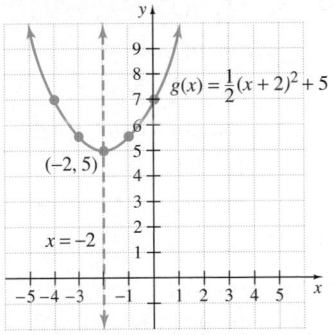

In general, the following holds.

GRAPH OF A QUADRATIC FUNCTION

The graph of a quadratic function written in the form $f(x) = a(x - h)^2 + k$ is a parabola with vertex (h, k). If $a > 0$, the parabola opens upward, and if $a < 0$, the parabola opens downward. The axis of symmetry is the line whose equation is $x = h$.

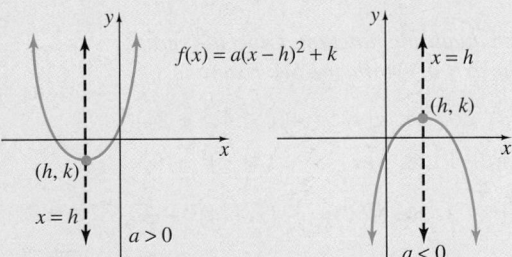

GRAPHING CALCULATOR EXPLORATIONS

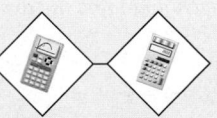

Use a graphing calculator to graph the first function of each pair that follows. Then use its graph to predict the graph of the second function. Check your prediction by graphing both on the same set of axes.

1. $F(x) = \sqrt{x}$; $G(x) = \sqrt{x} + 1$

2. $g(x) = x^3$; $H(x) = x^3 - 2$

3. $H(x) = |x|$; $f(x) = |x - 5|$

4. $h(x) = x^3 + 2$; $g(x) = (x - 3)^3 + 2$

5. $f(x) = |x + 4|$; $F(x) = |x + 4| + 3$

6. $G(x) = \sqrt{x} - 2$; $g(x) = \sqrt{x - 4} - 2$

MENTAL MATH

State the vertex of the graph of each quadratic function.

1. $f(x) = x^2$

2. $f(x) = -5x^2$

3. $g(x) = (x - 2)^2$

4. $g(x) = (x + 5)^2$

5. $f(x) = 2x^2 + 3$

6. $h(x) = x^2 - 1$

7. $g(x) = (x + 1)^2 + 5$

8. $h(x) = (x - 10)^2 - 7$

Exercise Set 10.6

Sketch the graph of each quadratic function. Label the vertex, and sketch and label the axis of symmetry. See Examples 1 through 5.

1. $f(x) = x^2 - 1$ **2.** $g(x) = x^2 + 3$

3. $h(x) = x^2 + 5$ **4.** $h(x) = x^2 - 4$

5. $g(x) = x^2 + 7$ **6.** $f(x) = x^2 - 2$

7. $f(x) = (x - 5)^2$ **8.** $g(x) = (x + 5)^2$

9. $h(x) = (x + 2)^2$ **10.** $H(x) = (x - 1)^2$

11. $G(x) = (x + 3)^2$ **12.** $f(x) = (x - 6)^2$

13. $f(x) = (x - 2)^2 + 5$ **14.** $g(x) = (x - 6)^2 + 1$

15. $h(x) = (x + 1)^2 + 4$ **16.** $G(x) = (x + 3)^2 + 3$

17. $g(x) = (x + 2)^2 - 5$ **18.** $h(x) = (x + 4)^2 - 6$

Sketch the graph of each quadratic function. Label the vertex, and sketch and label the axis of symmetry. See Examples 6 and 7.

19. $g(x) = -x^2$ **20.** $f(x) = 5x^2$

21. $h(x) = \dfrac{1}{3}x^2$ **22.** $g(x) = -3x^2$

23. $H(x) = 2x^2$ **24.** $f(x) = -\dfrac{1}{4}x^2$

Sketch the graph of each quadratic function. Label the vertex, and sketch and label the axis of symmetry. See Example 8.

25. $f(x) = 2(x - 1)^2 + 3$ **26.** $g(x) = 4(x - 4)^2 + 2$

27. $h(x) = -3(x + 3)^2 + 1$ **28.** $f(x) = -(x - 2)^2 - 6$

29. $H(x) = \dfrac{1}{2}(x - 6)^2 - 3$ **30.** $G(x) = \dfrac{1}{5}(x + 4)^2 + 3$

Sketch the graph of each quadratic function. Label the vertex, and sketch and label the axis of symmetry.

31. $f(x) = -(x - 2)^2$ **32.** $g(x) = -(x + 6)^2$

33. $F(x) = -x^2 + 4$ **34.** $H(x) = -x^2 + 10$

35. $F(x) = 2x^2 - 5$ **36.** $g(x) = \dfrac{1}{2}x^2 - 2$

37. $h(x) = (x - 6)^2 + 4$ **38.** $f(x) = (x - 5)^2 + 2$

39. $F(x) = \left(x + \dfrac{1}{2}\right)^2 - 2$ **40.** $H(x) = \left(x + \dfrac{1}{2}\right)^2 - 3$

41. $F(x) = \dfrac{3}{2}(x + 7)^2 + 1$ **42.** $g(x) = -\dfrac{3}{2}(x - 1)^2 - 5$

43. $f(x) = \dfrac{1}{4}x^2 - 9$ **44.** $H(x) = \dfrac{3}{4}x^2 - 2$

45. $G(x) = 5\left(x + \dfrac{1}{2}\right)^2$ **46.** $F(x) = 3\left(x - \dfrac{3}{2}\right)^2$

47. $h(x) = -(x - 1)^2 - 1$

48. $f(x) = -3(x + 2)^2 + 2$

49. $g(x) = \sqrt{3}(x + 5)^2 + \dfrac{3}{4}$

50. $G(x) = \sqrt{5}(x - 7)^2 - \dfrac{1}{2}$

51. $h(x) = 10(x + 4)^2 - 6$

52. $h(x) = 8(x + 1)^2 + 9$

53. $f(x) = -2(x - 4)^2 + 5$

54. $G(x) = -4(x + 9)^2 - 1$

Write the equation of the parabola that has the same shape as $f(x) = 5x^2$ but with the following vertex.

55. $(2, 3)$ **56.** $(1, 6)$

57. $(-3, 6)$ **58.** $(4, -1)$

The shifting properties covered in this section apply to the graphs of all functions. Given the accompanying graph of $y = f(x)$, sketch the graph of each of the following.

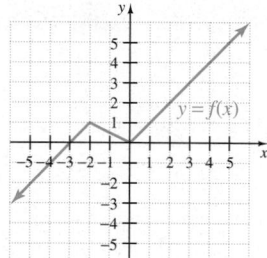

59. $y = f(x) + 1$ **60.** $y = f(x) - 2$

61. $y = f(x - 3)$ **62.** $y = f(x + 3)$

63. $y = f(x + 2) + 2$ **64.** $y = f(x - 1) + 1$

REVIEW EXERCISES

Add the proper constant to each binomial so that the resulting trinomial is a perfect square trinomial. See Section 10.1.

65. $x^2 + 8x$ **66.** $y^2 + 4y$

67. $z^2 - 16z$ **68.** $x^2 - 10x$

69. $y^2 + y$ **70.** $z^2 - 3z$

Solve by completing the square. See Section 10.1.

71. $x^2 + 4x = 12$ **72.** $y^2 + 6y = -5$

73. $z^2 + 10z - 1 = 0$ **74.** $x^2 + 14x + 20 = 0$

75. $z^2 - 8z = 2$ **76.** $y^2 - 10y = 3$

10.7 FURTHER GRAPHING OF QUADRATIC FUNCTIONS

CD-ROM   SSM

SSG   Video

▶ **OBJECTIVES**

1. Write quadratic functions in the form $y = a(x - h)^2 + k$.
2. Derive a formula for finding the vertex of a parabola.
3. Find the minimum or maximum value of a quadratic function.

1 We know that the graph of a quadratic function is a parabola. If a quadratic function is written in the form

$$f(x) = a(x - h)^2 + k$$

we can easily find the vertex (h, k) and graph the parabola. To write a quadratic function in this form, complete the square. (See Section 10.1 for a review of completing the square.)

Example 1   Graph $f(x) = x^2 - 4x - 12$. Find the vertex and any intercepts.

Solution   The graph of this quadratic function is a parabola. To find the vertex of the parabola, we will write the function in the form $y = (x - h)^2 + k$. To do this, we complete the square on the binomial $x^2 - 4x$. To simplify our work, we let $f(x) = y$.

$$y = x^2 - 4x - 12 \qquad \text{Let } f(x) = y.$$
$$y + 12 = x^2 - 4x \qquad \begin{array}{l}\text{Add 12 to both sides to get the} \\ x\text{-variable terms alone.}\end{array}$$

Now we add the square of half of -4 to both sides.

$$\frac{1}{2}(-4) = -2 \quad \text{and} \quad (-2)^2 = 4$$

$$y + 12 + 4 = x^2 - 4x + 4 \qquad \text{Add 4 to both sides.}$$
$$y + 16 = (x - 2)^2 \qquad \text{Factor the trinomial.}$$
$$y = (x - 2)^2 - 16 \qquad \text{Subtract 16 from both sides.}$$
$$f(x) = (x - 2)^2 - 16 \qquad \text{Replace } y \text{ with } f(x).$$

From this equation, we can see that the vertex of the parabola is $(2, -16)$, a point in quadrant IV, and the axis of symmetry is the line $x = 2$.

Notice that $a = 1$. Since $a > 0$, the parabola opens upward. This parabola opening upward with vertex $(2, -16)$ will have two x-intercepts and one y-intercept. (See the Helpful Hint after this example.)

x-intercepts: let y or $f(x) = 0$	y-intercept: let $x = 0$
$f(x) = x^2 - 4x - 12$	$f(x) = x^2 - 4x - 12$
$0 = x^2 - 4x - 12$	$f(0) = 0^2 - 4 \cdot 0 - 12$
$0 = (x - 6)(x + 2)$	$= -12$
$0 = x - 6 \quad \text{or} \quad 0 = x + 2$	
$6 = x \qquad\qquad -2 = x$	

The two x-intercepts are $(6, 0)$ and $(-2, 0)$. The y-intercept is $(0, -12)$. The sketch of $f(x) = x^2 - 4x - 12$ is shown.

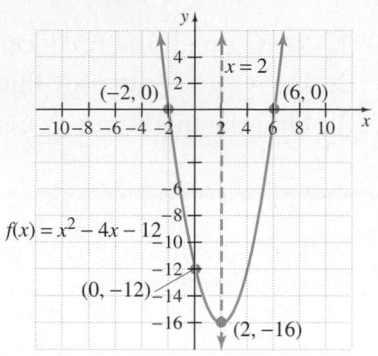

HELPFUL HINT

Parabola Opens Upward
Vertex in I or II: no x-intercept
Vertex in III or IV: 2 x-intercepts

Parabola Opens Downward
Vertex in I or II: 2 x-intercepts
Vertex in III or IV: no x-intercept.

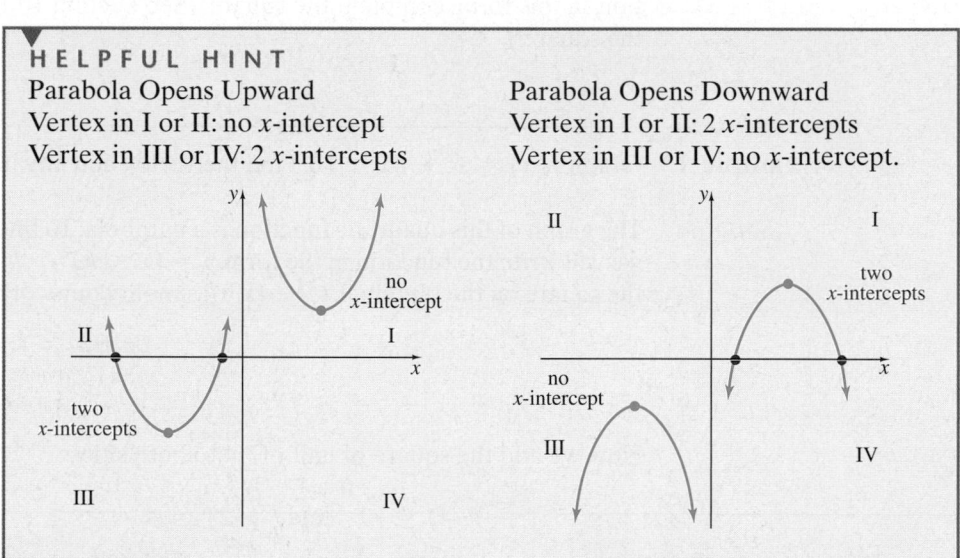

Example 2   Graph $f(x) = 3x^2 + 3x + 1$. Find the vertex and any intercepts.

Solution   Replace $f(x)$ with y and complete the square on x to write the equation in the form $y = a(x - h)^2 + k$.

$$y = 3x^2 + 3x + 1 \qquad \text{Replace } f(x) \text{ with } y.$$

$$y - 1 = 3x^2 + 3x \qquad \text{Isolate } x\text{-variable terms.}$$

Factor 3 from the terms $3x^2 + 3x$ so that the coefficient of x^2 is 1.

$$y - 1 = 3(x^2 + x) \qquad \text{Factor out 3.}$$

The coefficient of x in the parentheses above is 1. Then $\frac{1}{2}(1) = \frac{1}{2}$ and $\left(\frac{1}{2}\right)^2 = \frac{1}{4}$.

Since we are adding $\frac{1}{4}$ inside the parentheses, we are really adding $3\left(\frac{1}{4}\right)$, so we *must* add $3\left(\dfrac{1}{4}\right)$ to the left side.

$$y - 1 + 3\left(\frac{1}{4}\right) = 3\left(x^2 + x + \frac{1}{4}\right)$$

$$y - \frac{1}{4} = 3\left(x + \frac{1}{2}\right)^2 \qquad \textit{Simplify the left side and factor the right side.}$$

$$y = 3\left(x + \frac{1}{2}\right)^2 + \frac{1}{4} \qquad \textit{Add } \frac{1}{4} \textit{ to both sides.}$$

$$f(x) = 3\left(x + \frac{1}{2}\right)^2 + \frac{1}{4} \qquad \textit{Replace y with } f(x).$$

Then $a = 3$, $h = -\dfrac{1}{2}$, and $k = \dfrac{1}{4}$. This means that the parabola opens upward with vertex $\left(-\dfrac{1}{2}, \dfrac{1}{4}\right)$ and that the axis of symmetry is the line $x = -\dfrac{1}{2}$.

To find the y-intercept, let $x = 0$. Then

$$f(0) = 3(0)^2 + 3(0) + 1 = 1$$

Thus the y-intercept is $(0, 1)$.

This parabola has no x-intercepts since the vertex is in the second quadrant and opens upward. Use the vertex, axis of symmetry, and y-intercept to sketch the parabola.

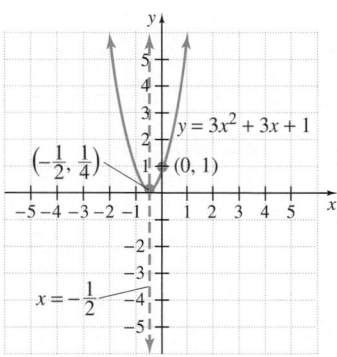

Example 3   Graph $f(x) = -x^2 - 2x + 3$. Find the vertex and any intercepts.

Solution   We write $f(x)$ in the form $a(x - h)^2 + k$ by completing the square. First we replace $f(x)$ with y.

$$f(x) = -x^2 - 2x + 3$$

$$y = -x^2 - 2x + 3$$

$$y - 3 = -x^2 - 2x \qquad \textit{Subtract 3 from both sides to get the x-variable terms alone.}$$

$$y - 3 = -1(x^2 + 2x) \qquad \textit{Factor −1 from the terms } -x^2 - 2x.$$

The coefficient of x is 2. Then $\dfrac{1}{2}(2) = 1$ and $1^2 = 1$. We add 1 to the right side inside the parentheses and add $-1(1)$ to the left side.

$$y - 3 - 1(1) = -1(x^2 + 2x + 1)$$

$$y - 4 = -1(x + 1)^2 \qquad \text{Simplify the left side and factor the right side.}$$

$$y = -1(x + 1)^2 + 4 \qquad \text{Add 4 to both sides.}$$

$$\underbrace{f(x) = -1(x + 1)^2 + 4} \qquad \text{Replace } y \text{ with } f(x).$$

> **HELPFUL HINT**
> This can be written as $f(x) = -1[x - (-1)]^2 + 4$. Notice that the vertex is $(-1, 4)$.

Since $a = -1$, the parabola opens downward with vertex $(-1, 4)$ and axis of symmetry $x = -1$.

To find the y-intercept, we let $x = 0$ and solve for y. Then

$$f(0) = -0^2 - 2(0) + 3 = 3$$

Thus, $(0, 3)$ is the y-intercept.

To find the x-intercepts, we let y or $f(x) = 0$ and solve for x.

$$f(x) = -x^2 - 2x + 3$$

$$0 = -x^2 - 2x + 3 \qquad \text{Let } f(x) = 0.$$

Now we divide both sides by -1 so that the coefficient of x^2 is 1.

$$\frac{0}{-1} = \frac{-x^2}{-1} - \frac{2x}{-1} + \frac{3}{-1} \qquad \text{Divide both sides by } -1.$$

$$0 = x^2 + 2x - 3 \qquad \text{Simplify.}$$

$$0 = (x + 3)(x - 1) \qquad \text{Factor.}$$

$$x + 3 = 0 \quad \text{or} \quad x - 1 = 0 \qquad \text{Set each factor equal to 0.}$$

$$x = -3 \text{ or} \qquad x = 1 \qquad \text{Solve.}$$

The x-intercepts are $(-3, 0)$ and $(1, 0)$. Use these points to sketch the parabola.

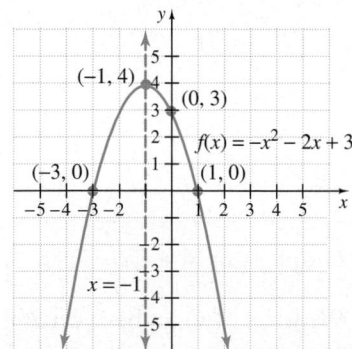

2 Recall from Section 6.8 that we introduced a formula for finding the vertex of a parabola. Now that we have practiced completing the square, we will show that the x-coordinate of the vertex of the graph of $f(x)$ or $y = ax^2 + bx + c$ can be found by the formula $x = \dfrac{-b}{2a}$. To do so, we complete the square on x and write the equation in the form $y = a(x - h)^2 + k$.

First, isolate the x-variable terms by subtracting c from both sides.

$$y = ax^2 + bx + c$$
$$y - c = ax^2 + bx$$

Next, factor a from the terms $ax^2 + bx$.

$$y - c = a\left(x^2 + \frac{b}{a}x\right)$$

Next, add the square of half of $\frac{b}{a}$, or $\left(\frac{b}{2a}\right)^2 = \frac{b^2}{4a^2}$, to the right side inside the parentheses. Because of the factor a, what we really added was $a\left(\frac{b^2}{4a^2}\right)$ and this must be added to the left side.

$$y - c + a\left(\frac{b^2}{4a^2}\right) = a\left(x^2 + \frac{b}{a}x + \frac{b^2}{4a^2}\right)$$

$$y - c + \frac{b^2}{4a} = a\left(x + \frac{b}{2a}\right)^2$$

Simplify the left side and factor the right side.

$$y = a\left(x + \frac{b}{2a}\right)^2 + c - \frac{b^2}{4a}$$

Add c to both sides and subtract $\frac{b^2}{4a}$ from both sides.

Compare this form with $f(x)$ or $y = a(x - h)^2 + k$ and see that h is $\frac{-b}{2a}$, which means that the x-coordinate of the vertex of the graph of $f(x) = ax^2 + bx + c$ is $\frac{-b}{2a}$.

VERTEX FORMULA

The graph of $f(x) = ax^2 + bx + c$, when $a \neq 0$, is a parabola with vertex

$$\left(\frac{-b}{2a}, f\left(\frac{-b}{2a}\right)\right)$$

Let's use this formula to find the vertex of the parabola we graphed in Example 1.

Example 4 Find the vertex of the graph of $f(x) = x^2 - 4x - 12$.

Solution In the quadratic function $f(x) = x^2 - 4x - 12$, notice that $a = 1, b = -4$, and $c = -12$. Then

$$\frac{-b}{2a} = \frac{-(-4)}{2(1)} = 2$$

The x-value of the vertex is 2. To find the corresponding $f(x)$ or y-value, find $f(2)$. Then

$$f(2) = 2^2 - 4(2) - 12 = 4 - 8 - 12 = -16$$

The vertex is $(2, -16)$. These results agree with our findings in Example 1.

3 The vertex of a parabola gives us some important information about its corresponding quadratic function. The quadratic function whose graph is a parabola that opens upward has a minimum value, and the quadratic function whose graph is a parabola that opens downward has a maximum value. The $f(x)$ or y-value of the vertex is the minimum or maximum value of the function.

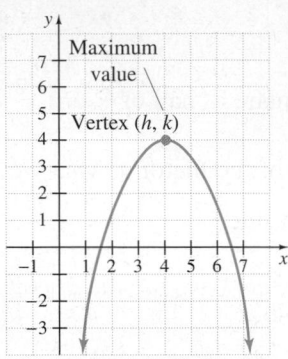 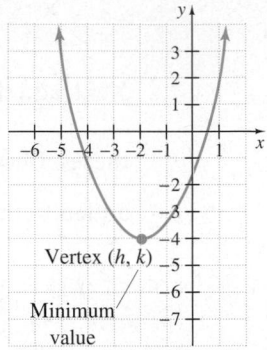

Example 5   **FINDING MAXIMUM HEIGHT**

A rock is thrown upward from the ground. Its height in feet above ground after t seconds is given by the function $f(t) = -16t^2 + 20t$. Find the maximum height of the rock and the number of seconds it took for the rock to reach its maximum height.

Solution   1. **UNDERSTAND.** The maximum height of the rock is the largest value of $f(t)$. Since the function $f(t) = -16t^2 + 20t$ is a quadratic function, its graph is a parabola. It opens downward since $-16 < 0$. Thus, the maximum value of $f(t)$ is the $f(t)$ or y-value of the vertex of its graph.

2. **TRANSLATE.** To find the vertex (h, k), notice that for $f(t) = -16t^2 + 20t$, $a = -16, b = 20$, and $c = 0$. We will use these values and the vertex formula

$$\left(\frac{-b}{2a}, \ f\left(\frac{-b}{2a}\right) \right)$$

3. **SOLVE.**

$$h = \frac{-b}{2a} = \frac{-20}{-32} = \frac{5}{8}$$

$$f\left(\frac{5}{8}\right) = -16\left(\frac{5}{8}\right)^2 + 20\left(\frac{5}{8}\right)$$

$$= -16\left(\frac{25}{64}\right) + \frac{25}{2}$$

$$= -\frac{25}{4} + \frac{50}{4} = \frac{25}{4}$$

4. **INTERPRET.** The graph of $f(t)$ is a parabola opening downward with vertex $\left(\frac{5}{8}, \frac{25}{4}\right)$. This means that the rock's maximum height is $\frac{25}{4}$ feet, or $6\frac{1}{4}$ feet, which was reached in $\frac{5}{8}$ second.

SPOTLIGHT ON DECISION MAKING

Suppose you are a member of a community theater group, the Slidell Players. For an upcoming performance of *Grease,* your group must decide on a ticket price. The graph shows the relationship between ticket price and box office receipts for past Slidell Players performances.

What ticket price would you suggest that the Slidell Players charge for its performance of *Grease*? Explain your reasoning.

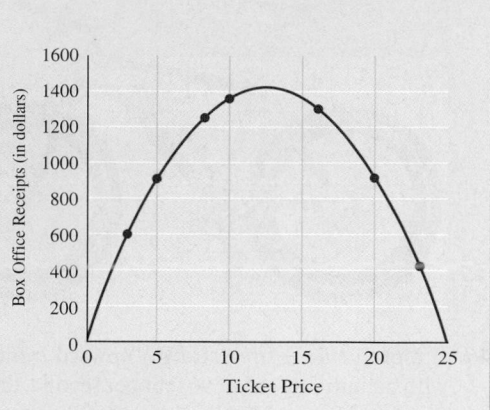

Exercise Set 10.7

Find the vertex of the graph of each quadratic function. See Examples 1 through 4.

1. $f(x) = x^2 + 8x + 7$

2. $f(x) = x^2 + 6x + 5$

3. $f(x) = -x^2 + 10x + 5$

4. $f(x) = -x^2 - 8x + 2$

5. $f(x) = 5x^2 - 10x + 3$

6. $f(x) = -3x^2 + 6x + 4$

7. $f(x) = -x^2 + x + 1$

8. $f(x) = x^2 - 9x + 8$

Match each function with its graph. See Examples 1 through 4.

A
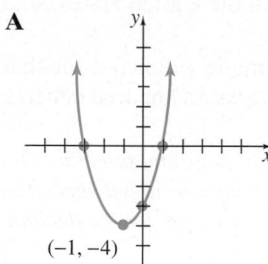
$(-1, -4)$

B
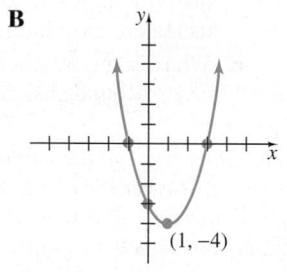
$(1, -4)$

C
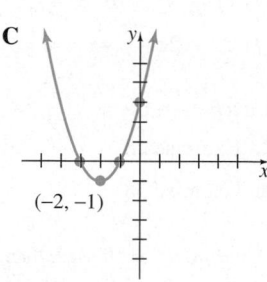
$(-2, -1)$

D
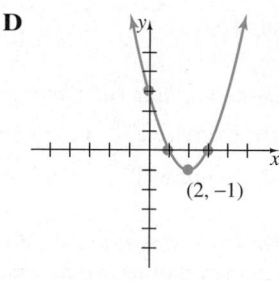
$(2, -1)$

9. $f(x) = x^2 - 4x + 3$

10. $f(x) = x^2 + 2x - 3$

11. $f(x) = x^2 - 2x - 3$

12. $f(x) = x^2 + 4x + 3$

Find the vertex of the graph of each quadratic function. Determine whether the graph opens upward or downward, find any intercepts, and sketch the graph. See Examples 1 through 3.

13. $f(x) = x^2 + 4x - 5$

14. $f(x) = x^2 + 2x - 3$

15. $f(x) = -x^2 + 2x - 1$

16. $f(x) = -x^2 + 4x - 4$

17. $f(x) = x^2 - 4$

18. $f(x) = x^2 - 1$

19. $f(x) = 4x^2 + 4x - 3$

20. $f(x) = 2x^2 - x - 3$

21. $f(x) = x^2 + 8x + 15$

22. $f(x) = x^2 + 10x + 9$

23. $f(x) = x^2 - 6x + 5$

24. $f(x) = x^2 - 4x + 3$

25. $f(x) = x^2 - 4x + 5$

26. $f(x) = x^2 - 6x + 11$

27. $f(x) = 2x^2 + 4x + 5$

28. $f(x) = 3x^2 + 12x + 16$

29. $f(x) = -2x^2 + 12x$

30. $f(x) = -4x^2 + 8x$

31. $f(x) = x^2 + 1$

32. $f(x) = x^2 + 4$

33. $f(x) = x^2 - 2x - 15$

34. $f(x) = x^2 - 4x + 3$

35. $f(x) = -5x^2 + 5x$

36. $f(x) = 3x^2 - 12x$

37. $f(x) = -x^2 + 2x - 12$

38. $f(x) = -x^2 + 8x - 17$

39. $f(x) = 3x^2 - 12x + 15$

40. $f(x) = 2x^2 - 8x + 11$

41. $f(x) = x^2 + x - 6$

42. $f(x) = x^2 + 3x - 18$

43. $f(x) = -2x^2 - 3x + 35$

44. $f(x) = 3x^2 - 13x - 10$

Solve. See Example 5.

45. The cost C in dollars of manufacturing x bicycles at Holladay's Production Plant is given by the function $C(x) = 2x^2 - 800x + 92,000$.

   a. Find the number of bicycles that must be manufactured to minimize the cost.

b. Find the minimum cost.

46. If a projectile is fired straight upward from the ground with an initial speed of 96 feet per second, then its height h in feet after t seconds is given by the equation

$$h(t) = -16t^2 + 96t$$

Find the maximum height of the projectile.

47. If Rheam Gaspar throws a ball upward with an initial speed of 32 feet per second, then its height h in feet after t seconds is given by the equation

$$h(t) = -16t^2 + 32t$$

Find the maximum height of the ball.

48. The Utah Ski Club sells calendars to raise money. The profit P, in cents, from selling x calendars is given by the equation $P(x) = 360x - x^2$.
  a. Find how many calendars must be sold to maximize profit.
  b. Find the maximum profit.

49. Find two numbers whose sum is 60 and whose product is as large as possible. [*Hint:* Let x and $60 - x$ be the two positive numbers. Their product can be described by the function $f(x) = x(60 - x)$.]

50. Find two numbers whose sum is 11 and whose product is as large as possible. (Use the hint for Exercise 49.)

51. Find two numbers whose difference is 10 and whose product is as small as possible. (Use the hint for Exercise 49.)

52. Find two numbers whose difference is 8 and whose product is as small as possible.

△ **53.** The length and width of a rectangle must have a sum of 40. Find the dimensions of the rectangle that will have the maximum area. (Use the hint for Exercise 49.)

△ **54.** The length and width of a rectangle must have a sum of 50. Find the dimensions of the rectangle that will have maximum area.

55. Methane is a gas produced by landfills, natural gas systems, and coal mining that contributes to the greenhouse effect and global warming. Methane emissions in the United States can be modeled by the quadratic function

$f(x) = -0.74x^2 + 8.66x + 159.07$, where $f(x)$ is the amount of methane produced in million metric tons and x is the number of years after 1990. (*Source:* based on data from the U.S. Environmental Protection Agency, 1993–1998)
  a. If this trend continues, what will U.S. emissions of methane be in 2004?
  b. In what year were methane emissions in the United States at their maximum? Round to the nearest whole year.
  c. Use the result of part b. to determine the maximum methane emissions level.

56. The number of inmates in custody in U.S. prisons and jails can be modeled by the quadratic function $p(x) = -716.2x^2 + 87,453.7x + 1,148,702$, where $p(x)$ is the number of inmates and x is the number of years after 1990. (*Source:* based on data from the Bureau of Justice Statistics, U.S. Department of Justice, 1990–1998)
  a. Will this function have a maximum or a minimum? How can you tell?
  b. According to this model, when will the number of prison inmates in custody in the United States be at its maximum/minimum?
  c. What is the number of inmates predicted for that year? Round answer to the nearest hundred inmates.

Find the vertex of the graph of each quadratic function. Determine whether the graph opens upward or downward, find the y-intercept, approximate the x-intercepts to one decimal place, and sketch the graph.

57. $f(x) = x^2 + 10x + 15$ **58.** $f(x) = x^2 - 6x + 4$

59. $f(x) = 3x^2 - 6x + 7$ **60.** $f(x) = 2x^2 + 4x - 1$

Use a graphing calculator to check each exercise.

61. Exercise 27   **62.** Exercise 28

63. Exercise 37   **64.** Exercise 38

Find the maximum or minimum value of each function. Approximate to two decimal places.

65. $f(x) = 2.3x^2 - 6.1x + 3.2$

66. $f(x) = 7.6x^2 + 9.8x - 2.1$

67. $f(x) = -1.9x^2 + 5.6x - 2.7$

68. $f(x) = -5.2x^2 - 3.8x + 5.1$

73. $f(x) = (x + 5)^2 + 2$ **74.** $f(x) = 2(x - 3)^2 + 2$

75. $f(x) = 3(x - 4)^2 + 1$ **76.** $f(x) = (x + 1)^2 + 4$

REVIEW EXERCISES

Sketch the graph of each function. See Sections 3.4 and 10.6.

69. $f(x) = x^2 + 2$ **70.** $f(x) = (x - 3)^2$

71. $g(x) = x + 2$ **72.** $h(x) = x - 3$

77. $f(x) = -(x - 4)^2 + \dfrac{3}{2}$

78. $f(x) = -2(x + 7)^2 + \dfrac{1}{2}$

10

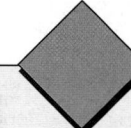

For additional Chapter Projects, visit the Real World Activities Website by going to http://www.prenhall.com/martin-gay.

CHAPTER PROJECT

Fitting a Quadratic Model to Data

Throughout the twentieth century, the eating habits of Americans changed noticeably. Americans started consuming less whole milk and butter, and started consuming more skim and low-fat milk and margarine. We also started eating more poultry and fish. In this project, you will have the opportunity to investigate trends in per capita consumption of poultry during the twentieth century. This project may be completed by working in groups or individually.

We will start by finding a quadratic model, $y = ax^2 + bx + c$, that has ordered pair solutions that correspond to the data for U.S. per capita consumption of poultry given in the table. To do so, substitute each data pair into the equation. Each time, the result is an equation in three unknowns: a, b,

and c. Because there are three pairs of data, we can form a system of three linear equations in three unknowns. Solving for the values of a, b, and c gives a quadratic model that represents the given data.

1. Write the system of equations that must be solved to find the values of a, b, and c needed for a quadratic model of the given data.
2. Solve the system of equations for $a, b,$ and c. Recall the various methods of solving linear systems used in Chapter 8. You might consider using matrices, Cramer's rule, or a graphing calculator to do so. Round to the nearest thousandth.
3. Write the quadratic model for the data. Note that the variable x represents the number of years after 1900.
4. In 1939, the actual U.S. per capita consumption of poultry was 12 pounds per person. Based on this information, how accurate do you think this model is for years other than those given in the table?
5. Use your model to estimate the per capita consumption of poultry in 1950.
6. According to the model, in what year was per capita consumption of poultry 50 pounds per person?
7. In what year was the per capita consumption of poultry at its lowest level? What was that level?
8. Who might be interested in a model like this and how would it be helpful?

U.S. PER CAPITA CONSUMPTION OF POULTRY (IN POUNDS)

Year	x	Poultry Consumption, y (in pounds)
1909	9	11
1969	69	33
1998	98	68

(*Source:* Economic research service, U.S. Department of Agriculture)

CHAPTER 10 VOCABULARY CHECK

Fill in each blank with one of the words or phrases listed below.

quadratic formula   quadratic   discriminant $\pm\sqrt{b}$

completing the square   quadratic inequality   jointly   directly   inversely

(h, k) $(0, k)$ $(h, 0)$ $\dfrac{-b}{2a}$

1. The _____ helps us find the number and type of solutions of a quadratic equation.

2. If $a^2 = b$, then $a =$ _____.

3. The graph of $f(x) = ax^2 + bx + c$ where a is not 0 is a parabola whose vertex has x-value of ____.

4. A(n) _____ is an inequality that can be written so that one side is a quadratic expression and the other side is 0.

5. The process of writing a quadratic equation so that one side is a perfect square trinomial is called _____.

6. The graph of $f(x) = x^2 + k$ has vertex _____.

7. The graph of $f(x) = (x - h)^2$ has vertex _____.

8. The graph of $f(x) = (x - h)^2 + k$ has vertex _____.

9. The formula $x = \dfrac{-b \pm \sqrt{b^2 - 4ac}}{2a}$ is called the _____.

10. A _____ equation is one that can be written in the form $ax^2 + bx + c = 0$ where $a, b,$ and c are real numbers and a is not 0.

11. In the equation $y = kx$, y varies _____ as x.

12. In the equation $y = \dfrac{k}{x}$, y varies _____ as x.

13. In the equation $y = kxz$, y varies _____ as x and z.

CHAPTER 10 HIGHLIGHTS

DEFINITIONS AND CONCEPTS	EXAMPLES
Section 10.1 Solving Quadratic Equations by Completing the Square	

DEFINITIONS AND CONCEPTS	EXAMPLES
Square Root Property If b is a real number and if $a^2 = b$, then $a = \pm\sqrt{b}$.	Solve $(x + 3)^2 = 14$. $x + 3 = \pm\sqrt{14}$ $x = -3 \pm \sqrt{14}$
To Solve a Quadratic Equation in x by Completing the Square **Step 1:** If the coefficient of x^2 is not 1, divide both sides of the equation by the coefficient of x^2. **Step 2:** Isolate the variable terms. **Step 3:** Complete the square by adding the square of half of the coefficient of x to both sides. **Step 4:** Write the resulting trinomial as the square of a binomial. **Step 5:** Apply the square root property and solve for x.	Solve $3x^2 - 12x - 18 = 0$. 1. $x^2 - 4x - 6 = 0$ 2. $\quad\;\; x^2 - 4x = 6$ 3. $\dfrac{1}{2}(-4) = -2$ and $(-2)^2 = 4$ $\quad\; x^2 - 4x + 4 = 6 + 4$ 4. $\quad\quad (x - 2)^2 = 10$ 5. $\quad\quad\; x - 2 = \pm\sqrt{10}$ $\quad\quad\quad\;\; x = 2 \pm \sqrt{10}$

(continued)

DEFINITIONS AND CONCEPTS	EXAMPLES

Section 10.2 Solving Quadratic Equations by the Quadratic Formula

A quadratic equation written in the form $ax^2 + bx + c = 0$ has solutions

$$x = \frac{-b \pm \sqrt{b^2 - 4ac}}{2a}$$

Solve $x^2 - x - 3 = 0$.

$$a = 1, b = -1, c = -3$$

$$x = \frac{-(-1) \pm \sqrt{(-1)^2 - 4(1)(-3)}}{2 \cdot 1}$$

$$x = \frac{1 \pm \sqrt{13}}{2}$$

Section 10.3 Solving Equations by Using Quadratic Methods

Substitution is often helpful in solving an equation that contains a repeated variable expression.

Solve $(2x + 1)^2 - 5(2x + 1) + 6 = 0$.
Let $m = 2x + 1$. Then

$$m^2 - 5m + 6 = 0 \quad \text{Let } m = 2x + 1.$$
$$(m - 3)(m - 2) = 0$$
$$m = 3 \quad \text{or} \quad m = 2$$
$$2x + 1 = 3 \quad \text{or} \quad 2x + 1 = 2 \quad \text{Substi-}$$
$$x = 1 \quad \text{or} \quad x = \frac{1}{2} \quad \begin{array}{l}\text{tute}\\ \text{back.}\end{array}$$

Section 10.4 Nonlinear Inequalities in One Variable

To Solve a Polynomial Inequality

Step 1: Write the inequality in standard form.
Step 2: Solve the related equation.
Step 3: Use solutions from Step 2 to separate the number line into regions.
Step 4: Use test points to determine whether values in each region satisfy the original inequality.
Step 5: Write the solution set as the union of regions whose test point value is a solution.

Solve $x^2 \geq 6x$.

1. $x^2 - 6x \geq 0$

2. $x^2 - 6x = 0$

$$x(x - 6) = 0$$
$$x = 0 \quad \text{or} \quad x = 6$$

3. $\begin{array}{ccc} A & B & C \end{array}$

$$\xleftarrow{\quad\;\;\underset{0}{|}\quad\;\;\underset{6}{|}\quad}\rightarrow$$

4.

Region	Test Point Value	$x^2 \geq 6x$	Result
A	-2	$(-2)^2 \geq 6(-2)$	True
B	1	$1^2 \geq 6(1)$	False
C	7	$7^2 \geq 6(7)$	True

5.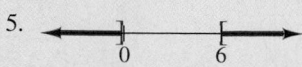

The solution set is $(-\infty, 0] \cup [6, \infty)$.

(continued)

DEFINITIONS AND CONCEPTS	EXAMPLES

Section 10.4 Nonlinear Inequalities in One Variable

To Solve a Rational Inequality

Step 1: Solve for values that make all denominators 0.

Step 2: Solve the related equation.

Step 3: Use solutions from Steps 1 and 2 to separate the number line into regions.

Step 4: Use test points to determine whether values in each region satisfy the original inequality.

Step 5: Write the solution set as the union of regions whose test point value is a solution.

Solve $\dfrac{6}{x-1} < -2$.

1. $x - 1 = 0$ Set denominator equal to 0.

    $x = 1$

2. $\dfrac{6}{x-1} = -2$

      $6 = -2(x-1)$ Multiply by $(x-1)$.

      $6 = -2x + 2$

      $4 = -2x$

    $-2 = x$

3.
```
      A      B      C
  ←───+──────+──────→
     -2      1
```

4. Only a test value from region B satisfies the original inequality.

5.
```
  ←───(──────)──────→
     -2      1
```

The solution set is $(-2, 1)$.

Section 10.5 Variation and Problem Solving

y **varies directly as** x, or y is **directly proportional to** x, if there is a nonzero constant k such that

$$y = kx$$

The circumference of a circle C varies directly as its radius r.

$$C = 2\pi r$$
$$\qquad \underset{k}{\uparrow}$$

y **varies inversely as** x, or y is **inversely proportional to** x, if there is a nonzero constant k such that

$$y = \frac{k}{x}$$

Pressure P varies inversely with volume V.

$$P = \frac{k}{V}$$

y **varies jointly as** x and z or y is **jointly proportional to** x and z if there is a nonzero constant k such that

$$y = kxz$$

The lateral surface area S of a cylinder varies jointly as its radius r and height h.

$$S = 2\pi rh$$
$$\qquad \underset{k}{\uparrow}$$

Section 10.6 Quadratic Functions and Their Graphs

Graph of a Quadratic Function

The graph of a quadratic function written in the form $f(x) = a(x - h)^2 + k$ is a parabola with vertex (h, k). If $a > 0$, the parabola opens upward; if $a < 0$, the parabola opens downward. The axis of symmetry is the line whose equation is $x = h$.

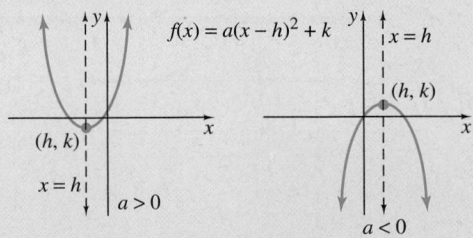

Graph $g(x) = 3(x - 1)^2 + 4$.

The graph is a parabola with vertex $(1, 4)$ and axis of symmetry $x = 1$. Since $a = 3$ is positive, the graph opens upward.

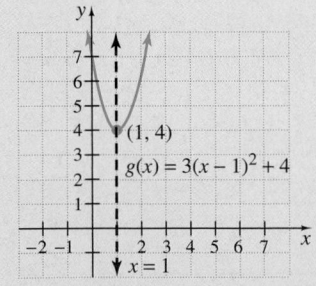

(continued)

DEFINITIONS AND CONCEPTS	EXAMPLES

Section 10.7 Further Graphing of Quadratic Functions

The graph of $f(x) = ax^2 + bx + c$, where $a \neq 0$, is a parabola with vertex

$$\left(\frac{-b}{2a}, f\left(\frac{-b}{2a}\right)\right)$$

Graph $f(x) = x^2 - 2x - 8$. Find the vertex and x- and y-intercepts.

$$\frac{-b}{2a} = \frac{-(-2)}{2 \cdot 1} = 1$$

$$f(1) = 1^2 - 2(1) - 8 = -9$$

The vertex is $(1, -9)$.

$$0 = x^2 - 2x - 8$$
$$0 = (x - 4)(x + 2)$$
$$x = 4 \quad \text{or} \quad x = -2$$

The x-intercepts are $(4, 0)$ and $(-2, 0)$.

$$f(0) = 0^2 - 2 \cdot 0 - 8 = -8$$

The y-intercept is $(0, -8)$.

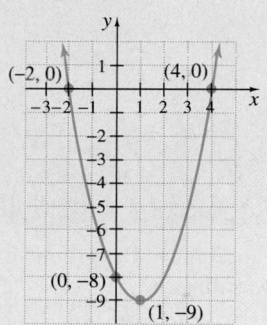

CHAPTER 10 REVIEW

(10.1) Solve by factoring.

1. $x^2 - 15x + 14 = 0$

2. $x^2 - x - 30 = 0$

3. $10x^2 = 3x + 4$ **4.** $7a^2 = 29a + 30$

Solve by using the square root property.

5. $4m^2 = 196$ **6.** $9y^2 = 36$

7. $(9n + 1)^2 = 9$ **8.** $(5x - 2)^2 = 2$

Solve by completing the square.

9. $z^2 + 3z + 1 = 0$ **10.** $x^2 + x + 7 = 0$

11. $(2x + 1)^2 = x$ **12.** $(3x - 4)^2 = 10x$

13. If P dollars are originally invested, the formula $A = P(1 + r)^2$ gives the amount A in an account paying interest rate r compounded annually after 2 years. Find the interest rate r such that $2500 increases to $2717 in

2 years. Round the result to the nearest hundredth of a percent.

14. Two ships leave a port at the same time and travel at the same speed. One ship is traveling due north and the other due east. In a few hours, the ships are 150 miles apart. How many miles has each ship traveled? Give an exact answer and a one-decimal-place approximation.

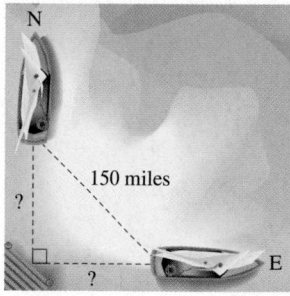

(10.2) *If the discriminant of a quadratic equation has the given value, determine the number and type of solutions of the equation.*

15. -8

16. 48

17. 100

18. 0

Solve by using the quadratic formula.

19. $x^2 - 16x + 64 = 0$

20. $x^2 + 5x = 0$

21. $x^2 + 11 = 0$

22. $2x^2 + 3x = 5$

23. $6x^2 + 7 = 5x$

24. $9a^2 + 4 = 2a$

25. $(5a - 2)^2 - a = 0$

26. $(2x - 3)^2 = x$

27. Cadets graduating from military school usually toss their hats high into the air at the end of the ceremony. One cadet threw his hat so that its distance $d(t)$ in feet above the ground t seconds after it was thrown was $d(t) = -16t^2 + 30t + 6$.

    a. Find the distance above the ground of the hat 1 second after it was thrown.

    b. Find the time it takes the hat to hit the ground. Give an exact time and a one-decimal-place approximation.

△ **28.** The hypotenuse of an isosceles right triangle is 6 centimeters longer than either of the legs. Find the length of the legs.

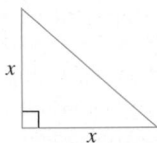

(10.3) *Solve each equation for the variable.*

29. $x^3 = 27$

30. $y^3 = -64$

31. $\dfrac{5}{x} + \dfrac{6}{x - 2} = 3$

32. $\dfrac{7}{8} = \dfrac{8}{x^2}$

33. $x^4 - 21x^2 - 100 = 0$

34. $5(x + 3)^2 - 19(x + 3) = 4$

35. $x^{2/3} - 6x^{1/3} + 5 = 0$

36. $x^{2/3} - 6x^{1/3} = -8$

37. $a^6 - a^2 = a^4 - 1$

38. $y^{-2} + y^{-1} = 20$

39. Two postal workers, Jerome Grant and Tim Bozik, can sort a stack of mail in 5 hours. Working alone, Tim can sort the mail in 1 hour less time than Jerome can. Find the time that each postal worker can sort the mail alone. Round the result to one decimal place.

40. A negative number decreased by its reciprocal is $-\dfrac{24}{5}$. Find the number.

(10.4) *Solve each inequality for x. Graph the solution set and write each solution set in interval notation.*

41. $2x^2 - 50 \le 0$

42. $\dfrac{1}{4}x^2 < \dfrac{1}{16}$

43. $(2x - 3)(4x + 5) \ge 0$

44. $(x^2 - 16)(x^2 - 1) > 0$

45. $\dfrac{x - 5}{x - 6} < 0$

46. $\dfrac{x(x + 5)}{4x - 3} \ge 0$

47. $\dfrac{(4x + 3)(x - 5)}{x(x + 6)} > 0$

48. $(x + 5)(x - 6)(x + 2) \le 0$

49. $x^3 + 3x^2 - 25x - 75 > 0$

50. $\dfrac{x^2 + 4}{3x} \le 1$

51. $\dfrac{(5x + 6)(x - 3)}{x(6x - 5)} < 0$

52. $\dfrac{3}{x - 2} > 2$

(10.5) *Solve each variation problem.*

53. A is directly proportional to B. If $A = 6$ when $B = 14$, find A when $B = 21$.

54. C is inversely proportional to D. If $C = 12$ when $D = 8$, find C when $D = 24$.

55. According to Boyle's law, the pressure exerted by a gas is inversely proportional to the volume, as long as the temperature stays the same. If a gas exerts a pressure of

1250 pounds per square inch when the volume is 2 cubic feet, find the volume when the pressure is 800 pounds per square inch.

△ **56.** The surface area of a sphere varies directly as the square of its radius. If the surface area is 36π square inches when the radius is 3 inches, find the surface area when the radius is 4 inches.

(10.6) Sketch the graph of each function. Label the vertex and the axis of symmetry.

57. $f(x) = x^2 - 4$

58. $g(x) = x^2 + 7$

59. $H(x) = 2x^2$

60. $h(x) = -\dfrac{1}{3}x^2$

61. $F(x) = (x - 1)^2$

62. $G(x) = (x + 5)^2$

63. $f(x) = (x - 4)^2 - 2$

64. $f(x) = -3(x - 1)^2 + 1$

(10.7) Sketch the graph of each function. Find the vertex and the intercepts.

65. $f(x) = x^2 + 10x + 25$ **66.** $f(x) = -x^2 + 6x - 9$

67. $f(x) = 4x^2 - 1$ **68.** $f(x) = -5x^2 + 5$

69. Find the vertex of the graph of $f(x) = -3x^2 - 5x + 4$. Determine whether the graph opens upward or downward, find the y-intercept, approximate the x-intercepts to one decimal place, and sketch the graph.

70. The function $h(t) = -16t^2 + 120t + 300$ gives the height in feet of a projectile fired from the top of a building in t seconds.

    a. When will the object reach a height of 350 feet? Round your answer to one decimal place.

    b. Explain why Part **a** has two answers.

71. Find two numbers whose product is as large as possible, given that their sum is 420.

72. Write an equation of a quadratic function whose graph is a parabola that has vertex $(-3, 7)$ and that passes through the origin.

CHAPTER 10 TEST

Solve each equation for the variable.

1. $5x^2 - 2x = 7$ **2.** $(x + 1)^2 = 10$

3. $m^2 - m + 8 = 0$ **4.** $u^2 - 6u + 2 = 0$

5. $7x^2 + 8x + 1 = 0$ **6.** $a^2 - 3a = 5$

7. $\dfrac{4}{x + 2} + \dfrac{2x}{x - 2} = \dfrac{6}{x^2 - 4}$

8. $x^4 - 8x^2 - 9 = 0$ **9.** $x^6 + 1 = x^4 + x^2$

10. $(x + 1)^2 - 15(x + 1) + 56 = 0$

Solve the equation for the variable by completing the square.

11. $x^2 - 6x = -2$ **12.** $2a^2 + 5 = 4a$

Solve each inequality for x. Graph the solution set and then write the solution set in interval notation.

13. $2x^2 - 7x > 15$ **14.** $(x^2 - 16)(x^2 - 25) > 0$

15. $\dfrac{5}{x + 3} < 1$ **16.** $\dfrac{7x - 14}{x^2 - 9} \le 0$

Graph each function. Label the vertex.

17. $f(x) = 3x^2$ **18.** $G(x) = -2(x - 1)^2 + 5$

Graph each function. Find and label the vertex, y-intercept, and x-intercepts (if any).

19. $h(x) = x^2 - 4x + 4$ **20.** $F(x) = 2x^2 - 8x + 9$

△ **21.** A 10-foot ladder is leaning against a house. The distance from the bottom of the ladder to the house is 4 feet less than the distance from the top of the ladder to the ground. Find how far the top of the ladder is from the ground. Give an exact answer and a one-decimal-place approximation.

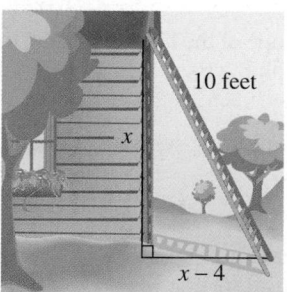

22. Dave and Sandy Hartranft can paint a room together in 4 hours. Working alone, Dave can paint the room in 2 hours less time than Sandy can. Find how long it takes Sandy to paint the room alone.

23. A stone is thrown upward from a bridge. The stone's height in feet, $s(t)$, above the water t seconds after the stone is thrown is a function given by the equation $s(t) = -16t^2 + 32t + 256$.

    a. Find the maximum height of the stone.

b. Find the time it takes the stone to hit the water. Round the answer to two decimal places.

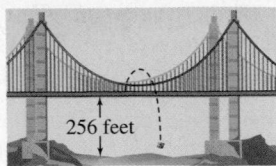

256 feet

△ **24.** Given the diagram shown, approximate to the nearest foot how many feet of walking distance a person saves by cutting across the lawn instead of walking on the sidewalk.

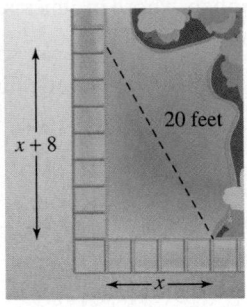

$x + 8$

20 feet

x

25. Suppose that W is inversely proportional to V. If $W = 20$ when $V = 12$, find W when $V = 15$.

26. Suppose that Q is jointly proportional to R and the square of S. If $Q = 24$ when $R = 3$ and $S = 4$, find Q when $R = 2$ and $S = 3$.

27. When an anvil is dropped into a gorge, the speed with which it strikes the ground is directly proportional to the square root of the distance it falls. An anvil that falls 400 feet hits the ground at a speed of 160 feet per second. Find the height of a cliff over the gorge if a dropped anvil hits the ground at a speed of 128 feet per second.

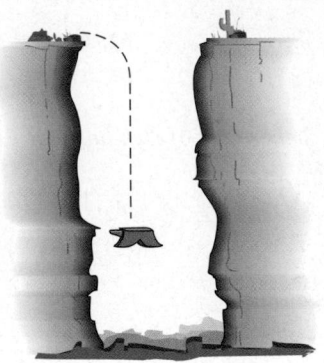

CHAPTER 10 CUMULATIVE REVIEW

1. Find the value of each expression when $x = 2$ and $y = -5$.

   a. $\dfrac{x - y}{12 + x}$ **b.** $x^2 - y$

2. Simplify each expression by combining like terms.

   a. $2x + 3x + 5 + 2$ **b.** $-5a - 3 + a + 2$

   c. $4y - 3y^2$ **d.** $2.3x + 5x - 6$

3. Find the slope of the line containing the points $(0, 3)$ and $(2, 5)$. Graph the line.

4. Solve the following system of equations by graphing.
$$\begin{cases} 2x + y = 7 \\ 2y = -4x \end{cases}$$

5. Solve the system:
$$\begin{cases} 7x - 3y = -14 \\ -3x + y = 6 \end{cases}$$

6. Solve the system:
$$\begin{cases} 3x - 2y = 2 \\ -9x + 6y = -6 \end{cases}$$

7. Albert and Louis live 15 miles away from each other. They decide to meet one day by walking toward one another. After 2 hours they meet. If Louis walks one mile per hour faster than Albert, find both walking speeds.

8. Simplify the following.

   a. $\left(\dfrac{-5x^2}{y^3}\right)^2$ **b.** $\dfrac{(x^3)^4 x}{x^7}$

   c. $\dfrac{(2x)^5}{x^3}$ **d.** $\dfrac{(a^2 b)^3}{a^3 b^2}$

9. Divide $\dfrac{3x^5 y^2 - 15x^3 y - x^2 y - 6x}{x^2 y}$.

10. If $P(x) = 2x^3 - 4x^2 + 5$

   a. Find $P(2)$ by substitution.

   b. Use synthetic division to find the remainder when $P(x)$ is divided by $x - 2$.

11. Solve $(5x - 1)(2x^2 + 15x + 18) = 0$.

12. Graph the quadratic function $f(x) = -x^2 + 2x - 3$ by plotting points.

13. Write the rational expression in lowest terms.
$$\dfrac{2x^2}{10x^3 - 2x^2}$$

14. Simplify.
$$\dfrac{x^{-1} + 2xy^{-1}}{x^{-2} - x^{-2}y^{-1}}$$

15. Solve for x: $\dfrac{x-5}{3} = \dfrac{x+2}{5}$

16. Simplify.

  a. $\sqrt{(-3)^2}$

  b. $\sqrt{x^2}$

  c. $\sqrt[4]{(x-2)^4}$

  d. $\sqrt[3]{(-5)^3}$

  e. $\sqrt[5]{(2x-7)^5}$

17. Use rational exponents to simplify. Assume that variables represent positive numbers.

  a. $\sqrt[6]{25}$ **b.** $\sqrt[8]{x^4}$

  c. $\sqrt[4]{r^2 s^6}$

18. Use the product rule to simplify.

  a. $\sqrt{25x^3}$ **b.** $\sqrt[3]{54x^6 y^8}$

  c. $\sqrt[4]{81z^{11}}$

19. Rationalize the denominator of each expression.

  a. $\dfrac{2}{\sqrt{5}}$ **b.** $\dfrac{2\sqrt{16}}{\sqrt{9x}}$

  c. $\sqrt[3]{\dfrac{1}{2}}$

20. Solve $\sqrt{2x+5} + \sqrt{2x} = 3$.

21. Find each quotient. Write in the form $a + bi$.

  a. $\dfrac{2+i}{1-i}$

  b. $\dfrac{7}{3i}$

22. Use the square root property to solve $(x+1)^2 = 12$.

23. Solve $x - \sqrt{x} - 6 = 0$.

24. Suppose that y varies directly as x. If y is 5 when x is 30, find the constant of variation and the direct variation equation.

The Number One Job in the United States

A Webmaster is responsible for creating and managing Web sites on the World Wide Web. According to the 1999 *Jobs Rated Almanac,* Webmaster is rated as the number one job in the United States based on factors such as income, stress, physical demands, potential growth, job security, and work environment.

Webmasters have technical, business, and visual design skills or understanding. Some Webmasters may focus on the computer programming aspects of developing Web sites, including graphics and Web site security issues. Others may focus on Web site content, such as company or organization history, product information, or other marketing topics. Webmasters should be comfortable with people management, project management, and strategic planning. Webmasters use math and problem-solving skills in tasks such as creating budgets, making hardware or software purchasing decisions, and compiling and analyzing Web site usage statistics.

 For more information about a career as a Webmaster or other Web professional, visit the World Organization of Webmasters Website by first going to www.prenhall.com/martin-gay.

In the Spotlight on Decision Making feature on page 707, you will have the opportunity to make a decision about modeling Website usage statistics as a Webmaster.

EXPONENTIAL AND LOGARITHMIC FUNCTIONS

11

In this chapter, we discuss two closely related functions: exponential and logarithmic functions. These functions are vital to applications in economics, finance, engineering, the sciences, education, and other fields. Models of tumor growth and learning curves are two examples of the uses of exponential and logarithmic functions.

11.1 THE ALGEBRA OF FUNCTIONS; COMPOSITE FUNCTIONS

CD-ROM SSM

SSG Video

▶ **OBJECTIVES**

1. Add, subtract, multiply, and divide functions.
2. Construct composite functions.

1 As we have seen in earlier chapters, it is possible to add, subtract, multiply, and divide functions. Although we have not stated it as such, the sums, differences, products, and quotients of functions are themselves functions. For example, if $f(x) = 3x$ and $g(x) = x + 1$, their product, $f(x) \cdot g(x) = 3x(x + 1) = 3x^2 + 3x$, is a new function. We can use the notation $(f \cdot g)(x)$ to denote this new function. Finding the sum, difference, product, and quotient of functions to generate new functions is called the **algebra of functions.**

ALGEBRA OF FUNCTIONS

Let f and g be functions. New functions from f and g are defined as follows.

Sum	$(f + g)(x) = f(x) + g(x)$
Difference	$(f - g)(x) = f(x) - g(x)$
Product	$(f \cdot g)(x) = f(x) \cdot g(x)$
Quotient	$\left(\dfrac{f}{g}\right)(x) = \dfrac{f(x)}{g(x)}, g(x) \neq 0$

Example 1 If $f(x) = x - 1$ and $g(x) = 2x - 3$, find

  a. $(f + g)(x)$

  b. $(f - g)(x)$

  c. $(f \cdot g)(x)$

  d. $\left(\dfrac{f}{g}\right)(x)$

Solution Use the algebra of functions and replace $f(x)$ by $x - 1$ and $g(x)$ by $2x - 3$. Then we simplify.

  a. $(f + g)(x) = f(x) + g(x)$

  $\qquad\qquad = (x - 1) + (2x - 3)$

  $\qquad\qquad = 3x - 4$

b. $(f - g)(x) = f(x) - g(x)$

$$= (x - 1) - (2x - 3)$$

$$= x - 1 - 2x + 3$$

$$= -x + 2$$

c. $(f \cdot g)(x) = f(x) \cdot g(x)$

$$= (x - 1)(2x - 3)$$

$$= 2x^2 - 5x + 3$$

d. $\left(\dfrac{f}{g}\right)(x) = \dfrac{f(x)}{g(x)} = \dfrac{x - 1}{2x - 3}$, where $x \neq \dfrac{3}{2}$

There is an interesting but not surprising relationship between the graphs of functions and the graphs of their sum, difference, product, and quotient. For example, the graph of $(f + g)(x)$ can be found by adding the graph of $f(x)$ to the graph of $g(x)$. We add two graphs by adding corresponding y-values.

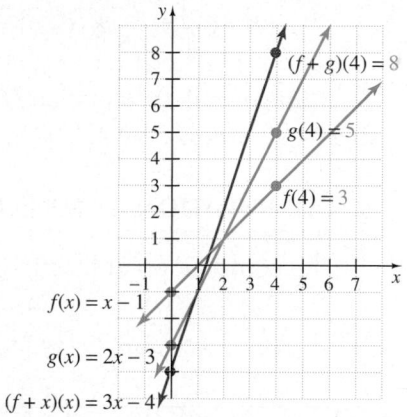

2 Another way to combine functions is called **function composition.** To understand this new way of combining functions, study the tables below. The first table shows degrees Celsius $C(x)$ as a function of degrees Fahrenheit x. The second table shows Kelvins $K(C)$ as a function of degrees Celsius C. (The Kelvin scale is a temperature scale devised by Lord Kelvin in 1848.) The function represented by the first table we will call C, and the second function we will call K.

$x = $ Degrees Fahrenheit (Input)	-31	-13	32	68	149	212
$C(x) = $ Degrees Celsius (Output)	-35	-25	0	20	65	100

$C = $ Degrees Celsius (Input)	-35	-25	0	20	65	100
$K(C) = $ Kelvins (Output)	238.15	248.15	273.15	293.15	338.15	373.15

Suppose that we want a table that shows a direct conversion from degrees Fahrenheit to Kelvins. In other words, suppose that a table is needed that shows Kelvins as a function of degrees Fahrenheit. This can easily be done because in the tables, the output of the first table $C(x)$ is the same as the input of the second table. If we use $C(x)$ to represent this, then we get the following table.

x = Degrees Fahrenheit (Input)	-31	-13	32	68	149	212
$K(C(x))$ = Kelvins (Output)	238.15	248.15	273.15	293.15	338.15	373.15

Since the output of the first table is used as the input of the second table, we write the new function as $K(C(x))$. The new function is formed from the composition of the other two functions. The mathematical symbol for this composition is $(K \circ C)(x)$. Thus, $(K \circ C)(x) = K(C(x))$.

It is possible to find an equation for the composition of the two functions C and K. In other words, we can find a function that converts degrees Fahrenheit directly to Kelvins. The function $C(x) = \dfrac{5}{9}(x - 32)$ converts degrees Fahrenheit to degrees Celsius, and the function $K(C) = C + 273.15$ converts degrees Celsius to Kelvins. Thus,

$$(K \circ C)(x) = K(C(x)) = K\left(\frac{5}{9}(x - 32)\right) = \frac{5}{9}(x - 32) + 273.15$$

In general, the notation $f(g(x))$ means "f composed with g" and can be written as $(f \circ g)(x)$. Also $g(f(x))$, or $(g \circ f)(x)$, means "g composed with f."

COMPOSITION OF FUNCTIONS

The composition of functions f and g is

$$(f \circ g)(x) = f(g(x))$$

> **HELPFUL HINT**
> $(f \circ g)(x)$ does not mean the same as $(f \cdot g)(x)$.
> $$(f \circ g)(x) = f(g(x)) \text{ while } (f \cdot g)(x) = f(x) \cdot g(x)$$

Example 2   If $f(x) = x^2$ and $g(x) = x + 3$, find each composition.

a. $(f \circ g)(2)$ and $(g \circ f)(2)$

b. $(f \circ g)(x)$ and $(g \circ f)(x)$

Solution   **a.** $(f \circ g)(2) = f(g(2))$

$\qquad\qquad\quad = f(5)$ Replace $g(2)$ with 5. [Since $g(x) = x + 3$,

$\qquad\qquad\quad = 5^2 = 25$ then $g(2) = 2 + 3 = 5$.]

$\qquad\quad (g \circ f)(2) = g(f(2))$

$\qquad\qquad\quad = g(4)$ Since $f(x) = x^2$, then $f(2) = 2^2 = 4$.

$\qquad\qquad\quad = 4 + 3 = 7$

b. $(f \circ g)(x) = f(g(x))$

$$= f(x + 3) \qquad \text{Replace } g(x) \text{ with } x + 3.$$

$$= (x + 3)^2 \qquad f(x + 3) = (x + 3)^2$$

$$= x^2 + 6x + 9 \qquad \text{Square } (x + 3).$$

$(g \circ f)(x) = g(f(x))$

$$= g(x^2) \qquad \text{Replace } f(x) \text{ with } x^2.$$

$$= x^2 + 3 \qquad g(x^2) = x^2 + 3$$

Example 3   If $f(x) = |x|$ and $g(x) = x - 2$, find each composition.

a. $(f \circ g)(x)$

b. $(g \circ f)(x)$

Solution   **a.** $(f \circ g)(x) = f(g(x)) = f(x - 2) = |x - 2|$

b. $(g \circ f)(x) = g(f(x)) = g(|x|) = |x| - 2$

> **HELPFUL HINT**
> In Examples 2 and 3, notice that $(g \circ f)(x) \neq (f \circ g)(x)$. In general, $(g \circ f)(x)$ *may* or *may not* equal $(f \circ g)(x)$.

Example 4   If $f(x) = 5x$, $g(x) = x - 2$, and $h(x) = \sqrt{x}$, write each function as a composition using two of the given functions.

a. $F(x) = \sqrt{x - 2}$

b. $G(x) = 5x - 2$

Solution   **a.** Notice the order in which the function F operates on an input value x. First, 2 is subtracted from x. This is the function $g(x) = x - 2$. Then the square root *of that result* is taken. The square root function is $h(x) = \sqrt{x}$. This means that $F = h \circ g$. To check, we find $h \circ g$.

$$(h \circ g)(x) = h(g(x)) = h(x - 2) = \sqrt{x - 2}$$

b. Notice the order in which the function G operates on an input value x. First, x is multiplied by 5, and then 2 is subtracted from the result. This means that $G = g \circ f$. To check, we find $g \circ f$.

$$(g \circ f)(x) = g(f(x)) = g(5x) = 5x - 2$$

GRAPHING CALCULATOR EXPLORATIONS

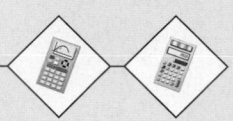

If $f(x) = \frac{1}{2}x + 2$ and $g(x) = \frac{1}{3}x^2 + 4$, then

$$(f + g)(x) = f(x) + g(x)$$

$$= \left(\frac{1}{2}x + 2\right) + \left(\frac{1}{3}x^2 + 4\right)$$

$$= \frac{1}{3}x^2 + \frac{1}{2}x + 6.$$

To visualize this addition of functions with a graphing calculator, graph

$$Y_1 = \frac{1}{2}x + 2, \qquad Y_2 = \frac{1}{3}x^2 + 4, \qquad Y_3 = \frac{1}{3}x^2 + \frac{1}{2}x + 6$$

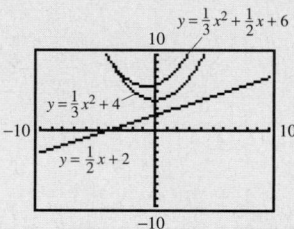

Use a TABLE feature to verify that for a given x value, $Y_1 + Y_2 = Y_3$. For example, verify that when $x = 0$, $Y_1 = 2$, $Y_2 = 4$, and $Y_3 = 2 + 4 = 6$.

Exercise Set 11.1

For the functions f and g, find ***a.*** $(f + g)(x)$, ***b.*** $(f - g)(x)$, ***c.*** $(f \cdot g)(x)$, *and* ***d.*** $\left(\frac{f}{g}\right)(x)$. *See Example 1.*

1. $f(x) = x - 7, g(x) = 2x + 1$

2. $f(x) = x + 4, g(x) = 5x - 2$

3. $f(x) = x^2 + 1, g(x) = 5x$

4. $f(x) = x^2 - 2, g(x) = 3x$

5. $f(x) = \sqrt{x}, g(x) = x + 5$

6. $f(x) = \sqrt[3]{x}, g(x) = x - 3$

7. $f(x) = -3x, g(x) = 5x^2$

8. $f(x) = 4x^3, g(x) = -6x$

If $f(x) = x^2 - 6x + 2$, $g(x) = -2x$, and $h(x) = \sqrt{x}$, find each composition. See Example 2.

9. $(f \circ g)(2)$

10. $(h \circ f)(-2)$

11. $(g \circ f)(-1)$

12. $(f \circ h)(1)$

13. $(g \circ h)(0)$

14. $(h \circ g)(0)$

Find $(f \circ g)(x)$ and $(g \circ f)(x)$. See Examples 2 and 3.

15. $f(x) = x^2 + 1, g(x) = 5x$

16. $f(x) = x - 3, g(x) = x^2$

17. $f(x) = 2x - 3, g(x) = x + 7$

18. $f(x) = x + 10, g(x) = 3x + 1$

19. $f(x) = x^3 + x - 2, g(x) = -2x$

20. $f(x) = -4x, g(x) = x^3 + x^2 - 6$

21. $f(x) = \sqrt{x}, g(x) = -5x + 2$

22. $f(x) = 7x - 1, g(x) = \sqrt[3]{x}$

If $f(x) = 3x$, $g(x) = \sqrt{x}$, and $h(x) = x^2 + 2$, write each function as a composition using two of the given functions. See Example 4.

23. $H(x) = \sqrt{x^2 + 2}$

24. $G(x) = \sqrt{3x}$

25. $F(x) = 9x^2 + 2$

26. $H(x) = 3x^2 + 6$

27. $G(x) = 3\sqrt{x}$

28. $F(x) = x + 2$

*Find $f(x)$ and $g(x)$ so that the given function
$h(x) = (f \circ g)(x)$.*

29. $h(x) = (x + 2)^2$

30. $h(x) = |x - 1|$

31. $h(x) = \sqrt{x + 5} + 2$

32. $h(x) = (3x + 4)^2 + 3$

33. $h(x) = \dfrac{1}{2x - 3}$

34. $h(x) = \dfrac{1}{x + 10}$

Given that $f(-1) = 4 \quad g(-1) = -4$
$\qquad\qquad\quad f(0) = 5 \quad g(0) = -3$
$\qquad\qquad\quad f(2) = 7 \quad g(2) = -1$
$\qquad\qquad\quad f(7) = 1 \quad g(7) = 4$

Find each function value.

35. $(f + g)(2)$

36. $(f - g)(7)$

37. $(f \circ g)(2)$

38. $(g \circ f)(2)$

39. $(f \cdot g)(7)$

40. $(f \cdot g)(0)$

41. $\left(\dfrac{f}{g}\right)(-1)$

42. $\left(\dfrac{g}{f}\right)(-1)$

Solve.

43. Business people are concerned with cost functions, revenue functions, and profit functions. Recall that the profit $P(x)$ obtained from x units of a product is equal to the revenue $R(x)$ from selling the x units minus the cost $C(x)$ of manufacturing the x units. Write an equation expressing this relationship among $C(x)$, $R(x)$, and $P(x)$.

44. Suppose the revenue $R(x)$ for x units of a product can be described by $R(x) = 25x$, and the cost $C(x)$ can be described by $C(x) = 50 + x^2 + 4x$. Find the profit $P(x)$ for x units.

REVIEW EXERCISES

Solve each equation for y. See Section 2.5.

45. $x = y + 2$

46. $x = y - 5$

47. $x = 3y$

48. $x = -6y$

49. $x = -2y - 7$

50. $x = 4y + 7$

11.2 INVERSE FUNCTIONS

CD-ROM

SSM

SSG

Video

▶ **OBJECTIVES**

1. Determine whether a function is a one-to-one function.
2. Use the horizontal line test to decide whether a function is a one-to-one function.
3. Find the inverse of a function.
4. Find the equation of the inverse of a function.
5. Graph functions and their inverses.
6. Determine whether two functions are inverses of each other.

1

In the next section, we begin a study of two new functions: exponential and logarithmic functions. As we learn more about these functions, we will discover that they share a special relation to each other: They are inverses of each other.

Before we study these functions, we need to learn about inverses. We begin by defining one-to-one functions.

Study the following table.

Degrees Fahrenheit (Input)	-31	-13	32	68	149	212
Degrees Celsius (Output)	-35	-25	0	20	65	100

Recall that since each degrees Fahrenheit (input) corresponds to exactly one degrees Celsius (output), this table of inputs and outputs does describe a function. Also notice that each output corresponds to a different input. This type of function is given a special name—a one-to-one function.

Does the set $f = \{(0, 1), (2, 2), (-3, 5), (7, 6)\}$ describe a one-to-one function? It is a function since each x-value corresponds to a unique y-value. For this particular function f, each y-value corresponds to a unique x-value. Thus, this function is also a **one-to-one function.**

ONE-TO-ONE FUNCTION

For a **one-to-one function,** each x-value (input) corresponds to only one y-value (output), and each y-value (output) corresponds to only one x-value (input).

Example 1 Determine whether each function described is one-to-one.

a. $f = \{(6, 2), (5, 4), (-1, 0), (7, 3)\}$

b. $g = \{(3, 9), (-4, 2), (-3, 9), (0, 0)\}$

c. $h = \{(1, 1), (2, 2), (10, 10), (-5, -5)\}$

d.

MINERAL (INPUT)	Talc	Gypsum	Diamond	Topaz	Stibnite
HARDNESS ON THE MOHS SCALE (OUTPUT)	1	2	10	8	2

e.

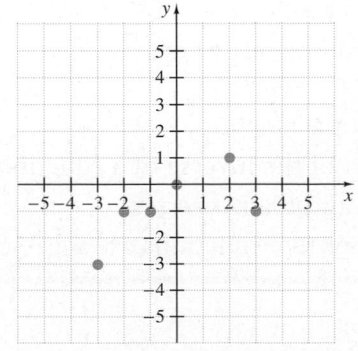

Solution **a.** f is one-to-one since each y-value corresponds to only one x-value.

b. g is not one-to-one because the y-value 9 in $(3, 9)$ and $(-3, 9)$ corresponds to two different x-values.

c. *h* is a one-to-one function since each *y*-value corresponds to only one *x*-value.

d. This table does not describe a one-to-one function since the output 2 corresponds to two different inputs, gypsum and stibnite.

e. This graph does not describe a one-to-one function since the *y*-value −1 corresponds to three different *x*-values, −2, −1, and 3.

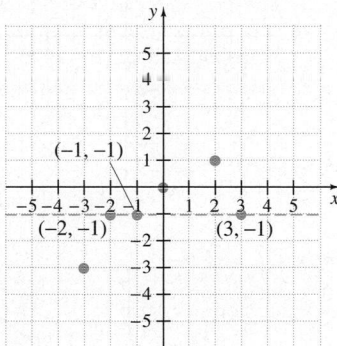

2 Recall that we recognize the graph of a function when it passes the vertical line test. Since every *x*-value of the function corresponds to exactly one *y*-value, each vertical line intersects the function's graph at most once. The graph shown next, for instance, is the graph of a function.

Is this function a *one-to-one* function? The answer is no. To see why not, notice that the *y*-value of the ordered pair (−3, 3), for example, is the same as the *y*-value of the ordered pair (3, 3). This function is therefore not one-to-one.

To test whether a graph is the graph of a one-to-one function, apply the vertical line test to see if it is a function, and then apply a similar **horizontal line test** to see if it is a one-to-one function.

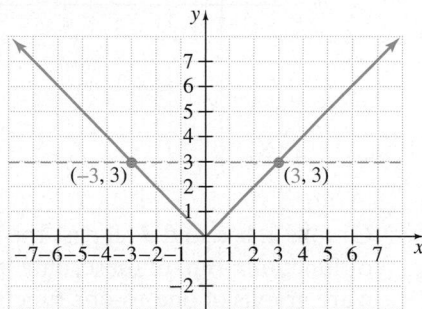

HORIZONTAL LINE TEST

If every horizontal line intersects the graph of a function at most once, then the function is a one-to-one function.

Example 2   Determine whether each graph is the graph of a one-to-one function.

a.

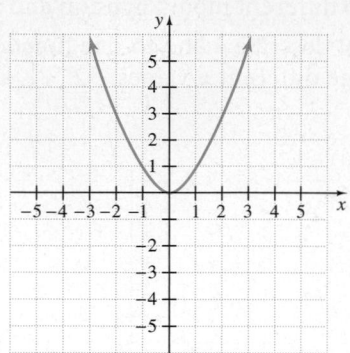

b.

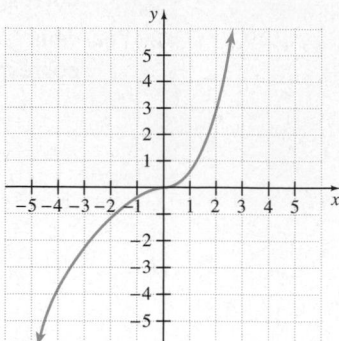

c.

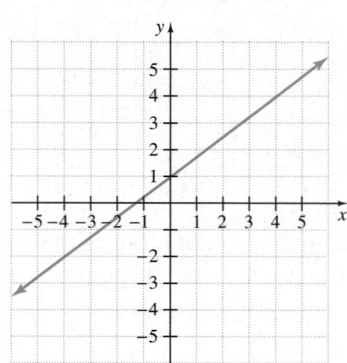

d.

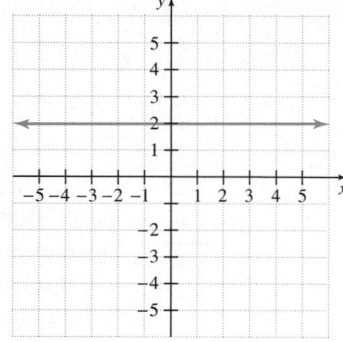

e.

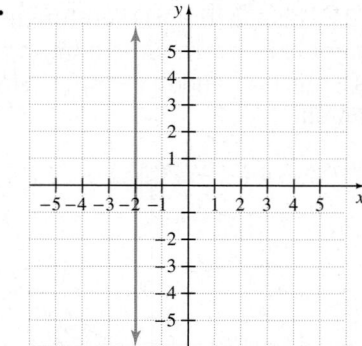

Solution   Graphs **a**, **b**, **c**, and **d** all pass the vertical line test, so only these graphs are graphs of functions. But, of these, only **b** and **c** pass the horizontal line test, so only **b** and **c** are graphs of one-to-one functions.

> **HELPFUL HINT**
> All linear equations are one-to-one functions except those whose graphs are horizontal or vertical lines. A vertical line does not pass the vertical line test and hence is not the graph of a function. A horizontal line is the graph of a function but does not pass the horizontal line test and hence is not the graph of a one-to-one function.

3 One-to-one functions are special in that their graphs pass both the vertical and horizontal line tests. They are special, too, in another sense: For each one-to-one function, we can find its **inverse function** by switching the coordinates of the ordered pairs of the function, or the inputs and the outputs. For example, the inverse of the one-to-one function

Degrees Fahrenheit (Input)	-31	-13	32	68	149	212
Degrees Celsius (Output)	-35	-25	0	20	65	100

is the function

Degrees Celsius (Input)	-35	-25	0	20	65	100
Degrees Fahrenheit (Output)	-31	-13	32	68	149	212

Notice that the ordered pair $(-31, -35)$ of the function, for example, becomes the ordered pair $(-35, -31)$ of its inverse.

Also, the inverse of the one-to-one function $f = \{(2, -3), (5, 10), (9, 1)\}$ is $\{(-3, 2), (10, 5), (1, 9)\}$. For a function f, we use the notation f^{-1}, read "f inverse," to denote its inverse function. Notice that since the coordinates of each ordered pair have been switched, the domain (set of inputs) of f is the range (set of outputs) of f^{-1}, and the range of f is the domain of f^{-1}. See the definition of inverse function.

INVERSE FUNCTION

The inverse of a one-to-one function f is the one-to-one function f^{-1} that consists of the set of all ordered pairs (y, x) where (x, y) belongs to f.

Example 3 Find the inverse of the one-to-one function.

$$f = \{(0, 1), (-2, 7), (3, -6), (4, 4)\}$$

Solution $f^{-1} = \{(1, 0), (7, -2), (-6, 3), (4, 4)\}$

    ↑   ↑   ↑   ↑   Switch coordinates
of each ordered pair.

HELPFUL HINT

The symbol f^{-1} is the single symbol used to denote the inverse of the function f. It is read as "f inverse." This symbol *does not mean* $\dfrac{1}{f}$.

4 If a one-to-one function f is defined as a set of ordered pairs, we can find f^{-1} by interchanging the x- and y-coordinates of the ordered pairs. If a one-to-one function f is given in the form of an equation, we can find f^{-1} by using a similar procedure.

FINDING THE INVERSE OF A ONE-TO-ONE FUNCTION $f(x)$

Step 1: Replace $f(x)$ with y.
Step 2: Interchange x and y.
Step 3: Solve the equation for y.
Step 4: Replace y with the notation $f^{-1}(x)$.

Example 4 Find an equation of the inverse of $f(x) = x + 3$.

Solution $f(x) = x + 3$

Step 1: $y = x + 3$ Replace $f(x)$ with y.
Step 2: $x = y + 3$ Interchange x and y.
Step 3: $x - 3 = y$ Solve for y.
Step 4: $f^{-1}(x) = x - 3$ Replace y with $f^{-1}(x)$.

The inverse of $f(x) = x + 3$ is $f^{-1}(x) = x - 3$. Notice that, for example,

$$f(1) = 1 + 3 = 4 \qquad \text{and} \qquad f^{-1}(4) = 4 - 3 = 1$$

Ordered pair: $(1, 4)$ Ordered pair: $(4, 1)$

The coordinates are
switched, as expected.

Example 5 Find the equation of the inverse of $f(x) = 3x - 5$. Graph f and f^{-1} on the same set of axes.

Solution $f(x) = 3x - 5$

Step 1: $y = 3x - 5$ Replace $f(x)$ with y.
Step 2: $x = 3y - 5$ Interchange x and y.
Step 3: $3y = x + 5$ Solve for y.
$$y = \frac{x + 5}{3}$$

Step 4: $f^{-1}(x) = \dfrac{x + 5}{3}$ Replace y with $f^{-1}(x)$.

Now we graph $f(x)$ and $f^{-1}(x)$ on the same set of axes. Both $f(x) = 3x - 5$ and

$f^{-1}(x) = \dfrac{x + 5}{3}$ are linear functions, so each graph is a line.

$$f(x) = 3x - 5 \qquad\qquad f^{-1}(x) = \dfrac{x + 5}{3}$$

x	$y = f(x)$
1	-2
0	-5
$\dfrac{5}{3}$	0

x	$y = f^{-1}(x)$
-2	1
-5	0
0	$\dfrac{5}{3}$

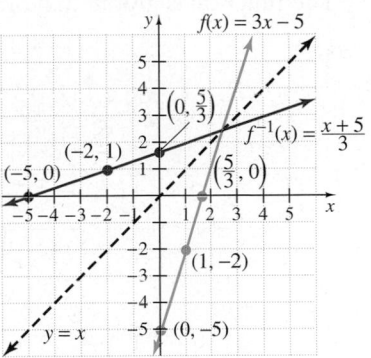

---

5   Notice that the graphs of f and f^{-1} in Example 5 are mirror images of each other, and the "mirror" is the dashed line $y = x$. This is true for every function and its inverse. For this reason, we say that *the graphs of f and f^{-1} are symmetric about the line $y = x$.*

To see why this happens, study the graph of a few ordered pairs and their switched coordinates.

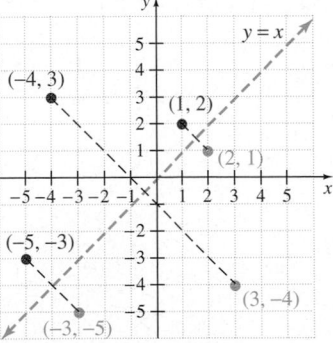

---

6   Notice also in the table of values in Example 5 that $f(0) = -5$ and $f^{-1}(-5) = 0$, as expected. Also, for example, $f(1) = -2$ and $f^{-1}(-2) = 1$. In words, we say that for some input x, the function f^{-1} takes the output of x, called $f(x)$, back to x.

$$x \rightarrow f(x) \quad \text{and} \quad f^{-1}(f(x)) \rightarrow x$$

$$f(0) = -5 \quad \text{and} \quad f^{-1}(-5) = 0$$
$$f(1) = -2 \quad \text{and} \quad f^{-1}(-2) = 1$$

In general,

If f is a one-to-one function, then the inverse of f is the function f^{-1} such that
$$(f^{-1} \circ f)(x) = x \quad \text{and} \quad (f \circ f^{-1})(x) = x$$

Example 6   Graph the inverse of each function.

Solution   The function is graphed in blue and the inverse is graphed in red.

a.

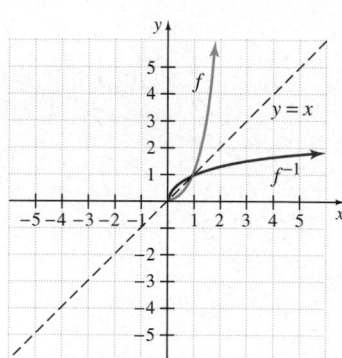

b.

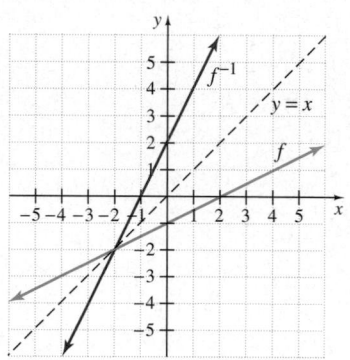

Example 7   Show that if $f(x) = 3x + 2$, then $f^{-1}(x) = \dfrac{x-2}{3}$.

Solution   See that $(f^{-1} \circ f)(x) = x$ and $(f \circ f^{-1})(x) = x$.

$$(f^{-1} \circ f)(x) = f^{-1}(f(x))$$

$$= f^{-1}(3x + 2) \qquad \text{Replace } f(x) \text{ with } 3x + 2.$$

$$= \frac{3x + 2 - 2}{3}$$

$$= \frac{3x}{3}$$

$$= x$$

$$(f \circ f^{-1})(x) = f(f^{-1}(x))$$

$$= f\left(\frac{x-2}{3}\right) \qquad \text{Replace } f^{-1}(x) \text{ with } \frac{x-2}{3}.$$

$$= 3\left(\frac{x-2}{3}\right) + 2$$

$$= x - 2 + 2$$

$$= x$$

GRAPHING CALCULATOR EXPLORATIONS

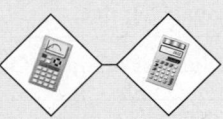

A graphing calculator can be used to visualize the results of Example 7. Recall that the graph of a function f and its inverse f^{-1} are mirror images of each other across the line $y = x$. To see this for the function from Example 7, use a square window and graph

the given function: $\quad Y_1 = 3x + 2$

its inverse: $\quad Y_2 = \dfrac{x - 2}{3}$

and the line: $\quad Y_3 = x$

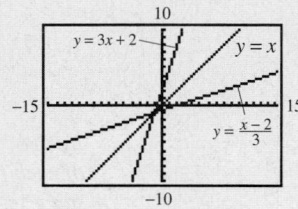

Exercises will follow in Exercise Set 11.2.

Exercise Set 11.2

Determine whether each function is a one-to-one function. If it is one-to-one, list the inverse function by switching coordinates, or inputs and outputs. See Examples 1 and 3.

1. $f = \{(-1, -1), (1, 1), (0, 2), (2, 0)\}$

2. $g = \{(8, 6), (9, 6), (3, 4), (-4, 4)\}$

3. $h = \{(10, 10)\}$

4. $r = \{(1, 2), (3, 4), (5, 6), (6, 7)\}$

5. $f = \{(11, 12), (4, 3), (3, 4), (6, 6)\}$

6. $g = \{(0, 3), (3, 7), (6, 7), (-2, -2)\}$

7.

Month of 1998 (Input)	January	February	March	April	May	June
Thousands of Houses on Sale at Month's End (Output)	282	277	281	285	282	287

(*Source:* U.S. Department of Housing and Urban Development)

8.

State (Input)	Washington	Ohio	Georgia	Colorado	California	Arizona
Electoral Votes (Output)	11	21	13	8	54	8

(*Source:* U.S. Bureau of the Census)

9.

State (Input)	California	Vermont	Virginia	Texas	South Dakota
Rank in Population (Output)	1	49	12	2	45

(*Source:* U.S. Bureau of the Census)

△ **10.**

Shape (Input)	Triangle	Pentagon	Quadrilateral	Hexagon	Decagon
Number of Sides (Output)	3	5	4	6	10

Given the one-to-one function $f(x) = x^3 + 2$, *find the following.* *[Hint: You do not need to find the equation for* $f^{-1}(x)$.]

11. a. $f(1)$ **12. a.** $f(0)$
   b. $f^{-1}(3)$ **b.** $f^{-1}(2)$

13. a. $f(-1)$ **14. a.** $f(-2)$
   b. $f^{-1}(1)$ **b.** $f^{-1}(-6)$

Determine whether the graph of each function is the graph of a one-to-one function. See Example 2.

15.

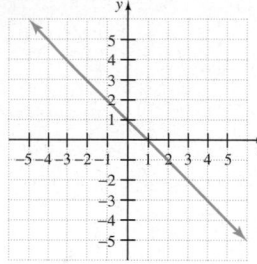

16.

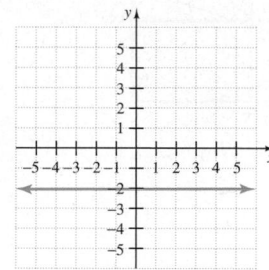

 17.

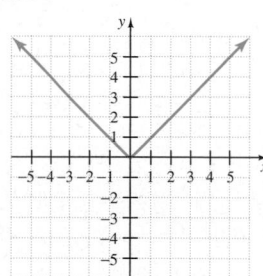

18.

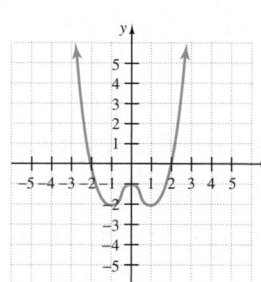

19.

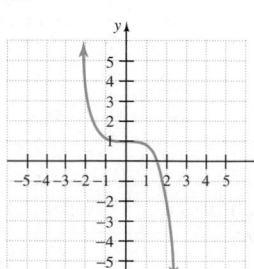

20.

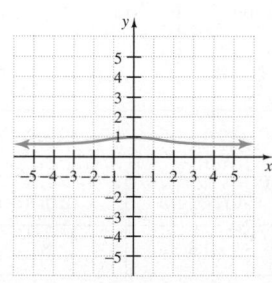

21.

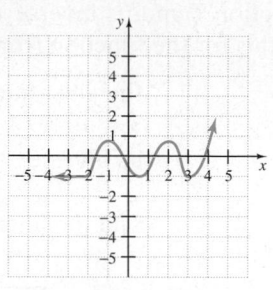

22.

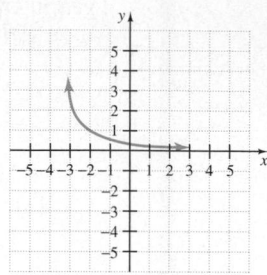

Each of the following functions is one-to-one. Find the inverse of each function and graph the function and its inverse on the same set of axes. See Examples 4 and 5.

23. $f(x) = x + 4$ **24.** $f(x) = x - 5$

◆ **25.** $f(x) = 2x - 3$ **26.** $f(x) = 4x + 9$

27. $f(x) = \dfrac{1}{2}x - 1$ **28.** $f(x) = -\dfrac{1}{2}x + 2$

29. $f(x) = x^3$ **30.** $f(x) = x^3 - 1$

Find the inverse of each one-to-one function. See Examples 4 and 5.

31. $f(x) = 5x + 2$ **32.** $f(x) = 6x - 1$

33. $f(x) = \dfrac{x - 2}{5}$ **34.** $f(x) = \dfrac{4x - 3}{2}$

35. $f(x) = \sqrt[3]{x}$ **36.** $f(x) = \sqrt[3]{x + 1}$

37. $f(x) = \dfrac{5}{3x + 1}$ **38.** $f(x) = \dfrac{7}{2x + 4}$

39. $f(x) = (x + 2)^3$ **40.** $f(x) = (x - 5)^3$

Graph the inverse of each function on the same set of axes. See Example 6.

41.

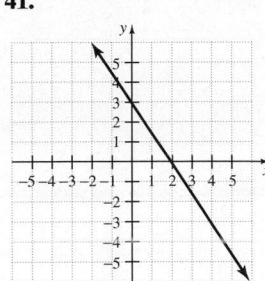

42.

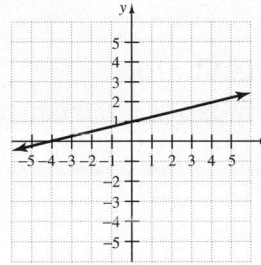

43.

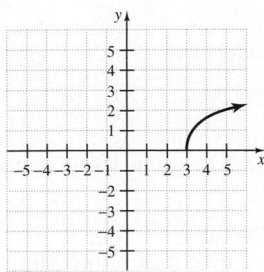

44.

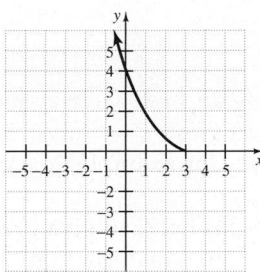

45.

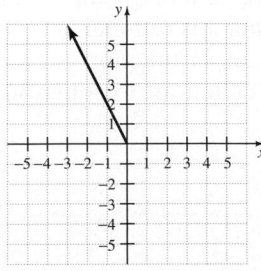

46.

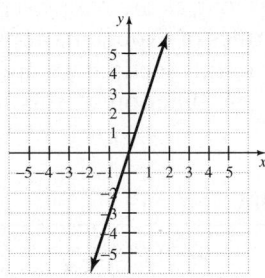

Solve. See Example 7.

47. If $f(x) = 2x + 1$, show that $f^{-1}(x) = \dfrac{x - 1}{2}$.

48. If $f(x) = 3x - 10$, show that $f^{-1}(x) = \dfrac{x + 10}{3}$.

49. If $f(x) = x^3 + 6$, show that $f^{-1}(x) = \sqrt[3]{x - 6}$.

50. If $f(x) = x^3 - 5$, show that $f^{-1}(x) = \sqrt[3]{x + 5}$.

For Exercises 51 and 52,

    a. Write the ordered pairs for $f(x)$ whose points are highlighted. (Include the points whose coordinates are given.)

    b. Write the corresponding ordered pairs for the inverse of f, f^{-1}.

    c. Graph the ordered pairs for f^{-1} found in Part **b.**

    d. Graph $f^{-1}(x)$ by drawing a smooth curve through the plotted points.

51.

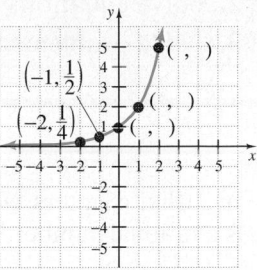

52.

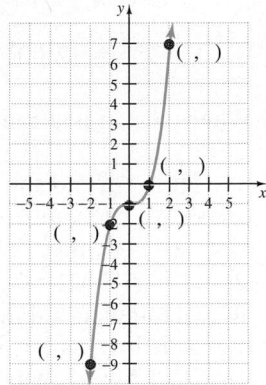

 *Find the inverse of each given one-to-one function. Then graph the function and its inverse on a square window.*

53. $f(x) = 3x + 1$ **54.** $f(x) = -2x - 6$

55. $f(x) = \sqrt[3]{x + 1}$ **56.** $f(x) = x^3 - 3$

REVIEW EXERCISES

Evaluate each of the following. See Section 9.2.

57. $25^{1/2}$ **58.** $49^{1/2}$

59. $16^{3/4}$ **60.** $27^{2/3}$

61. $9^{-3/2}$ **62.** $81^{-3/4}$

If $f(x) = 3^x$, find the following. In Exercises 65 and 66, give an exact answer and a two-decimal-place approximation. See Sections 3.3 and 9.2.

63. $f(2)$ **64.** $f(0)$

65. $f\left(\frac{1}{2}\right)$ **66.** $f\left(\frac{2}{3}\right)$

11.3 EXPONENTIAL FUNCTIONS

CD-ROM   SSM

SSG   Video

▶ **O B J E C T I V E S**

1. Graph exponential functions.
2. Solve equations of the form $b^x = b^y$.
3. Solve problems modeled by exponential equations.

1

In earlier chapters, we gave meaning to exponential expressions such as 2^x, where x is a rational number. For example,

$$2^3 = 2 \cdot 2 \cdot 2 \qquad \text{three factors, each factor is 2}$$

$$2^{3/2} = (2^{1/2})^3 = \sqrt{2} \cdot \sqrt{2} \cdot \sqrt{2} \qquad \text{three factors, each factor is } \sqrt{2}$$

When x is an irrational number (for example, $\sqrt{3}$), what meaning can we give to $2^{\sqrt{3}}$?

It is beyond the scope of this book to give precise meaning to 2^x if x is irrational. We can confirm your intuition and say that $2^{\sqrt{3}}$ is a real number, and since $1 < \sqrt{3} < 2$, then $2^1 < 2^{\sqrt{3}} < 2^2$. We can also use a calculator and approximate $2^{\sqrt{3}}$: $2^{\sqrt{3}} \approx 3.321997$. In fact, as long as the base b is positive, b^x is a real number for all real numbers x. Finally, the rules of exponents apply whether x is rational or irrational, as long as b is positive. In this section, we are interested in functions of the form $f(x) = b^x$, where $b > 0$. A function of this form is called an **exponential function.**

EXPONENTIAL FUNCTION

A function of the form

$$f(x) = b^x$$

is called an **exponential function** if $b > 0$, b is not 1, and x is a real number.

Next, we practice graphing exponential functions.

◆ **Example 1** Graph the exponential functions defined by $f(x) = 2^x$ and $g(x) = 3^x$ on the same set of axes.

Solution Graph each function by plotting points. Set up a table of values for each of the two functions.

$f(x) = 2^x$

x	0	1	2	3	-1	-2
$f(x)$	1	2	4	8	$\dfrac{1}{2}$	$\dfrac{1}{4}$

$g(x) = 3^x$

x	0	1	2	3	−1	−2
$g(x)$	1	3	9	27	$\dfrac{1}{3}$	$\dfrac{1}{9}$

If each set of points is plotted and connected with a smooth curve, the following graphs result.

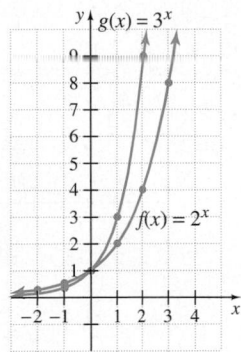

A number of things should be noted about the two graphs of exponential functions in Example 1. First, the graphs show that $f(x) = 2^x$ and $g(x) = 3^x$ are one-to-one functions since each graph passes the vertical and horizontal line tests. The y-intercept of each graph is $(0, 1)$, but neither graph has an x-intercept. From the graph, we can also see that the domain of each function is all real numbers and that the range is $(0, \infty)$. We can also see that as x-values are increasing, y-values are increasing also.

Example 2 Graph the exponential functions $y = \left(\dfrac{1}{2}\right)^x$ and $y = \left(\dfrac{1}{3}\right)^x$ on the same set of axes.

Solution As before, plot points and connect them with a smooth curve.

$y = \left(\dfrac{1}{2}\right)^x$

x	0	1	2	3	−1	−2
y	1	$\dfrac{1}{2}$	$\dfrac{1}{4}$	$\dfrac{1}{8}$	2	4

$y = \left(\dfrac{1}{3}\right)^x$

x	0	1	2	3	−1	−2
y	1	$\dfrac{1}{3}$	$\dfrac{1}{9}$	$\dfrac{1}{27}$	3	9

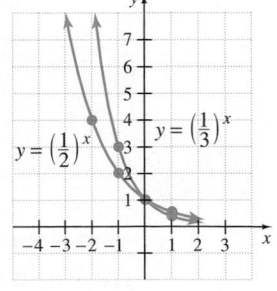

Each function in Example 2 again is a one-to-one function. The y-intercept of both is $(0, 1)$. The domain is the set of all real numbers, and the range is $(0, \infty)$.

Notice the difference between the graphs of Example 1 and the graphs of Example 2. An exponential function is always increasing if the base is greater than 1.

When the base is between 0 and 1, the graph is always decreasing. The following figures summarize these characteristics of exponential functions.

$$f(x) = b^x, \quad b > 0, \quad b \neq 1$$

- one-to-one function
- y-intercept $(0, 1)$
- no x-intercept
- domain: $(-\infty, \infty)$
- range: $(0, \infty)$

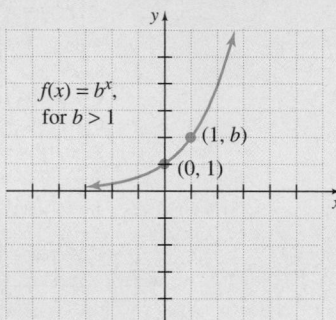

$f(x) = b^x$, for $b > 1$

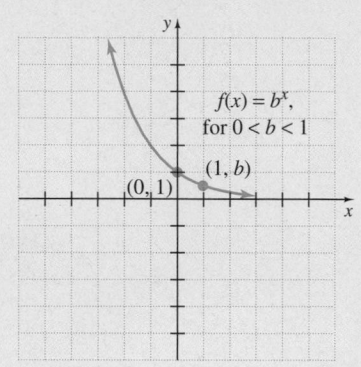

$f(x) = b^x$, for $0 < b < 1$

Example 3 Graph the exponential function $f(x) = 3^{x+2}$.

Solution As before, we find and plot a few ordered pair solutions. Then we connect the points with a smooth curve.

$y = 3^{x+2}$

x	0	-1	-2	-3	-4
y	9	3	1	$\dfrac{1}{3}$	$\dfrac{1}{9}$

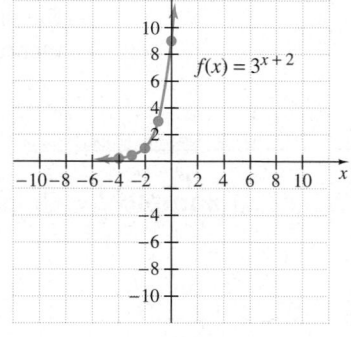

$f(x) = 3^{x+2}$

---

2 We have seen that an exponential function $y = b^x$ is a one-to-one function. Another way of stating this fact is a property that we can use to solve exponential equations.

UNIQUENESS OF b^x

Let $b > 0$ and $b \neq 1$. Then $b^x = b^y$ is equivalent to $x = y$.

Example 4 Solve each equation for x.

   a. $2^x = 16$ **b.** $9^x = 27$ **c.** $4^{x+3} = 8^x$

Solution **a.** We write 16 as a power of 2 and then use the uniqueness of b^x to solve.

$$2^x = 16$$
$$2^x = 2^4$$

Since the bases are the same and are nonnegative, by the uniqueness of b^x, we then have that the exponents are equal. Thus,

$$x = 4$$

The solution is 4.

b. Notice that both 9 and 27 are powers of 3.

$$9^x = 27$$
$$(3^2)^x = 3^3 \qquad \textit{Write 9 and 27 as powers of 3.}$$
$$3^{2x} = 3^3$$
$$2x = 3 \qquad \textit{Apply the uniqueness of } b^x.$$
$$x = \frac{3}{2} \qquad \textit{Divide by 2.}$$

To check, replace x with $\frac{3}{2}$ in the original expression, $9^x = 27$. The solution is $\frac{3}{2}$.

c. Write both 4 and 8 as powers of 2.

$$4^{x+3} = 8^x$$
$$(2^2)^{x+3} = (2^3)^x$$
$$2^{2x+6} = 2^{3x}$$
$$2x + 6 = 3x \qquad \textit{Apply the uniqueness of } b^x.$$
$$6 = x \qquad \textit{Subtract 2x from both sides.}$$

The solution is 6.

There is one major problem with the preceding technique. Often the two sides of an equation cannot easily be written as powers of a common base. We explore how to solve an equation such as $4 = 3^x$ with the help of **logarithms** later.

3 The bar graph here shows the increase in the number of cellular phone users. Notice that the graph of the exponential function $y = 6.052(1.378)^x$ approximates the heights of the bars.

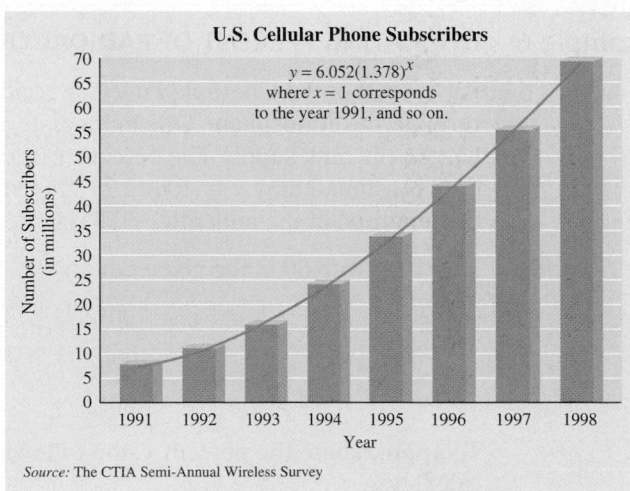

U.S. Cellular Phone Subscribers

$y = 6.052(1.378)^x$
where $x = 1$ corresponds
to the year 1991, and so on.

Number of Subscribers (in millions)

Year

Source: The CTIA Semi-Annual Wireless Survey

The graph above shows just one example of how the world abounds with patterns that can be modeled by exponential functions. To make these applications realistic, we use numbers that warrant a calculator. Another application of an exponential function has to do with interest rates on loans.

The exponential function defined by $A = P\left(1 + \dfrac{r}{n}\right)^{nt}$ models the dollars A accrued (or owed) after P dollars are invested (or loaned) at an annual rate of interest r compounded n times each year for t years. This function is known as the compound interest formula.

Example 5 USING THE COMPOUND INTEREST FORMULA

Find the amount owed at the end of 5 years if $1600 is loaned at a rate of 9% compounded monthly.

Solution We use the formula $A = P\left(1 + \dfrac{r}{n}\right)^{nt}$, with the following values.

$$P = 1600 \text{ (the amount of the loan)}$$
$$r = 9\% = 0.09 \text{ (the annual rate of interest)}$$
$$n = 12 \text{ (the number of times interest is compounded each year)}$$
$$t = 5 \text{ (the duration of the loan, in years)}$$

$$A = P\left(1 + \dfrac{r}{n}\right)^{nt} \qquad \text{compound interest formula}$$
$$= 1600\left(1 + \dfrac{0.09}{12}\right)^{12(5)} \qquad \text{Substitute known values.}$$
$$= 1600(1.0075)^{60}$$

To approximate A, use the $\boxed{y^x}$ or $\boxed{\wedge}$ key on your calculator.

$$\boxed{2505.0896}$$

Thus, the amount A owed is $2505.09.

Example 6 ESTIMATING PERCENT OF RADIOACTIVE MATERIAL

As a result of the Chernobyl nuclear accident, radioactive debris was carried through the atmosphere. One immediate concern was the impact that the debris had on the milk supply. The percent y of radioactive material in raw milk after t days is estimated by $y = 100(2.7)^{-0.1t}$. Estimate the expected percent of radioactive material in the milk after 30 days.

Solution Replace t with 30 in the given equation.

$$y = 100(2.7)^{-0.1t}$$
$$= 100(2.7)^{-0.1(30)} \qquad \text{Let } t = 30.$$
$$= 100(2.7)^{-3}$$

To approximate the percent y, the following keystrokes may be used on a scientific calculator.

$\boxed{2.7}\;\boxed{y^x}\;\boxed{3}\;\boxed{+/-}\;\boxed{=}\;\boxed{\times}\;\boxed{100}\;\boxed{=}$

or

$\boxed{2.7}\;\boxed{\wedge}\;\boxed{(-)}\;\boxed{3}\;\boxed{\times}\;\boxed{100}\;\boxed{\text{ENTER}}$

The display should read

$$5.0805263$$

Thus, approximately 5% of the radioactive material still remained in the milk supply after 30 days. ▬

GRAPHING CALCULATOR EXPLORATIONS

We can use a graphing calculator and its TRACE feature to solve Example 6 graphically.

To estimate the expected percent of radioactive material in the milk after 30 days, enter $Y_1 = 100(2.7)^{-0.1x}$. (The variable t in Example 6 is changed to x here to better accommodate our work on the graphing calculator.) The graph does not appear on a standard viewing window, so we need to determine an appropriate viewing window. Because it doesn't make sense to look at radioactivity *before* the Chernobyl nuclear accident, we use Xmin = 0. We are interested in finding the percent of radioactive material in the milk when $x = 30$, so we choose Xmax = 35 to leave enough space to see the graph at $x = 30$. Because the values of y are percents, it seems appropriate that $0 \le y \le 100$. (We also use Xscl = 1 and Yscl = 10.) Now we graph the function.

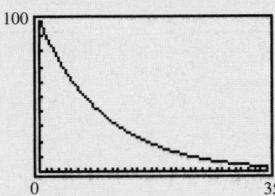

We can use the TRACE feature to obtain an approximation of the expected percent of radioactive material in the milk when $x = 30$. (A TABLE feature may also be used to approximate the percent.) To obtain a better approximation, let's use the ZOOM feature several times to zoom in near $x = 30$.

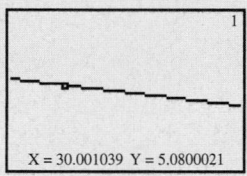

The percent of radioactive material in the milk 30 days after the Chernobyl accident was 5.08%, accurate to two decimal places.

Use a graphing calculator to find each percent. Approximate your solutions so that they are accurate to two decimal places.

1. Estimate the expected percent of radioactive material in the milk 2 days after the Chernobyl nuclear accident.

2. Estimate the expected percent of radioactive material in the milk 10 days after the Chernobyl nuclear accident.

3. Estimate the expected percent of radioactive material in the milk 15 days after the Chernobyl nuclear accident.

4. Estimate the expected percent of radioactive material in the milk 25 days after the Chernobyl nuclear accident.

Exercise Set 11.3

Graph each exponential function. See Examples 1 through 3.

1. $y = 4^x$

2. $y = 5^x$

3. $y = 2^x + 1$

4. $y = 3^x - 1$

5. $y = \left(\dfrac{1}{4}\right)^x$

6. $y = \left(\dfrac{1}{5}\right)^x$

7. $y = \left(\dfrac{1}{2}\right)^x - 2$

8. $y = \left(\dfrac{1}{3}\right)^x + 2$

9. $y = -2^x$

10. $y = -3^x$

11. $y = -\left(\dfrac{1}{4}\right)^x$

12. $y = -\left(\dfrac{1}{5}\right)^x$

13. $f(x) = 2^{x+1}$

14. $f(x) = 3^{x-1}$

15. $f(x) = 4^{x-2}$

16. $f(x) = 2^{x+3}$

17. Explain why the graph of an exponential function $y = b^x$ contains the point $(1, b)$.

18. Explain why an exponential function $y = b^x$ has a y-intercept of $(0, 1)$.

Solve each equation for x. See Example 4.

19. $3^x = 27$

20. $6^x = 36$

21. $16^x = 8$

22. $64^x = 16$

23. $32^{2x-3} = 2$

24. $9^{2x+1} = 81$

25. $\dfrac{1}{4} = 2^{3x}$

26. $\dfrac{1}{27} = 3^{2x}$

27. $5^x = 625$

28. $2^x = 64$

29. $4^x = 8$

30. $32^x = 4$

31. $27^{x+1} = 9$

32. $125^{x-2} = 25$

33. $81^{x-1} = 27^{2x}$

34. $4^{3x-7} = 32^{2x}$

Match each exponential equation with its graph.

35. $f(x) = \left(\dfrac{1}{2}\right)^x$

36. $f(x) = 2^x$

37. $f(x) = \left(\dfrac{1}{4}\right)^x$

38. $f(x) = 3^x$

A

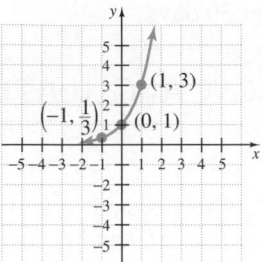

B

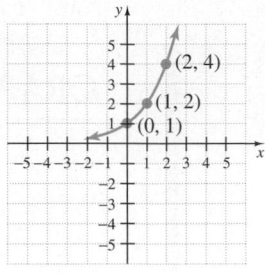

C

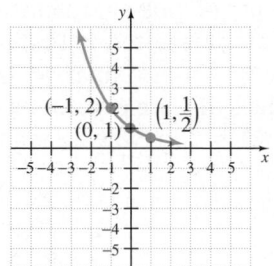

D

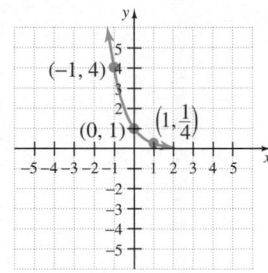

Solve. Unless otherwise indicated, round results to one decimal place. See Example 6.

39. One type of uranium has a daily radioactive decay rate of 0.4%. If 30 pounds of this uranium is available today, find how much will still remain after 50 days. Use $y = 30(2.7)^{-0.004t}$, and let t be 50.

40. The nuclear waste from an atomic energy plant decays at a rate of 3% each century. If 150 pounds of nuclear waste is disposed of, find how much of it will still remain after 10 centuries. Use $y = 150(2.7)^{-0.03t}$, and let t be 10.

41. The size of the rat population of a wharf area grows at a rate of 8% monthly. If there are 200 rats in January, find how many rats (rounded to the nearest whole) should be expected by next January. Use $y = 200(2.7)^{0.08t}$.

42. National Park Service personnel are trying to increase the size of the bison population of Theodore Roosevelt National Park. If 260 bison currently live in the park, and if the population's rate of growth is 2.5% annually, find how many bison (rounded to the nearest whole) there should be in 10 years. Use $y = 260(2.7)^{0.025t}$.

43. A rare isotope of a nuclear material is very unstable, decaying at a rate of 15% each second. Find how much isotope remains 10 seconds after 5 grams of the isotope is created. Use $y = 5(2.7)^{-0.15t}$.

44. An accidental spill of 75 grams of radioactive material in a local stream has led to the presence of radioactive debris decaying at a rate of 4% each day. Find how much debris still remains after 14 days. Use $y = 75(2.7)^{-0.04t}$.

45. Mexico City is growing at a rate of 0.7% annually. If there were 15,525,000 residents of Mexico City in 1994, find how many (to the nearest ten-thousand) are living in the city in 2000. Use $y = 15,525,000(2.7)^{0.007t}$.

46. An unusually wet spring has caused the size of the Cape Cod mosquito population to increase by 8% each day. If an estimated 200,000 mosquitoes are on Cape Cod on May 12, find how many thousands of mosquitoes will inhabit the Cape on May 25. Use $y = 200,000(2.7)^{0.08t}$.

Solve. Use $A = P\left(1 + \dfrac{r}{n}\right)^{nt}$. Round answers to two decimal places. See Example 5.

47. Find the amount Erica owes at the end of 3 years if $6000 is loaned to her at a rate of 8% compounded monthly.

48. Find the amount owed at the end of 5 years if $3000 is loaned at a rate of 10% compounded quarterly.

49. Find the total amount Janina has in a college savings account if $2000 was invested and earned 6% compounded semiannually for 12 years.

50. Find the amount accrued if $500 is invested and earns 7% compounded monthly for 4 years.

Use a graphing calculator to solve. Estimate each result to two decimal places.

51. Verify the results of Exercise 39.

52. From Exercise 39, estimate the number of pounds of uranium that will be available after 100 days.

53. From Exercise 39, estimate the number of pounds of uranium that will be available after 120 days.

54. Verify the results of Exercise 44.

55. From Exercise 44, estimate the amount of debris that remains after 10 days.

56. From Exercise 44, estimate the amount of debris that remains after 20 days.

Solve.

57. The world population is currently growing at a rate of 1.32% annually. In 1998, the midyear population of the world was 5,926,466,814 people. Predict the midyear world population (to the nearest million) in 2005. Use $y = 5,926,466,814(2.7)^{0.0132t}$, where t is the number of years after 1998. (*Source:* Based on data from the U.S. Bureau of the Census, International Data Base)

58. Retail revenue from shopping on the Internet is expected to grow at a rate of 64% per year. In 1997, a total of $2.4 billion in revenue was collected through Internet retail sales. To make the following predictions, use $y = 2.4(1.64)^t$, where t is the number of years after 1997.

(*Source:* Based on data from Forrester Research Inc.)

a. What level of retail revenues from Internet shopping is expected in 2001?

b. Predict the level of Internet shopping revenues in 2010.

59. Carbon dioxide (CO_2) is a greenhouse gas that contributes to global warming. Due to the combustion of fossil fuels, the amount of CO_2 in Earth's atmosphere has been increasing by 0.4% annually over the past century. In 1994, the concentration of CO_2 in the atmosphere was 358 parts per million by volume. To make the following predictions, use $y = 358(1.004)^t$, where t is the number of years after 1994. (*Source:* Based on data from the United Nations Environment Programme's Information Unit for Conventions)

a. Predict the concentration of CO_2 in the atmosphere in the year 2004.

b. Predict the concentration of CO_2 in the atmosphere in the year 2025.

The formula $y = 6.052(1.378)^x$ gives the number of cellular phone users y (in millions) in the United States for the years 1991 through 1998. In this formula, $x = 0$ corresponds to 1991, $x = 1$ corresponds to 1992, and so on. Use this formula to solve exercises 60 and 61.

60. Use this model to predict the number of cellular phone users in the year 2005.

61. Use this model to predict the number of cellular phone users in the year 2008.

REVIEW EXERCISES

Solve each equation. See Sections 2.2 and 6.6.

62. $5x - 2 = 18$

63. $3x - 7 = 11$

64. $3x - 4 = 3(x + 1)$

65. $2 - 6x = 6(1 - x)$

66. $x^2 + 6 = 5x$

67. $18 = 11x - x^2$

By inspection, find the value for x that makes each statement true.

68. $2^x = 8$

69. $3^x = 9$

70. $5^x = \dfrac{1}{5}$

71. $4^x = 1$

11.4 LOGARITHMIC FUNCTIONS

▶ **OBJECTIVES**

CD-ROM SSM

SSG Video

1. Write exponential equations with logarithmic notation and write logarithmic equations with exponential notation.
2. Solve logarithmic equations by using exponential notation.
3. Identify and graph logarithmic functions.

1 Since the exponential function $f(x) = 2^x$ is a one-to-one function, it has an inverse. We can create a table of values for f^{-1} by switching the coordinates in the accompanying table of values for $f(x) = 2^x$.

x	$y = f(x)$
-3	$\frac{1}{8}$
-2	$\frac{1}{4}$
-1	$\frac{1}{2}$
0	1
1	2
2	4
3	8

x	$y = f^{-1}(x)$
$\frac{1}{8}$	-3
$\frac{1}{4}$	-2
$\frac{1}{2}$	-1
1	0
2	1
4	2
8	3

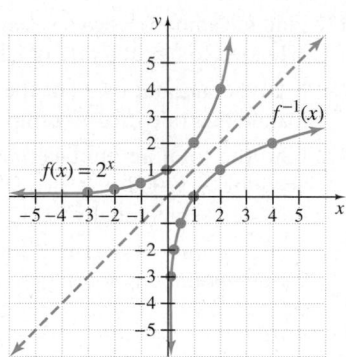

The graphs of $f(x)$ and its inverse are shown above. Notice that the graphs of f and f^{-1} are symmetric about the line $y = x$, as expected.

Now we would like to be able to write an equation for f^{-1}. To do so, we follow the steps for finding an inverse.

$$f(x) = 2^x$$

Step 1: Replace $f(x)$ by y. $y = 2^x$

Step 2: Interchange x and y. $x = 2^y$

Step 3: Solve for y.

At this point, we are stuck. To solve this equation for y, a new notation, the **logarithmic notation,** is needed. The symbol $\log_b x$ means "the power to which b is raised in order to produce a result of x."

$$\log_b x = y \quad \text{means} \quad b^y = x$$

We say that $\log_b x$ is "the logarithm of x to the base b" or "the log of x to the base b."

LOGARITHMIC DEFINITION

If $b > 0$ and $b \neq 1$, then

$$y = \log_b x \text{ means } x = b^y$$

for every $x > 0$ and every real number y.

Before returning to the function $x = 2^y$ and solving it for y in terms of x, let's practice using the new notation $\log_b x$.

It is important to be able to write exponential equations with logarithmic notation, and vice versa. The following table shows examples of both forms.

Logarithmic Equation	Corresponding Exponential Equation
$\log_3 9 = 2$	$3^2 = 9$
$\log_6 1 = 0$	$6^0 = 1$
$\log_2 8 = 3$	$2^3 = 8$
$\log_4 \dfrac{1}{16} = -2$	$4^{-2} = \dfrac{1}{16}$
$\log_8 2 = \dfrac{1}{3}$	$8^{1/3} = 2$

> **HELPFUL HINT**
> Notice that a *logarithm* is an *exponent*. In other words, $\log_3 9$ is the *power* that we raise 3 to in order to get 9.

Example 1   Write as an exponential equation.

    a. $\log_5 25 = 2$ **b.** $\log_6 \dfrac{1}{6} = -1$ **c.** $\log_2 \sqrt{2} = \dfrac{1}{2}$

Solution   **a.** $\log_5 25 = 2$ means $5^2 = 25$

    b. $\log_6 \dfrac{1}{6} = -1$ means $6^{-1} = \dfrac{1}{6}$

    c. $\log_2 \sqrt{2} = \dfrac{1}{2}$ means $2^{1/2} = \sqrt{2}$

Example 2   Write as a logarithmic equation.

    a. $9^3 = 729$ **b.** $6^{-2} = \dfrac{1}{36}$ **c.** $5^{1/3} = \sqrt[3]{5}$

Solution   **a.** $9^3 = 729$ means $\log_9 729 = 3$

    b. $6^{-2} = \dfrac{1}{36}$ means $\log_6 \dfrac{1}{36} = -2$

    c. $5^{1/3} = \sqrt[3]{5}$ means $\log_5 \sqrt[3]{5} = \dfrac{1}{3}$

Example 3   Find the value of each logarithmic expression.

    a. $\log_4 16$ **b.** $\log_{10} \dfrac{1}{10}$ **c.** $\log_9 3$

Solution   **a.** $\log_4 16 = 2$ because $4^2 = 16$ **b.** $\log_{10} \dfrac{1}{10} = -1$ because $10^{-1} = \dfrac{1}{10}$

    c. $\log_9 3 = \dfrac{1}{2}$ because $9^{1/2} = \sqrt{9} = 3$

> **HELPFUL HINT**
> Another method for evaluating logarithms such as those in Example 3 is to set the expression equal to x and then write them in exponential form to find x. For example:
>
> **a.** $\log_4 16 = x$ means $4^x = 16$. Since $4^2 = 16$, $x = 2$ or $\log_4 16 = 2$.
>
> **b.** $\log_{10} \frac{1}{10} = x$ means $10^x = \frac{1}{10}$. Since $10^{-1} = \frac{1}{10}$, $x = -1$ or $\log_{10} \frac{1}{10} = -1$.
>
> **c.** $\log_9 3 = x$ means $9^x = 3$. Since $9^{1/2} = 3$, $x = \frac{1}{2}$ or $\log_9 3 = \frac{1}{2}$.

2   The ability to interchange the logarithmic and exponential forms of a statement is often the key to solving logarithmic equations.

Example 4   Solve each equation for x.

     a. $\log_4 \frac{1}{4} = x$ **b.** $\log_5 x = 3$ **c.** $\log_x 25 = 2$

     d. $\log_3 1 = x$ **e.** $\log_b 1 = x$

Solution   **a.** $\log_4 \frac{1}{4} = x$ means $4^x = \frac{1}{4}$. Solve $4^x = \frac{1}{4}$ for x.

$$4^x = \frac{1}{4}$$
$$4^x = 4^{-1}$$

Since the bases are the same, by the uniqueness of b^x, we have that
$$x = -1$$

The solution is -1. To check, see that $\log_4 \frac{1}{4} = -1$, since $4^{-1} = \frac{1}{4}$.

b. $\log_5 x = 3$ means $5^3 = x$ or
$$x = 125$$
The solution is 125.

c. $\log_x 25 = 2$ means $x^2 = 25$ and $x > 0$ and $x \neq 1$.
$$x = 5$$
Even though $(-5)^2 = 25$, the base b of a logarithm must be positive. The solution is 5.

d. $\log_3 1 = x$ means $3^x = 1$. Either solve this equation by inspection or solve by writing 1 as 3^0 as shown.

$$3^x = 3^0 \qquad \text{Write 1 as } 3^0.$$
$$x = 0 \qquad \text{Apply the uniqueness of } b^x.$$

The solution is 0.

e. $\log_b 1 = x$ means $b^x = 1$ and $b > 0$ and $b \neq 1$.

$$b^x = b^0 \qquad \text{Write 1 as } b^0.$$

$$x = 0 \qquad \text{Apply the uniqueness of } b^x.$$

The solution is 0.

In Example 4e we proved an important property of logarithms. That is, $\log_b 1$ is always 0. This property as well as two important others are given next.

PROPERTIES OF LOGARITHMS

If b is a real number, $b > 0$, and $b \neq 1$, then

1. $\log_b 1 = 0$ 2. $\log_b b^x = x$ 3. $b^{\log_b x} = x$

To see that $\log_b b^x = x$, change the logarithmic form to exponential form. Then, $\log_b b^x = x$ means $b^x = b^x$. In exponential form, the statement is true, so in logarithmic form, the statement is also true.

Example 5 Simplify.

a. $\log_3 3^2$ **b.** $\log_7 7^{-1}$ **c.** $5^{\log_5 3}$ **d.** $2^{\log_2 6}$

Solution   **a.** From Property 2, $\log_3 3^2 = 2$.
b. From Property 2, $\log_7 7^{-1} = -1$.
c. From Property 3, $5^{\log_5 3} = 3$.
d. From Property 3, $2^{\log_2 6} = 6$.

3 Let us now return to the function $f(x) = 2^x$ and write an equation for its inverse, $f^{-1}(x)$. Recall our earlier work.

$$f(x) = 2^x$$

Step 1: Replace $f(x)$ by y. $y = 2^x$
Step 2: Interchange x and y. $x = 2^y$

Having gained proficiency with the notation $\log_b x$, we can now complete the steps for writing the inverse equation.

Step 3: Solve for y. $y = \log_2 x$
Step 4: Replace y with $f^{-1}(x)$. $f^{-1}(x) = \log_2 x$

Thus, $f^{-1}(x) = \log_2 x$ defines a function that is the inverse function of the function $f(x) = 2^x$. The function $f^{-1}(x)$ or $y = \log_2 x$ is called a **logarithmic function.**

LOGARITHMIC FUNCTION

If x is a positive real number, b is a constant positive real number, and b is not 1, then a **logarithmic function** is a function that can be defined by

$$f(x) = \log_b x$$

The domain of f is the set of positive real numbers, and the range of f is the set of real numbers.

We can explore logarithmic functions by graphing them.

Example 6   Graph the logarithmic function $y = \log_2 x$.

Solution   First we write the equation with exponential notation as $2^y = x$. Then we find some ordered pair solutions that satisfy this equation. Finally, we plot the points and connect them with a smooth curve. The domain of this function is $(0, \infty)$, and the range is all real numbers.

Since $x = 2^y$ is solved for x, we choose y-values and compute corresponding x-values.

If $y = 0$, $x = 2^0 = 1$
If $y = 1$, $x = 2^1 = 2$
If $y = 2$, $x = 2^2 = 4$
If $y = -1$, $x = 2^{-1} = \dfrac{1}{2}$

$x = 2^y$	y
1	0
2	1
4	2
$\dfrac{1}{2}$	-1

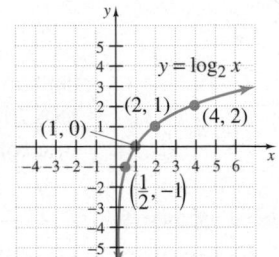

Example 7   Graph the logarithmic function $f(x) = \log_{1/3} x$.

Solution   Replace $f(x)$ with y, and write the result with exponential notation.

$$f(x) = \log_{1/3} x$$

$$y = \log_{1/3} x \qquad \text{Replace } f(x) \text{ with } y.$$

$$\left(\frac{1}{3}\right)^y = x \qquad \text{Write in exponential form.}$$

Now we can find ordered pair solutions that satisfy $\left(\dfrac{1}{3}\right)^y = x$, plot these points, and connect them with a smooth curve.

If $y = 0$, $x = \left(\dfrac{1}{3}\right)^0 = 1$

If $y = 1$, $x = \left(\dfrac{1}{3}\right)^1 = \dfrac{1}{3}$

If $y = -1$, $x = \left(\dfrac{1}{3}\right)^{-1} = 3$

If $y = -2$, $x = \left(\dfrac{1}{3}\right)^{-2} = 9$

$x = \left(\dfrac{1}{3}\right)^y$	y
1	0
$\dfrac{1}{3}$	1
3	-1
9	-2

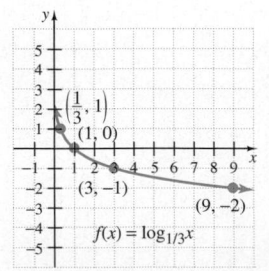

The domain of this function is $(0, \infty)$, and the range is the set of all real numbers.

The following figures summarize characteristics of logarithmic functions.

$$f(x) = \log_b x, b > 0, b \neq 1$$

- one-to-one function
- x-intercept $(1, 0)$
- no y-intercept

- domain: $(0, \infty)$
- range: $(-\infty, \infty)$

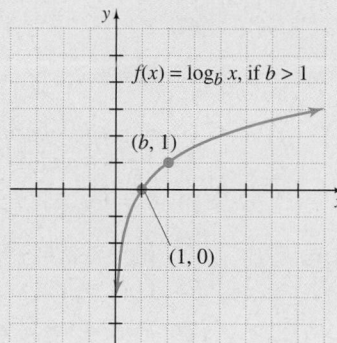

$f(x) = \log_b x$, if $b > 1$

$(b, 1)$

$(1, 0)$

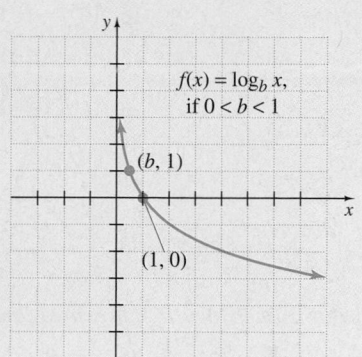

$f(x) = \log_b x$, if $0 < b < 1$

$(b, 1)$

$(1, 0)$

SPOTLIGHT ON DECISION MAKING

Suppose you are the Webmaster for a small but growing company. One of your duties is to ensure that your company's newly established Website can adequately handle the number of visitors to it. You decide to find a mathematical model for recent Website usage statistics to help predict future numbers of visitors. Ultimately, you would like to use this model to predict when your Website's server capacity must be expanded.

The first step in finding a model for the usage statistics is to decide what type of mathematical model to use: linear, quadratic, exponential, or logarithmic. The graph shows the number of visitors to your company's Website in each of the first five months since it was established. Use the graph to decide which type of mathematical model to use. Explain your reasoning.

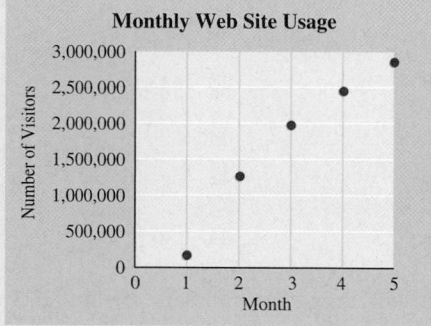

Month	1	2	3	4	5
Visitors	166,511	1,320,978	1,996,298	2,475,445	2,847,100

Exercise Set 11.4

Write each as an exponential equation. See Example 1.

1. $\log_6 36 = 2$

2. $\log_2 32 = 5$

3. $\log_3 \dfrac{1}{27} = -3$

4. $\log_5 \dfrac{1}{25} = -2$

5. $\log_{10} 1000 = 3$

6. $\log_{10} 10 = 1$

7. $\log_e x = 4$

8. $\log_e \dfrac{1}{e} = -1$

9. $\log_e \dfrac{1}{e^2} = -2$

10. $\log_e y = 7$

11. $\log_7 \sqrt{7} = \dfrac{1}{2}$

12. $\log_{11} \sqrt[4]{11} = \dfrac{1}{4}$

Write each as a logarithmic equation. See Example 2.

13. $2^4 = 16$

14. $5^3 = 125$

15. $10^2 = 100$

16. $10^4 = 10,000$

17. $e^3 = x$

18. $e^5 = y$

19. $10^{-1} = \dfrac{1}{10}$

20. $10^{-2} = \dfrac{1}{100}$

21. $4^{-2} = \dfrac{1}{16}$

22. $3^{-4} = \dfrac{1}{81}$

23. $5^{1/2} = \sqrt{5}$

24. $4^{1/3} = \sqrt[3]{4}$

Find the value of each logarithmic expression. See Example 3.

25. $\log_2 8$

26. $\log_3 9$

27. $\log_3 \dfrac{1}{9}$

28. $\log_2 \dfrac{1}{32}$

29. $\log_{25} 5$

30. $\log_8 \dfrac{1}{2}$

31. $\log_{1/2} 2$

32. $\log_{2/3} \dfrac{4}{9}$

33. $\log_7 1$

34. $\log_9 9$

35. $\log_2 2^4$

36. $\log_6 6^{-2}$

37. $\log_{10} 100$

38. $\log_{10} \dfrac{1}{10}$

39. $3^{\log_3 5}$

40. $5^{\log_5 7}$

41. $\log_3 81$

42. $\log_2 16$

43. $\log_4 \dfrac{1}{64}$

44. $\log_3 \dfrac{1}{9}$

45. Explain why negative numbers are not included as logarithmic bases.

46. Explain why 1 is not included as a logarithmic base.

Solve each equation for x. See Example 4.

47. $\log_3 9 = x$

48. $\log_2 8 = x$

49. $\log_3 x = 4$

50. $\log_2 x = 3$

51. $\log_x 49 = 2$

52. $\log_x 8 = 3$

53. $\log_2 \dfrac{1}{8} = x$

54. $\log_3 \dfrac{1}{81} = x$

55. $\log_3 \dfrac{1}{27} = x$

56. $\log_5 \dfrac{1}{125} = x$

57. $\log_8 x = \dfrac{1}{3}$

58. $\log_9 x = \dfrac{1}{2}$

59. $\log_4 16 = x$

60. $\log_2 16 = x$

61. $\log_{3/4} x = 3$

62. $\log_{2/3} x = 2$

63. $\log_x 100 = 2$

64. $\log_x 27 = 3$

Simplify. See Example 5.

65. $\log_5 5^3$

66. $\log_6 6^2$

67. $2^{\log_2 3}$

68. $7^{\log_7 4}$

69. $\log_9 9$

70. $\log_8 (8)^{-1}$

Graph each logarithmic function. Label any intercepts. See Examples 6 and 7.

71. $y = \log_3 x$

72. $y = \log_2 x$

73. $f(x) = \log_{1/4} x$

74. $f(x) = \log_{1/2} x$

75. $f(x) = \log_5 x$

76. $f(x) = \log_6 x$

77. $f(x) = \log_{1/6} x$

78. $f(x) = \log_{1/5} x$

Graph each function and its inverse function on the same set of axes. Label any intercepts.

79. $y = 4^x$; $y = \log_4 x$

80. $y = 3^x$; $y = \log_3 x$

81. $y = \left(\dfrac{1}{3}\right)^x$; $y = \log_{1/3} x$

82. $y = \left(\dfrac{1}{2}\right)^x$; $y = \log_{1/2} x$

83. The formula $\log_{10}(1 - k) = \dfrac{-0.3}{H}$ models the relationship between the half-life H of a radioactive material and its rate of decay k. Find the rate of decay of the iodine isotope I-131 if its half-life is 8 days. Round to 4 decimal places.

84. Explain why the graph of the function $y = \log_b x$ contains the point $(1, 0)$ no matter what b is.

85. $\text{Log}_3 10$ is between which two integers? Explain your answer.

REVIEW EXERCISES

Simplify each rational expression. See Section 7.1.

86. $\dfrac{x + 3}{3 + x}$

87. $\dfrac{x - 5}{5 - x}$

88. $\dfrac{x^2 - 8x + 16}{2x - 8}$

89. $\dfrac{x^2 - 3x - 10}{2 + x}$

Add or subtract as indicated. See Section 7.3 and 7.4.

90. $\dfrac{2}{x} + \dfrac{3}{x^2}$

91. $\dfrac{3x}{x + 3} + \dfrac{9}{x + 3}$

92. $\dfrac{m^2}{m + 1} - \dfrac{1}{m + 1}$

93. $\dfrac{5}{y + 1} - \dfrac{4}{y - 1}$

11.5 PROPERTIES OF LOGARITHMS

CD-ROM   SSM

SSG   Video

▶ **OBJECTIVES**

1. Use the product property of logarithms.
2. Use the quotient property of logarithms.
3. Use the power property of logarithms.
4. Use the properties of logarithms together.

In the previous section we explored some basic properties of logarithms. We now introduce and explore additional properties. Because a logarithm is an exponent, logarithmic properties are just restatements of exponential properties.

1 The first of these properties is called the **product property of logarithms,** because it deals with the logarithm of a product.

PRODUCT PROPERTY OF LOGARITHMS

If x, y, and b are positive real numbers and $b \neq 1$, then

$$\log_b xy = \log_b x + \log_b y$$

To prove this, let $\log_b x = M$ and $\log_b y = N$. Now write each logarithm with exponential notation.

$$\log_b x = M \quad \text{is equivalent to} \quad b^M = x$$

$$\log_b y = N \quad \text{is equivalent to} \quad b^N = y$$

Multiply the left sides and the right sides of the exponential equations, and we have that

$$xy = (b^M)(b^N) = b^{M+N}$$

If we write the equation $xy = b^{M+N}$ in equivalent logarithmic form, we have

$$\log_b xy = M + N$$

But since $M = \log_b x$ and $N = \log_b y$, we can write

$$\log_b xy = \log_b x + \log_b y \qquad \text{Let } M = \log_b x \text{ and } N = \log_b y.$$

In other words, the logarithm of a product is the sum of the logarithms of the factors. This property is sometimes used to simplify logarithmic expressions.

In the examples that follow, assume that variables represent positive numbers.

Example 1 Write each sum as a single logarithm.

a. $\log_{11} 10 + \log_{11} 3$ **b.** $\log_3 \dfrac{1}{2} + \log_3 12$ **c.** $\log_2 (x + 2) + \log_2 x$

Solution   In each case, both terms have a common logarithmic base.

a. $\log_{11} 10 + \log_{11} 3 = \log_{11} (10 \cdot 3)$ *Apply the product property.*

$$= \log_{11} 30$$

> **HELPFUL HINT**
> Check your logarithm properties. Make sure you understand that $\log_2 (x + 2)$ *is not* $\log_2 x + \log_2 2$.

b. $\log_3 \dfrac{1}{2} + \log_3 12 = \log_3 \left(\dfrac{1}{2} \cdot 12 \right) = \log_3 6$

c. $\log_2 (x + 2) + \log_2 x = \log_2 [(x + 2) \cdot x] = \log_2 (x^2 + 2x)$

2   The second property is the **quotient property of logarithms.**

QUOTIENT PROPERTY OF LOGARITHMS

If x, y, and b are positive real numbers and $b \neq 1$, then

$$\log_b \frac{x}{y} = \log_b x - \log_b y$$

The proof of the quotient property of logarithms is similar to the proof of the product property. Notice that the quotient property says that the logarithm of a quotient is the difference of the logarithms of the dividend and divisor.

Example 2 Write each difference as a single logarithm.

a. $\log_{10} 27 - \log_{10} 3$ **b.** $\log_5 8 - \log_5 x$ **c.** $\log_3 (x^2 + 5) - \log_3 (x^2 + 1)$

Solution   All terms have a common logarithmic base.

a. $\log_{10} 27 - \log_{10} 3 = \log_{10} \dfrac{27}{3} = \log_{10} 9$ *Apply the quotient property.*

b. $\log_5 8 - \log_5 x = \log_5 \dfrac{8}{x}$

c. $\log_3 (x^2 + 5) - \log_3 (x^2 + 1) = \log_3 \dfrac{x^2 + 5}{x^2 + 1}$

3   The third and final property we introduce is the **power property of logarithms.**

POWER PROPERTY OF LOGARITHMS

If x and b are positive real numbers, $b \neq 1$, and r is a real number, then

$$\log_b x^r = r \log_b x$$

Example 3   Use the power property to rewrite each expression.

a. $\log_5 x^3$ **b.** $\log_4 \sqrt{2}$

Solution   **a.** $\log_5 x^3 = 3 \log_5 x$

b. $\log_4 \sqrt{2} = \log_4 2^{1/2} = \dfrac{1}{2} \log_4 2$

---

4   Many times we must use more than one property of logarithms to simplify a logarithmic expression.

Example 4   Write as a single logarithm.

a. $2 \log_5 3 + 3 \log_5 2$ **b.** $3 \log_9 x - \log_9 (x + 1)$ **c.** $\log_4 25 + \log_4 3 - \log_4 5$

Solution   In each case, both terms have a common logarithmic base.

a. $2 \log_5 3 + 3 \log_5 2 = \log_5 3^2 + \log_5 2^3$ Apply the power property.

$\qquad\qquad\qquad\qquad\quad = \log_5 9 + \log_5 8$

$\qquad\qquad\qquad\qquad\quad = \log_5 (9 \cdot 8)$ Apply the product property.

$\qquad\qquad\qquad\qquad\quad = \log_5 72$

b. $3 \log_9 x - \log_9 (x + 1) = \log_9 x^3 - \log_9 (x + 1)$ Apply the power property.

$\qquad\qquad\qquad\qquad\qquad\quad = \log_9 \dfrac{x^3}{x + 1}$ Apply the quotient property.

c. Use both the product and quotient properties.

$\log_4 25 + \log_4 3 - \log_4 5 = \log_4 (25 \cdot 3) - \log_4 5$ Apply the product property.

$\qquad\qquad\qquad\qquad\qquad = \log_4 75 - \log_4 5$ Simplify.

$\qquad\qquad\qquad\qquad\qquad = \log_4 \dfrac{75}{5}$ Apply the quotient property.

$\qquad\qquad\qquad\qquad\qquad = \log_4 15$ Simplify.

Example 5 Write each expression as sums or differences of multiples of logarithms.

a. $\log_3 \dfrac{5 \cdot 7}{4}$ **b.** $\log_2 \dfrac{x^5}{y^2}$

Solution   **a.** $\log_3 \dfrac{5 \cdot 7}{4} = \log_3 (5 \cdot 7) - \log_3 4$ *Apply the quotient property.*

$$= \log_3 5 + \log_3 7 - \log_3 4 \qquad \textit{Apply the product property.}$$

b. $\log_2 \dfrac{x^5}{y^2} = \log_2 (x^5) - \log_2 (y^2)$ *Apply the quotient property.*

$$= 5 \log_2 x - 2 \log_2 y \qquad \textit{Apply the power property.}$$

> **HELPFUL HINT**
> Notice that we are not able to simplify further a logarithmic expression such as $\log_5 (2x - 1)$. None of the basic properties gives a way to write the logarithm of a difference in some equivalent form.

Example 6 If $\log_b 2 = 0.43$ and $\log_b 3 = 0.68$, use the properties of logarithms to evaluate.

a. $\log_b 6$ **b.** $\log_b 9$ **c.** $\log_b \sqrt{2}$

Solution   **a.** $\log_b 6 = \log_b (2 \cdot 3)$ *Write 6 as $2 \cdot 3$.*

$$= \log_b 2 + \log_b 3 \qquad \textit{Apply the product property.}$$

$$= 0.43 + 0.68 \qquad \textit{Substitute given values.}$$

$$= 1.11 \qquad \textit{Simplify.}$$

b. $\log_b 9 = \log_b 3^2$ *Write 9 as 3^2.*

$$= 2 \log_b 3$$

$$= 2(0.68) \qquad \textit{Substitute 0.68 for } \log_b 3.$$

$$= 1.36 \qquad \textit{Simplify.}$$

c. First, recall that $\sqrt{2} = 2^{1/2}$. Then

$$\log_b \sqrt{2} = \log_b 2^{1/2} \qquad \textit{Write } \sqrt{2} \textit{ as } 2^{1/2}.$$

$$= \frac{1}{2} \log_b 2 \qquad \textit{Apply the power property.}$$

$$= \frac{1}{2} (0.43) \qquad \textit{Substitute the given value.}$$

$$= 0.215 \qquad \textit{Simplify.}$$

A summary of the basic properties of logarithms that we have developed so far is given next.

PROPERTIES OF LOGARITHMS

If x, y, and b are positive real numbers, $b \neq 1$, and r is a real number, then

1. $\log_b 1 = 0$ 4. $\log_b xy = \log_b x + \log_b y$ *Product property*

2. $\log_b b^x = x$ 5. $\log_b \dfrac{x}{y} = \log_b x - \log_b y$ *Quotient property*

3. $b^{\log_b x} = x$ 6. $\log_b x^r = r \log_b x$ *Power property*

SPOTLIGHT ON DECISION MAKING

Suppose you are a quality assurance inspector for an electronics manufacturer. Your department has conducted reliability studies of a new model of CD player. Your studies show that the CD player's reliability can be described by the exponential function $R(t) = 2.7^{-(1/3)t}$, where the reliability R is the probability that the CD player is still working t years after it is manufactured.

The marketing department asks for your input in choosing a warranty period for the CD player. Popular warranty periods for similar competing CD players are 1 year, 2 years, and 3 years. Using the graph of the reliability for this CD player, which warranty period would you recommend? Explain your reasoning. What other factors would you want to consider?

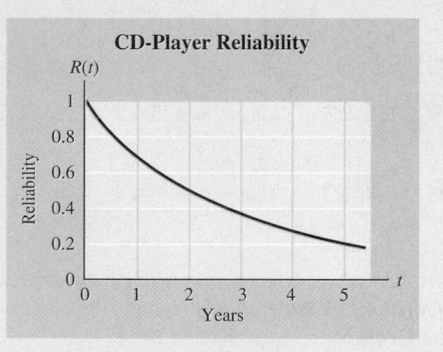

CD-Player Reliability

Exercise Set 11.5

Write each sum as a single logarithm. Assume that variables represent positive numbers. See Example 1.

1. $\log_5 2 + \log_5 7$
2. $\log_3 8 + \log_3 4$
3. $\log_4 9 + \log_4 x$
4. $\log_2 x + \log_2 y$
5. $\log_{10} 5 + \log_{10} 2 + \log_{10}(x^2 + 2)$
6. $\log_6 3 + \log_6 (x + 4) + \log_6 5$

Write each as a single logarithm. Assume that variables represent positive numbers. See Examples 2 and 4.

7. $\log_5 12 - \log_5 4$ 8. $\log_7 20 - \log_7 4$

9. $\log_2 x - \log_2 y$ 10. $\log_3 12 - \log_3 z$

11. $\log_4 2 + \log_4 10 - \log_4 5$
12. $\log_6 18 + \log_6 2 - \log_6 9$

Use the power property to rewrite each expression. See Example 3.

13. $\log_3 x^2$ 14. $\log_2 x^5$

15. $\log_4 5^{-1}$ 16. $\log_6 7^{-2}$

17. $\log_5 \sqrt{y}$ 18. $\log_5 \sqrt[3]{x}$

Write each as a single logarithm. Assume that variables represent positive numbers. See Example 4.

19. $2 \log_2 5$ 20. $3 \log_5 2$

21. $3 \log_5 x + 6 \log_5 z$

22. $2 \log_7 y + 6 \log_7 z$

23. $\log_{10} x - \log_{10} (x + 1) + \log_{10} (x^2 - 2)$

24. $\log_9 (4x) - \log_9 (x - 3) + \log_9 (x^3 + 1)$

25. $\log_4 5 + \log_4 7$

26. $\log_3 2 + \log_3 5$

27. $\log_3 8 - \log_3 2$

28. $\log_5 12 - \log_5 3$

29. $\log_7 6 + \log_7 3 - \log_7 4$

30. $\log_8 5 + \log_8 15 - \log_8 20$

31. $3 \log_4 2 + \log_4 6$

32. $2 \log_3 5 + \log_3 2$

33. $3 \log_2 x + \dfrac{1}{2} \log_2 x - 2 \log_2 (x + 1)$

34. $2 \log_5 x + \dfrac{1}{3} \log_5 x - 3 \log_5 (x + 5)$

35. $2 \log_8 x - \dfrac{2}{3} \log_8 x + 4 \log_8 x$

36. $5 \log_6 x - \dfrac{3}{4} \log_6 x + 3 \log_6 x$

Write each expression as a sum or difference of multiples of logarithms. Assume that variables represent positive numbers. See Example 5.

37. $\log_2 \dfrac{7 \cdot 11}{3}$

38. $\log_5 \dfrac{2 \cdot 9}{13}$

39. $\log_3 \dfrac{4y}{5}$

40. $\log_4 \dfrac{2}{9z}$

41. $\log_2 \dfrac{x^3}{y}$

42. $\log_5 \dfrac{x}{y^4}$

43. $\log_b \sqrt{7x}$

44. $\log_b \sqrt{\dfrac{3}{y}}$

45. $\log_7 \dfrac{5x}{4}$

46. $\log_9 \dfrac{7}{y}$

47. $\log_5 x^3(x + 1)$

48. $\log_2 y^3 z$

49. $\log_6 \dfrac{x^2}{x + 3}$

50. $\log_3 \dfrac{(x + 5)^2}{x}$

If $\log_b 3 = 0.5$ and $\log_b 5 = 0.7$, evaluate the following. See Example 6. If necessary, round to three decimal places.

51. $\log_b \dfrac{5}{3}$

52. $\log_b 25$

53. $\log_b 15$

54. $\log_b \dfrac{3}{5}$

55. $\log_b \sqrt[3]{5}$

56. $\log_b \sqrt[4]{3}$

Answer the following true or false.

57. $\log_2 x^3 = 3 \log_2 x$

58. $\log_3 (x + y) = \log_3 x + \log_3 y$

59. $\dfrac{\log_7 10}{\log_7 5} = \log_7 2$

60. $\log_7 \dfrac{14}{8} = \log_7 14 - \log_7 8$

61. $\dfrac{\log_7 x}{\log_7 y} = (\log_7 x) - (\log_7 y)$

62. $(\log_3 6) \cdot (\log_3 4) = \log_3 24$

If $\log_b 2 = 0.43$ and $\log_b 3 = 0.68$, evaluate the following.

63. $\log_b 8$

64. $\log_b 81$

65. $\log_b \dfrac{3}{9}$

66. $\log_b \dfrac{4}{32}$

67. $\log_b \sqrt{\dfrac{2}{3}}$

68. $\log_b \sqrt{\dfrac{3}{2}}$

REVIEW EXERCISES

69. Graph the functions $y = 10^x$ and $y = \log_{10} x$ on the same set of axes. See Section 11.4.

Evaluate each expression. See Section 11.4.

70. $\log_{10} 100$

71. $\log_{10} \dfrac{1}{10}$

72. $\log_7 7^2$

73. $\log_7 \sqrt{7}$

11.6   COMMON LOGARITHMS, NATURAL LOGARITHMS, AND CHANGE OF BASE

CD-ROM   SSM

SSG   Video

▶ **OBJECTIVES**

1. Identify common logarithms and approximate them by calculator.
2. Evaluate common logarithms of powers of 10.
3. Identify natural logarithms and approximate them by calculator.
4. Evaluate natural logarithms of powers of e.
5. Use the change of base formula.

In this section we look closely at two particular logarithmic bases. These two logarithmic bases are used so frequently that logarithms to their bases are given special names. **Common logarithms** are logarithms to base 10. **Natural logarithms** are logarithms to base e, which we introduce in this section. The work in this section is based on the use of the calculator, which has both the common "log" $\boxed{\text{LOG}}$ and the natural "log" $\boxed{\text{LN}}$ keys.

1   Logarithms to base 10, common logarithms, are used frequently because our number system is a base 10 decimal system. The notation $\log x$ means the same as $\log_{10} x$.

COMMON LOGARITHMS

$$\log x \ \text{means} \ \log_{10} x$$

Example 1   Use a calculator to approximate $\log 7$ to four decimal places.

Solution   Press the following sequence of keys.

$$\boxed{7}\ \boxed{\text{LOG}} \qquad \text{or} \qquad \boxed{\text{LOG}}\ \boxed{7}\ \boxed{\text{ENTER}}$$

To four decimal places,

$$\log 7 \approx 0.8451$$

2   To evaluate the common log of a power of 10, a calculator is not needed. According to the property of logarithms,

$$\log_b b^x = x$$

It follows that if b is replaced with 10, we have

$$\log 10^x = x$$

> **HELPFUL HINT**
> Remember that $\log 10^x$ means $\log_{10} 10^x = x$.

Example 2   Find the exact value of each logarithm.

a. $\log 10$ **b.** $\log 1000$ **c.** $\log \dfrac{1}{10}$ **d.** $\log \sqrt{10}$

Solution **a.** $\log 10 = \log 10^1 = 1$ **b.** $\log 1000 = \log 10^3 = 3$

c. $\log \dfrac{1}{10} = \log 10^{-1} = -1$ **d.** $\log \sqrt{10} = \log 10^{1/2} = \dfrac{1}{2}$

As we will soon see, equations containing common logs are useful models of many natural phenomena.

Example 3   Solve $\log x = 1.2$ for x. Give an exact solution, and then approximate the solution to four decimal places.

Solution   Remember that the base of a common log is understood to be 10.

> **HELPFUL HINT**
> The understood base is 10.

$$\log x = 1.2$$

$$10^{1.2} = x \qquad \text{Write with exponential notation.}$$

The exact solution is $10^{1.2}$. To four decimal places, $x \approx 15.8489$.

The Richter scale measures the intensity, or magnitude, of an earthquake. The formula for the magnitude R of an earthquake is $R = \log\left(\dfrac{a}{T}\right) + B$, where a is the amplitude in micrometers of the vertical motion of the ground at the recording station, T is the number of seconds between successive seismic waves, and B is an adjustment factor that takes into account the weakening of the seismic wave as the distance increases from the epicenter of the earthquake.

Example 4   **FINDING THE MAGNITUDE OF AN EARTHQUAKE**

Find an earthquake's magnitude on the Richter scale if a recording station measures an amplitude of 300 micrometers and 2.5 seconds between waves. Assume that B is 4.2. Approximate the solution to the nearest tenth.

Solution Substitute the known values into the formula for earthquake intensity.

$$R = \log\left(\frac{a}{T}\right) + B \qquad \text{Richter scale formula}$$

$$= \log\left(\frac{300}{2.5}\right) + 4.2 \qquad \text{Let } a = 300, T = 2.5, \text{ and } B = 4.2.$$

$$= \log(120) + 4.2$$

$$\approx 2.1 + 4.2 \qquad \text{Approximate } \log 120 \text{ by } 2.1.$$

$$= 6.3$$

This earthquake had a magnitude of 6.3 on the Richter scale.

3 **Natural logarithms** are also frequently used, especially to describe natural events; hence the label "natural logarithm." Natural logarithms are logarithms to the base e, which is a constant approximately equal to 2.7183. The number e is an irrational number, as is π. The notation $\log_e x$ is usually abbreviated to $\ln x$. (The abbreviation ln is read "el en.")

NATURAL LOGARITHMS

$$\ln x \text{ means } \log_e x$$

The graph of $y = \ln x$ is shown to the right.

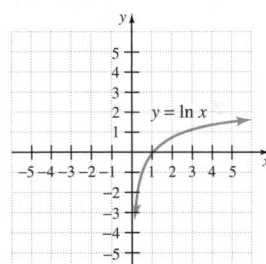

Example 5 Use a calculator to approximate ln 8 to four decimal places.

Solution Press the following sequence of keys.

$$\boxed{8}\ \boxed{\text{LN}} \qquad \text{or} \qquad \boxed{\text{LN}}\ \boxed{8}\ \boxed{\text{ENTER}}$$

To four decimal places,

$$\ln 8 \approx 2.0794$$

4 As a result of the property $\log_b b^x = x$, we know that $\log_e e^x = x$, or $\ln e^x = x$.

Example 6   Find the exact value of each natural logarithm.

a. $\ln e^3$ b. $\ln \sqrt[5]{e}$

Solution   a. $\ln e^3 = 3$ b. $\ln \sqrt[5]{e} = \ln e^{1/5} = \dfrac{1}{5}$

Example 7   Solve $\ln 3x = 5$. Give an exact solution, and then approximate the solution to four decimal places.

Solution   Remember that the base of a natural logarithm is understood to be e.

$$\ln 3x = 5$$

HELPFUL HINT
The understood base is e.

$$e^5 = 3x \qquad \text{Write with exponential notation.}$$

$$\frac{e^5}{3} = x \qquad \text{Solve for } x.$$

The exact solution is $\dfrac{e^5}{3}$. To four decimal places,

$$x \approx 49.4711.$$

Recall from Section 11.3 the formula $A = P\left(1 + \dfrac{r}{n}\right)^{nt}$ for compound interest, where n represents the number of compoundings per year. When interest is compounded continuously, the formula $A = Pe^{rt}$ is used, where r is the annual interest rate and interest is compounded continuously for t years.

Example 8   **FINDING FINAL LOAN PAYMENT**

Find the amount owed at the end of 5 years if $1600 is loaned at a rate of 9% compounded continuously.

Solution   Use the formula $A = Pe^{rt}$, where

$$P = \$1600 \text{ (the size of the loan)}$$

$$r = 9\% = 0.09 \text{ (the rate of interest)}$$

$$t = 5 \text{ (the 5-year duration of the loan)}$$

$$A = Pe^{rt}$$

$$= 1600e^{0.09(5)} \qquad \text{Substitute in known values.}$$

$$= 1600e^{0.45}$$

Now we can use a calculator to approximate the solution.

$$A \approx 2509.30$$

The total amount of money owed is $2509.30.

<u>5</u>   Calculators are handy tools for approximating natural and common logarithms. Unfortunately, some calculators cannot be used to approximate logarithms to bases other than e or 10—at least not directly. In such cases, we use the change of base formula.

CHANGE OF BASE

If a, b, and c are positive real numbers and neither b nor c is 1, then

$$\log_b a = \frac{\log_c a}{\log_c b}$$

Example 9   Approximate $\log_5 3$ to four decimal places.

Solution   Use the change of base property to write $\log_5 3$ as a quotient of logarithms to base 10.

$$\log_5 3 = \frac{\log 3}{\log 5}$$ Use the change of base property. In the change of base property, we let $a = 3$, $b = 5$, and $c = 10$.

$$\approx \frac{0.4771213}{0.69897}$$ Approximate logarithms by calculator.

$$\approx 0.6826062$$ Simplify by calculator.

To four decimal places, $\log_5 3 \approx 0.6826$.

Exercise Set 11.6

Use a calculator to approximate each logarithm to four decimal places. See Examples 1 and 5.

1. $\log 8$ **2.** $\log 6$ **3.** $\log 2.31$

4. $\log 4.86$ **5.** $\ln 2$ **6.** $\ln 3$

7. $\ln 0.0716$ **8.** $\ln 0.0032$ **9.** $\log 12.6$

10. $\log 25.9$ **11.** $\ln 5$ **12.** $\ln 7$

13. $\log 41.5$ **14.** $\ln 41.5$

15. Use a calculator and try to approximate $\log 0$. Describe what happens and explain why.

16. Use a calculator and try to approximate $\ln 0$. Describe what happens and explain why.

Find the exact value. See Examples 2 and 6.

17. $\log 100$ **18.** $\log 10,000$

19. $\log \left(\dfrac{1}{1000} \right)$ **20.** $\log \left(\dfrac{1}{100} \right)$

21. $\ln e^2$ **22.** $\ln e^4$

23. $\ln \sqrt[4]{e}$ **24.** $\ln \sqrt[5]{e}$

25. $\log 10^3$ **26.** $\ln e^5$

27. $\ln e^9$ **28.** $\log 10^7$

29. $\log 0.0001$ **30.** $\log 0.001$

31. $\ln \sqrt{e}$ **32.** $\log \sqrt{10}$

33. Without using a calculator, explain which of $\log 50$ or $\ln 50$ must be larger.

34. Without using a calculator, explain which of $\log 50^{-1}$ or $\ln 50^{-1}$ must be larger.

Solve each equation for x. Give an exact solution and a four-decimal-place approximation. See Examples 3 and 7.

35. $\log x = 1.3$ **36.** $\log x = 2.1$

37. $\log 2x = 1.1$ **38.** $\log 3x = 1.3$

39. $\ln x = 1.4$

40. $\ln x = 2.1$

41. $\ln (3x - 4) = 2.3$

42. $\ln (2x + 5) = 3.4$

43. $\log x = 2.3$

44. $\log x = 3.1$

45. $\ln x = -2.3$

46. $\ln x = -3.7$

47. $\log (2x + 1) = -0.5$

48. $\log (3x - 2) = -0.8$

49. $\ln 4x = 0.18$

50. $\ln 3x = 0.76$

Approximate each logarithm to four decimal places. See Example 9.

51. $\log_2 3$

52. $\log_3 2$

53. $\log_{1/2} 5$

54. $\log_{1/3} 2$

55. $\log_4 9$

56. $\log_9 4$

57. $\log_3 \dfrac{1}{6}$

58. $\log_6 \dfrac{2}{3}$

59. $\log_8 6$

60. $\log_6 8$

 *Use the formula $R = \log \left(\dfrac{a}{T} \right) + B$ to find the intensity R on the Richter scale of the earthquakes that fit the descriptions given. Round answers to one decimal place. See Example 4.*

61. Amplitude a is 200 micrometers, time T between waves is 1.6 seconds, and B is 2.1.

62. Amplitude a is 150 micrometers, time T between waves is 3.6 seconds, and B is 1.9.

63. Amplitude a is 400 micrometers, time T between waves is 2.6 seconds, and B is 3.1.

64. Amplitude a is 450 micrometers, time T between waves is 4.2 seconds, and B is 2.7.

Use the formula $A = Pe^{rt}$ to solve. See Example 8.

65. Find how much money Dana Jones has after 12 years if $1400 is invested at 8% interest compounded continuously.

66. Determine the size of an account in which $3500 earns 6% interest compounded continuously for 1 year.

67. Find the amount of money Barbara Mack owes at the end of 4 years if 6% interest is compounded continuously on her $2000 debt.

68. Find the amount of money for which a $2500 certificate of deposit is redeemable if it has been paying 10% interest compounded continuously for 3 years.

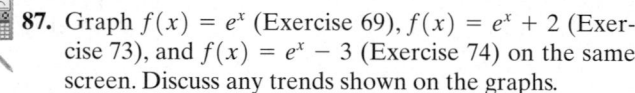 *Graph each function by finding ordered pair solutions, plotting the solutions, and then drawing a smooth curve through the plotted points.*

69. $f(x) = e^x$

70. $f(x) = e^{2x}$

71. $f(x) = e^{-3x}$

72. $f(x) = e^{-x}$

73. $f(x) = e^x + 2$

74. $f(x) = e^x - 3$

75. $f(x) = e^{x-1}$

76. $f(x) = e^{x+4}$

77. $f(x) = 3e^x$

78. $f(x) = -2e^x$

79. $f(x) = \ln x$

80. $f(x) = \log x$

81. $f(x) = -2 \log x$

82. $f(x) = 3 \ln x$

83. $f(x) = \log (x + 2)$

84. $f(x) = \log (x - 2)$

85. $f(x) = \ln x - 3$

86. $f(x) = \ln x + 3$

87. Graph $f(x) = e^x$ (Exercise 69), $f(x) = e^x + 2$ (Exercise 73), and $f(x) = e^x - 3$ (Exercise 74) on the same screen. Discuss any trends shown on the graphs.

88. Graph $f(x) = \ln x$ (Exercise 79), $f(x) = \ln x - 3$ (Exercise 85), and $f(x) = \ln x + 3$ (Exercise 86). Discuss any trends shown on the graphs.

REVIEW EXERCISES

Solve each equation for x. See Sections 2.3 and 6.6.

89. $6x - 3(2 - 5x) = 6$

90. $2x + 3 = 5 - 2(3x - 1)$

91. $2x + 3y = 6x$

92. $4x - 8y = 10x$

93. $x^2 + 7x = -6$

94. $x^2 + 4x = 12$

Solve each system of equations. See Section 4.1, 4.2, or 4.3.

95. $\begin{cases} x + 2y = -4 \\ 3x - y = 9 \end{cases}$

96. $\begin{cases} 5x + y = 5 \\ -3x - 2y = -10 \end{cases}$

11.7 EXPONENTIAL AND LOGARITHMIC EQUATIONS AND APPLICATIONS

CD-ROM

SSM

SSG

Video

▶ **OBJECTIVES**

1. Solve exponential equations.
2. Solve logarithmic equations.
3. Solve problems that can be modeled by exponential and logarithmic equations.

1

In Section 11.3 we solved exponential equations such as $2^x = 16$ by writing 16 as a power of 2 and applying the uniqueness of b^x.

$$2^x = 16$$
$$2^x = 2^4 \qquad \text{Write 16 as } 2^4.$$
$$x = 4 \qquad \text{Use the uniqueness of } b^x.$$

Solving the equation in this manner is possible since 16 is a power of 2. If solving an equation such as $2^x = a$ *number,* where the number is not a power of 2, we use logarithms. For example, to solve an equation such as $3^x = 7$, we use the fact that $f(x) = \log_b x$ is a one-to-one function. Another way of stating this fact is as a property of equality.

> **LOGARITHM PROPERTY OF EQUALITY**
>
> Let a, b, and c be real numbers such that $\log_b a$ and $\log_b c$ are real numbers and b is not 1. Then
>
> $$\log_b a = \log_b c \text{ is equivalent to } a = c$$

Example 1   Solve $3^x = 7$.

Solution   To solve, we use the logarithm property of equality and take the logarithm of both sides. For this example, we use the common logarithm.

$$3^x = 7$$
$$\log 3^x = \log 7 \qquad \text{Take the common log of both sides.}$$
$$x \log 3 = \log 7 \qquad \text{Apply the power property of logarithms.}$$
$$x = \frac{\log 7}{\log 3} \qquad \text{Divide both sides by log 3.}$$

The exact solution is $\dfrac{\log 7}{\log 3}$. If a decimal approximation is preferred,

$$\frac{\log 7}{\log 3} \approx \frac{0.845098}{0.4771213} \approx 1.7712 \text{ to four decimal places.}$$

The solution is $\dfrac{\log 7}{\log 3}$, or *approximately* 1.7712.

2   By applying the appropriate properties of logarithms, we can solve a broad variety of logarithmic equations.

Example 2   Solve $\log_4 (x - 2) = 2$.

Solution   Notice that $x - 2$ must be positive, so x must be greater than 2. With this in mind, we first write the equation with exponential notation.

$$\log_4 (x - 2) = 2$$
$$4^2 = x - 2$$
$$16 = x - 2$$
$$18 = x \qquad \text{Add 2 to both sides.}$$

Check   To check, we replace x with 18 in the original equation.

$$\log_4 (x - 2) = 2$$
$$\log_4 (18 - 2) \stackrel{?}{=} 2 \qquad \text{Let } x = 18.$$
$$\log_4 16 \stackrel{?}{=} 2$$
$$4^2 = 16 \qquad \text{True.}$$

The solution is 18.   ▬

Example 3   Solve $\log_2 x + \log_2 (x - 1) = 1$.

Solution   Notice that $x - 1$ must be positive, so x must be greater than 1. We use the product property on the left side of the equation.

$$\log_2 x + \log_2 (x - 1) = 1$$
$$\log_2 x(x - 1) = 1 \qquad \text{Apply the product property.}$$
$$\log_2 (x^2 - x) = 1$$

Next we write the equation with exponential notation and solve for x.

$$2^1 = x^2 - x$$
$$0 = x^2 - x - 2 \qquad \text{Subtract 2 from both sides.}$$
$$0 = (x - 2)(x + 1) \qquad \text{Factor.}$$
$$0 = x - 2 \quad \text{or} \quad 0 = x + 1 \qquad \text{Set each factor equal to 0.}$$
$$2 = x \qquad\qquad -1 = x$$

Recall that -1 cannot be a solution because x must be greater than 1. If we forgot this, we would still reject -1 after checking. To see this, we replace x with -1 in the original equation.

$$\log_2 x + \log_2 (x - 1) = 1$$
$$\log_2 (-1) + \log_2(-1 - 1) \stackrel{?}{=} 1 \qquad \text{Let } x = -1.$$

Because the logarithm of a negative number is undefined, -1 is rejected. Check to see that the solution is 2.   ▬

Example 4 Solve $\log (x + 2) - \log x = 2$.

Solution We use the quotient property of logarithms on the left side of the equation.

$$\log (x + 2) - \log x = 2$$

$$\log \frac{x + 2}{x} = 2 \qquad \text{Apply the quotient property.}$$

$$10^2 = \frac{x + 2}{x} \qquad \text{Write using exponential notation.}$$

$$100 = \frac{x + 2}{x} \qquad \text{Simplify.}$$

$$100x = x + 2 \qquad \text{Multiply both sides by } x.$$

$$99x = 2 \qquad \text{Subtract } x \text{ from both sides.}$$

$$x = \frac{2}{99} \qquad \text{Divide both sides by 99.}$$

Verify that the solution is $\frac{2}{99}$.

---

3   Logarithmic and exponential functions are used in a variety of scientific, technical, and business settings. A few examples follow.

Example 5   **ESTIMATING POPULATION SIZE**

The population size y of a community of lemmings varies according to the relationship $y = y_0 e^{0.15t}$. In this formula, t is time in months, and y_0 is the initial population at time 0. Estimate the population after 6 months if there were originally 5000 lemmings.

Solution We substitute 5000 for y_0 and 6 for t.

$$y = y_0 e^{0.15t}$$

$$= 5000 e^{0.15(6)} \qquad \text{Let } t = 6 \text{ and } y_0 = 5000.$$

$$= 5000 e^{0.9} \qquad \text{Multiply.}$$

Using a calculator, we find that $y \approx 12,298.016$. In 6 months the population will be approximately 12,300 lemmings.

Example 6   **DOUBLING AN INVESTMENT**

How long does it take an investment of \$2000 to double if it is invested at 5% interest compounded quarterly? The necessary formula is $A = P\left(1 + \frac{r}{n}\right)^{nt}$, where A is the accrued (or owed) amount, P is the principal invested, r is the annual rate of interest, n is the number of compounding periods per year, and t is the number of years.

Solution   We are given that $P = 2000$ and $r = 5\% = 0.05$. Compounding quarterly means 4 times a year, so $n = 4$. The investment is to double, so A must be 4000. Substitute these values and solve for t.

$$A = P\left(1 + \frac{r}{n}\right)^{nt}$$

$$4000 = 2000\left(1 + \frac{0.05}{4}\right)^{4t} \qquad \text{Substitute in known values.}$$

$$4000 = 2000(1.0125)^{4t} \qquad \text{Simplify } 1 + \frac{0.05}{4}.$$

$$2 = (1.0125)^{4t} \qquad \text{Divide both sides by 2000.}$$

$$\log 2 = \log 1.0125^{4t} \qquad \text{Take the logarithm of both sides.}$$

$$\log 2 = 4t(\log 1.0125) \qquad \text{Apply the power property.}$$

$$\frac{\log 2}{4 \log 1.0125} = t \qquad \text{Divide both sides by 4 log 1.0125.}$$

$$13.949408 \approx t \qquad \text{Approximate by calculator.}$$

Thus, it takes nearly 14 years for the money to double in value.

GRAPHING CALCULATOR EXPLORATIONS

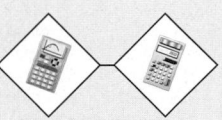

Use a graphing calculator to find how long it takes an investment of $1500 to triple if it is invested at 8% interest compounded monthly.

First, let $P = 1500$, $r = 0.08$, and $n = 12$ (for 12 months) in the formula

$$A = P\left(1 + \frac{r}{n}\right)^{nt}$$

Notice that when the investment has tripled, the accrued amount A is 4500. Thus,

$$4500 = 1500\left(1 + \frac{0.08}{12}\right)^{12t}$$

Determine an appropriate viewing window and enter and graph the equations

$$Y_1 = 1500\left(1 + \frac{0.08}{12}\right)^{12x}$$

and

$$Y_2 = 4500$$

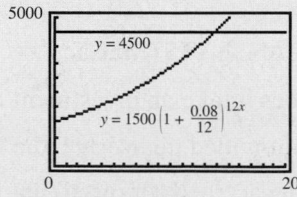

The point of intersection of the two curves is the solution. The x-coordinate tells how long it takes for the investment to triple.

Use a TRACE feature or an INTERSECT feature to approximate the coordinates of the point of intersection of the two curves. It takes approximately 13.78 years, or 13 years and 9 months, for the investment to triple in value to $4500.

Use this graphical solution method to solve each problem. Round each answer to the nearest hundredth.

1. Find how long it takes an investment of $5000 to grow to $6000 if it is invested at 5% interest compounded quarterly.

2. Find how long it takes an investment of $1000 to double if it is invested at 4.5% interest compounded daily. (Use 365 days in a year.)

3. Find how long it takes an investment of $10,000 to quadruple if it is invested at 6% interest compounded monthly.

4. Find how long it takes $500 to grow to $800 if it is invested at 4% interest compounded semiannually.

Exercise Set 11.7

Solve each equation. Give an exact solution, and also approximate the solution to four decimal places. See Example 1.

1. $3^x = 6$
2. $4^x = 7$
3. $3^{2x} = 3.8$
4. $5^{3x} = 5.6$
5. $2^{x-3} = 5$
6. $8^{x-2} = 12$
7. $9^x = 5$
8. $3^x = 11$
9. $4^{x+7} = 3$
10. $6^{x+3} = 2$
11. $7^{3x-4} = 11$
12. $5^{2x-6} = 12$
13. $e^{6x} = 5$
14. $e^{2x} = 8$

Solve each equation. See Examples 2 through 4.

15. $\log_2 (x + 5) = 4$
16. $\log_6 (x^2 - x) = 1$
17. $\log_3 x^2 = 4$
18. $\log_2 x^2 = 6$
19. $\log_4 2 + \log_4 x = 0$
20. $\log_3 5 + \log_3 x = 1$
21. $\log_2 6 - \log_2 x = 3$
22. $\log_4 10 - \log_4 x = 2$
23. $\log_4 x + \log_4 (x + 6) = 2$
24. $\log_3 x + \log_3 (x + 6) = 3$
25. $\log_5 (x + 3) - \log_5 x = 2$
26. $\log_6 (x + 2) - \log_6 x = 2$
27. $\log_3 (x - 2) = 2$
28. $\log_2 (x - 5) = 3$
29. $\log_4 (x^2 - 3x) = 1$
30. $\log_8 (x^2 - 2x) = 1$
31. $\ln 5 + \ln x = 0$
32. $\ln 3 + \ln (x - 1) = 0$
33. $3 \log x - \log x^2 = 2$
34. $2 \log x - \log x = 3$
35. $\log_2 x + \log_2 (x + 5) = 1$

36. $\log_4 x + \log_4 (x + 7) = 1$
37. $\log_4 x - \log_4 (2x - 3) = 3$
38. $\log_2 x - \log_2 (3x + 5) = 4$
39. $\log_2 x + \log_2 (3x + 1) = 1$
40. $\log_3 x + \log_3 (x - 8) = 2$

Solve. See Example 5.

41. The size of the wolf population at Isle Royale National Park increases at a rate of 4.3% per year. If the size of the current population is 83 wolves, find how many there should be in 5 years. Use $y = y_0 e^{0.043t}$ and round to the nearest whole.

42. The number of victims of a flu epidemic is increasing at a rate of 7.5% per week. If 20,000 persons are currently infected, find in how many days we can expect 45,000 to have the flu. Use $y = y_0 e^{0.075t}$ and round to the nearest whole. (*Hint:* Don't forget to convert your answer to days.)

43. The size of the population of Senegal is increasing at a rate of 2.6% per year. If 10,052,000 people lived in Senegal in 1999, find how many inhabitants there will be by 2005. Round to the nearest ten-thousand. Use $y = y_0 e^{0.026t}$.

44. In 1999, 1001 million people were citizens of India. Find how long it will take India's population to reach a size of 1500 million (that is, 1.5 billion) if the population size is growing at a rate of 1.7% per year. Use $y = y_0 e^{0.017t}$ and round to the nearest tenth. (*Source:* U.S. Bureau of the Census, International Data Base)

45. In 1999, Russia had a population of 146,394 thousand. At that time, Russia's population was declining at a rate of 0.5% per year. How long will it take for Russia's population to reach 120,000 thousand? Use $y = y_0 e^{-0.005t}$ and round to the nearest tenth. (*Source:* U.S. Bureau of the Census, International Data Base)

46. The population of Italy has been decreasing at a rate of 0.1% per year. If there were 56,735,000 people living in Italy in 1999, how many inhabitants will there be by 2020? Use $y = y_0 e^{-0.001t}$ and round to the nearest whole number. (*Source:* U.S. Bureau of the Census, International Data Base)

Use the formula $A = P\left(1 + \dfrac{r}{n}\right)^{nt}$ *to solve these compound interest problems. Round to the nearest tenth. See Example 6.*

47. Find how long it takes $600 to double if it is invested at 7% interest compounded monthly.

48. Find how long it takes $600 to double if it is invested at 12% interest compounded monthly.

49. Find how long it takes a $1200 investment to earn $200 interest if it is invested at 9% interest compounded quarterly.

50. Find how long it takes a $1500 investment to earn $200 interest if it is invested at 10% compounded semiannually.

51. Find how long it takes $1000 to double if it is invested at 8% interest compounded semiannually.

52. Find how long it takes $1000 to double if it is invested at 8% interest compounded monthly.

The formula $w = 0.00185h^{2.67}$ *is used to estimate the normal weight w of a boy h inches tall. Use this formula to solve the height–weight problems. Round to the nearest tenth.*

53. Find the expected weight of a boy who is 35 inches tall.

54. Find the expected weight of a boy who is 43 inches tall.

55. Find the expected height of a boy who weighs 85 pounds.

56. Find the expected height of a boy who weighs 140 pounds.

The formula $P = 14.7e^{-0.21x}$ *gives the average atmospheric pressure P, in pounds per square inch, at an altitude x, in miles above sea level. Use this formula to solve these pressure problems. Round answers to the nearest tenth.*

57. Find the average atmospheric pressure of Denver, which is 1 mile above sea level.

58. Find the average atmospheric pressure of Pikes Peak, which is 2.7 miles above sea level.

59. Find the elevation of a Delta jet if the atmospheric pressure outside the jet is 7.5 lb/in.^{2}.

60. Find the elevation of a remote Himalayan peak if the atmospheric pressure atop the peak is 6.5 lb/in.^{2}.

Psychologists call the graph of the formula $t = \dfrac{1}{c}\ln\left(\dfrac{A}{A - N}\right)$
the learning curve, since the formula relates time t passed, in

weeks, to a measure N of learning achieved, to a measure A of maximum learning possible, and to a measure c of an individual's learning style. Round to the nearest week.

61. Norman is learning to type. If he wants to type at a rate of 50 words per minute (N is 50) and his expected maximum rate is 75 words per minute (A is 75), find how many weeks it should take him to achieve his goal. Assume that c is 0.09.

62. An experiment with teaching chimpanzees sign language shows that a typical chimp can master a maximum of 65 signs. Find how many weeks it should take a chimpanzee to master 30 signs if c is 0.03.

63. Janine is working on her dictation skills. She wants to take dictation at a rate of 150 words per minute and believes that the maximum rate she can hope for is 210 words per minute. Find how many weeks it should take her to achieve the 150 words per minute level if c is 0.07.

64. A psychologist is measuring human capability to memorize nonsense syllables. Find how many weeks it should take a subject to learn 15 nonsense syllables if the maximum possible to learn is 24 syllables and c is 0.17.

Use a graphing calculator to solve each equation. For example, to solve Exercise 65, let $Y_1 = e^{0.3x}$ *and* $Y_2 = 8$, *and graph the equations. The x-value of the point of intersection is the solution. Round all solutions to two decimal places.*

65. $e^{0.3x} = 8$

66. $10^{0.5x} = 7$

67. $2\log(-5.6x + 1.3) = -x - 1$

68. $\ln(1.3x - 2.1) = -3.5x + 5$

69. Check Exercise 11.

70. Check Exercise 12.

71. Check Exercise 31.

72. Check Exercise 32.

REVIEW EXERCISES

If $x = -2$, $y = 0$, *and* $z = 3$, *find the value of each expression. See Section 1.4.*

73. $\dfrac{x^2 - y + 2z}{3x}$

74. $\dfrac{x^3 - 2y + z}{2z}$

75. $\dfrac{3z - 4x + y}{x + 2z}$

76. $\dfrac{4y - 3x + z}{2x + y}$

Find the inverse function of each one-to-one function. See Section 11.2.

77. $f(x) = 5x + 2$

78. $f(x) = \dfrac{x - 3}{4}$

11 CHAPTER PROJECT

 For additional Chapter Projects, visit the Real World Activities Website by going to http://www.prenhall.com/martin-gay.

Modeling Temperature

When a cold object is placed in a warm room, the object's temperature gradually rises until it becomes, or nearly becomes, room temperature. Similarly, if a hot object is placed in a cooler room, the object's temperature gradually falls to room temperature. The way in which a cold or hot object warms up or cools off is modeled by an exact mathematical relationship, known as Newton's law of cooling. This law relates the temperature of an object to the time elapsed since its warming or cooling began. In this project, you will have the opportunity to investigate this model of cooling and warming. This project may be completed by working in groups or individually.

To investigate Newton's law of cooling in this project, you will collect experimental data in one of two methods: Method 1, using a stopwatch and thermometer, or Method 2, using Texas Instruments' Calculator-Based Laboratory (CBL™) or Second Generation Calculator-Based Laboratory (CBL 2™).

Method 1 Materials

- Container of either cold or hot liquid
- Thermometer
- Stopwatch
- Graphing calculator with regression capabilities

Method 2 Materials

- Container of either cold or hot liquid
- A TI-82, TI-83, or TI-85 graphing calculator with unit-to-unit link cable
- CBL™ or CBL 2™ unit with temperature probe

DATA TABLE

Time, t	Temperature, T
0	

Steps for Collecting Data with Method 1:

a. Insert the thermometer into the liquid and allow a thermometer reading to register. Take a temperature reading T as you start the stopwatch (at $t = 0$) and record it in the accompanying data table.

b. Continue taking temperature readings at uniform intervals anywhere between 5 and 10 minutes long. At each reading use the stopwatch to measure the length of time that has elapsed since the temperature readings started with your first reading at $t = 0$. Record your time t and liquid temperature T in the data table. Gather data for six to twelve readings.

c. Plot the data from the data table. Plot t on the horizontal axis and T on the vertical axis.

Steps for Collecting Data with Method 2:

a. Enter the HEAT program appropriate for your calculator.

b. Prepare the CBL or CBL 2 and the graphing calculator. Insert the temperature probe into the liquid.

c. Start the HEAT program on the graphing calculator and follow its instructions to begin collecting data. The program will collect 36 temperature readings in degrees Celsius and plot them in real time with t on the horizontal axis and T on the vertical axis.

1. Which of the following mathematical models best fits the data you collected? Explain your reasoning. (Assume $a > 0$.)

   a. $T = ab^t + c$
   b. $T = ab^{-t} + c$
   c. $T = -ab^{-t} + c$
   d. $T = \ln(-ax + b) + c$
   e. $T = -\ln(-ax + b) + c$

2. What does the constant c represent in the model you chose? What is the value of c in this activity?

3. (Optional) Subtract the value of c from each of your observations of T. Enter the new

ordered pairs $(t, T - c)$ into a graphing calculator. Use the exponential or logarithmic regression feature to find a model for your experimental data. Graph the ordered pairs $(t, T - c)$ with the model you found. How well does the model fit the data? How does the model compare with your selection from Question 1?

Graphing Calculator Programs

TI-82 or TI-83 Program

```
PROGRAM:HEAT82
:PlotsOff
:Func
:FnOff
:AxesOn
:ClrDraw
:ClrList L3, L4
:-10→Ymin
:90→Ymax
:10→Yscl
:ClrHome
:{1, 0}→L1
:Send (L1)
:{1, 1, 1}→L1
:Send (L1)
:36→dim L3
:36→dim L4
:Disp "HOW MUCH TIME"
:Disp "BETWEEN POINTS"
:Disp "IN SECONDS?"
:Input T
:-2*T→Xmin
:36*T→Xmax
:T→Xscl
:seq(K, K, T, 36*T, T)→L3
:ClrHome
:Disp "PRESS ENTER"
:Disp "TO START"
:Pause
:ClrHome
:{3, T, -1, 0}→L1
:Send(L1)
:For (K, 1, 36, 1)
:Get (L4(K))
:Pt-On(L3(K), L4(K))
:End
:ClrHome
:Plot1(Scatter, L3, L4,·)
:DispGraph
:Stop
```

TI-85 Program

```
PROGRAM:HEAT85
:Func
:FnOff
:AxesOn
:ClDrw
:1→dimL L3:1→dimL L4
:-10→yMin
:90→yMax
:10→yScl
:Cl LCD
:{1, 0}→L1
:Outpt("CBLSEND", L1)
:{1, 1, 1}→L1
:Outpt ("CBLSEND", L1)
:36→dimL L3
:36→dimL L4
:Disp "HOW MUCH TIME"
:Disp "BETWEEN POINTS"
:Disp "IN SECONDS"
:Input T
:-2*T→xMin
:36*T→xMax
:T→xScl
:seq(K, K, T, 36*T, T)→L3
:ClLCD
:Disp "PRESS ENTER"
:Disp "TO START"
:Pause
:ClLCD
:{3, T, -1, 0}→L1
:Outpt ("CBLSEND", L1)
:For (K, 1, 36, 1)
:Input "CBLGET", L4(K)
:PtOn (L3(K), L4(K))
:End
:ClLCD
:Scatter L3, L4
:DispG
:Stop
```

CHAPTER 11 VOCABULARY CHECK

Fill in each blank with one of the words or phrases listed below.

inverse	common	composition	symmetric	exponential
vertical	logarithmic	natural	horizontal	

1. For each one-to-one function, we can find its _____ function by switching the coordinates of the ordered pairs of the function.

2. The _____ of functions f and g is $(f \circ g)(x) = f(g(x))$.

3. A function of the form $f(x) = b^x$ is called an _____ function if $b > 0$, b is not 1, and x is a real number.

4. The graphs of f and f^{-1} are _____ about the line $y = x$.

5. _____ logarithms are logarithms to base e.

6. _____ logarithms are logarithms to base 10.

7. To see whether a graph is the graph of a one-to-one function, apply the _____ line test to see if it is a function, and then apply the _____ line test to see if it is a one-to-one function.

8. A _____ function is a function that can be defined by $f(x) = \log_b x$ where x is a positive real number, b is a constant positive real number, and b is not 1.

CHAPTER 11 HIGHLIGHTS

DEFININTIONS AND CONCEPTS	EXAMPLES

Section 11.1 The Algebra of Functions; Composite Functions

Algebra of Functions

Sum	$(f + g)(x) = f(x) + g(x)$
Difference	$(f - g)(x) = f(x) - g(x)$
Product	$(f \cdot g)(x) = f(x) \cdot g(x)$
Quotient	$\left(\dfrac{f}{g}\right)(x) = \dfrac{f(x)}{g(x)}, g(x) \neq 0$

If $f(x) = 7x$ and $g(x) = x^2 + 1$,

$$(f + g)(x) = f(x) + g(x) = 7x + x^2 + 1$$

$$(f - g)(x) = f(x) - g(x) = 7x - (x^2 + 1)$$
$$= 7x - x^2 - 1$$

$$(f \cdot g)(x) = f(x) \cdot g(x) = 7x(x^2 + 1)$$
$$= 7x^3 + 7x$$

$$\left(\frac{f}{g}\right)(x) = \frac{f(x)}{g(x)} = \frac{7x}{x^2 + 1}$$

Composite Functions

The notation $(f \circ g)(x)$ means "f composed with g."

$$(f \circ g)(x) = f(g(x))$$

$$(g \circ f)(x) = g(f(x))$$

If $f(x) = x^2 + 1$ and $g(x) = x - 5$, find $(f \circ g)(x)$.

$$(f \circ g)(x) = f(g(x))$$
$$= f(x - 5)$$
$$= (x - 5)^2 + 1$$
$$= x^2 - 10x + 26$$

(continued)

DEFINITIONS AND CONCEPTS	EXAMPLES

Section 11.2 Inverse Functions

If f is a function, then f is a **one-to-one function** only if each y-value (output) corresponds to only one x-value (input).

Horizontal Line Test

If every horizontal line intersects the graph of a function at most once, then the function is a one-to-one function.

Determine whether each graph is a one-to-one function.

A
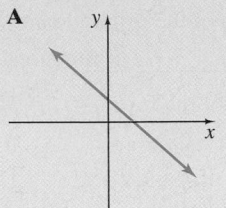

B
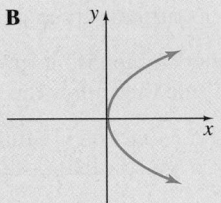

C
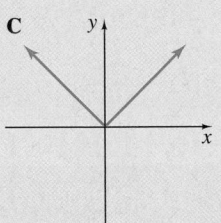

Graphs A and C pass the vertical line test, so only these are graphs of functions. Of graphs A and C, only graph A passes the horizontal line test, so only graph A is the graph of a one-to-one function.

The **inverse** of a one-to-one function f is the one-to-one function f^{-1} that is the set of all ordered pairs (b, a) such that (a, b) belongs to f.

To Find the Inverse of a One-to-One Function f(x)

Step 1: Replace $f(x)$ with y.
Step 2: Interchange x and y.
Step 3: Solve for y.
Step 4: Replace y with $f^{-1}(x)$.

Find the inverse of $f(x) = 2x + 7$.

$$y = 2x + 7 \qquad \text{Replace } f(x) \text{ with } y.$$

$$x = 2y + 7 \qquad \text{Interchange } x \text{ and } y.$$

$$2y = x - 7 \qquad \text{Solve for } y.$$

$$y = \frac{x - 7}{2}$$

$$f^{-1}(x) = \frac{x - 7}{2} \qquad \text{Replace } y \text{ with } f^{-1}(x).$$

The inverse of $f(x) = 2x + 7$ is
$$f^{-1}(x) = \frac{x - 7}{2}.$$

(continued)

DEFINITIONS AND CONCEPTS	EXAMPLES

Section 11.3 Exponential Functions

A function of the form $f(x) = b^x$ is an **exponential function,** where $b > 0, b \neq 1$, and x is a real number.

Graph the exponential function $y = 4^x$.

x	y
-2	$\dfrac{1}{16}$
-1	$\dfrac{1}{4}$
0	1
1	4
2	16

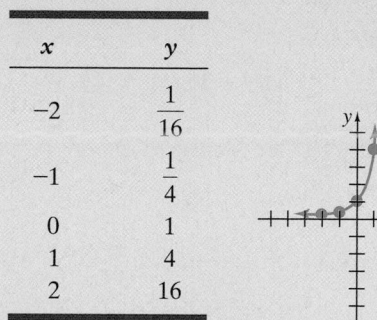

Uniqueness of b^x

If $b > 0$ and $b \neq 1$, then $b^x = b^y$ is equivalent to $x = y$.

Solve $2^{x+5} = 8$.

$$2^{x+5} = 2^3 \qquad \text{Write 8 as } 2^3.$$
$$x + 5 = 3 \qquad \text{Use the uniqueness of } b^x.$$
$$x = -2 \qquad \text{Subtract 5 from both sides.}$$

Section 11.4 Logarithmic Functions

Logarithmic Definition

If $b > 0$ and $b \neq 1$, then

$$y = \log_b x \quad \text{means} \quad x = b^y$$

for any positive number x and real number y.

Logarithmic Form	*Corresponding Exponential Statement*
$\log_5 25 = 2$	$5^2 = 25$
$\log_9 3 = \dfrac{1}{2}$	$9^{1/2} = 3$

Properties of Logarithms

If b is a real number, $b > 0$ and $b \neq 1$, then

$$\log_b 1 = 0, \quad \log_b b^x = x, \quad b^{\log_b x} = x$$

$$\log_5 1 = 0, \quad \log_7 7^2 = 2, \quad 3^{\log_3 6} = 6$$

Logarithmic Function

If $b > 0$ and $b \neq 1$, then a **logarithmic function** is a function that can be defined as

$$f(x) = \log_b x$$

The domain of f is the set of positive real numbers, and the range of f is the set of real numbers.

Graph $y = \log_3 x$.
Write $y = \log_3 x$ as $3^y = x$. Plot the ordered pair solutions listed in the table, and connect them with a smooth curve.

x	y
3	1
1	0
$\dfrac{1}{3}$	-1
$\dfrac{1}{9}$	-2

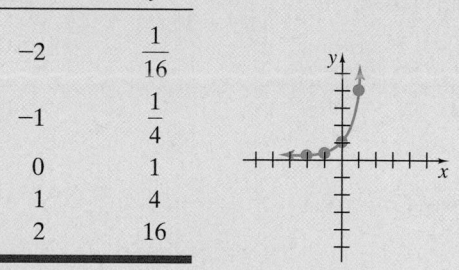

(continued)

DEFINITIONS AND CONCEPTS	EXAMPLES

Section 11.5 Properties of Logarithms

Let x, y, and b be positive numbers and $b \neq 1$.

Product Property

$$\log_b xy = \log_b x + \log_b y$$

Quotient Property

$$\log_b \frac{x}{y} = \log_b x - \log_b y$$

Power Property

$$\log_b x^r = r \log_b x$$

Write as a single logarithm.

$2 \log_5 6 + \log_5 x - \log_5 (y + 2)$

$= \log_5 6^2 + \log_5 x - \log_5 (y + 2)$ — Power property

$= \log_5 36 \cdot x - \log_5 (y + 2)$ — Product property

$= \log_5 \dfrac{36x}{y + 2}$ — Quotient property

Section 11.6 Common Logarithms, Natural Logarithms, and Change of Base

Common Logarithms

$$\log x \quad \text{means} \quad \log_{10} x$$

Natural Logarithms

$$\ln x \quad \text{means} \quad \log_e x$$

Continuously Compounded Interest Formula

$$A = Pe^{rt}$$

where r is the annual interest rate for P dollars invested for t years.

$\log 5 = \log_{10} 5 \approx 0.69897$

$\ln 7 = \log_e 7 \approx 1.94591$

Find the amount in an account at the end of 3 years if \$1000 is invested at an interest rate of 4% compounded continuously.

Here, $t = 3$ years, $P = \$1000$, and $r = 0.04$.

$A = Pe^{rt}$

$= 1000e^{0.04(3)}$

$\approx \$1127.50$

Section 11.7 Exponential and Logarithmic Equations and Applications

Logarithm Property of Equality

Let $\log_b a$ and $\log_b c$ be real numbers and $b \neq 1$. Then

$$\log_b a = \log_b c \quad \text{is equivalent to} \quad a = c$$

Solve $2^x = 5$.

$\log 2^x = \log 5$ — Log property of equality

$x \log 2 = \log 5$ — Power property

$x = \dfrac{\log 5}{\log 2}$ — Divide both sides by log 2.

$x \approx 2.3219$ — Use a calculator.

CHAPTER 11 REVIEW

(11.1) *If $f(x) = x - 5$ and $g(x) = 2x + 1$, find*

1. $(f + g)(x)$

2. $(f - g)(x)$

3. $(f \cdot g)(x)$

4. $\left(\dfrac{g}{f}\right)(x)$

If $f(x) = x^2 - 2$, $g(x) = x + 1$, and $h(x) = x^3 - x^2$, find each composition.

5. $(f \circ g)(x)$

6. $(g \circ f)(x)$

7. $(h \circ g)(2)$

8. $(f \circ f)(x)$

9. $(f \circ g)(-1)$

10. $(h \circ h)(2)$

(11.2) *Determine whether each function is a one-to-one function. If it is one-to-one, list the elements of its inverse.*

11. $h = \{(-9, 14), (6, 8), (-11, 12), (15, 15)\}$

12. $f = \{(-5, 5), (0, 4), (13, 5), (11, -6)\}$

13.

U.S. Region (Input)	West	Midwest	South	Northeast
Rank in Automobile Thefts (Output)	2	4	1	3

△ **14.**

Shape (Input)	Square	Triangle	Parallelogram	Rectangle
Number of Sides (Output)	4	3	4	4

Given that $f(x) = \sqrt{x + 2}$ is a one-to-one function, find the following.

15. a. $f(7)$

   b. $f^{-1}(3)$

16. a. $f(-1)$

   b. $f^{-1}(1)$

Determine whether each function is a one-to-one function.

17.

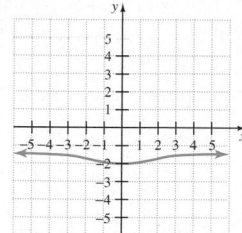

18.

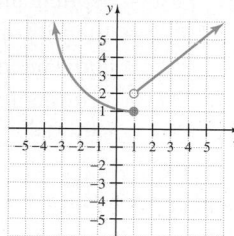

19.

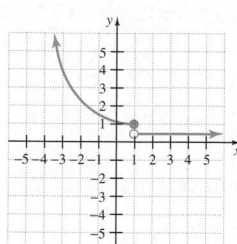

20.

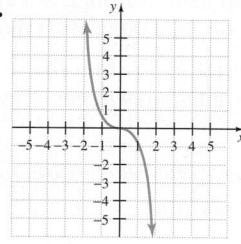

Find an equation defining the inverse function of the given one-to-one function.

21. $f(x) = x - 9$

22. $f(x) = x + 8$

23. $f(x) = 6x + 11$

24. $f(x) = 12x$

25. $f(x) = x^3 - 5$

26. $f(x) = \sqrt[3]{x + 2}$

27. $g(x) = \dfrac{12x - 7}{6}$

28. $r(x) = \dfrac{13}{2}x - 4$

On the same set of axes, graph the given one-to-one function and its inverse.

29. $g(x) = \sqrt{x}$

30. $h(x) = 5x - 5$

31. Find the inverse of the one-to-one function $f(x) = 2x - 3$. Then graph both $f(x)$ and $f^{-1}(x)$ with a square window.

(11.3) *Solve each equation for x.*

32. $4^x = 64$

33. $3^x = \dfrac{1}{9}$

34. $2^{3x} = \dfrac{1}{16}$

35. $5^{2x} = 125$

36. $9^{x+1} = 243$

37. $8^{3x-2} = 4$

Graph each exponential function.

38. $y = 3^x$

39. $y = \left(\dfrac{1}{3}\right)^x$

40. $y = 4 \cdot 2^x$

41. $y = 2^x + 4$

Use the formula $A = P\left(1 + \dfrac{r}{n}\right)^{nt}$ to solve the interest problems. In this formula,

A = amount accrued (or owed)

P = principal invested (or loaned)

r = rate of interest

n = number of compounding periods per year

t = time in years

42. Find the amount accrued if \$1600 is invested at 9% interest compounded semiannually for 7 years.

43. A total of $800 is invested in a 7% certificate of deposit for which interest is compounded quarterly. Find the value that this certificate will have at the end of 5 years.

44. Use a graphing calculator to verify the results of Exercise 40.

(11.4) *Write each equation with logarithmic notation.*

45. $49 = 7^2$

46. $2^{-4} = \dfrac{1}{16}$

Write each logarithmic equation with exponential notation.

47. $\log_{1/2} 16 = -4$

48. $\log_{0.4} 0.064 = 3$

Solve for x.

49. $\log_4 x = -3$

50. $\log_3 x = 2$

51. $\log_3 1 = x$

52. $\log_4 64 = x$

53. $\log_x 64 = 2$

54. $\log_x 81 = 4$

55. $\log_4 4^5 = x$

56. $\log_7 7^{-2} = x$

57. $5^{\log_5 4} = x$

58. $2^{\log_2 9} = x$

59. $\log_2 (3x - 1) = 4$

60. $\log_3 (2x + 5) = 2$

61. $\log_4 (x^2 - 3x) = 1$

62. $\log_8 (x^2 + 7x) = 1$

Graph each pair of equations on the same coordinate system.

63. $y = 2^x$ and $y = \log_2 x$

64. $y = \left(\dfrac{1}{2}\right)^x$ and $y = \log_{1/2} x$

(11.5) *Write each of the following as single logarithms.*

65. $\log_3 8 + \log_3 4$

66. $\log_2 6 + \log_2 3$

67. $\log_7 15 - \log_7 20$

68. $\log 18 - \log 12$

69. $\log_{11} 8 + \log_{11} 3 - \log_{11} 6$

70. $\log_5 14 + \log_5 3 - \log_5 21$

71. $2\log_5 x - 2\log_5 (x + 1) + \log_5 x$

72. $4\log_3 x - \log_3 x + \log_3 (x + 2)$

Use properties of logarithms to write each expression as a sum or difference of multiples of logarithms.

73. $\log_3 \dfrac{x^3}{x + 2}$

74. $\log_4 \dfrac{x + 5}{x^2}$

75. $\log_2 \dfrac{3x^2y}{z}$

76. $\log_7 \dfrac{yz^3}{x}$

If $\log_b 2 = 0.36$ and $\log_b 5 = 0.83$, find the following.

77. $\log_b 50$

78. $\log_b \dfrac{4}{5}$

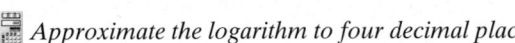 *(11.6)* *Use a calculator to approximate the logarithm to four decimal places.*

79. $\log 3.6$

80. $\log 0.15$

81. $\ln 1.25$

82. $\ln 4.63$

Find the exact value.

83. $\log 1000$

84. $\log \dfrac{1}{10}$

85. $\ln \dfrac{1}{e}$

86. $\ln e^4$

Solve each equation for x.

87. $\ln (2x) = 2$

88. $\ln (3x) = 1.6$

89. $\ln (2x - 3) = -1$

90. $\ln (3x + 1) = 2$

Use the formula $\ln \dfrac{I}{I_0} = -kx$ to solve radiation problems.
In this formula,

x = depth in millimeters
I = intensity of radiation
I_0 = initial intensity
k = a constant measure dependent on the material

Round answers to two decimal places.

91. Find the depth at which the intensity of the radiation passing through a lead shield is reduced to 3% of the original intensity if the value of k is 2.1.

92. If k is 3.2, find the depth at which 2% of the original radiation will penetrate.

Approximate the logarithm to four decimal places.

93. $\log_5 1.6$

94. $\log_3 4$

Use the formula $A = Pe^{rt}$ to solve the interest problems in which interest is compounded continuously. In this formula,

A = amount accrued (or owed)
P = principal invested (or loaned)
r = rate of interest
t = time in years

95. Bank of New York offers a 5-year, 6% continuously compounded investment option. Find the amount accrued if $1450 is invested.

96. Find the amount to which a $940 investment grows if it is invested at 11% compounded continuously for 3 years.

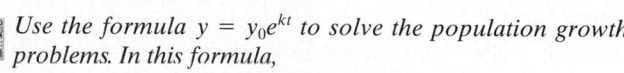 *(11.7) Solve each exponential equation for x. Give an exact solution and also approximate the solution to four decimal places.*

97. $3^{2x} = 7$

98. $6^{3x} = 5$

99. $3^{2x+1} = 6$

100. $4^{3x+2} = 9$

101. $5^{3x-5} = 4$

102. $8^{4x-2} = 3$

103. $2 \cdot 5^{x-1} = 1$

104. $3 \cdot 4^{x+5} = 2$

Solve the equation for x.

105. $\log_5 2 + \log_5 x = 2$

106. $\log_3 x + \log_3 10 = 2$

107. $\log(5x) - \log(x + 1) = 4$

108. $\ln(3x) - \ln(x - 3) = 2$

109. $\log_2 x + \log_2 2x - 3 = 1$

110. $-\log_6(4x + 7) + \log_6 x = 1$

Use the formula $y = y_0 e^{kt}$ to solve the population growth problems. In this formula,

$$y = \text{size of population}$$
$$y_0 = \text{initial count of population}$$
$$k = \text{rate of growth}$$
$$t = \text{time}$$

Round each answer to the nearest whole.

111. The population of mallard ducks in Nova Scotia is expected to grow at a rate of 6% per week during the spring migration. If 155,000 ducks are already in Nova Scotia, find how many are expected by the end of 4 weeks.

112. The population of Indonesia is growing at a rate of 1.5% per year. If the population in 1998 was 212,942,000, find the expected population by the year 2006. (*Source:* U.S. Bureau of the Census, International Data Base)

113. Japan is experiencing an annual growth rate of 0.2%. In 1998, the population of Japan was 125,932,000. How long will it take for the population to be 140,000,000? (*Source:* U.S. Bureau of the Census, International Data Base)

114. In 1998, Canada had a population of 30,675,000. How long will it take Canada to double in population if its growth rate is 1.1% annually? (*Source:* U.S. Bureau of the Census, International Data Base)

115. Egypt's population is increasing at a rate of 1.9% per year. How long will it take for its 1998 population of 66,050,000 to double in size? (*Source:* U.S. Bureau of the Census, International Data Base)

Use the compound interest equation $A = P\left(1 + \dfrac{r}{n}\right)^{nt}$ to solve the following. (See the directions for Exercises 42 and 43 for an explanation of this formula. Round answers to the nearest tenth.)

116. Find how long it will take a $5000 investment to grow to $10,000 if it is invested at 8% interest compounded quarterly.

117. An investment of $6000 has grown to $10,000 while the money was invested at 6% interest compounded monthly. Find how long it was invested.

Use a graphing calculator to solve each equation. Round all solutions to two decimal places.

118. $e^x = 2$

119. $10^{0.3x} = 7$

CHAPTER 11 TEST

If $f(x) = x$, $g(x) = x - 7$, and $h(x) = x^2 - 6x + 5$, find the following.

1. $(f \circ h)(0)$

2. $(g \circ f)(x)$

3. $(g \circ h)(x)$

On the same set of axes, graph the given one-to-one function and its inverse.

4. $f(x) = 7x - 14$

Determine whether the given graph is the graph of a one-to-one function.

5.

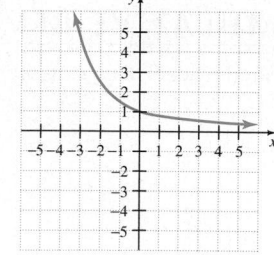

6.

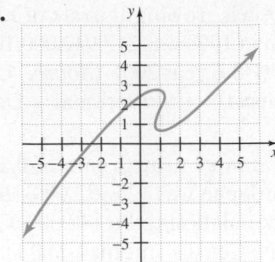

Determine whether each function is one-to-one. If it is one-to-one, find an equation or a set of ordered pairs that defines the inverse function of the given function.

7. $y = 6 - 2x$

8. $f = \{(0, 0), (2, 3), (-1, 5)\}$

9.

Word (Input)	Dog	Cat	House	Desk	Circle
First Letter of Word (Output)	d	c	h	d	c

Use the properties of logarithms to write each expression as a single logarithm.

10. $\log_3 6 + \log_3 4$

11. $\log_5 x + 3 \log_5 x - \log_5 (x + 1)$

12. Write the expression $\log_6 \dfrac{2x}{y^3}$ as the sum or difference of multiples of logarithms.

13. If $\log_b 3 = 0.79$ and $\log_b 5 = 1.16$, find the value of $\log_b \dfrac{3}{25}$.

14. Approximate $\log_7 8$ to four decimal places.

15. Solve $8^{x-1} = \dfrac{1}{64}$ for x. Give an exact solution.

16. Solve $3^{2x+5} = 4$ for x. Give an exact solution, and also approximate the solution to four decimal places.

Solve each logarithmic equation for x. Give an exact solution.

17. $\log_3 x = -2$

18. $\ln \sqrt{e} = x$

19. $\log_8 (3x - 2) = 2$

20. $\log_5 x + \log_5 3 = 2$

21. $\log_4 (x + 1) - \log_4 (x - 2) = 3$

22. Solve $\ln (3x + 7) = 1.31$ accurate to four decimal places.

23. Graph $y = \left(\dfrac{1}{2}\right)^x + 1$.

24. Graph the functions $y = 3^x$ and $y = \log_3 x$ on the same coordinate system.

Use the formula $A = P\left(1 + \dfrac{r}{n}\right)^{nt}$ to solve Exercises 25 and 26.

25. Find the amount in the account if $4000 is invested for 3 years at 9% interest compounded monthly.

26. Find how long it will take $2000 to grow to $3000 if the money is invested at 7% interest compounded semiannually. Round to the nearest whole.

Use the population growth formula $y = y_0 e^{kt}$ to solve Exercises 27 and 28.

27. The prairie dog population of the Grand Rapids area now stands at 57,000 animals. If the population is growing at a rate of 2.6% annually, find how many prairie dogs there will be in that area 5 years from now.

28. In an attempt to save an endangered species of wood duck, naturalists would like to increase the wood duck population from 400 to 1000 ducks. If the annual population growth rate is 6.2%, find how long it will take the naturalists to reach their goal. Round to the nearest whole year.

29. The formula $\log (1 + k) = \dfrac{0.3}{D}$ relates the doubling time D, in days, and the growth rate k for a population of mice. Find the rate at which the population is increasing if the doubling time is 56 days. Round to the nearest tenth of a percent.

30. Use a graphing calculator to approximate the solution of
$$e^{0.2x} = e^{-0.4x} + 2$$
to two decimal places.

CHAPTER 11 CUMULATIVE REVIEW

1. Name the property illustrated by each true statement.
 a. $3 \cdot y = y \cdot 3$
 b. $(x + 7) + 9 = x + (7 + 9)$
 c. $(b + 0) + 3 = b + 3$
 d. $2 \cdot (z \cdot 5) = 2 \cdot (5 \cdot z)$
 e. $-2 \cdot \left(-\dfrac{1}{2}\right) = 1$
 f. $-2 + 2 = 0$
 g. $-6 \cdot (y \cdot 2) = (-6 \cdot 2) \cdot y$

2. Solve: $\dfrac{x}{2} - 1 = \dfrac{2}{3}x - 3$

3. Simplify each expression.
 a. $\left(\dfrac{m}{n}\right)^7$
 b. $\left(\dfrac{x^3}{3y^5}\right)^4$

4. Perform each indicated operation. Write each result in standard decimal notation.
 a. $(8 \times 10^{-6})(7 \times 10^3)$

b. $\dfrac{12 \times 10^2}{6 \times 10^{-3}}$

d. $\sqrt[4]{-81}$

e. $\sqrt[3]{64x^3}$

5. Multiply: $(t + 2)$ by $(3t^2 - 4t + 2)$

6. Find $(x - 3)(x + 4)$ by the FOIL method.

7. Divide $3x^4 + 2x^3 - 8x + 6$ by $x^2 - 1$.

8. Factor $5(x + 3) + y(x + 3)$.

9. Factor $x^2 + 4x - 12$.

10. Factor $10x^2 - 13xy - 3y^2$.

11. Factor $x^2 - 25$.

12. Factor $30a^2b^3 + 55a^2b^2 - 35a^2b$.

13. Solve $3x^3 - 12x = 0$.

14. For a TV commercial, a piece of luggage is dropped from a cliff 256 feet above the ground to show the durability of the luggage. Neglecting air resistance, the height $h(t)$ in feet of the luggage above the ground after t seconds is given by the quadratic equation
$$h(t) = -16t^2 + 256$$
Find how long it takes for the luggage to hit the ground.

15. For the ICL Production Company, the rational function $C(x) = \dfrac{2.6x + 10,000}{x}$ describes the company's cost per disc of pressing x compact discs. Find the cost per disc for pressing

a. 100 compact discs

b. 1000 compact discs

16. Simplify the following expressions.

a. $\sqrt[4]{81}$

b. $\sqrt[5]{-243}$

c. $-\sqrt{25}$

17. Use rational exponents to write as a single radical.

a. $\sqrt{x} \cdot \sqrt[4]{x}$

b. $\dfrac{\sqrt{x}}{\sqrt[3]{x}}$

c. $\sqrt[3]{3} \cdot \sqrt{2}$

18. Multiply.

a. $\sqrt{3}(5 + \sqrt{30})$

b. $(\sqrt{5} - \sqrt{6})(\sqrt{7} + 1)$

c. $(7\sqrt{x} + 5)(3\sqrt{x} - \sqrt{5})$

d. $(4\sqrt{3} - 1)^2$

e. $(\sqrt{2x} - 5)(\sqrt{2x} + 5)$

19. Rationalize the denominator of $\dfrac{\sqrt[4]{x}}{\sqrt[4]{81y^5}}$.

20. Solve $\sqrt{4 - x} = x - 2$.

21. Solve $3x^2 - 9x + 8 = 0$ by completing the square.

22. Solve $\dfrac{3x}{x - 2} - \dfrac{x + 1}{x} = \dfrac{6}{x(x - 2)}$.

23. Solve $x^2 - 4x \leq 0$.

24. Graph $F(x) = (x - 3)^2 + 1$.

Working in Finance

A certified financial planner works with clients to develop a sound financial plan that helps the client reach his or her life goals. Financial planning is a growing field: As Baby Boomers approach retirement age and life spans lengthen overall, more and more people will seek professional assistance with financial management.

Financial planners work for investment firms, accounting firms, insurance companies, banks, credit counseling organizations, law firms, or in private practice. Through interviews and discussions, they assess clients' current financial positions. Planners then give advice on retirement planning, insurance needs, investment options, estate planning, tax strategies, and employee benefits. Financial planners may then help clients implement their new financial plans. Certified financial planners use math and problem-solving skills in such tasks as analyzing clients' current cash flow, estimating cash needs for future goals, and calculating investment returns.

 For more information about a career as a certified financial planner, visit the Certified Financial Planner Board of Standards Website by first going to www.prenhall.com/martin-gay.

In the Spotlight on Decision Making feature on page 761, you will have the opportunity to make a decision about reaching a client's retirement goals as a certified financial planner.

SEQUENCES, SERIES, AND THE BINOMIAL THEOREM

12

Having explored in some depth the concept of function, we turn now in this final chapter to *sequences*. In one sense, a sequence is simply an ordered list of numbers. In another sense, a sequence is itself a function. Phenomena modeled by such functions are everywhere around us. The starting place for all mathematics is the sequence of natural numbers: 1, 2, 3, 4, and so on.

Sequences lead us to *series*, which are a sum of ordered numbers. Through series we gain new insight, for example about the expansion of a binomial $(a + b)^n$, the concluding topic of this chapter.

739

12.1 SEQUENCES

CD-ROM   SSM

SSG   Video

▶ **OBJECTIVES**

1. Write the terms of a sequence given its general term.
2. Find the general term of a sequence.
3. Solve applications that involve sequences.

Suppose that a town's present population of 100,000 is growing by 5% each year. After the first year, the town's population will be

$$100{,}000 + 0.05(100{,}000) = 105{,}000$$

After the second year, the town's population will be

$$105{,}000 + 0.05(105{,}000) = 110{,}250$$

After the third year, the town's population will be

$$110{,}250 + 0.05(110{,}250) \approx 115{,}763$$

If we continue to calculate, the town's yearly population can be written as the **infinite sequence** of numbers

$$105{,}000, \ 110{,}250, \ 115{,}763, \ldots$$

If we decide to stop calculating after a certain year (say, the fourth year), we obtain the **finite sequence**

$$105{,}000, \ 110{,}250, \ 115{,}763, \ 121{,}551$$

SEQUENCES

An infinite sequence is a function whose domain is the set of natural numbers $\{1, 2, 3, 4, \ldots\}$.

A finite sequence is a function whose domain is the set of natural numbers $\{1, 2, 3, 4, \ldots, n\}$, where n is some natural number.

1

Given the sequence $2, 4, 8, 16, \ldots$, we say that each number is a **term** of the sequence. Because a sequence is a function, we could describe it by writing $f(n) = 2^n$, where n is a natural number. Instead, we use the notation

$$a_n = 2^n$$

Some function values are

$$a_1 = 2^1 = 2 \qquad \text{First term of the sequence}$$
$$a_2 = 2^2 = 4 \qquad \text{Second term}$$
$$a_3 = 2^3 = 8 \qquad \text{Third term}$$
$$a_4 = 2^4 = 16 \qquad \text{Fourth term}$$
$$a_{10} = 2^{10} = 1024 \qquad \text{Tenth term}$$

The nth term of the sequence a_n is called the **general term**.

Example 1   Write the first five terms of the sequence whose general term is given by

$$a_n = n^2 - 1$$

Solution   Evaluate a_n, where n is 1, 2, 3, 4, and 5.

$$a_n = n^2 - 1$$

$a_1 = 1^2 - 1 = 0$ Replace n with 1.

$a_2 = 2^2 - 1 = 3$ Replace n with 2.

$a_3 = 3^2 - 1 = 8$ Replace n with 3.

$a_4 = 4^2 - 1 = 15$ Replace n with 4.

$a_5 = 5^2 - 1 = 24$ Replace n with 5.

Thus, the first five terms of the sequence $a_n = n^2 - 1$ are 0, 3, 8, 15, and 24.

Example 2   If the general term of a sequence is given by $a_n = \dfrac{(-1)^n}{3n}$, find

    a. the first term of the sequence

    b. a_8

    c. the one-hundredth term of the sequence

    d. a_{15}

Solution   **a.** $a_1 = \dfrac{(-1)^1}{3(1)} = -\dfrac{1}{3}$ Replace n with 1.

    b. $a_8 = \dfrac{(-1)^8}{3(8)} = \dfrac{1}{24}$ Replace n with 8.

    c. $a_{100} = \dfrac{(-1)^{100}}{3(100)} = \dfrac{1}{300}$ Replace n with 100.

    d. $a_{15} = \dfrac{(-1)^{15}}{3(15)} = -\dfrac{1}{45}$ Replace n with 15.

2   Suppose we know the first few terms of a sequence and want to find a general term that fits the pattern of the first few terms.

Example 3   Find a general term a_n of the sequence whose first few terms are given.

    a. 1, 4, 9, 16, . . .

    b. $\dfrac{1}{1}, \dfrac{1}{2}, \dfrac{1}{3}, \dfrac{1}{4}, \dfrac{1}{5}, \ldots$

    c. $-3, -6, -9, -12, \ldots$

    d. $\dfrac{1}{2}, \dfrac{1}{4}, \dfrac{1}{8}, \dfrac{1}{16}, \ldots$

Solution **a.** These numbers are the squares of the first four natural numbers, so a general term might be $a_n = n^2$.

b. These numbers are the reciprocals of the first five natural numbers, so a general term might be $a_n = \dfrac{1}{n}$.

c. These numbers are the product of -3 and the first four natural numbers, so a general term might be $a_n = -3n$.

d. Notice that the denominators double each time.

$$\frac{1}{2}, \quad \frac{1}{2 \cdot 2}, \quad \frac{1}{2(2 \cdot 2)}, \quad \frac{1}{2(2 \cdot 2 \cdot 2)}$$

or

$$\frac{1}{2^1}, \quad \frac{1}{2^2}, \quad \frac{1}{2^3}, \quad \frac{1}{2^4}$$

We might then suppose that the general term is $a_n = \dfrac{1}{2^n}$. ∎

3 Sequences model many phenomena of the physical world, as illustrated by the following example.

Example 4 **FINDING A PUPPY'S WEIGHT GAIN**

The amount of weight, in pounds, a puppy gains in each month of its first year is modeled by a sequence whose general term is $a_n = n + 4$, where n is the number of the month. Write the first five terms of the sequence, and find how much weight the puppy should gain in its fifth month.

Solution Evaluate $a_n = n + 4$ when n is 1, 2, 3, 4, and 5.

$$a_1 = 1 + 4 = 5$$
$$a_2 = 2 + 4 = 6$$
$$a_3 = 3 + 4 = 7$$
$$a_4 = 4 + 4 = 8$$
$$a_5 = 5 + 4 = 9$$

The puppy should gain 9 pounds in its fifth month. ∎

SPOTLIGHT ON DECISION MAKING

Suppose you are considering two job offers. The first job offer pays $11.50 per hour and guarantees a $0.65-per-hour raise each year. The second job offer pays $10.75 per hour and guarantees a $1.10-per-hour raise each year. If one of your goals is to be earning at least $15 per hour in 5 years, which job offer would you accept? Explain. What other factors would you want to consider?

Exercise Set 12.1

Write the first five terms of each sequence whose general term is given. See Example 1.

1. $a_n = n + 4$

2. $a_n = 5 - n$

3. $a_n = (-1)^n$

4. $a_n = (-2)^n$

5. $a_n = \dfrac{1}{n + 3}$

6. $a_n = \dfrac{1}{7 - n}$

7. $a_n = 2n$

8. $a_n = -6n$

9. $a_n = -n^2$

10. $a_n = n^2 + 2$

11. $a_n = 2^n$

12. $a_n = 3^{n-2}$

13. $a_n = 2n + 5$

14. $a_n = 1 - 3n$

15. $a_n = (-1)^n n^2$

16. $a_n = (-1)^{n+1}(n - 1)$

Find the indicated term for each sequence whose general term is given. See Example 2.

17. $a_n = 3n^2; a_5$

18. $a_n = -n^2; a_{15}$

19. $a_n = 6n - 2; a_{20}$

20. $a_n = 100 - 7n; a_{50}$

21. $a_n = \dfrac{n + 3}{n}; a_{15}$

22. $a_n = \dfrac{n}{n + 4}; a_{24}$

23. $a_n = (-3)^n; a_6$

24. $a_n = 5^{n+1}; a_3$

25. $a_n = \dfrac{n - 2}{n + 1}; a_6$

26. $a_n = \dfrac{n + 3}{n + 4}; a_8$

27. $a_n = \dfrac{(-1)^n}{n}; a_8$

28. $a_n = \dfrac{(-1)^n}{2n}; a_{100}$

29. $a_n = -n^2 + 5; a_{10}$

30. $a_n = 8 - n^2; a_{20}$

31. $a_n = \dfrac{(-1)^n}{n + 6}; a_{19}$

32. $a_n = \dfrac{n - 4}{(-2)^n}; a_6$

Find a general term a_n for each sequence whose first four terms are given. See Example 3.

33. $3, 7, 11, 15$

34. $2, 7, 12, 17$

35. $-2, -4, -8, -16$

36. $-4, 16, -64, 256$

37. $\dfrac{1}{3}, \dfrac{1}{9}, \dfrac{1}{27}, \dfrac{1}{81}$

38. $\dfrac{2}{5}, \dfrac{2}{25}, \dfrac{2}{125}, \dfrac{2}{625}$

Solve. See Example 4.

39. The distance, in feet, that a Thermos dropped from a cliff falls in each consecutive second is modeled by a sequence whose general term is $a_n = 32n - 16$, where n is the number of seconds. Find the distance the Thermos falls in the second, third, and fourth seconds.

40. The population size of a culture of bacteria triples every hour such that its size is modeled by the sequence

$a_n = 50(3)^{n-1}$, where n is the number of the hour just beginning. Find the size of the culture at the beginning of the fourth hour and the size of the culture at the beginning of the first hour.

41. Mrs. Laser agrees to give her son Mark an allowance of $0.10 on the first day of his 14-day vacation, $0.20 on the second day, $0.40 on the third day, and so on. Write an equation of a sequence whose terms correspond to Mark's allowance. Find the allowance Mark will receive on the last day of his vacation.

42. A small theater has 10 rows with 12 seats in the first row, 15 seats in the second row, 18 seats in the third row, and so on. Write an equation of a sequence whose terms correspond to the seats in each row. Find the number of seats in the eighth row.

43. The number of cases of a new infectious disease is doubling every year such that the number of cases is modeled by a sequence whose general term is $a_n = 75(2)^{n-1}$, where n is the number of the year just beginning. Find how many cases there will be at the beginning of the sixth year. Find how many cases there were at the beginning of the first year.

44. A new college had an initial enrollment of 2700 students in 2000, and each year the enrollment increases by 150 students. Find the enrollment for each of 5 years, beginning with 2000.

45. An endangered species of sparrow had an estimated population of 800 in 2000, and scientists predict that its population will decrease by half each year. Estimate the population in 2004. Estimate the year the sparrow will be extinct.

46. A **Fibonacci sequence** is a special type of sequence in which the first two terms are 1 and each term thereafter is the sum of the two previous terms: $1, 1, 2, 3, 5, 8, \ldots$. Many plants and animals seem to grow according to a Fibonacci sequence, including pine cones, pineapple scales, nautilus shells, and certain flowers. Write the first 15 terms of the Fibonacci sequence.

Find the first five terms of each sequence. Round each term after the first to four decimal places.

47. $a_n = \dfrac{1}{\sqrt{n}}$

48. $\dfrac{\sqrt{n}}{\sqrt{n} + 1}$

49. $a_n = \left(1 + \dfrac{1}{n}\right)^n$

50. $a_n = \left(1 + \dfrac{0.05}{n}\right)^n$

REVIEW EXERCISES

Sketch the graph of each quadratic function. See Section 10.6.

51. $f(x) = (x - 1)^2 + 3$

52. $f(x) = (x - 2)^2 + 1$

53. $f(x) = 2(x + 4)^2 + 2$

54. $f(x) = 3(x - 3)^2 + 4$

12.2 ARITHMETIC AND GEOMETRIC SEQUENCES

CD-ROM   SSM

SSG   Video

▶ **OBJECTIVES**

1. Identify arithmetic sequences and their common differences.
2. Identify geometric sequences and their common ratios.

---

1   Find the first four terms of the sequence whose general term is $a_n = 5 + (n - 1)3$.

$$a_1 = 5 + (1 - 1)3 = 5 \qquad \text{Replace } n \text{ with 1.}$$
$$a_2 = 5 + (2 - 1)3 = 8 \qquad \text{Replace } n \text{ with 2.}$$
$$a_3 = 5 + (3 - 1)3 = 11 \qquad \text{Replace } n \text{ with 3.}$$
$$a_4 = 5 + (4 - 1)3 = 14 \qquad \text{Replace } n \text{ with 4.}$$

The first four terms are $5, 8, 11,$ and 14. Notice that the difference of any two successive terms is 3.

$$8 - 5 = 3$$
$$11 - 8 = 3$$
$$14 - 11 = 3$$

$$\underset{\substack{\uparrow \\ n\text{th} \\ \text{term}}}{a_n} - \underset{\substack{\uparrow \\ \text{previous} \\ \text{term}}}{a_{n-1}} = 3$$

Because the difference of any two successive terms is a constant, we call the sequence an **arithmetic sequence,** or an **arithmetic progression.** The constant difference d in successive terms is called the **common difference.** In this example, d is 3.

> ### ARITHMETIC SEQUENCE AND COMMON DIFFERENCE
>
> An **arithmetic sequence** is a sequence in which each term (after the first) differs from the preceding term by a constant amount d. The constant d is called the **common difference** of the sequence.

The sequence $2, 6, 10, 14, 18, \ldots$ is an arithmetic sequence. Its common difference is 4. Given the first term a_1 and the common difference d of an arithmetic sequence, we can find any term of the sequence.

Example 1   Write the first five terms of the arithmetic sequence whose first term is 7 and whose common difference is 2.

Solution

$$a_1 = 7$$
$$a_2 = 7 + 2 = 9$$
$$a_3 = 9 + 2 = 11$$
$$a_4 = 11 + 2 = 13$$
$$a_5 = 13 + 2 = 15$$

The first five terms are $7, 9, 11, 13, 15$.

Notice the general pattern of the terms in Example 1.

$$a_1 = 7$$
$$a_2 = 7 + 2 = 9 \quad \text{or} \quad a_2 = a_1 + d$$
$$a_3 = 9 + 2 = 11 \quad \text{or} \quad a_3 = a_2 + d = (a_1 + d) + d = a_1 + 2d$$
$$a_4 = 11 + 2 = 13 \quad \text{or} \quad a_4 = a_3 + d = (a_1 + 2d) + d = a_1 + 3d$$
$$a_5 = 13 + 2 = 15 \quad \text{or} \quad a_5 = a_4 + d = (a_1 + 3d) + d = a_1 + 4d$$

(subscript $- 1$) is multiplier

The pattern on the right suggests that the general term a_n of an arithmetic sequence is given by

$$a_n = a_1 + (n - 1)d$$

GENERAL TERM OF AN ARITHMETIC SEQUENCE

The general term a_n of an arithmetic sequence is given by

$$a_n = a_1 + (n - 1)d$$

where a_1 is the first term and d is the common difference.

Example 2 Consider the arithmetic sequence whose first term is 3 and common difference is -5.

a. Write an expression for the general term a_n.
b. Find the twentieth term of this sequence.

Solution **a.** Since this is an arithmetic sequence, the general term a_n is given by $a_n = a_1 + (n - 1)d$. Here, $a_1 = 3$ and $d = -5$, so

$$a_n = 3 + (n - 1)(-5) \quad \text{Let } a_1 = 3 \text{ and } d = -5.$$
$$= 3 - 5n + 5 \quad \text{Multiply.}$$
$$= 8 - 5n \quad \text{Simplify.}$$

b. $a_n = 8 - 5n$
$$a_{20} = 8 - 5 \cdot 20 \quad \text{Let } n = 20.$$
$$= 8 - 100 = -92$$

Example 3 Find the eleventh term of the arithmetic sequence whose first three terms are 2, 9, and 16.

Solution Since the sequence is arithmetic, the eleventh term is

$$a_{11} = a_1 + (11 - 1)d = a_1 + 10d$$

We know a_1 is the first term of the sequence, so $a_1 = 2$. Also, d is the constant difference of terms, so $d = a_2 - a_1 = 9 - 2 = 7$. Thus,

$$a_{11} = a_1 + 10d$$
$$= 2 + 10 \cdot 7 \quad \text{Let } a_1 = 2 \text{ and } d = 7.$$
$$= 72$$

Example 4   If the third term of an arithmetic progression is 12 and the eighth term is 27, find the fifth term.

Solution   We need to find a_1 and d to write the general term, which then enables us to find a_5, the fifth term. The given facts about terms a_3 and a_8 lead to a system of linear equations.

$$\begin{cases} a_3 = a_1 + (3-1)d \\ a_8 = a_1 + (8-1)d \end{cases} \quad \text{or} \quad \begin{cases} 12 = a_1 + 2d \\ 27 = a_1 + 7d \end{cases}$$

Next, we solve the system $\begin{cases} 12 = a_1 + 2d \\ 27 = a_1 + 7d \end{cases}$ by elimination. Multiply both sides of the second equation by -1 so that

$$\begin{cases} 12 = a_1 + 2d \\ -1(27) = -1(a_1 + 7d) \end{cases} \quad \begin{array}{c} \text{simplifies} \\ \text{to} \end{array} \quad \begin{cases} 12 = a_1 + 2d \\ -27 = -a_1 - 7d \end{cases}$$

$$-15 = -5d \qquad \text{Add the equations.}$$

$$3 = d \qquad \text{Divide both sides by } -5.$$

To find a_1, let $d = 3$ in $12 = a_1 + 2d$. Then

$$12 = a_1 + 2(3)$$

$$12 = a_1 + 6$$

$$6 = a_1$$

Thus, $a_1 = 6$ and $d = 3$, so

$$a_n = 6 + (n-1)(3)$$

$$= 6 + 3n - 3$$

$$= 3 + 3n$$

and

$$a_5 = 3 + 3 \cdot 5 = 18$$ ∎

Example 5   **FINDING SALARY**

Donna Theime has an offer for a job starting at $40,000 per year and guaranteeing her a raise of $1600 per year for the next 5 years. Write the general term for the arithmetic sequence that models Donna's potential annual salaries, and find her salary for the fourth year.

Solution   The first term, a_1, is 40,000, and d is 1600. So

$$a_n = 40{,}000 + (n-1)(1600) = 38{,}400 + 1600n$$

$$a_4 = 38{,}400 + 1600 \cdot 4 = 44{,}800$$

Her salary for the fourth year will be $44,800. ∎

2 We now investigate a **geometric sequence,** also called a **geometric progression.** In the sequence $5, 15, 45, 135, \ldots$, each term after the first is the *product* of 3 and the preceding term. This pattern of multiplying by a constant to get the next term defines a geometric sequence. The constant is called the **common ratio** because it is the ratio of any term (after the first) to its preceding term.

$$\frac{15}{5} = 3$$

$$\frac{45}{15} = 3$$

$$\frac{135}{45} = 3$$

$$\vdots$$

$$n\text{th term} \longrightarrow \frac{a_n}{a_{n-1}} = 3$$
$$\text{previous term} \longrightarrow$$

GEOMETRIC SEQUENCE AND COMMON RATIO

A **geometric sequence** is a sequence in which each term (after the first) is obtained by multiplying the preceding term by a constant r. The constant r is called the **common ratio** of the sequence.

The sequence $12, 6, 3, \dfrac{3}{2}, \ldots$ is geometric since each term after the first is the product of the previous term and $\dfrac{1}{2}$.

Example 6 Write the first five terms of a geometric sequence whose first term is 7 and whose common ratio is 2.

Solution
$$a_1 = 7$$
$$a_2 = 7(2) = 14$$
$$a_3 = 14(2) = 28$$
$$a_4 = 28(2) = 56$$
$$a_5 = 56(2) = 112$$

The first five terms are $7, 14, 28, 56,$ and 112.

Notice the general pattern of the terms in Example 6.

$$a_1 = 7$$
$$a_2 = 7(2) = 14 \qquad \text{or} \quad a_2 = a_1(r)$$
$$a_3 = 14(2) = 28 \qquad \text{or} \quad a_3 = a_2(r) = (a_1 \cdot r) \cdot r = a_1 r^2$$
$$a_4 = 28(2) = 56 \qquad \text{or} \quad a_4 = a_3(r) = (a_1 \cdot r^2) \cdot r = a_1 r^3$$
$$a_5 = 56(2) = 112 \qquad \text{or} \quad a_5 = a_4(r) = (a_1 \cdot r^3) \cdot r = a_1 r^4$$
$$\longrightarrow (\text{subscript} - 1) \text{ is power}$$

The pattern on the right on the previous page suggests that the general term of a geometric sequence is given by $a_n = a_1 r^{n-1}$.

GENERAL TERM OF A GEOMETRIC SEQUENCE

The general term a_n of a geometric sequence is given by

$$a_n = a_1 r^{n-1}$$

where a_1 is the first term and r is the common ratio.

Example 7   Find the eighth term of the geometric sequence whose first term is 12 and whose common ratio is $\frac{1}{2}$.

Solution   Since this is a geometric sequence, the general term a_n is given by

$$a_n = a_1 r^{n-1}$$

Here $a_1 = 12$ and $r = \frac{1}{2}$, so $a_n = 12\left(\frac{1}{2}\right)^{n-1}$. Evaluate a_n for $n = 8$.

$$a_8 = 12\left(\frac{1}{2}\right)^{8-1} = 12\left(\frac{1}{2}\right)^7 = 12\left(\frac{1}{128}\right) = \frac{3}{32}$$

Example 8   Find the fifth term of the geometric sequence whose first three terms are $2, -6,$ and 18.

Solution   Since the sequence is geometric and $a_1 = 2$, the fifth term must be $a_1 r^{5-1}$, or $2r^4$. We know that r is the common ratio of terms, so r must be $\frac{-6}{2}$, or -3. Thus,

$$a_5 = 2r^4$$
$$a_5 = 2(-3)^4 = 162$$

Example 9   If the second term of a geometric sequence is $\frac{5}{4}$ and the third term is $\frac{5}{16}$, find the first term and the common ratio.

Solution   Notice that $\frac{5}{16} \div \frac{5}{4} = \frac{1}{4}$, so $r = \frac{1}{4}$. Then

$$a_2 = a_1\left(\frac{1}{4}\right)^{2-1}$$

$$\frac{5}{4} = a_1\left(\frac{1}{4}\right)^1, \text{ or } a_1 = 5 \qquad \text{Replace } a_2 \text{ with } \frac{5}{4}.$$

The first term is 5.

Example 10   PREDICTING POPULATION OF A BACTERIAL CULTURE

The population size of a bacterial culture growing under controlled conditions is doubling each day. Predict how large the culture will be at the beginning of day 7 if it measures 10 units at the beginning of day 1.

Solution   Since the culture doubles in size each day, the population sizes are modeled by a geometric sequence. Here $a_1 = 10$ and $r = 2$. Thus,

$$a_n = a_1 r^{n-1} = 10(2)^{n-1} \quad \text{and} \quad a_7 = 10(2)^{7-1} = 640$$

The bacterial culture should measure 640 units at the beginning of day 7.   ▬

SPOTLIGHT ON DECISION MAKING

Suppose you are a research biologist studying a particular strain of bacteria that grows at a rate of 1.5 times per hour. For a particular experiment, you will start with a culture of 200 units of bacteria and will allow the culture to grow for 7 hours. Decide whether a culture dish that holds 5000 units will be large enough for this experiment. If not, would a dish that holds 10,000 units be a better choice?

Exercise Set 12.2

Write the first five terms of the arithmetic or geometric sequence whose first term, a_1, and common difference, d, or common ratio, r, are given. See Examples 1 and 6.

1. $a_1 = 4; d = 2$ **2.** $a_1 = 3; d = 10$

3. $a_1 = 6; d = -2$ **4.** $a_1 = -20; d = 3$

5. $a_1 = 1; r = 3$ **6.** $a_1 = -2; r = 2$

7. $a_1 = 48; r = \dfrac{1}{2}$ **8.** $a_1 = 1; r = \dfrac{1}{3}$

Find the indicated term of each sequence. See Examples 2 and 7.

9. The eighth term of the arithmetic sequence whose first term is 12 and whose common difference is 3

10. The twelfth term of the arithmetic sequence whose first term is 32 and whose common difference is -4

11. The fourth term of the geometric sequence whose first term is 7 and whose common ratio is -5

12. The fifth term of the geometric sequence whose first term is 3 and whose common ratio is 3

13. The fifteenth term of the arithmetic sequence whose first term is -4 and whose common difference is -4

14. The sixth term of the geometric sequence whose first term is 5 and whose common ratio is -4

Find the indicated term of each sequence. See Examples 3 and 8.

15. The ninth term of the arithmetic sequence $0, 12, 24, \ldots$

16. The thirteenth term of the arithmetic sequence $-3, 0, 3, \ldots$

17. The twenty-fifth term of the arithmetic sequence $20, 18, 16, \ldots$

18. The ninth term of the geometric sequence $5, 10, 20, \ldots$

19. The fifth term of the geometric sequence $2, -10, 50, \ldots$

20. The sixth term of the geometric sequence $\dfrac{1}{2}, \dfrac{3}{2}, \dfrac{9}{2}, \ldots$

Find the indicated term of each sequence. See Examples 4 and 9.

21. The eighth term of the arithmetic sequence whose fourth term is 19 and whose fifteenth term is 52

22. If the second term of an arithmetic sequence is 6 and the tenth term is 30, find the twenty-fifth term.

23. If the second term of an arithmetic progression is -1 and the fourth term is 5, find the ninth term.

24. If the second term of a geometric progression is 15 and the third term is 3, find a_1 and r.

25. If the second term of a geometric progression is $-\dfrac{4}{3}$ and the third term is $\dfrac{8}{3}$, find a_1 and r.

26. If the third term of a geometric sequence is 4 and the fourth term is -12, find a_1 and r.

27. Explain why 14, 10, and 6 may be the first three terms of an arithmetic sequence when it appears we are subtracting instead of adding to get the next term.

28. Explain why 80, 20, and 5 may be the first three terms of a geometric sequence when it appears we are dividing instead of multiplying to get the next term.

Given are the first three terms of a sequence that is either arithmetic or geometric. If the sequence is arithmetic, find a_1 and d. If a sequence is geometric, find a_1 and r.

29. 2, 4, 6

30. 8, 16, 24

31. 5, 10, 20

32. 2, 6, 18

33. $\dfrac{1}{2}, \dfrac{1}{10}, \dfrac{1}{50}$

34. $\dfrac{2}{3}, \dfrac{4}{3}, 2$

35. $x, 5x, 25x$

36. $y, -3y, 9y$

37. $p, p + 4, p + 8$

38. $t, t - 1, t - 2$

Find the indicated term of each sequence.

39. The twenty-first term of the arithmetic sequence whose first term is 14 and whose common difference is $\dfrac{1}{4}$

40. The fifth term of the geometric sequence whose first term is 8 and whose common ratio is -3

41. The fourth term of the geometric sequence whose first term is 3 and whose common ratio is $-\dfrac{2}{3}$

42. The fourth term of the arithmetic sequence whose first term is 9 and whose common difference is 5

43. The fifteenth term of the arithmetic sequence $\dfrac{3}{2}, 2, \dfrac{5}{2}, \ldots$

44. The eleventh term of the arithmetic sequence $2, \dfrac{5}{3}, \dfrac{4}{3}, \ldots$

45. The sixth term of the geometric sequence $24, 8, \dfrac{8}{3}, \ldots$

46. The eighteenth term of the arithmetic sequence $5, 2, -1, \ldots$

47. If the third term of an arithmetic sequence is 2 and the seventeenth term is -40, find the tenth term.

48. If the third term of a geometric sequence is -28 and the fourth term is -56, find a_1 and r.

Solve. See Examples 5 and 10.

49. An auditorium has 54 seats in the first row, 58 seats in the second row, 62 seats in the third row, and so on. Find the general term of this arithmetic sequence and the number of seats in the twentieth row.

50. A triangular display of cans in a grocery store has 20 cans in the first row, 17 cans in the next row, and so on, in an arithmetic sequence. Find the general term and the number of cans in the fifth row. Find how many rows there are in the display and how many cans are in the top row.

51. The initial size of a virus culture is 6 units, and it triples its size every day. Find the general term of the geometric sequence that models the culture's size.

52. A real estate investment broker predicts that a certain property will increase in value 15% each year. Thus, the yearly property values can be modeled by a geometric sequence whose common ratio r is 1.15. If the initial property value was $500,000, write the first four terms of the sequence and predict the value at the end of the third year.

53. A rubber ball is dropped from a height of 486 feet, and it continues to bounce one-third the height from which it last fell. Write out the first five terms of this geometric sequence and find the general term. Find how many bounces it takes for the ball to rebound less than 1 foot.

54. On the first swing, the length of the arc through which a pendulum swings is 50 inches. The length of each successive swing is 80% of the preceding swing. Determine whether this sequence is arithmetic or geometric. Find the length of the fourth swing.

55. Jose takes a job that offers a monthly starting salary of $4000 and guarantees him a monthly raise of $125 during his first year of training. Find the general term of this arithmetic sequence and his monthly salary at the end of his training.

56. At the beginning of Claudia Schaffer's exercise program, she rides 15 minutes on the Lifecycle. Each week she increases her riding time by 5 minutes. Write the general term of this arithmetic sequence, and find her riding time after 7 weeks. Find how many weeks it takes her to reach a riding time of 1 hour.

57. If a radioactive element has a half-life of 3 hours, then x grams of the element dwindles to $\dfrac{x}{2}$ grams after 3 hours. If a nuclear reactor has 400 grams of that radioactive element, find the amount of radioactive material after 12 hours.

  Write the first four terms of the arithmetic or geometric sequence whose first term, a_1, and common difference, d, or common ratio, r, are given.

58. $a_1 = \$3720, d = -\$268.50$

59. $a_1 = \$11{,}782.40, r = 0.5$

60. $a_1 = 26.8, r = 2.5$

61. $a_1 = 19.652; d = -0.034$

62. Describe a situation in your life that can be modeled by a geometric sequence. Write an equation for the sequence.

63. Describe a situation in your life that can be modeled by an arithmetic sequence. Write an equation for the sequence.

REVIEW EXERCISES

Evaluate. See Section 1.3.

64. $5(1) + 5(2) + 5(3) + 5(4)$

65. $\dfrac{1}{3(1)} + \dfrac{1}{3(2)} + \dfrac{1}{3(3)}$

66. $2(2 - 4) + 3(3 - 4) + 4(4 - 4)$

67. $3^0 + 3^1 + 3^2 + 3^3$

68. $\dfrac{1}{4(1)} + \dfrac{1}{4(2)} + \dfrac{1}{4(3)}$

69. $\dfrac{8 - 1}{8 + 1} + \dfrac{8 - 2}{8 + 2} + \dfrac{8 - 3}{8 + 3}$

12.3 SERIES

CD-ROM SSM SSG Video

▶ **OBJECTIVES**

1. Identify finite and infinite series and use summation notation.
2. Find partial sums.

1 A person who conscientiously saves money by saving first $100 and then saving $10 more each month than he saved the preceding month is saving money according to the arithmetic sequence

$$a_n = 100 + 10(n - 1)$$

Following this sequence, he can predict how much money he should save for any particular month. But if he also wants to know how much money *in total* he has saved, say, by the fifth month, he must find the *sum* of the first five terms of the sequence

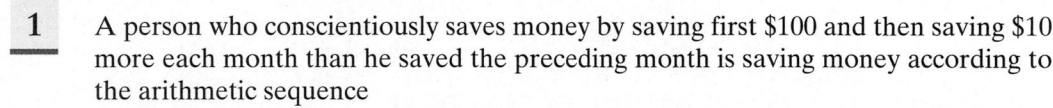

$$\underbrace{100}_{a_1} + \underbrace{100 + 10}_{a_2} + \underbrace{100 + 20}_{a_3} + \underbrace{100 + 30}_{a_4} + \underbrace{100 + 40}_{a_5}$$

A sum of the terms of a sequence is called a **series** (the plural is also "series"). As our example here suggests, series are frequently used to model financial and natural phenomena.

A series is a **finite series** if it is the sum of only the first k terms of the sequence, for some natural number k. A series is an **infinite series** if it is the sum of all the terms of the sequence. For example,

Sequence	*Series*	
$5, 9, 13$	$5 + 9 + 13$	Finite; k is 3.
$5, 9, 13, \ldots$	$5 + 9 + 13 + \cdots$	Infinite
$4, -2, 1, -\dfrac{1}{2}, \dfrac{1}{4}$	$4 + (-2) + 1 + \left(-\dfrac{1}{2}\right) + \left(\dfrac{1}{4}\right)$	Finite; k is 5.
$4, -2, 1, \ldots$	$4 + (-2) + 1 + \cdots$	Infinite
$3, 6, \ldots, 99$	$3 + 6 + \cdots + 99$	Finite; k is 33.

A shorthand notation for denoting a series when the general term of the sequence is known is called **summation notation.** The Greek uppercase letter **sigma,** Σ, is used to mean "sum." The expression $\displaystyle\sum_{n=1}^{5}(3n + 1)$ is read "the sum of $3n + 1$ as n goes from 1 to 5"; this expression means the sum of the first five terms of the sequence whose general term is $a_n = 3n + 1$. Often, the variable i is used instead of n in summation notation: $\displaystyle\sum_{i=1}^{5}(3i + 1)$. Whether we use n, i, k, or some other variable, the variable is called the **index of summation.** The notation $i = 1$ below the symbol Σ indicates the beginning value of i, and the number 5 above the symbol Σ indicates the ending value of i. Thus, the terms of the sequence are found by successively replacing i with the natural numbers 1, 2, 3, 4, 5. To find the sum, we write out the terms and then add.

$$\sum_{i=1}^{5}(3i + 1) = (3 \cdot 1 + 1) + (3 \cdot 2 + 1) + (3 \cdot 3 + 1)$$
$$+ (3 \cdot 4 + 1) + (3 \cdot 5 + 1)$$
$$= 4 + 7 + 10 + 13 + 16 = 50$$

Example 1 Evaluate.

a. $\displaystyle\sum_{i=0}^{6}\dfrac{i - 2}{2}$ **b.** $\displaystyle\sum_{i=3}^{5}2^i$

Solution **a.** $\displaystyle\sum_{i=0}^{6}\dfrac{i - 2}{2} = \dfrac{0 - 2}{2} + \dfrac{1 - 2}{2} + \dfrac{2 - 2}{2} + \dfrac{3 - 2}{2} + \dfrac{4 - 2}{2} + \dfrac{5 - 2}{2} + \dfrac{6 - 2}{2}$

$$= (-1) + \left(-\dfrac{1}{2}\right) + 0 + \dfrac{1}{2} + 1 + \dfrac{3}{2} + 2$$

$$= \dfrac{7}{2}, \text{ or } 3\dfrac{1}{2}$$

b. $\displaystyle\sum_{i=3}^{5}2^i = 2^3 + 2^4 + 2^5$

$$= 8 + 16 + 32$$

$$= 56$$

Example 2   Write each series with summation notation.

      a. $3 + 6 + 9 + 12 + 15$ **b.** $\dfrac{1}{2} + \dfrac{1}{4} + \dfrac{1}{8} + \dfrac{1}{16}$

Solution **a.** Since the *difference* of each term and the preceding term is 3, the terms correspond to the first five terms of the arithmetic sequence $a_n = a_1 + (n - 1)d$ with $a_1 = 3$ and $d = 3$. So $a_n = 3 + (n - 1)3$. Thus, in summation notation,

$$3 + 6 + 9 + 12 + 15 = \sum_{i=1}^{5} [3 + (i - 1)3]$$

      b. Since each term is the *product* of the preceding term and $\dfrac{1}{2}$, these terms correspond to the first four terms of the geometric sequence $a_n = a_1 r^{n-1}$. Here $a_1 = \dfrac{1}{2}$ and $r = \dfrac{1}{2}$, so $a_n = \left(\dfrac{1}{2}\right)\left(\dfrac{1}{2}\right)^{n-1} = \left(\dfrac{1}{2}\right)^{1+(n-1)} = \left(\dfrac{1}{2}\right)^{n}$. In summation notation,

$$\frac{1}{2} + \frac{1}{4} + \frac{1}{8} + \frac{1}{16} = \sum_{i=1}^{4} \left(\frac{1}{2}\right)^{i}$$

2   The sum of the first n terms of a sequence is a finite series known as a **partial sum,** S_n. Thus, for the sequence $a_1, a_2, \ldots, a_n$, the first three partial sums are

$$S_1 = a_1$$
$$S_2 = a_1 + a_2$$
$$S_3 = a_1 + a_2 + a_3$$

In general, S_n is the sum of the first n terms of a sequence.

$$S_n = \sum_{i=1}^{n} a_n$$

Example 3   Find the sum of the first three terms of the sequence whose general term is $a_n = \dfrac{n + 3}{2n}$.

Solution
$$S_3 = \sum_{i=1}^{3} \frac{i + 3}{2i} = \frac{1 + 3}{2 \cdot 1} + \frac{2 + 3}{2 \cdot 2} + \frac{3 + 3}{2 \cdot 3}$$

$$= 2 + \frac{5}{4} + 1 = 4\frac{1}{4}$$

The next example illustrates how these sums model real-life phenomena.

Example 4   **NUMBER OF BABY GORILLAS BORN**

The number of baby gorillas born at the San Diego Zoo is a sequence defined by $a_n = n(n - 1)$, where n is the number of years the zoo has owned gorillas. Find the *total* number of baby gorillas born in the *first 4 years*.

Solution   To solve, find the sum

$$S_4 = \sum_{i=1}^{4} i(i-1)$$

$$= 1(1-1) + 2(2-1) + 3(3-1) + 4(4-1)$$

$$= 0 + 2 + 6 + 12 = 20$$

There were 20 gorillas born in the first 4 years.

Exercise Set 12.3

Evaluate. See Example 1.

1. $\displaystyle\sum_{i=1}^{4} (i-3)$

2. $\displaystyle\sum_{i=1}^{5} (i+6)$

3. $\displaystyle\sum_{i=4}^{7} (2i+4)$

4. $\displaystyle\sum_{i=2}^{3} (5i-1)$

5. $\displaystyle\sum_{i=2}^{4} (i^2-3)$

6. $\displaystyle\sum_{i=3}^{5} i^3$

7. $\displaystyle\sum_{i=1}^{3} \left(\frac{1}{i+5}\right)$

8. $\displaystyle\sum_{i=2}^{4} \left(\frac{2}{i+3}\right)$

9. $\displaystyle\sum_{i=1}^{3} \frac{1}{6i}$

10. $\displaystyle\sum_{i=1}^{3} \frac{1}{3i}$

11. $\displaystyle\sum_{i=2}^{6} 3i$

12. $\displaystyle\sum_{i=3}^{6} -4i$

13. $\displaystyle\sum_{i=3}^{5} i(i+2)$

14. $\displaystyle\sum_{i=2}^{4} i(i-3)$

15. $\displaystyle\sum_{i=1}^{5} 2^i$

16. $\displaystyle\sum_{i=1}^{4} 3^{i-1}$

17. $\displaystyle\sum_{i=1}^{4} \frac{4i}{i+3}$

18. $\displaystyle\sum_{i=2}^{5} \frac{6-i}{6+i}$

Write each series with summation notation. See Example 2.

19. $1 + 3 + 5 + 7 + 9$

20. $4 + 7 + 10 + 13$

21. $4 + 12 + 36 + 108$

22. $5 + 10 + 20 + 40 + 80 + 160$

23. $12 + 9 + 6 + 3 + 0 + (-3)$

24. $5 + 1 + (-3) + (-7)$

25. $12 + 4 + \dfrac{4}{3} + \dfrac{4}{9}$

26. $80 + 20 + 5 + \dfrac{5}{4} + \dfrac{5}{16}$

27. $1 + 4 + 9 + 16 + 25 + 36 + 49$

28. $1 + (-4) + 9 + (-16)$

Find each partial sum. See Example 3.

29. Find the sum of the first two terms of the sequence whose general term is $a_n = (n+2)(n-5)$.

30. Find the sum of the first six terms of the sequence whose general term is $a_n = (-1)^n$.

31. Find the sum of the first two terms of the sequence whose general term is $a_n = n(n-6)$.

32. Find the sum of the first seven terms of the sequence whose general term is $a_n = (-1)^{n-1}$.

33. Find the sum of the first four terms of the sequence whose general term is $a_n = (n+3)(n+1)$.

34. Find the sum of the first five terms of the sequence whose general term is $a_n = \dfrac{(-1)^n}{2n}$.

35. Find the sum of the first four terms of the sequence whose general term is $a_n = -2n$.

36. Find the sum of the first five terms of the sequence whose general term is $a_n = (n-1)^2$.

37. Find the sum of the first three terms of the sequence whose general term is $a_n = -\dfrac{n}{3}$.

38. Find the sum of the first three terms of the sequence whose general term is $a_n = (n+4)^2$.

Solve. See Example 4.

39. A gardener is making a triangular planting with 1 tree in the first row, 2 trees in the second row, 3 trees in the third row, and so on for 10 rows. Write the sequence that describes the number of trees in each row. Find the total number of trees planted.

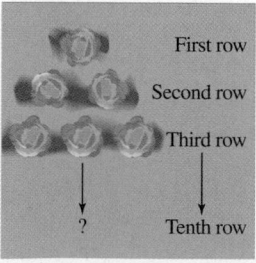

First row
Second row
Third row
?
Tenth row

40. Some surfers at the beach form a human pyramid with 2 surfers in the top row, 3 surfers in the second row, 4 surfers in the third row, and so on. If there are 6 rows in the pyramid, write the sequence that describes the number of surfers in each row of the pyramid. Find the total number of surfers.

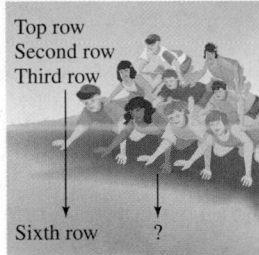

Top row
Second row
Third row

Sixth row   ?

41. A culture of fungus starts with 6 units and doubles every day. Write the general term of the sequence that describes the growth of this fungus. Find the number of fungus units there will be at the beginning of the fifth day.

42. A bacterial colony begins with 100 bacteria and doubles every 6 hours. Write the general term of the sequence describing the growth of the bacteria. Find the number of bacteria there will be after 24 hours.

43. A bacterial colony begins with 50 bacteria and doubles every 12 hours. Write the sequence that describes the growth of the bacteria. Find the number of bacteria there will be after 48 hours.

44. The number of otters born each year in a new aquarium forms a sequence whose general term is $a_n = (n - 1)(n + 3)$. Find the number of otters born in the third year, and find the total number of otters born in the first three years.

45. The number of opossums killed each month on a new highway forms the sequence whose general term is $a_n = (n + 1)(n + 2)$, where n is the number of the months. Find the number of opossums killed in the fourth month, and find the total number killed in the first four months.

46. In 1998 the population of an endangered fish was estimated by environmentalists to be decreasing each year. The size of the population in a given year is $24 - 4n$ thousand fish fewer than the previous year. Find the decrease in population in 2000, if year 1 is 1998. Find the total decrease in the fish population for the years 1998 through 2000.

47. The amount of decay in pounds of a radioactive isotope each year is given by the sequence whose general term is $a_n = 100(0.5)^n$, where n is the number of the year. Find the amount of decay in the fourth year, and find the total amount of decay in the first four years.

48. Susan has a choice between two job offers. Job A has an annual starting salary of $20,000 with guaranteed annual raises of $1200 for the next four years, whereas job B has an annual starting salary of $18,000 with guaranteed annual raises of $2500 for the next four years. Compare the fifth partial sums for each sequence to determine which job would pay Susan more money over the next 5 years.

49. A pendulum swings a length of 40 inches on its first swing. Each successive swing is $\frac{4}{5}$ of the preceding swing. Find the length of the fifth swing and the total length swung during the first five swings. (Round to the nearest tenth of an inch.)

50. Explain the difference between a sequence and a series.

51. a. Write the sum $\displaystyle\sum_{i=1}^{7} (i + i^2)$ without summation notation.

   b. Write the sum $\displaystyle\sum_{i=1}^{7} i + \sum_{i=1}^{7} i^2$ without summation notation.

   c. Compare the results of Parts **a.** and **b.**

   d. Do you think the following is true or false? Explain your answer.

$$\sum_{i=1}^{n} (a_n + b_n) = \sum_{i=1}^{n} a_n + \sum_{i=1}^{n} b_n$$

52. a. Write the sum $\displaystyle\sum_{i=1}^{6} 5i^3$ without summation notation.

   b. Write the expression $5 \cdot \displaystyle\sum_{i=1}^{6} i^3$ without summation notation.

   c. Compare the results of Parts **a.** and **b.**

   d. Do you think the following is true or false? Explain your answer.

$$\sum_{i=1}^{n} c \cdot a_n = c \cdot \sum_{i=1}^{n} a_n \text{ where } c \text{ is a constant}$$

REVIEW EXERCISES

Evaluate. See Section 1.3.

53. $\dfrac{5}{1 - \dfrac{1}{2}}$

54. $\dfrac{-3}{1 - \dfrac{1}{7}}$

55. $\dfrac{\dfrac{1}{3}}{1 - \dfrac{1}{10}}$

56. $\dfrac{\dfrac{6}{11}}{1 - \dfrac{1}{10}}$

57. $\dfrac{3(1 - 2^4)}{1 - 2}$

58. $\dfrac{2(1 - 5^3)}{1 - 5}$

59. $\dfrac{10}{2}(3 + 15)$

60. $\dfrac{12}{2}(2 + 19)$

12.4 PARTIAL SUMS OF ARITHMETIC AND GEOMETRIC SEQUENCES

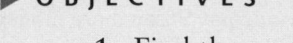

CD-ROM   SSM

SSG   Video

▶ **OBJECTIVES**

1. Find the partial sum of an arithmetic sequence.
2. Find the partial sum of a geometric sequence.
3. Find the sum of the terms of an infinite geometric sequence.

1 Partial sums S_n are relatively easy to find when n is small—that is, when the number of terms to add is small. But when n is large, finding S_n can be tedious. For a large n, S_n is still relatively easy to find if the addends are terms of an arithmetic sequence or a geometric sequence.

For an arithmetic sequence, $a_n = a_1 + (n - 1)d$ for some first term a_1 and some common difference d. So S_n, the sum of the first n terms, is

$$S_n = a_1 + (a_1 + d) + (a_1 + 2d) + \cdots + (a_1 + (n - 1)d)$$

We might also find S_n by "working backward" from the nth term a_n, finding the preceding term a_{n-1}, by subtracting d each time.

$$S_n = a_n + (a_n - d) + (a_n - 2d) + \cdots + (a_n - (n - 1)d)$$

Now add the left sides of these two equations and add the right sides.

$$2S_n = (a_1 + a_n) + (a_1 + a_n) + (a_1 + a_n) + \cdots + (a_1 + a_n)$$

The d terms subtract out, leaving n sums of the first term, a_1, and last term, a_n. Thus, we write

$$2S_n = n(a_1 + a_n)$$

or

$$S_n = \frac{n}{2}(a_1 + a_n)$$

PARTIAL SUM S_n OF AN ARITHMETIC SEQUENCE

The partial sum S_n of the first n terms of an arithmetic sequence is given by

$$S_n = \frac{n}{2}(a_1 + a_n)$$

where a_1 is the first term of the sequence and a_n is the nth term.

Example 1 Use the partial sum formula to find the sum of the first six terms of the arithmetic sequence $2, 5, 8, 11, 14, 17, \ldots$.

Solution Use the formula for S_n of an arithmetic sequence, replacing n with 6, a_1 with 2, and a_n with 17.

$$S_n = \frac{n}{2}(a_1 + a_n) = \frac{6}{2}(2 + 17) = 3(19) = 57$$

Example 2   Find the sum of the first 30 positive integers.

Solution   Because $1, 2, 3, \ldots, 30$ is an arithmetic sequence, use the formula for S_n with $n = 30$, $a_1 = 1$, and $a_n = 30$. Thus,

$$S_n = \frac{n}{2}(a_1 + a_n) = \frac{30}{2}(1 + 30) = 15(31) = 465$$

Example 3   **STACKING ROLLS OF CARPET**

Rolls of carpet are stacked in 20 rows with 3 rolls in the top row, 4 rolls in the next row, and so on, forming an arithmetic sequence. Find the total number of carpet rolls if there are 22 rolls in the bottom row.

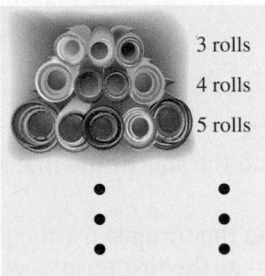

3 rolls
4 rolls
5 rolls

Solution   The list $3, 4, 5, \ldots, 22$ is the first 20 terms of an arithmetic sequence. Use the formula for S_n with $a_1 = 3$, $a_n = 22$, and $n = 20$ terms. Thus,

$$S_{20} = \frac{20}{2}(3 + 22) = 10(25) = 250$$

There are a total of 250 rolls of carpet.

2   We can also derive a formula for the partial sum S_n of the first n terms of a geometric series. If $a_n = a_1 r^{n-1}$, then

$$S_n = a_1 + a_1 r + a_1 r^2 + \cdots + a_1 r^{n-1}$$

$\uparrow$ $\uparrow$ $\uparrow$ $\uparrow$

1st   2nd   3rd nth
term   term   term   term

Multiply each side of the equation by $-r$.

$$-rS_n = -a_1 r - a_1 r^2 - a_1 r^3 - \cdots - a_1 r^n$$

Add the two equations.

$$S_n - rS_n = a_1 + (a_1 r - a_1 r) + (a_1 r^2 - a_1 r^2) + (a_1 r^3 - a_1 r^3) + \cdots - a_1 r^n$$

$$S_n - rS_n = a_1 - a_1 r^n$$

Now factor each side.

$$S_n(1 - r) = a_1(1 - r^n)$$

Solve for S_n by dividing both sides by $1 - r$. Thus,

$$S_n = \frac{a_1(1 - r^n)}{1 - r}$$

as long as r is not 1.

PARTIAL SUM S_n OF A GEOMETRIC SEQUENCE

The partial sum S_n of the first n terms of a geometric sequence is given by

$$S_n = \frac{a_1(1 - r^n)}{1 - r}$$

where a_1 is the first term of the sequence, r is the common ratio, and $r \neq 1$.

Example 4   Find the sum of the first six terms of the geometric sequence $5, 10, 20, 40, 80, 160$.

Solution   Use the formula for the partial sum S_n of the terms of a geometric sequence. Here, $n = 6$, the first term $a_1 = 5$, and the common ratio $r = 2$.

$$S_n = \frac{a_1(1 - r^n)}{1 - r}$$

$$S_6 = \frac{5(1 - 2^6)}{1 - 2} = \frac{5(-63)}{-1} = 315$$

Example 5   **FINDING AMOUNT OF DONATION**

A grant from an alumnus to a university specified that the university was to receive $800,000 during the first year and 75% of the preceding year's donation during each of the following 5 years. Find the total amount donated during the 6 years.

Solution   The donations are modeled by the first six terms of a geometric sequence. Evaluate S_n when $n = 6$, $a_1 = 800,000$, and $r = 0.75$.

$$S_6 = \frac{800,000[1 - (0.75)^6]}{1 - 0.75}$$

$$= \$2,630,468.75$$

The total amount donated during the 6 years is \$2,630,468.75.

3 Is it possible to find the sum of all the terms of an infinite sequence? Examine the partial sums of the geometric sequence $\frac{1}{2}, \frac{1}{4}, \frac{1}{8}, \ldots$.

$$S_1 = \frac{1}{2}$$

$$S_2 = \frac{1}{2} + \frac{1}{4} = \frac{3}{4}$$

$$S_3 = \frac{1}{2} + \frac{1}{4} + \frac{1}{8} = \frac{7}{8}$$

$$S_4 = \frac{1}{2} + \frac{1}{4} + \frac{1}{8} + \frac{1}{16} = \frac{15}{16}$$

$$S_5 = \frac{1}{2} + \frac{1}{4} + \frac{1}{8} + \frac{1}{16} + \frac{1}{32} = \frac{31}{32}$$

$$\vdots$$

$$S_{10} = \frac{1}{2} + \frac{1}{4} + \frac{1}{8} + \cdots + \frac{1}{2^{10}} = \frac{1023}{1024}$$

Even though each partial sum is larger than the preceding partial sum, we see that each partial sum is closer to 1 than the preceding partial sum. If n gets larger and larger, then S_n gets closer and closer to 1. We say that 1 is the **limit** of S_n and also that 1 is the sum of the terms of this infinite sequence. In general, if $|r| < 1$, the following formula gives the sum of the terms of an infinite geometric sequence.

SUM OF THE TERMS OF AN INFINITE GEOMETRIC SEQUENCE

The sum S_∞ of the terms of an infinite geometric sequence is given by

$$S_\infty = \frac{a_1}{1 - r}$$

where a_1 is the first term of the sequence, r is the common ratio, and $|r| < 1$. If $|r| \geq 1$, S_∞ does not exist.

What happens for other values of r? For example, in the following geometric sequence, $r = 3$.

$$6, 18, 54, 162, \ldots$$

Here, as n increases, the sum S_n increases also. This time, though, S_n does not get closer and closer to a fixed number but instead increases without bound.

Example 6   Find the sum of the terms of the geometric sequence $2, \frac{2}{3}, \frac{2}{9}, \frac{2}{27}, \ldots$.

Solution For this geometric sequence, $r = \dfrac{1}{3}$. Since $|r| < 1$, we may use the formula for S_∞ of a geometric sequence with $a_1 = 2$ and $r = \dfrac{1}{3}$.

$$S_\infty = \frac{a_1}{1-r} = \frac{2}{1 - \dfrac{1}{3}} = \frac{2}{\dfrac{2}{3}} = 3$$

The formula for the sum of the terms of an infinite geometric sequence can be used to write a repeating decimal as a fraction. For example,

$$0.33\overline{3} = \frac{3}{10} + \frac{3}{100} + \frac{3}{1000} + \cdots$$

This sum is the sum of the terms of an infinite geometric sequence whose first term a_1 is $\dfrac{3}{10}$ and whose common ratio r is $\dfrac{1}{10}$. Using the formula for S_∞,

$$S_\infty = \frac{a_1}{1-r} = \frac{\dfrac{3}{10}}{1 - \dfrac{1}{10}} = \frac{1}{3}$$

So $0.33\overline{3} = \dfrac{1}{3}$.

Example 7 DISTANCE TRAVELED BY A PENDULUM

On its first pass, a pendulum swings through an arc whose length is 24 inches. On each pass thereafter, the arc length is 75% of the arc length on the preceding pass. Find the total distance the pendulum travels before it comes to rest.

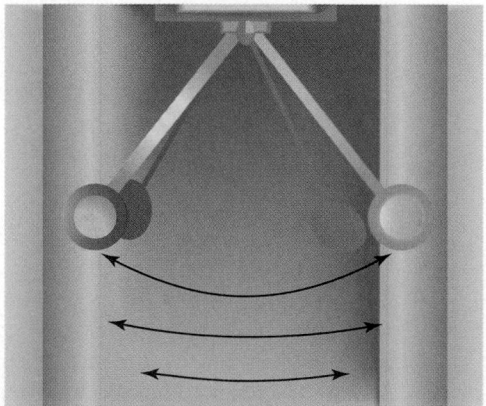

Solution We must find the sum of the terms of an infinite geometric sequence whose first term, a_1, is 24 and whose common ratio, r, is 0.75. Since $|r| < 1$, we may use the formula for S_∞.

$$S_\infty = \frac{a_1}{1-r} = \frac{24}{1 - 0.75} = \frac{24}{0.25} = 96$$

The pendulum travels a total distance of 96 inches before it comes to rest.

SPOTLIGHT ON DECISION MAKING

Suppose you are a certified financial planner. You are working with a 30-year-old client whose goal is to retire at age 65 with a sum of $500,000 to live off. This year she just started making the maximum annual contribution of $2000 to a Roth IRA (Individual Retirement Account) that pays 8% interest compounded annually.

You know that if she continues to make a $2000 contribution at the beginning of each year, by the end of the nth year, her account increases in value by the nth term of the geometric sequence

$a_n = 2000(1.08)^n$, which considers both the annual $2000 contribution and her earned interest. Decide whether your client will be able to reach her retirement goal by making only the maximum Roth IRA contribution each year until she retires, or if you should suggest an additional investment to help her reach her goal. (Note: Use a partial sum to find the value of the Roth IRA at the end of 35 years. To find a_1 of the geometric sequence, be sure to evaluate the equation for the nth term at $n = 1$.)

Exercise Set 12.4

Use the partial sum formula to find the partial sum of the given arithmetic or geometric sequence. See Examples 1 and 4.

1. Find the sum of the first six terms of the arithmetic sequence $1, 3, 5, 7, \ldots$.

2. Find the sum of the first seven terms of the arithmetic sequence $-7, -11, -15, \ldots$.

3. Find the sum of the first five terms of the geometric sequence $4, 12, 36, \ldots$.

4. Find the sum of the first eight terms of the geometric sequence $-1, 2, -4, \ldots$.

5. Find the sum of the first six terms of the arithmetic sequence $3, 6, 9, \ldots$.

6. Find the sum of the first four terms of the arithmetic sequence $-4, -8, -12, \ldots$.

7. Find the sum of the first four terms of the geometric sequence $2, \dfrac{2}{5}, \dfrac{2}{25}, \ldots$.

8. Find the sum of the first five terms of the geometric sequence $\dfrac{1}{3}, -\dfrac{2}{3}, \dfrac{4}{3}, \ldots$.

Solve. See Example 2.

9. Find the sum of the first ten positive integers.

10. Find the sum of the first eight negative integers.

11. Find the sum of the first four positive odd integers.

12. Find the sum of the first five negative odd integers.

Find the sum of the terms of each infinite geometric sequence. See Example 6.

13. $12, 6, 3, \ldots$

14. $45, 15, 5, \ldots$

15. $\dfrac{1}{10}, \dfrac{1}{100}, \dfrac{1}{1000}, \ldots$

16. $\dfrac{3}{5}, \dfrac{3}{20}, \dfrac{3}{80}, \ldots$

17. $-10, -5, -\dfrac{5}{2}, \ldots$

18. $-16, -4, -1, \ldots$

19. $2, -\dfrac{1}{4}, \dfrac{1}{32}, \ldots$

20. $-3, \dfrac{3}{5}, -\dfrac{3}{25}, \ldots$

21. $\dfrac{2}{3}, -\dfrac{1}{3}, \dfrac{1}{6}, \ldots$

22. $6, -4, \dfrac{8}{3}, \ldots$

Solve.

23. Find the sum of the first ten terms of the sequence $-4, 1, 6, \ldots, 41$ where 41 is the tenth term.

24. Find the sum of the first twelve terms of the sequence $-3, -13, -23, \ldots, -113$ where -113 is the twelfth term.

25. Find the sum of the first seven terms of the sequence $3, \dfrac{3}{2}, \dfrac{3}{4}, \ldots$.

26. Find the sum of the first five terms of the sequence $-2, -6, -18, \ldots$.

27. Find the sum of the first five terms of the sequence $-12, 6, -3, \ldots$.

28. Find the sum of the first four terms of the sequence $-\dfrac{1}{4}, -\dfrac{3}{4}, -\dfrac{9}{4}, \ldots$.

29. Find the sum of the first twenty terms of the sequence $\dfrac{1}{2}, \dfrac{1}{4}, 0, \ldots, -\dfrac{17}{4}$ where $-\dfrac{17}{4}$ is the twentieth term.

30. Find the sum of the first fifteen terms of the sequence $-5, -9, -13, \ldots, -61$ where -61 is the fifteenth term.

31. If a_1 is 8 and r is $-\dfrac{2}{3}$, find S_3.

32. If a_1 is 10, a_{18} is $\dfrac{3}{2}$, and d is $-\dfrac{1}{2}$, find S_{18}.

Solve. See Example 3.

33. Modern Car Company has come out with a new car model. Market analysts predict that 4000 cars will be sold in the first month and that sales will drop by 50 cars per month after that during the first year. Write out the first five terms of the sequence, and find the number of sold cars predicted for the twelfth month. Find the total predicted number of sold cars for the first year.

34. A company that sends faxes charges $3 for the first page sent and $0.10 less than the preceding page for each additional page sent. The cost per page forms an arithmetic sequence. Write the first five terms of this sequence, and use a partial sum to find the cost of sending a nine-page document.

35. Sal has two job offers: Firm *A* starts at $22,000 per year and guarantees raises of $1000 per year, whereas Firm *B* starts at $20,000 and guarantees raises of $1200 per year. Over a 10-year period, determine the more profitable offer.

36. The game of pool uses 15 balls numbered 1 to 15. In the variety called rotation, a player who sinks a ball receives as many points as the number on the ball. Use an arithmetic series to find the score of a player who sinks all 15 balls.

Solve. See Example 5.

37. A woman made $30,000 during the first year she owned her business and made an additional 10% over the previous year in each subsequent year. Find how much she made during her fourth year of business. Find her total earnings during the first four years.

38. In free fall, a parachutist falls 16 feet during the first second, 48 feet during the second second, 80 feet during the third second, and so on. Find how far she falls during the eighth second. Find the total distance she falls during the first 8 seconds.

39. A trainee in a computer company takes 0.9 times as long to assemble each computer as he took to assemble the preceding computer. If it took him 30 minutes to assemble the first computer, find how long it takes him to assemble the fifth computer. Find the total time he takes to assemble the first five computers (round to the nearest minute).

40. On a gambling trip to Reno, Carol doubled her bet each time she lost. If her first losing bet was $5 and she lost six consecutive bets, find how much she lost on the sixth bet. Find the total amount lost on these six bets.

Solve. See Example 7.

41. A ball is dropped from a height of 20 feet and repeatedly rebounds to a height that is $\dfrac{4}{5}$ of its previous height. Find the total distance the ball covers before it comes to rest.

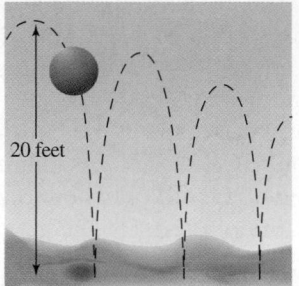

20 feet

42. A rotating flywheel coming to rest makes 300 revolutions in the first minute and in each minute thereafter makes $\dfrac{2}{5}$ as many revolutions as in the preceding minute. Find how many revolutions the wheel makes before it comes to rest.

Solve.

43. In the pool game of rotation, player *A* sinks balls numbered 1 to 9, and player *B* sinks the rest of the balls. Use arithmetic series to find each player's score (see Exercise 36).

44. A godfather deposited $250 in a savings account on the day his godchild was born. On each subsequent birthday he deposited $50 more than he deposited the previous year. Find how much money he deposited on his godchild's twenty-first birthday. Find the total amount deposited over the 21 years.

45. During the holiday rush a business can rent a computer system for $200 the first day, with the rental fee decreasing $5 for each additional day. Find the fee paid for 20 days during the holiday rush.

46. The spraying of a field with insecticide killed 6400 weevils the first day, 1600 the second day, 400 the third day, and so on. Find the total number of weevils killed during the first 5 days.

47. A college student humorously asks his parents to charge him room and board according to this geometric sequence: $0.01 for the first day of the month, $0.02 for the second day, $0.04 for the third day, and so on. Find the total room and board he would pay for 30 days.

48. Following its television advertising campaign, a bank attracted 80 new customers the first day, 120 the second day, 160 the third day, and so on, in an arithmetic sequence. Find how many new customers were attracted during the first 5 days following its television campaign.

49. Write $0.88\overline{8}$ as an infinite geometric series and use the formula for S_∞ to write it as a rational number.

50. Write $0.54\overline{54}$ as an infinite geometric series and use the formula S_∞ to write it as a rational number.

51. Explain whether the sequence $5, 5, 5, \ldots$ is arithmetic, geometric, neither, or both.

52. Describe a situation in everyday life that can be modeled by an infinite geometric series.

REVIEW EXERCISES

Evaluate. See Section 1.3.

53. $6 \cdot 5 \cdot 4 \cdot 3 \cdot 2 \cdot 1$

54. $8 \cdot 7 \cdot 6 \cdot 5 \cdot 4 \cdot 3 \cdot 2 \cdot 1$

55. $\dfrac{3 \cdot 2 \cdot 1}{2 \cdot 1}$

56. $\dfrac{5 \cdot 4 \cdot 3 \cdot 2 \cdot 1}{3 \cdot 2 \cdot 1}$

Multiply. See Section 5.5.

57. $(x + 5)^2$

58. $(x - 2)^2$

59. $(2x - 1)^3$

60. $(3x + 2)^3$

12.5   THE BINOMIAL THEOREM

CD-ROM   SSM

SSG   Video

▶ **OBJECTIVES**

1. Use Pascal's triangle to expand binomials.
2. Evaluate factorials.
3. Use the binomial theorem to expand binomials.
4. Find the nth term in the expansion of a binomial raised to a positive power.

In this section, we learn how to **expand** binomials of the form $(a + b)^n$ easily. Expanding a binomial such as $(a + b)^n$ means to write the factored form as a sum. First, we review the patterns in the expansions of $(a + b)^n$.

$$(a + b)^0 = 1 \qquad\qquad 1\ \text{term}$$

$$(a + b)^1 = a + b \qquad\qquad 2\ \text{terms}$$

$$(a + b)^2 = a^2 + 2ab + b^2 \qquad\qquad 3\ \text{terms}$$

$$(a + b)^3 = a^3 + 3a^2b + 3ab^2 + b^3 \qquad\qquad 4\ \text{terms}$$

$$(a + b)^4 = a^4 + 4a^3b + 6a^2b^2 + 4ab^3 + b^4 \qquad\qquad 5\ \text{terms}$$

$$(a + b)^5 = a^5 + 5a^4b + 10a^3b^2 + 10a^2b^3 + 5ab^4 + b^5 \qquad 6\ \text{terms}$$

Notice the following patterns.

1. The expansion of $(a + b)^n$ contains $n + 1$ terms. For example, for $(a + b)^3$, $n = 3$, and the expansion contains $3 + 1$ terms, or 4 terms.
2. The first term of the expansion of $(a + b)^n$ is a^n, and the last term is b^n.
3. The powers of a decrease by 1 for each term, whereas the powers of b increase by 1 for each term.
4. For each term of the expansion of $(a + b)^n$, the sum of the exponents of a and b is n. (For example, the sum of the exponents of $5a^4b$ is $4 + 1$, or 5, and the sum of the exponents of $10a^3b^2$ is $3 + 2$, or 5.)

1   There are patterns in the coefficients of the terms as well. Written in a triangular array, the coefficients are called **Pascal's triangle.**

$$
\begin{array}{lccccccccc}
(a + b)^0: & & & & & 1 & & & & n = 0 \\
(a + b)^1: & & & & 1 & & 1 & & & n = 1 \\
(a + b)^2: & & & 1 & & 2 & & 1 & & n = 2 \\
(a + b)^3: & & 1 & & 3 & & 3 & & 1 & n = 3 \\
(a + b)^4: & 1 & & 4 & & 6 & & 4 & & 1 \quad n = 4 \\
(a + b)^5: 1 & & 5 & & 10 & & 10 & & 5 & & 1 \quad n = 5
\end{array}
$$

Each row in Pascal's triangle begins and ends with 1. Any other number in a row is the sum of the two closest numbers above it. Using this pattern, we can write the next row, for $n = 6$, by first writing the number 1. Then we can add the consecutive numbers in the row for $n = 5$ and write each sum "between and below" the pair. We complete the row by writing a 1.

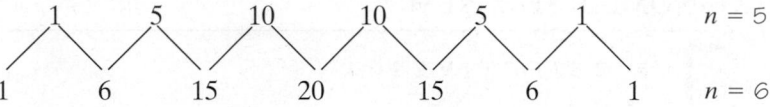

We can use Pascal's triangle and the patterns noted to expand $(a + b)^n$ without actually multiplying any terms.

Example 1   Expand $(a + b)^6$.

Solution   Using the $n = 6$ row of Pascal's triangle as the coefficients and following the patterns noted, $(a + b)^6$ can be expanded as

$$a^6 + 6a^5b + 15a^4b^2 + 20a^3b^3 + 15a^2b^4 + 6ab^5 + b^6$$

---

2   For a large n, the use of Pascal's triangle to find coefficients for $(a + b)^n$ can be tedious. An alternative method for determining these coefficients is based on the concept of a **factorial.**

The **factorial of n,** written $n!$ (read "n factorial"), is the product of the first n consecutive natural numbers.

FACTORIAL OF n: $n!$

If n is a natural number, then $n! = n(n - 1)(n - 2)(n - 3) \cdots 3 \cdot 2 \cdot 1$.
The factorial of 0, written 0!, is defined to be 1.

For example, $3! = 3 \cdot 2 \cdot 1 = 6$, $5! = 5 \cdot 4 \cdot 3 \cdot 2 \cdot 1 = 120$, and $0! = 1$.

Example 2   Evaluate each expression.

a. $\dfrac{5!}{6!}$

b. $\dfrac{10!}{7!3!}$

c. $\dfrac{3!}{2!1!}$

d. $\dfrac{7!}{7!0!}$

Solution   **a.** $\dfrac{5!}{6!} = \dfrac{5 \cdot 4 \cdot 3 \cdot 2 \cdot 1}{6 \cdot 5 \cdot 4 \cdot 3 \cdot 2 \cdot 1} = \dfrac{1}{6}$

b. $\dfrac{10!}{7!3!} = \dfrac{10 \cdot 9 \cdot 8 \cdot 7!}{7! \cdot 3 \cdot 2 \cdot 1} = \dfrac{10 \cdot 9 \cdot 8}{3 \cdot 2 \cdot 1} = 10 \cdot 3 \cdot 4 = 120$

c. $\dfrac{3!}{2!1!} = \dfrac{3 \cdot 2 \cdot 1}{2 \cdot 1 \cdot 1} = 3$

d. $\dfrac{7!}{7!0!} = \dfrac{7!}{7! \cdot 1} = 1$

> ▼ **HELPFUL HINT**
> We can use a calculator with a factorial key to evaluate a factorial. A calculator uses scientific notation for large results.

3   It can be proved, although we won't do so here, that the coefficients of terms in the expansion of $(a + b)^n$ can be expressed in terms of factorials. Following patterns 1 through 4 given earlier and using the factorial expressions of the coefficients, we have what is known as the **binomial theorem.**

BINOMIAL THEOREM

If n is a positive integer, then

$$(a + b)^n = a^n + \frac{n}{1!} a^{n-1}b^1 + \frac{n(n-1)}{2!} a^{n-2}b^2$$
$$+ \frac{n(n-1)(n-2)}{3!} a^{n-3}b^3 + \cdots + b^n$$

We call the formula for $(a + b)^n$ given by the binomial theorem the **binomial formula.**

Example 3   Use the binomial theorem to expand $(x + y)^{10}$.

Solution   Let $a = x$, $b = y$, and $n = 10$ in the binomial formula.

$$(x + y)^{10} = x^{10} + \frac{10}{1!}x^9 y + \frac{10 \cdot 9}{2!}x^8 y^2 + \frac{10 \cdot 9 \cdot 8}{3!}x^7 y^3 + \frac{10 \cdot 9 \cdot 8 \cdot 7}{4!}x^6 y^4$$

$$+ \frac{10 \cdot 9 \cdot 8 \cdot 7 \cdot 6}{5!}x^5 y^5 + \frac{10 \cdot 9 \cdot 8 \cdot 7 \cdot 6 \cdot 5}{6!}x^4 y^6$$

$$+ \frac{10 \cdot 9 \cdot 8 \cdot 7 \cdot 6 \cdot 5 \cdot 4}{7!}x^3 y^7$$

$$+ \frac{10 \cdot 9 \cdot 8 \cdot 7 \cdot 6 \cdot 5 \cdot 4 \cdot 3}{8!}x^2 y^8$$

$$+ \frac{10 \cdot 9 \cdot 8 \cdot 7 \cdot 6 \cdot 5 \cdot 4 \cdot 3 \cdot 2}{9!}xy^9 + y^{10}$$

$$= x^{10} + 10x^9 y + 45x^8 y^2 + 120x^7 y^3 + 210x^6 y^4 + 252x^5 y^5 + 210x^4 y^6$$

$$+ 120x^3 y^7 + 45x^2 y^8 + 10xy^9 + y^{10}$$

Example 4   Use the binomial theorem to expand $(x + 2y)^5$.

Solution   Let $a = x$ and $b = 2y$ in the binomial formula.

$$(x + 2y)^5 = x^5 + \frac{5}{1!}x^4(2y) + \frac{5 \cdot 4}{2!}x^3(2y)^2 + \frac{5 \cdot 4 \cdot 3}{3!}x^2(2y)^3$$

$$+ \frac{5 \cdot 4 \cdot 3 \cdot 2}{4!}x(2y)^4 + (2y)^5$$

$$= x^5 + 10x^4y + 40x^3y^2 + 80x^2y^3 + 80xy^4 + 32y^5$$

Example 5   Use the binomial theorem to expand $(3m - n)^4$.

Solution   Let $a = 3m$ and $b = -n$ in the binomial formula.

$$(3m - n)^4 = (3m)^4 + \frac{4}{1!}(3m)^3(-n) + \frac{4 \cdot 3}{2!}(3m)^2(-n)^2$$

$$+ \frac{4 \cdot 3 \cdot 2}{3!}(3m)(-n)^3 + (-n)^4$$

$$= 81m^4 - 108m^3n + 54m^2n^2 - 12mn^3 + n^4$$

4   Sometimes it is convenient to find a specific term of a binomial expansion without writing out the entire expansion. By studying the expansion of binomials, a pattern forms for each term. This pattern is most easily stated for the $(r + 1)$st term.

$(r + 1)$ST TERM IN A BINOMIAL EXPANSION

The $(r + 1)$st term of the expansion of $(a + b)^n$ is $\dfrac{n!}{r!(n - r)!}a^{n-r}b^r$.

Example 6   Find the eighth term in the expansion of $(2x - y)^{10}$.

Solution   Use the formula, with $n = 10$, $a = 2x$, $b = -y$, and $r + 1 = 8$. Notice that, since $r + 1 = 8$, $r = 7$.

$$\frac{n!}{r!(n - r)!}a^{n-r}b^r = \frac{10!}{7!3!}(2x)^3(-y)^7$$

$$= 120(8x^3)(-y^7)$$

$$= -960x^3y^7$$

Exercise Set 12.5

Use Pascal's triangle to expand the binomial. See Example 1.

1. $(m + n)^3$

2. $(x + y)^4$

3. $(c + d)^5$

4. $(a + b)^6$

5. $(y - x)^5$

6. $(q - r)^7$

7. Explain how to generate a row of Pascal's triangle.

8. Write the $n = 8$ row of Pascal's triangle.

Evaluate each expression. See Example 2.

9. $\dfrac{8!}{7!}$

10. $\dfrac{6!}{0!}$

11. $\dfrac{7!}{5!}$

12. $\dfrac{8!}{5!}$

13. $\dfrac{10!}{7!2!}$

14. $\dfrac{9!}{5!3!}$

15. $\dfrac{8!}{6!0!}$

16. $\dfrac{10!}{4!6!}$

Use the binomial formula to expand each binomial. See Examples 3 through 5.

17. $(a + b)^7$

18. $(x + y)^8$

19. $(a + 2b)^5$

20. $(x + 3y)^6$

21. $(q + r)^9$

22. $(b + c)^6$

23. $(4a + b)^5$

24. $(3m + n)^4$

25. $(5a - 2b)^4$

26. $(m - 4)^6$

27. $(2a + 3b)^3$

28. $(4 - 3x)^5$

29. $(x + 2)^5$

30. $(3 + 2a)^4$

Find the indicated term. See Example 6.

31. The fifth term of the expansion of $(c - d)^5$

32. The fourth term of the expansion of $(x - y)^6$

33. The eighth term of the expansion of $(2c + d)^7$

34. The tenth term of the expansion of $(5x - y)^9$

35. The fourth term of the expansion of $(2r - s)^5$

36. The first term of the expansion of $(3q - 7r)^6$

37. The third term of the expansion of $(x + y)^4$

38. The fourth term of the expansion of $(a + b)^8$

39. The second term of the expansion of $(a + 3b)^{10}$

40. The third term of the expansion of $(m + 5n)^7$

REVIEW EXERCISES

Sketch the graph of each function. Decide whether each function is one-to-one. See Sections 3.3, 10.6, and 11.2.

41. $f(x) = |x|$

42. $g(x) = 3(x - 1)^2$

43. $H(x) = 2x + 3$

44. $F(x) = -2$

45. $f(x) = x^2 + 3$

46. $h(x) = -(x + 1)^2 - 4$

12

For additional Chapter Projects, visit the Real World Activities
Website by going to http://www.prenhall.com/martin-gay.

CHAPTER PROJECT

Modeling College Tuition

Annual college tuition has steadily increased since 1970. According to the College Board, by the 1999–2000 academic year, the average annual tuition at a public four-year university had increased to $3356. Similarly, average annual tuition at private four-year universities had grown to $15,380.

Over the past few years, annual tuition at four-year public universities has been increasing at an average rate of 4.5% per year. Over the same time period, annual tuition at four-year private universities has been increasing at an average rate of $804.30 per year. In this project, you will have the opportunity to model and investigate the trend in increasing tuition at public and private universities. This project may be completed by working in groups or individually.

1. Using the information given in the introductory paragraphs, decide whether the sequence of public university tuitions is arithmetic or geometric.
2. Using the information given in the introductory paragraphs, decide whether the sequence of private university tuitions is arithmetic or geometric.
3. Find the general term of the sequence that describes the pattern of average annual tuition for four-year public universities. Let $n = 1$ represent the 1999–2000 academic year.
4. Find the general term of the sequence that describes the pattern of average annual tuition for four-year private universities. Let $n = 1$ represent the 1999–2000 academic year.

5. Assuming that the rate of tuition increase remains the same, use the general term equation from Question 3 to find the average annual tuition at a four-year public university for the 2002–2003 academic year.
6. Assuming that the rate of tuition increase remains the same, use the general term equation from Question 4 to find the average annual tuition at a four-year private university for the 2003–2004 academic year.
7. Use partial sums to find the average cost of a four-year college education at a public university for a student who started college in the 1999–2000 academic year.
8. Use partial sums to find the average cost of a four-year college education at a private university for a student who starts college in the 2002–2003 academic year. (Hint: One way to do this is to find S_7 and subtract S_3 from it. If you use this method, explain why this gives the desired sum.)
9. (Optional) Use newspapers or news magazines to find a situation that can be modeled by a sequence. Briefly describe the situation, and decide whether it is an arithmetic or a geometric sequence. Find an equation of the general term of the sequence.

Academic Year	n
1999–2000	$n = 1$
2000–2001	$n = 2$
2001–2002	$n = 3$
2002–2003	$n = 4$
2003–2004	$n = 5$
2004–2005	$n = 6$
2005–2006	$n = 7$

CHAPTER 12 VOCABULARY CHECK

Fill in each blank with one of the words or phrases listed below.

general term	common difference	infinite sequence	common ratio
Pascal's triangle	finite sequence	factorial of *n*	arithmetic sequence
geometric sequence	series		

1. A(n) _____ is a function whose domain is the set of natural numbers $\{1, 2, 3, \ldots, n\}$, where n is some natural number.
2. The _____, written $n!$, is the product of the first n consecutive natural numbers.
3. A(n) _____ is a function whose domain is the set of natural numbers.
4. A(n) _____ is a sequence in which each term (after the first) is obtained by multiplying the preceding term by a constant amount r. The constant r is called the _____ of the sequence.
5. A sum of the terms of a sequence is called a _____.
6. The nth term of the sequence a_n is called the _____.
7. A(n) _____ is a sequence in which each term (after the first) differs from the preceding term by a constant amount d. The constant d is called the _____ of the sequence.
8. A triangular array of the coefficients of the terms of the expansions of $(a + b)^n$ is called _____.

CHAPTER 12 HIGHLIGHTS

DEFINITIONS AND CONCEPTS	EXAMPLES
Section 12.1 Sequences	
An **infinite sequence** is a function whose domain is the set of natural numbers $\{1, 2, 3, 4, \ldots\}$.	*Infinite Sequence* $2, 4, 6, 8, 10, \ldots$
A **finite sequence** is a function whose domain is the set of natural numbers $\{1, 2, 3, 4, \ldots, n\}$, where n is some natural number.	*Finite Sequence* $1, -2, 3, -4, 5, -6$
The notation a_n, where n is a natural number, is used to denote a sequence.	Write the first four terms of the sequence whose general term is $a_n = n^2 + 1$. $a_1 = 1^2 + 1 = 2$ $a_2 = 2^2 + 1 = 5$ $a_3 = 3^2 + 1 = 10$ $a_4 = 4^2 + 1 = 17$
Section 12.2 Arithmetic and Geometric Sequences	
An **arithmetic sequence** is a sequence in which each term differs from the preceding term by a constant amount d, called the **common difference.**	*Arithmetic Sequence* $5, 8, 11, 14, 17, 20, \ldots$ Here, $a_1 = 5$ and $d = 3$.
The **general term** a_n of an arithmetic sequence is given by $$a_n = a_1 + (n - 1)d$$ where a_1 is the first term and d is the common difference.	The general term is $a_n = a_1 + (n - 1)d$ or $a_n = 5 + (n - 1)3$

(continued)

DEFINITIONS AND CONCEPTS	EXAMPLES

Section 12.2 Arithmetic and Geometric Sequences

A **geometric sequence** is a sequence in which each term is obtained by multiplying the preceding term by a constant r, called the **common ratio**.

The **general term** a_n of a geometric sequence is given by

$$a_n = a_1 r^{n-1}$$

where a_1 is the first term and r is the common ratio.

Geometric Sequence

$$12, -6, 3, -\frac{3}{2}, \ldots$$

Here $a_1 = 12$ and $r = -\frac{1}{2}$.

The general term is

$$a_n = a_1 r^{n-1} \text{ or}$$

$$a_n = 12\left(-\frac{1}{2}\right)^{n-1}$$

Section 12.3 Series

A sum of the terms of a sequence is called a **series**.

A shorthand notation for denoting a series is called **summation notation**:

index of summation $\longrightarrow$ $\displaystyle\sum_{i=1}^{4}$ $\longrightarrow$ Greek letter sigma used to mean sum

Sequence	*Series*	
3, 7, 11, 15	$3 + 7 + 11 + 15$	finite
$3, 7, 11, 15, \ldots$	$3 + 7 + 11 + 15 + \cdots$	infinite

$$\sum_{i=1}^{4} 3^i = 3^1 + 3^2 + 3^3 + 3^4$$

$$= 3 + 9 + 27 + 81$$

$$= 120$$

Section 12.4 Partial Sums of Arithmetic and Geometric Sequences

Partial sum, S_n, of the first n terms of an arithmetic sequence:

$$S_n = \frac{n}{2}(a_1 + a_n)$$

where a_1 is the first term and a_n is the nth term.

Partial sum, S_n, of the first n terms of a geometric sequence:

$$S_n = \frac{a_1(1 - r^n)}{1 - r}$$

where a_1 is the first term, r is the common ratio, and $r \neq 1$.

Sum of the terms of an infinite geometric sequence:

$$S_\infty = \frac{a_1}{1 - r}$$

where a_1 is the first term, r is the common ratio, and $|r| < 1$. (If $|r| \geq 1$, S_∞ does not exist.)

The sum of the first five terms of the arithmetic sequence

$$12, 24, 36, 48, 60, \ldots \text{ is}$$

$$S_n = \frac{5}{2}(12 + 60) = 180$$

The sum of the first five terms of the geometric sequence

$$15, 30, 60, 120, 240, \ldots \text{ is}$$

$$S_5 = \frac{15(1 - 2^5)}{1 - 2} = 465$$

The sum of the terms of the infinite geometric sequence

$$1, \frac{1}{3}, \frac{1}{9}, \frac{1}{27}, \ldots \text{ is}$$

$$S_\infty = \frac{1}{1 - \frac{1}{3}} = \frac{3}{2}$$

(continued)

DEFINITIONS AND CONCEPTS	EXAMPLES

Section 12.5 The Binomial Theorem

The **factorial of n**, written $n!$, is the product of the first n consecutive natural numbers.

$$5! = 5 \cdot 4 \cdot 3 \cdot 2 \cdot 1 = 120$$

Binomial Theorem

If n is a positive integer, then

$$(a + b)^n = a^n + \frac{n}{1!}a^{n-1}b^1 + \frac{n(n-1)}{2!}a^{n-2}b^2$$

$$+ \frac{n(n-1)(n-2)}{3!}a^{n-3}b^3 + \cdots + b^n$$

Expand $(3x + y)^4$.

$$(3x + y)^4 = (3x)^4 + \frac{4}{1!}(3x)^3(y)^1$$

$$+ \frac{4 \cdot 3}{2!}(3x)^2(y)^2 + \frac{4 \cdot 3 \cdot 2}{3!}(3x)^1 y^3 + y^4$$

$$= 81x^4 + 108x^3 y + 54x^2 y^2 + 12xy^3 + y^4$$

CHAPTER 12 REVIEW

(12.1) Find the indicated term(s) of the given sequence.

1. The first five terms of the sequence $a_n = -3n^2$

2. The first five terms of the sequence $a_n = n^2 + 2n$

3. The one-hundredth term of the sequence $a_n = \frac{(-1)^n}{100}$

4. The fiftieth term of the sequence $a_n = \frac{2n}{(-1)^2}$

5. The general term a_n of the sequence $\frac{1}{6}, \frac{1}{12}, \frac{1}{18}, \ldots$

6. The general term a_n of the sequence $-1, 4, -9, 16, \ldots$

Solve the following applications.

7. The distance in feet that an olive falling from rest in a vacuum will travel during each second is given by an arithmetic sequence whose general term is $a_n = 32n - 16$, where n is the number of the second. Find the distance the olive will fall during the fifth, sixth, and seventh seconds.

8. A culture of yeast doubles every day in a geometric progression whose general term is $a_n = 100(2)^{n-1}$, where n is the number of the day just ending. Find how many days it takes the yeast culture to measure at least 10,000. Find the original measure of the yeast culture.

9. The Centers for Disease Control and Prevention (CDC) reported that a new type of virus infected approximately 450 people during 1999, the year it was first discovered. The CDC predicts that during the next decade the virus will infect three times as many people each year as the year before. Write out the first five terms of this geometric sequence, and predict the number of infected people there will be in 2003.

10. The first row of an amphitheater contains 50 seats, and each row thereafter contains 8 additional seats. Write the first ten terms of this arithmetic progression, and find the number of seats in the tenth row.

(12.2)

11. Find the first five terms of the geometric sequence whose first term is -2 and whose common ratio is $\frac{2}{3}$.

12. Find the first five terms of the arithmetic sequence whose first term is 12 and whose common difference is -1.5.

13. Find the thirtieth term of the arithmetic sequence whose first term is -5 and whose common difference is 4.

14. Find the eleventh term of the arithmetic sequence whose first term is 2 and whose common difference is $\frac{3}{4}$.

15. Find the twentieth term of the arithmetic sequence whose first three terms are 12, 7, and 2.

16. Find the sixth term of the geometric sequence whose first three terms are 4, 6, and 9.

17. If the fourth term of an arithmetic sequence is 18 and the twentieth term is 98, find the first term and the common difference.

18. If the third term of a geometric sequence is -48 and the fourth term is 192, find the first term and the common ratio.

19. Find the general term of the sequence $\frac{3}{10}, \frac{3}{100}, \frac{3}{1000}, \ldots$

20. Find a general term that satisfies the terms shown for the sequence $50, 58, 66, \ldots$

Determine whether each of the following sequences is arithmetic, geometric, or neither. If a sequence is arithmetic, find a_1 and d. If a sequence is geometric, find a_1 and r.

21. $\dfrac{8}{3}, 4, 6, \ldots$

22. $-10.5, -6.1, -1.7$

23. $7x, -14x, 28x$

24. $3x^2, 9x^4, 81x^8, \ldots$

Solve the following applications.

25. To test the bounce of a racquetball, the ball is dropped from a height of 8 feet. The ball is judged "good" if it rebounds at least 75% of its previous height with each bounce. Write out the first six terms of this geometric sequence (round to the nearest tenth). Determine if a ball is "good" that rebounds to a height of 2.5 feet after the fifth bounce.

26. A display of oil cans in an auto parts store has 25 cans in the bottom row, 21 cans in the next row, and so on, in an arithmetic progression. Find the general term and the number of cans in the top row.

27. Suppose that you save $1 the first day of a month, $2 the second day, $4 the third day, continuing to double your savings each day. Write the general term of this geometric sequence and find the amount you will save on the tenth day. Estimate the amount you will save on the thirtieth day of the month, and check your estimate with a calculator.

28. On the first swing, the length of an arc through which a pendulum swings is 30 inches. The length of the arc for each successive swing is 70% of the preceding swing. Find the length of the arc for the fifth swing.

29. Rosa takes a job that has a monthly starting salary of $900 and guarantees her a monthly raise of $150 during her 6-month training period. Find the general term of this sequence and her salary at the end of her training.

30. A sheet of paper is $\dfrac{1}{512}$-inch thick. By folding the sheet in half, the total thickness will be $\dfrac{1}{256}$ inch. A second fold produces a total thickness of $\dfrac{1}{128}$ inch. Estimate the thickness of the stack after 15 folds, and then check your estimate with a calculator.

(12.3) *Write out the terms and find the sum for each of the following.*

31. $\displaystyle\sum_{i=1}^{5} (2i - 1)$

32. $\displaystyle\sum_{i=1}^{5} i(i + 2)$

33. $\displaystyle\sum_{i=2}^{4} \dfrac{(-1)^i}{2i}$

34. $\displaystyle\sum_{i=3}^{5} 5(-1)^{i-1}$

Find the partial sum of the given sequence.

35. S_4 of the sequence $a_n = (n - 3)(n + 2)$

36. S_6 of the sequence $a_n = n^2$

37. S_5 of the sequence $a_n = -8 + (n - 1)3$

38. S_3 of the sequence $a_n = 5(4)^{n-1}$

Write the sum with Σ notation.

39. $1 + 3 + 9 + 27 + 81 + 243$

40. $6 + 2 + (-2) + (-6) + (-10) + (-14) + (-18)$

41. $\dfrac{1}{4} + \dfrac{1}{16} + \dfrac{1}{64} + \dfrac{1}{256}$

42. $1 + \left(-\dfrac{3}{2}\right) + \dfrac{9}{4}$

Solve.

43. A yeast colony begins with 20 yeast and doubles every 8 hours. Write the sequence that describes the growth of the yeast, and find the total yeast after 48 hours.

44. The number of cranes born each year in a new aviary forms a sequence whose general term is $a_n = n^2 + 2n - 1$. Find the number of cranes born in the fourth year and the total number of cranes born in the first four years.

45. Harold has a choice between two job offers. Job A has an annual starting salary of $39,500 with guaranteed annual raises of $2200 for the next four years, whereas job B has an annual starting salary of $41,000 with guaranteed annual raises of $1400 for the next four years. Compare the salaries for the fifth year under each job offer.

46. A sample of radioactive waste is decaying such that the amount decaying in kilograms during year n is $a_n = 200(0.5)^n$. Find the amount of decay in the third year, and the total amount of decay in the first three years.

(12.4) *Find the partial sum of the given sequence.*

47. The sixth partial sum of the sequence $15, 19, 23, \ldots$.

48. The ninth partial sum of the sequence $5, -10, 20, \ldots$.

49. The sum of the first 30 odd positive integers

50. The sum of the first 20 positive multiples of 7

51. The sum of the first 20 terms of the sequence $8, 5, 2, \ldots$.

52. The sum of the first eight terms of the sequence $\dfrac{3}{4}, \dfrac{9}{4}, \dfrac{27}{4}, \ldots$

53. S_4 if $a_1 = 6$ and $r = 5$

54. S_{100} if $a_1 = -3$ and $d = -6$

Find the sum of each infinite geometric sequence.

55. $5, \dfrac{5}{2}, \dfrac{5}{4}, \ldots$

56. $18, -2, \dfrac{2}{9}, \ldots$

57. $-20, -4, -\dfrac{4}{5}, \ldots$

58. $0.2, 0.02, 0.002, \ldots$

Solve.

59. A frozen yogurt store owner cleared $20,000 the first year he owned his business and made an additional 15% over the previous year in each subsequent year. Find how much he made during his fourth year of business. Find his total earnings during the first 4 years (round to the nearest dollar).

60. On his first morning in a television assembly factory, a trainee takes 0.8 times as long to assemble each television as he took to assemble the one before. If it took him 40 minutes to assemble the first television, find how long it takes him to assemble the fourth television. Find the total time he takes to assemble the first four televisions (round to the nearest minute).

61. During the harvest season a farmer can rent a combine machine for $100 the first day, with the rental fee decreasing $7 for each additional day. Find how much the farmer pays for the rental on the seventh day. Find how much total rent the farmer pays for 7 days.

62. A rubber ball is dropped from a height of 15 feet and rebounds 80% of its previous height after each bounce. Find the total distance the ball travels before it comes to rest.

63. After a pond was sprayed once with insecticide, 1800 mosquitoes were killed the first day, 600 the second day, 200 the third day, and so on. Find the total number of mosquitoes killed during the first 6 days after the spraying (round to the nearest unit).

64. See Exercise 63. Find the day on which the insecticide is no longer effective, and find the total number of mosquitoes killed (round to the nearest mosquito).

65. Use the formula S_∞ to write $0.5\overline{55}$ as a fraction.

66. A movie theater has 27 seats in the first row, 30 seats in the second row, 33 seats in the third row, and so on. Find the total number of seats in the theater if there are 20 rows.

(12.5) Use Pascal's triangle to expand each binomial.

67. $(x + z)^5$ **68.** $(y - r)^6$

69. $(2x + y)^4$ **70.** $(3y - z)^4$

Use the binomial formula to expand the following.

71. $(b + c)^8$ **72.** $(x - w)^7$

73. $(4m - n)^4$ **74.** $(p - 2r)^5$

Find the indicated term.

75. The fourth term of the expansion of $(a + b)^7$

76. The eleventh term of the expansion of $(y + 2z)^{10}$

CHAPTER 12 TEST

Find the indicated term(s) of the given sequence.

1. The first five terms of the sequence $a_n = \dfrac{(-1)^n}{n + 4}$

2. The first five terms of the sequence $a_n = \dfrac{3}{(-1)^n}$

3. The eightieth term of the sequence $a_n = 10 + 3(n - 1)$

4. The two-hundredth term of the sequence $a_n = (n + 1)(n - 1)(-1)^n$

5. The general term of the sequence $\dfrac{2}{5}, \dfrac{2}{25}, \dfrac{2}{125}, \ldots$

6. The general term of the sequence $-9, 18, -27, 36, \ldots$

Find the partial sum of the given sequence.

7. S_5 of the sequence $a_n = 5(2)^{n-1}$

8. S_{30} of the sequence $a_n = 18 + (n - 1)(-2)$

9. S_∞ of the sequence $a_1 = 24$ and $r = \dfrac{1}{6}$

10. S_∞ of the sequence $\dfrac{3}{2}, -\dfrac{3}{4}, \dfrac{3}{8}, \ldots$

11. $\displaystyle\sum_{i=1}^{4} i(i - 2)$ **12.** $\displaystyle\sum_{i=2}^{4} 5(2)^i(-1)^{i-1}$

Expand the binomial by using Pascal's triangle.

13. $(a - b)^6$ **14.** $(2x + y)^5$

Expand the binomial by using the binomial formula.

15. $(y + z)^8$ **16.** $(2p + r)^7$

Solve the following applications.

17. The population of a small town is growing yearly according to the sequence defined by $a_n = 250 + 75(n - 1)$, where n is the number of the year just beginning. Predict the population at the beginning of the tenth year. Find the town's initial population.

18. A gardener is making a triangular planting with one shrub in the first row, three shrubs in the second row, five shrubs in the third row, and so on, for eight rows. Write the finite series of this sequence, and find the total number of shrubs planted.

19. A pendulum swings through an arc of length 80 centimeters on its first swing. On each successive swing, the length of the arc is $\frac{3}{4}$ the length of the arc on the preceding swing. Find the length of the arc on the fourth swing, and find the total arc length for the first four swings.

20. See Exercise 19. Find the total arc length before the pendulum comes to rest.

21. A parachutist in free-fall falls 16 feet during the first second, 48 feet during the second second, 80 feet during the third second, and so on. Find how far he falls during the tenth second. Find the total distance he falls during the first 10 seconds.

22. Use the formula S_∞ to write $0.42\overline{42}$ as a fraction.

CHAPTER 12 CUMULATIVE REVIEW

1. Is the relation $x = y^2$ also a function?

2. Graph $x = 2$.

3. Find the slope of the line $y = 2$.

4. Find the equation of the horizontal line containing the point $(2, 3)$.

5. Graph the union of $x + \frac{1}{2} y \geq -4$ or $y \leq -2$.

6. Use synthetic division to divide $x^4 - 2x^3 - 11x^2 + 5x + 34$ by $x + 2$.

7. Solve $-1 \leq \frac{2x}{3} + 5 \leq 2$.

8. Solve $|y| = 0$.

9. Solve $\left|2x - \frac{1}{10}\right| < -13$.

10. Simplify the following.
   a. $\sqrt{50}$
   b. $\sqrt[3]{24}$
   c. $\sqrt{26}$
   d. $\sqrt[4]{32}$

11. Find the interest rate r if $2000 compounded annually grows to $2420 in 2 years.

12. Solve $(x - 3)^2 - 3(x - 3) - 4 = 0$.

13. Solve $\frac{5}{x + 1} < -2$.

14. A rock is thrown upward from the ground. Its height in feet above ground after t seconds is given by the function $f(t) = -16t^2 + 20t$. Find the maximum height of the rock and the number of seconds it took for the rock to reach its maximum height.

15. If $f(x) = x^2$ and $g(x) = x + 3$, find each composition.
   a. $(f \circ g)(2)$ and $(g \circ f)(2)$
   b. $(f \circ g)(x)$ and $(g \circ f)(x)$

16. Find the inverse of the one-to-one function.
   $f = \{(0, 1), (-2, 7), (3, -6), (4, 4)\}$

17. Solve each equation for x.
   a. $2^x = 16$
   b. $9^x = 27$
   c. $4^{x+3} = 8^x$

18. Simplify.
   a. $\log_3 3^2$
   b. $\log_7 7^{-1}$
   c. $5^{\log_5 3}$
   d. $2^{\log_2 6}$

19. Write each sum as a single logarithm.
   a. $\log_{11} 10 + \log_{11} 3$
   b. $\log_3 \frac{1}{2} + \log_3 12$
   c. $\log_2 (x + 2) + \log_2 x$

20. Find the amount owed at the end of 5 years if $1600 is loaned at a rate of 9% compounded continuously.

21. Solve $3^x = 7$.

22. Solve $\log_4 (x - 2) = 2$.

23. Write the first five terms of the sequence whose general term is given by $a_n = n^2 - 1$.

24. Find the eleventh term of the arithmetic sequence whose first three terms are 2, 9, and 16.

25. Evaluate.

    a. $\displaystyle\sum_{i=0}^{6} \frac{i - 2}{2}$ **b.** $\displaystyle\sum_{i=3}^{5} 2^i$

26 Find the sum of the first 30 positive integers.

Designing Your World

Schools, homes, hospitals, airports, auditoriums, community centers, jails, theaters, day-care centers, and office buildings are just some of the types of structures designed by architects to be safe, economical, and functional.

To become a licensed architect, a person must have a professional degree in architecture, complete a three-year internship, and pass the Architect Registration Examination. An architecture degree typically includes courses in building design, computer-aided design and drafting (CADD), physics and other physical sciences, architectural history, and mathematics. Architects need solid computer and communication skills. They also need a good understanding of geometry and spatial relationships to visualize a building during the design process.

 For more information about a career in architecture, visit The American Institute of Architects Website by first going to www.prenhall.com/martin-gay.

In the Spotlight on Decision Making feature on page 805, you will have the opportunity to make a decision as an architect about redesigning a bridge arch in the shape of a half-ellipse.

CONIC SECTIONS

In Chapter 10, we analyzed some of the important connections between a parabola and its equation. Parabolas are interesting in their own right but are more interesting still because they are part of a collection of curves known as conic sections. This chapter is devoted to quadratic equations in two variables and their conic section graphs: the parabola, circle, ellipse, and hyperbola.

13.1 THE PARABOLA AND THE CIRCLE

CD-ROM SSM

SSG Video

▶ **OBJECTIVES**

1. Graph parabolas of the form $x = a(y - k)^2 + h$ and $y = a(x - h)^2 + k$.
2. Use the distance formula and the midpoint formula.
3. Graph circles of the form $(x - h)^2 + (y - k)^2 = r^2$.
4. Write the equation of a circle, given its center and radius.
5. Find the center and the radius of a circle, given its equation.

Conic sections derive their name because each conic section is the intersection of a right circular cone and a plane. The circle, parabola, ellipse, and hyperbola are the conic sections.

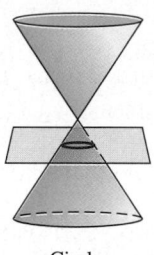
Circle

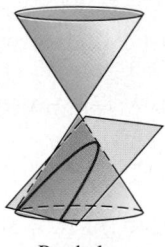
Parabola

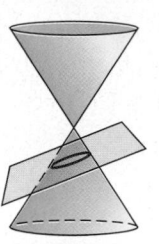
Ellipse

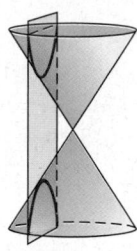
Hyperbola

1 Thus far, we have seen that $f(x)$ or $y = a(x - h)^2 + k$ is the equation of a parabola that opens upward if $a > 0$ or downward if $a < 0$. Parabolas can also open left or right, or even on a slant. Equations of these parabolas are not functions of x, of course, since a parabola opening any way other than upward or downward fails the vertical line test. In this section, we introduce parabolas that open to the left and to the right. Parabolas opening on a slant will not be developed in this book.

Just as $y = a(x - h)^2 + k$ is the equation of a parabola that opens upward or downward, $x = a(y - k)^2 + h$ is the equation of a parabola that opens to the right or to the left. The parabola opens to the right if $a > 0$ and to the left if $a < 0$. The parabola has vertex (h, k), and its axis of symmetry is the line $y = k$.

PARABOLAS

$$y = a(x - h)^2 + k$$

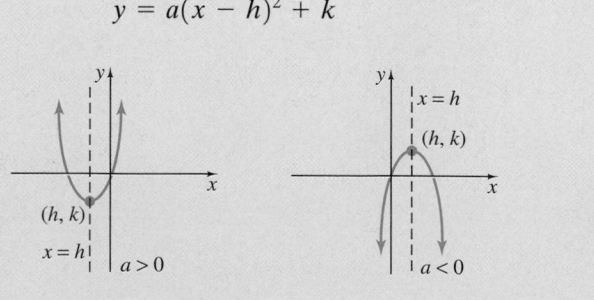

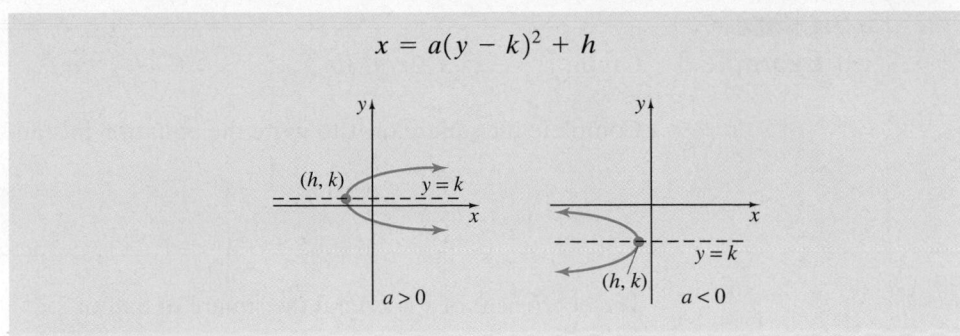

$$x = a(y - k)^2 + h$$

The equations $y = a(x - h)^2 + k$ and $x = a(y - k)^2 + h$ are called **standard forms.**

Example 1 Graph the parabola $x = 2y^2$.

Solution Written in standard form, the equation $x = 2y^2$ is $x = 2(y - 0)^2 + 0$ with $a = 2$, $h = 0$, and $k = 0$. Its graph is a parabola with vertex $(0, 0)$, and its axis of symmetry is the line $y = 0$. Since $a > 0$, this parabola opens to the right. The table shows a few more ordered pair solutions of $x = 2y^2$. Its graph is also shown.

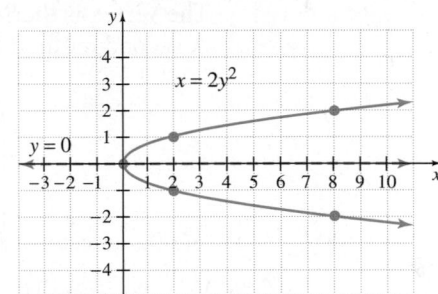

x	y
8	-2
2	-1
0	0
2	1
8	2

Example 2 Graph the parabola $x = -3(y - 1)^2 + 2$.

Solution The equation $x = -3(y - 1)^2 + 2$ is in the form $x = a(y - k)^2 + h$ with $a = -3$, $k = 1$, and $h = 2$. Since $a < 0$, the parabola opens to the left. The vertex (h, k) is $(2, 1)$, and the axis of symmetry is the line $y = 1$. When $y = 0$, $x = -1$, so the x-intercept is $(-1, 0)$. Again, we obtain a few ordered pair solutions and then graph the parabola.

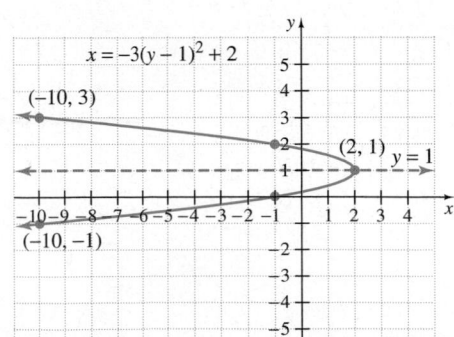

x	y
2	1
-1	0
-1	2
-10	3
-10	-1

Example 3 Graph $y = -x^2 - 2x + 15$.

Solution Complete the square on x to write the equation in standard form.

$$y - 15 = -x^2 - 2x \qquad \text{Subtract 15 from both sides.}$$

$$y - 15 = -1(x^2 + 2x) \qquad \text{Factor } -1 \text{ from the terms } -x^2 - 2x.$$

The coefficient of x is 2. Find the square of half of 2.

$$\frac{1}{2}(2) = 1 \quad \text{and} \quad 1^2 = 1$$

$$y - 15 - 1(1) = -1(x^2 + 2x + 1) \qquad \text{Add } -1(1) \text{ to both sides.}$$

$$y - 16 = -1(x + 1)^2 \qquad \begin{array}{l}\text{Simplify the left side and}\\\text{factor the right side.}\end{array}$$

$$y = -(x + 1)^2 + 16 \qquad \text{Add 16 to both sides.}$$

The equation is now in standard form $y = a(x - h)^2 + k$ with $a = -1$, $h = -1$, and $k = 16$.

The vertex is then (h, k), or $(-1, 16)$.

A second method for finding the vertex is by using the formula $\dfrac{-b}{2a}$.

$$x = \frac{-(-2)}{2(-1)} = \frac{2}{-2} = -1$$

$$y = -(-1)^2 - 2(-1) + 15 = -1 + 2 + 15 = 16$$

Again, we see that the vertex is $(-1, 16)$, and the axis of symmetry is the vertical line $x = -1$. The y-intercept is $(0, 15)$. Now we can use a few more ordered pair solutions to graph the parabola.

x	y
-1	16
0	15
-2	15
1	12
-3	12
3	0
-5	0

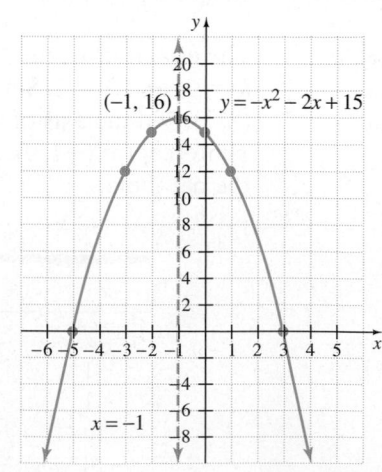

Example 4   Graph $x = 2y^2 + 4y + 5$.

Solution   Notice that this equation is quadratic in y, so its graph is a parabola that opens to the left or the right. We can complete the square on y or we can use the formula $\dfrac{-b}{2a}$ to find the vertex.

Since the equation is quadratic in y, the formula gives us the y-value of the vertex.

$$y = \frac{-4}{2 \cdot 2} = \frac{-4}{4} = -1$$

$$x = 2(-1)^2 + 4(-1) + 5 = 2 \cdot 1 - 4 + 5 = 3$$

The vertex is $(3, -1)$, and the axis of symmetry is the line $y = -1$. The parabola opens to the right since $a > 0$. The x-intercept is $(5, 0)$.

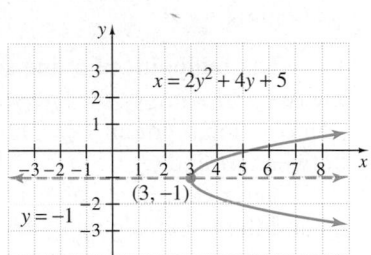

2   Another conic section is the circle. Before we review the circle, we need a formula to calculate the distance between points of the Cartesian coordinate system. To find the distance between two points, we use the distance formula, which is derived from the Pythagorean theorem.

To find the distance d between two points (x_1, y_1) and (x_2, y_2) as shown to the left, notice that the length of leg a is $x_2 - x_1$ and that the length of leg b is $y_2 - y_1$.

Thus, the Pythagorean theorem tells us that

$$d^2 = a^2 + b^2$$

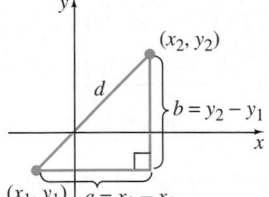

or

$$d^2 = (x_2 - x_1)^2 + (y_2 - y_1)^2$$

or

$$d = \sqrt{(x_2 - x_1)^2 + (y_2 - y_1)^2}$$

This formula gives us the distance between any two points on the real plane.

DISTANCE FORMULA

The distance d between two points (x_1, y_1) and (x_2, y_2) is given by

$$d = \sqrt{(x_2 - x_1)^2 + (y_2 - y_1)^2}$$

Example 5   Find the distance between $(2, -5)$ and $(1, -4)$. Give an exact distance and a three-decimal-place approximation.

Solution   To use the distance formula, it makes no difference which point we call (x_1, y_1) and which point we call (x_2, y_2). We will let $(x_1, y_1) = (2, -5)$ and $(x_2, y_2) = (1, -4)$.

$$d = \sqrt{(x_2 - x_1)^2 + (y_2 - y_1)^2}$$
$$= \sqrt{(1 - 2)^2 + [-4 - (-5)]^2}$$
$$= \sqrt{(-1)^2 + (1)^2}$$
$$= \sqrt{1 + 1}$$
$$= \sqrt{2} \approx 1.414$$

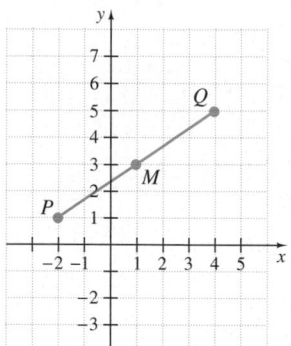

The distance between the two points is exactly $\sqrt{2}$ units, or approximately 1.414 units.

The **midpoint** of a line segment is the **point** located exactly halfway between the two endpoints of the line segment. On the graph to the left, the point M is the midpoint of line segment PQ. Thus, the distance between M and P equals the distance between M and Q.

The x-coordinate of M is at half the distance between the x-coordinates of P and Q, and the y-coordinate of M is at half the distance between the y-coordinates of P and Q. That is, the x-coordinate of M is the average of the x-coordinates of P and Q; the y-coordinate of M is the average of the y-coordinates of P and Q.

MIDPOINT FORMULA

The midpoint of the line segment whose endpoints are (x_1, y_1) and (x_2, y_2) is the point with coordinates

$$\left(\frac{x_1 + x_2}{2}, \frac{y_1 + y_2}{2} \right)$$

Example 6   Find the midpoint of the line segment that joins points $P(-3, 3)$ and $Q(1, 0)$.

Solution   Use the midpoint formula. It makes no difference which point we call (x_1, y_1) or which point we call (x_2, y_2). Let $(x_1, y_1) = (-3, 3)$ and $(x_2, y_2) = (1, 0)$.

$$\text{midpoint} = \left(\frac{x_1 + x_2}{2}, \frac{y_1 + y_2}{2} \right)$$
$$= \left(\frac{-3 + 1}{2}, \frac{3 + 0}{2} \right)$$
$$= \left(\frac{-2}{2}, \frac{3}{2} \right)$$
$$= \left(-1, \frac{3}{2} \right)$$

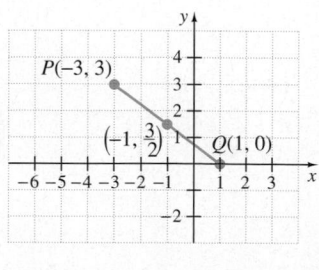

The midpoint of the segment is $\left(-1, \dfrac{3}{2} \right)$.

3

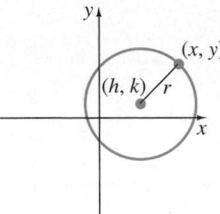

Another conic section is the **circle.** A circle is the set of all points in a plane that are the same distance from a fixed point called the **center.** The distance is called the **radius** of the circle. To find a standard equation for a circle, let (h, k) represent the center of the circle, and let (x, y) represent any point on the circle. The distance between (h, k) and (x, y) is defined to be the circle's radius, r units. We can find this distance r by using the distance formula.

$$r = \sqrt{(x - h)^2 + (y - k)^2}$$

$$r^2 = (x - h)^2 + (y - k)^2 \qquad \text{Square both sides.}$$

CIRCLE

The graph of $(x - h)^2 + (y - k)^2 = r^2$ is a circle with center (h, k) and radius r.

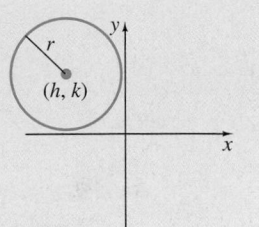

The equation $(x - h)^2 + (y - k)^2 = r^2$ is called **standard form.**

If an equation can be written in the standard form

$$(x - h)^2 + (y - k)^2 = r^2$$

then its graph is a circle, which we can draw by graphing the center (h, k) and using the radius r.

Example 7   Graph $x^2 + y^2 = 4$.

Solution   The equation can be written in standard form as

$$(x - 0)^2 + (y - 0)^2 = 2^2$$

The center of the circle is $(0, 0)$, and the radius is 2. Its graph is shown.

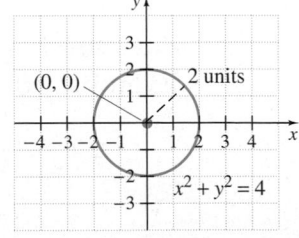

HELPFUL HINT
Notice the difference between the equation of a circle and the equation of a parabola. The equation of a circle contains both x^2-and y^2-terms on the same side of the equation with equal coefficients. The equation of a parabola has either an x^2-term or a y^2-term but not both.

Example 8   Graph $(x + 1)^2 + y^2 = 8$.

Solution   The equation can be written as $(x + 1)^2 + (y - 0)^2 = 8$ with $h = -1$, $k = 0$, and $r = \sqrt{8}$. The center is $(-1, 0)$, and the radius is $\sqrt{8} = 2\sqrt{2} \approx 2.8$.

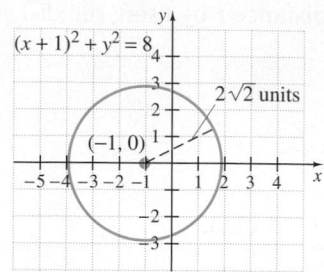

4   Since a circle is determined entirely by its center and radius, this information is all we need to write the equation of a circle.

Example 9   Find an equation of the circle with center $(-7, 3)$ and radius 10.

Solution   Using the given values $h = -7$, $k = 3$, and $r = 10$, we write the equation

$$(x - h)^2 + (y - k)^2 = r^2$$

or

$$[x - (-7)]^2 + (y - 3)^2 = 10^2 \qquad \text{Substitute the given values.}$$

or

$$(x + 7)^2 + (y - 3)^2 = 100$$

5   To find the center and the radius of a circle from its equation, write the equation in standard form. To write the equation of a circle in standard form, we complete the square on both x and y.

Example 10   Graph $x^2 + y^2 + 4x - 8y = 16$.

Solution   Since this equation contains x^2 and y^2 terms on the same side of the equation with equal coefficients, its graph is a circle. To write the equation in standard form, group the terms involving x and the terms involving y, and then complete the square on each variable.

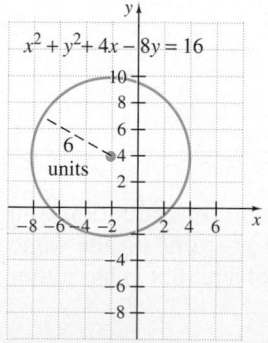

$$(x^2 + 4x) + (y^2 - 8y) = 16$$

Thus, $\frac{1}{2}(4) = 2$ and $2^2 = 4$. Also, $\frac{1}{2}(-8) = -4$ and $(-4)^2 = 16$. Add 4 and then 16 to both sides.

$$(x^2 + 4x + 4) + (y^2 - 8y + 16) = 16 + 4 + 16$$
$$(x + 2)^2 + (y - 4)^2 = 36 \qquad \text{Factor.}$$

This circle has the center $(-2, 4)$ and radius 6, as shown.

GRAPHING CALCULATOR EXPLORATIONS

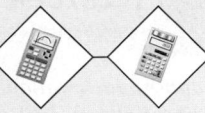

To graph an equation such as $x^2 + y^2 = 25$ with a graphing calculator, we first solve the equation for y.

$$x^2 + y^2 = 25$$
$$y^2 = 25 - x^2$$
$$y = \pm\sqrt{25 - x^2}$$

The graph of $y = \sqrt{25 - x^2}$ will be the top half of the circle, and the graph of $y = -\sqrt{25 - x^2}$ will be the bottom half of the circle.

To graph, press $\boxed{Y=}$ and enter $Y_1 = \sqrt{25 - x^2}$ and $Y_2 = -\sqrt{25 - x^2}$. Insert parentheses around $25 - x^2$ so that $\sqrt{25 - x^2}$ and not $\sqrt{25} - x^2$ is graphed.

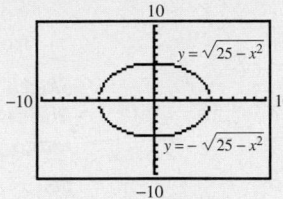

The graph does not appear to be a circle because we are currently using a standard window and the screen is rectangular. This causes the tick marks on the x-axis to be farther apart than the tick marks on the y-axis and, thus, creates the distorted circle. If we want the graph to appear circular, we must define a square window by using a feature of the graphing calculator or by redefining the window to show the x-axis from -15 to 15 and the y-axis from -10 to 10. Using a square window, the graph appears as follows.

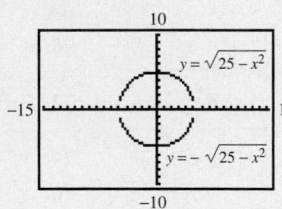

Use a graphing calculator to graph each circle.

1. $x^2 + y^2 = 55$ **2.** $x^2 + y^2 = 20$

3. $5x^2 + 5y^2 = 50$ **4.** $6x^2 + 6y^2 = 105$

5. $2x^2 + 2y^2 - 34 = 0$ **6.** $4x^2 + 4y^2 - 48 = 0$

7. $7x^2 + 7y^2 - 89 = 0$ **8.** $3x^2 + 3y^2 - 35 = 0$

MENTAL MATH

The graph of each equation is a parabola. Determine whether the parabola opens upward, downward, to the left, or to the right.

1. $y = x^2 - 7x + 5$ **2.** $y = -x^2 + 16$ **3.** $x = -y^2 - y + 2$

4. $x = 3y^2 + 2y - 5$ **5.** $y = -x^2 + 2x + 1$ **6.** $x = -y^2 + 2y - 6$

Exercise Set 13.1

The graph of each equation is a parabola. Find the vertex of the parabola and sketch its graph. See Examples 1 through 4.

1. $x = 3y^2$ **2.** $x = -2y^2$
3. $x = (y - 2)^2 + 3$ **4.** $x = (y - 4)^2 - 1$
5. $y = 3(x - 1)^2 + 5$ **6.** $x = -4(y - 2)^2 + 2$
7. $x = y^2 + 6y + 8$ **8.** $x = y^2 - 6y + 6$
9. $y = x^2 + 10x + 20$ **10.** $y = x^2 + 4x - 5$
11. $x = -2y^2 + 4y + 6$ **12.** $x = 3y^2 + 6y + 7$

Find the distance between each pair of points. Approximate the distances in Exercises 21 and 22 to two decimal places. See Example 5.

13. $(5, 1)$ and $(8, 5)$ **14.** $(2, 3)$ and $(14, 8)$
15. $(-3, 2)$ and $(1, -3)$
16. $(3, -2)$ and $(-4, 1)$
17. $(-9, 4)$ and $(-8, 1)$
18. $(-5, -2)$ and $(-6, -6)$
19. $(0, -\sqrt{2})$ and $(\sqrt{3}, 0)$
20. $(-\sqrt{5}, 0)$ and $(0, \sqrt{7})$
21. $(1.7, -3.6)$ and $(-8.6, 5.7)$
22. $(9.6, 2.5)$ and $(-1.9, -3.7)$
23. $(2\sqrt{3}, \sqrt{6})$ and $(-\sqrt{3}, 4\sqrt{6})$
24. $(5\sqrt{2}, -4)$ and $(-3\sqrt{2}, -8)$

Find the midpoint of the line segment whose endpoints are given. See Example 6.

25. $(6, -8), (2, 4)$ **26.** $(3, 9), (7, 11)$
27. $(-2, -1), (-8, 6)$ **28.** $(-3, -4), (6, -8)$
29. $(7, 3), (-1, -3)$ **30.** $(-2, 5), (-1, 6)$
31. $\left(\dfrac{1}{2}, \dfrac{3}{8}\right), \left(-\dfrac{3}{2}, \dfrac{5}{8}\right)$ **32.** $\left(-\dfrac{2}{5}, \dfrac{7}{15}\right), \left(-\dfrac{2}{5}, -\dfrac{4}{15}\right)$
33. $(\sqrt{2}, 3\sqrt{5}), (\sqrt{2}, -2\sqrt{5})$
34. $(\sqrt{8}, -\sqrt{12}), (3\sqrt{2}, 7\sqrt{3})$
35. $(4.6, -3.5), (7.8, -9.8)$
36. $(-4.6, 2.1), (-6.7, 1.9)$

The graph of each equation is a circle. Find the center and the radius, and then sketch. See Examples 7, 8, and 10.

37. $x^2 + y^2 = 9$ **38.** $x^2 + y^2 = 25$
39. $x^2 + (y - 2)^2 = 1$ **40.** $(x - 3)^2 + y^2 = 9$
41. $(x - 5)^2 + (y + 2)^2 = 1$
42. $(x + 3)^2 + (y + 3)^2 = 4$
43. $x^2 + y^2 + 6y = 0$ **44.** $x^2 + 10x + y^2 = 0$

45. $x^2 + y^2 + 2x - 4y = 4$
46. $x^2 + 6x - 4y + y^2 = 3$
47. $x^2 + y^2 - 4x - 8y - 2 = 0$
48. $x^2 + y^2 - 2x - 6y - 5 = 0$

Write an equation of the circle with the given center and radius. See Example 9.

49. $(2, 3); 6$ **50.** $(-7, 6); 2$
51. $(0, 0); \sqrt{3}$ **52.** $(0, -6); \sqrt{2}$
53. $(-5, 4); 3\sqrt{5}$ **54.** the origin; $4\sqrt{7}$

55. If you are given a list of equations of circles and parabolas and none are in standard form, explain how you would determine which is an equation of a circle and which is an equation of a parabola. Explain also how you would distinguish the upward or downward parabolas from the left-opening or right-opening parabolas.

Sketch the graph of each equation. If the graph is a parabola, find its vertex. If the graph is a circle, find its center and radius.

56. $x = y^2 + 2$ **57.** $x = y^2 - 3$
58. $y = (x + 3)^2 + 3$ **59.** $y = (x - 2)^2 - 2$
60. $x^2 + y^2 = 49$ **61.** $x^2 + y^2 = 1$
62. $x = (y - 1)^2 + 4$ **63.** $x = (y + 3)^2 - 1$
64. $(x + 3)^2 + (y - 1)^2 = 9$
65. $(x - 2)^2 + (y - 2)^2 = 16$
66. $x = -2(y + 5)^2$ **67.** $x = -(y - 1)^2$
68. $x^2 + (y + 5)^2 = 5$ **69.** $(x - 4)^2 + y^2 = 7$
70. $y = 3(x - 4)^2 + 2$ **71.** $y = 5(x + 5)^2 + 3$
72. $2x^2 + 2y^2 = \dfrac{1}{2}$ **73.** $\dfrac{x^2}{8} + \dfrac{y^2}{8} = 2$
74. $y = x^2 - 2x - 15$ **75.** $y = x^2 + 7x + 6$
76. $x^2 + y^2 + 6x + 10y - 2 = 0$
77. $x^2 + y^2 + 2x + 12y - 12 = 0$
78. $x = y^2 + 6y + 2$ **79.** $x = y^2 + 8y - 4$
80. $x^2 + y^2 - 8y + 5 = 0$ **81.** $x^2 - 10y + y^2 + 4 = 0$
82. $x = -2y^2 - 4y$ **83.** $x = -3y^2 + 30y$
84. $\dfrac{x^2}{3} + \dfrac{y^2}{3} = 2$ **85.** $5x^2 + 5y^2 = 25$
86. $y = 4x^2 - 40x + 105$ **87.** $y = 5x^2 - 20x + 16$

Solve.

88. Two surveyors need to find the distance across a lake. They place a reference pole at point A in the diagram. Point B is 3 meters east and 1 meter north of the reference point *A*. Point *C* is 19 meters east and 13 meters north of point *A*. Find the distance across the lake, from *B* to *C*.

100 meters and the maximum height of the arch is 40 meters.

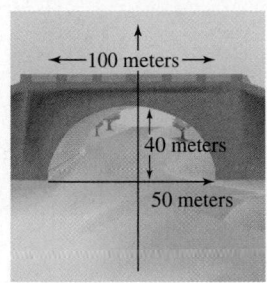

△ **89.** Determine whether the triangle with vertices $(2, 6)$, $(0, -2)$, and $(5, 1)$ is an isosceles triangle.

△ **90.** Cindy Brown, an architect, is drawing plans on grid paper for a circular pool with a fountain in the middle. The paper is marked off in centimeters, and each centimeter represents 1 foot. On the paper, the diameter of the "pool" is 20 centimeters, and "fountain" is the point $(0, 0)$.

   a. Sketch the architect's drawing. Be sure to label the axes.

   b. Write an equation that describes the circular pool.

   c. Cindy plans to place a circle of lights around the fountain such that each light is 5 feet from the center of the fountain. Write an equation for the circle of lights and sketch the circle on your drawing.

91. A bridge constructed over a bayou has a supporting arch in the shape of a parabola. Find an equation of the parabolic arch if the length of the road over the arch is

Use a graphing calculator to verify each exercise. Use a square viewing window.

92. Exercise 84.　　　　**93.** Exercise 85.
94. Exercise 86.　　　　**95.** Exercise 87.

REVIEW EXERCISES

Graph each equation. See Section 3.4.

96. $y = 2x + 5$　　　　**97.** $y = -3x + 3$
98. $y = 3$　　　　　　　**99.** $x = -2$

Rationalize each denominator and simplify if possible. See Section 9.5.

100. $\dfrac{1}{\sqrt{3}}$　　　　**101.** $\dfrac{\sqrt{5}}{\sqrt{8}}$

102. $\dfrac{4\sqrt{7}}{\sqrt{6}}$　　　**103.** $\dfrac{10}{\sqrt{5}}$

13.2 THE ELLIPSE AND THE HYPERBOLA

CD-ROM

SSM

SSG

Video

▶ **OBJECTIVES**

  1. Define and graph an ellipse.
  2. Define and graph a hyperbola.

1

An **ellipse** can be thought of as the set of points in a plane such that the sum of the distances of those points from two fixed points is constant. Each of the two fixed points is called a **focus**. (The plural of focus is **foci**.) The point midway between the foci is called the **center**.

   An ellipse may be drawn by hand by using two thumbtacks, a piece of string, and a pencil. Secure the two thumbtacks in a piece of cardboard, for example, and tie each end of the string to a tack. Use your pencil to pull the string tight and draw the ellipse. The two thumbtacks are the foci of the drawn ellipse.

 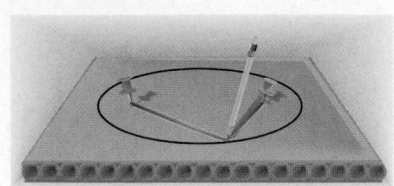 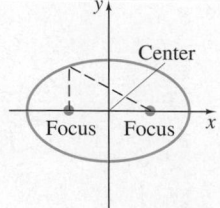

ELLIPSE WITH CENTER (0, 0)

The graph of an equation of the form $\dfrac{x^2}{a^2} + \dfrac{y^2}{b^2} = 1$ is an ellipse with center $(0, 0)$. The x-intercepts are $(a, 0)$ and $(-a, 0)$, and the y-intercepts are $(0, b)$, and $(0, -b)$.

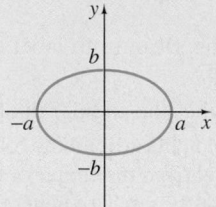

The **standard form** of an ellipse with center $(0, 0)$ is $\dfrac{x^2}{a^2} + \dfrac{y^2}{b^2} = 1$.

Example 1 Graph $\dfrac{x^2}{9} + \dfrac{y^2}{16} = 1$.

Solution The equation is of the form $\dfrac{x^2}{a^2} + \dfrac{y^2}{b^2} = 1$, with $a = 3$ and $b = 4$, so its graph is an ellipse with center $(0, 0)$, x-intercepts $(3, 0)$ and $(-3, 0)$, and y-intercepts $(0, 4)$ and $(0, -4)$.

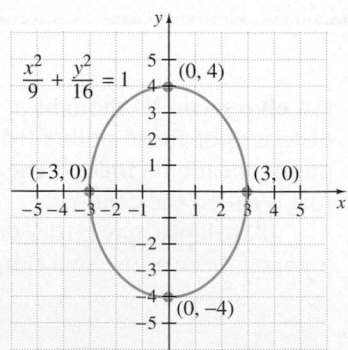

Example 2   Graph the equation $4x^2 + 16y^2 = 64$.

Solution   Although this equation contains a sum of squared terms in x and y on the same side of an equation, this is not the equation of a circle since the coefficients of x^2 and y^2 are not the same. The graph of this equation is an ellipse. Since the standard form of the equation of an ellipse has 1 on one side, divide both sides of this equation by 64.

$$4x^2 + 16y^2 = 64$$

$$\frac{4x^2}{64} + \frac{16y^2}{64} = \frac{64}{64} \qquad \text{Divide both sides by 64.}$$

$$\frac{x^2}{16} + \frac{y^2}{4} = 1 \qquad \text{Simplify.}$$

We now recognize the equation of an ellipse with $a = 4$ and $b = 2$. This ellipse has center $(0, 0)$, x-intercepts $(4, 0)$ and $(-4, 0)$, and y-intercepts $(0, 2)$ and $(0, -2)$.

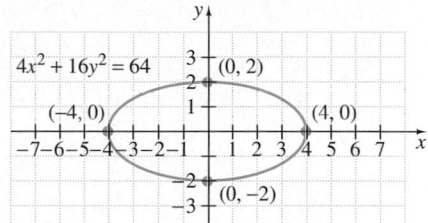

The center of an ellipse is not always $(0, 0)$, as shown in the next example.

Example 3   Graph $\dfrac{(x + 3)^2}{25} + \dfrac{(y - 2)^2}{36} = 1$.

Solution   The center of this ellipse is found in a way that is similar to finding the center of a circle. This ellipse has center $(-3, 2)$. Notice that $a = 5$ and $b = 6$. To find four points on the graph of the ellipse, first graph the center, $(-3, 2)$. Since $a = 5$, count 5 units right and then 5 units left of the point with coordinates $(-3, 2)$. Next, since $b = 6$, start at $(-3, 2)$ and count 6 units up and then 6 units down to find two more points on the ellipse.

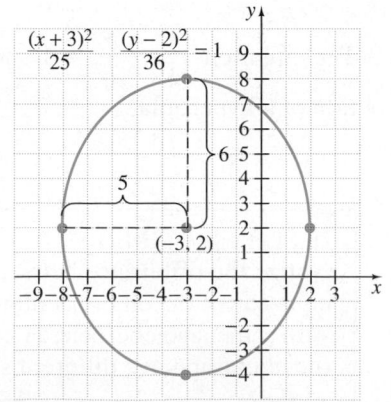

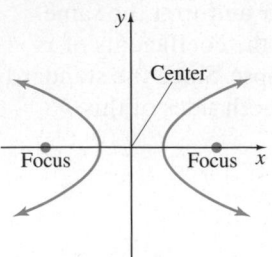

2 The final conic section is the **hyperbola.** A hyperbola is the set of points in a plane such that the absolute value of the difference of the distances from two fixed points is constant. Each of the two fixed points is called a **focus.** The point midway between the foci is called the **center.**

Using the distance formula, we can show that the graph of $\dfrac{x^2}{a^2} - \dfrac{y^2}{b^2} = 1$ is a hyperbola with center $(0, 0)$ and x-intercepts $(a, 0)$ and $(-a, 0)$. Also, the graph of $\dfrac{y^2}{b^2} - \dfrac{x^2}{a^2} = 1$ is a hyperbola with center $(0, 0)$ and y-intercepts $(0, b)$ and $(0, -b)$.

HYPERBOLA WITH CENTER $(0, 0)$

The graph of an equation of the form $\dfrac{x^2}{a^2} - \dfrac{y^2}{b^2} = 1$ is a hyperbola with center $(0, 0)$ and x-intercepts $(a, 0)$ and $(-a, 0)$.

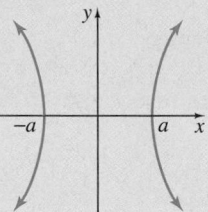

The graph of an equation of the form $\dfrac{y^2}{b^2} - \dfrac{x^2}{a^2} = 1$ is a hyperbola with center $(0, 0)$ and y-intercepts $(0, b)$ and $(0, -b)$.

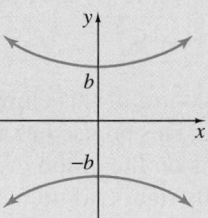

The equations $\dfrac{x^2}{a^2} - \dfrac{y^2}{b^2} = 1$ and $\dfrac{y^2}{b^2} - \dfrac{x^2}{a^2} = 1$ are the **standard forms** for the equation of a hyperbola.

> **HELPFUL HINT**
> Notice the difference between the equation of an ellipse and a hyperbola. The equation of the ellipse contains x^2 and y^2 terms on the same side of the equation with same-sign coefficients. For a hyperbola, the coefficients on the same side of the equation have different signs.

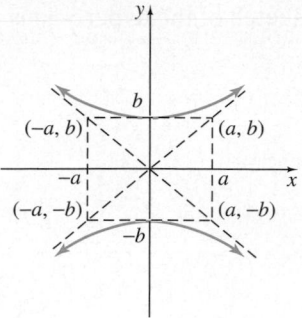

Graphing a hyperbola such as $\dfrac{y^2}{b^2} - \dfrac{x^2}{a^2} = 1$ is made easier by recognizing one of its important characteristics. Examining the figure to the left, notice how the sides of the branches of the hyperbola extend indefinitely and seem to approach the dashed lines in the figure. These dashed lines are called the **asymptotes** of the hyperbola.

To sketch these lines, or asymptotes, draw a rectangle with vertices (a, b), $(-a, b)$, $(a, -b)$, and $(-a, -b)$. The asymptotes of the hyperbola are the extended diagonals of this rectangle.

Example 4   Sketch the graph of $\dfrac{x^2}{16} - \dfrac{y^2}{25} = 1$.

Solution   This equation has the form $\dfrac{x^2}{a^2} - \dfrac{y^2}{b^2} = 1$, with $a = 4$ and $b = 5$. Thus, its graph is a hyperbola that opens to the left and right. It has center $(0, 0)$ and x-intercepts $(4, 0)$ and $(-4, 0)$. To aid in graphing the hyperbola, we first sketch its asymptotes. The extended diagonals of the rectangle with corners $(4, 5)$, $(4, -5)$, $(-4, 5)$, and $(-4, -5)$ are the asymptotes of the hyperbola. Then we use the asymptotes to aid in sketching the hyperbola.

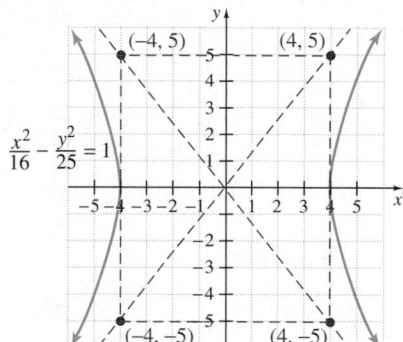

Example 5   Sketch the graph of the equation $4y^2 - 9x^2 = 36$.

Solution   Since this is a difference of squared terms in x and y on the same side of the equation, its graph is a hyperbola, as opposed to an ellipse or a circle. The standard form of the equation of a hyperbola has a 1 on one side, so divide both sides of the equation by 36.

$$4y^2 - 9x^2 = 36$$

$$\frac{4y^2}{36} - \frac{9x^2}{36} = \frac{36}{36} \qquad \text{Divide both sides by 36.}$$

$$\frac{y^2}{9} - \frac{x^2}{4} = 1 \qquad \text{Simplify.}$$

The equation is of the form $\dfrac{y^2}{b^2} - \dfrac{x^2}{a^2} = 1$, with $a = 2$ and $b = 3$, so the hyperbola is centered at $(0, 0)$ with y-intercepts $(0, 3)$ and $(0, -3)$. The sketch of the hyperbola is shown.

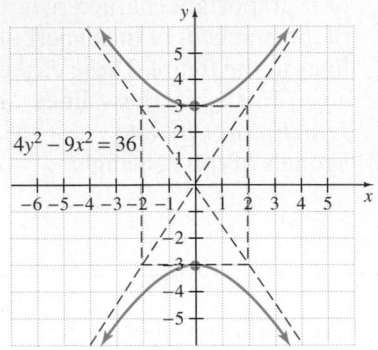

Following is a summary of conic sections.

CONIC SECTIONS

	Standard Form	*Graph*

Parabola $y = a(x - h)^2 + k$

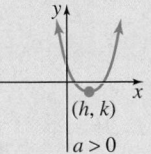

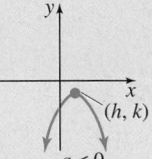

Parabola $x = a(y - k)^2 + h$

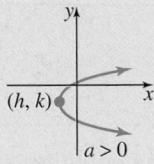

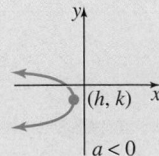

Circle $(x - h)^2 + (y - k)^2 = r^2$

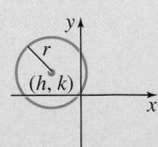

Ellipse $\dfrac{x^2}{a^2} + \dfrac{y^2}{b^2} = 1$

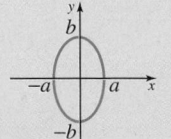

Hyperbola	$\dfrac{x^2}{a^2} - \dfrac{y^2}{b^2} = 1$	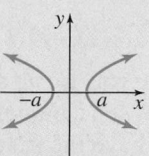
Hyperbola	$\dfrac{y^2}{b^2} - \dfrac{x^2}{a^2} = 1$	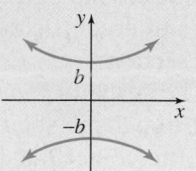

GRAPHING CALCULATOR EXPLORATIONS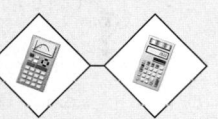

To find the graph of an ellipse by using a graphing calculator, use the same procedure as for graphing a circle. For example, to graph $x^2 + 3y^2 = 22$, first solve for y.

$$3y^2 = 22 - x^2$$

$$y^2 = \frac{22 - x^2}{3}$$

$$y = \pm\sqrt{\frac{22 - x^2}{3}}$$

Next press the $\boxed{Y =}$ key and enter $Y_1 = \sqrt{\dfrac{22 - x^2}{3}}$ and $Y_2 = -\sqrt{\dfrac{22 - x^2}{3}}$. (Insert two sets of parentheses in the radicand to get $\sqrt{((22 - x^2)/3)}$ so that the desired graph is obtained.) The graph appears as follows.

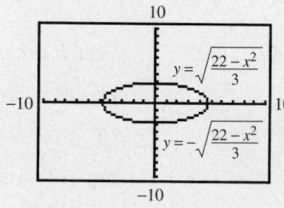

Use a graphing calculator to graph each ellipse.

1. $10x^2 + y^2 = 32$ **2.** $x^2 + 6y^2 = 35$

3. $20x^2 + 5y^2 = 100$ **4.** $4y^2 + 12x^2 = 48$

5. $7.3x^2 + 15.5y^2 = 95.2$ **6.** $18.8x^2 + 36.1y^2 = 205.8$

SPOTLIGHT ON DECISION MAKING

Suppose you are an astronomer. You know that the orbits of stars, planets, comets, asteroids, and satellites all have the shape of one of the conic sections. *Eccentricity* is a measure used to describe the shape and elongation of an orbital path. The table shows ranges of eccentricities for the different types of conic sections.

For each of the following comets known to pass through our solar system, decide what type of orbit the comet has based on its eccentricity *e*. Describe how the shape of the comet's orbit affects how often it passes through our solar system. (For more exercises on eccentricity, see Exercise Set 13.2.)

a. Spacewatch (1997 P2), $e = 1.02851919$
b. Whipple, $e = 0.25871336$
c. Lee (1999 H1), $e = 0.99973749$
d. Giacobini-Zinner, $e = 0.70647162$
e. Tabur (1997 N1), $e = 1.00004712$

Conic Section	Eccentricity e
Circle	$e = 0$
Ellipse	$0 < e < 1$
Parabola	$e = 1$
Hyperbola	$e > 1$

Exercise Set 13.2

Sketch the graph of each equation. See Examples 1 and 2.

1. $\dfrac{x^2}{4} + \dfrac{y^2}{25} = 1$

2. $\dfrac{x^2}{9} + y^2 = 1$

3. $\dfrac{x^2}{16} + \dfrac{y^2}{9} = 1$

4. $x^2 + \dfrac{y^2}{4} = 1$

5. $9x^2 + 4y^2 = 36$

6. $x^2 + 4y^2 = 16$

7. $4x^2 + 25y^2 = 100$

8. $36x^2 + y^2 = 36$

Sketch the graph of each equation. See Example 3.

9. $\dfrac{(x + 1)^2}{36} + \dfrac{(y - 2)^2}{49} = 1$

10. $\dfrac{(x - 3)^2}{9} + \dfrac{(y + 3)^2}{16} = 1$

11. $\dfrac{(x - 1)^2}{4} + \dfrac{(y - 1)^2}{25} = 1$

12. $\dfrac{(x + 3)^2}{16} + \dfrac{(y + 2)^2}{4} = 1$

Sketch the graph of each equation. See Examples 4 and 5.

13. $\dfrac{x^2}{4} - \dfrac{y^2}{9} = 1$

14. $\dfrac{x^2}{36} - \dfrac{y^2}{36} = 1$

15. $\dfrac{y^2}{25} - \dfrac{x^2}{16} = 1$

16. $\dfrac{y^2}{25} - \dfrac{x^2}{49} = 1$

Sketch the graph of each equation. See Example 5.

17. $x^2 - 4y^2 = 16$

18. $4x^2 - y^2 = 36$

19. $16y^2 - x^2 = 16$

20. $4y^2 - 25x^2 = 100$

21. If you are given a list of equations of circles, parabolas, ellipses, and hyperbolas, explain how you could distinguish the different conic sections from their equations.

Identify whether each equation, when graphed, will be a parabola, circle, ellipse, or hyperbola. Sketch the graph of each equation.

22. $(x - 7)^2 + (y - 2)^2 = 4$

23. $y = x^2 + 4$

24. $y = x^2 + 12x + 36$

25. $\dfrac{x^2}{4} + \dfrac{y^2}{9} = 1$ **26.** $\dfrac{y^2}{9} - \dfrac{x^2}{9} = 1$

27. $\dfrac{x^2}{16} - \dfrac{y^2}{4} = 1$ **28.** $\dfrac{x^2}{16} + \dfrac{y^2}{4} = 1$

29. $x^2 + y^2 = 16$

30. $x = y^2 + 4y - 1$ **31.** $x = -y^2 + 6y$

32. $9x^2 - 4y^2 = 36$ **33.** $9x^2 + 4y^2 = 36$

34. $\dfrac{(x - 1)^2}{49} + \dfrac{(y + 2)^2}{25} = 1$

35. $y^2 = x^2 + 16$

36. $\left(x + \dfrac{1}{2}\right)^2 + \left(y - \dfrac{1}{2}\right)^2 = 1$

37. $y = -2x^2 + 4x - 3$

The orbits of stars, planets, comets, asteroids, and satellites all have the shape of one of the conic sections. Astronomers use a measure called eccentricity *to describe the shape and elongation of an orbital path. For the circle and ellipse, eccentricity e is calculated with the formula* $e = \dfrac{c}{d}$, *where* $c^2 = |a^2 - b^2|$ *and d is the larger value of a or b. For a hyperbola, eccentricity e is calculated with the formula* $e = \dfrac{c}{d}$, *where* $c^2 = a^2 + b^2$ *and the value of d is equal to a if the hyperbola has x-intercepts or equal to b if the hyperbola has y-intercepts. (For more information about eccentricity, see the Spotlight on Decision Making in this section.)*

A $\dfrac{x^2}{36} - \dfrac{y^2}{13} = 1$ **B** $\dfrac{x^2}{4} + \dfrac{y^2}{4} = 1$

C $\dfrac{x^2}{25} + \dfrac{y^2}{16} = 1$ **D** $\dfrac{y^2}{25} - \dfrac{x^2}{39} = 1$

E $\dfrac{x^2}{17} + \dfrac{y^2}{81} = 1$ **F** $\dfrac{x^2}{36} + \dfrac{y^2}{36} = 1$

G $\dfrac{x^2}{16} - \dfrac{y^2}{65} = 1$ **H** $\dfrac{x^2}{144} + \dfrac{y^2}{140} = 1$

38. Identify the type of conic section represented by each of the equations A–H.

39. For each of the equations A–H, identify the values of a^2 and b^2.

40. For each of the equations A–H, calculate the value of c^2 and c.

41. For each of the equations A–H, find the value of d.

42. For each of the equations A–H, calculate the eccentricity e.

43. What do you notice about the values of e for the equations you identified as ellipses?

44. What do you notice about the values of e for the equations you identified as circles?

45. What do you notice about the values of e for the equations you identified as hyperbolas?

Solve.

46. A planet's orbit about the Sun can be described as an ellipse. Consider the Sun as the origin of a rectangular coordinate system. Suppose that the x-intercepts of the elliptical path of the planet are $\pm130{,}000{,}000$ and that the y-intercepts are $\pm125{,}000{,}000$. Write the equation of the elliptical path of the planet.

47. Comets orbit the Sun in elongated ellipses. Consider the Sun as the origin of a rectangular coordinate system. Suppose that the equation of the path of the comet is

$$\dfrac{(x - 1{,}782{,}000{,}000)^2}{3.42 \cdot 10^{23}} + \dfrac{(y - 356{,}400{,}000)^2}{1.368 \cdot 10^{22}} = 1$$

Find the center of the path of the comet.

48. Use a graphing calculator to verify Exercise 5.

49. Use a graphing calculator to verify Exercise 6.

REVIEW EXERCISES

Solve each inequality. See Section 8.3.

50. $x < 5$ and $x < 1$

51. $x < 5$ or $x < 1$

52. $2x - 1 \geq 7$ or $-3x \leq -6$

53. $2x - 1 \geq 7$ and $-3x \leq -6$

Perform the indicated operations. See Sections 5.1, 5.3, and 5.4.

54. $(2x^3)(-4x^2)$

55. $2x^3 - 4x^3$

56. $-5x^2 + x^2$

57. $(-5x^2)(x^2)$

58. $-5x^2 - x^2$

A Look Ahead

Example

Sketch the graph of $\dfrac{(x - 2)^2}{25} - \dfrac{(y - 1)^2}{9} = 1$.

Solution
This hyperbola has center $(2, 1)$. Notice that $a = 5$ and $b = 3$.

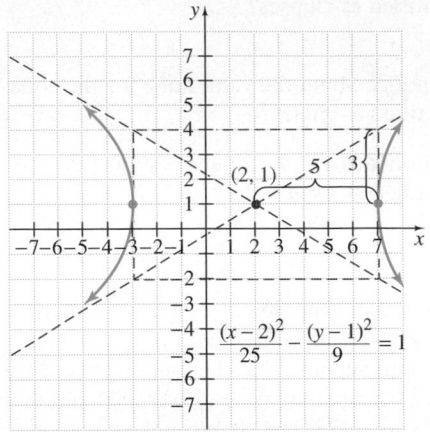

Sketch the graph of each equation. See the preceding example.

59. $\dfrac{(x-1)^2}{4} - \dfrac{(y+1)^2}{25} = 1$

60. $\dfrac{(x+2)^2}{9} - \dfrac{(y-1)^2}{4} = 1$

61. $\dfrac{y^2}{16} - \dfrac{(x+3)^2}{9} = 1$

62. $\dfrac{(y+4)^2}{4} - \dfrac{x^2}{25} = 1$

63. $\dfrac{(x+5)^2}{16} - \dfrac{(y+2)^2}{25} = 1$

64. $\dfrac{(x-3)^2}{9} - \dfrac{(y-2)^2}{4} = 1$

13.3 SOLVING NONLINEAR SYSTEMS OF EQUATIONS

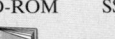

CD-ROM   SSM

SSG   Video

▶ **O B J E C T I V E S**

1. Solve a nonlinear system by substitution.
2. Solve a nonlinear system by elimination.

In Chapter 4, we used graphing, substitution, and elimination methods to find solutions of systems of linear equations in two variables. We now apply these same methods to nonlinear systems of equations in two variables. A **nonlinear system of equations** is a system of equations at least one of which is not linear. Since we will be graphing the equations in each system, we are interested in real number solutions only.

1   First, nonlinear systems are solved by the substitution method.

Example 1   Solve the system.

$$\begin{cases} x^2 - 3y = 1 \\ x - y = 1 \end{cases}$$

Solution   We can solve this system by substitution if we solve one equation for one of the variables. Solving the first equation for x is not the best choice since doing so introduces a radical. Also, solving for y in the first equation introduces a fraction. We solve the second equation for y.

$$x - y = 1 \qquad \text{Second equation}$$

$$x - 1 = y \qquad \text{Solve for } y.$$

Replace y with $x - 1$ in the first equation, and then solve for x.

$$x^2 - 3y = 1 \qquad \text{First equation}$$

$$x^2 - 3(x - 1) = 1 \qquad \text{Replace } y \text{ with } x - 1.$$

$$x^2 - 3x + 3 = 1$$

$$x^2 - 3x + 2 = 0$$

$$(x - 2)(x - 1) = 0$$

$$x = 2 \quad \text{or} \quad x = 1$$

Let $x = 2$ and then let $x = 1$ in the equation $y = x - 1$ to find corresponding y-values.

Let $x = 2$.   Let $x = 1$.

$$y = x - 1 \qquad\qquad y = x - 1$$

$$y = 2 - 1 = 1 \qquad\qquad y = 1 - 1 = 0$$

The solutions are $(2, 1)$ and $(1, 0)$. Check both solutions in both equations. Both solutions satisfy both equations, so both are solutions of the system. The graph of each equation in the system is shown next. Intersections of the graphs are at $(2, 1)$ and $(1, 0)$.

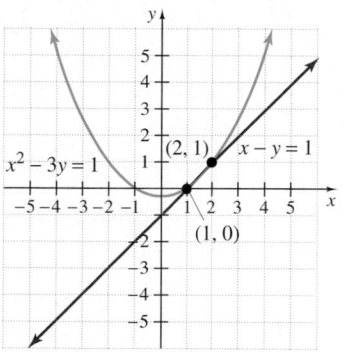

Example 2   Solve the system.

$$\begin{cases} y = \sqrt{x} \\ x^2 + y^2 = 6 \end{cases}$$

Solution   This system is ideal for substitution since y is expressed in terms of x in the first equation. Notice that if $y = \sqrt{x}$, then both x and y must be nonnegative if they are real numbers. Substitute $\sqrt{x}$ for y in the second equation, and solve for x.

$$x^2 + y^2 = 6$$

$$x^2 + (\sqrt{x})^2 = 6 \qquad \text{Let } y = \sqrt{x}.$$

$$x^2 + x = 6$$

$$x^2 + x - 6 = 0$$

$$(x + 3)(x - 2) = 0$$

$$x = -3 \quad \text{or} \quad x = 2$$

The solution -3 is discarded because we have noted that x must be nonnegative. To see this, let $x = -3$ in the first equation. Then let $x = 2$ in the first equation to find a corresponding y-value.

Let $x = -3$.

$$y = \sqrt{x}$$

$$y = \sqrt{-3} \qquad \text{not a real number}$$

Let $x = 2$.

$$y = \sqrt{x}$$

$$y = \sqrt{2}$$

Since we are interested only in real number solutions, the only solution is $(2, \sqrt{2})$. Check to see that this solution satisfies both equations. The graph of each equation in the system is shown next.

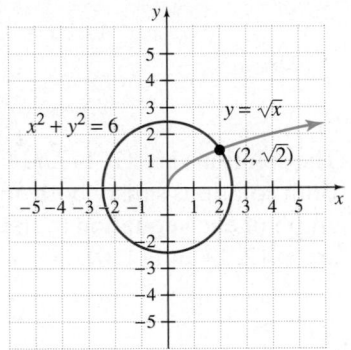

Example 3   Solve the system.

$$\begin{cases} x^2 + y^2 = 4 \\ x + y = 3 \end{cases}$$

Solution   We use the substitution method and solve the second equation for x.

$$x + y = 3 \qquad \textit{Second equation}$$

$$x = 3 - y$$

Now we let $x = 3 - y$ in the first equation.

$$x^2 + y^2 = 4 \qquad \textit{First equation}$$

$$(3 - y)^2 + y^2 = 4 \qquad \textit{Let } x = 3 - y.$$

$$9 - 6y + y^2 + y^2 = 4$$

$$2y^2 - 6y + 5 = 0$$

By the quadratic formula, where $a = 2$, $b = -6$, and $c = 5$, we have

$$y = \frac{6 \pm \sqrt{(-6)^2 - 4 \cdot 2 \cdot 5}}{2 \cdot 2} = \frac{6 \pm \sqrt{-4}}{4}$$

Since $\sqrt{-4}$ is not a real number, there is no real solution. Graphically, the circle and the line do not intersect, as shown.

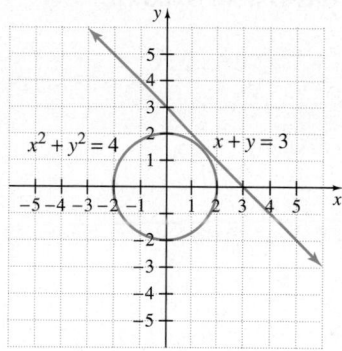

2   Some nonlinear systems may be solved by the elimination method.

Example 4   Solve the system.

$$\begin{cases} x^2 + 2y^2 = 10 \\ x^2 - y^2 = 1 \end{cases}$$

Solution   We will use the elimination, or addition, method to solve this system. To eliminate x^2 when we add the two equations, multiply both sides of the second equation by -1. Then

$$\begin{cases} x^2 + 2y^2 = 10 \\ (-1)(x^2 - y^2) = -1 \cdot 1 \end{cases} \quad \text{is equivalent to} \quad \begin{cases} x^2 + 2y^2 = 10 \\ -x^2 + y^2 = -1 \end{cases}$$

$$3y^2 = 9 \qquad \text{Add.}$$
$$y^2 = 3 \qquad \text{Divide both}$$
$$y = \pm\sqrt{3} \qquad \text{sides by 3.}$$

To find the corresponding x-values, we let $y = \sqrt{3}$ and $y = -\sqrt{3}$ in either original equation. We choose the second equation.

Let $y = \sqrt{3}$.	Let $y = -\sqrt{3}$.
$x^2 - y^2 = 1$	$x^2 - y^2 = 1$
$x^2 - (\sqrt{3})^2 = 1$	$x^2 - (-\sqrt{3})^2 = 1$
$x^2 - 3 = 1$	$x^2 - 3 = 1$
$x^2 = 4$	$x^2 = 4$
$x = \pm\sqrt{4} = \pm2$	$x = \pm\sqrt{4} = \pm2$

The solutions are $(2, \sqrt{3})$, $(-2, \sqrt{3})$, $(2, -\sqrt{3})$, and $(-2, -\sqrt{3})$. Check all four ordered pairs in both equations of the system. The graph of each equation in this system is shown.

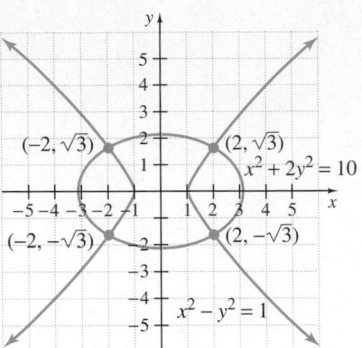

Exercise Set 13.3

Solve each nonlinear system of equations for real solutions. See Examples 1 through 4.

1. $\begin{cases} x^2 + y^2 = 25 \\ 4x + 3y = 0 \end{cases}$

2. $\begin{cases} x^2 + y^2 = 25 \\ 3x + 4y = 0 \end{cases}$

3. $\begin{cases} x^2 + 4y^2 = 10 \\ y = x \end{cases}$

4. $\begin{cases} 4x^2 + y^2 = 10 \\ y = x \end{cases}$

5. $\begin{cases} y^2 = 4 - x \\ x - 2y = 4 \end{cases}$

6. $\begin{cases} x^2 + y^2 = 4 \\ x + y = -2 \end{cases}$

7. $\begin{cases} x^2 + y^2 = 9 \\ 16x^2 - 4y^2 = 64 \end{cases}$

8. $\begin{cases} 4x^2 + 3y^2 = 35 \\ 5x^2 + 2y^2 = 42 \end{cases}$

9. $\begin{cases} x^2 + 2y^2 = 2 \\ x - y = 2 \end{cases}$

10. $\begin{cases} x^2 + 2y^2 = 2 \\ x^2 - 2y^2 = 6 \end{cases}$

11. $\begin{cases} y = x^2 - 3 \\ 4x - y = 6 \end{cases}$

12. $\begin{cases} y = x + 1 \\ x^2 - y^2 = 1 \end{cases}$

13. $\begin{cases} y = x^2 \\ 3x + y = 10 \end{cases}$

14. $\begin{cases} 6x - y = 5 \\ xy = 1 \end{cases}$

15. $\begin{cases} y = 2x^2 + 1 \\ x + y = -1 \end{cases}$

16. $\begin{cases} x^2 + y^2 = 9 \\ x + y = 5 \end{cases}$

17. $\begin{cases} y = x^2 - 4 \\ y = x^2 - 4x \end{cases}$

18. $\begin{cases} x = y^2 - 3 \\ x = y^2 - 3y \end{cases}$

19. $\begin{cases} 2x^2 + 3y^2 = 14 \\ -x^2 + y^2 = 3 \end{cases}$

20. $\begin{cases} 4x^2 - 2y^2 = 2 \\ -x^2 + y^2 = 2 \end{cases}$

21. $\begin{cases} x^2 + y^2 = 1 \\ x^2 + (y + 3)^2 = 4 \end{cases}$

22. $\begin{cases} x^2 + 2y^2 = 4 \\ x^2 - y^2 = 4 \end{cases}$

23. $\begin{cases} y = x^2 + 2 \\ y = -x^2 + 4 \end{cases}$

24. $\begin{cases} x = -y^2 - 3 \\ x = y^2 - 5 \end{cases}$

25. $\begin{cases} 3x^2 + y^2 = 9 \\ 3x^2 - y^2 = 9 \end{cases}$

26. $\begin{cases} x^2 + y^2 = 25 \\ x = y^2 - 5 \end{cases}$

27. $\begin{cases} x^2 + 3y^2 = 6 \\ x^2 - 3y^2 = 10 \end{cases}$

28. $\begin{cases} x^2 + y^2 = 1 \\ y = x^2 - 9 \end{cases}$

29. $\begin{cases} x^2 + y^2 = 36 \\ y = \dfrac{1}{6}x^2 - 6 \end{cases}$

30. $\begin{cases} x^2 + y^2 = 16 \\ y = -\dfrac{1}{4}x^2 + 4 \end{cases}$

31. How many real solutions are possible for a system of equations whose graphs are a circle and a parabola? Draw diagrams to illustrate each possibility.

32. How many real solutions are possible for a system of equations whose graphs are an ellipse and a line? Draw diagrams to illustrate each possibility.

33. The sum of the squares of two numbers is 130. The difference of the squares of the two numbers is 32. Find the two numbers.

34. The sum of the squares of two numbers is 20. Their product is 8. Find the two numbers.

△ **35.** During the development stage of a new rectangular keypad for a security system, it was decided that the area of the rectangle should be 285 square centimeters and the perimeter should be 68 centimeters. Find the dimensions of the keypad.

△ **36.** A rectangular holding pen for cattle is to be designed so that its perimeter is 92 feet and its area is 525 feet. Find the dimensions of the holding pen.

*Recall that in business, a demand function expresses the quantity of a commodity demanded as a function of the commodity's unit price. A supply function expresses the quantity of a commodity supplied as a function of the commodity's unit price. When the quantity produced and supplied is equal to the quantity demanded, then we have what is called **market equilibrium**.*

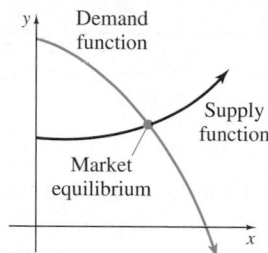

37. The demand function for a certain compact disc is given by the function

$$p = -0.01x^2 - 0.2x + 9$$

and the corresponding supply function is given by

$$p = 0.01x^2 - 0.1x + 3$$

where p is in dollars and x is in thousands of units. Find the equilibrium quantity and the corresponding price by solving the system consisting of the two given equations.

38. The demand function for a certain style of picture frame is given by the function

$$p = -2x^2 + 90$$

and the corresponding supply function is given by

$$p = 9x + 34$$

where p is in dollars and x is in thousands of units. Find the equilibrium quantity and the corresponding price by solving the system consisting of the two given equations.

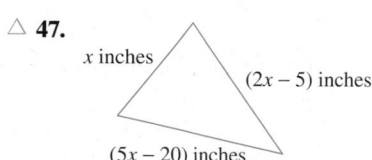

Use a graphing calculator to verify the results of each exercise.

39. Exercise 3.

40. Exercise 4.

41. Exercise 23.

42. Exercise 24.

REVIEW EXERCISES

Graph each inequality in two variables. See Section 3.7.

43. $x > -3$

44. $y \leq 1$

45. $y < 2x - 1$

46. $3x - y \leq 4$

Find the perimeter of each geometric figure. See Section 5.3.

△ **47.**

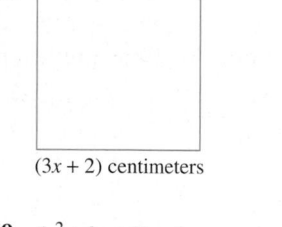

x inches

$(2x - 5)$ inches

$(5x - 20)$ inches

△ **48.**

$(3x + 2)$ centimeters

△ **49.** $(x^2 + 3x + 1)$ meters

x^2 meters

△ **50.**

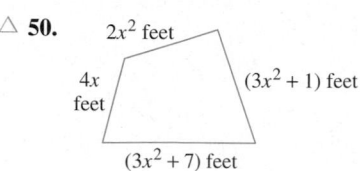

$2x^2$ feet

$4x$ feet

$(3x^2 + 1)$ feet

$(3x^2 + 7)$ feet

13.4 NONLINEAR INEQUALITIES AND SYSTEMS OF INEQUALITIES

▶ **OBJECTIVES**

CD-ROM   SSM

SSG   Video

1. Graph a nonlinear inequality.
2. Graph a system of nonlinear inequalities.

---

1 We can graph a nonlinear inequality in two variables such as $\frac{x^2}{9} + \frac{y^2}{16} \leq 1$ in a way similar to the way we graphed a linear inequality in two variables in Section 3.7. First, we graph the related equation $\frac{x^2}{9} + \frac{y^2}{16} = 1$. The graph of the equation is our boundary. Then, using test points, we determine and shade the region whose points satisfy the inequality.

Example 1   Graph $\frac{x^2}{9} + \frac{y^2}{16} \leq 1$.

Solution   First, graph the equation $\frac{x^2}{9} + \frac{y^2}{16} = 1$. Sketch a solid curve since the graph of $\frac{x^2}{9} + \frac{y^2}{16} \leq 1$ includes the graph of $\frac{x^2}{9} + \frac{y^2}{16} = 1$. The graph is an ellipse, and it divides the plane into two regions, the "inside" and the "outside" of the ellipse. To determine which region contains the solutions, select a test point in either region and determine whether the coordinates of the point satisfy the inequality. We choose $(0, 0)$ as the test point.

$$\frac{x^2}{9} + \frac{y^2}{16} \leq 1$$

$$\frac{0^2}{9} + \frac{0^2}{16} \leq 1 \qquad \text{Let } x = 0 \text{ and } y = 0.$$

$$0 \leq 1 \qquad \text{True.}$$

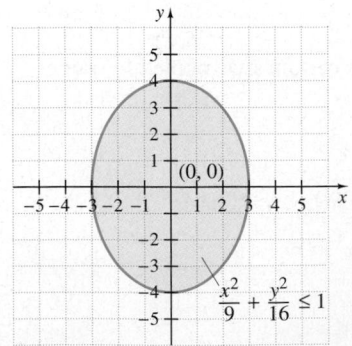

Since this statement is true, the solution set is the region containing $(0, 0)$. The graph of the solution set includes the points on and inside the ellipse, as shaded in the figure.

Example 2   Graph $4y^2 > x^2 + 16$.

Solution   The related equation is $4y^2 = x^2 + 16$. Subtract x^2 from both sides and divide both sides by 16, and we have $\dfrac{y^2}{4} - \dfrac{x^2}{16} = 1$, which is a hyperbola. Graph the hyperbola as a dashed curve since the graph of $4y^2 > x^2 + 16$ does *not* include the graph of $4y^2 = x^2 + 16$. The hyperbola divides the plane into three regions. Select a test point in each region—not on a boundary line—to determine whether that region contains solutions of the inequality.

Test region A with (0, 4)	*Test region B with (0, 0)*	*Test region C with (0, −4)*
$4y^2 > x^2 + 16$	$4y^2 > x^2 + 16$	$4y^2 > x^2 + 16$
$4(4)^2 > 0^2 + 16$	$4(0)^2 > 0^2 + 16$	$4(-4)^2 > 0^2 + 16$
$64 > 16$ True.	$0 > 16$ False.	$64 > 16$ True.

The graph of the solution set includes the shaded regions A and C only, not the boundary.

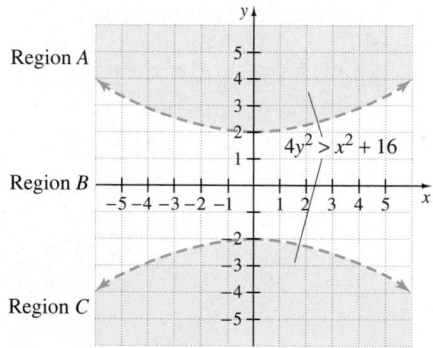

---

2   In Section 3.7 we graphed systems of linear inequalities. Recall that the graph of a system of inequalities is the intersection of the graphs of the inequalities.

Example 3   Graph the system.

$$\begin{cases} x \le 1 - 2y \\ y \le x^2 \end{cases}$$

Solution   We graph each inequality on the same set of axes. The intersection is shown in the third graph. It is the darkest shaded region along with its boundary lines. The coordinates of the points of intersection can be found by solving the related system.

$$\begin{cases} x = 1 - 2y \\ y = x^2 \end{cases}$$

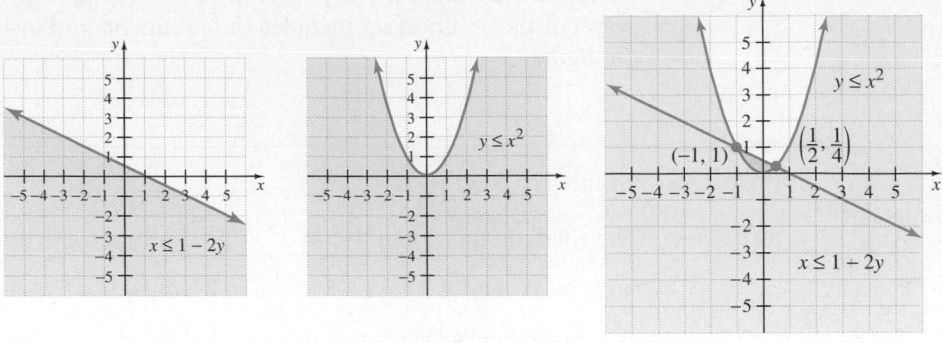

Example 4 Graph the system.

$$\begin{cases} x^2 + y^2 < 25 \\ \dfrac{x^2}{9} - \dfrac{y^2}{25} < 1 \\ y < x + 3 \end{cases}$$

Solution We graph each inequality. The graph of $x^2 + y^2 < 25$ contains points "inside" the circle that has center $(0, 0)$ and radius 5. The graph of $\dfrac{x^2}{9} - \dfrac{y^2}{25} < 1$ is the region between the two branches of the hyperbola with x-intercepts -3 and 3 and center $(0, 0)$. The graph of $y < x + 3$ is the region "below" the line with slope 1 and y-intercept $(0, 3)$. The graph of the solution set of the system is the intersection of all the graphs, the darkest shaded region shown. The boundary of this region is not part of the solution.

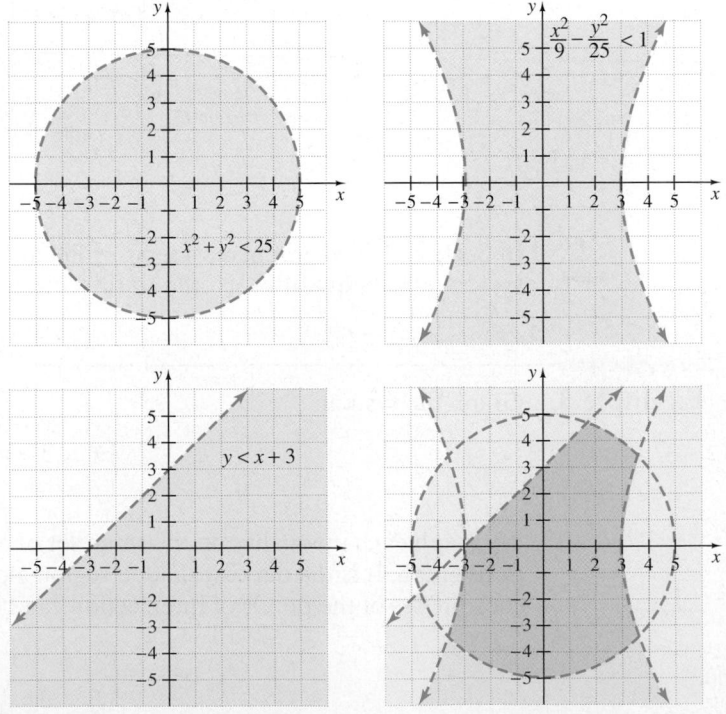

SPOTLIGHT ON DECISION MAKING

Suppose you are an architect. You have just designed a bridge over a two-lane road. Spanning the road is an arch in the shape of a half-ellipse that is 40 feet wide at the base of the arch and is 15 feet tall at the center of the arch. A colleague has just pointed out that the bridge must have a 13-foot clearance for vehicles on the road. The road is 22 feet wide, and its center line falls directly beneath the highest point of the arch. Decide whether your current bridge design will allow 13-foot-tall vehicles to pass on the road beneath it or if your bridge must be redesigned. Explain your reasoning. (*Hint:* Envision a 13-foot-tall semi truck passing under the bridge in the right-hand lane of the road. Try checking whether points along the top of the truck would fall within the ellipse that defines the arch of the bridge. Remember that the truck can drive within any portion of the right-hand lane.)

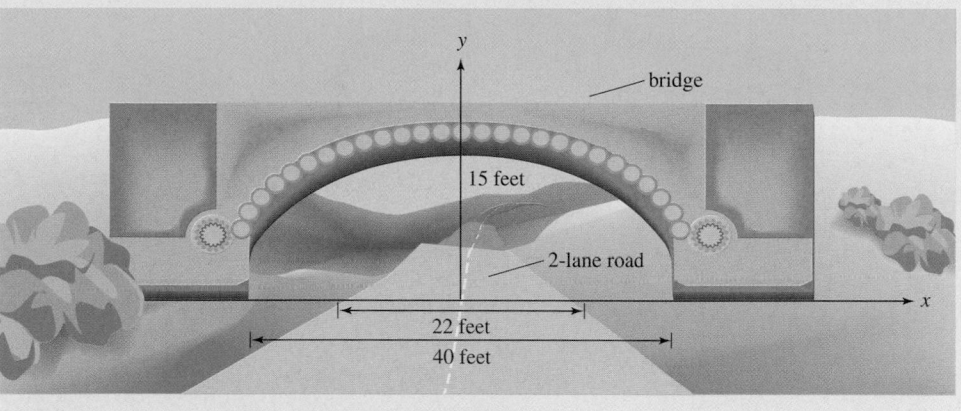

Exercise Set 13.4

Graph each inequality. See Examples 1 and 2.

1. $y < x^2$

2. $y < -x^2$

3. $x^2 + y^2 \geq 16$

4. $x^2 + y^2 < 36$

5. $\dfrac{x^2}{4} - y^2 < 1$

6. $x^2 - \dfrac{y^2}{9} \geq 1$

7. $y > (x - 1)^2 - 3$

8. $y > (x + 3)^2 + 2$

9. $x^2 + y^2 \leq 9$

10. $x^2 + y^2 > 4$

11. $y > -x^2 + 5$

12. $y < -x^2 + 5$

13. $\dfrac{x^2}{4} + \dfrac{y^2}{9} \leq 1$

14. $\dfrac{x^2}{25} + \dfrac{y^2}{4} \geq 1$

15. $\dfrac{y^2}{4} - x^2 \leq 1$

16. $\dfrac{y^2}{16} - \dfrac{x^2}{9} > 1$

17. $y < (x - 2)^2 + 1$

18. $y > (x - 2)^2 + 1$

19. $y \leq x^2 + x - 2$

20. $y > x^2 + x - 2$

21. Discuss how graphing a linear inequality such as $x + y < 9$ is similar to graphing a nonlinear inequality such as $x^2 + y^2 < 9$.

22. Discuss how graphing a linear inequality such as $x + y < 9$ is different from graphing a nonlinear inequality such as $x^2 + y^2 < 9$.

Graph each system. See Examples 3 and 4.

23. $\begin{cases} 2x - y < 2 \\ y \leq -x^2 \end{cases}$

24. $\begin{cases} x - 2y > 4 \\ y > -x^2 \end{cases}$

25. $\begin{cases} 4x + 3y \geq 12 \\ x^2 + y^2 < 16 \end{cases}$

26. $\begin{cases} 3x - 4y \leq 12 \\ x^2 + y^2 < 16 \end{cases}$

27. $\begin{cases} x^2 + y^2 \leq 9 \\ x^2 + y^2 \geq 1 \end{cases}$

28. $\begin{cases} x^2 + y^2 \geq 9 \\ x^2 + y^2 \geq 16 \end{cases}$

29. $\begin{cases} y > x^2 \\ y \geq 2x + 1 \end{cases}$

30. $\begin{cases} y \leq -x^2 + 3 \\ y \leq 2x - 1 \end{cases}$

31. $\begin{cases} x > y^2 \\ y > 0 \end{cases}$

32. $\begin{cases} x < (y + 1)^2 + 2 \\ x + y \ge 3 \end{cases}$

 **33.** $\begin{cases} x^2 + y^2 > 9 \\ \quad y > x^2 \end{cases}$

34. $\begin{cases} x^2 + y^2 \le 9 \\ \quad y < x^2 \end{cases}$

35. $\begin{cases} \dfrac{x^2}{4} + \dfrac{y^2}{9} \ge 1 \\ x^2 + y^2 \ge 4 \end{cases}$

36. $\begin{cases} x^2 + (y - 2)^2 \ge 9 \\ \dfrac{x^2}{4} + \dfrac{y^2}{25} < 1 \end{cases}$

37. $\begin{cases} x^2 - y^2 \ge 1 \\ \quad y \ge 0 \end{cases}$

38. $\begin{cases} x^2 - y^2 \ge 1 \\ \quad x \ge 0 \end{cases}$

39. $\begin{cases} x + y \ge 1 \\ 2x + 3y < 1 \\ \quad x > -3 \end{cases}$

40. $\begin{cases} x - y < -1 \\ 4x - 3y > 0 \\ \quad y > 0 \end{cases}$

41. $\begin{cases} x^2 - y^2 < 1 \\ \dfrac{x^2}{16} + y^2 \le 1 \\ \quad x \ge -2 \end{cases}$

42. $\begin{cases} x^2 - y^2 \ge 1 \\ \dfrac{x^2}{16} + \dfrac{y^2}{4} \le 1 \\ \quad y \ge 1 \end{cases}$

43. Graph the system.

$$\begin{cases} y \le x^2 \\ y \ge x + 2 \\ x \ge 0 \\ y \ge 0 \end{cases}$$

REVIEW EXERCISES

Determine which graph is the graph of a function. See Section 3.3.

44.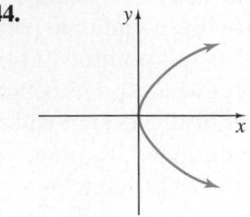

45.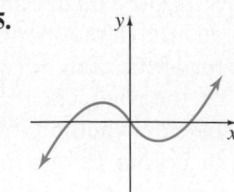

46.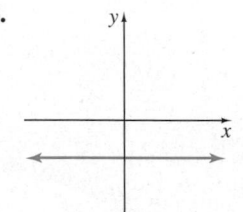

47.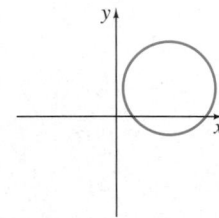

Find each function value if $f(x) = 3x^2 - 2$. *See Section 3.3.*

48. $f(-1)$

49. $f(-3)$

50. $f(a)$

51. $f(b)$

13 CHAPTER PROJECT

For additional Chapter Projects, visit the Real World Activities Website by going to http://www.prenhall.com/martin-gay.

Modeling Conic Sections

In this project, you will have the opportunity to construct and investigate a model of an ellipse. You will need two thumbtacks or nails, graph paper, cardboard, tape, string, a pencil, and a ruler. This project may be completed by working in groups or individually.

Follow these steps, answering any questions as you go.

1. Draw an *x*-axis and a *y*-axis on the graph paper as shown in Figure 1.
2. Place the graph paper on the cardboard and attach it with tape.

3. Locate two points on the *x*-axis, each about $1\frac{1}{2}$ inches from the origin and on opposite sides of the origin (see Figure 1). Insert thumbtacks (or nails) at each of these locations.
4. Fasten a 9-inch piece of string to the thumbtacks as shown in Figure 2. Use your pencil to draw and keep the string taut while you carefully move the pencil in a path all around the thumbtacks.
5. Using the grid of the graph paper as a guide, find an approximate equation of the ellipse you drew.
6. Experiment by moving the tacks closer together or farther apart and drawing new ellipses. What do you observe?

7. Write a paragraph explaining why the figure drawn by the pencil is an ellipse. How might you use the same materials to draw a circle?
8. (Optional) Choose one of the ellipses you drew with the string and pencil. Use a ruler to draw any four tangent lines to the ellipse. (A line is tangent to the ellipse if it intersects, or just touches, the ellipse at only one point. See Figure 3.) Extend the tangent lines to yield four points of intersection among the tangents. Use a straightedge to draw a line connecting each pair of opposite points of intersection. What do you observe? Repeat with a different ellipse. Can you make a conjecture about the relationship among the lines that connect opposite points of intersection?

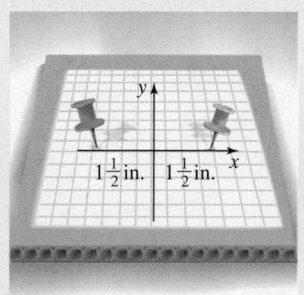

Figure 1

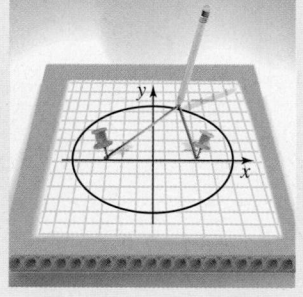

Figure 2

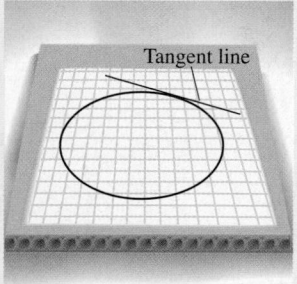

Figure 3

CHAPTER 13 VOCABULARY CHECK

Fill in each blank with one of the words or phrases listed below.

| circle | midpoint | radius | distance |
| center | ellipse | hyperbola | nonlinear system of equations |

1. The _____ formula is $d = \sqrt{(x_2 - x_1)^2 + (y_2 - y_1)^2}$.

2. A(n) _____ is the set of all points in a plane that are the same distance from a fixed point, called the _____.

3. A _____ is a system of equations at least one of which is not linear.

4. A(n) _____ is the set of points on a plane such that the sum of the distances of those points from two fixed points is a constant.

5. In a circle, the distance from the center to a point of the circle is called its _____.

6. A(n) _____ is the set of points in a plane such that the absolute value of the difference of the distance from two fixed points is constant.

7. The _____ formula is $\left(\dfrac{x_1 + x_2}{2}, \dfrac{y_1 + y_2}{2} \right)$.

CHAPTER 13 HIGHLIGHTS

DEFINITIONS AND CONCEPTS	EXAMPLES

Section 13.1 The Parabola and the Circle

Parabolas

$$y = a(x - h)^2 + k$$

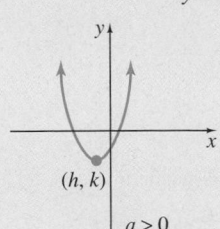

$a > 0$

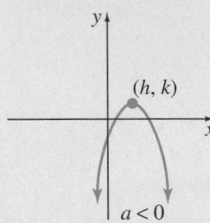

$a < 0$

$$x = a(y - k)^2 + h$$

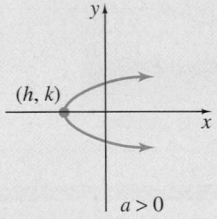

$a > 0$

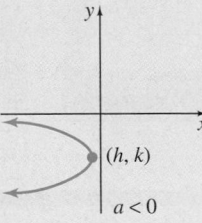

$a < 0$

Graph $\quad x = 3y^2 - 12y + 13$.

$$x - 13 = 3y^2 - 12y$$

$$x - 13 + 3(4) = 3(y^2 - 4y + 4) \quad \text{Add } 3(4)$$

$$x = 3(y - 2)^2 + 1 \quad \text{to both sides.}$$

Since $a = 3$, this parabola opens to the right with vertex $(1, 2)$. Its axis of symmetry is $y = 2$. The x-intercept is $(13, 0)$.

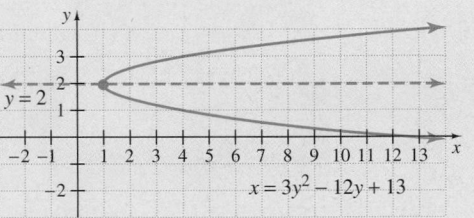

Distance Formula

The distance d between two points (x_1, y_1) and (x_2, y_2) is given by

$$d = \sqrt{(x_2 - x_1)^2 + (y_2 - y_1)^2}$$

Find the distance between points $(-1, 6)$ and $(-2, -4)$. Let $(x_1, y_1) = (-1, 6)$ and $(x_2, y_2) = (-2, -4)$.

$$d = \sqrt{(x_2 - x_1)^2 + (y_2 - y_1)^2}$$
$$= \sqrt{(-2 - (-1))^2 + (-4 - 6)^2}$$
$$= \sqrt{1 + 100} = \sqrt{101}$$

Midpoint Formula

The midpoint of the line segment whose endpoints are (x_1, y_1) and (x_2, y_2) is the point with coordinates

$$\left(\frac{x_1 + x_2}{2}, \frac{y_1 + y_2}{2} \right)$$

Find the midpoint of the line segment whose endpoints are $(-1, 6)$ and $(-2, -4)$.

$$\left(\frac{-1 + (-2)}{2}, \frac{6 + (-4)}{2} \right)$$

The midpoint is

$$\left(-\frac{3}{2}, 1 \right)$$

Circle

The graph of $(x - h)^2 + (y - k)^2 = r^2$ is a circle with center (h, k) and radius r.

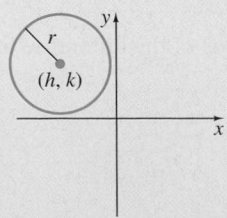

Graph $x^2 + (y + 3)^2 = 5$.
This equation can be written as

$$(x - 0)^2 + (y + 3)^2 = 5 \text{ with } h = 0,$$
$$k = -3, \text{ and } r = \sqrt{5}.$$

The center of this circle is $(0, -3)$, and the radius is $\sqrt{5}$.

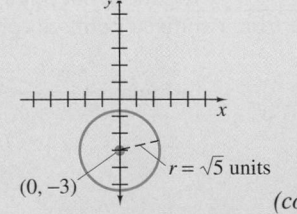

$(0, -3)$ $\quad r = \sqrt{5}$ units

(continued)

DEFINITIONS AND CONCEPTS	EXAMPLES

Section 13.2 The Ellipse and the Hyperbola

Ellipse with Center (0, 0)

The graph of an equation of the form $\dfrac{x^2}{a^2} + \dfrac{y^2}{b^2} = 1$ is an ellipse with center $(0, 0)$. The x-intercepts are $(a, 0)$ and $(-a, 0)$, and the y-intercepts are $(0, b)$ and $(0, -b)$.

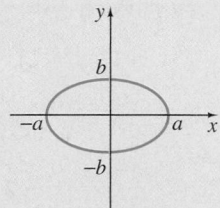

Graph $4x^2 + 9y^2 = 36$.

$$\frac{x^2}{9} + \frac{y^2}{4} = 1 \qquad \text{Divide by 36.}$$

$$\frac{x^2}{3^2} + \frac{y^2}{2^2} = 1$$

The ellipse has center $(0, 0)$, x-intercepts $(3, 0)$ and $(-3, 0)$, and y-intercepts $(0, 2)$ and $(0, -2)$.

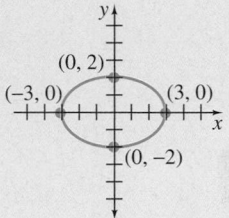

Hyperbola with Center (0, 0)

The graph of an equation of the form $\dfrac{x^2}{a^2} - \dfrac{y^2}{b^2} = 1$ is a hyperbola with center $(0, 0)$ and x-intercepts $(a, 0)$ and $(-a, 0)$.

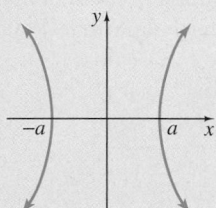

Graph $\dfrac{x^2}{9} - \dfrac{y^2}{4} = 1$. Here $a = 3$ and $b = 2$.

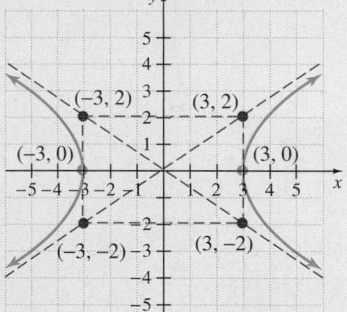

The graph of an equation of the form $\dfrac{y^2}{b^2} - \dfrac{x^2}{a^2} = 1$ is a hyperbola with center $(0, 0)$ and y-intercepts $(0, b)$ and $(0, -b)$.

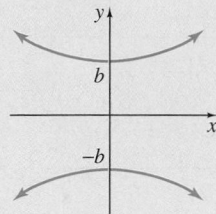

Section 13.3 Solving Nonlinear Systems of Equations

A **nonlinear system of equations** is a system of equations at least one of which is not linear. Both the substitution method and the elimination method may be used to solve a nonlinear system of equations.

Solve the nonlinear system $\begin{cases} y = x + 2 \\ 2x^2 + y^2 = 3 \end{cases}$

(continued)

DEFINITIONS AND CONCEPTS	EXAMPLES

Section 13.3 Solving Nonlinear Systems of Equations

Substitute $x + 2$ for y in the second equation.

$$2x^2 + y^2 = 3$$

$$2x^2 + (x + 2)^2 = 3$$

$$2x^2 + x^2 + 4x + 4 = 3$$

$$3x^2 + 4x + 1 = 0$$

$$(3x + 1)(x + 1) = 0$$

$$x = -\frac{1}{3}, x = -1$$

If $x = -\frac{1}{3}$, $y = x + 2 = -\frac{1}{3} + 2 = \frac{5}{3}$.

If $x = -1$, $y = x + 2 = -1 + 2 = 1$.

The solutions are $\left(-\frac{1}{3}, \frac{5}{3}\right)$ and $(-1, 1)$.

Section 13.4 Nonlinear Inequalities and Systems of Inequalities

The graph of a system of inequalities is the intersection of the graphs of the inequalities.

Graph the system $\begin{cases} x \geq y^2 \\ x + y \leq 4 \end{cases}$

The graph of the system is the darkest shaded region along with its boundary lines.

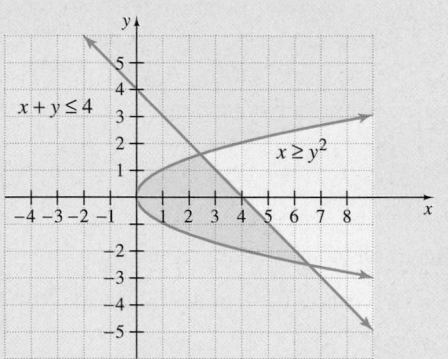

CHAPTER 13 REVIEW

(13.1) *Find the distance between each pair of points. For Exercises 7 and 8, round the distance to two decimal places.*

1. $(-6, 3)$ and $(8, 4)$

2. $(3, 5)$ and $(8, 9)$

3. $(-4, -6)$ and $(-1, 5)$

4. $(-1, 5)$ and $(2, -3)$

5. $(-\sqrt{2}, 0)$ and $(0, -4\sqrt{6})$

6. $(-\sqrt{5}, -\sqrt{11})$ and $(-\sqrt{5}, -3\sqrt{11})$

7. $(7.4, -8.6)$ and $(-1.2, 5.6)$

8. $(2.3, 1.8)$ and $(10.7, -9.2)$

Find the midpoint of the line segment whose endpoints are given.

9. $(2, 6)$ and $(-12, 4)$

10. $(-3, 8)$ and $(11, 24)$

11. $(-6, -5)$ and $(-9, 7)$

12. $(4, -6)$ and $(-15, 2)$

13. $\left(0, -\dfrac{3}{8}\right)$ and $\left(\dfrac{1}{10}, 0\right)$

14. $\left(\dfrac{3}{4}, -\dfrac{1}{7}\right)$ and $\left(-\dfrac{1}{4}, -\dfrac{3}{7}\right)$

15. $(\sqrt{3}, -2\sqrt{6})$ and $(\sqrt{3}, -4\sqrt{6})$

16. $(-5\sqrt{3}, 2\sqrt{7})$ and $(-3\sqrt{3}, 10\sqrt{7})$

Write an equation of the circle with the given center and radius.

17. center $(-4, 4)$, radius 3

18. center $(5, 0)$, radius 5

19. center $(-7, -9)$, radius $\sqrt{11}$

20. center $(0, 0)$, radius $\dfrac{7}{2}$

Sketch the graph of the equation. If the graph is a circle, find its center. If the graph is a parabola, find its vertex.

21. $x^2 + y^2 = 7$

22. $x = 2(y - 5)^2 + 4$

23. $x = -(y + 2)^2 + 3$

24. $(x - 1)^2 + (y - 2)^2 = 4$

25. $y = -x^2 + 4x + 10$

26. $x = -y^2 - 4y + 6$

27. $x = \dfrac{1}{2}y^2 + 2y + 1$

28. $y = -3x^2 + \dfrac{1}{2}x + 4$

29. $x^2 + y^2 + 2x + y = \dfrac{3}{4}$

30. $x^2 + y^2 + 3y = \dfrac{7}{4}$

31. $4x^2 + 4y^2 + 16x + 8y = 1$

32. $3x^2 + 6x + 3y^2 = 9$

33. $y = x^2 + 6x + 9$

34. $x = y^2 + 6y + 9$

35. Write an equation of the circle centered at $(5.6, -2.4)$ with diameter 6.2.

(13.2) *Sketch the graph of each equation.*

36. $x^2 + \dfrac{y^2}{4} = 1$

37. $x^2 - \dfrac{y^2}{4} = 1$

38. $\dfrac{y^2}{4} - \dfrac{x^2}{16} = 1$

39. $\dfrac{y^2}{4} + \dfrac{x^2}{16} = 1$

40. $\dfrac{x^2}{5} + \dfrac{y^2}{5} = 1$

41. $\dfrac{x^2}{5} - \dfrac{y^2}{5} = 1$

42. $-5x^2 + 25y^2 = 125$

43. $4y^2 + 9x^2 = 36$

44. $\dfrac{(x - 2)^2}{4} + (y - 1)^2 = 1$

45. $\dfrac{(x + 3)^2}{9} + \dfrac{(y - 4)^2}{25} = 1$

46. $x^2 - y^2 = 1$

47. $36y^2 - 49x^2 = 1764$

48. $y^2 = x^2 + 9$

49. $x^2 = 4y^2 - 16$

50. $100 - 25x^2 = 4y^2$

Sketch the graph of each equation. Identify whether each equation, when graphed, will be a parabola, circle, ellipse, or hyperbola.

51. $y = x^2 + 4x + 6$

52. $y^2 = x^2 + 6$

53. $y^2 + x^2 = 4x + 6$

54. $y^2 + 2x^2 = 4x + 6$

55. $x^2 + y^2 - 8y = 0$

56. $x - 4y = y^2$

57. $x^2 - 4 = y^2$

58. $x^2 = 4 - y^2$

59. $6(x - 2)^2 + 9(y + 5)^2 = 36$

60. $36y^2 = 576 + 16x^2$

61. $\dfrac{x^2}{16} - \dfrac{y^2}{25} = 1$

62. $3(x - 7)^2 + 3(y + 4)^2 = 1$

Use a graphing calculator to verify the results of each exercise.

63. Exercise 39.

64. Exercise 40.

65. Exercise 51.

66. Exercise 58.

(13.3) Solve each system of equations.

67. $\begin{cases} y = 2x - 4 \\ y^2 = 4x \end{cases}$

68. $\begin{cases} x^2 + y^2 = 4 \\ x - y = 4 \end{cases}$

69. $\begin{cases} y = x + 2 \\ y = x^2 \end{cases}$

70. $\begin{cases} y = x^2 - 5x + 1 \\ y = -x + 6 \end{cases}$

71. $\begin{cases} 4x - y^2 = 0 \\ 2x^2 + y^2 = 16 \end{cases}$

72. $\begin{cases} x^2 + 4y^2 = 16 \\ x^2 + y^2 = 4 \end{cases}$

73. $\begin{cases} x^2 + y^2 = 10 \\ 9x^2 + y^2 = 18 \end{cases}$

74. $\begin{cases} x^2 + 2y = 9 \\ 5x - 2y = 5 \end{cases}$

75. $\begin{cases} y = 3x^2 + 5x - 4 \\ y = 3x^2 - x + 2 \end{cases}$

76. $\begin{cases} x^2 - 3y^2 = 1 \\ 4x^2 + 5y^2 = 21 \end{cases}$

△ **77.** Find the length and the width of a room whose area is 150 square feet and whose perimeter is 50 feet.

78. What is the greatest number of real solutions possible for a system of two equations whose graphs are an ellipse and a hyperbola?

(13.4) Graph the inequality or system of inequalities.

79. $y \leq -x^2 + 3$

80. $x^2 + y^2 < 9$

81. $x^2 - y^2 < 1$

82. $\dfrac{x^2}{4} + \dfrac{y^2}{9} \geq 1$

83. $\begin{cases} 2x \leq 4 \\ x + y \geq 1 \end{cases}$

84. $\begin{cases} 3x + 4y \leq 12 \\ x - 2y > 6 \end{cases}$

85. $\begin{cases} y > x^2 \\ x + y \geq 3 \end{cases}$

86. $\begin{cases} x^2 + y^2 \leq 16 \\ x^2 + y^2 \geq 4 \end{cases}$

87. $\begin{cases} x^2 + y^2 < 4 \\ x^2 - y^2 \leq 1 \end{cases}$

88. $\begin{cases} x^2 + y^2 < 4 \\ y \geq x^2 - 1 \\ x \geq 0 \end{cases}$

CHAPTER 13 TEST

1. Find the distance between the points $(-6, 3)$ and $(-8, -7)$.

2. Find the distance between the points $(-2\sqrt{5}, \sqrt{10})$ and $(-\sqrt{5}, 4\sqrt{10})$.

3. Find the midpoint of the line segment whose endpoints are $(-2, -5)$ and $(-6, 12)$.

4. Find the midpoint of the line segment whose endpoints are $\left(-\dfrac{2}{3}, -\dfrac{1}{5}\right)$ and $\left(-\dfrac{1}{3}, \dfrac{4}{5}\right)$.

Sketch the graph of each equation.

5. $x^2 + y^2 = 36$

6. $x^2 - y^2 = 36$

7. $16x^2 + 9y^2 = 144$

8. $y = x^2 - 8x + 16$

9. $x^2 + y^2 + 6x = 16$

10. $x = y^2 + 8y - 3$

11. $\dfrac{(x - 4)^2}{16} + \dfrac{(y - 3)^2}{9} = 1$

12. $y^2 - x^2 = 1$

Solve each system.

13. $\begin{cases} x^2 + y^2 = 169 \\ 5x + 12y = 0 \end{cases}$

14. $\begin{cases} x^2 + y^2 = 26 \\ x^2 - y^2 = 24 \end{cases}$

15. $\begin{cases} y = x^2 - 5x + 6 \\ y = 2x \end{cases}$

16. $\begin{cases} x^2 + 4y^2 = 5 \\ y = x \end{cases}$

Graph the solution of each system.

17. $\begin{cases} 2x + 5y \geq 10 \\ y \geq x^2 + 1 \end{cases}$

18. $\begin{cases} \dfrac{x^2}{4} + y^2 \leq 1 \\ x + y > 1 \end{cases}$

19. $\begin{cases} x^2 + y^2 > 1 \\ \dfrac{x^2}{4} - y^2 \geq 1 \end{cases}$

20. $\begin{cases} x^2 + y^2 \geq 4 \\ x^2 + y^2 < 16 \\ y \geq 0 \end{cases}$

21. Which graph best resembles the graph of
$x = a(y - k)^2 + h$ if $a > 0, h < 0,$ and $k > 0$?

A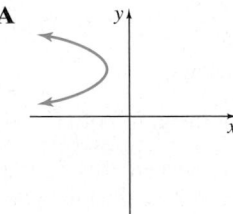

B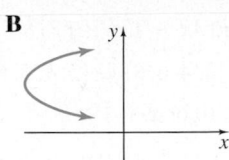

C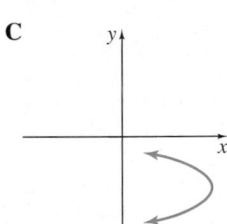

D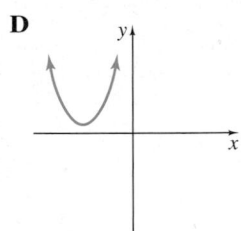

22. A bridge has an arch in the shape of a half-ellipse. If the equation of the ellipse, measured in feet, is $100x^2 + 225y^2 = 22{,}500$, find the height of the arch from the road and the width of the arch.

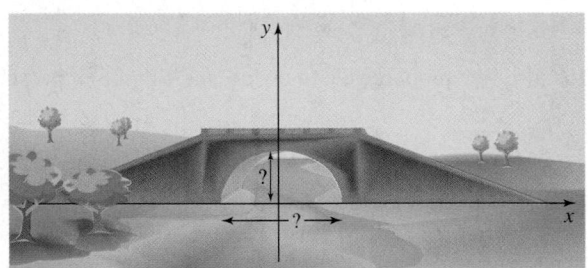

CHAPTER 13 CUMULATIVE REVIEW

1. Graph $x = -2y$ by plotting intercepts.

2. Divide $2x^2 - x - 10$ by $x + 2$.

3. Use the remainder theorem and synthetic division to find $P(4)$ if

$P(x) = 4x^6 - 25x^5 + 35x^4 + 17x^2.$

4. Simplify each complex fraction.

a. $\dfrac{\dfrac{2x}{27y^2}}{\dfrac{6x^2}{9}}$

b. $\dfrac{\dfrac{5x}{x + 2}}{\dfrac{10}{x - 2}}$

c. $\dfrac{\dfrac{x}{y^2} + \dfrac{1}{y}}{\dfrac{y}{x^2} + \dfrac{1}{x}}$

Solve.

5. Find the intersection: $\{2, 4, 6, 8\} \cap \{3, 4, 5, 6\}$

6. Solve $2x \geq 0$ and $4x - 1 \leq -9$

7. Find the union: $\{2, 4, 6, 8\} \cup \{3, 4, 5, 6\}$

8. Solve $5x - 3 \leq 10$ or $x + 1 \geq 5$.

Solve:

9. $|5w + 3| = 7$

10. $\left|\dfrac{x}{2} - 1\right| = 11$

11. $|3x + 2| = |5x - 8|$

12. $|m - 6| < 2$

13. $|2x + 9| + 5 > 3$

14. Find the cube roots.

   a. $\sqrt[3]{1}$

   b. $\sqrt[3]{-64}$

   c. $\sqrt[3]{\dfrac{8}{125}}$

   d. $\sqrt[3]{x^6}$

   e. $\sqrt[3]{-8x^9}$

15. Multiply.

   a. $z^{2/3}(z^{1/3} - z^5)$

   b. $(x^{1/3} - 5)(x^{1/3} + 2)$

16. Use the quotient rule to divide, and simplify if possible.

   a. $\dfrac{\sqrt{20}}{\sqrt{5}}$

   b. $\dfrac{\sqrt{50x}}{2\sqrt{2}}$

   c. $\dfrac{7\sqrt[3]{48x^4y^8}}{\sqrt[3]{6y^2}}$

17. Add or subtract as indicated.

   a. $\dfrac{\sqrt{45}}{4} - \dfrac{\sqrt{5}}{3}$

   b. $\sqrt[3]{\dfrac{7x}{8}} + 2\sqrt[3]{7x}$

18. Rationalize the denominator of $\sqrt{\dfrac{7x}{3y}}$.

19. Solve $\sqrt{2x - 3} = 9$.

20. Find the following powers of i.

   a. i^7

   b. i^{20}

   c. i^{46}

   d. i^{-12}

21. Solve $p^2 + 2p = 4$ by completing the square.

22. Solve $\dfrac{1}{4}m^2 - m + \dfrac{1}{2} = 0$.

23. Solve $p^4 - 3p^2 - 4 = 0$.

24. Solve $\dfrac{x + 2}{x - 3} \leq 0$.

25. Graph $g(x) = \dfrac{1}{2}(x + 2)^2 + 5$. Find the vertex and the axis of symmetry.

26. Find the vertex of the graph of $f(x) = x^2 - 4x - 12$.

27. Find the distance between $(2, -5)$ and $(1, -4)$. Give an exact distance and a three-decimal-place approximation.

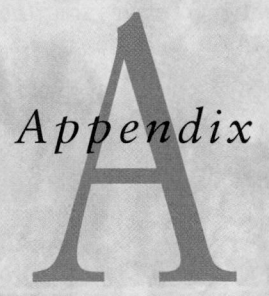
OPERATIONS ON DECIMALS

To **add** or **subtract** decimals, write the numbers vertically with decimal points lined up. Add or subtract as with whole numbers and place the decimal point in the answer directly below the decimal points in the problem.

Example 1   Add: $5.87 + 23.279 + 0.003$

Solution
$$
\begin{array}{r}
5.87 \\
23.279 \\
+\,0.003 \\
\hline
29.152
\end{array}
$$

Example 2   Subtract: $32.15 - 11.237$

Solution
$$
\begin{array}{r}
{}^{1}\ {}^{11}\ {}^{4}\ {}^{10} \\
3\ \overset{1}{\cancel{2}} . \overset{11}{\cancel{1}}\ \overset{4}{\cancel{5}}\ \overset{10}{\cancel{0}} \\
-\ 1\ 1\ .\ 2\ \ 3\ \ 7 \\
\hline
2\ 0\ .\ 9\ \ 1\ \ 3
\end{array}
$$

To **multiply** decimals, multiply the numbers as if they were whole numbers. The decimal point in the product is placed so that the number of decimal places in the product is the same as the sum of the number of decimal places in the factors.

Example 3   Multiply: 0.072×3.5

Solution
$$
\begin{array}{r}
0.072 \\
\times\ \ \ 3.5 \\
\hline
360 \\
216\ \ \\
\hline
0.2520
\end{array}
$$

0.072   *3 decimal places*
× 3.5   *1 decimal place*
0.2520   *4 decimal places*

To **divide** decimals, move the decimal point in the divisor to the right of the last digit. Move the decimal point in the dividend the same number of places that the decimal point in the divisor was moved. The decimal point in the quotient lies directly above the moved decimal point in the dividend.

Example 4 Divide: $9.46 \div 0.04$

Solution

$$
\begin{array}{r}
236.5 \\
04.\overline{)946.0} \\
-8 \\
\hline
14 \\
-12 \\
\hline
26 \\
-24 \\
\hline
20 \\
-20 \\
\end{array}
$$

Appendix A Exercise Set

Perform the indicated operations.

1. $9.076 + 8.004$

2.
$$
\begin{array}{r}
6.3 \\
\times\ 0.05 \\
\end{array}
$$

3.
$$
\begin{array}{r}
27.004 \\
-14.2 \\
\end{array}
$$

4.
$$
\begin{array}{r}
0.0036 \\
7.12 \\
32.502 \\
+0.05 \\
\end{array}
$$

5.
$$
\begin{array}{r}
107.92 \\
+3.04 \\
\end{array}
$$

6. $7.2 \div 4$

7. $10 - 7.6$

8. $40 \div 0.25$

9. $126.32 - 97.89$

10.
$$
\begin{array}{r}
3.62 \\
7.11 \\
12.36 \\
4.15 \\
+2.29 \\
\end{array}
$$

11.
$$
\begin{array}{r}
3.25 \\
\times\ 70 \\
\end{array}
$$

12.
$$
\begin{array}{r}
26.014 \\
-\ 7.8 \\
\end{array}
$$

13. $8.1 \div 3$

14.
$$
\begin{array}{r}
1.2366 \\
0.005 \\
15.17 \\
+\ 0.97 \\
\end{array}
$$

15. $55.405 - 6.1711$

16. $8.09 + 0.22$

17. $60 \div 0.75$

18. $20 - 12.29$

19. $7.612 \div 100$

20.
$$
\begin{array}{r}
8.72 \\
1.12 \\
14.86 \\
3.98 \\
+\ 1.99 \\
\end{array}
$$

21. $12.312 \div 2.7$

22. $0.443 \div 100$

23.
$$
\begin{array}{r}
569.2 \\
71.25 \\
+\ \ 8.01 \\
\end{array}
$$

24. $3.706 - 2.91$

25. $768 - 0.17$

26. $63 \div 0.28$

27. $12 + 0.062$

28. $0.42 + 18$

29. $76 - 14.52$

30. $1.1092 \div 0.47$

31. $3.311 \div 0.43$

32. $7.61 + 0.0004$

33.
$$
\begin{array}{r}
762.12 \\
89.7 \\
+\ 11.55 \\
\end{array}
$$

34. $444 \div 0.6$

35. $23.4 - 0.821$

36. $3.7 + 5.6$

37. $476.12 - 112.97$

38. $19.872 \div 0.54$

39. $0.007 + 7$

40.
$$
\begin{array}{r}
51.77 \\
+\ 3.6 \\
\end{array}
$$

REVIEW OF ANGLES, LINES, AND SPECIAL TRIANGLES

The word **geometry** is formed from the Greek words, **geo**, meaning earth, and **metron**, meaning measure. Geometry literally means to measure the earth.

This section contains a review of some basic geometric ideas. It will be assumed that fundamental ideas of geometry such as point, line, ray, and angle are known. In this appendix, the notation $\angle 1$ is read "angle 1" and the notation $m\angle 1$ is read "the measure of angle 1."

We first review types of angles.

ANGLES

A **right angle** is an angle whose measure is 90°. A right angle can be indicated by a square drawn at the vertex of the angle, as shown below.

An angle whose measure is more than 0° but less than 90° is called an **acute angle**.

An angle whose measure is greater than 90° but less than 180° is called an **obtuse angle**.

An angle whose measure is 180° is called a **straight angle**.

Two angles are said to be **complementary** if the sum of their measures is 90°. Each angle is called the **complement** of the other.

Two angles are said to be **supplementary** if the sum of their measures is 180°. Each angle is called the **supplement** of the other.

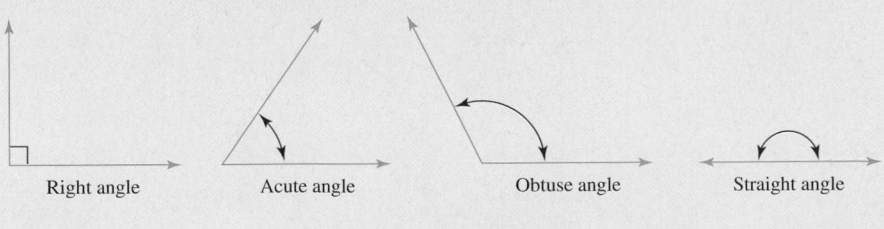

Right angle   Acute angle   Obtuse angle   Straight angle

(continued)

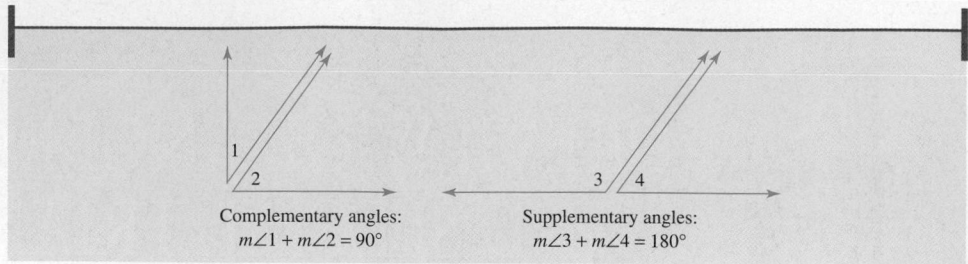

Complementary angles:
$m\angle 1 + m\angle 2 = 90°$

Supplementary angles:
$m\angle 3 + m\angle 4 = 180°$

Example 1 If an angle measures 28°, find its complement.

Solution Two angles are complementary if the sum of their measures is 90°. The complement of a 28° angle is an angle whose measure is $90° - 28° = 62°$. To check, notice that $28° + 62° = 90°$.

Plane is an undefined term that we will describe. A plane can be thought of as a flat surface with infinite length and width, but no thickness. A plane is two dimensional. The arrows in the following diagram indicate that a plane extends indefinitely and has no boundaries.

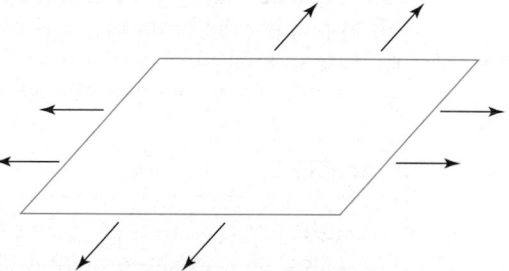

Figures that lie on a plane are called **plane figures**. (See the description of common plane figures in Appendix C.) Lines that lie in the same plane are called **coplanar**.

LINES

Two lines are **parallel** if they lie in the same plane but never meet.
Intersecting lines meet or cross in one point.
Two lines that form right angles when they intersect are said to be
  perpendicular.

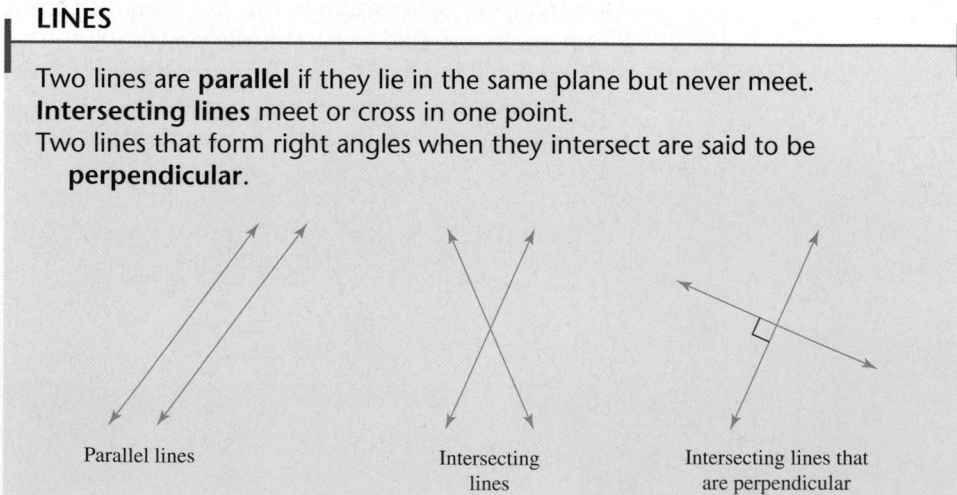

Parallel lines

Intersecting
lines

Intersecting lines that
are perpendicular

Two intersecting lines form **vertical angles**. Angles 1 and 3 are vertical angles. Also angles 2 and 4 are vertical angles. It can be shown that **vertical angles have equal measures**.

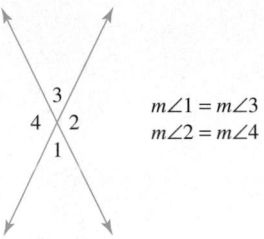

$$m\angle 1 = m\angle 3$$
$$m\angle 2 = m\angle 4$$

Adjacent angles have the same vertex and share a side. Angles 1 and 2 are adjacent angles. Other pairs of adjacent angles are angles 2 and 3, angles 3 and 4, and angles 4 and 1.

A **transversal** is a line that intersects two or more lines in the same plane. Line l is a transversal that intersects lines m and n. The eight angles formed are numbered and certain pairs of these angles are given special names.

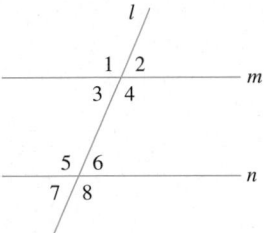

Corresponding angles: $\angle 1$ and $\angle 5$, $\angle 3$ and $\angle 7$, $\angle 2$ and $\angle 6$, and $\angle 4$ and $\angle 8$.

Exterior angles: $\angle 1, \angle 2, \angle 7$, and $\angle 8$.

Interior angles: $\angle 3, \angle 4, \angle 5$, and $\angle 6$.

Alternate interior angles: $\angle 3$ and $\angle 6$, $\angle 4$ and $\angle 5$.

These angles and parallel lines are related in the following manner.

PARALLEL LINES CUT BY A TRANSVERSAL

1. If two parallel lines are cut by a transversal, then
   a. **corresponding angles are equal** and
   b. **alternate interior angles are equal.**

2. If corresponding angles formed by two lines and a transversal are equal, then the lines are parallel.

3. If alternate interior angles formed by two lines and a transversal are equal, then the lines are parallel.

Example 2   Given that lines *m* and *n* are parallel and that the measure of angle 1 is 100°, find the measures of angles 2, 3, and 4.

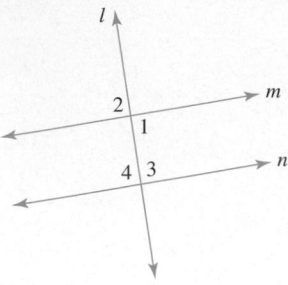

Solution $m\angle 2 = 100°$, since angles 1 and 2 are vertical angles.
$m\angle 4 = 100°$, since angles 1 and 4 are alternate interior angles.
$m\angle 3 = 180° - 100° = 80°$, since angles 4 and 3 are supplementary angles.

A **polygon** is the union of three or more coplanar line segments that intersect each other only at each end point, with each end point shared by exactly two segments.

A **triangle** is a polygon with three sides. The sum of the measures of the three angles of a triangle is 180°. In the following figure, $m\angle 1 + m\angle 2 + m\angle 3 = 180°$.

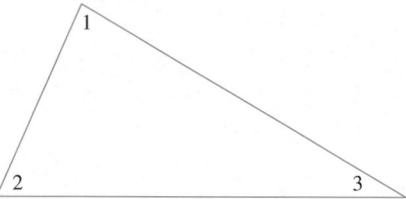

Example 3   Find the measure of the third angle of the triangle shown.

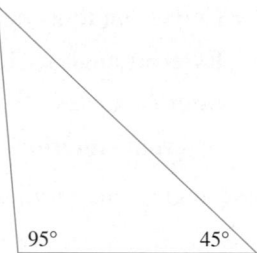

Solution   The sum of the measures of the angles of a triangle is 180°. Since one angle measures 45° and the other angle measures 95°, the third angle measures $180° - 45° - 95° = 40°$.

Two triangles are **congruent** if they have the same size and the same shape. In congruent triangles, the measures of corresponding angles are equal and the lengths of corresponding sides are equal. The following triangles are congruent.

Corresponding angles are equal: $m\angle 1 = m\angle 4$, $m\angle 2 = m\angle 5$, and $m\angle 3 = m\angle 6$. Also, lengths of corresponding sides are equal: $a = x$, $b = y$, and $c = z$.

Any one of the following may be used to determine whether two triangles are congruent.

CONGRUENT TRIANGLES

1. If the measures of two angles of a triangle equal the measures of two angles of another triangle and the lengths of the sides between each pair of angles are equal, the triangles are congruent.

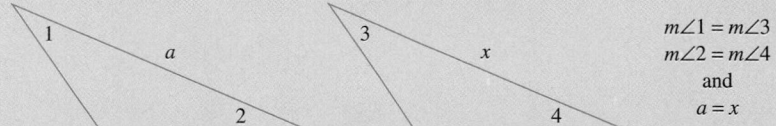

$$m\angle 1 = m\angle 3$$
$$m\angle 2 = m\angle 4$$
and
$$a = x$$

2. If the lengths of the three sides of a triangle equal the lengths of corresponding sides of another triangle, the triangles are congruent.

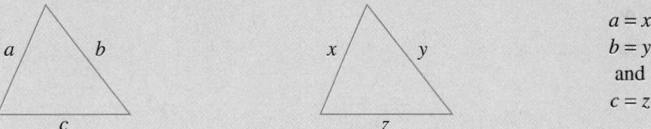

$$a = x$$
$$b = y$$
and
$$c = z$$

3. If the lengths of two sides of a triangle equal the lengths of corresponding sides of another triangle, and the measures of the angles between each pair of sides are equal, the triangles are congruent.

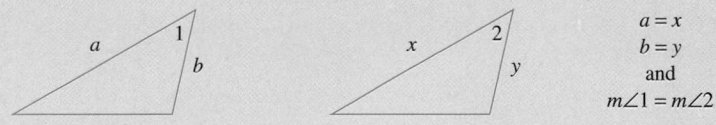

$$a = x$$
$$b = y$$
and
$$m\angle 1 = m\angle 2$$

Two triangles are **similar** if they have the same shape. In similar triangles, the measures of corresponding angles are equal and corresponding sides are in proportion. The following triangles are similar. (All similar triangles drawn in this appendix will be oriented the same.)

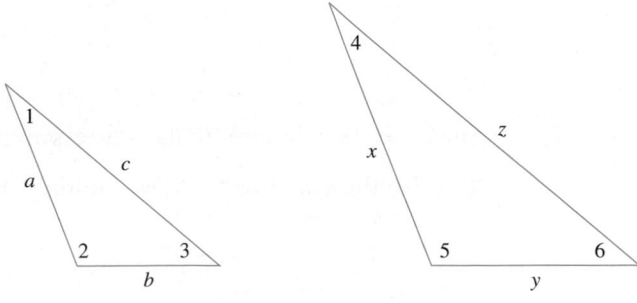

Corresponding angles are equal: $m\angle 1 = m\angle 4$, $m\angle 2 = m\angle 5$, and $m\angle 3 = m\angle 6$. Also, corresponding sides are proportional: $\dfrac{a}{x} = \dfrac{b}{y} = \dfrac{c}{z}$.

Any one of the following may be used to determine whether two triangles are similar.

SIMILAR TRIANGLES

1. If the measures of two angles of a triangle equal the measures of two angles of another triangle, the triangles are similar.

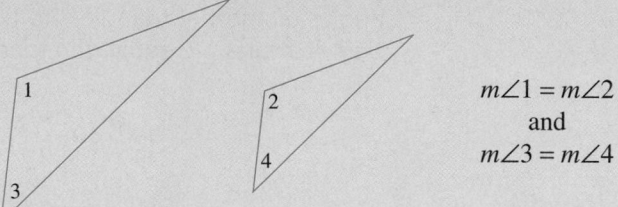

$m\angle 1 = m\angle 2$
and
$m\angle 3 = m\angle 4$

2. If three sides of one triangle are proportional to three sides of another triangle, the triangles are similar.

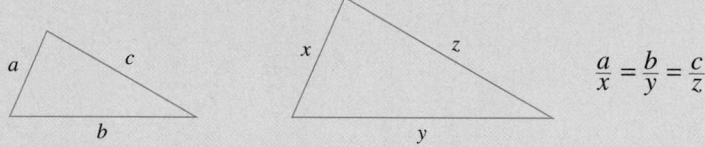

$\dfrac{a}{x} = \dfrac{b}{y} = \dfrac{c}{z}$

3. If two sides of a triangle are proportional to two sides of another triangle and the measures of the included angles are equal, the triangles are similar.

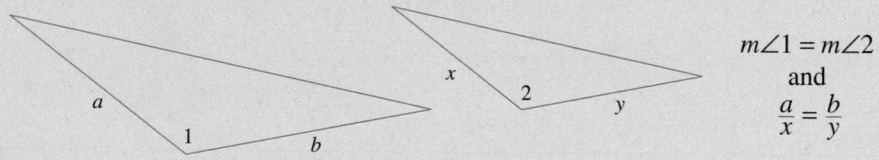

$m\angle 1 = m\angle 2$
and
$\dfrac{a}{x} = \dfrac{b}{y}$

Example 4 Given that the following triangles are similar, find the missing length x.

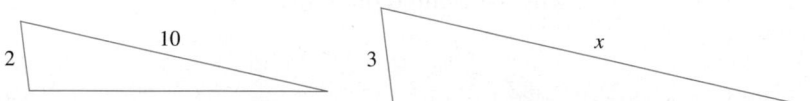

Solution Since the triangles are similar, corresponding sides are in proportion. Thus, $\dfrac{2}{3} = \dfrac{10}{x}$.

To solve this equation for x, we multiply both sides by the LCD, $3x$.

$$3x\left(\frac{2}{3}\right) = 3x\left(\frac{10}{x}\right)$$
$$2x = 30$$
$$x = 15$$

The missing length is 15 units.

A **right triangle** contains a right angle. The side opposite the right angle is called the **hypotenuse**, and the other two sides are called the **legs**. The **Pythagorean theorem** gives a formula that relates the lengths of the three sides of a right triangle.

THE PYTHAGOREAN THEOREM

If a and b are the lengths of the legs of a right triangle, and c is the length of the hypotenuse, then $a^2 + b^2 = c^2$.

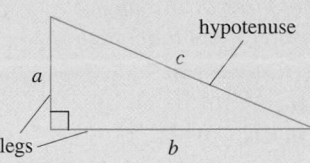

Example 5   Find the length of the hypotenuse of a right triangle whose legs have lengths of 3 centimeters and 4 centimeters.

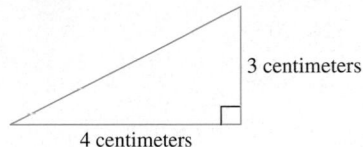

Solution   Because we have a right triangle, we use the Pythagorean theorem. The legs have lengths of 3 centimeters and 4 centimeters, so let $a = 3$ and $b = 4$ in the formula.

$$a^2 + b^2 = c^2$$
$$3^2 + 4^2 = c^2$$
$$9 + 16 = c^2$$
$$25 = c^2$$

Since c represents a length, we assume that c is positive. Thus, if c^2 is 25, c must be 5. The hypotenuse has a length of 5 centimeters.

Appendix B Exercise Set

Find the complement of each angle. See Example 1.

1. $19°$

2. $65°$

3. $70.8°$

4. $45\frac{2}{3}°$

5. $11\frac{1}{4}°$

6. $19.6°$

Find the supplement of each angle.

7. $150°$

8. $90°$

9. $30.2°$

10. $81.9°$

11. $79\frac{1}{2}°$

12. $165\frac{8}{9}°$

13. If lines m and n are parallel, find the measures of angles 1 through 7. See Example 2.

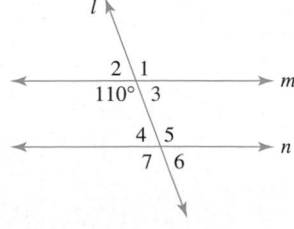

14. If lines *m* and *n* are parallel, find the measures of angles 1 through 5. See Example 2.

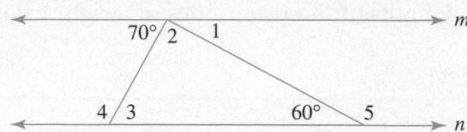

In each of the following, the measures of two angles of a triangle are given. Find the measure of the third angle. See Example 3.

15. 11°, 79°   **16.** 8°, 102°
17. 25°, 65°   **18.** 44°, 19°
19. 30°, 60°   **20.** 67°, 23°

In each of the following, the measure of one angle of a right triangle is given. Find the measures of the other two angles.

21. 45°   **22.** 60°
23. 17°   **24.** 30°
25. $39\frac{3}{4}°$ **26.** 72.6°

Given that each of the following pairs of triangles is similar, find the missing lengths. See Example 4.

27.

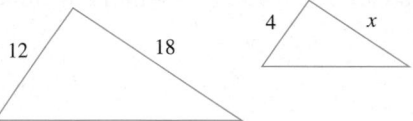

28.

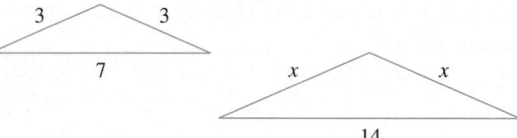

29.

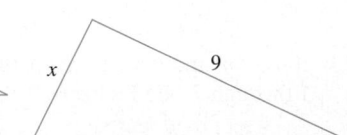

30.

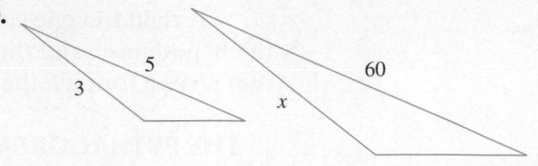

Use the Pythagorean theorem to find the missing lengths in the right triangles. See Example 5.

31.

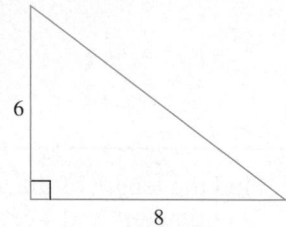

32.

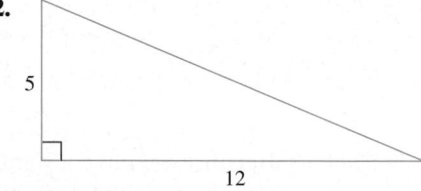

33.

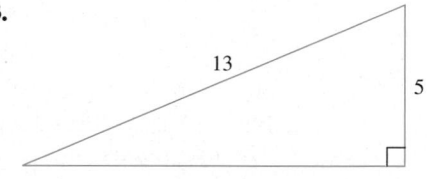

34.

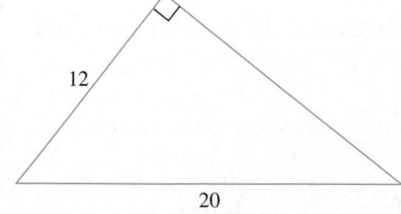

REVIEW OF GEOMETRIC FIGURES

Plane figures have length and width but no thickness or depth.		
Name	*Description*	*Figure*
POLYGON	Union of three or more coplanar line segments that intersect with each other only at each end point, with each end point shared by two segments.	
TRIANGLE	Polygon with three sides (sum of measures of the three angles is 180°).	
SCALENE TRIANGLE	Triangle with no sides of equal length.	
ISOSCELES TRIANGLE	Triangle with two sides of equal length.	
EQUILATERAL TRIANGLE	Triangle with all sides of equal length.	
RIGHT TRIANGLE	Triangle that contains a right angle.	leg, hypotenuse, leg

Plane figures have length and width but no thickness or depth.

Name	Description	Figure
QUADRILATERAL	Polygon with four sides (sum of measures of the four angles is 360°).	
TRAPEZOID	Quadrilateral with exactly one pair of opposite sides parallel.	
ISOSCELES TRAPEZOID	Trapezoid with legs of equal length.	
PARALLELOGRAM	Quadrilateral with both pairs of opposite sides parallel and equal in length.	
RHOMBUS	Parallelogram with all sides of equal length.	
RECTANGLE	Parallelogram with four right angles.	
SQUARE	Rectangle with all sides of equal length.	
CIRCLE	All points in a plane the same distance from a fixed point called the **center**.	

Solids have length, width, and depth.

Name	Description	Figure
RECTANGULAR SOLID	A solid with six sides, all of which are rectangles.	
CUBE	A rectangular solid whose six sides are squares.	
SPHERE	All points the same distance from a fixed point, called the center.	
RIGHT CIRCULAR CYLINDER	A cylinder consisting of two circular bases that are perpendicular to its altitude.	
RIGHT CIRCULAR CONE	A cone with a circular base that is perpendicular to its altitude.	

REVIEW OF VOLUME AND SURFACE AREA

A **convex solid** is a set of points, *S*, not all in one plane, such that for any two points *A* and *B* in *S*, all points between *A* and *B* are also in *S*. In this appendix, we will find the volume and surface area of special types of solids called polyhedrons. A solid formed by the intersection of a finite number of planes is called a **polyhedron**. The box below is an example of a polyhedron.

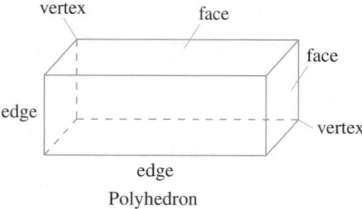

Polyhedron

Each of the plane regions of the polyhedron is called a **face** of the polyhedron. If the intersection of two faces is a line segment, this line segment is an **edge** of the polyhedron. The intersections of the edges are the **vertices** of the polyhedron.

Volume is a measure of the space of a solid. The volume of a box or can, for example, is the amount of space inside. Volume can be used to describe the amount of juice in a pitcher or the amount of concrete needed to pour a foundation for a house.

The volume of a solid is the number of **cubic units** in the solid. A cubic centimeter and a cubic inch are illustrated.

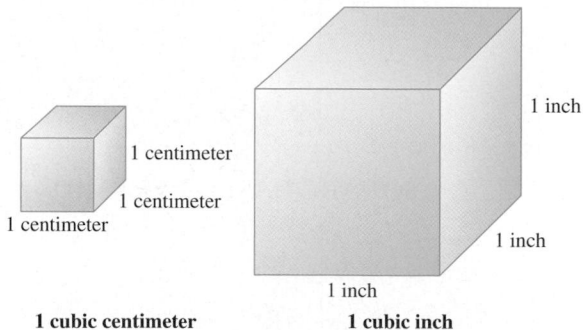

1 cubic centimeter   **1 cubic inch**

The **surface area** of a polyhedron is the sum of the areas of the faces of the polyhedron. For example, each face of the cube to the left above has an area of 1 square centimeter. Since there are 6 faces of the cube, the sum of the areas of the faces is

6 square centimeters. Surface area can be used to describe the amount of material needed to cover or form a solid. Surface area is measured in square units.

Formulas for finding the volumes, V, and surface areas, SA, of some common solids are given next.

VOLUME AND SURFACE AREA FORMULAS OF COMMON SOLIDS

Solid	*Formulas*

RECTANGULAR SOLID

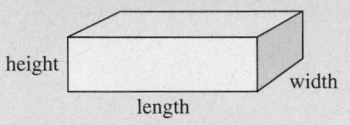

$V = lwh$
$SA = 2lh + 2wh + 2lw$
where h = height, w = width, l = length

CUBE

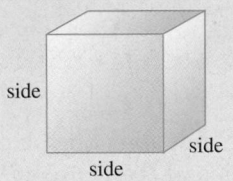

$V = s^3$
$SA = 6s^2$
where s = side length

SPHERE

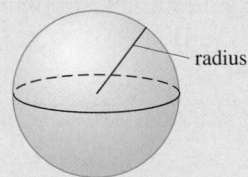

$V = \dfrac{4}{3}\pi r^3$

$SA = 4\pi r^2$
where r = radius

CIRCULAR CYLINDER

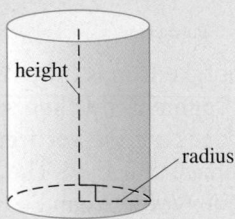

$V = \pi r^2 h$
$SA = 2\pi rh + 2\pi r^2$
where h = height, r = radius

CONE

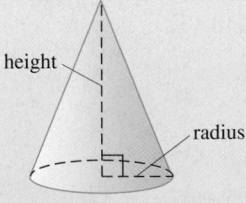

$V = \dfrac{1}{3}\pi r^2 h$

$SA = \pi r\sqrt{r^2 + h^2} + \pi r^2$
where h = height, r = radius

SQUARE-BASED PYRAMID

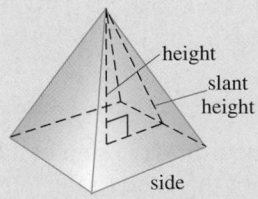

$V = \dfrac{1}{3}s^2 h$

$SA = B + \dfrac{1}{2}pl$

where B = area of base; p = perimeter of base, h = height, s = side, l = slant height

> **HELPFUL HINT**
> Volume is measured in cubic units. Surface area is measured in square units.

Example 1   Find the volume and surface area of a rectangular box that is 12 inches long, 6 inches wide, and 3 inches high.

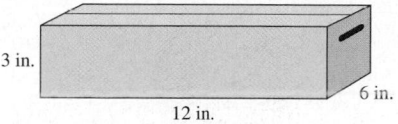

3 in.

12 in.

6 in.

Solution   Let $h = 3$ in., $l = 12$ in., and $w = 6$ in.

$V = lwh$

$V = 12 \text{ inches} \cdot 6 \text{ inches} \cdot 3 \text{ inches} = 216$ cubic inches

The volume of the rectangular box is 216 cubic inches.

$SA = 2lh + 2wh + 2lw$

$\quad = 2(12 \text{ in.})(3 \text{ in.}) + 2(6 \text{ in.})(3 \text{ in.}) + 2(12 \text{ in.})(6 \text{ in.})$

$\quad = 72 \text{ sq. in.} + 36 \text{ sq. in.} + 144 \text{ sq. in.}$

$\quad = 252 \text{ sq. in.}$

The surface area of the rectangular box is 252 square inches.

Example 2   Find the volume and surface area of a ball of radius 2 inches. Give the exact volume and surface area and then use the approximation $\dfrac{22}{7}$ for π.

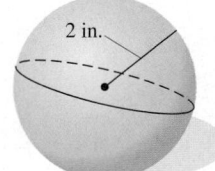

2 in.

Solution

$V = \dfrac{4}{3}\pi r^3$ Formula for volume of a sphere.

$V = \dfrac{4}{3} \cdot \pi (2 \text{ in.})^3$ Let $r = 2$ inches.

$\quad = \dfrac{32}{3}\pi$ cu. in.   Simplify.

$\quad \approx \dfrac{32}{3} \cdot \dfrac{22}{7}$ cu. in.   Approximate π with $\dfrac{22}{7}$.

$\quad = \dfrac{704}{21}$ or $33\dfrac{11}{21}$ cu. in.

The volume of the sphere is exactly $\dfrac{32}{3}\pi$ cubic inches or approximately $33\dfrac{11}{21}$ cubic inches.

$SA = 4\pi r^2$ Formula for surface area.

$SA = 4 \cdot \pi (2 \text{ in.})^2$ Let $r = 2$ inches.

$\quad = 16\pi$ sq. in.   Simplify.

$\quad \approx 16 \cdot \dfrac{22}{7}$ sq. in.   Approximate π with $\dfrac{22}{7}$.

$\quad = \dfrac{352}{7}$ or $50\dfrac{2}{7}$ sq. in.

The surface area of the sphere is exactly 16π square inches or approximately $50\dfrac{2}{7}$ square inches.

Appendix D Exercise Set

Find the volume and surface area of each solid. See Examples 1 and 2. For formulas that contain π, give an exact answer and then approximate using $\frac{22}{7}$ for π.

 1.
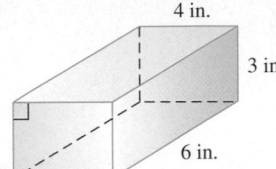
4 in.
3 in.
6 in.

2.
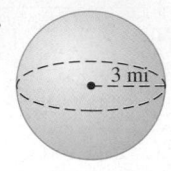
3 mi

3.
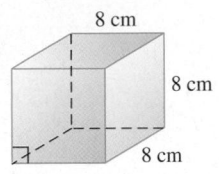
8 cm
8 cm
8 cm

4.
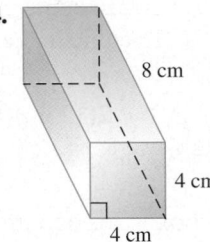
8 cm
4 cm
4 cm

5. (For surface area, use 3.14 for π and round off the answer to two decimal places.)

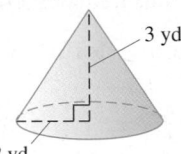
3 yd
2 yd

6.
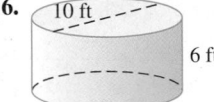
10 ft
6 ft

 7.
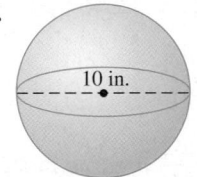
10 in.

8. Find the volume only.
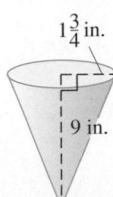
$1\frac{3}{4}$ in.
9 in.

9.
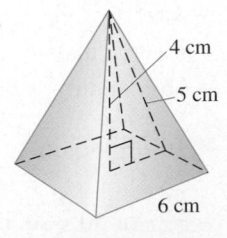
4 cm
5 cm
6 cm

10.
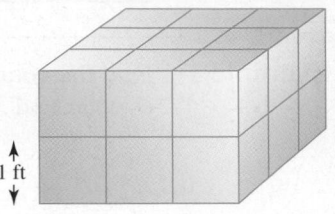
1 ft

Solve.

11. Find the volume of a cube with edges of $1\frac{1}{3}$ inches.

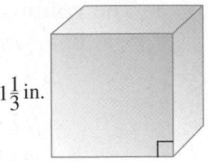
$1\frac{1}{3}$ in.

12. A water storage tank is in the shape of a cone with the pointed end down. If the radius is 14 ft and the depth of the tank is 15 ft, approximate the volume of the tank in cubic feet. Use $\frac{22}{7}$ for π.

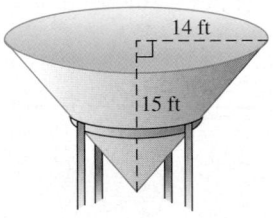
14 ft
15 ft

13. Find the surface area of a rectangular box 2 ft by 1.4 ft by 3 ft.

14. Find the surface area of a box in the shape of a cube that is 5 ft on each side.

15. Find the volume of a pyramid with a square base 5 in. on a side and a height of 1.3 in.

16. Approximate to the nearest hundredth the volume of a sphere with a radius of 2 cm. Use 3.14 for π.

17. A paperweight is in the shape of a square-based pyramid 20 cm tall. If an edge of the base is 12 cm, find the volume of the paperweight.

18. A bird bath is made in the shape of a hemisphere (half-sphere). If its radius is 10 in., approximate the volume. Use $\dfrac{22}{7}$ for π.

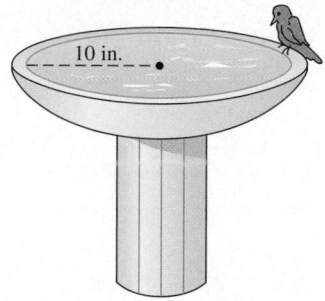

10 in.

19. Find the exact surface area of a sphere with a radius of 7 in.

20. A tank is in the shape of a cylinder 8 ft tall and 3 ft in radius. Find the exact surface area of the tank.

21. Find the volume of a rectangular block of ice 2 ft by $2\dfrac{1}{2}$ ft by $1\dfrac{1}{2}$ ft.

22. Find the capacity (volume in cubic feet) of a rectangular ice chest with inside measurements of 3 ft by $1\dfrac{1}{2}$ ft by $1\dfrac{3}{4}$ ft.

23. An ice cream cone with a 4-cm diameter and 3-cm depth is filled exactly level with the top of the cone. Approximate how much ice cream (in cubic centimeters) is in the cone. Use $\dfrac{22}{7}$ for π.

24. A child's toy is in the shape of a square-based pyramid 10 in. tall. If an edge of the base is 7 in., find the volume of the toy.

MEAN, MEDIAN, AND MODE

It is sometimes desirable to be able to describe a set of data, or a set of numbers, by a single "middle" number. Three such **measures of central tendency** are the mean, the median, and the mode.

The most common measure of central tendency is the mean (sometimes called the arithmetic mean or the average). The **mean** of a set of data items, denoted by $\bar{x}$, is the sum of the items divided by the number of items.

Example 1   Seven students in a psychology class conducted an experiment on mazes. Each student was given a pencil and asked to successfully complete the same maze. The timed results are below.

STUDENT	Ann	Thanh	Carlos	Jesse	Melinda	Ramzi	Dayni
TIME (SECONDS)	13.2	11.8	10.7	16.2	15.9	13.8	18.5

a. Who completed the maze in the shortest time? Who completed the maze in the longest time?
b. Find the mean.
c. How many students took longer than the mean time? How many students took shorter than the mean time?

Solution   **a.** Carlos completed the maze in 10.7 seconds, the shortest time. Dayni completed the maze in 18.5 seconds, the longest time.
b. To find the mean, $\bar{x}$, find the sum of the data items and divide by 7, the number of items.

$$\bar{x} = \frac{13.2 + 11.8 + 10.7 + 16.2 + 15.9 + 13.8 + 18.5}{7} = \frac{100.1}{7} = 14.3$$

c. Three students, Jesse, Melinda, and Dayni, had times longer than the mean time. Four students, Ann, Thanh, Carlos, and Ramzi, had times shorter than the mean time.

Two other measures of central tendency are the median and the mode.

The **median** of an ordered set of numbers is the middle number. If the number of items is even, the median is the mean of the two middle numbers. The **mode** of a set of numbers is the number that occurs most often. It is possible for a data set to have no mode or more than one mode.

Example 2   Find the median and the mode of the following list of numbers. These numbers were high temperatures for fourteen consecutive days in a city in Montana.

$$76, 80, 85, 86, 89, 87, 82, 77, 76, 79, 82, 89, 89, 92$$

Solution   First, write the numbers in order.

$$76, 76, 77, 79, 80, 82, 82, 85, 86, 87, 89, 89, 89, 92$$

two middle numbers

mode

Since there are an even number of items, the median is the mean of the two middle numbers.

$$\text{median} = \frac{82 + 85}{2} = 83.5$$

The mode is 89, since 89 occurs most often.

Appendix E Exercise Set

For each of the following data sets, find the mean, the median, and the mode. If necessary, round the mean to one decimal place.

1. 21, 28, 16, 42, 38

2. 42, 35, 36, 40, 50

3. 7.6, 8.2, 8.2, 9.6, 5.7, 9.1

4. 4.9, 7.1, 6.8, 6.8, 5.3, 4.9

5. 0.2, 0.3, 0.5, 0.6, 0.6, 0.9, 0.2, 0.7, 1.1

6. 0.6, 0.6, 0.8, 0.4, 0.5, 0.3, 0.7, 0.8, 0.1

7. 231, 543, 601, 293, 588, 109, 334, 268

8. 451, 356, 478, 776, 892, 500, 467, 780

Ten tall buildings in the United States are listed below. Use this table for Exercises 9–12.

Building	Height (feet)
Sears Tower, Chicago, IL	1454
One World Trade Center (1972), New York, NY	1368
One World Trade Center (1973), New York, NY	1362
Empire State, New York, NY	1250
Amoco, Chicago, IL	1136
John Hancock Center, Chicago, IL	1127
First Interstate World Center, Los Angeles, CA	1107
Chrysler, New York, NY	1046
NationsBank Tower, Atlanta, GA	1023
Texas Commerce Tower, Houston, TX	1002

9. Find the mean height for the five tallest buildings.
10. Find the median height for the five tallest buildings.
11. Find the median height for the ten tallest buildings.
12. Find the mean height for the ten tallest buildings.

During an experiment, the following times (in seconds) were recorded: 7.8, 6.9, 7.5, 4.7, 6.9, 7.0.

13. Find the mean. Round to the nearest tenth.
14. Find the median.   **15.** Find the mode.

In a mathematics class, the following test scores were recorded for a student: 86, 95, 91, 74, 77, 85.

16. Find the mean. Round to the nearest hundredth.
17. Find the median.   **18.** Find the mode.

The following pulse rates were recorded for a group of fifteen students: 78, 80, 66, 68, 71, 64, 82, 71, 70, 65, 70, 75, 77, 86, 72.

19. Find the mean.   **20.** Find the median.
21. Find the mode.

22. How many rates were higher than the mean?
23. How many rates were lower than the mean?
24. Have each student in your algebra class take his/her pulse rate. Record the data and find the mean, the median, and the mode.

Find the missing numbers in each list of numbers. (These numbers are not necessarily in numerical order)

25. __, __, 16, 18, __
The mode is 21.
The mean is 20.

26. __, __, __, __, 40
The mode is 35.
The median is 37.
The mean is 38.

INTRODUCTION TO USING A GRAPHING UTILITY

Appendix **F**

The Viewing Window and Interpreting Window Settings

In this appendix, we will use the term **graphing utility** to mean a graphing calculator or a computer software graphing package. All graphing utilities graph equations by plotting points on a screen. While plotting several points can be slow and sometimes tedious for us, a graphing utility can quickly and accurately plot hundreds of points. How does a graphing utility show plotted points? A computer or calculator screen is made up of a grid of small rectangular areas called **pixels**. If a pixel contains a point to be plotted, the pixel is turned "on"; otherwise, the pixel remains "off." The graph of an equation is then a collection of pixels turned "on." The graph of $y = 3x + 1$ from a graphing calculator is shown in Figure A-1. Notice the irregular shape of the line caused by the rectangular pixels.

$y = 3x + 1$

Figure A-1

The portion of the coordinate plane shown on the screen in Figure A-1 is called the **viewing window** or the **viewing rectangle.** Notice the x-axis and the y-axis on the graph. While tick marks are shown on the axes, they are not labeled. This means that from this screen alone, we do not know how many units each tick mark represents. To see what each tick mark represents and the minimum and maximum values on the axes, check the *window setting* of the graphing utility. It defines the viewing window. The window of the graph of $y = 3x + 1$ shown in Figure A-1 has the following setting (Figure A-2):

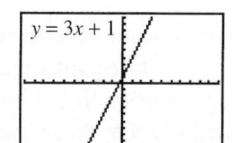

Figure A-2

Xmin $= -10$	The minimum x-value is -10.
Xmax $= 10$	The maximum x-value is 10.
Xscl $= 1$	The x-axis scale is 1 unit per tick mark.
Ymin $= -10$	The minimum y-value is -10.
Ymax $= 10$	The maximum y-value is 10.
Yscl $= 1$	The y-axis scale is 1 unit per tick mark.

By knowing the scale, we can find the minimum and the maximum values on the axes simply by counting tick marks. For example, if both the Xscl (x-axis scale) and the Yscl are 1 unit per tick mark on the graph in Figure A-3, we can count the tick marks and find that the minimum x-value is -10 and the maximum x-value is 10. Also, the minimum y-value is -10 and the maximum y-value is 10. If the Xscl (x-axis

839

scale) changes to 2 units per tick mark (shown in Figure A-4), by counting tick marks, we see that the minimum x-value is now -20 and the maximum x-value is now 20.

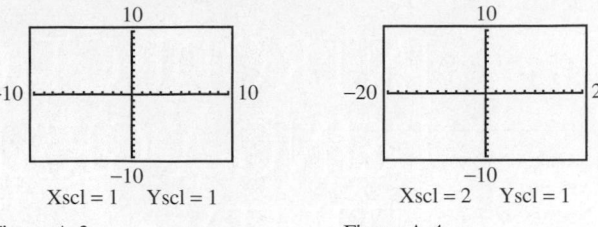

Figure A-3   Figure A-4

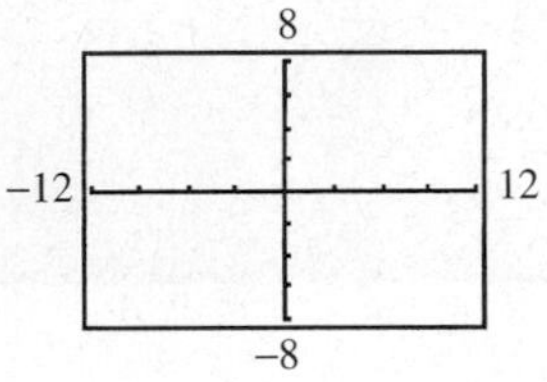

It is also true that if we know the Xmin and the Xmax values, we can calculate the Xscl by the displayed axes. For example, the Xscl of the graph in Figure A-5 must be 3 units per tick mark for the maximum and minimum x-values to be as shown. Also, the Yscl of that graph must be 2 units per tick mark for the maximum and minimum y-values to be as shown.

Figure A-5

We will call the viewing window in Figure A-3 a *standard* viewing window or rectangle. Although a standard viewing window is sufficient for much of this text, special care must be taken to ensure that all key features of a graph are shown. Figures A-6, A-7, and A-8 show the graph of $y = x^2 + 11x - 1$ on three different viewing windows. Note that certain viewing windows for this equation are misleading.

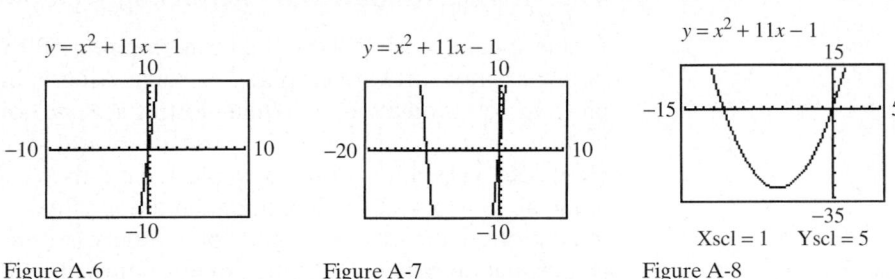

Figure A-6   Figure A-7   Figure A-8

How do we ensure that all distinguishing features of the graph of an equation are shown? It helps to know about the equation that is being graphed. For example, the equation $y = x^2 + 11x - 1$ is not a linear equation and its graph is not a line. This equation is a quadratic equation and, therefore, its graph is a parabola. By knowing this information, we know that the graph shown in Figure A-6, although correct, is misleading. Of the three viewing rectangles shown, the graph in Figure A-8 is best because it shows more of the distinguishing features of the parabola. Properties of equations needed for graphing will be studied in this text.

Viewing Window and Interpreting Window Settings Exercise Set

In Exercises 1–4, determine whether all ordered pairs listed will lie within a standard viewing rectangle.

1. $(-9, 0), (5, 8), (1, -8)$
2. $(4, 7), (0, 0), (-8, 9)$
3. $(-11, 0), (2, 2), (7, -5)$
4. $(3, 5), (-3, -5), (15, 0)$

In Exercises 5–10, choose an Xmin, Xmax, Ymin, and Ymax so that all ordered pairs listed will lie within the viewing rectangle.

5. $(-90, 0), (55, 80), (0, -80)$
6. $(4, 70), (20, 20), (-18, 90)$
7. $(-11, 0), (2, 2), (7, -5)$

8. $(3, 5), (-3, -5), (15, 0)$

9. $(200, 200), (50, -50), (70, -50)$

10. $(40, 800), (-30, 500), (15, 0).$

Write the window setting for each viewing window shown. Use the following format:

Xmin =	Ymin =
Xmax =	Ymax =
Xscl =	Yscl =

15.
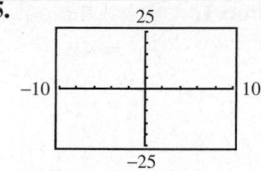

16.
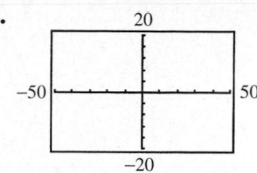

11.
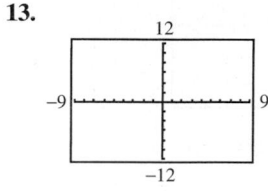

12.
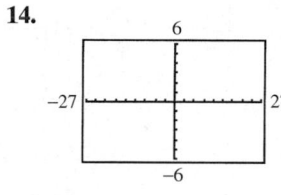

17.
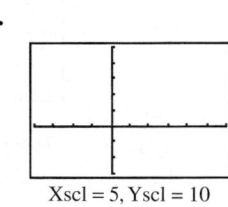
Xscl = 1, Yscl = 3

18.
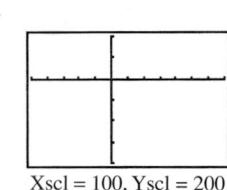
Xscl = 10, Yscl = 2

13.
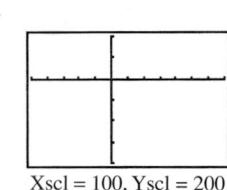

14.

19.
Xscl = 5, Yscl = 10

20.
Xscl = 100, Yscl = 200

Graphing Equations and Square Viewing Window

In general, the following steps may be used to graph an equation on a standard viewing window.

GRAPHING AN EQUATION IN x AND y WITH A GRAPHIING UTILITY ON A STANDARD VIEWING WINDOW

Step 1: Solve the equation for y.

Step 2: Use your graphing utility and enter the equation in the form
$Y = $ *expression involving x*

Step 3: Activate the graphing utility.

Special care must be taken when entering the *expression involving x* in Step 2. You must be sure that the graphing utility you are using interprets the expression as you want it to. For example, let's graph $3y = 4x$. To do so,

Step 1:   Solve the equation for y.

$$3y = 4x$$

$$\frac{3y}{3} = \frac{4x}{3}$$

$$y = \frac{4}{3}x$$

Step 2:   Using your graphing utility, enter the expression $\frac{4}{3}x$ after the

Y = prompt. In order for your graphing utility to correctly interpret the expression, you may need to enter $(4/3)x$ or $(4 \div 3)x$.

Step 3:   Activate the graphing utility. The graph should appear as in Figure A-9.

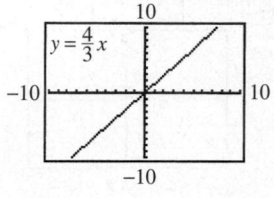

Figure A-9

Distinguishing features of the graph of a line include showing all the intercepts of the line. For example, the window of the graph of the line in Figure A-10 does not show both intercepts of the line, but the window of the graph of the same line in Figure A-11 does show both intercepts. Notice the notation below each graph. This is a shorthand notation of the range setting of the graph. This notation means [Xmin, Xmax] by [Ymin, Ymax].

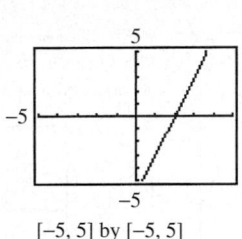

[−5, 5] by [−5, 5]

Figure A-10

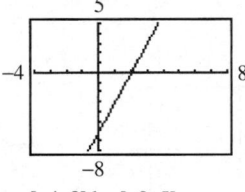

[−4, 8] by [−8, 5]

Figure A-11

On a standard viewing window, the tick marks on the y-axis are closer together than the tick marks on the x-axis. This happens because the viewing window is a rectangle, and so 10 equally spaced tick marks on the positive y-axis will be closer together than 10 equally spaced tick marks on the positive x-axis. This causes the appearance of graphs to be distorted.

For example, notice the different appearances of the same line graphed using different viewing windows. The line in Figure A-12 is distorted because the tick marks along the x-axis are farther apart than the tick marks along the y-axis. The graph of the same line in Figure A-13 is not distorted because the viewing rectangle has been selected so that there is equal spacing between tick marks on both axes.

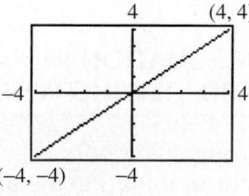

Figure A-12

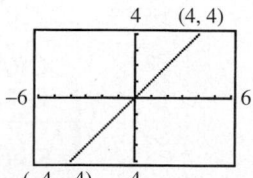

Figure A-13

We say that the line in Figure A-13 is graphed on a *square* setting. Some graphing utilities have a built-in program that, if activated, will automatically provide a square setting. A square setting is especially helpful when we are graphing perpendicular lines, circles, or when a true geometric perspective is desired. Some examples of square screens are shown in Figures A-14 and A-15.

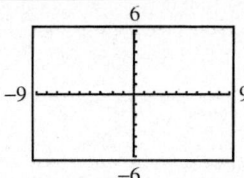

Figure A-14

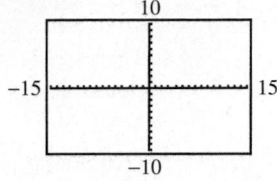

Figure A-15

Other features of a graphing utility such as Trace, Zoom, Intersect, and Table are discussed in appropriate Graphing Calculator Explorations in this text.

Graphing Equations and Square Viewing Window Exercise Set

Graph each linear equation in two variables, using the two different range settings given. Determine which setting shows all intercepts of a line.

1. $y = 2x + 12$

    Setting A: $[-10, 10]$ by $[-10, 10]$

    Setting B: $[-10, 10]$ by $[-10, 15]$

2. $y = -3x + 25$

    Setting A: $[-5, 5]$ by $[-30, 10]$

    Setting B: $[-10, 10]$ by $[-10, 30]$

3. $y = -x - 41$

    Setting A: $[-50, 10]$ by $[-10, 10]$

    Setting B: $[-50, 10]$ by $[-50, 15]$

4. $y = 6x - 18$

    Setting A: $[-10, 10]$ by $[-20, 10]$

    Setting B: $[-10, 10]$ by $[-10, 10]$

5. $y = \dfrac{1}{2}x - 15$

    Setting A: $[-10, 10]$ by $[-20, 10]$

    Setting B: $[-10, 35]$ by $[-20, 15]$

6. $y = -\dfrac{2}{3}x - \dfrac{29}{3}$

    Setting A: $[-10, 10]$ by $[-10, 10]$

    Setting B: $[-15, 5]$ by $[-15, 5]$

The graph of each equation is a line. Use a graphing utility and a standard viewing window to graph each equation.

7. $3x = 5y$ **8.** $7y = -3x$ **9.** $9x - 5y = 30$

10. $4x + 6y = 20$ **11.** $y = -7$ **12.** $y = 2$

13. $x + 10y = -5$ **14.** $x - 5y = 9$

Graph the following equations using the square window setting given. Some keystrokes that may be helpful are given.

15. $y = \sqrt{x}$ $[-12, 12]$ by $[-8, 8]$

    Suggested keystrokes: $\sqrt{}\, x$

16. $y = \sqrt{2x}$ $[-12, 12]$ by $[-8, 8]$

    Suggested keystrokes: $\sqrt{}\, (2x)$

17. $y = x^2 + 2x + 1$ $[-15, 15]$ by $[-10, 10]$

    Suggested keystrokes: $x \wedge 2 + 2x + 1$

18. $y = x^2 - 5$ $[-15, 15]$ by $[-10, 10]$

    Suggested keystrokes: $x \wedge 2 - 5$

19. $y = |x|$ $[-9, 9]$ by $[-6, 6]$

    Suggested keystrokes: $ABS\,(x)$

20. $y = |x - 2|$ $[-9, 9]$ by $[-6, 6]$

    Suggested keystrokes: $ABS\,(x - 2)$

Graph each line. Use a standard viewing window; then, if necessary, change the viewing window so that all intercepts of each line show.

21. $x + 2y = 30$ **22.** $1.5x - 3.7y = 40.3$

Appendix G

TABLE OF SQUARES AND SQUARE ROOTS

n	n^2	$\sqrt{n}$	n	n^2	$\sqrt{n}$
1	1	1.000	31	961	5.568
2	4	1.414	32	1024	5.657
3	9	1.732	33	1089	5.745
4	16	2.000	34	1156	5.831
5	25	2.236	35	1225	5.916
6	36	2.449	36	1296	6.000
7	49	2.646	37	1369	6.083
8	64	2.828	38	1444	6.164
9	81	3.000	39	1521	6.245
10	100	3.162	40	1600	6.325
11	121	3.317	41	1681	6.403
12	144	3.464	42	1764	6.481
13	169	3.606	43	1849	6.557
14	196	3.742	44	1936	6.633
15	225	3.873	45	2025	6.708
16	256	4.000	46	2116	6.782
17	289	4.123	47	2209	6.856
18	324	4.243	48	2304	6.928
19	361	4.359	49	2401	7.000
20	400	4.472	50	2500	7.071
21	441	4.583	51	2601	7.141
22	484	4.690	52	2704	7.211
23	529	4.796	53	2809	7.280
24	576	4.899	54	2916	7.348
25	625	5.000	55	3025	7.416
26	676	5.099	56	3136	7.483
27	729	5.196	57	3249	7.550
28	784	5.292	58	3364	7.616
29	841	5.385	59	3481	7.681
30	900	5.477	60	3600	7.746

n	n^2	$\sqrt{n}$	n	n^2	$\sqrt{n}$
61	3721	7.810	81	6561	9.000
62	3844	7.874	82	6724	9.055
63	3969	7.937	83	6889	9.110
64	4096	8.000	84	7056	9.165
65	4225	8.062	85	7225	9.220
66	4356	8.124	86	7396	9.274
67	4489	8.185	87	7569	9.327
68	4624	8.246	88	7744	9.381
69	4761	8.307	89	7921	9.434
70	4900	8.367	90	8100	9.487
71	5041	8.426	91	8281	9.539
72	5184	8.485	92	8464	9.592
73	5329	8.544	93	8649	9.644
74	5476	8.602	94	8836	9.695
75	5625	8.660	95	9025	9.747
76	5776	8.718	96	9216	9.798
77	5929	8.775	97	9409	9.849
78	6084	8.832	98	9604	9.899
79	6241	8.888	99	9801	9.950
80	6400	8.944	100	10000	10.000

ANSWERS TO SELECTED EXERCISES

■ **CHAPTER 1   REVIEW OF REAL NUMBERS**

Exercise Set 1.2 **1.** $<$ **3.** $>$ **5.** $=$ **7.** $<$ **9.** $32 < 212$ **11.** $44{,}300 > 34{,}611$ **13.** true **15.** false **17.** false
19. true **21.** $30 \le 45$ **23.** $8 < 12$ **25.** $5 \ge 4$ **27.** $15 \ne -2$ **29.** $535; -8$ **31.** $-433{,}853$ **33.** $350; -126$ **35.** 1988
37. 1988, 1989, 1990, 1991 **39.** $6068 \ge 2649$ **41.** whole, integers, rational, real **43.** integers, rational, real
45. natural, whole, integers, rational, real **47.** rational, real **49.** irrational, real **51.** false **53.** true **55.** true
57. true **59.** false **61.** $>$ **63.** $>$ **65.** $<$ **67.** $<$ **69.** $>$ **71.** $=$ **73.** $<$ **75.** $<$ **77.** $-0.04 > -26.7$ **79.** sun
81. sun **83.** $20 \le 25$ **85.** $6 > 0$ **87.** $-12 < -10$ **89.** Answers may vary.

Mental Math **1.** $\dfrac{3}{8}$ **3.** $\dfrac{5}{7}$ **5.** numerator; denominator

Exercise Set 1.3 **1.** $3 \cdot 11$ **3.** $2 \cdot 7 \cdot 7$ **5.** $2 \cdot 2 \cdot 5$ **7.** $3 \cdot 5 \cdot 5$ **9.** $3 \cdot 3 \cdot 5$ **11.** $\dfrac{1}{2}$ **13.** $\dfrac{2}{3}$ **15.** $\dfrac{3}{7}$ **17.** $\dfrac{3}{5}$ **19.** $\dfrac{3}{8}$ **21.** $\dfrac{1}{2}$
23. $\dfrac{6}{7}$ **25.** 15 **27.** $\dfrac{1}{6}$ **29.** $\dfrac{25}{27}$ **31.** $\dfrac{11}{20}$ sq. mi **33.** $\dfrac{3}{5}$ **35.** 1 **37.** $\dfrac{1}{3}$ **39.** $\dfrac{9}{35}$ **41.** $\dfrac{21}{30}$ **43.** $\dfrac{4}{18}$ **45.** $\dfrac{16}{20}$ **47.** $\dfrac{23}{21}$ **49.** $1\dfrac{2}{3}$
51. $\dfrac{5}{66}$ **53.** $\dfrac{7}{5}$ **55.** $\dfrac{1}{5}$ **57.** $\dfrac{3}{8}$ **59.** $\dfrac{1}{9}$ **61.** $\dfrac{5}{7}$ **63.** $\dfrac{65}{21}$ **65.** $\dfrac{2}{5}$ **67.** $\dfrac{9}{7}$ **69.** $\dfrac{3}{4}$ **71.** $\dfrac{17}{3}$ **73.** $\dfrac{7}{26}$ **75.** 1 **77.** $\dfrac{1}{5}$ **79.** $5\dfrac{1}{6}$
81. $\dfrac{17}{18}$ **83.** $55\dfrac{1}{4}$ ft **85.** $5\dfrac{8}{25}$ m **87.** Answers may vary. **89.** $3\dfrac{3}{8}$ mi **91.** $\dfrac{3}{4}$ **93.** $\dfrac{49}{200}$

Calculator Explorations **1.** 125 **3.** 59,049 **5.** 30 **7.** 9857

Mental Math **1.** multiply **3.** subtract

Exercise Set 1.4 **1.** 243 **3.** 27 **5.** 1 **7.** 5 **9.** $\dfrac{1}{125}$ **11.** $\dfrac{16}{81}$ **13.** 49 **15.** 16 **17.** 1.44 **19.** 17 **21.** 20 **23.** 10
25. 21 **27.** 45 **29.** 0 **31.** $\dfrac{2}{7}$ **33.** 30 **35.** 2 **37.** $\dfrac{7}{18}$ **39.** $\dfrac{27}{10}$ **41.** $\dfrac{7}{5}$ **43.** no **45. a.** 64 **b.** 43 **c.** 19 **d.** 22 **47.** 9
49. 1 **51.** 1 **53.** 11 **55.** 45 **57.** 27 **59.** 132 **61.** $\dfrac{37}{18}$ **63.** 16, 64, 144, 256 **65.** solution **67.** not a solution
69. not a solution **71.** solution **73.** not a solution **75.** $x + 15$ **77.** $x - 5$ **79.** $3x + 22$ **81.** $1 + 2 = 9 \div 3$
83. $3 \ne 4 \div 2$ **85.** $5 + x = 20$ **87.** $13 - 3x = 13$ **89.** $\dfrac{12}{x} = \dfrac{1}{2}$ **91.** Answers may vary. **93.** 28 m **95.** 12,000 sq. ft
97. 6.5% **99.** $27.75

Mental Math **1.** negative **3.** 0 **5.** negative

Exercise Set 1.5 **1.** 9 **3.** -14 **5.** 1 **7.** -12 **9.** -5 **11.** -12 **13.** -4 **15.** 7 **17.** -2 **19.** 0 **21.** -19 **23.** 31
25. -47 **27.** -2.1 **29.** -8 **31.** 38 **33.** -13.1 **35.** $\dfrac{2}{8} = \dfrac{1}{4}$ **37.** $-\dfrac{3}{16}$ **39.** $-\dfrac{13}{10}$ **41.** -8 **43.** -59 **45.** -9 **47.** 5

A1

49. 11 **51.** −18 **53.** 19 **55.** −0.7 **57.** Tues **59.** 7° **61.** 1° **63.** −6° **65.** −654 ft **67.** −$218.8 million **69.** −17

71. −6 **73.** 2 **75.** 0 **77.** −6 **79.** Answers may vary. **81.** −2 **83.** 0 **85.** $-\dfrac{2}{3}$ **87.** Answers may vary.

89. negative **91.** positive **93.** yes **95.** no

Exercise Set 1.6 **1.** −10 **3.** −5 **5.** 19 **7.** $\dfrac{1}{6}$ **9.** 2 **11.** −11 **13.** 11 **15.** 5 **17.** 37 **19.** −6.4 **21.** −71 **23.** 0

25. 4.1 **27.** $\dfrac{2}{11}$ **29.** $-\dfrac{11}{12}$ **31.** 8.92 **33.** 13 **35.** −5 **37.** −1 **39.** −23 **41.** Answers may vary. **43.** −26 **45.** −24

47. 3 **49.** −45 **51.** −4 **53.** 13 **55.** 6 **57.** 9 **59.** −9 **61.** −7 **63.** $\dfrac{7}{5}$ **65.** 21 **67.** $\dfrac{1}{4}$ **69.** January, −22° **71.** −12°

73. 49° **75.** 100° **77.** 23 yd loss **79.** 384 B.C. **81.** $-2\dfrac{3}{8}$ points **83.** 22,965 ft **85.** 130° **87.** 30° **89.** not a solution

91. not a solution **93.** solution **95.** true **97.** false **99.** negative, −30,387

Calculator Explorations **1.** 38 **3.** −441 **5.** $163.\overline{3}$ **7.** 54,499 **9.** 15,625

Mental Math **1.** positive **3.** negative **5.** positive

Exercise Set 1.7 **1.** −24 **3.** −2 **5.** 50 **7.** −12 **9.** 42 **11.** −18 **13.** $\dfrac{3}{10}$ **15.** $\dfrac{2}{3}$ **17.** −7 **19.** 0.14 **21.** −800

23. −28 **25.** 25 **27.** $-\dfrac{8}{27}$ **29.** −121 **31.** $-\dfrac{1}{4}$ **33.** 0.84 **35.** −30 **37.** 90 **39.** true **41.** false **43.** 16 **45.** −1

47. 25 **49.** −49 **51.** $\dfrac{1}{9}$ **53.** $\dfrac{3}{2}$ **55.** $-\dfrac{1}{14}$ **57.** $-\dfrac{11}{3}$ **59.** $\dfrac{1}{0.2}$ **61.** −6.3 **63.** −9 **65.** 4 **67.** −4 **69.** 0 **71.** −5

73. undefined **75.** 3 **77.** −15 **79.** $-\dfrac{18}{7}$ **81.** $\dfrac{20}{27}$ **83.** −1 **85.** $-\dfrac{9}{2}$ **87.** −4 **89.** 16 **91.** −3 **93.** $-\dfrac{16}{7}$ **95.** 2

97. $\dfrac{6}{5}$ **99.** −5 **101.** $\dfrac{3}{2}$ **103.** −21 **105.** 41 **107.** −134 **109.** 3 **111.** −1 **113.** −$498 million **115.** Answers may vary.

117. 1, −1 **119.** positive **121.** not possible **123.** negative **125.** $-2 + \dfrac{-15}{3}$; −7 **127.** $2[-5 + (-3)]$; −16 **129.** yes

131. yes **133.** yes

Exercise Set 1.8 **1.** $16 + x$ **3.** $y \cdot (-4)$ **5.** yx **7.** $13 + 2x$ **9.** $x \cdot (yz)$ **11.** $(2 + a) + b$ **13.** $(4a) \cdot b$
15. $a + (b + c)$ **17.** $17 + b$ **19.** $24y$ **21.** y **23.** $26 + a$ **25.** $-72x$ **27.** s **29.** Answers may vary. **31.** $4x + 4y$
33. $9x - 54$ **35.** $6x + 10$ **37.** $28x - 21$ **39.** $18 + 3x$ **41.** $-2y + 2z$ **43.** $-21y - 35$ **45.** $5x + 20m + 10$
47. $-4 + 8m - 4n$ **49.** $-5x - 2$ **51.** $-r + 3 + 7p$ **53.** $3x + 4$ **55.** $-x + 3y$ **57.** $6r + 8$ **59.** $-36x - 70$
61. $-16x - 25$ **63.** $4(1 + y)$ **65.** $11(x + y)$ **67.** $-1(5 + x)$ **69.** $30(a + b)$ **71.** commutative property of
multiplication **73.** associative property of addition **75.** distributive property **77.** associative property of multiplication
79. identity property of addition **81.** distributive property **83.** commutative and associative properties of multiplication
85. $-8; \dfrac{1}{8}$ **87.** $-x; \dfrac{1}{x}$ **89.** $-2x; \dfrac{1}{2x}$ **91.** no **93.** yes **95.** Answers may vary.

Exercise Set 1.9 **1.** approx. 7.8 million **3.** 2002 **5.** PGA/LPGA tours **7.** Major League Baseball, NBA
9. approx. $15 million **11.** France **13.** France, United States, Spain, Italy **15.** 34 million **17.** approx. $142 million
19. *Snow White and the Seven Dwarfs* **21.** Answers may vary. **23.** 1994 **25.** 1989, 1990 **27.** approx. 54%
29. 1994 **31.** approx. 59 beats per minute **33.** approx. 26 beats per minute **35.** 20 students **37.** 1985
39. Answers may vary. **41.** 4 million **43.** 69 million **45.** 1960 **47.** 12 million **49.** 30° north, 90° west
51. Answers may vary.

Chapter 1 Review **1.** < **3.** > **5.** < **7.** = **9.** > **11.** $4 \geq -3$ **13.** $0.03 < 0.3$ **15. a.** $\{1, 3\}$ **b.** $\{0, 1, 3\}$

c. $\{-6, 0, 1, 3\}$ **d.** $\left\{-6, 0, 1, 1\dfrac{1}{2}, 3, 9.62\right\}$ **e.** $\{\pi\}$ **f.** $\left\{-6, 0, 1, 1\dfrac{1}{2}, 3, \pi, 9.62\right\}$ **17.** Friday **19.** $2 \cdot 2 \cdot 3 \cdot 3$ **21.** $\dfrac{12}{25}$

23. $\dfrac{13}{10}$ **25.** $9\dfrac{3}{8}$ **27.** 15 **29.** $\dfrac{7}{12}$ **31.** $A = \dfrac{34}{121}$ sq. in.; $P = 2\dfrac{4}{11}$ in. **33.** $2\dfrac{15}{16}$ lb **35.** $11\dfrac{5}{16}$ lb **37.** Odera **39.** $3\dfrac{7}{8}$ lb

41. 16 **43.** $\dfrac{4}{49}$ **45.** 70 **47.** 37 **49.** $\dfrac{18}{7}$ **51.** $20 - 12 = 2 \cdot 4$ **53.** 18 **55.** 5 **57.** 63° **59.** no **61.** $\dfrac{-2}{3}$ **63.** 7

65. −17 **67.** −5 **69.** 3.9 **71.** −14 **73.** 5 **75.** −19 **77.** 15 **79.** $51 **81.** $-\dfrac{1}{6}$ **83.** −48 **85.** 3 **87.** −36

89. undefined **91.** undefined **93.** −5 **95.** commutative property of addition **97.** distributive property
99. associative property of addition **101.** distributive property **103.** multiplicative inverse
105. commutative property of addition **107.** 7 million **109.** number of subscribers is increasing **111.** Miami, 1.8%
113. New York, Cleveland, Houston, Chicago, Boston, Dallas/Fort Worth, San Francisco

Chapter 1 Test **1.** $|−7| > 5$ **2.** $(9 + 5) \geq 4$ **3.** −5 **4.** −11 **5.** −14 **6.** −39 **7.** 12 **8.** −2 **9.** undefined **10.** −8

11. $-\dfrac{1}{3}$ **12.** $4\dfrac{5}{8}$ **13.** $\dfrac{51}{40}$ **14.** −32 **15.** −48 **16.** 3 **17.** 0 **18.** > **19.** > **20.** > **21.** = **22.** $2221 < 10{,}993$

23. a. {1, 7} **b.** {0, 1, 7} **c.** {−5, −1, 0, 1, 7} **d.** $\left\{-5, -1, 0, \dfrac{1}{4}, 1, 7, 11.6\right\}$ **e.** $\{\sqrt{7}, 3\pi\}$

f. $\left\{-5, -1, 0, \dfrac{1}{4}, 1, 7, 11.6, \sqrt{7}, 3\pi\right\}$ **24.** 40 **25.** 12 **26.** 22 **27.** −1 **28.** associative property

29. commutative property **30.** distributive property **31.** multiplicative inverse **32.** 9 **33.** −3 **34.** second down
35. yes **36.** 17° **37.** $650 million **38.** loss of $420 **39.** $8 billion **40.** $25 billion **41.** $5.5 billion **42.** 1996
43. Indiana, 25.2 million tons **44.** Texas, 5 million tons **45.** 16 million tons **46.** 3 million tons

■ CHAPTER 2   EQUATIONS, INEQUALITIES, AND PROBLEM SOLVING

Mental Math **1.** −7 **3.** 1 **5.** 17 **7.** like **9.** unlike **11.** like

Exercise Set 2.1 **1.** $15y$ **3.** $13w$ **5.** $-7b - 9$ **7.** $-m - 6$ **9.** $5y - 20$ **11.** $7d - 11$ **13.** $-3x + 2y - 1$
15. $2x + 14$ **17.** Answers may vary. **19.** $10x - 3$ **21.** $-4x - 9$ **23.** $5x^2$ **25.** $4x - 3$ **27.** $8x - 53$ **29.** −8
31. $7.2x - 5.2$ **33.** $k - 6$ **35.** $0.9m + 1$ **37.** $-12y + 16$ **39.** $x + 5$ **41.** −11 **43.** $1.3x + 3.5$ **45.** $x + 2$

47. $-15x + 18$ **49.** $2k + 10$ **51.** $-3x + 5$ **53.** $2x - 4$ **55.** $\dfrac{3}{4}x + 12$ **57.** $-2 + 12x$ **59.** $-4m - 3$ **61.** $8(x + 6)$

63. $x - 10$ **65.** $\dfrac{7x}{6}$ **67.** $7x - 7$ **69.** $(18x - 2)$ ft **71.** balanced **73.** balanced **75.** $(15x + 23)$ in. **77.** $\dfrac{8}{5}$ **79.** $\dfrac{1}{2}$
81. −9 **83.** x **85.** y **87.** x **89.** $5b^2c^3 + b^3c^2$ **91.** $5x^2 + 9x$ **93.** $-7x^2y$

Mental Math **1.** 2 **3.** 12 **5.** 17 **7.** 9 **9.** 2 **11.** −5

Exercise Set 2.2 **1.** 3 **3.** −2 **5.** −14 **7.** 0.5 **9.** −3 **11.** −0.7 **13.** 3 **15.** 11 **17.** 0 **19.** −3 **21.** 16 **23.** −4
25. 0 **27.** 12 **29.** 10 **31.** −12 **33.** 3 **35.** −2 **37.** 0 **39.** Answers may vary. **41.** 10 **43.** −20 **45.** 0 **47.** −5

49. 0 **51.** $-\dfrac{3}{2}$ **53.** −21 **55.** $\dfrac{11}{2}$ **57.** 1 **59.** $-\dfrac{1}{4}$ **61.** 12 **63.** −30 **65.** $\dfrac{9}{10}$ **67.** −30 **69.** 2 **71.** −2 **73.** 23
75. $20 - p$ **77.** $(10 - x)$ ft **79.** $(180 - x)°$ **81.** $n + 284$ **83.** $(m - 60)$ ft **85.** $(173 - 3x)°$ **87.** 250 ml
89. Answers may vary. **91.** $4x + 12$ **93.** $x + 21$ **95.** −2.95 **97.** 0.02 **99.** $7x - 12$ **101.** $12z + 44$ **103.** 1

Calculator Explorations **1.** solution **3.** not a solution **5.** solution

Exercise Set 2.3 **1.** 1 **3.** $\dfrac{9}{2}$ **5.** $\dfrac{3}{2}$ **7.** 0 **9.** 2 **11.** −5 **13.** 10 **15.** 18 **17.** 1 **19.** 50 **21.** 0.2 **23.** all real numbers
25. no solution **27.** no solution **29.** Answers may vary. **31.** Answers may vary. **33.** 4 **35.** −4 **37.** 3 **39.** −2

41. 4 **43.** $\dfrac{7}{3}$ **45.** no solution **47.** $\dfrac{9}{5}$ **49.** $\dfrac{4}{19}$ **51.** 1 **53.** no solution **55.** $\dfrac{7}{2}$ **57.** −17 **59.** $\dfrac{19}{6}$ **61.** all real numbers

63. 3 **65.** 13 **67.** 15.3 **69.** −0.2 **71.** $2x + \dfrac{1}{5} = 3x - \dfrac{4}{5}; 1$ **73.** $2x + 7 = x + 6; -1$ **75.** $3x - 6 = 2x + 8; 14$

77. $\dfrac{1}{3}x = \dfrac{5}{6}; \dfrac{5}{2}$ **79.** $x - 4 = 2x; -4$ **81.** $\dfrac{x}{4} + \dfrac{1}{2} = \dfrac{3}{4}; 1$ **83.** 4 cm, 4 cm, 4 cm, 8 cm, 8 cm **85.** $\dfrac{5}{4}$ **87.** Midway **89.** 145

91. −1 **93.** $\dfrac{1}{5}$ **95.** $(6x - 8)$ m **97.** $-\dfrac{7}{8}$ **99.** no solution **101.** 0

Exercise Set 2.4 **1.** 1 **3.** −25 **5.** $-\dfrac{3}{4}$ **7.** −16 **9.** governor of Nebraska: $65,000; governor of Washington: $130,000

11. 1st piece: 5 in.; 2nd piece: 10 in.; 3rd piece: 25 in. **13.** 172 mi **15.** 1st angle: 37.5°; 2nd angle: 37.5°; 3rd angle: 105°
17. Brown: 66,362; Randall: 53,074 **19.** 45°, 135° **21.** Belgium: 32; France: 33; Spain: 34 **23.** 0.2 in./yr

25. height: 34 in.; diameter: 49 in.   **27.** Spurs: 78; Knicks: 77   **29.** $2\frac{1}{2}$ **31.** 17 Democratic governors; 31 Republican governors
33. 5 ft, 12 ft   **35.** 58°, 60°, 62°   **37.** 350 mi   **39.** Answers may vary.   **41.** Texas and Florida
43. Hawaii: $37.9 million; Pennsylvania: $23 million   **45.** 15 ft by 24 ft   **47.** Answers may vary.   **49.** Answers may vary.
51. c   **53.** −10   **55.** −9   **57.** −15   **59.** $5(-x) = x + 60$ **61.** $50 - (x + 9) = 0$

Exercise Set 2.5   **1.** $h = 3$ **3.** $h = 3$ **5.** $h = 20$ **7.** $c = 12$ **9.** $r = 2.5$ **11.** $T = 3$ **13.** $h = 15$ **15.** $h = \dfrac{f}{5g}$

17. $W = \dfrac{V}{LH}$ **19.** $y = 7 - 3x$ **21.** $R = \dfrac{A - P}{PT}$ **23.** $A = \dfrac{3V}{h}$ **25.** $a = P - b - c$ **27.** $\dfrac{S - 2\pi r^2}{2\pi r}$ **29.** 131 ft

31. 7.5 hr   **33.** −10°C   **35.** 96 piranhas   **37.** 137.5 mi   **39.** 2 bags   **41.** 6.25 hr   **43.** 800 cu. ft   **45.** one 16-in. pizza

47. 4.65 min   **49.** 2.25 hr   **51.** −109.3°F   **53.** 500 sec or $8\frac{1}{3}$ min   **55.** 33,493,333,333 cu. mi   **57.** 10.7   **59.** 44.3 sec

61. −40°   **63.** It multiplies the volume by 8.   **65.** $\dfrac{9}{x + 5}$ **67.** $3(x + 4)$ **69.** $2(10 + 4x)$ **71.** $3(x - 12)$

Mental Math   **1.** no   **3.** yes

Exercise Set 2.6   **1.** 1.2   **3.** 0.225   **5.** 0.0012   **7.** 75%   **9.** 200%   **11.** 12.5%   **13.** 38%   **15.** 54%   **17.** 136.8°
19. Answers may vary.   **21.** 4%   **23.** 14.4°   **25.** 49,950   **27.** 11.2   **29.** 55%   **31.** 180   **33.** 4.6   **35.** 50   **37.** 30%
39. $39 decrease; $117 sale price   **41.** 243 ft   **43.** 55.40%   **45.** 54 people   **47.** No, many people use several medications.
49. 31 men   **51.** 27%, 5%, 35%, 2%   **53.** 0.15%   **55.** 23.6 million   **57.** 75% increase   **59.** 24%; no   **61.** 166,567%
63. 16%   **65.** 12   **67.** Answers may vary.   **69.** 11%   **71.** 7.7%   **73.** 19.3%   **75.** Answers may vary.   **77.** 6   **79.** 208
81. −55

Mental Math   **1.** $x > 2$ **3.** $x \geq 8$

Exercise Set 2.7   **1.** $\xleftarrow{\hspace{1cm}}$ -3 , $(-\infty, -3)$ **3.** $\xleftarrow{\hspace{1cm}}$ 0.3 , $[0.3, \infty)$ **5.** $\xleftarrow{\hspace{1cm}}$ 5 , $(5, \infty)$

7. $\xleftarrow{\hspace{1cm}}$ -2 5 , $(-2, 5)$ **9.** $\xleftarrow{\hspace{1cm}}$ -1 5 , $(-1, 5)$ **11.** $\xleftarrow{\hspace{1cm}}$ -3 , $(-\infty, -3)$

13. $\xleftarrow{\hspace{1cm}}$ -5 , $[-5, \infty)$ **15.** $\xleftarrow{\hspace{1cm}}$ -2 , $[-2, \infty)$ **17.** $\xleftarrow{\hspace{1cm}}$ 11 , $(-\infty, 11]$

19. $\xleftarrow{\hspace{1cm}}$ $\frac{8}{3}$, $\left[\frac{8}{3}, \infty\right)$ **21.** $\xleftarrow{\hspace{1cm}}$ -13 , $(-13, \infty)$ **23.** $\xleftarrow{\hspace{1cm}}$ 7 , $(-\infty, 7]$

25. $\xleftarrow{\hspace{1cm}}$ 0 , $(-\infty, \infty)$ **27.** $\xleftarrow{\hspace{1cm}}$ 0 , $\emptyset$ **29.** Answers may vary.   **31.** $\emptyset$; Answers may vary.

33. $\xleftarrow{\hspace{1cm}}$ -1 , $(-1, \infty)$ **35.** $\xleftarrow{\hspace{1cm}}$ 2 , $(-\infty, 2]$ **37.** $\xleftarrow{\hspace{1cm}}$ 2 , $(2, \infty)$

39. $\xleftarrow{\hspace{1cm}}$ -8 , $[-8, \infty)$ **41.** $\xleftarrow{\hspace{1cm}}$ -1 , $(-\infty, -1]$ **43.** $\xleftarrow{\hspace{1cm}}$ 0 , $(0, \infty)$

45. $\xleftarrow{\hspace{1cm}}$ -2 , $(-2, \infty)$ **47.** $\xleftarrow{\hspace{1cm}}$ -2 , $(-\infty, -2]$ **49.** $\xleftarrow{\hspace{1cm}}$ -8 , $(-\infty, -8]$

51. $\xleftarrow{\hspace{1cm}}$ $-\frac{3}{5}$, $\left[-\frac{3}{5}, \infty\right)$ **53.** $\xleftarrow{\hspace{1cm}}$ -9 , $[-9, \infty)$ **55.** $\xleftarrow{\hspace{1cm}}$ 38 , $(38, \infty)$

57. $\xleftarrow{\hspace{1cm}}$ 0 , $[0, \infty)$ **59.** $\xleftarrow{\hspace{1cm}}$ -5 , $(-\infty, -5]$ **61.** $\xleftarrow{\hspace{1cm}}$ $\frac{1}{4}$, $\left(-\infty, \frac{1}{4}\right)$

63. $\xleftarrow{\hspace{1cm}}$ -1 , $(-\infty, -1]$ **65.** $\xleftarrow{\hspace{1cm}}$ $-\frac{79}{3}$, $\left[-\frac{79}{3}, \infty\right)$ **67.** $\xleftarrow{\hspace{1cm}}$ -15 , $(-\infty, -15)$

69. $x < 200$ recommended; $200 \leq x \leq 240$ borderline; $x > 240$ high   **71.** $x > -10$ **73.** 86 people   **75.** 35 cm

77. 10%   **79.** 193   **81.** 8   **83.** 1   **85.** $\dfrac{16}{49}$ **87.** 51 million   **89.** 1996   **91.** $\xleftarrow{\hspace{1cm}}$ 1 , $x > 1$

93. 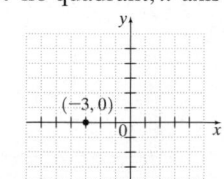, $x < \dfrac{5}{8}$ **95.** , $x \le 0$

Chapter 2 Review   **1.** $6x$ **3.** $4x - 2$ **5.** $3n - 18$ **7.** $-6x + 7$ **9.** $3x - 7$ **11.** 4   **13.** -6 **15.** -9 **17.** 5; 5
19. $10 - x$ **21.** $(175 - x)°$ **23.** 4   **25.** -1 **27.** -1 **29.** 6   **31.** 2   **33.** no solution   **35.** $\dfrac{3}{4}$ **37.** 20   **39.** 0   **41.** $\dfrac{20}{7}$
43. $\dfrac{23}{7}$ **45.** 102   **47.** 3   **49.** 1052 ft   **51.** 307; 955   **53.** $w = 9$ **55.** $m = \dfrac{y - b}{x}$ **57.** $x = \dfrac{2y - 7}{5}$ **59.** $\pi = \dfrac{C}{D}$
61. 15 m   **63.** 1 hr and 20 min   **65.** 93.5   **67.** 70%   **69.** 1280   **71.** 6%   **73.** 120 travelers   **75.** 14.3%
77. , $(-\infty, -4]$ **79.** , $(-17, \infty)$ **81.** , $(-\infty, 4]$
83. , $(-\infty, 1)$ **85.** , $(2, \infty)$ **87.** score must be less than 83

Chapter 2 Test   **1.** $y - 10$ **2.** $5.9x + 1.2$ **3.** $-2x + 10$ **4.** $-15y + 1$ **5.** -5 **6.** 8   **7.** $\dfrac{7}{10}$ **8.** 0   **9.** 27   **10.** $-\dfrac{19}{6}$
11. 3   **12.** $\dfrac{3}{11}$ **13.** 0.25   **14.** $\dfrac{25}{7}$ **15.** 21   **16.** 7 gal   **17.** \$8500 @ 10%; \$17,000 @ 12%   **18.** $2\dfrac{1}{2}$ hr   **19.** $x = 6$
20. $h = \dfrac{V}{\pi r^2}$ **21.** $y = \dfrac{3x - 10}{4}$ **22.** , $x < -2$ **23.** , $x < 4$
24. , $x > \dfrac{2}{5}$ **25.** , $(-\infty, 4]$ **26.** 81.3%   **27.** \$5.9314 billion   **28.** 24.12°   **29.** 17%
30. 13%   **31.** about 5980

Chapter 2 Cumulative Review   **1. a.** 11, 112   **b.** 0, 11, 112   **c.** $-3, -2, 0, 11, 112$ **d.** $-3, -2, 0, \dfrac{1}{4}, 11, 112$ **e.** $\sqrt{2}$
f. all numbers in the given set; Sec. 1.2, Ex. 5   **2. a.** 4   **b.** 5   **c.** 0; Sec. 1.2, Ex. 7
3. a. $2 \cdot 2 \cdot 2 \cdot 5$ **b.** $3 \cdot 3 \cdot 7$; Sec. 1.3, Ex. 1   **4.** $\dfrac{8}{20}$; Sec. 1.3, Ex. 6   **5.** 54; Sec. 1.4, Ex. 4   **6.** 2 is a solution; Sec. 1.4, Ex. 7
7. -3; Sec. 1.5, Ex. 2   **8.** 2; Sec. 1.5, Ex. 4   **9. a.** 10   **b.** $\dfrac{1}{2}$ **c.** $2x$ **d.** -6; Sec. 1.5, Ex. 10   **10. a.** 9.9   **b.** $-\dfrac{4}{5}$
c. $\dfrac{2}{15}$; Sec. 1.6, Ex. 2   **11. a.** 52°   **b.** 118°; Sec. 1.6, Ex. 8   **12. a.** -0.06 **b.** $-\dfrac{7}{15}$; Sec. 1.7, Ex. 3   **13. a.** 6   **b.** -12
c. $-\dfrac{8}{15}$; Sec. 1.7, Ex. 7   **14. a.** $5 + x$ **b.** $x \cdot 3$; Sec. 1.8, Ex. 1   **15. a.** $8(2 + x)$ **b.** $7(s + t)$; Sec. 1.8, Ex. 5
16. $-2x - 1$; Sec. 2.1, Ex. 7   **17.** 17; Sec. 2.2, Ex. 1   **18.** 8; Sec. 2.2, Ex. 4   **19.** 140; Sec. 2.2, Ex. 7   **20.** 2; Sec. 2.3, Ex. 1
21. 10; Sec. 2.4, Ex. 1   **22.** $\dfrac{V}{wh} = l$; Sec. 2.5, Ex. 4   **23. a.** 0.35   **b.** 0.895   **c.** 1.5; Sec. 2.6, Ex. 1
24. 87.5%; Sec. 2.6, Ex. 5   **25.** ; $(-\infty, -10]$; Sec. 2.7, Ex. 4

■ CHAPTER 3   GRAPHS AND FUNCTIONS

Mental Math   **1.** $(5, 2)$ **3.** $(3, -1)$ **5.** $(-5, -2)$ **7.** $(-1, 0)$
Exercise Set 3.1   **1.** quadrant I   **3.** no quadrant, x-axis   **5.** quadrant IV   **7.** no quadrant, x-axis

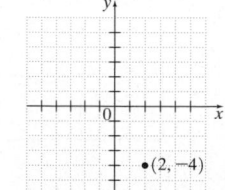
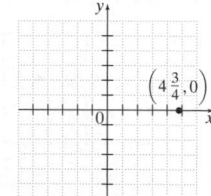

9. no quadrant, origin   **11.** no quadrant, y-axis

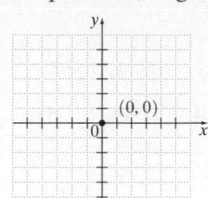

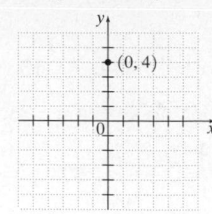

13. $A(0, 0)$; $B\left(3\dfrac{1}{2}, 0\right)$; $C(3, 2)$; $D(-1, 3)$; $E(-2, -2)$; $F(0, -1)$; $G(2, -1)$

15. 26 units

17. a. $(1991, 1.14)$, $(1992, 1.13)$, $(1993, 1.11)$, $(1994, 1.11)$, $(1995, 1.15)$, $(1996, 1.23)$, $(1997, 1.23)$, $(1998, 1.06)$, $(1999, 1.17)$

b.

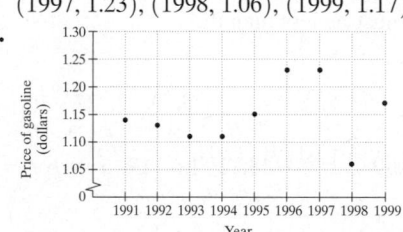

19. a. $(1994, 578)$, $(1995, 613)$, $(1996, 654)$, $(1997, 675)$, $(1998, 717)$

b.

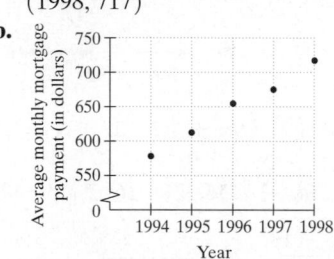

c. Average monthly mortgage payment increases each year.

21. yes; no; yes   **23.** no; yes; yes   **25.** no; yes; yes   **27.** yes; no   **29.** no; no   **31.** yes; yes   **33.** yes; yes
35. $(-4, -2)$, $(4, 0)$ **37.** $(0, 9)$, $(3, 0)$ **39.** $(11, -7)$; Answers may vary; Ex. $(2, -7)$

41. $(0, 2)$, $(6, 0)$, $(3, 1)$ **43.** $(0, -12)$, $(5, -2)$, $(-3, -18)$ **45.** $\left(0, \dfrac{5}{7}\right)$, $\left(\dfrac{5}{2}, 0\right)$, $(-1, 1)$

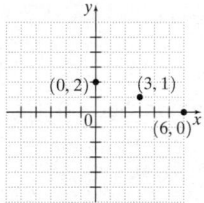

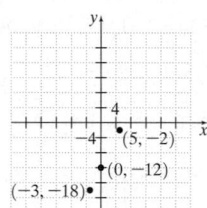

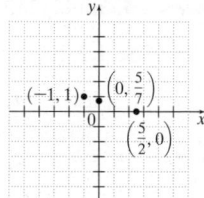

47. $(3, 0)$, $(3, -0.5)$, $\left(3, \dfrac{1}{4}\right)$ **49.** $(0, 0)$, $(-5, 1)$, $(10, -2)$ **51.** Answers may vary.

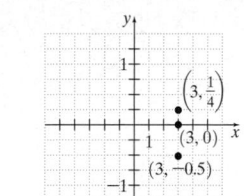

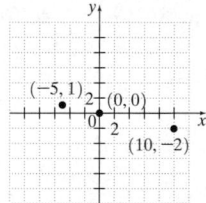

53. a. 13,000; 21,000; 29,000

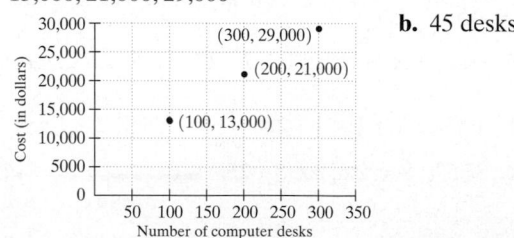

b. 45 desks

55. a. 29.219; 45.599; 54.699   **b.** 1977
57. In 1995, there were 670 Target stores.   **59.** year 6: 66 stores; year 7: 60 stores; year 8: 55 stores   **61.** $a = b$

63. quadrant IV   **65.** quadrants II or III   **67.** $y = 5 - x$ **69.** $y = -\dfrac{1}{2}x + \dfrac{5}{4}$ **71.** $y = -2x$ **73.** $y = \dfrac{1}{3}x - 2$

Graphing Calculator Explorations

1.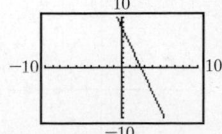

3.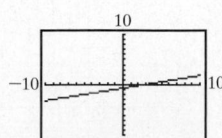

5.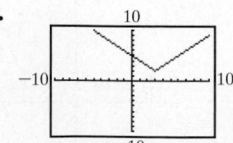

7.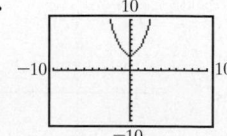

Exercise Set 3.2   **1.** yes   **3.** yes   **5.** no   **7.** yes

9. 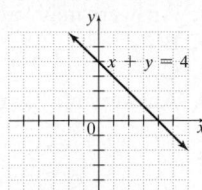   **11.** 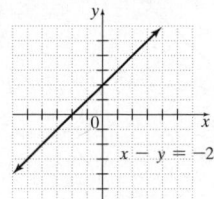   **13.** 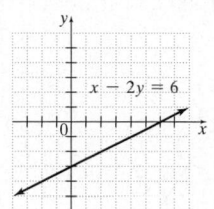   **15.**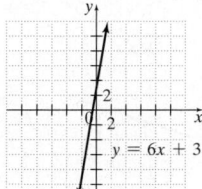

17. 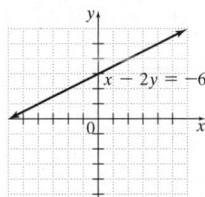   **19.** 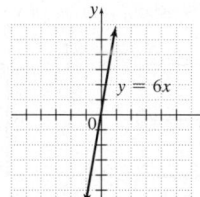   **21.** 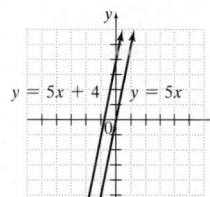   **23.**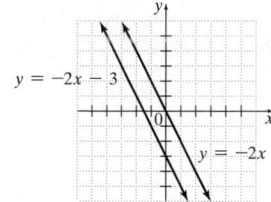

25. 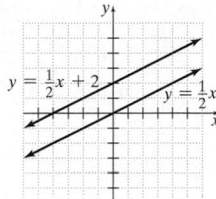   **27.** c   **29.** d   **31.** 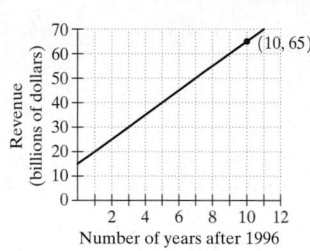 ; 65 billion dollars

33. 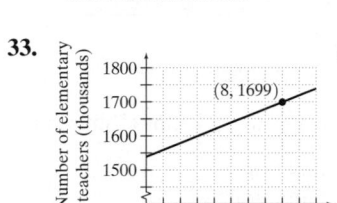   ; 1699 thousand teachers   **35.** linear 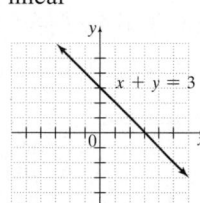   **37.** linear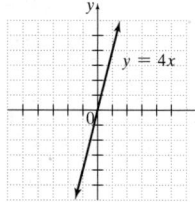

39. linear 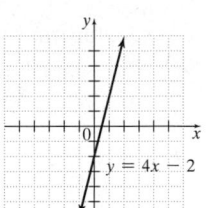   **41.** not linear 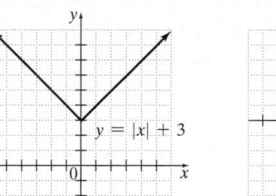   **43.** linear 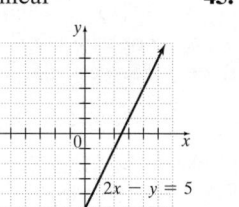   **45.** not linear 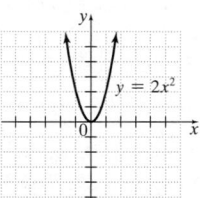   **47.** not linear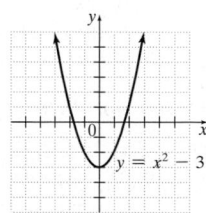

49. linear 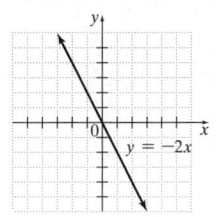   **51.** linear 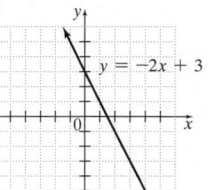   **53.** not linear 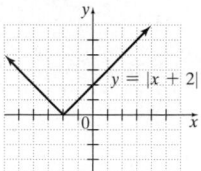   **55.** not linear 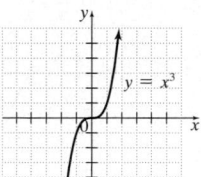   **57.** not linear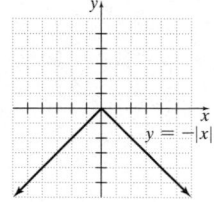

59. linear

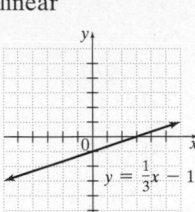

61. linear

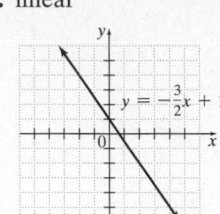

63. a.

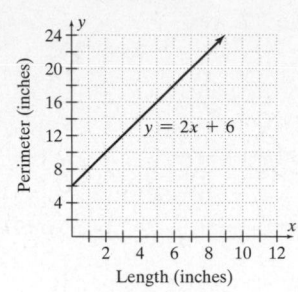

b. 14 in.

65. 1991
67. Answers may vary.
69. B
71. C
73. Answers may vary.

75. $y = 3x + 5$

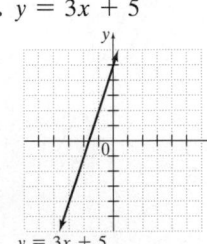

77. $y = x^2 + 2$

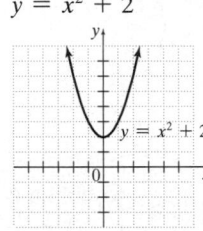

79.

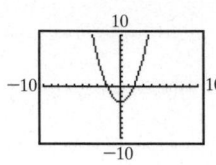

81.

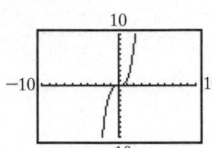

83. $(4, -1)$ **85.** -5 **87.** $-\dfrac{1}{10}$

Graphing Calculator Explorations   **1.**

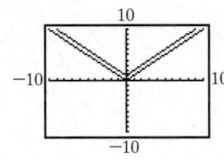

3.

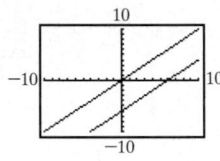

5.

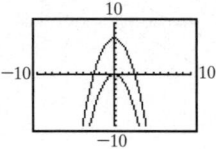

Exercise Set 3.3   **1.** domain: $\{-1, 0, -2, 5\}$; range: $\{7, 6, 2\}$; function   **3.** domain: $\{-2, 6, -7\}$; range: $\{4, -3, -8\}$; not a function   **5.** domain: $\{1\}$; range: $\{1, 2, 3, 4\}$; not a function   **7.** domain: $\left\{\dfrac{3}{2}, 0\right\}$; range: $\left\{\dfrac{1}{2}, -7, \dfrac{4}{5}\right\}$; not a function

9. domain: $\{-3, 0, 3\}$; range: $\{-3, 0, 3\}$; function   **11.** domain: $\{-1, 1, 2, 3\}$; range: $\{2, 1\}$; function
13. domain: {Colorado, Alaska, Delaware, Illinois, Connecticut, Texas}; range: $\{6, 1, 20, 30\}$; function
15. domain: $\{32°, 104°, 212°, 50°\}$; range: $\{0°, 40°, 10°, 100°\}$; function   **17.** domain: $\{2, -1, 5, 100\}$; range: $\{0\}$; function
19. function   **21.** Answers may vary.   **23.** function   **25.** not a function   **27.** function   **29.** 5:20 A.M.
31. Answers may vary.   **33.** domain: $[0, \infty)$; range: $(-\infty, \infty)$; not a function   **35.** domain: $[-1, 1]$; range: $(-\infty, \infty)$; not a function   **37.** domain: $(-\infty, \infty)$; range: $(-\infty, -3] \cup [3, \infty)$; not a function   **39.** domain: $[2, 7]$; range $[1, 6]$; not a function   **41.** domain: $\{-2\}$; range: $(-\infty, \infty)$; not a function   **43.** domain: $(-\infty, \infty)$; range: $(-\infty, 3]$; function
45. Answers may vary.   **47.** yes   **49.** no   **51.** yes   **53.** yes   **55.** 15   **57.** 38   **59.** 7   **61.** 3   **63. a.** 0   **b.** 1   **c.** -1
65. a. 246   **b.** 6   **c.** $\dfrac{9}{2}$ **67. a.** -5 **b.** -5 **c.** -5 **69. a.** 5.1   **b.** 15.5   **c.** 9.533   **71.** $(1, -10)$ **73.** $f(-1) = -2$
75. $-4, 0$ **77.** infinite number   **79. a.** \$13.4 billion   **b.** \$13.672 billion   **81.** \$34.374 billion   **83.** $f(x) = x + 7$
85. 25π sq. cm   **87.** 2744 cu. in.   **89.** 166.38 cm   **91.** 163.2 mg
93. a. 91.4; The per capita consumption of poultry was 91.4 pounds in 1997.   **b.** 106.7 lb

95. $5, -5, 6$

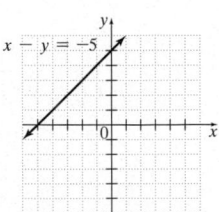

97. $2, \dfrac{8}{7}, \dfrac{12}{7}$

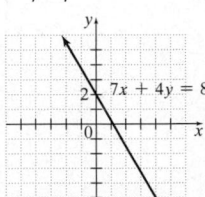

99. $0, 0, -6$

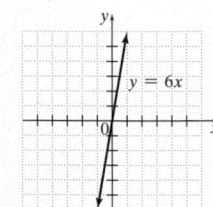

101. yes; 170 m

103. a. $-3s + 12$
b. $-3r + 12$
105. a. 132
b. $a^2 - 12$

Graphing Calculator Explorations

1. $y = \dfrac{x}{3.5}$

3. $y = -\dfrac{5.78}{2.31}x + \dfrac{10.98}{2.31}$

5. $y = |x| + 3.78$

7. $y = 5.6x^2 + 7.7x + 1.5$

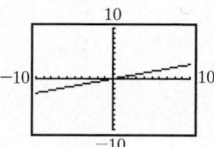 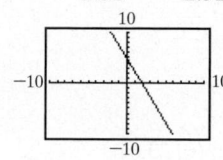 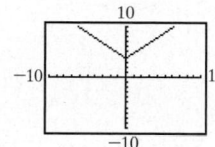 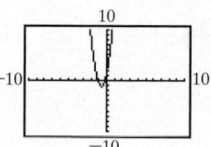

Exercise Set 3.4   1. 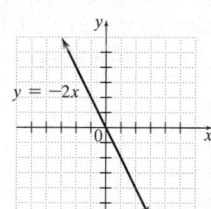   **3.** 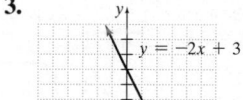   **5.** 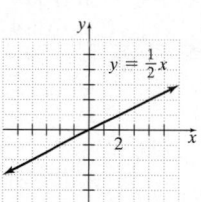   **7.**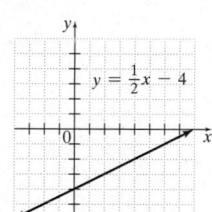

9. C   **11.** D   **13.** 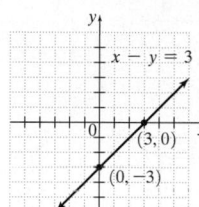   **15.** 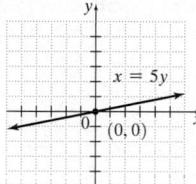   **17.** 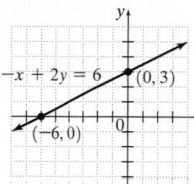   **19.**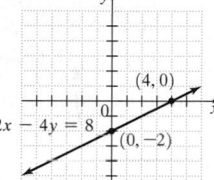

21. Answers may vary.   **23.** 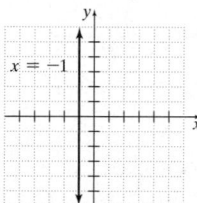   **25.** 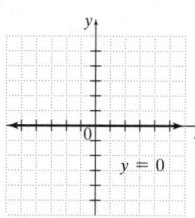   **27.**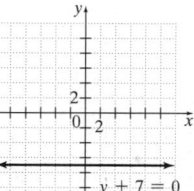

29. C   **31.** A   **33.** The vertical line $x = 0$ has y-intercepts.

35. 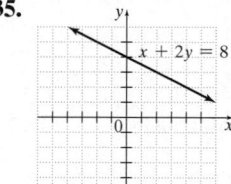   **37.** 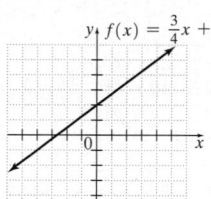   **39.** 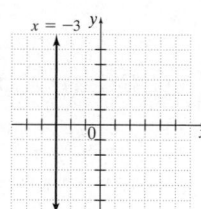   **41.**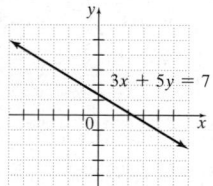

43. 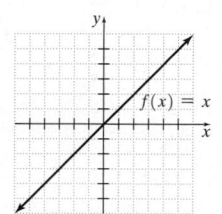   **45.** 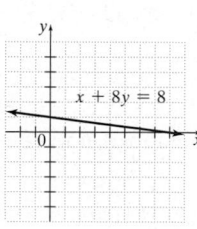   **47.** 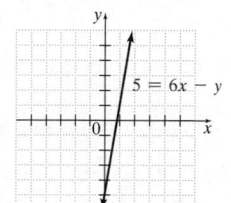   **49.**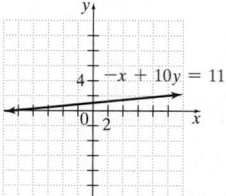

51. 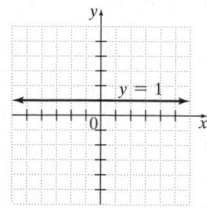   **53.** 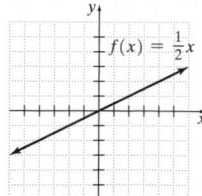   **55.** 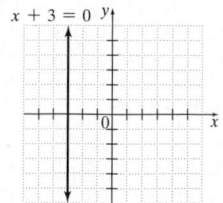   **57.**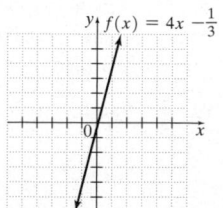

59.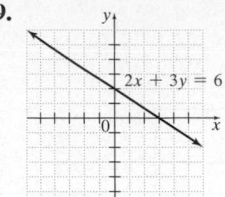

61. a. $(0, 500)$; If no tables are produced, 500 chairs can be produced
b. $(750, 0)$; If no chairs are produced, 750 tables can be produced
c. 466 chairs

63. a. $64
b.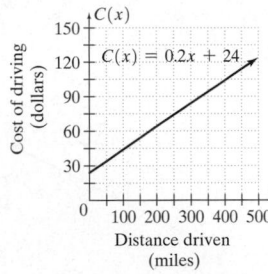

c. The line moves upward from left to right.

65. a. $2243.20
b. 2007
c. Answers may vary.

69.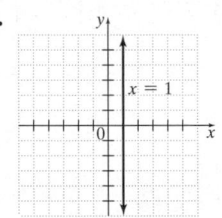

67. a. $(0, 23.6)$
b. In 1995 (0 years after 1995), U.S. farm expenses for livestock feed were 23.6 billion dollars.

71. 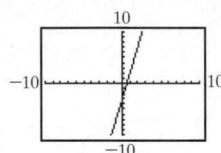   **73.**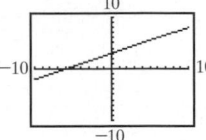

75. $\dfrac{3}{2}$ **77.** 6   **79.** $-\dfrac{6}{5}$

Graphing Calculator Explorations   **1.** 18.4   **3.** -1.5

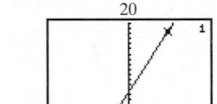   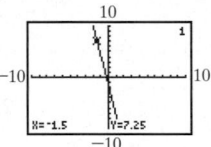

5. $14.0; 4.2, -9.4$

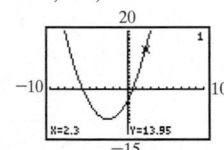   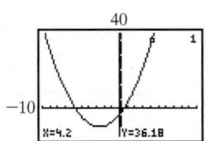   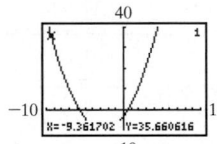

Mental Math   **1.** upward   **3.** horizontally

Exercise Set 3.5   **1.** $\dfrac{9}{5}$ **3.** $-\dfrac{7}{2}$ **5.** $-\dfrac{5}{6}$ **7.** $\dfrac{1}{3}$ **9.** $-\dfrac{4}{3}$ **11.** 0   **13.** undefined   **15.** 2   **17.** -1 **19.** $\dfrac{3}{5}$ **21.** 12.5%
23. 40%   **25.** 0.02   **27.** Every 1 year, there are/should be 15 million more internet users.   **29.** It costs $0.36 per 1 mile to own and operate a compact car.   **31.** l_2 **33.** l_2 **35.** l_2 **37. a.** $l_1: -2, l_2: -1, l_3: -\dfrac{2}{3}$ **b.** lesser   **39.** $m = -2, b = 6$

41. $m = 5, b = 10$ **43.** $m = -\dfrac{3}{4}, b = -\dfrac{3}{2}$ **45.** $m = -\dfrac{1}{4}, b = 0$ **47.** D   **49.** C   **51.** 0   **53.** undefined   **55.** 0

57. Answers may vary.   **59.** $m = 1, b = 2$ **61.** $m = \dfrac{4}{7}, b = -4$ **63.** $m = \dfrac{1}{2}, b = \dfrac{7}{2}$

65. slope is undefined, no y-intercept   **67.** $m = \dfrac{1}{7}, b = 0$ **69.** slope is undefined, no y-intercept   **71.** $m = 0, b = -\dfrac{11}{2}$

73. parallel   **75.** perpendicular   **77.** neither   **79.** Answers may vary.   **81. a.** $28,559.40   **b.** $m = 1054.7$; The annual average income increases $1054.70 every year   **c.** $b = 23,285.9$; At year $x = 0$, or 1991, the annual average income was $23,285.90.   **83. a.** $m = 7.6, b = 113$ **b.** The number of people employed as paralegals increases 7.6 thousand for every 1 year.   **c.** There were 113 thousand paralegals employed in 1996.
85. a. The yearly cost of tuition increases $72.90 every 1 year.   **b.** The yearly cost of tuition in 1990 was $785.20.

87. 1   **89.** -1 **91.** $\dfrac{3}{4}$ **93. a.** $(6, 20)$ **b.** $(10, 13)$ **c.** $-\dfrac{7}{4}$ or -1.75 yd per sec   **d.** $\dfrac{3}{2}$ or 1.5 yd per sec

95. 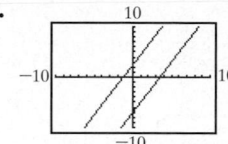 **97. a.** 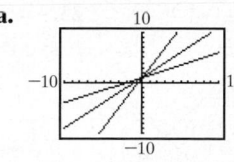 **b.** 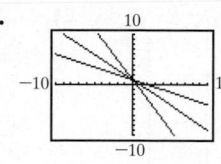 **c.** true   **99.** $y = -3x - 30$
101. $y = -8x - 23$

Mental Math   **1.** $m = -4, b = 12$ **3.** $m = 5, b = 0$ **5.** $m = \frac{1}{2}, b = 6$ **7.** parallel   **9.** neither

Exercise Set 3.6   **1.** $y = -x + 1$ **3.** $y = 2x + \frac{3}{4}$ **5.** $y = \frac{2}{7}x$

7. 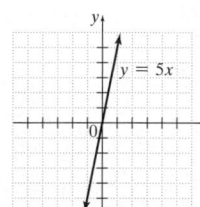   **9.** 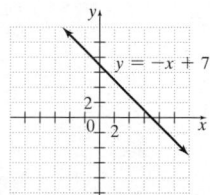   **11.**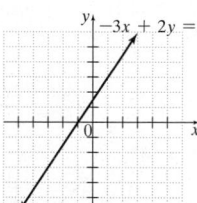

13. $y = 3x - 1$ **15.** $y = -2x - 1$
17. $y = \frac{1}{2}x + 5$ **19.** $y = -\frac{9}{10}x - \frac{27}{10}$
21. $2x + y = 3$ **23.** $2x - 3y = -7$
25. $f(x) = 3x - 6$ **27.** $f(x) = -2x + 1$
29. $f(x) = -\frac{1}{2}x - 5$ **31.** $f(x) = \frac{1}{3}x - 7$

33. Answers may vary.   **35.** -2 **37.** 2   **39.** -2 **41.** $y = -4$ **43.** $x = 4$ **45.** $y = 5$ **47.** $f(x) = 4x - 4$
49. $f(x) = -3x + 1$ **51.** $f(x) = -\frac{3}{2}x - 6$ **53.** $2x - y = -7$ **55.** $f(x) = -x + 7$ **57.** $x + 2y = 22$
59. $2x + 7y = -42$ **61.** $4x + 3y = -20$ **63.** $x = -2$ **65.** $x + 2y = 2$ **67.** $y = 12$ **69.** $8x - y = 47$ **71.** $x = 5$
73. $f(x) = -\frac{3}{8}x - \frac{29}{4}$ **75. a.** $y = 740x + 3280$ **b.** 9940 vehicles   **77. a.** $P(x) = 12{,}000x + 18{,}000$ **b.** $102,000
c. end of the ninth yr   **79. a.** $y = 4625x + 109{,}900$ **b.** $174,650   **81. a.** $y = 16.6x + 225$ **b.** 357.8 thousand people
83. 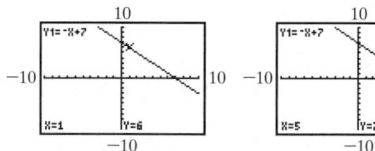   **85.** 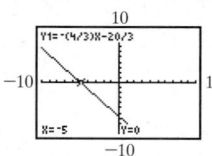   **87.** 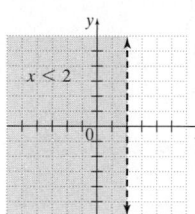, $(-\infty, 14]$

89. $\left[\frac{7}{2}, \infty\right)$

91. $\left(-\infty, -\frac{1}{4}\right)$ **93.** $-4x + y = 4$ **95.** $2x + y = -23$ **97.** $3x - 2y = -13$

Exercise Set 3.7   **1.** 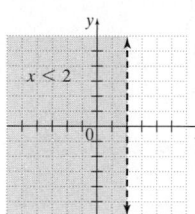   **3.** 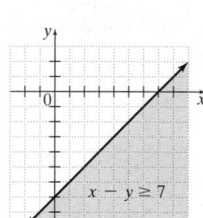   **5.** 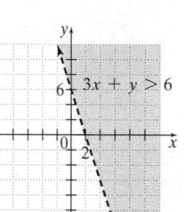   **7.**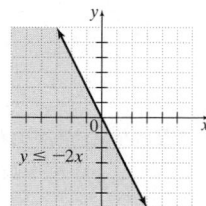

9. 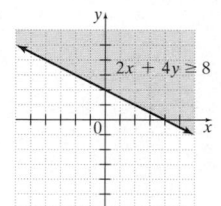   **11.** 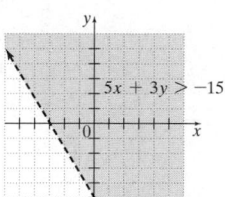   **13.** Answers may vary.   **15.**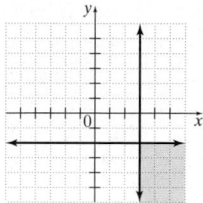

17.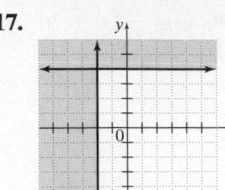

19.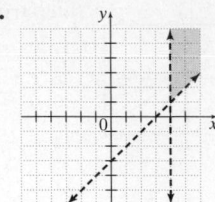

21.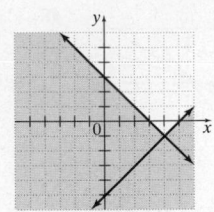

23.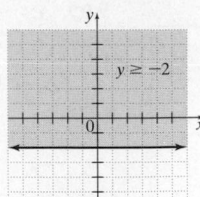

25.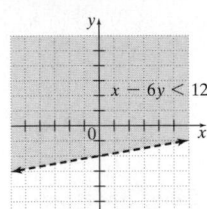

27.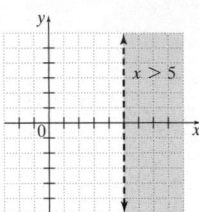

29.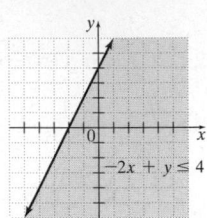

31.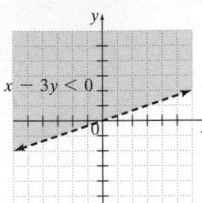

33.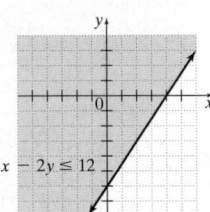

35.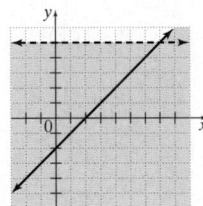

37.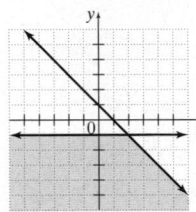

39.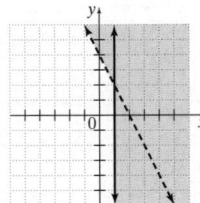

41.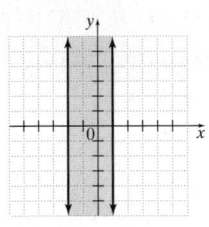

43.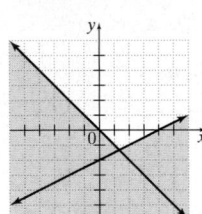

45.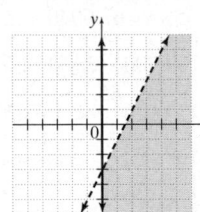

47. D   **49.** A
51. $x \geq 2$ **53.** $y \leq -3$
55. $y > 4$ **57.** $x < 1$

59. $0 \leq x \leq 20$ and $y \geq 10$.

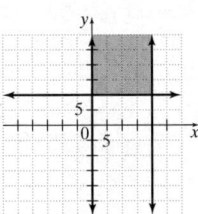

61.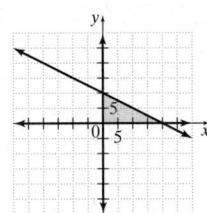

63. 9 **65.** 25 **67.** -16 **69.** $\dfrac{4}{49}$

71. domain: $(-\infty, -2] \cup [2, \infty)$;
range: $(-\infty, \infty)$; not a function

Chapter 3 Review   **1.** A: Quadrant IV; B: Quadrant II;
C: no quadrant, y-axis; D: Quadrant III

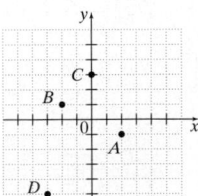

3. no, yes

5. a. $(8.00, 1)$, $(7.50, 10)$, $(6.50, 25)$,
$(5.00, 50)$, $(2.00, 100)$

b.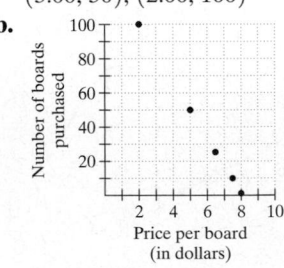

7. linear

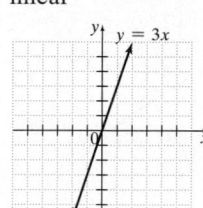

9. linear

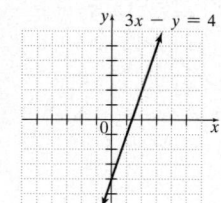

11. nonlinear

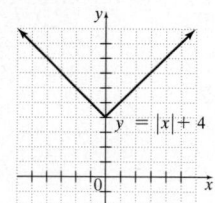

13. linear

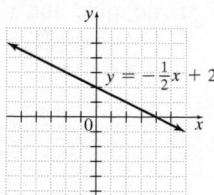

15. linear

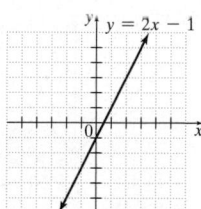

17. linear

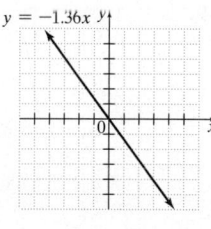

19. domain: $\left\{-\dfrac{1}{2}, 6, 0, 25\right\}$; range: $\left\{\dfrac{3}{4} \text{ or } 0.75, -12, 25\right\}$; function

21. domain: $\{2, 4, 6, 8\}$; range: $\{2, 4, 5, 6\}$; not a function

23. domain: $(-\infty, \infty)$; range: $(-\infty, -1] \cup [1, \infty)$; not a function

25. domain: $(-\infty, \infty)$; range: $\{4\}$; function

27. -3 **29.** 18   **31.** -3 **33.** 381 lb   **35.** 0   **37.** $-2, 4$

39.

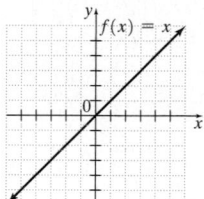

41.

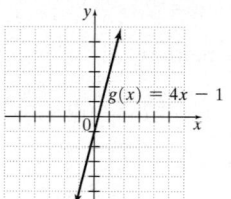

43. A   **45.** D   **47.**

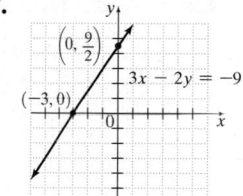

49.

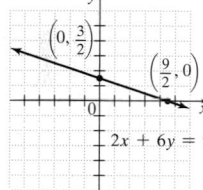

51.

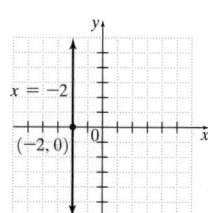

53.

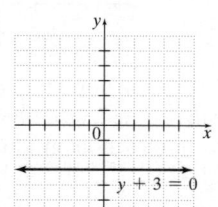

55. -3 **57.** $\dfrac{5}{2}$ **59.** $m = \dfrac{2}{5}, b = -\dfrac{4}{3}$ **61.** 0

59. Every 1 year, 1.24 million more persons have a bachelors degree or higher.

65. l_2

67. a. $m = 0.3$; The cost increases by \$0.30 for each additional mile driven.

b. $b = 42$; The cost for 0 miles driven is \$42.

69. parallel   **71.**

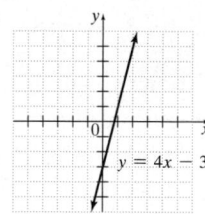

73.

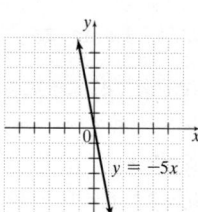

75. $x = -2$ **77.** $y = 5$ **79.** $2x - y = 12$

81. $11x + y = -52$ **83.** $y = -5$ **85.** $f(x) = -x - 2$

87. $f(x) = -\dfrac{3}{2}x - 8$ **89.** $f(x) = -\dfrac{3}{2}x - 1$

91. a. $y = \dfrac{17}{22}x + 43$ **b.** 52 million

93.

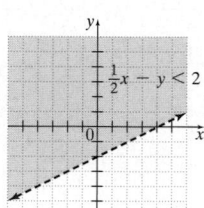

95.

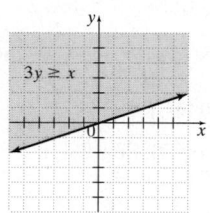

97.

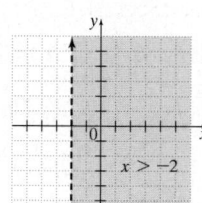

99.

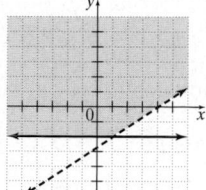

Chapter 3 Test   **1.** A: Quadrant IV;   **2.** $(-6, -3)$ **3.**   **4.**
B: no quadrant, x-axis;
C: Quadrant II

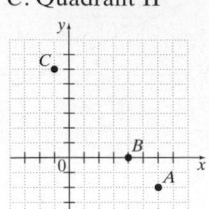

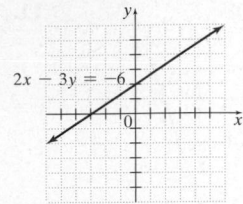

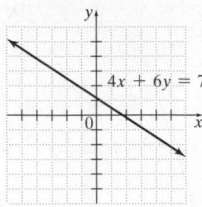

5.   **6.**   **7.** $-\dfrac{3}{2}$ **9.**   **10.**

8. $m = -\dfrac{1}{4}, b = \dfrac{2}{3}$

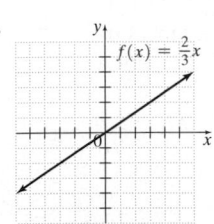

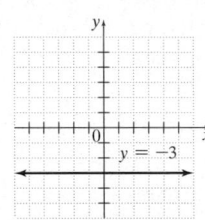

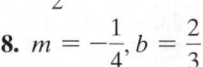

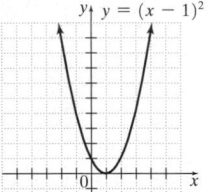

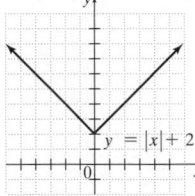

11. $y = -8$ **12.** $x = -4$ **13.** $y = -2$ **14.** $3x + y = 11$ **15.** $5x - y = 2$

16. $f(x) = -\dfrac{1}{2}x$ **17.** $f(x) = -\dfrac{1}{3}x + \dfrac{5}{3}$ **18.** $f(x) = -\dfrac{1}{2}x - \dfrac{1}{2}$ **19.** neither

20.   **21.**   **22.**   **23.**

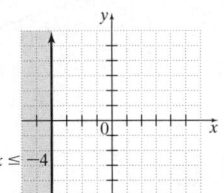

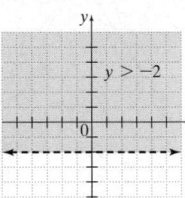

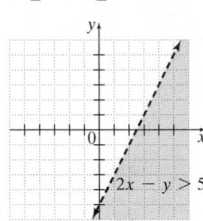

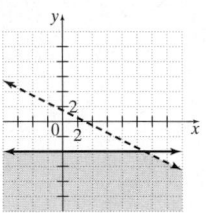

24. domain: $(-\infty, \infty)$; range: $\{5\}$; function   **25.** domain: $\{-2\}$; range: $(-\infty, \infty)$; not a function
26. domain: $(-\infty, \infty)$; range: $[0, \infty)$; function   **27.** domain: $(-\infty, \infty)$; range: $(-\infty, \infty)$; function
28. a. \$22,892 **b.** \$28,016 **c.** 2008 **d.** The average yearly earnings for high school graduates increases \$732 per year.
e. The average yearly earnings for a high school graduate in 1996 was \$21,428.
29. Every 1 year, 71 million more movie tickets are sold.

Chapter 3 Cumulative Review   **1. a.** $<$ **b.** $>$ **c.** $>$; Sec. 1.2, Ex. 1   **2.** $\dfrac{2}{39}$; Sec. 1.3, Ex. 3   **3.** $\dfrac{8}{3}$; Sec. 1.4, Ex. 3

4. a. -19 **b.** 30   **c.** -0.5 **d.** $-\dfrac{4}{5}$ **e.** 6.7   **f.** $\dfrac{1}{40}$; Sec. 1.5, Ex. 6   **5.** -6; Sec. 1.6, Ex. 4   **6. a.** -6 **b.** -24

c. $\dfrac{3}{4}$; Sec. 1.7, Ex. 10   **7. a.** $22 + x$ **b.** $-21x$; Sec. 1.8, Ex. 3   **8. a.** -3 **b.** 22   **c.** 1   **d.** -1 **e.** $\dfrac{1}{7}$; Sec. 2.1, Ex. 1

9. -1.6; Sec. 2.2, Ex. 2   **10.** 6; Sec. 2.2, Ex. 5   **11.** $3x + 3$; Sec. 2.2, Ex. 10   **12.** 0; Sec. 2.3, Ex. 4

13. 46 Democratic senators; 54 Republican senators; Sec. 2.4, Ex. 3   **14.** 40 ft; Sec. 2.5, Ex. 2   **15.** $\dfrac{y - b}{m} = x$; Sec. 2.5, Ex. 5

16. a. 73% **b.** 139% **c.** 25%; Sec. 2.6, Ex. 2   **17.** 800; Sec. 2.6, Ex. 6   **18.**  ; Sec. 2.7, Ex. 1

19. a. solution   **b.** not a solution   **c.** solution; Sec. 3.1, Ex. 3   **20. a.** linear   **b.** linear   **c.** not linear

d. linear; Sec. 3.2, Ex. 1   **21.** yes; Sec. 3.3, Ex. 3   **22. a.** $\left(0, \dfrac{3}{7}\right)$ **b.** $(0, -3.2)$; Sec. 3.4, Ex. 3   **23.** $\dfrac{2}{3}$; Sec. 3.5, Ex.3

24. $y = \dfrac{1}{4}x - 3$; Sec. 3.6, Ex. 1   **25.**   ; Sec. 3.7, Ex. 1

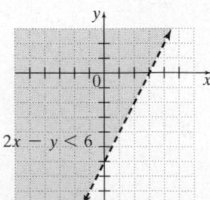

■ CHAPTER 4   SOLVING SYSTEMS OF LINEAR EQUATIONS

Calculator Explorations   **1.** $(0.37, 0.23)$ **3.** $(0.03, -1.89)$

Mental Math   **1.** 1 solution, $(-1, 3)$ **3.** infinite number of solutions   **5.** no solution   **7.** 1 solution, $(3, 2)$

Exercise Set 4.1   **1. a.** no   **b.** yes   **c.** no   **3. a.** no   **b.** yes   **c.** no   **5. a.** yes   **b.** yes   **c.** yes   **7.** Answers may vary.

9. $(2, 3)$; consistent; independent

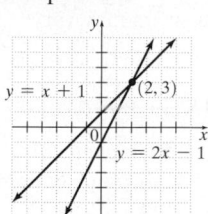

11. $(1, -2)$; consistent; independent

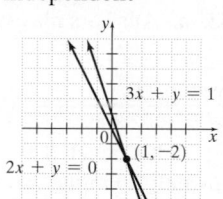

13. $(-2, 1)$; consistent; independent

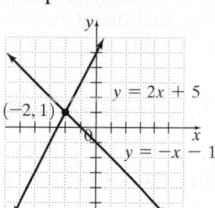

15. $(4, 2)$; consistent; independent

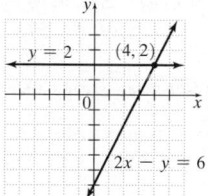

17. no solution; inconsistent; independent

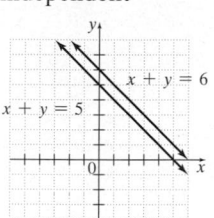

19. infinite number of solutions; consistent; dependent

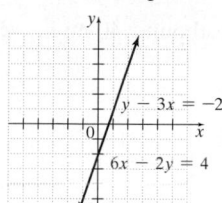

21. $(0, -1)$; consistent; independent

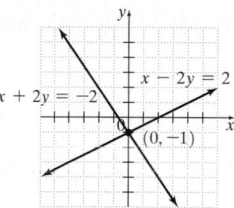

23. $(4, -3)$; consistent; independent

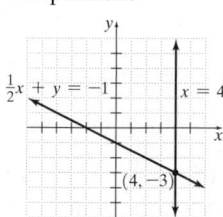

25. $(-5, -7)$; consistent; independent

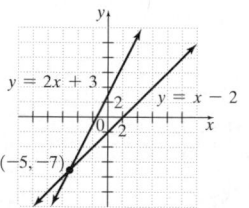

27. $(5, 2)$; consistent; independent

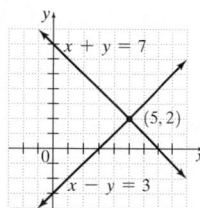

29. Answers may vary.
31. intersecting, one solution
33. parallel, no solution
35. identical lines, infinite number of solutions
37. intersecting, one solution
39. intersecting, one solution
41. identical lines, infinite number of solutions

43. parallel, no solution
45. Answers may vary.
47. 1984, 1988

49. a. $(4, 9)$ **b.**

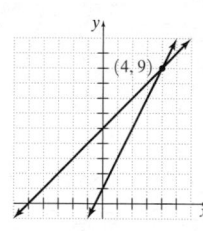

; yes   **51.** -1 **53.** 3   **55.** -7

Exercise Set 4.2   **1.** $(2, 1)$ **3.** $(-3, 9)$ **5.** $(4, 2)$ **7.** $(10, 5)$ **9.** $(2, 7)$ **11.** $(-2, 4)$ **13.** $(-2, -1)$ **15.** no solution

17. infinite number of solutions   **19.** $(3, -1)$ **21.** $(3, 5)$ **23.** $\left(\dfrac{2}{3}, -\dfrac{1}{3}\right)$ **25.** $(-1, -4)$ **27.** $(-6, 2)$ **29.** $(2, 1)$

31. no solution   **33.** $\left(-\dfrac{1}{5}, \dfrac{43}{5}\right)$ **35.** Answers may vary.   **37.** $(1, -3)$

39. a. $(21, 18)$ **b.** Answers may vary.   **c.**

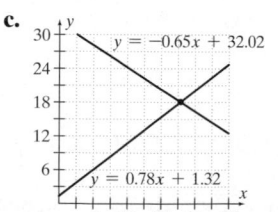

41. $(-2.6, 1.3)$
43. $(3.28, 2.11)$
45. $-6x - 4y = -12$
47. $-12x + 3y = 9$
49. $5n$
51. $-15b$

Exercise Set 4.3   **1.** $(1, 2)$ **3.** $(2, -3)$ **5.** $(6, 0)$ **7.** no solution   **9.** $\left(2, -\dfrac{1}{2}\right)$ **11.** $(6, -2)$

13. infinite number of solutions   **15.** $(-2, -5)$ **17.** $(5, -2)$ **19.** $(-7, 5)$ **21.** $\left(\dfrac{12}{11}, -\dfrac{4}{11}\right)$ **23.** no solution   **25.** $\left(\dfrac{3}{2}, 3\right)$

27. $(1, 6)$ **29.** infinite number of solutions   **31.** $(-2, 0)$ **33.** Answers may vary.   **35.** $(2, 5)$ **37.** $(-3, 2)$ **39.** $(0, 3)$

41. $(5, 7)$ **43.** $\left(\dfrac{1}{3}, 1\right)$ **45.** infinite number of solutions   **47.** $(-8.9, 10.6)$ **49. a.** $(12, 555)$ **b.** Answers may vary.

c. 1993 to 1997   **51. a.** $b = 15$ **b.** any real number except 15   **53. a.** Answers may vary.   **b.** Answers may vary.

55. $2x + 6 = x - 3$ **57.** $20 - 3x = 2$ **59.** $4(n + 6) = 2n$

Exercise Set 4.4   **1.** c   **3.** b   **5.** a   **7.** $\begin{cases} x + y = 15 \\ x - y = 7 \end{cases}$ **9.** $\begin{cases} x + y = 6500 \\ x = y + 800 \end{cases}$ **11.** 33 and 50   **13.** 14 and -3

15. Cooper: 686 points; Swoopes: 585 points   **17.** child's ticket: \$18; adult's ticket: \$29   **19.** quarters: 53; nickels: 27

21. IBM: \$43.51; GA Financial: \$107.76   **23.** still water: 6.5 mph; current: 2.5 mph   **25.** still air: 455 mph; wind: 65 mph

27. 12% solution: $7\dfrac{1}{2}$ oz; 4% solution: $4\dfrac{1}{2}$ oz   **29.** \$4.95 beans: 113 lb; \$2.65 beans; 87 lb   **31.** $60°, 30°$ **33.** $20°, 70°$

35. 20% solution: 10 l; 70% solution: 40 l   **37.** number sold at \$9.50: 23; number sold at \$7.50: 67

39. width: 9 ft; length: 15 ft   **41.** $4\dfrac{1}{2}$ hr   **43.** westbound: 80 mph; eastbound: 40 mph   **45.** b

47.
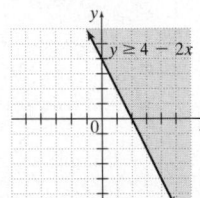

49.
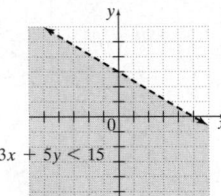

Chapter 4 Review   **1. a.** no   **b.** yes   **c.** no   **3. a.** no   **b.** no   **c.** yes

5. $(3, -1)$ **7.** $(-3, -2)$ **9.** $\left(\dfrac{1}{2}, \dfrac{1}{2}\right)$ **11.** intersecting, one solution

13. identical, infinite number of solutions

15. $(-1, 4)$ **17.** $(3, -2)$

19. no solution; inconsistent

21. $(3, 1)$ **23.** $(8, -6)$

25. $(-6, 2)$ **27.** $(3, 7)$

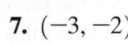
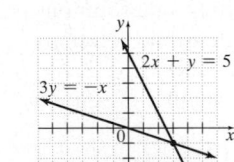
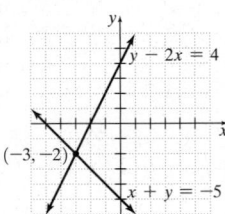
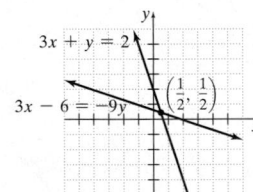

29. infinite number of solutions; dependent   **31.** $\left(2, -2\dfrac{1}{2}\right)$ **33.** $(-6, 15)$ **35.** $(-3, 1)$ **37.** -6 and 22

39. ship: 21.1 mph; current: 3.2 mph   **41.** width: 1.15 ft; length: 1.85 ft   **43.** one egg: \$0.40; one strip of bacon: \$0.65

Chapter 4 Test   **1.** false   **2.** false   **3.** true   **4.** false   **5.** no   **6.** yes

7. $(-4, 2)$ **8.** $(-4, 1)$ **9.** $\left(\dfrac{1}{2}, -2\right)$ **10.** $(4, -2)$ **11.** $\left(2, \dfrac{1}{2}\right)$ **12.** $(4, -5)$ **13.** $(7, 2)$ **14.** $(5, -2)$

15. 20 \$1.00 bills; 42 \$5.00 bills   **16.** \$1225 at 5%; \$2775 at 9%

17. Texas: 226 thousand; Missouri: 110 thousand

Chapter 4 Cumulative Review   **1. a.** $<$ **b.** $=$ **c.** $>$; Sec. 1.2, Ex. 6   **2. a.** $\dfrac{1}{2}$ **b.** 9; Sec. 1.6, Ex. 6   **3. a.** $-\dfrac{39}{5}$

b. 2; Sec. 1.7, Ex. 9   **4. a.** $5x + 7$ **b.** $-4a - 1$ **c.** $4y - 3y^2$ **d.** $7.3x - 6$; Sec. 2.1, Ex. 4   **5.** -3; Sec. 2.2, Ex. 3

6. -11; Sec. 2.2, Ex. 6   **7.** $\dfrac{16}{3}$; Sec. 2.3, Ex. 2   **8.** $\dfrac{P-2l}{2}=w$; Sec. 2.5, Ex. 6

9.
x	y
-1	-3
0	0
-3	-9
; Sec. 3.1, Ex. 5   **10. a.** 5   **b.** 1   **c.** 35   **d.** -2; Sec. 3.3, Ex. 8

11. 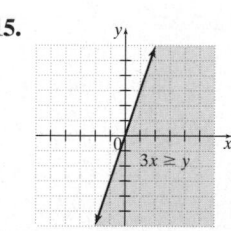 ; Sec. 3.4, Ex. 1   **12.** slope: $\dfrac{3}{4}$; y-intercept: $(0,-1)$; Sec. 3.5, Ex. 6

13. a. parallel   **b.** neither; Sec. 3.5, Ex. 10

14. $f(x)=\dfrac{5}{8}x-\dfrac{5}{2}$; Sec. 3.6, Ex. 5

15. ; Sec. 3.7, Ex. 2   **16. a.** solution   **b.** not a solution; Sec. 4.1, Ex. 1

17. one solution; Sec. 4.1, Ex. 6   **18.** $(4,2)$; Sec. 4.2, Ex. 1

19. $\left(6,\dfrac{1}{2}\right)$; Sec. 4.2, Ex. 2   **20.** $(6,1)$; Sec. 4.3, Ex. 1

21. $\left(-\dfrac{5}{4},-\dfrac{5}{2}\right)$; Sec. 4.3, Ex. 6   **22.** 29 and 8; Sec. 4.4, Ex. 1

23. 5% saline solution: 2.5 l; 25% saline solution: 7.5 l; Sec. 4.4, Ex. 4

■ CHAPTER 5   EXPONENTS AND POLYNOMIALS

Mental Math   **1.** base: 3; exponent: 2   **3.** base: -3; exponent: 6   **5.** base: 4; exponent: 2
7. base: 5; exponent: 1; base: 3; exponent: 4   **9.** base: 5; exponent: 1; base: x; exponent: 2

Exercise Set 5.1   **1.** 49   **3.** -5 **5.** -16 **7.** 16   **9.** $\dfrac{1}{27}$ **11.** 112   **13.** Answers may vary.   **15.** 4   **17.** 135   **19.** 150

21. $\dfrac{32}{5}$ **23.** 343 cu. m   **25.** volume   **27.** x^7 **29.** $(-3)^{12}$ **31.** $15y^5$ **33.** $-24z^{20}$ **35.** p^7q^7 **37.** $\dfrac{m^9}{n^9}$ **39.** $x^{10}y^{15}$

41. $\dfrac{4x^2z^2}{y^{10}}$ **43.** x^2 **45.** 4   **47.** p^6q^5 **49.** $\dfrac{y^3}{2}$ **51.** 1   **53.** -2 **55.** 2   **57.** $-\dfrac{27a^6}{b^9}$ **59.** x^{39} **61.** $\dfrac{z^{14}}{625}$ **63.** $7776m^4n^3$

65. -25 **67.** $\dfrac{1}{64}$ **69.** $81x^2y^2$ **71.** 1   **73.** 40   **75.** b^6 **77.** a^9 **79.** $-16x^7$ **81.** $64a^3$ **83.** $36x^2y^2z^6$ **85.** $\dfrac{y^{15}}{8x^{12}}$ **87.** x

89. $2x^2y$ **91.** $243x^7y^{24}$ **93.** Answers may vary.   **95.** $20x^5$ sq. ft   **97.** $25\pi y^2$ sq. cm   **99.** $27y^{12}$ cu. ft   **101.** $-2x+7$

103. $2y-10$ **105.** $-x-4$ **107.** $-x+5$ **109.** x^{9a} **111.** a^{5b} **113.** x^{5a} **115.** $x^{5a^2}y^{5ab}z^{5ac}$

Calculator Explorations   **1.** 5.31 EE 3   **3.** 6.6 EE -9 **5.** 1.5 EE 13   **7.** 8.15×10^{19}

Mental Math   **1.** $\dfrac{5}{x^2}$ **3.** y^6 **5.** $4y^3$

Exercise Set 5.2   **1.** $\dfrac{1}{64}$ **3.** $\dfrac{7}{x^3}$ **5.** -64 **7.** $\dfrac{5}{6}$ **9.** p^3 **11.** $\dfrac{q^4}{p^5}$ **13.** $\dfrac{1}{x^3}$ **15.** z^3 **17.** $\dfrac{4}{3}$ **19.** $\dfrac{1}{9}$ **21.** $-p^4$ **23.** -2

25. x^4 **27.** p^4 **29.** m^{11} **31.** r^6 **33.** $\dfrac{1}{x^{15}y^9}$ **35.** $\dfrac{1}{x^4}$ **37.** $\dfrac{1}{a^2}$ **39.** $4k^3$ **41.** $3m$ **43.** $-\dfrac{4a^5}{b}$ **45.** $-\dfrac{6x}{7y^2}$ **47.** $\dfrac{a^{30}}{b^{12}}$

49. $\dfrac{1}{x^{10}y^6}$ **51.** $\dfrac{z^2}{4}$ **53.** $\dfrac{1}{32x^5}$ **55.** $\dfrac{49a^4}{b^6}$ **57.** $a^{24}b^8$ **59.** x^9y^{19} **61.** $-\dfrac{y^8}{8x^2}$ **63.** $\dfrac{27}{x^6z^3}$ cu. in.   **65.** 7.8×10^4

67. 1.67×10^{-6} **69.** 6.35×10^{-3} **71.** 1.16×10^6 **73.** 2.0×10^7 **75.** 9.3×10^7 **77.** 1.2×10^8

79. 0.0000000008673   **81.** 0.033   **83.** 20,320   **85.** 6,250,000,000,000,000,000   **87.** 9,460,000,000,000   **89.** 0.000036

91. 0.0000000000000000028   **93.** 0.0000005   **95.** 200,000   **97.** 1.512×10^{10} cu. ft   **99.** Answers may vary.   **101.** 100

103. -394.5 **105.** 1.3 sec   **107.** $-2y-18$ **109.** $-7x+2$ **111.** $-3z-14$ **113.** a^m **115.** $27y^{6z}$ **117.** y^{5a}

119. $\dfrac{1}{z^{6a+4}}$

Graphing Calculator Explorations   **1.** x^3-4x^2+7x-8 **3.** $-2.1x^2-3.2x-1.7$ **5.** $7.69x^2-1.26x+5.3$

Exercise Set 5.3   **1.** 0   **3.** 2   **5.** 3   **7.** degree 1; binomial   **9.** degree 2; trinomial   **11.** degree 3; monomial
13. degree 3; none of these   **15.** Answers may vary.   **17.** 57   **19.** 499   **21.** 1   **23.** 989 ft   **25.** 477 ft   **27.** $6y$

29. $11x - 3$ **31.** $xy + 2x - 1$ **33.** $18y^2 - 17$ **35.** $3x^2 - 3xy + 6y^2$ **37.** $x^2 - 4x + 8$ **39.** $y^2 + 3$
41. $-2x^2 + 5x$ **43.** $-2x^2 - 4x + 15$ **45.** $4x - 13$ **47.** $x^2 + 2$ **49.** $12x^3 + 8x + 8$ **51.** $7x^3 + 4x^2 + 8x - 10$
53. $-18y^2 + 11yx + 14$ **55.** $-x^3 + 8a - 12$ **57.** $5x^2 - 9x - 3$ **59.** $-3x^2 + 3$ **61.** $8xy^2 + 2x^3 + 3x^2 - 3$
63. $7y^2 - 3$ **65.** $5x^2 + 22x + 16$ **67.** $-q^4 + q^2 - 3q + 5$ **69.** $15x^2 + 8x - 6$ **71.** $x^4 - 7x^2 + 5$
73. $4x^{2y} + 2x^y - 11$ **75.** $(x^2 + 7x + 4)$ ft **77.** $(3y^2 + 4y + 11)$ m **79.** 404 ft per sec **81.** 384 ft; 386 ft; 356 ft
83. a. 284 ft **b.** 536 ft **c.** 756 ft **d.** 944 ft **85.** 19 sec **87. a.** \$1783.05 **b.** \$3515.05 **c.** \$6274.30
d. No, $f(x)$ is not linear. **89.** A **91.** D **93.** $4x^2 + 7x + x^2 + 5x; 5x^2 + 12x$ **95.** $15x - 10$ **97.** $-2x^2 + 10x - 12$
99. a. $2a - 3$ **b.** $-2x - 3$ **c.** $2x + 2h - 3$ **101. a.** $4a$ **b.** $-4x$ **c.** $4x + 4h$ **103. a.** $4a - 1$ **b.** $-4x - 1$
c. $4x + 4h - 1$

Mental Math **1.** $10xy$ **3.** x^7 **5.** $18x^3$

Exercise Set 5.4 **1.** $4a^2 - 8a$ **3.** $7x^3 + 14x^2 - 7x$ **5.** $6x^4 - 3x^3$ **7.** $x^2 + 3x$ **9.** $a^2 + 5a - 14$
11. $4y^2 - 16y + 16$ **13.** $30x^2 - 79xy + 45y^2$ **15.** $4x^4 - 20x^2 + 25$ **17.** $x^2 + 5x + 6$ **19.** $x^3 - 5x^2 + 13x - 14$
21. $x^4 + 5x^3 - 3x^2 - 11x + 20$ **23.** $10a^3 - 27a^2 + 26a - 12$ **25.** $x^3 + 6x^2 + 12x + 8$ **27.** $8y^3 - 36y^2 + 54y - 27$
29. $2x^3 + 10x^2 + 11x - 3$ **31.** $x^5 - 4x^3 - 7x^2 - 45x + 63$ **33. a.** 25; 13 **b.** 324; 164 **c.** no; Answers may vary.
35. $2a^2 + 8a$ **37.** $6x^3 - 9x^2 + 12x$ **39.** $15x^2 + 37xy + 18y^2$ **41.** $x^3 + 7x^2 + 16x + 12$ **43.** $49x^2 + 56x + 16$
45. $-6a^4 + 4a^3 - 6a^2$ **47.** $x^3 + 10x^2 + 33x + 36$ **49.** $a^3 + 3a^2 + 3a + 1$ **51.** $x^2 + 2xy + y^2$ **53.** $x^2 - 13x + 42$
55. $3a^3 + 6a$ **57.** $-4y^3 - 12y^2 + 44y$ **59.** $25x^2 - 1$ **61.** $5x^3 - x^2 + 16x + 16$ **63.** $8x^3 - 60x^2 + 150x - 125$
65. $32x^3 + 48x^2 - 6x - 20$ **67.** $49x^2y^2 - 14xy + y^2$ **69.** $5y^4 - 16y^3 - 4y^2 - 7y - 6$
71. $6x^4 - 8x^3 - 7x^2 + 22x - 12$ **73.** $(4x^2 - 25)$ sq. yd **75.** $(6x^2 - 4x)$ sq. in. **77.** $(x^2 + 6x + 5)$ sq. units
79. a. $a^2 - b^2$ **b.** $4x^2 - 9y^2$ **c.** $16x^2 - 49$ **d.** Answers may vary. **81.** $16p^2$ **83.** $49m^4$ **85.** \$3500 **87.** \$500
89. There is a loss in value each year.

Graphing Calculator Explorations **1.** $x^2 - 16$ **3.** $9x^2 - 42x + 49$ **5.** $5x^3 - 14x^2 - 13x - 2$

Mental Math **1.** false **3.** false

Exercise Set 5.5 **1.** $x^2 + 7x + 12$ **3.** $x^2 + 5x - 50$ **5.** $5x^2 + 4x - 12$ **7.** $4y^2 - 25y + 6$ **9.** $6x^2 + 13x - 5$
11. $x^2 - 4x + 4$ **13.** $4x^2 - 4x + 1$ **15.** $9a^2 - 30a + 25$ **17.** $25x^2 + 90x + 81$ **19.** Answers may vary.

21. $a^2 - 49$ **23.** $9x^2 - 1$ **25.** $9x^2 - \dfrac{1}{4}$ **27.** $81x^2 - y^2$ **29.** $(4x^2 + 4x + 1)$ sq. ft **31.** $a^2 + 9a + 20$

33. $a^2 + 14a + 49$ **35.** $12a^2 - a - 1$ **37.** $x^2 - 4$ **39.** $9a^2 + 6a + 1$ **41.** $4x^2 + 3xy - y^2$ **43.** $x^3 - 3x^2 - 17x + 3$

45. $4a^2 - 12a + 9$ **47.** $25x^2 - 36z^2$ **49.** $x^2 - 8x + 15$ **51.** $x^2 - \dfrac{1}{9}$ **53.** $a^2 + 8a - 33$ **55.** $x^2 - 4x + 4$

57. $6b^2 - b - 35$ **59.** $49p^2 - 64$ **61.** $\dfrac{1}{9}a^4 - 49$ **63.** $15x^4 - 5x^3 + 10x^2$ **65.** $4r^2 - 9s^2$ **67.** $9x^2 - 42xy + 49y^2$

69. $16x^2 - 25$ **71.** $x^2 + 8x + 16$ **73.** $a^2 - \dfrac{1}{4}y^2$ **75.** $\dfrac{1}{25}x^2 - y^2$ **77.** $3a^3 + 2a^2 + 1$ **79.** $(3x^2 - 9)$ sq. units

81. $(24x^2 - 32x + 8)$ sq. m **83.** $16b^2 + 32b + 16$ **85.** $4s^2 - 12s + 8$ **87.** $x^2 + 8x + 16 + 2xy - 8y + y^2$

89. $x^2 + 2xy + y^2 - 9$ **91.** $a^2 - 6a + 9 - b^2$ **93.** $\dfrac{1}{3}$ **95.** 1

Mental Math **1.** a^2 **3.** a^2 **5.** k^3 **7.** p^5 **9.** k^2

Exercise Set 5.6 **1.** $2a + 4$ **3.** $3ab + 4$ **5.** $2y + \dfrac{3y}{x} - \dfrac{2y}{x^2}$ **7.** $x^2 + 2x + 1$ **9.** $(x^4 + 2x^2 - 6)$ m **11.** $x + 1$

13. $2x - 8$ **15.** $x - \dfrac{1}{2}$ **17.** $2x^2 - \dfrac{1}{2}x + 5$ **19.** $(3x - 7)$ in. **21.** $\dfrac{5b^5}{2a^3}$ **23.** $x^3y^3 - 1$ **25.** $a + 3$ **27.** $2x + 5$

29. $4y - 6y^2$ **31.** $2x + 23 + \dfrac{130}{x - 5}$ **33.** $10x + 3y - 6x^2y^2$ **35.** $2x + 4$ **37.** $y + 5$ **39.** $2x + 3$

41. $2x^2 - 8x + 38 - \dfrac{156}{x + 4}$ **43.** $3x + 3 - \dfrac{1}{x - 1}$ **45.** $-2x^3 + 3x^2 - x + 4$ **47.** $3x^3 + 5x + 4 - \dfrac{2x}{x^2 - 2}$

49. $x - \dfrac{5}{3x^2}$ **51.** $(3x^3 + x - 4)$ ft **53.** $(2x + 5)$ m **55.** 4 **57.** 372 **59.** Answers may vary.

61. $3x^2 + 10x + 8 + \dfrac{4}{x - 2}$ **63.** $=$ **65.** $=$ **67.** $x \geq 4$ **69.** $x \leq 17$ **71.** $2x^2 + \dfrac{1}{2}x - 5$

73. $2x^3 + \dfrac{9}{2}x^2 + 10x + 21 + \dfrac{42}{x - 2}$ **75.** $3x^4 - 2x$

Exercise Set 5.7   **1.** $x + 8$ **3.** $x - 1$ **5.** $x^2 - 5x - 23 - \dfrac{41}{x - 2}$ **7.** $4x + 8 + \dfrac{7}{x - 2}$ **9.** 3   **11.** 73   **13.** -8

15. $x^2 + \dfrac{2}{x - 3}$ **17.** $6x + 7 + \dfrac{1}{x + 1}$ **19.** $2x^3 - 3x^2 + x - 4$ **21.** $3x - 9 + \dfrac{12}{x + 3}$

23. $3x^2 - \dfrac{9}{2}x + \dfrac{7}{4} + \dfrac{47}{8(x - 1/2)}$ **25.** $3x^2 + 3x - 3$ **27.** $3x^2 + 4x - 8 + \dfrac{20}{x + 1}$ **29.** $x^2 + x + 1$ **31.** $x - 6$

33. 1   **35.** -133 **37.** 3   **39.** $-\dfrac{187}{81}$ **41.** $\dfrac{95}{32}$ **43.** Answers may vary.   **45.** $(x + 3)(x^2 + 4) = x^3 + 3x^2 + 4x + 12$

47. 0   **49.** $x^3 + 2x^2 + 7x + 28$ **51.** $(x - 1)$ m   **53.** $-12a^3 + 16a$ **55.** $4y^3 - 32y^2 - 16y$

57. $-36x^2y^2z - 63x^2y^3z - 18xy$ **59.** $-42s^3r^2 - 63s^2r^3 - 63s^2r^2 - 56sr$ **61.** \$103 million   **63.** The Eagles (1994)

Chapter 5 Review   **1.** base: 3; exponent: 2   **3.** base: 5; exponent: 4   **5.** 36   **7.** -65 **9.** 1   **11.** 8   **13.** $-10x^5$ **15.** $\dfrac{b^4}{16}$

17. $\dfrac{x^6y^6}{4}$ **19.** $40a^{19}$ **21.** $\dfrac{3}{64}$ **23.** $\dfrac{1}{x}$ **25.** 5   **27.** 1   **29.** $6a^6b^9$ **31.** $\dfrac{1}{49}$ **33.** $\dfrac{2}{x^4}$ **35.** 125   **37.** $\dfrac{17}{16}$ **39.** $8q^3$ **41.** $\dfrac{s^4}{r^3}$

43. $-\dfrac{3x^3}{4r^4}$ **45.** $\dfrac{x^{15}}{8}$ **47.** $\dfrac{a^2b^2c^4}{9}$ **49.** $\dfrac{5}{x^3}$ **51.** c^4 **53.** $\dfrac{1}{x^6y^{13}}$ **55.** $-\dfrac{15}{16}$ **57.** a^{11m} **59.** $27x^3y^{6z}$ **61.** 2.7×10^{-4}

63. 8.08×10^7 **65.** 3.2667×10^7 **67.** 867,000   **69.** 0.00086   **71.** 100,000,000,000,000,000,000   **73.** 0.016

75. a. $3, -2, 1, -1, -6; 3, 2, 2, 2, 0$ **b.** 3   **77.** $12x - 6x^2 - 6x^2y$ **79.** $4x^2 + 8y + 6$ **81.** $8x^2 + 2b - 22$

83. $12x^2y - 7xy + 3$ **85.** $x^3 + x - 2xy^2 - y - 7$ **87.** 58   **89.** $(6x^2y - 12x + 12)$ cm   **91.** $-56xz^2$ **93.** $-12x^2a^2y^4$

95. $9x - 63$ **97.** $54a - 27$ **99.** $-32y^3 + 48y$ **101.** $-3a^3b - 3a^2b - 3ab^2$ **103.** $42b^4 - 28b^2 + 14b$

105. $6x^2 - 11x - 10$ **107.** $42a^2 + 11a - 3$ **109.** $x^6 + 2x^5 + x^2 + 3x + 2$ **111.** $x^6 + 8x^4 + 16x^2 - 16$

113. $15y^3 - 3y^2 + 6y$ **115.** $12a^2 + 25a - 7$ **117.** $x^2 - 10x + 25$ **119.** $16x^2 + 16x + 4$ **121.** $x^3 - 3x^2 + 4$

123. $25x^2 - 1$ **125.** $a^2 - 4b^2$ **127.** $16a^4 - 4b^2$ **129.** $\dfrac{x^4b^9}{3y^6}$ **131.** $1 + \dfrac{x}{2y} - \dfrac{9}{4xy}$ **133.** $3x^3 + 9x^2 + 2x + 6 - \dfrac{2}{x - 3}$

135. $3x^2 + 2x - 1$ **137.** $3x^2 + 6x + 24 + \dfrac{44}{x - 2}$ **139.** $x^4 - x^3 + x^2 - x + 1 - \dfrac{2}{x + 1}$

141. $3x^3 + 13x^2 + 51x + 204 + \dfrac{814}{x - 4}$ **143.** 3043   **145.** $(x^3 + 2x^2 - 6)$ mi

Chapter 5 Test   **1.** 32   **2.** 81   **3.** -81 **4.** $\dfrac{1}{64}$ **5.** $-15x^{11}$ **6.** y^5 **7.** $\dfrac{1}{r^5}$ **8.** $\dfrac{y^{14}}{x^2}$ **9.** $\dfrac{1}{6xy^8}$ **10.** 5.63×10^5

11. 8.63×10^{-5} **12.** 0.0015   **13.** 62,300   **14.** 0.036   **15. a.** $4, 7, 1, -2; 3, 3, 4, 0$ **b.** 4   **16.** $-2x^2 + 12xy + 11$

17. $16x^3 + 7x^2 - 3x - 13$ **18.** $-3x^3 + 5x^2 + 4x + 5$ **19.** $x^3 + 8x^2 + 3x - 5$ **20.** $3x^3 + 22x^2 + 41x + 14$

21. $6x^4 - 9x^3 + 21x^2$ **22.** $3x^2 + 16x - 35$ **23.** $9x^2 - 49$ **24.** $16x^2 - 16x + 4$ **25.** $64x^2 + 48x + 9$

26. $x^4 - 81b^2$ **27.** 1001 ft; 985 ft; 857 ft; 601 ft   **28.** $\dfrac{2}{x^2yz}$ **29.** $\dfrac{x}{2y} + \dfrac{1}{4} - \dfrac{7}{8y}$ **30.** $x + 2$ **31.** $9x^2 - 6x + 4 - \dfrac{16}{3x + 2}$

32. 2917 thousand cases   **33.** $4x^3 - 15x^2 + 47x - 142 + \dfrac{425}{x + 3}$ **34.** 91   **35.** $x^2 - 4y^2$ **36. a.** 960 ft   **b.** 953.44 ft

Chapter 5 Cumulative Review   **1. a.** true   **b.** true   **c.** false   **d.** true; Sec. 1.2, Ex. 2   **2. a.** $\dfrac{64}{25}$ **b.** $\dfrac{1}{20}$

c. $\dfrac{5}{4}$; Sec. 1.3, Ex. 4   **3. a.** 9   **b.** 125   **c.** 16   **d.** 7   **e.** $\dfrac{9}{49}$; Sec. 1.4, Ex. 1   **4. a.** -10 **b.** -21 **c.** -12; Sec. 1.5, Ex. 3

5. -12; Sec. 1.6, Ex. 3   **6. a.** $\dfrac{1}{22}$ **b.** $\dfrac{16}{3}$ **c.** $-\dfrac{1}{10}$ **d.** $-\dfrac{13}{9}$; Sec. 1.7, Ex. 5   **7. a.** $(5 + 4) + 6$ **b.** $-1 \cdot (2 \cdot 5)$; Sec. 1.8, Ex. 2

8. a. Alaska Village Electric   **b.** American Electric Power   **c.** American Electric Power: 5¢ per kilowatt-hour;
Green Mountain Power: 11¢ per kilowatt-hour; Montana Power Co.: 6¢ per kilowatt-hour;
Alaska Village Electric: 42¢ per kilowatt-hour   **d.** 37¢; Sec. 1.9, Ex. 1

9. a. $5x + 10$ **b.** $-2y - 0.6z + 2$ **c** $-x - y + 2z - 6$; Sec. 2.1, Ex. 5   **10.** -1.6; Sec. 2.2, Ex. 2

11. -10; Sec. 2.3, Ex. 8   **12.** $\dfrac{5F - 160}{9} = C$; Sec. 2.5, Ex. 7   **13.** 144; Sec. 2.6, Ex. 3   **14.** $\longleftrightarrow$; Sec. 2.7, Ex. 2

15. a. $(0, 12)$ **b.** $(2, 6)$ **c.** $(-1, 15)$; Sec. 3.1, Ex. 4

16. 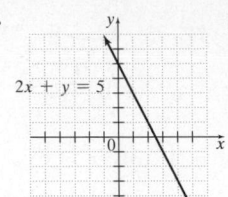   ; Sec. 3.2, Ex. 2   **17.** no; Sec. 3.3, Ex. 4   **18.** 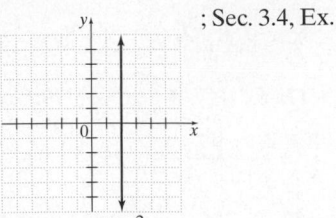   ; Sec. 3.4, Ex. 6

19. undefined slope; Sec. 3.5, Ex. 8   **20.** $y = 3$; Sec. 3.6, Ex. 7
21. 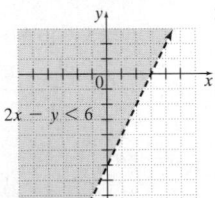   ; Sec. 3.7, Ex. 1   **22.** $-6x^7$; Sec. 5.1, Ex. 4   **23.** 4; Sec. 5.3, Ex. 3
24. $4x^2 - 4xy + y^2$; Sec. 5.4, Ex. 3

■ CHAPTER 6   FACTORING POLYNOMIALS

Mental Math   **1.** $2 \cdot 7$ **3.** $2 \cdot 5$ **5.** 3   **7.** 3

Exercise Set 6.1   **1.** 4   **3.** 6   **5.** y^2 **7.** xy^2 **9.** 4   **11.** $4y^3$ **13.** $3x^3$ **15.** $9x^2y$ **17.** $15(2x - 1)$ **19.** $6cd(4d^2 - 3c)$
21. $-6a^3x(4a - 3)$ **23.** $4x(3x^2 + 4x - 2)$ **25.** $5xy(x^2 - 3x + 2)$ **27.** Answers may vary.   **29.** $(x + 2)(y + 3)$
31. $(y - 3)(x - 4)$ **33.** $(x + y)(2x - 1)$ **35.** $(x + 3)(5 + y)$ **37.** $(y - 4)(2 + x)$ **39.** $(y - 2)(3x + 8)$
41. $(y + 3)(y^2 + 1)$ **43.** $12x^3 - 2x; 2x(6x^2 - 1)$ **45.** $200x + 25\pi; 25(8x + \pi)$ **47.** $3(x - 2)$ **49.** $2x(16y - 9x)$
51. $4(x - 2y + 1)$ **53.** $(x + 2)(8 - y)$ **55.** $-8x^8y^5(5y + 2x)$ **57.** $-3(x - 4)$ **59.** $6x^3y^2(3y - 2 + x^2)$
61. $(x - 2)(y^2 + 1)$ **63.** $(y + 3)(5x + 6)$ **65.** $(x - 2y)(4x - 3)$ **67.** $42yz(3x^3 + 5y^3z^2)$ **69.** $(3 - x)(5 + y)$
71. $2(3x^2 - 1)(2y - 7)$ **73.** Answers may vary.   **75.** factored   **77.** not factored   **79.** $(n^3 - 6)$ units
81. a. 850 million   **b.** 1855 million   **c.** $5(12x^2 - 17x + 156)$ **83.** $x^2 + 7x + 10$ **85.** $a^2 - 15a + 56$ **87.** 2, 6
89. $-1, -8$ **91.** $-2, 5$ **93.** $-8, 3$

Mental Math   **1.** $+5$ **3.** -3 **5.** $+2$

Exercise Set 6.2   **1.** $(x + 6)(x + 1)$ **3.** $(x + 8)(x + 1)$ **5.** $(x - 5)(x - 3)$ **7.** $(x - 9)(x - 1)$ **9.** prime
11. $(x - 6)(x + 3)$ **13.** prime   **15.** $(x + 5y)(x + 3y)$ **17.** $(x - y)(x - y)$ **19.** $(x - 4y)(x + y)$
21. $2(z + 8)(z + 2)$ **23.** $2x(x - 5)(x - 4)$ **25.** $7(x + 3y)(x - y)$ **27.** product; sum   **29.** $(x + 12)(x + 3)$
31. $(x - 2)(x + 1)$ **33.** $(r - 12)(r - 4)$ **35.** $(x - 7)(x + 3)$ **37.** $(x + 5y)(x + 2y)$ **39.** prime
41. $2(t + 8)(t + 4)$ **43.** $x(x - 6)(x + 4)$ **45.** $(x - 9)(x - 7)$ **47.** $(x + 2y)(x - y)$ **49.** $3(x - 18)(x - 2)$
51. $(x - 24)(x + 6)$ **53.** $6x(x + 4)(x + 5)$ **55.** $2t^3(t - 4)(t - 3)$ **57.** $5xy(x - 8y)(x + 3y)$ **59.** $4y(x^2 + x - 3)$
61. $2b(a - 7b)(a - 3b)$ **63.** 8; 16   **65.** 6; 26   **67.** 5; 8; 9   **69.** 3; 4   **71.** Answers may vary.   **73.** $2x^2 + 11x + 5$
75. $15y^2 - 17y + 4$ **77.** $9a^2 + 23a - 12$
79. 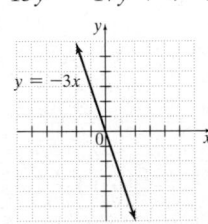   **81.**

83. $2y(x + 5)(x + 10)$
85. $-12y^3(x^2 + 2x + 3)$
87. $(x + 1)(y - 5)(y + 3)$

Mental Math   **1.** yes   **3.** no   **5.** yes

Exercise Set 6.3   **1.** $(2x + 3)(x + 5)$ **3.** $(2x + 1)(x - 5)$ **5.** $(2y + 3)(y - 2)$ **7.** $(4a - 3)^2$ **9.** $(9r - 8)(4r + 3)$
11. $(5x + 1)(2x + 3)$ **13.** $3(7x + 5)(x - 3)$ **15.** $2(2x - 3)(3x + 1)$ **17.** $x(4x + 3)(x - 3)$ **19.** $(x + 11)^2$
21. $(x - 8)^2$ **23.** $(4y - 5)^2$ **25.** $(xy - 5)^2$ **27.** Answers may vary.   **29.** $(2x + 11)(x - 9)$ **31.** $(2x - 7)(2x + 3)$
33. $(6x - 7)(5x - 3)$ **35.** $(4x - 9)(6x - 1)$ **37.** $(3x - 4y)^2$ **39.** $(x - 7y)^2$ **41.** $(2x + 5)(x + 1)$ **43.** prime
45. $(5 - 2y)(2 + y)$ **47.** $(4x + 3y)^2$ **49.** $2y(4x - 7)(x + 6)$ **51.** $(3x - 2)(x + 1)$ **53.** $(xy + 2)^2$ **55.** $(7y + 3x)^2$
57. $3(x^2 - 14x + 21)$ **59.** $(7a - 6)(6a - 1)$ **61.** $(6x - 7)(3x + 2)$ **63.** $(5p - 7q)^2$ **65.** $(5x + 3)(3x - 5)$

67. $(7t + 1)(t - 4)$ **69.** $a^2 + 2ab + b^2$ **71.** 2; 14   **73.** 5; 13   **75.** 2   **77.** 4; 5   **79.** $x^2 - 4$ **81.** $a^3 + 27$
83. $y^3 - 125$ **85.** \$75,000 and above   **87.** Answers may vary.   **89.** $-3xy^2(4x - 5)(x + 1)$
91. $-2pq(p - 3q)(15p + q)$ **93.** $(y - 1)^2(4x^2 + 10x + 25)$

Calculator Explorations

16	14	16
16	14	16
2.89	0.89	2.89
171.61	169.61	171.61
1	−1	1

Mental Math   **1.** 1^2 **3.** 9^2 **5.** 3^2 **7.** 1^3 **9.** 2^3

Exercise Set 6.4   **1.** $(x + 2)(x - 2)$ **3.** $(y + 7)(y - 7)$ **5.** $(5y - 3)(5y + 3)$ **7.** $(11 - 10x)(11 + 10x)$
9. $3(2x - 3)(2x + 3)$ **11.** $(13a - 7b)(13a + 7b)$ **13.** $(xy - 1)(xy + 1)$ **15.** $(x^2 + 3)(x^2 - 3)$
17. $(7a^2 + 4)(7a^2 - 4)$ **19.** $(x^2 + y^5)(x^2 - y^5)$ **21.** $(x + 6)$ **23.** $(a + 3)(a^2 - 3a + 9)$ **25.** $(2a + 1)(4a^2 - 2a + 1)$
27. $5(k + 2)(k^2 - 2k + 4)$ **29.** $(xy - 4)(x^2y^2 + 4xy + 16)$ **31.** $(x + 5)(x^2 - 5x + 25)$
33. $3x(2x - 3y)(4x^2 + 6xy + 9y^2)$ **35.** $(2x + y)$ **37.** $(x - 2)(x + 2)$ **39.** $(9 - p)(9 + p)$ **41.** $(2r - 1)(2r + 1)$
43. $(3x - 4)(3x + 4)$ **45.** prime   **47.** $(3 - t)(9 + 3t + t^2)$ **49.** $8(r - 2)(r^2 + 2r + 4)$ **51.** $(t - 7)(t^2 + 7t + 49)$
53. $(x - 13y)(x + 13y)$ **55.** $(xy - z)(xy + z)$ **57.** $(xy + 1)(x^2y^2 - xy + 1)$ **59.** $(s - 4t)(s^2 + 4st + 16t^2)$
61. $2(3r - 2)(3r + 2)$ **63.** $x(3y - 2)(3y + 2)$ **65.** $25y^2(y - 2)(y + 2)$ **67.** $xy(x - 2y)(x + 2y)$
69. $4s^3t^3(2s^3 + 25t^3)$ **71.** $xy^2(27xy - 1)$ **73. a.** 777 ft   **b.** 441 ft   **c.** 7 sec   **d.** $(29 + 4t)(29 - 4t)$
75. Answers may vary.   **77.** $4x^3 + 2x^2 - 1 + \dfrac{3}{x}$ **79.** $2x + 1$ **81.** $3x + 4 - \dfrac{2}{x + 3}$ **83.** $(a - 2 - b)(a + 2 + b)$
85. $(x - 2)^2(x + 1)(x + 3)$

Exercise Set 6.5   **1.** $(a + b)^2$ **3.** $(a - 3)(a + 4)$ **5.** $(a + 2)(a - 3)$ **7.** $(x + 1)^2$ **9.** $(x + 1)(x + 3)$
11. $(x + 3)(x + 4)$ **13.** $(x + 4)(x - 1)$ **15.** $(x + 5)(x - 3)$ **17.** $(x - 6)(x + 5)$ **19.** $2(x - 7)(x + 7)$
21. $(x + 3)(x + y)$ **23.** $(x + 8)(x - 2)$ **25.** $4x(x + 7)(x - 2)$ **27.** $2(3x + 4)(2x + 3)$ **29.** $(2a - b)(2a + b)$
31. $(5 - 2x)(4 + x)$ **33.** prime   **35.** $(4x - 5)(x + 1)$ **37.** $4(t^2 + 9)$ **39.** $(x + 1)(a + 2)$ **41.** $4a(3a^2 - 6a + 1)$
43. prime   **45.** $(5p - 7q)^2$ **47.** $(5 - 2y)(25 + 10y + 4y^2)$ **49.** $(5 - x)(6 + x)$ **51.** $(7 - x)(2 + x)$
53. $3x^2y(x + 6)(x - 4)$ **55.** $5xy^2(x - 7y)(x - y)$ **57.** $3xy(4x^2 + 81)$ **59.** $(x - y - z)(x - y + z)$
61. $(s + 4)(3r - 1)$ **63.** $(4x - 3)(x - 2y)$ **65.** $6(x + 2y)(x + y)$ **67.** $(x + 3)(y - 2)(y + 2)$
69. $(5 + x)(x + y)$ **71.** $(7t - 1)(2t - 1)$ **73.** $(3x + 5)(x - 1)$ **75.** $(x + 12y)(x - 3y)$ **77.** $(1 - 10ab)(1 + 2ab)$
79. $(x - 3)(x + 3)(x - 1)(x + 1)$ **81.** $(x - 4)(x + 4)(x^2 + 2)$ **83.** $(x - 15)(x - 8)$ **85.** $2x(3x - 2)(x - 4)$
87. $(3x - 5y)(9x^2 + 15xy + 25y^2)$ **89.** $(xy + 2z)(x^2y^2 - 2xyz + 4z^2)$ **91.** $2xy(1 - 6x)(1 + 6x)$
93. $(x - 2)(x + 2)(x + 6)$ **95.** $2a^2(3a + 5)$ **97.** $(a^2 + 2)(a + 2)$ **99.** $(x - 2)(x + 2)(x + 7)$ **101.** Answers may vary.
103. 6   **105.** −2   **107.** $\dfrac{1}{5}$ **109.** 8 in.   **111.** $(-2, 0); (4, 0); (0, 2); (0, -2)$ **113.** $(2, 0); (4, 0); (0, 4)$

Mental Math   **1.** 3, 7   **3.** −8, −6   **5.** −1, 3

Exercise Set 6.6   **1.** 2, −1   **3.** 0, −6   **5.** $-\dfrac{3}{2}, \dfrac{5}{4}$ **7.** $\dfrac{7}{2}, -\dfrac{2}{7}$ **9.** $(x - 6)(x + 1) = 0$ **11.** 9, 4   **13.** −4, 2   **15.** 8, −4
17. $\dfrac{7}{3}, -2$ **19.** $\dfrac{8}{3}, -9$ **21.** $x^2 - 12x + 35 = 0$ **23.** 0, 8, 4   **25.** $\dfrac{3}{4}$ **27.** $0, \dfrac{1}{2}, -\dfrac{1}{2}$ **29.** $0, \dfrac{1}{2}, -\dfrac{3}{8}$ **31.** 0, −7   **33.** −5, 4
35. −5, 6   **37.** $-\dfrac{4}{3}, 5$ **39.** $-\dfrac{3}{2}, -\dfrac{1}{2}, 3$ **41.** −5, 3   **43.** 0, 16   **45.** $-\dfrac{9}{2}, \dfrac{8}{3}$ **47.** $\dfrac{5}{4}, \dfrac{11}{3}$ **49.** $-\dfrac{2}{3}, \dfrac{1}{6}$ **51.** $-\dfrac{5}{3}, \dfrac{1}{2}$ **53.** $\dfrac{17}{2}$
55. $2, -\dfrac{4}{5}$ **57.** −4, 3   **59.** $\dfrac{1}{2}, -\dfrac{1}{2}$ **61.** −2, −11   **63.** 3   **65.** −3   **67.** $\dfrac{3}{4}, -\dfrac{4}{3}$ **69.** $0, \dfrac{5}{2}, -1$ **71.** $\left(-\dfrac{4}{3}, 0\right); (1, 0)$
73. $(-2, 0); (5, 0)$ **75.** $(-6, 0); \left(\dfrac{1}{2}, 0\right)$ **77.** E   **79.** B   **81.** C

83. a. 300; 304; 276; 216; 124; 0; −156

85. $\dfrac{47}{45}$ **87.** $\dfrac{17}{60}$ **89.** $\dfrac{15}{8}$ **91.** $\dfrac{7}{10}$ **93.** $0, \dfrac{1}{2}$

95. $0, -15$ **97.** $0, 13$

b. 5 sec **c.** 304 ft

d. 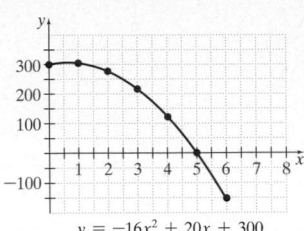 ; Answers may vary.

$y = -16x^2 + 20x + 300$

Graphing Calculator Explorations **1.** −3.562, 0.562 **3.** −0.874, 2.787 **5.** −0.465, 1.910

Exercise Set 6.7 **1.** width $= x$; length $= x + 4$ **3.** x and $x + 2$ if x is an odd integer **5.** base $= x$; height $= 4x + 1$
7. 11 units **9.** 15 cm, 13 cm, 22 cm, 70 cm **11.** base $= 16$ mi; height $= 6$ mi **13.** 5 sec **15.** length $= 5$ cm; width $= 6$ cm
17. 54 diagonals **19.** 10 sides **21.** −12 or 11 **23.** slow boat: 8 mph; fast boat: 15 mph **25.** 13 and 7 **27.** 5 in.
29. 12 mm, 16 mm, 20 mm **31.** 10 km **33.** 36 ft **35.** 6.25 sec **37.** 10 sec **39.** width: 7 ft; length: 13 ft
41. 10-in. square tier **43.** 20% **45.** length: 15 mi; width: 8 mi **47.** 105 units **49.** boom length: 36 ft; height of mainsail: 100 ft
51. D **53.** A **55.** C **57.** Answers may vary. Ex.: $f(x) = x^2 - 13x + 42$
59. Answers may vary. Ex.: $f(x) = x^2 - x - 12$ **61.** $(-4, 0), (0, 0), (3, 0)$; function
63. $(-5, 0), (5, 0), (0, -4)$; function **65.** Answers may vary.

Mental Math **1.** upward **3.** downward

Exercise Set 6.8 **1. a.** domain: $(-\infty, \infty)$; range: $(-\infty, 5]$ **b.** x-intercepts: $(-2, 0), (6, 0)$; y-intercept: $(0, 5)$ **c.** $(0, 5)$
d. There is no such point. **e.** $-2, 6$ **f.** between $x = -2$ and $x = 6$ **g.** $-2, 6$ **3. a.** domain: $(-\infty, \infty)$; range: $[-4, \infty)$
b. x-intercepts: $(-3, 0), (1, 0)$; y-intercept: $(0, -3)$ **c.** There is no such point. **d.** $(-1, -4)$ **e.** $-3, 1$ **f.** $x < -3$ or $x > 1$
g. $-3, 1$ **5. a.** domain: $(-\infty, \infty)$; range: $(-\infty, \infty)$ **b.** x-intercepts: $(-2, 0), (0, 0), (2, 0)$; y-intercept: $(0, 0)$
c. There is no such point. **d.** There is no such point. **e.** $-2, 0, 2$ **f.** between $x = -2$ and 0; $x > 2$ **g.** $-2, 0, 2$

7. 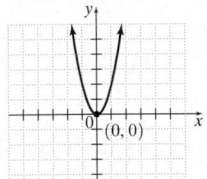 **9.** 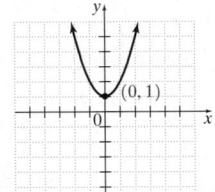 **11.**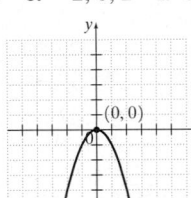

13. $(-4, -9)$
15. $(-1, 1)$
17. $(5, 30)$
19. 2
21. 0
23. 1 x-intercept; 1 y-intercept

25. 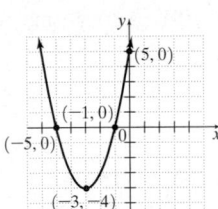 **27.** 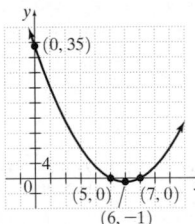 **29.** 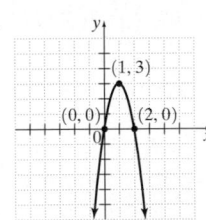 **31.**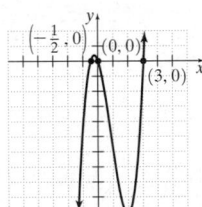

33. 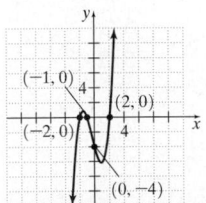 **35.** no **37.** 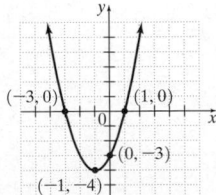 **39.** 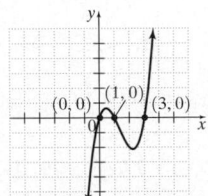 **41.**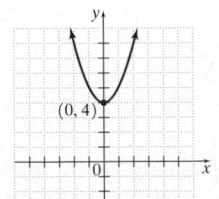

43. 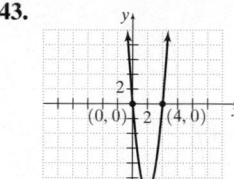   **45.** 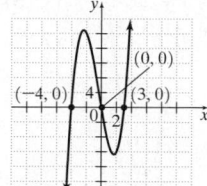   **47.** 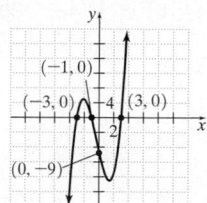   **49.**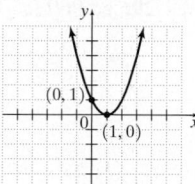

51. 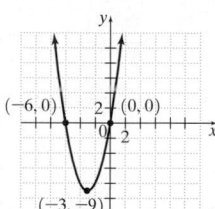   **53.** 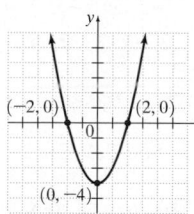   **55.** 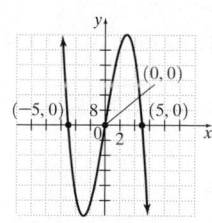   **57.**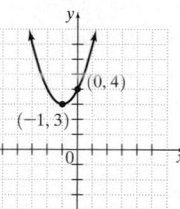

59. 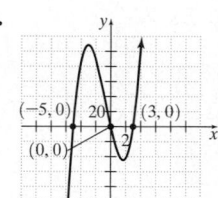   **61.** 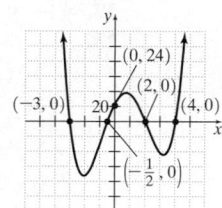   **63.** 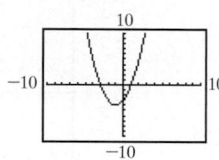   **65.**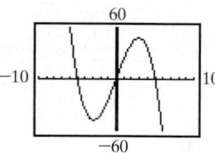

67. 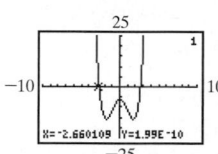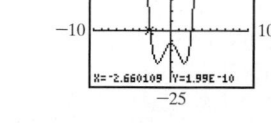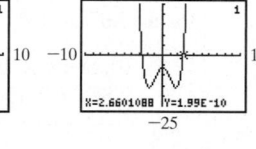 $; x = -2.7, x = 2.7$ **69.** $-\dfrac{9}{20}$ **71.** $\dfrac{a^{13}}{b^2}$ **73.** $\dfrac{4y^7}{5x^4}$

Chapter 6 Review   **1.** $2x - 5$ **3.** $4x(5x + 3)$ **5.** $-2x^2y(4x - 3y)$ **7.** $(x + 1)(5x - 1)$ **9.** $(2x - 1)(3x + 5)$
11. $(x + 4)(x + 2)$ **13.** prime   **15.** $(x + 4)(x - 2)$ **17.** $(x + 5y)(x + 3y)$ **19.** $2(3 - x)(12 + x)$
21. $(2x - 1)(x + 6)$ **23.** $(2x + 3)(2x - 1)$ **25.** $(6x - y)(x - 4y)$ **27.** $(2x + 3y)(x - 13y)$
29. $(6x + 5y)(3x - 4y)$ **31.** $(2x - 3)(2x + 3)$ **33.** prime   **35.** $(2x + 3)(4x^2 - 6x + 9)$
37. $2(3 - xy)(9 + 3xy + x^2y^2)$ **39.** $(2x - 1)(2x + 1)(4x^2 + 1)$ **41.** $(2x - 3)(x + 4)$ **43.** $(x - 1)(x + 3)$
45. $2xy(2x - 3y)$ **47.** $(5x + 3)(25x^2 - 15x + 9)$ **49.** $(x + 7 - y)(x + 7 + y)$ **51.** $-6, 2$
53. $-\dfrac{1}{5}, -3$ **55.** $-4, 6$ **57.** $2, 8$ **59.** $-\dfrac{2}{7}, \dfrac{3}{8}$ **61.** $-\dfrac{2}{5}$ **63.** 3 **65.** $0, -\dfrac{7}{4}, 3$ **67.** 36 yd

69. a. 17.5 sec and 10 sec; The rocket reaches a height of 2800 ft on its way up and on its way back down.   **b.** 27.5 sec
71. domain: $(-\infty, \infty)$; range: $(-\infty, 4]$ **73.** $(-1, 4)$
75. 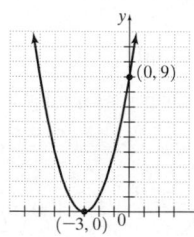   **77.** 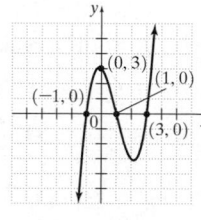   **79.** 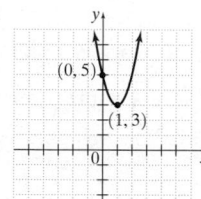   **81.**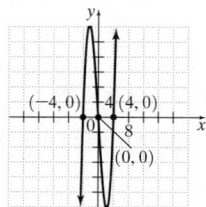

Chapter 6 Test   **1.** $3x(3x + 1)(x + 4)$ **2.** prime   **3.** prime   **4.** $(y - 12)(y + 4)$ **5.** $(3a - 7)(a + b)$
6. $(3x - 2)(x - 1)$ **7.** prime   **8.** $(x + 12y)(x + 2y)$ **9.** $x^4(26x^2 - 1)$ **10.** $5x(10x^2 + 2x - 7)$
11. $5(6 - x)(6 + x)$ **12.** $(4x - 1)(16x^2 + 4x + 1)$ **13.** $(6t + 5)(t - 1)$ **14.** $(y - 2)(y + 2)(x - 7)$
15. $x(1 - x)(1 + x)(1 + x^2)$ **16.** $-xy(y^2 + x^2)$ **17.** $-7, 2$ **18.** $-7, 1$ **19.** $0, \dfrac{3}{2}, -\dfrac{4}{3}$ **20.** $0, 3, -3$ **21.** $0, -4$
22. $-3, 5$ **23.** $-3, 8$ **24.** $0, \dfrac{5}{2}$ **25.** width: 6 ft; length: 11 ft   **26.** 17 ft   **27.** 8 and 9

28. $(x + 2y)(x - 2y)$ **29. a.** 960 ft   **b.** 953.44 ft   **c.** 11 sec   **30.**   **31.**

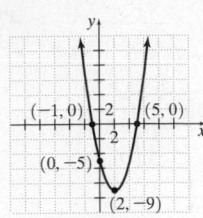

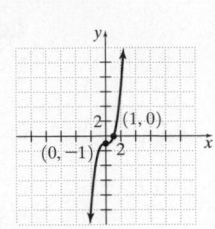

Chapter 6 Cumulative Review   **1. a.** $9 \le 11$ **b.** $8 > 1$ **c.** $3 \ne 4$; Sec. 1.2, Ex. 3   **2. a.** $\dfrac{6}{7}$ **b.** $\dfrac{11}{27}$ **c.** $\dfrac{22}{5}$; Sec. 1.3, Ex. 2

3. $\dfrac{14}{3}$; Sec. 1.4, Ex. 5   **4. a.** -12 **b.** -9; Sec. 1.5, Ex. 7   **5. a.** -24 **b.** -2 **c.** 50; Sec. 1.7, Ex. 1

6. a. $4x$ **b.** $11y^2$ **c.** $8x^2 - x$; Sec. 2.1, Ex. 3   **7.** shorter: 2 ft; longer: 8 ft; Sec. 2.4, Ex. 2

8.   ; Sec. 3.2, Ex. 4   **9.** no; Sec. 3.3, Ex. 4

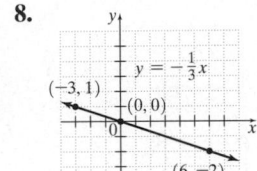

10.   ; Sec. 3.4, Ex. 6   **11.** 0; Sec. 3.5, Ex. 9

12. $y = 3$; Sec. 3.6, Ex. 7

13.   ; Sec. 3.7, Ex. 4

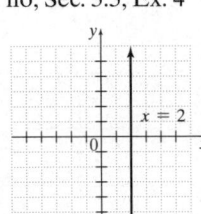

14. a. 250   **b.** 1; Sec. 5.1, Ex. 2   **15. a.** $\dfrac{1}{9}$ **b.** $\dfrac{2}{x^3}$ **c.** $\dfrac{3}{4}$ **d.** $\dfrac{1}{16}$; Sec. 5.2, Ex. 1   **16. a.** 3.67×10^8 **b.** 3.0×10^{-6}

c. 2.052×10^{10} **d.** 8.5×10^{-4}; Sec. 5.2, Ex. 5   **17.** 4; Sec. 5.3, Ex. 3   **18.** $6x^2 - 11x - 10$; Sec. 5.4, Ex. 2

19. $9y^2 + 6y + 1$; Sec. 5.5, Ex. 4   **20. a.** x^3 **b.** y; Sec. 6.1, Ex. 2   **21.** $(x + 3)(x + 4)$; Sec. 6.2, Ex. 1

22. $(4x - 1)(2x - 5)$; Sec. 6.3, Ex. 2   **23.** $(5a + 3b)(5a - 3b)$; Sec. 6.4, Ex. 2b   **24.** 3, -1; Sec. 6.6, Ex. 1

■ CHAPTER 7   RATIONAL EXPRESSIONS

Graphing Calculator Explorations   **1.** $\{x \mid x \text{ is a real number and } x \ne -2, x \ne 2\}$

3. $\left\{x \mid x \text{ is a real number and } x \ne -4, x \ne \dfrac{1}{2}\right\}$

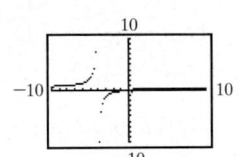

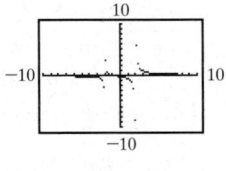

Exercise Set 7.1   **1.** $\dfrac{10}{3}, -8, -\dfrac{7}{3}$ **3.** $-\dfrac{17}{48}, \dfrac{2}{7}, -\dfrac{3}{8}$ **5.** $\{x \mid x \text{ is a real number}\}$ **7.** $\{t \mid t \text{ is a real number and } t \ne 0\}$

9. $\{x \mid x \text{ is a real number and } x \ne 7\}$ **11.** $\{x \mid x \text{ is a real number and } x \ne -2, x \ne 0, x \ne 1\}$ **13.** $\{x \mid x \text{ is a real number}$

and $x \ne 2, x \ne -2\}$ **15.** Answers may vary.   **17.** $\dfrac{5x^2}{9}$ **19.** $\dfrac{x^4}{2y^2}$ **21.** 1   **23.** $-\dfrac{1}{1 + x}$ **25.** $\dfrac{4}{3}$ **27.** -2 **29.** $\dfrac{x + 1}{x - 3}$

31. $\dfrac{2(x + 3)}{x - 3}$ **33.** $\dfrac{3}{x}$ **35.** $\dfrac{x + 1}{x^2 + 1}$ **37.** $\dfrac{1}{2(q - 1)}$ **39.** $x - 4$ **41.** $-x^2 - 5x - 25$ **43.** $\dfrac{4x^2 + 6x + 9}{2}$ **45.** 400 mg

47. no; $B \approx 24$ **49. a.** \$37.5 million   **b.** \$85.7 million   **c.** \$48.2 million   **51.** a, c; Answers may vary.

53. $4, 2, 1, \dfrac{1}{2}, \dfrac{1}{4}; -\dfrac{1}{4}, -\dfrac{1}{2}, -1, -2, -4;$

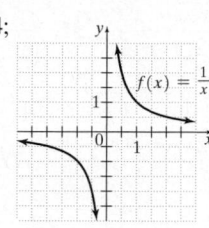

55. a. \$200 million **b.** \$500 million **c.** \$300 million

57.

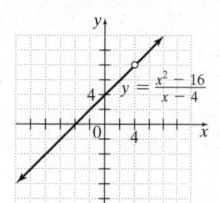

59.

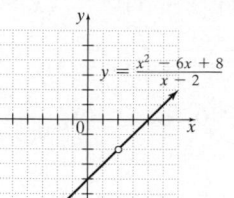

61. $\dfrac{2}{27}$ **63.** $\dfrac{7}{4}$ **65.** $\dfrac{40}{273}$ **67.** $\dfrac{64}{75}$

Mental Math **1.** $\dfrac{2x}{3y}$ **3.** $\dfrac{5y^2}{7x^2}$ **5.** $\dfrac{9}{5}$

Exercise Set 7.2 **1.** $\dfrac{21}{4y}$ **3.** x^4 **5.** $-\dfrac{b^2}{6}$ **7.** $\dfrac{x^2}{10}$ **9.** $\dfrac{1}{3}$ **11.** 1 **13.** $\dfrac{x+5}{x}$ **15.** $\dfrac{2}{9x^2(x-5)}$ sq. ft **17.** x^4 **19.** $\dfrac{12}{y^6}$

21. $x(x+4)$ **23.** $\dfrac{3(x+1)}{x^3(x-1)}$ **25.** $m^2 - n^2$ **27.** $-\dfrac{x+2}{x-3}$ **29.** $-\dfrac{x+2}{x-3}$ **31.** Answers may vary. **33.** $\dfrac{1}{6b^4}$ **35.** $\dfrac{9}{7x^2y^7}$

37. $\dfrac{5}{6}$ **39.** $\dfrac{3x}{8}$ **41.** $\dfrac{3}{2}$ **43.** $\dfrac{3x+4y}{2(x+2y)}$ **45.** $-2(x+3)$ **47.** $\dfrac{2(x+2)}{x-2}$ **49.** $\dfrac{(a+5)(a+3)}{(a+2)(a+1)}$ **51.** $-\dfrac{1}{x}$ **53.** $\dfrac{2(x+3)}{x-4}$

55. $-(x^2+2)$ **57.** -1 **59.** $4x^3(x-3)$ **61.** 1440 **63.** $411{,}972$ sq. yd **65.** 73 **67.** 3424.8 mph **69.** $\dfrac{(a+b)^2}{a-b}$

71. $\dfrac{3x+5}{x^2+4}$ **73.** $\dfrac{4}{x-2}$ **75.** $\dfrac{a-b}{6(a^2+ab+b^2)}$ **77.** 1 **79.** $-\dfrac{10}{9}$ **81.** $-\dfrac{1}{5}$

83.

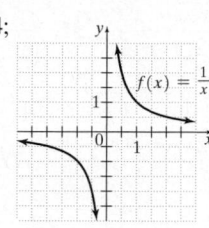

$x - 2y = 6$

85. $\dfrac{x}{2}$ **87.** $\dfrac{5a(2a+b)(3a-2b)}{b^2(a-b)(a+2b)}$

Mental Math **1.** 1 **3.** $\dfrac{7x}{9}$ **5.** $\dfrac{1}{9}$ **7.** $\dfrac{7-10y}{5}$

Exercise Set 7.3 **1.** $\dfrac{a+9}{13}$ **3.** $\dfrac{y+10}{3+y}$ **5.** $\dfrac{3m}{n}$ **7.** $\dfrac{5x+7}{x-3}$ **9.** $\dfrac{1}{2}$ **11.** 4 **13.** $x+5$ **15.** $x+4$ **17.** 1 **19.** $\dfrac{12}{x^3}$

21. $-\dfrac{5}{x+4}$ **23.** $\dfrac{x-2}{x+y}$ **25.** 4 **27.** 3 **29.** $\dfrac{1}{a+5}$ **31.** $\dfrac{1}{x-6}$ **33.** $\dfrac{20}{x-2}$ m **35.** Answers may vary. **37.** 33

39. $4x^3$ **41.** $8x(x+2)$ **43.** $6(x+1)^2$ **45.** $8-x$ or $x-8$ **47.** $40x^3(x-1)^2$ **49.** $(2x+1)(2x-1)$

51. $(2x-1)(x+4)(x+3)$ **53.** Answers may vary. **55.** $\dfrac{6x}{4x^2}$ **57.** $\dfrac{24b^2}{12ab^2}$ **59.** $\dfrac{18}{2(x+3)}$ **61.** $\dfrac{9ab+2b}{5b(a+2)}$

63. $\dfrac{x^2+x}{(x+4)(x+2)(x+1)}$ **65.** $\dfrac{18y-2}{30x^2-60}$ **67.** $\dfrac{15x(x-7)}{3x(2x+1)(x-7)(x-5)}$ **69.** $-\dfrac{5}{x-2}$ **71.** $\dfrac{7+x}{x-2}$

73. $95{,}304$ Earth days **75.** Answers may vary. **77.** $0, -5$ **79.** $1, 5$ **81.** $\dfrac{3}{10}$ **83.** $\dfrac{58}{45}$

Mental Math **1.** D **3.** A

Exercise Set 7.4 **1.** $\dfrac{5}{x}$ **3.** $\dfrac{75a+6b^2}{5b}$ **5.** $\dfrac{6x+5}{2x^2}$ **7.** $\dfrac{21}{2(x+1)}$ **9.** $\dfrac{17x+30}{2(x-2)(x+2)}$ **11.** $\dfrac{35x-6}{4x(x-2)}$

13. $\dfrac{5+10y-y^3}{y^2(2y+1)}$ **15.** Answers may vary. **17.** $-\dfrac{2}{x-3}$ **19.** $-\dfrac{1}{x^2-1}$ **21.** $\dfrac{1}{x-2}$ **23.** $\dfrac{5+2x}{x}$ **25.** $\dfrac{6x-7}{x-2}$

27. $-\dfrac{y+4}{y+3}$ **29.** $\left(\dfrac{90x-40}{x}\right)^\circ$ **31.** 2 **33.** $3x^3 - 4$ **35.** $\dfrac{x+2}{(x+3)^2}$ **37.** $\dfrac{9b-4}{5b(b-1)}$ **39.** $\dfrac{2+m}{m}$ **41.** $\dfrac{10}{1-2x}$

43. $\dfrac{15x - 1}{(x + 1)^2(x - 1)}$　**45.** $\dfrac{x^2 - 3x - 2}{(x - 1)^2(x + 1)}$　**47.** $\dfrac{a + 2}{2(a + 3)}$　**49.** $\dfrac{x - 10}{2(x - 2)}$　**51.** $\dfrac{-3 - 2y}{(y - 1)(y - 2)}$　**53.** $\dfrac{-5x + 23}{(x - 3)(x - 2)}$

55. $\dfrac{12x - 32}{(x + 2)(x - 2)(x - 3)}$　**57.** $\dfrac{6x + 9}{(3x - 2)(3x + 2)}$　**59.** $\dfrac{2x^2 - 2x - 46}{(x + 1)(x - 6)(x - 5)}$　**61.** $\dfrac{2x - 16}{(x - 4)(x + 4)}$ in.　**63.** C

65. B　**67.** 10　**69.** 2　**71.** $\dfrac{25a}{9(a - 2)}$　**73.** $\dfrac{x + 4}{(x - 2)(x - 1)}$　**75.** $\dfrac{2x^2 + 9x - 18}{6x^2}$　**77.** $\dfrac{4}{3}$　**79.** $\dfrac{4a^2}{9(a - 1)}$　**81.** 4

83.　　　　**85.**　　　　**87.** Answers may vary.　**89.** $\dfrac{1}{2}$　**91.** $\dfrac{3}{7}$　**93.** $m = 2$

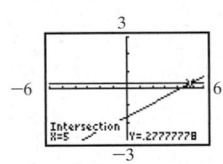　　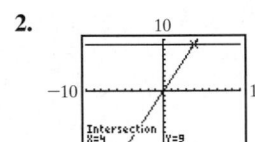

95. $m = -\dfrac{1}{3}$　**97.** $\dfrac{3}{2x}$　**99.** $\dfrac{4 - 3x}{x^2}$　**101.** $\dfrac{1 - 3x}{x^3}$

Exercise Set 7.5　**1.** $\dfrac{5}{6}$　**3.** $\dfrac{8}{5}$　**5.** 4　**7.** $\dfrac{7}{13}$　**9.** $\dfrac{4}{x}$　**11.** $\dfrac{9x - 18}{9x^2 - 4}$　**13.** $\dfrac{1 - x}{1 + x}$　**15.** $\dfrac{xy^2}{x^2 + y^2}$　**17.** $\dfrac{2b^2 + 3a}{b^2 - ab}$　**19.** $\dfrac{x}{x^2 - 1}$

21. $\dfrac{x + 1}{x + 2}$　**23.** $\dfrac{10}{69}$　**25.** $\dfrac{2(x + 1)}{2x - 1}$　**27.** $\dfrac{x(x + 1)}{6}$　**29.** $\dfrac{x}{2 - 3x}$　**31.** $-\dfrac{y}{x + y}$　**33.** $-\dfrac{2x^3}{y(x - y)}$　**35.** $\dfrac{2x + 1}{y}$

37. $\dfrac{x - 3}{9}$　**39.** $\dfrac{1}{x + 2}$　**41.** $\dfrac{x}{5x - 10}$　**43.** $\dfrac{x - 2}{2x - 1}$　**45.** $-\dfrac{x^2 + 4}{4x}$　**47.** $\dfrac{x - 3y}{x + 3y}$　**49.** $\dfrac{1 + a}{1 - a}$　**51.** $\dfrac{x^2 + 6xy}{2y}$

53. $\dfrac{5a}{2a + 4}$　**55.** $5xy^2 + 2x^2y$　**57.** $\dfrac{xy}{2x + 5y}$　**59.** $\dfrac{xy}{x + y}$　**61.** $x^2 + x$　**63.** $\dfrac{770a}{770 - s}$　**65.** $\dfrac{13}{24}$　**67.** $\dfrac{R_1 R_2}{R_2 + R_1}$　**69.** 12 hr

71. a. $\dfrac{1}{a + h}$　b. $\dfrac{1}{a}$　c. $\dfrac{\dfrac{1}{a + h} - \dfrac{1}{a}}{h}$　d. $\dfrac{-1}{a(a + h)}$　**73.** a. $\dfrac{3}{a + h + 1}$　b. $\dfrac{3}{a + 1}$　c. $\dfrac{\dfrac{3}{a + h + 1} - \dfrac{3}{a + 1}}{h}$

d. $\dfrac{-3}{(a + h + 1)(a + 1)}$　**75.** $\dfrac{x^2 y^2}{4}$　**77.** $-9x^3 y^4$　**79.** $\dfrac{x - 1}{x}$　**81.** $2x$　**83.** $3a^2 + 4a + 4$

Calculator Explorations　**1.**　　　　**2.**

3.　　　　　　**4.**　　　**5.**

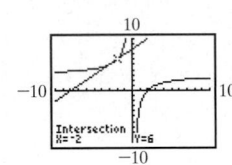　　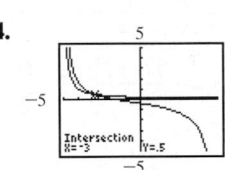　　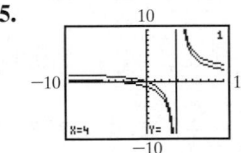

6.

Mental Math　**1.** 10　**3.** 36

Exercise Set 7.6　**1.** 30　**3.** 0　**5.** $-5, 2$　**7.** 5　**9.** 3　**11.** $100°, 80°$　**13.** $22.5°, 67.5°$　**15.** $\dfrac{1}{4}$　**17.** no solution　**19.** $6, -4$

21. 5　**23.** $-\dfrac{10}{9}$　**25.** $\dfrac{11}{14}$　**27.** no solution　**29.** expression; $\dfrac{3 + 2x}{3x}$　**31.** equation; $x = 3$　**33.** expression; $\dfrac{x - 1}{x(x + 1)}$

35. equation; no solution　**37.** Answers may vary.　**39.** $-\dfrac{21}{11}$　**41.** 1　**43.** $\dfrac{9}{2}$　**45.** no solution　**47.** $-\dfrac{9}{4}$　**49.** $-\dfrac{3}{2}, 4$

51. -2　**53.** $\dfrac{12}{5}$　**55.** $-2, 8$　**57.** $\dfrac{1}{5}$　**59.** $R = \dfrac{D}{T}$　**61.** $y = \dfrac{3x + 6}{5x}$　**63.** $b = -\dfrac{3a^2 + 2a - 4}{6}$　**65.** $B = \dfrac{2A}{H}$

67. $r = \dfrac{C}{2\pi}$ **69.** $a = \dfrac{bc}{c+b}$ **71.** $n = \dfrac{m^2 - 3p}{2}$ **73.** $\dfrac{17}{4}$ **75.** $\dfrac{1}{16}, \dfrac{1}{3}$ **77.** $-\dfrac{1}{5}, 1$ **79.** -0.17 **81.** 0.42
83. $(2, 0), (0, -2)$ **85.** $(-4, 0), (-2, 0), (3, 0), (0, 4)$

Exercise Set 7.7 **1.** $\dfrac{2}{15}$ **3.** $\dfrac{5}{6}$ **5.** $\dfrac{5}{12}$ **7.** $\dfrac{1}{10}$ **9.** $\dfrac{7}{20}$ **11.** $\dfrac{19}{18}$ **13.** Answers may vary. **15.** 4 **17.** $\dfrac{50}{9}$ **19.** $\dfrac{21}{4}$ **21.** 30
23. 7 **25.** $-\dfrac{1}{3}$ **27.** -3 **29.** $\dfrac{14}{9}$ **31.** 5 **33.** 123 lb **35.** 165 calories **37.** 3833 women **39.** 9 gal **41.** 7800 people
43. $182\dfrac{6}{7}$ cal **45.** 110 oz for $5.79 **47.** 8 oz for $0.90 **49.** 4-pack **51.** gallon **53.** 530 megawatts **55.** 1,484,000 people
57. yes; Answers may vary. **59.** a **61.** $m = 3$; upward **63.** $m = -\dfrac{9}{5}$; downward **65.** $m = 0$; horizontal

Exercise Set 7.8 **1.** 2 **3.** -3 **5.** $2\dfrac{2}{9}$ hr **7.** $1\dfrac{1}{2}$ min **9.** $\dfrac{12}{x}, \dfrac{18}{x+1}$; 6 mph **11.** 1st portion: 10 mph; cooldown: 8 mph
13. $x = 6$ **15.** $x = 5$ **17.** 2 **19.** $108.00 **21.** 63 mph **23.** $y = 21.25$ **25.** $y = 5\dfrac{5}{7}$ ft **27.** $y = 37\dfrac{1}{2}$ ft **29.** 5 **31.** 217 mph
33. 8 mph **35.** 3 hr **37.** 20 hr **39.** $26\dfrac{2}{3}$ ft **41.** $1\dfrac{1}{5}$ hr **43.** $5\dfrac{1}{4}$ hr **45.** first pump: 28 min; second pump: 84 min
47. 38 yr; 42 yr **53.** **55.** **57.**
49. Answers may vary.
51. 3.75 min

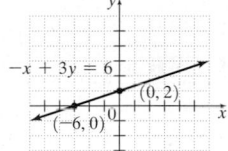

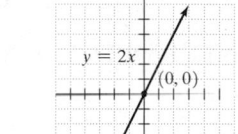

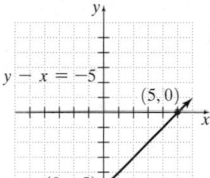

Chapter 7 Review **1.** $\{x \mid x \text{ is a real number and } x \neq 5\}$ **3.** $\{x \mid x \text{ is a real number and } x \neq 0, x \neq -8\}$ **5.** $\dfrac{x^2}{3}$ **7.** $\dfrac{9m^2p}{5}$
9. $\dfrac{1}{5}$ **11.** $\dfrac{1}{x-1}$ **13.** $\dfrac{2(x-3)}{x-4}$ **15. a.** $119 **b.** $77 **c.** decrease **17.** $-\dfrac{9x^2}{8}$ **19.** $-\dfrac{2x(2x+5)}{(x-6)^2}$ **21.** $\dfrac{4x}{3y}$ **23.** $\dfrac{2}{3}$
25. $\dfrac{x}{x+6}$ **27.** $\dfrac{3(x+2)}{3x+y}$ **29.** $-\dfrac{2(2x+3)}{y-2}$ **31.** $\dfrac{5x+2}{3x-1}$ **33.** $\dfrac{2x+1}{2x^2}$ **35.** $(x-8)(x+8)(x+3)$
37. $\dfrac{3x^2 + 4x - 15}{(x+2)^2(x+3)}$ **39.** $\dfrac{-2x+10}{(x-3)(x-1)}$ **41.** $\dfrac{-2x-2}{x+3}$ **43.** $\dfrac{x-4}{3x}$ **45.** $\dfrac{x^2+2x-3}{(x+2)^2}$ **47.** $\dfrac{29x}{12(x-1)}; \dfrac{3xy}{5(x-1)}$
49. $\dfrac{2x}{x-3}$ **51.** $\dfrac{5}{3a^2}$ **53.** $\dfrac{2x^2+1}{x+2}$ **55. a.** $\dfrac{3}{a+h}$ **b.** $\dfrac{3}{a}$ **c.** $\dfrac{\dfrac{3}{a+h} - \dfrac{3}{a}}{h}$ **d.** $\dfrac{-3}{a(a+h)}$ **57.** 30 **59.** 3, -4
61. no solution **63.** 5 **65.** $-6, 1$ **67.** $y = \dfrac{560 - 8x}{7}$ **69.** $\dfrac{2}{3}$ **71.** $x = 500$ **73.** $c = 50$ **75.** no solution
77. no solution **79.** 15 oz for $1.63 **81.** $33.75 **83.** 3 **85.** 30 mph; 20 mph **87.** $17\dfrac{1}{2}$ hr **89.** $x = 15$ **91.** $x = 15$

Chapter 7 Test **1.** $\{x \mid x \text{ is a real number and } x \neq 0, x \neq 6\}$ **2. a.** $115 **b.** $103 **3.** $\dfrac{3}{5}$ **4.** $\dfrac{1}{x-10}$ **5.** $\dfrac{1}{x+6}$
6. $\dfrac{1}{x^2 - 3x + 9}$ **7.** $\dfrac{2m(m+2)}{m-2}$ **8.** $\dfrac{a+2}{a+5}$ **9.** $-\dfrac{1}{x+y}$ **10.** $\dfrac{(x-6)(x-7)}{(x+7)(x+2)}$ **11.** 15 **12.** $\dfrac{y-2}{4}$ **13.** $-\dfrac{1}{2x+5}$
14. $\dfrac{3a-4}{(a-3)(a+2)}$ **15.** $\dfrac{3}{x-1}$ **16.** $\dfrac{2(x+5)}{x(y+5)}$ **17.** $\dfrac{x^2+2x+35}{(x+9)(x+2)(x-5)}$ **18.** $\dfrac{4y^2+13y-15}{(y+4)(y+5)(y+1)}$
19. $\dfrac{30}{11}$ **20.** -6 **21.** no solution **22.** no solution **23.** $\dfrac{xz}{2y}$ **24.** $\dfrac{b^2-a^2}{2b^2}$ **25.** $\dfrac{5y^2-1}{y+2}$ **26.** 18 bulbs **27.** 5 or 1
28. 30 mph **29.** $6\dfrac{2}{3}$ hr **30.** 6 oz for $1.19 **31.** $x = 12$

Chapter 7 Cumulative Review   **1. a.** $\dfrac{15}{x} = 4$ **b.** $12 - 3 = x$ **c.** $4x + 17 = 21$; Sec. 1.4, Ex. 9

2. a. function   **b.** not a function   **c.** function; Sec. 3.3, Ex. 2

3. 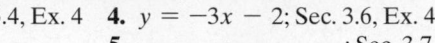 ; Sec. 3.4, Ex. 4   **4.** $y = -3x - 2$; Sec. 3.6, Ex. 4

5. 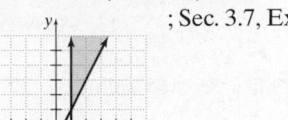 ; Sec. 3.7, Ex. 3   **6. a.** 4^7 **b.** x^7 **c.** y^4 **d.** y^{12}
e. $(-5)^{15}$; Sec. 5.1, Ex. 3

7. a. $\dfrac{27}{8}$ **b.** x^3 **c.** $\dfrac{q^9}{p^4}$; Sec. 5.2, Ex. 2

8. $27a^3 + 27a^2b + 9ab^2 + b^3$; Sec. 5.4, Ex. 5

9. a. $t^2 + 4t + 4$ **b.** $p^2 - 2qp + q^2$ **c.** $4x^2 + 20x + 25$ **d.** $x^4 - 14x^2y + 49y^2$; Sec. 5.5, Ex. 5

10. a. 4   **b.** 1   **c.** 3; Sec. 6.1, Ex. 1   **11.** $-3a(3a^4 - 6a + 1)$; Sec. 6.1, Ex. 5   **12.** $3(m + 2)(m - 10)$; Sec. 6.2, Ex. 7

13. $(3x + 2)(x + 3)$; Sec. 6.3, Ex. 1   **14.** $(x + 6)^2$; Sec. 6.3, Ex. 7   **15.** prime polynomial; Sec. 6.4, Ex. 5

16. $(x + 2)(x^2 - 2x + 4)$; Sec. 6.4, Ex. 6   **17.** $(2x + 3)(x + 1)(x - 1)$; Sec. 6.5, Ex. 2

18. $3(2m + n)(2m - n)$; Sec. 6.5, Ex. 3   **19.** $-\dfrac{1}{2}, 4$; Sec. 6.6, Ex. 3   **20.** $(1, 0), (4, 0)$; Sec. 6.6, Ex. 8

21. base: 6 m; height: 10 m; Sec. 6.7, Ex. 3   **22.** $\dfrac{2}{x(x + 1)}$; Sec. 7.2, Ex. 6   **23.** 14-ounce box; Sec. 7.7, Ex. 5

■ CHAPTER 8   TRANSITIONS TO INTERMEDIATE ALGEBRA

Exercise Set 8.1   **1.** $-3, -8$ **3.** -2 **5.** $\dfrac{1}{4}, -\dfrac{2}{3}$ **7.** $1, 9$ **9.** 0   **11.** 5   **13.** $\dfrac{3}{5}, -1$ **15.** no solution   **17.** $\dfrac{1}{8}$ **19.** 0

21. $6, -3$ **23.** 0   **25.** $\dfrac{2}{5}, -\dfrac{1}{2}$ **27.** $\dfrac{3}{4}, -\dfrac{1}{2}$ **29.** 29   **31.** -8 **33.** $-\dfrac{1}{3}, 0$ **35.** $-\dfrac{7}{8}$ **37.** $\dfrac{31}{4}$ **39.** 1   **41.** $-7, 4$ **43.** $4, 6$

45. $-\dfrac{1}{2}$ **47. a.** incorrect   **b.** correct   **c.** correct   **d.** incorrect   **49.** $K = -11$ **51.** $K = 24$ **53.** -4.86 **55.** 1.53

57. $\dfrac{8}{x}$ **59.** $8x$ **61.** $3x + 2$

Exercise Set 8.2   **1.** $4y$ **3.** $3z + 3$ **5.** $(15x + 30)$ cents   **7.** $10x + 3$ **9.** -5 **11.** $45, 225$
13. approximately 658.59 million acres   **15.** 51,700 homes   **17.** 20%   **19.** 117 automobile loans
21. Dallas/Ft. Worth, 60.5 million; Atlanta, 68.2 million; Chicago, 70.3 million   **23.** B767-300ER, 216 seats;
B737-200, 112 seats   **25.** $430.00   **27. a.** 49,057 telephone company operators   **b.** Answers may vary.   **29.** 5 years
31. 17 million returns   **33.** square: each side, 18 cm; triangle: each side, 24 cm   **35.** length, 14 cm; width, 6 cm
37. width: 8.4 m; height: 47 m   **39.** $64°, 32°, 84°$ **41.** $80°, 100°$ **43.** $15°, 75°$ **45.** $40°, 140°$ **47.** $75, 76, 77$
49. Fallon's zip code is 89406; Fernley's zip code is 89408; Gardnerville Ranchos' zip code is 89410
51. any three consecutive integers   **53.** 25 skateboards   **55.** 800 books   **57.** Answers may vary.   **59.** 11 million trees
61. -11 and -6 or 6 and 11   **63.** 75 ft   **65.** 105 units   **67.** 12 cm and 9 cm   **69.** 2 in.

71. ; $[0, 5]$ **73.** ; $\left(-\dfrac{1}{2}, \dfrac{3}{2}\right)$

Exercise Set 8.3   **1.** $\{2, 3, 4, 5, 6, 7\}$ **3.** $\{4, 6\}$ **5.** $\{\ldots, -2, -1, 0, 1, \ldots\}$ **7.** $\{5, 7\}$
9. $\{x \mid x$ is an odd integer or $x = 2$ or $x = 4\}$ **11.** $\{2, 4\}$ **13.** ; $(-2, 5)$ **15.** ; $[6, \infty)$

17. ; $(-\infty, -3]$ **19.** ; $(11, 17)$ **21.** ; $[1, 4]$

23. ; $\left[-3, \dfrac{3}{2}\right]$ **25.** ; $[-21, -9]$ **27.** ; $(-\infty, -1) \cup (0, \infty)$

29. ; $[2, \infty)$ **31.** ; $(-\infty, \infty)$ **33.** Answers may vary.   **35.** ; $(-1, 2)$

37. ; $(-\infty, \infty)$ **39.** ; $[-1, \infty)$ **41.** ; $[-5, \infty)$ **43.** ; $\left[\dfrac{3}{2}, 6\right]$

45. $\left(\dfrac{5}{4}, \dfrac{11}{4}\right)$ **47.** $\varnothing$ **49.** $(-7, \infty)$ **51.** $\left(-5, \dfrac{5}{2}\right)$

53. $\left(0, \dfrac{14}{3}\right]$ **55.** $(-\infty, -3]$ **57.** $(-\infty, 1] \cup \left(\dfrac{29}{7}, \infty\right)$

59. $\varnothing$ **61.** $\left[-\dfrac{1}{2}, \dfrac{3}{2}\right)$ **63.** $\left(-\dfrac{4}{3}, \dfrac{7}{3}\right)$ **65.** $(6, 12)$

67. $-20.2° \le F \le 95°$ **69.** $67 \le$ final score ≤ 94 **71.** $1993, 1994, 1995$ **73.** -12 **75.** -4 **77.** $-7, 7$ **79.** 0
81. $(6, \infty)$ **83.** $[3, 7]$ **85.** $(-\infty, -1)$

Mental Math   **1.** 7 **3.** -5 **5.** -6 **7.** 12

Exercise Set 8.4   **1.** $7, -7$ **3.** $4.2, -4.2$ **5.** $7, -2$ **7.** $8, 4$ **9.** $5, -5$ **11.** $3, -3$ **13.** 0 **15.** no solution   **17.** $\dfrac{1}{5}$

19. $|x| = 5$ **21.** $9, -\dfrac{1}{2}$ **23.** $-\dfrac{5}{2}$ **25.** Answers may vary.   **27.** $4, -4$ **29.** 0 **31.** no solution   **33.** $0, \dfrac{14}{3}$ **35.** $2, -2$

37. no solution   **39.** $7, -1$ **41.** no solution   **43.** no solution   **45.** $-\dfrac{1}{8}$ **47.** $\dfrac{1}{2}, -\dfrac{5}{6}$ **49.** $2, -\dfrac{12}{5}$ **51.** $3, -2$ **53.** $-8, \dfrac{2}{3}$

55. no solution   **57.** 4 **59.** $13, -8$ **61.** $3, -3$ **63.** $8, -7$ **65.** $2, 3$ **67.** $2, -\dfrac{10}{3}$ **69.** $\dfrac{3}{2}$ **71.** no solution

73. Answers may vary.   **75.** 13% **77.** \$1.088 billion   **79.** Answers may vary.   **81.** $\varnothing$

Exercise Set 8.5   **1.** $[-4, 4]$ **3.** $(1, 5)$ **5.** $(-5, -1)$

7. $[-10, 3]$ **9.** $[-5, 5]$ **11.** $\varnothing$ **13.** $[0, 12]$

15. $(-\infty, -3) \cup (3, \infty)$ **17.** $(-\infty, -24] \cup [4, \infty)$

19. $(-\infty, -4) \cup (4, \infty)$ **21.** $(-\infty, \infty)$ **23.** $\left(-\infty, \dfrac{2}{3}\right) \cup (2, \infty)$

25. $\{0\}$ **27.** $\left(-\infty, -\dfrac{3}{8}\right) \cup \left(-\dfrac{3}{8}, \infty\right)$ **29.** $|x| < 7$ **31.** $|x| \le 5$

33. $[-2, 2]$ **35.** $(-\infty, -1) \cup (1, \infty)$ **37.** $(-5, 11)$

39. $(-\infty, 4) \cup (6, \infty)$ **41.** $\varnothing$ **43.** $(-\infty, \infty)$

45. $[-2, 9]$ **47.** $(-\infty, -11] \cup [1, \infty)$ **49.** $(-\infty, 0) \cup (0, \infty)$

51. $(-\infty, \infty)$ **53.** $\left[-\dfrac{1}{2}, 1\right]$ **55.** $(-\infty, -3) \cup (0, \infty)$

57. $\varnothing$ **59.** $(-\infty, \infty)$ **61.** $\left(-\dfrac{2}{3}, 0\right)$ **63.** $(-\infty, \infty)$

65. $(-\infty, -1) \cup (1, \infty)$ **67.** $(-\infty, -12) \cup (0, \infty)$

69. $(-\infty, -6) \cup (0, \infty)$ **71.** $\left(-\dfrac{31}{5}, \dfrac{11}{5}\right)$ **73.** $[-1, 8]$

75. $\left[-\dfrac{23}{8}, \dfrac{17}{8}\right]$ **77.** $(-2, 5)$ **79.** $5, -2$ **81.** $(-\infty, -7] \cup [17, \infty)$ **83.** $-\dfrac{9}{4}$ **85.** $(-2, 1)$ **87.** $2, \dfrac{4}{3}$

89. no solution   **91.** $\dfrac{19}{2}, -\dfrac{17}{2}$ **93.** $\left(-\infty, -\dfrac{25}{3}\right) \cup \left(\dfrac{35}{3}, \infty\right)$ **95.** Answers may vary.   **97.** $3.45 < x < 3.55$ **99.** -1.5
101. 0

Exercise Set 8.6 **1.** $(-2, 5, 1)$ **3.** $(-2, 3, -1)$ **5.** $\{(x, y, z)|x - 2y + z = -5\}$ **7.** no solution

9. Answers may vary; One possibility is: $\begin{cases} 3x & = -3 \\ 2x + 4y & = 6 \\ x - 3y + z = -11 \end{cases}$ **11.** $(0, 0, 0)$ **13.** $(-3, -35, -7)$ **15.** $(6, 22, -20)$

17. no solution **19.** $(3, 2, 2)$ **21.** $\{(x, y, z)|x + 2y - 3z = 4\}$ **23.** $(-3, -4, -5)$ **25.** $(12, 6, 4)$ **27.** $(1, 1, -1)$
29. two units of Mix A; 3 units of Mix B; 1 unit of Mix C **31.** 204 free throws; 154 two-point field goals;
58 three-point field goals **33.** $20°, 50°, 110°$ **35.** $a = 3, b = 4, c = -1$ **37.** $x = 60, y = 55, z = 65$

39. $a = 12\frac{5}{6}, b = -161\frac{1}{6}, c = 1430$; 1900 students in 2005 **41.** $5x + 5z = 10$ **43.** $-5y + 2z = 2$ **45.** $(1, 1, 0, 2)$

47. $(1, -1, 2, 3)$

Exercise Set 8.7 **1.** $(2, -1)$ **3.** $(-4, 2)$ **5.** no solution **7.** $\{(x, y)|x - y = 3\}$ **9.** $(-2, 5, -2)$ **11.** $(1, -2, 3)$
13. $(4, -3)$ **15.** $(2, 1, -1)$ **17.** $(9, 9)$ **19.** no solution **21.** no solution **23.** $(1, -4, 3)$ **25.** Answers may vary.
27. function **29.** not a function **31.** -13 **33.** -36 **35.** 0

Exercise Set 8.8 **1.** 26 **3.** -19 **5.** 0 **7.** $(1, 2)$ **9.** $\{(x, y)|3x + y = 1\}$ **11.** $(9, 9)$ **13.** 8 **15.** 0 **17.** 54

19. $(-2, 0, 5)$ **21.** $(6, -2, 4)$ **23.** 16 **25.** 15 **27.** $\frac{13}{6}$ **29.** 0 **31.** 56 **33.** 0 **35.** 5 **37.** $(-3, -2)$ **39.** no solution

41. $(-2, 3, -1)$ **43.** $(3, 4)$ **45.** $(-2, 1)$ **47.** $\{(x, y, z)|x - 2y + z = -3\}$ **49.** $(0, 2, -1)$
51. $+ \ - \ + \ -$ **53.** $6x - 18$ **55.** $9x - 15$
    $- \ + \ - \ +$ **57.** **59.** **61.** -125 **63.** 24
    $+ \ - \ + \ -$
    $- \ + \ - \ +$

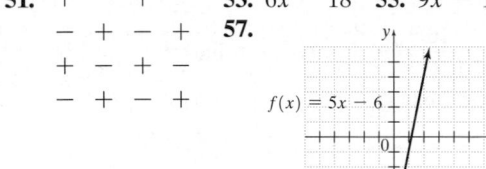
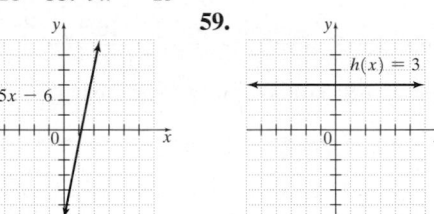

Chapter 8 Review **1.** 3 **3.** $-\frac{45}{14}$ **5.** $-4, 9$ **7.** 6 **9.** all real numbers **11.** no solution **13.** $-3, 5$

15. $-\frac{1}{2}, \frac{1}{3}$ **17.** 32 **19.** 8 **21.** $-6, 5$ **23.** 2 **25.** -7 **27.** \$22,896 **29.** No such odd integers exist.

31. 358 mi **33.** $-\frac{15}{2}, 7$ **35.** 5 sec **37.** $\left[-2, -\frac{9}{5}\right)$ **39.** $\left(-\frac{3}{5}, 0\right)$ **41.** $\left[-\frac{4}{3}, \frac{7}{6}\right]$ **43.** $(-\infty, \infty)$ **45.** $(5, \infty)$ **47.** $5, 11$

49. $-1, \frac{11}{3}$ **51.** $-\frac{1}{6}$ **53.** no solution **55.** $1, 5$ **57.** no solution **59.** $-10, -\frac{4}{3}$ **61.** $\underset{-4 \qquad 1}{\longleftrightarrow}; (-\infty, -4] \cup [1, \infty)$

63. $\underset{-3 \qquad 3}{\longleftrightarrow}; (-3, 3)$ **65.** $\underset{0}{\longleftrightarrow}; (-\infty, \infty)$ **67.** $\underset{-\frac{1}{2} \quad 2}{\longleftrightarrow}; \left(-\frac{1}{2}, 2\right)$ **69.** $\underset{0}{\longleftrightarrow}; \varnothing$

71. $(2, 0, -3)$ **73.** $(-1, 2, 0)$ **75.** $(5, 3, 0)$ **77.** $(3, 1, 1)$ **79.** 30 lb of creme-filled; 5 lb of chocolate-covered nuts;
10 lb of chocolate-covered raisins **81.** Two sides, 22 cm each; third side, 29 cm **83.** $(-3, 1)$ **85.** $\left(-\frac{2}{3}, 3\right)$

87. $\left(\frac{5}{4}, \frac{5}{8}\right)$ **89.** $(1, 3)$ **91.** $(1, 2, 3)$ **93.** $(3, -2, 5)$ **95.** $(1, 1, -2)$ **97.** -17 **99.** 34 **101.** $\left(-\frac{2}{3}, 3\right)$ **103.** $(-3, 1)$

105. no solution **107.** $(1, 2, 3)$ **109.** $(2, 1, 0)$ **111.** no solution

Chapter 8 Test **1.** 10 **2.** 1 **3.** $-3, 5$ **4.** $-7, 1$ **5.** 12 **6.** $-\frac{80}{29}$ **7.** $1, \frac{2}{3}$ **8.** no solution **9.** $\left(\frac{3}{2}, 5\right]$

10. $(-\infty, -2) \cup \left(\frac{4}{3}, \infty\right)$ **11.** $[5, \infty)$ **12.** $[4, \infty)$ **13.** $[-3, -1)$ **14.** $(-\infty, \infty)$ **15.** 34 **16.** -6 **17.** $(-1, -2, 4)$

18. no solution **19.** $\left(\frac{7}{2}, -10\right)$ **20.** $(2, -1)$ **21.** $(3, 6)$ **22.** $(3, -1, 2)$ **23.** $(5, 0, -4)$ **24.** $\{(x, y)|x - y = -2\}$

25. $(5, -3)$ **26.** $(-1, -1, 0)$ **27.** no solution **28.** 53 double rooms, 27 single rooms **29.** 800 packages
30. 211,468 people **31.** approximately 8 dogs **32.** more than 850 sunglasses **33.** $-3, 8$

Chapter 8 Cumulative Review   **1. a.** 5   **b.** 1   **c.** 35   **d.** -2; Sec 3.3, Ex. 8

2.

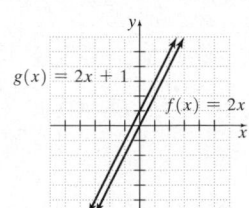

; Sec. 3.4, Ex. 1   **3.** slope: $\dfrac{3}{4}$; y-intercept: $(0, -1)$; Sec 3.5, Ex. 6

4. a. parallel   **b.** neither; Sec. 3.5, Ex. 10

5. $f(x) = \dfrac{5}{8}x - \dfrac{5}{2}$; Sec. 3.6, Ex. 5

6.

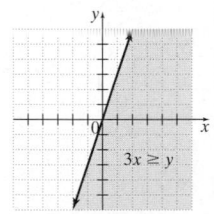

; Sec. 3.7, Ex. 2   **7.** 1; Sec. 8.1, Ex. 1   **8.** $-\dfrac{1}{6}$, 3; Sec. 8.1, Ex. 5   **9.** $\{4, 6\}$; Sec. 8.3, Ex. 1

10. $(-\infty, 4)$; Sec. 8.3, Ex. 2   **11.** $(-3, 2)$; Sec. 8.3, Ex. 4

12. $\{2, 3, 4, 5, 6, 8\}$; Sec. 8.3, Ex. 6   **13.** $(-\infty, \infty)$; Sec. 8.3, Ex. 8

14. $2, -2$; Sec. 8.4, Ex. 1   **15.** $24, -20$; Sec. 8.4, Ex. 3   **16.** $-1, 1$; Sec. 8.4, Ex. 4

17. 4; Sec. 8.4, Ex. 9   **18.** $[-3, 3]$; Sec. 8.5, Ex. 1   **19.** $(4, 8)$; Sec. 8.5, Ex. 2

20. $(-\infty, \infty)$; Sec. 8.5, Ex. 6   **21.** 25 cm, 62 cm, 62 cm; Sec. 8.2, Ex. 4

22. $(-4, 2, -1)$; Sec. 8.6, Ex. 1   **23.** $(-1, 2)$; Sec. 8.7, Ex. 1

24. a. -2 **b.** -10; Sec. 8.8, Ex. 1

■ CHAPTER 9   RATIONAL EXPONENTS, RADICALS, AND COMPLEX NUMBERS

Exercise Set 9.1   **1.** 10   **3.** $\dfrac{1}{2}$ **5.** 0.01   **7.** -6 **9.** x^5 **11.** $4y^3$ **13.** 2.646   **15.** 6.164   **17.** 14.142   **19.** 4   **21.** $\dfrac{1}{2}$

23. -1 **25.** x^4 **27.** $-3x^3$ **29.** -2 **31.** not a real number   **33.** -2 **35.** x^4 **37.** $2x^2$ **39.** $9x^2$ **41.** $4x^2$ **43.** 8

45. -8 **47.** $2|x|$ **49.** x **51.** $|x - 5|$ **53.** $|x + 2|$ **55.** -11 **57.** $2x$ **59.** y^6 **61.** $5ab^{10}$ **63.** $-3x^4y^3$ **65.** a^4b

67. $-2x^2y$ **69.** $\dfrac{5}{7}$ **71.** $\dfrac{x}{2y}$ **73.** $-\dfrac{z^7}{3x}$ **75.** $\dfrac{x}{2}$ **77.** $\sqrt{3}$ **79.** -1 **81.** -3 **83.** $\sqrt{7}$

85. $[0, \infty)$ **87.** $[3, \infty)$ **89.** $(-\infty, \infty)$ **91.** $(-\infty, \infty)$; $0, 1, -1, 2, -2$

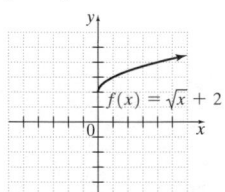   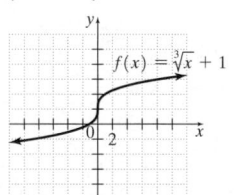   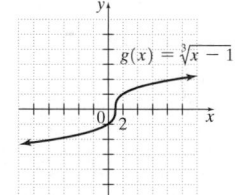

93. Answers may vary.   **95.**   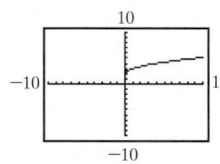   **97.**   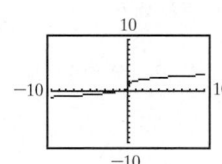   **99.** $-32x^{15}y^{10}$

101. $-60x^7y^{10}z^5$

103. $\dfrac{x^9y^5}{2}$

Exercise Set 9.2   **1.** 7   **3.** 3   **5.** $\dfrac{1}{2}$ **7.** 13   **9.** $2\sqrt[3]{m}$ **11.** $3x^2$ **13.** -3 **15.** -2 **17.** 8   **19.** 16   **21.** not a real number

23. $\sqrt[5]{(2x)^3}$ **25.** $\sqrt[3]{(7x + 2)^2}$ **27.** $\dfrac{64}{27}$ **29.** $\dfrac{1}{16}$ **31.** $\dfrac{1}{16}$ **33.** not a real number   **35.** $\dfrac{1}{x^{1/4}}$ **37.** $a^{2/3}$ **39.** $\dfrac{5x^{3/4}}{7}$

41. Answers may vary.   **43.** $a^{7/3}$ **45.** x **47.** $3^{5/8}$ **49.** $y^{1/6}$ **51.** $8u^3$ **53.** $-b$ **55.** $27x^{2/3}$ **57.** $y - y^{7/6}$

59. $2x^{5/3} - 2x^{2/3}$ **61.** $4x^{2/3} - 9$ **63.** $x^{8/3}(1 + x^{2/3})$ **65.** $x^{1/5}(x^{1/5} - 3)$ **67.** $x^{-1/3}(5 + x)$ **69.** $\sqrt{x}$ **71.** $\sqrt[3]{2}$ **73.** $2\sqrt{x}$

75. $\sqrt{xy}$ **77.** $\sqrt[15]{y^{11}}$ **79.** $\sqrt[12]{b^5}$ **81.** $\sqrt{a}$ **83.** $\sqrt[6]{432}$ **85.** $\sqrt[15]{343y^5}$ **87.** $\dfrac{t^{1/2}}{u^{1/2}}$ **89.** 1316 calories

91. a. \$43,340 million   **b.** \$63,374 million   **93.** $x^{3/8}$ **95.** $y^{1/4}$ **97.** 1.8206   **99.** 22.0515   **101.** $4 \cdot 5$ **103.** $9 \cdot 5$

105. $8 \cdot 7$ **107.** $8 \cdot 10$

Exercise Set 9.3 **1.** $\sqrt{14}$ **3.** 2 **5.** $\sqrt[3]{36}$ **7.** $\sqrt{6x}$ **9.** $\sqrt{\dfrac{14}{xy}}$ **11.** $\sqrt[4]{20x^3}$ **13.** $\dfrac{\sqrt{6}}{7}$ **15.** $\dfrac{\sqrt{2}}{7}$ **17.** $\dfrac{\sqrt[4]{x^3}}{2}$ **19.** $\dfrac{\sqrt[3]{4}}{3}$
21. $\dfrac{\sqrt[4]{8}}{x^2}$ **23.** $\dfrac{\sqrt[3]{2x}}{3y^4\sqrt[3]{3}}$ **25.** $\dfrac{x\sqrt{y}}{10}$ **27.** $\dfrac{\sqrt{5x}}{2y}$ **29.** $-\dfrac{z^2\sqrt[3]{z}}{3x}$ **31.** $4\sqrt{2}$ **33.** $4\sqrt[3]{3}$ **35.** $25\sqrt{3}$ **37.** $2\sqrt{6}$ **39.** $10x^2\sqrt{x}$
41. $2y^2\sqrt[3]{2y}$ **43.** $a^2b\sqrt[4]{b^3}$ **45.** $y^2\sqrt{y}$ **47.** $5ab\sqrt{b}$ **49.** $-2x^2\sqrt[5]{y}$ **51.** $x^4\sqrt[3]{50x^2}$ **53.** $-4a^4b^3\sqrt{2b}$ **55.** $3x^3y^4\sqrt{xy}$
57. $5r^3s^4$ **59.** $\sqrt{2}$ **61.** 2 **63.** 10 **65.** x^2y **67.** $24m^2$ **69.** $\dfrac{15x\sqrt{2x}}{2}$ or $\dfrac{15x}{2}\sqrt{2x}$ **71.** $2a^2\sqrt[4]{2}$

73. a. 20π sq. cm **b.** 211.57 sq. ft **75. a.** 3.8 times **b.** 2.9 times **c.** Answers may vary. **77.** $48x^2$
79. $3x - 2$ **81.** $-72y^4$ **83.** $x + 2$ **85.** $4x^2 + 4x + 1$
Mental Math **1.** $6\sqrt{3}$ **3.** $3\sqrt{x}$ **5.** $12\sqrt[3]{x}$

Exercise Set 9.4 **1.** $-2\sqrt{2}$ **3.** $10x\sqrt{2x}$ **5.** $17\sqrt{2} - 15\sqrt{5}$ **7.** $-\sqrt[3]{2x}$ **9.** $5b\sqrt{b}$ **11.** $\dfrac{31\sqrt{2}}{15}$ **13.** $\dfrac{\sqrt[3]{11}}{3}$ **15.** $\dfrac{5\sqrt{5x}}{9}$
17. $14 + \sqrt{3}$ **19.** $7 - 3y$ **21.** $6\sqrt{3} - 6\sqrt{2}$ **23.** $-23\sqrt[3]{5}$ **25.** $2b\sqrt{b}$ **27.** $20y\sqrt{2y}$ **29.** $2y\sqrt[3]{2x}$ **31.** $6\sqrt[3]{11} - 4\sqrt{11}$
33. $4x\sqrt[4]{x^3}$ **35.** $\dfrac{2\sqrt{3}}{3}$ **37.** $\dfrac{5x\sqrt[3]{x}}{7}$ **39.** $\dfrac{5\sqrt{7}}{2x}$ **41.** $\dfrac{\sqrt[3]{2}}{6}$ **43.** $\dfrac{14x\sqrt[3]{2x}}{9}$ **45.** $15\sqrt{3}$ in. **47.** $\sqrt{35} + \sqrt{21}$
49. $7 - 2\sqrt{10}$ **51.** $3\sqrt{x} - x\sqrt{3}$ **53.** $6x - 13\sqrt{x} - 5$ **55.** $\sqrt[3]{a^2} + \sqrt[3]{a} - 20$ **57.** $6\sqrt{2} - 12$ **59.** $2 + 2x\sqrt{3}$
61. $-16 - \sqrt{35}$ **63.** $x - y^2$ **65.** $3 + 2x\sqrt{3} + x^2$ **67.** $5x - 3\sqrt{15x} - 3\sqrt{10x} + 9\sqrt{6}$ **69.** $2\sqrt[3]{2} - \sqrt[3]{4}$
71. $-4\sqrt[6]{x^5} + \sqrt[3]{x^2} + 8\sqrt[3]{x} - 4\sqrt{x} + 7$ **73. a.** $22\sqrt{5}$ ft **b.** 150 sq. ft **75.** Answers may vary. **77.** $x - 7$
79. $\dfrac{7}{x + y}$ **81.** $2a - 3$ **83.** $\dfrac{-2 + \sqrt{3}}{3}$
Mental Math **1.** $\sqrt{2} - x$ **3.** $5 + \sqrt{a}$ **5.** $7\sqrt{5} - 8\sqrt{x}$

Exercise Set 9.5 **1.** $\dfrac{\sqrt{14}}{7}$ **3.** $\dfrac{\sqrt{5}}{5}$ **5.** $\dfrac{\sqrt[3]{6}}{2}$ **7.** $\dfrac{4\sqrt[3]{9}}{3}$ **9.** $\dfrac{3\sqrt{2x}}{4x}$ **11.** $\dfrac{3\sqrt[3]{2x}}{2x}$ **13.** $\dfrac{2\sqrt{x}}{x}$ **15.** $\dfrac{3\sqrt{3a}}{a}$ **17.** $\dfrac{3\sqrt[3]{4}}{2}$
19. $\dfrac{2\sqrt{21}}{7}$ **21.** $\dfrac{\sqrt{10xy}}{5y}$ **23.** $\dfrac{3\sqrt[4]{2}}{2}$ **25.** $\dfrac{2\sqrt[4]{9x}}{3x^2}$ **27.** $\dfrac{5\sqrt[5]{4ab^4}}{2ab^3}$ **29.** $\dfrac{5}{\sqrt{15}}$ **31.** $\dfrac{6}{\sqrt{10}}$ **33.** $\dfrac{2x}{7\sqrt{x}}$ **35.** $\dfrac{5y}{\sqrt[3]{100xy}}$
37. $\dfrac{2}{\sqrt{10}}$ **39.** $\dfrac{2x}{11\sqrt{2x}}$ **41.** $\dfrac{7}{2\sqrt[3]{49}}$ **43.** $\dfrac{3x^2}{10\sqrt[3]{9x}}$ **45.** $\dfrac{6x^2y^3}{\sqrt{6z}}$ **47.** Answers may vary. **49.** $-2(2 + \sqrt{7})$
51. $\dfrac{7(3 + \sqrt{x})}{9 - x}$ **53.** $-5 + 2\sqrt{6}$ **55.** $\dfrac{2a + 2\sqrt{a} + \sqrt{ab} + \sqrt{b}}{4a - b}$ **57.** $-\dfrac{8(1 - \sqrt{10})}{9}$ **59.** $\dfrac{x - \sqrt{xy}}{x - y}$ **61.** $\dfrac{5 + 3\sqrt{2}}{7}$
63. $\dfrac{-7}{12 + 6\sqrt{11}}$ **65.** $\dfrac{3}{10 + 5\sqrt{7}}$ **67.** $\dfrac{x - 9}{x - 3\sqrt{x}}$ **69.** $\dfrac{1}{3 + 2\sqrt{2}}$ **71.** $\dfrac{x - 1}{x - 2\sqrt{x} + 1}$ **73.** $r = \dfrac{\sqrt{A\pi}}{2\pi}$

75. Answers may vary. **77.** 5 **79.** $-\dfrac{1}{2}, 6$ **81.** 2, 6

Graphing Calculator Explorations **1.** 3.19 **3.** no solution **5.** 3.23

Exercise Set 9.6 **1.** 8 **3.** 7 **5.** no solution **7.** 7 **9.** 6 **11.** $-\dfrac{9}{2}$ **13.** 29 **15.** 4 **17.** -4 **19.** no solution
21. 7 **23.** 9 **25.** 50 **27.** no solution **29.** $\dfrac{15}{4}$ **31.** 13 **33.** 5 **35.** -12 **37.** 9 **39.** -3 **41.** 1
43. 1 **45.** $\dfrac{1}{2}$ **47.** 0, 4 **49.** $\dfrac{37}{4}$ **51.** Answers may vary. **53.** $3\sqrt{5}$ ft **55.** $2\sqrt{10}$ m **57.** $2\sqrt{131}$ m ≈ 22.9 m
59. $\sqrt{100.84}$ mm ≈ 10.0 mm **61.** 17 ft **63.** 13 ft **65.** 14,657,415 sq. mi **67.** 100 ft **69.** 1 **71.** 2743 deliveries
73. a. Answers may vary. **b.** Answers may vary. **75.** not a function **77.** not a function **79.** not a function
81. $-\dfrac{20 + 16y}{3y}$ **83.** $\dfrac{x + y}{x - y}$ **85.** $-1, 2$ **87.** $-8, -6, 0, 2$
Mental Math **1.** $9i$ **3.** $i\sqrt{7}$ **5.** -4 **7.** $8i$
Exercise Set 9.7 **1.** $2i\sqrt{6}$ **3.** $-6i$ **5.** $24i\sqrt{7}$ **7.** $-3\sqrt{6}$ **9.** $-\sqrt{14}$ **11.** $-5\sqrt{2}$ **13.** $4i$ **15.** $i\sqrt{3}$ **17.** $2\sqrt{2}$
19. $6 - 4i$ **21.** $-2 + 6i$ **23.** $-2 - 4i$ **25.** $18 + 12i$ **27.** 7 **29.** $12 - 16i$ **31.** $-4i$ **33.** $\dfrac{28}{25} - \dfrac{21}{25}i$

35. $4 + i$ **37.** $\dfrac{17}{13} + \dfrac{7}{13}i$ **39.** 63 **41.** $2 - i$ **43.** 20 **45.** 10 **47.** 2 **49.** $-5 + \dfrac{16}{3}i$ **51.** $17 + 144i$

53. $\dfrac{3}{5} - \dfrac{1}{5}i$ **55.** $5 - 10i$ **57.** $\dfrac{1}{5} - \dfrac{8}{5}i$ **59.** $8 - i$ **61.** 1 **63.** i **65.** $-i$ **67.** -1 **69.** $1 - i$ **71.** 0

73. $2 + 3i$ **75.** $2 + i\sqrt{2}$ **77.** $\dfrac{1}{2} - \dfrac{\sqrt{3}}{2}i$ **79.** Answers may vary.   **81.** $6 - 6i$ **83.** yes   **85.** $33°$

87. $5x^3 - 10x^2 + 17x - 34 + \dfrac{70}{x + 2}$ **89.** 5 **91.** 11 **93.** 16.7%

Chapter 9 Review   **1.** 9 **3.** -2 **5.** $-\dfrac{1}{7}$ **7.** -6 **9.** $-a^2b^3$ **11.** $2ab^2$ **13.** $\dfrac{x^6}{6y}$ **15.** $|-x|$ **17.** -27 **19.** $-x$

21. $5|(x - y)^5|$ **23.** $-x$ **25.** $(-\infty, \infty); -2, -1, 0, 1, 2$ **27.** $-\dfrac{1}{3}$ **29.** $-\dfrac{1}{4}$ **31.** $\dfrac{1}{4}$ **33.** $\dfrac{343}{125}$

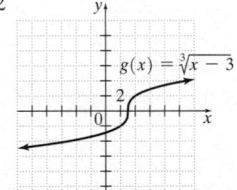

35. not a real number   **37.** $5^{1/5}x^{2/5}y^{3/5}$ **39.** $5\sqrt[3]{xy^2z^5}$ **41.** $a^{13/6}$ **43.** $\dfrac{1}{a^{9/2}}$ **45.** a^4b^6 **47.** $\dfrac{b^{5/6}}{49a^{1/4}c^{5/3}}$ **49.** 4.472

51. 5.191 **53.** -26.246 **55.** $\sqrt[6]{1372}$ **57.** $2\sqrt{6}$ **59.** $2x$ **61.** $2\sqrt{15}$ **63.** $3\sqrt[3]{6}$ **65.** $6x^3\sqrt{x}$ **67.** $\dfrac{p^8\sqrt{p}}{11}$ **69.** $\dfrac{y\sqrt[4]{xy^2}}{3}$

71. a. $\dfrac{5}{\sqrt{\pi}}$ m or $\dfrac{5\sqrt{\pi}}{\pi}$ m   **b.** 5.75 in.   **73.** $xy\sqrt{2y}$ **75.** $3a\sqrt[4]{2a}$ **77.** $\dfrac{3\sqrt{2}}{4x}$ **79.** $-4ab\sqrt[4]{2b}$ **81.** $x - 6\sqrt{x} + 9$

83. $4x - 9y$ **85.** $\sqrt[3]{a^2} + 4\sqrt[3]{a} + 4$ **87.** $a + 64$ **89.** $\dfrac{\sqrt{3x}}{6}$ **91.** $\dfrac{2x^2\sqrt{2x}}{y}$ **93.** $-\dfrac{10 + 5\sqrt{7}}{3}$ **95.** $-5 + 2\sqrt{6}$

97. $\dfrac{6}{\sqrt{2y}}$ **99.** $\dfrac{4x^3}{y\sqrt{2x}}$ **101.** $\dfrac{x - 25}{-3\sqrt{x} + 15}$ **103.** no solution   **105.** no solution   **107.** 16 **109.** $\sqrt{241}$ **111.** 4.24 ft

113. $-i\sqrt{6}$ **115.** $-\sqrt{10}$ **117.** $-13 - 3i$ **119.** $10 + 4i$ **121.** $1 + 5i$ **123.** 87 **125.** $-\dfrac{1}{3} + \dfrac{1}{3}i$

Chapter 9 Test   **1.** $6\sqrt{6}$ **2.** $-x^{16}$ **3.** $\dfrac{1}{5}$ **4.** 5 **5.** $\dfrac{4x^2}{9}$ **6.** $-a^6b^3$ **7.** $\dfrac{8a^{1/3}c^{2/3}}{b^{5/12}}$ **8.** $a^{7/12} - a^{7/3}$ **9.** $|4xy|$ or $4|xy|$

10. -27 **11.** $\dfrac{3\sqrt{y}}{y}$ **12.** $\dfrac{8 - 6\sqrt{x} + x}{8 - 2x}$ **13.** $\dfrac{\sqrt[3]{b^2}}{b}$ **14.** $\dfrac{6 - x^2}{8(\sqrt{6} - x)}$ **15.** $-x\sqrt{5x}$ **16.** $4\sqrt{3} - \sqrt{6}$

17. $x + 2\sqrt{x} + 1$ **18.** $\sqrt{6} - 4\sqrt{3} + \sqrt{2} - 4$ **19.** -20 **20.** 23.685 **21.** 0.019 **22.** $2, 3$ **23.** no solution   **24.** 6

25. $i\sqrt{2}$ **26.** $-2i\sqrt{2}$ **27.** $-3i$ **28.** 40 **29.** $7 + 24i$ **30.** $-\dfrac{3}{2} + \dfrac{5}{2}i$ **31.** $\dfrac{5\sqrt{2}}{2}$

32. $[-2, \infty)$;   ; $0, 1, 2, 3$ **33.** 27 mph   **34.** 360 ft

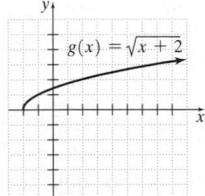

Chapter 9 Cumulative Review   **1. a.** $-\dfrac{39}{5}$ **b.** 2; Sec. 1.7, Ex. 9

2.

x	y
-1	-3
0	0
-3	-9

; Sec 3.1, Ex. 5

3. a. domain: $\{2, 0, 3\}$; range: $\{3, 4, -1\}$
   b. domain: $\{-4, -3, -2, -1, 0, 1, 2, 3\}$; range: $\{1\}$
   c. domain: $\{$Erie, Escondido, Gary, Miami, Waco$\}$;
   range: $\{104, 109, 117, 359\}$; Sec. 3.3, Ex. 1

4.

; Sec. 3.4, Ex. 7   **5.** $\frac{2}{3}$; Sec. 3.5, Ex. 5

6. one; Sec. 4.1, Ex. 6

7. $\left(6, \frac{1}{2}\right)$; Sec. 4.2, Ex. 2   **8.** no solution; Sec. 4.3, Ex. 3

9. 5% saline solution: 2.5 L; 25% saline solution: 7.5 L; Sec. 4.4, Ex. 4

10. a. -4 **b.** 11; Sec. 5.3, Ex. 4   **11.** $2x^2 - x + 4$; Sec. 5.6, Ex. 1

12. $2x^2 + 5x + 2 + \frac{7}{x - 3}$; Sec. 5.7, Ex. 1   **13.** $(x + 2y)(x + 3y)$; Sec. 6.2, Ex. 6

14. a. $x^2 - 2x + 4$ **b.** $\frac{2}{y - 5}$; Sec. 7.1, Ex. 6   **15.** $\frac{3y^9}{160}$; Sec. 7.2, Ex. 4   **16.** 1; Sec. 7.3, Ex. 2

17. $\frac{x(3x - 1)}{(x + 1)^2(x - 1)}$; Sec. 7.4, Ex. 7   **18. a.** $\frac{x(x - 2)}{2(x + 2)}$ **b.** $\frac{x^2}{y^2}$; Sec. 7.5, Ex. 2   **19.** -5; Sec. 7.6, Ex. 1

20. 15 yd; Sec. 7.8, Ex. 4   **21. a.** $\frac{1}{8}$ **b.** $\frac{1}{9}$; Sec. 9.2, Ex. 3   **22.** $\frac{x - 4}{5(\sqrt{x} - 2)}$; Sec. 9.5, Ex. 7

■ CHAPTER 10   QUADRATIC EQUATIONS AND FUNCTIONS

Graphing Calculator Explorations   **1.** $-1.27, 6.27$ **3.** $-1.10, 0.90$ **5.** no real solution

Exercise Set 10.1   **1.** $-4, 4$ **3.** $-\sqrt{7}, \sqrt{7}$ **5.** $-3\sqrt{2}, 3\sqrt{2}$ **7.** $-\sqrt{10}, \sqrt{10}$ **9.** $-8, -2$ **11.** $6 - 3\sqrt{2}, 6 + 3\sqrt{2}$

13. $\frac{3 - 2\sqrt{2}}{2}, \frac{3 + 2\sqrt{2}}{2}$ **15.** $-3i, 3i$ **17.** $-\sqrt{6}, \sqrt{6}$ **19.** $-2i\sqrt{2}, 2i\sqrt{2}$ **21.** $1 - 4i, 1 + 4i$ **23.** $-7 - \sqrt{5}, -7 + \sqrt{5}$

25. $-3 - 2i\sqrt{2}, -3 + 2i\sqrt{2}$ **27.** $x^2 + 16x + 64 = (x + 8)^2$ **29.** $z^2 - 12z + 36 = (z - 6)^2$

31. $p^2 + 9p + \frac{81}{4} = \left(p + \frac{9}{2}\right)^2$ **33.** $x^2 + x + \frac{1}{4} = \left(x + \frac{1}{2}\right)^2$ **35.** $-8x, 8x$ **37.** $-5z, 5z$ **39.** $-5, -3$

41. $-3 - \sqrt{7}, -3 + \sqrt{7}$ **43.** $\frac{-1 - \sqrt{5}}{2}, \frac{-1 + \sqrt{5}}{2}$ **45.** $-1 - \sqrt{6}, -1 + \sqrt{6}$ **47.** $\frac{6 - \sqrt{30}}{3}, \frac{6 + \sqrt{30}}{3}$

49. $\frac{3 - \sqrt{11}}{2}, \frac{3 + \sqrt{11}}{2}$ **51.** $-4, \frac{1}{2}$ **53.** $-1, 5$ **55.** $-4 - \sqrt{15}, -4 + \sqrt{15}$ **57.** $\frac{-3 - \sqrt{21}}{3}, \frac{-3 + \sqrt{21}}{3}$ **59.** $-1, \frac{5}{2}$

61. $-1 - i, -1 + i$ **63.** $3 - \sqrt{6}, 3 + \sqrt{6}$ **65.** $-2 - i\sqrt{2}, -2 + i\sqrt{2}$ **67.** $\frac{-15 - 7\sqrt{5}}{10}, \frac{-15 + 7\sqrt{5}}{10}$

69. $\frac{1 - i\sqrt{47}}{4}, \frac{1 + i\sqrt{47}}{4}$ **71.** $-5 - i\sqrt{3}, -5 + i\sqrt{3}$ **73.** $-4, 1$ **75.** $\frac{2 - i\sqrt{2}}{2}, \frac{2 + i\sqrt{2}}{2}$ **77.** $\frac{-3 - \sqrt{69}}{6}, \frac{-3 + \sqrt{69}}{6}$

79. 20%   **81.** 11%   **83.** Answers may vary.   **85.** simple   **87.** 8.11 sec   **89.** 6.73 sec   **91.** 6 in.   **93.** $\frac{27\sqrt{2}}{2}$ in.

95. 2.828 thousand units   **97.** $\frac{7}{5}$ **99.** $\frac{1}{5}$ **101.** $5 - 10\sqrt{3}$ **103.** $\frac{3 - 2\sqrt{7}}{4}$ **105.** $2\sqrt{7}$ **107.** $\sqrt{13}$

Exercise Set 10.2   **1.** $-6, 1$ **3.** $-\frac{3}{5}, 1$ **5.** 3   **7.** $\frac{-7 - \sqrt{33}}{2}, \frac{-7 + \sqrt{33}}{2}$ **9.** $\frac{1 - \sqrt{57}}{8}, \frac{1 + \sqrt{57}}{8}$

11. $\frac{7 - \sqrt{85}}{6}, \frac{7 + \sqrt{85}}{6}$ **13.** $1 - \sqrt{3}, 1 + \sqrt{3}$ **15.** $-\frac{3}{2}, 1$ **17.** $\frac{3 - \sqrt{11}}{2}, \frac{3 + \sqrt{11}}{2}$ **19.** Answers may vary.

21. $\frac{3 - i\sqrt{87}}{8}, \frac{3 + i\sqrt{87}}{8}$ **23.** $-2 - \sqrt{11}, -2 + \sqrt{11}$ **25.** $\frac{-5 - i\sqrt{5}}{10}, \frac{-5 + i\sqrt{5}}{10}$ **27.** Answers may vary.

29. two real solutions   **31.** one real solution   **33.** two real solutions   **35.** two complex but not real solutions

37. $\frac{-5 - \sqrt{17}}{2}, \frac{-5 + \sqrt{17}}{2}$ **39.** $\frac{5}{2}, 1$ **41.** $\frac{3 - \sqrt{29}}{2}, \frac{3 + \sqrt{29}}{2}$ **43.** $\frac{-1 - \sqrt{19}}{6}, \frac{-1 + \sqrt{19}}{6}$

45. $-3 - 2i, -3 + 2i$ **47.** $\frac{-1 - i\sqrt{23}}{4}, \frac{-1 + i\sqrt{23}}{4}$ **49.** 1   **51.** $\frac{19 - \sqrt{345}}{2}, \frac{19 + \sqrt{345}}{2}$ **53.** 14 ft

55. $2 + 2\sqrt{2}$ cm, $2 + 2\sqrt{2}$ cm, $4 + 2\sqrt{2}$ cm   **57.** width: $-5 + 5\sqrt{17}$ ft; length: $5 + 5\sqrt{17}$ ft

59. a. $50\sqrt{2}$ m   **b.** 5000 sq. m   **61.** $\frac{1 + \sqrt{5}}{2}$ **63.** 8.9 sec   **65.** 2.8 sec   **67.** Sunday to Monday   **69.** Wednesday

71. 32; yes   **73. a.** $3056 million   **b.** 2005
75. 8.9 sec: 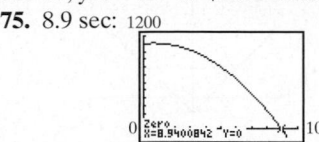 ; 2.8 sec: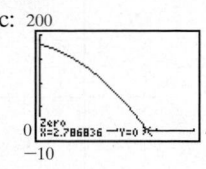

77. two real solutions   **79.** $\dfrac{11}{5}$ **81.** 15

83. $(x^2 + 5)(x + 2)(x - 2)$
85. $(z + 3)(z - 3)(z + 2)(z - 2)$
87. $\dfrac{\sqrt{3}}{3}$ **89.** $\dfrac{-\sqrt{2} - i\sqrt{2}}{2}, \dfrac{-\sqrt{2} + i\sqrt{2}}{2}$

91. $\dfrac{\sqrt{3} - \sqrt{11}}{4}, \dfrac{\sqrt{3} + \sqrt{11}}{4}$

Exercise Set 10.3 **1.** 2   **3.** 16   **5.** 1, 4   **7.** $3 - \sqrt{7}, 3 + \sqrt{7}$ **9.** $\dfrac{3 - \sqrt{57}}{4}, \dfrac{3 + \sqrt{57}}{4}$ **11.** $\dfrac{1 - \sqrt{29}}{2}, \dfrac{1 + \sqrt{29}}{2}$

13. $-2, 2, -2i, 2i$ **15.** $-\dfrac{1}{2}, \dfrac{1}{2}, -i\sqrt{3}, i\sqrt{3}$ **17.** $-3, 3, -2, 2$ **19.** $125, -8$ **21.** $-\dfrac{4}{5}, 0$ **23.** $-\dfrac{1}{8}, 27$ **25.** $-\dfrac{2}{3}, \dfrac{4}{3}$

27. $-\dfrac{1}{125}, \dfrac{1}{8}$ **29.** $-\sqrt{2}, \sqrt{2}, -\sqrt{3}, \sqrt{3}$ **31.** $\dfrac{-9 - \sqrt{201}}{6}, \dfrac{-9 + \sqrt{201}}{6}$ **33.** 2, 3   **35.** 3   **37.** 27, 125   **39.** $1, -3i, 3i$

41. $\dfrac{1}{8}, -8$ **43.** $-\dfrac{1}{2}, \dfrac{1}{3}$ **45.** 4   **47.** -3 **49.** $-\sqrt{5}, \sqrt{5}, -2i, 2i$ **51.** $-3, \dfrac{3 - 3i\sqrt{3}}{2}, \dfrac{3 + 3i\sqrt{3}}{2}$ **53.** 6, 12

55. $-\dfrac{1}{3}, \dfrac{1}{3}, -\dfrac{i\sqrt{6}}{3}, \dfrac{i\sqrt{6}}{3}$ **57.** 5 mph, then 4 mph   **59.** inlet pipe, 15.5 hr; hose, 16.5 hr   **61.** 55 mph, 66 mph

63. 8.5 hr.   **65.** 12 or -8 **67. a.** $(x - 6)$ **b.** 300 cu. cm $= (x - 6) \cdot (x - 6) \cdot 3$ cu. cm   **c.** 16 cm by 16 cm
69. a. 281.0 ft per sec   **b.** 284.8 ft per sec   **c.** Papis: 191.6 mph; Montoya: 194.2 mph   **71.** Answers may vary.
73. $(-\infty, 3]$ **75.** $(-5, \infty)$ **77.** domain: $\{x|x \text{ is a real number}\}$; range: $\{y|y \text{ is a real number}\}$; function
79. domain: $\{x|x \text{ is a real number}\}$; range: $\{y|y \geq -1\}$; function

Exercise Set 10.4   **1.** ; $(-\infty, -5) \cup (-1, \infty)$ **3.** ; $[-4, 3]$ **5.** ; $[2, 5]$

7. ; $\left(-5, -\dfrac{1}{3}\right)$ **9.** ; $(2, 4) \cup (6, \infty)$

11. ; $(-\infty, -4] \cup [0, 1]$ **13.** ; $(-\infty, -3) \cup (-2, 2) \cup (3, \infty)$

15. ; $(-7, 2)$ **17.** ; $(-1, \infty)$ **19.** ; $(-\infty, -1] \cup (4, \infty)$

21. Answers may vary.   **23.** ; $(-\infty, 2) \cup \left(\dfrac{11}{4}, \infty\right)$ **25.** ; $(0, 2] \cup [3, \infty)$

27. ; $(-\infty, -7) \cup (8, \infty)$ **29.** ; $\left[-\dfrac{5}{4}, \dfrac{3}{2}\right]$ **31.** ; $(-\infty, 0) \cup (1, \infty)$

33. ; $(-\infty, -4] \cup [4, 6]$ **35.** ; $\left(-\infty, -\dfrac{2}{3}\right] \cup \left[\dfrac{3}{2}, \infty\right)$

37. ; $\left(-4, -\dfrac{3}{2}\right) \cup \left(\dfrac{3}{2}, \infty\right)$ **39.** ; $(-\infty, -5] \cup [-1, 1] \cup [5, \infty)$

41. ; $\left(-\infty, -\dfrac{5}{3}\right) \cup \left(\dfrac{7}{2}, \infty\right)$ **43.** ; $(0, 10)$ **45.** ; $(-\infty, -4) \cup [5, \infty)$

47. ; $(-\infty, -6] \cup (-1, 0] \cup (7, \infty)$ **49.** ; $(-\infty, 1) \cup (2, \infty)$

51. ; $(-\infty, -8] \cup (-4, \infty)$ **53.** ; $(-\infty, 0] \cup \left(5, \dfrac{11}{2}\right]$

55. ; $(0, \infty)$ **57.** Any number less than -1 or between 0 and 1.   **59.** x is between 2 and 11.

61. 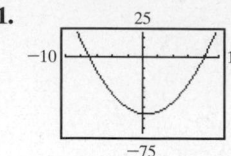   **63.** 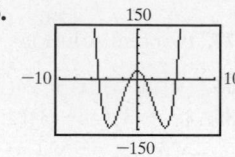   **65.** 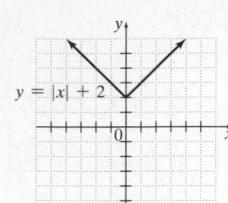   **67.**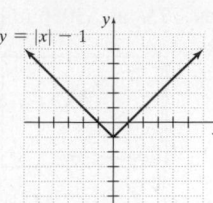

69. 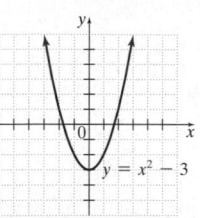   **71.**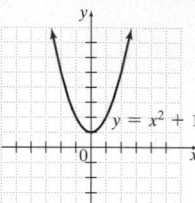

Exercise Set 10.5   **1.** $k = \dfrac{1}{5}; y = \dfrac{1}{5}x$ **3.** $k = \dfrac{3}{2}; y = \dfrac{3}{2}x$ **5.** $k = 14; y = 14x$ **7.** $k = 0.25; y = 0.25x$ **9.** 4.05 lb

11. $P = 566{,}222$ tons   **13.** $k = 30; y = \dfrac{30}{x}$ **15.** $k = 700; y = \dfrac{700}{x}$ **17.** $k = 2; y = \dfrac{2}{x}$ **19.** $k = 0.14; y = \dfrac{0.14}{x}$

21. 54 mph   **23.** 72 amps   **25.** divided by 4   **27.** $x = kyz$ **29.** $r = kst^3$ **31.** 22.5 tons   **33.** 15π cu. in.
35. 90 hp   **37.** 800 millibars   **39.** multiplied by 2   **41.** multiplied by 4

43. 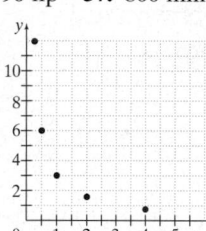   **45.**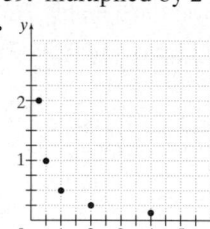

47. $C = 12\pi$ cm; $A = 36\pi$ sq. cm
49. $C = 14\pi$ m; $A = 49\pi$ sq. m
51. 6
53. 2
55. $\dfrac{1}{5}$
57. $\dfrac{5}{11}$

Graphing Calculator Explorations   **1.** 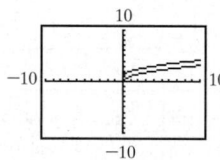   **3.** 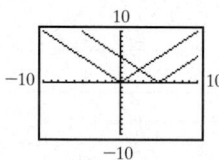   **5.**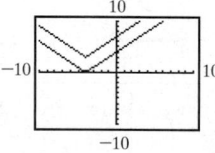

Mental Math   **1.** $(0, 0)$ **3.** $(2, 0)$ **5.** $(0, 3)$ **7.** $(-1, 5)$

Exercise Set 10.6   **1.** 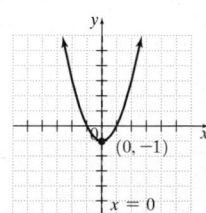   **3.** 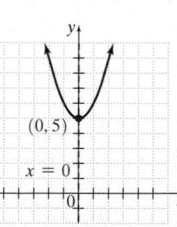   **5.** 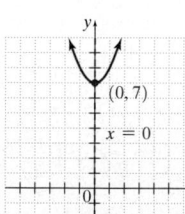   **7.**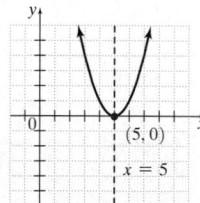

9. 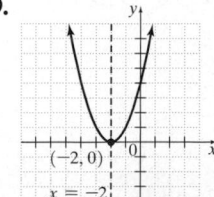   **11.** 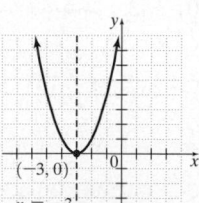   **13.** 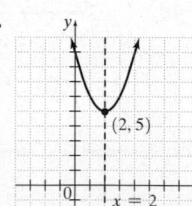   **15.**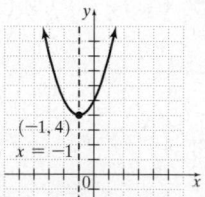

17.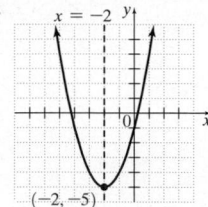

19.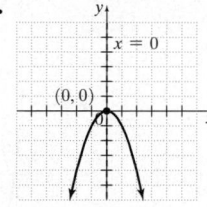

21.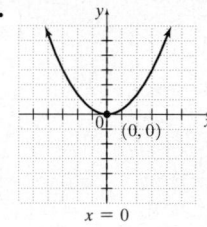

23.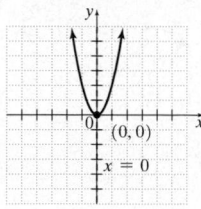

25.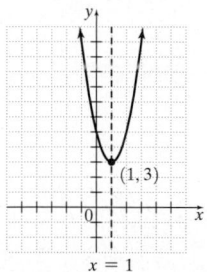

27.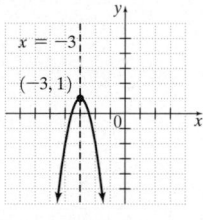

29.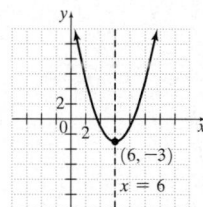

31.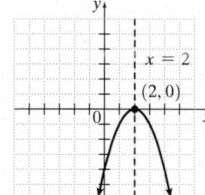

33.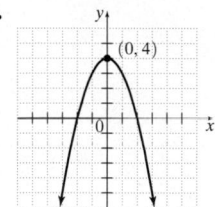

35.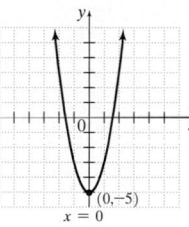

37.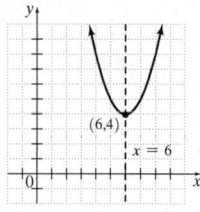

39.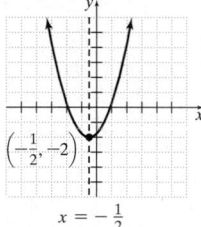

41.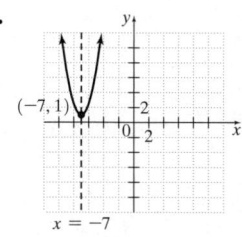

43.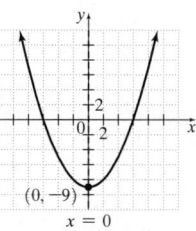

45.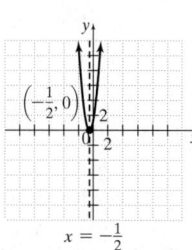

47.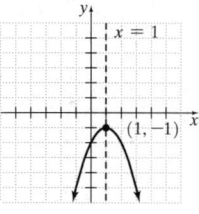

49.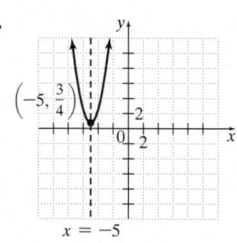

51.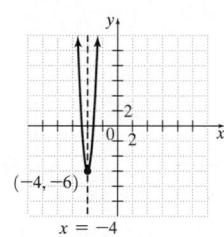

53.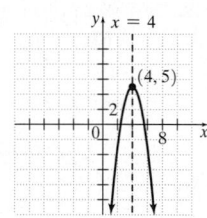

55. $f(x) = 5(x - 2)^2 + 3$

57. $f(x) = 5(x + 3)^2 + 6$

59.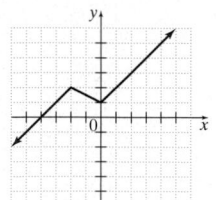

61.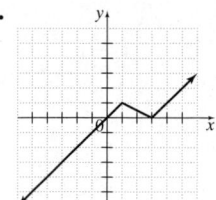

63.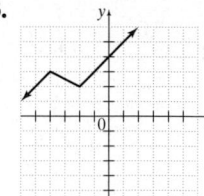

65. $x^2 + 8x + 16$

67. $z^2 - 16z + 64$

69. $y^2 + y + \dfrac{1}{4}$

71. $-6, 2$

73. $-5 - \sqrt{26}, -5 + \sqrt{26}$

75. $4 - 3\sqrt{2}, 4 + 3\sqrt{2}$

Exercise Set 10.7   **1.** $(-4, -9)$ **3.** $(5, 30)$ **5.** $(1, -2)$ **7.** $\left(\dfrac{1}{2}, \dfrac{5}{4}\right)$ **9.** D   **11.** B

13. 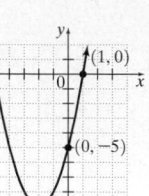   **15.** 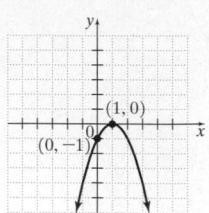   **17.** 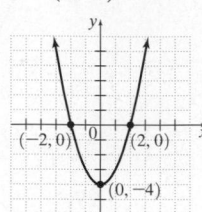   **19.**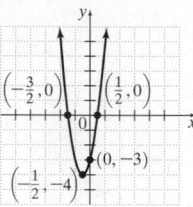

21. 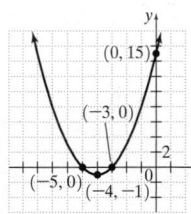   **23.** 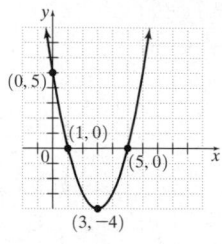   **25.** 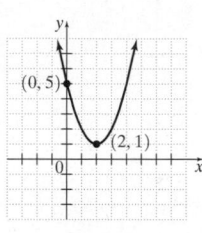   **27.**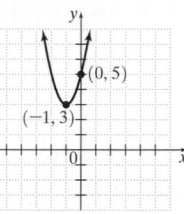

29. 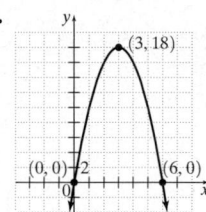   **31.** 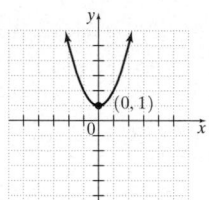   **33.** 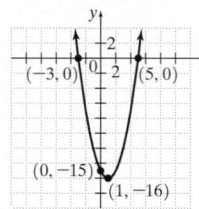   **35.**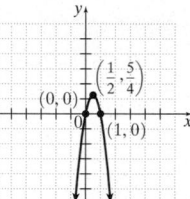

37. 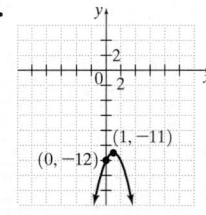   **39.** 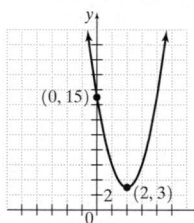   **41.** 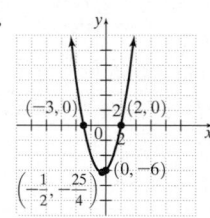   **43.**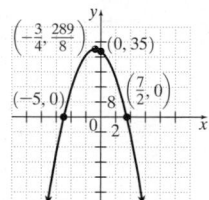

45. a. 200 bicycles   **b.** \$12,000   **47.** 16 ft   **49.** 30 and 30   **51.** $5, -5$ **53.** length, 20 units; width, 20 units
55. a. 135.27 million metric tons   **b.** 1996   **c.** 184.39 million metric tons

57. 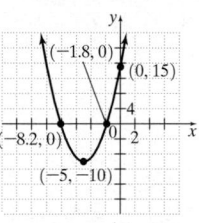   **59.** 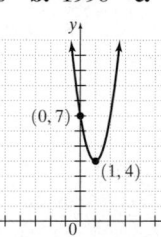   **61.** 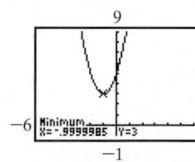   **63.**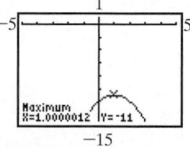

65. -0.84 **67.** 1.43   **69.** 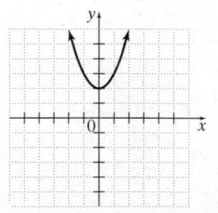   **71.** 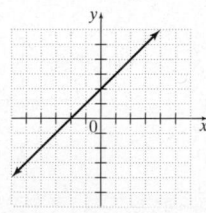   **73.**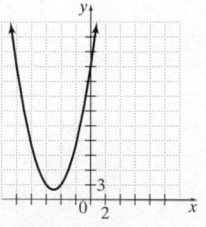

75. 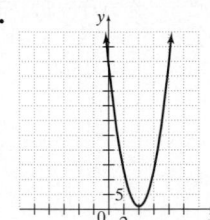   **77.**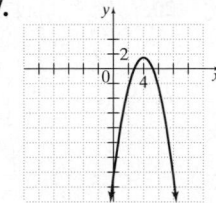

Chapter 10 Review **1.** $14, 1$ **3.** $\dfrac{4}{5}, -\dfrac{1}{2}$ **5.** $-7, 7$ **7.** $-\dfrac{4}{9}, \dfrac{2}{9}$ **9.** $\dfrac{-3 - \sqrt{5}}{2}, \dfrac{-3 + \sqrt{5}}{2}$ **11.** $\dfrac{-3 - i\sqrt{7}}{8}, \dfrac{-3 + i\sqrt{7}}{8}$

13. 4.25% **15.** two complex but not real solutions   **17.** two real solutions   **19.** 8 **21.** $-i\sqrt{11}, i\sqrt{11}$

23. $\dfrac{5 - i\sqrt{143}}{12}, \dfrac{5 + i\sqrt{143}}{12}$ **25.** $\dfrac{21 - \sqrt{41}}{50}, \dfrac{21 + \sqrt{41}}{50}$ **27. a.** 20 ft   **b.** $\dfrac{15 + \sqrt{321}}{16}$ sec; 2.1 sec

29. $3, \dfrac{-3 + 3i\sqrt{3}}{2}, \dfrac{-3 - 3i\sqrt{3}}{2}$ **31.** $\dfrac{2}{3}, 5$ **33.** $-5, 5, -2i, 2i$ **35.** $1, 125$ **37.** $-1, 1, -i, i$ **39.** Jerome: 10.5 hr; Tim: 9.5 hr

41. ⟵[———]⟶; $[-5, 5]$ **43.** ⟵——] [——⟶; $\left(-\infty, -\dfrac{5}{4}\right] \cup \left[\dfrac{3}{2}, \infty\right)$ **45.** ⟵——( )——⟶; $(5, 6)$

47. ⟵——) (——) (——⟶; $(-\infty, -6) \cup \left(-\dfrac{3}{4}, 0\right) \cup (5, \infty)$ **49.** ⟵——( )——(——⟶; $(-5, -3) \cup (5, \infty)$

51. ⟵—( )—( )—⟶; $\left(-\dfrac{6}{5}, 0\right) \cup \left(\dfrac{5}{6}, 3\right)$ **53.** 9 **55.** 3.125 cu. ft

57. 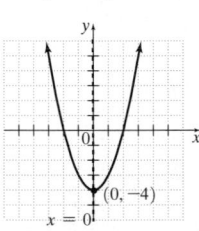   **59.** 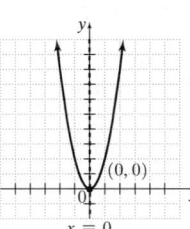   **61.** 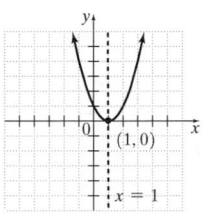   **63.**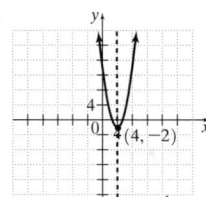

65. 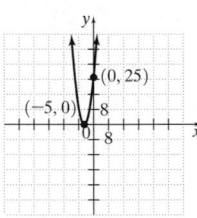   **67.** 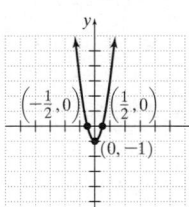   **69.** 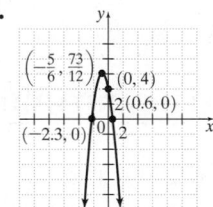   **71.** 210 and 210

Chapter 10 Test **1.** $\dfrac{7}{5}, -1$ **2.** $-1 - \sqrt{10}, -1 + \sqrt{10}$ **3.** $\dfrac{1 + i\sqrt{31}}{2}, \dfrac{1 - i\sqrt{31}}{2}$ **4.** $3 - \sqrt{7}, 3 + \sqrt{7}$ **5.** $-\dfrac{1}{7}, -1$

6. $\dfrac{3 + \sqrt{29}}{2}, \dfrac{3 - \sqrt{29}}{2}$ **7.** $-2 - \sqrt{11}, -2 + \sqrt{11}$ **8.** $-3, 3, -i, i$ **9.** $-1, 1, -i, i$ **10.** $6, 7$ **11.** $3 - \sqrt{7}, 3 + \sqrt{7}$

12. $\dfrac{2 - i\sqrt{6}}{2}, \dfrac{2 + i\sqrt{6}}{2}$ **13.** ⟵——) (——⟶; $\left(-\infty, -\dfrac{3}{2}\right) \cup (5, \infty)$

14. ⟵——) (——) (——⟶; $(-\infty, -5) \cup (-4, 4) \cup (5, \infty)$ **15.** ⟵——) (——⟶; $(-\infty, -3) \cup (2, \infty)$

16. ⟵——) [——⟶; $(-\infty, -3) \cup [2, 3)$

17. 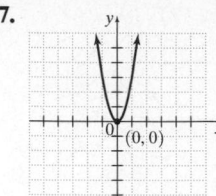 **18.** 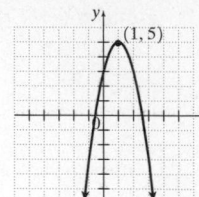 **19.** 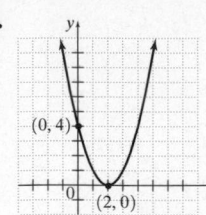 **20.**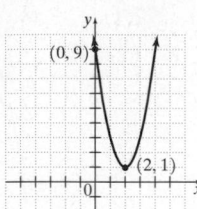

21. $(2 + \sqrt{46})$ ft ≈ 8.8 ft   **22.** $(5 + \sqrt{17})$ hr ≈ 9.12 hr   **23. a.** 272 ft   **b.** 5.12 sec   **24.** 7 ft   **25.** 16   **26.** 9
27. 256 ft

Chapter 10 Cumulative Review   **1. a.** $\dfrac{1}{2}$ **b.** 9; Sec. 1.6, Ex. 6   **2. a.** $5x + 7$ **b.** $-4a - 1$ **c.** $4y - 3y^2$

d. $7.3x - 6$; Sec. 2.1, Ex. 4

3. slope: 1; 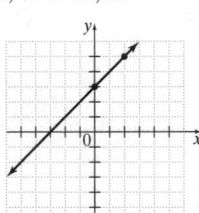 ; Sec. 3.5, Ex. 1

4. no solution; Sec. 4.1, Ex. 3   **5.** $(-2, 0)$; Sec. 4.2, Ex. 3
6. infinite number of solutions; Sec. 4.3, Ex. 4
7. Albert: 3.25 mph; Louis: 4.25 mph; Sec. 4.4, Ex. 3

8. a. $\dfrac{25x^4}{y^6}$ **b.** x^6 **c.** $32x^2$ **d.** a^3b; Sec. 5.1, Ex. 10

9. $3x^3y - 15x - 1 - \dfrac{6}{xy}$; Sec. 5.6, Ex. 2

10. a. 5   **b.** 5; Sec. 5.7, Ex. 3   **11.** $-6, -\dfrac{3}{2}, \dfrac{1}{5}$; Sec. 6.6, Ex. 6

12. 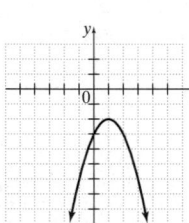 ; Sec. 6.8, Ex. 2   **13.** $\dfrac{1}{5x - 1}$; Sec. 7.1, Ex. 3b   **14.** $\dfrac{xy + 2x^3}{y - 1}$; Sec. 7.5, Ex. 3

15. $\dfrac{31}{2}$; Sec. 7.7, Ex. 3

16. a. 3   **b.** $|x|$ **c.** $|x - 2|$ **d.** -5 **e.** $2x - 7$; Sec. 9.1, Ex. 5
17. a. $\sqrt[3]{5}$ **b.** $\sqrt{x}$ **c.** $\sqrt{rs^3}$; Sec. 9.2, Ex. 7
18. a. $5x\sqrt{x}$ **b.** $3x^2y^2\sqrt[3]{2y^2}$ **c.** $3z^2\sqrt[4]{z^3}$; Sec. 9.3, Ex. 4

19. a. $\dfrac{2\sqrt{5}}{5}$ **b.** $\dfrac{8\sqrt{x}}{3x}$ **c.** $\dfrac{\sqrt[3]{4}}{2}$; Sec. 9.5, Ex. 1   **20.** $\dfrac{2}{9}$; Sec. 9.6, Ex. 5   **21. a.** $\dfrac{1}{2} + \dfrac{3}{2}i$ **b.** $-\dfrac{7}{3}i$; Sec. 9.7, Ex. 5

22. $-1 + 2\sqrt{3}, -1 - 2\sqrt{3}$; Sec. 10.1, Ex. 3   **23.** 9; Sec. 10.3, Ex. 1   **24.** $\dfrac{1}{6}$; $y = \dfrac{1}{6}x$; Sec. 10.5, Ex. 1

■ CHAPTER 11   EXPONENTIAL AND LOGARITHMIC FUNCTIONS

Exercise Set 11.1   **1. a.** $3x - 6$ **b.** $-x - 8$ **c.** $2x^2 - 13x - 7$ **d.** $\dfrac{x - 7}{2x + 1}$, where $x \neq -\dfrac{1}{2}$ **3. a.** $x^2 + 5x + 1$

b. $x^2 - 5x + 1$ **c.** $5x^3 + 5x$ **d.** $\dfrac{x^2 + 1}{5x}$, where $x \neq 0$ **5. a.** $\sqrt{x} + x + 5$ **b.** $\sqrt{x} - x - 5$ **c.** $x\sqrt{x} + 5\sqrt{x}$

d. $\dfrac{\sqrt{x}}{x + 5}$, where $x \neq -5$ **7. a.** $5x^2 - 3x$ **b.** $-5x^2 - 3x$ **c.** $-15x^3$ **d.** $-\dfrac{3}{5x}$, where $x \neq 0$ **9.** 42   **11.** -18 **13.** 0
15. $(f \circ g)(x) = 25x^2 + 1$; $(g \circ f)(x) = 5x^2 + 5$ **17.** $(f \circ g)(x) = 2x + 11$; $(g \circ f)(x) = 2x + 4$
19. $(f \circ g)(x) = -8x^3 - 2x - 2$; $(g \circ f)(x) = -2x^3 - 2x + 4$ **21.** $(f \circ g)(x) = \sqrt{-5x + 2}$; $(g \circ f)(x) = -5\sqrt{x} + 2$
23. $H(x) = (g \circ h)(x)$ **25.** $F(x) = (h \circ f)(x)$ **27.** $G(x) = (f \circ g)(x)$ **29.** Answers may vary.
31. Answers may vary.   **33.** Answers may vary.   **35.** 6   **37.** 4   **39.** 48   **41.** -1 **43.** $P(x) = R(x) - C(x)$

45. $y = x - 2$ **47.** $y = \dfrac{x}{3}$ **49.** $y = -\dfrac{x + 7}{2}$

Exercise Set 11.2   **1.** one-to-one; $f^{-1} = \{(-1, -1), (1, 1), (2, 0), (0, 2)\}$ **3.** one-to-one; $h^{-1} = \{(10, 10)\}$
5. one-to-one; $f^{-1} = \{(12, 11), (3, 4), (4, 3), (6, 6)\}$ **7.** not one-to-one

9. one-to-one;

Rank in Population (Input)	1	49	12	2	45
State (Output)	CA	VT	VA	TX	SD

11. a. 3 **b.** 1   **13. a.** 1 **b.** -1 **15.** one-to-one   **17.** not one-to-one   **19.** one-to-one   **21.** not one-to-one

23. $f^{-1}(x) = x - 4$ **25.** $f^{-1}(x) = \dfrac{x + 3}{2}$ **27.** $f^{-1}(x) = 2x + 2$ **29.** $f^{-1}(x) = \sqrt[3]{x}$

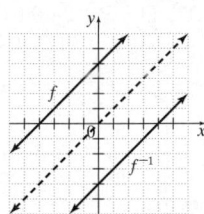 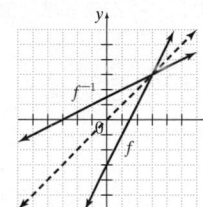 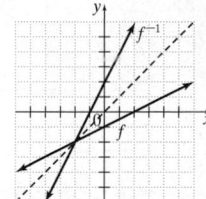 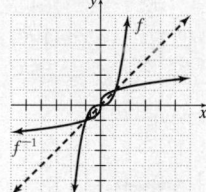

31. $f^{-1}(x) = \dfrac{x - 2}{5}$ **33.** $f^{-1}(x) = 5x + 2$ **35.** $f^{-1}(x) = x^3$ **37.** $f^{-1}(x) = \dfrac{5 - x}{3x}$ **39.** $f^{-1}(x) = \sqrt[3]{x} - 2$

41. 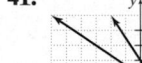 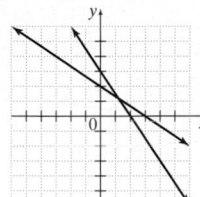   **43.** 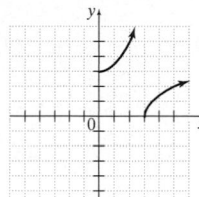   **45.** 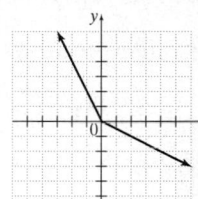   **47.** $(f \circ f^{-1})(x) = x; (f^{-1} \circ f)(x) = x$
49. $(f \circ f^{-1})(x) = x; (f^{-1} \circ f)(x) = x$

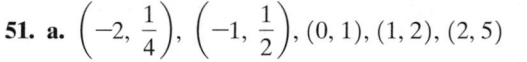

51. a. $\left(-2, \dfrac{1}{4}\right), \left(-1, \dfrac{1}{2}\right), (0, 1), (1, 2), (2, 5)$ **c.** 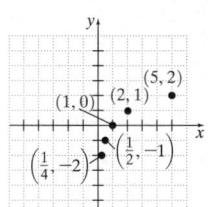   **d.**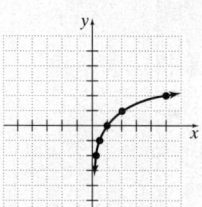
b. $\left(\dfrac{1}{4}, -2\right), \left(\dfrac{1}{2}, -1\right), (1, 0), (2, 1), (5, 2)$

53. $f^{-1}(x) = \dfrac{x - 1}{3}$; 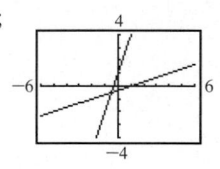   **55.** $f^{-1}(x) = x^3 - 1$; 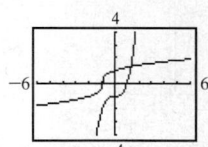   **57.** 5 **59.** 8 **61.** $\dfrac{1}{27}$
63. 9 **65.** $3^{1/2} \approx 1.73$

Graphing Calculator Explorations   **1.** 81.98%; 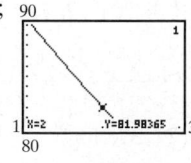   **3.** 22.54%;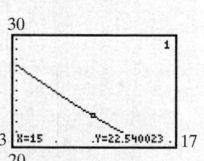

Exercise Set 11.3   **1.** 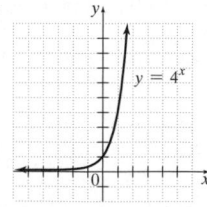   **3.** 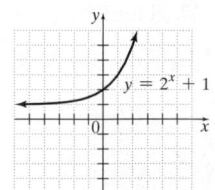   **5.** 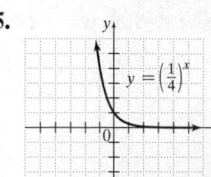   **7.**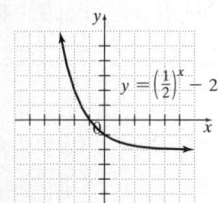

9. 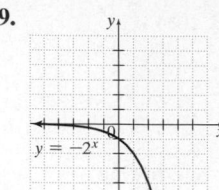 **11.** 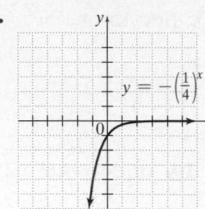 **13.** 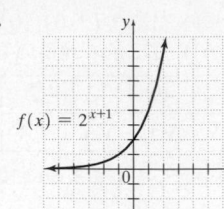 **15.**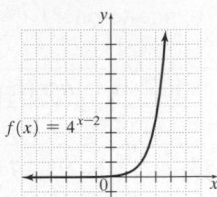

17. Answers may vary. **19.** 3 **21.** $\dfrac{3}{4}$ **23.** $\dfrac{8}{5}$ **25.** $-\dfrac{2}{3}$ **27.** 4 **29.** $\dfrac{3}{2}$ **31.** $-\dfrac{1}{3}$ **33.** -2 **35.** C **37.** D **39.** 24.6 lb

41. 519 rats **43.** 1.1 g **45.** 16,190,000 residents **47.** $7621.42 **49.** $4065.59

51. 24.60 lb; 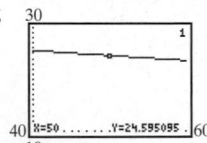   **53.** 18.62 lb; 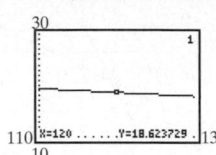   **55.** 50.41 g;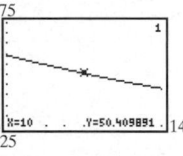

57. 6,496,000 people **59. a.** 372.6 parts per million by volume **b.** 405.2 parts per million by volume
61. 1410 cellular phone users **63.** 6 **65.** no solution **67.** 2, 9 **69.** 2 **71.** 0

Exercise Set 11.4 **1.** $6^2 = 36$ **3.** $3^{-3} = \dfrac{1}{27}$ **5.** $10^3 = 1000$ **7.** $e^4 = x$ **9.** $e^{-2} = \dfrac{1}{e^2}$ **11.** $7^{1/2} = \sqrt{7}$ **13.** $\log_2 16 = 4$

15. $\log_{10} 100 = 2$ **17.** $\log_e x = 3$ **19.** $\log_{10} \dfrac{1}{10} = -1$ **21.** $\log_4 \dfrac{1}{16} = -2$ **23.** $\log_5 \sqrt{5} = \dfrac{1}{2}$ **25.** 3 **27.** -2 **29.** $\dfrac{1}{2}$

31. -1 **33.** 0 **35.** 4 **37.** 2 **39.** 5 **41.** 4 **43.** -3 **45.** Answers may vary. **47.** 2 **49.** 81 **51.** 7 **53.** -3

55. -3 **57.** 2 **59.** 2 **61.** $\dfrac{27}{64}$ **63.** 10 **65.** 3 **67.** 3 **69.** 1

71. 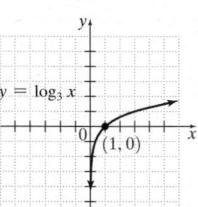 **73.** 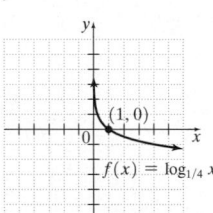 **75.** 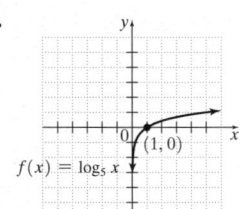 **77.**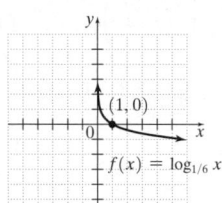

79. 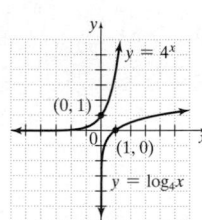 **81.** 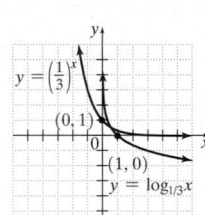 **83.** 0.0827 **85.** 2 and 3 **87.** -1
89. $x - 5$ **91.** 3 **93.** $\dfrac{y - 9}{y^2 - 1}$

Exercise Set 11.5 **1.** $\log_5 14$ **3.** $\log_4 9x$ **5.** $\log_{10}(10x^2 + 20)$ **7.** $\log_5 3$ **9.** $\log_2 \dfrac{x}{y}$ **11.** $\log_4 4$, or 1 **13.** $2 \log_3 x$

15. $-1 \log_4 5 = -\log_4 5$ **17.** $\dfrac{1}{2} \log_5 y$ **19.** $\log_2 25$ **21.** $\log_5 x^3 z^6$ **23.** $\log_{10} \dfrac{x^3 - 2x}{x + 1}$ **25.** $\log_4 35$ **27.** $\log_3 4$

29. $\log_7 \dfrac{9}{2}$ **31.** $\log_4 48$ **33.** $\log_2 \dfrac{x^{7/2}}{(x + 1)^2}$ **35.** $\log_8 x^{16/3}$ **37.** $\log_2 7 + \log_2 11 - \log_2 3$ **39.** $\log_3 4 + \log_3 y - \log_3 5$

43. $\dfrac{1}{2} \log_b 7 + \dfrac{1}{2} \log_b x$ **45.** $\log_7 5 + \log_7 x - \log_7 4$ **47.** $3 \log_5 x + \log_5(x + 1)$ **49.** $2 \log_6 x - \log_6(x + 3)$

51. 0.2 **53.** 1.2 **55.** 0.233 **57.** true **59.** false **61.** false **63.** 1.29 **65.** -0.68 **67.** -0.125

69.

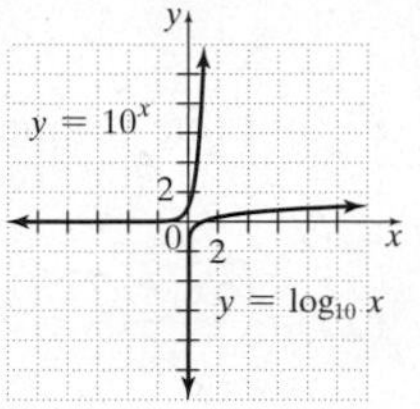

71. -1 **73.** $\frac{1}{2}$

Exercise Set 11.6 **1.** 0.9031 **3.** 0.3636 **5.** 0.6931 **7.** -2.6367 **9.** 1.1004 **11.** 1.6094 **13.** 1.6180 **15.** Answers may vary. **17.** 2 **19.** -3 **21.** 2 **23.** $\frac{1}{4}$ **25.** 3 **27.** 9 **29.** -4 **31.** $\frac{1}{2}$ **33.** Answers may vary. **35.** $10^{1.3} \approx 19.9526$ **37.** $\frac{10^{1.1}}{2} \approx 6.2946$ **39.** $e^{1.4} \approx 4.0552$ **41.** $\frac{4 + e^{2.3}}{3} \approx 4.6581$ **43.** $10^{2.3} \approx 199.5262$ **45.** $e^{-2.3} \approx 0.1003$ **47.** $\frac{10^{-0.5} - 1}{2} \approx -0.3419$ **49.** $\frac{e^{0.18}}{4} \approx 0.2993$ **51.** 1.5850 **53.** -2.3219 **55.** 1.5850 **57.** -1.6309 **59.** 0.8617 **61.** 4.2 **63.** 5.3 **65.** \$3656.38 **67.** \$2542.50

69.

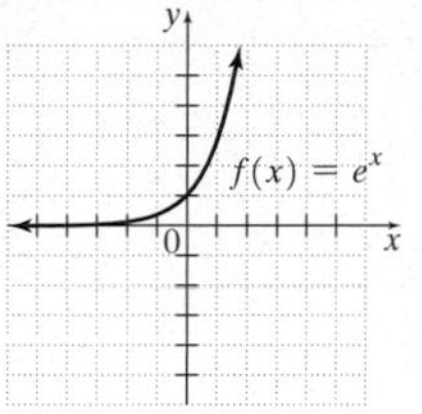

71.

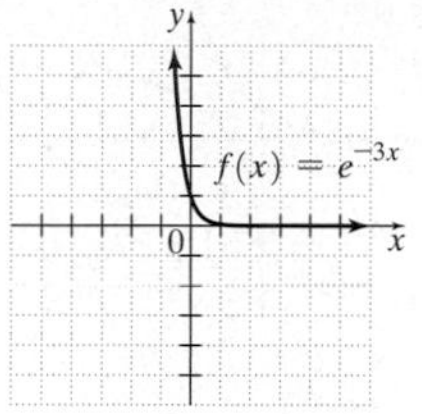

73.

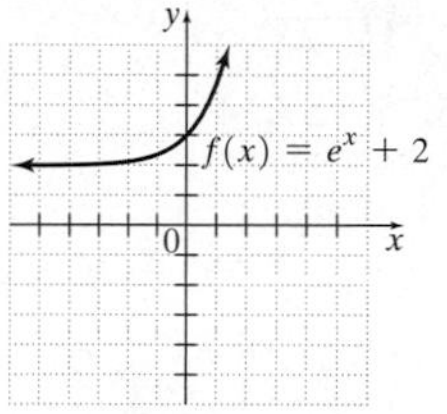

75.

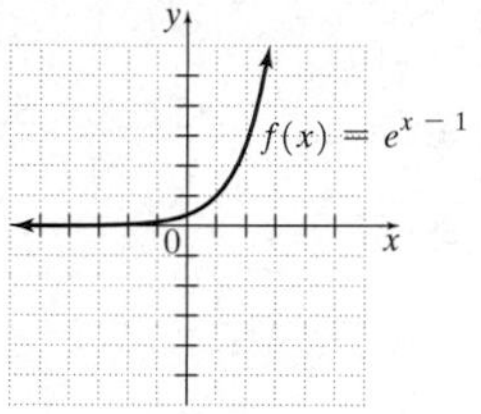

77.

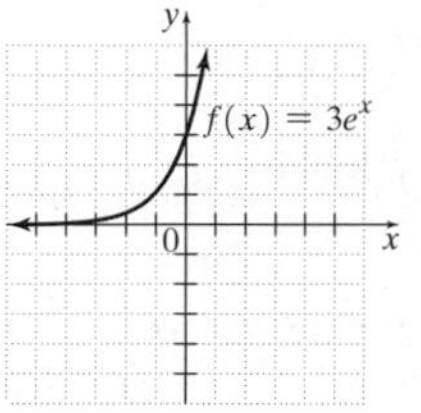

79.

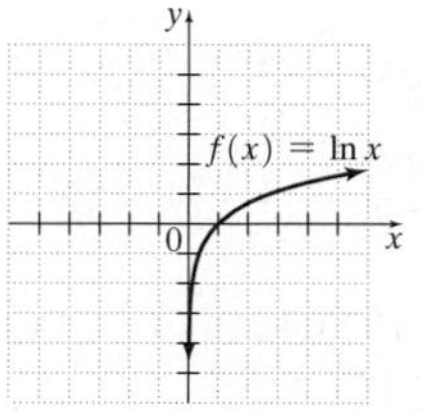

81.

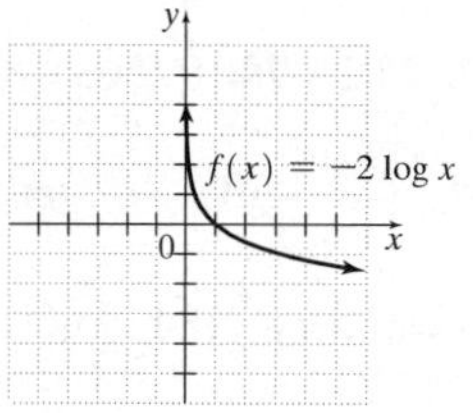

83.

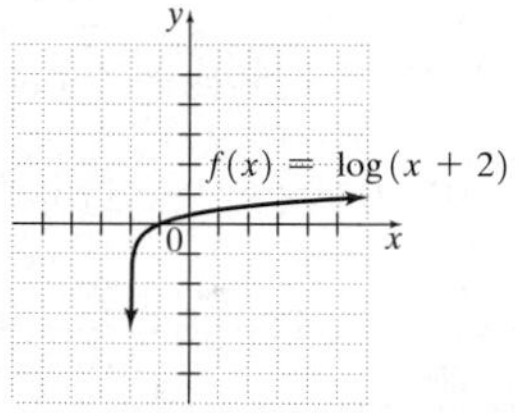

85.

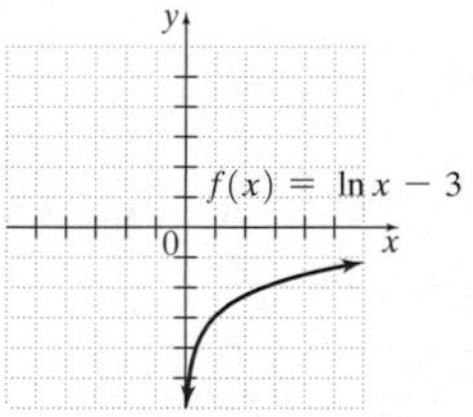

87.

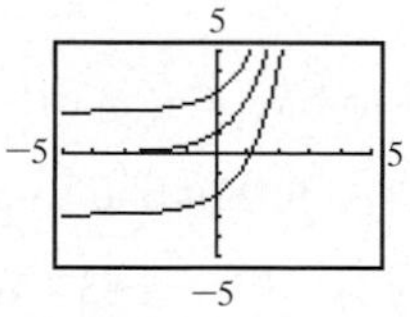

Answers may vary.

89. $\frac{4}{7}$ **91.** $x = \frac{3y}{4}$ **93.** $-6, -1$ **95.** $(2, -3)$

Graphing Calculator Explorations **1.** 3.67 years, or 3 years and 8 months **3.** 23.16 years, or 23 years and 2 months

Exercise Set 11.7 **1.** $\frac{\log 6}{\log 3}$; 1.6309 **3.** $\frac{\log 3.8}{2 \log 3}$; 0.6076 **5.** $3 + \frac{\log 5}{\log 2}$; 5.3219 **7.** $\frac{\log 5}{\log 9}$; 0.7325 **9.** $\frac{\log 3}{\log 4} - 7$; -6.2075 **11.** $\frac{1}{3}\left(4 + \frac{\log 11}{\log 7}\right)$; 1.7441 **13.** $\frac{\ln 5}{6}$; 0.2682 **15.** 11 **17.** 9, -9 **19.** $\frac{1}{2}$ **21.** $\frac{3}{4}$ **23.** 2 **25.** $\frac{1}{8}$ **27.** 11 **29.** 4, -1 **31.** $\frac{1}{5}$ **33.** 100 **35.** $\frac{-5 + \sqrt{33}}{2}$ **37.** $\frac{192}{127}$ **39.** $\frac{2}{3}$ **41.** 103 wolves **43.** 11,750,000 inhabitants **45.** 39.8 yr **47.** 10 yr **49.** 1.7 yr **51.** 8.8 yr **55.** 55.7 in. **57.** 11.9 lb/sq. in. **59.** 3.2 mi **61.** 12 weeks **63.** 18 weeks **65.** 6.93 **67.** $-3.68, 0.19$ **69.** 1.74 **71.** 0.2 **73.** $-\frac{5}{3}$ **75.** $\frac{17}{4}$ **77.** $f^{-1}(x) = \frac{x - 2}{5}$

Chapter 11 Review   **1.** $3x - 4$ **3.** $2x^2 - 9x - 5$ **5.** $x^2 + 2x - 1$ **7.** 18   **9.** -2
11. one-to-one; $h^{-1} = \{(14, -9), (8, 6), (12, -11), (15, 15)\}$

13. one-to-one;

Rank in Automobile Thefts (Input)	2	4	1	3
US Region (Output)	W	Midwest	S	NE

15. a. 3   **b.** 7   **17.** not one-to-one

19. not one-to-one   **21.** $f^{-1}(x) = x + 9$ **23.** $f^{-1}(x) = \dfrac{x - 11}{6}$ **25.** $f^{-1}(x) = \sqrt[3]{x + 5}$ **27.** $g^{-1}(x) = \dfrac{6x + 7}{12}$

29. 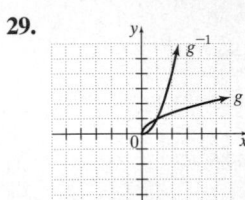   **31.** $f^{-1}(x) = \dfrac{x + 3}{2}$ 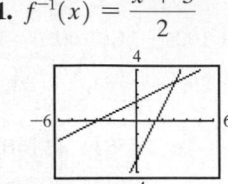   **33.** -2 **35.** $\dfrac{3}{2}$ **37.** $\dfrac{8}{9}$ **39.**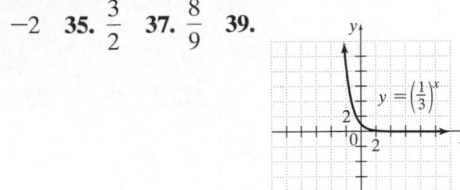

41. 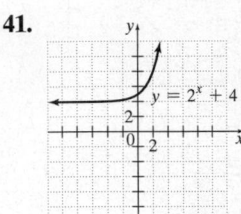   **43.** \$1131.82   **45.** $\log_7 49 = 2$ **63.**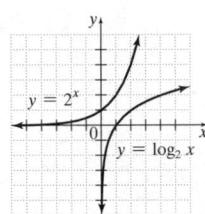
47. $\left(\dfrac{1}{2}\right)^{-4} = 16$ **49.** $\dfrac{1}{64}$
51. 0   **53.** 8   **55.** 5   **57.** 4
59. $\dfrac{17}{3}$ **61.** $-1, 4$

65. $\log_3 32$ **67.** $\log_7 \dfrac{3}{4}$ **69.** $\log_{11} 4$ **71.** $\log_5 \dfrac{x^3}{(x + 1)^2}$ **73.** $3 \log_3 x - \log_3(x + 2)$

75. $\log_2 3 + 2 \log_2 x + \log_2 y - \log_2 z$ **77.** 2.02   **79.** 0.5563   **81.** 0.2231   **83.** 3   **85.** -1 **87.** $\dfrac{e^2}{2}$ **89.** $\dfrac{e^{-1} + 3}{2}$

91. 1.67 mm   **93.** 0.2920   **95.** \$1957.30   **97.** $\dfrac{\log 7}{2 \log 3}$; 0.8856   **99.** $\dfrac{1}{2}\left(\dfrac{\log 6}{\log 3} - 1\right)$; 0.3155   **101.** $\dfrac{1}{3}\left(\dfrac{\log 4}{\log 5} + 5\right)$; 1.9538

103. $-\dfrac{\log 2}{\log 5} + 1$; 0.5693   **105.** $\dfrac{25}{2}$ **107.** no solution   **109.** $2\sqrt{2}$ **111.** 197,044 ducks   **113.** 53 yr   **115.** 36 yr
117. 8.5 yr   **119.** 2.82

Chapter 11 Test   **1.** 5   **2.** $x - 7$ **3.** $x^2 - 6x - 2$

4. 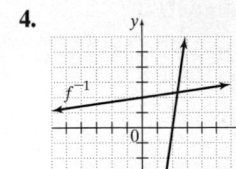   **5.** one-to-one   **6.** not one-to-one   **7.** one-to-one; $f^{-1}(x) = \dfrac{-x + 6}{2}$

8. one-to-one; $f^{-1} = \{(0, 0), (3, 2), (5, -1)\}$ **9.** not one-to-one   **10.** $\log_3 24$

11. $\log_5 \dfrac{x^4}{x + 1}$ **12.** $\log_6 2 + \log_6 x - 3 \log_6 y$ **13.** -1.53 **14.** 1.0686   **15.** -1

16. $\dfrac{1}{2}\left(\dfrac{\log 4}{\log 3} - 5\right)$; -1.8691 **17.** $\dfrac{1}{9}$ **18.** $\dfrac{1}{2}$ **19.** 22   **20.** $\dfrac{25}{3}$ **21.** $\dfrac{43}{21}$ **22.** -1.0979

23. 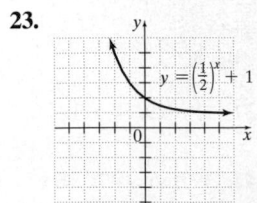   **24.** 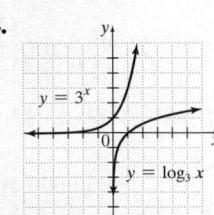   **25.** \$5234.58   **26.** 6 yr   **27.** 64,913 prairie dogs   **28.** 15 yr
29. 1.2%   **30.** 3.95;

Chapter 11 Cumulative Review   **1. a.** commutative property of multiplication   **b.** associative property of addition
c. identity element for addition   **d.** commutative property of multiplication   **e.** multiplicative inverse property
f. additive inverse property   **g.** commutative and associative properties of multiplication; Sec. 1.8, Ex. 6   **2.** 12; Sec. 2.3, Ex. 3

3. a. $\dfrac{m^7}{n^7}$ **b.** $\dfrac{x^{12}}{81 y^{20}}$; Sec. 5.1, Ex. 7   **4. a.** 0.056   **b.** 200,000; Sec. 5.2, Ex. 7   **5.** $3t^3 + 2t^2 - 6t + 4$; Sec. 5.4, Ex. 4

6. $x^2 + x - 12$; Sec. 5.5, Ex. 1   **7.** $3x^2 + 2x + 3 + \dfrac{-6x + 9}{x^2 - 1}$; Sec. 5.6, Ex. 5   **8.** $(x + 3)(5 + y)$; Sec. 6.1, Ex. 7

9. $(x - 2)(x + 6)$; Sec. 6.2, Ex. 3   **10.** $(2x - 3y)(5x + y)$; Sec. 6.3, Ex. 4   **11.** $(x + 5)(x - 5)$; Sec. 6.4, Ex. 1
12. $5a^2b(2b - 1)(3b + 7)$; Sec. 6.5, Ex. 5   **13.** $0, -2, 2$; Sec. 6.6, Ex. 5   **14.** 4 sec; Sec. 6.7, Ex. 1   **15. a.** $102.60
b. $12.60; Sec. 7.1, Ex. 1   **16. a.** 3   **b.** -3 **c.** -5 **d.** not a real number   **e.** $4x$; Sec. 9.1, Ex. 4   **17. a.** $\sqrt[4]{x^3}$ **b.** $\sqrt[6]{x}$
c. $\sqrt[6]{72}$; Sec. 9.2, Ex. 8   **18. a.** $5\sqrt{3} + 3\sqrt{10}$ **b.** $\sqrt{35} + \sqrt{5} - \sqrt{42} - \sqrt{6}$ **c.** $21x - 7\sqrt{5x} + 15\sqrt{x} - 5\sqrt{5}$

d. $49 - 8\sqrt{3}$ **e.** $2x - 25$; Sec. 9.4, Ex. 3   **19.** $\dfrac{\sqrt[4]{xy^3}}{3y^2}$; Sec. 9.5, Ex. 3   **20.** 3; Sec. 9.6, Ex. 4

21. $\dfrac{9 + i\sqrt{15}}{6}, \dfrac{9 - i\sqrt{15}}{6}$; Sec. 10.1, Ex. 8   **22.** $\dfrac{-1 + \sqrt{33}}{4}, \dfrac{-1 - \sqrt{33}}{4}$; Sec. 10.3, Ex. 2   **23.** $[0, 4]$; Sec. 10.4, Ex. 2

24. 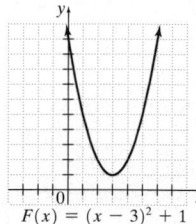 ; Sec. 10.6, Ex. 2

$F(x) = (x - 3)^2 + 1$

■ CHAPTER 12   SEQUENCES, SERIES, AND THE BINOMIAL THEOREM

Exercise Set 12.1   **1.** $5, 6, 7, 8, 9$ **3.** $-1, 1, -1, 1, -1$ **5.** $\dfrac{1}{4}, \dfrac{1}{5}, \dfrac{1}{6}, \dfrac{1}{7}, \dfrac{1}{8}$ **7.** $2, 4, 6, 8, 10$ **9.** $-1, -4, -9, -16, -25$

11. $2, 4, 8, 16, 32$ **13.** $7, 9, 11, 13, 15$ **15.** $-1, 4, -9, 16, -25$ **17.** 75   **19.** 118   **21.** $\dfrac{6}{5}$ **23.** 729   **25.** $\dfrac{4}{7}$ **27.** $\dfrac{1}{8}$

29. -95 **31.** $-\dfrac{1}{25}$ **33.** $a_n = 4n - 1$ **35.** $a_n = -2^n$ **37.** $a_n = \dfrac{1}{3^n}$ **39.** 48 ft, 80 ft, and 112 ft

41. $a_n = 0.10(2)^{n-1}$; $819.20   **43.** 2400 cases; 75 cases   **45.** 50 sparrows in 2004: extinct in 2010
47. $1, 0.7071, 0.5774, 0.5, 0.4472$ **49.** $2, 2.25, 2.3704, 2.4414, 2.4883$ **51.** 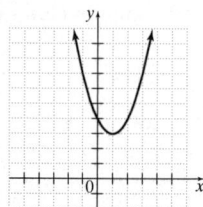   **53.**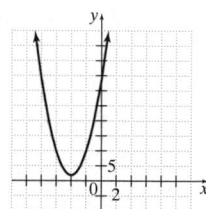

Exercise Set 12.2   **1.** $4, 6, 8, 10, 12$ **3.** $6, 4, 2, 0, -2$ **5.** $1, 3, 9, 27, 81$ **7.** $48, 24, 12, 6, 3$ **9.** 33   **11.** -875 **13.** -60
15. 96   **17.** -28 **19.** 1250   **21.** 31   **23.** 20   **25.** $a_1 = \dfrac{2}{3}; r = -2$ **27.** Answers may vary.   **29.** $a_1 = 2; d = 2$

31. $a_1 = 5; r = 2$ **33.** $a_1 = \dfrac{1}{2}; r = \dfrac{1}{5}$ **35.** $a_1 = x; r = 5$ **37.** $a_1 = p; d = 4$ **39.** 19   **41.** $-\dfrac{8}{9}$ **43.** $\dfrac{17}{2}$ **45.** $\dfrac{8}{81}$

47. -19 **49.** $a_n = 4n + 50$; 130 seats   **51.** $a_n = 6(3)^{n-1}$ **53.** $486, 162, 54, 18, 6; a_n = \dfrac{486}{3^{n-1}}$; 6 bounces

55. $a_n = 4000 + 125(n - 1)$ or $a_n = 3875 + 125n$; $5375   **57.** 25 g   **59.** $11,782.40, $5891.20, $2945.60, $1472.80

61. $19.652, 19.618, 19.584, 19.55$ **63.** Answers may vary.   **65.** $\dfrac{11}{18}$ **67.** 40   **69.** $\dfrac{907}{495}$

Exercise Set 12.3   **1.** -2 **3.** 60   **5.** 20   **7.** $\dfrac{73}{168}$ **9.** $\dfrac{11}{36}$ **11.** 60   **13.** 74   **15.** 62   **17.** $\dfrac{241}{35}$ **19.** $\displaystyle\sum_{i=1}^{5} (2i - 1)$

21. $\displaystyle\sum_{i=1}^{4} 4(3)^{i-1}$ **23.** $\displaystyle\sum_{i=1}^{6} (-3i + 15)$ **25.** $\displaystyle\sum_{i=1}^{4} \dfrac{4}{3^{i-2}}$ **27.** $\displaystyle\sum_{i=1}^{7} i^2$ **29.** -24 **31.** -13 **33.** 82   **35.** -20 **37.** -2
39. $1, 2, 3, \ldots, 10$; 55 trees   **41.** $a_n = 6(2)^{n-1}$; 96 units   **43.** $a_n = 50(2)^n$; n represents the number of 12-hour periods;
800 bacteria   **45.** 30 opossums; 68 opossums   **47.** 6.25 lb; 93.75 lb   **49.** 16.4 in.; 134.5 in.

51. a. $2 + 6 + 12 + 20 + 30 + 42 + 56$ **b.** $1 + 2 + 3 + 4 + 5 + 6 + 7 + 1 + 4 + 9 + 16 + 25 + 36 + 49$

c. Answers may vary. **d.** True; answers may vary. **53.** 10 **55.** $\dfrac{10}{27}$ **57.** 45 **59.** 90

Exercise Set 12.4 **1.** 36 **3.** 484 **5.** 63 **7.** 2.496 **9.** 55 **11.** 16 **13.** 24 **15.** $\dfrac{1}{9}$ **17.** -20 **19.** $\dfrac{16}{9}$ **21.** $\dfrac{4}{9}$ **23.** 185

25. $\dfrac{381}{64}$ **27.** $-\dfrac{33}{4}$, or -8.25 **29.** $-\dfrac{75}{2}$ **31.** $\dfrac{56}{9}$ **33.** 4000, 3950, 3900, 3850, 3800: 3450 cars; 44,700 cars

35. Firm A (Firm A, \$265,000; Firm B, \$254,000) **37.** \$39,930; \$139,230 **39.** 20 min; 123 min **41.** 180 ft

43. Player A, 45 points; Player B, 75 points **45.** \$3050 **47.** \$10,737,418.23 **49.** $\dfrac{8}{10} + \dfrac{8}{100} + \dfrac{8}{1000} + \cdots; \dfrac{8}{9}$

51. Answers may vary. **53.** 720 **55.** 3 **57.** $x^2 + 10x + 25$ **59.** $8x^3 - 12x^2 + 6x - 1$

Exercise Set 12.5 **1.** $m^3 + 3m^2n + 3mn^2 + n^3$ **3.** $c^5 + 5c^4d + 10c^3d^2 + 10c^2d^3 + 5cd^4 + d^5$

5. $y^5 - 5y^4x + 10y^3x^2 - 10y^2x^3 + 5yx^4 - x^5$ **7.** Answers may vary. **9.** 8 **11.** 42 **13.** 360 **15.** 56

17. $a^7 + 7a^6b + 21a^5b^2 + 35a^4b^3 + 35a^3b^4 + 21a^2b^5 + 7ab^6 + b^7$ **19.** $a^5 + 10a^4b + 40a^3b^2 + 80a^2b^3 + 80ab^4 + 32b^5$

21. $q^9 + 9q^8r + 36q^7r^2 + 84q^6r^3 + 126q^5r^4 + 126q^4r^5 + 84q^3r^6 + 36q^2r^7 + 9qr^8 + r^9$

23. $1024a^5 + 1280a^4b + 640a^3b^2 + 160a^2b^3 + 20ab^4 + b^5$ **25.** $625a^4 - 1000a^3b + 600a^2b^2 - 160ab^3 + 16b^4$

27. $8a^3 + 36a^2b + 54ab^2 + 27b^3$ **29.** $x^5 + 10x^4 + 40x^3 + 80x^2 + 80x + 32$ **31.** $5cd^4$ **33.** d^7 **35.** $-40r^2s^3$

37. $6x^2y^2$ **39.** $30a^9b$ **41.** not one-to-one **43.** one-to-one **45.** not one-to-one

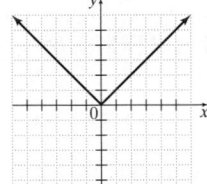 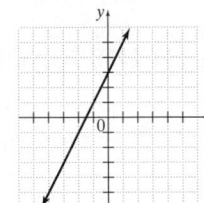 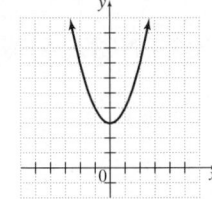

Chapter 12 Review **1.** $-3, -12, -27, -48, -75$ **3.** $\dfrac{1}{100}$ **5.** $a_n = \dfrac{1}{6n}$ **7.** 144 ft, 176 ft, 208 ft

9. 450, 1350, 4050, 12,150, 36,450; 36,450 infected people in 2003 **11.** $-2, -\dfrac{4}{3}, \dfrac{8}{9}, -\dfrac{16}{27}, \dfrac{32}{81}$ **13.** 111 **15.** -83

17. $a_1 = 3; d = 5$ **19.** $a_n = \dfrac{3}{10^n}$ **21.** $a_1 = \dfrac{8}{3}, r = \dfrac{3}{2}$ **23.** $a_1 = 7x, r = -2$ **25.** 8, 6, 4.5, 3.4, 2.5, 1.9; good

27. $a_n = 2^{n-1}, \$512, \$536,870,912$ **29.** $a_n = 900 + (n-1)150$ or $a_n = 150n + 750; \$1650/month$

31. $1 + 3 + 5 + 7 + 9 = 25$ **33.** $\dfrac{1}{4} - \dfrac{1}{6} + \dfrac{1}{8} = \dfrac{5}{24}$ **35.** -4 **37.** -10 **39.** $\displaystyle\sum_{i=1}^{6} 3^{i-1}$ **41.** $\displaystyle\sum_{i=1}^{4} \dfrac{1}{4^i}$

43. $a_n = 20(2)^n$; n represents the number of 8-hour periods; 1280 yeast **45.** Job A, \$48,300; Job B, \$46,600 **47.** 150

49. 900 **51.** -410 **53.** 936 **55.** 10 **57.** -25 **59.** \$30,418; \$99,868 **61.** \$58; \$553 **63.** 2696 mosquitoes **65.** $\dfrac{5}{9}$

67. $x^5 + 5x^4z + 10x^3z^2 + 10x^2z^3 + 5xz^4 + z^5$ **69.** $16x^4 + 32x^3y + 24x^2y^2 + 8xy^3 + y^4$

71. $b^8 + 8b^7c + 28b^6c^2 + 56b^5c^3 + 70b^4c^4 + 56b^3c^5 + 28b^2c^6 + 8bc^7 + c^8$

73. $256m^4 - 256m^3n + 96m^2n^2 - 16mn^3 + n^4$ **75.** $35a^4b^3$

Chapter 12 Test **1.** $-\dfrac{1}{5}, \dfrac{1}{6}, -\dfrac{1}{7}, \dfrac{1}{8}, -\dfrac{1}{9}$ **2.** $-3, 3, -3, 3, -3$ **3.** 247 **4.** 39,999 **5.** $a_n = \dfrac{2}{5}\left(\dfrac{1}{5}\right)^{n-1}$ **6.** $a_n = (-1)^n 9n$

7. 155 **8.** -330 **9.** $\dfrac{144}{5}$ **10.** 1 **11.** 10 **12.** -60 **13.** $a^6 - 6a^5b + 15a^4b^2 - 20a^3b^3 + 15a^2b^4 - 6ab^5 + b^6$

14. $32x^5 + 80x^4y + 80x^3y^2 + 40x^2y^3 + 10xy^4 + y^5$

15. $y^8 + 8y^7z + 28y^6z^2 + 56y^5z^3 + 70y^4z^4 + 56y^3z^5 + 28y^2z^6 + 8yz^7 + z^8$

16. $128p^7 + 448p^6r + 672p^5r^2 + 560p^4r^3 + 280p^3r^4 + 84p^2r^5 + 14pr^6 + r^7$ **17.** 925 people; 250 people initially

18. $1 + 3 + 5 + 7 + 9 + 11 + 13 + 15; 64$ shrubs **19.** 33.75 cm, 218.75 cm **20.** 320 cm **21.** 304 ft; 1600 ft **22.** $\dfrac{14}{33}$

Chapter 12 Cumulative Review   **1.** no; Sec. 3.3, Ex. 4
2.

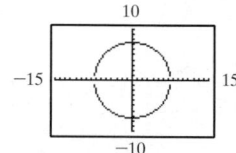

; Sec. 3.4, Ex. 6   **3.** 0; Sec. 3.5, Ex. 9   **5.**

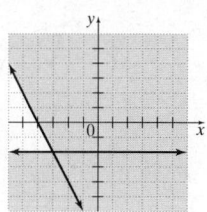

; Sec. 3.7, Ex. 4

4. $y = 3$; Sec. 3.6, Ex. 7

6. $x^3 - 4x^2 - 3x + 11 + \dfrac{12}{x + 2}$; Sec. 5.7, Ex. 2   **7.** $\left[-9, -\dfrac{9}{2}\right]$; Sec. 8.3, Ex. 5   **8.** 0; Sec. 8.4, Ex. 5   **9.** $\varnothing$; Sec. 8.5, Ex. 4

10. a. $5\sqrt{2}$ **b.** $2\sqrt[3]{3}$ **c.** $\sqrt{26}$ **d.** $2\sqrt[4]{2}$; Sec. 9.3, Ex. 3   **11.** 10%; Sec. 10.1, Ex. 9   **12.** 2, 7; Sec. 10.3, Ex. 4

13. $\left(-\dfrac{7}{2}, -1\right)$; Sec. 10.4, Ex. 5   **14.** $\dfrac{25}{4}$ ft; $\dfrac{5}{8}$ sec; Sec. 10.7, Ex. 5   **15. a.** 25; 7   **b.** $x^2 + 6x + 9$; $x^2 + 3$; Sec. 11.1, Ex. 2

16. $f^{-1} = \{(1, 0), (7, -2), (-6, 3), (4, 4)\}$; Sec. 11.2, Ex. 3   **17. a.** 4   **b.** $\dfrac{3}{2}$ **c.** 6; Sec. 11.3, Ex. 4   **18. a.** 2   **b.** -1 **c.** 3

d. 6; Sec. 11.4, Ex. 5   **19. a.** $\log_{11} 30$ **b.** $\log_3 6$ **c.** $\log_2 (x^2 + 2x)$; Sec. 11.5, Ex. 1   **20.** $2509.30; Sec. 11.6, Ex. 8

21. $\dfrac{\log 7}{\log 3} \approx 1.7712$; Sec. 11.7, Ex. 1   **22.** 18; Sec. 11.7, Ex. 2   **23.** 0, 3, 8, 15, 24; Sec. 12.1, Ex. 1   **24.** 72; Sec. 12.2, Ex. 3

25. a. $\dfrac{7}{2}$ **b.** 56; Sec. 12.3, Ex. 1   **26.** 465; Sec. 12.4, Ex. 2

■ CHAPTER 13   CONIC SECTIONS

Graphing Calculator Explorations
1.

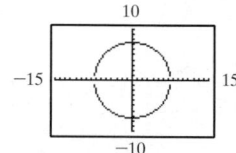

3.

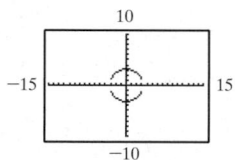

5.

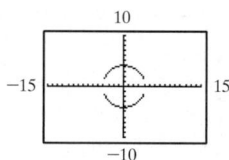

7.

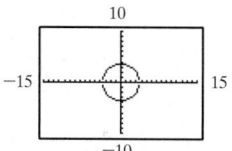

Mental Math   **1.** upward   **3.** to the left   **5.** downward

Exercise Set 13.1   **1.**

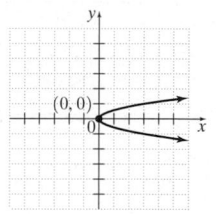

3.

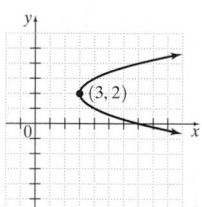

5.

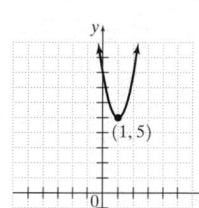

7.

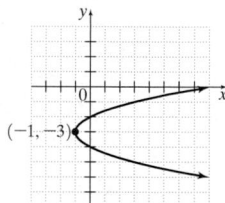

9.

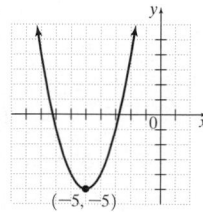

11.

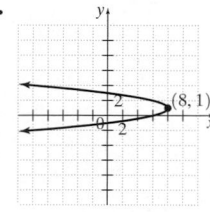

13. 5 units   **15.** $\sqrt{41}$ units   **17.** $\sqrt{10}$ units
19. $\sqrt{5}$ units   **21.** 13.88 units   **23.** 9 units   **25.** $(4, -2)$
27. $\left(-5, \dfrac{5}{2}\right)$ **29.** $(3, 0)$ **31.** $\left(-\dfrac{1}{2}, \dfrac{1}{2}\right)$ **33.** $\left(\sqrt{2}, \dfrac{\sqrt{5}}{2}\right)$
35. $(6.2, -6.65)$

37.

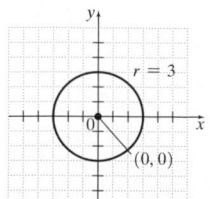

39.

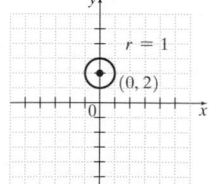

41.

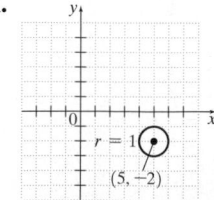

43.

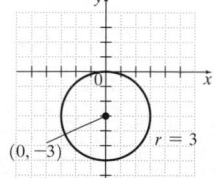

45.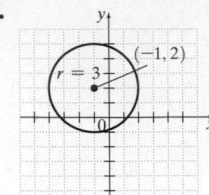

47.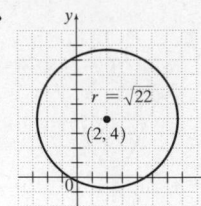

49. $(x - 2)^2 + (y - 3)^2 = 36$
51. $x^2 + y^2 = 3$
53. $(x + 5)^2 + (y - 4)^2 = 45$
55. Answers may vary.

57.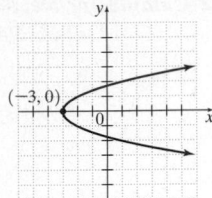

59.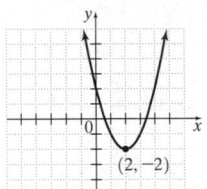

61.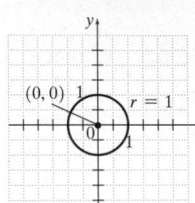

63.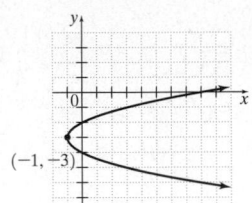

65.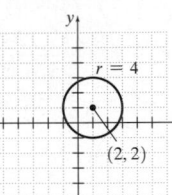

67.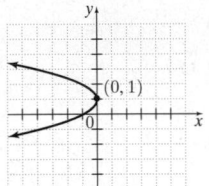

69.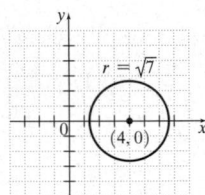

71.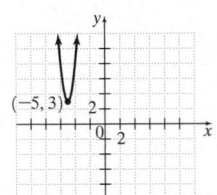

73.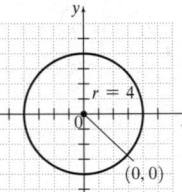

75.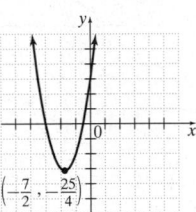

77.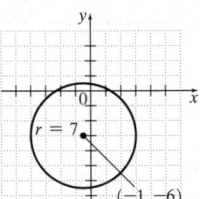

79.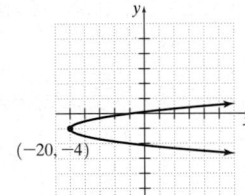

81.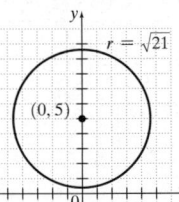

83.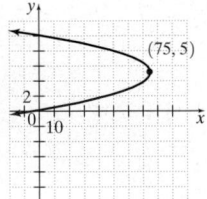

85.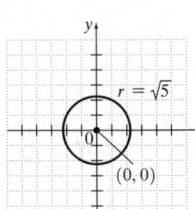

87.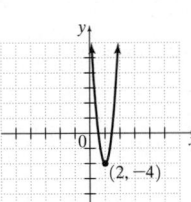

89. Yes, it is.   **91.** $y = -\dfrac{2}{125}x^2 + 40$

93.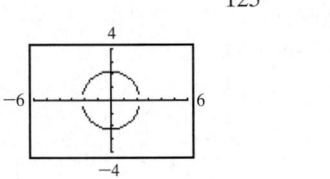

95.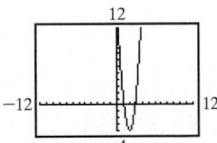

97.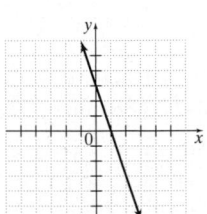

99.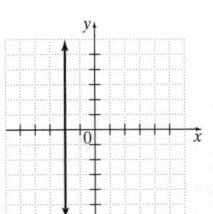

101. $\dfrac{\sqrt{10}}{4}$ **103.** $2\sqrt{5}$

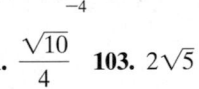

Graphing Calculator Explorations   **1.**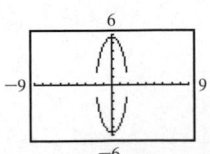

3.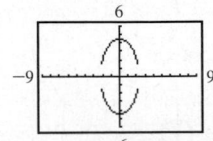

5.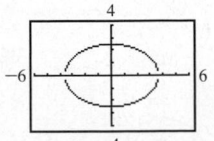

Exercise Set 13.2   **1.** 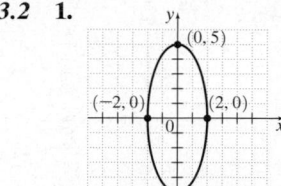 **3.** 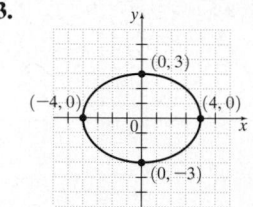 **5.** 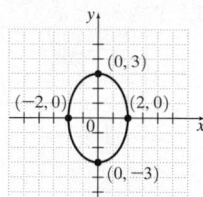 **7.**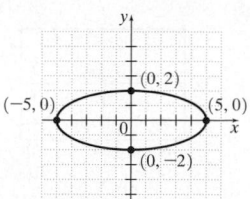

9. 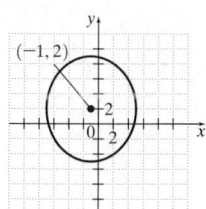 **11.** 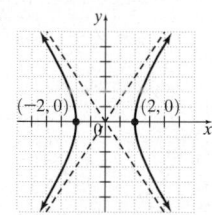 **13.** 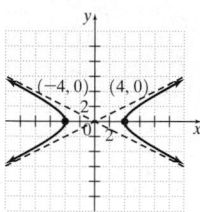 **15.**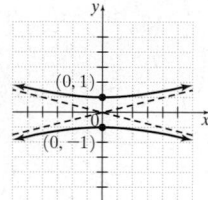

17. 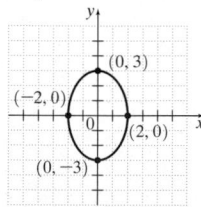 **19.** 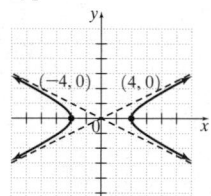 **21.** Answers may vary.   **23.** parabola

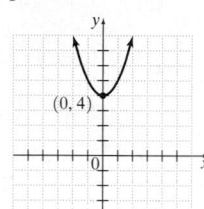

25. ellipse   **27.** hyperbola   **29.** circle   **31.** parabola

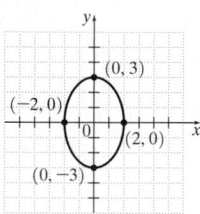 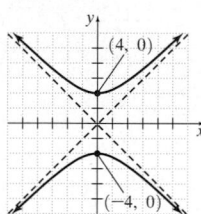 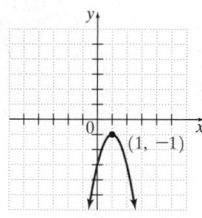 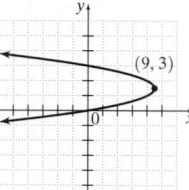

33. ellipse   **35.** hyperbola   **37.** parabola   **39.** A: 36, 13; B: 4, 4; C: 25, 16; D: 39, 25; E: 17, 81; F: 36, 36; G: 16, 65; H: 144, 140

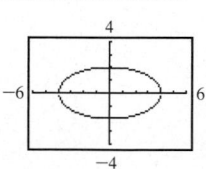 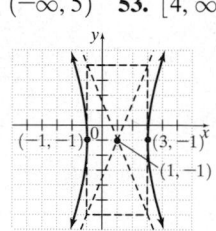 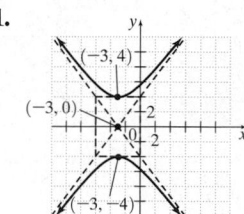

41. A: 6; B: 2; C: 5; D: 5; E: 9; F: 6; G: 4; H: 12
43. greater than zero and less than one
45. greater than one
47. (1,782,000,000, 356,400,000)

49. 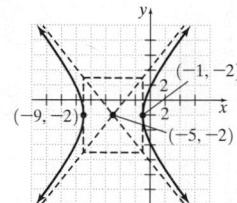 **51.** $(-\infty, 5)$ **53.** $[4, \infty)$ **55.** $-2x^3$ **57.** $-5x^4$
59. 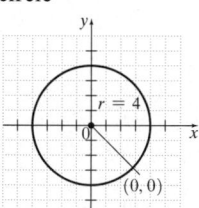 **61.** **63.**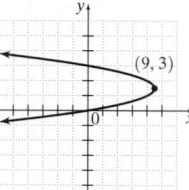

Exercise Set 13.3   **1.** $(3, -4), (-3, 4)$ **3.** $(\sqrt{2}, \sqrt{2}), (-\sqrt{2}, -\sqrt{2})$ **5.** $(4, 0), (0, -2)$
7. $(-\sqrt{5}, -2), (-\sqrt{5}, 2), (\sqrt{5}, -2), (\sqrt{5}, 2)$ **9.** no solution   **11.** $(1, -2), (3, 6)$ **13.** $(2, 4), (-5, 25)$ **15.** no solution
17. $(1, -3)$ **19.** $(-1, -2), (-1, 2), (1, -2), (1, 2)$ **21.** $(0, -1)$ **23.** $(-1, 3), (1, 3)$ **25.** $(\sqrt{3}, 0), (-\sqrt{3}, 0)$
27. no solution   **29.** $(-6, 0), (6, 0), (0, -6)$ **31.** 0, 1, 2, 3, or 4   **33.** 9 and 7; 9 and -7; -9 and 7; -9 and -7
35. 15 cm by 19 cm   **37.** 15 thousand compact discs; price: $3.75

39. 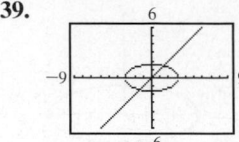   **41.** 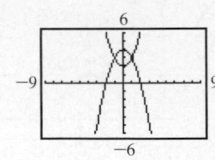   **43.** 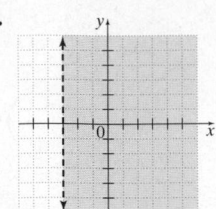   **45.**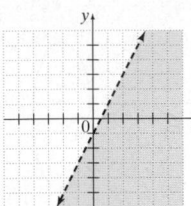

47. $(8x - 25)$ in.   **49.** $(4x^2 + 6x + 2)$ m

Exercise Set 13.4   **1.** 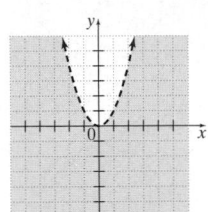   **3.** 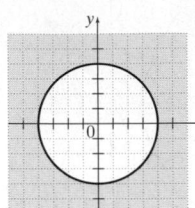   **5.** 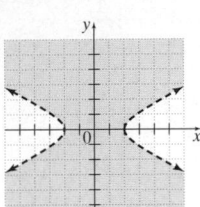   **7.**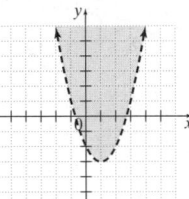

9. 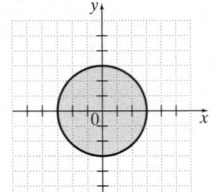   **11.** 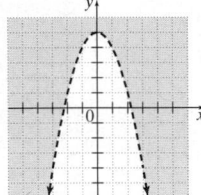   **13.** 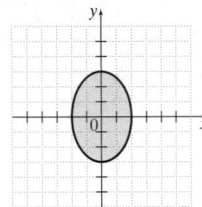   **15.**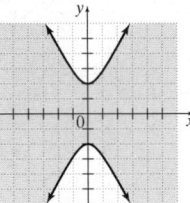

17. 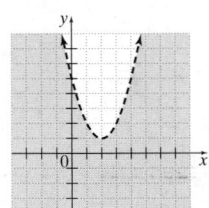   **19.** 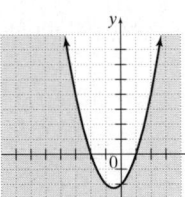   **21.** Answers may vary.   **23.**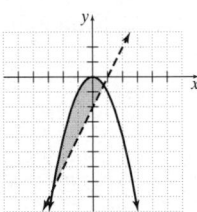

25. 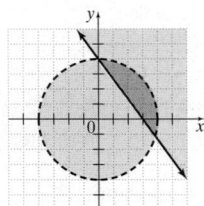   **27.** 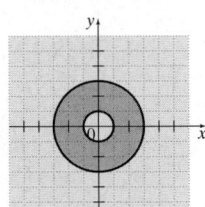   **29.** 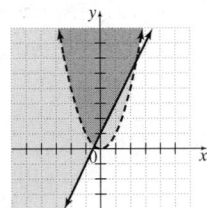   **31.**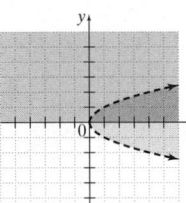

33. 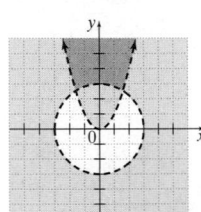   **35.** 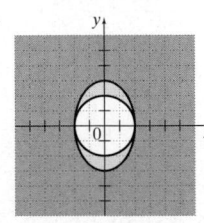   **37.** 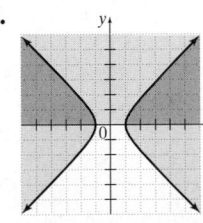   **39.**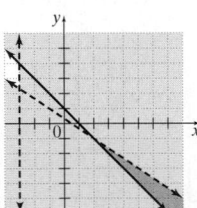

41. 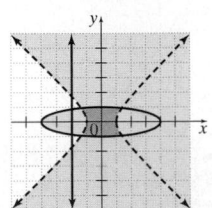   **43.** 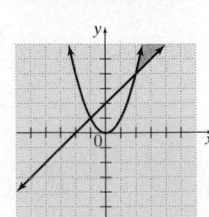   **45.** function   **47.** not a function   **49.** 25   **51.** $3b^2 - 2$

Chapter 13 Review   **1.** $\sqrt{197}$ units   **3.** $\sqrt{130}$ units   **5.** $7\sqrt{2}$ units   **7.** 16.60 units   **9.** $(-5, 5)$ **11.** $\left(-\dfrac{15}{2}, 1\right)$

13. $\left(\dfrac{1}{20}, -\dfrac{3}{16}\right)$ **15.** $(\sqrt{3}, -3\sqrt{6})$ **17.** $(x + 4)^2 + (y - 4)^2 = 9$ **19.** $(x + 7)^2 + (y + 9)^2 = 11$

21.

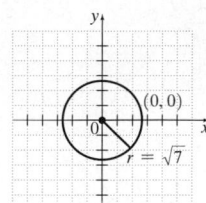

23.

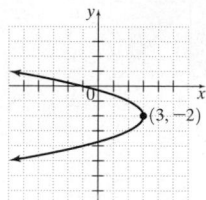

25.

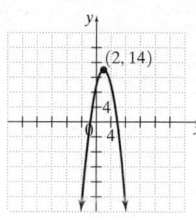

27.

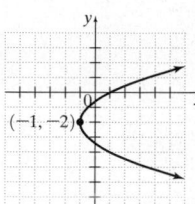

29.

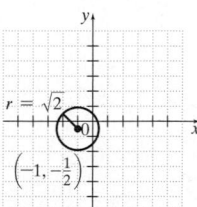

31.

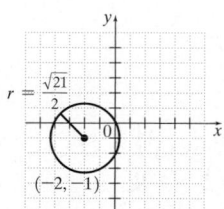

33.

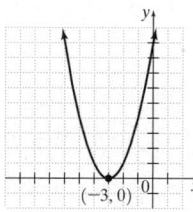

35. $(x - 5.6)^2 + (y + 2.4)^2 = 9.61$

37.

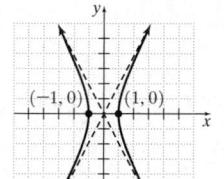

39.

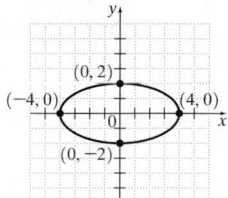

41.

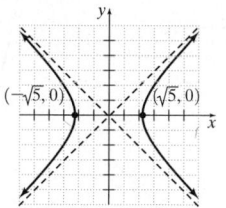

43.

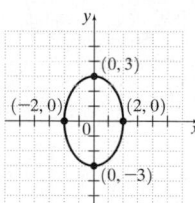

45.

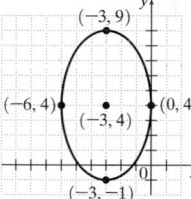

47.

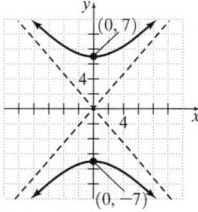

49.

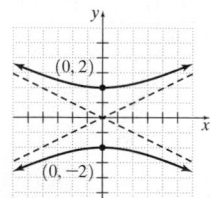

51.

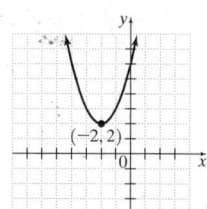

53.

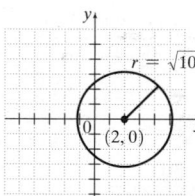

55.

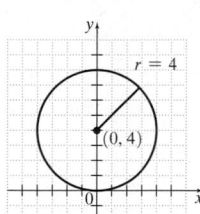

57.

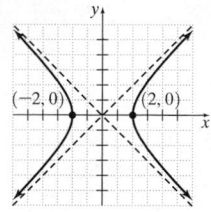

59.

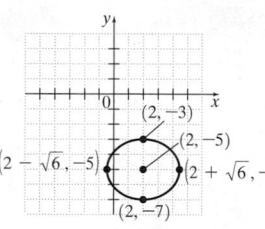

61.

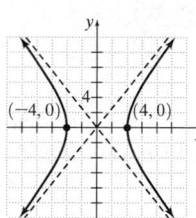

63.

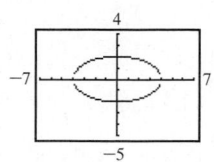

65.

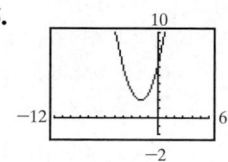

67. $(1, -2), (4, 4)$
69. $(-1, 1), (2, 4)$
71. $(2, 2\sqrt{2}), (2, -2\sqrt{2})$
73. $(-1, 3), (-1, -3), (1, 3), (1, -3)$
75. $(1, 4)$
77. 15 ft by 10 ft

79. 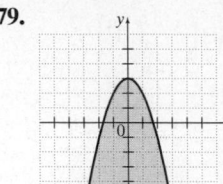   **81.** 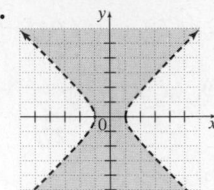   **83.** 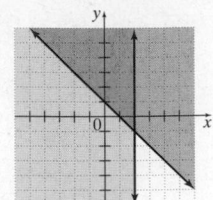   **85.**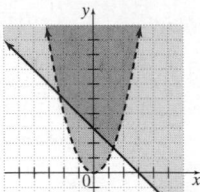

87.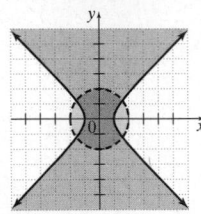

Chapter 13 Test   **1.** $2\sqrt{26}$ units   **2.** $\sqrt{95}$ units   **3.** $\left(-4, \dfrac{7}{2}\right)$ **4.** $\left(-\dfrac{1}{2}, \dfrac{3}{10}\right)$

5. 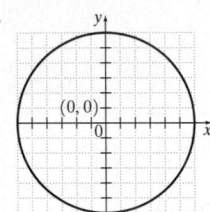   **6.** 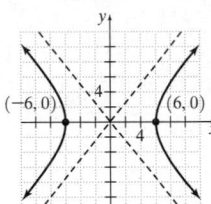   **7.** 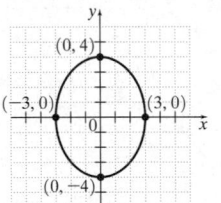   **8.** 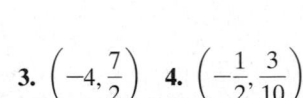   **9.**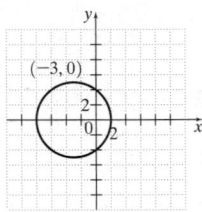

10. 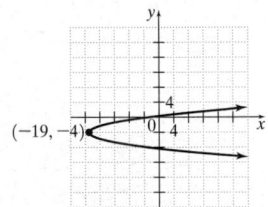   **11.** 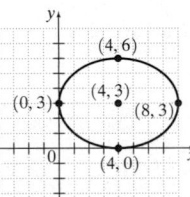   **12.**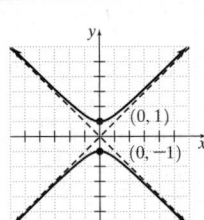

13. $(-12, 5), (12, -5)$
14. $(-5, -1), (-5, 1), (5, -1), (5, 1)$
15. $(6, 12), (1, 2)$
16. $(1, 1), (-1, -1)$

17. 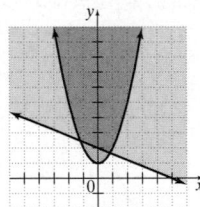   **18.** 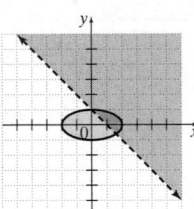   **19.** 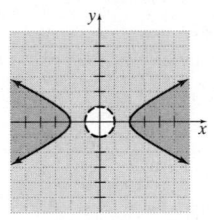   **20.**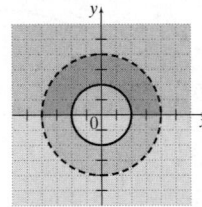

21. B
22. height: 10 ft;
width: 30 ft

Chapter 13 Cumulative Review   **1.** 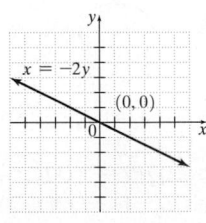 ; Sec. 3.4, Ex. 5   **2.** $2x - 5$; Sec. 5.6, Ex. 3
3. 16; Sec. 5.7, Ex. 4
4. a. $\dfrac{1}{9xy^2}$ **b.** $\dfrac{x(x-2)}{2(x+2)}$ **c.** $\dfrac{x^2}{y^2}$; Sec. 7.5, Ex. 1
5. $\{4, 6\}$; Sec. 8.3, Ex. 1
6. $\varnothing$; Sec. 8.3, Ex. 3
7. $\{2, 3, 4, 5, 6, 8\}$; Sec. 8.3, Ex. 6

8. $\left(-\infty, \dfrac{13}{5}\right] \cup [4, \infty)$; Sec. 8.3, Ex. 7   **9.** $-2, \dfrac{4}{5}$; Sec. 8.4, Ex. 2   **10.** $24, -20$; Sec. 8.4, Ex. 3   **11.** $\dfrac{3}{4}, 5$; Sec. 8.4, Ex. 8

12. $(4, 8)$; Sec. 8.5, Ex. 2   **13.** $(-\infty, \infty)$; Sec. 8.5, Ex. 6   **14. a.** 1   **b.** -4 **c.** $\dfrac{2}{5}$ **d.** x^2 **e.** $-2x^3$; Sec. 9.1, Ex. 3

15. a. $z - z^{17/3}$ **b.** $x^{2/3} - 3x^{1/3} - 10$; Sec. 9.2, Ex. 5   **16. a.** 2   **b.** $\dfrac{5}{2}\sqrt{x}$ **c.** $14xy^2\sqrt[3]{x}$; Sec. 9.3, Ex. 5

17. a. $\dfrac{5\sqrt{5}}{12}$ **b.** $\dfrac{5\sqrt[3]{7x}}{2}$; Sec. 9.4, Ex. 2 **18.** $\dfrac{\sqrt{21xy}}{3y}$; Sec. 9.5, Ex. 2 **19.** 42; Sec. 9.6, Ex. 1

20. a. $-i$ **b.** 1 **c.** -1 **d.** 1; Sec. 9.7, Ex. 6 **21.** $-1 + \sqrt{5}, -1 - \sqrt{5}$; Sec. 10.1, Ex. 5

22. $2 + \sqrt{2}, 2 - \sqrt{2}$; Sec. 10.2, Ex. 3 **23.** $2, -2, i, -i$; Sec. 10.3, Ex. 3 **24.** $[-2, 3)$; Sec. 10.4, Ex. 4

25. ; Sec. 10.6, Ex. 8 **26.** $(2, -16)$; Sec. 10.7, Ex. 4

27. $\sqrt{2} \approx 1.414$; Sec. 13.1, Ex. 5

■ APPENDIX A   OPERATIONS ON DECIMALS

Appendix A Exercise Set **1.** 17.08 **3.** 12.804 **5.** 110.96 **7.** 2.4 **9.** 28.43 **11.** 227.5 **13.** 2.7 **15.** 49.2339 **17.** 80 **19.** 0.07612 **21.** 4.56 **23.** 648.46 **25.** 767.83 **27.** 12.062 **29.** 61.48 **31.** 7.7 **33.** 863.37 **35.** 22.579 **37.** 363.15 **39.** 7.007

■ APPENDIX B   REVIEW OF ANGLES, LINES, AND SPECIAL TRIANGLES

Appendix B Exercise Set **1.** $71°$ **3.** $19.2°$ **5.** $78\frac{3}{4}°$ **7.** $30°$ **9.** $149.8°$ **11.** $100\frac{1}{2}°$

13. $m\angle 1 = m\angle 5 = m\angle 7 = 110°; m\angle 2 = m\angle 3 = m\angle 4 = m\angle 6 = 70°$ **15.** $90°$ **17.** $90°$ **19.** $90°$ **21.** $45°, 90°$

23. $73°, 90°$ **25.** $50\frac{1}{4}°, 90°$ **27.** $x = 6$ **29.** $x = 4.5$ **31.** 10 **33.** 12

■ APPENDIX D   REVIEW OF VOLUME AND SURFACE AREA

Appendix D Exercise Set **1.** $V = 72$ cu. in.; $SA = 108$ sq. in. **3.** $V = 512$ cu. cm; $SA = 384$ sq. cm

5. $V = 4\pi$ cu. yd $\approx 12\frac{4}{7}$ cu. yd; $SA = (2\sqrt{13}\pi + 4\pi)$ sq. yd ≈ 35.20 sq. yd

7. $V = \dfrac{500}{3}\pi$ cu. in. $\approx 523\frac{17}{21}$ cu. in.; $SA = 100\pi$ sq. in. $\approx 314\frac{2}{7}$ sq. in. **9.** $V = 48$ cu. cm; $SA = 96$ sq. cm

11. $2\frac{10}{27}$ cu. in. **13.** 26 sq. ft **15.** ≈ 10.83 cu. in. **17.** 960 cu. cm **19.** 196π sq. in. **21.** $7\frac{1}{2}$ cu. ft **23.** $12\frac{4}{7}$ cu. cm

■ APPENDIX E   MEAN, MEDIAN, AND MODE

Appendix E Exercise Set **1.** mean: 29, median: 28, no mode **3.** mean: 8.1, median: 8.2, mode: 8.2

5. mean: 0.6, median: 0.6, mode: 0.2 and 0.6 **7.** mean: 370.9, median: 313.5, no mode **9.** 1314 ft **11.** 1131.5 ft **13.** 6.8

15. 6.9 **17.** 85.5 **19.** 73 **21.** 70 and 71 **23.** 9 **25.** 21, 21, 24

APPENDIX F   INTRODUCTION TO USING A GRAPHING UTILITY

Viewing Window and Interpreting Window Settings Exercise Set

1. yes **3.** no **5.** Answers may vary. **7.** Answers may vary. **9.** Answers may vary.

11. Xmin = −12   Ymin = −12 **13.** Xmin = −9   Ymin = −12 **15.** Xmin = −10   Ymin = −25
      Xmax = 12   Ymax = 12   Xmax = 9   Ymax = 12   Xmax = 10   Ymax = 25
      Xscl = 3   Yscl = 3   Xscl = 1   Yscl = 2   Xscl = 2   Yscl = 5

17. Xmin = −10   Ymin = −30 **19.** Xmin = −20   Ymin = −30
      Xmax = 10   Ymax = 30   Xmax = 30   Ymax = 50
      Xscl = 1   Yscl = 3   Xscl = 5   Yscl = 10

Graphing Equations and Square Viewing Window Exercise Set

1. Setting B   **3.** Setting B   **5.** Setting B

7.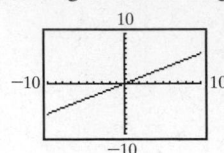

9.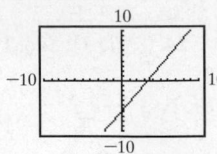

11.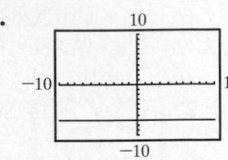

13.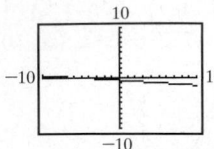

15.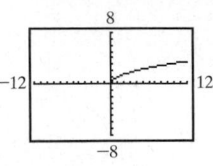

17.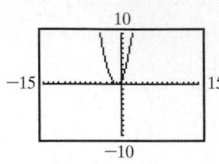

19.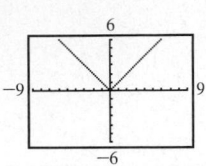

21.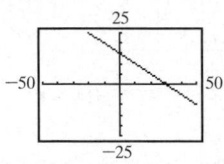

SUBJECT INDEX

Photo Credits

Chapter 1 CO Bob Daemmrich/The Image Works, (p. 5) Frank Johnston/PhotoDisc, Inc., (p. 11) Tim Davis/Photo Researchers, Inc., (p. 17) John Chumack/Photo Researchers, Inc., (p. 28) Didier Klein/Vandystadt/Allsport Photography (USA), Inc., (p. 40) © Richard T. Nowitz/Corbis, (p. 40) Emory Kristof/NGS Image Collection, (p. 47) N. Gellette/Liaison Agency, Inc., (p. 55) David Young-Wolfe/PhotoEdit

Chapter 2 CO PhotoEdit, (p. 95) AP/Wide World Photos, (p. 106) Bruce Hoertel/Liaison Agency, Inc., (p. 108) Brian K. Diggs/AP/Wide World Photos, (p. 110) Nasa/Science Source/Photo Researchers, Inc., (p. 110) Mark Lennihan/AP/Wide World Photos, (p. 112) Bachmann/Stock Boston, (p. 114) Sean Reid/Alaska Stock, (p. 120) Mt. Stromlo and Siding Spring Observatories, Australian National University/Science Photo Library/Photo Researchers, Inc., (p. 120) Norbert Wu/Stone, (p. 121) John Elk III/Stock Boston, (p. 125) Jonathan Nourok/PhotoEdit, (p. 146) Jean-Claude LeJeune/Stock Boston

Chapter 3 CO © Bernard Boutrit/Woodfin Camp/PictureQuest, (p. 168) Michael Newman/PhotoEdit, (p. 171) Amy C. Etra/PhotoEdit, (p. 198) John Neubauer/PhotoEdit, (p. 213) Bill Bachmann/Photo Researchers, Inc., (p. 218) M. K. Denny/PhotoEdit, (p. 218) Spencer Grant/PhotoEdit, (p. 235) Photo Courtesy of Motorola, Inc.

Chapter 4 CO Nicola Sutton/PhotoDisc, Inc., (p. 258) David Young-Wolfe/PhotoEdit, (p. 267) Doug Densinger/Allsport Photography (USA), Inc., (p. 267) Elsa Hasch/Allsport Photography (USA), Inc., (p. 268) J. Scott Applewhite/AP/Wide World Photos, (p. 276) Donna McWilliam/AP/Wide World Photos

Chapter 5 CO Michael Newman/PhotoEdit, (p. 297) Reuters/Jack Newton/Archive Photos

Chapter 6 CO Jonathan Nourok/PhotoEdit, (p. 389) PhotoDisc, Inc.

Chapter 7 CO Carl J. Single/The Image Works, (p. 416) Chuck Keeler/The Image Works, (p. 422) Tebo Photography, (p. 424) Rich Pedroncelli/AP/Wide World Photos, (p. 424) UPI/Corbis, (p. 443) Gary Benson/Gary J. Benson Photography, (p. 460) © Charles O'Rear/CORBIS, (p. 475) Keith Brofsky/PhotoDisc, Inc.

Chapter 8 CO Robert Brenner/PhotoEdit, (p. 485) Peter J. Schulz/Liaison Agency, Inc., (p. 486) Michael Newman/PhotoEdit, (p. 488) Jeremy Woodhouse/PhotoDisc, Inc., (p. 520) Tony Gutierrez/AP/Wide World Photos

Chapter 9 CO David Young-Wolff/PhotoEdit, (p. 564) John Henley/The Stock Market, (p. 587) Steve Gottlieb/FPG International LLC

Chapter 10 CO Henley & Savage/The Stock Market, (p. 615) Tony Freeman/PhotoEdit, (p. 635) Tim Flach/Stone, (p. 635) Arthur S. Aubry Photography/PhotoDisc, Inc., (p. 646) Richard A. Cooke III/Stone, (p. 650) Mary Teresa Giancoli, (p. 666) Jody Dole/The Image Bank, (p. 666) Simon Fraser/Northumbrian Environmental Management, Ltd./Science Photo Library/Photo Researchers, Inc., (p. 672) AP/Wide World Photos

Chapter 11 CO Gary Landsman/The Stock Market, (p. 701) PhotoDisc, Inc.

Chapter 12 CO Steve Smith/FPG International LLC

Chapter 13 CO Telegraph Colour Library/FPG International LLC

COMMON GRAPHS AND MODELS

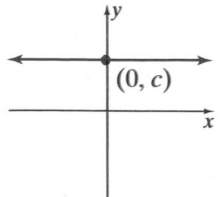

Horizontal Line;
Zero Slope
$y = c$

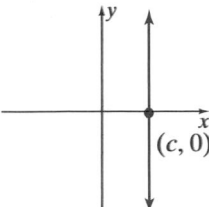

Vertical Line;
Undefined Slope
$x = c$

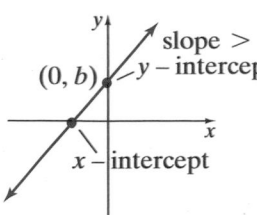

Linear Function;
Positive Slope
$y = mx + b; m > 0$

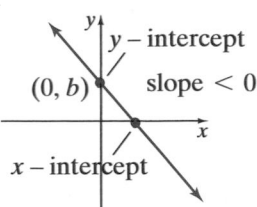

Linear Function;
Negative Slope
$y = mx + b; m < 0$

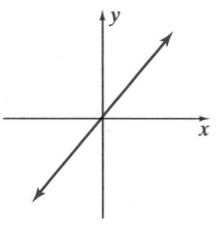

$y = x$

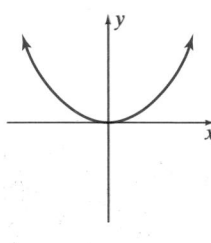

$y = x^2$

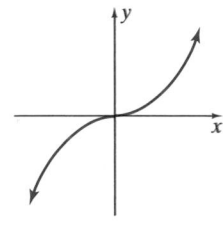

$y = x^3$

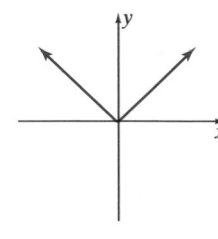

$y = |x|$

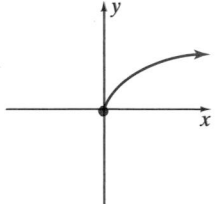

$y = \sqrt{x}; x \geq 0$

Quadratic Function
$y = ax^2 + bx + c; a \neq 0$
Parabola opens upward if $a > 0$
Parabola opens downward if $a < 0$

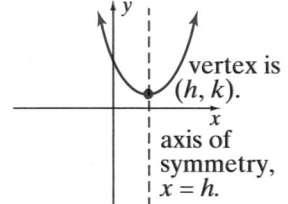

Quadratic Function
$y = a(x - h)^2 + k; a \neq 0$
Parabola opens upward if $a > 0$
Parabola opens downward if $a < 0$

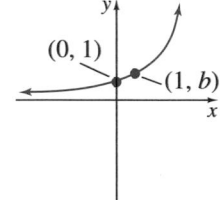

Exponential
Function
$y = b^x$ for $b > 1$

SYSTEMS OF LINEAR EQUATIONS

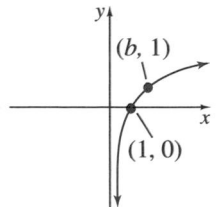

Logarithmic Function
$y = \log_b x$ for $b > 1$

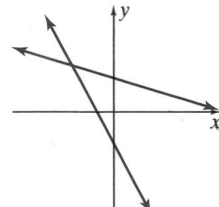

Independent and
consistent; one solution

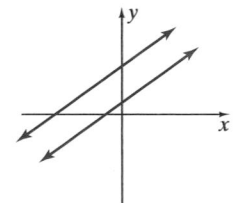

Independent and
inconsistent, no solution

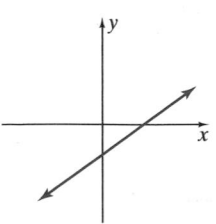

Dependent and consistent;
infinitely many solutions